# Dissection Experience

**Customizable images enliven presentations and quizzes for lecture or lab.**

Veins of the Abdomen and Pelvis

- Inferior vena cava
- Renal v.
- R. gonadal v.
- Common iliac v.

6. Which muscle's origin is highlighted in red?

a. Infraspinatus
b. *Subscapularis*
c. Teres minor
d. Teres major
e. Supraspinatus

*Bonus:* Which of the above muscles is not part of the rotator cuff?

**Layered cadaver dissections not available anywhere else!**

# www.aprevealed.com

**Portable cadavers can replace or enhance the laboratory experience.**

**Imaging**
Correlate dissected anatomy with radiologic images like X-ray, MRI, and CT scans.

**Self-Test**
Gauge proficiency with quizzes and simulated practical exams.

**Search**
Type in a term to quickly locate any structure in the program.

# Seeley's
# Essentials of
# Anatomy
# &Physiology

*seventh edition*

**Cinnamon VanPutte**
*Southwestern Illinois College*

**Jennifer Regan**
*Jones County Junior College*

**Andy Russo**
*University of Iowa*

Mc
Graw
Hill **Higher Education**

Boston   Burr Ridge, IL   Dubuque, IA   New York   San Francisco   St. Louis
Bangkok   Bogotá   Caracas   Kuala Lumpur   Lisbon   London   Madrid   Mexico City
Milan   Montreal   New Delhi   Santiago   Seoul   Singapore   Sydney   Taipei   Toronto

SEELEY'S ESSENTIALS OF ANATOMY AND PHYSIOLOGY, SEVENTH EDITION

ISBN    978–0–07–352563–1
MHID    0–07–352563–4

Publisher: *Michelle Watnick*
Senior Sponsoring Editor: *James F. Connely*
Director of Development: *Kristine Tibbetts*
Senior Developmental Editor: *Kathleen R. Loewenberg*
Marketing Manager: *Lynn M. Breithaupt*
Lead Project Manager: *Peggy J. Selle*
Senior Production Supervisor: *Laura Fuller*
Senior Media Project Manager: *Tammy Juran*
Senior Designer: *David W. Hash*
Cover/Interior Designer: *Jamie E. O'Neal*
(USE) Cover Image: *©Peter Sterling/Taxi/Getty Images*
Senior Photo Research Coordinator: *John C. Leland*
Photo Research: *Jerry Marshall/pictureresearching.com*
Supplement Producer: *Mary Jane Lampe*
Compositor: *Aptara®, Inc.*
Typeface: *10/12  Minion*
Printer: *Quebecor World Dubuque, IA*

The credits section for this book begins on page C-1 and is considered an extension of the copyright page.

**Library of Congress Cataloging-in-Publication Data**

VanPutte, Cinnamon L.
   Seeley's essentials of anatomy and physiology. – 7th ed. /  Cinnamon L. VanPutte, Jennifer L. Regan, Andrew F. Russo.
      p. cm.
   Rev. ed. of: Essentials of anatomy & physiology / Rod R. Seeley, Trent D. Stephens, Philip Tate. 6th ed. c2007.
   Includes index.
   ISBN  978-0-07-352563-1 — ISBN  0-07-352563-4 (hard copy : alk. paper)  1.  Human physiology. 2.  Human anatomy.  I. Regan, Jennifer L. II. Russo, Andrew F.
III. Seeley, Rod R. Essentials of anatomy & physiology. IV. Title. V. Title: Essentials of anatomy and physiology.
   QP34.5.S418 2010
   612–dc22
                                                                                                              2008039050

www.mhhe.com

SEELEY'S ESSENTIALS OF ANATOMY AND PHYSIOLOGY, SEVENTH EDITION

ISBN    978–0–07–352563–1
MHID    0–07–352563–4

Publisher: *Michelle Watnick*
Senior Sponsoring Editor: *James F. Connely*
Director of Development: *Kristine Tibbetts*
Senior Developmental Editor: *Kathleen R. Loewenberg*
Marketing Manager: *Lynn M. Breithaupt*
Lead Project Manager: *Peggy J. Selle*
Senior Production Supervisor: *Laura Fuller*
Senior Media Project Manager: *Tammy Juran*
Senior Designer: *David W. Hash*
Cover/Interior Designer: *Jamie E. O'Neal*
(USE) Cover Image: *©Peter Sterling/Taxi/Getty Images*
Senior Photo Research Coordinator: *John C. Leland*
Photo Research: *Jerry Marshall/pictureresearching.com*
Supplement Producer: *Mary Jane Lampe*
Compositor: *Aptara®, Inc.*
Typeface: *10/12 Minion*
Printer: *Quebecor World Dubuque, IA*

The credits section for this book begins on page C 1 and is considered an extension of the copyright page.

**Library of Congress Cataloging-in-Publication Data**

VanPutte, Cinnamon L.
   Seeley's essentials of anatomy and physiology. – 7th ed. / Cinnamon L. VanPutte, Jennifer L. Regan, Andrew F. Russo.
      p. cm.
   Rev. ed. of: Essentials of anatomy & physiology / Rod R. Seeley, Trent D. Stephens, Philip Tate. 6th ed. c2007.
   Includes index.
   ISBN 978–0–07–352563–1 — ISBN 0–07–352563–4 (hard copy : alk. paper)  1. Human physiology. 2. Human anatomy.  I. Regan, Jennifer L. II. Russo, Andrew F.
   III. Seeley, Rod R. Essentials of anatomy & physiology. IV. Title. V. Title: Essentials of anatomy and physiology.
   QP34.5.S418 2010
   612–dc22
                                                                                                    2008039050

www.mhhe.com

# Seeley's
# Essentials of
# Anatomy
# &Physiology
### seventh edition

**Cinnamon VanPutte**
*Southwestern Illinois College*

**Jennifer Regan**
*Jones County Junior College*

**Andy Russo**
*University of Iowa*

Mc Graw Hill **Higher Education**

Boston   Burr Ridge, IL   Dubuque, IA   New York   San Francisco   St. Louis
Bangkok   Bogotá   Caracas   Kuala Lumpur   Lisbon   London   Madrid   Mexico City
Milan   Montreal   New Delhi   Santiago   Seoul   Singapore   Sydney   Taipei   Toronto

## DEDICATION

This text is dedicated to our families. Without their uncompromising
support and love, this effort would not have been possible.
Our spouses and children have been more than patient while we've spent
many nights at the computer surrounded by mountains of books.
We also want to acknowledge and dedicate this edition to the previous
authors as we continue the standard of excellence that they have set
for so many years. For each of us, authoring this text is a culmination of our
passion for teaching and represents an opportunity to pass knowledge on
to students beyond our own classrooms; this has all been made possible
by the support and mentorship we in turn have received from our
teachers, colleagues, friends, and family.

# About the Authors

## Andrew F. Russo
**Professor of Molecular
Physiology and Biophysics
University of Iowa**

Andrew has 20 years of classroom experience with human physiology, neurobiology, molecular biology, and cell biology courses at the University of Iowa. He is a recipient of the Collegiate Teaching Award and is currently the course director for Medical Cell Biology and Director of the Biosciences Graduate Program. He is also a member of several professional societies, including the American Physiological Society and the Society for Neuroscience. Andrew received his Ph.D. in Biochemistry from the University of California at Berkeley. His research interests are focused on the molecular neurobiology of migraine. His decision to join the author team for *Seeley's Essentials of Human Anatomy and Physiology* is the culmination of a passion for teaching that began in graduate school. He is excited about the opportunity to hook students' interest in learning by presenting cutting-edge clinical and scientific advances. Andy is married to Maureen, a physical therapist, and has three daughters Erilynn, Becky, and Colleen, now in college and graduate school. He enjoys all types of outdoor sports, especially bicycling, skiing, ultimate Frisbee and, before moving to Iowa, bodyboard surfing.

## Cinnamon L. VanPutte
**Associate Professor of Biology
Southwestern Illinois College**

Cinnamon has been teaching biology and human anatomy and physiology for almost two decades. At Southwestern Illinois College she is a full-time faculty member and the coordinator for the anatomy and physiology courses. Cinnamon is an active member of several professional societies, including the Human Anatomy & Physiology Society. Her Ph.D. in Zoology, with an emphasis in endocrinology, is from Texas A&M University. She worked in Dr. Duncan MacKenzie's lab, where she was indoctrinated in the major principles of physiology and the importance of critical thinking. The critical thinking component of *Seeley's Essentials of Human Anatomy and Physiology* epitomizes Cinnamon's passion for the field of human anatomy and physiology; she is committed to maintaining this tradition of excellence. Cinnamon loves to spend time helping to coach her daughter, Savannah, in soccer, watching her son, Ethan, doing karate, and taking walks with her husband, Robb.

## Jennifer L. Regan
**Instructor
Jones County Junior College**

For the past ten years, Jennifer has taught introductory biology, human anatomy and physiology, and genetics at the university and community college level. She has received the Instructor of the Year Award at both the departmental and college level while teaching at USM. In addition, she has been recognized for her dedication to teaching by student organizations such as the Alliance for Graduate Education in Mississippi and Increasing Minority Access to Graduate Education. Jennifer has dedicated much of her career to improving lecture and laboratory instruction at her institutions. Critical thinking and lifelong learning are two characteristics Jennifer hopes to instill in her students. She appreciates the Seeley approach to learning and is excited about contributing to further development of the textbook. She received her Ph.D. in biology at the University of Houston, under the direction of Edwin H. Bryant and Lisa M. Meffert. She is an active member of several professional organizations, including the Human Anatomy and Physiology Society. During her free time, Jennifer enjoys spending time with her husband, Hobbie, and two sons, Patrick and Nicholas.

# Brief Contents

Chapter 1    The Human Organism    1

Chapter 2    The Chemical Basis of Life    20

Chapter 3    Cell Structures and Their Functions    42

Chapter 4    Tissues    71

Chapter 5    Integumentary System    95

Chapter 6    Skeletal System: Bones and Joints    111

Chapter 7    Muscular System    151

Chapter 8    Nervous System    192

Chapter 9    Senses    238

Chapter 10   Endocrine System    262

Chapter 11   Blood    291

Chapter 12   Heart    312

Chapter 13   Blood Vessels and Circulation    344

Chapter 14   The Lymphatic System and Immunity    379

Chapter 15   Respiratory System    406

Chapter 16   Digestive System    437

Chapter 17   Nutrition, Metabolism, and Body Temperature Regulation    472

Chapter 18   Urinary System and Fluid Balance    496

Chapter 19   Reproductive System    526

Chapter 20   Development, Heredity, and Aging    558

# Contents

Preface   xi
Guided Tour   xvi

## Chapter 1
### The Human Organism   1

Anatomy   2

Physiology   2

Structural and Functional Organization
   of the Human Body   2
   Chemical Level   2
   Cell Level   2
   Tissue Level   3
   Organ Level   3
   Organ System Level   3
   Organism Level   5

Characteristics of Life   6

Homeostasis   7
   Negative Feedback   7
   Positive Feedback   7

Terminology and the Body Plan   10
   Body Positions   10
   Directional Terms   10
   Body Parts and Regions   10
   Planes   12
   Body Cavities   15
   Serous Membranes   16

## Chapter 2
### The Chemical Basis
### of Life   20

Basic Chemistry   21
   Matter, Mass, and Weight   21
   Elements and Atoms   21
   Atomic Structure   22
   Electrons and Chemical Bonding   22
   Hydrogen Bonds   24
   Molecules and Compounds   25
   Dissociation   25

Chemical Reactions   26
   Classification of Chemical
      Reactions   26
   Reversible Reactions   28
   Energy and Chemical Reactions   28
   Rate of Chemical Reactions   29

Acids and Bases   30
   The pH Scale   30
   Salts   31
   Buffers   31

Inorganic Chemistry   31
   Oxygen and Carbon Dioxide   31
   Water   31

Organic Chemistry   32
   Carbohydrates   32
   Lipids   33
   Proteins   35
   Nucleic Acids: DNA and RNA   38
   Adenosine Triphosphate   38

## Chapter 3
### Cell Structures and Their
### Functions   42

Cell Structures   43

Functions of the Cell   44

Cell Membrane   44

Movement Through the Cell
   Membrane   45
   Diffusion   45
   Osmosis   47
   Carrier-Mediated Transport
      Mechanisms   49
   Endocytosis and
      Exocytosis   51

Organelles   53
   Nucleus   53
   Ribosomes   53
   Rough and Smooth Endoplasmic
      Reticulum   54
   Golgi Apparatus   54
   Secretory Vesicles   54
   Lysosomes and Peroxisomes   54
   Mitochondria   56
   Cytoskeleton   57
   Centrioles   57
   Cilia, Flagella, and
      Microvilli   57

Whole-Cell Activity   59
   Gene Expression   59
   Cell Life Cycle   61
   Differentiation   63

Systems Pathology: Cystic
   Fibrosis   66

   Apoptosis   66

Cellular Aspects of Aging   67

## Chapter 4
### Tissues   71

Epithelial Tissue   72
   Functions of Epithelia   72
   Classification of Epithelia   72
   Structural and Functional
      Relationships   76
   Glands   78

Connective Tissue   79
   Functions of Connective Tissue   80
   Classification of Connective
      Tissue   80

Muscle Tissue   84

Nervous Tissue   87

Membranes   87
   Mucous Membranes   88
   Serous Membranes   88
   Synovial Membranes   88

Tissue Damage and Inflammation   89
   Chronic Inflammation   89

Tissue Repair   90

Effects of Aging on Tissues   92

## Chapter 5
### Integumentary System   95

Functions of the Integumentary
   System   96

Skin   96
   Epidermis   96
   Dermis   97
   Skin Color   98

Hypodermis   100

Accessory Skin Structures   100
   Hair   100
   Glands   101
   Nails   101

Physiology of the Integumentary
   System   102
   Protection   102
   Sensation   102
   Vitamin D Production   102
   Temperature Regulation   102
   Excretion   104

Integumentary System as a Diagnostic Aid    104

Burns    104

Skin Cancer    105

*Systems Pathology: Burns    106*

Effects of Aging on the Integumentary System    108

## Chapter 6
### Skeletal System: Bones and Joints    111

Functions of the Skeletal System    112

Extracellular Matrix    112

General Features of Bone    113
   Structure of a Long Bone    113
   Histology of Bone    113
   Bone Ossification    115
   Bone Growth    116
   Bone Remodeling    117
   Bone Repair    119

Bone and Calcium Homeostasis    119

General Considerations of Bone Anatomy    121

Axial Skeleton    122
   Skull    122
   Vertebral Column    127
   Thoracic Cage    130

Appendicular Skeleton    131
   Pectoral Girdle    131
   Upper Limb    132
   Pelvic Girdle    135
   Lower Limb    136

Joints    140
   Fibrous Joints    140
   Cartilaginous Joints    140
   Synovial Joints    141
   Types of Movement    143

Effects of Aging on the Skeletal System and Joints    145

*Systems Pathology: Osteoporosis    146*

## Chapter 7
### Muscular System    151

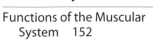

Functions of the Muscular System    152

Characteristics of Skeletal Muscle    152
   Skeletal Muscle Structure    153
   Excitability of Muscle Fibers    153
   Nerve Supply    157
   Muscle Contraction    158

Smooth Muscle and Cardiac Muscle    165

Skeletal Muscle Anatomy    166
   General Principles    166
   Nomenclature    167

Muscles of the Head and Neck    167
Trunk Muscles    172
Upper Limb Muscles    176
Lower Limb Muscles    182

Effects of Aging on Skeletal Muscle    187

*Systems Pathology: Duchenne Muscular Dystrophy    188*

## Chapter 8
### Nervous System    192

Functions of the Nervous System    193

Divisions of the Nervous System    193

Cells of the Nervous System    194
   Neurons    194
   Neuroglia    195
   Myelin Sheaths    195
   Organization of Nervous Tissue    196

Electrical Signals and Neural Pathways    198
   Resting Membrane Potential    198
   Action Potentials    199
   The Synapse    201
   Reflexes    203
   Neuronal Pathways    204

Central and Peripheral Nervous Systems    206

Spinal Cord    206
   Spinal Cord Reflexes    208

Spinal Nerves    208

Brain    211
   Brainstem    211
   Cerebellum    213
   Diencephalon    213
   Cerebrum    214

Sensory Functions    215
   Ascending Tracts    215
   Sensory Areas of the Cerebral Cortex    216

Motor Functions    216
   Motor Areas of the Cerebral Cortex    217
   Descending Tracts    217
   Basal Nuclei    218
   Cerebellum    218

Other Brain Functions    220
   Communication Between the Right and Left Hemispheres    220
   Speech    220
   Brain Waves and Consciousness    220
   Memory    221
   Limbic System    221

Meninges, Ventricles, and Cerebrospinal Fluid    222
   Meninges    222
   Ventricles    222
   Cerebrospinal Fluid    222

Cranial Nerves    224

Autonomic Nervous System    227
   Anatomy of the Sympathetic Division    227
   Anatomy of the Parasympathetic Division    227
   Autonomic Neurotransmitters    229
   Functions of the Sympathetic and Parasympathetic Divisions of the ANS    229

Enteric Nervous System    230

Effects of Aging on the Nervous System    230

*Systems Pathology: Stroke    232*

## Chapter 9
### Senses    238

General Senses    239
   Pain    240

Special Senses    241

Olfaction    241
   Neuronal Pathways for Olfaction    241

Taste    242
   Neuronal Pathways for Taste    242

Vision    242
   Accessory Structures of the Eye    243
   Anatomy of the Eye    244
   Functions of the Eye    249
   Neuronal Pathways for Vision    249

Hearing and Balance    251
   Anatomy and Function of the Ear    251
   Hearing    254
   Neuronal Pathways for Hearing    255
   Balance    255
   Neuronal Pathways for Balance    256

Effects of Aging on the Senses    258

## Chapter 10
### Endocrine System    262

Functions of the Endocrine System    263

Chemical Signals    263

Hormones    265
   Chemistry of Hormones    265

Receptors    266
   Receptor Types for Hormones    266
   Receptor Responses    266

Regulation of Hormone Secretion    268

Endocrine Glands and Their Hormones    269
   Pituitary and Hypothalamus    269
   Thyroid Gland    275
   Parathyroid Glands    276
   Adrenal Glands    278
   Pancreas, Insulin, and Diabetes    281

Testes and Ovaries   285
Thymus   285
Pineal Gland   285

*Systems Pathology: Graves Disease (Hyperthyroidism)   286*

Other Hormones   286

Effects of Aging on the Endocrine System   288

# Chapter 11
## Blood   291

Functions of Blood   292

Composition of Blood   292

Plasma   292

Formed Elements   294
Production of Formed Elements   294
Red Blood Cells   294
White Blood Cells   298
Platelets   299

Preventing Blood Loss   299
Vascular Spasm   299
Platelet Plugs   300
Blood Clotting   301

Blood Grouping   303
ABO Blood Group   303
Rh Blood Group   305

Diagnostic Blood Tests   306
Type and Crossmatch   306
Complete Blood Count   306
Differential White Blood Count   307
Clotting   307
Blood Chemistry   308

# Chapter 12
## Heart   312

Functions of the Heart   313

Size, Form, and Location of the Heart   313

Anatomy of the Heart   314
Pericardium   314
External Anatomy   314
Heart Chambers and Internal Anatomy   316
Heart Valves   318
Route of Blood Flow Through the Heart   320
Blood Supply to the Heart   321

Histology of the Heart   322
Heart Wall   322
Cardiac Muscle   322

Electrical Activity of the Heart   324
Action Potentials in Cardiac Muscle   324
Conduction System of the Heart   325
Electrocardiogram   326

Cardiac Cycle   328

Heart Sounds   331

Regulation of Heart Function   332
Intrinsic Regulation of the Heart   332
Extrinsic Regulation of the Heart   332

*Systems Pathology: Myocardial Infarction   338*

Effects of Aging on the Heart   340

# Chapter 13
## Blood Vessels and Circulation   344

Functions of the Circulatory System   345

General Features of Blood Vessel Structure   345
Arteries   345
Capillaries   347
Veins   347

Blood Vessels of the Pulmonary Circulation   348

Blood Vessels of the Systemic Circulation: Arteries   348
Aorta   348
Arteries of the Head and Neck   352
Arteries of the Upper Limbs   353
Thoracic Aorta and Its Branches   354
Abdominal Aorta and Its Branches   354
Arteries of the Pelvis   356
Arteries of the Lower Limbs   356

Blood Vessels of the Systemic Circulation: Veins   356
Veins of the Head and Neck   356
Veins of the Upper Limbs   356
Veins of the Thorax   356
Veins of the Abdomen and Pelvis   359
Veins of the Lower Limbs   360

The Physiology of Circulation   361
Blood Pressure   362
Pressure and Resistance   363
Pulse Pressure   364
Capillary Exchange   365

Control of Blood Flow in Tissues   366
Local Control of Blood Flow   366
Nervous and Hormonal Control of Blood Flow   367

Regulation of Arterial Pressure   368
Baroreceptor Reflexes   368
Chemoreceptor Reflexes   369
Hormonal Mechanisms   371
Summary of Regulatory Mechanisms   373

Effects of Aging on the Blood Vessels   373

# Chapter 14
## The Lymphatic System and Immunity   379

Functions of the Lymphatic System   380

Anatomy of the Lymphatic System   380
Lymphatic Capillaries and Vessels   380
Lymphatic Organs   380
Overview of the Lymphatic System   384

Immunity   386

Innate Immunity   386
Physical Barriers   386
Chemical Mediators   386
White Blood Cells   386
Inflammatory Response   387

Adaptive Immunity   388
Origin and Development of Lymphocytes   389
Activation and Multiplication of Lymphocytes   390
Antibody-Mediated Immunity   392
Cell-Mediated Immunity   395

Acquired Immunity   397
Active Natural Immunity   398
Active Artificial Immunity   398
Passive Natural Immunity   398
Passive Artificial Immunity   398

Overview of Immune Interactions   398

Immunotherapy   398

*Systems Pathology: Systemic Lupus Erythematosus   400*

Effects of Aging on the Lymphatic System and Immunity   400

# Chapter 15
## Respiratory System   406

Functions of the Respiratory System   407

Anatomy of the Respiratory System   407
Nose   407
Pharynx   409
Larynx   409
Trachea   410
Bronchi   412
Lungs   413
Pleural Cavities   414
Lymphatic Supply   415

Ventilation and Respiratory Volumes   416
Changing Thoracic Volume   416
Pressure Changes and Airflow   417
Lung Recoil   419
Changing Alveolar Volume   419
Respiratory Volumes and Capacities   420

Gas Exchange    421
    Respiratory Membrane Thickness    422
    Surface Area    422
    Partial Pressure    422

Gas Transport in the Blood    424
    Oxygen Transport    424
    Carbon Dioxide Transport and
        Blood pH    424

Rhythmic Breathing    424
    Respiratory Areas in the Brainstem    424
    Generation of Rhythmic
        Breathing    426
    Nervous Control of Breathing    426
    Chemical Control of Breathing    427
    Effect of Exercise on Breathing    429

Systems Pathology: Asthma    430

Respiratory Adaptations to
    Exercise    430

Effects of Aging on the Respiratory
    System    433

## Chapter 16
### Digestive System    437

Functions of the Digestive
    System    438

Anatomy and Histology of the Digestive
    System    438
    Peritoneum    439

Oral Cavity, Pharynx, and
    Esophagus    440
    Anatomy of the Oral Cavity    440
    Saliva    443
    Mastication    444
    Pharynx    444
    Esophagus    444
    Swallowing    444

Stomach    446
    Anatomy of the Stomach    446
    Secretions of the Stomach    448
    Regulation of Stomach Secretions    448
    Movement in the Stomach    450

Small Intestine    451
    Anatomy of the Small Intestine    451
    Secretions of the Small Intestine    452
    Movement in the Small Intestine    453
    Absorption in the Small Intestine    453

Liver and Pancreas    453
    Anatomy of the Liver    453
    Functions of the Liver    455
    Anatomy of the Pancreas    457
    Functions of the Pancreas    458

Large Intestine    459
    Anatomy of the Large Intestine    459
    Functions of the Large Intestine    460

Digestion, Absorption, and
    Transport    460
    Carbohydrates    461

Lipids    462
Proteins    464
Water and Minerals    465

Effects of Aging on the Digestive
    System    465

Systems Pathology: Diarrhea    466

## Chapter 17
### Nutrition, Metabolism, and Body Temperature Regulation    472

Nutrition    473
    Nutrients    473
    Kilocalories    474
    Carbohydrates    474
    Lipids    474
    Proteins    477
    Vitamins    477
    Minerals    479
    Daily Values    479

Metabolism    481
    Regulation of Metabolism    482
    Carbohydrate Metabolism    483
    Lipid Metabolism    486
    Protein Metabolism    486
    Metabolic States    486
    Metabolic Rate    487

Body Temperature Regulation    491

## Chapter 18
### Urinary System and Fluid Balance    496

Functions of the Urinary
    System    497

Anatomy of the Kidneys    497
    Arteries and Veins    502

Urine Formation    502
    Filtration    503
    Tubular Reabsorption    504
    Tubular Secretion    508

Regulation of Urine Concentration and
    Volume    509
    Hormonal Mechanisms    510
    Anatomy and Histology of the Ureters,
        Urinary Bladder, and Urethra    513

Urine Movement    514

Body Fluid Compartments    514
    Composition of the Fluid in the Body
        Fluid Compartments    516
    Exchange Between Body Fluid
        Compartments    516

Regulation of Extracellular Fluid
    Composition    516
    Thirst Regulation    516
    Ion Concentration Regulation    517

Regulation of Acid-Base Balance    518
    Buffers    518
    Respiratory System    519
    Kidneys    519
    Acidosis and Alkalosis    519

Systems Pathology: Acute Renal
    Failure    520

## Chapter 19
### Reproductive System    526

Functions of the Reproductive
    System    527

Formation of Gametes    527

Male Reproductive System    528
    Scrotum    528
    Testes    530
    Spermatogenesis    530
    Ducts    531
    Penis    532
    Glands    534
    Secretions    535

Physiology of Male Reproduction    535
    Regulation of Sex Hormone
        Secretion    535
    Puberty in Males    537
    Effects of Testosterone    537
    Male Sexual Behavior and the Male
        Sex Act    537
    Infertility in Males    539

Female Reproductive System    539
    Ovaries    540
    Oogenesis and Fertilization    540
    Uterine Tubes    543
    Uterus    543
    Vagina    543
    External Genitalia    544
    Mammary Glands    544

Physiology of Female
    Reproduction    545
    Puberty in Females    545
    Menstrual Cycle    546
    Menopause    548
    Female Sexual Behavior and the Female
        Sex Act    549
    Infertility in Females    549

Effects of Aging on the Reproductive
    System    552

Systems Pathology: Benign Uterine
    Tumors    553

## Chapter 20
### Development, Heredity, and Aging    558

Prenatal Development    559
    Fertilization    559
    Early Cell Division    559
    Blastocyst    561

Implantation of the Blastocyst and
    Development of the Placenta   561
Maternal Hormonal Changes   563
Formation of the Germ Layers   564
Neural Tube and Neural Crest
    Formation   565
Formation of the General Body
    Structure   566
Development of the Organ
    Systems   566
Growth of the Fetus   567

Parturition   572

The Newborn   574
Respiratory and Circulatory
    Changes   574
Digestive Changes   576

Lactation   576

First Year Following Birth   577

Life Stages   577
The Aging Process   577
Death   578

Genetics   579
Chromosomes   579
Genes   579
Genetic Disorders   582
Genetic Counseling   582

# Appendices

A   Table of Measurements   A-1

B   Some Reference Laboratory
    Values   A-2

C   Solution Concentrations   A-7

D   Answers to Critical Thinking
    Questions   A-8

E   Answers to Predict
    Questions   A-18

**Glossary   G-1**

**Credits   C-1**

**Index   I-1**

# Preface

*Seeley's Essentials of Anatomy and Physiology* is designed to help students develop a solid, basic understanding of essential concepts in anatomy and physiology without an encyclopedic presentation of detail. Our goal as authors is to offer a textbook that provides enough information to allow students to understand basic concepts, and from that knowledge, make reasonable predictions and analyses. We have taken great care to select critically important information and present it in a way that maximizes understanding.

We have provided explanations of how the systems respond to physical activity, disease, and aging, with a special focus on how regulatory systems control functions so that homeostasis is maintained. We have included timely and interesting examples to demonstrate the application of knowledge, often in a clinical context. For example, fundamental information is presented to allow students to understand the normal structure and function of the heart, how the heart reacts to changes in activity levels, and how the heart responds to age-related changes. It is also important for students to predict the response of the heart to blood loss or heart attack. Enough additional material is included to help students understand the effects of transfusions, treatments that reduce damage to heart muscle and methods that help the heart continue to pump blood.

We've designed the figures to support the explanations in the text. Illustrations of structures are accurate, attractive, and clear. They reflect a contemporary design and are coordinated so that colors and styles of structures in multiple figures are consistent with one another throughout the book. Many of the figures depict not only structures but also their relationships to surrounding structures. Care has been taken so that figure labels and legends are consistent with the terms used in the body of the text.

## WHAT SETS THIS BOOK APART?

### Clarity—*Just Enough Information Presented Concisely*

*Seeley's Essentials of Anatomy and Physiology* is written in succinct, understandable language. In this seventh edition in particular, we have taken measures to make the content more accessible to readers and improve the clarity of important illustrations. For students to comprehend and retain what they are reading, basic content must be presented and explained logically without confusing digressions. Relevant examples are used often to help illuminate main points, and readers can count on being able to easily pull key concepts out of sections without having to suffer through long analogies and conversational-like writing that veers them off course. Given the important material and the amount of terminology to be learned in an introductory anatomy and physiology course, this textbook offers just enough concise content, presented in a clear, logical style.

### Emphasis on Critical Thinking—*Building a Knowledge Base for Solving Problems*

An emphasis on critical thinking is integrated throughout this textbook. This approach can be found in questions embedded in the narrative; in clinical material that is designed to bridge concepts explained in the text with real-life applications and scenarios; in end-of-chapter questions that go beyond rote memorization; and in a visual program that presents material in understandable, relevant images.

### Ease of Comprehension—*Carefully Designed Pedagogy Supports Key Themes*

The pedagogical framework of *Seeley's Essentials of Anatomy and Physiology* provides a multi-pronged approach to learning. This well-structured foundation is centered on clarity, ease of comprehension, and critical thinking. Knowledge- and comprehension-level questions are balanced with questions that require more complex reasoning in both the narrative of the text and the end-of-chapter exercises. Clinical information is presented in several ways that support and restate concepts learned in the textual material.

### Exceptional Art Program—*Simplicity, Consistency, and Logic Underscore Visuals*

The text's visual program is precisely planned to enhance comprehension in a number of ways: tables combined with illustrations; relevant photos side-by-side with drawings; cadaver photos where most relevant and appropriate; numbered step-by-step Process Figures; intuitive homeostasis figures that include a "Start" icon, color-coded directional arrows, and helpful illustrations; and instructional flow diagrams.

### Clinical Emphasis—*Case Studies Bring Relevance to the Reader*

When problems in structure and/or function of the human body occur, this is often the best time to comprehend how the two are related. *Seeley's Essentials of Anatomy and Physiology* has always had a strong emphasis on clinical material. A Case in Point readings, untitled Clinical Asides, and Clinical Impact boxes all work to provide a thorough clinical education that fully supports the surrounding textual material.

# SEVENTH EDITION CHANGES— *WHAT'S NEW?*

The seventh edition of *Seeley's Essentials of Anatomy and Physiology* is the result of extensive analysis of the text and evaluation of input from anatomy and physiology instructors who have thoroughly reviewed chapters. We are grateful to these professionals and have used their constructive comments in our continuing efforts to enhance the strengths of this book. This edition brings a new team of authors along with a fresh look at what students need in the one-semester anatomy and physiology course. All of the beloved features listed previously have been retained, but the text is now tighter and flows more logically, and the art has been enhanced and customized to better fit the one-semester course, aligning perfectly with accompanying explanations.

## Improved Art—*Visualizing Relationships Between Structure and Function*

The effectiveness of the art has been improved in several ways in this edition. Many figures have been enhanced for clarity. Some of the figures have been heavily revised to help students more easily understand concepts:

- Homeostasis figures have been improved by making the explanations within boxes more succinct, reducing the number of boxes and arrows, and adding helpful illustrations. These changes contribute to a cleaner, less intimidating look to these important figures. We've kept the "Start Here" icon and color-coded arrows to make it easier to follow the events occurring when a variable increases or decreases. These simplified flow charts succinctly map out key homeostatic events, giving students a quick and effective summary of the mechanisms described in the text.
- Most of the **Process Figures have been made more clear and easier to follow.** When appropriate, additional numbered circles have been added to the corresponding explanations, so that readers can easily make connections between the text and the parts of the figure. These process figures provide well-organized, self-contained visual explanations of how physiological mechanisms work. They help students learn processes by combining detailed illustrations with parallel descriptions of the major features of each step.
- Many **new tables have been added** to the seventh edition.
- **Over forty figures have been improved** in this edition to *present more current information, achieve consistency, coordinate better with the text, or improve comprehension* (a complete listing of changes is available on the textbook's accompanying website).
- Colored **marginal tabs now visually indicate the different body systems** and help facilitate easier navigation of the textbook.

The end result of all of these art changes is a beautiful, accurate, up-to-date art program that has been scrutinized and approved by our peers and fully supports the hallmarks of clarity, ease of comprehension, and critical thinking that this text emphasizes.

## Refined and Updated Textual Copy— *Organizing Information in a Logical Sequence*

Textual changes to this edition were made always with the guiding principle of "Will it enhance the learning experience for the student?" in mind. We have kept the well-received order of topics in each chapter. Additionally, we have added new research findings, implemented suggestions from reviewers who teach anatomy and physiology, and carefully examined the existing text to improve the book in the following ways:

- **Lists of diseases** formerly found in the Clinical Focus boxes **have been transferred to succinct tables** following the appropriate boxed readings.
- The **writing style has been tightened** to present clearer explanations.
- **Terminology has been made more consistent** throughout the text and the illustrations.
- **New information** applicable to the discipline has been researched, validated, and included where appropriate.
- **Factual data have been checked and updated,** or corrected, if necessary.
- Several **explanations have been expanded** to make the topics easier for students to understand.
- **Predict Questions and Critical Thinking Questions have been updated.**

# ACKNOWLEDGMENTS

In today's world, no textbook is brought to fruition through the work of the authors alone. Without the support of friends, family, and colleagues, it would not have been possible for us to complete our work on this text. The final product is truly a team effort. We want to express sincere gratitude to the staff of McGraw-Hill for their help and encouragement. We sincerely appreciate Publisher Michelle Watnick, Developmental Editor Kathy Loewenberg, and Sponsoring Editor Jim Connely for their hours of work, suggestions, patience and undying encouragement. We also thank Project Manager Peggy Selle, Photo Editor John Leland, Production Supervisor Laura Fuller, and Designer David Hash for their hours spent turning a manuscript into a book; Media Project Manager Tammy Juran for her assistance in building the various products that support our text; and Marketing Manager Lynn Breithaupt for her enthusiasm in promoting this book. A very special thank you goes to Kennie Harris for her strict attention to detail and her dedication in helping mold us into a fine-tuned author machine. The McGraw-Hill employees with whom we have worked are extremely professional, but more than that, they are completely dedicated to their role as part of the textbook team.

Thanks are gratefully offered to Copy Editor Deb DeBord for carefully polishing our words.

We also thank the many illustrators who worked on the development and execution of the illustration program and who provided photographs and photomicrographs for the seventh edition of *Essentials of Anatomy and Physiology*. The art program for this text represents a monumental effort, and we appreciate their contribution to the overall appearance and pedagogical value of the illustrations.

Finally, we sincerely thank the reviewers and the teachers who have provided us with excellent constructive criticism. The remuneration they received represents only a token payment for their efforts. To conscientiously review a textbook requires a true commitment and dedication to excellence in teaching. Their helpful criticisms and suggestions for improvement were significant contributions that we greatly appreciate. We acknowledge them by name in the next section.

*Cinnamon VanPutte*

*Jennifer Regan*

*Andy Russo*

## REVIEWERS

Deborah Allen
*Jefferson College*

Jerry M. Allen
*Somerset Community College*

Lynn Atkins
*University of Maine*

Nahel W. Awadallah
*Sampson Community College*

Mahmoud Bishr
*Penn Valley Community College*

A. Braganza
*Chabot College*

Jocelyn Cash
*Central Piedmont Community College*

Lu Anne Clark
*Lansing Community College*

David S. Clayton
*Richmond Community College*

Sonya Conway
*Northern Illinois University*

Janie Corbitt
*Central Georgia Technical College*

Paula deDiego
*Mount Wachusett Community College*

Bridget A. Falkenstein
*Sierra College*

Bradley J. Fillmore
*Chadron State College*

Amy D. Goode
*Illinois Central College*

Eric S. Green
*Salt Lake Community College*

Barbara R. Heard
*Atlantic Cape Community College*

Murray S. Jensen
*University of Minnesota*

Thomas Jordan
*Pima Community College*

Susanne J. Kalup
*Westmoreland County Community College*

Kamal Kamal
*Valencia Community College–West*

Jean M. Krueger-Watson
*Clark College*

Claire M. Leonard
*William Paterson University*

Jennifer M. Lucas
*Montgomery Community College*

M. J. Malachowski
*City College of San Francisco*

Angela K. Mick
*Glendale Community College*

Robert K. Moats, II
*Ohio University–Chillicothe*

Payman Nasr
*Kent State University*

David Pounds
*The North Carolina School of the Arts*

Melissa A. Reed
*Pitt Community College*

Julie C. Rutherford
*Concordia College*

Leba W. Sarkis
*Aims Community College*

Gerald Schafer
*Mission College, Thailand*

Marialana Speidel
*Jefferson College*

Marty D. Spranger
*Wayne State University*

Mary Elizabeth Torrano
*American River College*

Jorge Vasquez-Kool
*Johnston Community College*

Nicholas J. D. Wright
*Eastern New Mexico University*

Steven A. Woodley
*Clark Community College*

Mark D. Young
*West Kentucky Community and Technical College*

Paul L. Zillgitt
*University of Wisconsin–Waukesha*

# TEACHING AND LEARNING SUPPLEMENTS

McGraw-Hill offers various tools and technology products to support the seventh edition of *Seeley's Essentials of Anatomy and Physiology.* Students can order supplemental study materials by contacting their campus bookstore. Instructors can obtain teaching aids by calling the McGraw-Hill Customer Service Department at 1–800–338–3987, visiting our A&P catalog at **www.mhhe.com/ap**, or contacting their local McGraw-Hill sales representative.

## Anatomy & Physiology | REVEALED® 2.0 Student Tutorial

Anatomy & Physiology | REVEALED® 2.0 is a unique multimedia study aid designed to help students learn and review human anatomy using digital cadaver specimens. Dissections, animations, imaging, and self-tests all work together as an exceptional tool for the study of structure and function. Anatomy & Physiology | REVEALED® 2.0 includes

- Integumentary system
- Skeletal and muscular systems
- Nervous system
- Cardiovascular, respiratory, and, lymphatic, systems
- Digestive, urinary, reproductive, and endocrine systems
- Expanded physiology content
- Histology material

A new online version of Anatomy & Physiology | REVEALED® is now available! Visit **www.mhhe.com/aprevealed** for more information.

## Correlated Website

The website that accompanies *Seeley's Essentials of Anatomy and Physiology* at **www.mhhe.com/seeleyess7** allows instructors to browse, select, and export files containing artwork from the text in multiple formats to create customized classroom presentations, visually based tests and quizzes, dynamic course website content, or printed support materials. The digital assets on the website are all available for teaching presentations:

- **Art**   Full-color digital files of all the illustrations in the book and unlabeled versions of the same artwork can be readily incorporated into lecture presentations, exams, or custom-made classroom materials. In addition, all files are pre-inserted into blank PowerPoint slides for easy lecture presentations.
- **Photos**   Digital files of instructionally significant photographs from the text can be reproduced for multiple classroom uses.
- **Tables**   Every table that appears in the text is available to instructors in electronic form.

- **Animations**   Numerous full-color animations illustrating physiological processes are provided. Harness the visual impact of processes in motion by importing these files into classroom presentations or online course materials.

Students will benefit from **online tutoring** (a 24-hour tutorial service moderated by qualified instructors), **practice quizzing, animation quizzing,** and other **study tools,** all correlated by chapter. Help with difficult concepts is only a click away!

## Online Computerized Test Bank

A comprehensive bank of test questions is provided within a computerized test bank powered by McGraw-Hill's flexible electronic testing program EZ Test Online. EZ Test Online allows instructors to create and access paper or online tests or quizzes in an easy-to-use program anywhere, at any time without installing the testing software. Now, with EZ Test Online, instructors can select questions from multiple McGraw-Hill test banks, or author their own, and then either print the test for paper distribution or give the test online. Visit **www.eztestonline.com** to learn more about creating and managing tests, online scoring and reporting, and support resources.

## Laboratory Manual

The *Laboratory Manual* to accompany *Seeley's Essentials of Anatomy and Physiology,* authored by Kevin Patton of St. Charles Community College and St. Louis University School of Medicine, divides the material typically covered in anatomy and physiology labs into forty-three subunits. Because of these subunits, instructors may design a laboratory course that is integrated with the emphasis and flow of their course lecture material. Basic content is introduced first, and gradually more complex activities are presented. This laboratory manual also contains coloring and labeling exercises, optional computer exercises, boxed hints, safety alerts, separate lab reports, and a full-color histology mini-reference. Access to an online comprehensive instructor's manual containing answers to the lab reports is provided to instructors.

## Electronic Books

If you, or your students, are ready for an alternative version of the traditional textbook, McGraw-Hill offers innovative and inexpensive electronic textbooks. By purchasing E-books from McGraw-Hill, students can save as much as 50% on selected titles delivered on an easy-to-use, advanced E-book platform.

E-books from McGraw-Hill are *smart, interactive, searchable, and portable.* Each title comes with a powerful suite of built-in tools that allow detailed searching, highlighting, note taking, and student-to-student or instructor-to-student note sharing. In addition, media-rich E-books incorporate relevant animations and videos into the textbook content for a true multimedia learning experience.

E-books will help students study smarter and quickly find the information they need while saving money. Instructors contact your McGraw-Hill sales representative to discuss E-book packaging options.

## Physiology Interactive Lab Simulations (Ph.I.L.S.) 3.0

This unique student study tool is the perfect way to reinforce key physiology concepts with powerful lab experiments. Created by Dr. Phil Stephens of Villanova University, the program offers thirty-seven laboratory simulations that may be used to supplement or substitute for wet labs. Students can adjust variables, view outcomes, make predictions, draw conclusions, and print lab reports. The easy-to-use software offers the flexibility to change the parameters of the lab experiment—there is no limit to the number of times an experiment can be repeated.

## "MediaPhys" Tutorial

This physiology study aid offers detailed explanations, high-quality illustrations, and amazing animations to provide a thorough introduction to the world of physiology. MediaPhys is filled with interactive activities and quizzes to help reinforce physiology concepts that are often difficult for students to understand.

## e-Instruction

The Classroom Performance System (CPS) is an interactive system that allows instructors to administer in-class questions electronically. Students answer questions via hand-held remote control keypads, and their individual responses are logged into a grade book. Aggregated responses can be displayed in graphical form. Using this immediate feedback, instructors can quickly determine if students understand the lecture topic or if more clarification is needed. CPS promotes student participation, class productivity, and individual student accountability.

## Course Delivery Systems

With help from our partners WebCT, BlackBoard, Angel, Top-Class, eCollege, and other course management systems, professors can take complete control over their course content. Course cartridges containing website content, online testing, and powerful student tracking features are readily available for use within these platforms.

# Guided Tour

## A SOUND LEARNING SYSTEM

*Seeley's Essentials of Anatomy & Physiology* is designed to help students learn in a systematic fashion. Simple facts are the building blocks for developing explanations of more complex concepts. The clear and logical text discussion is presented within a supporting framework of learning aids that help organize studying, reinforce learning, and promote problem-solving skills.

"The Seeley Essentials textbook has a logical flow of information—topics build well on each other. Students learn the introductory information first, and then get the details and see how the current information ties into previously presented material. It does a great job of introducing, explaining details, summarizing and "linking" information, and building a great picture of human anatomy and physiology." —Bradley J. Fillmore, *Chadron State College*

"The writing style is unusually 'reader friendly.'"—Claire M. Leonard, *William Paterson University*

The pizza this boy is eating will activate all aspects of his digestive system. Pizza contains most of the major nutrients: carbohydrates, proteins, and fat. The body's ability to break down, separate, and appropriately distribute these nutrients is amazing.

# 16 Chapter

## Digestive System

## Chapter Outline and Objectives

*An outline that combines chapter topics with their associated learning outcomes is presented at the beginning of each chapter to introduce key concepts students will encounter in their reading.*

### Chapter Outline and Objectives

**Functions of the Digestive System (p. 438)**
1. List the major functions of the digestive system.

**Anatomy and Histology of the Digestive System (p. 438)**
2. Describe the general histology of the digestive tract.

**Oral Cavity, Pharynx, and Esophagus (p. 440)**
3. Describe the structure of a tooth.
4. Describe the major salivary glands. Compare their structures and functions.
5. Describe mastication and swallowing.

**Stomach (p. 446)**
6. Outline the anatomical and physiological characteristics of the stomach.
7. Describe the stomach secretions, their functions, and their regulation.
8. Describe gastric movements and stomach emptying and how they are regulated.

**Small Intestine (p. 451)**
9. List the characteristics of the small intestine that account for its large surface area.

10. Describe the secretions and movements that occur in the small intestine.

**Liver and Pancreas (p. 453)**
11. Describe the anatomy, histology, and ducts of the liver and pancreas.
12. Describe the major functions of the liver and pancreas, and explain how they are regulated.

**Large Intestine (p. 459)**
13. List the parts of the large intestine, and describe its anatomy and histology.
14. Describe the major functions of the large intestine, and explain how movement is regulated.

**Digestion, Absorption, and Transport (p. 460)**
15. Describe the digestion, absorption, and transport of carbohydrates, lipids, and proteins.
16. Discuss water movement into and out of the digestive tract.

**Effects of Aging on the Digestive System (p. 465)**
17. Describe the effects of aging on the digestive system.

437

## Predict Questions

*These innovative critical thinking exercises encourage students to become active learners as they read. Predict Questions challenge readers to use their understanding of new concepts to solve a problem. Answers to Predict Questions are given at the end of the text, allowing students to evaluate their responses and discover the logic used in arriving at the correct answer.*

**Figure 10.11  Thyroid and Parathyroid Glands**

(*a*) Anterior view of the thyroid gland. (*b*) The four small parathyroid glands are embedded in the posterior surface of the thyroid gland. (*c*) Three-dimensional interpretive drawing of thyroid follicles and parafollicular cells. (*d*) Light micrograph of thyroid and parathyroid tissue.

This causes the thyroid gland to enlarge, a condition called a **goiter** (goy'ter). One type of goiter develops if iodine in the diet is too low. As less $T_3$ and $T_4$ is synthesized and secreted, TSH-releasing hormone and TSH secretion increase above normal levels and cause dramatic enlargement of the thyroid gland.

**Predict 3**

*In people with Graves disease (hyperthyroidism), the immune system produces a large amount of a protein that is so much like TSH that it binds to cells of the thyroid gland and acts like TSH. Unlike TSH, however, the secretion of this protein does not respond to negative feedback. Predict the effect of this abnormal protein on the structure and function of the thyroid gland and the release of hormones from the hypothalamus and anterior pituitary gland.*

In addition to secreting thyroid hormones, the thyroid gland secretes a hormone called **calcitonin** (kal-si-tō'nin) from the C cells (see figure 10.11*c*). Calcitonin is secreted if the blood concentration of $Ca^{2+}$ becomes too high, and it causes $Ca^{2+}$ levels to decrease to their normal range (figure 10.13). Calcitonin binds to membrane-bound receptors of osteoclasts and reduces the rate of $Ca^{2+}$ resorption (breakdown) from bone by inhibiting the osteoclasts. Calcitonin may prevent blood $Ca^{2+}$ levels from becoming overly elevated following a meal that contains a high concentration of $Ca^{2+}$.

Calcitonin helps prevent elevated blood $Ca^{2+}$ levels, but a lack of calcitonin secretion does not result in a prolonged increase in those levels. Other mechanisms controlling blood $Ca^{2+}$ levels compensate for the lack of calcitonin secretion.

### Parathyroid Glands

Four tiny **parathyroid** (par-ă-thī'royd) **glands** are embedded in the posterior wall of the thyroid gland (see figure 10.11*b,d*). The

produced in one part of a cell, such as the cell membrane, and then travel to another part of the same cell, where they bind to receptors in either the cytoplasm or the nucleus. **Intercellular chemical signals** are released from one cell and carried in the intercellular fluid; they bind to receptors that are found in some cells, but usually not in all cells, of the body. Intercellular chemical signals can be categorized into six types, depending on the source of the chemical signals or their targets (table 10.1).

**Autocrine** (aw'tō-krin; *autos,* self) **chemical signals** are released by cells and have a local effect on the same cell type from which they are released. Examples include chemicals such as the eicosanoids (*eicosa,* refers to 20-carbon atoms) that are released from smooth muscle cells and from platelets in response to inflammation. These chemicals cause the relaxation of blood vessel smooth muscle cells and the aggregation of platelets. As a result, the blood vessels dilate, and blood clots.

**Paracrine** (par'ă-krin; *para,* alongside) **chemical signals** are released by cells that produce effects on other cell types near the cells from which they are released, without being transported in blood. For example, cells in the pancreas release a peptide called somatostatin, which functions locally to inhibit the secretion of insulin from other cells of the pancreas.

**Hormones** (hōr'mōnz) and **neurohormones** are intercellular chemical signals secreted into the circulatory system (see "Hormones" later in this chapter). They are carried in the blood to the organs they control, where they bind to receptor molecules and produce a response.

**Neurotransmitters** and **neuromodulators** are intercellular chemical signals, secreted by nerve cells, that play important roles in nervous system function (see chapter 8).

## Vocabulary Aids

*Learning anatomy and physiology, is, in many ways, like learning a new language. Mastering the terminology is vital to building a knowledge base. Key terms are set in boldface where they are defined in the chapter (with acceptable alternate terms in italics), and most terms are included in the glossary at the end of the book. Pronunciation guides are included for difficult words.*

*Because knowing the origin of a term can enhance understanding and retention, the meanings of the words or root words that form a term are given when they are relevant. Furthermore, a handy list of prefixes, suffixes, and combining forms is printed on the inside of the back cover as a quick reference to help students identify commonly used root words.*

# Guided Tour

## INSTRUCTIVE ARTWORK MAKES THE DIFFERENCE

A picture is worth a thousand words—especially when learning anatomy and physiology. Because words alone cannot convey the nuances of anatomy or the intricacies of physiology, the seventh edition of *Seeley's Essentials of Anatomy and Physiology* employs a dynamic program of full-color illustrations and photographs that support and further clarify the text explanations. Brilliantly rendered and carefully reviewed for accuracy and consistency, the precisely labeled illustrations and photos provide concrete, visual reinforcement of the topics discussed throughout the text.

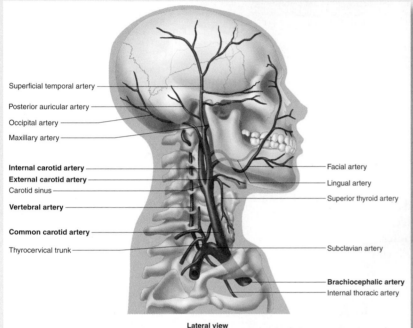

**Lateral view**

## Realistic Anatomical Art

*The anatomical figures in* Seeley's Essentials of Anatomy and Physiology *have been carefully rendered to convey realistic, three-dimensional detail. Richly textured bones and artfully shaded muscles and vessels lend a sense of realism to the figures which helps readers envision the appearance of actual structures within the body.*

*The colors used to represent various anatomical structures have been applied consistently throughout the book. This reliable color pattern promotes continuity between figures and allows students to identify the structures in each illustration.*

*Reference diagrams orient students to the perspective of important or complex illustrations.*

## Multi-level Perspective

*Illustrations depicting complex structures or processes combine macroscopic and microscopic views to help students see the relationships between increasingly detailed drawings.*

"As soon as I received a copy of this book, I requested to change the text that I was using to Seeley for the following academic year. This book discusses all topics in an interesting style, and addresses all learning methods. The feedback from my students can be summarized in one phrase 'easy on the eyes.'"—Nahel W. Awadallah, *Sampson Community College*

# Guided Tour

## Combination Art

*Drawings are often paired with photographs to enhance the visualization of structures.*

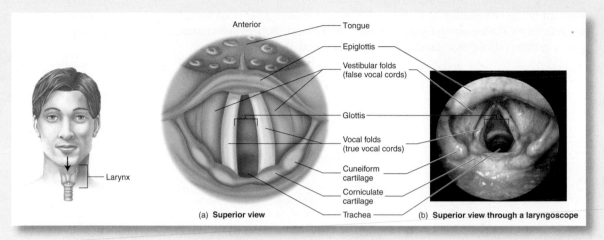

- Anterior
- Tongue
- Epiglottis
- Vestibular folds (false vocal cords)
- Glottis
- Vocal folds (true vocal cords)
- Cuneiform cartilage
- Corniculate cartilage
- Larynx
- Trachea

(a) **Superior view**  (b) **Superior view through a laryngoscope**

## Histology Micrographs

*Light micrographs, as well as scanning and transmission electron micrographs, are used in conjunction with illustrations to present a true picture of anatomy and physiology from the cellular level.*

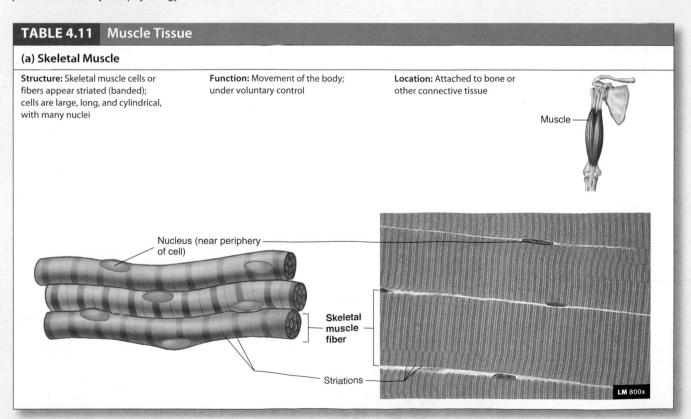

| TABLE 4.11 | Muscle Tissue |
| --- | --- |

### (a) Skeletal Muscle

**Structure:** Skeletal muscle cells or fibers appear striated (banded); cells are large, long, and cylindrical, with many nuclei

**Function:** Movement of the body; under voluntary control

**Location:** Attached to bone or other connective tissue

- Muscle
- Nucleus (near periphery of cell)
- Skeletal muscle fiber
- Striations

LM 800x

# Guided Tour

## SPECIALIZED FIGURES CLARIFY TOUGH CONCEPTS

Studying anatomy and physiology does not have to be an intimidating task mired in memorization. *Seeley's Essentials of Anatomy and Physiology* uses two special types of illustrations to help students not only learn the steps involved in specific processes, but also apply the knowledge as they predict outcomes in similar situations. Process Figures organize the key occurrences of physiological processes in an easy-to-follow format. Homeostasis figures summarize the mechanisms of homeostasis by diagramming how a given system regulates a parameter within a narrow range of values.

## Process Figures

*Process Figures break down physiological processes into a series of smaller steps, allowing readers to track the key occurrences and learn them as they work through the illustration.*

*Indicators within the artwork correspond to the numbered steps along the side. These colored circles help students zero in on the site where the action described in each step takes place.*

*Process Figures and Homeostasis Figures are identified next to the figure number. The accompanying caption provides additional explanation.*

**1 Resting membrane potential.** Na⁺ channels (*pink*) and some, but not all, K⁺ channels (*purple*) are closed. A few K⁺ diffuse down their concentration gradient through the open K⁺ channels, making the inside of the cell membrane negatively charged compared to the outside.

**2 Depolarization.** Na⁺ channels are open. A few Na⁺ diffuse down their concentration gradient through the open Na⁺ channels, making inside of the cell membrane positively charged compared to the outside.

**3 Repolarization.** Na⁺ channels are closed, and Na⁺ movement into the cells stops. More K⁺ channels open. K⁺ movement out of the cell increases, making the inside of the cell membrane negatively charged compared to the outside, once again.

**PROCESS Figure 7.4  Ion Channels and the Action Potential**

Step 1 illustrates the status of Na⁺ and K⁺ channels in a resting cell. Steps 2 and 3 show how the channels open and close to produce an action potential. Next to each step, the charge difference across the plasma membrane is illustrated.

# Guided Tour

## Homeostasis Figures

These specialized flow charts diagram the mechanisms that body systems use to maintain homeostasis. This edition features a new design for these important figures.

Changes caused by an increase of a variable toward the outside of its normal range are shown in the boxes outlined in green.

Pancreatic islets detect an increase in blood glucose and secrete insulin.

Insulin stimulates glucose uptake by most tissues and promotes glycogen storage in skeletal muscle and liver. Excess glucose is stored as fat.

Blood glucose level increases: **Homeostasis Disturbed**

Blood glucose level decreases: **Homeostasis Restored**

The normal range for a given value is represented by the graphs in the center of each figure.

Start at the yellow icon and follow the green arrows to learn about the chain of events triggered by an **increase** in the variable, or the red arrows for events resulting from a **decrease** in the variable.

Blood glucose (normal range)

Start here

Blood glucose (normal range)

Blood glucose level decreases: **Homeostasis Disturbed**

Blood glucose level increases: **Homeostasis Restored**

Small accompanying illustrations remind students of the organs or structures involved in the process.

Pancreatic islets detect a decrease in blood glucose and do not secrete insulin.

Decreased insulin results in decreased glucose uptake, increased glycogen breakdown by the liver and skeletal muscle, and increased glucose synthesis.

Changes caused by a decrease of a variable toward the outside of its normal range are shown in boxes outlined in red.

"The homeostasis figures in Chapters 12 and 13 are fantastic. I always emphasize in my classes that homeostasis is the reason our bodies function the way they do. The advantages of having figures like this is that there is a clear start/stop point and the figures clearly represent the mechanisms responsible for maintaining the feedback necessary for regulating homeostasis."—Melissa Reed, *Pitt Community College*

# Guided Tour

## CLINICAL CONTENT PUTS KNOWLEDGE INTO PRACTICE

*Seeley's Essentials of Anatomy and Physiology* provides clinical examples to promote interest and demonstrate relevance, but clinical information is used primarily to illustrate the application of basic knowledge. Exposure to clinical case studies is especially beneficial if students are planning on using their knowledge of anatomy and physiology in a health-related career.

### A CASE IN POINT
#### Venous Thrombosis

Harry Leggs is a 55-year-old college professor who teaches a night class in a small town about 50 miles from his home. One evening, as he walked to his car after teaching the class, Harry became aware of a slight pain in his right leg. By the time he reached home about 90 minutes later, the calf of his right leg was swollen. When he extended his knee and plantar flexed his foot, the pain increased. Harry knew this was a serious condition, so he drove to the emergency room. There, technicians performed a Doppler test to monitor the flow of blood through the blood vessels in Harry's right leg. The test confirmed that a thrombus had formed in one of the deep veins of the leg.

Harry's symptoms stemmed from several events. Small proteins released from fibrinogen during clot formation caused the activation of kinins that stimulate pain receptors (see chapter 14). Also, the blocked vein caused edema and led to tissue ischemia, a decrease in blood flow to tissues, which may resulted in tissue damage and pain. Edema occurred because of the increase in venous pressure inferior to the blocked vein (see figure 13.24).

A major danger associated with a deep venous thrombosis is that portions of the thrombus may form emboli. The emboli are likely to be transported in the blood to the lungs, where they form pulmonary emboli. Pulmonary emboli are life-threatening because they may be large or numerous enough to block blood flow to substantial areas of the lungs.

Harry was admitted to the hospital for a few days. His movements were restricted, and he was given an anticoagulant. Even after his release, he continued taking the anticoagulant for a few weeks until the clot had dissolved.

### A Case in Point

*These brief case studies explore relevant issues of clinical interest and explain how material just presented in the text can be used to understand important anatomical and physiological concepts, particularly in a clinical setting.*

"Overall, the text chapters are clear, easy to read, and very concise. They give the students exactly what they need to know. The case studies are great. I encourage my students to read and apply them to help understand the content of the chapter."
—Mahmoud Bishr, *Penn Valley College*

## CLINICAL IMPACT

### Blood Vessels Used for Coronary Bypass Surgery

The great saphenous vein is often surgically removed and used in coronary bypass surgery. Portions of the saphenous vein are grafted to create a route for blood flow that bypasses blocked portions of the coronary arteries. The circulation interrupted by the removal of the saphenous vein flows through other veins of the lower limb instead. The internal thoracic artery is also used for coronary bypasses. The distal end of the artery is freed and attached to a coronary artery at a point that bypasses the blocked portion. This technique appears to be better because the internal thoracic artery does not become blocked as fast as the saphenous vein. However, due to the length of the saphenous vein and the fact that more than one segment can be removed for use in bypass surgery, surgeons use the saphenous vein.

### Clinical Impact

*The topics of these boxed essays include pathologies, current research, sports medicine, exercise physiology, pharmacology, and various clinical applications.*

# Guided Tour

## Systems Pathology

*These in-depth boxes explore a specific disorder or condition related to a particular body system. Presented in a modified case study format, each Systems Pathology box begins with a patient history followed by background information about the featured topic.*

### SYSTEMS PATHOLOGY

#### Duchenne Muscular Dystrophy

A couple became concerned about their 3-year-old son when they noticed that he was much weaker than other boys his age. The differences appeared to become more obvious as time passed. Eventually, the boy had difficulty sitting, standing, and walking. He seemed clumsy and fell often. He had difficulty climbing stairs, and he often rose from sitting on the floor to a standing position by using his hands and arms to climb up his legs. His muscles appeared poorly developed. The couple took their son to a physician, and after several tests, they were informed that he had Duchenne muscular dystrophy.

#### Background Information

**Duchenne muscular dystrophy (DMD)** is usually identified in children around 3 years of age when their parents notice slow motor development with progressive weakness and muscle wasting (atrophy). Typically, muscular weakness begins in the pelvic girdle, causing a waddling gait. Rising from the floor by climbing up the legs is characteristic and is caused by weakness of the lumbar and gluteal muscles (figure 7A). Within 3 to 5 years, the muscles of the shoulder girdle become involved. The replacement of muscle with connective tissue contributes to muscular atrophy and shortened, inflexible muscles called contractures. The contractures limit movements and can cause severe deformities of the skeleton. By 10 to 12 years of age, people with DMD are usually unable to walk, and few live beyond age 20. There is no effective treatment to prevent the progressive deterioration of muscles in DMD. Therapy primarily involves exercises to help strengthen muscles and prevent contractures.

**Figure 7A**
DMD is characterized by progressive muscle weakness. Rising from the floor to a standing position becomes difficult, and the patient often has to use his or her arms to push against the thighs while rising.

Duchenne muscular dystrophy results from an abnormal gene on the X chromosome and is therefore a sex-linked (X-linked) condition. The gene is carried by females, but DMD affects males almost exclusively, at a frequency of 1 in 3000. The DMD gene is responsible for producing a protein called **dystrophin**, which plays a role in attaching myofibrils to other proteins in the cell membrane and regulating their activity. In a normal individual, dystrophin is thought to protect muscle cells against mechanical stress. In DMD patients, part of the [gene is] missing, and the protein it produces is nonfunctional, resulting [in pro-]gressive muscular weakness and muscle contractures.

*A System Interactions table at the end of every box summarizes how the condition affects each body system.*

| System Interactions | Effects of Duchenne Muscular Dystrophy on Other Systems |
|---|---|
| **System** | **Interactions** |
| Skeletal | Shortened, inflexible muscles (contractures) cause severe deformities of the skeletal system. Kyphoscoliosis, severe curvature of the spinal column laterally and anteriorly, can be so severe that normal respiratory movements are impaired. Surgery is sometimes required to prevent contractures from making it impossible for the individual to sit in a wheelchair. |
| Nervous | Some degree of mental retardation occurs in a large percentage of people with DMD, although the specific cause is unknown. Research suggests that dystrophin is important in the formation of synapses between nerve cells. |
| Cardiovascular | DMD affects cardiac muscle; consequently, heart failure occurs in a large number of people with advanced DMD. Cardiac involvement becomes serious in as many as 95% of cases and is one of the leading causes of death for individuals with DMD. |
| Lymphatic and Immune | No obvious direct effects occur to the lymphatic system, but damaged muscle fibers are phagocytized by macrophages. |
| Respiratory | Deformity of the thorax and increasing weakness of the respiratory muscles results in inadequate respiratory movements, which cause an increase in respiratory infections, such as pneumonia. Insufficient movement of air into and out of the lungs due to weak respiratory muscles is a major contributing factor in many deaths. |
| Digestive | DMD affects smooth muscle tissue, and the reduced ability of smooth muscle to contract can cause disorders of the digestive system. Digestive system disorders include an enlarged colon diameter, and twisting of the small intestine, resulting in intestinal obstruction, cramping, and reduced absorption of nutrients. |
| Urinary | Reduced smooth muscle function and wheelchair dependency increase the frequency of urinary tract infections. |

**Predict 6**

*A boy with advanced Duchenne muscular dystrophy developed pulmonary edema (accumulation of fluid in the lungs) and pneumonia caused by a bacterial infection. His physician diagnosed the condition in the following way: The pulmonary edema was the result of heart failure, and the increased fluid in the lungs provided a site where bacteria could invade and grow. The fact that the boy could not breathe deeply or cough effectively made the condition worse. How would the muscle tissues in a boy with advanced DMD with heart failure and ineffective respiratory movements differ from the muscle tissues in a boy with less advanced DMD?*

*Every Systems Pathology box includes a Predict Question specific to the case study.*

### DISEASES AND DISORDERS  Muscular System

| CONDITION | DESCRIPTION |
|---|---|
| Cramps | Painful, spastic contractions of a muscle; usually due to a buildup of lactic acid |
| Fibromyalgia (fī-brō-mī-al'ja) | Non-life-threatening, chronic, widespread pain in muscles with no known cure; also known as chronic muscle pain syndrome |
| Hypertrophy | Enlargement of a muscle due to an increased number of myofibrils, as occurs with increased muscle use |
| Atrophy | Decrease in muscle size due to a decreased number of myofilaments; can occur due to disuse of a muscle, as in paralysis |
| Muscular dystrophy | Group of genetic disorders in which all types of muscle degenerate and atrophy |
|    Duchenne muscular dystrophy | See Systems Pathology |
|    Myotonic muscular dystrophy | Muscles are weak and fail to relax following forceful contractions; affects the hands most severely; dominant trait in 1/20,000 births |
| Myasthenia gravis | See A Case in Point, earlier in this chapter |
| Tendinitis (ten-di-ni'tis) | Inflammation of a tendon or its attachment point due to overuse of the muscle |
| Go to **www.mhhe.com/seeleyess7** for additional information on these pathologies. | |

## New! Diseases and Disorders Tables

*Examples of diseases and/or disorders related to a particular system are collected in one easy-to-read table for students, helping them appreciate the clinical possibilities related to that body system.*

# Guided Tour

## STUDY FEATURES ENSURE SUCCESS

A carefully devised set of learning aids at the end of each chapter assists students in reviewing the chapter content, evaluating their grasp of key concepts, and utilizing what they've learned.

---

### ⎍⎍ SUMMARY

Chemistry is the study of the composition and structure of substances and the reactions they undergo.

**Basic Chemistry**  (p. 21)

*Matter, Mass, and Weight*
1. Matter is anything that occupies space and has mass.
2. Mass is the amount of matter in an object, and weight results from the gravitational attraction between the earth and an object.

*Elements and Atoms*
1. An element is the simplest type of matter having unique chemical and physical properties.
2. An atom is the smallest particle of an element that has the chemical characteristics of that element. An element is composed of only one kind of atom.

*Atomic Structure*
1. Atoms consist of neutrons, positively charged protons, and negatively charged electrons.
2. An atom is electrically neutral because the number of protons equals the number of electrons.
3. Protons and neutrons are in the nucleus, and electrons can be represented by an electron cloud around the nucleus.
4. The atomic number is the unique number of protons in each atom of an element. The mass number is the number of protons and neutrons.

*Electrons and Chemical Bonding*
1. An ionic bond results when an electron is transferred from one atom to another.
2. A covalent bond results when a pair of electrons is shared between atoms. A polar covalent bond is an unequal sharing of electron pairs.

*Hydrogen Bonds*
A hydrogen bond is the weak attraction between the oppositely charged

*Energy and Chemical Reactions*
1. Energy is the capacity to do work. Potential energy is stored energy that could do work, and kinetic energy does work by causing the movement of an object.
2. Energy exists in chemical bonds as potential energy.
3. Energy is released in chemical reactions when the products contain less potential energy than the reactants. The energy can be lost as heat, used to synthesize molecules, or used to do work.
4. Energy must be added in reactions when the products contain more potential energy than the reactants.
5. Energy can be neither created nor destroyed, but one type of energy can be changed into another.

*Rate of Chemical Reactions*
1. The rate of a chemical reaction increases when the concentra reactants increases, the temperature increases, or a catalyst is
2. A catalyst (enzyme) increases the rate of a chemical reactic without being altered permanently.

**Acids and Bases**  (p. 30)

Acids are proton (hydrogen ion) donors, and bases are proton a

*The pH Scale*
1. A neutral solution has an equal number of $H^+$ and $OH^-$ a of 7.0.
2. An acidic solution has more $H^+$ than $OH^-$ and a pH less t
3. A basic solution has fewer $H^+$ than $OH^-$ and a pH greater

*Salts*
A salt forms when an acid reacts with a base.

*Buffers*
Buffers are chemicals that resist changes in pH when acids or b added.

2. Lipids provide energy (fats), serve as structural components (phospholipids), and regulate physiological processes (steroids).
3. The building blocks of triglycerides (fats) are glycerol and fatty acids.
4. Fatty acids can be saturated (have only single covalent bonds between carbon atoms) or unsaturated (have one or more double covalent bonds between carbon atoms).

*Proteins*
1. Proteins regulate chemical reactions (enzymes), serve as structural components, and cause muscle contraction.
2. The building blocks of proteins are amino acids.
3. Denaturation of proteins disrupts hydrogen bonds, which changes the shape of proteins and makes them nonfunctional.
4. Enzymes are specific, bind to reactants according to the lock-and-key model, and function by lowering activation energy.

*Nucleic Acids: DNA and RNA*
1. The basic unit of nucleic acids is the nucleotide, which is a monosaccharide with an attached phosphate and organic base.
2. DNA nucleotides contain the monosaccharide deoxyribose and the organic bases adenine, thymine, guanine, and cytosine. DNA occurs as a double strand of joined nucleotides and is the genetic material of cells.
3. RNA nucleotides are composed of the monosaccharide ribose. The organic bases are the same as for DNA, except that thymine is replaced with uracil.

*Adenosine Triphosphate*
ATP stores energy, which can be used in cell processes.

### ⎍⎍ REVIEW AND COMPREHENSION

1. Define chemistry. Why is an understanding of chemistry important to the study of human anatomy and physiology?
2. Define matter. What is the difference between mass and weight?
3. Define element and atom. How many different kinds of atoms are present in a specific element?
4. List the components of an atom, and explain how they are organized to form an atom. Compare the charges of the subatomic particles.
5. Define the atomic number and the mass number of an element.
6. Distinguish among ionic, covalent, polar covalent, and hydrogen bonds. Define ion.
7. What is the difference between a molecule and a compound?
8. What happens to ionic and covalent compounds when they dissolve in water?
9. Define chemical reaction. Describe synthesis, decomposition, and exchange reactions, giving an example of each.
10. What is meant by the equilibrium condition in a reversible reaction?
11. Define potential energy and kinetic energy.
12. Give an example of a chemical reaction that releases energy and an example of a chemical reaction that requires the input of energy.
13. Name three ways that the rate of chemical reactions can be increased.
14. What is an acid and what is a base? Describe the pH scale.
15. Define a salt. What is a buffer, and why are buffers important?
16. Distinguish between inorganic and organic chemistry.
17. Why is oxygen necessary for human life? Where does the carbon dioxide we breathe out come from?
18. List four functions that water performs in the human body.
19. Name the four major types of organic molecules. Give a function for each.
20. Describe the action of enzymes in terms of activation energy and the lock-and-key model.

---

## Chapter Summary

*The summary outline briefly states the important facts and concepts covered in each chapter to provide a convenient "big picture" of the chapter content.*

## Review and Comprehension

*These practice questions cover the main points presented in the chapter. Completing these exercises helps students gauge their mastery of the material.*

## Answers to Predict Questions

*The Predict Questions that appear throughout the text are answered in Appendix E, allowing students to evaluate their responses and understand the logic used to arrive at the correct answer.*

---

### ⎍⎍ CRITICAL THINKING

1. If an atom of iodine (I) gains an electron, what is the charge of the resulting ion? Write the symbol for this ion.
2. For each of the following chemical equations, determine if a synthesis reaction, a decomposition reaction, or dissociation has taken place:
   a. $HCl \rightarrow H^+ + Cl^-$
   b. Glucose + Fructose → Sucrose (table sugar)
   c. $2 H_2O \rightarrow 2 H_2 + O_2$
3. In terms of the energy in chemical bonds, explain why eating food is necessary for increasing muscle mass.
4. Given that the hydrogen ion concentration in a solution is based on the following reversible reaction,

$$CO_2 + H_2O \rightleftharpoons H^+ + HCO_3^-$$

what happens to the pH of the solution when $NaHCO_3$ (sodium bicarbonate) is added to the solution? (*Hint*: The sodium bicarbonate dissociates to form $Na^+$ and $HCO_3^-$.)

5. A mixture of chemicals is warmed slightly. As a consequence, although little heat is added, the solution becomes very hot. Explain what happens to make the solution hot.
6. Two solutions, when mixed together at room temperature, produce a chemical reaction. However, when the solutions are boiled and allowed to cool to room temperature before mixing, no chemical reaction takes place. Explain.

*Answers in Appendix D*

Visit this textbook's website at **www.mhhe.com/seeleyess7** for practice quizzes, animations, interactive learning exercises, and other study tools.

McGraw-Hill offers a study CD that features interactive cadaver dissection. *Anatomy & Physiology Revealed* includes cadaver photos that allow you to peel away layers of the human body to reveal structures beneath the surface. This program also includes animations, radiologic imaging, audio pronunciations, and practice quizzing.

## Critical Thinking

*These innovative exercises encourage students to apply chapter concepts to solve a problem. Answering these questions helps students build a working knowledge of anatomy and physiology while developing reasoning skills. Answers are provided in Appendix D.*

## Website Study Support

*Practice quizzes, animation quizzes, flashcards, and 24-hour tutor assistance are all available on the website accompanying this book at* **www.mhhe.com/seeleyess7.**

# List of Clinical Impact Essays

**Chapter 1**  Cadavers and the Law   4
Humors and Homeostasis   8

**Chapter 2**  Ions and Electricity   25
Clinical Applications of Atomic Particles   27
Acidosis and Alkalosis   30

**Chapter 3**  Carbohydrate and Lipid Disorders   56
Relationships Between Cell Structure
    and Cell Function   59
Cancer   65

**Chapter 5**  Adaptive Advantages of Skin Color   99
Acne   103

**Chapter 6**  Bone Fractures   120
Carpal Tunnel Syndrome   135

**Chapter 7**  Acetylcholine Antagonists   159

**Chapter 8**  Spinal Cord Injury   209
Radial Nerve Damage   211
Biofeedback and Meditation   229
Autonomic Dysfunctions   230

**Chapter 9**  Corneal Transplants   245
Color Blindness   251

**Chapter 10**  Lipid- and Water-Soluble Hormones
    in Medicine   266
Hormones and Stress   282

**Chapter 11**  Stem Cells and Cancer Therapy   296
Hemoglobin-Based Oxygen Carriers   297
Clinical Importance of Activating
    Platelets   300
The Danger of Unwanted Clots   302
Erythrocytosis   307
Anemia   308

**Chapter 12**  Disorders of the Pericardium   316
Heart Attack   322
Fibrillation of the Heart   326
The Consequences of an Incompetent
    Bicuspid Valve   331
Consequences of Heart Failure   333
Treatment and Prevention of Heart
    Disease   336

**Chapter 13**  Varicose Veins   348
Blood Vessels Used for Coronary Bypass
    Surgery   361
Hypertension   363
Edema   365
Circulatory Shock   369

**Chapter 14**  Ruptured Spleen   384
Treating Viral Infections and Cancer
    with Interferons   387
Inhibiting and Stimulating Immunity   392
Use of Monoclonal Antibodies   393

**Chapter 15**  Establishing Airflow   411
Pneumothorax   420
Effects of High Altitude and Emphysema   429

**Chapter 16**  Peritonitis   439
Dietary Fiber   444
Hypertrophic Pyloric Stenosis   446
Lactose Intolerance   453
Oral Rehydration Therapy   462
Cystic Fibrosis   463
High- and Low-Density Lipids   464

**Chapter 17**  Fatty Acids and Blood Clotting   476
Free Radicals and Antioxidants   479
Enzymes and Disease   482
Starvation and Obesity   489
Too Hot or Too Cold   493

**Chapter 18**  Alcohol and Caffeine   510
Diabetes Insipidus   511

**Chapter 19**  Descent of the Testes   528
Circumcision   534
Anabolic Steroids   537
Male Pattern Baldness   538
Cancer of the Cervix   543
Cancer of the Breast   544
Amenorrhea   548
Control of Pregnancy   550

**Chapter 20**  In Vitro Fertilization and
    Embryo Transfer   562
Human Genome Project   583

The dancer in the photo is perfectly balanced, yet a slight movement in any direction would cause him to adjust his position. The human body adjusts its balance among all parts through a process called homeostasis.

# 1 Chapter

# The Human Organism

## Chapter Outline and Objectives

**Anatomy (p. 2)**

1. Define anatomy, and explain the importance of understanding the relationship between structure and function.

**Physiology (p. 2)**

2. Define physiology, and state two major goals of physiology.

**Structural and Functional Organization of the Human Body (p. 2)**

3. Describe the six levels of organization of the body, and give the major characteristics of each level.
4. List 11 organ systems, and give the major functions of each.

**Characteristics of Life (p. 6)**

5. List six characteristics of life.

**Homeostasis (p. 7)**

6. Define homeostasis, and explain why it is important.
7. Diagram a negative-feedback mechanism and a positive-feedback mechanism, and describe their relationships to homeostasis.

**Terminology and the Body Plan (p. 10)**

8. Describe a person in the anatomical position.
9. Define the directional terms for the human body, and use them to locate specific body structures.
10. Define the parts and regions of the body.
11. Name and describe the three major planes of the body and the body organs.
12. Describe the major trunk cavities.
13. Describe the serous membranes, and give their functions.

Human anatomy and physiology is the study of the structure and function of the human body. The human body has many intricate parts with coordinated functions maintained by a complex system of checks and balances. The coordinated function of all the parts of the human body allows us to detect stimuli, respond to stimuli, and perform many other actions.

Knowing the structure and function of the human body allows us to understand how the body responds to a stimulus. For example, eating a candy bar results in an increase in blood sugar (the stimulus). Knowledge of the pancreas allows us to predict that the pancreas will secrete insulin (the response). Insulin moves into blood vessels and is transported to cells, where it increases the movement of sugar from the blood into the cells, providing them with a source of energy. As glucose moves into cells, blood sugar levels decrease.

Knowing human anatomy and physiology also provides the basis for understanding disease. For example, in one type of diabetes mellitus, the pancreas does not secrete adequate amounts of insulin. Without adequate insulin, not enough sugar moves into cells, depriving them of a needed source of energy and causing them to malfunction.

The study of human anatomy and physiology is important for students who plan a career in the health sciences because health professionals need a sound knowledge of structure and function in order to perform their duties. In addition, understanding anatomy and physiology prepares all of us to evaluate recommended treatments, critically review advertisements and reports in the popular literature, and rationally discuss the human body with health professionals and nonprofessionals.

# ANATOMY

**Anatomy** (ă-nat′ŏ-mē) is the scientific discipline that investigates the structure of the body. The word *anatomy* means to dissect, or cut apart and separate, the parts of the body for study. Anatomy covers a wide range of studies, including the structure of body parts, their microscopic organization, and the processes by which they develop. In addition, anatomy examines the relationship between the structure of a body part and its function. Just as the structure of a hammer makes it well suited for pounding nails, the structure of body parts allows them to perform specific functions effectively. For example, bones can provide strength and support because bone cells secrete a hard, mineralized substance. Understanding the relationship between structure and function makes it easier to understand and appreciate anatomy.

Two basic approaches to the study of anatomy are systemic anatomy and regional anatomy. **Systemic anatomy** is the study of the body by systems, such as the cardiovascular, nervous, skeletal, and muscular systems. It is the approach taken in this and most introductory textbooks. **Regional anatomy** is the study of the organization of the body by areas. Within each region, such as the head, abdomen, or arm, all systems are studied simultaneously. This is the approach taken in most medical and dental schools.

Anatomists have two general ways to examine the internal structures of a living person: surface anatomy and anatomical imaging. **Surface anatomy** is the study of external features, such as bony projections, which serve as landmarks for locating deeper structures (for examples, see chapters 6 and 7). **Anatomical imaging** involves the use of x-rays, ultrasound, magnetic resonance imaging (MRI), and other technologies to create pictures of internal structures. Both surface anatomy and anatomical imaging provide important information for diagnosing disease.

# PHYSIOLOGY

**Physiology** (fiz-ē-ol′ō-jē; the study of nature) is the scientific discipline that deals with the processes or functions of living things. It is important in physiology to recognize structures as dynamic rather than fixed and unchanging. The major goals of physiology are (1) to understand and predict the body's responses to stimuli and (2) to understand how the body maintains conditions within a narrow range of values in the presence of a continually changing environment. **Human physiology** is the study of a specific organism, the human, whereas **cellular physiology** and **systemic physiology** are subdivisions that emphasize specific organizational levels.

# STRUCTURAL AND FUNCTIONAL ORGANIZATION OF THE HUMAN BODY

The body can be studied at six structural levels: chemical, cell, tissue, organ, organ system, and organism (figure 1.1).

## Chemical Level

The structural and functional characteristics of all organisms are determined by their chemical makeup. The **chemical** level of organization involves how atoms interact and combine into molecules. The function of a molecule is intimately related to its structure. For example, collagen molecules are strong, ropelike fibers that give skin structural strength and flexibility. With old age, the structure of collagen changes, and the skin becomes fragile and more easily torn. A brief overview of chemistry is presented in chapter 2.

## Cell Level

**Cells** are the basic structural and functional units of organisms, such as plants and animals. Molecules can combine to form **organelles** (or′gă-nelz; little organs), which are the small structures that make up cells. For example, the nucleus contains the cell's hereditary information, and mitochondria manufacture adenosine triphosphate (ATP), a molecule cells use for a source of energy. Although cell types differ in their structure and function, they have many characteristics in common. Knowledge of these characteristics

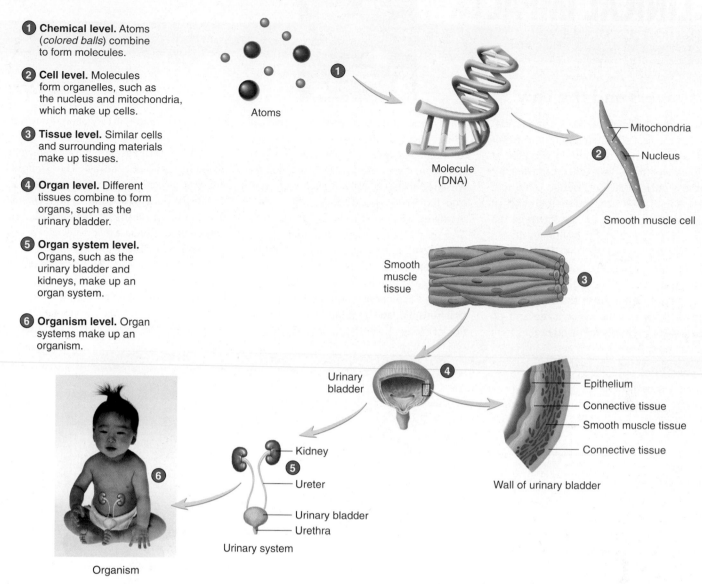

1. **Chemical level.** Atoms (*colored balls*) combine to form molecules.

2. **Cell level.** Molecules form organelles, such as the nucleus and mitochondria, which make up cells.

3. **Tissue level.** Similar cells and surrounding materials make up tissues.

4. **Organ level.** Different tissues combine to form organs, such as the urinary bladder.

5. **Organ system level.** Organs, such as the urinary bladder and kidneys, make up an organ system.

6. **Organism level.** Organ systems make up an organism.

Atoms

Molecule (DNA)

Mitochondria

Nucleus

Smooth muscle cell

Smooth muscle tissue

Urinary bladder

Epithelium

Connective tissue

Smooth muscle tissue

Connective tissue

Wall of urinary bladder

Kidney

Ureter

Urinary bladder

Urethra

Urinary system

Organism

**PROCESS Figure 1.1   Levels of Organization for the Human Body**

and their variations is essential to a basic understanding of anatomy and physiology. The cell is discussed in chapter 3.

## Tissue Level

A **tissue** (tish′ū) is a group of similar cells and the materials surrounding them. The characteristics of the cells and surrounding materials determine the functions of the tissue. The many tissues that make up the body are classified into four primary types: epithelial, connective, muscle, and nervous. Tissues are discussed in chapter 4.

## Organ Level

An **organ** (ōr′găn; a tool) is composed of two or more tissue types that together perform one or more common functions. The

urinary bladder, skin, stomach, and heart are examples of organs (figure 1.2).

## Organ System Level

An **organ system** is a group of organs classified as a unit because of a common function or set of functions. For example, the urinary system consists of the kidneys, ureter, urinary bladder, and urethra. The kidneys produce urine, which is transported by the ureters to the urinary bladder, where it is stored until eliminated from the body by passing through the urethra. In this text, we consider 11 major organ systems: integumentary, skeletal, muscular, lymphatic, respiratory, digestive, nervous, endocrine, cardiovascular, urinary, and reproductive (figure 1.3).

The coordinated activity of the organ systems is necessary for normal function. For example, the digestive system takes in and

# CLINICAL IMPACT

## Cadavers and the Law

The study of human bodies is the foundation of medical education, and for much of history, anatomists have used the bodies of people who have died, called cadavers, for these studies. However, public sentiment has often made it difficult for anatomists to obtain human bodies for dissection. In the early 1800s, the benefits of human dissection for training physicians had become very apparent, and the need for cadavers increased beyond the ability to acquire them legally. Thus arose the resurrectionists, or body snatchers. For a fee and no questions asked, they removed bodies from graves and provided them to medical schools. Because the bodies were not easy to obtain and were not always in the best condition, two enterprising men named William Burke and William Hare went one step further. Over a period of time, they murdered 17 people in Scotland and sold their bodies to a medical school. When discovered, Hare testified against Burke and went free. Burke was convicted, hanged, and publicly dissected. Discovery of Burke's activities so outraged the public that sensible laws regulating the acquisition of cadavers were soon passed, and this dark chapter in the history of anatomy was closed.

Today it is quite simple to donate your body for scientific study. The Uniform Anatomical Gift Act allows individuals to donate their organs or entire cadaver by putting a notation on their driver's license. You only need to contact a medical school or private agency to file the forms that give them the rights to your cadaver. Once the donor dies, the family of the deceased usually pays only the transportation costs for the remains. After dissection, the body is cremated, and the cremains can be returned to the family.

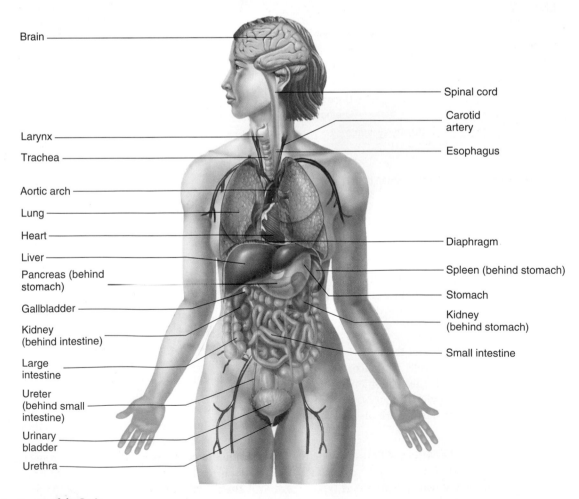

Brain

Larynx

Trachea

Aortic arch

Lung

Heart

Liver

Pancreas (behind stomach)

Gallbladder

Kidney (behind intestine)

Large intestine

Ureter (behind small intestine)

Urinary bladder

Urethra

Spinal cord

Carotid artery

Esophagus

Diaphragm

Spleen (behind stomach)

Stomach

Kidney (behind stomach)

Small intestine

**Figure 1.2** Major Organs of the Body

**Integumentary System**

Provides protection, regulates temperature, prevents water loss, and helps produce vitamin D. Consists of skin, hair, nails, and sweat glands.

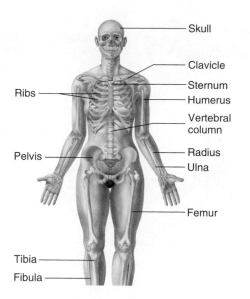

**Skeletal System**

Provides protection and support, allows body movements, produces blood cells, and stores minerals and fat. Consists of bones, associated cartilages, ligaments, and joints.

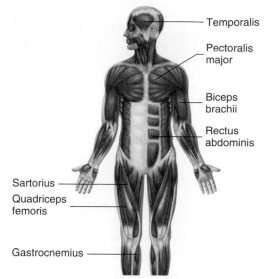

**Muscular System**

Produces body movements, maintains posture, and produces body heat. Consists of muscles attached to the skeleton by tendons.

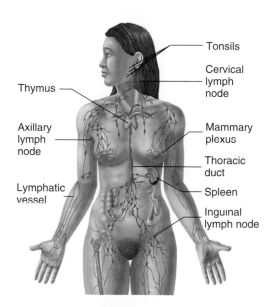

**Lymphatic System**

Removes foreign substances from the blood and lymph, combats disease, maintains tissue fluid balance, and absorbs fats from the digestive tract. Consists of the lymphatic vessels, lymph nodes, and other lymphatic organs.

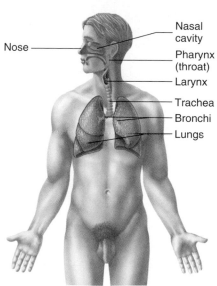

**Respiratory System**

Exchanges oxygen and carbon dioxide between the blood and air and regulates blood pH. Consists of the lungs and respiratory passages.

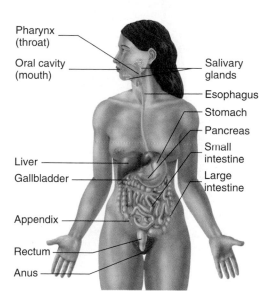

**Digestive System**

Performs the mechanical and chemical processes of digestion, absorption of nutrients, and elimination of wastes. Consists of the mouth, esophagus, stomach, intestines, and accessory organs.

**Figure 1.3  Organ Systems of the Body**

processes food, which is carried by the blood of the cardiovascular system to the cells of the other systems. These cells use the food and produce waste products that are carried by the blood to the kidneys of the urinary system, which removes waste products from the blood. Because the organ systems are so interrelated, dysfunction in one organ system can have profound effects on other systems. For example, a heart attack can result in inadequate circulation of blood. Consequently, the organs of other systems, such as the brain and kidneys, can malfunction. Throughout this text, Systems Pathology essays consider the interactions of the organ systems.

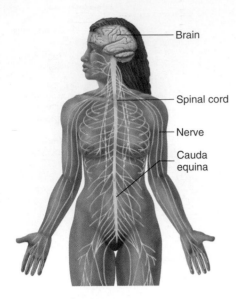

## Nervous System

A major regulatory system that detects sensations and controls movements, physiological processes, and intellectual functions. Consists of the brain, spinal cord, nerves, and sensory receptors.

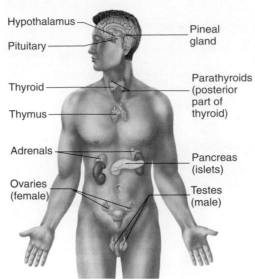

## Endocrine System

A major regulatory system that influences metabolism, growth, reproduction, and many other functions. Consists of glands, such as the pituitary, that secrete hormones.

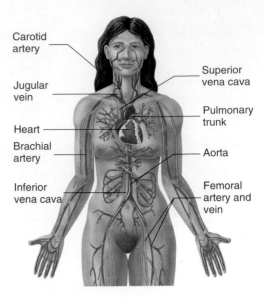

## Cardiovascular System

Transports nutrients, waste products, gases, and hormones throughout the body; plays a role in the immune response and the regulation of body temperature. Consists of the heart, blood vessels, and blood.

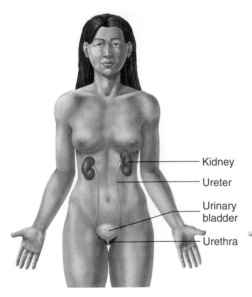

## Urinary System

Removes waste products from the blood and regulates blood pH, ion balance, and water balance. Consists of the kidneys, urinary bladder, and ducts that carry urine.

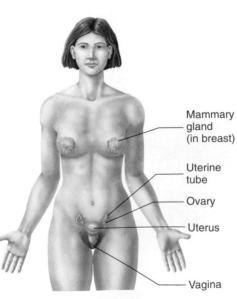

## Female Reproductive System

Produces oocytes and is the site of fertilization and fetal development; produces milk for the newborn; produces hormones that influence sexual function and behaviors. Consists of the ovaries, vagina, uterus, mammary glands, and associated structures.

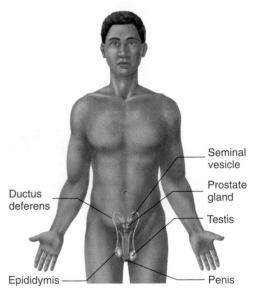

## Male Reproductive System

Produces and transfers sperm cells to the female and produces hormones that influence sexual functions and behaviors. Consists of the testes, accessory structures, ducts, and penis.

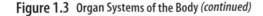

**Figure 1.3** Organ Systems of the Body *(continued)*

## Organism Level

An **organism** is any living thing considered as a whole, whether composed of one cell, such as a bacterium, or trillions of cells, such as a human. The human organism is a complex of organ systems that are mutually dependent on one another (figure 1.3).

# CHARACTERISTICS OF LIFE

Humans share many characteristics with other organisms, and much of our knowledge about the human body has come from studying other organisms. For example, the study of bacteria (single-celled organisms) has provided much information about human cells, and open heart surgery techniques were perfected on other mammals before being attempted on humans.

The most important common feature of all organisms is life. Essential characteristics of life are organization, metabolism, responsiveness, growth, development, and reproduction.

1. **Organization** refers to the specific interrelationships among the parts of an organism and how those parts interact to perform specific functions. Living things are highly organized. All organisms are composed of one or more cells. Cells, in turn, are composed of highly specialized organelles, which depend on the precise functions of large molecules. Disruption of this organized state can result in loss of function and death.

2. **Metabolism** (mĕ-tab′ō-lizm) is the ability to use energy to perform vital functions, such as growth, movement, and reproduction. Plants capture energy from sunlight, and humans obtain energy from food.

3. **Responsiveness** is the ability of an organism to sense changes in the environment and make the adjustments that help maintain its life. Responses include movement toward food or water and away from danger or poor environmental conditions. Organisms can also make adjustments that maintain their internal environment. For example, if body temperature increases in a hot environment, sweat glands produce sweat, which can lower body temperature back toward the normal level.

4. **Growth** refers to an increase in size of all or part of the organism. It can result from an increase in cell number, cell size, or the amount of substance surrounding cells. For example, bones become larger as the number of bone cells increases and they become surrounded by bone matrix.

5. **Development** includes the changes an organism undergoes through time; it begins with fertilization and ends at death. The greatest developmental changes occur before birth, but many changes continue after birth, and some continue throughout life. Development usually involves growth, but it also involves differentiation. Differentiation is change in cell structure and function from generalized to specialized. For example, following fertilization, generalized cells specialize to become specific cell types, such as skin, bone, muscle, or nerve cells. These differentiated cells form the tissues and organs.

6. **Reproduction** is the formation of new cells or new organisms. Without reproduction of cells, growth and tissue repair are impossible. Without reproduction of the organism, the species becomes extinct.

# HOMEOSTASIS

**Homeostasis** (hō′mē-ō-stā′sis; *homeo-*, the same) is the existence and maintenance of a relatively constant environment within the body. Most body cells are surrounded by a small amount of fluid, and normal cell functions depend on the maintenance of the cells' fluid environment within a narrow range of conditions, including temperature, volume, and chemical content. These conditions are

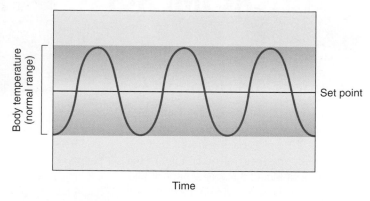

**Figure 1.4** Homeostasis

Homeostasis is the maintenance of a variable, such as body temperature, around an ideal normal value, or set point. The value of the variable fluctuates around the set point to establish a normal range of values.

called **variables** because their values can change. For example, body temperature is a variable that can increase in a hot environment or decrease in a cold environment.

Homeostatic mechanisms, such as sweating or shivering, normally maintain body temperature near an ideal normal value, or **set point** (figure 1.4). Note that these mechanisms are not able to maintain body temperature precisely at the set point. Instead, body temperature increases and decreases slightly around the set point, producing a **normal range** of values. As long as body temperatures remain within this normal range, homeostasis is maintained.

The organ systems help control the internal environment so that it remains relatively constant. For example, the digestive, respiratory, cardiovascular, and urinary systems function together so that each cell in the body receives adequate oxygen and nutrients and so that waste products do not accumulate to a toxic level. If the fluid surrounding cells deviates from homeostasis, the cells do not function normally and may even die. Disease disrupts homeostasis and sometimes results in death. Modern medicine attempts to understand disturbances in homeostasis and works to reestablish a normal range of values.

## Negative Feedback

Most systems of the body are regulated by **negative-feedback mechanisms,** which maintain homeostasis. *Negative* means that any deviation from the set point is made smaller or is resisted. Negative feedback does not prevent variation but maintains variation within a normal range.

The maintenance of normal blood pressure is an example of a negative-feedback mechanism. Normal blood pressure is important because it helps move blood from the heart to tissues. The blood supplies the tissues with oxygen and nutrients and removes waste products. Thus, normal blood pressure is required to ensure that tissue homeostasis is maintained.

Many negative-feedback mechanisms, such as the one that maintains normal blood pressure, have three components: (1) A **receptor** (rē-sep′tŏr, rē-sep′tōr) monitors the value of a variable, such as blood pressure; (2) a **control center,** such as part of the brain, establishes the set point around which the variable is maintained;

## Humors and Homeostasis

The idea that the body maintains a balance (homeostasis) can be traced back to ancient Greece. Early physicians believed that the body supported four juices, or humors: the red juice of blood, the yellow juice of bile, the white juice secreted from the nose and lungs, and a black juice in the pancreas. They also thought that health resulted from a proper balance of these juices and that an excess of any one of them caused disease. Normally, they believed, the body would attempt to heal itself by expelling the excess juice, as when mucus runs from the nose of a person with a cold. This belief led to the practice of bloodletting to restore the body's normal balance of juices. Typically, physicians used sharp instruments to puncture the larger, external vessels, but sometimes they applied leeches, blood-eating organisms, to the skin.

Tragically, in the eighteenth and nineteenth centuries, bloodletting went to extremes. During this period, a physician might recommend bloodletting, but barbers conducted the actual procedure. In fact, the traditional red-and-white-striped barber pole originated as a symbol for bloodletting. The brass basin on top of the pole represented the bowl for leeches, and the bowl on the bottom represented the basin for collecting blood. The stripes represented the bandages used as tourniquets, and the pole itself stood for the wooden staff patients gripped during the procedure. The fact that bloodletting did not improve the patient's condition was taken as evidence that not enough blood had been removed to restore a healthy balance of the body's juices. Thus, the obvious solution was to let still more blood, undoubtedly causing many deaths. Eventually, the failure of this approach became obvious, and the practice was abandoned.

The modern term for bloodletting is **phlebotomy** (fle-bot′ō-mē), but it is practiced in a controlled setting and removes only small volumes of blood, usually for laboratory testing. There are some diseases in which bloodletting is still useful—for example, **polycythemia** (pol′ē-sī-thē′mē-ă), an overabundance of red blood cells. However, bloodletting in these patients does not continue until the patient faints or dies. Fortunately, we now understand more about how the body maintains homeostasis.

and (3) an **effector** (ē-fek′tŏr), such as the heart, can change the value of the variable. Blood pressure depends in part on contraction (beating) of the heart: As the heart rate increases, blood pressure increases, and as the heart rate decreases, blood pressure decreases.

The receptors that monitor blood pressure are located within large blood vessels near the heart. If blood pressure increases slightly, the receptors detect that change and send the information to the control center in the brain. The control center causes the heart rate to decrease and blood vessels to dilate, resulting in a decrease in blood pressure. If blood pressure decreases slightly, the receptors inform the control center, which increases the heart rate and constricts blood vessels, producing an increase in blood pressure. As a result, blood pressure is maintained within a normal range (figure 1.5).

### Predict 1

*Donating a pint of blood reduces blood volume, which results in a decrease in blood pressure (just as the air pressure in a tire decreases as air is let out). What effect does donating blood have on the heart rate? What would happen if a negative-feedback mechanism did not return the value of a variable, such as blood pressure, to its normal range?*

## Positive Feedback

**Positive-feedback mechanisms** are *not* homeostatic and are rare in healthy individuals. *Positive* implies that when a value deviates from normal, the system's response is to make the deviation even greater. Positive feedback therefore usually creates a cycle leading away from homeostasis, and in some cases, results in death.

An example of positive feedback is inadequate delivery of blood to cardiac (heart) muscle. Contraction of cardiac muscle generates blood pressure and moves blood through the blood vessels to the tissues. A system of blood vessels on the outside of the heart provides cardiac muscle with a blood supply sufficient to allow normal contractions to occur. In effect, the heart pumps blood to itself. Just as with other tissues, blood pressure must be maintained to ensure adequate delivery of blood to the cardiac muscle. Following extreme blood loss, blood pressure decreases to the point that the delivery of blood to cardiac muscle is inadequate. As a result, cardiac muscle homeostasis is disrupted, and cardiac muscle does not function normally. The heart pumps less blood, which causes the blood pressure to drop even further. The additional decrease in blood pressure further reduces blood delivery to cardiac muscle, and the heart pumps even less blood, which again decreases the blood pressure (figure 1.6). The process continues until the blood pressure is too low to sustain the cardiac muscle, the heart stops beating, and death results.

Following a moderate amount of blood loss (e.g., after donating a pint of blood), negative-feedback mechanisms result in an increase in heart rate that restores blood pressure. However, if blood loss is severe, negative-feedback mechanisms may not be able to maintain homeostasis, and the positive-feedback effect of an ever-decreasing blood pressure can develop.

A basic principle to remember is that many disease states result from the failure of negative-feedback mechanisms to maintain homeostasis. The purpose of medical therapy is to overcome

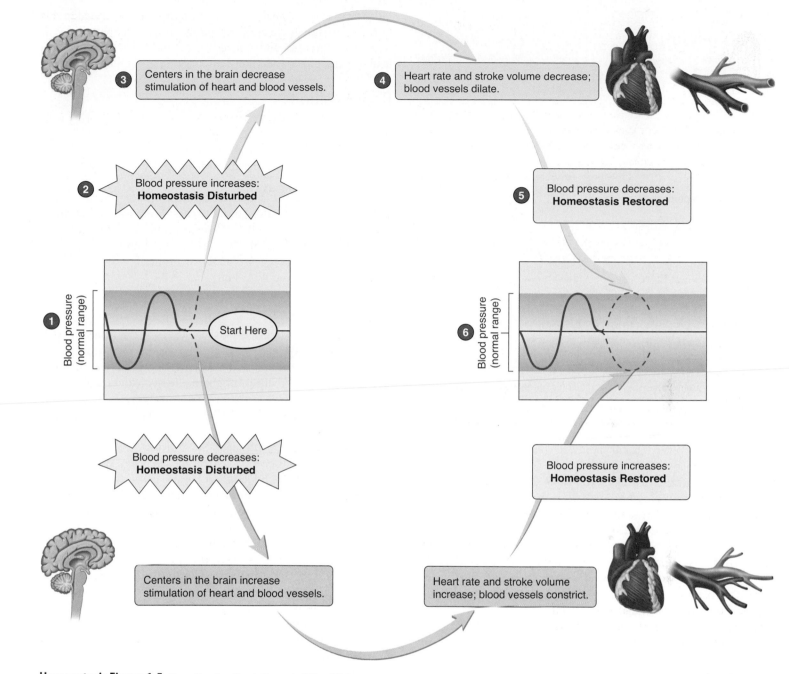

## Homeostasis Figure 1.5  Negative-Feedback Control of Blood Pressure

Throughout this book, all homeostasis figures have the same format as shown here. The changes caused by the increase of a variable outside the normal range are shown in the *green boxes,* and the changes caused by a decrease are shown in the *red boxes*. To help you learn how to interpret homeostasis figures, some of the steps in this figure are numbered. (1) Blood pressure is within its normal range. (2) Blood pressure increases outside the normal range, which causes homeostasis to be disturbed. (3) The blood pressure control center in the brain responds to the change in blood pressure. (4) The control center causes heart rate and stroke volume to decrease and blood vessels to dilate. (5) These changes cause blood pressure to decrease. (6) Blood pressure returns to its normal range, and homeostasis is restored. Observe the responses to a decrease in blood pressure outside its normal range by following the *red arrows*.

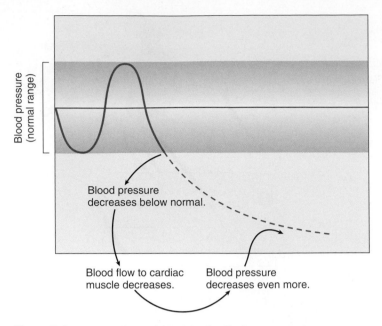

**Figure 1.6  Example of Harmful Positive Feedback**

A decrease in blood pressure outside the normal range causes decreased blood flow to the heart. The heart is unable to pump enough blood to maintain blood pressure, and blood flow to the cardiac muscle decreases. Thus, the heart's ability to pump decreases further, and blood pressure decreases even more.

illness by aiding negative-feedback mechanisms. For example, a transfusion can reverse a constantly decreasing blood pressure and restore homeostasis.

A few positive-feedback mechanisms do operate in the body under normal conditions, but in all cases they are eventually limited in some way. Birth is an example of a normally occurring positive-feedback mechanism. Near the end of pregnancy, the uterus is stretched by the baby's large size. This stretching, especially around the opening of the uterus, stimulates contractions of the uterine muscles. The uterine contractions push the baby against the opening of the uterus, stretching it further. This stimulates additional contractions, which result in additional stretching. This positive-feedback sequence ends only when the baby is delivered from the uterus and the stretching stimulus is eliminated.

**Predict 2**

*Is the sensation of thirst associated with a negative- or a positive-feedback mechanism? Explain. (Hint: What is being regulated when you become thirsty?)*

# TERMINOLOGY AND THE BODY PLAN

When you begin to study anatomy and physiology, the number of new words may seem overwhelming. Learning is easier and more interesting if you pay attention to the origin, or **etymology** (et′ĕ-mol′o-jē), of new words. Most of the terms are derived from Latin or Greek. For example, *anterior* in Latin means "to go before." Therefore, the anterior surface of the body is the one that goes before when we are walking.

Words are often modified by adding a prefix or suffix. For example, the suffix *-itis* means an inflammation, so *appendicitis* is an

inflammation of the appendix. As new terms are introduced in this text, their meanings are often explained. The glossary and the list of word roots, prefixes, and suffixes on the inside back cover of the textbook also provide additional information about the new terms.

## Body Positions

The **anatomical position** refers to a person standing erect with the face directed forward, the upper limbs hanging to the sides, and the palms of the hands facing forward (figure 1.7). A person is **supine** when lying face upward and **prone** when lying face downward.

The position of the body can affect the description of body parts relative to each other. In the anatomical position, the elbow is above the hand, but in the supine or prone position, the elbow and hand are at the same level. To avoid confusion, relational descriptions are always based on the anatomical position, no matter the actual position of the body.

## Directional Terms

Directional terms describe parts of the body relative to each other (figure 1.7 and table 1.1). It is important to become familiar with these directional terms as soon as possible because you will see them repeatedly throughout the text. *Right* and *left* are retained as directional terms in anatomical terminology. *Up* is replaced by **superior,** *down* by **inferior,** *front* by **anterior,** and *back* by **posterior.**

As previously mentioned, the word *anterior* means that which goes before; the word **ventral** means belly. Therefore, the anterior surface of the human body is also called the ventral surface, or belly, because the belly "goes first" when we are walking. The word *posterior* means that which follows, and **dorsal** means back. Thus, the posterior surface of the body is the dorsal surface, or back, which follows as we are walking.

**Proximal** means nearest, whereas **distal** means distant. These terms are used to refer to linear structures, such as the limbs, in which one end is near another structure and the other end is farther away. Each limb is attached at its proximal end to the body, and the distal end, such as the hand, is farther away.

**Medial** means toward the midline, and **lateral** means away from the midline. The nose is located in a medial position on the face, and the eyes are lateral to the nose. The term **superficial** refers to a structure close to the surface of the body, and **deep** is toward the interior of the body. For example, the skin is superficial to muscle and bone.

**Predict 3**

*Provide the correct directional term for the following statement:*
*When a boy is standing on his head, his nose is _____ to his mouth.*

## Body Parts and Regions

Health professionals use a number of terms when referring to different regions or parts of the body. Figure 1.8 shows the anatomical terms, with the common terms in parentheses. The central region of the body consists of the **head, neck,** and **trunk.** The trunk can be divided into the **thorax** (chest), **abdomen** (region between the thorax and pelvis), and **pelvis** (the inferior

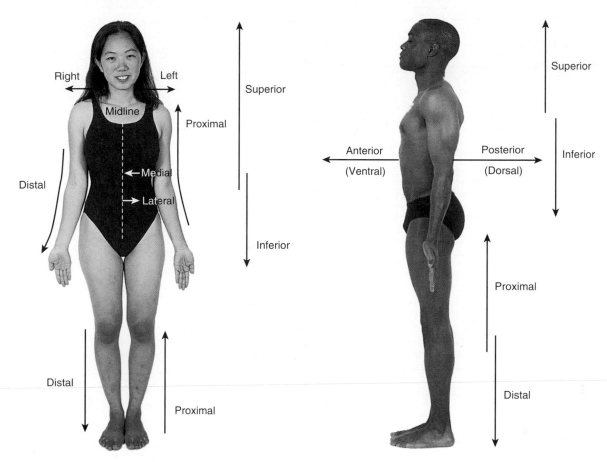

**Figure 1.7 Directional Terms**

All directional terms are in relation to the body in the anatomical position: a person standing erect with the face directed forward, the arms hanging to the sides, and the palms of the hands facing forward.

| TABLE 1.1 | Directional Terms for the Human Body | | |
|---|---|---|---|
| **Terms** | **Etymology** | **Definition*** | **Example** |
| Right | | Toward the body's right side | The right ear |
| Left | | Toward the body's left side | The left ear |
| Inferior | Lower | Below | The nose is inferior to the forehead. |
| Superior | Higher | Above | The mouth is superior to the chin. |
| Anterior | To go before | Toward the front of the body | The teeth are anterior to the throat. |
| Posterior | *Posterus,* following | Toward the back of the body | The brain is posterior to the eyes. |
| Dorsal | *Dorsum,* back | Toward the back (synonymous with *posterior*) | The spine is dorsal to the breastbone. |
| Ventral | *Venter,* belly | Toward the belly (synonymous with *anterior*) | The navel is ventral to the spine. |
| Proximal | *Proximus,* nearest | Closer to a point of attachment | The elbow is proximal to the wrist. |
| Distal | *di + sto,* to be distant | Farther from a point of attachment | The knee is distal to the hip. |
| Lateral | *Latus,* side | Away from the midline of the body | The nipple is lateral to the breastbone. |
| Medial | *Medialis,* middle | Toward the middle or midline of the body | The bridge of the nose is medial to the eye. |
| Superficial | *Superficialis,* surface | Toward or on the surface | The skin is superficial to muscle. |
| Deep | *Deop,* deep | Away from the surface, internal | The lungs are deep to the ribs. |

*All directional terms refer to a human in the anatomical position.

**Head**
(cephalic)
— Frontal (forehead)
— Orbital (eye)
— Nasal (nose)
— Oral (mouth)

**Neck** — Cervical

Otic (ear)
Buccal (cheek)
Mental (chin)

Clavicular (collarbone)

**Trunk**

**Thoracic**
(thorax)
— Pectoral (chest)
— Sternal (breastbone)
— Mammary (breast)

Abdominal (abdomen)
Umbilical (navel)

Pelvic (pelvis)
Inguinal (groin)
Pubic (genital)

Axillary (armpit)
Brachial (arm)
Antecubital (front of elbow)
Antebrachial (forearm)

**Upper limb**

Carpal (wrist)
Palmar (palm)
Digital (fingers)

Manual (hand)

Coxal (hip)
Femoral (thigh)
Patellar (kneecap)
Crural (leg)

**Lower limb**

Talus (ankle)
Dorsum (top of foot)
Digital (toes)

Pedal (foot)

(a)

**Figure 1.8  Body Parts and Regions**

The anatomical and common (*in parentheses*) names are indicated for some parts and regions of the body. (*a*) Anterior view.

end of the trunk associated with the hips). The upper limb is divided into the arm, forearm, wrist, and hand. The **arm** extends from the shoulder to the elbow, and the **forearm** extends from the elbow to the wrist. The lower limb is divided into the thigh, leg, ankle, and foot. The **thigh** extends from the hip to the knee, and the **leg** extends from the knee to the ankle. Note that, contrary to popular usage, the terms *arm* and *leg* refer to only a part of the respective limb.

The abdomen is often subdivided superficially into four sections, or **quadrants,** by two imaginary lines—one horizontal and one vertical—that intersect at the navel (figure 1.9*a*). The quadrants formed are the right-upper, left-upper, right-lower, and left-lower quadrants. In addition to these quadrants, the abdomen is sometimes subdivided into **regions** by four imaginary lines—two horizontal and two vertical. These four lines create an imaginary tic-tac-toe figure on the abdomen, resulting in nine regions: epigastric (ep-i-gas′trik), right and left hypochondriac (hī-pō-kon′drē-ak), umbilical (ŭm-bil′i-kăl), right and left lumbar (lŭm′bar), hypogastric (hī-pō-gas′trik), and right and left iliac (il′ē-ak) (figure 1.9*b*). Clinicians use the quadrants or regions as reference points for locating the underlying organs. For example, the appendix is in the right-lower quadrant, and the pain of an acute appendicitis is usually felt there.

*Predict 4*

*Using figures 1.2 and 1.9a, determine in which quadrant each of the following organs is located: spleen, gallbladder, kidneys, most of the stomach, and most of the liver.*

## A CASE IN POINT
### Epigastric Pain

Wilby Hurtt has pain in the epigastric region (Figure 1.9*b*), which is most noticeable following meals and at night when he is lying in bed. He probably has gastroesophageal reflux disease (GERD), in which stomach acid improperly moves into the esophagus, damaging and irritating its lining. Epigastric pain, however, can have many causes and should be evaluated by a physician. For example, gallstones, stomach or small intestine ulcers, inflammation of the pancreas, and heart disease can also cause epigastric pain.

## Planes

At times, it is conceptually useful to discuss the body in reference to a series of planes (imaginary flat surfaces) passing through it (figure 1.10). Sectioning the body is a way to "look inside" and observe the body's structures. A **sagittal** (saj′i-tăl) **plane** runs

(b)

**Figure 1.8** Body Parts and Regions *(continued)*

(b) Posterior view.

Labels (posterior view, top to bottom):
- Cranial (skull)
- Occipital (base of skull)
- Nuchal (back of neck)
- Acromial (point of shoulder)
- Scapular (shoulder blade)
- Dorsal (back)
- Vertebral (spinal column)
- Olecranon (point of elbow)
- Lumbar (loin)
- Sacral (between hips)
- Dorsum (back of hand)
- Gluteal (buttock)
- Perineal (perineum)
- Popliteal (hollow behind knee)
- Sural (calf)
- Plantar (sole)
- Calcaneal (heel)
- Trunk
- Upper limb
- Lower limb

**Figure 1.9** Subdivisions of the Abdomen

Lines are superimposed over internal organs to demonstrate the relationship of the organs to the subdivisions. (a) Abdominal quadrants consist of four subdivisions. (b) Abdominal regions consist of nine subdivisions.

(a) Abdominal quadrants:
- Right-upper quadrant
- Left-upper quadrant
- Right-lower quadrant
- Left-lower quadrant

(b) Abdominal regions:
- Right hypochondriac region
- Epigastric region
- Left hypochondriac region
- Right lumbar region
- Umbilical region
- Left lumbar region
- Right iliac region
- Hypogastric region
- Left iliac region

Cerebrum

Cerebellum
Brainstem

Nasal cavity

Spinal cord

Tongue
Pharynx (throat)

Vertebral
column

Trachea

(b) Sagittal section of the head

Sagittal
plane

Transverse,
or horizontal,
plane

Liver

Stomach

Large
intestine

Spleen

Kidney

Vertebra

Spinal
cord

Kidney

(c) Transverse section through the abdomen

Frontal, or
coronal, plane

Skin
Fat
Hip muscle

Coxal bone
(hipbone)

Femur
(thighbone)

Thigh muscles

(a)

(d) Frontal section through the right hip

# Figure 1.10  Planes of Section of the Body

(a) Planes of section through the body are indicated by "glass" sheets.
Also shown are actual sections through (b) the head (viewed from the
right), (c) the abdomen (*inferior view*), and (d) the hip (*anterior view*).

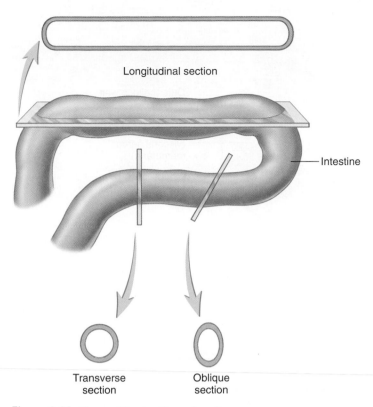

Longitudinal section

Intestine

Transverse
section

Oblique
section

**Figure 1.11  Planes of Section Through an Organ**

Planes of section through the small intestine are indicated by "glass" sheets. The views of the small intestine after sectioning are also shown. Although the small intestine is basically a tube, the sections appear quite different in shape.

vertically through the body and separates it into right and left parts. The word *sagittal* literally means the flight of an arrow and refers to the way the body would be split by an arrow passing anteriorly to posteriorly. A **median plane** is a sagittal plane that passes through the midline of the body, dividing it into equal right and left halves. A **transverse** (trans-vers′) **plane,** or *horizontal plane,* runs parallel to the surface of the ground, dividing the body into superior and inferior parts. A **frontal plane,** or *coronal* (kōr′ŏ-năl, kō-rō′nal; crown) *plane,* runs vertically from right to left and divides the body into anterior and posterior parts.

Organs are often sectioned to reveal their internal structure (figure 1.11). A cut through the long axis of the organ is a **longitudinal section,** and a cut at a right angle to the long axis is a **transverse section,** or *cross section.* If a cut is made across the long axis at other than a right angle, it is called an **oblique section.**

## Body Cavities

The body contains many cavities. Some of these cavities, such as the nasal cavity, open to the outside of the body, and some do not. The trunk contains three large cavities that do not open to the outside of the body: the thoracic cavity, the abdominal cavity, and the pelvic cavity (figure 1.12). The **thoracic cavity** is surrounded by the rib cage and is separated from the abdominal cavity by the muscular diaphragm. It is divided into right and left parts by a median structure called the **mediastinum** (me′dē-as-tī′nŭm; wall). The mediastinum is a partition containing the heart, the thymus, the trachea, the esophagus, and other structures. The two lungs are located on each side of the mediastinum.

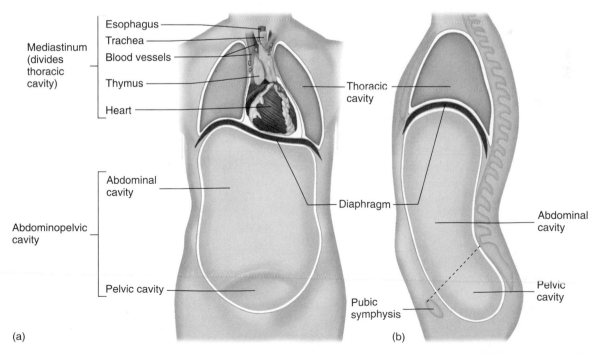

Esophagus
Trachea
Blood vessels
Mediastinum (divides thoracic cavity)
Thymus
Heart
Abdominal cavity
Abdominopelvic cavity
Pelvic cavity
Thoracic cavity
Diaphragm
Pubic symphysis
Abdominal cavity
Pelvic cavity

(a)                                    (b)

**Figure 1.12  Trunk Cavities**

(*a*) Anterior view showing the major trunk cavities. The diaphragm separates the thoracic cavity from the abdominal cavity. The mediastinum, which includes the heart, is a partition of organs dividing the thoracic cavity. (*b*) Sagittal section of the trunk cavities viewed from the left. The *dashed line* shows the division between the abdominal and pelvic cavities. The mediastinum has been removed to show the thoracic cavity.

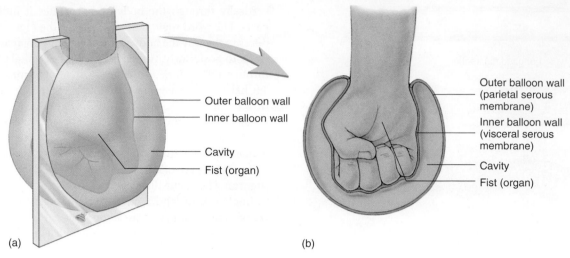

**Figure 1.13**  Serous Membranes

(a) Fist pushing into a balloon. A "glass" sheet indicates the location of a cross section through the balloon. (b) Interior view produced by the section in (a). The fist represents an organ, and the walls of the balloon represent the serous membranes. The inner wall of the balloon represents a visceral serous membrane in contact with the fist (organ). The outer wall of the balloon represents a parietal serous membrane.

The **abdominal cavity** is bounded primarily by the abdominal muscles and contains the stomach, the intestines, the liver, the spleen, the pancreas, and the kidneys. The **pelvic** (pel′vik) **cavity** is a small space enclosed by the bones of the pelvis and contains the urinary bladder, part of the large intestine, and the internal reproductive organs. The abdominal and pelvic cavities are not physically separated and sometimes are called the **abdominopelvic** (ab-dom′i-nō-pel′vik) **cavity.**

## Serous Membranes

**Serous** (sēr′ŭs) **membranes** line the trunk cavities and cover the organs of these cavities. To understand the relationship between serous membranes and an organ, imagine pushing your fist into an inflated balloon. The inner balloon wall in contact with your fist (organ) represents the **visceral** (vis′er-ăl; organ) **serous membrane,** and the outer part of the balloon wall represents the **parietal** (pă-rī′ĕ-tăl; wall) **serous membrane** (figure 1.13). The cavity, or space, between the visceral and parietal serous membranes is normally filled with a thin, lubricating film of serous fluid produced by the membranes. As an organ rubs against another organ or against the body wall, the serous fluid and smooth serous membranes reduce friction.

The thoracic cavity contains three serous membrane–lined cavities: a pericardial cavity and two pleural cavities. The **pericardial** (per-i-kar′dē-ăl; around the heart) **cavity** surrounds the heart (figure 1.14a). The visceral pericardium covers the heart, which is contained within a connective tissue sac lined with the parietal pericardium. The pericardial cavity, which contains pericardial fluid, is located between the visceral pericardium and the parietal pericardium.

A **pleural** (ploor′ăl; associated with the ribs) **cavity** surrounds each lung, which is covered by visceral pleura (figure 1.14b). Parietal pleura lines the inner surface of the thoracic wall, the lateral surfaces of the mediastinum, and the superior surface of the

diaphragm. The pleural cavity is located between the visceral pleura and the parietal pleura and contains pleural fluid.

The abdominopelvic cavity contains a serous membrane–lined cavity called the **peritoneal** (per′i-tō-nē′ăl; to stretch over) **cavity** (figure 1.14c). Visceral peritoneum covers many of the organs of the abdominopelvic cavity. Parietal peritoneum lines the wall of the abdominopelvic cavity and the inferior surface of the diaphragm. The peritoneal cavity is located between the visceral peritoneum and the parietal peritoneum and contains peritoneal fluid.

The serous membranes can become inflamed—usually as a result of an infection. **Pericarditis** (per′i-kar-dī′tis) is inflammation of the pericardium, **pleurisy** (ploor′i-sē) is inflammation of the pleura, and **peritonitis** (per′i-tō-nī′tis) is inflammation of the peritoneum.

### A CASE IN POINT
#### Peritonitis

May Day is rushed to the hospital emergency room. Earlier that day, she experienced diffuse abdominal pain, but no fever. Then the pain became more intense and shifted to her right-lower quadrant. She also developed a fever. The examining physician concludes that May Day has **appendicitis,** an inflammation of the appendix that is usually caused by an infection. The appendix is a small, wormlike sac attached to the large intestine. The outer surface of the appendix is visceral peritoneum. An infection of the appendix can rupture its wall, releasing bacteria into the peritoneal cavity and causing peritonitis. Appendicitis is the most common cause of emergency abdominal surgery in children, and it often leads to peritonitis if not treated. May has her appendix removed, is treated with antibiotics, and makes a full recovery.

**Mesenteries** (mes′en-ter-ēz), which consist of two layers of peritoneum fused together (figure 1.14c), connect the visceral

(a)

- Parietal pericardium
- Visceral pericardium
- Pericardial cavity containing pericardial fluid
- Heart

(b)

- Parietal pleura
- Visceral pleura
- Pleural cavity containing pleural fluid
- Lung
- Diaphragm

(c)

- Parietal peritoneum
- Visceral peritoneum
- Peritoneal cavity containing peritoneal fluid
- Retroperitoneal organs
- Mesentery
- Retroperitoneal organs
- Organs surrounded by visceral peritoneum
- Mesenteries

**Figure 1.14  Location of Serous Membranes**

(a) Frontal section showing the parietal pericardium (*blue*), visceral pericardium (*red*), and pericardial cavity. (b) Frontal section showing the parietal pleura (*blue*), visceral pleura (*red*), and pleural cavities. (c) Sagittal section through the abdominopelvic cavity showing the parietal peritoneum (*blue*), visceral peritoneum (*red*), peritoneal cavity, mesenteries (*purple*), and retroperitoneal organs.

peritoneum of some abdominopelvic organs to the parietal peritoneum on the body wall or to the visceral peritoneum of other abdominopelvic organs. The mesenteries anchor the organs to the body wall and provide a pathway for nerves and blood vessels to reach the organs. Other abdominopelvic organs are more closely attached to the body wall and do not have mesenteries. Parietal peritoneum covers these other organs, which are said to be **retro-**

**peritoneal** (re′trō-per′i-tō-nē′ăl; *retro,* behind). The retroperitoneal organs include the kidneys, the adrenal glands, the pancreas, parts of the intestines, and the urinary bladder (figure 1.14*c*).

**Predict 5**

*Explain how an organ can be located within the abdominopelvic cavity but not be within the peritoneal cavity.*

---

## SUMMARY

Knowledge of anatomy and physiology can be used to predict the body's responses to stimuli when healthy or diseased.

### Anatomy    (p. 2)

1. Anatomy is the study of the structures of the body.
2. Systemic anatomy is the study of the body by organ systems. Regional anatomy is the study of the body by areas.
3. Surface anatomy uses superficial structures to locate deeper structures, and anatomical imaging is a noninvasive method for examining deep structures.

### Physiology    (p. 2)

Physiology is the study of the processes and functions of the body.

### Structural and Functional Organization of the Human Body    (p. 2)

1. The human body can be organized into six levels: chemical, cell, tissue, organ, organ system, and organism.
2. The 11 organ systems are the integumentary, skeletal, muscular, lymphatic, respiratory, digestive, nervous, endocrine, cardiovascular, urinary, and reproductive systems (see figure 1.3).

## Characteristics of Life   (p. 6)

The characteristics of life are organization, metabolism, responsiveness, growth, development, and reproduction.

## Homeostasis   (p. 7)

1. Homeostasis is the condition in which body functions, body fluids, and other factors of the internal environment are maintained within a range of values suitable to support life.
2. Negative-feedback mechanisms maintain homeostasis.
3. Positive-feedback mechanisms make deviations from normal even greater. Although a few positive-feedback mechanisms normally exist in the body, most positive-feedback mechanisms are harmful.

## Terminology and the Body Plan   (p. 10)

### Body Positions

1. A human standing erect with the face directed forward, the arms hanging to the sides, and the palms facing forward is in the anatomical position.
2. A person lying face upward is supine and face downward is prone.

### Directional Terms

Directional terms always refer to the anatomical position, regardless of the body's actual position (see table 1.1).

### Body Parts and Regions

1. The body can be divided into the head, neck, trunk, upper limbs, and lower limbs.
2. The abdomen can be divided superficially into four quadrants or nine regions, which are useful for locating internal organs or describing the location of a pain.

### Planes

1. A sagittal plane divides the body into left and right parts, a transverse plane divides the body into superior and inferior parts, and a frontal plane divides the body into anterior and posterior parts.
2. A longitudinal section divides an organ along its long axis, a transverse section cuts an organ at a right angle to the long axis, and an oblique section cuts across the long axis at an angle other than a right angle.

### Body Cavities

1. The thoracic cavity is bounded by the ribs and the diaphragm. The mediastinum divides the thoracic cavity into two parts.
2. The abdominal cavity is bounded by the diaphragm and the abdominal muscles.
3. The pelvic cavity is surrounded by the pelvic bones.

### Serous Membranes

1. The trunk cavities are lined by serous membranes. The parietal part of a serous membrane lines the wall of the cavity, and the visceral part covers the internal organs.
2. The serous membranes secrete fluid that fills the space between the parietal and visceral membranes. The serous membranes protect organs from friction.
3. The pericardial cavity surrounds the heart, the pleural cavities surround the lungs, and the peritoneal cavity surrounds certain abdominal and pelvic organs.
4. Mesenteries are parts of the peritoneum that hold the abdominal organs in place and provide a passageway for blood vessels and nerves to organs.
5. Retroperitoneal organs are found "behind" the parietal peritoneum. The kidneys, the adrenal glands, the pancreas, parts of the intestines, and the urinary bladder are examples of retroperitoneal organs.

 REVIEW AND COMPREHENSION

1. Define anatomy, surface anatomy, anatomical imaging, and physiology.
2. List six structural levels at which the body can be studied.
3. Define tissue. What are the four primary tissue types?
4. Define organ and organ system. What are the 11 organ systems of the body and their functions?
5. Name six characteristics of life.
6. What does the term *homeostasis* mean? If a deviation from homeostasis occurs, what kind of mechanism restores homeostasis?
7. Describe a negative-feedback mechanism in terms of receptor, control center, and effector. Give an example of a negative-feedback mechanism in the body.
8. Define positive feedback. Why are positive-feedback mechanisms generally harmful? Give an example of a harmful and a beneficial positive-feedback mechanism in the body.
9. Why is knowledge of the etymology of anatomical and physiological terms useful?
10. Describe the anatomical position. Why is it important to remember the anatomical position when using directional terms?
11. Define and give an example of the following directional terms: *inferior, superior, anterior, posterior, dorsal, ventral, proximal, distal, lateral, medial, superficial,* and *deep.*
12. List the subdivisions of the trunk, the upper limbs, and the lower limbs.
13. Describe the four-quadrant and nine-region methods of subdividing the abdomen. What is the purpose of these methods?
14. Define the sagittal, median, transverse, and frontal planes of the body.
15. Define the longitudinal, transverse, and oblique sections of an organ.
16. Define the following cavities: thoracic, abdominal, pelvic, and abdominopelvic. What is the mediastinum?
17. What is the difference between the visceral and parietal layers of a serous membrane? What function do serous membranes perform?
18. Name the serous membranes associated with the heart, lungs, and abdominopelvic organs.
19. Define mesentery. What does the term *retroperitoneal* mean? Give an example of a retroperitoneal organ.

 CRITICAL THINKING

1. A male has lost blood as a result of a gunshot wound. Even though the bleeding has been stopped, his blood pressure is low and dropping, and his heart rate is elevated. Following a blood transfusion, his blood pressure increases and his heart rate decreases. Which of the following statement(s) is (are) consistent with these observations?

a. Negative-feedback mechanisms can be inadequate without medical intervention.

b. The transfusion interrupted a positive-feedback mechanism.

c. The increased heart rate after the gunshot wound and before the transfusion is a result of a positive-feedback mechanism.

d. Both a and b are correct.

e. Answers a, b, and c are correct.

2. During physical exercise, the respiration rate increases. Two students are discussing the mechanisms involved. Student A claims they are positive-feedback mechanisms, and student B claims they are negative-feedback mechanisms. Do you agree with student A or student B, and why?

3. Complete the following statements using the correct directional terms for the human body.

a. The navel is _____ to the nose.

b. The heart is _____ to the breastbone (sternum).

c. The forearm is _____ to the arm.

d. The ear is _____ to the brain.

4. Describe, using as many directional terms as you can, the relationship between your kneecap and your heel.

5. According to "Dear Abby," a wedding band should be worn closest to the heart, and an engagement ring should be worn as a "guard" on the outside. Should a woman's wedding band be worn proximal or distal to her engagement ring?

6. In which quadrant and region is the pancreas located? In which quadrant and region is the urinary bladder located?

7. During pregnancy, which would increase more in size, the mother's abdominal cavity or her pelvic cavity? Explain.

8. A bullet enters the left side of a male, passes through the left lung, and lodges in the heart. Name in order the serous membranes and their cavities through which the bullet passes.

9. Can a kidney be removed without cutting through parietal peritoneum? Explain.

*Answers in Appendix D*

Visit this textbook's website at **www.mhhe.com/seeleyess7** for practice quizzes, animations, interactive learning exercises, and other study tools.

McGraw-Hill offers a study CD that features interactive cadaver dissection. ***Anatomy & Physiology Revealed*** includes cadaver photos that allow you to peel away layers of the human body to reveal structures beneath the surface. This program also includes animations, radiologic imaging, audio pronunciations, and practice quizzing.

The human body is made of many chemicals. Understanding the interactions between and the behavior of these chemicals is essential for understanding human physiology. For example, interactions between water molecules explain why sweating helps you feel cooler.

# Chapter 2

# The Chemical Basis of Life

## Chapter Outline and Objectives

### Basic Chemistry (p. 21)

1. Define matter, mass, and weight.
2. Define element and atom.
3. Name the subatomic particles of an atom, and describe how they are organized.
4. Describe two types of chemical bonds.
5. Define hydrogen bond, and explain its importance.
6. Distinguish between a molecule and a compound.
7. Describe the process of dissociation.

### Chemical Reactions (p. 26)

8. Using symbols, explain synthesis, decomposition, and exchange reactions.
9. Explain how reversible reactions produce chemical equilibrium.
10. Distinguish between chemical reactions that release energy and those that take in energy.
11. List the factors that affect the rate of chemical reactions.

### Acids and Bases (p. 30)

12. Describe the pH scale and its relationship to acidic and basic solutions.
13. Explain why buffers are important.

### Inorganic Chemistry (p. 31)

14. Distinguish between inorganic and organic molecules.
15. State the importance of oxygen, carbon dioxide, and water for living organisms.

### Organic Chemistry (p. 32)

16. Describe four important types of organic molecules and their functions in the human body.
17. Explain how enzymes work.

Chemicals make up the body's structures, and the interactions of chemicals with one another are responsible for the body's functions. The processes of nerve impulse generation, digestion, muscle contraction, and metabolism can all be described in chemical terms. Many abnormal conditions and their treatments can also be explained in chemical terms, even though their symptoms appear as malfunctions in organ systems. For example, Parkinson disease, which causes uncontrolled shaking movements, results from a shortage of a chemical called dopamine in certain nerve cells in the brain. It can be treated by giving patients another chemical, which brain cells convert to dopamine.

A basic knowledge of chemistry is essential for understanding anatomy and physiology.

**Chemistry** is the scientific discipline concerned with the atomic composition and structure of substances and the reactions they undergo. This chapter outlines some basic chemical principles and emphasizes their relationship to humans. Refer back to this chapter when chemical phenomena are discussed later in the text.

# BASIC CHEMISTRY

## Matter, Mass, and Weight

All living and nonliving things are composed of **matter,** which is anything that occupies space and has mass. **Mass** is the amount of matter in an object, and **weight** is the gravitational force acting on an object of a given mass. For example, the weight of an apple results from the force of gravity "pulling" on the apple's mass.

### Predict 1

*The difference between mass and weight can be illustrated by considering an astronaut. How would an astronaut's mass and weight in outer space compare with his mass and weight on the earth's surface?*

The international unit for mass is the **kilogram (kg),** which is the mass of a platinum-iridium cylinder kept at the International Bureau of Weights and Measurements in France. The mass of all other objects is compared with this cylinder. For example, a 2.2-lb lead weight and 1 liter (L) (1.06 qt) of water each has a mass of approximately 1 kg. An object with 1/1000 the mass of the standard kilogram cylinder is said to have a mass of 1 **gram (g).**

## Elements and Atoms

An **element** is the simplest type of matter having unique chemical properties. A list of the elements commonly found in the human body appears in table 2.1. About 96% of the body's weight results from the elements oxygen, carbon, hydrogen, and nitrogen. Many other elements also play important roles in the human body. For example, calcium helps form bones, and sodium ions are essential for neuronal activity. Some of these elements are present in only trace amounts but are still essential for life.

An **atom** (at'ŏm; indivisible) is the smallest particle of an element that has the chemical characteristics of that element. An element is composed of atoms of only one kind. For example, the element carbon is composed of only carbon atoms, and the element oxygen is composed of only oxygen atoms.

| TABLE 2.1 | Common Elements in the Human Body | | | | |
|---|---|---|---|---|---|
| Element | Symbol | Atomic Number | Mass Number | Percent in Human Body by Weight | Percent in Human Body by Number of Atoms |
| Hydrogen | H | 1 | 1 | 9.5 | 63.0 |
| Carbon | C | 6 | 12 | 18.5 | 9.5 |
| Nitrogen | N | 7 | 14 | 3.3 | 1.4 |
| Oxygen | O | 8 | 16 | 65.0 | 25.5 |
| Sodium | Na | 11 | 23 | 0.2 | 0.3 |
| Phosphorus | P | 15 | 31 | 1.0 | 0.22 |
| Sulfur | S | 16 | 32 | 0.3 | 0.05 |
| Chlorine | Cl | 17 | 35 | 0.2 | 0.03 |
| Potassium | K | 19 | 39 | 0.4 | 0.06 |
| Calcium | Ca | 20 | 40 | 1.5 | 0.31 |
| Iron | Fe | 26 | 56 | Trace | Trace |
| Iodine | I | 53 | 127 | Trace | Trace |

An element, or an atom of that element, is often represented by a symbol. Usually the symbol is the first letter or letters of the element's name—for example, C for carbon, H for hydrogen, and Ca for calcium. Occasionally, the symbol is taken from the Latin, Greek, or Arabic name for the element; for example, the symbol for sodium is Na, from the Latin word *natrium*.

## Atomic Structure

The characteristics of matter result from the structure, organization, and behavior of atoms. Atoms are composed of subatomic particles, some of which have an electrical charge. The three major types of subatomic particles are neutrons, protons, and electrons. **Neutrons** (noo'tronz) have no electrical charge, **protons** (prō'tonz) have positive charges, and **electrons** (e-lek'tronz) have negative charges. The positive charge of a proton is equal in magnitude to the negative charge of an electron. The number of protons and number of electrons in each atom are equal, and the individual charges cancel each other. Therefore, each atom is electrically neutral.

Protons and neutrons form the **nucleus** at the center of the atom, and electrons move around the nucleus (figure 2.1). The nucleus accounts for 99.97% of an atom's mass, but only 1-ten-trillionth of its volume. Most of the volume of an atom is occupied by the electrons. Although it is impossible to know precisely where any given electron is located at any particular moment, the region where electrons are most likely to be found can be represented by an **electron cloud** (figure 2.1).

Each element is uniquely defined by the number of protons in the atoms of that element. For example, only hydrogen atoms have one proton, only carbon atoms have six protons, and only oxygen atoms have eight protons (figure 2.2; see table 2.1). The number of protons in each atom is called the **atomic number,** and because the number of electrons and number of protons are equal, the atomic number is also the number of electrons.

Protons and neutrons have about the same mass, and they are responsible for most of the mass of atoms. Electrons, on the other hand, have very little mass. The **mass number** of an element is the number of protons plus the number of neutrons in each atom. For example, the mass number for carbon is 12 because it has 6 protons and 6 neutrons.

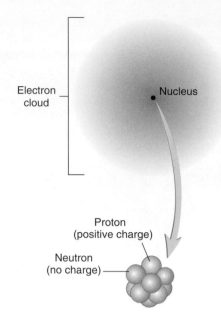

**Figure 2.1  Model of an Atom**

The tiny, dense nucleus consists of positively charged protons and uncharged neutrons. Most of the volume of an atom is occupied by rapidly moving, negatively charged electrons, which can be represented as an electron cloud.

*The atomic number of fluorine is 9, and the mass number is 19. What is the number of protons, neutrons, and electrons in an atom of fluorine?*

## Electrons and Chemical Bonding

The chemical behavior of an atom is determined largely by its outermost electrons. **Chemical bonding** occurs when the outermost electrons are transferred, or shared, between atoms. Two major types of chemical bonding are ionic bonding and covalent bonding.

### Ionic Bonding

An atom is electrically neutral because it has an equal number of protons and electrons. If an atom loses or gains electrons, the numbers of protons and electrons are no longer equal, and a

**Figure 2.2  Hydrogen, Carbon, and Oxygen Atoms**

Within the nucleus of an atom are specific numbers of positively charged protons ($p^+$) and uncharged neutrons ($n^0$). The negatively charged electrons ($e^-$) are around the nucleus. The atoms depicted here are electrically neutral because the number of protons and number of electrons within each atom are equal.

**Figure 2.3  Ionic Bonding**

(a) A sodium atom loses an electron to become a smaller, positively charged ion, and a chlorine atom gains an electron to become a larger, negatively charged ion. The attraction between the oppositely charged ions results in ionic bonding and the formation of sodium chloride. (b) The $Na^+$ and $Cl^-$ are organized to form a cube-shaped array. (c) A photomicrograph of salt crystals reflects the cubic arrangement of the ions.

charged particle called an **ion** (ī'on) is formed. After an atom loses an electron, it has one more proton than it has electrons and is positively charged. For example, a sodium atom (Na) can lose an electron to become a positively charged sodium ion ($Na^+$) (figure 2.3a). After an atom gains an electron, it has one more electron than it has protons and is negatively charged. For example, a chlorine atom (Cl) can accept an electron to become a negatively charged chloride ion ($Cl^-$). Because oppositely charged ions are attracted to each other, positively charged ions tend to remain close to negatively charged ions, forming an **ionic** (ī-on'ik) **bond.** Thus, $Na^+$ and $Cl^-$ are held together by ionic bonding to form an array of ions called sodium chloride (NaCl), or table salt (figure 2.3b,c).

Ions are denoted by using the symbol of the atom from which the ion was formed and adding a plus (+) or minus (−) superscript to indicate the ion's charge. For example, a sodium ion is $Na^+$, and a chloride ion is $Cl^-$. If more than one electron has been lost or gained, a number is used with the plus or minus sign. Thus, $Ca^{2+}$ is a calcium ion formed by the loss of two electrons. Table 2.2 lists some ions commonly found in the body.

**Predict 3**

*If an iron (Fe) atom loses three electrons, what is the charge of the resulting ion? Write the symbol for this ion.*

| TABLE 2.2 | | Important Ions in the Human Body |
|---|---|---|
| **Ion** | **Symbol** | **Significance\*** |
| Calcium | $Ca^{2+}$ | Part of bones and teeth; functions in blood clotting, muscle contraction, release of neurotransmitters |
| Sodium | $Na^+$ | Membrane potentials, water balance |
| Potassium | $K^+$ | Membrane potentials |
| Hydrogen | $H^+$ | Acid-base balance |
| Hydroxide | $OH^-$ | Acid-base balance |
| Chloride | $Cl^-$ | Water balance |
| Bicarbonate | $HCO_3^-$ | Acid-base balance |
| Ammonium | $NH_4^+$ | Acid-base balance |
| Phosphate | $PO_4^{3-}$ | Part of bones and teeth; functions in energy exchange, acid-base balance |
| Iron | $Fe^{2+}$ | Red blood cell function |
| Magnesium | $Mg^{2+}$ | Necessary for enzymes |
| Iodide | $I^-$ | Present in thyroid hormones |

\*The ions are part of the structures or play important roles in the processes listed.

The two hydrogen atoms do not interact because they are too far apart.

The positively charged nucleus of each hydrogen atom begins to attract the electron of the other.

A covalent bond forms when the electrons are shared between the nuclei because the electrons are equally attracted to each nucleus.

**Figure 2.4  Covalent Bonding**

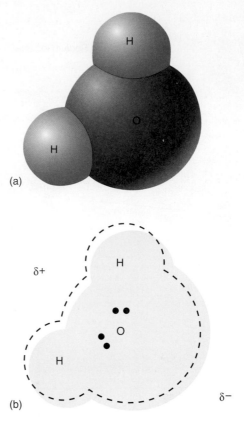

(a)

(b)

**Figure 2.5  Polar Covalent Bonds**

(*a*) A water molecule forms when two hydrogen atoms form covalent bonds with an oxygen atom. (*b*) Electron pairs (*indicated by the black dots*) are shared between the hydrogen atoms and oxygen. The *dashed outline* shows the expected location of the electron cloud if the electrons are shared equally. The electrons are shared unequally, as shown by the electron cloud (*yellow*) not coinciding with the *dashed outline*. Consequently, the oxygen side of the molecule has a slight negative charge (*indicated by* $\delta^-$), and the hydrogen side of the molecule has a slight positive charge (*indicated by* $\delta^+$).

## Covalent Bonding

A **covalent bond** forms when atoms share one or more pairs of electrons. The resulting combination of atoms is called a **molecule** (mol′ĕ-kūl). An example is the covalent bond between two hydrogen atoms to form a hydrogen molecule (figure 2.4). Each hydrogen atom has one electron. As the atoms get closer together, the positively charged nucleus of each atom begins to attract the electron of the other atom. At an optimal distance, the two nuclei mutually attract the two electrons, and each electron is shared by both nuclei. The two hydrogen atoms are now held together by a covalent bond.

The sharing of one pair of electrons by two atoms results in a **single covalent bond.** A single line between the symbols of the atoms involved (for example, H—H) represents a single covalent bond. A **double covalent bond** results when two atoms share two pairs of electrons. When a carbon atom combines with two oxygen atoms to form carbon dioxide, two double covalent bonds are formed. Double covalent bonds are indicated by a double line between the atoms (O=C=O).

Electrons can be shared unequally in covalent bonds. When there is an unequal, asymmetrical sharing of electrons, the bond is called a **polar covalent bond** because the unequal sharing of electrons results in one end (pole) of the molecule having a partial electrical charge opposite to that of the other end. For example, two hydrogen atoms can share their electrons with an oxygen

atom to form a water molecule ($H_2O$), as shown in figure 2.5. However, the hydrogen atoms do not share the electrons equally with the oxygen atom, and the electrons tend to spend more time around the oxygen atoms than around the hydrogen atoms. Molecules with this asymmetrical electrical charge are called **polar molecules.**

When there is an equal sharing of electrons between atoms, the bond is called a **nonpolar covalent bond.** Molecules with a symmetrical electrical charge are called **nonpolar molecules.**

## Hydrogen Bonds

A polar molecule has a positive "end" and a negative "end". The positive end of one polar molecule can be weakly attracted to the negative end of another polar molecule. Although this attraction is called a **hydrogen bond,** it is not a chemical bond because electrons are not transferred, or shared, between the atoms of the different polar molecules. The attraction between molecules resulting from hydrogen bonds is much weaker than in ionic or covalent bonds. For example, the positively charged hydrogen of one water molecule is weakly attracted to a negatively charged oxygen of

## Ions and Electricity

Ions that dissociate in water are sometimes called **electrolytes** (ē-lek′trō-lītz) because they have the capacity to conduct an electrical current, which is the flow of charged particles. An electrocardiogram (ECG) is a recording of electrical currents produced by the heart. These currents can be detected by electrodes on the surface of the body because the ions in the body fluids conduct electrical currents.

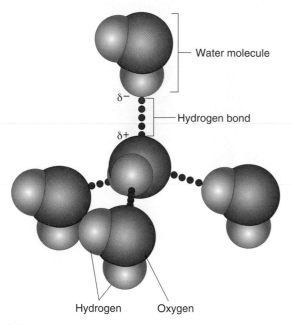

**Figure 2.6 Hydrogen Bonds**
The positive hydrogen part of one water molecule (δ⁺) forms a hydrogen bond (*red dotted line*) with the negative oxygen part of another water molecule (δ⁻). As a result, hydrogen bonds hold the water molecules together.

another water molecule (figure 2.6). Thus, the water molecules are held together by hydrogen bonds.

Hydrogen bonds also play an important role in determining the shape of complex molecules. The bonds can occur between different polar parts of a single large molecule to hold the molecule in its normal three-dimensional shape (see "Proteins" and "Nucleic Acids: DNA and RNA" later in this chapter).

## Molecules and Compounds

A molecule is formed when two or more atoms chemically combine to form a structure that behaves as an independent unit. Sometimes the atoms that combine are of the same type, as when two hydrogen atoms combine to form a hydrogen molecule. But more typically, a molecule consists of two or more different types of atoms, such as two hydrogen atoms and an oxygen atom combining to form water. Thus, a glass of water consists of a collection of individual water molecules positioned next to one another.

A **compound** (kom′pownd; to place together) is a substance resulting from the chemical combination of two or more *different* types of atoms. Water is an example of a substance that is a compound and a molecule. Not all molecules are compounds. For example, a hydrogen molecule is not a compound because it does not consist of different types of atoms.

Some compounds are molecules and some are not. (Remember that, to be a molecule, a structure must be an independent unit.) Covalent compounds, in which different types of atoms are held together by covalent bonds, are molecules because the sharing of electrons results in distinct units. On the other hand, ionic compounds, in which ions are held together by the force of attraction between opposite charges, are not molecules because they do not consist of distinct units. Sodium chloride is an example of a substance that is a compound but not a molecule. A piece of sodium chloride does not consist of individual sodium chloride molecules positioned next to one another. Instead, it is an organized array of individual $Na^+$ and individual $Cl^-$ in which each charged ion is surrounded by several ions of the opposite charge (see figure 2.3b).

Molecules and compounds can be represented by the symbols of the atoms forming the molecule or compound plus subscripts denoting the quantity of each type of atom present. For example, glucose (a sugar) can be represented as $C_6H_{12}O_6$, indicating that glucose is composed of 6 carbon, 12 hydrogen, and 6 oxygen atoms.

## Dissociation

When ionic compounds dissolve in water, their ions **dissociate** (di-sō′sē-āt′), or separate, from each other because the positively charged ions are attracted to the negative ends of the water molecules, and the negatively charged ions are attracted to the positive ends of the water molecules. For example, when sodium chloride dissociates in water, the $Na^+$ and $Cl^-$ separate, and water molecules surround and isolate the ions, keeping them in solution (figure 2.7).

When molecules dissolve in water, the molecules usually remain intact even though they are surrounded by water molecules. Thus, in a glucose solution, glucose molecules are surrounded by water molecules.

## Figure 2.7  Dissociation

Sodium chloride (table salt) dissociates in water. The positively charged Na$^+$ are attracted to the negatively charged oxygen (*red*) end of the water molecule, and the negatively charged Cl$^-$ are attracted to the positively charged hydrogen (*blue*) end of the water molecule.

# CHEMICAL REACTIONS

In a **chemical reaction,** atoms, ions, molecules, or compounds interact either to form or to break chemical bonds. The substances that enter into a chemical reaction are called the **reactants,** and the substances that result from the chemical reaction are called the **products.**

## Classification of Chemical Reactions

For our purposes, chemical reactions can be classified as synthesis, decomposition, or exchange reactions.

### Synthesis Reactions

When two or more reactants combine to form a larger, more complex product, the process is called a **synthesis reaction,** represented symbolically as

$$A + B \rightarrow AB$$

Examples of synthesis reactions include the synthesis of the complex molecules of the human body from the basic "building blocks" obtained in food and the synthesis of **adenosine triphosphate (ATP)** (ă-den′ō-sēn trī-fos′făt) molecules. In ATP, *A* stands for adenosine, *T* stands for *tri-* (or three), and *P* stands for a phosphate

group (PO$_4^{3-}$). Thus, ATP consists of adenosine and three phosphate groups. ATP is synthesized when adenosine diphosphate (ADP), which has two (*di-*) phosphate groups, combines with a phosphate group to form the larger ATP molecule. The phosphate group that reacts with ADP is often denoted as P$_i$, where the *i* indicates that the phosphate group is associated with an inorganic substance (see "Inorganic Chemistry" later in this chapter).

$$\begin{array}{ccccc} \text{A-P-P} & + & \text{P}_i & \rightarrow & \text{A-P-P-P} \\ \text{(ADP)} & & \text{(Phosphate} & & \text{(ATP)} \\ & & \text{group)} & & \end{array}$$

### Decomposition Reactions

In a **decomposition reaction,** reactants are broken down into smaller, less complex products. A decomposition reaction is the reverse of a synthesis reaction and can be represented in this way:

$$AB \rightarrow A + B$$

Examples of decomposition reactions include the breakdown of food molecules into basic building blocks and the breakdown of ATP to ADP and a phosphate group.

$$\begin{array}{ccccc} \text{A-P-P-P} & \rightarrow & \text{A-P-P} & + & \text{P}_i \\ \text{(ATP)} & & \text{(ADP)} & & \text{(Phosphate group)} \end{array}$$

# CLINICAL IMPACT

## Clinical Applications of Atomic Particles

Protons, neutrons, and electrons are responsible for the chemical properties of atoms. They also have other properties that can be useful in a clinical setting. For example, some of these properties have enabled the development of methods for examining the inside of the body.

**Isotopes** (ī′sō-tōpz; *isos*, equal + *topos*, part) are two or more forms of the same element that have the same number of protons and electrons but a different number of neutrons. Thus, isotopes have the same atomic number (i.e., number of protons) but different mass numbers (i.e., sum of the protons and neutrons). For example, hydrogen and its isotope deuterium each have an atomic number of 1 because they both have 1 proton. However, hydrogen has no neutrons, whereas deuterium has 1 neutron. Therefore, the mass number of hydrogen is 1, and that of deuterium is 2. Water made with deuterium is called heavy water because of the weight of the "extra" neutron. Because isotopes of the same atom have the same number of electrons, they are very similar in their chemical behavior. The nuclei of some isotopes are stable and do not change. Radioactive isotopes, however, have unstable nuclei that lose neutrons or protons. Several different kinds of radiation can be produced when neutrons and protons, or the products formed by their breakdown, are released from the nucleus of the isotope.

The radiation given off by some radioactive isotopes can penetrate and destroy tissues. Rapidly dividing cells are more sensitive to radiation than are slowly dividing cells. Radiation is used to treat cancerous (malignant) tumors because cancer cells divide rapidly. If the treatment is effective, few healthy cells are destroyed, but the cancerous cells are killed.

Radioactive isotopes are also used in medical diagnosis. The radiation can be detected, and the movement of the radioactive isotopes throughout the body can be traced. For example, the thyroid gland normally takes up iodine and uses it in the formation of thyroid hormones. Radioactive iodine can be used to determine if iodine uptake is normal in the thyroid gland.

Radiation can be produced in ways other than by changing the nucleus of atoms. X-rays are a type of radiation formed when electrons lose energy by moving from a higher energy state to a lower one. Health professionals use x-rays to examine bones to determine if they are broken and x-rays of teeth to see if they have caries (cavities). Mammograms, which are low-energy radiographs (x-ray films) of the breast, can reveal tumors because the tumors are slightly denser than normal tissue.

Computers can be used to analyze a series of radiographs, each made at a slightly different body location. The computer assembles these radiographic "slices" through the body to form a three-dimensional image. A **computed tomography** (tō-mog′ră-fē) **(CT) scan** is an example of this technique (figure 2A). CT scans are used to detect tumors and other abnormalities in the body.

**Magnetic resonance imaging (MRI)** is another method for looking into the body (figure 2B). The patient is placed into a very powerful magnetic field, which aligns the hydrogen nuclei. Radiowaves given off by the hydrogen nuclei are monitored, and a computer uses these data to make an image of the body. Because MRI detects hydrogen, it is very effective for visualizing soft tissues that contain a lot of water. MRI technology can reveal tumors and other abnormalities.

**Figure 2A**

CT scan of the brain with iodine injection, showing three brain tumors (ovals) that have metastasized (spread) to the brain from cancer in the large intestine.

**Figure 2B**

Colorized MRI brain scan showing a stroke. The whitish area in the lower right part of the MRI is blood that has leaked into the surrounding tissue.

## Exchange Reactions

An **exchange reaction** is a combination of a decomposition reaction and a synthesis reaction. In decomposition, the reactants are broken down. In synthesis, the products of the decomposition reaction are combined. The symbolic representation of an exchange reaction is

$$AB + CD \rightarrow AC + BD$$

An example of an exchange reaction is the reaction of hydrochloric acid (HCl) with sodium hydroxide (NaOH) to form table salt (NaCl) and water ($H_2O$):

$$HCl + NaOH \rightarrow NaCl + H_2O$$

## Reversible Reactions

A **reversible reaction** is a chemical reaction that can proceed from reactants to products and from products to reactants. When the rate of product formation is equal to the rate of reactant formation, the reaction is said to be at **equilibrium.** At equilibrium, the amount of the reactants relative to the amount of products remains constant.

The following analogy may help clarify the concept of reversible reactions and equilibrium. Imagine a trough containing water. The trough is divided into two compartments by a partition, but the partition contains holes that allow the water to move freely between the compartments. Because the water can move in either direction, this is like a reversible reaction. Let the amount of water in the left compartment represent the amount of reactant, and the amount of water in the right compartment represent the amount of product. At equilibrium, the amount of reactant relative to the amount of product in each compartment is always the same because the partition allows water to pass between the two compartments until the level of water is the same in both compartments. If the amount of reactant is increased by adding water to the left compartment, water flows from the left compartment through the partition to the right compartment until the level of water is the same in both. Thus, the amounts of reactant and product are once again equal. Unlike this analogy, however, the amount of reactants relative to the products in most reversible reactions is not one to one. Depending on the specific reversible reaction, there can be one part reactant to two parts product, two parts reactant to one part product, or many other possibilities.

An important reversible reaction in the human body occurs when carbon dioxide ($CO_2$) and water ($H_2O$) form hydrogen ions ($H^+$) and bicarbonate ions ($HCO_3^-$). The reversibility of the reaction is indicated by two arrows pointing in opposite directions:

$$CO_2 + H_2O \rightleftarrows H^+ + HCO_3^-$$

If $CO_2$ is added to $H_2O$, the amount of $CO_2$ relative to the amount of $H^+$ increases. However, the reaction of $CO_2$ with $H_2O$ produces more $H^+$, and the amount of $CO_2$ relative to the amount of $H^+$ returns to equilibrium. Conversely, adding $H^+$ results in the formation of more $CO_2$, and the equilibrium is restored.

Maintaining a constant level of $H^+$ in body fluids is necessary for the nervous system to function properly. This level can be maintained, in part, by controlling blood $CO_2$ levels. For example, slowing the respiration rate causes blood $CO_2$ levels to increase, which causes an increase in $H^+$ concentration in the blood.

**Predict 4**

*If the respiration rate increases, $CO_2$ is removed from the blood. What effect does this have on blood $H^+$ levels?*

## Energy and Chemical Reactions

**Energy** is defined as the capacity to do **work**—that is, to move matter. Energy can be subdivided into potential energy and kinetic energy. **Potential energy** is stored energy that could do work but is not doing so. For example, a coiled spring has potential energy. It could push against an object and move the object, but as long as the spring does not uncoil, no work is accomplished. **Kinetic** (ki-net′ik; of motion) **energy** is energy caused by the movement of an object and is the form of energy that actually does work. An uncoiling spring pushing an object and causing it to move is an example. When potential energy is released, it becomes kinetic energy, thus doing work.

Potential and kinetic energy exist in many different forms: chemical energy, mechanical energy, heat energy, electrical energy, and electromagnetic (radiant) energy. Here we examine how chemical energy and mechanical energy play important roles in the body.

The **chemical energy** of a substance is a form of potential energy stored in chemical bonds. Consider two balls attached by a relaxed spring. In order to push the balls together and compress the spring, energy must be put into this system. As the spring is compressed, potential energy increases. When the compressed spring expands, potential energy decreases. In the same way, similarly charged particles, such as two negatively charged electrons or two positively charged nuclei, repel each other. As similarly charged particles move closer together, their potential energy increases, much like compressing a spring, and as they move farther apart, their potential energy decreases. Chemical bonding is a form of potential energy because of the charges and positions of the subatomic particles bound together.

Chemical reactions are important because of the products they form and the energy changes that result as the relative position of subatomic particles changes. If the products of a chemical reaction contain less potential energy than the reactants, energy is released. For example, food molecules contain more potential energy than waste products. The difference in potential energy between food and waste products is used by the human body to drive activities such as growth, repair, movement, and heat production.

An example of a reaction that releases energy is the breakdown of ATP to ADP and a phosphate group (figure 2.8a). The phosphate group is attached to the ADP molecule by a covalent bond, which has potential energy. After the breakdown of ATP, some of that energy is released as heat, and some is available to cells for activities such as synthesizing new molecules or contracting muscles:

$$ATP \rightarrow ADP + P_i + Energy \text{ (used by cells)}$$

According to the law of conservation of energy, the total energy of the universe is constant. Therefore, energy is neither created nor destroyed. However, one type of energy can be changed into another. Potential energy is converted into kinetic energy. Since the conversion between energy states is not 100%

**Figure 2.8**  **Energy and Chemical Reactions**

In the two reactions depicted here, the upper shelf represents a higher energy level, and the lower shelf represents a lower energy level. (*a*) Energy is released as a result of the breakdown of ATP. (*b*) The input of energy is required for the synthesis of ATP.

efficient, heat energy is released. For example, as a spring is released, its potential energy is converted to mechanical energy and heat energy. **Mechanical energy** is energy resulting from the position or movement of objects. Many of the activities of the human body, such as moving a limb, breathing, or circulating blood, involve mechanical energy.

**Predict 5**

*Why does body temperature increase during exercise?*

If the products of a chemical reaction contain more energy than the reactants (figure 2.8*b*), energy must be added from another source. The energy released during the breakdown of food molecules is the source of energy for this kind of reaction in the body. The energy from food molecules is used to synthesize molecules such as ATP, fats, and proteins:

$$\text{ADP} + \text{P}_i + \text{Energy (from food molecules)} \rightarrow \text{ATP}$$

## Rate of Chemical Reactions

The rate at which a chemical reaction proceeds is influenced by several factors, including how easily the substances react with one another, their concentrations, the temperature, and the presence of a catalyst.

### Reactants

Reactants differ from one another in their ability to undergo chemical reactions. For example, iron corrodes much more rapidly than does stainless steel. For this reason, during the refurbishment

of the Statue of Liberty, in the 1980s the iron bars forming the statue's skeleton were replaced with stainless steel bars.

### Concentration

Within limits, the greater the concentration of the reactants, the greater the rate at which a chemical reaction will occur because, as the concentration increases, the reacting molecules are more likely to come in contact with one another. For example, the normal concentration of oxygen inside cells enables it to come in contact with other molecules, producing the chemical reactions necessary for life. If the oxygen concentration decreases, the rate of chemical reactions decreases. A decrease in oxygen in cells can impair cell function and even result in cell death.

### Temperature

The rate of chemical reactions also increases when the temperature is increased. For example, reactions occur throughout the body at a faster rate when a person has a fever of only a few degrees. The result is increased activity in most organ systems, such as increased heart and respiratory rates. By contrast, the rate of reactions decreases when body temperature drops. The clumsy movement of very cold fingers results largely from the reduced rate of chemical reactions in cold muscle tissue.

### Catalysts

At normal body temperatures, most chemical reactions would take place too slowly to sustain life if not for substances called

## Acidosis and Alkalosis

The normal pH range for human blood is 7.35 to 7.45. If blood pH drops below 7.35, a condition called **acidosis** (as-i-dō′sis) results. The nervous system is depressed, and the individual becomes disoriented and possibly comatose. If blood pH rises above 7.45, **alkalosis** (al-kă-lō′sis) results. The nervous system becomes overexcitable, and the individual can be extremely nervous or have convulsions. Both acidosis and alkalosis can result in death.

catalysts. A **catalyst** (kat′ă-list) increases the rate of a chemical reaction, without itself being permanently changed or depleted. An enzyme is a protein molecule that acts as a catalyst. Many of the chemical reactions that occur in the body require enzymes. We consider enzymes in greater detail later in this chapter.

# ACIDS AND BASES

The body has many molecules and compounds, called acids and bases, that can alter body functions. An **acid** is a proton donor. Because a hydrogen atom without its electron is a proton, any substance that releases hydrogen ions in water is an acid. For example, hydrochloric acid (HCl) in the stomach forms hydrogen ions ($H^+$) and chloride ions ($Cl^-$):

$$HCl \rightarrow H^+ + Cl^-$$

A **base** is a proton acceptor. For example, sodium hydroxide (NaOH) forms sodium ions ($Na^+$) and hydroxide ions ($OH^-$). It is a base because the $OH^-$ is a proton acceptor that binds with a $H^+$ to form water.

## The pH Scale

The **pH scale** indicates the $H^+$ concentration of a solution (figure 2.9). The scale ranges from 0 to 14. A **neutral solution** has an equal number of $H^+$ and $OH^-$ and thus a pH of 7.0. An **acidic solution** has a greater concentration of $H^+$ than $OH^-$ and thus a pH less than 7.0. A **basic**, or *alkaline* (al′kă-līn), **solution** has fewer $H^+$ than $OH^-$ and thus a pH greater than 7.0.

As the pH value becomes smaller, the solution becomes more acidic; as the pH value becomes larger, the solution becomes more basic. A change of one unit on the pH scale represents a 10-fold change in the $H^+$ concentration. For example, a solution with a pH of 6.0 has 10 times more $H^+$ than a solution with a pH of 7.0. Thus, small changes in pH represent large changes in $H^+$ concentration.

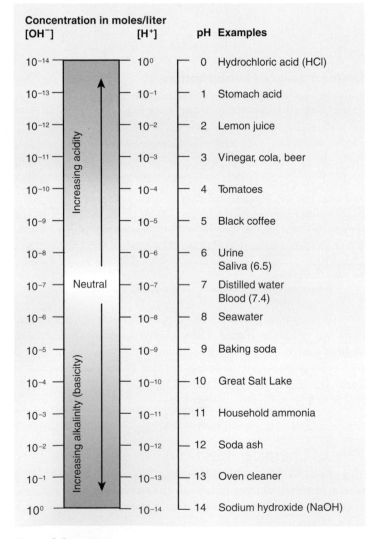

**Figure 2.9 pH Scale**

A pH of 7 is neutral. Values less than 7 are acidic (the lower the number, the more acidic). Values greater than 7 are basic (the higher the number, the more basic). This scale shows some representative fluids and their approximate pH values.

## Salts

A **salt** is a compound consisting of a positive ion other than $H^+$ and a negative ion other than $OH^-$. Salts are formed by the reaction of an acid and a base. For example, hydrochloric acid (HCl) combines with sodium hydroxide (NaOH) to form the salt sodium chloride (NaCl):

$$\text{HCl} \quad + \quad \text{NaOH} \quad \rightarrow \quad \text{NaCl} \quad + \quad H_2O$$
$$\text{(Acid)} \qquad \text{(Base)} \qquad \text{(Salt)} \qquad \text{(Water)}$$

## Buffers

The chemical behavior of many molecules changes as the pH of the solution in which they are dissolved changes. The survival of an organism depends on its ability to maintain homeostasis by keeping body fluid pH within a narrow range. One way normal body fluid pH is maintained is through the use of buffers. A **buffer**

**Figure 2.10  Buffers**

(a) The addition of an acid to a nonbuffered solution results in an increase of $H^+$ and a decrease in pH. (b) The addition of an acid to a buffered solution results in a much smaller change in pH. The added $H^+$ bind to the buffer (symbolized by the letter B).

(bŭf′er) is a chemical that resists changes in pH when either an acid or a base is added to a solution containing the buffer. When an acid is added to a buffered solution, the buffer binds to the $H^+$, preventing these ions from causing a decrease in the pH of the solution (figure 2.10).

> **Predict 6**
>
> *If a base is added to a solution, will the pH of the solution increase or decrease? If the solution is buffered, what response from the buffer prevents the change in pH?*

# INORGANIC CHEMISTRY

Early scientists believed that inorganic substances came from nonliving sources and that organic substances were extracted from living organisms. As the science of chemistry developed, however, it became apparent that the body also contains inorganic substances and that organic substances can be manufactured in the laboratory. As currently defined, **inorganic chemistry** deals with those substances that do not contain carbon, whereas **organic chemistry** is the study of carbon-containing substances. These definitions have a few exceptions. For example, carbon dioxide and carbon monoxide are classified as inorganic molecules, even though they contain carbon.

Inorganic substances play many vital roles in human anatomy and physiology. Examples include the oxygen and other gases we breathe, the calcium phosphate that makes up our bones, and the metals that are required for protein functions, such as iron in hemoglobin and zinc in alcohol dehydrogenase. In the next sections, we discuss the important roles of oxygen, carbon dioxide, and water—all inorganic molecules—in the body.

## Oxygen and Carbon Dioxide

**Oxygen ($O_2$)** is an inorganic molecule consisting of two oxygen atoms bound together by a double covalent bond. About 21% of the gas in the atmosphere is oxygen, and it is essential for most living organisms. Humans require oxygen in the final step of a series of chemical reactions in which energy is extracted from food molecules (see chapter 17).

**Carbon dioxide ($CO_2$)** consists of one carbon atom bound to two oxygen atoms. Each oxygen atom is bound to the carbon atom by a double covalent bond. Carbon dioxide is produced when food molecules, such as glucose, are metabolized within the cells of the body. Once carbon dioxide is produced, it is eliminated from the cell as a metabolic by-product, transferred to the lungs by the blood, and exhaled during respiration. If carbon dioxide is allowed to accumulate within cells, it becomes toxic.

## Water

**Water ($H_2O$)** is an inorganic molecule that consists of one atom of oxygen joined by polar covalent bonds to two atoms of hydrogen. Water has many important roles in humans and all living organisms:

1. *Stabilizing body temperature.* Water can absorb large amounts of heat and remain at a stable temperature. Blood, which is mostly water, is warmed deep in the body and then flows to the surface, where the heat is released. In addition,

water evaporation in the form of sweat results in significant heat loss from the body.

2. *Providing protection.* Water is an effective lubricant. For example, tears protect the surface of the eye from the rubbing of the eyelids. Water also forms a fluid cushion around organs, which helps protect them from damage. The fluid that surrounds the brain is an example.

3. *Facilitating chemical reactions.* Most of the chemical reactions necessary for life do not take place unless the reacting molecules are dissolved in water. For example, sodium chloride must dissociate in water into $Na^+$ and $Cl^-$ before those ions can react with other ions. Water also directly participates in many chemical reactions. For example, during the digestion of food, large molecules and water react to form smaller molecules.

4. *Transporting substances.* Many substances dissolve in water and can be moved from place to place as the water moves. For example, blood transports nutrients, gases, and waste products within the body.

# ORGANIC CHEMISTRY

Carbon's ability to form covalent bonds with other atoms makes possible the formation of the large, diverse, complicated molecules necessary for life. Carbon atoms bound together by covalent bonds constitute the "backbone" of many large molecules. Two mechanisms that allow the formation of a wide variety of molecules are variation in the length of the carbon chains and the combination of the atoms bound to the backbone. For example, proteins have thousands of carbon atoms bound by covalent bonds to one another and to other atoms, such as nitrogen, sulfur, hydrogen, and oxygen.

The four major groups of organic molecules essential to living organisms are carbohydrates, lipids, proteins, and nucleic acids. Each of these groups has specific structural and functional characteristics (table 2.3).

## Carbohydrates

**Carbohydrates** are composed of carbon, hydrogen, and oxygen atoms. In most carbohydrates, for each carbon atom there are two hydrogen atoms and one oxygen atom. Note that this two-to-one ratio is the same as in water ($H_2O$). The molecules are called carbohydrates because each carbon (*carbo-*) is combined with the same atoms that form water (*hydrated*). For example, the chemical formula for glucose is $C_6H_{12}O_6$.

The smallest carbohydrates are **monosaccharides** (mon-ō-sak'ă-rīdz; one sugar), or simple sugars. Glucose (blood sugar) and fructose (fruit sugar) are important monosaccharide energy sources for many of the body's cells. Larger carbohydrates are formed by chemically binding monosaccharides together. For this reason, monosaccharides are considered the building blocks of

| TABLE 2.3 | Important Organic Molecules and Their Functions in the Body | | | |
|---|---|---|---|---|
| **Molecule** | **Elements** | **Building Blocks** | **Function** | **Examples** |
| Carbohydrate | C, H, O | Monosaccharides | Energy | Monosaccharides can be used as energy sources. Glycogen (a polysaccharide) is an energy-storage molecule. |
| Lipid | C, H, O (P, N in some) | Glycerol and fatty acids (for fats) | Energy | Fats can be stored and broken down later for energy; per unit of weight, fats yield twice as much energy as carbohydrates. |
| | | | Structure | Phospholipids and cholesterol are important components of cell membranes. |
| | | | Regulation | Steroid hormones regulate many physiological processes (e.g., estrogen and testosterone are responsible for many of the differences between males and females). |
| Protein | C, H, O, N (S in most) | Amino acids | Regulation | Enzymes control the rate of chemical reactions. Hormones regulate many physiological processes (e.g., insulin affects glucose transport into cells). |
| | | | Structure | Collagen fibers form a structural framework in many parts of the body. |
| | | | Energy | Proteins can be broken down for energy; per unit of weight, they yield the same energy as carbohydrates. |
| | | | Contraction | Actin and myosin in muscle are responsible for muscle contraction. |
| | | | Transport | Hemoglobin transports oxygen in the blood. |
| | | | Protection | Antibodies and complement protect against microorganisms and other foreign substances. |
| Nucleic acid | C, H, O, N, P | Nucleotides | Regulation | DNA directs the activities of the cell. |
| | | | Heredity | Genes are pieces of DNA that can be passed from one generation to the next. |
| | | | Gene expression | RNA is involved in gene expression. |

**Figure 2.11** **Carbohydrates**

(a) Glucose and fructose are monosaccharides that combine to form the disaccharide sucrose. (b) Glycogen is a polysaccharide formed by combining many glucose molecules.

carbohydrates. **Disaccharides** (dī-sak′ă-rīdz; two sugars) are formed when two monosaccharides are joined by a covalent bond. For example, glucose and fructose combine to form the disaccharide sucrose (table sugar) (figure 2.11a). **Polysaccharides** (pol-ē-sak′ă-rīdz; many sugars) consist of many monosaccharides bound in long chains. Glycogen, or animal starch, is a polysaccharide of glucose (figure 2.11b). When cells containing glycogen need energy, the glycogen is broken down into individual glucose molecules, which can be used as energy sources. Plant starch, also a polysaccharide of glucose, can be ingested and broken down into glucose. Cellulose, another polysaccharide of glucose, is an important structural component of plant cell walls. Humans cannot digest cellulose, however, and it is eliminated in the feces, where the cellulose fibers provide bulk.

## Lipids

**Lipids** are substances that dissolve in nonpolar solvents, such as alcohol or acetone, but not in polar solvents, such as water. Lipids are composed mainly of carbon, hydrogen, and oxygen, but other

elements, such as phosphorus and nitrogen, are minor components of some lipids. Lipids contain a lower proportion of oxygen to carbon than do carbohydrates.

Fats, phospholipids, and steroids are examples of lipids. **Fats** are important energy-storage molecules; they also pad and insulate the body. The building blocks of fats are **glycerol** (glis′er-ol) and **fatty acids** (figure 2.12). Glycerol is a 3-carbon molecule with a **hydroxyl** (hī-drok′sil) **group** (—OH) attached to each carbon atom, and fatty acids consist of a carbon chain with a **carboxyl** (kar-bok′sil) **group** attached at one end. A carboxyl group consists of both an oxygen atom and a hydroxyl group attached to a carbon atom (—COOH):

$$
\begin{array}{ccc}
\text{O} & & \text{O} \\
\parallel & & \parallel \\
\text{—C—OH} & \text{or} & \text{HO—C—}
\end{array}
$$

The carboxyl group is responsible for the acidic nature of the molecule because it releases hydrogen ions into solution. **Triglycerides** (trī-glis′er-īdz) are the most common type of fat molecules. Triglycerides have three fatty acids bound to a glycerol molecule.

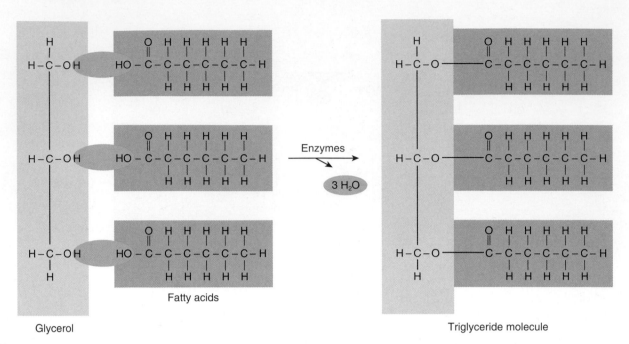

**Figure 2.12** **Triglyceride**

One glycerol molecule and three fatty acids are combined to produce a triglyceride.

Fatty acids differ from one another according to the length and degree of saturation of their carbon chains. Most naturally occurring fatty acids contain 14–18 carbon atoms. A fatty acid is **saturated** if it contains only single covalent bonds between the carbon atoms (figure 2.13a). Sources of saturated fats include beef, pork, whole milk, cheese, butter, eggs, coconut oil, and palm oil. The carbon chain is **unsaturated** if it has one or more double covalent bonds (figure 2.13b). Because the double covalent bonds can occur anywhere along the carbon chain, many types of unsaturated fatty acids with an equal degree of unsaturation are possible. **Monounsaturated fats,** such as olive and peanut oils, have one double covalent bond between carbon atoms. **Polyunsaturated fats,** such as safflower, sunflower, corn, and fish oils, have two or more double covalent bonds between carbon atoms. Unsaturated

fats are the best type of fats in the diet because, unlike saturated fats, they do not contribute to the development of cardiovascular disease.

**Phospholipids** are similar to triglycerides, except that one of the fatty acids bound to the glycerol is replaced by a molecule containing phosphorus (figure 2.14). A phospholipid is polar at the end of the molecule to which the phosphate is bound and nonpolar at the other end. The polar end of the molecule is attracted to water and is said to be **hydrophilic** (water-loving). The nonpolar end is repelled by water and is said to be **hydrophobic** (water-fearing). Phospholipids are important structural components of cell membranes (see chapter 3).

The **eicosanoids** (ī′kō-să-noydz) are a group of important chemicals derived from fatty acids. Eicosanoids are made in most cells

**Figure 2.13** **Fatty Acids**

(a) Palmitic acid is a saturated fatty acid; it contains no double bonds between the carbons. (b) Linolenic acid is an unsaturated fatty acid; note the three double bonds between the carbons.

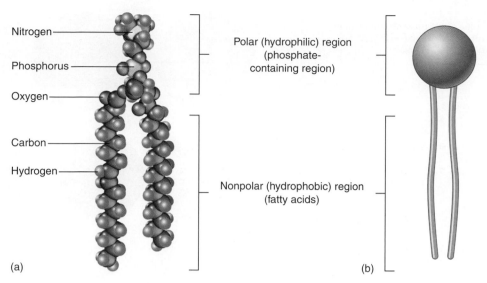

**Figure 2.14** Phospholipids

(*a*) Molecular model of a phospholipid. (*b*) Simplified depiction of a phospholipid.

and are important regulatory molecules. Among their numerous effects is their role in the response of tissues to injuries. One example of eicosanoids is **prostaglandins** (pros′tă-glan′dinz), which have been implicated in regulating the secretion of some hormones, blood clotting, some reproductive functions, and many other processes. Many of the therapeutic effects of aspirin and other anti-inflammatory drugs result from their ability to inhibit prostaglandin synthesis.

**Steroids** are composed of carbon atoms bound together into four ringlike structures. Cholesterol is an important steroid because other steroid molecules are synthesized from it. For example, bile salts, which increase fat absorption in the intestines, are derived from cholesterol, as are the reproductive hormones estrogen, progesterone, and testosterone (figure 2.15). In addition, cholesterol is an important component of cell membranes. Although high levels of cholesterol in the blood increase the risk of cardiovascular disease, a certain amount of cholesterol is vital for normal function.

## Proteins

All **proteins** contain carbon, hydrogen, oxygen, and nitrogen, and most have some sulfur. The building blocks of proteins are **amino**

**Figure 2.15** Steroids

Steroids are four-ringed molecules that differ from one another according to the groups attached to the rings. Cholesterol, the most common steroid, can be modified to produce other steroids.

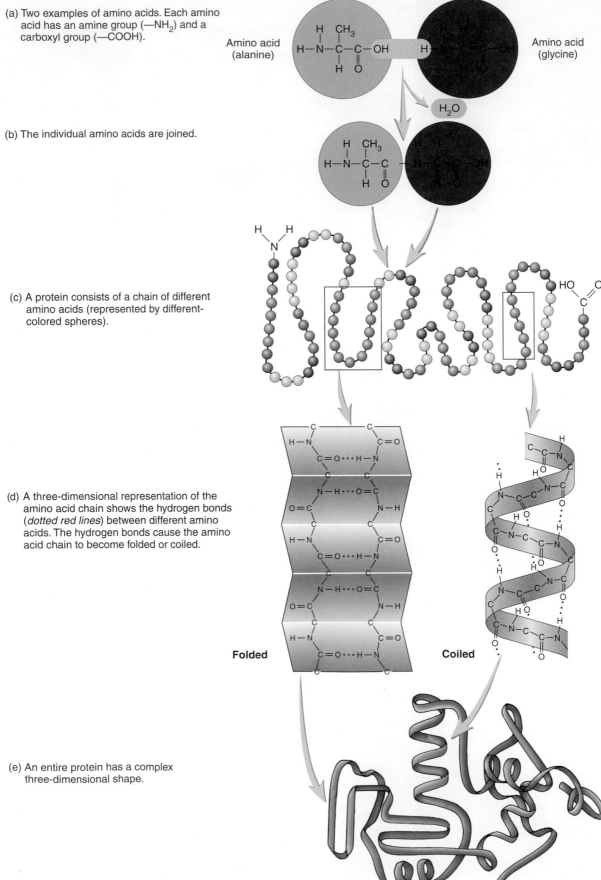

(a) Two examples of amino acids. Each amino acid has an amine group (—NH₂) and a carboxyl group (—COOH).

Amino acid (alanine)

Amino acid (glycine)

(b) The individual amino acids are joined.

(c) A protein consists of a chain of different amino acids (represented by different-colored spheres).

(d) A three-dimensional representation of the amino acid chain shows the hydrogen bonds (*dotted red lines*) between different amino acids. The hydrogen bonds cause the amino acid chain to become folded or coiled.

**Folded**

**Coiled**

(e) An entire protein has a complex three-dimensional shape.

**Figure 2.16** Protein Structure

**Figure 2.17  Activation Energy and Enzymes**

(a) Activation energy is needed to change ATP to ADP. The upper shelf represents a higher energy level, and the lower shelf represents a lower energy level. The "wall" extending above the upper shelf represents the activation energy. Even though energy is given up when moving from the upper to the lower shelf, the activation energy "wall" must be overcome before the reaction can proceed. (b) The enzyme lowers the activation energy, making it easier for the reaction to proceed.

(ă-mē′nō) **acids,** which are organic acids containing an **amine** (ă-mĕn′) **group** (–NH₂) and a carboxyl group (figure 2.16a). There are 20 basic types of amino acids. Humans can synthesize 12 of them from simple organic molecules, but the remaining 8 so-called essential amino acids must be obtained in the diet.

A protein consists of many amino acids joined together to form a chain (figure 2.16b,c). Although there are only 20 amino acids, they can combine to form numerous types of proteins with unique structures and functions. Different proteins have different kinds and numbers of amino acids. Hydrogen bonds between amino acids in the chain cause the chain to fold or coil into a specific three-dimensional shape (figure 2.16d,e). The ability of proteins to perform their functions depends on their shape. If the hydrogen bonds that maintain the shape of the protein are broken, the protein becomes nonfunctional. This change in shape is called **denaturation,** and it can be caused by abnormally high temperatures or changes in pH.

Proteins perform many important functions. For example, enzymes are proteins that regulate the rate of chemical reactions, structural proteins provide the framework for many of the body's tissues, and muscles contain proteins that are responsible for muscle contraction.

### Enzymes

An **enzyme** (en′zīm) is a protein catalyst that increases the rate at which a chemical reaction proceeds without the enzyme being permanently changed. Enzymes increase the rate of chemical reactions by lowering the **activation energy,** which is the energy necessary to start a chemical reaction. For example, heat in the form of a spark is required to start the reaction between oxygen and gasoline. Most of the chemical reactions that occur in the body have high activation energies, which are decreased by enzymes (figure 2.17). The lowered activation energies enable reactions to proceed at rates that sustain life.

Consider an analogy in which paper clips represent amino acids and your hands represent enzymes. Paper clips in a box only occasionally join together. Using your hands, however, you can rapidly make a chain of paper clips. In a similar fashion, enzymes can quickly join amino acids into a chain, forming a protein. An enzyme allows the rate of a chemical reaction to take place more than a million times faster than it would without the enzyme.

The three-dimensional shape of enzymes is critical for their normal function. According to the **lock-and-key model** of enzyme action, the shapes of an enzyme and those of the reactants allow the enzyme to bind easily to the reactants. Bringing the reactants very close to one another reduces the activation energy for the reaction. Because the enzyme and the reactants must fit together, enzymes are very specific for the reactions they control, and each enzyme controls only one type of chemical reaction. After the reaction takes place, the enzyme is released and can be used again (figure 2.18).

The body's chemical events are regulated primarily by mechanisms that control either the concentration or the activity of

Molecule A    Molecule B

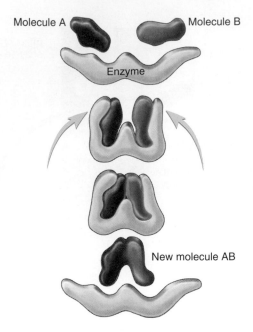

Enzyme

New molecule AB

**Figure 2.18** Enzyme Action

The enzyme brings two reacting molecules together. This is possible because the reacting molecules "fit" the shape of the enzyme (lock-and-key model). After the reaction, the unaltered enzyme can be used again.

enzymes. The rate at which enzymes are produced in cells or whether the enzymes are in an active or inactive form determines the rate of each chemical reaction.

## Nucleic Acids: DNA and RNA

**Deoxyribonucleic** (dē-oks'ē-rī'bō-noo-klē'ik) **acid (DNA)** is the genetic material of cells, and copies of DNA are transferred from one generation of cells to the next. DNA contains the information that determines the structure of proteins. **Ribonucleic** (rī'bō-noo-klē'ik) **acid (RNA)** is structurally related to DNA, and three types of RNA also play important roles in gene expression or protein synthesis. In chapter 3, we explore the means by which DNA and RNA direct the functions of the cell.

The **nucleic** (noo-klē'ik, noo-klā'ik) **acids** are large molecules composed of carbon, hydrogen, oxygen, nitrogen, and phosphorus. Both DNA and RNA consist of basic building blocks called **nucleotides** (noo'klē-ō-tīdz). Each nucleotide is composed of a sugar (monosaccharide) to which a nitrogenous organic base and a phosphate group are attached (figure 2.19). The sugar is deoxyribose for DNA, ribose for RNA. The organic bases are thymine (thī'mēn, thī'min), cytosine (sī'tō-sēn), and uracil (ūr'ă-sil), which are single-ringed molecules, and adenine (ad'ĕ-nēn) and guanine (gwahn'ēn), which are double-ringed molecules.

DNA has two strands of nucleotides joined together to form a twisted, ladderlike structure called a double helix. The sides of the ladder are formed by covalent bonds between the sugar molecules and phosphate groups of adjacent nucleotides. The rungs of the ladder are formed by the bases of the nucleotides of one side connected to the bases of the other side by hydrogen bonds. Each

nucleotide of DNA contains one of the organic bases: adenine, thymine, cytosine, or guanine. Adenine binds only to thymine because the structure of these organic bases allows two hydrogen bonds to form between them. Cytosine binds only to guanine because the structure of these organic bases allows three hydrogen bonds to form between them.

The sequence of organic bases in DNA molecules stores genetic information. Each DNA molecule consists of millions of organic bases, and their sequence ultimately determines the type and sequence of amino acids found in protein molecules. Because enzymes are proteins, DNA structure determines the rate and type of chemical reactions that occur in cells by controlling enzyme structure. Therefore, the information contained in DNA ultimately defines all cellular activities. Other proteins, such as collagen, that are coded by DNA determine many of the structural features of humans.

RNA has a structure similar to a single strand of DNA. Like DNA, four different nucleotides make up the RNA molecule, and the organic bases are the same, except that thymine is replaced with uracil. Uracil can bind only to adenine.

## Adenosine Triphosphate

**Adenosine triphosphate** (ă-den'ō-sēn trī-fos'fāt) **(ATP)** is an especially important organic molecule found in all living organisms. It consists of adenosine (the sugar ribose with the organic base adenine) and three phosphate groups (figure 2.20). Adenosine diphosphate (ADP) has only two phosphate groups. The potential energy stored in the covalent bond between the second and third phosphate groups is important to living organisms because it provides the energy used in nearly all of the chemical reactions within cells.

ATP is often called the energy currency of cells because it is capable of both storing and providing energy. The concentration of ATP is maintained within a narrow range of values, and essentially all energy-requiring chemical reactions stop when the quantity of ATP becomes inadequate.

### A CASE IN POINT
#### Cyanide Poisoning

Although Justin Hale was rescued from his burning home, he suffered from cyanide poisoning. Inhalation of smoke released by the burning of rubber and plastic during household fires is the most common cause of cyanide poisoning. Cyanide compounds can be lethal to humans because they interfere with the production of ATP in mitochondria (see chapter 3). Without adequate ATP, cells malfunction and can die. The heart and brain are especially susceptible to cyanide poisoning.

Cyanide poisoning by inhalation or absorption through the skin can also occur in certain manufacturing industries, and cyanide gas was used during the Holocaust to kill people. Deliberate suicide by ingesting cyanide is rare but has been made famous by the "suicide capsules" in spy movies. In 1982, seven people in the Chicago area died after taking Tylenol that someone had laced with cyanide. Subsequent copycat tamperings occurred and led to the widespread use of tamper-proof capsules and packaging.

1. The building blocks of nucleic acids are nucleotides, which consist of a phosphate group, a sugar, and a nitrogen base.

2. The phosphate groups connect the sugars to form two strands of nucleotides (*purple columns*).

3. Hydrogen bonds (*dotted red lines*) between the nucleotides join the two nucleotide strands together. Adenine binds to thymine, and cytosine binds to guanine.

4. The two nucleotide strands coil to form a double-stranded helix.

| Cytosine (C) | | Guanine (G) |
| Thymine (T) | | Adenine (A) |

**Figure 2.19** Structure of DNA

**Figure 2.20** Structure of ATP

SUMMARY

Chemistry is the study of the composition and structure of substances and the reactions they undergo.

# Basic Chemistry    (p. 21)

## Matter, Mass, and Weight

1. Matter is anything that occupies space and has mass.
2. Mass is the amount of matter in an object, and weight results from the gravitational attraction between the earth and an object.

## Elements and Atoms

1. An element is the simplest type of matter having unique chemical and physical properties.
2. An atom is the smallest particle of an element that has the chemical characteristics of that element. An element is composed of only one kind of atom.

## Atomic Structure

1. Atoms consist of neutrons, positively charged protons, and negatively charged electrons.
2. An atom is electrically neutral because the number of protons equals the number of electrons.
3. Protons and neutrons are in the nucleus, and electrons can be represented by an electron cloud around the nucleus.
4. The atomic number is the unique number of protons in each atom of an element. The mass number is the number of protons and neutrons.

## Electrons and Chemical Bonding

1. An ionic bond results when an electron is transferred from one atom to another.
2. A covalent bond results when a pair of electrons is shared between atoms. A polar covalent bond is an unequal sharing of electron pairs.

## Hydrogen Bonds

A hydrogen bond is the weak attraction between the oppositely charged regions of polar molecules. Hydrogen bonds are important in determining the three-dimensional structure of large molecules.

## Molecules and Compounds

1. A molecule is two or more atoms chemically combined to form a structure that behaves as an independent unit.
2. A compound is two or more different types of atoms chemically combined. A compound can be a molecule (covalent compound) or an organized array of ions (ionic compound).

## Dissociation

Dissociation is the separation of ions in an ionic compound by polar water molecules.

# Chemical Reactions    (p. 26)

## Classification of Chemical Reactions

1. A synthesis reaction is the combination of reactants to form a new, larger product.
2. A decomposition reaction is the breakdown of larger reactants into smaller products.
3. An exchange reaction is a combination of a decomposition reaction, in which reactants are broken down, and a synthesis reaction, in which the products of the decomposition reaction recombine.

## Reversible Reactions

1. In a reversible reaction, the reactants can form products, or the products can form reactants.
2. The amount of reactants relative to products is constant at equilibrium.

# Energy and Chemical Reactions

1. Energy is the capacity to do work. Potential energy is stored energy that could do work, and kinetic energy does work by causing the movement of an object.
2. Energy exists in chemical bonds as potential energy.
3. Energy is released in chemical reactions when the products contain less potential energy than the reactants. The energy can be lost as heat, used to synthesize molecules, or used to do work.
4. Energy must be added in reactions when the products contain more potential energy than the reactants.
5. Energy can be neither created nor destroyed, but one type of energy can be changed into another.

## Rate of Chemical Reactions

1. The rate of a chemical reaction increases when the concentration of the reactants increases, the temperature increases, or a catalyst is present.
2. A catalyst (enzyme) increases the rate of a chemical reaction without being altered permanently.

# Acids and Bases    (p. 30)

Acids are proton (hydrogen ion) donors, and bases are proton acceptors.

## The pH Scale

1. A neutral solution has an equal number of $H^+$ and $OH^-$ and a pH of 7.0.
2. An acidic solution has more $H^+$ than $OH^-$ and a pH less than 7.0.
3. A basic solution has fewer $H^+$ than $OH^-$ and a pH greater than 7.0.

## Salts

A salt forms when an acid reacts with a base.

## Buffers

Buffers are chemicals that resist changes in pH when acids or bases are added.

# Inorganic Chemistry    (p. 31)

1. Inorganic chemistry is mostly concerned with non-carbon-containing substances but does include such carbon-containing substances as carbon dioxide and carbon monoxide.
2. Some inorganic chemicals play important roles in the body.

## Oxygen and Carbon Dioxide

1. Oxygen is involved with the extraction of energy from food molecules.
2. Carbon dioxide is a by-product of the breakdown of food molecules.

## Water

1. Water stabilizes body temperature.
2. Water provides protection by acting as a lubricant or a cushion.
3. Water is necessary for many chemical reactions.
4. Water transports many substances.

# Organic Chemistry    (p. 32)

Organic molecules contain carbon atoms bound together by covalent bonds.

## Carbohydrates

1. Carbohydrates provide the body with energy.
2. Monosaccharides are the building blocks that form more complex carbohydrates, such as disaccharides and polysaccharides.

## Lipids

1. Lipids are substances that dissolve in nonpolar solvents, such as alcohol or acetone, but not in polar solvents, such as water. Fats, phospholipids, and steroids are examples of lipids.

2. Lipids provide energy (fats), serve as structural components (phospholipids), and regulate physiological processes (steroids).
3. The building blocks of triglycerides (fats) are glycerol and fatty acids.
4. Fatty acids can be saturated (have only single covalent bonds between carbon atoms) or unsaturated (have one or more double covalent bonds between carbon atoms).

### Proteins

1. Proteins regulate chemical reactions (enzymes), serve as structural components, and cause muscle contraction.
2. The building blocks of proteins are amino acids.
3. Denaturation of proteins disrupts hydrogen bonds, which changes the shape of proteins and makes them nonfunctional.
4. Enzymes are specific, bind to reactants according to the lock-and-key model, and function by lowering activation energy.

### Nucleic Acids: DNA and RNA

1. The basic unit of nucleic acids is the nucleotide, which is a monosaccharide with an attached phosphate and organic base.
2. DNA nucleotides contain the monosaccharide deoxyribose and the organic bases adenine, thymine, guanine, and cytosine. DNA occurs as a double strand of joined nucleotides and is the genetic material of cells.
3. RNA nucleotides are composed of the monosaccharide ribose. The organic bases are the same as for DNA, except that thymine is replaced with uracil.

### Adenosine Triphosphate

ATP stores energy, which can be used in cell processes.

# REVIEW AND COMPREHENSION

1. Define chemistry. Why is an understanding of chemistry important to the study of human anatomy and physiology?
2. Define matter. What is the difference between mass and weight?
3. Define element and atom. How many different kinds of atoms are present in a specific element?
4. List the components of an atom, and explain how they are organized to form an atom. Compare the charges of the subatomic particles.
5. Define the atomic number and the mass number of an element.
6. Distinguish among ionic, covalent, polar covalent, and hydrogen bonds. Define ion.
7. What is the difference between a molecule and a compound?
8. What happens to ionic and covalent compounds when they dissolve in water?
9. Define chemical reaction. Describe synthesis, decomposition, and exchange reactions, giving an example of each.
10. What is meant by the equilibrium condition in a reversible reaction?
11. Define potential energy and kinetic energy.
12. Give an example of a chemical reaction that releases energy and an example of a chemical reaction that requires the input of energy.
13. Name three ways that the rate of chemical reactions can be increased.
14. What is an acid and what is a base? Describe the pH scale.
15. Define a salt. What is a buffer, and why are buffers important?
16. Distinguish between inorganic and organic chemistry.
17. Why is oxygen necessary for human life? Where does the carbon dioxide we breathe out come from?
18. List four functions that water performs in the human body.
19. Name the four major types of organic molecules. Give a function for each.
20. Describe the action of enzymes in terms of activation energy and the lock-and-key model.

# CRITICAL THINKING

1. If an atom of iodine (I) gains an electron, what is the charge of the resulting ion? Write the symbol for this ion.
2. For each of the following chemical equations, determine if a synthesis reaction, a decomposition reaction, or dissociation has taken place:
   a. $HCl \rightarrow H^+ + Cl^-$
   b. Glucose + Fructose $\rightarrow$ Sucrose (table sugar)
   c. $2\,H_2O \rightarrow 2\,H_2 + O_2$
3. In terms of the energy in chemical bonds, explain why eating food is necessary for increasing muscle mass.
4. Given that the hydrogen ion concentration in a solution is based on the following reversible reaction,

   $$CO_2 + H_2O \rightleftarrows H^+ + HCO_3^-$$

   what happens to the pH of the solution when $NaHCO_3$ (sodium bicarbonate) is added to the solution? (*Hint:* The sodium bicarbonate dissociates to form $Na^+$ and $HCO_3^-$.)
5. A mixture of chemicals is warmed slightly. As a consequence, although little heat is added, the solution becomes very hot. Explain what happens to make the solution hot.
6. Two solutions, when mixed together at room temperature, produce a chemical reaction. However, when the solutions are boiled and allowed to cool to room temperature before mixing, no chemical reaction takes place. Explain.

*Answers in Appendix D*

Visit this textbook's website at **www.mhhe.com/seeleyess7** for practice quizzes, animations, interactive learning exercises, and other study tools.

McGraw-Hill offers a study CD that features interactive cadaver dissection. **Anatomy & Physiology Revealed** includes cadaver photos that allow you to peel away layers of the human body to reveal structures beneath the surface. This program also includes animations, radiologic imaging, audio pronunciations, and practice quizzing.

3D fluorescence microscopy image of a HeLa cell (human cancer cell line) in the anaphase stage of mitosis. The chromosomes have migrated to their respective spindle poles, which are moving apart. A cleavage furrow has formed in preparation for division. Red: microtubules; green: microtubule motor (dynein); light blue: centromeres; dark blue: chromosomes. Mag: 1800x at 24 mm × 36 mm.

# Chapter 3

# Cell Structures and Their Functions

## Chapter Outline and Objectives

**Cell Structure (p. 43)**

1. Explain how the structures of a cell contribute to its function.

**Functions of the Cell (p. 44)**

2. List the four main functions of a cell.

**Cell Membrane (p. 44)**

3. Describe the structure of the cell membrane.

**Movement Through the Cell Membrane (p. 45)**

4. Define diffusion and concentration gradient.

5. Explain the role of osmosis and that of osmotic pressure in controlling the movement of water across the cell membrane. Compare hypotonic, isotonic, and hypertonic solutions.

6. Define carrier-mediated transport, and compare the processes of facilitated diffusion, active transport, and secondary active transport.

7. Describe endocytosis and exocytosis.

**Organelles (p. 53)**

8. Describe the structure and function of the nucleus and the nucleoli.

9. Compare the structure and function of rough and smooth endoplasmic reticulum.

10. Describe the roles of the Golgi apparatuses and secretory vesicles in secretion.

11. Explain the role of lysosomes and peroxisomes in digesting material taken into cells by phagocytosis.

12. Describe the structure and function of mitochondria.

13. Describe the structure and function of the cytoskeleton.

14. Describe the structure and function of centrioles.

15. Compare the structure and function of cilia, flagella, and microvilli.

**Whole-Cell Activity (p. 59)**

16. Describe the process of gene expression.

17. Explain what is accomplished during mitosis.

18. Define differentiation, and explain how different cell types develop.

**Cellular Aspects of Aging (p. 67)**

19. List the major theories of aging.

The cell is the basic living unit of all organisms. The simplest organisms consist of single cells, whereas humans are composed of trillions of cells. An average-sized cell is one-fifth the size of the smallest dot you can make on a sheet of paper with a sharp pencil! But despite their extremely small size, cells are complex living structures. Cells have many characteristics in common; however, most cells are also specialized to perform specific functions. The human body is made up of many populations of specialized cells. The coordinated functions of these populations are critical to the survival of humans and all other complex organisms.

The study of cells is an important link between the study of chemistry in chapter 2 and the study of tissues in chapter 4. Knowledge of chemistry makes it possible to understand cells because cells are composed of chemicals, and those chemicals are responsible for many of the cells' characteristics. Cells, in turn, determine the form and functions of the body tissues. In addition, a great many diseases and other human disorders have a cellular basis. This chapter considers the structures of cells and how cells perform the activities necessary for life.

# CELL STRUCTURE

Each cell is a highly organized unit. Within cells, specialized structures called **organelles** (or′gă-nelz; little organs) perform specific functions (figure 3.1 and table 3.1). The nucleus is an organelle containing the cell's genetic material. The living material surrounding the nucleus is called **cytoplasm** (sī′tō-plazm), and it contains many types of organelles. The cytoplasm is enclosed by the cell membrane, or plasma membrane.

The number and type of organelles within each cell determine the cell's specific structure and functions. For example, cells secreting large amounts of protein contain well-developed organelles that synthesize and secrete protein, whereas muscle cells contain proteins and organelles that enable them to contract. The following sections describe the structure and main functions of the major organelles in cells.

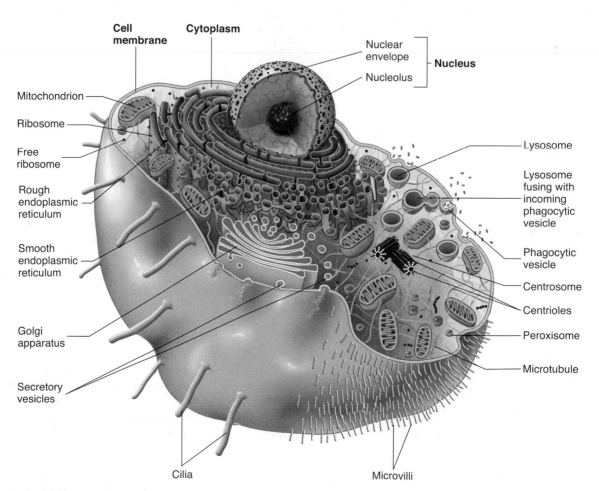

## Figure 3.1  Generalized Cell

This generalized cell shows the major organelles contained in cells. However, no single cell contains all of these organelle types. Furthermore, one kind of cell may contain many organelles of one type, whereas another kind of cell may contain very few.

## TABLE 3.1    Organelles and Their Locations and Functions

| Organelles | Location | Function(s) |
|---|---|---|
| Nucleus | Often near center of the cell | Contains genetic material of cell (DNA) and nucleoli; site of ribosome and messenger RNA synthesis |
| Nucleolus | In the nucleus | Site of ribosomal RNA synthesis and ribosomal subunit assembly |
| Ribosomes | In cytoplasm | Site of protein synthesis |
| Rough endoplasmic reticulum (rough ER) | In cytoplasm | Has many ribosomes attached; site of protein synthesis |
| Smooth endoplasmic reticulum (smooth ER) | In cytoplasm | Site of lipid synthesis; participates in detoxification |
| Golgi apparatus | In cytoplasm | Modifies protein structure and packages proteins in secretory vesicles |
| Secretory vesicle | In cytoplasm | Contains materials produced in the cell; formed by the Golgi apparatus; secreted by exocytosis |
| Lysosome | In cytoplasm | Contains enzymes that digest material taken into the cell |
| Mitochondrion | In cytoplasm | Site of aerobic respiration and the major site of ATP synthesis |
| Microtubule | In cytoplasm | Supports cytoplasm; assists in cell division and forms components of cilia and flagella |
| Centrioles | In cytoplasm | Facilitate the movement of chromosomes during cell division |
| Cilia | On cell surface with many on each cell | Move substances over surfaces of certain cells |
| Flagella | On sperm cell surface with one per cell | Propel sperm cells |
| Microvilli | Extensions of cell surface with many on each cell | Increase surface area of certain cells |

# FUNCTIONS OF THE CELL

Cells are the basic units of all living organisms, including humans. Cells are the smallest units that have all the characteristics of life. Our body cells perform several important functions:

1. *Cell metabolism and energy use.* The chemical reactions that occur within cells are collectively called cell metabolism. Energy released during metabolism is used for cell activities, such as the synthesis of new molecules, muscle contraction, and heat production, which helps maintain body temperature.
2. *Synthesis of molecules.* Cells synthesize various types of molecules, including proteins, nucleic acids, and lipids. The different cells of the body do not all produce the same molecules. Therefore, the structural and functional characteristics of cells are determined by the types of molecules they produce.
3. *Communication.* Cells produce and receive chemical and electrical signals that allow them to communicate with one another. For example, nerve cells communicate with one another and with muscle cells, causing muscle cells to contract.
   *Reproduction and inheritance.* Each cell contains a copy of the genetic information of the individual. Specialized cells (sperm cells and oocytes) transmit that genetic information to the next generation.

# CELL MEMBRANE

The **cell membrane,** or *plasma* (plaz′mă) *membrane,* is the outermost component of a cell. The cell membrane encloses the cytoplasm and forms the boundary between material inside the cell and material outside it. Substances outside the cell are called **extracellular** substances, and those inside the cell are called **intracellular** substances. Besides enclosing the cell, the cell membrane supports the cell contents, acts as a selective barrier that determines what moves into and out of the cell, and plays a role in communication between cells. The major molecules that make up the cell membrane are phospholipids and proteins. In addition, the membrane contains other molecules, such as cholesterol and carbohydrates.

Studies of the arrangement of molecules in the cell membrane have given rise to a model of its structure called the **fluid-mosaic model** (figure 3.2). The phospholipids form a double layer of molecules. The polar, phosphate-containing ends of the phospholipids are hydrophilic (water-loving) and therefore face the extracellular and intracellular fluids of the cell. The nonpolar, fatty acid ends of the phospholipids are hydrophobic (water-fearing) and therefore face away from the fluid on either side of the membrane, toward the center of the double layer of phospholipids. The double layer of phospholipids forms a lipid barrier between the inside and outside of the cell.

The double layer of phospholipid molecules has a liquid quality. Cholesterol within the phospholipid membrane gives it

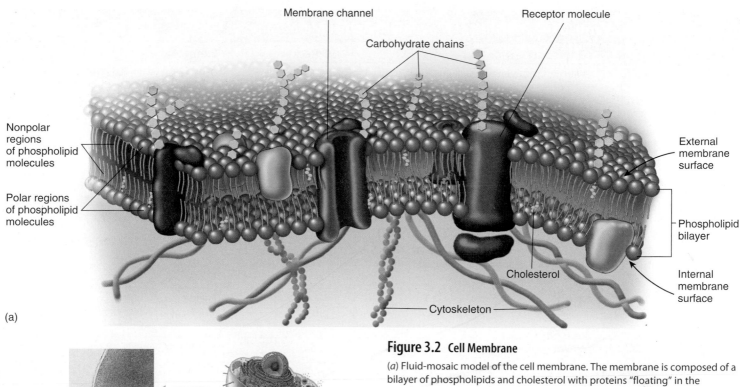

Membrane channel

Receptor molecule

Carbohydrate chains

Nonpolar regions of phospholipid molecules

Polar regions of phospholipid molecules

External membrane surface

Phospholipid bilayer

Cholesterol

Internal membrane surface

Cytoskeleton

(a)

(b)

TEM 100,000x

### Figure 3.2  Cell Membrane

(*a*) Fluid-mosaic model of the cell membrane. The membrane is composed of a bilayer of phospholipids and cholesterol with proteins "floating" in the membrane. The nonpolar hydrophobic region of each phospholipid molecule is directed toward the center of the membrane, and the polar hydrophilic region is directed toward the fluid environment either outside or inside the cell. (*b*) Transmission electron micrograph of a cell membrane, with the membrane indicated by the *red arrows*. Proteins at either surface of the lipid bilayer stain more readily than the lipid bilayer does and give the membrane the appearance of having three parts: The two dark, outer parts are proteins and the phospholipid heads, and the lighter, central part is the phospholipid tails and cholesterol.

added strength and flexibility. Protein molecules "float" among the phospholipid molecules and, in some cases, extend from the inner to the outer surface of the cell membrane. Carbohydrates may be bound to some protein molecules, modifying their functions. The proteins function as membrane channels, carrier molecules, receptor molecules, enzymes, or structural supports in the membrane. **Membrane channels** and carrier molecules are involved with the movement of substances through the cell membrane. **Receptor molecules** are part of an intercellular communication system that enables cell recognition and coordination of the activities of cells. For example, a nerve cell can release a chemical messenger that moves to a muscle cell and temporarily binds to a receptor on the muscle cell membrane. The binding acts as a signal that triggers a response, such as contraction of the muscle cell.

## MOVEMENT THROUGH THE CELL MEMBRANE

Cell membranes are **selectively permeable,** meaning that they allow some substances, but not others, to pass into or out of the cells. Intracellular material has a different composition than extracellular material, and the cell's survival depends on maintaining the difference. Substances such as enzymes, glycogen, and potassium ions

($K^+$) are found at higher concentrations intracellularly, whereas $Na^+$, $Ca^{2+}$, and $Cl^-$ are found in greater concentrations extracellularly. In addition, nutrients must enter cells continually, and waste products must exit. Because of the permeability characteristics of cell membranes and their ability to transport certain molecules, cells are able to maintain proper intracellular concentrations of molecules. Rupture of the membrane, alteration of its permeability characteristics, or inhibition of transport processes can disrupt the normal intracellular concentration of molecules and lead to cell death.

Table 3.2 lists the specific types of movement across cell membranes and we discuss each method in detail in the following sections.

## Diffusion

A **solution** is generally composed of one or more substances, called **solutes,** dissolved in the predominant liquid or gas, which is called the **solvent.** Solutes, such as ions or molecules, tend to move from an area of higher concentration to an area of lower concentration of that solute in solution. This process is called **diffusion** (figure 3.3, steps 1 and 2). Examples of diffusion include the distribution of smoke or perfume throughout a room in which there are no air currents and the gradual spread of a dye or salt throughout a beaker of still water.

## TABLE 3.2   Types and Characteristics of Movement Across Membranes

| Type | Transport | Requires ATP | Examples |
|---|---|---|---|
| Diffusion | With the concentration gradient through the lipid portion of the cell membrane or through membrane channels | No | Oxygen, carbon dioxide, chloride ions, urea |
| Osmosis | With the concentration gradient (for water) through the lipid portion of the cell membrane or through membrane channels | No | Water |
| Carrier-mediated transport mechanisms | | | |
| Facilitated diffusion | With the concentration gradient by carrier molecules | No | Glucose in most cells |
| Active transport | Against the concentration gradient* by carrier molecules | Yes | $Na^+$, $K^+$, $Ca^{2+}$, and $H^+$; amino acids |
| Secondary active transport | Against the concentration gradient by carrier molecules; the energy for secondary active transport of one substance comes from the concentration gradient of another | Yes | Glucose, amino acids |
| Endocytosis | Movement into cells by vesicles | Yes | Ingestion of particles by phagocytosis or receptor-mediated endocytosis and liquids by pinocytosis |
| Exocytosis | Movement out of cells by vesicles | Yes | Secretion of proteins |

*Active transport normally moves substances *against* their concentration gradient, but it can also move substances *with* their concentration gradient.

Diffusion results from the constant random motion of all solutes in a solution. More solute particles occur in an area of higher concentration than in one of lower concentration. Because particles move randomly, the chances are greater that solute particles will move from the higher toward the lower concentration than from the lower to the higher concentration. At equilibrium, the net movement of solutes stops, although the random motion continues, and the movement of solutes in any one direction is balanced by an equal movement in the opposite direction (figure 3.3, step 3).

A **concentration gradient** is the difference in the concentration of a solute in a solvent between two points divided by the distance between the two points. The concentration gradient is said to be steeper when the concentration difference is large and/or the distance is small. When we say that a substance moves down (or with) the concentration gradient, we mean that solutes are diffusing from a higher toward a lower concentration of solutes. Speaking of movement up (or against) a concentration gradient describes the movement of solutes from a lower toward a higher concentration of solutes. This second type of movement does not occur by diffusion and requires energy in order to occur.

In the body, diffusion is an important means of transporting substances through the extracellular and intracellular fluids. In addition, substances, such as nutrients and some waste products, can diffuse into and out of the cell. The normal intracellular concentrations of many substances depend on diffusion. For example, if the extracellular concentration of oxygen is reduced,

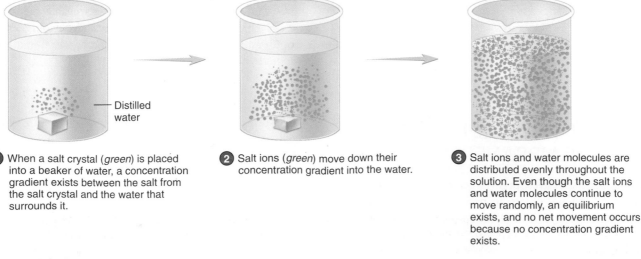

1 When a salt crystal (*green*) is placed into a beaker of water, a concentration gradient exists between the salt from the salt crystal and the water that surrounds it.

2 Salt ions (*green*) move down their concentration gradient into the water.

3 Salt ions and water molecules are distributed evenly throughout the solution. Even though the salt ions and water molecules continue to move randomly, an equilibrium exists, and no net movement occurs because no concentration gradient exists.

**PROCESS Figure 3.3** Diffusion

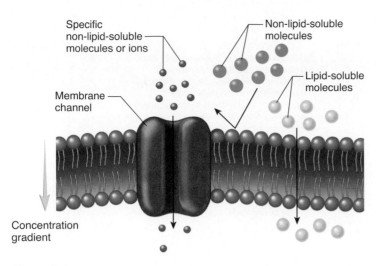

**Figure 3.4** Diffusion Through the Cell Membrane

Non-lipid-soluble molecules (*red*) diffuse through membrane channels. Lipid-soluble molecules (*yellow*) diffuse directly through the cell membrane.

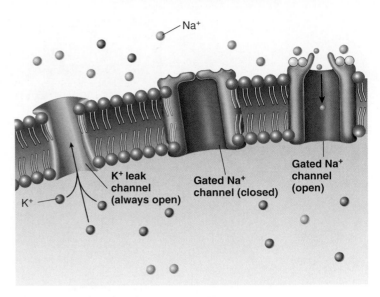

**Figure 3.5** Leak and Gated Membrane Channels

In this example, the $K^+$ leak channel (*purple*) is always open, allowing $K^+$ to diffuse across the cell membrane. The gated $Na^+$ channel (*pink*) regulates the movement of $Na^+$ across the membrane by opening and closing.

not enough oxygen diffuses into the cell, and the cell cannot function normally.

**Predict 1**

*Urea is a toxic waste produced inside liver cells. It diffuses from those cells into the blood and is eliminated from the body by the kidneys. What would happen to the intracellular and extracellular concentrations of urea if the kidneys stopped functioning?*

The phospholipid bilayer acts as a barrier to most water-soluble substances. However, certain small, water-soluble molecules can diffuse between the phospholipid molecules of cell membranes. Other water-soluble molecules can diffuse across the cell membrane only by passing through cell membrane **channels** (figure 3.4). Molecules that are lipid-soluble, such as oxygen, carbon dioxide, and steroids, pass readily through the phospholipid bilayer.

Cell membrane channels consist of large protein molecules that extend from one surface of cell membranes to the other (figure 3.5). There are several channel types, each of which allows only certain molecules to pass through. The size, shape, and charge of a molecule all determine whether it can pass through each kind of channel. For example, $Na^+$ pass through $Na^+$ channels, and $K^+$ and $Cl^-$ pass through $K^+$ and $Cl^-$ channels, respectively. Rapid movement of water across the cell membrane also occurs through membrane channels.

In addition, cell membrane channels differ in the degree to which ions pass through them. Some channels constantly allow ions to pass through. These channels are called **leak channels.** Other channels limit the movement of ions across the membrane by opening and closing. These channels are called **gated channels** (figure 3.5).

## Osmosis

**Osmosis** (os-mō′sis) is the diffusion of water (a solvent) across a selectively permeable membrane, such as the cell membrane, from a region of higher water concentration to one of lower water concentration (figure 3.6; see table 3.2). Osmosis is important to cells because large volume changes caused by water movement can disrupt normal cell functions. Osmosis occurs when the cell membrane

is less permeable, selectively permeable, or not permeable to solutes *and* a concentration gradient for water exists across the cell membrane. Water diffuses from a solution with a higher water concentration across the cell membrane into a solution with a lower water concentration. The ability to predict the direction of water movement across the cell membrane depends on knowing which solution on either side of the membrane has the highest water concentration.

The concentration of a solution, however, is not expressed in terms of water, but in terms of solute concentration. For example, if sugar solution A is more concentrated than sugar solution B, then solution A has more sugar (solute) than solution B. As the concentration of a solution increases, the amount of water (solvent) proportionately decreases. Water diffuses from the less concentrated solution, which has fewer solute molecules and more water molecules, into the more concentrated solution, which has more solute molecules and fewer water molecules. In other words, water diffuses toward areas of high solute concentration and dilutes those solutes.

**Osmotic pressure** is the force required to prevent the movement of water across a selectively permeable membrane. Thus, osmotic pressure is a measure of the tendency of water to move by osmosis across a selectively permeable membrane. It can be measured by placing a solution into a tube that is closed at one end by a selectively permeable membrane and immersing the tube in distilled water (see figure 3.6, step 1). Water molecules move by osmosis through the membrane into the tube, forcing the solution to move up the tube (see figure 3.6, step 2). As the solution rises, its weight produces **hydrostatic pressure** (see figure 3.6, step 3), which moves water out of the tube back into the distilled water surrounding the tube. Net movement of water into the tube stops when the hydrostatic pressure in the tube causes water to move out of the tube at the same rate that it diffuses into the tube by osmosis. The osmotic pressure of the solution in the tube is equal to the hydrostatic pressure that prevents net movement of water into the tube.

The greater the concentration of a solution, the greater its osmotic pressure, and the greater the tendency for water to move

* Because the tube contains salt ions (*green and red spheres*) as well as water molecules (*blue spheres*), there is proportionately less water in the tube than in the beaker, which contains only water. The water molecules diffuse with their concentration gradient into the tube (*blue arrows*). Because the salt ions cannot leave the tube, the total fluid volume inside the tube increases, and fluid moves up the glass tube (*black arrow*) as a result of osmosis.

3% salt solution

Selectively permeable membrane

Distilled water

Salt solution rising

Water

Weight of water column

The solution stops rising when the weight of the water column prevents further movement of water into the tube by osmosis.

Osmosis

**1** The end of a tube containing a 3% salt solution (*green*) is closed at one end with a selectively permeable membrane, which allows water molecules to pass through but retains the salt ions within the tube.

**2** The tube is immersed in distilled water. Water moves into the tube by osmosis (see inset above*). The concentration of salt in the tube decreases as water rises in the tube (*lighter green color*).

**3** Water moves by osmosis into the tube until the weight of the column of water in the tube (hydrostatic pressure) prevents further movement of water into the tube. The hydrostatic pressure that prevents net movement of water into the tube is equal to the osmotic pressure of the solution in the tube.

**PROCESS Figure 3.6  Osmosis**

into the solution. This occurs because water moves from less concentrated solutions (less solute, more water) into more concentrated solutions (more solute, less water). The greater the concentration of a solution, the greater the tendency for water to move into the solution, and so greater the osmotic pressure must be to prevent that movement.

When placed into a solution, a cell may swell, remain unchanged, or shrink, depending on the concentration gradient between the solution and the cell's cytoplasm. A **hypotonic** (hī′pō-ton′ik; *hypo,* under) solution usually has a lower concentration of

solutes and a higher concentration of water than the cytoplasm of the cell. Thus, the solution has less tone, or osmotic pressure, than the cell. Water moves by osmosis into the cell, causing it to swell. If the cell swells enough, it can rupture, a process called **lysis** (lī′sis) (figure 3.7*a*). When a cell is immersed in an **isotonic** (ī′sō-ton′ik; *iso,* equal) solution, the concentrations of various solutes and water are the same on both sides of the cell membrane. The cell therefore neither shrinks nor swells (figure 3.7*b*). When a cell is immersed in a **hypertonic** (hi′per-ton′ik; *hyper,* above) solution, the solution usually has a higher concentration of solutes and

Red blood cell

H₂O

Hypotonic solution

Isotonic solution

Hypertonic solution

(a) When a red blood cell is placed in a hypotonic solution (one having a low solute concentration), water enters the cell by osmosis (*black arrows*), causing the cell to swell or even burst (lyse; *puff of red in lower part of cell* ).

(b) When a red blood cell is placed in an isotonic solution (one having a concentration of solutes equal to that inside the cell). water moves into and out of the cell at the same rate (*black arrows*). No net water movement occurs, and the cell shape remains normal.

(c) When a red blood cell is placed in a hypertonic solution (one having a high solute concentration), water moves by osmosis out of the cell and into the solution (*black arrows*), resulting in shrinkage (crenation).

**Figure 3.7**  Effects of Hypotonic, Isotonic, and Hypertonic Solutions on Red Blood Cells

a lower concentration of water than the cytoplasm of the cell. Water moves by osmosis from the cell into the hypertonic solution, resulting in cell shrinkage, or **crenation** (krē-nā′shŭn) (figure 3.7c). In general, solutions injected into blood vessels or into tissues must be isotonic because swelling or shrinking disrupts normal cell function and can lead to cell death.

## Carrier-Mediated Transport Mechanisms

Many nutrient molecules, such as amino acids and glucose, cannot enter the cell by diffusion, and many substances produced in cells, such as proteins, cannot leave the cell by diffusion. **Carrier molecules,** which are proteins within the cell membrane, are involved in **carrier-mediated transport mechanisms,** which move large, water-soluble molecules or electrically charged ions across the cell membrane. After a molecule to be transported binds to a specific carrier molecule on one side of the membrane, the three-dimensional shape of the carrier molecule changes, and the transported molecule is moved to the opposite side of the cell membrane. The transported molecule is then released by the carrier molecule, which resumes its original shape and is available to transport another molecule. Carrier-mediated transport mechanisms exhibit **specificity;** that is, only specific molecules are transported by the carriers. There are three kinds of carrier-mediated transport: facilitated diffusion, active transport, and secondary active transport.

### Facilitated Diffusion

**Facilitated diffusion** is a carrier-mediated transport process that moves substances into or out of cells from a higher to a lower concentration of that substance (figure 3.8; see table 3.2). Because movement is with the concentration gradient, metabolic energy in the form of ATP is not required.

**Extracellular fluid**

Carrier molecule

Concentration gradient

**Cytoplasm**

Transported molecule (glucose)

**1** The carrier molecule binds with a molecule, such as glucose, on the outside of the cell membrane.

**2** The carrier molecule changes shape and releases the molecule on the inside of the cell membrane.

**PROCESS Figure 3.8**  Facilitated Diffusion

**Extracellular fluid**

① Three sodium ions (Na⁺) and adenosine triphosphate (ATP) bind to the sodium–potassium (Na⁺–K⁺) pump.

Na⁺

**Cytoplasm**

Na⁺–K⁺ pump

ATP

①

ATP binding site on Na⁺–K⁺ pump

② The ATP breaks down to adenosine diphosphate (ADP) and a phosphate (P) and releases energy. That energy is used to power the shape change in the Na⁺–K⁺ pump. Phosphate remains bound to the Na⁺–K⁺ pump–ATP binding site.

③

Na⁺

K⁺

③ The Na⁺–K⁺ pump changes shape, and the Na⁺ are transported across the membrane.

Na⁺–K⁺ pump changes shape (requires energy).

P

Breakdown of ATP (releases energy)

②

ADP

④

Na⁺

K⁺ ⑤

④ The Na⁺ diffuse away from the Na⁺–K⁺ pump.

⑤ Two potassium ions (K⁺) bind to the Na⁺–K⁺ pump.

⑥ The phosphate is released from the Na⁺–K⁺ pump binding site.

⑥

P

Na⁺–K⁺ pump resumes original shape.

⑦ The Na⁺–K⁺ pump changes shape, transporting K⁺ across the membrane, and the K⁺ diffuse away from the pump. The Na⁺–K⁺ pump can again bind to Na⁺ and ATP.

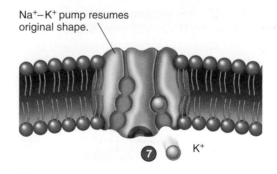

⑦     K⁺

**PROCESS Figure 3.9** Active Transport: Sodium-Potassium Pump

*Predict 2*

*The transport of glucose into most cells occurs by facilitated diffusion. Because diffusion occurs from a higher to a lower concentration, glucose cannot accumulate within these cells at a higher concentration than exists outside the cell. Once glucose enters a cell, it is rapidly converted to other molecules, such as glucose phosphate or glycogen. What effect does this conversion have on the cell's ability to transport glucose?*

## Active Transport

**Active transport** is a carrier-mediated process that moves substances across the cell membrane from regions of lower concentration to those of higher concentration against a concentration gradient (see table 3.2). Consequently, active transport processes accumulate substances on one side of the cell membrane at concentrations many times greater than those on the other side. Active transport requires energy in the form of ATP; if ATP is not available, active transport stops. One example of active transport is the movement of various amino acids from the small intestine into the blood.

In some cases, the active transport mechanism can exchange one substance for another. For example, the **sodium-potassium pump** moves Na⁺ out of cells and K⁺ into cells (figure 3.9). The result is a higher concentration of Na$^+$ outside the cell and a higher concentration of K$^+$ inside the cell. The concentration gradients for Na$^+$ and K$^+$, established by the sodium-potassium pump, are essential in maintaining the resting membrane potential (see chapter 8).

**Cystic fibrosis** is a genetic disorder that affects the active transport of Cl$^-$ into cells, as described in the Systems Pathology essay later in this chapter.

## A CASE IN POINT
### Addison Disease

Lowe Blood experienced unexplained weight loss, fatigue, and low blood pressure. After running several tests, his physician concluded that he had **Addison disease**, which involves decreased aldosterone production by the adrenal cortex. When aldosterone production falls too low, certain kidney cells are unable to transport Na$^+$. This results in Na$^+$ and water loss in the urine, causing blood volume and blood pressure to drop.

### Secondary Active Transport

**Secondary active transport** involves the active transport of one substance, such as Na$^+$, across the cell membrane, establishing a concentration gradient. The diffusion of that transported substance down its concentration gradient provides the energy to transport a second substance, such as glucose, across the cell membrane (figure 3.10). In **cotransport,** the diffusing substance moves in the same direction as the transported substance; in **countertransport,** the diffusing substance moves in the opposite direction as the transported substance.

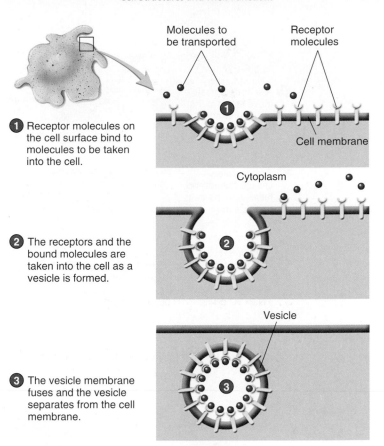

**PROCESS Figure 3.11** Receptor-Mediated Endocytosis

1. Receptor molecules on the cell surface bind to molecules to be taken into the cell.
2. The receptors and the bound molecules are taken into the cell as a vesicle is formed.
3. The vesicle membrane fuses and the vesicle separates from the cell membrane.

## Endocytosis and Exocytosis

Large water-soluble molecules, small pieces of matter, and even whole cells can be transported across cell membranes in **vesicles,** which are membrane-bound sacs. Because of the fluid nature of membranes, vesicles and cell membranes can fuse, allowing the contents of the vesicles to cross the cell membrane.

**Endocytosis** (en'dō-sī-tō'sis) is the uptake of material through the cell membrane by the formation of a vesicle (see table 3.2). The cell membrane invaginates to form a vesicle containing the material to be taken into the cell. The vesicle then moves into the cytoplasm.

Endocytosis usually exhibits specificity. The cell membrane contains specific receptor molecules that bind to specific substances. When a specific substance binds to the receptor molecule, endocytosis is triggered, and the substance is transported into the cell. This process is called **receptor-mediated endocytosis** (figure 3.11). Cholesterol and growth factors are examples of molecules that can be taken into a cell by receptor-mediated endocytosis. Bacterial phagocytosis is also receptor-mediated.

The term **phagocytosis** (fag'ō-sī-tō'sis; cell-eating) is often used for endocytosis when solid particles are ingested. A part of the cell membrane extends around a particle and fuses so that the particle is surrounded by the membrane. That part of the membrane then "pinches off" to form a vesicle containing the particle. The vesicle is now within the cytoplasm of the cell, and the cell membrane is left intact. White blood cells and some other cell

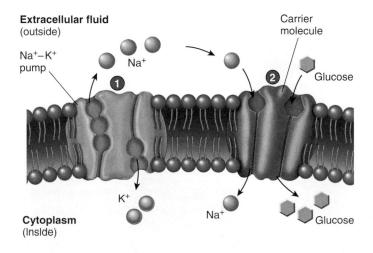

1. A sodium–potassium (Na$^+$–K$^+$) pump maintains a concentration of Na$^+$ that is higher outside the cell than inside.

2. Na$^+$ move back into the cell by a carrier molecule that also moves glucose. The concentration gradient for Na$^+$ provides the energy required to move glucose, by cotransport, against its concentration gradient.

**PROCESS Figure 3.10** Secondary Active Transport

① A secretory vesicle moves toward the cell membrane.

② The secretory vesicle membrane fuses with the cell membrane.

③ The secretory vesicle's contents are released into the extracellular fluid.

(a)

Cell membrane

Secretory vesicle

Vesicle contents

Secretory vesicle fused to cell membrane

Released contents of secretory vesicle

TEM 30,000x

(b)

**PROCESS Figure 3.12  Exocytosis**

(a) Diagram of exocytosis. (b) Transmission electron micrograph of exocytosis.

(a)

Nuclear pores

Ribosomes

Nucleus

Outer membrane

Space

Inner membrane

**Nuclear envelope**

Nucleolus

Nuclear envelope

Interior of nucleus

Nucleolus

Chromatin

TEM 20,000x

(b)

Outer membrane of nuclear envelope

Inner membrane of nuclear envelope

Nuclear pores

SEM 50,000x   (c)

**Figure 3.13  Nucleus**

(a) The nuclear envelope consists of inner and outer membranes, which become fused at the nuclear pores. The nucleolus is a condensed region of the nucleus not bounded by a membrane and consisting mostly of RNA and protein. (b) Transmission electron micrograph of the nucleus. (c) Scanning electron micrograph showing the membranes of the nuclear envelope and the nuclear pores.

types phagocytize bacteria, cell debris, and foreign particles. Phagocytosis is an important means by which white blood cells take up and destroy harmful substances that have entered the body. **Pinocytosis** (pin′ō-sī-tō′sis; cell-drinking) is distinguished from phagocytosis in that much smaller vesicles are formed, and they contain liquid rather than particles.

In some cells, membrane-bound sacs called **secretory vesicles** accumulate materials for *release* from the cell. The secretory vesicles move to the cell membrane, where the membrane of the vesicle fuses with the cell membrane, and the material in the vesicle is eliminated from the cell. This process is called **exocytosis** (ek′sō-sī-tō′sis; *exo,* outside) (figure 3.12; see table 3.2). Examples of exocytosis are the secretion of digestive enzymes by the pancreas and of mucus by the salivary glands.

To sum up, endocytosis results in the uptake of materials by cells, and exocytosis allows the release of materials from cells. Vesicle formation for both endocytosis and exocytosis requires energy in the form of ATP.

# ORGANELLES
## Nucleus

The **nucleus** (noo′klē-ŭs; a little nut or the stone of a fruit) is a large organelle usually located near the center of the cell (see figure 3.1). All cells of the body have a nucleus at some point in their life cycle, although some cells, such as red blood cells, lose their nuclei as they mature. Other cells, such as skeletal muscle cells, contain more than one nucleus.

The nucleus is bounded by a **nuclear envelope,** which consists of outer and inner membranes with a narrow space between them (figure 3.13). At many points on the surface of the nucleus, the inner and outer membranes come together to form **nuclear pores,** through which materials can pass into or out of the nucleus.

Human cells contain 23 pairs of **chromosomes** (krō′mō′-somz), which consist of DNA and proteins. During most of a cell's life, the chromosomes are loosely coiled and collectively called **chromatin** (figure 3.14; see figure 3.13*b*). When a cell prepares to divide, the chromosomes become tightly coiled and are visible when viewed with a microscope (see "Cell Life Cycle" later in this chapter). The genes that influence the structural and functional features of every individual are portions of DNA molecules. These sections of DNA molecules determine the structure of proteins. By determining the structure of proteins, genes direct cell structure and function.

**Nucleoli** (noo-klē′ō-li; sing. nucleolus, little nucleus) are diffuse bodies with no surrounding membrane that are found within the nucleus (see figure 3.13). There are usually one to several nucleoli per cell. The subunits of ribosomes, cytoplasmic

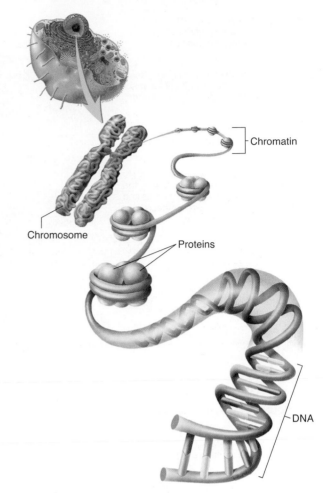

**Figure 3.14  Structure of a Chromosome**

organelles, are formed within a nucleolus. Proteins produced in the cytoplasm move through the nuclear pores into the nucleus and to the nucleolus. These proteins are joined to **ribosomal ribonucleic** (rī′bō-noo-klē′ik) **acid (rRNA),** produced within the nucleolus, to form large and small ribosomal subunits (figure 3.15). The ribosomal subunits then move from the nucleus through the nuclear pores into the cytoplasm, where one large and one small subunit join to form a ribosome during protein synthesis.

## Ribosomes

**Ribosomes** (rī′bō-sōmz) are the organelles where proteins are produced (see "Gene Expression" later in this chapter). Free ribosomes are not attached to any other organelles in the cytoplasm,

1. Ribosomal proteins, produced in the cytoplasm, are transported through nuclear pores into the nucleolus.

2. rRNA, most of which is produced in the nucleolus, is assembled with ribosomal proteins to form small and large ribosomal subunits.

3. The small and large ribosomal subunits leave the nucleolus and the nucleus through nuclear pores.

4. The small and large subunits, now in the cytoplasm, combine with each other and with mRNA during protein synthesis.

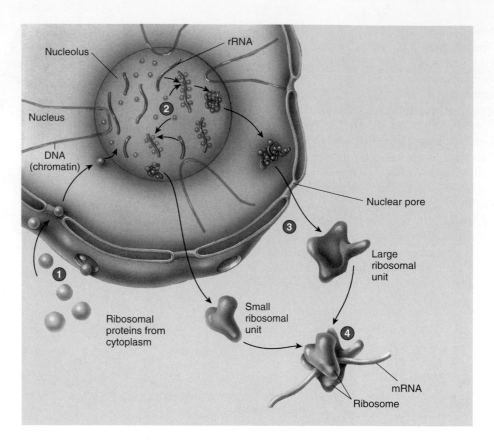

**PROCESS Figure 3.15** Production of Ribosomes

whereas other ribosomes are attached to a network of membranes called the endoplasmic reticulum.

## Rough and Smooth Endoplasmic Reticulum

The **endoplasmic reticulum** (en′dō-plas′mik re-tik′ū-lŭm; *rete,* a network) **(ER)** is a series of membranes forming sacs and tubules that extends from the outer nuclear membrane into the cytoplasm (figure 3.16). **Rough ER** is ER with ribosomes attached to it. A large amount of rough ER in a cell indicates that it is synthesizing large amounts of protein for export from the cell. On the other hand, ER without ribosomes is called **smooth ER.** Smooth ER is a site for lipid synthesis and participates in detoxification of chemicals within cells. In skeletal muscle cells, the smooth ER stores calcium ions.

## Golgi Apparatus

The **Golgi** (gol′jē) **apparatus** consists of closely packed stacks of curved, membrane-bound sacs (figure 3.17). It collects, modifies, packages, and distributes proteins and lipids manufactured by the ER. For example, proteins produced at the ribosomes enter the Golgi apparatus from the ER. In some cases, the Golgi apparatus chemically modifies the proteins by attaching carbohydrate or lipid molecules to them. The proteins then are packaged into membrane sacs, called secretory vesicles, that pinch off from the margins of the Golgi apparatus (see the next section). The Golgi apparatus is present in larger numbers and is most highly developed in cells that secrete protein, such as those of the salivary glands or the pancreas.

## Secretory Vesicles

A **vesicle** (ves′i-kl; a bladder) is a small, membrane-bound sac that transports or stores materials within cells. Secretory vesicles pinch off from the Golgi apparatus and move to the surface of the cell (figure 3.17). Their membranes then fuse with the cell membrane, and the contents of the vesicles are released to the exterior of the cell. In many cells, secretory vesicles accumulate in the cytoplasm and are released to the exterior when the cell receives a signal. For example, nerve cells release substances called neurotransmitters from secretory vesicles to communicate with other cells. Also, secretory vesicles containing the hormone insulin remain in the cytoplasm of pancreatic cells until rising blood glucose levels stimulate their release.

## Lysosomes and Peroxisomes

**Lysosomes** (lī′sō-sōmz) are membrane-bound vesicles formed from the Golgi apparatus. They contain a variety of enzymes that function as intracellular digestive systems. Vesicles formed by endocytosis may fuse with lysosomes (figure 3.18). The enzymes within the lysosomes break down the materials in the endocytotic vesicle. For example, white blood cells phagocytize bacteria. Then enzymes within lysosomes destroy the phagocytized bacteria.

**Peroxisomes** (per-ok′si-sōmz) are small, membrane-bound vesicles containing enzymes that break down fatty acids, amino acids, and hydrogen peroxide ($H_2O_2$). Hydrogen peroxide is a by-product of fatty acid and amino acid breakdown and can be toxic to a cell. The enzymes in peroxisomes break down hydrogen peroxide to water and oxygen. Cells active in detoxification, such as liver and kidney cells, have many peroxisomes.

(a)

(b)

**Figure 3.16  Endoplasmic Reticulum**

(*a*) The endoplasmic reticulum is continuous with the nuclear envelope and can exist as either rough endoplasmic reticulum (with ribosomes) or smooth endoplasmic reticulum (without ribosomes). (*b*) Transmission electron micrograph of the rough endoplasmic reticulum.

(a)

(b)  Mitochondrion

**Figure 3.17  Golgi Apparatus**

(*a*) The Golgi apparatus is composed of flattened, membranous sacs and resembles a stack of dinner plates or pancakes. (*b*) Transmission electron micrograph of the Golgi apparatus.

## Carbohydrate and Lipid Disorders

Some diseases result from nonfunctional lysosomal enzymes. For example, **Pompe disease** is caused by the inability of lysosomal enzymes to break down the carbohydrate glycogen produced in certain cells. Glycogen accumulates in large amounts in the heart, liver, and skeletal muscle cells; the accumulation in heart muscle cells often leads to heart failure. Specialists are best suited to treat Pompe disease, although there is no cure. However, clinical trials have shown that replacement of the nonfunctional enzymes can reduce the symptoms of Pompe disease.

**Lipid-storage disorders** are a group of hereditary diseases characterized by the accumulation of large amounts of lipids in cells that lack the enzymes necessary to break down the lipids. The accumulation of lipids leads to cell and tissue damage in the brain, spleen, liver, and bone marrow. **Tay-Sachs disease,** one such lipid-storage disorder, affects the nervous system. Tay-Sachs disease results from a lack of lysosomal enzymes that normally break down specific lipids in nerve cells of the brain. Buildup of these lipids in the brain leads to neurological problems, such as blindness, deafness, and paralysis, within the first several months of life. There is no successful treatment for Tay-Sachs disease; children with this disease usually die before age 4.

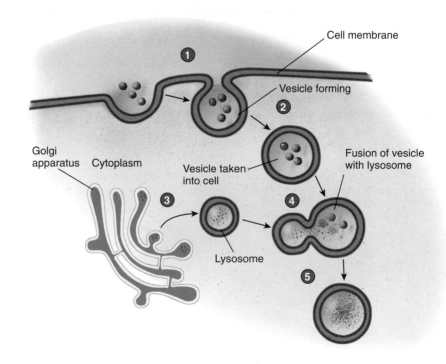

1. A vesicle forms around material outside the cell.

2. The vesicle is pinched off from the cell membrane and becomes a separate vesicle inside the cell.

3. A lysosome is pinched off the Golgi apparatus.

4. The lysosome fuses with the vesicle.

5. The enzymes from the lysosome mix with the material in the vesicle, and the enzymes digest the material.

**PROCESS Figure 3.18** Action of Lysosomes

## Mitochondria

**Mitochondria** (mī′tō-kon′drē-ă; sing. mitochondrion) are small organelles with inner and outer membranes separated by a space (figure 3.19; see figure 3.1). The outer membranes have a smooth contour, but the inner membranes have numerous folds, called **cristae** (kris′tē), which project like shelves into the interior of the mitochondria.

Mitochondria are the major sites of adenosine triphosphate (ATP) production within cells. ATP is the main energy source for most chemical reactions within the cell, and cells with a large energy requirement have more mitochondria than cells that require less energy. Mitochondria carry out aerobic respiration (discussed in greater detail in chapter 17), which requires oxygen to allow the reactions that break down food molecules to produce

**Figure 3.19** Mitochondrion

(*a*) Typical mitochondrion structure. (*b*) Transmission electron micrograph of mitochondria in longitudinal and cross sections.

ATP. Cells that carry out extensive active transport, described earlier in this chapter, contain many mitochondria. When muscles enlarge as a result of exercise, the mitochondria increase in number within the muscle cells and provide the additional ATP required for muscle contraction.

## Cytoskeleton

The **cytoskeleton** (sī-tō-skel′ĕ-ton) consists of proteins that support the cell, hold organelles in place, and enable the cell to change shape. These protein structures are microtubules, microfilaments, and intermediate filaments (figure 3.20).

 **Microtubules** are hollow structures formed from protein subunits. They perform a variety of roles, such as helping support the cytoplasm of cells, assisting in cell division, and forming essential components of certain organelles, such as cilia and flagella.

 **Microfilaments** are small fibrils formed from protein subunits that structurally support the cytoplasm. Some microfilaments are involved with cell movement. For example, microfilaments in muscle cells enable the cells to shorten, or contract.

 **Intermediate filaments** are fibrils formed from protein subunits that are smaller in diameter than microtubules but larger in diameter than microfilaments. They provide mechanical support to the cell.

## Centrioles

The **centrosome** (sen′trō-sōm) is a specialized zone of cytoplasm close to the nucleus, where microtubule formation occurs. It contains two **centrioles** (sen′trē-ōlz), which are normally oriented perpendicular to each other. Each centriole is a small, cylindrical organelle composed of nine triplets; each triplet consists of three parallel microtubules joined together (figure 3.21). Additional microtubules, extending from the area of the centrioles, play an important role in cell division, as we learn later in this chapter.

## Cilia, Flagella, and Microvilli

**Cilia** (sĭl′ē-ă; sing. cilium, an eyelash) project from the surface of cells (see figure 3.1). They vary in number from none to thousands per cell and are capable of moving. Cilia are cylindrical structures that extend from the cell. Cilia are composed of microtubules, organized in a pattern similar to that of centrioles, which are enclosed by the cell membrane. Cilia are numerous on surface cells that line the respiratory tract. Their coordinated movement transports mucus, in which dust particles are embedded, upward and away from the lungs. This action helps keep the lungs clear of debris.

**Figure 3.20  Cytoskeleton**

(*a*) Diagram of the cytoskeleton. (*b*) Scanning electron micrograph of the cytoskeleton.

**Figure 3.21  Centriole**

(*a*) Structure of a centriole, which is composed of nine triplets of microtubules. Each triplet contains one complete microtubule fused to two incomplete microtubules. (*b*) Transmission electron micrograph of a pair of centrioles, which are normally located together near the nucleus. One is shown in cross section and one in longitudinal section.

## Relationships Between Cell Structure and Cell Function

Each cell is well adapted for the functions it performs, and the abundance of organelles in each cell reflects those functions. For example, epithelial cells lining the larger-diameter respiratory passages secrete mucus and transport it toward the throat, where it is either swallowed or expelled from the body by coughing. Particles of dust and other debris suspended in the air become trapped in the mucus. The production and transport of mucus from the respiratory passages keeps these passages clean. Cells of the respiratory system have abundant rough ER, Golgi apparatuses, secretory vesicles, and cilia. The ribosomes on the rough ER are the sites where proteins, a major component of mucus, are produced. The Golgi apparatuses package the proteins and other components of mucus into secretory vesicles, which move to the surface of the epithelial cells. The contents of the secretory vesicles are released onto the surface of the epithelial cells. Cilia on the cell surface then propel the mucus toward the throat.

In people who smoke, prolonged exposure of the respiratory epithelium to the irritation of tobacco smoke causes the respiratory epithelial cells to change in structure and function. The cells flatten and form several layers of epithelial cells. These flattened epithelial cells no longer contain abundant rough ER, Golgi apparatuses, secretory vesicles, or cilia. The altered respiratory epithelium is adapted to protect the underlying cells from irritation, but it can no longer secrete mucus, nor does it have cilia to move the mucus toward the throat to clean the respiratory passages. Extensive replacement of normal epithelial cells in respiratory passages is associated with chronic inflammation of the respiratory passages (bronchitis), which is common in people who smoke heavily. Coughing is a common manifestation of inflammation and mucus accumulation in the respiratory passages.

Flagella (flă-jel′ă; sing. flagellum, a whip) have a structure similar to that of cilia but are much longer, and they usually occur only one per cell. Sperm cells each have one flagellum, which propels the sperm cell.

Microvilli (mī′krō-vil′ī; mikros, small + villus, shaggy hair) are specialized extensions of the cell membrane that are supported by microfilaments (see figure 3.1), but they do not actively move as cilia and flagella do. Microvilli are numerous on cells that have them and they increase the surface area of those cells. They are abundant on the surface of cells that line the intestine, kidney, and other areas in which absorption is an important function.

### Predict 3

*List the organelles that are plentiful in cells that (a) synthesize and secrete proteins, (b) actively transport substances into cells, and (c) ingest foreign substances by endocytosis. Explain the function of each organelle you list.*

## WHOLE-CELL ACTIVITY

A cell's characteristics are ultimately determined by the types of proteins it produces. The proteins produced are in turn determined by the genetic information in the nucleus. In order to understand how a cell functions, we must consider the relationship between genes and proteins. For example, the transport of many food molecules into the cell requires cell membrane proteins. Amino acids are assembled to synthesize proteins, including the transport proteins of the cell membrane. Information contained in DNA within the nucleus determines which amino acids are combined at ribosomes to form proteins.

The human body is composed of trillions of cells, with many different characteristics. Each human begins life as a single cell. Through cell division and cell differentiation, the cells that make up the human body are formed. The following sections illustrate the whole-cell activities that determine the characteristics of a functioning cell and the growth and maintenance of the human body.

## Gene Expression

DNA contains the information that directs gene expression. This process is called **gene expression.** The proteins produced in a cell include those that serve as structural components inside the cell, structural proteins secreted to the outside of the cell, and enzymes that regulate chemical reactions in the cell. DNA influences the structural and functional characteristics of the entire organism because it directs protein synthesis. Whether an individual has blue eyes, brown hair, or other inherited traits is determined ultimately by DNA.

A DNA molecule consists of nucleotides joined together to form two nucleotide strands (see figures 2.19 and 3.23). The two strands are connected and resemble a ladder that is twisted around its long axis. The nucleotides function as chemical "letters" that form chemical "words." A **gene** is a sequence of nucleotides (making a word) that provides a chemical set of instructions for making a specific protein. Each DNA molecule contains many different genes.

Recall from chapter 2 that proteins consist of amino acids. The unique structural and functional characteristics of different proteins are determined by the kinds, numbers, and arrangement of their amino acids. The nucleotide sequence of a gene determines the amino acid sequence of a specific protein.

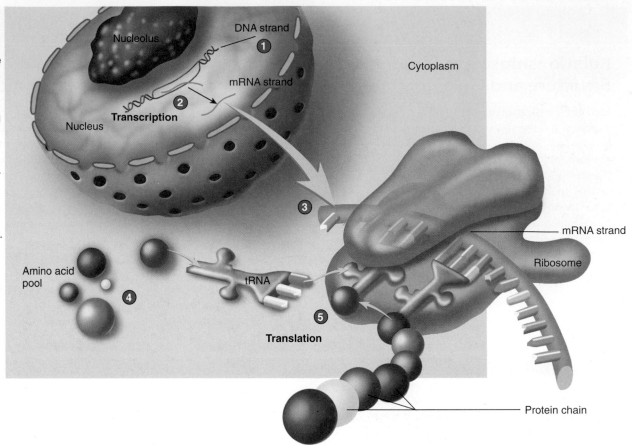

**①** DNA contains the information necessary to produce proteins.

**②** Transcription of one DNA strand results in mRNA, which is a copy of the information in the DNA strand needed to make a protein.

**③** The mRNA leaves the nucleus and goes to a ribosome.

**④** Amino acids, the building blocks of proteins, are carried to the ribosome by tRNAs.

**⑤** In the process of translation, the information contained in mRNA is used to determine the number, kinds, and arrangement of amino acids in the protein (see fig. 3.24).

**PROCESS Figure 3.22  Overview of Gene Expression**

Gene expression involves two steps—transcription and translation (figure 3.22). This process can be illustrated with an analogy. Suppose a chef wants a cake recipe that is found only in a cookbook in the library. Because the book cannot be checked out, the chef makes a copy, or **transcription,** of the recipe. Later in the kitchen, the information contained in the copied recipe is used to make the cake. The changing of something from one form to another (from recipe to cake) is called **translation.**

In terms of this analogy, DNA (the cookbook) contains many genes (recipes) for making different proteins (food items). DNA, however, is too large a molecule to pass through the nuclear pores to the ribosomes (kitchen) where the proteins are made. Just as a book stays in the library, DNA remains in the nucleus. Therefore, through transcription, the cell makes a copy of the gene necessary to make a particular protein. The copy, which is called **messenger RNA (mRNA),** travels from the nucleus to the ribosomes in the cytoplasm, where the information in the copy is used to construct a protein by means of translation. Of course, the actual ingredients are needed to turn a recipe into a cake. The ingredients necessary to synthesize a protein are amino acids. Specialized molecules, called **transfer RNAs (tRNAs),** carry the amino acids to the ribosome (figure 3.22).

In summary, gene expression involves transcription (making a copy of a gene) and translation (converting that copied information into a protein). Next, we consider the details of transcription and translation.

### Transcription

The first step in gene expression, transcription, takes place in the nucleus of the cell (figure 3.22, steps 1 and 2). DNA determines the structure of mRNA through transcription. The double strands of a DNA segment separate, and DNA nucleotides pair with RNA nucleotides (figure 3.23). Each nucleotide of DNA contains one of the following organic bases: thymine, adenine, cytosine, or guanine; each nucleotide of mRNA contains uracil, adenine, cytosine, or guanine. The number and sequence of nucleotides in the DNA serve as a template to determine the number and sequence of nucleotides in the mRNA. DNA nucleotides pair only with specific RNA nucleotides: DNA's thymine with RNA's adenine, DNA's adenine with RNA's uracil, DNA's cytosine with RNA's guanine, and DNA's guanine with RNA's cytosine.

After the DNA nucleotides pair up with the RNA nucleotides, an enzyme catalyzes reactions that form chemical bonds between the RNA nucleotides to form a long mRNA segment. Once the mRNA segment has been transcribed, portions of the mRNA molecule may be removed.

The information in mRNA is carried in groups of three nucleotides called **codons,** which specify a particular amino acid.

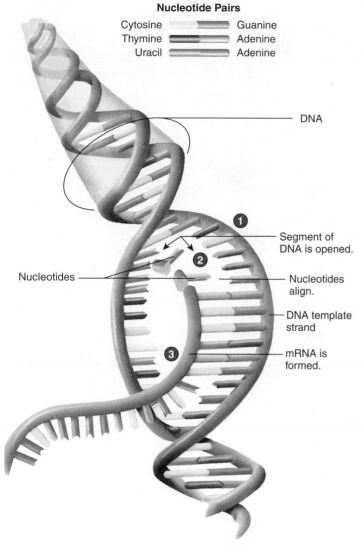

**Nucleotide Pairs**

| | |
|---|---|
| Cytosine | Guanine |
| Thymine | Adenine |
| Uracil | Adenine |

DNA

Segment of DNA is opened.

Nucleotides

Nucleotides align.

DNA template strand

mRNA is formed.

**1** The strands of the DNA molecule separate from each other. One DNA strand serves as a template for mRNA synthesis.

**2** Nucleotides that will form mRNA pair with DNA nucleotides according to the base-pair combinations shown in the key at the top of the figure. Thus, the sequence of nucleotides in the template DNA strand (*purple*) determines the sequence of nucleotides in mRNA (*gray*).

An enzyme (not shown) joins the nucleotides of mRNA together.

**3** As nucleotides are added, an mRNA molecule is formed.

**PROCESS Figure 3.23   Formation of mRNA by Transcription of DNA**

For example, the nucleotide sequence uracil, cytosine, and adenine (UCA) specifies the amino acid serine. There are 64 possible mRNA codons, but only 20 amino acids. As a result, more than 1 codon can specify the same amino acid. For example, CGA, CGG, CGU, and CGC code for the amino acid arginine; UUU and UUC code for phenylalanine. Some codons do not specify a particular amino acid but perform other functions. For example, UAA does not code for an amino acid. It is called a stop codon because it acts as a signal to end the translation process.

## Translation

Translation, the synthesis of proteins based on the information in mRNA, occurs at ribosomes (see figure 3.22, steps 3–5). The mRNA molecules produced by transcription pass through the nuclear pores to the ribosomes. Ribosomes consist of small and large subunits, which combine with mRNA during translation. The process of translation requires two types of RNA in addition to the mRNA: tRNA and **ribosomal RNA (rRNA).** There is one type of tRNA for each mRNA codon. The **anticodon,** a series of three nucleotides of tRNA, pairs with the codon of the mRNA. An amino acid is bound to another part of the tRNA. This ensures that the correct amino acid is matched with the codon of the mRNA. For example, the tRNA that pairs with the UUU codon of mRNA has the anticodon AAA and has the amino acid phenylalanine bound to it.

During translation, a ribosome binds to an mRNA. The ribosome aligns the mRNA with tRNA molecules so that the anticodons of tRNA can pair with the appropriate codons on the mRNA (figure 3.24, steps 1 and 2). An enzyme associated with the ribosomes causes the formation of a **peptide bond** between the amino acids bound to the tRNAs (figure 3.24, step 3). The ribosome moves down the mRNA one codon at a time, releasing one of the tRNA and allowing the next tRNA to move into position. As the process continues, a **polypeptide chain** is formed. Translation ends when the ribosome reaches the stop codon on the mRNA (figure 3.24, step 4). The polypeptide chain is released and becomes folded to form the three-dimensional structure of the protein molecule (see figure 2.16). A protein can consist of a single polypeptide chain or two or more polypeptide chains that are joined after each chain is produced on separate ribosomes.

**Predict 4**

*Explain how changing one nucleotide within a DNA molecule of a cell can change the structure of a protein produced by that cell.*

## Cell Life Cycle

A human body is composed of trillions of cells; however, it began as just one cell. During growth and development, cell division allowed for a dramatic increase in the number of cells. The same process allows for the replacement and repair of damaged tissue. The cell life cycle includes a nondividing phase, called **interphase,** and cell division. During interphase, the DNA (located in chromosomes in the cell's nucleus) is replicated. The two strands of DNA separate from each other, and each strand serves as a template for the production of a new strand of DNA (figure 3.25). Nucleotides in the DNA of each template strand pair with nucleotides that are subsequently joined by enzymes to form a new strand of DNA. The sequence of nucleotides in the DNA template determines the sequence of nucleotides in the new strand of DNA because adenine pairs with thymine, and cytosine pairs with guanine. The two new strands of DNA combine with the two template strands to form two double strands of DNA.

① To start protein synthesis, a ribosome binds to mRNA. The ribosome also has two binding sites for tRNA, one of which is occupied by a tRNA with its amino acid. Note that the codon of mRNA and the anticodon of tRNA are aligned and joined. The other tRNA binding site is open.

② By occupying the open tRNA binding site, the next tRNA is properly aligned with mRNA and with the other tRNA.

③ An enzyme within the ribosome catalyzes a synthesis reaction to form a peptide bond between the amino acids. Note that the amino acids are now associated with only one of the tRNAs.

④ The ribosome shifts position by three nucleotides. The tRNA without the amino acid is released from the ribosome, and the tRNA with the amino acids takes its position. A tRNA binding site is left open by the shift. Additional amino acids can be added by repeating steps 2 through 4. Eventually, a stop codon in the mRNA ends the addition of amino acids to the protein (polypeptide), which is released from the ribosome.

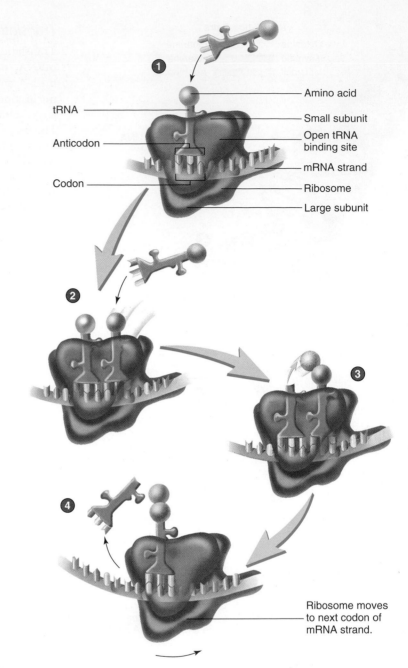

**PROCESS Figure 3.24** Translation of mRNA to Produce a Protein

At the end of interphase, a cell has two complete sets of genetic material. The DNA is dispersed throughout the nucleus as thin threads called chromatin (figure 3.26, step 1; see also figures 3.13b and 3.14).

**Cell division** is the formation of two daughter cells from a single parent cell. The new cells necessary for growth and tissue repair are formed through mitosis (discussed next), and the sex cells necessary for reproduction are formed through meiosis (discussed in chapter 19).

Each cell of the human body, except for sex cells, contains 46 chromosomes. Sex cells have half the number of chromosomes as other cells (see "Meiosis" in chapter 19). The 46 chromosomes are the **diploid** (dip′loyd) number of chromosomes and are organized to form 23 pairs of chromosomes. Of the 23 pairs, 1 pair is the sex chromosomes, which consist of 2 **X chromosomes** if the person is a female or an X chromosome and a **Y chromosome** if the person is a male. The remaining 22 pairs of chromosomes are called **autosomes** (aw′tō-sōmz). The sex chromosomes determine

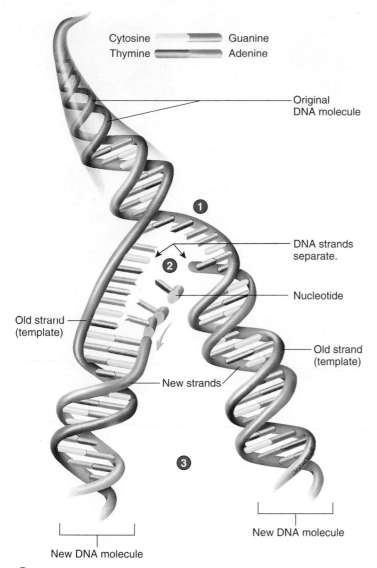

The strands of the DNA molecule separate from each other.

Each old strand (*dark purple*) functions as a template on which a new strand (*light purple*) is formed. The base-pairing relationship between nucleotides determines the sequence of nucleotides in the newly formed strands.

Two identical DNA molecules are produced.

**PROCESS Figure 3.25  Replication of DNA**

Replication during interphase produces two identical molecules of DNA.

the individual's sex, and the autosomes determine most other characteristics.

## Mitosis

Most cells of the body, except those that give rise to sex cells, divide by **mitosis** (mī-tō'sis). During mitosis, a cell divides to form two daughter cells with the same amount and type of DNA as the parent cell. Because DNA determines the structure and function of cells, the daughter cells, which have the same DNA as

the parent cell, can have the same structure and perform the same functions as the parent cell.

For convenience, mitosis is divided into four stages: prophase, metaphase, anaphase, and telophase (figure 3.26, steps 2–5). Although each stage represents certain major events, the process of mitosis is continuous. Learning each of the stages is helpful, but the most important concept to understand is how each of the two cells produced by mitosis obtains the same number and type of chromosomes as the parent cell.

- *Prophase.* During **prophase** (figure 3.26, step 2), the chromatin condenses to form visible chromosomes. After interphase, each chromosome is made up of two separate but genetically identical strands of chromatin, called **chromatids** (krō'mă-tidz), which are linked at one point by a specialized region called the **centromere** (sen'trō-mēr; *kentron,* center + *meros,* part). Replication of the genetic material during interphase results in the two identical chromatids of each chromosome. Also during prophase, microtubules called spindle fibers extend from the centrioles to the centromeres (see figures 3.1 and 3.21; figure 3.26, step 2). The centrioles divide and migrate to each pole of the cell. In late prophase, the nucleolus and nuclear envelope disappear.
- *Metaphase.* In **metaphase** (figure 3.26, step 3), the chromosomes align near the center of the cell.
- *Anaphase.* At the beginning of **anaphase** (figure 3.26, step 4), the chromatids separate. When this happens, each chromatid is then called a chromosome. At this point, two identical sets of 46 chromosomes are present in the cell. Each of the two sets of 46 chromosomes is moved by the spindle fibers toward the centriole at one of the poles of the cell. At the end of anaphase, each set of chromosomes has reached an opposite pole of the cell, and the cytoplasm begins to divide.
- *Telophase.* During **telophase** (tel'ō-fāz) (figure 3.26, step 5), the chromosomes in each of the daughter cells become organized to form two separate nuclei. The chromosomes begin to unravel and resemble the genetic material during interphase.

Following telophase, the cytoplasm of the two cells completes division, and two separate daughter cells are produced (figure 3.26, step 6).

## Differentiation

A sperm cell and an oocyte unite to form a single cell, and a new individual begins. The single cell formed during fertilization divides by mitosis to form two cells, which divide to form four cells, and so on (see chapter 20). The trillions of cells that ultimately make up the body of an adult, as a result, stem from that single cell. Therefore, all the cells in an individual's body contain the same amount and type of DNA. But even though the genetic information contained in cells is identical, not all cells look and function alike. Bone cells, for example, do not look like or function the same as muscle cells, nerve cells, or red blood cells (figure 3.27).

1 **Interphase** is the time between cell divisions. DNA is found as thin threads of chromatin in the nucleus. DNA replication occurs during interphase.

2 In **prophase**, the chromatin condenses into chromosomes. Each chromosome consists of two chromatids joined at the centromere. The centrioles move to the opposite ends of the cell, and the nucleolus and the nuclear envelope disappear.

3 In **metaphase**, the chromosomes align in the center of the cell in association with the spindle fibers.

4 In **anaphase**, the chromatids separate to form two sets of identical chromosomes. The chromosomes, assisted by the spindle fibers, move toward the centrioles at each end of the cell.

5 In **telophase**, the chromosomes disperse, the nuclear envelopes and the nucleoli form, and the cytoplasm begins to divide to form two cells.

6 Mitosis is complete, and a new interphase begins. The chromosomes have unraveled to become chromatin. Cell division has produced two daughter cells, each with DNA that is identical to the DNA of the parent cell.

**PROCESS Figure 3.26** Cell Life Cycle

# CLINICAL IMPACT

## Cancer

A **tumor** (too′mōr) is any swelling that occurs within the body, usually involving cell proliferation. Tumors can be either **malignant** (mă-lig′nănt; with malice or intent to cause harm) or **benign** (bē-nīn′; kind).

Benign tumors are usually less dangerous than malignant ones. They are not inclined to spread, but they may increase in size. As a benign tumor enlarges, it can compress surrounding tissues and impair their functions. In some cases (e.g., brain tumors), the results are fatal.

Malignant tumors can spread by local growth and expansion or by **metastasis** (mē-tas′tă-sis; moving to another place), which results when tumor cells separate from the main neoplasm and are carried by the lymphatic system or blood vessels to a new site, where a second tumor forms.

**Cancer** (kan′ser) refers to a malignant, spreading tumor and the illness that results from it. Cancers lack the normal growth control that is exhibited by most other adult tissues. Cancer results when a cell or group of cells breaks away from the normal control of growth and differentiation. This breaking loose involves the genetic machinery and can be induced by viruses, environmental toxins, and other causes. The illness associated with cancer usually occurs as the tumor invades and destroys the healthy surrounding tissues, eliminating their functions.

Promising anticancer therapies are being developed in which the cells responsible for immune responses can be stimulated to recognize tumor cells and destroy them. A major advantage in such anticancer treatments is that the cells of the immune system can specifically attack the tumor cells and not other, healthy tissues. Other therapies currently under investigation include techniques to starve a tumor to death by cutting off its blood supply. Drugs that can inhibit blood vessel development are currently in use (bevacizumab, Avastin) or under investigation (thalidomide).

### Predict 5

*Cancer cells divide continuously. The normal mechanisms that regulate whether cell division occurs or ceases do not function properly in cancer cells. Cancer cells, such as breast cancer cells, do not look like normal, mature cells. Explain.*

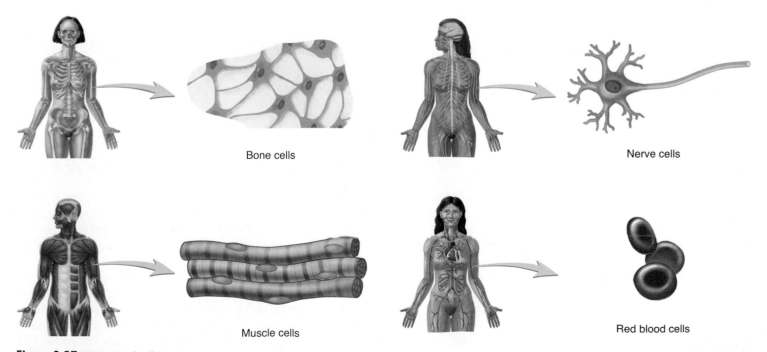

Bone cells

Nerve cells

Muscle cells

Red blood cells

**Figure 3.27** **Diversity of Cell Types**

# SYSTEMS PATHOLOGY

## Cystic Fibrosis

Tim is a 5-year-old white male. He is small for his age and has had frequent bouts of pulmonary infections all his life. Tim always seems to have a "runny nose." None of the infections have been very serious, mostly just irritating. This time, however, his congestion becomes so extreme that he is unable to breathe and is rushed to the hospital. There, a series of tests demonstrates that Tim suffers from cystic fibrosis.

### Background Information

**Cystic fibrosis** is a genetic disorder that occurs at a rate of approximately 1 per 2000 births and currently affects 33,000 people in the United States. It is the most common lethal genetic disorder among whites. The diagnosis is based on the following signs: recurrent respiratory disease, increased $Na^+$ in the sweat, and high levels of unabsorbed fats in the stool. Approximately 98% of all cases of cystic fibrosis are diagnosed before the patient is 18 years old.

At the molecular level, cystic fibrosis results from an abnormality in $Cl^-$ channels. There are three types of cystic fibrosis. (1) In about 70% of cases, a defective channel protein fails to reach the cell membrane from its site of production inside the cell. (2) In a less common type, the channel protein is incorporated into the cell membrane but fails to bind ATP. (3) In the final category, the channel protein is incorporated into the cell membrane and ATP is bound to the channel protein, but the channel does not open. The result of any of these defects is that $Cl^-$ do not exit cells at a normal rate.

Normally, as $Cl^-$ move out of cells that line body tubes, such as ducts or respiratory passages, water follows by osmosis. In cystic fibrosis, $Cl^-$ do not exit these cells at normal rates, and therefore less water moves into the tubes. With less water present, the mucus produced by the cells lining those tubes is thick, and the cells' cilia cannot readily move the mucus over the cell surface. As a result, the tubes become clogged with mucus and lose much of their normal function.

The most critical effects of cystic fibrosis, accounting for 90% of deaths, are on the respiratory system. Cystic fibrosis also affects the secretory cells lining the ducts of the pancreas, sweat glands, and salivary glands.

In normal lungs, ciliated cells move a thin, fluid layer of mucus. In people with cystic fibrosis, the viscous mucus resists movement by cilia and accumulates in the lung passages. The accumulated mucus obstructs the passageways and increases the likelihood of infections. Infections are frequent and may result in chronic pneumonia. Chronic coughing occurs as the affected person attempts to remove the mucus.

At one time, cystic fibrosis was fatal during early childhood, but many patients are now surviving into young adulthood because of modern medical treatment. Currently, approximately 80% of people with cystic fibrosis live past age 20. Pulmonary therapy consists of supporting and enhancing existing respiratory functions, and infections are treated with antibiotics.

The buildup of thick mucus in the pancreatic and hepatic ducts blocks them so that pancreatic digestive enzymes and bile salts cannot reach the small intestine. As a result, fats and fat-soluble vitamins, which require bile salts for absorption and which cannot be adequately digested without pancreatic enzymes, are not taken up by intestinal cells in normal amounts. The patient suffers from deficiencies of vitamins A, D, E, and K, which result in conditions such as night blindness, skin disorders, rickets, and excessive bleeding. Therapy includes administering the missing vitamins to the patient and reducing dietary fat intake.

The process by which cells develop with specialized structures and functions is called **differentiation**. During differentiation of a cell, some portions of DNA are active, but others are inactive. The active and inactive sections of DNA differ with each cell type. For example, the portion of DNA responsible for the structure and function of a bone cell is different from that responsible for the structure and function of a muscle cell. Differentiation, then, results from the selective activation and inactivation of segments of DNA. The mechanisms that determine which portions of DNA are active in any one cell type are not fully understood, but the resulting differentiation produces the many cell types that function together to make a person. Eventually, as cells differentiate and mature, the rate at which they divide slows or even stops.

## Apoptosis

**Apoptosis** (ăp′op-tō′sis), or *programmed cell death,* is a normal process by which cell numbers within various tissues are adjusted and controlled. In the developing fetus, apoptosis removes extra tissue, such as cells between the developing fingers and toes. In some adult tissues, apoptosis eliminates excess cells to maintain a constant number of cells within the tissue. Damaged or potentially dangerous cells, virus-infected cells, and potential cancer cells are also eliminated by apoptosis.

Apoptosis is regulated by specific genes. The proteins coded for by those genes initiate events within the cell that ultimately lead to the cell's death. As apoptosis begins, the chromatin within the nucleus condenses and fragments. This is followed by fragmentation of the nucleus and finally by death and fragmentation of the cell. Specialized cells called macrophages phagocytize the cell fragments.

| System Interactions | Effects of Cystic Fibrosis on Other Systems |
|---|---|
| **System** | **Interactions** |
| Integumentary | Cystic fibrosis is characterized by increased perspiration with abnormally high quantities of $Na^+$ in the sweat, which can lead to decreased blood $Na^+$ levels. A number of skin rashes and other disorders can develop as a result of the abnormal perspiration. |
| Nervous | Night blindness can develop as a result of vitamin A deficiency caused by insufficient absorption of the vitamin in the digestive tract. |
| Endocrine | Diabetes mellitus resulting from decreased production of the hormone insulin may develop because mucus blocks the pancreatic duct so that pancreatic digestive enzymes are retained within the pancreas, where they destroy the pancreatic tissues (pancreatic islets) that produce insulin. |
| Cardiovascular | Blood vessels can become fragile, resulting in excessive bleeding. Decreased blood clotting results from insufficient vitamin K absorption from the digestive tract. Red blood cell membranes become fragile because of inadequate vitamin E absorption. |
| Respiratory | Viscous mucus blocks the respiratory passages and inhibits respiration. Recurrent respiratory infections also occur. Decreased airflow into and out of the lungs results in reduced oxygen flow to the tissues. Respiratory complications account for most deaths. |
| Digestive | Pancreatic ducts and ducts from the liver and salivary glands are blocked with thick mucus. Fats and the fat-soluble vitamins (A, D, E, and K) are poorly absorbed. Deficiencies in fat-soluble vitamins affect many other systems. The intestine can become impacted with dehydrated stool. Gallstones can form in the gallbladder or liver ducts. |
| Reproductive | Reproductive ability is greatly decreased. In 95% of males with cystic fibrosis, living sperm cells are absent in the semen. Viscous secretions in the male or female reproductive tract decrease fertility. |

Future treatments for cystic fibrosis could include the development of drugs that correct or assist $Cl^-$ transport. Alternatively, the disease may someday be cured through gene therapy—that is, inserting a functional copy of the defective gene into patients' cells.

**Predict 6**

*Predict the effect of cystic fibrosis on the concentrations of $Cl^-$ inside and outside the cell. In normal muscle and nerve cells at rest, many $K^+$ channels are open and $K^+$ tend to flow out of the cell down their concentration gradient. How is this flow of $K^+$ affected in the cells of people with cystic fibrosis?*

# CELLULAR ASPECTS OF AGING

We are all familiar with the outward signs of aging, such as wrinkled skin, gray hair, and reduced vision. A number of cellular structures or events appear to be involved in causing these effects. The major theories that attempt to explain how aging occurs concentrate on molecules within the cell, such as lipids, proteins, and nucleic acids. It is estimated that at least 35% of the factors affecting aging are genetic.

1. *Cellular clock.* One theory of aging suggests the existence of a cellular clock that, after a certain passage of time or a certain number of cell divisions, results in the death of a given cell line.

2. *Death genes.* Another theory suggests that there are "death genes," which turn on late in life, or sometimes prematurely, causing cells to deteriorate and die.

3. *DNA damage.* Other theories suggest that, through time, DNA is damaged, resulting in cell degeneration and death.

4. *Free radicals.* DNA is also susceptible to direct damage, resulting in mutations that may result in cellular dysfunction and, ultimately, cell death. One of the major sources of DNA damage is apparently **free radicals,** which are atoms or molecules with an unpaired electron.

5. *Mitochondrial damage.* Mitochondrial DNA may be more sensitive to free-radical damage than is nuclear DNA. Mitochondrial DNA damage may result in loss of proteins critical to mitochondrial function. Because the mitochondria are the source of ATP, loss of mitochondrial function could lead to the loss of energy critical to cell function and, ultimately, to cell death. One proposal suggests that reduced caloric intake may reduce free-radical damage to mitochondria.

# SUMMARY

## Cell Structure    (p. 43)

1. Cells are highly organized units containing organelles, which perform specific functions.
2. The nucleus contains genetic material, and cytoplasm is the living material between the nucleus and the cell membrane.

## Functions of the Cell    (p. 44)

1. Cells are the basic unit of life.
2. Cells protect and support the body, as well as provide for cell metabolism, communication, and inheritance.

## Cell Membrane    (p. 44)

1. The cell membrane forms the outer boundary of the cell. It determines what enters and leaves the cell.
2. The cell membrane is composed of a double layer of phospholipid molecules in which proteins float. The proteins function as membrane channels, carrier molecules, receptor molecules, enzymes, and structural components of the membrane.

## Movement Through the Cell Membrane    (p. 45)

### Diffusion

1. Diffusion is the movement of a solute from an area of higher concentration to an area of lower concentration within a solvent. At equilibrium, the distribution of molecules is uniform.
2. A concentration gradient is the concentration of a solute at one point in a solvent minus the concentration of that solute at another point in the solvent divided by the distance between the points.
3. Lipid-soluble molecules pass through the cell membrane readily by dissolving in the lipid portion of the membrane. Small molecules and ions can pass through membrane channels.

### Osmosis

1. Osmosis is the diffusion of water across a selectively permeable membrane.
2. Osmotic pressure is the force required to prevent movement of water across a selectively permeable membrane.
3. In a hypotonic solution, cells swell (and can undergo lysis); in an isotonic solution, cells neither swell nor shrink; in a hypertonic solution, cells shrink and undergo crenation.

### Carrier-Mediated Transport Mechanisms

1. Carrier-mediated transport is the movement of a substance across a membrane by means of a carrier molecule. The substances transported tend to be large, water-soluble molecules or ions.
2. Facilitated diffusion moves substances from a higher to a lower concentration and does not require energy in the form of ATP.
3. Active transport can move substances from a lower to a higher concentration and requires ATP.
4. Secondary active transport uses the energy of one substance moving down its concentration gradient to move another substance across the cell membrane. In cotransport, both substances move in the same direction; in countertransport, they move in opposite directions.

### Endocytosis and Exocytosis

1. Endocytosis is the movement of materials into cells by the formation of a vesicle. Receptor-mediated endocytosis involves cell receptors attaching to molecules, which are then transported into the cell. Phagocytosis is the movement of solid material into cells. Pinocytosis is similar to phagocytosis, except that the material ingested is much smaller and is in solution.
2. Exocytosis is the secretion of materials from cells by vesicle formation.

## Organelles    (p. 53)

### Nucleus

1. The nuclear envelope consists of two separate membranes that form nuclear pores at many points on the surface of the nucleus.
2. DNA and associated proteins are found inside the nucleus as chromatin. DNA is the hereditary material of the cell and controls cell activities.
3. Nucleoli consist of RNA and proteins and are the sites of ribosomal subunit assembly.

### Ribosomes

1. Ribosomes are the sites of protein synthesis.
2. A ribosome is composed of one large and one small subunit.

### Rough and Smooth Endoplasmic Reticulum

1. Rough ER is ER with ribosomes attached. It is a major site of protein synthesis.
2. Smooth ER does not have ribosomes attached and is a major site of lipid synthesis.

### Golgi Apparatus

The Golgi apparatus is a series of closely packed membrane sacs that collect, modify, package, and distribute proteins and lipids produced by the ER.

### Secretory Vesicles

Secretory vesicles are membrane-bound sacs that carry substances from the Golgi apparatus to the cell membrane, where the vesicle contents are released.

### Lysosomes and Peroxisomes

Lysosomes and peroxisomes are membrane-bound sacs containing enzymes. Within the cell, lysosomes break down phagocytized material. Peroxisomes break down fatty acids, amino acids, and hydrogen peroxide.

### Mitochondria

Mitochondria are the major sites for the production of ATP, which cells use as an energy source. Mitochondria carry out aerobic respiration (requires oxygen).

### Cytoskeleton

1. The cytoskeleton supports the cytoplasm and organelles and is involved with cell movements.
2. The cytoskeleton is composed of microtubules, microfilaments, and intermediate filaments.

### Centrioles

Centrioles, located in the centrosome, are made of microtubules. They facilitate chromosome movement during cell division.

### Cilia, Flagella, and Microvilli

1. Cilia move substances over the surface of cells.
2. Flagella are much longer than cilia and propel sperm cells.
3. Microvilli increase the surface area of cells and thus aid in absorption.

## Whole-Cell Activity    (p. 59)

A cell's characteristics are ultimately determined by the types of proteins it produces, which are determined by the genetic information in the nucleus. Understanding how genetic information is used in the cell and distributed to daughter cells is important for understanding basic cellular activity.

### Gene Expression

1. Cell activity is regulated by enzymes (proteins), and DNA controls enzyme production.
2. During transcription, the sequence of nucleotides in DNA (a gene) determines the sequence of nucleotides in mRNA; the mRNA moves through the nuclear pores to ribosomes.
3. During translation, the sequence of codons in mRNA is used at ribosomes to produce proteins. Anticodons of tRNA bind to the codons of mRNA, and the amino acids carried by tRNA are joined to form a protein.

### Cell Life Cycle

1. The life cycle of a cell consists of a series of events that produce new cells for growth and for tissue repair.
2. The two phases of the cell life cycle are interphase and cell division.
3. DNA replicates during interphase, the nondividing phase of the cell cycle.
4. Cell division occurs through mitosis, which is divided into four stages:
   • Prophase—each chromosome consists of two chromatids joined at the centromere.
   • Metaphase—chromosomes align at the center of the cell.
   • Anaphase—chromatids separate at the centromere and migrate to opposite poles.
   • Telophase—the two new nuclei assume their normal structure, and cell division is completed, producing two new daughter cells.

### Differentiation

Differentiation, the process by which cells develop specialized structures and functions, results from the selective activation and inactivation of DNA sections.

### Apoptosis

Apoptosis is the programmed death of cells. Apoptosis regulates the number of cells within various tissues of the body.

## Cellular Aspects of Aging (p. 67)

Aging may be due to the presence of "cellular clocks," the function of "death genes," DNA damage, free radicals, or mitochondrial damage.

## REVIEW AND COMPREHENSION

1. Define cytoplasm and organelle.
2. List the functions of a cell.
3. Describe the structure of the cell membrane. What functions does it perform?
4. Define solution, solute, solvent, diffusion, and concentration gradient.
5. How do lipid-soluble molecules, small molecules that are not lipid-soluble, and large molecules that are not lipid-soluble cross the cell membrane?
6. Define osmosis and osmotic pressure.
7. What happens to cells that are placed in isotonic solutions? In hypertonic or hypotonic solutions? What are crenation and lysis?
8. What is carrier-mediated transport? How are facilitated diffusion and active transport similar, and how are they different?
9. How does secondary active transport work? Define cotransport and countertransport.
10. Describe receptor-mediated endocytosis, phagocytosis, pinocytosis, and exocytosis. What do they accomplish?
11. Describe the structure of the nucleus and the nuclear envelope. Name the structures in the nucleus, and give their functions.
12. Where are ribosomes assembled, and what kinds of molecules are found in them?
13. What is endoplasmic reticulum? Compare the functions of rough and smooth endoplasmic reticulum.
14. Describe the Golgi apparatus, and state its function.
15. Where are secretory vesicles produced? What are their contents, and how are they released?
16. What are the functions of lysosomes and peroxisomes?
17. Describe the structure and function of mitochondria.
18. Name the components of the cytoskeleton, and give their functions.
19. Describe the structure and function of centrioles.
20. Describe the structure and function of cilia, flagella, and microvilli.
21. Describe how proteins are synthesized and how the structure of DNA determines the structure of proteins.
22. Define autosome, sex chromosome, and diploid number.
23. How do the sex chromosomes of males and females differ?
24. Describe what happens during interphase and each stage of mitosis. What kinds of tissues undergo mitosis?
25. Define cell differentiation. In general terms, how does differentiation occur?
26. List the principal theories of aging.

## CRITICAL THINKING

1. The body of a man was found floating in the salt water of Grand Pacific Bay, which has a concentration slightly greater than that of body fluids. When examined during an autopsy, the cells in his lung tissue were clearly swollen. Choose the most logical conclusion.
   a. He probably drowned in the bay.
   b. He may have been murdered elsewhere.
   c. He did not drown.

2. Patients with kidney failure can be kept alive by dialysis, which removes toxic waste products from the blood. In a dialysis machine, blood flows past one side of a selectively permeable dialysis membrane, and dialysis fluid flows on the other side of the membrane. Small substances, such as ions, glucose, and urea, can pass through the dialysis membrane, but larger substances, such as proteins, cannot. If you wanted to use a dialysis machine to remove

only the toxic waste product urea from blood, what could you use for the dialysis fluid?

a. a solution that is isotonic and contains only protein
b. a solution that is isotonic and contains the same concentration of substances as blood, except for having no urea in it
c. distilled water
d. blood

3. Secretory vesicles fuse with the cell membrane to release their contents to the outside of the cell. In this process, the membrane of the secretory vesicle becomes part of the cell membrane. Because small pieces of membrane are continually added to the cell membrane, we would expect the cell membrane to become larger and larger as secretion continues. However, the cell membrane stays the same size. Explain how this happens.

4. Suppose that a cell has the following characteristics: many mitochondria, well-developed rough ER, well-developed Golgi apparatuses, and numerous vesicles. Predict the major function of the cell. Explain how each characteristic supports your prediction.

5. The proteins (hemoglobin) in red blood cells normally organize relative to one another, forming "stacks" of proteins that are, in part, responsible for the normal shape of red blood cells. In sickle-cell anemia, proteins inside red blood cells do not stack normally. Consequently, the red blood cells become sickle-shaped and plug up small blood vessels. Sickle-cell anemia is hereditary and results from changing one nucleotide for a different nucleotide within the gene that is responsible for producing the protein. Explain how this change results in an abnormally functioning protein.

*Answers in Appendix D*

Visit this textbook's website at **www.mhhe.com/seeleyess7** for practice quizzes, animations, interactive learning exercises, and other study tools.

McGraw-Hill offers a study CD that features interactive cadaver dissection. ***Anatomy & Physiology Revealed*** includes cadaver photos that allow you to peel away layers of the human body to reveal structures beneath the surface. This program also includes animations, radiologic imaging, audio pronunciations, and practice quizzing.

The human body is constructed in layers of increasing complexity. The simplest "layer" is the cell. Cells combine into groups with similar shapes and functions. These groups of cells work together and are called tissues. Tissues, such as the one shown in the picture, provide a framework for most body parts.

# 4 Chapter

# Tissues

## Chapter Outline and Objectives

**Epithelial Tissue (p. 72)**

1. List the characteristics of epithelial tissue.
2. Name the major classifications of epithelial tissue, and give an example of each.
3. Explain the function of the following features of epithelial tissue: cell layers, cell shapes, free cell surfaces, and connections between cells.
4. Define and categorize glands.

**Connective Tissue (p. 79)**

5. Describe the classification of connective tissue, and give examples of each major type.

**Muscle Tissue (p. 84)**

6. Name the three types of muscle tissue, and list their functions.

**Nervous Tissue (p. 87)**

7. State the functions of nervous tissue, and describe a neuron.

**Membranes (p. 87)**

8. List the structural and functional characteristics of mucous, serous, and synovial membranes.

**Tissue Damage and Inflammation (p. 89)**

9. Describe the process of inflammation in response to tissue damage, and explain why inflammation protects the body.

**Tissue Repair (p. 90)**

10. Describe the major events involved in tissue repair.

**Effects of Aging on Tissues (p. 92)**

11. Describe the age-related changes that occur in cells and in extracellular matrix.

In some ways, the human body is like a complex machine. For example, all of the parts of an automobile cannot be made from a single type of material. Metal, capable of withstanding the heat of the engine, cannot be used for windows or tires. Similarly, the many parts of the human body are made of collections of specialized cells and the materials surrounding them. Muscle cells that contract to produce body movements have a different structure and function than epithelial cells that protect, secrete, or absorb.

A **tissue** (tish'ū) is a group of cells with similar structure and function that have similar extracellular substances located between them. The microscopic study of tissue structure is called **histology** (his-tol'ō-je; histo-, tissue + -ology, study). Knowledge of tissue structure and function is important in understanding how individual cells are organized to form tissues and how tissues are organized to form organs, organ systems, and the complete organism. The structure of each tissue type is related to its function, and the structure of the tissues in an organ is related to the organ's function.

Changes in tissues can result in development, growth, aging, trauma, or disease. For example, skeletal muscles enlarge because skeletal muscle cells increase in size in response to exercise. Reduced elasticity of blood vessel walls in aging people results from gradual changes in connective tissue. Many tissue abnormalities, including cancer, result from changes in tissues that can be identified by microscopic examination.

The four basic tissue types are epithelial, connective, muscle, and nervous. This chapter emphasizes epithelial and connective tissues. Later chapters consider muscle and nervous tissues in more detail.

# EPITHELIAL TISSUE

**Epithelium** (ep-i-thē'lē-ŭm; pl. epithelia, ep-i-thē'lē-ă; epi, on + thele, covering or lining), or epithelial tissue, covers internal and external surfaces throughout the body. It also forms most glands. Surfaces of the body include the outer layer of the skin and the lining of cavities, such as the digestive tract, respiratory passages, and blood vessels. Epithelium consists almost entirely of cells with very little extracellular material between them. Although there are some exceptions, most epithelia have a **free surface,** which is not in contact with other cells, and a **basal surface** adjacent to a basement membrane, which attaches the epithelial cells to underlying tissues (figure 4.1). Epithelium may consist of a single layer of cells or multiple layers of epithelial cells between the free surface and the basement membrane.

The **basement membrane** is secreted partly by epithelial cells and partly by the cells of the underlying tissues. It consists of a meshwork of protein molecules with other molecules bound to them. Substances that cross the epithelium must also cross the basement membrane. It can function as a filter and as a barrier to the movement of cells. For example, if some epithelial cells are converted to cancer cells, the basement membrane can, for some time, help prevent the spread of the cancer into the underlying tissues.

Blood vessels do not extend from the underlying tissues into epithelium, so gases and nutrients that reach the epithelium must diffuse across the basement membrane from the underlying tissues, where blood vessels are abundant. Waste products produced by the epithelial cells diffuse across the basement membrane to blood vessels.

## Functions of Epithelia

The major functions of epithelia are

1. *Protecting underlying structures.* Examples include the outer layer of the skin and the epithelium of the oral cavity, which protect the underlying structures from abrasion.

2. *Acting as a barrier.* Epithelium prevents many substances from moving through it. For example, the epithelium of the skin acts as a barrier to water and reduces water loss from the body. The epithelium of the skin also prevents many toxic molecules and microorganisms from entering the body.

3. *Permitting the passage of substances.* Epithelium allows many substances to move through the epithelial layer. For example, oxygen and carbon dioxide are exchanged between the air and blood by diffusion through the epithelium in the lungs.

4. *Secreting substances.* Sweat glands, mucous glands, and the enzyme-secreting portion of the pancreas are all composed of epithelial cells.

5. *Absorbing substances.* The cell membranes of certain epithelial tissues contain carrier proteins (see chapter 3) that regulate the absorption of materials. For example, the epithelial cells of the intestine absorb digested food molecules, vitamins, and ions.

## Classification of Epithelia

Epithelia are classified according to the number of cell layers and the shape of the cells (table 4.1). **Simple epithelium** consists of a single layer of cells. **Stratified epithelium** consists of more than one layer of epithelial cells, with some cells sitting on top of others. Categories of epithelium based on cell shape are **squamous** (skwā'mŭs; flat), **cuboidal** (cubelike), and **columnar** (tall and thin). In most cases, each epithelium is given two names, such as simple squamous, simple columnar, or stratified squamous epithelium. When epithelium is stratified, it is named according to the shape of the cells at the free surface.

**Simple squamous epithelium** is a single layer of thin, flat cells (table 4.2a). Some substances easily pass through this thin layer of cells, but other substances do not. For example, the respiratory passages end as small sacs called **alveoli** (al-vē'o-lī; sing. alveolus, hollow sac). The alveoli consist of simple squamous

Lung
Pleura
Free surface
Epithelial cells with little extracellular matrix
Nucleus
Basement membrane
Connective tissue
Capillary
LM 640x
Surface view
LM 640x
Cross-sectional view

**Figure 4.1** Characteristics of Epithelium

Surface and cross-sectional views of epithelium illustrate the following characteristics: little extracellular material between cells, a free surface, and a basement membrane attaching epithelial cells to underlying tissues. The capillaries in connective tissue do not penetrate the basement membrane. Nutrients, oxygen, and waste products diffuse across the basement membrane between the capillaries and the epithelial cells.

| TABLE 4.1 | Classification of Epithelia | |
|---|---|
| **Number of Layers** | **Cell Shape** |
| Simple (one layer) | Squamous<br>Cuboidal<br>Columnar |
| Pseudostratified (a modified form of simple epithelium) | Columnar |
| Stratified (more than one layer) | Squamous<br>Cuboidal<br>Columnar |
| Transitional (a type of stratified epithelium) | Roughly cuboidal to columnar when not stretched and squamouslike when stretched |

epithelium that allows oxygen from the air to diffuse into the body and carbon dioxide to diffuse out of the body into the air. Simple squamous epithelium in the filtration membranes of the kidneys forms thin barriers through which small molecules, but not large ones, can pass. Small molecules, including water from blood, are filtered through these barriers as a major step in urine formation. Large molecules, such as proteins and blood cells, remain in the blood vessels of the kidneys.

Simple squamous epithelium also prevents abrasion between organs in the pericardial, pleural, and peritoneal cavities (see chapter 1). The outer surfaces of organs are covered with simple squamous epithelium that secretes a slippery fluid. The fluid lubricates the surfaces between the organs, preventing damage from friction when the organs rub against one another.

**Simple cuboidal epithelium** is a single layer of cubelike cells (table 4.2b) that carry out active transport, facilitated diffusion, or secretion. Epithelial cells that secrete molecules such as proteins contain organelles that synthesize them. These cells have a greater volume than simple squamous epithelial cells and contain more cell organelles. The kidney tubules have large portions of their walls composed of simple cuboidal epithelium. These cells secrete waste products into the tubules and reabsorb useful materials from the tubules as urine is formed. Some cuboidal epithelial cells have cilia that move mucus over the free surface or microvilli that increase the surface area for secretion and absorption.

## TABLE 4.2    Simple Epithelium

### (a) Simple Squamous Epithelium

**Structure:** Single layer of flat, often hexagonal cells; the nuclei appear as bumps when viewed in cross section because the cells are so flat

**Function:** Diffusion, filtration, some secretion, and some protection against friction

**Location:** Lining of blood vessels and the heart, lymphatic vessels, alveoli of the lungs, portions of the kidney tubules, lining of serous membranes of body cavities (pleural, pericardial, peritoneal)

Lung alveoli

Free surface

Nucleus

Basement membrane

**Simple squamous epithelial cell**

LM 640x

### (b) Simple Cuboidal Epithelium

**Structure:** Single layer of cube-shaped cells; some cells have microvilli (kidney tubules) or cilia (terminal bronchioles of the lungs)

**Function:** Active transport and facilitated diffusion result in secretion and absorption by cells of the kidney tubules; secretion by cells of glands and choroid plexuses; movement of particles embedded in mucus out of the terminal bronchioles by ciliated cells

**Location:** Kidney tubules, glands and their ducts, choroid plexuses of the brain, lining of terminal bronchioles of the lungs, and surfaces of the ovaries

Kidney

Free surface

Nucleus

**Simple cuboidal epithelial cell**

Basement membrane

LM 640x

## (c) Simple Columnar Epithelium

**Structure:** Single layer of tall, narrow cells; some cells have cilia (bronchioles of lungs, auditory tubes, uterine tubes, and uterus) or microvilli (intestines)

**Function:** Movement of particles out of the bronchioles of the lungs by ciliated cells; partially responsible for the movement of oocytes through the uterine tubes by ciliated cells; secretion by cells of the glands, the stomach, and the intestine; absorption by cells of the intestine

**Location:** Glands and some ducts, bronchioles of lungs, auditory tubes, uterus, uterine tubes, stomach, intestines, gallbladder, bile ducts, and ventricles of the brain

Lining of stomach and intestines

Free surface

Goblet cell containing mucus

Nucleus

**Simple columnar epithelial cell**

Basement membrane

LM 640x

## (d) Pseudostratified Columnar Epithelium

Structure: Single layer of cells; some cells are tall and thin and reach the free surface, and others do not; the nuclei of these cells are at different levels and appear stratified; the cells are almost always ciliated and are associated with goblet cells that secrete mucus onto the free surface

Function: Synthesize and secrete mucus onto the free surface and move mucus (or fluid) that contains foreign particles over the surface of the free surface and from passages

Location: Lining of nasal cavity, nasal sinuses, auditory tubes, pharynx, trachea, and bronchi of lungs

Trachea

Bronchus

Cilia

Free surface

Goblet cell containing mucus

**Pseudostratified columnar epithelial cell**

Nucleus

Basement membrane

LM 413x

**Simple columnar epithelium** is a single layer of tall, thin cells (table 4.2c). These large cells contain organelles that enable them to perform complex functions. For example, the simple columnar epithelium of the small intestine produces and secretes mucus and digestive enzymes. **Mucus** (mū′kŭs) is a clear, viscous (thick) fluid. The mucus protects the lining of the intestine, and the digestive enzymes complete the process of digesting food. The columnar cells then absorb the digested foods by active transport, facilitated diffusion, or simple diffusion.

**Pseudostratified columnar epithelium** is a special type of simple epithelium (table 4.2d). The prefix *pseudo-* means false, so this type of epithelium appears stratified but is not. It consists of one layer of cells, with all the cells attached to the basement membrane. But it looks like two or more layers of cells because some of the cells are tall and reach the free surface, whereas others are short and do not reach the free surface. Pseudostratified columnar epithelium lines some glands and ducts, the auditory tubes, and some of the respiratory passages, such as the nasal cavity, nasal sinuses, pharynx, trachea, and bronchi. Pseudostratified columnar epithelium secretes mucus, which covers its free surface. Cilia located on the free surface move the mucus and the debris that accumulates in it. For example, cilia of the respiratory passages move mucus toward the throat, where it is swallowed.

**Stratified squamous epithelium** forms a thick epithelium because it consists of several layers of cells (table 4.3a). The deepest cells are cuboidal or columnar and are capable of dividing and producing new cells. As these newly formed cells are pushed to the surface, they become flat and thin. As the cells flatten, the cytoplasm of the epithelial cells is replaced by a protein called keratin, and the cells die. One type of stratified squamous epithelium forms the outer layer of the skin and is called keratinized squamous epithelium (see chapter 5). The dead cells provide protection against abrasion, form a barrier that prevents microorganisms and toxic chemicals from entering the body, and reduces the loss of water from the body. If cells at the surface are damaged or rubbed away, they are replaced by cells formed in the deeper layers.

In contrast, stratified squamous epithelium of the mouth is composed of living cells with a moist surface. This nonkeratinized (moist) stratified squamous epithelium also provides protection against abrasion and acts as a mechanical barrier, preventing microorganisms from entering the body. Water, however, can move across it more readily than across the skin.

**Stratified cuboidal epithelium** consists of more than one layer of cuboidal epithelial cells. This epithelial type is relatively rare and is found in sweat gland ducts, ovarian follicular cells, and the salivary glands. It functions in absorption, secretion, and protection.

**Stratified columnar epithelium** consists of more than one layer of epithelial cells, but only the surface cells are columnar. The deeper layers are irregular or cuboidal in shape. Like stratified cuboidal epithelium, stratified columnar epithelium is relatively rare. It is found in the mammary gland ducts, the larynx, and a portion of the male urethra. This epithelium carries out secretion, protection, and some absorption.

**Transitional epithelium** is a special type of stratified epithelium that can be greatly stretched (table 4.3b). In the unstretched state, transitional epithelium consists of five or more layers of cuboidal or columnar cells that often are dome-shaped at the free surface. As transitional epithelium is stretched, the cells change to a low cuboidal or squamous shape, and the number of cell layers decreases. Transitional epithelium lines cavities that can expand greatly, such as the urinary bladder. It also protects underlying structures from the caustic effects of urine.

## Structural and Functional Relationships
### Cell Layers and Cell Shapes

The number of cell layers and the shape of the cells in a specific type of epithelium reflect the function the epithelium performs. Two important functions are controlling the passage of materials through the epithelium and protecting the underlying tissues. Simple epithelium, with its single layer of cells, is found in organs that primarily function to move materials. Examples include the diffusion of gases across the wall of the alveoli of the lungs, filtration of fluid across the filtration membranes in the kidneys, secretion from glands, and nutrient absorption by the intestines. The movement of materials through a stratified epithelium is hindered by its many layers. Stratified epithelium is well adapted for its protective function. As the outer cell layers are damaged, they are replaced by cells from deeper layers. Stratified squamous epithelium is found in areas of the body where abrasion can occur, such as in the skin, anal canal, and vagina.

Differences in function are also reflected in cell shape. Cells are normally flat and thin when the function is diffusion, such as in the alveoli of the lungs, or filtration, such as in the kidney tubules. Cells with the major function of secretion or absorption are usually cuboidal or columnar. They are larger because they contain more organelles, which are responsible for the function of the cell. The stomach, for example, is lined with simple columnar epithelium. These cells contain many **secretory vesicles** (ves′i-klz) filled with mucus. The large amounts of mucus produced by the simple columnar epithelium protect the stomach lining against the digestive enzymes and acid produced in the stomach. An ulcer, or irritation in the stomach's epithelium and underlying tissue, can develop if this protective mechanism fails. Simple cuboidal epithelial cells that secrete or absorb molecules, as occurs in the kidney tubules, contain many mitochondria, which produce the ATP required for active transport.

> **Predict 1**
>
> *Explain the consequences of having (a) nonkeratinized stratified epithelium rather than simple columnar epithelium lining the digestive tract and (b) nonkeratinized stratified squamous epithelium rather than keratinized stratified squamous epithelium at the surface of the skin.*

The shape and number of layers of epithelial cells can change if they are subjected to long-term irritation or other abnormal conditions. People who smoke cigarettes eventually experience changes in the epithelium of the larger respiratory passages. The delicate pseudostratified columnar epithelium, which performs a cleaning function by moving mucus and debris from the passageways, is replaced by stratified squamous epithelium, which

## TABLE 4.3    Stratified Epithelium

### (a) Stratified Squamous Epithelium

**Structure:** Several layers of cells that are cuboidal in the basal layer and progressively flattened toward the surface; the epithelium can be nonkeratinized (moist) or keratinized; in nonkeratinized stratified squamous epithelium, the surface cells retain a nucleus and cytoplasm; in keratinized stratified epithelium, the cytoplasm of cells at the surface is replaced by a protein called keratin, and the cells are dead

**Function:** Protects against abrasion, forms a barrier against infection, and reduces loss of water from the body

**Location:** Keratinized—outer layer of the skin; nonkeratinized—mouth, throat, larynx, esophagus, anus, vagina, inferior urethra, and corneas

- Skin
- Cornea
- Mouth
- Esophagus

- Free surface
- **Nonkeratinized stratified squamous epithelial cell**
- Nuclei
- Basement membrane

LM 286x

### (b) Transitional Epithelium

**Structure:** Stratified cells that appear cuboidal when the organ or tube is not stretched and squamous when the organ or tube is stretched by fluid

**Function:** Accommodates fluctuations in the volume of fluid in an organ or a tube; protects against the caustic effects of urine

**Location:** Lining of urinary bladder, ureters, and superior urethra

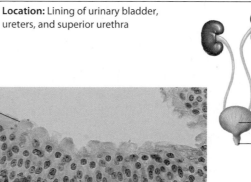

- Ureter
- Urinary bladder
- Urethra

- Free surface
- **Transitional epithelial cell**
- Nucleus
- Basement membrane

LM 413x

Tissue not stretched

- Free surface
- **Transitional epithelial cell**
- Nucleus
- Basement membrane

LM 413x

Tissue stretched

is more resistant to irritation but does not perform a cleaning function. Also, lung cancer most often results from changes in epithelial cells in the lung passageways of smokers. The changes in cell structure are used to identify the cancer.

## A CASE IN POINT
### Detecting Cancer

Wanna Wonder was wary about having a Pap test. She insisted on knowing why she had to endure such a procedure. Her physician explained that the test could detect cervical cancer. Using a swab, a physician collects cells from the cervix of the uterus. The cells are stained and examined with a microscope. Compared with normal cells, cancer cells have variable sizes and shapes, with large, unusually shaped nuclei. If uterine cancer is detected early, it can be treated before it spreads to other areas of the body.

### Free Surfaces

Most epithelia have a free surface that is not in contact with other cells and faces away from underlying tissues. The characteristics of the free surface reflect its functions. The free surface can be smooth or lined with microvilli or cilia. A smooth free surface reduces friction as material moves across it. For example, the lining of blood vessels is simple squamous epithelium with a smooth surface, which reduces friction as blood flows through the vessels. **Microvilli** are cylindrical extensions of the cell membrane that increase the free surface area (see chapter 3). Normally, many microvilli cover the free surface of each cell involved in absorption or secretion, such as the cells lining the small intestine. **Cilia** (see chapter 3) propel materials along the free surface of cells. The nasal cavity and trachea are lined with pseudostratified columnar ciliated epithelium. Intermixed with the ciliated cells are specialized mucus-producing cells called **goblet cells** (see table 4.2*d*). Dust and other materials are trapped in the mucus that covers the epithelium, and movement of the cilia propels the mucus with its entrapped particles to the back of the throat, where it is swallowed or coughed up. The constant movement of mucus helps keep the respiratory passages clean.

### Cell Connections

Epithelial cells are connected to one another in several ways (figure 4.2). **Tight junctions** bind adjacent cells together and form permeability barriers. Tight junctions prevent the passage of materials between epithelial cells because they completely surround each cell, similar to the way a belt surrounds the waist. Materials that pass through the epithelial layer must pass through the cells, so those cells regulate what materials can cross. Tight junctions are found in the lining of the intestines and in most other simple epithelia. **Desmosomes** (dez'mō-sōmz; *desmos,* a band + *soma,* body) are mechanical links that bind cells together. Modified desmosomes, called **hemidesmosomes** (hem-ē-dez'mō-sōmz; *hemi,* one-half), also anchor cells to the basement membrane. Many desmosomes are found in epithelia subjected to stress, such as the stratified squamous epithelium of the skin. **Gap junctions** are small channels that allow small molecules and ions

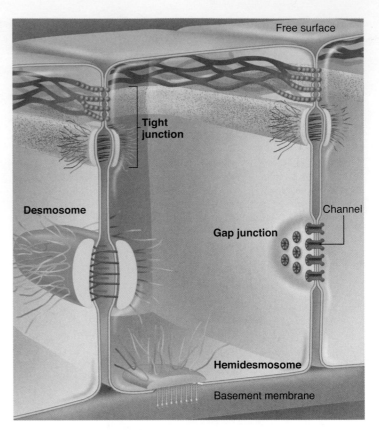

### Figure 4.2 Cell Connections

Desmosomes anchor cells to one another, and hemidesmosomes anchor cells to the basement membrane. Tight junctions bind adjacent cells together. Gap junctions have channels that allow adjacent cells to communicate with each other. Few cells have all of these types of connections.

to pass from one epithelial cell to an adjacent one. Most epithelial cells are connected to one another by gap junctions, and researchers believe that molecules or ions moving through the gap junctions act as communication signals to coordinate the activities of the cells.

## Glands

A **gland** is a structure that secretes substances onto a surface, into a cavity, or into the blood. Most glands are composed primarily of epithelium and are multicellular. But sometimes single goblet cells are classified as unicellular glands because they secrete mucus onto epithelial surfaces. Glands with ducts are called **exocrine** (ek'sō-krin; *exo,* outside + *krino,* to separate) glands (figure 4.3). The exocrine glands can be **simple,** with ducts that have no branches, or **compound,** with ducts that have many branches. The end of a duct can be **tubular.** Some tubular glands are straight, and others are coiled. Some glands have ends that are expanded into a saclike structure called an **acinus** (as'i-nŭs; grapelike) or an **alveolus** (al-vē'ō-lŭs; small cavity). Some compound glands have acini and tubules (tubuloacinar or tubuloalveolar glands) that secrete substances. Secretions from exocrine glands pass through the ducts onto a surface or into an organ. For

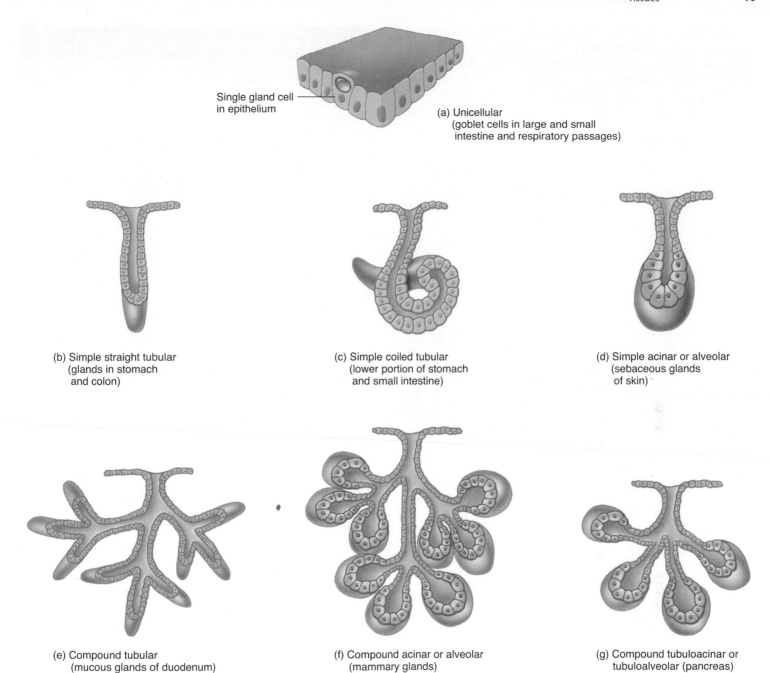

Single gland cell
in epithelium

(a) Unicellular
(goblet cells in large and small
intestine and respiratory passages)

(b) Simple straight tubular
(glands in stomach
and colon)

(c) Simple coiled tubular
(lower portion of stomach
and small intestine)

(d) Simple acinar or alveolar
(sebaceous glands
of skin)

(e) Compound tubular
(mucous glands of duodenum)

(f) Compound acinar or alveolar
(mammary glands)

(g) Compound tubuloacinar or
tubuloalveolar (pancreas)

**Figure 4.3  Structure of Exocrine Glands**

Based on the shapes of their ducts, exocrine glands may be called simple or compound. Based on their secretory units, they may be classed as tubular, acinar, or alveolar.

example, sweat from sweat glands and oil from sebaceous glands flow onto the skin surface.

   **Endocrine** (en′dō-krin; *endo,* within) glands have no ducts and empty their secretions into the blood. These secretions, called **hormones** (hōr′mōnz), are carried by the blood to other parts of the body. Endocrine glands include the thyroid gland and the insulin-secreting portions of the pancreas. We discuss endocrine glands more fully in chapter 10.

# CONNECTIVE TISSUE

**Connective tissue** is found throughout the body. It is usually characterized by large amounts of extracellular material that separates cells from one another. The extracellular material, or **extracellular matrix** (mā′triks), has three major components: (1) protein fibers, (2) ground substance consisting of nonfibrous protein and other molecules, and (3) fluid.

Three types of protein fibers help form most connective tissues. **Collagen** (kol′lă-jen; glue-producing) **fibers,** which resemble microscopic ropes, are flexible but resist stretching. **Reticular** (rē-tik′ū-lăr) **fibers** are very fine, short collagen fibers that branch to form a supporting network. **Elastic fibers** have a structure similar to that of coiled metal bed springs; after being stretched, they can recoil to their original shape.

**Ground substance** is the shapeless background against which cells and collagen fibers can be seen when using a light microscope. Although ground substance appears shapeless, the molecules within it are highly structured. **Proteoglycans** (prō′tē-ō-glī′kanz; *proteo,* protein + *glycan,* polysaccharide) resemble the limbs of pine trees, with proteins forming the branches and polysaccharides forming the pine needles. This structure enables proteoglycans to trap large quantities of water between the polysaccharides.

Connective tissue cells are named according to their functions. Cells whose names contain the suffix *-blast* (germ) produce the matrix; cells ending in *-cyte* (cell) maintain it; and cells ending in *-clast* (break) break it down for remodeling. For example, **fibroblasts** (fī′brō-blass; *fibra,* fiber) are cells that form fibers and ground substance in the extracellular matrix of fibrous connective tissue, and **fibroyctes** are cells that maintain it. **Osteoblasts** (os′tē-ō-blasts; *osteo,* bone) form bone, **osteocytes** (os′tē-ō-sītz) maintain bone, and **osteoclasts** (os′tē-ō-klasts) break down bone.

Also found in connective tissue are cells associated with the immune system. **Macrophages** (mak′rō-fāj-ez; *makros,* large + *phago,* to eat) are large white blood cells that are capable of moving about and ingesting foreign substances, including microorganisms, in the connective tissue. **Mast cells** are nonmotile cells that release chemicals, such as histamine, that promote inflammation.

## Functions of Connective Tissue

Connective tissue performs the following major functions:

1. *Enclosing and separating other tissues.* Sheets of connective tissue form capsules around organs, such as the liver and the kidneys. Connective tissue also forms layers that separate tissues and organs. For example, connective tissues separate muscles, arteries, veins, and nerves from one another.
2. *Connecting tissues to one another.* Tendons are strong cables, or bands, of connective tissue that attach muscles to bone, and ligaments are connective tissue bands that hold bones together.
3. *Supporting and moving parts of the body.* Bones of the skeletal system provide rigid support for the body, and semirigid cartilage supports structures, such as the nose, the ears, and the surfaces of joints. Joints between bones allow one part of the body to move relative to other parts.
4. *Storing compounds.* Adipose tissue (fat) stores high-energy molecules, and bones store minerals, such as calcium and phosphate.
5. *Cushioning and insulating.* Adipose tissue cushions and protects the tissues it surrounds and provides an insulating layer beneath the skin that helps conserve heat.

| TABLE 4.4 | Classification of Connective Tissues |
|---|---|
| Loose (areolar) connective tissue | |
| Adipose tissue | |
| Dense connective tissue<br>    Dense collagenous connective tissue<br>        Collagen fibers arranged in the same direction<br>        Collagen fibers arranged in many directions<br>    Dense elastic connective tissue<br>        Elastic fibers arranged in the same direction<br>        Elastic fibers arranged in many directions | |
| Cartilage<br>    Hyaline cartilage<br>    Fibrocartilage<br>    Elastic cartilage | |
| Bone | |
| Blood | |

6. *Transporting.* Blood transports gases, nutrients, enzymes, hormones, and cells of the immune system throughout the body.
7. *Protecting.* Cells of the immune system and blood provide protection against toxins and tissue injury, as well as against microorganisms. Bones protect underlying structures from injury.

## Classification of Connective Tissue

During embryonic development, embryonic connective tissue gives rise to six major categories of connective tissue (table 4.4). The extracellular matrix predominates in nearly all types of connective tissue and is largely responsible for the functional characteristics of each connective tissue type.

**Loose,** or *areolar* (a-re′ō-lar) **connective tissue** has extracellular matrix consisting mostly of collagen fibers and a few elastic fibers. These fibers are widely separated from one another (table 4.5). The most common cells in loose connective tissue are the fibroblasts, which are responsible for producing the matrix. Loose connective tissue is widely distributed throughout the body and is the loose packing material that fills the spaces between glands, muscles, and nerves. The basement membranes of epithelia often rest on loose connective tissue, and it attaches the skin to underlying tissues.

**Adipose** (ad′i-pōs; fat) tissue has an extracellular matrix composed of collagen and elastic fibers, but it is not a typical connective tissue because there is very little of this extracellular matrix. The individual cells are large and closely packed together (table 4.6). Adipose cells are filled with lipids and store energy. Adipose tissue also pads and protects parts of the body and acts as a thermal insulator.

**Dense connective tissue** has an extracellular matrix that consists of densely packed fibers produced by fibroblasts. There are two major subcategories of dense connective tissue: dense collagenous and dense elastic. **Dense collagenous connective tissue** has an extracellular matrix consisting mostly of collagen

## TABLE 4.5    Loose Connective Tissue

**Structure:** A fine network of fibers (mostly collagen fibers with a few elastic fibers) with spaces between the fibers; fibroblasts, macrophages, and lymphocytes are located in the spaces

**Function:** Loose packing, support, and nourishment for the structures with which it is associated

**Location:** Widely distributed throughout the body; substance on which epithelial basement membranes rest; packing between glands, muscles, and nerves; attaches the skin to underlying tissues

Elastic fiber
Nucleus
Collagen fiber

LM 400x

Epidermis
Dermis
Skin
Loose connective tissue with fat
Muscle

## TABLE 4.6    Connective Tissue: Adipose Tissue

**Structure:** Little extracellular matrix surrounding cells; the adipocytes, or fat cells, are so full of lipid that the cytoplasm is pushed to the periphery of the cell

**Function:** Packing material, thermal insulator, energy storage, and protection of organs against injury from being bumped or jarred

**Location:** Predominantly in subcutaneous areas, mesenteries, renal pelves, around kidneys, attached to the surface of the colon, mammary glands, and in loose connective tissue that penetrates into spaces and crevices

Nucleus

Adipocytes or fat cells

LM 100x

Adipose tissue

Mammary gland

fibers (table 4.7a). Structures made up of dense collagenous connective tissue include tendons, which attach muscle to bone; many ligaments, which attach bones to other bones; and much of the dermis, which is the connective tissue of the skin. Dense collagenous connective tissue also forms many capsules that surround organs, such as the liver and kidneys. In tendons and ligaments, the collagen fibers are oriented in the same direction, but in the dermis and in organ capsules, they are oriented in many different directions.

**Predict 2**

*In tendons, collagen fibers are oriented parallel to the length of the tendon. In the skin, collagen fibers are oriented in many directions. What are the functional advantages of the fiber arrangements in tendons and in the skin?*

**Dense elastic connective tissue** has abundant elastic fibers among its collagen fibers. The elastic fibers allow the tissue to stretch and recoil. Examples include the dense elastic connective

## TABLE 4.7    Dense Connective Tissue

### (a) Dense Collagenous Connective Tissue

**Structure:** Matrix composed of collagen fibers running in somewhat the same direction in tendons and ligaments; collagen fibers run in several directions in the dermis of the skin and in organ capsules

**Function:** Withstand great pulling forces exerted in the direction of fiber orientation due to great tensile strength and stretch resistance

**Location:** Tendons (attach muscle to bone) and ligaments (attach bones to each other); also found in the dermis of the skin, organ capsules, and the outer layer of many blood vessels

Nucleus of fibroblast

Collagen fibers

LM 165x

Ligament

Tendon

### (b) Dense Elastic Connective Tissue

**Structure:** Matrix composed of collagen fibers and elastin fibers running in somewhat the same direction in elastic ligaments; elastic fibers run in connective tissue of blood vessel walls

**Function:** Capable of stretching and recoiling like a rubber band with strength in the direction of fiber orientation

**Location:** Elastic ligaments between the vertebrae and along the dorsal aspect of the neck (nucha) and in the vocal cords; also found in elastic connective tissue of blood vessel walls

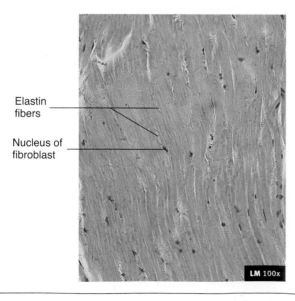

Elastin fibers

Nucleus of fibroblast

LM 100x

Base of tongue

Vocal folds (true vocal cords)

Vestibular fold (false vocal cord)

tissue in the vocal cords (table 4.7*b*), in elastic ligaments, and in the walls of large arteries. The elastic fibers are oriented in the same direction in elastic ligaments and in the vocal cords, but they are oriented in many different directions in the walls of arteries.

## A CASE IN POINT
### Marfan Syndrome

Izzy Taller is a 6-foot, 10-inch-tall high school student. The basketball coach convinces Izzy to try out for the basketball team, which requires a physical exam. The physician determines that Izzy's limbs, fingers, and toes are disproportionately long in relation to the rest of his body; he has an abnormal heart sound, indicating a problem with a heart valve; and he has poor vision because the lenses of his eyes are positioned abnormally. The physician diagnoses Izzy's condition as Marfan syndrome, an autosomal-dominant (see chapter 20) genetic disorder that results in the production of an abnormal protein called fibrillin-1. This protein is necessary for the normal formation and maintenance of elastic fibers and the regulation of a growth factor called TGFß. Because elastic fibers are found in connective tissue throughout the body, several body systems can be affected. The major cause of death in Marfan syndrome is rupture of the aorta due to weakening of the connective tissue in its wall. Izzy Taller should not join the basketball team.

**Cartilage** (kar′ti-lij; gristle) is composed of cartilage cells, or **chondrocytes** (kon′drō-sītz), located in spaces called **lacunae**

(lă-koo′nē; small spaces) within an extensive matrix (table 4.8*a*). Collagen in the matrix gives cartilage flexibility and strength. Cartilage is resilient because the proteoglycans of the matrix trap water, which makes the cartilage relatively rigid and enables it to spring back after being compressed. Cartilage provides support, but if bent or slightly compressed, it resumes its original shape. Cartilage heals slowly after an injury because blood vessels do not penetrate it. Thus, the cells and nutrients necessary for tissue repair do not easily reach the damaged area.

**Hyaline** (hī′ă-lin; clear or glassy) **cartilage** (see table 4.8*a*) is the most abundant type of cartilage and has many functions. It covers the ends of bones where they come together to form joints. In joints, hyaline cartilage forms smooth, resilient surfaces that can withstand repeated compression. Hyaline cartilage also forms the cartilage rings of the respiratory tract, the nasal cartilages, and the costal cartilages, which attach the ribs to the sternum (breastbone).

**Fibrocartilage** (table 4.8*b*) has more collagen than does hyaline cartilage, and bundles of collagen fibers can be seen in the matrix. In addition to withstanding compression, it is able to resist pulling or tearing forces. Fibrocartilage is found in the disks between the vertebrae (bones of the back) and in some joints, such as the knee and temporomandibular (jaw) joints.

**Elastic cartilage** (table 4.8*c*) contains elastic fibers in addition to collagen and proteoglycans. The elastic fibers appear as coiled fibers among bundles of collagen fibers. Elastic cartilage is

| TABLE 4.8 | Connective Tissue: Cartilage |
|---|---|

### (a) Hyaline Cartilage

**Structure:** Collagen fibers are small and evenly dispersed in the matrix, making the matrix appear transparent; the cartilage cells, or chondrocytes, are found in spaces, or lacunae, within the firm but flexible matrix

**Function:** Allows growth of long bones; provides rigidity with some flexibility in the trachea, bronchi, ribs, and nose; forms rugged, smooth, yet somewhat flexible articulating surfaces; forms the embryonic skeleton

**Location:** Growing long bones, cartilage rings of the respiratory system, costal cartilage of ribs, nasal cartilages, articulating surface of bones, and the embryonic skeleton

Chondrocyte in a lacuna
Nucleus
Matrix
LM 240x

Bone
Hyaline cartilage

*continued*

## TABLE 4.8    *continued*

### (b) Fibrocartilage

Structure: Collagen fibers similar to those in hyaline cartilage; the fibers are more numerous than in other cartilages and are arranged in thick bundles

Function: Somewhat flexible and capable of withstanding considerable pressure; connects structures subjected to great pressure

Location: Intervertebral disks, pubic symphysis, and articular disks (e.g., knees and temporomandibular [jaw] joints)

Chondrocyte in lacuna
Nucleus
Collagen fibers in matrix

LM 240x

Intervertebral disk

### (c) Elastic Cartilage

Structure: Similar to hyaline cartilage, but matrix also contains elastin fibers

Function: Provides rigidity with even more flexibility than hyaline cartilage because elastic fibers return to their original shape after being stretched

Location: External ears, epiglottis, and auditory tubes

Elastic fibers in matrix

Chondrocytes in lacunae

Nucleus

LM 240x

able to recoil to its original shape when bent. The external ear, epiglottis, and auditory tube contain elastic cartilage.

*Predict 3*

*Scars consist of dense connective tissue made of collagen fibers. Vitamin C is required for collagen synthesis. Predict the effect of scurvy, a nutritional disease caused by vitamin C deficiency, on wound healing.*

**Bone** is a hard connective tissue that consists of living cells and a mineralized matrix (table 4.9). Bone cells, or osteocytes (*osteo,* bone), are located within lacunae. The strength and rigidity of the mineralized matrix enables bones to support and protect other tissues and organs. The two types of bone are **compact bone** and **spongy bone** (see chapter 6).

**Blood** is unique because the matrix is liquid, enabling blood cells to move through blood vessels (table 4.10). Some blood cells even leave the blood and wander into other tissues. The liquid matrix enables blood to flow rapidly through the body, carrying food, oxygen, waste products, and other materials. Blood is discussed more fully in chapter 11.

## MUSCLE TISSUE

The main characteristic of **muscle tissue** is its ability to contract, or shorten, making movement possible. Muscle contraction results from contractile proteins located within the muscle cells (see chapter 7). The length of muscle cells is greater than the

## TABLE 4.9    Bone

**Structure:** Hard, bony matrix predominates; many osteocytes (not seen in this bone preparation) are located within lacunae; the matrix is organized into layers called lamellae

**Function:** Provides great strength and support and protects internal organs, such as the brain; bone also provides attachment sites for muscles and ligaments; the joints of bones allow movements

**Location:** All bones of the body

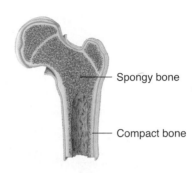

Lacuna

Central canal

Matrix organized into lamellae

LM 240x

Spongy bone

Compact bone

## TABLE 4.10    Connective Tissue: Blood

**Structure:** Blood cells and a fluid matrix

**Function:** Transports oxygen, carbon dioxide, hormones, nutrients, waste products, and other substances; protects the body from infections and is involved in temperature regulation

**Location:** Within the blood vessels; white blood cells frequently leave the blood vessels and enter the interstitial spaces

White blood cells

Red blood cells

LM 400x

Red blood cell

White blood cell

diameter. Muscle cells are sometimes called **muscle fibers** because they often resemble tiny threads.

The three types of muscle tissue are skeletal, cardiac, and smooth. **Skeletal muscle** is what we normally think of as "muscle" (table 4.11*a*). It is the meat of animals and constitutes about 40% of a person's body weight. As the name implies, skeletal muscle attaches to the skeleton and enables the body to move. Skeletal muscle is described as voluntary (under conscious control) because a person can purposefully cause skeletal muscle contraction to achieve specific body movements. However, the nervous system can cause skeletal muscles to contract without conscious involvement, as occurs during reflex movements and the maintenance of muscle tone. Skeletal muscle cells tend to be long and cylindrical, with several nuclei per cell. Some skeletal muscle cells extend the length of an entire muscle. Skeletal muscle cells are **striated** (strī′āt-ed), or banded, because of the arrangement of contractile proteins within the cells (see chapter 7).

## TABLE 4.11    Muscle Tissue

### (a) Skeletal Muscle

**Structure:** Skeletal muscle cells or fibers appear striated (banded); cells are large, long, and cylindrical, with many nuclei

**Function:** Movement of the body; under voluntary control

**Location:** Attached to bone or other connective tissue

Muscle

Nucleus (near periphery of cell)

**Skeletal muscle fiber**

Striations

LM 800x

### (b) Cardiac Muscle

**Structure:** Cardiac muscle cells are cylindrical and striated and have a single nucleus; they are branched and connected to one another by intercalated disks, which contain gap junctions

**Function:** Pumps the blood; under involuntary (unconscious) control

**Location:** In the heart

Nucleus

**Cardiac muscle cell**

Intercalated disks (special junctions between cells)

Striations

LM 800x

## (c) Smooth Muscle

**Structure:** Smooth muscle cells are tapered at each end, are not striated, and have a single nucleus

**Function:** Regulates the size of organs, forces fluid through tubes, controls the amount of light entering the eye, and produces "goose flesh" in the skin; under involuntary (unconscious) control

**Location:** In hollow organs, such as the stomach and intestine

Wall of stomach
Wall of colon
Wall of small intestine

Nucleus

Smooth muscle cell

LM 800x

---

**Cardiac muscle** is the muscle of the heart; it is responsible for pumping blood (table 4.11*b*). Cardiac muscle is under involuntary (unconscious) control, although a person can learn to influence the heart rate by using techniques such as meditation and biofeedback. Cardiac muscle cells are cylindrical but much shorter than skeletal muscle cells. Cardiac muscle cells are striated and usually have one nucleus per cell. They are often branched and connected to one another by **intercalated** (in-ter′kǎ-lā-ted) **disks.** The intercalated disks, which contain specialized gap junctions, are important in coordinating the contractions of the cardiac muscle cells (see chapter 12).

**Smooth muscle** forms the walls of hollow organs (except the heart); it is also found in the skin and the eyes (table 4.11*c*). Smooth muscle is responsible for a number of functions, such as moving food through the digestive tract and emptying the urinary bladder. Like cardiac muscle, smooth muscle is controlled involuntarily. Smooth muscle cells are tapered at each end, have a single nucleus, and are not striated.

### Predict 4

*Make a table that summarizes the characteristics of the three major muscle types. The muscle types should form a column at the left side of the table, and the characteristics of muscle should form a row at the top of the table.*

## NERVOUS TISSUE

**Nervous tissue** forms the brain, spinal cord, and nerves. It is responsible for coordinating and controlling many body activities.

For example, the conscious control of skeletal muscles and the unconscious regulation of cardiac muscle are accomplished by nervous tissue. Awareness of ourselves and the external environment, emotions, reasoning skills, and memory are other functions performed by nervous tissue. Many of these functions depend on the ability of nervous tissue cells to communicate with one another and with the cells of other tissues by means of electrical signals called **action potentials.**

Nervous tissue consists of neurons and support cells. The **neuron** (noor′on), or *nerve cell,* is responsible for conducting action potentials. It is composed of three parts: a cell body, dendrites, and an axon (table 4.12). The **cell body** contains the nucleus and is the site of general cell functions. **Dendrites** (den′drītz; relating to a tree) and **axons** (ak′sonz) are nerve cell processes (extensions). Dendrites usually receive stimuli leading to electrical changes that either increase or decrease action potentials in the neuron's axon. Action potentials usually originate at the base of an axon where it joins the cell body and travel to the end of the axon. **Neuroglia** (noo-rog′lē-ǎ; *glia,* glue) are the support cells of the nervous system; they nourish, protect, and insulate the neurons. We consider nervous tissue in greater detail in chapter 8.

## MEMBRANES

A **membrane** is a thin sheet or layer of tissue that covers a structure or lines a cavity. Most membranes consist of epithelium and the connective tissue on which the epithelium rests. The **skin,** or *cutaneous* (kū-tā′nē-ŭs, skin) *membrane,* is the external

## TABLE 4.12   Nervous Tissue

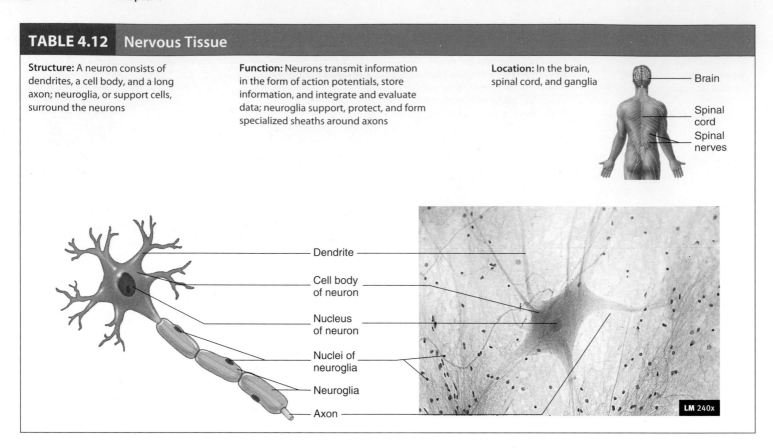

**Structure:** A neuron consists of dendrites, a cell body, and a long axon; neuroglia, or support cells, surround the neurons

**Function:** Neurons transmit information in the form of action potentials, store information, and integrate and evaluate data; neuroglia support, protect, and form specialized sheaths around axons

**Location:** In the brain, spinal cord, and ganglia

Brain

Spinal cord

Spinal nerves

Dendrite

Cell body of neuron

Nucleus of neuron

Nuclei of neuroglia

Neuroglia

Axon

LM 240x

---

membrane. It is composed of stratified squamous epithelium and dense connective tissue (see chapter 5). The three major categories of internal membranes are mucous, serous, and synovial membranes.

## Mucous Membranes

**Mucous** (mū′kŭs) **membranes** consist of various kinds of epithelium resting on a thick layer of loose connective tissue. They line cavities that open to the outside of the body, such as the digestive, respiratory, and reproductive tracts (figure 4.4a). Many, but not all, mucous membranes have mucous glands, which secrete mucus. The functions of mucous membranes vary, depending on their location, but they include protection, absorption, and secretion. For example, the stratified squamous epithelium of the oral cavity (mouth) performs a protective function, whereas the simple columnar epithelium of the intestine absorbs nutrients and secretes digestive enzymes and mucus. Mucous membrane also lines the nasal passages. When it becomes inflamed, we experience the "runny nose" characteristic of the common cold or an allergy.

## Serous Membranes

**Serous** (sēr′ŭs; producing watery secretion) **membranes** consist of simple squamous epithelium resting on a delicate layer of loose connective tissue. Serous membranes line the trunk cavities and cover the organs within these cavities (figure 4.4b).

Serous membranes secrete serous fluid, which covers the surface of the membranes. The smooth surface of the epithelial cells of the serous membranes combined with the lubricating qualities of the serous fluid prevent damage from abrasion when organs in the thoracic or abdominopelvic cavities rub against one another. Serous membranes are named according to their location: The **pleural** (ploor′ăl; a rib or the side) **membranes** are associated with the lungs; the **pericardial** (per-i-kar′dē-ăl; around the heart) **membranes** are associated with the heart; and the **peritoneal** (per′i-tō-nē′ăl; to stretch over) **membranes** are located in the abdominopelvic cavity (see figure 1.14). When the suffix -*itis* is added to the name of a structure, it means that the structure is inflamed. Thus, **pericarditis** (per′i-kar-dī′tis) and **peritonitis** (per′i-tō-nī′tis) refer to inflammation of the pericardial membranes and peritoneal membranes, respectively. **Pleurisy** (ploor′i-sē) is inflammation of the pleural membranes.

## Synovial Membranes

**Synovial** (si-nō′vē-ăl) **membranes** are made up of only connective tissue. They line the inside of joint cavities (the space where bones come together within a movable joint) (figure 4.4c). Synovial membranes produce **synovial fluid,** which makes the joint very slippery, thereby reducing friction and allowing smooth movement within the joint. Synovial and other connective tissue membranes are discussed in chapter 6.

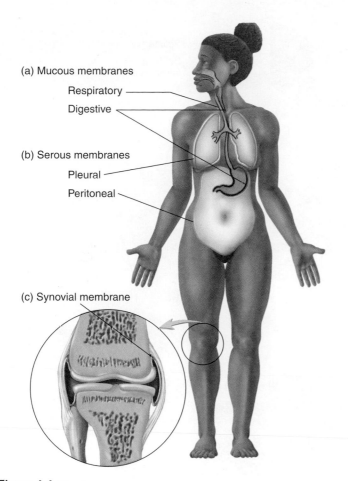

(a) Mucous membranes
Respiratory
Digestive

(b) Serous membranes
Pleural
Peritoneal

(c) Synovial membrane

**Figure 4.4  Membranes**

(a) Mucous membranes line cavities that open to the outside and often contain mucous glands, which secrete mucus. (b) Serous membranes line cavities that do not open to the exterior, do not contain glands, but do secrete serous fluid. (c) Synovial membranes line the cavities that surround synovial joints.

# TISSUE DAMAGE AND INFLAMMATION

**Inflammation** (*flamma*, flame) occurs when tissues are damaged. For example, viruses infect epithelial cells of the upper respiratory tract to produce inflammation and the symptoms of the common cold. Inflammation can also result from the immediate and painful events that follow trauma, such as closing your finger in a car door or cutting yourself with a knife. Figure 4.5 illustrates the stages of the inflammatory response. Inflammation mobilizes the body's defenses and isolates and destroys microorganisms, foreign materials, and damaged cells so that tissue repair can proceed (see chapter 14). Inflammation produces five major symptoms: redness, heat, swelling, pain, and disturbance of function. Although unpleasant, the processes producing the symptoms are usually beneficial.

Following an injury, substances called **chemical mediators** are released or activated in the injured tissues and adjacent blood vessels. The mediators include **histamine** (his′tă-mēn) and **prostaglandins** (pros-tă-glan′dinz). Some mediators cause dilation of blood vessels, which produces redness and heat, similar to what

occurs when a person blushes. Dilation of blood vessels is beneficial because it increases the speed with which blood cells and other substances important for fighting infections and repairing the injury are brought to the injury site.

Chemical mediators of inflammation also increase the permeability of blood vessels, allowing materials and blood cells to move out of the vessels and into the tissue, where they can deal directly with the injury. **Edema** (e-dē′mă), or swelling, of the tissues results when water, proteins, and other substances from the blood move into the tissues. One of the proteins, fibrin, forms a fibrous network that "walls off" the site of injury from the rest of the body. This mechanism can help prevent the spread of infectious agents. One type of blood cell that enters the tissues is the **neutrophil** (noo′trō-fil), a phagocytic white blood cell that fights infections by ingesting bacteria. Neutrophils die after ingesting a small number of bacteria; the mixture of dead neutrophils, other cells, and fluid that can accumulate is called **pus.**

Pain associated with inflammation is produced in several ways. Nerve cell endings are stimulated by direct damage and by some chemical mediators to produce pain sensations. In addition, the increased pressure in the tissue caused by edema and pus accumulation can cause pain.

Pain, limitation of movement resulting from edema, and tissue destruction all contribute to the disturbance of function, which can be adaptive because it warns the person to protect the injured area from further damage.

## Chronic Inflammation

Chronic, or prolonged, inflammation results when the agent responsible for an injury is not removed or something else interferes with the healing process. Infections of the lungs or kidneys usually result in a brief period of inflammation followed by repair. However, prolonged infections, or prolonged exposure to irritants, can result in chronic inflammation. Chronic inflammation caused by irritants, such as silica in the lungs, or abnormal immune responses can result in the replacement of normal tissue by fibrous connective tissue. Chronic inflammation of the stomach or small intestine may cause ulcers. The loss of normal tissue leads to the loss of normal organ function. Consequently, chronic inflammation of organs, such as the lungs, liver, or kidneys, can lead to death.

When the inflammatory response lasts longer or is more intense than is desirable, drugs are sometimes used to suppress the symptoms by inhibiting the synthesis, release, or actions of the chemical mediators of inflammation. For example, medications called antihistamines suppress the effects of histamine released in people with hay fever. Aspirin and related drugs, such as ibuprofen and naproxen, are effective anti-inflammatory agents that relieve pain by preventing the synthesis of prostaglandins and related substances.

**Predict 5**

*In some injuries, tissues are so severely damaged that cells die and blood vessels are destroyed. For injuries such as these, where do the signs of inflammation, such as redness, heat, edema, and pain, occur?*

**1** A splinter in the skin causes damage and introduces bacteria. Chemical mediators of inflammation are released or activated in injured tissues and adjacent blood vessels. Some blood vessels rupture, causing bleeding.

**2** Chemical mediators cause capillaries to dilate and the skin to become red. Chemical mediators also increase capillary permeability, and fluid leaves the capillaries, producing swelling *(arrows)*.

**3** White blood cells (e.g., neutrophils) leave the dilated blood vessels and move to the site of bacterial infection, where they begin to phagocytize bacteria and other debris.

**PROCESS Figure 4.5** **Inflammation**

# TISSUE REPAIR

**Tissue repair** is the substitution of viable cells for dead cells. Tissue repair can occur by regeneration or replacement. In **regeneration,** the new cells are the same type as those that were destroyed, and normal function is usually restored. In **replacement,** a new type of tissue develops that eventually causes scar production and the loss of some tissue function. The tissues involved and the severity of the wound determine the type of tissue repair that dominates.

Regeneration can completely repair some tissues, such as the skin and the mucous membrane of the intestine. In these cases,

regeneration is accomplished primarily by stem cells. **Stem cells** are self-renewing, undifferentiated cells that continue to divide throughout life. With each division, there is a daughter stem cell and a second cell that can undergo differentiation. The differentiated cells are the same cell types as the dead cells. Regeneration can also involve division of differentiated cells in connective tissue and glands, such as the liver and pancreas. These cells do not normally divide, but retain the ability to divide after an injury.

Replacement is the predominant repair mechanism in some tissues. In the adult brain, heart, and skeletal muscle there are relatively few stem cells and the mature neurons, cardiac muscle,

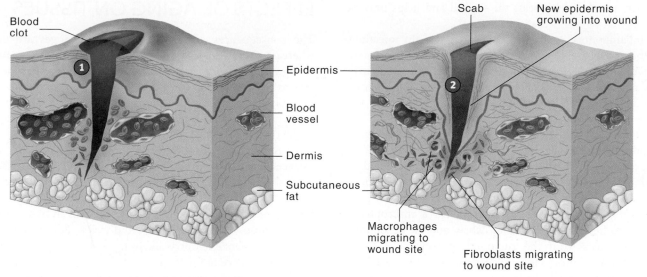

1 Fresh wound cuts through the epithelium (epidermis) and underlying connective tissue (dermis), and a clot forms.

2 Approximately 1 week after the injury, a scab is present, and epithelium (new epidermis) is growing into the wound.

3 Approximately 2 weeks after the injury, the epithelium has grown completely into the wound, and fibroblasts have formed granulation tissue.

4 Approximately 1 month after the injury, the wound has completely closed, the scab has been sloughed, and the granulation tissue is being replaced by new connective tissue.

**PROCESS Figure 4.6  Tissue Repair**

and skeletal muscle do not divide. While these cells cannot divide, they can recover from a limited amount of damage. For example, if the axon of a neuron is damaged, the neuron can grow a new axon, but it will die if the cell body is sufficiently damaged. If these cells are killed, they are often replaced by connective tissue.

Although neurons cannot form additional neurons, a small population of stem cells has been found in the adult brain. These stem cells can divide and form new neurons. It may be possible to develop treatments for some brain injuries that stimulate the stem cells. Researchers have identified a class of chemicals, called growth factors, that stimulate stem cells to divide and make injured

neurons recover more rapidly. The new neurons may be incorporated with other functional neurons of the central nervous system.

In addition to the type of cells involved, the severity of an injury can influence whether repair is by regeneration or replacement. Generally, the more severe the injury, the greater the likelihood that repair involves replacement.

Repair of the skin is an illustration of tissue repair (figure 4.6). When the edges of a wound are close together, the wound fills with blood, and a clot forms (see chapter 11). The **clot** contains the threadlike protein fibrin, which binds the edges of the wound together and stops the bleeding. The surface of the clot dries to form a **scab,** which seals the wound and helps prevent infection.

An inflammatory response is activated to fight infectious agents in the wound and to help the repair process (see figure 4.5). Dilation of blood vessels brings blood cells and other substances to the injury area, and increased blood vessel permeability allows them to enter the tissue. The area is "walled off" by the fibrin, and neutrophils enter the tissue from the blood.

The epithelium at the edge of the wound undergoes regeneration and migrates under the scab while the inflammatory response proceeds. Eventually, the epithelial cells from the edges meet, and the epithelium is restored. After the epithelium is repaired, the scab is sloughed off (shed).

A second type of phagocytic cell, called a macrophage, removes the dead neutrophils, cellular debris, and the decomposing clot. Fibroblasts from the surrounding connective tissue migrate into the area, producing collagen and other extracellular matrix components. Capillaries grow from blood vessels at the edge of the wound and revascularize the area. The result is the replacement of the clot by **granulation tissue,** a delicate, granular-appearing connective tissue that consists of fibroblasts, collagen, and capillaries. Eventually, normal connective tissue replaces the granulation tissue. Sometimes a large amount of granulation tissue persists as a scar, which at first is bright red because numerous blood vessels are present. The scar turns from red to white as collagen accumulates and the blood vessels decrease in number.

When the wound edges are far apart, the clot may not completely close the gap, and it takes much longer for the epithelial cells to regenerate and cover the wound. With increased tissue damage, the degree of inflammation is greater, there is more cell debris for the phagocytes to remove, and the risk of infection increases. Much more granulation tissue forms, and **wound contracture,** a result of the contraction of fibroblasts in the granulation tissue, pulls the edges of the wound closer together. Although wound contracture reduces the size of the wound and speeds healing, it can lead to disfiguring and debilitating scars.

**Predict 6**

*Explain why it is advisable to suture large wounds.*

# EFFECTS OF AGING ON TISSUES

The consequences of some age-related changes are obvious, whereas others are subtle. For example, the appearance of skin changes as people age, and athletic performance declines, after approximately 30–35 years. With advanced age, the number of neurons and muscle cells decreases substantially. Reduced visual acuity, smell, taste, and touch occur, and the functional capacities of the respiratory and cardiovascular systems decline.

At the tissue level, age-related changes affect cells and the extracellular matrix produced by them. In general, cells divide more slowly in older than in younger people. The rate of red blood cell synthesis declines in the elderly. Injuries in the very young heal more rapidly and more completely than in older people, in part, because of the more rapid cell division. For example, a fracture in an infant's femur is likely to heal quickly and eventually leave no evidence in the bone. A similar fracture in an adult heals more slowly, and a scar, seen in radiographs of the bone, is likely to persist throughout life.

The consequences of changes in the extracellular matrix are important. Collagen fibers become more irregular in structure, even though they may increase in number. As a consequence, connective tissues with abundant collagen, such as tendons and ligaments, become less flexible and more fragile. Elastic fibers fragment, bind to $Ca^{2+}$, and become less elastic, causing elastic connective tissues to become less elastic. The reduced flexibility and elasticity of connective tissue is responsible for increased wrinkling of the skin, as well as an increased tendency for older people's bones to break.

The walls of arteries become less elastic because of changes in collagen and elastic fibers. Atherosclerosis results as plaques form in the walls of blood vessels, which contain collagen fibers, lipids, and calcium deposits. These changes result in reduced blood supply to tissues and increased susceptibility to blockage and rupture of arteries.

However, it is increasingly evident that many of the cell losses and functional declines can be slowed by physical and mental exercises.

---

# ⟋⋏⟍ SUMMARY

1. A tissue is a group of cells with similar structure and function, along with the extracellular substances located between the cells.
2. Histology is the study of tissues.

## Epithelial Tissue   (p. 72)

Epithelial tissue covers surfaces; it usually has a basement membrane, little extracellular material, and no blood vessels.

### Functions of Epithelia

General functions of epithelia include protecting underlying structures, acting as a barrier, permitting the passage of substances, secreting substances, and absorbing substances.

### Classification of Epithelia

1. Epithelia are classified according to the number of cell layers and the shape of the cells.
2. Simple epithelium has one layer of cells, whereas stratified epithelium has more than one.
3. Pseudostratified columnar epithelium is simple epithelium that appears to have two or more cell layers.
4. Transitional epithelium is stratified epithelium that can be greatly stretched.

### Structural and Functional Relationships

1. Simple epithelium is involved with diffusion, secretion, or absorption. Stratified epithelium serves a protective role. Squamous

cells function in diffusion or filtration. Cuboidal or columnar cells, which contain more organelles, secrete or absorb.
2. A smooth, free surface reduces friction. Microvilli increase surface area, and cilia move materials over the cell surface.
3. Tight junctions bind adjacent cells together and form a permeability barrier.
4. Desmosomes mechanically bind cells together, and hemidesmosomes mechanically bind cells to the basement membrane.
5. Gap junctions allow intercellular communication.

### Glands
1. A gland is a single cell or a multicellular structure that secretes.
2. Exocrine glands have ducts, and endocrine glands do not.

## Connective Tissue    (p. 79)
1. Connective tissue has an extracellular matrix consisting of protein fibers, ground substance, and fluid.
2. Collagen fibers are flexible but resist stretching; reticular fibers form a fiber network; and elastic fibers recoil.
3. Connective tissue cells that are blast cells form the matrix, cyte cells maintain it, and clast cells break it down.

### Functions of Connective Tissue
Connective tissues enclose and separate other tissues, connect tissues to one another, help support and move body parts, store compounds, cushion and insulate the body, transport substances, and protect against toxins and injury.

### Classification of Connective Tissue
1. Embryonic connective tissue gives rise to six major categories of connective tissue.
2. Loose, or areolar, connective tissue is the "packing material" of the body; it fills the spaces between organs and holds them in place.
3. Adipose tissue, or fat, stores energy. Adipose tissue also pads and protects parts of the body and acts as a thermal insulator.
4. Dense connective tissue has a matrix consisting of either densely packed collagen fibers (in tendons, ligaments, and the dermis of the skin) or densely packed elastic fibers (in elastic ligaments and the walls of arteries).
5. Cartilage provides support and is found in structures such as the disks between the vertebrae, the external ear, and the costal cartilages.
6. Bone has a mineralized matrix and forms most of the skeleton of the body.
7. Blood has a liquid matrix and is found in blood vessels.

## Muscle Tissue    (p. 84)
1. Muscle tissue is specialized to shorten, or contract.
2. The three types of muscle tissue are skeletal, cardiac, and smooth muscle.

## Nervous Tissue    (p. 87)
1. Nervous tissue is specialized to conduct action potentials (electrical signals).
2. Neurons conduct action potentials, and neuroglia support the neurons.

## Membranes    (p. 87)
### Mucous Membranes
Mucous membranes line cavities that open to the outside of the body (digestive, respiratory, and reproductive tracts). They contain glands and secrete mucus.

### Serous Membranes
Serous membranes line trunk cavities that do not open to the outside of the body (pleural, pericardial, and peritoneal cavities). They do not contain glands but do secrete serous fluid.

### Synovial Membranes
Synovial membranes line joint cavities and secrete a lubricating fluid.

## Tissue Damage and Inflammation    (p. 89)
1. Inflammation isolates and destroys harmful agents.
2. Inflammation produces redness, heat, swelling, pain, and disturbance of function.

### Chronic Inflammation
Chronic inflammation results when the agent causing injury is not removed or something else interferes with the healing process.

## Tissue Repair    (p. 90)
1. Tissue repair is the substitution of viable cells for dead cells by regeneration or replacement. In regeneration, stem cells, which can divide throughout life, and other dividing cells regenerate new cells of the same type as those that were destroyed. In replacement, the destroyed cells are replaced by different cell types, which causes scar formation.
2. Tissue repair involves clot formation, inflammation, the formation of granulation tissue, and the regeneration or replacement of tissues. In severe wounds, wound contracture can occur.

## Effects of Aging on Tissues    (p. 92)
1. Cells divide more slowly as people age. Injuries heal more slowly.
2. Extracellular matrix containing collagen and elastic fibers becomes less flexible and less elastic. Consequently, skin wrinkles, elasticity in arteries is reduced, and bones break more easily.

## REVIEW AND COMPREHENSION

1. Define tissue and histology.
2. In what areas of the body is epithelium located? What are four characteristics of epithelial tissue?
3. Explain how epithelial tissue is classified according to the number of cell layers and the shape of the cells. What are pseudostratified columnar and transitional epithelium?
4. What kinds of functions does a single layer of epithelium perform? A stratified layer? Give an example of each.
5. Contrast the functions performed by squamous cells with those of cuboidal or columnar cells. Give an example of each.
6. What is the function of an epithelial free surface that is smooth, one that has microvilli, and one that has cilia?
7. Name the ways in which epithelial cells may be connected to one another, and give the function for each way.
8. Define gland. Distinguish between an exocrine and an endocrine gland.
9. Explain the differences among connective tissue cells that are termed blast, cyte, and clast cells.
10. What are the functions of connective tissues?
11. What are the six major connective tissue types? How are they used to classify connective tissue?

12. Describe loose or areolar connective tissue, and give an example.

13. How is adipose tissue different from other connective tissues? List the functions of adipose tissue.

14. Describe dense collagenous connective tissue and dense elastic connective tissue, and give two examples of each.

15. Describe the components of cartilage. Give an example of hyaline cartilage, fibrocartilage, and elastic cartilage.

16. Describe the components of bone.

17. Describe the cell types and matrix of blood, and list the functions of blood.

18. Functionally, what is unique about muscle? Which of the muscle types is under voluntary control? What tasks does each type perform?

19. Functionally, what is unique about nervous tissue? What do neurons and neuroglia accomplish? What is the difference between an axon and a dendrite?

20. Compare mucous and serous membranes according to the type of cavity they line and their secretions. Name the serous membranes associated with the lungs, heart, and abdominopelvic organs.

21. What is the function of the inflammatory response? Name the five symptoms of inflammation, and explain how each is produced.

22. Define tissue repair. What is the difference between regeneration and replacement?

23. Describe the process of tissue repair when the edges of a wound are close together versus when they are far apart.

24. Describe the effect of aging on cell division and the formation of extracellular matrix.

 ## CRITICAL THINKING

1. What types of epithelium are likely to be found lining the trachea of a heavy smoker? Predict the changes that are likely to occur after he or she stops smoking for 1 or 2 years.

2. The blood-brain barrier is a specialized epithelium in capillaries that prevents many materials from passing from the blood into the brain. What kind of cell connections would you expect to find in the blood-brain barrier?

3. One of the functions of the pancreas is to secrete digestive enzymes that are carried by ducts to the small intestine. How many cell layers and what cell shape, cell surface, and type of cell-to-cell connections would you expect to be present in the epithelium that produces the digestive enzymes?

4. Explain the consequences:
   a. if simple columnar epithelium replaced the nonkeratinized stratified squamous epithelium that lines the mouth
   b. if tendons were composed of dense elastic connective tissue instead of dense collagenous connective tissue
   c. if bones were made entirely of elastic cartilage

5. Some dense connective tissue has elastic fibers in addition to collagen fibers. This enables a structure to stretch and then recoil to its original shape. Examples are certain ligaments that hold together the vertebrae (bones of the back). When the back is bent (flexed), the ligaments are stretched. How does the elastic nature of these ligaments help the back function? How are the fibers in the ligaments organized?

6. The aorta is a large blood vessel attached to the heart. When the heart beats, blood is ejected into the aorta, which expands to accept the blood. The wall of the aorta is constructed with dense connective tissue that has elastic fibers. How do you think the fibers are arranged?

7. Antihistamines block the effect of a chemical mediator, histamine, which is released during the inflammatory response. When could taking an antihistamine be harmful, and when could it be beneficial?

8. Granulation tissue and scars consist of dense irregular collagenous connective tissue. Vitamin C is required for collagen synthesis. Predict the effect of scurvy, which is a nutritional disease caused by vitamin C deficiency, on wound healing.

*Answers in Appendix D*

Visit this textbook's website at **www.mhhe.com/seeleyess7** for practice quizzes, animations, interactive learning exercises, and other study tools.

McGraw-Hill offers a study CD that features interactive cadaver dissection. **Anatomy & Physiology Revealed** includes cadaver photos that allow you to peel away layers of the human body to reveal structures beneath the surface. This program also includes animations, radiologic imaging, audio pronunciations, and practice quizzing.

Playing in the sun is good for your body in more ways than one. First and foremost, it's fun! But you may not be aware that, while you're frolicking in the waves, sunlight triggers chemicals in your skin to produce vitamin D, which is important for bone health. However, excessive sun exposure can lead to premature skin aging and possibly skin cancer.

# 5 Chapter

# Integumentary System

## Chapter Outline and Objectives

### Functions of the Integumentary System (p. 96)
1. Describe the general functions of the integumentary system.

### Skin (p. 96)
2. Describe the structure and function of the epidermis.
3. Describe the epidermal strata, and relate them to the process of keratinization.
4. Explain how melanin, blood, carotene, and collagen affect skin color.
5. Describe the structure and discuss the function of the dermis.

### Hypodermis (p. 100)
6. Describe the structure and discuss the function of the hypodermis.

### Accessory Skin Structures (p. 100)
7. Describe the structure of a hair, and discuss the phases of hair growth.

8. Name the glands of the skin, and describe the secretions they produce.
9. Describe the parts of a nail, and explain how nails grow.

### Physiology of the Integumentary System (p. 102)
10. Discuss the specific functions of skin, hair, glands, and nails.

### Integumentary System as a Diagnostic Aid (p. 104)
11. Explain how the integumentary system can be used as a diagnostic aid.

### Burns (p. 104)
12. Classify burns on the basis of the amount of skin damage produced.

### Skin Cancer (p. 105)
13. Name and define the types of skin cancer.

### Effects of Aging on the Integumentary System (p. 108)
14. List the changes the integumentary system undergoes with age.

The **integumentary** (in-teg-ū-men'tă-rē) **system** consists of the skin and accessory structures, such as hair, glands, and nails. Integument means covering, and the integumentary system is familiar to most people because it covers the outside of the body and is easily observed. In addition, humans are concerned with the appearance of the integumentary system. Skin without blemishes is considered attractive, whereas acne is a source of embarrassment for many teenagers. The development of wrinkles and the graying or loss of hair are signs of aging that some people find unattractive.

Because of these feelings, much time, effort, and money are spent on changing the appearance of the integumentary system; people apply lotion to their skin, color their hair, and trim their nails. They also try to prevent sweating by using antiperspirants and to reduce or mask body odor by washing and by using deodorants and perfumes.

The appearance of the integumentary system can indicate physiological imbalances in the body. Some disorders, such as acne or warts, affect just the integumentary system. Other disorders affect different parts of the body but are reflected in the integumentary system, which provides useful signs for diagnosis. For example, reduced blood flow through the skin during a heart attack can cause a person to look pale, whereas increased blood flow as a result of fever can cause a flushed appearance. Also, some diseases cause skin rashes, such as those characteristic of measles, chickenpox, and allergic reactions. In addition, the integumentary system and the other systems often interact in complex ways in both healthy and diseased states (see Systems Pathology, "Burns," later in this chapter).

# FUNCTIONS OF THE INTEGUMENTARY SYSTEM

Although we are often concerned with how the integumentary system looks, it has many important functions that go beyond appearance. Major functions of the integumentary system include

1. *Protection.* The skin provides protection against abrasion and ultraviolet light. It also prevents microorganisms from entering the body and reduces water loss, thus preventing dehydration.
2. *Sensation.* The integumentary system has sensory receptors that can detect heat, cold, touch, pressure, and pain.
3. *Vitamin D production.* When exposed to ultraviolet light, the skin produces a molecule that can be transformed into vitamin D, an important regulator of calcium homeostasis.
4. *Temperature regulation.* The amount of blood flow beneath the skin's surface and the activity of sweat glands in the skin both help regulate body temperature.
5. *Excretion.* Small amounts of waste products are lost through the skin and in gland secretions.

# SKIN

The skin is made up of two major tissue layers. The **epidermis** (ep-i-der'mis; upon the dermis) is the most superficial layer of skin. It is a layer of epithelial tissue that rests on the **dermis** (der'mis), a layer of dense connective tissue (figure 5.1). The thickness of the epidermis and dermis varies, depending on location, but on average the dermis is 10 to 20 times thicker than the epidermis. The epidermis prevents water loss and resists abrasion. The dermis is responsible for most of the skin's structural strength. The skin rests on the **hypodermis** (hī-pō-der'mis; under the dermis), which is a layer of connective tissue. The hypodermis is not part of the skin, but it does connect the skin to underlying muscle or bone. To give an analogy, if the hypodermis is the foundation on which a house rests, the dermis forms most of the house and the epidermis is its roof.

## Epidermis

The epidermis is stratified squamous epithelium; in its deepest layers, new cells are produced by mitosis. As new cells form, they push older cells to the surface, where they slough, or flake off. The outermost cells protect the cells underneath, and the deeper, replicating cells replace cells lost from the surface. During their movement, the cells change shape and chemical composition. This process is called **keratinization** (ker'ă-tin-i-zā'shŭn) because the cells become filled with the protein **keratin** (ker'ă-tin), which makes them hard. As keratinization proceeds, epithelial cells eventually die and produce an outer layer of dead, hard cells that resists abrasion and forms a permeability barrier.

Although keratinization is a continuous process, distinct layers called **strata** (stra'tă; layer) can be seen in the epidermis (figure 5.2). The deepest stratum, the **stratum basale** (bā'să-lē; a base), consists of cuboidal or columnar cells that undergo mitotic divisions about every 19 days. One daughter cell becomes a new stratum basale cell and can divide again. The other daughter cell is pushed toward the surface, a journey that takes about 40–56 days. As cells move to the surface, changes in the cells produce intermediate strata.

The **stratum corneum** (kōr'nē-ŭm) is the most superficial stratum of the epidermis. It consists of dead squamous cells filled with keratin. Keratin gives the stratum corneum its structural strength. The stratum corneum cells are also coated and surrounded by lipids, which help prevent fluid loss through the skin.

**Predict 1**

*What kinds of substances could easily pass through the skin by diffusion? What kinds would have difficulty?*

The stratum corneum is composed of 25 or more layers of dead squamous cells joined by desmosomes (see chapter 4). Eventually, the desmosomes break apart, and the cells are sloughed from the skin. Excessive stratum corneum cells sloughed from the surface of the scalp are called dandruff. In skin subjected to friction, the number of layers in the stratum corneum greatly increases, producing a thickened area called a **callus** (kal'ŭs; hard skin). Over a bony prominence, the stratum corneum can thicken to form a cone-shaped structure called a **corn.**

**Figure 5.1  Skin and Hypodermis**

The skin, consisting of the epidermis and the dermis, is connected by the hypodermis to underlying structures. Note the accessory structures (hairs, glands, and arrector pili), some of which project into the hypodermis, and the large amount of fat in the hypodermis.

## Dermis

The dermis is composed of dense collagenous connective tissue containing fibroblasts, fat cells, and macrophages. Nerves, hair follicles, smooth muscles, glands, and lymphatic vessels extend into the dermis (see figure 5.1).

Collagen and elastic fibers are responsible for the structural strength of the dermis. In fact, the dermis is the part of an animal hide from which leather is made. The collagen fibers of the dermis are oriented in many different directions and can resist stretch. However, more collagen fibers are oriented in some directions than in others. This produces **cleavage lines,** or *tension lines,* in the skin, and the skin is most resistant to stretch along these lines (figure 5.3). It is important for surgeons to be aware of cleavage lines. An incision made across the cleavage lines is likely to gap and produce considerable scar tissue, but an incision made parallel with the lines tends to gap less and produce less scar tissue (see chapter 4). If the skin is overstretched for any reason, the dermis can be damaged, leaving lines that are visible through the epidermis. These lines, called **stretch marks,** can develop on the abdomen and breasts of a woman during pregnancy or on the skin of athletes who have quickly increased muscle size by intense weight training.

The upper part of the dermis has projections called **dermal papillae** (pă-pil′e; nipple), which extend toward the epidermis (see figure 5.2). The dermal papillae contain many blood vessels that supply the overlying epidermis with nutrients, remove waste products, and help regulate body temperature. The dermal papillae in the palms of the hands, the soles of the feet, and the tips of the digits are arranged in parallel, curving ridges that shape the overlying epidermis into fingerprints and footprints. The ridges increase friction and improve the grip of the hands and feet.

### A CASE IN POINT

#### Injections

Howey Stickum, a student nurse, learns three ways to give injections. An **intradermal injection** is administered by drawing the skin taut and inserting a small needle at a shallow angle into the dermis; an example is the tuberculin skin test. A **subcutaneous injection** is achieved by pinching the skin to form a "tent" and inserting a short needle into the fatty tissue of the hypodermis; an example is an insulin injection. An **intramuscular injection** is accomplished by inserting a long needle at a 90-degree angle to the skin into a muscle deep to the hypodermis. Intramuscular injections are used for most vaccines and certain antibiotics.

Integumentary

(a)

(b) Dermal papillae

**Figure 5.2  Epidermis and Dermis**

(*a*) The epidermis rests on the dermis. Dermal papillae project toward the epidermis. (*b*) Photomicrograph of the epidermis resting on the dermis. Note the strata of the epidermis and the papillae of the dermis.

An incision made across cleavage lines can gap, increasing the time needed for healing and resulting in increased scar tissue formation.

An incision made parallel to cleavage lines results in less gapping, faster healing, and less scar tissue.

**Figure 5.3  Cleavage Lines**

The orientation of collagen fibers produces cleavage lines, or tension lines, in the skin.

## Skin Color

Factors that determine skin color include pigments in the skin, blood circulating through the skin, and the thickness of the stratum corneum. **Melanin** (mel′ă-nin; black) is the group of pigments primarily responsible for skin, hair, and eye color. Most melanin molecules are brown to black pigments, but some are yellowish or reddish. Melanin provides protection against ultraviolet light from the sun.

Melanin is produced by **melanocytes** (mel′ă-nō-sītz; *melano*, black + *kytos*, cell), which are irregularly shaped cells with many long processes that extend between the epithelial cells of the deep part of the epidermis (figure 5.4). The Golgi apparatuses of the melanocytes package melanin into vesicles called **melanosomes** (mel′ă-nō-sōmz), which move into the cell processes of the melanocytes. Epithelial cells phagocytize the tips of the melanocyte cell processes, thereby acquiring melanosomes. Although all the epithelial cells of the epidermis can contain melanin, only the melanocytes produce it.

Large amounts of melanin form freckles or moles in some regions of the skin, as well as darkened areas in the genitalia, the nipples, and the circular areas around the nipples. Other areas, such as the lips, palms of the hands, and soles of the feet, contain less melanin.

Melanin production is determined by genetic factors, exposure to light, and hormones. Genetic factors are responsible for the amounts of melanin produced in different races. Since all races have about the same number of melanocytes, racial variations in skin color are determined by the amount, kind, and distribution of melanin. Although many genes are responsible for skin color, a single mutation can prevent the manufacture of melanin. For example, **albinism** (al′bi-nizm) is a recessive genetic trait that causes a deficiency or an absence of melanin. Albinos have fair skin, white hair, and unpigmented irises in the eyes.

Exposure to ultraviolet light—for example, in sunlight—stimulates melanocytes to increase melanin production. The result is a suntan.

Certain hormones, such as estrogen and melanocyte-stimulating hormone, cause an increase in melanin production during pregnancy in the mother, darkening the nipples, the

# CLINICAL IMPACT

## Adaptive Advantages of Skin Color

The evolution of skin color in humans is intriguing because it helps explain certain modern health problems. During human evolution, the skeletal system of our ancestors changed, resulting in an upright posture and the ability to walk and run greater distances. Excess heat is produced as a result of increased physical activity, however, which can cause overheating. An increase in the number of sweat glands in the skin and a reduction in the amount of hair covering the skin help eliminate the excess heat.

With the reduction in hair over most of the body, prolonged exposure of the skin to sunlight can be harmful in two ways. First, it promotes the development of skin cancer (see "Skin Cancer" later in this chapter) by damaging DNA; second, it decreases the levels of the B vitamin folate in the blood by breaking it down through a photochemical reaction. Low folate levels are known to increase the risk of abnormal

development of the fetal nervous system (see chapter 20). On the other hand, exposure to ultraviolet light from the sun stimulates the production of vitamin D (see "Vitamin D Production" later in this chapter). Vitamin D promotes the uptake of calcium from the small intestine, which is important for the normal development of the skeletal system in the fetus and in children. Inadequate quantities of vitamin D can result in rickets, a condition in which the bones are soft, weak, and easily broken. Thus, increased skin pigmentation protects against skin cancer and abnormal development of the nervous system but impairs skeletal system development.

The optimal amount of melanin in the skin should be large enough to protect against the harmful effects of ultraviolet light but small enough to allow ultraviolet light to stimulate vitamin D production. Ultraviolet light intensity is high in the tropics but diminishes toward the poles. The

skin color of populations is a genetic adaptation to their different exposure to ultraviolet light. Dark-skinned people in the tropics have more melanin, which provides protection against ultraviolet light, but can still produce vitamin D year-round. Light-skinned people at higher latitudes have less melanin, which increases the body's ability to produce vitamin D while providing ultraviolet light protection.

When people migrate from the regions where their ancestors evolved, cultural adaptations help them adjust to their changed ultraviolet light environment. For example, clothing and portable shade, such as tents or umbrellas, provide protection against ultraviolet light, and eating vitamin D–rich foods, such as fish, provides vitamin D. Still, light-skinned people who move to southern climates are more likely to develop skin cancer, and dark-skinned people who move to northern climates are more likely to develop rickets.

---

pigmented circular areas around the nipples, and the genitalia even more. The cheekbones, forehead, and chest can also darken, resulting in "the mask of pregnancy," and a dark line of pigmentation can appear on the midline of the abdomen.

Blood flowing through the skin imparts a reddish hue, and when blood flow increases, the red color intensifies. Examples include blushing and the redness resulting from the inflammatory response. A decrease in blood flow, as occurs in shock, can make

① Melanosomes are produced by the Golgi apparatus of the melanocyte.

② Melanosomes move into the melanocyte cell processes.

③ Epithelial cells phagocytize the tips of the melanocyte cell processes.

④ The melanosomes, which were produced inside the melanocytes, have been transferred to epithelial cells and are now inside them.

Epithelial cell
Melanocyte
Melanosomes
Nucleus
Golgi apparatus

**PROCESS Figure 5.4  Melanin Transfer to Epithelial Cells**

Melanocytes make melanin, which is packaged into melanosomes and transferred to many epithelial cells.

the skin appear pale. A decrease in the blood oxygen content produces a bluish color of the skin, called **cyanosis** (sī-ă-nō′sis; dark blue). Birthmarks are congenital (present at birth) disorders of the blood vessels (capillaries) in the dermis.

**Carotene** (kar′ō-tēn) is a yellow pigment found in plants such as squash and carrots. Humans normally ingest carotene and use it as a source of vitamin A. Carotene is lipid-soluble; when consumed, it accumulates in the lipids of the stratum corneum and in the fat cells of the dermis and hypodermis. If large amounts of carotene are consumed, the skin can become quite yellowish.

The location of pigments and other substances in the skin affects the color produced. If a dark pigment is located in the dermis or hypodermis, light reflected off the dark pigment can be scattered by collagen fibers of the dermis to produce a blue color. The deeper within the dermis or hypodermis any dark pigment is located, the bluer the pigment appears because of the light-scattering effect of the overlying tissue. This effect causes the blue color of tattoos, bruises, and some superficial blood vessels.

**Predict 2**

> Explain the differences in skin color between (a) the palms of the hands and the lips, (b) the palms of the hands of a person who does heavy manual labor and one who does not, and (c) the anterior and posterior surfaces of the forearm.

## HYPODERMIS

Just as a house rests on a foundation, the skin rests on the hypodermis, which attaches it to underlying bone and muscle and supplies it with blood vessels and nerves (see figure 5.1). The hypodermis, which is not part of the skin, is sometimes called *subcutaneous* (sŭb-koo-tā′nē-ŭs; under the skin) *tissue.* The hypodermis is loose connective tissue that contains about half the body's stored fat, although the amount and location vary with age, sex, and diet. Fat in the hypodermis functions as padding and insulation, and it is responsible for some of the differences in appearance between men and women as well as between individuals of the same sex.

The hypodermis can be used to estimate total body fat. The skin and hypodermis are pinched at selected locations, and the thickness of the fold is measured. The thicker the fold, the greater the amount of total body fat. The percentage of body fat varies in the population, but on average women have higher total body fat than do males. The acceptable percentage of body fat varies from 21% to 30% for females and from 13% to 25% for males. A body fat percentage above the acceptable range is an indicator of obesity.

## ACCESSORY SKIN STRUCTURES

The accessory skin structures are hair, glands, and nails.

### Hair

In humans, **hair** is found everywhere on the skin, except on the palms, the soles, the lips, the nipples, parts of the genitalia, and the distal segments of the fingers and toes.

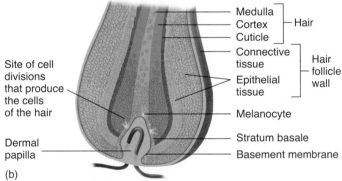

**Figure 5.5  Hair Follicle**

(a) Hair within a hair follicle. (b) Enlargement of the hair bulb and hair follicle wall.

Each hair arises from a **hair follicle,** an extension of the epidermis that originates deep in the dermis (figure 5.5a). The **shaft** of the hair protrudes above the surface of the skin, whereas the **root** and **hair bulb** are below the surface. A hair has a hard **cortex,** which surrounds a softer center, the **medulla** (me-dool′ă). The cortex is covered by the **cuticle** (kū′ti-kl; skin), a single layer of overlapping cells that holds the hair in the hair follicle. The hair follicle can play an important role in repair of the skin. If the surface epidermis is damaged, the epithelial cells within the hair follicle can divide and serve as a source of new epithelial cells.

Hair is produced in the hair bulb, which rests on a dermal papilla (figure 5.5b). Blood vessels within the papilla supply the hair bulb with the nourishment needed to produce the hair. Hair is produced in cycles. During the growth stage, it is formed by epithelial cells within the hair bulb. These cells, like the cells of the stratum basale in the skin, divide and undergo keratinization. The hair grows longer as these cells are added to the base of the hair within the hair bulb. Thus, the hair root and shaft consist of columns of dead keratinized epithelial cells. During the resting stage, growth stops, and the hair is held in the hair follicle. When the next growth stage begins, a new hair is formed, and the old hair falls out. The duration of each stage depends on the individual hair. Eyelashes grow for about 30 days and rest for 105 days, whereas scalp hairs grow for 3 years and rest for 1–2 years. The loss of hair normally means that the hair is being replaced because the old hair falls out of the hair follicle when the new hair begins to grow. In some men, however, a permanent loss of hair results in "pattern baldness." Although many of the hair follicles are lost, some remain and produce a very short, transparent hair, which for practical purposes is invisible. These changes occur when male sex hormones act on the hair follicles of men who have the genetic predisposition for pattern baldness.

Hair color is determined by varying amounts and types of melanin. The production and distribution of melanin by melanocytes occurs in the hair bulb by the same method as in the skin. With age, the amount of melanin in hair can decrease, causing the hair color to become faded, or the hair can contain no melanin and be white. Gray hair is usually a mixture of unfaded, faded, and white hairs.

**Predict 3**

*Marie Antoinette's hair supposedly turned white overnight after she heard she would be sent to the guillotine. Explain why you believe or disbelieve this story.*

Associated with each hair follicle are smooth muscle cells called the **arrector** (ă-rek′tōr; that which raises) **pili** (pī′li; hair) (figure 5.5*a*). Contraction of the arrector pili causes the hair to become more perpendicular to the skin's surface, or to "stand on end," and it produces a raised area of skin called "goose flesh."

## Glands

The major glands of the skin are the **sebaceous** (sē-bā′shŭs) **glands** and the **sweat glands** (figure 5.6). Sebaceous glands are simple, branched acinar glands (see chapter 4). Most are connected by a duct to the superficial part of a hair follicle. They produce **sebum,** an oily, white substance rich in lipids. The sebum lubricates the hair and the surface of the skin, which prevents drying and protects against some bacteria.

There are two kinds of sweat glands: eccrine and apocrine. **Eccrine** (ek′rin) **sweat glands** are simple, coiled, tubular glands located in almost every part of the skin but most numerous in the palms and soles. They produce a secretion that is mostly water with a few salts. Eccrine sweat glands have ducts that open onto the surface of the skin through sweat pores. When the body temperature starts to rise above normal levels, the sweat glands produce sweat, which evaporates and cools the body. Sweat can also be released in the palms, soles, armpits, and other places because of emotional stress. Emotional sweating is used in lie detector (polygraph) tests because sweat gland activity usually increases when a person tells a lie. Such tests can detect even small amounts

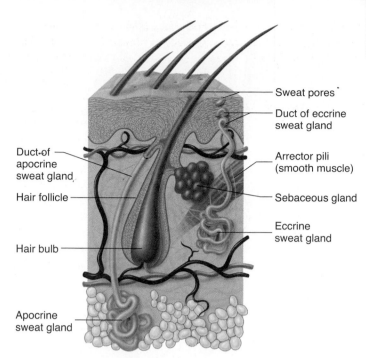

**Figure 5.6   Glands of the Skin**
Sebaceous glands and apocrine sweat glands empty into a hair follicle. Eccrine sweat glands empty onto the surface of the skin.

of sweat because the salt solution conducts electricity and lowers the electrical resistance of the skin.

**Apocrine** (ap′ō-krin) **sweat glands** are simple, coiled, tubular glands that produce a thick secretion rich in organic substances. They open into hair follicles, but only in the armpits and genitalia. Apocrine sweat glands become active at puberty because of the influence of sex hormones. The organic secretion, which is essentially odorless when released, is quickly broken down by bacteria into substances responsible for what is commonly known as body odor.

## Nails

The **nail** is a thin plate, consisting of layers of dead stratum corneum cells that contain a very hard type of keratin. The visible part of the nail is the **nail body,** and the part of the nail covered by skin is the **nail root** (figure 5.7). The **cuticle,** or *eponychium* (ep-ō-nik′ē-ŭm), is stratum corneum that extends onto the nail body. The nail root extends distally from the **nail matrix.** The nail also attaches to the underlying **nail bed,** which is located distal to the nail matrix. The nail matrix and bed are epithelial tissue with a stratum basale that gives rise to the cells that form the nail. The nail matrix is thicker than the nail bed and produces most of the nail. A small part of the nail matrix, the **lunula** (loo′noo-lă; moon), can be seen through the nail body as a whitish, crescent-shaped area at the base of the nail. Cell production within the nail matrix causes the nail to grow. Unlike hair, nails grow continuously and do not have a resting stage.

(a)                                                                    (b)

**Figure 5.7** Nail

(a) Dorsal view of the exterior nail. (b) Lateral view of a sagittal section through the nail. Most of the epidermis is absent from the nail bed.

# PHYSIOLOGY OF THE INTEGUMENTARY SYSTEM

## Protection

The integumentary system performs many protective functions.

1. The intact skin plays an important role in reducing water loss because its lipids act as a barrier to the diffusion of water.
2. The skin prevents microorganisms and other foreign substances from entering the body. Secretions from skin glands also produce an environment unsuitable for some microorganisms.
3. The stratified squamous epithelium of the skin protects underlying structures against abrasion.
4. Melanin absorbs ultraviolet light and protects underlying structures from its damaging effects.
5. Hair provides protection in several ways: The hair on the head acts as a heat insulator, eyebrows keep sweat out of the eyes, eyelashes protect the eyes from foreign objects, and hair in the nose and ears prevents the entry of dust and other materials.
6. The nails protect the ends of the fingers and toes from damage and can be used in defense.

## Sensation

Receptors in the epidermis and dermis can detect pain, heat, cold, and pressure (see chapter 9). Although hair does not have a nerve supply, sensory receptors around the hair follicle can detect the movement of a hair.

## Vitamin D Production

When the skin is exposed to ultraviolet light, a precursor molecule of **vitamin D** is formed. The precursor is carried by the blood to the liver, where it is modified, and then to the kidneys, where the precursor is modified further to form active vitamin D. If exposed to enough ultraviolet light, humans can produce all the vitamin D they need. However, many people need to ingest vitamin D as well because clothing and indoor living reduce their exposure to ultraviolet light. Fatty fish (and fish oils) and vitamin D–fortified milk are the best sources of vitamin D. Eggs, butter, and liver contain small amounts of vitamin D but are not considered significant sources because too large a serving size is necessary to meet the daily vitamin D requirement. Adequate levels of vitamin D are necessary because vitamin D stimulates the intestines to absorb calcium and phosphate, the substances necessary for normal bone growth (see chapter 6) and normal muscle function (see chapter 7).

## Temperature Regulation

Body temperature normally is maintained at about 37°C (98.6°F). Regulation of body temperature is important because the rate of chemical reactions within the body can be increased or decreased by changes in body temperature. Even slight changes in temperature can make enzymes operate less efficiently and disrupt the normal rates of chemical changes in the body.

Exercise, fever, and an increase in environmental temperature tend to raise body temperature. In order to maintain homeostasis, the body must rid itself of excess heat. Blood vessels in the dermis dilate and enable more blood to flow within the skin, thus transferring heat from deeper tissues to the skin (figure 5.8a), where the heat is lost by radiation (infrared energy), convection (air movement), or conduction (direct contact with an object). Sweat that spreads over the surface of the skin and evaporates also carries away heat and reduces body temperature.

If body temperature begins to drop below normal, heat can be conserved by the constriction of dermal blood vessels, which reduces blood flow to the skin (figure 5.8b). Thus, less heat is transferred from deeper structures to the skin, and heat loss is reduced. However, with smaller amounts of warm blood flowing through the skin, the skin temperature decreases. If the skin temperature drops below about 15°C (59°F), dermal blood vessels dilate.

### Predict 4

*You may have noticed that, on cold winter days, people's noses and ears turn red. Can you explain the advantage of this response?*

# CLINICAL IMPACT

## Acne

**Acne** (ak′nē) is inflammation of the hair follicles and sebaceous glands. Four factors are believed to be responsible: hormones, sebum, abnormal production of cells, and the bacterium *Propionibacterium acnes*. During puberty, hormones, especially testosterone, stimulate the sebaceous glands, and sebum production increases. Because both the testes and the ovaries produce testosterone, the effect is seen in males and females.

The lesions of acne begin with the overproduction of epidermal cells in the hair follicle. These cells are shed from the wall of the hair follicle, and they stick to one another to form a mass of cells mixed with sebum that blocks the hair follicle. An accumulation of sebum behind the blockage produces a whitehead. A blackhead develops when the accumulating mass of cells and sebum pushes through the opening of

the hair follicle. Although there is general agreement that dirt is not responsible for the black color, the exact cause is disputed. A pimple results if the wall of the hair follicle ruptures, forming an entry into the surrounding tissue. *P. acnes* and other bacteria stimulate an inflammatory response, which results in the formation of a red pimple filled with pus. If tissue damage is extensive, scarring occurs.

**1** Blood vessel dilation results in increased blood flow toward the surface of the skin.

**2** Increased blood flow beneath the epidermis results in increased heat loss (*gold arrows*).

Blood vessel dilates (vasodilation).

Heat loss across the epidermis increases.

Epidermis

(a)

**3** Blood vessel constriction results in decreased blood flow toward the surface of the skin.

**4** Decreased blood flow beneath the epidermis results in decreased heat loss.

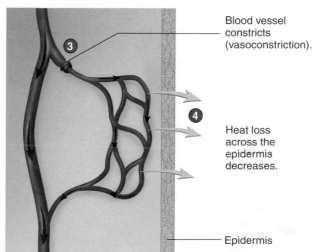

Blood vessel constricts (vasoconstriction).

Heat loss across the epidermis decreases.

Epidermis

(b)

**PROCESS Figure 5.8** Heat Exchange in the Skin

## Excretion

**Excretion** is the removal of waste products from the body. In addition to water and salts, sweat contains small amounts of waste products, such as urea, uric acid, and ammonia. Even though the body can lose large amounts of sweat, the sweat glands do not play a significant role in the excretion of waste products.

## INTEGUMENTARY SYSTEM AS A DIAGNOSTIC AID

The integumentary system is useful in diagnosis because it is observed easily and often reflects events occurring in other parts of the body. For example, cyanosis, a bluish color to the skin caused by decreased blood oxygen content, is an indication of impaired circulatory or respiratory function. A yellowish skin color, called **jaundice** (jawn′dis), can occur when the liver is damaged by a disease, such as viral hepatitis. Normally, the liver secretes bile pigments, which are products of the breakdown of worn-out red blood cells, into the intestine. Bile pigments are yellow, and their buildup in the blood and tissues can indicate impaired liver function.

Rashes and lesions in the skin can be symptoms of problems elsewhere in the body. For example, scarlet fever results when bacteria infecting the throat release a toxin into the blood that causes a reddish rash on the skin. The development of a rash can also indicate an allergic reaction to foods or to drugs, such as penicillin.

The condition of the skin, hair, and nails is affected by nutritional status. In vitamin A deficiency, the skin produces excess keratin and assumes a characteristic sandpaper texture, whereas in iron-deficiency anemia the nails lose their normal contour and become flat or concave (spoon-shaped).

Hair concentrates many substances that can be detected by laboratory analysis, and comparison of a patient's hair with a "normal" hair can be useful in certain diagnoses. For example, lead poisoning results in high levels of lead in the hair. However, the use of hair analysis to determine the general health or nutritional status of an individual is unreliable.

## BURNS

A **burn** is injury to a tissue caused by heat, cold, friction, chemicals, electricity, or radiation. Burns are classified according to their depth (figure 5.9). In **partial-thickness burns,** part of the stratum basale remains viable, and regeneration of the epidermis occurs from within the burn area, as well as from the edges of the burn. Partial-thickness burns are divided into first- and second-degree burns.

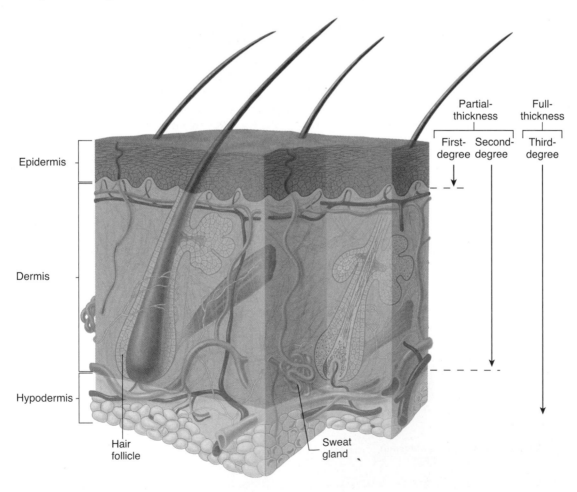

**Figure 5.9 Burns**

Burns are classified based on the extent of skin damage. Partial-thickness burns are subdivided into first-degree burns (damage to only the epidermis) and second-degree burns (damage to the epidermis and part of the dermis). Full-thickness, or third-degree, burns destroy the epidermis, dermis, and sometimes deeper tissues.

*First-degree burns* involve only the epidermis and are red and painful. Slight **edema** (e-dē′mă), or swelling, may be present. They can be caused by sunburn or brief exposure to very hot or very cold objects, and they heal without scarring in about a week.

*Second-degree burns* damage both the epidermis and the dermis. If dermal damage is minimal, symptoms include redness, pain, edema, and blisters. Healing takes about 2 weeks, and no scarring results. However, if the burn goes deep into the dermis, the wound appears red, tan, or white; can take several months to heal; and might scar. In all second-degree burns, the epidermis regenerates from epithelial tissue in hair follicles and sweat glands, as well as from the edges of the wound.

In **full-thickness burns,** or *third-degree burns,* the epidermis and the dermis are completely destroyed, and recovery occurs from the edges of the burn wound. Third-degree burns often are surrounded by areas of first- and second-degree burns. Although the first- and second-degree burn areas are painful, the region of third-degree burn is usually painless because sensory receptors in the epidermis and dermis have been destroyed. Third-degree burns appear white, tan, brown, black, or deep cherry red.

Deep partial-thickness and full-thickness burns take a long time to heal, and they form scar tissue with disfiguring and debilitating wound contracture (see chapter 4). To prevent these complications and to speed healing, skin grafts are often performed. In a procedure called a split skin graft, the epidermis and part of the dermis are removed from another part of the body and placed over the burn. Interstitial fluid from the burn nourishes the graft until blood vessels can grow into the graft and supply it with nourishment. Meanwhile, the donor tissue produces new epidermis from epithelial tissue in the hair follicles and sweat glands in the same manner as in superficial second-degree burns.

When it is not possible or practical to move skin from one part of the body to a burn site, physicians sometimes use artificial skin or grafts from human cadavers. However, these techniques are often unsatisfactory because the body's immune system recognizes the graft as a foreign substance and rejects it. A solution to this problem is to grow some of the burn victim's own skin in the laboratory. A piece of healthy skin from the burn victim is removed and placed in a flask with nutrients and hormones that stimulate rapid growth. The new skin that results consists only of epidermis and does not contain glands or hair.

# SKIN CANCER

Skin cancer is the most common type of cancer. Although chemicals and radiation (x-rays) are known to induce cancer, the development of skin cancer most often is associated with exposure to ultraviolet (UV) light from the sun. Consequently, most skin cancers develop on the face, neck, or hands. The group of people most likely to have skin cancer are fair-skinned (they have less protection from the sun) or older than 50 (they have had long exposure to the sun).

There are three main types of skin cancer. **Basal cell carcinoma** (kar-si-nō′mă), the most frequent type, begins with cells in the stratum basale and extends into the dermis to produce an open ulcer (figure 5.10*a*). Surgical removal or radiation therapy cures this type of cancer. Fortunately, there is little danger that this type of cancer will spread, or metastasize, to other areas of the body. **Squamous cell carcinoma** develops from cells immediately superficial to the stratum basale. Normally, these cells undergo little or no cell division, but in squamous cell carcinoma, the cells continue to divide as they produce keratin. The typical result is a nodular, keratinized tumor confined to the epidermis (figure 5.10*b*). If untreated, the tumor can invade the dermis, metastasize, and cause death. **Malignant melanoma** (mel′ă-nō′mă) is a rare form of skin cancer that arises from melanocytes, usually in a preexisting mole. A mole is an aggregation, or "nest," of melanocytes. The melanoma can appear as a large, flat, spreading lesion or as a deeply pigmented nodule (figure 5.10*c*). Metastasis is common, and unless diagnosed and treated early in development, this cancer is often fatal.

Limiting exposure to the sun and using sunscreens that block ultraviolet light can reduce the likelihood of developing skin cancer. Ultraviolet light is classified into two types based on their wavelengths: UVA has a longer wavelength than UVB. Exposure to UVA causes most tanning of the skin but is associated with the development of malignant melanoma. Exposure to UVB causes most burning of the skin and is associated with the development of basal cell and squamous cell carcinomas. It is advisable to use sunscreens that effectively block both UVA and UVB.

(a)

(b)

(c)

**Figure 5.10  Cancer of the Skin**
(*a*) Basal cell carcinoma. (*b*) Squamous cell carcinoma. (*c*) Malignant melanoma.

# SYSTEMS PATHOLOGY

## Burns

Sam is a 23-year-old man who often stayed up late watching TV or reading until he fell asleep. Sam was also a chain smoker. One night, he took several sleeping pills. Unfortunately, he fell asleep before putting out his cigarette, which started a fire. As a result, Sam was severely burned, receiving both full-thickness and partial-thickness burns (figure 5A). He was rushed to the emergency room and later transferred to a burn unit.

For the first day after his accident, Sam's condition was critical because he had gone into shock. Administration of large volumes of intravenous fluid stabilized his condition. As part of his treatment, Sam was also given a high-protein, high-calorie diet.

A week later, dead tissue was removed from Sam's most serious burns (figure 5B), and skin grafts were performed. Despite the use of topical antimicrobial drugs and sterile bandages, however, some of the burns became infected. An additional complication was the development of a venous thrombosis (blood clot) in Sam's leg.

Although the burns were painful and the treatment was prolonged, Sam made a full recovery. He no longer smokes.

## Background Information

When large areas of skin are severely burned, the resulting systemic effects can be life-threatening. Within minutes of a major burn injury, there is increased permeability of capillaries, which are the small blood vessels in which fluid, gases, nutrients, and waste products are normally exchanged between the blood and tissues. This increased permeability occurs at the burn site and throughout the body. As a result, fluid and ions are lost from the burn wound and into tissue spaces. The loss of fluid decreases blood volume, which decreases the heart's ability to pump blood. The resulting decrease in blood delivery to tissues can cause tissue damage, shock, and even death. Treatment consists of administering intravenous fluid at a faster rate than it leaks out of the capillaries. Although this fluid replacement can reverse the shock and prevent death, fluid continues to leak into tissue spaces, causing pronounced edema (swelling).

Typically, after 24 hours, capillary permeability returns to normal, and the amount of intravenous fluid administered can be greatly decreased. How burns cause capillary permeability to change is not well understood. It is clear that, following a burn, immunological and metabolic changes occur that affect not only capillaries but the rest of the body as well. For example, chemical mediators (see chapter 4), which are released in response to the tissue damage, contribute to changes in capillary permeability throughout the body.

Substances released from the burn may also play a role in causing cells to function abnormally. Burn injuries result in an almost immediate hypermetabolic state, which persists until wound closure. Two

Full-thickness burn

Partial-thickness burn

**Figure 5A**
Full-thickness and partial-thickness burns

**Figure 5B**
Patient in a burn unit

other factors contributing to the increased metabolism are (1) a resetting of the temperature control center in the brain to a higher temperature and (2) hormones released by the endocrine system (e.g., epinephrine and norepinephrine from the adrenal glands, which can increase cell metabolism). Compared with a normal body temperature of approximately 37°C (98.6°F), a typical burn patient may have a body temperature of 38.5°C (101.3°F), despite the higher loss of water by evaporation from the burn.

In severe burns, the increased metabolic rate can result in loss of as much as 30–40% of the patient's preburn weight. To help compensate, treatment may include doubling or tripling the patient's caloric

## System Interactions — Effects of Burns on Other Systems

| System | Interactions |
| --- | --- |
| Skeletal | Red bone marrow replaces red blood cells destroyed in the burned skin. |
| Muscular | Loss of muscle mass results from the hypermetabolic state caused by the burn. |
| Nervous | Pain is sensed in the partial-thickness burns. The temperature-regulatory center in the brain is set to a higher temperature, contributing to increased body temperature. Abnormal $K^+$ concentrations disturb normal nervous system activity: Elevated levels are caused by the release of $K^+$ from damaged cells; low levels can be caused by the rapid loss of these ions in fluid from the burn. |
| Endocrine | Increased secretion of epinephrine and norepinephrine from the adrenal glands in response to the injury contributes to increased body temperature by increasing metabolism. |
| Cardiovascular | Increased capillary permeability causes decreased blood volume, resulting in decreased blood delivery to tissues, edema, and shock. The heart's pumping effectiveness is impaired by ion imbalance and substances released from the burn. Increased blood clotting causes venous thrombosis. Preferential delivery of blood to the injury promotes healing. |
| Lymphatic and Immune | Inflammation increases in response to tissue damage. Later, depression of the immune system can result in infection. |
| Respiratory | Edema causes airway obstruction. Respiration rate increases due to increased metabolism and lactic acid buildup. |
| Digestive | Decreased blood delivery as a result of the burn causes degeneration of the intestinal lining and liver. Bacteria from the intestine can cause systemic infections. The liver releases blood-clotting factors in response to the injury. Increased nutrients necessary to support increased metabolism and repair of the integumentary system are absorbed. |
| Urinary | The kidneys compensate for the increased fluid loss caused by the burn by greatly reducing or even stopping urine production. Decreased blood volume causes decreased blood flow to the kidneys, which reduces urine output but can damage kidney tissue. Hemoglobin, released from red blood cells damaged in the burned skin, can decrease urine production by blocking fluid movement from the blood into the kidney. |

intake. In addition, the need for protein, which is necessary for tissue repair, is greater.

Normal skin maintains homeostasis by preventing microorganisms from entering the body. Because burns damage and sometimes completely destroy the skin, microorganisms can cause infections. For this reason, burn patients are maintained in an aseptic (sterile) environment, which attempts to prevent the entry of microorganisms into the wound. They are also given antimicrobial drugs, which kill microorganisms or suppress their growth. **Debridement** (dā-brēd-mon′), the removal of dead tissue from the burn, helps prevent infections by cleaning the wound and removing tissue in which infections could develop. Skin grafts, performed within a week of the injury, also help close the wound and prevent the entry of microorganisms.

Despite these efforts, however, infections are still the major cause of death for burn victims. Depression of the immune system during the first or second week after the injury contributes to the high infection rate. First, the thermally altered tissue is recognized as a foreign substance, which stimulates the immune system. Then, the immune system is overwhelmed as immune system cells become less effective and the production of the chemicals that normally provide resistance to infections decreases (see chapter 14). The greater the magnitude of the burn, the greater the depression of the immune system, and the greater the risk of infection.

**Venous thrombosis** (throm-bō′sis), the development of a clot in a vein, is another complication of burns. Blood normally forms a clot when exposed to damaged tissue, such as at a burn site, but clotting can also occur elsewhere, such as in veins, where clots can block blood flow, resulting in tissue destruction. The concentration of chemicals that cause blood clotting (called clotting factors) increases for two reasons: Loss of fluid from the burn patient concentrates the chemicals, and the liver releases an increased amount of clotting factors.

*Predict 5*

*When Sam was first admitted to the burn unit, the nurses carefully monitored his urine output. Why does that make sense in light of his injuries?*

## DISEASES AND DISORDERS  Skin

| CONDITION | DESCRIPTION |
|---|---|
| Ringworm | Fungal infection that produces patchy scaling and inflammatory response in the skin |
| Eczema and dermatitis | Inflammatory conditions of the skin caused by allergy, infection, poor circulation, or exposure to chemical or environmental factors |
| Psoriasis | Chronic skin disease characterized by thicker than normal epidermal layer (stratum corneum) that sloughs to produce large, silvery scales; bleeding may occur if the scales are scraped away |
| **Bacterial Infections** | |
| Impetigo | Small blisters containing pus; easily rupture to form a thick, yellowish crust; usually affects children |
| Decubitus ulcers (bedsores or pressure sores) | Develop in people who are bedridden or confined to a wheelchair; compression of tissue and reduced circulation result in destruction of the hypodermis and skin, which later become infected by bacteria, forming ulcers |
| **Viral Infections** | |
| Rubeola (measles) | Skin lesions; caused by a virus contracted through the respiratory tract; may develop into pneumonia or infect the brain, causing damage |
| Rubella (German measles) | Skin lesions; usually mild viral disease contracted through the respiratory tract; may be dangerous if contracted during pregnancy because the virus can cross the placenta and damage the fetus |
| Chicken pox | Skin lesions; usually mild viral disease contracted through the respiratory tract |
| Shingles | Painful skin lesions that can recur when the dormant virus is activated by trauma, stress, or another illness; caused by the chickenpox virus after childhood infection |
| Cold sores (fever blisters) | Skin lesions; caused by herpes simplex I virus; transmitted by oral or respiratory routes; lesions recur |
| Genital herpes | Genital lesions; caused by herpes simplex II virus; transmitted by sexual contact |

Go to **www.mhhe.com/seeleyess7** for additional information on these pathologies.

# EFFECTS OF AGING ON THE INTEGUMENTARY SYSTEM

As the body ages, the skin is more easily damaged because the epidermis thins and the amount of collagen in the dermis decreases. Skin infections are more likely, and repair of the skin occurs more slowly. A decrease in the number of elastic fibers in the dermis and loss of fat from the hypodermis cause the skin to sag and wrinkle. A decrease in the activity of sweat glands and in the blood supply to the dermis results in reduced ability to regulate body temperature. The skin becomes drier as sebaceous gland activity decreases. The number of melanocytes generally decreases, but in some areas the number of melanocytes increases to produce **age spots.** Note that age spots are different from **freckles,** which are caused by increased melanin production. Gray or white hair also results because of a decrease in or a lack of melanin production. Skin that is exposed to sunlight shows signs of aging more rapidly than nonexposed skin, so avoiding overexposure to sunlight and using sunscreen is advisable.

 SUMMARY

The integumentary system consists of the skin, hair, glands, and nails.

## Functions of the Integumentary System  (p. 96)

The integumentary system protects us from the external environment. Other functions include sensation, vitamin D production, temperature regulation, and excretion of small amounts of waste products.

## Skin  (p. 96)

### Epidermis
1. The epidermis is stratified squamous epithelium divided into strata.
   - New cells are produced in the stratum basale.
   - The stratum corneum consists of many layers of dead squamous cells containing keratin. The most superficial layers are sloughed.
2. Keratinization is the transformation of stratum basale cells into stratum corneum cells.
   - Structural strength results from keratin inside the cells and from desmosomes, which hold the cells together.
   - Lipids surrounding the stratum corneum cells help prevent fluid loss.

### Dermis
1. The dermis is dense connective tissue.
2. Collagen and elastic fibers provide structural strength, and the blood vessels of the papillae supply the epidermis with nutrients.

### Skin Color
1. Melanocytes produce melanin, which is responsible for different skin colors. Melanin production is determined genetically but can be modified by exposure to ultraviolet light and by hormones.
2. Carotene, a plant pigment ingested as a source of vitamin A, can cause the skin to appear yellowish.

3. Increased blood flow produces a red skin color, whereas decreased blood flow causes a pale skin color. Decreased blood oxygen results in the blue skin color of cyanosis.
4. Scattering of light by collagen produces a bluish color.

## Hypodermis   (p. 100)

1. The hypodermis, which is not part of the skin, is loose connective tissue that attaches the skin to underlying tissues.
2. About half of the body's fat is stored in the hypodermis.

## Accessory Skin Structures   (p. 100)

### Hair

1. Hairs are columns of dead, keratinized epithelial cells. Each hair consists of a shaft (above the skin), a root (below the skin), and a hair bulb (site of hair cell formation).
2. Hairs have a growth phase and a resting phase.
3. Contraction of the arrector pili, which are smooth muscles, causes hair to "stand on end" and produces "goose flesh."

### Glands

1. Sebaceous glands produce sebum, which oils the hair and the surface of the skin.
2. Eccrine sweat glands produce sweat, which cools the body.
3. Apocrine sweat glands produce an organic secretion that causes body odor when broken down by bacteria.

### Nails

1. The nail consists of the nail body and the nail root.
2. The nail matrix produces the nail, which is composed of stratum corneum cells containing hard keratin.

## Physiology of the Integumentary System   (p. 102)

### Protection

The skin reduces water loss, prevents the entry of microorganisms, and provides protection against abrasion and ultraviolet light; hair and nails also perform protective functions.

### Sensation

The skin contains sensory receptors for pain, heat, cold, and pressure.

### Vitamin D Production

1. Ultraviolet light stimulates the production of a precursor molecule in the skin that is modified by the liver and kidneys into vitamin D.
2. Vitamin D increases calcium uptake in the intestines.

### Temperature Regulation

1. Through dilation and constriction of blood vessels, the skin controls heat loss from the body.
2. Evaporation of sweat cools the body.

### Excretion

Skin glands remove small amounts of waste products but are not important in excretion.

## Integumentary System as a Diagnostic Aid   (p. 104)

The integumentary system is easily observed and often reflects events occurring in other parts of the body (e.g., cyanosis, jaundice, rashes).

## Burns   (p. 104)

1. Partial-thickness burns damage only the epidermis (first-degree burn) or the epidermis and the dermis (second-degree burn).
2. Full-thickness burns (third-degree burns) destroy the epidermis, the dermis, and usually underlying tissues.

## Skin Cancer   (p. 105)

1. Basal cell carcinoma involves the cells of the stratum basale and is readily treatable.
2. Squamous cell carcinoma involves the cells immediately superficial to the stratum basale and can metastasize.
3. Malignant melanoma involves melanocytes, can metastasize, and is often fatal.

## Effects of Aging on the Integumentary System   (p. 108)

1. Blood flow to the skin is reduced, the skin becomes thinner, and elasticity is lost.
2. Sweat and sebaceous glands are less active, and the number of melanocytes decreases.

## REVIEW AND COMPREHENSION

1. Name the components of the integumentary system.
2. What kind of tissue is the epidermis? In which stratum of the epidermis are new cells formed? From which stratum are they sloughed?
3. Define keratinization. What structural changes does keratinization produce to make the skin resistant to abrasion and water loss?
4. What type of tissue is the dermis? What is responsible for its structural strength? How does the dermis supply the epidermis with blood?
5. Name the cells that produce melanin. What happens to the melanin after it is produced? What is the function of melanin?
6. Describe the factors that determine the amount of melanin produced in the skin.
7. How do melanin, blood, carotene, and collagen affect skin color?
8. What type of tissue is the hypodermis, and what are its functions?
9. What is a hair follicle? Define the root, shaft, and hair bulb of a hair. What kinds of cells are found in a hair?
10. Why is a hair follicle important in the repair of skin?
11. What part of a hair is the site of hair growth? What are the stages of hair growth?
12. What happens when the arrector pili of the skin contract?
13. What secretion do the sebaceous glands produce? What is the function of the secretion?
14. Which glands of the skin are responsible for cooling the body? Which glands are involved in producing body odor?
15. Name the parts of a nail. Where are the cells that make up the nail produced, and what kind of cells make up a nail? What is the lunula? Describe nail growth.
16. How do the components of the integumentary system provide protection?
17. List the types of sensations detected by receptors in the skin.
18. Describe the production of vitamin D by the body. What is the function of vitamin D?

19. How does the integumentary system help regulate body temperature?

20. Name the substances excreted by skin glands. Is the skin an important site of excretion?

21. Why is the skin a useful diagnostic aid? Give three examples of how the skin functions as a diagnostic aid.

22. Define the different categories of burns. How is repair accomplished after each type?

23. What is the most common cause of skin cancer? Describe three types of skin cancer and the risks of each type.

24. What changes occur in the skin as a result of aging?

# CRITICAL THINKING

1. A woman has stretch marks on her abdomen, yet she states that she has never been pregnant. Is this possible?

2. It has been several weeks since Tom has competed in a tennis match. After the match, he discovers that a blister has formed beneath an old callus on his foot and the callus has fallen off. When he examines the callus, it appears yellow. Can you explain why?

3. The lips are muscular folds forming the anterior boundary of the oral cavity. A mucous membrane covers the lips internally, and the skin of the face covers them externally. The red part of the lips (called the vermillion border) is covered by keratinized epithelium that is a transition between the epithelium of the mucous membrane and the facial skin. The vermillion border can become chapped (dry and cracked), whereas the mucous membrane and the facial skin do not. Propose as many reasons as you can to explain why the vermillion border is more prone to drying than the mucous membrane or the facial skin.

4. Pulling on hair can be quite painful, but cutting hair is not painful. Explain.

5. Given what you know about the cause of acne, propose some ways to prevent or treat it.

6. Consider the following statement: Dark-skinned children are more susceptible to rickets (insufficient calcium in the bones) than light-skinned children. Defend or refute this statement.

7. Harry, a light-skinned man, jogs on a cool day. What color would you expect his skin to be (a) after going outside and just before starting to run, (b) during the run, and (c) 5 minutes after the run?

*Answers in Appendix D*

Visit this textbook's website at **www.mhhe.com/seeleyess7** for practice quizzes, animations, interactive learning exercises, and other study tools.

McGraw-Hill offers a study CD that features interactive cadaver dissection. ***Anatomy & Physiology Revealed*** includes cadaver photos that allow you to peel away layers of the human body to reveal structures beneath the surface. This program also includes animations, radiologic imaging, audio pronunciations, and practice quizzing.

Have you ever had a radiograph taken? If you've had a radiograph like the one shown in the picture, you probably know how important bones are to homeostasis. Bones allow our bodies to move, protect our vital organs, and store extra calcium. Without healthy bones, our bodies cannot maintain proper homeostasis.

# 6 Chapter

# Skeletal System Bones and Joints

## Chapter Outline and Objectives

### Functions of the Skeletal System (p. 112)
1. List the functions of the skeletal system.

### Extracellular Matrix (p. 112)
2. Describe the components of the extracellular matrix, and state the function of each.

### General Features of Bone (p. 113)
3. Describe the structure of compact and spongy bone.
4. Outline the processes of bone ossification, growth, remodeling, and repair.

### Bone and Calcium Homeostasis (p. 119)
5. Explain the role of bone in calcium homeostasis.

### General Considerations of Bone Anatomy (p. 121)
6. List and define the major features of a typical bone.

### Axial Skeleton (p. 122)
7. Describe the main features of the skull from the lateral, frontal, internal, and inferior views.

8. Describe the shape of the vertebral column, and list its divisions. Describe the general features of a vertebra and the differences among vertebrae from each region of the vertebral column.
9. List the bones of the thoracic cage, including the three types of ribs.

### Appendicular Skeleton (p. 131)
10. Name and describe the bones of the pectoral girdle and upper limb.
11. Name and describe the bones of the pelvic girdle and lower limb.

### Joints (p. 140)
12. List and describe the various types of joints.
13. Describe the major types of joint movement.

### Effects of Aging on the Skeletal System and Joints (p. 145)
14. Describe the effects of aging on bones and joints.

Sitting, standing, walking, picking up a pencil, and taking a breath all involve the skeletal system. Without the skeletal system, there would be no rigid framework to support the soft tissues of the body and no system of joints and levers to allow the body to move. The skeletal system consists of bones, such as those shown in figure 6.1, as well as their associated connective tissues, which include cartilage, tendons, and ligaments. The term *skeleton* is derived from a Greek word meaning dried. But the skeleton is far from being dry and nonliving. Rather, the skeletal system consists of dynamic, living tissues that are able to grow, detect pain stimuli, adapt to stress, and undergo repair after injury.

A joint, or an articulation, is a place where two bones come together. Many joints are movable, although some of them allow only limited movement; others allow no apparent movement. The structure of a given joint is directly correlated to its degree of movement.

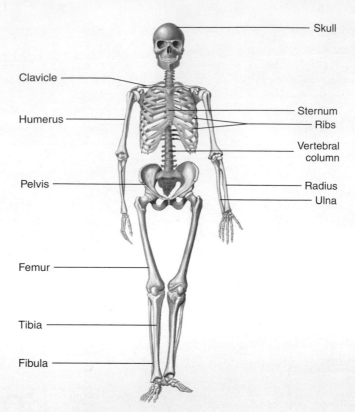

**Figure 6.1** Major Bones of the Skeletal System

# FUNCTIONS OF THE SKELETAL SYSTEM

The skeleton is usually thought of as the framework of the body, but the skeletal system has many other functions in addition to support. The major functions of the skeletal system include

1. *Support.* Rigid, strong bone is well suited for bearing weight and is the major supporting tissue of the body. Cartilage provides firm yet flexible support within certain structures, such as the nose, external ear, thoracic cage, and trachea. Ligaments are strong bands of fibrous connective tissue that attach to bones and hold them together.
2. *Protection.* Bone is hard and protects the organs it surrounds. For example, the skull encloses and protects the brain, and the vertebrae surround the spinal cord. The rib cage protects the heart, lungs, and other organs of the thorax.
3. *Movement.* Skeletal muscles attach to bones by tendons, which are strong bands of connective tissue. Contraction of the skeletal muscles moves the bones, producing body movements. Joints, where two or more bones come together, allow movement between bones. Smooth cartilage covers the ends of bones within some joints, allowing the bones to move freely. Ligaments allow some movement between bones but prevent excessive movement.

4. *Storage.* Some minerals in the blood—principally, calcium and phosphorus—are taken into bone and stored. Should blood levels of these minerals decrease, the minerals are released from bone into the blood. Adipose tissue is also stored within bone cavities. If needed, the fats are released into the blood and used by other tissues as a source of energy.
5. *Blood cell production.* Many bones contain cavities filled with red bone marrow, which produces blood cells and platelets (see chapter 11).

# EXTRACELLULAR MATRIX

The bone, cartilage, tendons, and ligaments of the skeletal system are all connective tissues. Their characteristics are largely determined by the composition of their extracellular matrix. The matrix always contains collagen, proteoglycan (ground substance), and other organic molecules, as well as water and minerals. But the types and quantities of these substances differ in each type of connective tissue.

**Collagen** (kol′lă-jen; *koila*, glue + *-gen*, producing) is a tough, ropelike protein. **Proteoglycans** (prō′tē-ō-glī′kanz; *proteo*, protein + *glycan*, polysaccharide) are large molecules consisting of polysaccharides attached to core proteins, much as the needles of a pine tree are attached to the tree's branches. The proteoglycans form large aggregates, much as pine branches combine to form a whole tree. Proteoglycans can attract and retain large amounts of water between their polysaccharide "needles."

The extracellular matrix of **tendons** and **ligaments** contains large amounts of collagen fibers, making these structures very tough, like ropes or cables. The extracellular matrix of **cartilage** (kar′ti-lij) contains collagen and proteoglycans. Collagen makes cartilage tough, whereas the water-filled proteoglycans make it smooth and resilient. As a result, cartilage is relatively rigid, but it springs back to its original shape after being bent or slightly compressed. It is an excellent shock absorber.

The extracellular matrix of bone contains collagen and minerals, including calcium and phosphate. The ropelike collagen fibers, like the reinforcing steel bars in concrete, lend flexible strength to the bone. The mineral component, like the concrete itself, gives the bone compression (weight-bearing) strength. Most of the mineral in bone is in the form of calcium phosphate crystals called **hydroxyapatite** (hī-drok′sē-ap-ă-tīt).

### Predict 1

*What would a bone be like if all of the minerals were removed? What would it be like if all of the collagen were removed?*

### A CASE IN POINT
**Brittle Bone Disease**

May Trix is a 10-year-old girl who has a history of numerous broken bones. At first, physicians suspected she was a victim of child abuse, but eventually they determined that she has **brittle bone disease,** or *osteogenesis imperfecta,* which literally means imperfect bone formation. May is short for her age, and her limbs are short and bowed. Her vertebral column is also abnormally curved. Brittle bone disease is a rare disorder caused by any one of a number of faulty genes that results in either too little collagen formation or a poor quality of collagen. As a result, the bone matrix has decreased flexibility and is more easily broken than normal bone.

## GENERAL FEATURES OF BONE

There are four categories of bone, based on their shape: long, short, flat, and irregular. **Long bones** are longer than they are wide. Most of the bones of the upper and lower limbs are long bones. **Short bones** are approximately as broad as they are long; examples are the bones of the wrist and ankle. **Flat bones** have a relatively thin, flattened shape. Examples of flat bones are certain skull bones, the ribs, the scapulae (shoulder blades), and the sternum. **Irregular bones** include the vertebrae and facial bones, which have shapes that do not fit readily into the other three categories.

### Structure of a Long Bone

A long bone serves as a useful model for illustrating the parts of a typical bone (figure 6.2). Each long bone consists of a central shaft, called the **diaphysis** (dī-af′i-sis; growing between), and two ends, each called an **epiphysis** (e-pif′i-sis; growing upon). A thin layer of **articular** (ar-tik′ū-lăr; joint) **cartilage** covers the ends of the epiphyses where the bone articulates (joins) with other bones. A long bone that is still growing has an **epiphyseal plate,** or *growth plate,* composed of cartilage, between each epiphysis and the diaphysis (figure 6.2a). The epiphyseal plate is where the bone grows in length. When bone growth stops, the cartilage of each epiphyseal plate is replaced by bone and becomes an **epiphyseal line** (figure 6.2b).

Bones contain cavities, such as the large **medullary cavity** in the diaphysis, as well as smaller cavities in the epiphyses of long bones and in the interior of other bones. These spaces are filled with soft tissue called **marrow. Yellow marrow** consists mostly of adipose tissue. **Red marrow** consists of blood-forming cells and is the only site of blood formation in adults (see chapter 11). Children's bones have proportionately more red marrow than do adult bones because, as a person ages, red marrow is mostly replaced by yellow marrow. In adults, red marrow is confined to the bones in the central axis of the body and in the most proximal epiphyses of the limbs.

Most of the outer surface of bone is covered by dense connective tissue called the **periosteum** (per-ē-os′tē-ŭm; *peri,* around + *osteon,* bone), which contains blood vessels and nerves (figure 6.2c). The surface of the medullary cavity is lined with a thinner connective tissue membrane, the **endosteum** (en-dos′tē-ŭm; *endo,* inside).

## Histology of Bone

The periosteum and endosteum contain **osteoblasts** (os′tē-ō-blasts; bone-forming cells), which function in the formation of bone, as well as in the repair and remodeling of bone. When osteoblasts become surrounded by matrix, they are referred to as **osteocytes** (os′tē-ō-sītz; bone cells).

Bone is formed in thin sheets of extracellular matrix called **lamellae** (lă-mel′ē; plates), with osteocytes located between the lamellae within spaces called **lacunae** (lă-koo′nē; a hollows) (figure 6.3). Cell processes extend from the osteocytes across the extracellular matrix of the lamellae within tiny canals called **canaliculi** (kan-ă-lik′ū-lī; sing. canaliculus, little canal).

There are two major types of bone, based on their histological structure. **Compact bone** is mostly solid matrix and cells. **Spongy bone,** or *cancellous* (k an′sě-lŭs) *bone,* consists of a lacy network of bone with many small, marrow-filled spaces.

### Compact Bone

Compact bone (figure 6.3) forms most of the diaphysis of long bones and the thinner surfaces of all other bones. Most of the lamellae of compact bone are organized into sets of concentric rings, with each set surrounding a **central canal,** or *haversian* (ha-ver′shan), *canal.* Blood vessels that run parallel to the long axis of the bone are contained within the central canals. Each central canal, with the lamellae and osteocytes surrounding it, is called an **osteon** (os′tē-on), or *haversian system.* Each osteon, seen in cross section, looks like a microscopic target, with the central canal as the "bull's eye" (figure 6.3). Osteocytes, located in lacunae, are connected to one another by cell processes in canaliculi. The canaliculi give the osteon the appearance of having tiny cracks within the lamellae.

Nutrients leave the blood vessels of the central canals and diffuse to the osteocytes through the canaliculi. Waste products diffuse in the opposite direction. The blood vessels in the central

Skeletal

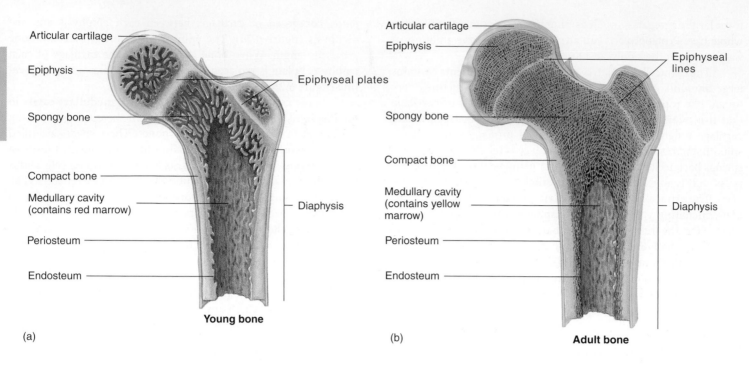

**Young bone**

(a)

**Adult bone**

(b)

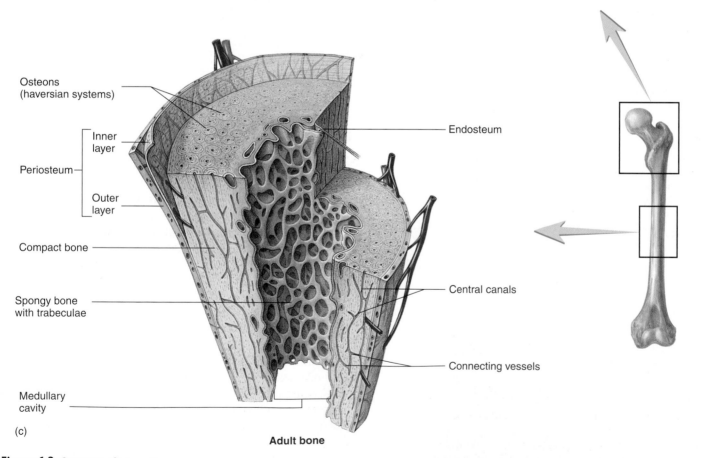

**Adult bone**

(c)

## Figure 6.2  Structure of a Long Bone

(a) Young long bone (the femur) showing the epiphysis, epiphyseal plates, and diaphysis. (b) Adult long bone with epiphyseal lines. (c) Internal features of a portion of the diaphysis in (a).

**Figure 6.3  Structure of Bone Tissue**

(*a*) Photomicrograph of compact bone. (*b*) Fine structure of compact bone.

canals, in turn, are connected to blood vessels in the periosteum and endosteum.

## Spongy Bone

Spongy bone, so called because of its appearance, is located mainly in the epiphyses of long bones. It forms the interior of all other bones. Spongy bone consists of delicate interconnecting rods or plates of bone called **trabeculae** (tră-bek′ū-lē; beams), which resemble the beams or scaffolding of a building (figure 6.4*a*). Like scaffolding, the trabeculae add strength to a bone without the added weight that would be present if the bone were solid mineralized matrix. The spaces between the trabeculae are filled with marrow. Each trabecula consists of several lamellae with osteocytes between them (figure 6.4*b*). Usually, no blood vessels penetrate the trabeculae, and the trabeculae have no central canals. Nutrients exit vessels in the marrow and pass by diffusion through canaliculi to the osteocytes of the trabeculae.

## Bone Ossification

**Ossification** (os′i-fi-kā′shŭn; *os,* bone + *facio,* to make) is the formation of bone by osteoblasts. After an osteoblast becomes completely surrounded by bone matrix, it becomes a mature bone cell, or osteocyte. In the fetus, bones develop by two processes, each involving the formation of bone matrix on preexisting connective tissue (figure 6.5). Bone formation that occurs within connective tissue membranes is called intramembranous ossification, and bone formation that occurs inside cartilage is called endochondral ossification. Both types of bone formation result in compact and spongy bone.

**Intramembranous** (in′tră-mem′bră-nŭs) **ossification** occurs when osteoblasts begin to produce bone in connective tissue membranes. This occurs primarily in the bones of the skull. Osteoblasts line up on the surface of connective tissue fibers and begin depositing bone matrix to form trabeculae. The process begins in areas called **ossification centers** (figure 6.5*a*), and the trabeculae radiate out from the centers. Usually, two or more ossification centers exist in each flat skull bone, and the skull bones result from fusion of these centers as they enlarge. The trabeculae are constantly remodeled after their initial formation, and they may enlarge or be replaced by compact bone.

The bones at the base of the skull and most of the remaining skeletal system develop through the process of **endochondral ossification** from cartilage models. The cartilage models have the

Skeletal

(a)

(b)

**Figure 6.4  Spongy Bone**

(*a*) Beams of bone, the trabeculae, surround spaces in the bone. In life, the spaces are filled with red or yellow bone marrow and with blood vessels. (*b*) Transverse section of a trabecula.

(a)

(b)

**Figure 6.5  Bone Formation in a Fetus**

(*a*) Intramembranous ossification occurs in a 12-week-old fetus at ossification centers in the flat bones of the skull (*yellow*). Endochondral ossification occurs in the bones forming the inferior part of the skull (*blue*). (*b*) Radiograph of an 18-week-old fetus, showing intramembranous and endochondral ossification. Intramembranous ossification occurs at ossification centers in the flat bones of the skull. Endochondral ossification has formed bones in the diaphyses of long bones. The epiphyses are still cartilage at this stage of development.

general shape of the mature bone (figure 6.6, step 1). During endochondral ossification, cartilage cells, called **chondrocytes,** increase in number, enlarge, and die. Then the cartilage matrix becomes calcified (figure 6.6, step 2). As this process is occurring in the center of the cartilage model, blood vessels accumulate in the perichondrium. The presence of blood vessels in the outer surface of future bone causes some of the unspecified connective tissue cells on the surface to become osteoblasts. These osteoblasts then produce a collar of bone around part of the outer surface of the diaphysis, and the perichondrium becomes periosteum in that area. Blood vessels also grow into the center of the diaphyses, bringing in osteoblasts and stimulating ossification. The center part of the diaphysis, where bone first begins to appear, is called the **primary ossification center** (figure 6.6, step 3). Osteoblasts invade spaces in the center of the bone left by the dying cartilage cells. Some of the calcified cartilage matrix is removed by cells called **osteoclasts** (os′tē-ō-klastz; bone-eating cells), and the osteoblasts line up on the remaining calcified matrix and begin to form bone trabeculae. As the bone develops, it is constantly changing. A medullary cavity forms in the center of the diaphysis as

osteoclasts remove bone and calcified cartilage, which are replaced by bone marrow. Later, **secondary ossification centers** form in the epiphyses (figure 6.6, step 4).

## Bone Growth

Bone growth occurs by the deposition of new bone lamellae onto existing bone or other connective tissue. As osteoblasts deposit new bone matrix on the surface of bones between the periosteum and the existing bone matrix, the bone increases in width, or diameter. This process is called appositional growth. Growth in the length of a bone, which is the major source of increased height in an individual, occurs in the epiphyseal plate. This type

**1** A cartilage model, with the general shape of the mature bone, is produced by chondrocytes. A perichondrium surrounds most of the cartilage model.

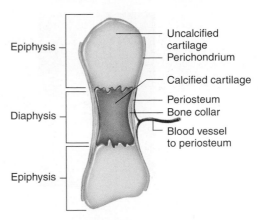

**2** A bone collar is produced, and the perichondrium of the diaphysis becomes the periosteum. The chondrocytes enlarge, and cartilage is calcified.

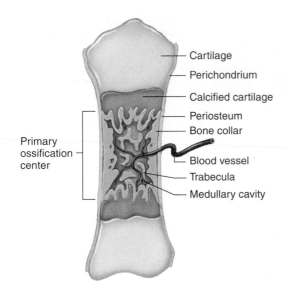

**3** A primary ossification center forms as blood vessels and osteoblasts invade the calcified cartilage. The osteoblasts lay down bone matrix, forming trabeculae.

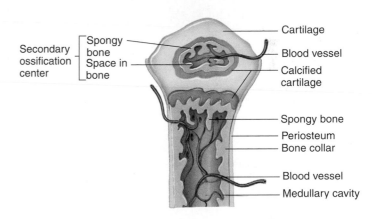

**4** Secondary ossification centers form in the epiphyses of long bones.

**PROCESS Figure 6.6** Endochondral Ossification of a Long Bone

of bone growth occurs through endochondral ossification (figure 6.7). Chondrocytes increase in number on the epiphyseal side of the epiphyseal plate. They line up in columns parallel to the long axis of the bone, causing the bone to elongate. Then the chondrocytes enlarge and die. The cartilage matrix becomes calcified. Much of the cartilage that forms around the enlarged cells is removed by osteoclasts, and the dying chondrocytes are replaced by osteoblasts. The osteoblasts start forming bone by depositing bone lamellae on the surface of the calcified cartilage. This process produces bone on the diaphyseal side of the epiphyseal plate.

**Predict 2**

*Describe the appearance of an adult if cartilage growth did not occur in the long bones during childhood.*

## Bone Remodeling

**Bone remodeling** involves the removal of existing bone by osteoclasts and the deposition of new bone by osteoblasts. Bone remodeling occurs in all bone. Remodeling is responsible for changes in bone shape, the adjustment of bone to stress, bone repair, and calcium ion regulation in the body fluids. Remodeling is also involved in bone growth when newly formed spongy bone in the epiphyseal plate forms compact bone. A long bone increases in length and diameter as new bone is deposited on the outer surface and growth occurs at the epiphyseal plate. At the same time, bone is removed from the inner, medullary surface of the bone. As the bone diameter increases, the thickness of the compact bone relative to the medullary cavity tends to remain fairly constant. If the size of the medullary cavity did not also increase as bone size

Skeletal

Femur

Patella

Epiphysis
of tibia

Epiphyseal
plate

Diaphysis
of tibia

Epiphyseal side

Epiphyseal
plate

1 New cartilage is produced on
the epiphyseal side of the plate
as the chondrocytes divide and
form stacks of cells.

2 Chondrocytes mature and
enlarge.

3 Matrix is calcified, and
chondrocytes die.

4 The cartilage on the diaphyseal
side of the plate is replaced
by bone.

LM 400x

(a)

(b)        Diaphyseal side

Length of bone
increases.

Thickness of
epiphyseal plate
remains unchanged.

Epiphyseal
plate

Chondrocytes
divide and enlarge.

Bone is added
to diaphysis.

Bone of
diaphysis

Calcified cartilage
is replaced by bone.

(c)

## PROCESS Figure 6.7  Endochondral Bone Growth

(a) Location of the epiphyseal plate in a long bone. (b) Photomicrograph of an epiphyseal plate, demonstrating chondrocyte division and enlargement and the areas of calcification and ossification. (c) As the chondrocytes of the epiphyseal plate divide and align in columns, the cartilage expands toward the epiphysis, and the bone elongates. At the same time, the older cartilage is calcified and then replaced by bone, which is remodeled, resulting in expansion of the medullary cavity of the diaphysis. The net result is an epiphyseal plate that remains uniform in thickness through time but is constantly moving toward the epiphysis, resulting in elongation of the bone.

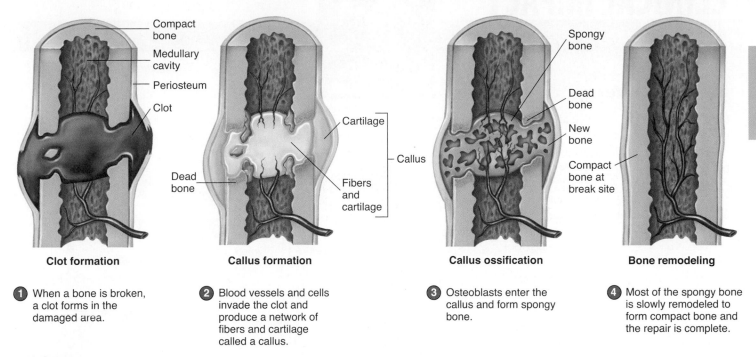

**Clot formation**

**Callus formation**

**Callus ossification**

**Bone remodeling**

**1** When a bone is broken, a clot forms in the damaged area.

**2** Blood vessels and cells invade the clot and produce a network of fibers and cartilage called a callus.

**3** Osteoblasts enter the callus and form spongy bone.

**4** Most of the spongy bone is slowly remodeled to form compact bone and the repair is complete.

**PROCESS Figure 6.8**    **Bone Repair**

increased, the compact bone of the diaphysis would become thick and very heavy.

Because bone is the major storage site for calcium in the body, bone remodeling is important to maintain blood calcium levels within normal limits. Calcium is removed from bones when blood calcium levels decrease, and it is deposited when dietary calcium is adequate. This removal and deposition is under hormonal control (see "Bone and Calcium Homeostasis" later in this chapter).

If too much bone is deposited, the bones become thick or develop abnormal spurs or lumps that can interfere with normal function. Too little bone formation or too much bone removal, as occurs in osteoporosis, weakens the bones and makes them susceptible to fracture (see Systems Pathology, "Osteoporosis").

## Bone Repair

Sometimes a bone is broken and needs to be repaired. When this occurs, blood vessels in the bone are also damaged. The vessels bleed, and a clot forms in the damaged area (figure 6.8, step 1). Two to three days after the injury, blood vessels and cells from surrounding tissues begin to invade the clot. Some of these cells produce a fibrous network of connective tissue between the broken bones, which holds the bone fragments together and fills the gap between them. Other cells produce islets of cartilage in the fibrous network. The network of fibers and islets of cartilage between the two bone fragments is called a callus (figure 6.8, step 2).

Osteoblasts enter the callus and begin forming spongy bone (figure 6.8, step 3). Spongy bone formation in the callus is usually complete 4–6 weeks after the injury. Immobilization of

the bone is critical up to this time because movement can refracture the delicate new matrix. Subsequently, the spongy bone is slowly remodeled to form compact and spongy bone, and the repair is complete (figure 6.8, step 4). Although immobilization at a fracture point is critical during the early stages of bone healing, complete immobilization is not good for the bone, the muscles, or the joints. Not long ago, it was common practice to immobilize a bone completely for as long as 10 weeks. But we now know that, if a bone is immobilized for as little as 2 weeks, the muscles associated with that bone may lose as much as half their strength. Furthermore, if a bone is completely immobilized, it is not subjected to the normal mechanical stresses that help it form. Bone matrix is reabsorbed, and the strength of the bone decreases. In experimental animals, complete immobilization of the back for 1 month resulted in up to a threefold decrease in vertebral compression strength. Modern therapy attempts to balance bone immobilization with enough exercise to keep muscle and bone from decreasing in size and strength and to maintain joint mobility. These goals are accomplished by limiting the amount of time a cast is left on the patient and by using "walking casts," which allow some stress on the bone and some movement. Total healing of the fracture may require several months. If a bone heals properly, the healed region can be even stronger than the adjacent bone.

## BONE AND CALCIUM HOMEOSTASIS

Bone is the major storage site for calcium in the body, and movement of calcium into and out of bone helps determine blood calcium levels, which is critical for normal muscle and nervous

# CLINICAL IMPACT

## Bone Fractures

Bone fractures can be classified as **open** (or *compound*) if the bone protrudes through the skin and **closed,** (or *simple*) if the skin is not perforated. Figure 6A illustrates some of the different types of fractures. If the fracture totally separates the two bone fragments, it is called **complete;** if it doesn't, it is called **incomplete.** An incomplete fracture that occurs on the convex side of the curve of a bone is called a **greenstick fracture.** A **comminuted** (kom'i-nū-ted; broken into small pieces) fracture is one in which the bone breaks into more than two fragments. An **impacted** fracture occurs when one of the fragments of one part of the bone is driven into the spongy bone of another fragment.

Fractures can also be classified according to the direction of the fracture line as **linear** (parallel to the long axis), **transverse** (at right angles to the long axis), or **oblique,** or **spiral** (at an angle other than a right angle to the long axis).

**Figure 6A**

Types of bone fractures.

**1** PTH indirectly stimulates osteoclasts to break down bone and release Ca²⁺ into the blood. Calcitonin inhibits osteoclasts. Osteoblasts remove Ca²⁺ from the blood to make bone (*blue arrows*).

**2** In the kidneys, PTH increases Ca²⁺ reabsorption from the urine.

**3** In the kidneys, PTH also promotes the formation of active vitamin D (*green arrows*), which increases Ca²⁺ absorption from the small intestine.

**PROCESS Figure 6.9  Calcium Homeostasis**

## TABLE 6.1  Named Bones in the Adult Human Skeleton

| Bones | | | Number | Bones | Number |
|---|---|---|---|---|---|
| **Axial Skeleton** | | | | ***Thoracic Cage*** | |
| ***Skull*** | | | | Ribs | 24 |
| Braincase | | | | Sternum (3 parts, sometimes considered 3 bones) | 1 |
| | Paired | Parietal | 2 | TOTAL THORACIC CAGE | 25 |
| | | Temporal | 2 | TOTAL AXIAL SKELETON | 80 |
| | Unpaired | Frontal | 1 | | |
| | | Occipital | 1 | **Appendicular Skeleton** | |
| | | Sphenoid | 1 | ***Pectoral Girdle*** | |
| | | Ethmoid | 1 | Scapula | 2 |
| Face | | | | Clavicle | 2 |
| | Paired | Maxilla | 2 | ***Upper Limb*** | |
| | | Zygomatic | 2 | Humerus | 2 |
| | | Palatine | 2 | Ulna | 2 |
| | | Nasal | 2 | Radius | 2 |
| | | Lacrimal | 2 | Carpal bones | 16 |
| | | Inferior nasal concha | 2 | Metacarpal bones | 10 |
| | Unpaired | Mandible | 1 | Phalanges | 28 |
| | | Vomer | 1 | TOTAL GIRDLE AND UPPER LIMB | 64 |
| | | TOTAL SKULL | 22 | | |
| ***Auditory Ossicles*** | Malleus | | 2 | ***Pelvic Girdle*** | |
| | Incus | | 2 | Coxal bone | 2 |
| | Stapes | | 2 | ***Lower Limb*** | |
| | TOTAL | | 6 | Femur | 2 |
| ***Hyoid*** | | | 1 | Tibia | 2 |
| ***Vertebral Column*** | | | | Fibula | 2 |
| Cervical vertebrae | | | 7 | Patella | 2 |
| Thoracic vertebrae | | | 12 | Tarsal bones | 14 |
| Lumbar vertebrae | | | 5 | Metatarsal bones | 10 |
| Sacrum | | | 1 | Phalanges | 28 |
| Coccyx | | | 1 | TOTAL GIRDLE AND LOWER LIMB | 62 |
| | | TOTAL VERTEBRAL COLUMN | 26 | TOTAL APPENDICULAR SKELETON | 126 |
| | | | | TOTAL BONES | 206 |

system function. Calcium ($Ca^{2+}$) moves into bone as osteoblasts build new bone and out of bone as osteoclasts break down bone (figure 6.9). When osteoblast and osteoclast activity is balanced, the movements of calcium into and out of a bone are equal.

When blood calcium levels are too low, osteoclast activity increases, osteoclasts release calcium from bone into the blood, and blood calcium levels increase. Conversely, if blood calcium levels are too high, osteoclast activity decreases, osteoblasts remove calcium from the blood to produce new bone, and blood calcium levels decrease.

Calcium homeostasis is maintained by two hormones. **Parathyroid hormone (PTH),** secreted from the parathyroid glands when blood calcium levels are too low, stimulates increased bone breakdown and increased blood calcium levels by indirectly stimulating osteoclast activity. PTH also increases calcium uptake from the urine in the kidney. Additionally, PTH stimulates the kidneys to form active vitamin D, which increases calcium absorption from the small intestine. Decreasing blood calcium levels stimulate PTH secretion.

**Calcitonin** (kal-si-tō′nin), secreted from the thyroid gland when blood calcium levels are too high, decreases osteoclast activity and thus decreases blood calcium levels. Increasing blood calcium levels stimulate calcitonin secretion. PTH and calcitonin are described more fully in chapter 10.

## GENERAL CONSIDERATIONS OF BONE ANATOMY

It is traditional to list 206 bones in the average adult skeleton (table 6.1), although the actual number varies from person to person and decreases with age as some bones fuse.

Anatomists use several common terms to describe the features of bones (table 6.2). For example, a hole in a bone is called a **foramen** (fō-rā′men; pl. foramina, fō-rā′min-ă; *foro*, to pierce). A foramen usually exists in a bone because some structure, such as a nerve or blood vessel, passes through the bone at that point. If the hole is elongated into a tunnel-like passage through the bone, it is

Skeletal

## TABLE 6.2    Anatomical Terms for Features of Bones

| Term | Description |
|---|---|
| **Major Features** | |
| Body, shaft | Main portion |
| Head | Enlarged (often rounded) end |
| Neck | Constricted area between head and body |
| Condyle | Smooth, rounded articular surface |
| Facet | Small, flattened articular surface |
| Crest | Prominent ridge |
| Process | Prominent projection |
| Tubercle, or tuberosity | Knob or enlargement |
| Trochanter | Large tuberosity found only on proximal femur |
| Epicondyle | Enlargement near or above a condyle |
| **Openings or Depressions** | |
| Foramen | Hole |
| Canal, meatus | Tunnel |
| Fissure | Cleft |
| Sinus | Cavity |
| Fossa | Depression |

called a **canal** or a **meatus** (mē-ā′tus; a passage). A depression in a bone is called a **fossa** (fos′ă). A lump on a bone is called a **tubercle** (too′ber-kl; a knob) or a **tuberosity** (too′ber-os′i-tē), and a projection from a bone is called a **process.** Most tubercles and processes are sites of muscle attachment on the bone. Increased muscle pull, as occurs when a person lifts weights to build up muscle mass, can increase the size of some tubercles. The smooth, rounded end of a bone, where it forms a joint with another bone, is called a **condyle** (kon′dīl; knuckle).

The bones of the skeleton are divided into axial and appendicular portions (figure 6.10).

# AXIAL SKELETON

The axial skeleton is composed of the skull, the vertebral column, and the thoracic cage.

## Skull

The 22 bones of the skull are divided into those of the braincase and those of the face (see table 6.1). The **braincase,** which encloses the cranial cavity, consists of 8 bones that immediately surround and protect the brain; 14 **facial bones** form the structure of the face. Thirteen of the facial bones are rather solidly connected to form the bulk of the face. The mandible, however, forms a freely movable joint with the rest of the skull. There are also three auditory ossicles (os′i-klz) in each middle ear (six total).

Many students studying anatomy never see the individual bones of the skull. Even if they do, it makes more sense from a functional, or clinical, perspective to study most of the bones as they appear together in the intact skull because many of the anatomical features of the skull cannot be fully appreciated by examining the separate bones. For example, several ridges on the skull cross more than one bone, and several foramina are located between bones rather than within a single bone. For these reasons, it is more relevant to think of the skull, excluding the mandible, as a single unit. The major features of the intact skull are therefore described from four views.

### Lateral View

The **parietal bones** (pă-rī′ĕ-tăl; wall) and **temporal** (tem′pō-răl) **bones** form a large portion of the side of the head (figure 6.11). (The word *temporal* refers to time, and the temporal bone is so named because the hairs of the temples turn white, indicating the passage of time.) These two bones join each other on the side of the head at the **squamous** (skwā′mŭs) **suture.** A suture is a joint uniting bones of the skull. Anteriorly, the parietal bone is joined to the **frontal** (forehead) **bone** by the **coronal** (kōr′ō-năl; *corona,* crown) **suture,** and posteriorly it is joined to the **occipital** (oksip′i-tăl; back of the head) bone by the **lambdoid** (lam′doyd) **suture.** A prominent feature of the temporal bone is a large opening, the **external auditory canal,** a canal that enables sound waves to reach the eardrum. The **mastoid** (mas′toyd) **process** of the temporal bone can be seen and felt as a prominent lump just posterior to the ear. Important neck muscles involved in rotation of the head attach to the mastoid process.

Part of the **sphenoid** (sfē′noyd) **bone** can be seen immediately anterior to the temporal bone. Although it appears to be two small, paired bones on the sides of the skull, the sphenoid bone is actually a single bone that extends completely across the skull. It resembles a butterfly, with its body in the center of the skull and its wings extending to the sides of the skull. Anterior to the sphenoid bone is the **zygomatic** (zī-gō-mat′ik) **bone,** or cheekbone, which can be easily felt. The **zygomatic arch,** which consists of joined processes of the temporal and zygomatic bones, forms a bridge across the side of the face and provides a major attachment site for a muscle moving the mandible.

The **maxilla** (mak-sil′ă; jawbone) forms the upper jaw, and the **mandible** (man′di-bl; jaw) forms the lower jaw. The maxilla articulates by sutures to the temporal bone. The maxilla contains the superior set of teeth, and the mandible contains the inferior teeth.

### Frontal View

The major structures seen from the frontal view are the frontal bone, the zygomatic bones, the maxillae, and the mandible (figure 6.12a). The teeth are very prominent in this view. Many bones of the face can be easily felt through the skin (figure 6.12b).

From this view, the most prominent openings into the skull are the **orbits** (ōr′bitz; eye sockets) and the **nasal cavity.** The orbits are cone-shaped fossae, so named because the eyes rotate within them. The bones of the orbits provide both protection for the eyes and attachment points for the muscles that move the eyes.

**Axial Skeleton**                    **Appendicular Skeleton**                    **Axial Skeleton**

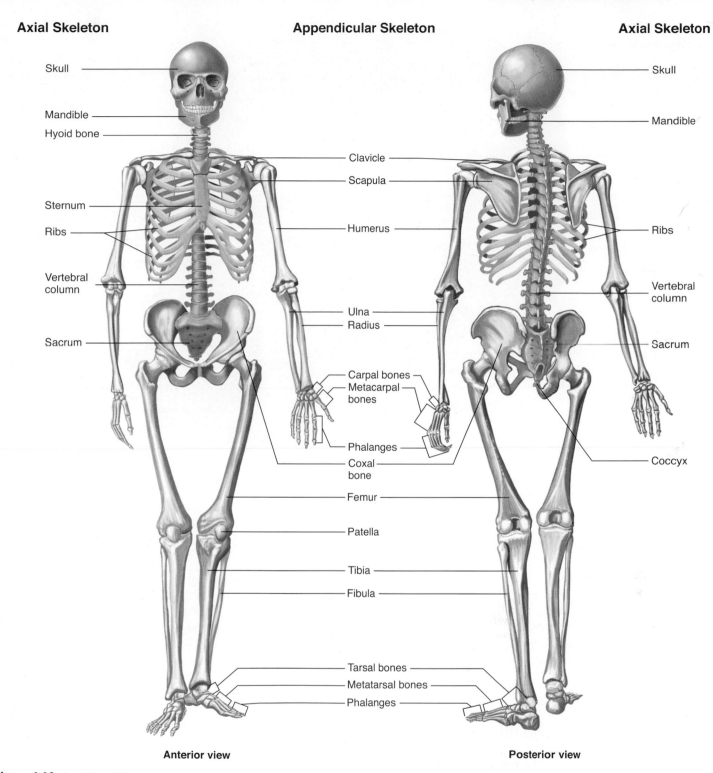

Skull

Mandible
Hyoid bone

Sternum

Ribs

Vertebral
column

Sacrum

Clavicle
Scapula

Humerus

Ulna
Radius

Carpal bones
Metacarpal
bones

Phalanges
Coxal
bone

Femur

Patella

Tibia

Fibula

Tarsal bones
Metatarsal bones
Phalanges

Skull

Mandible

Ribs

Vertebral
column

Sacrum

Coccyx

**Anterior view**                                    **Posterior view**

**Figure 6.10**  Complete Skeleton

Bones of the axial skeleton are listed in the far left- and right-hand columns; bones of the appendicular skeleton are listed in the center. (The skeleton is not shown in the anatomical position.)

The orbit is a good example of why it is valuable to study the skull as an intact structure. No fewer than seven bones come together to form the orbit, and for the most part, the contribution of each bone to the orbit cannot be appreciated when the bones are examined individually.

Each orbit has several openings through which structures communicate with other cavities (figure 6.12a). The largest of these are the **superior** and **inferior orbital fissures.** They provide openings through which nerves and blood vessels communicate with the orbit or pass to the face. The optic nerve, for the sense of

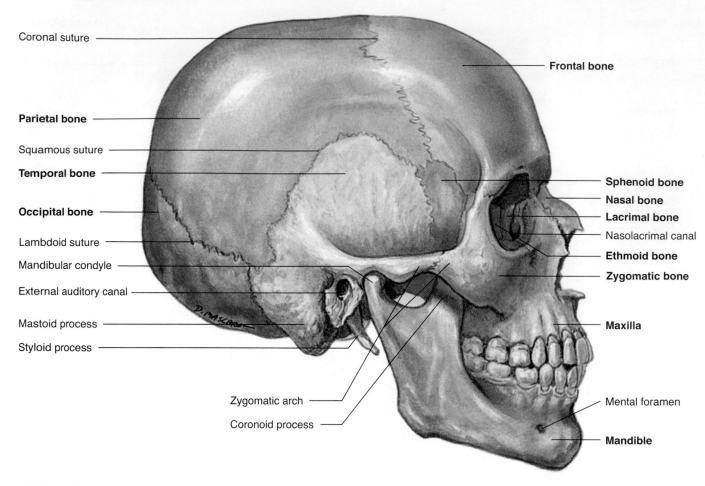

Coronal suture

Frontal bone

Parietal bone

Squamous suture

Temporal bone

Sphenoid bone

Nasal bone

Lacrimal bone

Occipital bone

Nasolacrimal canal

Ethmoid bone

Lambdoid suture

Zygomatic bone

Mandibular condyle

External auditory canal

Mastoid process

Maxilla

Styloid process

Zygomatic arch

Mental foramen

Coronoid process

Mandible

**Figure 6.11  Skull as Seen from the Right Side**                    **Lateral view**

vision, passes from the eye through the **optic foramen** and enters the cranial cavity. The **nasolacrimal** (nā-zō-lak′ri-măl; *nasus,* nose + *lacrima,* tear) **canal** (see figure 6.11) passes from the orbit into the nasal cavity. It contains a duct that carries tears from the eyes to the nasal cavity. A small **lacrimal** (lak′ri-măl) **bone** can be seen in the orbit just above the opening of this canal.

*Predict 3*

*Why does your nose run when you cry?*

The nasal cavity is divided into right and left halves by a **nasal septum** (sep′tŭm; wall) (figure 6.12a). The bony part of the nasal septum consists primarily of the **vomer** (vō′mer) inferiorly and the **perpendicular plate** of the **ethmoid** (eth′moyd; sieve-shaped) **bone** superiorly. The anterior part of the nasal septum is formed by cartilage.

The external part of the nose is formed mostly of cartilage. The bridge of the nose is formed by the **nasal bones.**

Each of the lateral walls of the nasal cavity has three bony shelves, called the **nasal conchae** (kon′kē; resembling a conch shell). The inferior nasal concha is a separate bone, and the middle and superior conchae are projections from the ethmoid bone. The conchae increase the surface area in the nasal cavity.

The increased surface area of the overlying epithelium facilitates moistening and warming of the air inhaled through the nose (see chapter 15).

Several of the bones associated with the nasal cavity have large cavities within them, called the **paranasal** (par-ă-nā′săl; *para,* alongside) **sinuses** (figure 6.13), which open into the nasal cavity. The sinuses decrease the weight of the skull and act as reso-nating chambers during voice production. Compare a normal voice with the voice of a person who has a cold and whose sinuses are "stopped up." The sinuses are named for the bones where they are located and include the frontal, maxillary, ethmoidal, and sphenoidal sinuses.

The skull has additional sinuses, called the **mastoid air cells,** which are located inside the mastoid processes of the temporal bone. These air cells open into the middle ear instead of into the nasal cavity. An auditory tube connects the middle ear to the naso-pharynx (upper part of throat).

## Interior of the Cranial Cavity

When the floor of the cranial cavity is viewed from above with the roof cut away (figure 6.14), it can be divided roughly into three cranial fossae (anterior, middle, and posterior), which are formed

**Figure 6.12  Skull and Face**

(a) Frontal view of the skull. (b) Bony landmarks of the face.

as the developing skull conforms to the shape of the brain. The bones forming the floor of the cranial cavity, from anterior to posterior, are the frontal, ethmoid, sphenoid, temporal, and occipital bones. Several foramina can be seen in the floor of the middle fossa. These allow nerves and blood vessels to pass through the skull. For example, the foramen rotundum and foramen ovale transmit important nerves to the face. A major artery to the meninges (the membranes around the brain) passes through the foramen spinosum. The internal carotid artery passes through the carotid canal, and the internal jugular vein passes through the jugular foramen

(see chapter 13). The large **foramen magnum,** through which the spinal cord joins the brain, is located in the posterior fossa. The central region of the sphenoid bone is modified into a structure resembling a saddle, the **sella turcica** (sel′ă tŭr′sĭ-kă; Turkish saddle), which contains the pituitary gland.

### Base of Skull Viewed from Below

Many of the same foramina that are visible in the interior of the skull can also be seen in the base of the skull, when viewed from

Skeletal

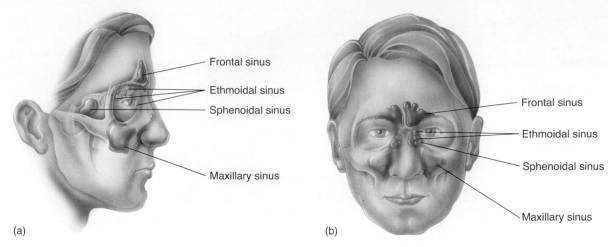

Frontal sinus
Ethmoidal sinus
Sphenoidal sinus

Maxillary sinus

(a)

Frontal sinus
Ethmoidal sinus
Sphenoidal sinus

Maxillary sinus

(b)

**Figure 6.13  Paranasal Sinuses**

(*a*) Lateral view. (*b*) Anterior view.

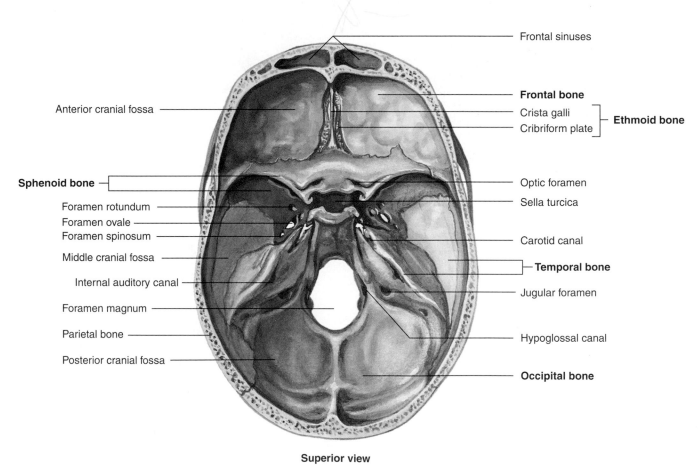

Frontal sinuses

**Frontal bone**
Crista galli
Cribriform plate    } **Ethmoid bone**

Anterior cranial fossa

**Sphenoid bone**

Foramen rotundum
Foramen ovale
Foramen spinosum
Middle cranial fossa
Internal auditory canal
Foramen magnum
Parietal bone
Posterior cranial fossa

Optic foramen
Sella turcica

Carotid canal

**Temporal bone**

Jugular foramen

Hypoglossal canal

**Occipital bone**

**Superior view**

**Figure 6.14  Floor of the Cranial Cavity**

The roof of the skull has been removed, and the floor is viewed from above.

below, with the mandible removed (figure 6.15). Other specialized structures, such as processes for muscle attachments, can also be seen. The foramen magnum is located in the occipital bone near the center of the skull base. **Occipital condyles** (ok-sip′i-tăl kon′dĭlz), the smooth points of articulation between the skull and the vertebral column, are located beside the foramen magnum.

Two long, pointed **styloid** (stī′loyd; stylus or pen-shaped) **processes** project from the inferior surface of the temporal bone. The muscles involved in moving the tongue, the hyoid bone, and the pharynx (throat) originate from this process. The **mandibular fossa,** where the mandible articulates with the temporal bone, is anterior to the mastoid process.

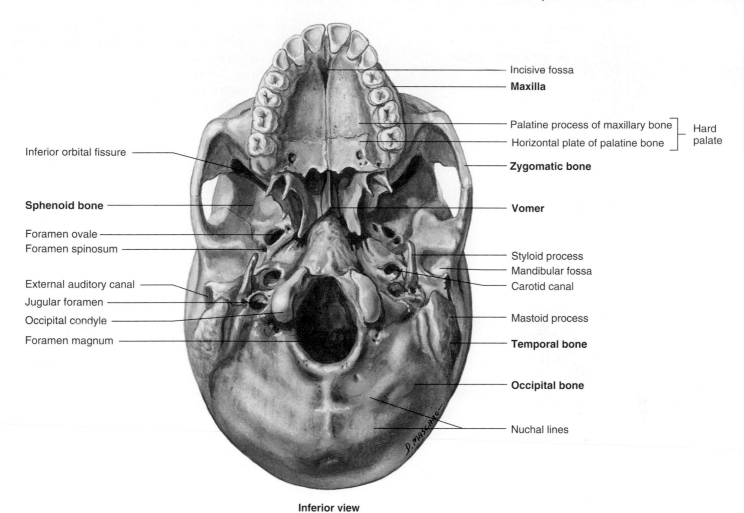

Incisive fossa

**Maxilla**

Palatine process of maxillary bone ⎤
Horizontal plate of palatine bone ⎦ Hard palate

Inferior orbital fissure

**Zygomatic bone**

**Sphenoid bone**

**Vomer**

Foramen ovale

Foramen spinosum

Styloid process

Mandibular fossa

Carotid canal

External auditory canal

Jugular foramen

Occipital condyle

Mastoid process

Foramen magnum

**Temporal bone**

**Occipital bone**

Nuchal lines

**Inferior view**

**Figure 6.15** **Base of the Skull as Viewed from Below**
The mandible has been removed.

The **hard palate** (pal′ăt) forms the floor of the nasal cavity and the roof of the mouth. The anterior two-thirds of the hard palate is formed by the maxillae, the posterior one-third by the **palatine** (pal′ă-tīn) **bones.** The connective tissue and muscles that make up the **soft palate** extend posteriorly from the hard, or bony, palate. The hard and soft palates separate the nasal cavity and nasopharynx from the mouth, enabling us to chew and breathe at the same time.

### Hyoid Bone

The **hyoid bone** (figure 6.16) is an unpaired, U-shaped bone. It is not part of the skull (see table 6.1) and has no direct bony attachment to the skull. Muscles and ligaments attach it to the skull. The hyoid bone provides an attachment for some tongue muscles, and it is an attachment point for important neck muscles that elevate the larynx (voicebox) during speech or swallowing.

## Vertebral Column

The **vertebral column,** or backbone, is the central axis of the skeleton, extending from the base of the skull to slightly past the end

Body

**Anterior view**

Body

**Lateral view**
(from the left side)

**Figure 6.16** **Hyoid Bone**

Skeletal

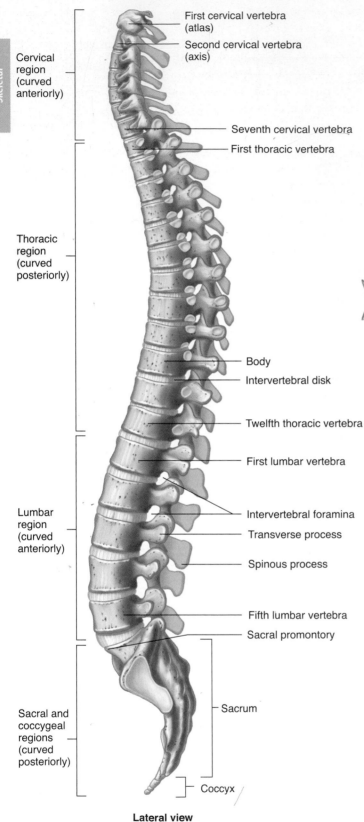

Cervical region (curved anteriorly)

First cervical vertebra (atlas)

Second cervical vertebra (axis)

Seventh cervical vertebra

First thoracic vertebra

Thoracic region (curved posteriorly)

Body

Intervertebral disk

Twelfth thoracic vertebra

First lumbar vertebra

Lumbar region (curved anteriorly)

Intervertebral foramina

Transverse process

Spinous process

Fifth lumbar vertebra

Sacral promontory

Sacral and coccygeal regions (curved posteriorly)

Sacrum

Coccyx

**Lateral view**

**Figure 6.17** Vertebral Column

Complete column viewed from the left side.

of the pelvis. In adults, it usually consists of 26 individual bones, grouped into five regions (figure 6.17; see table 6.1): 7 **cervical** (ser′vĭ-kal; neck) **vertebrae** (ver′tĕ-brē; *verto,* to turn), 12 **thoracic** (thō-ras′ik) **vertebrae,** 5 **lumbar** (lŭm′bar) **vertebrae,** 1 **sacral** (sā′krăl) **bone,** and 1 **coccygeal** (kok-sij′ē-ăl) **bone.** The adult sacral and coceygeal bones fuse from 5 and 3–4 individual bones, respectively. For convenience, each of the five regions is identified by a letter, and the vertebrae within each region are numbered: C1–C7, T1–T12, L1–L5, S, and CO. You can remember the number of vertebrae in each region by remembering meal times: 7, 12, and 5.

The adult vertebral column has four major curvatures. The cervical region curves anteriorly, the thoracic region curves posteriorly, the lumbar region curves anteriorly, and the sacral and coccygeal regions together curve posteriorly.

Abnormal vertebral curvatures are not uncommon. **Kyphosis** (kī-fō′sis) is an abnormal posterior curvature of the spine, mostly in the upper thoracic region, resulting in a hunchback condition. **Lordosis** (lōr-dō′sis; curving forward) is an abnormal anterior curvature of the spine, mainly in the lumbar region, resulting in a swayback condition. **Scoliosis** (skō-lē-ō′sis) is an abnormal lateral curvature of the spine.

The vertebral column performs the following five major functions: (1) supports the weight of the head and trunk; (2) protects the spinal cord; (3) allows spinal nerves to exit the spinal cord; (4) provides a site for muscle attachment; and (5) permits movement of the head and trunk.

## General Plan of the Vertebrae

Each vertebra consists of a body, an arch, and various processes (figure 6.18). The weight-bearing portion of each vertebra is the **body.** The vertebral bodies are separated by **intervertebral disks** (see figure 6.17), which are formed by fibrocartilage. The **vertebral arch** surrounds a large opening called the **vertebral foramen.** The vertebral foramina of all the vertebrae form the **vertebral canal,** where the spinal cord is located. The vertebral canal protects the spinal cord from injury. Each vertebral arch consists of two **pedicles** (ped′ĭ-klz), which extend from the body to the

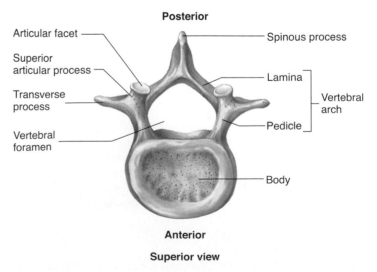

Posterior

Articular facet

Superior articular process

Transverse process

Vertebral foramen

Spinous process

Lamina

Vertebral arch

Pedicle

Body

Anterior

**Superior view**

**Figure 6.18** Vertebra

transverse process of each vertebra, and two **laminae** (lam′i-nē; thin plates), which extend from the transverse processes to the spinous process. A **transverse process** extends laterally from each side of the arch, between the pedicle and lamina, and a single **spinous process** projects dorsally from where the two laminae meet. The spinous processes can be seen and felt as a series of lumps down the midline of the back (see figure 6.24b). The transverse and spinous processes provide attachment sites for the muscles that move the vertebral column. Spinal nerves exit the spinal cord through the **intervertebral foramina,** which are formed by notches in the pedicles of adjacent vertebrae (see figure 6.17). Each vertebra has a superior and inferior **articular process** where the vertebrae articulate with each other. Each articular process has a smooth "little face" called an **articular facet** (fas′et).

## Regional Differences in Vertebrae

The **cervical vertebrae** (figure 6.19a–c) have very small bodies, except for the atlas, which has no body. Because the cervical vertebrae are relatively delicate and have small bodies, dislocations and fractures are more common in this area than in other regions of the vertebral column. Each of the transverse processes has a transverse foramen through which the vertebral arteries pass toward the brain. Several of the cervical vertebrae also have partly split spinous processes. The first cervical vertebra (figure 6.19a) is called the **atlas** because it holds up the head, as Atlas in classical mythology held up the world. Movement between the atlas and the occipital bone is responsible for a "yes" motion of the head. It also allows a slight tilting of the head from side to side. The second cervical vertebra (figure 6.19b) is called the **axis** because a considerable amount of rotation occurs at this vertebra, as in shaking the head "no." This rotation occurs around a process called the **dens** (denz), which protrudes superiorly from the axis.

The **thoracic vertebrae** (figure 6.19d) possess long, thin spinous processes that are directed inferiorly. The thoracic vertebrae also have extra articular facets on their lateral surfaces that articulate with the ribs.

Transverse process
Posterior arch
Vertebral foramen
Superior articular facet (articulates with occipital condyle)
Facet for dens
Transverse foramen
Anterior arch

**Atlas (first cervical vertebra), superior view**

(a)

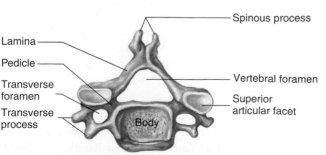

Dens

Body

Dens (articulates with atlas)

**Axis (second cervical vertebra), superior view**

**Lateral view**

(b)

Lamina
Pedicle
Transverse foramen
Transverse process
Body
Spinous process
Vertebral foramen
Superior articular facet

**Cervical vertebra, superior view**

(c)

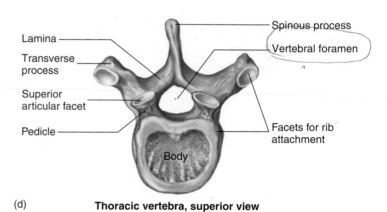

Lamina
Transverse process
Superior articular facet
Pedicle
Body
Spinous process
Vertebral foramen
Facets for rib attachment

(d)    **Thoracic vertebra, superior view**

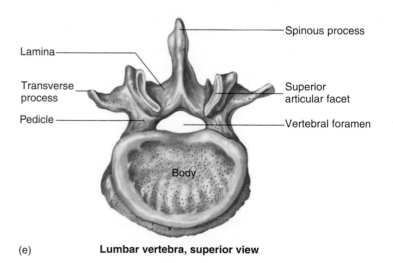

Lamina
Transverse process
Pedicle
Body
Spinous process
Superior articular facet
Vertebral foramen

(e)    **Lumbar vertebra, superior view**

**Figure 6.19  Regional Differences in Vertebrae**
The posterior portion lies at the top of each illustration.

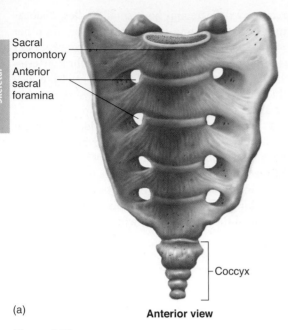

Sacral promontory

Anterior sacral foramina

Coccyx

(a)        **Anterior view**

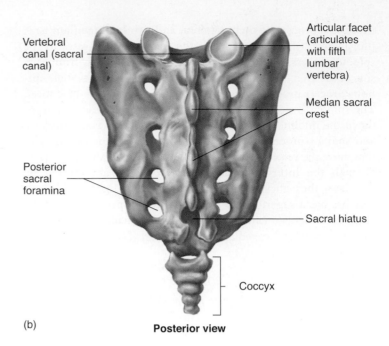

Vertebral canal (sacral canal)

Articular facet (articulates with fifth lumbar vertebra)

Median sacral crest

Posterior sacral foramina

Sacral hiatus

Coccyx

(b)        **Posterior view**

**Figure 6.20** Sacrum

The **lumbar vertebrae** (figure 6.19e) have large, thick bodies and heavy, rectangular transverse and spinous processes. Because the lumbar vertebrae have massive bodies and carry a large amount of weight, ruptured intervertebral disks are more common in this area than in other regions of the column. The superior articular facets of the lumbar vertebrae face medially, whereas the inferior articular facets face laterally. This arrangement tends to "lock" adjacent lumbar vertebrae together, giving the lumbar part of the vertebral column more strength. The articular facets in other regions of the vertebral column have a more "open" position, allowing for more rotational movement but less stability than in the lumbar region.

The five sacral vertebrae are fused into a single bone called the **sacrum** (figure 6.20). The spinous processes of the first four sacral vertebrae form the **median sacral crest.** The spinous process of the fifth vertebra does not form, leaving a **sacral hiatus** (hī-ā-tŭs) at the inferior end of the sacrum, which is often the site of "caudal" anesthetic injections given just before childbirth. The anterior edge of the body of the first sacral vertebra bulges to form the **sacral promontory** (prom'on-tō-rē) (see figure 6.17), a landmark that can be felt during a vaginal examination. It is used as a reference point to determine if the pelvic openings are large enough to allow for normal vaginal delivery of a baby.

The **coccyx** (kok'siks), or tailbone, usually consists of four more-or-less fused vertebrae. The vertebrae of the coccyx do not have the typical structure of most other vertebrae. They consist of extremely reduced vertebral bodies, without the foramina or processes, usually fused into a single bone. The coccyx is easily broken when a person falls by sitting down hard on a solid surface or in women during childbirth.

## Thoracic Cage

The **thoracic cage,** or *rib cage,* protects the vital organs within the thorax and prevents the collapse of the thorax during respiration.

It consists of the thoracic vertebrae, the ribs with their associated cartilages, and the sternum (figure 6.21).

### *Ribs and Costal Cartilages*

The 12 pairs of ribs can be divided into true ribs and false ribs. The **true ribs,** ribs 1–7, attach directly to the sternum by means of costal cartilages. The **false ribs,** ribs 8–12, do not attach directly to the sternum. Ribs 8–10 attach to the sternum by a common cartilage; ribs 11 and 12 do not attach at all to the sternum and are called **floating ribs.**

### A CASE IN POINT
#### Rib Fractures

Han D. Mann's ladder fell as he was working on his roof, and he landed on the ladder with his chest. Three ribs were fractured on his right side. It was difficult for Han to cough, laugh, or even breathe without severe pain in the right side of his chest. The middle ribs are those most commonly fractured, and the portion of each rib that forms the lateral wall of the thorax is the weakest and most commonly broken. The pain from rib fractures occurs because the broken ends move during respiration and other chest movements, stimulating pain receptors. Broken rib ends can damage internal organs, such as the lungs, spleen, liver, and diaphragm. Fractured ribs are not often dislocated, but dislocated ribs may have to be set for proper healing to occur. Binding the chest to limit movement can facilitate healing and lessen pain.

### *Sternum*

The **sternum** (ster'nŭm), or breastbone (figure 6.21), is divided into three parts: the **manubrium** (mă-nū'brē-ŭm; handle), the **body,** and the **xiphoid** (zif'oyd, zī'foyd; sword) **process.** The sternum

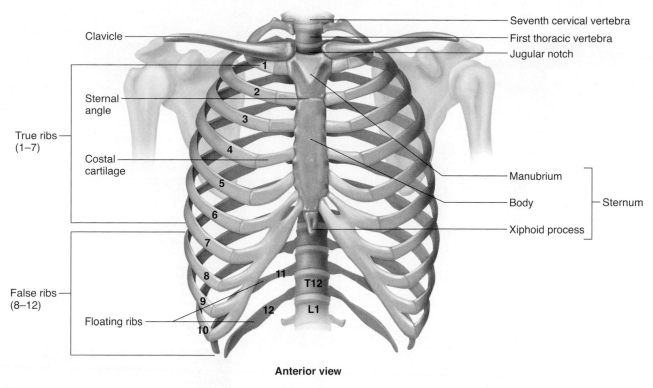

**Anterior view**

**Figure 6.21** Thoracic Cage

resembles a sword, with the manubrium forming the handle, the body forming the blade, and the xiphoid process forming the tip. At the superior end of the sternum, a depression called the **jugular notch** is located between the ends of the clavicles where they articulate with the sternum. A slight elevation, called the **sternal angle,** can be felt at the junction of the manubrium and the body of the sternum. This junction is an important landmark because it identifies the location of the second rib. This identification allows the ribs to be counted; for example, it can help a health professional locate the apex of the heart, which is between the fifth and sixth ribs.

The xiphoid process is another important landmark of the sternum. During cardiopulmonary resuscitation (CPR), it is very important to place the hands over the body of the sternum rather than over the xiphoid process. Pressure applied to the xiphoid process can drive it into an underlying abdominal organ, such as the liver, causing internal bleeding.

# APPENDICULAR SKELETON

The **appendicular** (ap′en-dik′ū-lăr; appendage) skeleton consists of the bones of the upper and lower limbs, as well as the girdles, which attach the limbs to the axial skeleton.

## Pectoral Girdle

The **pectoral** (pek′tō-răl) **girdle,** or *shoulder girdle* (figure 6.22), consists of four bones, two scapulae and two clavicles, which attach the upper limb to the body. The **scapula** (skap′ū-lă), or *shoulder blade,* is a flat, triangular bone with three large

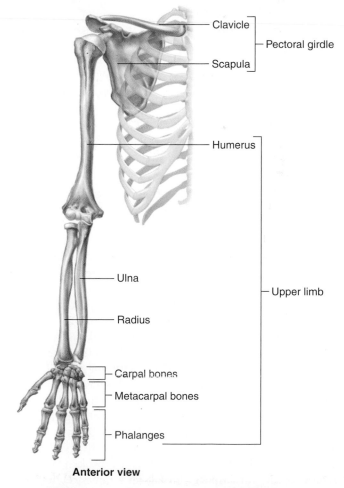

**Anterior view**

**Figure 6.22** Bones of the Pectoral Girdle and Right Upper Limb

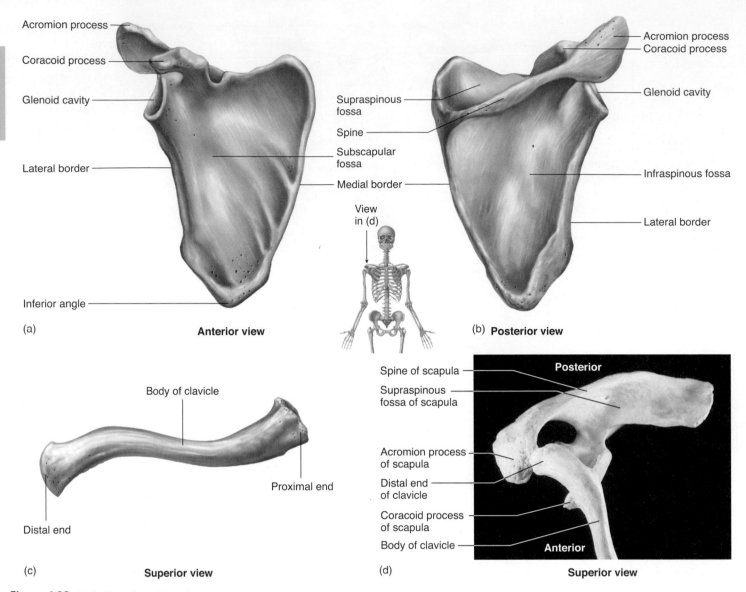

**Figure 6.23** **Right Scapula and Clavicle**

(a) Right scapula, anterior view. (b) Right scapula, posterior view. (c) Right clavicle, superior view. (d) Photograph of the right scapula and clavicle from a superior view, showing the relationship between the distal end of the clavicle and the acromion process of the scapula.

fossae where muscles extending to the arm are attached (figures 6.23 and 6.24; see figure 6.10). A fourth fossa, the **glenoid** (glen′oyd) **cavity,** is where the head of the humerus connects to the scapula. A ridge, called the **spine,** runs across the posterior surface of the scapula. A projection, called the **acromion** (ă-krō′mē-on; *akron,* tip + *omos,* shoulder) **process,** extends from the scapular spine to form the point of the shoulder. The **clavicle** (klav′i-kl), or *collarbone,* articulates with the scapula at the acromion process. The proximal end of the clavicle is attached to the sternum, providing the only bony attachment of the scapula to the remainder of the skeleton. The clavicle is the first bone to begin ossification in the fetus. This relatively brittle bone may be fractured in the newborn during delivery. The bone remains slender in children and may be broken as a child attempts to take the impact of a fall on an outstretched hand. The clavicle is thicker in adults and is less vulnerable to fracture.

Even though it is the first bone to begin ossification, it is the last to complete ossification. The **coracoid** (kōr′ă-koyd) **process** curves below the clavicle and provides for the attachment of arm and chest muscles.

## Upper Limb

The upper limb consists of the bones of the arm, forearm, wrist, and hand (see figure 6.22).

### Arm

The **arm** is the region between the shoulder and the elbow; it contains the **humerus** (hū′mer-ŭs; shoulder) (figure 6.25). The proximal end of the humerus has a smooth, rounded **head,** which attaches the humerus to the scapula at the glenoid cavity. Around the edge of the humeral head is the anatomical neck. When the

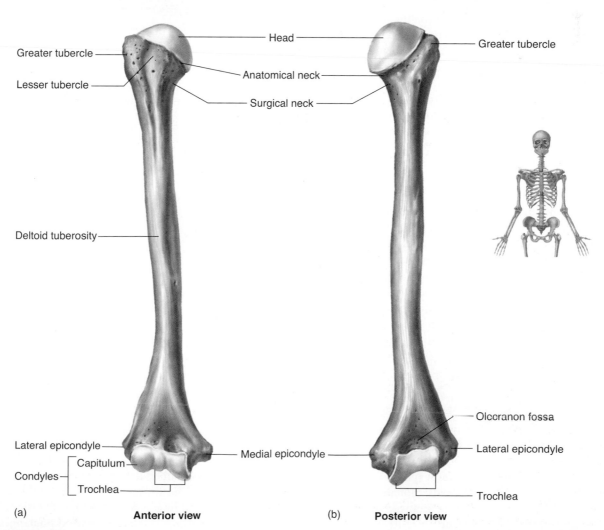

**Figure 6.24  Surface Anatomy of the Pectoral Girdle and Thoracic Cage**

(*a*) Bones of the pectoral girdle and the anterior thorax. (*b*) Bones of the scapula and the posterior vertebral column.

**Figure 6.25  Right Humerus**

joint needs to be surgically replaced, this neck is not easily accessible. A more accessible site for surgical removal is at the surgical neck, located at the proximal end of the humeral shaft. Lateral to the head are two tubercles, a **greater tubercle** and a **lesser tubercle.** Muscles originating on the scapula attach to the greater and lesser tubercles and hold the humerus to the scapula. Approximately one-third of the way down the shaft of the humerus, on the lateral surface, is the **deltoid tuberosity,** where the deltoid muscle attaches. The size of the deltoid tuberosity can increase as the result of frequent and powerful pulls from the deltoid muscle. For example, in bodybuilders, the deltoid muscle and the deltoid tuberosity enlarge substantially. Anthropologists, examining ancient human remains, can use the presence of enlarged deltoid tuberosities as evidence that a person was engaged in lifting heavy objects during life. If the humerus of a person exhibits an unusually large deltoid tuberosity for her age, it may indicate, in some societies, that she was a slave and was required to lift heavy loads. The distal end of the humerus is modified into specialized condyles that connect the humerus to the forearm bones. **Epicondyles** (ep′i-kon′dīlz; *epi,* upon) on the distal end of the humerus, just lateral to the condyles, provide attachment sites for forearm muscles.

## Forearm

The **forearm** has two bones: the **ulna** (ŭl′nă) on the medial (little finger) side of the forearm and the **radius** on the lateral (thumb) side (figure 6.26). The proximal end of the ulna forms a **trochlear notch** that fits tightly over the end of the humerus, forming most of the elbow joint. Just proximal to the trochlear notch is an extension of the ulna, called the **olecranon** (ō-lek′ră-non; elbow) **process,** which can be felt as the point of the elbow (see figure 6.28). Just distal to the trochlear notch is a **coronoid** (kōr′ŏ-noyd) **process,** which helps complete the "grip" of the ulna on the distal end of the humerus. The distal end of the ulna forms a head, which articulates with the bones of the wrist, and a **styloid process** is located on its medial side. The ulnar head can be seen as a prominent lump on the posterior ulnar side of the wrist. The proximal end of the radius has a head by which the radius articulates with both the humerus and the ulna. The radius does not attach as firmly to the humerus as the ulna does. The radial head rotates against the humerus and ulna. Just distal to the radial head is a **radial tuberosity,** where one of the arm muscles, the biceps brachii, attaches. The distal end of the radius articulates with the wrist bones. A styloid process is located on the lateral side of the distal end of the radius. The radial and ulnar styloid processes provide attachment sites for ligaments of the wrist.

## Wrist

The **wrist** is a relatively short region between the forearm and the hand; it is composed of eight **carpal** (kar′păl; wrist) **bones** (figure 6.27). These eight bones are the scaphoid (skaf′oyd), lunate (lū′nāt), triquetrum (trī-kwē′trŭm), pisiform (pis′i-fŏrm), trapezium (tra-pē′zē-ŭm), trapezoid (trap′ě-zoyd), capitate (kap′i-tāt), and hamate (ha′māt). The carpal bones are arranged in two rows of four bones each and form a slight curvature that is concave anteriorly and convex posteriorly. A number of mnemonics have

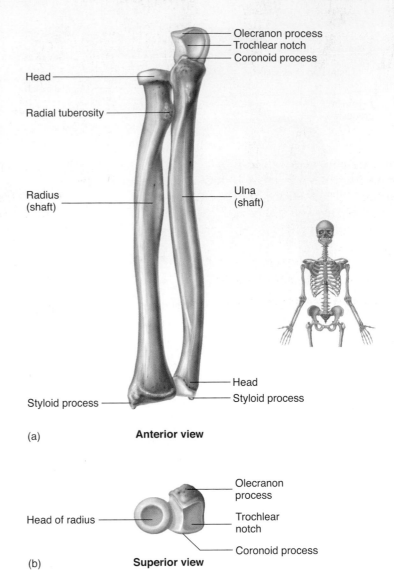

**Figure 6.26  Right Ulna and Radius**

(*a*) Anterior view of the right ulna and radius. (*b*) Proximal ends of the right ulna and radius.

been developed to help students remember the carpal bones. The following one allows students to remember them in order from lateral to medial for the proximal row (top) and from medial to lateral (by the thumb) for the distal row: So Long Top Part, Here Comes The Thumb—that is, Scaphoid, Lunate, Triquetrum, Pisiform, Hamate, Capitate, Trapezoid, and Trapezium.

## Hand

Five **metacarpal** (met′ă-kar′păl) **bones** are attached to the carpal bones and form the bony framework of the hand (figure 6.27). The metacarpal bones are aligned with the five **digits:** the thumb and fingers. They are numbered 1 to 5, from the thumb to the little finger. The ends, or heads, of the five metacarpal bones associated with the thumb and fingers form the knuckles (figure 6.28). Each finger consists of three small bones called **phalanges** (fă-lan′jēz; sing. phalanx, fā′langks), named after the Greek phalanx, a wedge

# CLINICAL IMPACT

## Carpal Tunnel Syndrome

The bones and ligaments on the anterior side of the wrist form a **carpal tunnel,** which does not have much "give." Tendons and nerves pass from the forearm through the carpal tunnel to the hand. Fluid and connective tissue can accumulate in the carpal tunnel as a result of inflammation associated with overuse or trauma. The inflammation can also cause the tendons in the carpal tunnel to enlarge. The accumulated fluid and enlarged tendons can apply pressure to a major nerve passing through the tunnel. The pressure on this nerve causes **carpal tunnel syndrome,** characterized by tingling, burning, and numbness in the hand.

Treatments for carpal tunnel syndrome vary, depending on the severity of the syndrome. Mild cases can be treated nonsurgically with either anti-inflammatory medications or stretching exercises. However, if symptoms have lasted for more than 6 months, surgery is recommended. Surgical techniques involve cutting the carpal ligament to enlarge the carpal tunnel and ease pressure on the nerve.

---

of soldiers holding their spears, tips outward, in front of them. The phalanges of each finger are called proximal, middle, and distal, according to their position in the digit. The thumb has two phalanges, proximal and distal. The digits are also numbered 1 to 5, starting from the thumb.

## Pelvic Girdle

The **pelvic girdle** is the place where the lower limbs attach to the body (figure 6.29). The right and left **coxal** (kok′sul) **bones,** or hipbones, join each other anteriorly and the **sacrum** posteriorly to form a ring of bone called the **pelvic girdle.** The **pelvis** (pel′vis; basin) includes the pelvic girdle and the coccyx (figure 6.30). The sacrum and coccyx form part of the pelvis but are also part of the axial skeleton. Each coxal bone is formed by three bones fused to one another to form a single bone (figure 6.31). The **ilium** (il′ē-ŭm) is the most superior, the **ischium** (is′kē-ŭm) is inferior and posterior, and the **pubis** (pū′bis) is inferior and anterior. An **iliac crest** can be seen along the superior margin of each ilium, and an **anterior superior iliac spine,** an important hip landmark, is located at the anterior end of the iliac crest. The coxal bones join each other anteriorly at the **pubic** (pū′bik) **symphysis** and join the sacrum posteriorly at the **sacroiliac** (sā-krō-il′ē-ak) **joints** (see figure 6.30). The **acetabulum** (as-ĕ-tab′ū-lŭm; vinegar cup) is the socket of the hip joint. The **obturator** (ob′too-rā-tŏr) **foramen** is the large hole in each coxal bone that is closed off by muscles and other structures.

**Carpal bones** (proximal row): Scaphoid, Lunate, Triquetrum, Pisiform — Radius, Ulna — **Carpal bones** (distal row): Hamate, Capitate, Trapezoid, Trapezium

**Metacarpal bones**

1  2  3  4  5

Digits

**Proximal phalanx** of thumb

**Distal phalanx** of thumb

**Proximal phalanx** of finger

**Middle phalanx** of finger

**Distal phalanx** of finger

**Posterior view**

**Figure 6.27  Bones of the Right Wrist and Hand**

Heads of metacarpal bones (knuckles)

Head of ulna

Olecranon process

Acromion process

Medial border of scapula

Olecranon process

**Figure 6.28  Surface Anatomy Showing Bones of the Pectoral Girdle and Upper Limb**

The male pelvis can be distinguished from the female pelvis because it is usually larger and more massive, but the female pelvis tends to be broader (figure 6.32; table 6.3). Both the inlet and the outlet of the female pelvis are larger than those of the male pelvis, and the subpubic angle is greater in the female (figure 6.32*a,b*). The increased size of these openings helps accommodate the fetus

during childbirth. The **pelvic inlet** is formed by the pelvic brim and the sacral promontory. The **pelvic outlet** is bounded by the ischial spines, the pubic symphysis, and the coccyx (figure 6.32*c*).

## Lower Limb

The lower limb consists of the bones of the thigh, leg, ankle, and foot (see figure 6.29).

### Thigh

The **thigh** is the region between the hip and the knee (figure 6.33*a*). It contains a single bone called the **femur.** The **head** of the femur articulates with the acetabulum of the coxal bone. At the distal end

| TABLE 6.3 | Differences Between Male and Female Pelvic Girdles |
|---|---|
| **Area** | **Description of Difference** |
| General | Female pelvis somewhat lighter in weight and wider laterally but shorter superiorly to inferiorly and less funnel-shaped; less obvious muscle attachment points in female than in male |
| Sacrum | Broader in female, with the inferior portion directed more posteriorly; the sacral promontory projects less anteriorly in female |
| Pelvic inlet | Heart-shaped in male; oval in female |
| Pelvic outlet | Broader and more shallow in female |
| Subpubic angle | Less than 90 degrees in male; 90 degrees or more in female |
| Ilium | More shallow and flared laterally in female |
| Ischial spines | Farther apart in female |
| Ischial tuberosities | Turned laterally in female and medially in male (not shown in figure 6.31) |

**Figure 6.29**   Bones of the Pelvic Girdle and Right Lower Limb

Sacrum · Coxal bone — Pelvic girdle

Femur

Patella

Tibia — Lower limb

Fibula

Tarsal bones · Metatarsal bones · Phalanges

**Anterior view**

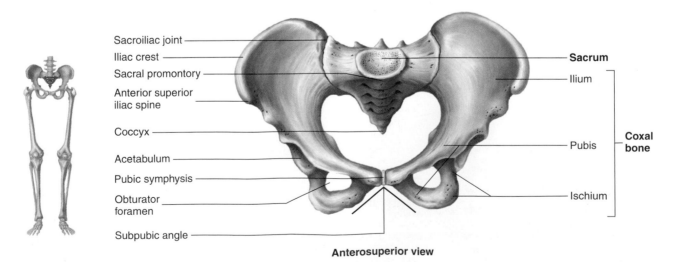

Sacroiliac joint · Iliac crest · Sacral promontory · Anterior superior iliac spine · Coccyx · Acetabulum · Pubic symphysis · Obturator foramen · Subpubic angle

Sacrum · Ilium · Pubis — Coxal bone · Ischium

**Anterosuperior view**

**Figure 6.30**   Pelvis

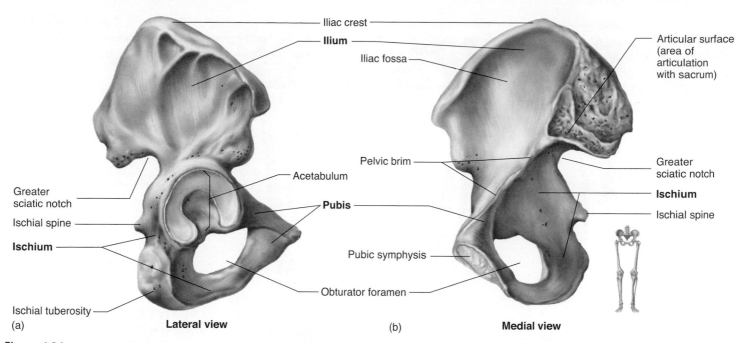

**Figure 6.31** Right Coxal Bone

**Figure 6.32** Comparison of the Male Pelvis to the Female Pelvis

(a) In a male, the pelvic inlet (*red dashed line*) and outlet (*blue dashed line*) are small, and the subpubic angle is less than 90 degrees. (b) In a female, the pelvic inlet (*red dashed line*) and outlet (*blue dashed line*) are larger, and the subpubic angle is 90 degrees or greater. (c) A midsagittal section through the pelvis shows the pelvic inlet (*red arrow* and *red dashed line*) and the pelvic outlet (*blue arrow* and *blue dashed line*).

(a)

**Anterior view**        **Posterior view**

(b)

**Anterior view**

**Figure 6.33** Bones of the Thigh

(a) Right femur. (b) Patella.

of the femur, the **condyles** articulate with the tibia. **Epicondyles,** located medial and lateral to the condyles, are points of ligament attachment. The femur can be distinguished from the humerus by its long neck, located between the head and the **trochanters** (trō′kan-terz). A "broken hip" is usually a break of the femoral neck. A broken hip is difficult to repair and often requires pinning to hold the femoral head to the shaft. A major complication can occur if the blood vessels between the femoral head and the acetabulum are damaged. If this occurs, the femoral head may degenerate from lack of nourishment. The trochanters are points of muscle attachment. The **patella** (pa-tel′ă), or kneecap (figure 6.33b), is located within the major tendon of the anterior thigh muscles and enables the tendon to bend over the knee.

## Leg

The **leg** is the region between the knee and the ankle (figure 6.34). It contains two bones, called the **tibia** (tib′ē-ă; shinbone) and the **fibula** (fib′ū-lă). The tibia is the larger of the two and is the major weight-bearing bone of the leg. The rounded condyles of the femur rest on the flat condyles on the proximal end of the tibia.

Just distal to the condyles of the tibia, on its anterior surface, is the **tibial tuberosity,** where the muscles of the anterior thigh attach. The fibula does not articulate with the femur, but its head is attached to the proximal end of the tibia. The distal ends of the tibia and fibula form a partial socket that articulates with a bone of the ankle (the talus). A prominence can be seen on each side of the ankle (figure 6.34). These are the **medial malleolus** (mal-ē′ō-lŭs) of the tibia and the **lateral malleolus** of the fibula.

## Ankle

The **ankle** consists of seven **tarsal** (tar′săl; the sole of the foot) **bones** (figure 6.35). The tarsal bones are the **talus** (tā′lŭs; ankle bone), **calcaneus** (kal-kā′nē-ŭs; heel), **cuboid** (kū′boyd), and **navicular** (nă-vik′yū-lăr), and the medial, intermediate, and lateral **cuneiforms** (kū′nē-i-fōrmz). The talus articulates with the tibia and fibula to form the ankle joint, and the calcaneus forms the heel (figure 6.36). A mnemonic for the distal row is **MILC**—that is, **M**edial, **I**ntermediate, and **L**ateral cuneiforms and the **C**uboid. A mnemonic for the proximal three bones is **N**o **T**hanks **C**ow—that is, **N**avicular, **T**alus, and **C**alcaneus.

Lateral condyle

Head

Medial condyle

Tibial tuberosity

Fibula

Tibia

Lateral malleolus

Medial malleolus

**Anterior view**

**Figure 6.34  Bones of the Leg**

The right tibia and fibula are shown.

Calcaneus

Talus

Cuboid

Navicular

Medial cuneiform

Intermediate cuneiform

Lateral cuneiform

**Tarsal bones**

**Metatarsal bones**

5

4

3   2   1

Digits

**Proximal phalanx**

**Middle phalanx**

**Distal phalanx**

Proximal phalanx of great toe

Distal phalanx of great toe

(a)

**Superior view**

Navicular

Fibula

Tibia

Talus

Cuboid

Calcaneus

Phalanges

Metatarsal bones

Tarsal bones

(b)

**Medial view**

**Figure 6.35  Bones of the Right Foot**

**Figure 6.36** Surface Anatomy Showing Bones of the Lower Limb

## Foot

The **metatarsal** (met′ă-tar′săl) **bones** and **phalanges** of the foot are arranged and numbered in a manner very similar to the metacarpal bones and phalanges of the hand (see figure 6.35). The metatarsal bones are somewhat longer than the metacarpal bones, whereas the phalanges of the foot are considerably shorter than those of the hand.

There are three primary **arches** in the foot, formed by the positions of the tarsal bones and metatarsal bones, and held in place by ligaments. Two longitudinal arches extend from the heel to the ball of the foot, and a transverse arch extends across the foot. The arches function similarly to the springs of a car, allowing the foot to give and spring back.

# JOINTS

A **joint,** or an *articulation,* is a place where two bones come together. A joint is usually considered movable, but that is not always the case. Many joints exhibit limited movement, and others are completely, or almost completely, immovable.

One method of classifying joints is a functional classification. Based on the degree of motion, a joint may be called a **synarthrosis** (sin′ar-thrō′sis; nonmovable joint), an **amphiarthrosis** (am′fi-ar-thrō′sis; slightly movable joint), or a **diarthrosis** (dī-ar-thrō′sis; freely movable joint). However, functional classification is somewhat restrictive and is not used in this text. Instead, we use a structural classification whereby joints are classified according to the type of connective tissue that binds the bones together and whether there is a fluid-filled joint capsule. The three major structural classes of joints are fibrous, cartilaginous, and synovial.

## Fibrous Joints

**Fibrous joints** consist of two bones that are united by fibrous tissue and that exhibit little or no movement. Joints in this group are further subdivided on the basis of structure as sutures, syndesmoses, or gomphoses. **Sutures** (soo′choorz) are fibrous joints between the bones of the skull (see figure 6.11). In a newborn, some parts of the sutures are quite wide and called **fontanels** (fon′tă-nelz′), or

(a)    **Lateral view**

(b)    **Superior view**

**Figure 6.37** Fetal Skull Showing Fontanels and Sutures

soft spots (figure 6.37). They allow flexibility in the skull during the birth process, as well as growth of the head after birth. **Syndesmoses** (sin′dez-mō′sēz) are fibrous joints in which the bones are separated by some distance and held together by ligaments. An example is the fibrous membrane connecting most of the distal parts of the radius and ulna. **Gomphoses** (gom-fō′sēz) consist of pegs fitted into sockets and held in place by ligaments. The joint between a tooth and its socket is a gomphosis.

## Cartilaginous Joints

**Cartilaginous joints** unite two bones by means of cartilage. Only slight movement can occur at these joints. Examples are the cartilage in the epiphyseal plates of growing long bones and the cartilages

**Figure 6.38  Structure of a Synovial Joint**

between the ribs and the sternum. The cartilage of some cartilaginous joints, where much strain is placed on the joint, may be reinforced by additional collagen fibers. This type of cartilage, called **fibrocartilage** (see chapter 4), forms joints such as the intervertebral disks.

## Synovial Joints

**Synovial** (si-nō′vē-ăl) **joints** are freely movable joints that contain fluid in a cavity surrounding the ends of articulating bones. Most joints that unite the bones of the appendicular skeleton are synovial joints, whereas many of the joints that unite the bones of the axial skeleton are not. This pattern reflects the greater mobility of the appendicular skeleton compared to that of the axial skeleton.

Several features of synovial joints are important to their function (figure 6.38). The articular surfaces of bones within synovial joints are covered with a thin layer of **articular cartilage,** which provides a smooth surface where the bones meet. The **joint cavity** is filled with fluid. The cavity is enclosed by a **joint capsule,** which helps hold the bones together and allows for movement. Portions of the fibrous part of the joint capsule may be thickened to form ligaments. In addition, ligaments and tendons outside the joint capsule contribute to the strength of the joint.

A **synovial membrane** lines the joint cavity everywhere except over the articular cartilage. The membrane produces **synovial fluid,** which is a complex mixture of polysaccharides, proteins, fat, and cells. Synovial fluid forms a thin, lubricating film covering the surfaces of the joint. In certain synovial joints, the synovial membrane may extend as a pocket, or sac, called a **bursa**

(ber′să; pocket). Bursae are located between structures that rub together, such as where a tendon crosses a bone; they reduce friction, which could damage the structures involved. Inflammation of a bursa, often resulting from abrasion, is called **bursitis.** A synovial membrane may extend as a **tendon sheath** along some tendons associated with joints (figure 6.38).

### Types of Synovial Joints

Synovial joints are classified according to the shape of the adjoining articular surfaces (figure 6.39). **Plane joints,** or *gliding joints,* consist of two opposed flat surfaces that glide over each other. Examples of these joints are the articular facets between vertebrae. **Saddle joints** consist of two saddle-shaped articulating surfaces oriented at right angles to each other. Movement in these joints can occur in two planes. The joint between the metacarpal bone and the carpal bone (trapezium) of the thumb is a saddle joint. **Hinge joints** permit movement in one plane only. They consist of a convex cylinder of one bone applied to a corresponding concavity of the other bone. Examples are the elbow and knee joints (figure 6.40a,b). The flat condylar surface of the knee joint is modified into a concave surface by shock-absorbing fibrocartilage pads called **menisci** (mĕ-nis′sī). **Pivot joints** restrict movement to rotation around a single axis. Each pivot joint consists of a cylindrical bony process that rotates within a ring composed partly of bone and partly of ligament. The rotation that occurs between the axis and atlas when shaking the head "no" is an example. The articulation between the proximal ends of the ulna and radius is also a pivot joint.

Skeletal

Plane

Saddle

Hinge

Pivot

Ball-and-socket

Ellipsoid

| Class and Example of Joint | Structures Joined | Movement |
|---|---|---|
| **Plane** | | |
| Acromioclavicular | Acromion process of scapula and clavicle | Slight |
| Carpometacarpal | Carpals and metacarpals 2–5 | Multiple axes as a group |
| Costovertebral | Ribs and vertebrae | Slight |
| Intercarpal | Between carpal bones | Slight |
| Intermetatarsal | Between metatarsal bones | Slight |
| Intertarsal | Between tarsal bones | Slight |
| Intervertebral | Between articular processes of adjacent vertebrae | Slight |
| Sacroiliac | Between sacrum and coxal bone (complex joint with several planes and synchondroses) | Slight |
| Tarsometatarsal | Tarsal bones and metatarsal bones | Slight |
| **Saddle** | | |
| Carpometacarpal pollicis | Carpal and metacarpal of thumb | Two axes |
| Intercarpal | Between carpal bones | Slight |
| Sternoclavicular | Manubrium of sternum and clavicle | Slight |
| **Hinge** | | |
| Cubital (elbow) | Humerus, ulna, and radius | One axis |
| Knee | Femur and tibia | One axis |
| Interphalangeal | Between phalanges | One axis |
| Talocrural (ankle) | Talus, tibia, and fibula | Multiple axes; one predominates |
| **Pivot** | | |
| Medial atlantoaxial | Atlas and axis | Rotation |
| Proximal radioulnar | Radius and ulna | Rotation |
| Distal radioulnar | Radius and ulna | Rotation |
| **Ball-and-Socket** | | |
| Coxal (hip) | Coxal bone and femur | Multiple axes |
| Humeral (shoulder) | Scapula and humerus | Multiple axes |
| **Ellipsoid** | | |
| Atlantooccipital | Atlas and occipital bone | Two axes |
| Metacarpophalangeal (knuckles) | Metacarpal bones and phalanges | Two axes |
| Metatarsophalangeal (ball of foot) | Metatarsal bones and phalanges | Two axes |
| Radiocarpal (wrist) | Radius and carpal bones | Multiple axes |
| Temporomandibular | Mandible and temporal bone | Multiple axes; one predominates |

**Figure 6.39  Types of Synovial Joints**

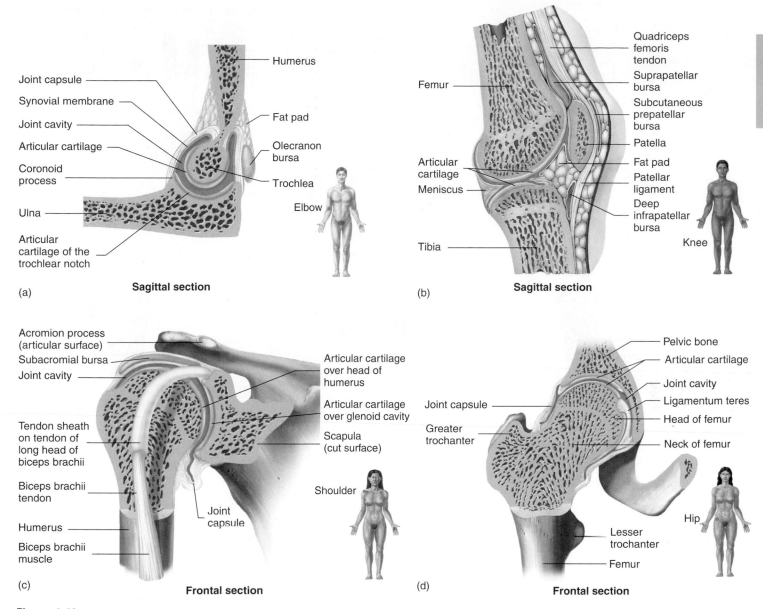

**Figure 6.40  Examples of Synovial Joints**
(*a*) Elbow. (*b*) Knee. (*c*) Shoulder. (*d*) Hip.

**Ball-and-socket joints** consist of a ball (head) at the end of one bone and a socket in an adjacent bone into which a portion of the ball fits. This type of joint allows a wide range of movement in almost any direction. Examples are the shoulder and hip joints (figure 6.40*c,d*). **Ellipsoid** (ē-lip′soyd) **joints,** or *condyloid* (kon′di-loyd) *joints,* are elongated ball-and-socket joints. The shape of the joint limits its range of movement nearly to that of a hinge motion, but in two planes. Examples of ellipsoid joints are the joint between the occipital condyles of the skull and the atlas of the vertebral column and the joints between the metacarpal bones and phalanges.

## Types of Movement

The types of movement occurring at a given joint are related to the structure of that joint. Some joints are limited to only one type of movement, whereas others permit movement in several directions.

All the movements are described relative to the anatomical position. Because most movements are accompanied by movements in the opposite direction, they are often illustrated in pairs (figure 6.41).

**Flexion** and **extension** are common opposing movements that may be defined in a number of ways, although each definition has an exception. The literal definitions are to bend (flex) and to straighten (extend). But we have chosen to use definitions with more utility and fewer exceptions. Thus, we say that flexion moves a part of the body in the anterior direction from the frontal plane, whereas extension moves a part in the posterior direction from the frontal plane (figure 6.41*a,b*). An exception is the knee, where flexion moves the leg in a posterior direction and extension moves it in an anterior direction.

Movement of the foot toward the plantar surface (sole of the foot), as when standing on the toes, is commonly called **plantar flexion.** Movement of the foot toward the shin, as when walking on the heels, is called **dorsiflexion.**

Skeletal

**Figure 6.41  Types of Movement**

(*a*) Flexion and extension of the elbow. (*b*) Flexion and extension of the neck. (*c*) Abduction and adduction of the fingers. (*d*) Pronation and supination of the hand. (*e*) Medial and lateral rotation of the arm. (*f*) Circumduction of the arm.

## A CASE IN POINT
### Dislocated Shoulder

The shoulder joint is the most commonly dislocated joint in the body. Loosh Holder dislocated his shoulder joint while playing basketball. As a result of a "charging" foul, Loosh was knocked backward and fell. As he broke his fall with his extended right arm, the head of the right humerus was forced out of the glenoid cavity. While being helped up from the floor, Loosh felt severe pain in his shoulder, his right arm sagged, and he could not move his arm at the shoulder. Most dislocations result in stretching of the joint capsule and movement of the humeral head to the inferior, anterior side of the glenoid cavity. The dislocated humeral head is moved back to its normal position by carefully pulling it laterally over the inferior lip of the glenoid cavity and then superiorly into the glenoid cavity. Once the shoulder joint capsule has been stretched by a shoulder dislocation, the shoulder joint may be predisposed to future dislocations. Some individuals have hereditary "loose" joints and are more likely to experience a dislocated shoulder.

**Abduction** (ab-dŭk′shun; to take away) is movement away from the median or midsagittal plane; **adduction** (to bring together) is movement toward the median plane (figure 6.41*c*). Moving the legs away from the midline of the body, as in the outward movement of "jumping jacks," is abduction, and bringing the legs back together is adduction.

**Pronation** (prō-nā′shŭn) and **supination** (soo′pi-nā′shun) are best demonstrated with the elbow flexed at a 90-degree angle. When the elbow is flexed, pronation is rotation of the forearm so that the palm is down, and supination is rotation of the forearm so that the palm faces up (figure 6.41*d*).

**Eversion** (ē-ver′zhŭn) is turning the foot so that the plantar surface (bottom of the foot) faces laterally; **inversion** (in-ver′zhŭn) is turning the foot so that the plantar surface faces medially.

**Rotation** is the turning of a structure around its long axis, as in shaking the head "no." Rotation of the arm can best be demonstrated with the elbow flexed (figure 6.41*e*) so that rotation is not confused with supination and pronation of the forearm. With the elbow flexed, medial rotation of the arm brings the forearm against the anterior surface of the abdomen, and lateral rotation moves it away from the body.

**Circumduction** (ser-kŭm-dŭk′shŭn) occurs at freely movable joints, such as the shoulder. In circumduction, the arm moves so that it describes a cone with the shoulder joint at the apex (figure 6.41*f*).

In addition to the movements pictured in figure 6.41, several other movement types have been identified:

- **Protraction** (prō-trak′shŭn) is a movement in which a structure, such as the mandible, glides anteriorly.
- In **retraction** (rē-trak′shŭn), the structure glides posteriorly.
- **Elevation** is movement of a structure in a superior direction. Closing the mouth involves elevation of the mandible.
- **Depression** is movement of a structure in an inferior direction. Opening the mouth involves depression of the mandible.
- **Excursion** is movement of a structure to one side, as in moving the mandible from side to side.
- **Opposition** is a movement unique to the thumb and little finger. It occurs when the tips of the thumb and little finger

are brought toward each other across the palm of the hand. The thumb can also oppose the other digits.
- **Reposition** returns the digits to the anatomical position.

Most movements that occur in the course of normal activities are combinations of movements. A complex movement can be described by naming the individual movements involved.

When the bones of a joint are forcefully pulled apart and the ligaments around the joint are pulled or torn, a *sprain* results. A *separation* exists when the bones remain apart after injury to a joint. A *dislocation* is when the end of one bone is pulled out of the socket in a ball-and-socket, ellipsoid, or pivot joint.

**Hyperextension** is usually defined as an abnormal, forced extension of a joint beyond its normal range of motion. For example, if a person falls and attempts to break the fall by putting out a hand, the force of the fall directed into the hand and wrist may cause hyperextension of the wrist, which may result in sprained joints or broken bones. Some health professionals, however, define hyperextension as the normal movement of a structure into the space posterior to the anatomical position.

**Predict 4**

*What combination of movements at the shoulder and elbow joints allows a person to perform a crawl stroke in swimming?*

# EFFECTS OF AGING ON THE SKELETAL SYSTEM AND JOINTS

The most significant age-related changes in the skeletal system affect the joints as well as the quality and quantity of bone matrix. The bone matrix in an older bone is more brittle than in a younger bone because decreased collagen production results in relatively more mineral and less collagen fibers. With aging, the amount of matrix also decreases because the rate of matrix formation by osteoblasts becomes slower than the rate of matrix breakdown by osteoclasts.

Bone mass is at its highest around age 30, and men generally have denser bones than women because of the effects of testosterone and greater body weight. Race also affects bone mass. African-Americans and Hispanics have higher bone masses than caucasians and Asians. After age 35, both men and women experience a loss of bone of 0.3–0.5% a year. This loss can increase 10-fold in women after menopause, when they can lose bone mass at a rate of 3–5% a year for approximately 5–7 years (see Systems Pathology, "Osteoporosis").

Significant loss of bone increases the likelihood of bone fractures. For example, loss of trabeculae greatly increases the risk of fractures of the vertebrae. In addition, loss of bone and the resulting fractures can cause deformity, loss of height, pain, and stiffness. Loss of bone from the jaws can also lead to tooth loss.

A number of changes occur within many joints as a person ages. Changes in synovial joints have the greatest effect and often present major problems for elderly people. With use, the cartilage covering articular surfaces can wear down. When a person is young, production of new, resilient matrix compensates for the wear. As a person ages, the rate of replacement declines, and the matrix becomes more rigid, thus adding to its rate of wear. The production rate of lubricating synovial fluid

# SYSTEMS PATHOLOGY

## Osteoporosis

Betty is a 65-year-old woman who has smoked heavily for 50 years. She does not exercise, seldom goes outside, has a poor diet, and is slightly underweight. While attending a family picnic, Betty tripped on a lawn sprinkler and fell. She was unable to stand because of severe hip pain, so she was rushed to the hospital, where a radiograph revealed that she had osteoporosis and a fracture in the neck of her femur (figure 6B).

Because of the location of the fracture and the osteoporosis, Betty's orthopedic surgeon suggested that she undergo hip-replacement surgery. Betty had a successful hip transplant and then began physical therapy. She was also placed on a diet and exercise program designed to improve the strength of her bones. In addition, she was prescribed alendronate therapy (see following discussion) and advised to quit smoking.

### Background Information

**Osteoporosis,** or porous bone, results from reduction in the overall quantity of bone matrix. It occurs when the rate of bone reabsorption exceeds the rate of bone formation. The loss of bone mass makes bones so porous and weakened that they become deformed and prone to fracture (figure 6C). The occurrence of osteoporosis increases with age. In both men and women, bone mass starts to decrease at about age 40 and continually decreases thereafter. Women can eventually lose approximately one-half, and men one-quarter, of their spongy bone. Osteoporosis is 2.5 times more common in women than in men.

In postmenopausal women, decreased production of the female sex hormone estrogen can cause osteoporosis. The degeneration occurs mostly in spongy bone, especially in the vertebrae of the spine and the bones of the forearm. Collapse of the vertebrae can cause a decrease in height or, in more severe cases, kyphosis in the upper back.

Other conditions that result in decreased estrogen can also cause osteoporosis. Examples include removal of the ovaries before menopause, extreme exercise to the point of amenorrhea (lack of menstrual flow), anorexia nervosa (self-starvation), and cigarette smoking.

In males, reduction in testosterone levels can cause loss of bone tissue. However, decreasing testosterone levels are usually less of a problem for men than decreasing estrogen levels are for women for two reasons. First, because males have denser bones than females, loss of some bone tissue has less of an effect. Second, testosterone levels generally don't decrease significantly until after age 65, and even then the rate of decrease is often slow.

Inadequate dietary intake or absorption of calcium can contribute to osteoporosis. Absorption of calcium from the small intestine decreases with age, and individuals with osteoporosis often have

Coxal bone
(hipbone)

Break

Femur
(thighbone)

### Figure 6B

Radiograph of a broken hip. A "broken hip" is usually a break in the femur in the hip region. (In this case, the break is at the base of the neck.)

Normal bone

Osteoporotic bone

### Figure 6C

Photomicrograph of normal bone and osteoporotic bone.

| System Interactions | Effects of Osteoporosis on Other Systems |
| --- | --- |
| **System** | **Interactions** |
| Integumentary | Little exposure to the sun results in lowered vitamin D production and decreased calcium absorption. |
| Muscular | Muscle atrophy and weakness make it difficult to maintain balance, and thus falls and injuries are more likely. |
| Nervous | Pain sensations following injury and during rehabilitation help prevent further injury. |
| Endocrine | Calcitonin is used to treat osteoporosis. |
| Cardiovascular | Blood clotting following injury starts the process of tissue repair. Blood cells are carried to the injury site to fight infections and remove cell debris. Blood vessels grow into the recovering tissue, providing nutrients and removing waste products. |
| Lymphatic and Immune | Immune cells resist possible infection following surgery, such as hip-replacement surgery, and release chemicals that promote tissue repair. |
| Respiratory | Excessive smoking lowers estrogen levels, which increases bone loss. |
| Digestive | Inadequate calcium and vitamin D in the diet can contribute to inadequate calcium absorption by the digestive system and osteoporosis. |
| Reproductive | Decreased estrogen following menopause contributes to osteoporosis. |

insufficient intake of calcium, vitamin D, and vitamin C. Drugs that interfere with the uptake or use of calcium can also increase the risk of osteoporosis.

Finally, osteoporosis can result from inadequate exercise or disuse caused by fractures or paralysis. Significant amounts of bone are lost after 8 weeks of immobilization.

Treatments for osteoporosis are designed to reduce bone loss or increase bone formation (or both). Increased dietary calcium and vitamin D can increase calcium uptake and promote bone formation. Daily doses of 1000–1500 mg of calcium and 800 IU (20 μg) of vitamin D are recommended. Exercise, such as walking or using light weights, also appears to be effective not only in reducing bone loss but also in increasing bone mass.

In postmenopausal women, estrogen replacement therapy decreases osteoclast activity, which reduces bone loss, but does not result in an increase in bone mass because osteoclast activity still exceeds osteoblast activity. Estrogen therapy also apparently increases the risk of breast cancer, strokes, heart attacks, and Alzheimer disease and is no longer recommended for treating

osteoporosis. Calcitonin (Miacalcin), which inhibits osteoclast activity, is now available as a nasal spray. Calcitonin can be used to treat osteoporosis in men and women and has been shown to produce a slight increase in bone mass. Another drug, alendronate (Fosamax), belongs to a class called the bisphosphonates, which bind to hydroxyapatite and inhibit bone breakdown by osteoclasts. Alendronate increases bone mass and reduces fracture rates even more effectively than calcitonin. Slow-releasing sodium fluoride (Slow Fluoride) in combination with calcium citrate (Citracal) also appears to increase bone mass.

Early diagnosis of osteoporosis can lead to more preventative treatments. Instruments that measure the absorption of photons (particles of light) by bone are currently used; of these, dual-energy x-ray absorptiometry (DEXA) is considered the best.

### Predict 5

*What advice should Betty give to her granddaughter so that the granddaughter will be less likely to develop osteoporosis when she is Betty's age?*

## DISEASES AND DISORDERS  Skeletal System

| CONDITION | DESCRIPTION |
|---|---|
| Tumors | May be malignant or benign and cause a range of bone defects |
| **SKELETAL DISORDERS** | |
| **Growth and Developmental Disorders** | |
| Gigantism | Abnormally increased body size due to excessive growth at the epiphyseal plates |
| Dwarfism | Abnormally small body size due to improper growth at the epiphyseal plates |
| Osteogenesis imperfecta | Brittle bones that fracture easily due to insufficient or abnormal collagen |
| Rickets | Growth retardation due to nutritional deficiencies in minerals ($Ca^{2+}$) or vitamin D; results in bones that are soft, weak, and easily broken |
| **Bacterial Infections** | |
| Osteomyelitis | Bone inflammation often due to a bacterial infection that may lead to complete destruction of the bone |
| Tuberculosis | Typically, a lung bacterium that can also affect bone |
| **Decalcification** | |
| Osteomalacia | Softening of adult bones due to calcium depletion; often caused by vitamin D deficiency |
| Osteoporosis | Reduction in overall quantity of bone tissue; see Systems Pathology |
| **JOINT DISORDERS** | |
| Arthritis | Inflammation of a joint; causes include infectious agents, metabolic disorders, trauma, and immune disorders |
| Rheumatoid arthritis | General connective tissue autoimmune disease |
| Degenerative joint disease (osteoarthritis) | Gradual degeneration of a joint with advancing age; can be delayed with exercise |
| Gout | Increased production and accumulation of uric acid crystals in tissues, including joint capsules |
| **Bursitis and bunions** | |
| Bursitis | Inflammation of a bursa |
| Bunion | Bursitis over the joint at the base of the great toe; irritated by tight shoes |
| Joint replacement | Replacement of painful joints with artificial joints |

Go to **www.mhhe.com/seeleyess7** for additional information on these pathologies.

also declines with age, further contributing to the wear of the articular cartilage. Many people also experience arthritis, an inflammatory degeneration of the joints, with advancing age. In addition, the ligaments and tendons surrounding a joint shorten and become less flexible with age, resulting in decreased range of motion. Furthermore, many older people are less physically active, which causes the joints to become less flexible and decreases their range of motion.

The most effective preventative measure against the effects of aging on the skeletal system is the combination of increasing physical activity and taking dietary calcium and vitamin D supplements. Intensive exercise can even reverse loss of bone matrix.

## SUMMARY

The skeletal system consists of bone, cartilage, tendons, and ligaments.

### Functions of the Skeletal System  (p. 112)

1. The skeletal system provides the major support for the body.
2. Bone protects internal organs.
3. Joints allow movement between bones.
4. Bones store and release minerals as needed by the body.
5. Bone marrow gives rise to blood cells and platelets.

### Extracellular Matrix  (p. 112)

1. Bone, cartilage, tendons, and ligaments are connective tissues.

2. Varying amounts of collagen, proteoglycan, organic molecules, water, and minerals in the matrix determine the characteristics of connective tissue.

### General Features of Bone  (p. 113)

There are four categories of bone: long, short, flat, and irregular.

#### Structure of a Long Bone

Long bones consist of a diaphysis (shaft), epiphyses (ends), and epiphyseal (growth) plates. The diaphysis contains a medullary cavity, which is filled with marrow, and the end of the epiphysis is covered by articular cartilage.

### Histology of Bone

1. Osteoblasts are bone-forming cells.
2. Osteocytes are bone cells located between thin sheets of extracellular matrix called lamellae.
3. Compact bone tissue consists of osteons, which are composed of osteocytes organized into lamellae surrounding central canals.
4. Spongy bone tissue consists of trabeculae without central canals.

### Bone Ossification

1. Intramembranous ossification occurs within connective tissue membranes.
2. Endochondral ossification occurs within cartilage.

### Bone Growth

Bone elongation occurs at the epiphyseal plate as chondrocytes proliferate, enlarge, die, and are replaced by bone.

### Bone Remodeling

Bone remodeling consists of removal of existing bone by osteoclasts and deposition of new bone by osteoblasts.

### Bone Repair

During bone repair, cells move into the damaged area and form a callus, which is replaced by bone.

## Bone and Calcium Homeostasis   (p. 119)

1. Osteoclasts remove calcium from bone, causing blood calcium levels to increase.
2. Osteoblasts deposit calcium into bone, causing blood calcium levels to decrease.
3. Parathyroid hormone increases bone breakdown, whereas calcitonin decreases bone breakdown.

## General Considerations of Bone Anatomy   (p. 121)

There are 206 bones in the average adult skeleton.

## Axial Skeleton   (p. 122)

The axial skeleton includes the skull, vertebral column, and thoracic cage.

### Skull

1. The skull consists of 22 bones: 8 forming the braincase and 14 facial bones. The hyoid bone and 6 auditory ossicles are associated with the skull.
2. From a lateral view, the parietal, temporal, and sphenoid bones can be seen.
3. From a frontal view, the orbits and nasal cavity can be seen, as well as associated bones and structures, such as the frontal bone, zygomatic bone, maxilla, and mandible.
4. The interior of the cranial cavity contains three fossae with several foramina.
5. Seen from below, the base of the skull reveals numerous foramina and other structures, such as processes for muscle attachment.

### Vertebral Column

1. The vertebral column contains 7 cervical, 12 thoracic, and 5 lumbar vertebrae, plus 1 sacral bone and 1 coccygeal bone.
2. Each vertebra consists of a body, an arch, and processes.
3. Regional differences in vertebrae are as follows: Cervical vertebrae have transverse foramina; thoracic vertebrae have long spinous processes and attachment sites for the ribs; lumbar vertebrae have rectangular transverse and spinous processes, and the position of their facets limits rotation; the sacrum is a single, fused bone; the coccyx is four or fewer fused vertebrae.

### Thoracic Cage

1. The thoracic cage consists of the thoracic vertebrae, the ribs, and the sternum.
2. There are 12 pairs of ribs: 7 true and 5 false (2 of the false ribs are also called floating ribs).
3. The sternum consists of the manubrium, the body, and the xiphoid process.

## Appendicular Skeleton   (p. 131)

The appendicular skeleton consists of the bones of the upper and lower limbs and their girdles.

### Pectoral Girdle

The pectoral girdle includes the scapulae and clavicles.

### Upper Limb

The upper limb consists of the arm (humerus), forearm (ulna and radius), wrist (eight carpal bones), and hand (five metacarpal bones, three phalanges in each finger, and two phalanges in the thumb).

### Pelvic Girdle

The pelvic girdle is made up of the two coxal bones. Each coxal bone consists of an ilium, an ischium, and a pubis. The coxal bones, sacrum, and coccyx form the pelvis.

### Lower Limb

The lower limb includes the thigh (femur), leg (tibia and fibula), ankle (seven tarsal bones), and foot (metatarsal bones and phalanges, similar to the bones in the hand).

## Joints   (p. 140)

A joint is a place where bones come together.

### Fibrous Joints

Fibrous joints consist of bones united by fibrous connective tissue. They allow little or no movement.

### Cartilaginous Joints

Cartilaginous joints consist of bones united by cartilage, and they exhibit slight movement.

### Synovial Joints

1. Synovial joints consist of articular cartilage over the uniting bones, a joint cavity lined by a synovial membrane and containing synovial fluid, and a joint capsule. They are highly movable joints.
2. Synovial joints can be classified as plane, saddle, hinge, pivot, ball-and-socket, or ellipsoid.

### Types of Movement

The major types of movement are flexion/extension, abduction/adduction, pronation/supination, eversion/inversion, rotation, circumduction, protraction/retraction, elevation/depression, excursion, and opposition/reposition.

## Effects of Aging on the Skeletal System and Joints   (p. 145)

1. Bone matrix becomes more brittle and decreases in total amount during aging.
2. Joints lose articular cartilage and become less flexible with age.
3. Prevention measures include exercise and calcium and vitamin D supplements.

# REVIEW AND COMPREHENSION

1. What are the primary functions of the skeletal system?
2. Name the major types of fibers and molecules in the extracellular matrix of the skeletal system. How do they contribute to the functions of tendons, ligaments, cartilage, and bones?
3. Define diaphysis, epiphysis, epiphyseal plate, medullary cavity, articular cartilage, periosteum, and endosteum.
4. Describe the structure of compact bone. How do nutrients reach the osteocytes in compact bone?
5. Describe the structure of spongy bone. What are trabeculae? How do nutrients reach osteocytes in trabeculae?
6. Define and describe intramembranous and endochondral ossification.
7. How do bones grow in diameter? How do long bones grow in length?
8. What is accomplished by bone remodeling? How does bone repair occur?
9. Define the axial skeleton and the appendicular skeleton.
10. Name the bones of the braincase and the face.
11. Give the locations of the paranasal sinuses. What are their functions?
12. What is the function of the hard palate?
13. Through what foramen does the brain connect to the spinal cord?
14. How do the vertebrae protect the spinal cord? Where do spinal nerves exit the vertebral column?
15. Name and give the number of each type of vertebra. Describe the characteristics that distinguish the different types of vertebrae from one another.
16. What is the function of the thoracic cage? Name the parts of the sternum. Distinguish among true, false, and floating ribs.
17. Name the bones that make up the pectoral girdle, arm, forearm, wrist, and hand. How many phalanges are in each finger and in the thumb?
18. Define the pelvic girdle. What bones fuse to form each coxal bone? Where and with what bones do the coxal bones articulate?
19. Name the bones of the thigh, leg, ankle, and foot.
20. Define joint, or articulation. Name and describe the differences among the three major classes of joints.
21. Describe the structure of a synovial joint. How do the different parts of the joint permit joint movement?
22. On what basis are synovial joints classified? Describe the different types of synovial joints, and give examples of each. What movements does each type of joint allow?
23. Describe and give examples of flexion/extension, abduction/adduction, and supination/pronation.

# CRITICAL THINKING

1. A 12-year-old boy fell while playing basketball. The physician explained that the head (epiphysis) of the femur was separated from the shaft (diaphysis). Although the bone was set properly, by the time the boy was 16, it was apparent that the injured lower limb was shorter than the normal one. Explain why this difference occurred.
2. Justin Time leaped from his hotel room to avoid burning to death in a fire. If he landed on his heels, what bone was likely to fracture? Unfortunately for Justin, a 240-pound fireman, Hefty Stomper, ran by and stepped heavily on the distal part of Justin's foot (not the toes). What bones now could be broken?
3. One day while shopping, Ms. Wantta Bargain picked up her 3-year-old son, Somm, by his right wrist and lifted him into a shopping cart. She heard a clicking sound, and Somm immediately began to cry and hold his elbow. Given that lifting the child caused a separation at the elbow, which is more likely: separation of the radius and humerus or separation of the ulna and humerus?
4. Why are women knock-kneed more commonly than men?
5. A skeleton was discovered in a remote mountain area. The coroner determined that, not only was the skeleton human but it was female. Explain.

*Answers in Appendix D*

Visit this textbook's website at **www.mhhe.com/seeleyess7** for practice quizzes, animations, interactive learning exercises, and other study tools.

McGraw-Hill offers a study CD that features interactive cadaver dissection. **Anatomy & Physiology Revealed** includes cadaver photos that allow you to peel away layers of the human body to reveal structures beneath the surface. This program also includes animations, radiologic imaging, audio pronunciations, and practice quizzing.

The human body is capable of performing work, such as lifting these weights. However, the body can perform this function only by using the muscular system. Skeletal muscles work together with bones to form a pulley and lever system. Muscles control every motion the body makes.

# 7 Chapter

# Muscular System

## Chapter Outline and Objectives

### Functions of the Muscular System (p. 152)
1. List the functions of the muscular system.

### Characteristics of Skeletal Muscle (p. 152)
2. Describe the microscopic structure of a muscle, and produce diagrams that illustrate the arrangement of myofibrils, sarcomeres, and myofilaments.
3. Describe a resting membrane potential and an action potential.
4. Describe a neuromuscular junction.
5. Describe the events that occur in muscle contraction and relaxation.
6. Define muscle twitch, tetanus, recruitment, and summation.
7. Distinguish between aerobic and anaerobic respiration.
8. Compare the mechanisms involved in the major types of fatigue.
9. Distinguish between isometric and isotonic contractions.
10. Define muscle tone.
11. Distinguish between fast-twitch and slow-twitch fibers, and explain the function for which each type is best adapted.

### Smooth Muscle and Cardiac Muscle (p. 165)
12. Distinguish among skeletal, smooth, and cardiac muscle.

### Skeletal Muscle Anatomy (p. 166)
13. Define the following and give an example of each: origin, insertion, agonist, antagonist, synergist, and prime mover.
14. Describe various facial expressions, and list the major muscles involved in each.
15. Describe mastication, tongue movement, and swallowing, and list the muscles or groups of muscles involved in each.
16. Describe the muscles of the trunk and the actions they produce.
17. Describe the movements of the arm, forearm, and hand, and list the muscle groups involved in each movement.
18. Describe the movements of the thigh, leg, and foot, and list the muscle groups involved in each movement.

### Effects of Aging on Skeletal Muscle (p. 187)
19. Describe the effects of aging on skeletal muscle.

A runner rounds the last corner of the track and sprints for the finish line. Her arms and legs are pumping as she tries to reach her maximum speed. Her heart is beating rapidly, and her breathing is rapid, deep, and regular. Blood is shunted away from her digestive organs, and a greater volume is delivered to her skeletal muscles to maximize their oxygen supply. These actions are accomplished by muscle tissue, the most abundant tissue of the body and one of the most adaptable.

You don't have to be running for the muscular system to be at work. Even when you aren't "moving," postural muscles keep you sitting or standing upright, respiratory muscles keep you breathing, the heart continually pumps blood to all parts of your body, and blood vessels constrict or relax to direct blood to organs where it is needed.

# FUNCTIONS OF THE MUSCULAR SYSTEM

Movement within the body is accomplished in various ways: by cilia or flagella on the surface of certain cells, by the force of gravity, or by the contraction of muscles. But most of the body's movement results from the last-named—muscle contraction. As described in chapter 4, there are three types of muscle tissue: skeletal, cardiac, and smooth (figure 7.1). This chapter deals primarily with the structure and function of skeletal muscle; cardiac and smooth muscle are described only briefly. Following are the major functions of the muscular system:

1. *Movement of the body.* Contraction of skeletal muscles is responsible for the overall movements of the body, such as walking, running, and manipulating objects with the hands.
2. *Maintenance of posture.* Skeletal muscles constantly maintain tone, which keeps us sitting or standing erect.
3. *Respiration.* Muscles of the thorax carry out the movements necessary for respiration.
4. *Production of body heat.* When skeletal muscles contract, heat is given off as a by-product. This released heat is critical to the maintenance of body temperature.
5. *Communication.* Skeletal muscles are involved in all aspects of communication, including speaking, writing, typing, gesturing, and facial expressions.
6. *Constriction of organs and vessels.* The contraction of smooth muscle within the walls of internal organs and vessels causes those structures to constrict. This constriction can help propel and mix food and water in the digestive tract, propel secretions from organs, and regulate blood flow through vessels.
7. *Contraction of the heart.* The contraction of cardiac muscle causes the heart to beat, propelling blood to all parts of the body.

# CHARACTERISTICS OF SKELETAL MUSCLE

**Skeletal muscle,** with its associated connective tissue, constitutes approximately 40% of body weight. Skeletal muscle is so named because most of the muscles are attached to the skeletal system. It

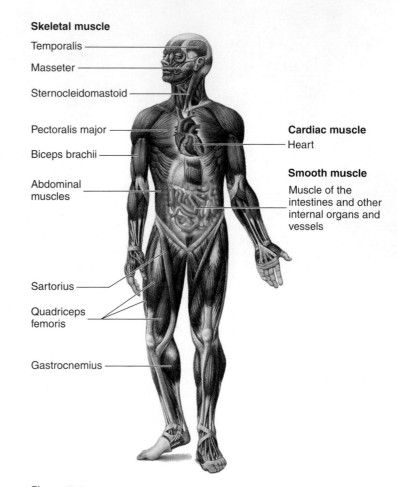

**Figure 7.1  Overview of the Muscular System**

is also called *striated muscle* because transverse bands, or striations, can be seen in the muscle under the microscope.

Skeletal muscle has four major functional characteristics: contractility, excitability, extensibility, and elasticity.

1. **Contractility** (kon-trak-til′i-tē) is the ability of skeletal muscle to shorten with force. When skeletal muscles contract, they cause the structures to which they are attached to move. Skeletal muscles shorten forcefully during contraction, but they lengthen passively. Either

gravity or the contraction of an opposing muscle produces a force that pulls on the shortened muscle, causing it to lengthen.

2. **Excitability** (ek-sī′tă-bil′i-tē) is the capacity of skeletal muscle to respond to a stimulus. Normally, the stimulus is from nerves that we consciously control.

3. **Extensibility** (eks-ten′sĭ-bil′i-tē) means that skeletal muscles stretch. After a contraction, skeletal muscles can be stretched to their normal resting length and beyond to a limited degree.

4. **Elasticity** (ĕ-las-tis′i-tē) is the ability of skeletal muscles to recoil to their original resting length after they have been stretched.

## Skeletal Muscle Structure

### Connective Tissue Coverings of Muscle

Each skeletal muscle is surrounded by a connective tissue sheath called the **epimysium** (ep-i-mis′ē-ŭm), or *muscular fascia* (fash′ē-ă) (figure 7.2*a*). A muscle is composed of numerous visible bundles called **muscle fasciculi** (fă-sik′ū-lī), which are surrounded by loose connective tissue called the **perimysium** (per′i-mis′ē-ŭm). A fasciculus is composed of several muscle cells, which are called **muscle fibers** (figure 7.2*b*). Each muscle fiber is surrounded by loose connective tissue called **endomysium** (en′dō-mis′ē-ŭm).

### Muscle Fiber Structure

Muscle contraction is much easier to understand when we consider the structure of a muscle fiber. A muscle fiber is a single cylindrical fiber, with several nuclei located at its periphery. The largest human muscle fibers are up to 30 cm long and 0.15 mm in diameter. Such giant cells may contain several thousand nuclei. The cell membrane of the muscle fiber is called the **sarcolemma** (sar′kō-lem′ă; *sarco*, flesh) (figure 7.2*b*). The multiple nuclei of the muscle fiber are located just deep to the sarcolemma. Along the surface of the sarcolemma are many tubelike invaginations, called **transverse tubules,** or **T tubules,** which occur at regular intervals along the muscle fiber and extend inward into it. The T tubules are associated with a highly organized smooth endoplasmic reticulum called the **sarcoplasmic reticulum** (re-tik′ū-lŭm). T tubules connect the sarcolemma to the sarcoplasmic reticulum. The sarcoplasmic reticulum has a relatively high concentration of $Ca^{2+}$, which play a major role in muscle contraction.

Inside each muscle fiber is the cytoplasm, called the **sarcoplasm** (sar′kō-plazm). It contains numerous **myofibrils** mī-ō-fī′brilz; *myo*, muscle), threadlike structures that extend from one end of the muscle fiber to the other (figure 7.2*c*). Myofibrils consist of two major kinds of protein fibers: **actin** (ak′tin) **myofilaments** (mī-ō-fil′ă-ments) and **myosin** (mī′ō-sin) myofilaments (figure 7.2*d*). The actin and myosin myofilaments are arranged into highly ordered, repeating units along the myofibril called **sarcomeres** (sar′kō-mērz) (figure 7.2*c,d*).

### Actin and Myosin Myofilaments

Actin myofilaments, or thin filaments, are made up of three components: actin, troponin, and tropomyosin. The actin myofilaments, which resemble two minute strands of pearls twisted together, have attachment sites for the myosin myofilaments (figure 7.2*e*). **Troponin** (trō′pō-nin) molecules are attached at specific intervals along the actin myofilaments. These molecules have binding sites for $Ca^{2+}$. **Tropomyosin** (trō-pō-mī′ō-sin) filaments are located along the groove between the twisted strands of actin myofilament subunits. The tropomyosin filaments block the myosin myofilament binding sites on the actin myofilaments in an unstimulated muscle. In other words, if no $Ca^{2+}$ are present, the tropomyosin filaments cover the attachment sites on the actin myofilament. However, when $Ca^{2+}$ are present, they bind to troponin, which causes the tropomyosin filaments to expose the attachement sites on the actin myofilaments.

Myosin myofilaments, or thick myofilaments, resemble bundles of minute golf clubs (figure 7.2*f*). The parts of the myosin molecule that resemble golf club heads are referred to as **myosin heads.** The myosin heads have three important properties: (1) The heads can bind to attachment sites on the actin myofilaments; (2) they can bend and straighten during contraction; and (3) they can break down ATP, releasing energy.

### Sarcomeres

The sarcomere is the basic structural and functional unit of skeletal muscle because it is the smallest portion of skeletal muscle capable of contracting. Each sarcomere extends from one Z disk to another Z disk. Each **Z disk** is a network of protein fibers forming an attachment site for actin myofilaments. The arrangement of the actin and myosin myofilaments in sarcomeres gives the myofibril a banded appearance (figure 7.3). A light **I band,** which consists only of actin myofilaments, spans each Z disk and ends at the myosin myofilaments. A darker, central region in each sarcomere, called an **A band,** extends the length of the myosin myofilaments. The actin and myosin myofilaments overlap for some distance at both ends of the A band. In the center of each sarcomere is a second light zone, called the **H zone,** which consists only of myosin myofilaments. The myosin myofilaments are anchored in the center of the sarcomere at a dark-staining band, called the **M line.** The alternating I bands and A bands of the sarcomeres are responsible for the striations in skeletal muscle fibers observed through the microscope (see table 4.11*a*). It is the close association of the sarcomeres, the T tubules, and the sarcoplasmic reticulum that enables a nerve stimulus to initiate contraction of the muscle fiber.

## Excitability of Muscle Fibers

Muscle fibers, like other cells of the body, have electrical properties. This section describes the electrical properties of skeletal muscle fibers, and later sections illustrate their role in contraction.

The inside of most cell membranes is negatively charged compared to the outside of the cell membrane; in other words, the

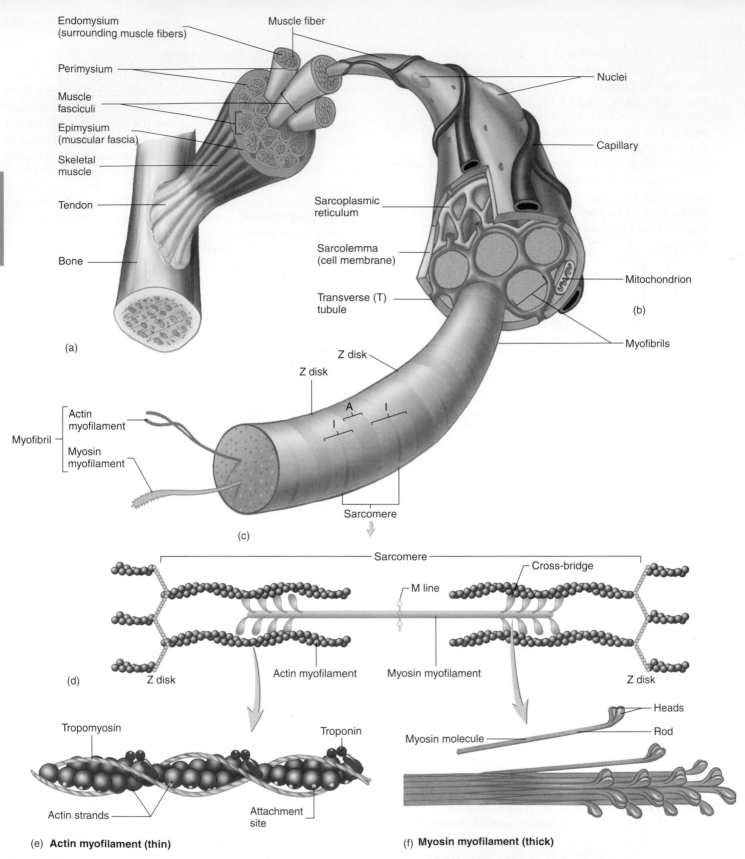

**Figure 7.2** Structure of a Muscle

(a) Part of a muscle attached by a tendon to a bone. A muscle is composed of muscle fasciculi, each surrounded by perimysium. The fasciculi are composed of bundles of individual muscle fibers (muscle cells), each surrounded by endomysium. The entire muscle is surrounded by a connective tissue sheath called epimysium, or muscular fascia. (b) Enlargement of one muscle fiber containing several myofibrils. (c) A myofibril extended out the end of the muscle fiber, showing the banding patterns of the sarcomeres. (d) A single sarcomere of a myofibril is composed mainly of actin myofilaments and myosin myofilaments. The Z disks anchor the actin myofilaments, and the myosin myofilaments are held in place by the M line. (e) Part of an actin myofilament is enlarged. (f) Part of a myosin myofilament is enlarged.

**Figure 7.3  Skeletal Muscle**

(*a*) Electron micrograph of skeletal muscle, showing several sarcomeres in a muscle fiber. (*b*) Diagram of two adjacent sarcomeres, depicting the structures responsible for the banding pattern.

cell membrane is **polarized** (figure 7.4, step 1). The charge difference, called the **resting membrane potential,** occurs because there is an uneven distribution of ions across the cell membrane. The resting membrane potential develops for three reasons: (1) The concentration of $K^+$ inside the cell membrane is higher than that outside the cell membrane; (2) the concentration of $Na^+$ outside the cell membrane is higher than that inside the cell membrane; and (3) the cell membrane is more permeable to $K^+$ than it is to $Na^+$. Recall from chapter 3 the different types of ion channels: nongated, or leak, channels, which are always open, and chemically gated channels, which are closed until a chemical, such as a neurotransmitter, binds to them and stimulates them to open. Thus, because excitable cells have many $K^+$ leak channels, $K^+$ leak out of the cell faster than $Na^+$ leak into the cell. In other words, some $K^+$ channels are open, whereas other ion channels, such as those for $Na^+$, are closed. In addition, negatively charged molecules, such as proteins, are in essence "trapped" inside the cell because the cell membrane is impermeable to them. For these reasons, the inside of the cell membrane is more negatively charged than the outside of the cell membrane.

A few $K^+$ are able to diffuse down their concentration gradient from inside to just outside the cell membrane. Because $K^+$ are positively charged, their movement from inside the cell to the outside causes the inside of the cell membrane to become even more negatively charged compared to the outside of the cell membrane.

Potassium ions diffuse down their concentration gradient only until the charge difference across the cell membrane is great enough to prevent any additional diffusion of $K^+$ out of the cell. The resting membrane potential is an equilibrium in which the tendency for $K^+$ to diffuse out of the cell is opposed by the negative charges inside the cell, which tend to attract the positively charged $K^+$ into the cell. At rest, the sodium-potassium pump transports $K^+$ from outside the cell to the inside and transports $Na^+$ from inside the cell to the outside. The active transport of $Na^+$ and $K^+$ by the sodium-potassium pump maintains the uneven distribution of $Na^+$ and $K^+$ across the cell membrane (see chapter 3).

In a stimulated muscle fiber or nerve cell, the stimulation causes $Na^+$ channels to open quickly and the membrane to

Muscular

**1 Resting membrane potential.**
Na⁺ channels (*pink*) and some,
but not all, K⁺ channels (*purple*)
are closed. A few K⁺ diffuse down
their concentration gradient
through the open K⁺ channels,
making the inside of the cell
membrane negatively charged
compared to the outside.

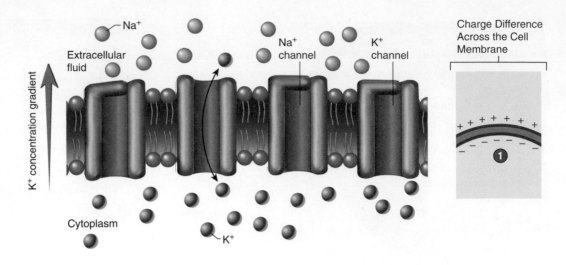

**2 Depolarization.** Na⁺ channels are
open. A few Na⁺ diffuse down their
concentration gradient through the
open Na⁺ channels, making the
inside of the cell membrane
positively charged compared to
the outside.

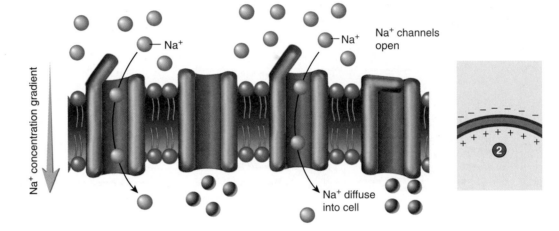

**3 Repolarization.** Na⁺ channels are
closed, and Na⁺ movement into
the cells stops. More K⁺ channels
open. K⁺ movement out of the cell
increases, making the inside of
the cell membrane negatively
charged compared to the outside,
once again.

**PROCESS Figure 7.4  Ion Channels and the Action Potential**

Step 1 illustrates the status of Na⁺ and K⁺ channels in a resting cell. Steps 2 and 3 show how the channels open and close to produce an action potential. Next to each
step, the charge difference across the plasma membrane is illustrated.

become very permeable to Na$^+$ for a brief time (figure 7.4, step 2). Because the Na$^+$ concentration is much greater outside the cell than inside and the charge inside the cell membrane is negative, a few positively charged Na$^+$ quickly diffuse down their concentration gradient and toward the negative charges into the cell, causing the inside of the cell membrane to become more positive than the outside of the cell. This change is called **depolarization.** Near the end of depolarization, Na$^+$ channels close, and additional K$^+$ channels open (figure 7.4, step 3). Consequently, the tendency for Na$^+$ to enter the cell decreases, and the tendency for K$^+$ to leave the cell increases. These changes cause the inside of the cell membrane to become more negative than the outside once again. Additional K$^+$ channels then close as the charge across the cell membrane returns to its resting condition (figure 7.4, step 1). The change back to the resting membrane potential is called **repolarization.** The rapid depolarization and repolarization of the cell membrane is called an **action potential.** In a muscle fiber, an action potential results in muscle contraction. The resting membrane potential and action potential are described in more detail in chapter 8.

## Nerve Supply

Skeletal muscle fibers do not contract unless they are stimulated by motor neurons. **Motor neurons** are specialized nerve cells that stimulate muscles to contract. Motor neurons generate action potentials that travel to skeletal muscle fibers. Axons of these neurons enter muscles and send out branches to several muscle fibers. Each branch forms a junction with a muscle fiber, called a **neuromuscular junction** (figure 7.5). A more general term, **synapse** (sin'aps), refers to the cell-to-cell junction between a nerve cell and either another nerve cell or an effector cell, such as a muscle or gland cell. Neuromuscular junctions are located near the center of a muscle fiber. A single motor neuron and all the skeletal muscle fibers it innervates constitute a **motor unit.** A motor unit in a small, precisely controlled muscle, such as in the hand, may have only one or a few muscle fibers per unit, whereas the motor units of large thigh muscles may have as many as 1000 muscle fibers per motor unit. Therefore, the fewer fibers there are in the motor units of a muscle, the greater control you have over that muscle. Many motor units constitute a single muscle.

**Figure 7.5  Neuromuscular Junction**

(a) In a neuromuscular junction, several branches of an axon junction with a single muscle fiber. (b) Photomicrograph of neuromuscular junctions.

1. An action potential arrives at the presynaptic terminal, causing Ca²⁺ channels to open.

2. Calcium ions (Ca²⁺) enter the presynaptic terminal and initiate the release of a neurotransmitter, acetylcholine (ACh), from synaptic vesicles into the presynaptic cleft.

3. Diffusion of ACh across the synaptic cleft and binding of ACh to ACh receptors on the postsynaptic muscle fiber membrane opens Na⁺ channels.

4. Sodium ions (Na⁺) diffuse down their concentration gradient, which results in depolarization of the muscle fiber membrane; once threshold has been reached, a postsynaptic action potential results.

**PROCESS Figure 7.6  Function of the Neuromuscular Junction**

ACh is released in response to an action potential at the neuromuscular junction.

A neuromuscular junction is formed by a cluster of enlarged axon terminals resting in indentations of the muscle fiber's cell membrane. An enlarged axon terminal is the **presynaptic terminal;** the space between the presynaptic terminal and the muscle fiber membrane is the **synaptic cleft;** and the muscle fiber membrane is the **postsynaptic membrane.** Each presynaptic terminal contains many small vesicles, called **synaptic vesicles.** These vesicles contain **acetylcholine** (as-e-til-kō′lēn), or **ACh,** which functions as a **neurotransmitter,** a molecule released by a presynaptic nerve cell that stimulates or inhibits a postsynaptic cell.

When an action potential reaches the presynaptic terminal, it causes Ca²⁺ channels to open. Ca²⁺ enter the presynaptic terminal and cause several synaptic vesicles to release acetylcholine into the synaptic cleft by exocytosis (figure 7.6). The acetylcholine diffuses across the synaptic cleft and binds to acetylcholine receptor sites on the Na⁺ channels in the muscle fiber cell membrane. The

combination of acetylcholine with its receptor opens Na⁺ channels and therefore makes the cell membrane more permeable to Na⁺. The resulting movement of Na⁺ into the muscle fiber initiates an action potential, which travels along the length of the muscle fiber and causes it to contract. The acetylcholine released into the synaptic cleft between the neuron and the muscle fiber is rapidly broken down by an enzyme, **acetylcholinesterase** (as′e-til-kō-lin-es′ter-ās). This enzymatic breakdown ensures that one action potential in the neuron yields only one action potential in the skeletal muscle fibers of that motor unit and only one contraction of each muscle fiber.

### A CASE IN POINT

#### Myasthenia Gravis

Les Strong was 45 years old when he first noticed muscle weakness in his face. While shaving, he could not tighten the left side of his face. Within a few days, he noticed the same problem on the right side of his face. He also experienced ptosis (drooping) of the left eyelid and slightly slurred speech. These changes worried him, and he scheduled an appointment with his physician. After a thorough examination, Les's physician referred him to a neurologist, who thought Les's muscular problems might be due to a condition called myasthenia gravis.

**Myasthenia gravis** is an autoimmune disorder in which antibodies are formed against acetylcholine receptors. As a result, acetylcholine receptors in the postsynaptic membranes of skeletal muscles are destroyed. With fewer receptors, acetylcholine is less likely to stimulate muscle contraction, resulting in muscle weakness and fatigue. In Les's case, a blood test revealed the presence of acetycholine receptor antibodies, confirming the diagnosis of myasthenia gravis. Les was prescribed the anticholinesterase drug neostigmine (nē′ō-stig′min), which greatly reduced his symptoms and allowed him to continue a relatively normal life. Anticholinesterase drugs inhibit acetylcholinesterase activity, preventing the breakdown of acetylcholine. Consequently, acetylcholine levels in the synapse remain elevated, which increases the stimulation of the reduced number of functional acetylcholine receptors.

## Muscle Contraction

Contraction of skeletal muscle tissue occurs as actin and myosin myofilaments slide past one another, causing the sarcomeres to shorten. Many sarcomeres joined end to end form myofibrils. Shortening of the sarcomeres causes myofibrils to shorten, thereby causing the entire muscle to shorten.

The sliding of actin myofilaments past myosin myofilaments during contraction is called the **sliding filament model** of muscle contraction. During contraction, neither the actin nor the myosin fibers shorten. The H zones and I bands shorten during contraction, but the A bands do not change in length (figure 7.7).

During muscle relaxation, sarcomeres lengthen. This lengthening requires an opposing force, such as that produced by other muscles or by gravity.

Table 7.1 traces the sequence of events that occur as a skeletal muscle contracts. Action potentials produced in skeletal muscle fibers at the neuromuscular junction travel along the sarcolemma and the T tubule membranes. The action potentials cause the membranes of the sarcoplasmic reticulum adjacent to

# CLINICAL IMPACT

## Acetylcholine Antagonists

Anything that affects the production, release, or degradation of acetylcholine or its ability to bind to its receptor on the muscle cell membrane can affect the transmission of action potentials across the neuromuscular junction. For example, some insecticides contain organophosphates that bind to and inhibit the function of acetylcholinesterase. As a result, acetylcholine is not degraded and accumulates in the synaptic cleft, where it acts as a constant stimulus to the muscle fiber. Insects exposed to such insecticides die partly because their respiratory muscles contract and cannot relax—a condition called **spastic paralysis** (spas′tik pa-ral′i-sis), which is followed by fatigue in the muscles.

Humans respond similarly to these insecticides. The skeletal muscles responsible for respiration cannot undergo their normal cycle of contraction and relaxation. Instead, they remain in a state of spastic paralysis until they become fatigued. Patients die of respiratory failure. Other organic poisons, such as **curare** (koo-ră′rē), the substance originally used by South American Indians in poison arrows, bind to the acetylcholine receptors on the muscle cell membrane and prevent acetylcholine from binding to them. Curare does not allow activation of the receptors; therefore, the muscle is incapable of contracting in response to nervous stimulation, a condition called **flaccid** (flak′sid, flas′id) **paralysis.** Curare is not a poison to which people are commonly exposed, but it has been used to investigate the role of acetylcholine in the neuromuscular synapse and is sometimes administered in small doses to relax muscles during certain kinds of surgery.

| TABLE 7.1 | Summary of Skeletal Muscle Contraction (see figures 7.7 and 7.8) |
|---|---|

1. An action potential travels along an axon membrane to a neuromuscular junction.
2. Calcium channels open and $Ca^{2+}$ enter the presynaptic terminal.
3. Acetylcholine is released from the synaptic vesicles in the presynaptic terminal of the neuron.
4. Acetylcholine diffuses across the synaptic cleft and binds to receptor sites on $Na^+$ channels in the muscle cell postsynaptic membrane, causing $Na^+$ channels to open.
5. Sodium ions diffuse into the muscle cell, initiating an action potential in the muscle cell, which travels along the sarcolemma and T tubule membranes.
6. Action potentials in the T tubules cause the sarcoplasmic reticulum to release $Ca^{2+}$.
7. On actin myofilaments, $Ca^{2+}$ bind to troponin molecules, causing the tropomyosin molecules to move into grooves along the actin myofilament, which exposes myosin attachment sites.
8. ATP molecules, bound to the myosin heads, are broken down to ADP and P, releasing energy, which is briefly stored in the myosin heads.
9. The heads of myosin myofilaments bind to exposed attachment sites on the actin myofilaments, forming cross-bridges. The P are released from the myosin heads.
10. The heads of the myosin myofilaments bend, causing the actin myofilaments to slide over the surface of the myosin myofilaments.
11. ADP molecules are released from the myosin head.
12. ATP molecules bind to the myosin heads and are broken down to ADP and P. Energy is stored in the myosin heads. The actin myofilament is released, and the myosin heads bend back to their resting position.
13. As long as $Ca^{2+}$ remain attached to troponin, and as long as ATP remains available, steps 8 through 11 are repeated, and the muscle continues to contract.

the T tubules to become more permeable to $Ca^{2+}$, and $Ca^{2+}$ diffuse into the sarcoplasm. The $Ca^{2+}$ bind to troponin molecules attached to the actin myofilaments (figure 7.8). This binding causes tropomyosin molecules to move into a groove along the actin molecule, exposing myosin attachment sites on the actin myofilament. The exposed attachment sites on the actin myofilament bind to the heads of the myosin myofilaments to form **cross-bridges** between the actin and myosin myofilaments.

Energy for muscle contraction is supplied to the muscles in the form of adenosine triphosphate (ATP), a high-energy molecule produced from the energy that is released during the metabolism of food (see chapters 3 and 17). The energy is released as ATP breaks down to adenosine diphosphate (ADP) and phosphate (P). During muscle contraction, the energy released from ATP is briefly stored in the myosin head. This energy is used to move the heads of the myosin myofilaments toward the center of the sarcomere, causing the actin myofilaments to slide past the myosin myofilaments. In the process, ADP and P are released from the myosin heads.

As a new ATP molecule attaches to the head of the myosin molecule, the cross-bridge is released, the ATP breaks down to ADP and P (which both remain bound to the myosin head), and the myosin head reassumes its original position, where it can attach to the next site. As long as $Ca^{2+}$ remain attached to troponin, and as long as ATP remains available, the cycle of cross-bridge formation, movement, and release repeats (table 7.1). A new ATP must bind to myosin before the cross-bridge can be released. After a person dies, ATP is not available, and the cross-bridges that have formed are not released, causing the muscles to become rigid. This condition is called **rigor mortis** (rig′er mōr′tis; stiffness + death).

Part of the energy from ATP involved in muscle contraction is required for the formation and movement of the cross-bridges, and part is released as heat. The heat released during muscle

**Relaxed muscle**

**1** Actin and myosin myofilaments in a relaxed muscle (*right*) and a contracted muscle (*#4 below*) are the same length. Myofilaments do not change length during muscle contraction.

Myosin myofilament     Actin myofilament     Sarcomere

**Contracting muscle**

**2** During contraction, actin myofilaments at each end of the sarcomere slide past the myosin myofilaments toward each other. As a result, the Z disks are brought closer together, and the sarcomere shortens.

Actin myofilaments move toward each other.     Sarcomere shortens as Z disks move toward each other.

**Contracting muscle**

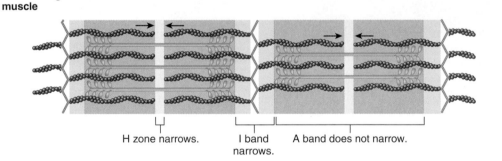

**3** As the actin myofilaments slide over the myosin myofilaments, the H zones (*yellow*) and the I bands (*blue*) narrow. The A bands, which are equal to the length of the myosin myofilaments, do not narrow, because the length of the myosin myofilaments does not change.

H zone narrows.     I band narrows.     A band does not narrow.

**Fully contracted muscle**

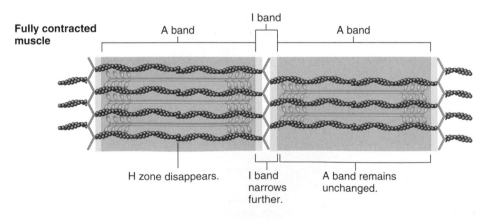

**4** In a fully contracted muscle, the ends of the actin myofilaments overlap, and the H zone disappears.

H zone disappears.     I band narrows further.     A band remains unchanged.

**PROCESS Figure 7.7** Sarcomere Shortening

contraction increases body temperature, which explains why a person becomes warmer during exercise. Shivering, a type of generalized muscle contraction, is one of the body's mechanisms for dealing with cold. The muscle movement involved in shivering produces heat, which raises the body temperature.

Muscle relaxation occurs as $Ca^{2+}$ are actively transported back into the sarcoplasmic reticulum (a process that requires ATP). As a consequence, the attachment sites on the actin molecules are once again covered by tropomyosin so that cross-bridges cannot reform.

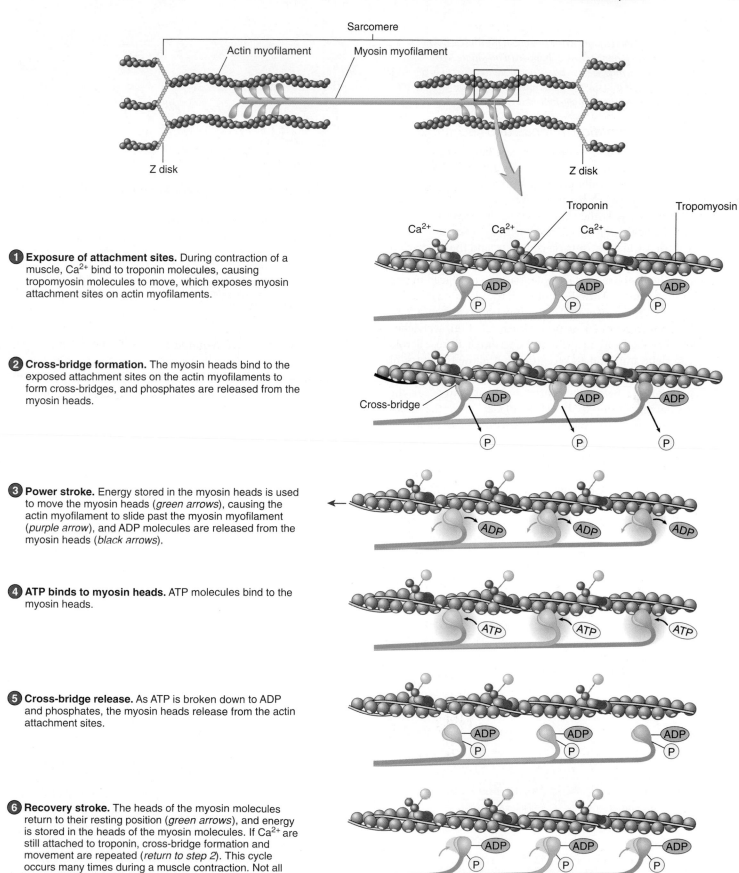

**1** **Exposure of attachment sites.** During contraction of a muscle, $Ca^{2+}$ bind to troponin molecules, causing tropomyosin molecules to move, which exposes myosin attachment sites on actin myofilaments.

**2** **Cross-bridge formation.** The myosin heads bind to the exposed attachment sites on the actin myofilaments to form cross-bridges, and phosphates are released from the myosin heads.

**3** **Power stroke.** Energy stored in the myosin heads is used to move the myosin heads (*green arrows*), causing the actin myofilament to slide past the myosin myofilament (*purple arrow*), and ADP molecules are released from the myosin heads (*black arrows*).

**4** **ATP binds to myosin heads.** ATP molecules bind to the myosin heads.

**5** **Cross-bridge release.** As ATP is broken down to ADP and phosphates, the myosin heads release from the actin attachment sites.

**6** **Recovery stroke.** The heads of the myosin molecules return to their resting position (*green arrows*), and energy is stored in the heads of the myosin molecules. If $Ca^{2+}$ are still attached to troponin, cross-bridge formation and movement are repeated (*return to step 2*). This cycle occurs many times during a muscle contraction. Not all cross-bridges form and release simultaneously.

**PROCESS Figure 7.8** Breakdown of ATP and Cross-Bridge Movement During Muscle Contraction

Muscular

**Predict 1**

*What happens if inadequate ATP is present in the muscle fiber before a stimulus is applied? What happens if adequate ATP is present within the muscle fiber, but action potentials occur at a frequency so great that Ca²⁺ are not transported back into the sarcoplasmic reticulum between individual action potentials?*

## Muscle Twitch, Summation, Tetanus, and Recruitment

A **muscle twitch** is the contraction of a muscle fiber in response to a stimulus. Because most muscle fibers are grouped into motor units, a muscle twitch usually involves all the muscle fibers in a motor unit. A muscle twitch has three phases (figure 7.9). The **lag phase,** or *latent phase,* is the time between the application of a stimulus and the beginning of contraction. The **contraction phase** is the time during which the muscle contracts, and the **relaxation phase** is the time during which the muscle relaxes.

During the lag phase, action potentials are produced in one or more motor neurons. An action potential travels along the axon of a motor neuron to a neuromuscular junction (see figure 7.5). Once the stimulus reaches the neuromuscular junction, acetylcholine must be released from the presynaptic terminal, diffuse across the synaptic cleft, and bind to receptors that allow the entry of Na⁺, which initiates an action potential on the postsynaptic membrane (see figure 7.6). Before the contraction phase can occur, the action potential must result in the release of Ca²⁺ from the sarcoplasmic reticulum and the formation of cross-bridges (see figure 7.8, steps 1–2).

The contraction phase results from cross-bridge movement and cycling (see figure 7.8, steps 3–6), which increases the tension produced by the muscle fibers (figure 7.9).

During the relaxation phase, Ca²⁺ are actively transported back into the sarcoplasmic reticulum. As Ca²⁺ diffuse away from the troponin molecules, tropomyosin molecules once again block the attachment sites. Cross-bridge formation is prevented, and the tension produced by the muscle fibers decreases (figure 7.9).

The strength of muscle contractions vary from weak to strong. For example, the force muscles generate to lift a feather is

**Figure 7.10** Multiple-Wave Summation

Stimuli 1–5 are of increasing frequency. For each stimulus, the *up arrow* indicates the start of stimulation, and the *down arrow* indicates the end of stimulation. Stimulus frequency 1 produces successive muscle twitches with complete relaxation between stimuli. Stimuli frequencies 2–4 do not allow complete relaxation between stimuli. Stimulus frequency 5 causes tetanus— no relaxation between stimuli.

much less than the force required to lift a 25-pound weight. The force of contraction a muscle produces is increased in two ways: (1) Summation involves increasing the force of contraction of the muscle fibers within the muscle, and (2) recruitment involves increasing the number of muscle fibers contracting.

In **summation,** the force of contraction of individual muscle fibers is increased by rapidly stimulating them. When stimulus frequency, which is the number of times a motor neuron is stimulated per second, is low, there is time for complete relaxation of muscle fibers between muscle twitches (figure 7.10, stimulus frequency 1). As stimulus frequency increases (figure 7.10, stimuli frequencies 2–4), there is not enough time between contractions for muscle fibers to relax completely. Thus, one contraction *summates,* or is added onto, a previous contraction. As a result, the overall force of contraction increases. **Tetanus** (tet′a-nus; convulsive tension) is a sustained contraction that occurs when the frequency of stimulation is so rapid that no relaxation occurs (figure 7.10, stimulus frequency 5). The increased force of contraction produced in summation and tetanus occurs because Ca²⁺ build up in myofibrils, which promotes cross-bridge formation and cycling. The buildup of Ca²⁺ occurs because the rapid production of action potentials in muscle fibers causes Ca²⁺ to be released from the sarcoplasmic reticulum faster than they are actively transported back into the sarcoplasmic reticulum.

In **recruitment,** the number of muscle fibers contracting is increased by increasing the number of motor units stimulated, and the muscle contracts with more force. When only a few motor units are stimulated, a small force of contraction is produced because only a small number of muscle fibers are contracting. As the number of motor units stimulated increases, more muscle fibers are stimulated to contract, and the force of contraction increases. Maximum force of contraction is produced in a given muscle when all the motor units of that muscle are stimulated (recruited).

If all the motor units in a muscle could be stimulated simultaneously, the resulting motion would be quick and jerky. However, because the motor units are recruited gradually, some are stimulated and held in tetanus while additional motor units are

**Figure 7.9** Phases of a Muscle Twitch

Hypothetical muscle twitch in a single muscle fiber. After the stimulus, there is a short lag phase, followed by a contraction phase and a relaxation phase.

recruited; thus, contractions are slow, smooth, and sustained. In the same way, smooth relaxation of muscle occurs because some motor units are held in tetanus while other motor units relax.

## Energy Requirements for Muscle Contraction

The energy needed for muscle contraction must come from ATP. Even resting muscle fibers require fairly large amounts of ATP for cell maintenance. It is therefore necessary for muscle fibers to produce ATP constantly. Most of the ATP required for muscle contraction is produced through aerobic (with oxygen) respiration; a smaller amount comes from anaerobic (without oxygen) respiration (see chapter 17).

**Aerobic respiration** requires $O_2$ and breaks down glucose to produce ATP, $CO_2$, and $H_2O$. It takes place in mitochondria located within the muscle fiber sarcoplasm between the myofibrils. Like all cells with a high metabolic rate, muscle fibers depend on large amounts of $O_2$ and thus contain large numbers of mitochondria.

**Anaerobic respiration,** which does not require $O_2$, results in the breakdown of glucose to yield ATP and lactic acid. Anaerobic respiration occurs in the cytoplasm of cells. It is important when $O_2$ availability is limited. Some of the lactic acid that results can diffuse out of the muscle fibers into the blood, but some remains in the muscle fibers, where it causes irritation and short-term pain. (Muscle pain that lasts for several days, however, indicates mechanical injury to the muscle.)

Aerobic respiration is much more efficient than anaerobic respiration. The metabolism of a glucose molecule by aerobic respiration theoretically can produce approximately 18 times as much ATP as is produced by anaerobic respiration. In addition, aerobic respiration can use a greater variety of nutrient molecules to produce ATP than can anaerobic respiration. For example, aerobic respiration can use fatty acids and amino acids to generate ATP.

Although aerobic respiration produces more ATP molecules than does anaerobic respiration, anaerobic respiration can occur faster. For example, during intense exercise, such as a sprint, smaller amounts of ATP are produced by anaerobic respiration. By using many glucose molecules, anaerobic respiration can rapidly produce much ATP, but the process can proceed for only a short time.

Muscle fibers cannot stockpile ATP in preparation for periods of activity. However, they can store another high-energy molecule, **creatine** (krē′a-tēn) **phosphate.** Creatine phosphate provides a means of storing energy that can be used rapidly to help maintain an adequate amount of ATP in a contracting muscle fiber. During periods of inactivity, as excess ATP is produced in the muscle fiber, the energy contained in the ATP is used to synthesize creatine phosphate. During periods of activity, the small reserves of ATP in the cell are used first. Then the energy stored in creatine phosphate is accessed quickly to produce ATP, which is used in muscle contraction and to restore ATP reserves. Figure 7.11 summarizes how aerobic and anaerobic respiration and creatine

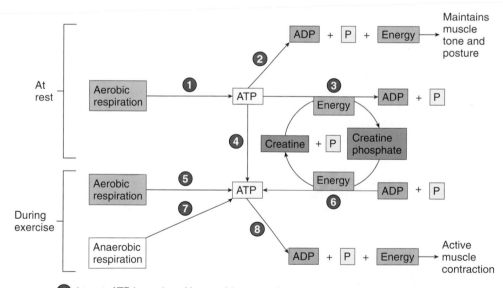

1. At rest, ATP is produced by aerobic respiration.

2. Small amounts of ATP are used in muscle contractions that maintain muscle tone and posture.

3. Excess ATP is used to produce creatine phosphate, an energy-storage molecule.

4. As exercise begins, ATP already in the cell is used first.

5. During moderate exercise, aerobic respiration provides most of the ATP necessary for muscle contraction.

6. Energy stored in creatine phosphate can also be used to produce ATP.

7. During times of extreme exercise, anaerobic respiration provides small amounts of ATP that can sustain muscle contraction for brief periods.

8. Throughout the time of exercise, ATP from all of these sources (4–7) provides energy for active muscle contraction.

**PROCESS Figure 7.11   Fate of ATP in Resting and Exercising Muscle**

phosphate production interact to keep the muscles supplied with the ATP they need.

Resting muscles or muscles undergoing long-term exercise, such as long-distance running, depend primarily on aerobic respiration for ATP synthesis. Although some glucose is used as an energy source, fatty acids are a more important energy source during both sustained exercise and resting conditions. During short periods of intense exercise, such as sprinting, ATP is used up more quickly than it can be produced by aerobic respiration. Once the ATP reserves begin to decrease, the energy stored in creatine phosphate is used to maintain ATP levels in the contracting muscle fiber. Once the creatine phosphate stores are depleted, anaerobic respiration becomes important. Anaerobic respiration rapidly metabolizes available glucose to provide enough ATP to support intense muscle contraction for up to 2–3 minutes. During intense exercise, glycogen stored in muscle fibers can be broken down to glucose, which can then be used to produce more ATP. Anaerobic metabolism is ultimately limited by depletion of glucose and buildup of lactic acid within the muscle fiber.

After intense exercise, the respiration rate and volume remain elevated for a time, even though the muscles are no longer actively contracting. This increased respiration provides the $O_2$ to pay back the *oxygen deficit*. The *recovery oxygen consumption* is the amount of $O_2$ needed in chemical reactions that occur to (1) convert lactic acid to glucose, (2) replenish the depleted ATP and creatine phosphate stores in muscle fibers, and (3) replenish $O_2$ stores in the lungs, blood, and muscles. After the lactic acid produced by anaerobic respiration is converted to glucose and creatine phosphate levels are restored, respiration rate returns to normal. The magnitude of the oxygen deficit depends on the intensity of the exercise, the length of time it was sustained, and the physical condition of the individual. The metabolic capacity of an individual in poor physical condition is much lower than that of a well-trained athlete. With exercise and training, a person's ability to carry out both aerobic and anaerobic activities is enhanced.

### Predict 2

*After a 10-mile run with a sprint at the end, a runner continues to breathe heavily for a time. Indicate the type of respiration that is producing energy during the run, during the sprint, and after the run.*

## Fatigue

Fatigue is a state of reduced work capacity. **Muscular fatigue** results when muscle fibers use ATP faster than they produce it and cannot remove lactic acid before it builds up. As a consequence, ATP falls to levels too low for muscle fibers to produce their maximum force of contraction, and contractions become weaker and weaker.

An example of muscle fatigue occurs when a runner collapses on the track and must be helped off. The runner's muscles can no longer function regardless of how determined the runner is. Under conditions of extreme muscular fatigue, muscles may become incapable of either contracting or relaxing. This condition, called **physiological contracture,** occurs when there is too little ATP to bind to myosin myofilaments. Because binding of

ATP to the myosin heads is necessary for cross-bridge release between the actin and myosin, the cross-bridges between the actin and myosin myofilaments cannot be broken, and the muscle cannot relax.

The most common type of fatigue, **psychological fatigue,** involves the central nervous system rather than the muscles themselves. The muscles are still capable of contracting, but the individual "perceives" that continued muscle contraction is impossible. A determined burst of activity in a tired runner in response to pressure from a competitor is an example of how psychological fatigue can be overcome.

## Types of Muscle Contractions

Muscle contractions are classified as either isometric or isotonic. In **isometric** (equal distance) **contractions,** the length of the muscle does not change, but the amount of tension increases during the contraction process. Isometric contractions are responsible for the constant length of the body's postural muscles, such as the muscles of the back. On the other hand, in **isotonic** (equal tension) **contractions,** the amount of tension produced by the muscle is constant during contraction, but the length of the muscle decreases. Movements of the arms or fingers are predominantly isotonic contractions. Most muscle contractions are a combination of isometric and isotonic contractions in which the muscles shorten and the degree of tension increases.

**Concentric** (kon-sen′trik) **contractions** are isotonic contractions in which muscle tension increases as the muscle shortens. Many common movements are produced by concentric muscle contractions. **Eccentric** (ek-sen′trik) **contractions** are isotonic contractions in which tension is maintained in a muscle, but the opposing resistance causes the muscle to lengthen. Eccentric contractions are used when a person slowly lowers a heavy weight. Substantial force is produced in muscles during eccentric contractions, and muscles can be injured during repetitive eccentric contractions, as sometimes occurs in the hamstring muscles when a person runs downhill.

## Muscle Tone

**Muscle tone** is the constant tension produced by body muscles over long periods of time. Muscle tone is responsible for keeping the back and legs straight, the head in an upright position, and the abdomen from bulging. Muscle tone depends on a small percentage of all the motor units in a muscle being stimulated at any point in time, causing their muscle fibers to contract tetanically and out of phase with one another.

## Slow-Twitch and Fast-Twitch Fibers

Muscle fibers are sometimes classified as either slow-twitch or fast-twitch fibers (table 7.2). This classification is based on differences in the rod portion of the myosin myofilament (see figure 7.2). Slow-twitch fibers contain type I myosin as the predominant or even exclusive type. Fast-twitch fibers contain either type IIa or type IIb myosin myofilaments. Each of these three myosin types is the product of a different myosin gene. Fast-twitch muscle fibers contract quickly and fatigue quickly, whereas slow-twitch muscle fibers contract more slowly and are more resistant

| TABLE 7.2 | Characteristics of Skeletal Muscle Fiber Types | | |
|---|---|---|---|
| | Slow-Twitch (Type I) | Fast-Twitch (Type IIa) | Fast-Twitch (Type IIb) |
| Fiber Diameter | Smallest | Intermediate | Largest |
| Myoglobin Content | High | High | Low |
| Mitochondria | Many | Many | Few |
| Metabolism | High aerobic capacity | High anaerobic capacity; intermediate aerobic capacity | Highest anaerobic capacity |
| Fatigue Resistance | High | Intermediate | Low |
| Myosin Head Activity | Slow | Fast | Fast |
| Glycogen Concentration | Low | High | High |
| Functions | Maintenance of posture; endurance activities | Endurance activities in endurance-trained muscles | Rapid, intense movement of short duration (sprinting) |

to fatigue. Type IIb fibers can contract 10 times faster than type I fibers, and type IIa fibers contract at an intermediate speed. Type IIa fibers are more fatigue-resistant than type IIb fibers. Fast-twitch muscles have larger stores of glycogen and are well adapted for performing anaerobic respiration, whereas slow-twitch muscles contain more mitochondria and are better suited for aerobic respiration.

The white meat of a chicken's breast is composed mainly of fast-twitch fibers. The muscles are adapted to contract rapidly for a short time but fatigue quickly. Chickens normally do not fly long distances. They spend most of their time walking. Ducks, on the other hand, fly for much longer periods and over greater distances. The red, or dark, meat of a chicken's leg or a duck's breast is composed of slow-twitch fibers. The darker appearance is due partly to a richer blood supply and partly to the presence of **myoglobin,** which stores oxygen temporarily. Myoglobin can continue to release oxygen in a muscle even when a sustained contraction has interrupted the continuous flow of blood.

Humans exhibit no clear separation of slow-twitch and fast-twitch muscle fibers in individual muscles. Most muscles have both types of fibers, although the number of each type varies in a given muscle. The large postural muscles contain more slow-twitch fibers, whereas muscles of the upper limb contain more fast-twitch fibers.

Average, healthy, active adults have roughly equal numbers of slow- and fast-twitch fibers in their muscles and over three times as many type IIa as type IIb fibers. In fact, athletes who are able to perform a variety of anaerobic and aerobic exercises tend to have a balanced mixture of fast-twitch and slow-twitch muscle fibers. However, a world-class sprinter may have over 80% type II fibers, with type IIa slightly predominating, whereas a world-class endurance athlete may have 95% type I fibers.

The ratio of muscle fiber types in a person's body apparently has a large hereditary component but can also be considerably influenced by training. Exercise increases the blood supply to muscles, the number of mitochondria per muscle fiber, and the number of myofibrils and myofilaments, thus causing muscle fibers to enlarge, or **hypertrophy** (hī-per′trō-fē). With weight training, type IIb myosin myofilaments can be replaced by type IIa myosin myofilaments as muscles enlarge. Muscle nuclei quit expressing type IIb genes and begin expressing type IIa genes, which are more resistant to fatigue. If the exercise stops, the type IIa genes turn off, and the type IIb genes turn back on. Vigorous exercise programs can cause a limited number of type I myofilaments to be replaced by type IIa myofilaments.

The number of cells in a skeletal muscle remains somewhat constant following birth. Enlargement of muscles after birth is primarily the result of an increase in the size of the existing muscle fibers. As people age, the number of muscle fibers actually decreases. However, there are undifferentiated cells just below the endomysium called **satellite cells.** When stimulated, satellite cells can differentiate and develop into a limited number of new, functional muscle fibers. These cells are stimulated by the destruction of existing muscle fibers, such as by injury or disease or during intensive strength training.

# SMOOTH MUSCLE AND CARDIAC MUSCLE

Smooth muscle cells are small and spindle-shaped, usually with one nucleus per cell (table 7.3). They contain less actin and myosin than do skeletal muscle cells, and the myofilaments are not organized into sarcomeres. As a result, smooth muscle cells are

## TABLE 7.3    Comparison of Muscle Types

| | Skeletal Muscle | Cardiac Muscle | Smooth Muscle |
|---|---|---|---|
| Location | Attached to bone | Heart | Walls or hollow organs, blood vessels, and glands |
| Appearance | | | |
| Cell Shape | Long, cylindrical | Branched | Spindle-shaped |
| Nucleus | Multiple, peripheral | Usually single, central | Single, central |
| Special Features | ———— | Intercalated disks | Cell-to-cell attachments |
| Striations | Yes | Yes | No |
| Autorhythmic | No | Yes | Yes |
| Control | Voluntary | Involuntary | Involuntary |
| Function | Move the whole body | Contract heart to propel blood through the body | Compress organs, ducts, tubes, and so on |

not striated. Smooth muscle cells contract more slowly than skeletal muscle cells when stimulated by neurotransmitters from the nervous system and do not develop an oxygen deficit. The resting membrane potential of some smooth muscle cells fluctuates between slow depolarization and repolarization phases. As a result, smooth muscle cells can periodically and spontaneously generate action potentials that cause the smooth muscle cells to contract. The resulting periodic spontaneous contraction of smooth muscle is called **autorhythmicity.** Smooth muscle is under involuntary control, whereas skeletal muscle is under voluntary motor control. Some hormones, such as those that regulate the digestive system, can stimulate smooth muscle to contract.

Smooth muscle cells are organized to form layers. Most of those cells have gap junctions, specialized cell-to-cell contacts (see chapter 4), that allow action potentials to spread to all the smooth muscle cells in a tissue. Thus, all the smooth muscle cells tend to function as a unit and contract at the same time.

Cardiac muscle shares some characteristics with both smooth and skeletal muscle (table 7.3). Cardiac muscle cells are long, striated, and branching, with usually only one nucleus per cell. The actin and myosin myofilaments are organized into sarcomeres, but the distribution of myofilaments is not as uniform as in skeletal muscle. As a result, cardiac muscle cells are striated, but not as distinctly striated as skeletal muscle. When stimulated by neurotransmitters, the rate of cardiac muscle contraction is between that of smooth and skeletal muscle. Cardiac muscle contraction is autorhythmic. Cardiac muscle exhibits limited anaerobic respiration. Instead, it continues to contract at a level that can be sustained by aerobic respiration and consequently does not fatigue.

Cardiac muscle cells are connected to one another by **intercalated** (in-ter′ka-lā-ted) **disks.** Intercalated disks are specialized structures that include tight junctions and gap junctions and that facilitate action potential conduction between the cells. This cell-to-cell connection allows cardiac muscle cells to function as a unit. As a result, an action potential in one cardiac muscle cell can stimulate action potentials in adjacent cells, causing all to contract together. As with smooth muscle, cardiac muscle is under involuntary control and is influenced by hormones, such as epinephrine.

# SKELETAL MUSCLE ANATOMY

## General Principles

Most muscles extend from one bone to another and cross at least one joint. At each end, the muscle is connected to the bone by a **tendon.** Some broad, sheetlike tendons are called **aponeuroses** (ap′ō-noo-rō′sēz). A **retinaculum** (ret-i-nak′ū-lum; bracelet) is a band of connective tissue that holds down the tendons at each wrist and ankle. Muscle contraction causes most body movements by pulling one of the bones toward the other across the movable joint. Some muscles are not attached to bone at both ends. For example, some facial muscles attach to the skin, which moves as the muscles contract.

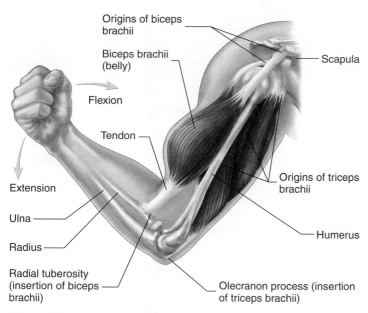

Origins of biceps brachii

Biceps brachii (belly)

Flexion

Scapula

Tendon

Extension

Origins of triceps brachii

Ulna

Radius

Humerus

Radial tuberosity (insertion of biceps brachii)

Olecranon process (insertion of triceps brachii)

**Figure 7.12 Muscle Attachment**

Muscles are attached to bones by tendons. The biceps brachii has two origins that originate on the scapula. The triceps brachii has three origins that originate on the scapula and humerus. The biceps tendon inserts onto the radial tuberosity and onto nearby connective tissue. The triceps brachii inserts onto the olecranon of the ulna.

The two points of attachment of each muscle are its origin and insertion. The **origin,** also called the *head,* is the most stationary end of the muscle. The **insertion** is the end of the muscle attached to the bone undergoing the greatest movement. Origins are usually, but not always, proximal or medial to the insertion of a given muscle. The part of the muscle between the origin and the insertion is the **belly** (figure 7.12). Some muscles have multiple origins; for example, the biceps brachii has two, and the triceps brachii has three.

Muscles are typically grouped so that the action of one muscle or group of muscles is opposed by that of another muscle or group of muscles. For example, the biceps brachii flexes the elbow, and the triceps brachii extends the elbow. A muscle that accomplishes a certain movement, such as flexion, is called the **agonist** (ag′ō-nist). A muscle acting in opposition to an agonist is called an **antagonist** (an-tag′ō-nist). For example, when flexing the elbow, the biceps brachii is the agonist, whereas the triceps brachii, which relaxes and stretches to allow the elbow to bend, is the antagonist. When extending the elbow, the muscles' roles are reversed; the triceps brachii is the agonist, and the biceps brachii is the antagonist.

Muscles also tend to function in groups to accomplish specific movements. For example, the deltoid, biceps brachii, and pectoralis major all help flex the shoulder. Furthermore, many muscles are members of more than one group, depending on the type of movement being produced. For example, the anterior part of the deltoid muscle functions with the flexors of the shoulder, whereas the posterior part functions with the extensors of the shoulder. Members of a group of muscles working together

to produce a movement are called **synergists** (sin′er-jistz). For example, the biceps brachii and the brachialis are synergists in elbow flexion. Among a group of synergists, if one muscle plays the major role in accomplishing the desired movement, it is called the **prime mover.** The brachialis is the prime mover in flexing the elbow. **Fixators** are muscles that hold one bone in place relative to the body while a usually more distal bone is moved. For example, the muscles of the scapula act as fixators to hold the scapula in place while other muscles contract to move the humerus.

## Nomenclature

Most muscles have descriptive names (figure 7.13). Some muscles are named according to their location, such as the pectoralis (chest) muscles. Other muscles are named according to their origin and insertion, such as the brachioradialis (*brachio,* arm) muscle, which extends from the arm to the radius. Some muscles are named according to the number of origins, such as the biceps (*bi,* two + *ceps,* head) brachii, which has two origins, and some according to their function, such as the flexor digitorum, which flexes the digits (fingers). Other muscles are named according to their size (vastus, large), their shape (deltoid, triangular), or the orientation of their fasciculi (rectus, straight). Recognizing the descriptive nature of muscle names makes learning those names much easier.

Examining surface anatomy can also be a great help in understanding muscle anatomy. We have pointed out certain muscles of the upper and lower limbs that can be seen on the surface of the body, and figure 7.13 shows the most superficial muscles. Some muscles are especially well developed in bodybuilders (figure 7.14).

## Muscles of the Head and Neck

The muscles of the head and neck include those involved in forming facial expressions, chewing, moving the tongue, swallowing, producing sounds, moving the eyes, and moving the head and neck.

### Facial Expression

Several muscles act on the skin around the eyes and eyebrows (figure 7.15 and table 7.4). The **occipitofrontalis** (ok-sip′i-tō-frŭn-tā′lis) raises the eyebrows. The occipital and frontal portions of the muscle are connected by the epicranial aponeurosis. The **orbicularis oculi** (ōr-bik′ū-la′ris, circular + ok′ū-lī, eye) encircle the eyes, tightly close the eyelids, and cause "crow's feet" wrinkles in the skin at the lateral corners of the eyes.

Several other muscles function in moving the lips and the skin surrounding the mouth (figure 7.15). The **orbicularis oris** (ōr′is; mouth), which encircles the mouth, and the **buccinator** (buk′sĭ-nā′tōr; *bucca,* cheek) are sometimes called the kissing muscles because they pucker the mouth. The buccinator also flattens the cheeks as in whistling or blowing a trumpet and is therefore sometimes called the trumpeter's muscle. Smiling is accomplished primarily by the **zygomaticus** (zī′gō-mat′i-kŭs) muscles, which elevate the upper lip and corner of the mouth. Sneering is accomplished by the **levator labii superioris**

Muscular

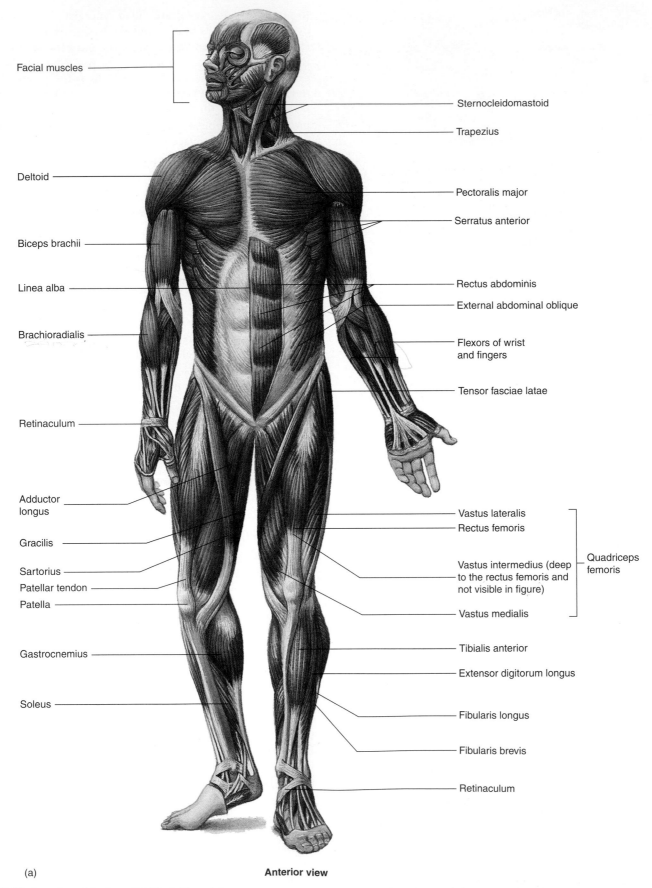

Facial muscles

Sternocleidomastoid

Trapezius

Deltoid

Pectoralis major

Serratus anterior

Biceps brachii

Linea alba

Rectus abdominis

External abdominal oblique

Brachioradialis

Flexors of wrist
and fingers

Tensor fasciae latae

Retinaculum

Adductor
longus

Vastus lateralis

Rectus femoris

Gracilis

Vastus intermedius (deep
to the rectus femoris and
not visible in figure)

Quadriceps
femoris

Sartorius

Patellar tendon

Patella

Vastus medialis

Gastrocnemius

Tibialis anterior

Extensor digitorum longus

Soleus

Fibularis longus

Fibularis brevis

Retinaculum

(a)                                    **Anterior view**

## Figure 7.13  Overview of the Superficial Body Musculature

*Red* is muscle; *white* is connective tissue, such as tendons, aponeuroses, and retinacula.

Muscular

Sternocleidomastoid

Seventh cervical vertebra

Infraspinatus

Teres minor

Teres major

Triceps brachii

Extensors
of the wrist
and fingers

Hamstring
muscles

Semitendinosus

Biceps femoris

Semimembranosus

Fibularis longus

Fibularis brevis

Trapezius

Deltoid

Latissimus dorsi

External abdominal
oblique

Gluteus medius

Gluteus maximus

Adductor magnus

Iliotibial tract

Gracilis

Gastrocnemius

Soleus

Calcaneal tendon
(Achilles tendon)

(b)

**Posterior view**

**Figure 7.13** **Overview of the Superficial Body Musculature** (*continued*)

**Anterior view**

**Lateral view**

## Figure 7.14 Bodybuilders

Name as many muscles as you can in these photos. Compare these photos to the labeled muscles in figure 7.13.

(le-vā′ter lā′bē-ī soo-pēr′ē-ōr′is) because the muscle elevates one side of the upper lip. Frowning and pouting are largely performed by the **depressor anguli oris** (dē-pres′ŏr an′gū-lī ōr′ŭs), which depresses the corner of the mouth.

*Harry Wolf, a notorious flirt, on seeing Sally Gorgeous, raises his eyebrows, winks, whistles, and smiles. Name the facial muscles he uses to carry out this communication. Sally, thoroughly displeased with this exhibition, frowns and sneers in disgust. What muscles does she use?*

### Mastication

The four pairs of muscles for chewing, or **mastication** (mas-ti-kā′shŭn), are some of the strongest muscles in the body (table 7.5). The **temporalis** (tem′pŏ-rā′lis) and **masseter** (ma-sē′ter) muscles (figure 7.15) can be easily seen and felt on the side of the head during mastication. The **pterygoid** (ter′ĭ-goyd) muscles, consisting of two pairs, are deep to the mandible.

### Tongue and Swallowing Muscles

The tongue is very important in mastication and speech. It moves food around in the mouth and, with the buccinator muscle, holds the food in place while the teeth grind the food. The tongue pushes food up to the palate and back toward the pharynx to initiate swallowing. The tongue consists of a mass of **intrinsic muscles,** which are located entirely within the tongue and change its shape. The **extrinsic muscles** are located outside the tongue but are attached to and move the tongue (figure 7.16 and table 7.6).

Swallowing involves a number of structures and their associated muscles, including the hyoid muscles, soft palate, pharynx (throat), and larynx (voicebox). The **hyoid** (hi′oyd) **muscles** are divided into a suprahyoid group (superior to the hyoid bone) and an infrahyoid group (inferior to the hyoid bone) (figure 7.16 and table 7.6). When the suprahyoid muscles hold the hyoid bone in place from above, the infrahyoid muscles can elevate the larynx. To observe this effect, place your hand on your larynx (Adam's apple) and swallow.

The muscles of the soft palate close the posterior opening to the nasal cavity during swallowing, preventing food and liquid from entering the nasal cavity. When we swallow, muscles elevate the pharynx and larynx and then constrict the pharynx. Specifically, the **pharyngeal** (fă-rin′jē-ăl) **elevators** elevate the pharynx, and the **pharyngeal constrictors** constrict the pharynx from superior to inferior, forcing the food into the esophagus. Pharyngeal muscles also open the auditory tube, which connects the middle ear to the pharynx. Opening the auditory tube equalizes the pressure between the middle ear and the atmosphere. This is why it is sometimes helpful to chew gum or swallow when ascending or descending a mountain in a car or changing altitude in an airplane.

### Neck Muscles

The deep neck muscles (figure 7.17 and table 7.7) include neck flexors, located along the anterior surfaces of the vertebral bodies,

Muscular

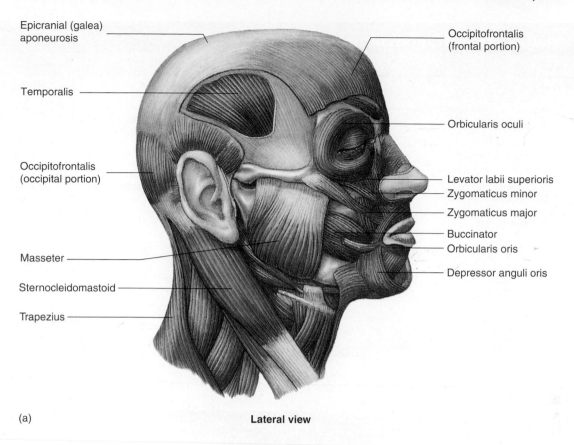

Epicranial (galea) aponeurosis

Temporalis

Occipitofrontalis (occipital portion)

Masseter

Sternocleidomastoid

Trapezius

Occipitofrontalis (frontal portion)

Orbicularis oculi

Levator labii superioris
Zygomaticus minor
Zygomaticus major
Buccinator
Orbicularis oris
Depressor anguli oris

(a)                    **Lateral view**

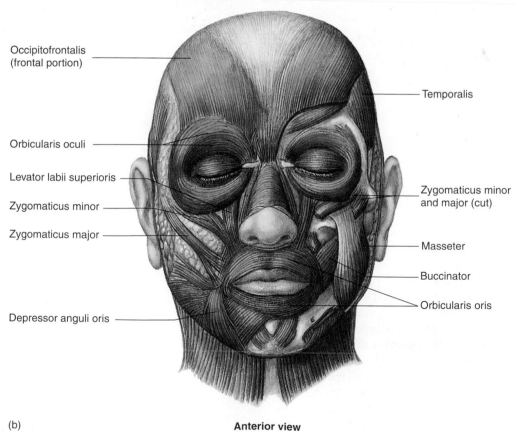

Occipitofrontalis (frontal portion)

Orbicularis oculi

Levator labii superioris

Zygomaticus minor

Zygomaticus major

Depressor anguli oris

Temporalis

Zygomaticus minor and major (cut)

Masseter

Buccinator

Orbicularis oris

(b)                    **Anterior view**

**Figure 7.15**  Muscles of Facial Expression

Muscular

## TABLE 7.4  Muscles of Facial Expression (see figure 7.15)

| Muscle | Origin | Insertion | Action |
|---|---|---|---|
| Buccinator (buk′sĭ-nā′tōr) | Maxilla and mandible | Orbicularis oris at angle of mouth | Retracts angle of mouth; flattens cheek |
| Depressor anguli oris (dē-pres′ōr an′gŭ-lī ōr′ŭs) | Lower border of mandible | Skin of lip near angle of mouth | Depresses angle of mouth |
| Levator labii superioris (le-vā′ter lā′bē-ī soo-pēr′ē-ōr′is) | Maxilla | Skin and orbicularis oris of upper lip | Elevates upper lip |
| Occipitofrontalis (ok-sip′i-tō-frŭn′tā′lis) | Occipital bone | Skin of eyebrow and nose | Moves scalp; elevates eyebrows |
| Orbicularis oculi (ōr-bik′ū-lā′ris ok′ū-lī) | Maxilla and frontal bones | Circles orbit and inserts near origin | Closes eye |
| Orbicularis oris (ōr-bik′ū-lā′ris ōr′is) | Nasal septum, maxilla, and mandible | Fascia and other muscles of lips | Closes lip |
| Zygomaticus major (zī′gō-mat′i-kŭs) | Zygomatic bone | Angle of mouth | Elevates and abducts upper lip |
| Zygomaticus minor (zī′gō-mat′i-kŭs) | Zygomatic bone | Orbicularis oris of upper lip | Elevates and abducts upper lip |

## TABLE 7.5  Muscles of Mastication (see figure 7.15)

| Muscle | Origin | Insertion | Action |
|---|---|---|---|
| Temporalis (tem′pŏ-rā′lis) | Temporal fossa | Anterior portion of mandibular ramus and coronoid process | Elevates and retracts mandible; involved in excursion |
| Masseter (mă-sē′ter) | Zygomatic arch | Lateral side of mandibular ramus | Elevates and protracts mandible; involved in excursion |
| Lateral pterygoid (ter′i-goyd) (not shown in illustration) | Lateral pterygoid plate and greater wing of sphenoid | Condylar process of mandible and articular disk | Protracts and depresses mandible; involved in excursion |
| Medial pterygoid (not shown in illustration) | Lateral pterygoid plate of sphenoid and tuberosity of maxilla | Medial surface of mandible | Protracts and elevates mandible; involved in excursion |

and neck extensors, located posteriorly. Rotation and lateral flexion of the head are accomplished by lateral and posterior neck muscles. The **sternocleidomastoid** (ster′nō-klī′dō-mas′toyd) muscle (see figure 7.15a), the prime mover of the lateral muscle group, is easily seen on the anterior and lateral sides of the neck. Contraction of only one sternocleidomastoid muscle rotates the head. Contraction of both sternocleidomastoids flexes the neck or extends the head, depending on what the other neck muscles are doing. **Torticollis** (tōr′ti-kol′is; a twisted neck), or wry neck, may result from injury to one of the sternocleidomastoid muscles. It is sometimes caused by damage to a baby's neck muscles during a difficult birth and usually can be corrected by exercising the muscle.

**Predict 4**

*Shortening of the right sternocleidomastoid muscle rotates the head in which direction?*

## Trunk Muscles

Trunk muscles include those that move the vertebral column, the thorax and abdominal wall, and the pelvic floor.

### Muscles Moving the Vertebral Column

In humans, the back muscles are very strong to maintain erect posture. The **erector spinae** (ē-rek′tōr spī′nē) group of muscles on each side of the back are primarily responsible for keeping the back straight and the body erect (table 7.8 and figure 7.17). **Deep back muscles,** located between the spinous and transverse processes of adjacent vertebrae, are responsible for several movements of the vertebral column, including extension, lateral flexion, and rotation. When the deep back muscles are stretched abnormally or torn, muscle strains and sprains of lumbar vertebral ligaments can occur, resulting in low back pain. Treatments

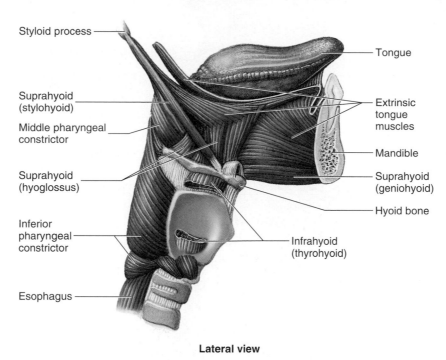

Styloid process

Tongue

Suprahyoid
(stylohyoid)

Extrinsic
tongue
muscles

Middle pharyngeal
constrictor

Mandible

Suprahyoid
(hyoglossus)

Suprahyoid
(geniohyoid)

Hyoid bone

Inferior
pharyngeal
constrictor

Infrahyoid
(thyrohyoid)

Esophagus

**Lateral view**

**Figure 7.16  Tongue and Swallowing Muscles**

Muscles of the tongue, hyoid, pharynx, and larynx as seen from the right.

**TABLE 7.6    Tongue and Swallowing Muscles (see figure 7.16)**

| Muscle | Origin | Insertion | Action |
|---|---|---|---|
| Tongue muscles | | | |
| Intrinsic (not shown) | Inside tongue | Inside tongue | Changes shape of tongue |
| Extrinsic | Bones around oral cavity or soft palate | Onto tongue | Moves tongue |
| Hyoid muscles | | | |
| Suprahyoid (e.g., geniohyoid, stylohyoid, and hyoglossus) | Base of skull, mandible | Hyoid bone | Elevates or stabilizes hyoid |
| Infrahyoid (e.g., thyrohyoid) | Sternum, larynx | Hyoid bone | Depresses or stabilizes hyoid |
| Soft palate (not shown) | Skull or soft palate | Palate, tongue, or pharynx | Moves soft palate, tongue, or pharynx |
| Pharyngeal muscles | | | |
| Elevators (not shown) | Soft palate and auditory tube | Pharynx | Elevate pharynx |
| Constrictors Superior (not shown) Middle Inferior | Larynx and hyoid | Pharynx | Constrict pharynx |

include anti-inflammatory medication and RICE (*r*est, *i*ce, *c*ompression, and *e*levation). Low back exercises can also help the problem.

*Thoracic Muscles*

The muscles of the thorax (figure 7.18 and table 7.9) are involved almost entirely in the process of breathing. The **external intercostals** (in′ter-kos′tŭlz; between ribs) elevate the ribs during inspiration. The **internal intercostals** contract during forced expiration, depressing the ribs. However, the major movement produced in the thorax during quiet breathing is accomplished by the dome-shaped **diaphragm** (dī′ă-fram). When it contracts, the dome is flattened, causing the volume of the thoracic cavity to increase, resulting in inspiration.

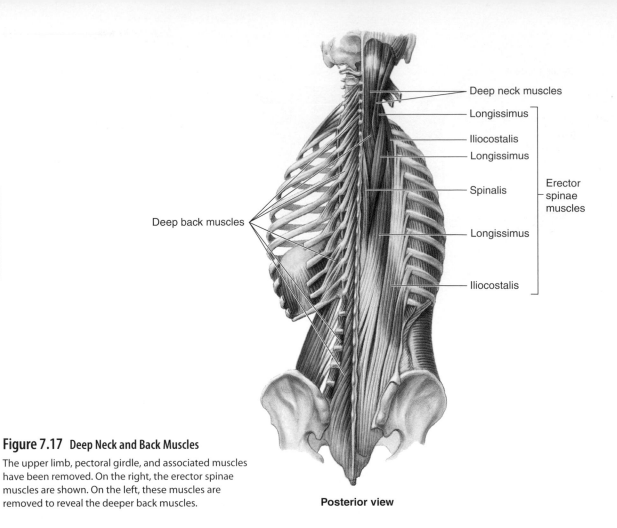

**Figure 7.17  Deep Neck and Back Muscles**

The upper limb, pectoral girdle, and associated muscles have been removed. On the right, the erector spinae muscles are shown. On the left, these muscles are removed to reveal the deeper back muscles.

**Posterior view**

| TABLE 7.7 | Neck Muscles (see figures 7.13, 7.15, 7.17, and 7.21) | | |
|---|---|---|---|
| **Muscle** | **Origin** | **Insertion** | **Action** |
| Deep neck muscles | | | |
|   Flexors (not shown) | Anterior side of vertebrae | Base of skull | Flex head and neck |
|   Extensors | Posterior side of vertebrae | Base of skull | Extend head and neck |
| Sternocleidomastoid (ster′nō-klī′dō-mas′toyd) | Manubrium of sternum and medial part of clavicle | Mastoid process and nuchal line of skull | Individually rotate head; together flex neck or extend head |
| Trapezius (tra-pē′zē-ŭs) | Posterior surface of skull and upper vertebral column (C7–T12) | Clavicle, acromion process, and scapular spine | Extends head and neck |

| TABLE 7.8 | Muscles Acting on the Vertebral Column (see figure 7.17) | | |
|---|---|---|---|
| **Muscle** | **Origin** | **Insertion** | **Action** |
| Superficial | | | |
|   Erector spinae (ĕ-rek′tōr spī′nē) divides into three columns: | Sacrum, ilium, vertebrae, and ribs | Ribs, vertebrae, and skull | Extends vertebral column |
|     Iliocostalis (il′ē-ō-kos-tā′lis) | | | |
|     Longissimus (lon-gis′i-mŭs) | | | |
|     Spinalis (spī-nā′lis) | | | |
| Deep back muscles | Vertebrae | Vertebrae | Extend vertebral column and help bend vertebral column laterally |

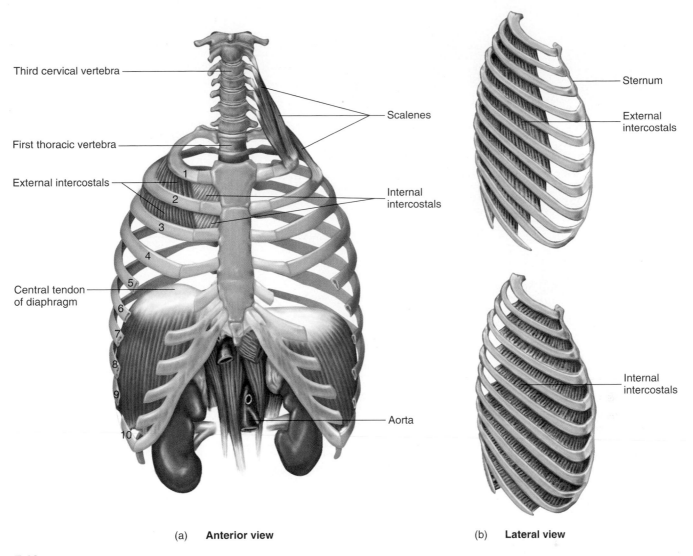

(a)    **Anterior view**                    (b)    **Lateral view**

**Figure 7.18**  Muscles of the Thorax

(*a*) Anterior view shows a few selected intercostal muscles and the diaphragm. (*b*) Lateral view shows the external and internal intercostals.

| TABLE 7.9 | Muscles of the Thorax (see figures 7.18 and 7.21) | | |
|---|---|---|---|
| **Muscle** | **Origin** | **Insertion** | **Action** |
| Scalenes (skā′lēnz) | Cervical vertebrae | First and second ribs | Inspiration; elevate ribs |
| External intercostals (in′ter-kos′tŭlz) | Inferior edge of each rib | Superior edge of next rib below origin | Inspiration; elevate ribs |
| Internal intercostals (in′ter-kos′tŭlz) | Superior edge of each rib | Inferior edge of next rib above origin | Forced expiration; depress ribs |
| Diaphragm (dī′ă-fram) | Inferior ribs, sternum, and lumbar vertebrae | Central tendon of diaphragm | Inspiration; depress floor of thorax |

## Abdominal Wall Muscles

The muscles of the anterior abdominal wall (figure 7.19 and table 7.10) flex and rotate the vertebral column, compress the abdominal cavity, and hold in and protect the abdominal organs. In a relatively muscular person with little fat, a vertical indentation, extending from the sternum through the navel to the pubis, is visible. This tendinous area of the abdominal wall, called the **linea alba** (lin′ē-ă al′bă; white line), consists of white connective tissue rather than muscle. On each side of the linea alba is the **rectus abdominis** (rek′tŭs ab-dom′ĭ-nis; *rectus*, straight) muscle. **Tendinous intersections** cross the rectus abdominis at three or more locations, causing the abdominal wall of a lean, well-muscled

Muscular

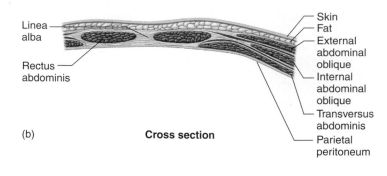

**Rectus abdominis (sheath removed)**

**Rectus abdominis (covered by sheath)**

**Linea alba**

**Umbilicus**

**External abdominal oblique**

**External abdominal oblique**

**Internal abdominal oblique**

**Transversus abdominis**

**Tendinous intersection**

(a)     **Anterior view**

**Linea alba**

**Rectus abdominis**

(b)     **Cross section**

**Skin**
**Fat**
**External abdominal oblique**
**Internal abdominal oblique**
**Transversus abdominis**
**Parietal peritoneum**

**Figure 7.19**   Muscles of the Anterior Abdominal Wall

(a) In this anterior view, windows reveal the various muscle layers. (b) A cross-sectional view of the muscle layers.

| TABLE 7.10 | Muscles of the Abdominal Wall (see figures 7.13, 7.19, 7.21) | | |
|---|---|---|---|
| **Muscle** | **Origin** | **Insertion** | **Action** |
| Rectus abdominis (rek'tŭs ab-dom'i-nis) | Pubic crest and pubic symphysis | Xiphoid process and inferior ribs | Flexes vertebral column; compresses abdomen |
| External abdominal oblique | Ribs 5–12 | Iliac crest, inguinal ligament, and fascia of rectus abdominis | Compresses abdomen; flexes and rotates vertebral column |
| Internal abdominal oblique | Iliac crest, inguinal ligament, and lumbar fascia | Ribs 10–12 and fascia of rectus abdominis | Compresses abdomen; flexes and rotates vertebral column |
| Transversus abdominis (trans-ver'sŭs ab-dom'in-is) | Costal cartilages 7–12, lumbar fascia, iliac crest, and inguinal ligament | Xiphoid process, fascia of rectus abdominis, and pubic tubercle | Compresses abdomen |

person to appear segmented. Lateral to the rectus abdominis are three layers of muscle. From superficial to deep, these muscles are the **external abdominal oblique,** the **internal abdominal oblique,** and the **transversus abdominis** (trans-ver'sŭs ab-dom'in-is) muscles. The fasciculi of these three muscle layers are oriented in different directions. When these muscles contract, they flex and rotate the vertebral column or compress the abdominal contents.

### Pelvic Floor and Perineal Muscles

The pelvis is a ring of bone with an inferior opening that is closed by a muscular floor through which the anus and the openings of the urinary tract and reproductive tract penetrate. Most of the **pelvic floor,** also referred to as the *pelvic diaphragm,* is formed by the **levator ani** (le-vā'ter ā'nī) muscle. The area inferior to the

pelvic floor is the **perineum** (per'i-nē'ŭm), which contains a number of muscles associated with the male or female reproductive structures (figure 7.20 and table 7.11). Several of these muscles help regulate urination and defecation.

## Upper Limb Muscles

The muscles of the upper limb include those that attach the limb and pectoral girdle to the body and those in the arm, forearm, and hand.

### Scapular Movements

The upper limb is primarily connected to the body by muscles. The muscles that attach the scapula to the thorax and move the scapula include the **trapezius** (tra-pē'zē-ŭs), the **levator scapulae**

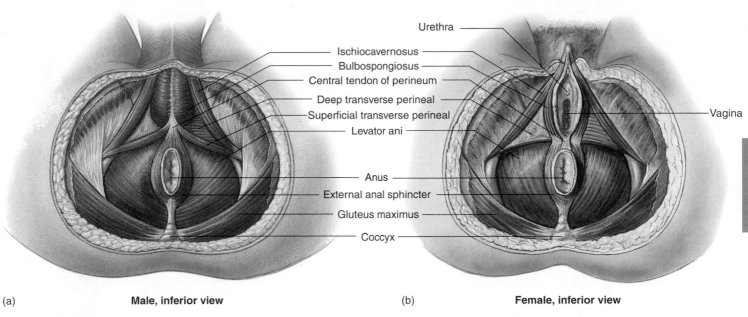

Urethra

Ischiocavernosus
Bulbospongiosus
Central tendon of perineum
Deep transverse perineal
Superficial transverse perineal
Levator ani

Vagina

Anus
External anal sphincter
Gluteus maximus
Coccyx

(a)    **Male, inferior view**

(b)    **Female, inferior view**

**Figure 7.20** Muscles of the Pelvic Floor and Perineum

| TABLE 7.11 | Muscles of the Pelvic Floor and Perineum (see figure 7.20) | | |
|---|---|---|---|
| **Muscle** | **Origin** | **Insertion** | **Action** |
| **Pelvic Floor** | | | |
| Levator ani (le′vā-ter ā′nī) | Posterior pubis and ischial spine | Sacrum and coccyx | Elevates anus; supports pelvic viscera |
| **Perineum** | | | |
| Bulbospongiosus (bul′bō-spŭn′jē-ō′sŭs) | Male—central tendon of perineum | Dorsal surface of penis and bulb of penis | Constricts urethra; erects penis |
| | Female—central tendon of perineum | Base of clitoris | Erects clitoris |
| Ischiocavernosus (ish′ē-ō-kav′er-nō′sŭs) | Ischial ramus | Corpus cavernosum | Compresses base of penis or clitoris |
| External anal sphincter | Coccyx | Central tendon of perineum | Keeps orifice of anal canal closed |
| Transverse perinei (pĕr′i-nē′ī) | | | |
| Deep | Ischial ramus | Midline connective tissue | Supports pelvic floor |
| Superficial | Ischial ramus | Central tendon of perineum | Fixes central tendon |

(le-vā′ter skap′ū-lē), the **rhomboids** (rom′boydz), the **serratus** (ser-ā′tŭs; serrated) **anterior,** and the **pectoralis** (pek′tō-ra′lis) **minor** (figure 7.21 and table 7.12). These muscles act as fixators to hold the scapula firmly in position when the muscles of the arm contract. The scapular muscles also move the scapula into different positions, thereby increasing the range of movement of the upper limb. The trapezius forms the upper line from each shoulder to the neck. The origin of the serratus anterior from the first eight or nine ribs can be seen along the lateral thorax.

### Arm Movements

The arm is attached to the thorax by the **pectoralis major** and **latissimus dorsi** (lă-tis′i-mŭs dōr′sī) muscles (figure 7.22a and table 7.13; see figure 7.21c). The pectoralis major adducts the arm and flexes the shoulder. It can also extend the shoulder from a flexed position. The latissimus dorsi medially rotates and adducts the arm and powerfully extends the shoulder. Because a swimmer uses these three motions during the power stroke of the crawl, the latissimus dorsi is often called the swimmer's muscle.

Muscular

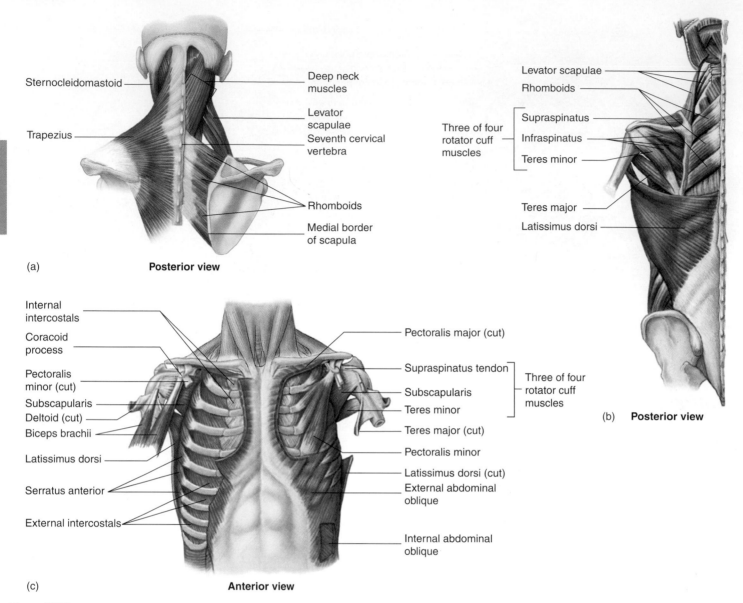

**(a)  Posterior view**

**(b)  Posterior view**

**(c)  Anterior view**

**Figure 7.21  Muscles of the Shoulder**

(a) Posterior view of the neck and upper shoulder. The left side shows the superficial muscles. On the right, the superficial muscles are removed to show the deep muscles. (b) Posterior view of the thoracic region, with the trapezius and deltoid muscles removed. (c) Anterior view of the thoracic region.

| TABLE 7.12 | Muscles Acting on the Scapula (see figures 7.13, 7.21, and 7.22) | | |
|---|---|---|---|
| **Muscle** | **Origin** | **Insertion** | **Action** |
| Levator scapulae (le-vā′ter skap′ū-lē) | Transverse processes of C1–C4 | Superior angle of scapula | Elevates, retracts, and rotates scapula; laterally flexes neck |
| Pectoralis minor (pek′tō-ra′lis) | Ribs 3–5 | Coracoid process of scapula | Depresses scapula or elevates ribs |
| Rhomboids (rom′boydz) Major Minor | Spinous processes of T1–T4 Spinous processes of T1–T4 | Medial border of scapula Medial border of scapula | Retracts, rotates, and fixes scapula Retracts, slightly elevates, rotates, and fixes scapula |
| Serratus anterior (ser-ā′tŭs) | Ribs 1–9 | Medial border of scapula | Rotates and protracts scapula; elevates ribs |
| Trapezius (tra-pē′zē-ŭs) | Posterior surface of skull and spinous processes of C7–T12 | Clavicle, acromion process, and scapular spine | Elevates, depresses, retracts, rotates, and fixes scapula; extends head and neck |

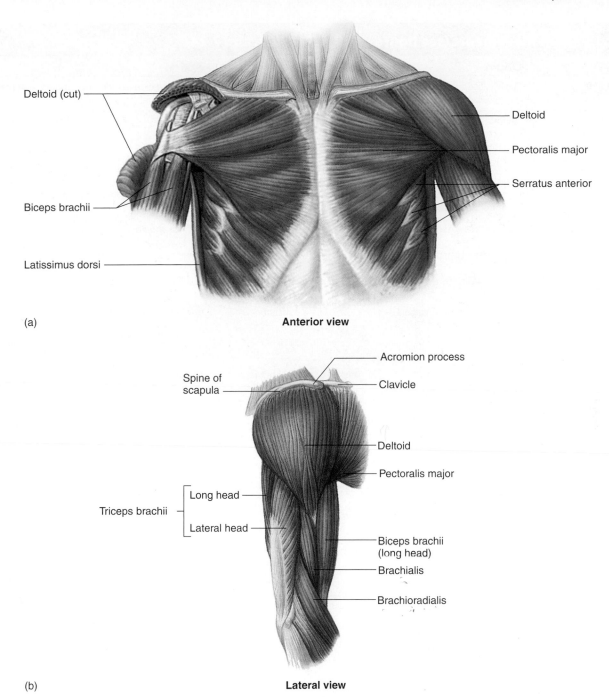

(a)    **Anterior view**

(b)    **Lateral view**

**Figure 7.22** Arm Muscles

Another group of four muscles, called the **rotator cuff muscles,** attaches the humerus to the scapula and forms a cuff or cap over the proximal humerus (table 7.13; see figure 7.21*b,c*). These muscles stabilize the joint by holding the head of the humerus in the glenoid cavity during arm movements, especially abduction. A rotator cuff injury involves damage to one or more of these muscles or their tendons. The **deltoid** (del′toyd) muscle attaches the humerus to the scapula and clavicle and is the major abductor of the upper limb. The pectoralis major forms the upper chest, and

the deltoid forms the rounded mass of the shoulder (see figure 7.24). The deltoid is a common site for administering injections.

## Forearm Movements

The arm can be divided into anterior and posterior compartments. The **triceps brachii** (trī′seps brā′kē-ī; three heads, arm), the primary extensor of the elbow, occupies the posterior compartment (figure 7.22*b* and table 7.14). The anterior compartment

**TABLE 7.13       Arm Movements (see figures 7.13, 7.21, 7.22, and 7.24)**

| Muscle | Origin | Insertion | Action |
|---|---|---|---|
| Deltoid (del′toyd) | Clavicle, acromion process, and scapular spine | Deltoid tuberosity | Flexes and extends shoulder; abducts and medially and laterally rotates arm |
| Latissimus dorsi (lă-tis′i-mŭs dōr′sī) | Spinous processes of T7–L5, sacrum and iliac crest, and inferior angle of scapula in some people | Medial crest of intertubercular groove | Extends shoulder; adducts and medially rotates arm |
| Pectoralis major (pek′tō-rā′lis) | Clavicle, sternum, superior six costal cartilages, and abdominal muscles | Lateral crest of intertubercular groove | Flexes shoulder; extends shoulder from flexed position; adducts and medially rotates arm |
| Teres major (ter′ēz) | Lateral border of scapula | Medial crest of intertubercular groove | Extends shoulder; adducts and medially rotates arm |
| **Rotator Cuff** | | | |
| Infraspinatus (in′fră-spī-nā′tŭs) | Infraspinous fossa of scapula | Greater tubercle of humerus | Stabilizes and extends shoulder and laterally rotates arm |
| Subscapularis (sŭb′skap-ū-lār′is) | Subscapular fossa of scapula | Lesser tubercle of humerus | Stabilizes and extends shoulder and medially rotates arm |
| Supraspinatus (sŭ′pră-spī-nā′tŭs) | Supraspinous fossa of scapula | Greater tubercle of humerus | Stabilizes shoulder and abducts arm |
| Teres minor (te′rēz) | Lateral border of scapula | Greater tubercle of humerus | Stabilizes and extends shoulder; adducts and laterally rotates arm |

**TABLE 7.14       Arm Muscles (see figures 7.13, 7.21, 7.22, and 7.24)**

| Muscle | Origin | Insertion | Action |
|---|---|---|---|
| Biceps brachii (bī′seps brā′kē-ī) | Long head—supraglenoid tubercle Short head—coracoid process | Radial tuberosity and aponeurosis of biceps brachii | Flexes elbow; supinates forearm; flexes shoulder |
| Brachialis (brā′kē-al′is) | Anterior surface of humerus | Coronoid process of ulna | Flexes elbow |
| Triceps brachii (trī′seps brā′kē-ī) | Long head—lateral border of scapula Lateral head—lateral and posterior surface of humerus Medial head—posterior humerus | Olecranon process of ulna | Extends elbow; extends shoulder; adducts arm |

is occupied mostly by the **biceps** (bī′seps) **brachii** and the **brachialis** (brā′kē-ăl-is), the primary flexors of the elbow. The **brachioradialis** (brā′kē-ō-rā′dē-al′is), which is actually a posterior forearm muscle, helps flex the elbow.

## Supination and Pronation

Supination of the forearm, or turning the flexed forearm so that the palm is up, is accomplished by the **supinator** (soo′pi-nā-ter) (figure 7.23 and table 7.15) and the biceps brachii, which tends to supinate the forearm while flexing the elbow. Pronation, turning the forearm so that the palm is down, is a function of two **pronator** (prō-nā′ter) muscles.

## Wrist and Finger Movements

The 20 muscles of the forearm can also be divided into anterior and posterior groups. Only a few of these muscles, the most superficial, appear in table 7.15 and figure 7.23. Most of the anterior forearm muscles are responsible for flexion of the wrist and fingers, whereas most of the posterior forearm muscles cause extension. A strong band of fibrous connective tissue, the retinaculum (figure 7.23b), covers the flexor and extensor tendons and holds them in place around the wrist so that they do not "bowstring" during muscle contraction. Because the retinaculum does not stretch as a result of pressure, this characteristic is a contributing factor in carpal tunnel syndrome (see chapter 6).

Muscular

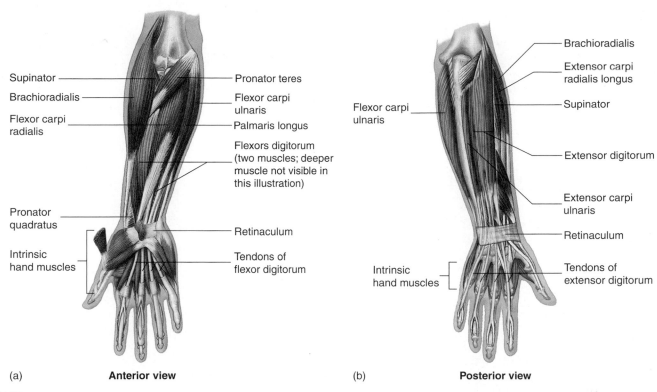

**Figure 7.23** Muscles of the Forearm

(a) Anterior view. The flexor retinaculum has been removed. (b) Posterior view.

| TABLE 7.15 | Forearm Muscles (see figure 7.23) | | |
|---|---|---|---|
| **Muscle** | **Origin** | **Insertion** | **Action** |
| **Anterior Forearm** | | | |
| Palmaris longus (pawl-mār′is lon′gus) | Medial epicondyle of humerus | Aponeurosis over palm | Tightens skin of palm |
| Flexor carpi radialis (kar′pī rā′dē-a-lĭs) | Medial epicondyle of humerus | Second and third metacarpal bones | Flexes and abducts wrist |
| Flexor carpi ulnaris (kar′pī ŭl-nār′is) | Medial epicondyle of humerus and ulna | Pisiform | Flexes and abducts wrist |
| Flexor digitorum profundus (dij′i-tōr′ŭm prō-fŭn′dŭs) (not shown) | Ulna | Distal phalanges of digits 2–5 | Flexes fingers and wrist |
| Flexor digitorum superficialis (sū′per-fish′ē-a′lis) | Medial epicondyle of humerus, coronoid process, and radius | Middle phalanges of digits 2–5 | Flexes fingers and wrist |
| Pronator | | | |
| Quadratus (prō′nă-tōr kwah-drā′tŭs) | Distal ulna | Distal radius | Pronates forearm |
| Teres (prō′nă-tōr te′rēz) | Medial epicondyle of humerus and coronoid process of ulna | Radius | Pronates forearm |
| **Posterior Forearm** | | | |
| Brachioradialis (brā′kē-ō-rā′dē-a′lis) | Lateral supracondylar ridge of humerus | Styloid process of radius | Flexes forearm |
| Extensor carpi radialis brevis (kar′pī rā′dē-a′lis brev′ıs) (not shown) | Lateral epicondyle of humerus | Base of third metacarpal bone | Extends and abducts wrist |
| Extensor carpi radialis longus (lon′gus) | Lateral supracondylar ridge of humerus | Base of second metacarpal bone | Extends and abducts wrist |
| Extensor carpi ulnaris (kar′pī ŭl-nār′is) | Lateral epicondyle of humerus and ulna | Base of fifth metacarpal bone | Extends and adducts wrist |
| Extensor digitorum (dij′i-tōr′ŭm) | Lateral epicondyle of humerus | Extensor tendon expansion over phalanges of digits 2–5 | Extends fingers and wrist |
| Supinator (sū′pi-nā′tōr) | Lateral epicondyle of humerus and ulna | Radius | Supinates forearm (and hand) |

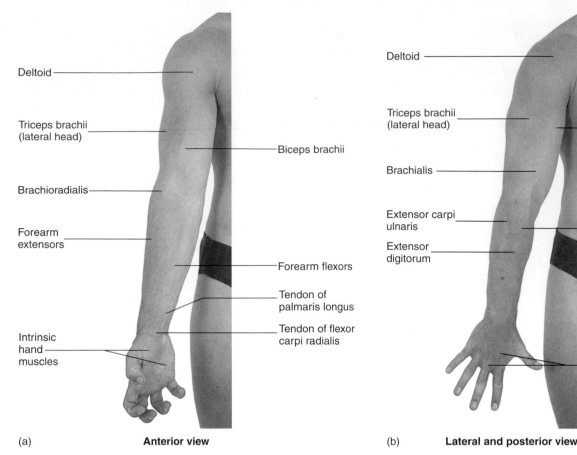

(a)    **Anterior view**

(b)    **Lateral and posterior view**

**Figure 7.24** Surface Anatomy, Muscles of the Upper Limb

The **flexor carpi** (kar′pī) muscles flex the wrist, and the **extensor carpi** muscles extend the wrist. The tendon of the flexor carpi radialis serves as a landmark for locating the radial pulse (figure 7.24a). The tendons of the wrist extensors are visible on the posterior surface of the forearm (figure 7.24b). Forceful, repeated contraction of the wrist extensor muscles, as occurs in a tennis backhand, may result in inflammation and pain where the extensor muscles attach to the lateral humeral epicondyle. This condition is sometimes referred to as "tennis elbow." Flexion of the fingers is the function of the **flexor digitorum** (dij′i-tōr′ŭm; flexor of the digits, or fingers). Extension of the fingers is accomplished by the **extensor digitorum.** The tendons of this muscle are very visible on the dorsal surface of the hand (figure 7.24b). The thumb has its own set of flexors, extensors, adductors, and abductors. The little finger has some similar muscles.

Nineteen muscles, called **intrinsic hand muscles,** are located within the hand. **Interossei** (in′ter-os′ē-ī; between bones) muscles, located between the metacarpal bones, are responsible for abduction and adduction of the fingers. Other intrinsic hand muscles are involved in the many possible movements of the thumb and fingers. These muscles account for the fleshy masses at the base of the thumb and little finger and the fleshy region between the metacarpal bones of the thumb and index finger.

## Lower Limb Muscles

The muscles of the lower limb include those located in the hip, the thigh, the leg, and the foot.

### Thigh Movements

Several hip muscles originate on the coxal bone and insert onto the femur (figure 7.25 and table 7.16). The anterior muscle, the **iliopsoas** (il′ē-ō-sō′ŭs), flexes the hip (figure 7.25a). The posterior and lateral hip muscles consist of the **gluteal muscles** and the tensor fasciae latae. The **tensor fasciae latae** (ten′sōr fa′shē-ă la′tē) is so named because it tenses a thick band of fascia on the lateral side of the thigh called the iliotibial tract. By so doing, it helps steady the femur on the tibia when a person is standing. The **gluteus** (glū′tē-us; buttock) **maximus,** which extends the hip and abducts and laterally rotates the thigh, contributes most of the mass that can be seen as the buttocks. The **gluteus medius,** which abducts and medially rotates the thigh, creates a smaller mass just superior and lateral to the maximus (see figure 7.26b). The gluteus maximus functions optimally to extend the hip when the thigh is flexed at a 45-degree angle. The gluteus medius is a common site for injections in the buttocks because the sciatic nerve lies deep to the gluteus maximus and could be damaged during an injection.

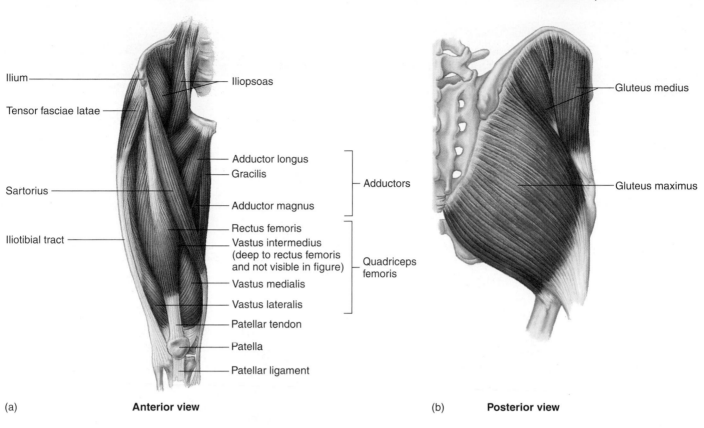

Ilium

Tensor fasciae latae

Sartorius

Iliotibial tract

Iliopsoas

Adductor longus
Gracilis

Adductor magnus

— Adductors

Rectus femoris
Vastus intermedius
(deep to rectus femoris
and not visible in figure)

Vastus medialis

Vastus lateralis

Quadriceps
femoris

Patellar tendon

Patella

Patellar ligament

(a)    **Anterior view**

Gluteus medius

Gluteus maximus

(b)    **Posterior view**

Muscular

Ischial tuberosity

Hamstring
muscles

Semitendinosus

Biceps femoris

Semimembranosus

Tibia

Fibula

(c)    **Posterior view**

**Figure 7.25  Muscles of the Hip and Thigh**

(*a*) Anterior view. The vastus intermedius is labeled to allow for a complete listing of the quadriceps femoris muscles, but the muscle lies deep to the rectus femoris and cannot be seen in the figure. (*b*) Posterior view of the hip muscles. (*c*) Posterior view of the thigh muscles.

| TABLE 7.16 | Muscles Moving the Thigh (see figure 7.25) | | |
|---|---|---|---|
| Muscle | Origin | Insertion | Action |
| Iliopsoas (il′ē-ō-sō′ŭs) | Iliac fossa and vertebrae T12–L5 | Lesser trochanter of femur and hip capsule | Flexes hip |
| Gluteus maximus (glŭ′tē-ŭs mak′si-mŭs) | Posterior surface of ilium, sacrum, and coccyx | Gluteal tuberosity of femur and iliotibial tract | Extends hip; abducts and laterally rotates thigh |
| Gluteus medius (glŭ′tē-ŭs mē′dē-ŭs) | Posterior surface of ilium | Greater trochanter of femur | Abducts and medially rotates thigh |
| Gluteus minimus (glŭ′tē-ŭs min′i-mŭs) (not shown) | Posterior surface of ilium | Greater trochanter of femur | Abducts and medially rotates thigh |
| Tensor fasciae latae (ten′sōr fa′shē-ē lā′tē) | Anterior superior iliac spine | Through lateral fascia of thigh to lateral condyle of tibia | Steadies femur on tibia through iliotibial tract when standing; flexes hip; medially rotates and abducts thigh |

(a)        **Anterior view**
(b)        **Posterolateral view**

**Figure 7.26** Surface Anatomy of the Lower Limb

In addition to the hip muscles, some of the thigh muscles also attach to the coxal bone and can move the thigh. There are three groups of thigh muscles: the anterior thigh muscles, which flex the hip; the posterior thigh muscles, which extend the hip; and the medial thigh muscles, which adduct the thigh.

*Predict 5*

*Consider a sprinter's stance and a bicyclist's racing posture, and explain why these athletes use these postures.*

## Leg Movements

The anterior thigh muscles are the **quadriceps femoris** (kwah′dri-seps fe-mōr′is; four muscles) and the **sartorius** (sar-tōr′ē-ŭs) (table 7.17; figures 7.25a and 7.26a). The quadriceps femoris muscles are the primary extensors of the knee. They have a common insertion, the patellar tendon, on and around the patella. The patellar ligament is an extension of the patellar tendon onto the tibial tuberosity. The patellar ligament is tapped with a rubber

## TABLE 7.17    Leg Movements (see figures 7.13, 7.25, and 7.27)

| Muscle | Origin | Insertion | Action |
|---|---|---|---|
| **Anterior Compartment** | | | |
| Quadriceps femoris (kwah′dri-seps fem′ō-ris) | | | |
|     Rectus femoris (rek′tŭs fem′ō-ris) | Ilium | Tibial tuberosity via patellar ligament | Extends knee; flexes hip |
|     Vastus lateralis (vas′tus lat-er-ā′lis) | Greater trochanter and linea aspera of femur | Tibial tuberosity via patellar ligament | Extends knee |
|     Vastus medialis (vas′tus mē′dē-ā′lis) | Linea aspera of femur | Tibial tuberosity via patellar ligament | Extends knee |
|     Vastus intermedius (vas′tŭs in′ter-mē′dē-ŭs) | Body of femur | Tibial tuberosity via patellar ligament | Extends knee |
| Sartorius (sar-tōr′ē-ŭs) | Anterior superior iliac spine | Medial side of tibial tuberosity | Flexes hip and knee; laterally rotates thigh |
| **Medial Compartment** | | | |
| Adductor longus (a′dŭk-ter lon′gŭs) | Pubis | Linea aspera of femur | Adducts and laterally rotates thigh; flexes hip |
| Adductor magnus (a′dŭk-ter mag′nŭs) | Pubis and ischium | Femur | Adducts and laterally rotates thigh; extends knee |
| Gracilis (gras′i-lis) | Pubis near symphysis | Tibia | Adducts thigh; flexes knee |
| **Posterior Compartment (Hamstring Muscles)** | | | |
| Biceps femoris (bi′seps fem′ō-ris) | Long head—ischial tuberosity<br>Short head—femur | Head of fibula | Flexes knee; laterally rotates leg; extends hip |
| Semimembranosus (se′mē-mem′bră-nō′sŭs) | Ischial tuberosity | Medial condyle of tibia and collateral ligament | Flexes knee; medially rotates leg; extends hip |
| Semitendinosus (se′mē-ten′di-nō′sŭs) | Ischial tuberosity | Tibia | Flexes knee; medially rotates leg; extends hip |

hammer when testing the knee-jerk reflex in a physical examination (see figure 8.19). Health professionals often use one of the quadriceps muscles, the vastus lateralis, as an intermuscular injection site. The sartorius, the longest muscle in the body, is called the "tailor's muscle" because it flexes the hip and knee and rotates the thigh laterally for sitting cross-legged, as tailors used to sit while sewing.

The posterior thigh muscles are called **hamstring muscles,** and they are responsible for flexing the knee (table 7.17; see figures 7.13*b* and 7.25*c*). Their tendons are easily felt and seen on the medial and lateral posterior aspect of a slightly bent knee (figure 7.26*b*). The hamstrings are so named because these tendons in hogs or pigs could be used to suspend hams during curing. Animals such as wolves often bring down their prey by biting through the hamstrings, thus preventing the prey animal from running. "To hamstring" someone is therefore to render him or her helpless. A "pulled hamstring" results from tearing one or more of these muscles or their tendons, usually where the tendons attach to the coxal bone.

The medial thigh muscles, the **adductor** (a′dŭk-ter) **muscles,** are primarily involved, as the name implies, in adducting the thigh (table 7.17).

### Ankle and Toe Movements

The 13 muscles in the leg, with tendons extending into the foot, can be divided into three groups: anterior, posterior, and lateral. As with the forearm, only the most superficial muscles are illustrated in figure 7.27 and listed in table 7.18. The anterior muscles (figure 7.27*a*) are extensor muscles involved in dorsiflexion (extension) of the foot and extension of the toes.

The superficial muscles of the posterior compartment of the leg (figure 7.27*b*), the **gastrocnemius** (gas′trok-nē′mē-us) and the **soleus** (sō′lē-ŭs), form the bulge of the calf (posterior leg; figure 7.27*b*). They join to form the common **calcaneal** (kal-kā′nē-ăl; heel) **tendon,** or *Achilles tendon.* These muscles are flexors and are involved in plantar flexion of the foot. The deep muscles of the posterior compartment plantar flex and invert the foot and flex the toes.

The lateral muscles of the leg, called the **fibularis** (fib-ū-lā′ris) **muscles** (figure 7.27*c*), are primarily everters (turning the lateral side of the foot outward) of the foot, but they also aid in plantar flexion.

The 20 muscles located within the foot, called the **intrinsic foot muscles,** flex, extend, abduct, and adduct the toes. They are arranged in a manner similar to the intrinsic muscles of the hand.

**Anterior view**
(a)

Soleus
Gastrocnemius
Fibularis longus
Fibularis brevis
Soleus
Tibialis anterior
Extensor digitorum longus
Extensor hallucis longus
Fibularis tertius

**Posterior view**
(b)

Two heads of gastrocnemius
Gastrocnemius
Soleus
Calcaneal tendon (Achilles tendon)

**Lateral view**
(c)

Gastrocnemius
Soleus
Fibularis longus (cut)
Tibialis anterior
Fibularis brevis
Extensor digitorum longus
Fibularis tertius
Tendon of fibularis longus (cut)

**Posterior view**
(d)

Gastrocnemius
Soleus
Calcaneal (Achilles) tendon

**Figure 7.27  Superficial Muscles of the Leg**

(a–c) Anterior, posterior, and lateral views. (d) Surface anatomy.

## TABLE 7.18    Muscles of the Leg Acting on the Leg, Ankle, and Foot (see figures 7.13, 7.26, and 7.27)

| Muscle | Origin | Insertion | Action |
|---|---|---|---|
| **Anterior Compartment** | | | |
| Extensor digitorum longus (dij'i-tōr'ŭm lon'gŭs) | Lateral condyle of tibia and fibula | Four tendons to phalanges of four lateral toes | Extends four lateral toes; dorsiflexes and everts foot |
| Extensor hallucis longus (hal'i-sis lon'gŭs) | Middle fibula and interosseous membrane | Distal phalanx of great toe | Extends great toe; dorsiflexes and inverts foot |
| Tibialis anterior (tib'ē-a'lis) | Tibia and interosseous membrane | Medial cuneiform and first metatarsal bone | Dorsiflexes and inverts foot |
| Fibularis tertius (peroneus tertius) (per'ō-nē'ŭs ter'shē-ŭs) | Fibula and interosseous membrane | Fifth metatarsal bone | Dorsiflexes and everts foot |
| **Posterior Compartment** | | | |
| *Superficial* | | | |
| Gastrocnemius (gas'trok-nē'mē-ŭs) | Medial and lateral condyles of femur | Through calcaneal (Achilles) tendon to calcaneus | Plantar flexes foot; flexes leg |
| Soleus (sō'lē-ŭs) | Fibula and tibia | Through calcaneal tendon to calcaneus | Plantar flexes foot |
| *Deep* | | | |
| Flexor digitorum longus (dij'i-tōr'ŭm lon'gŭs) (not shown) | Tibia | Four tendons to distal phalanges of four lateral toes | Flexes four lateral toes; plantar flexes and inverts foot |
| Flexor hallucis longus (hal'i-sis lon'gŭs) (not shown) | Fibula | Distal phalanx of great toe | Flexes great toe; plantar flexes and inverts foot |
| Tibialis posterior (tib'ē-a'lis) (not shown) | Tibia, interosseous membrane, and fibula | Navicular, cuneiforms, cuboid, and second through fourth metatarsal bones | Plantar flexes and inverts foot |
| **Lateral Compartment** | | | |
| Fibularis brevis (peroneus brevis) (fib-ū-lā'ris brev'is) | Fibula | Fifth metatarsal bone | Everts and plantar flexes foot |
| Fibularis longus (peroneus longus) (fib-ū-lā'ris lon'gŭs) | Fibula | Medial cuneiform and first metatarsal bone | Everts and plantar flexes foot |

## A CASE IN POINT
### Groin Pull

Lowe Kikker was playing soccer on a very cold October day. He raced up to kick the ball, just as a player for the other team kicked the ball from the opposite direction. Initially, the ball did not move. Lowe felt a sudden, sharp pain in the medial side of his right thigh and fell to the ground, holding the injured area. His coach and team trainer ran onto the field. As the trainer pressed on the center part of his medial thigh, Lowe felt severe pain. He was able to stand and was helped off the field. On the sideline, the trainer applied ice to the adductor region of Lowe's right thigh. For the next few days, Lowe was told to rest, along with icing the region, compressing the muscles, and elevating the thigh (RICE). He was also given anti-inflammatory drugs. After a couple days of rest, Lowe began massage therapy. He missed the next game but was able to play in the finals. Lowe's injury was considered a grade 1 injury because there was pain and tenderness in the medial thigh muscles but no swelling. Grade 2

injury, in which the muscle is partially torn, is more painful and results in swelling of the area. Grade 3 injury, which involves a complete muscle tear, causes so much pain and swelling that the injured person cannot even walk. A groin pull involves one or more of the adductor muscles, most commonly the adductor longus. The damage usually occurs at the musculotendon junction, near its insertion.

## EFFECTS OF AGING ON SKELETAL MUSCLE

Aging skeletal muscle undergoes several changes that reduce muscle mass, increase the time a muscle takes to contract in response to nervous stimuli, reduce stamina, and increase recovery time. Loss of muscle fibers begins as early as 25 years

## Duchenne Muscular Dystrophy

A couple became concerned about their 3-year-old son when they noticed that he was much weaker than other boys his age. The differences appeared to become more obvious as time passed. Eventually, the boy had difficulty sitting, standing, and walking. He seemed clumsy and fell often. He had difficulty climbing stairs, and he often rose from sitting on the floor to a standing position by using his hands and arms to climb up his legs. His muscles appeared poorly developed. The couple took their son to a physician, and after several tests, they were informed that he had Duchenne muscular dystrophy.

### Background Information

**Duchenne muscular dystrophy (DMD)** is usually identified in children around 3 years of age when their parents notice slow motor development with progressive weakness and muscle wasting (atrophy). Typically, muscular weakness begins in the pelvic girdle, causing a waddling gait. Rising from the floor by climbing up the legs is characteristic and is caused by weakness of the lumbar and gluteal muscles (figure 7A). Within 3 to 5 years, the muscles of the shoulder girdle become involved. The replacement of muscle with connective tissue contributes to muscular atrophy and shortened, inflexible muscles called contractures. The contractures limit movements and can cause severe deformities of the skeleton. By 10 to 12 years of age, people with DMD are usually unable to walk, and few live beyond age 20. There is no effective treatment to prevent the progressive deterioration of muscles in DMD. Therapy primarily involves exercises to help strengthen muscles and prevent contractures.

**Figure 7A**

DMD is characterized by progressive muscle weakness. Rising from the floor to a standing position becomes difficult, and the patient often has to use his or her arms to push against the thighs while rising.

Duchenne muscular dystrophy results from an abnormal gene on the X chromosome and is therefore a sex-linked (X-linked) condition. The gene is carried by females, but DMD affects males almost exclusively, at a frequency of 1 in 3000. The DMD gene is responsible for producing a protein called **dystrophin,** which plays a role in attaching myofibrils to other proteins in the cell membrane and regulating their activity. In a normal individual, dystrophin is thought to protect muscle cells against mechanical stress. In DMD patients, part of the gene is missing, and the protein it produces is nonfunctional, resulting in progressive muscular weakness and muscle contractures.

of age, and by age 80 the muscle mass has been reduced by approximately 50%. Weight-lifting exercises help slow the loss of muscle mass but do not prevent the loss of muscle fibers. In addition, fast-twitch muscle fibers decrease in number more rapidly than slow-twitch fibers. Most of the loss of strength and speed is due to the loss of muscle fibers, particularly fast-twitch muscle fibers. The surface area of the neuromuscular junction decreases, and as a result, action potentials in neurons stimulate action potentials in muscle cells more slowly; thus, fewer action potentials are produced in muscle fibers. The number of motor neurons also decreases, and the remaining neurons innervate more muscle fibers. This decreases the number of motor units in skeletal muscle, with a greater number of muscle fibers for each neuron, which may result in less precise muscle control. Aging is also associated with a decrease in the density of capillaries in skeletal muscles so that a longer recovery period is required after exercise.

Many of the age-related changes in skeletal muscle can be slowed dramatically if people remain physically active. As people age, they often assume a sedentary lifestyle. Studies show that elderly people who are sedentary can become stronger and more mobile in response to exercise.

| System Interactions | Effects of Duchenne Muscular Dystrophy on Other Systems |
|---|---|
| **System** | **Interactions** |
| Skeletal | Shortened, inflexible muscles (contractures) cause severe deformities of the skeletal system. Kyphoscoliosis, severe curvature of the spinal column laterally and anteriorly, can be so severe that normal respiratory movements are impaired. Surgery is sometimes required to prevent contractures from making it impossible for the individual to sit in a wheelchair. |
| Nervous | Some degree of mental retardation occurs in a large percentage of people with DMD, although the specific cause is unknown. Research suggests that dystrophin is important in the formation of synapses between nerve cells. |
| Cardiovascular | DMD affects cardiac muscle; consequently, heart failure occurs in a large number of people with advanced DMD. Cardiac involvement becomes serious in as many as 95% of cases and is one of the leading causes of death for individuals with DMD. |
| Lymphatic and Immune | No obvious direct effects occur to the lymphatic system, but damaged muscle fibers are phagocytized by macrophages. |
| Respiratory | Deformity of the thorax and increasing weakness of the respiratory muscles results in inadequate respiratory movements, which cause an increase in respiratory infections, such as pneumonia. Insufficient movement of air into and out of the lungs due to weak respiratory muscles is a major contributing factor in many deaths. |
| Digestive | DMD affects smooth muscle tissue, and the reduced ability of smooth muscle to contract can cause disorders of the digestive system. Digestive system disorders include an enlarged colon diameter, and twisting of the small intestine, resulting in intestinal obstruction, cramping, and reduced absorption of nutrients. |
| Urinary | Reduced smooth muscle function and wheelchair dependency increase the frequency of urinary tract infections. |

### Predict 6

A boy with advanced Duchenne muscular dystrophy developed pulmonary edema (accumulation of fluid in the lungs) and pneumonia caused by a bacterial infection. His physician diagnosed the condition in the following way: The pulmonary edema was the result of heart failure, and the increased fluid in the lungs provided a site where bacteria could invade and grow. The fact that the boy could not breathe deeply or cough effectively made the condition worse. How would the muscle tissues in a boy with advanced DMD with heart failure and ineffective respiratory movements differ from the muscle tissues in a boy with less advanced DMD?

## DISEASES AND DISORDERS Muscular System

| CONDITION | DESCRIPTION |
|---|---|
| Cramps | Painful, spastic contractions of a muscle; usually due to a buildup of lactic acid |
| Fibromyalgia (fī-brō-mī-al′ja) | Non-life-threatening, chronic, widespread pain in muscles with no known cure; also known as chronic muscle pain syndrome |
| Hypertrophy | Enlargement of a muscle due to an increased number of myofibrils, as occurs with increased muscle use |
| Atrophy | Decrease in muscle size due to a decreased number of myofilaments; can occur due to disuse of a muscle, as in paralysis |
| Muscular dystrophy | Group of genetic disorders in which all types of muscle degenerate and atrophy |
|     Duchenne muscular dystrophy | See Systems Pathology |
|     Myotonic muscular dystrophy | Muscles are weak and fail to relax following forceful contractions; affects the hands most severely; dominant trait in 1/20,000 births |
| Myasthenia gravis | See A Case in Point, earlier in this chapter |
| Tendinitis (ten-di-nī′tis) | Inflammation of a tendon or its attachment point due to overuse of the muscle |
| Go to **www.mhhe.com/seeleyess7** for additional information on these pathologies. | |

# SUMMARY

## Functions of the Muscular System    (p. 152)

The muscular system produces body movement, maintains posture, causes respiration, produces body heat, performs movements involved in communication, constricts organs and vessels, and pumps blood.

## Characteristics of Skeletal Muscle    (p. 152)

Skeletal muscle has contractility, excitability, extensibility, and elasticity.

### Skeletal Muscle Structure

1. Muscle fibers are organized into fasciculi, and fasciculi are organized into muscles by associated connective tissue.
2. Each skeletal muscle fiber is a single cell containing numerous myofibrils.
3. Myofibrils are composed of actin and myosin myofilaments.
4. Sarcomeres are joined end to end to form myofibrils.

### Excitability of Muscle Fibers

1. Cell membranes have a negative charge on the inside relative to a positive charge outside. This is called the resting membrane potential.
2. Action potentials are a brief reversal of the membrane charge. They are carried rapidly along the cell membrane.
3. Sodium ions ($Na^+$) move into cells during depolarization, and $K^+$ move out of cells during repolarization.

### Nerve Supply

1. Motor neurons carry action potentials to skeletal muscles, where the neuron and muscle fibers form neuromuscular junctions.
2. Neurons release acetylcholine, which binds to receptors on muscle cell membranes, stimulates an action potential in the muscle cell, and causes the muscle to contract.

### Muscle Contraction

1. Action potentials are carried along T tubules to the sarcoplasmic reticulum, where they cause the release of calcium ions.
2. Calcium ions, released from the sarcoplasmic reticulum, bind to the actin myofilaments, exposing attachment sites.
3. Myosin forms cross-bridges with the exposed actin attachment sites.
4. The myosin molecules bend, causing the actin molecules to slide past; this is the sliding filament model. The H and I bands shorten; the A bands do not.
5. This process requires ATP breakdown.
6. A muscle twitch is the contraction of a muscle fiber in response to a stimulus; it consists of a lag phase, a contraction phase, and a relaxation phase.
7. Tetanus occurs when stimuli occur so rapidly that a muscle does not relax between twitches.
8. Small contraction forces are generated when small numbers of motor units are recruited, and greater contraction forces are generated when large numbers of motor units are recruited.
9. Energy is produced by aerobic (with oxygen) and anaerobic (without oxygen) respiration.
10. After intense exercise, the rate of aerobic respiration remains elevated to repay the oxygen deficit.
11. Muscular fatigue occurs as ATP is depleted during muscle contraction. Physiological contracture occurs in extreme fatigue when a muscle can neither contract nor relax.
12. Muscles contract either isometrically (tension increases, but muscle length stays the same) or isotonically (tension remains the same, but muscle length decreases).
13. Muscle tone consists of a small percentage of muscle fibers contracting tetanically and is responsible for posture.
14. Muscles contain a combination of slow-twitch and fast-twitch fibers.
15. Slow-twitch fibers are better suited for aerobic respiration, and fast-twitch fibers are adapted for anaerobic respiration.
16. Sprinters have more fast-twitch fibers, whereas distance runners have more slow-twitch fibers.

## Smooth Muscle and Cardiac Muscle    (p. 165)

1. Smooth muscle is not striated, has one nucleus per cell, contracts more slowly than skeletal muscle, can be autorhythmic, and is under involuntary control.
2. Cardiac muscle is striated, usually has one nucleus per cell, has intercalated disks, is autorhythmic, and is under involuntary control.

## Skeletal Muscle Anatomy    (p. 166)

### General Principles

1. Most muscles have an origin on one bone, have an insertion onto another, and cross at least one joint.
2. A muscle causing a specific movement is an agonist. A muscle causing the opposite movement is an antagonist.
3. Muscles working together are synergists.
4. A prime mover is the muscle of a synergistic group that is primarily responsible for the movement.

### Nomenclature

Muscles are named according to their location, origin and insertion, number of heads, or function.

### Muscles of the Head and Neck

1. Muscles of facial expression are associated primarily with the mouth and eyes.
2. Four pairs of muscles are involved in mastication.
3. Tongue movements involve intrinsic and extrinsic muscles.
4. Swallowing involves the suprahyoid and infrahyoid muscles, plus muscles of the soft palate, pharynx, and larynx.
5. Neck muscles move the head.

### Trunk Muscles

1. The erector spinae muscles hold the body erect.
2. Intercostal muscles and the diaphragm are involved in breathing.
3. Muscles of the abdominal wall flex and rotate the vertebral column, compress the abdominal cavity, and hold in and protect the abdominal organs.
4. Muscles form the floor of the pelvis.

### Upper Limb Muscles

1. The upper limb is attached to the body primarily by muscles.
2. Arm movements are accomplished by pectoral, rotator cuff, and deltoid muscles.
3. The elbow is flexed and extended by anterior and posterior arm muscles, respectively.
4. Supination and pronation of the forearm are accomplished by supinators and pronators in the forearm.
5. Movements of the wrist and fingers are accomplished by most of the 20 forearm muscles and 19 intrinsic muscles in the hand.

### Lower Limb Muscles

1. Hip muscles flex and extend the hip and abduct the thigh.
2. Thigh muscles flex and extend the hip and adduct the thigh. They also flex and extend the knee.
3. Muscles of the leg and foot are similar to those of the forearm and hand.

## Effects of Aging on Skeletal Muscle    (p. 187)

Aging is associated with decreased muscle mass, slower reaction time, reduced stamina, and increased recovery time.

# REVIEW AND COMPREHENSION

1. List the seven major functions of the muscular system.
2. Define contractility, excitability, extensibility, and elasticity.
3. List the connective tissue layers associated with muscles.
4. What are fasciculi?
5. What is a muscle fiber?
6. Explain the relevance of the structural relationship among sarcomeres, T tubules, and the sarcoplasmic reticulum.
7. What is a sarcomere?
8. Describe the composition of a myofibril. Describe the structure of actin and myosin myofilaments.
9. Explain the resting membrane potential and how it is produced.
10. Describe the production of an action potential.
11. What is a neuromuscular junction? What happens there?
12. Describe the sliding filament model of muscle contraction.
13. Explain how an action potential results in muscle contraction.
14. Define muscle twitch, tetanus, and recruitment.
15. Describe the two ways energy is produced in skeletal muscle.
16. Explain fatigue.
17. Compare isometric, isotonic, concentric, and eccentric contraction.
18. What is muscle tone?
19. Compare slow-twitch and fast-twitch muscle fibers.
20. How do smooth muscles and cardiac muscles differ from skeletal muscles?
21. Define origin, insertion, agonist, antagonist, synergist, prime mover, and fixator.
22. Describe the muscles of facial expression.
23. What is mastication? What muscles are involved?
24. What are the intrinsic and extrinsic tongue muscles?
25. What muscles are involved in swallowing?
26. What muscles are involved in respiration?
27. Describe the functions of the muscles of the anterior abdominal wall.
28. What is primarily responsible for attaching the upper limb to the body?
29. Describe, by muscle groups, movements of the arm, forearm, and hand.
30. Describe, by muscle groups, movements of the thigh, leg, and foot.

# CRITICAL THINKING

1. Bob Canner improperly canned some home-grown vegetables and contracted botulism poisoning after eating them. Botulism results from a toxin, produced by bacteria, that prevents skeletal muscles from contracting. Symptoms include difficulty in swallowing and breathing. Eventually, Bob died of respiratory failure because his respiratory muscles relaxed and would not contract. Assuming that botulism toxin affects the neuromuscular junction, propose as many ways as you can to explain how botulism toxin produced the observed symptoms.

2. Harvey Leche milked cows by hand each morning before school. One morning, he overslept and had to hurry to get to school on time. While he was milking the cows as fast as he could, his hands became very tired, and then for a short time he could neither release his grip nor squeeze harder. Explain what happened.

3. A researcher was investigating the fast-twitch versus slow-twitch composition of muscle tissue in the gastrocnemius muscle (in the calf of the leg) of athletes. Describe the general differences this researcher would see when comparing the muscles from athletes who were outstanding in the following events: 100-m dash, weight lifting, and 10,000-m run.

4. Describe an exercise routine that would build up each of the following groups of muscles: anterior arm, posterior arm, anterior forearm, anterior thigh, posterior leg, and abdomen.

5. Sherri Speedster started a 100-m dash but fell to the ground in pain. Examination of her right lower limb revealed the following symptoms: The knee was held in a slightly flexed position, but she could not flex it voluntarily; she could extend the knee with difficulty, but this caused her considerable pain; and there was considerable pain and bulging of the muscles in the posterior thigh. Explain the nature of her injury.

*Answers in Appendix D*

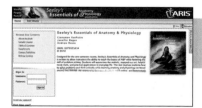

Visit this textbook's website at **www.mhhe.com/seeleyess7** for practice quizzes, animations, interactive learning exercises, and other study tools.

McGraw-Hill offers a study CD that features interactive cadaver dissection. ***Anatomy & Physiology Revealed*** includes cadaver photos that allow you to peel away layers of the human body to reveal structures beneath the surface. This program also includes animations, radiologic imaging, audio pronunciations, and practice quizzing.

This image of the brain shows that the nervous system is very active. In fact, the nervous system is the control center of the body. It is through the brain, spinal cord, and nerves that we are able to detect and respond to the world around us.

# Chapter  8

# Nervous System

## Chapter Outline and Objectives

**Functions of the Nervous System (p. 193)**

1. List the functions of the nervous system.

**Divisions of the Nervous System (p. 193)**

2. List the divisions of the nervous system, and describe the characteristics of each.

**Cells of the Nervous System (p. 194)**

3. Describe the structure of neurons and the function of their components. Describe the location, structure, and general function of neuroglia.

4. Define and describe the structure of a nucleus, a ganglion, a nerve tract, and a nerve.

**Electrical Signals and Neural Pathways (p. 198)**

5. Describe a resting membrane potential, and explain how an action potential is generated and propagated. Compare the roles of leak and gated ion channels.

6. Describe the structure and function of a synapse.

7. List the parts of a reflex arc, and describe its function.

8. Describe a converging and a diverging circuit and the role of summation in neural pathways.

**Central and Peripheral Nervous Systems (p. 206)**

9. Compare and contrast the central and peripheral nervous systems.

**Spinal Cord (p. 206)**

10. Describe the relationship between the spinal cord and the spinal nerves.

11. Describe a cross section of the spinal cord.

**Spinal Nerves (p. 208)**

12. Define plexus, and describe the three primary plexuses, including their branches.

**Brain (p. 211)**

13. List the parts of the brain.

14. List the parts of the brainstem, and state their functions.

15. State where the cerebellum is located.

16. List the parts of the diencephalon, and state their functions.

17. List the lobes of the cerebrum, and state a function for each.

**Sensory Functions (p. 215)**

18. List the major ascending tracts, and state a function for each.

19. Describe the sensory and association areas of the cerebral cortex, and discuss their interactions.

**Motor Functions (p. 216)**

20. Describe the motor area of the cerebral cortex, and discuss how it interacts with other parts of the frontal lobe.

21. Distinguish between upper and lower motor neurons and between direct and indirect tracts.

22. Discuss how the basal nuclei and cerebellum regulate motor functions.

**Other Brain Functions (p. 220)**

23. Discuss the right and left cerebral hemispheres and speech.

24. Compare and contrast the features of sensory, short-term, and the two types of long-term memory.

**Meninges and Cerebrospinal Fluid (p. 222)**

25. Describe the three meningeal layers surrounding the central nervous system, the four ventricles of the brain, and the origin and circulation of the cerebrospinal fluid.

**Cranial Nerves (p. 224)**

26. List the various types of cranial nerves, and briefly describe their functions.

**Autonomic Nervous System (p. 227)**

27. Contrast the structure of the autonomic nervous system and the somatic nervous system.

28. Name the two divisions of the autonomic nervous system, and describe the anatomical and neurotransmitter differences between them.

29. Compare and contrast the general functions of the parasympathetic and sympathetic nervous systems.

**Enteric Nervous System (p. 230)**

30. Discuss how the enteric nervous system can act independently of the CNS.

**Effects of Aging on the Nervous System (p. 230)**

31. Describe the changes that occur in the nervous system with advancing age.

A man is walking down a street on a beautiful autumn day, enjoying the warmth of the sun. He passes a produce stand. The apples look good to him, and he decides that a big, juicy one would taste wonderful. He selects just the right one and picks it up. At the register, he reaches into his pocket, feels the coins there, and pulls out a few. He counts out the correct change, pays for the apple, and puts the rest of the change back into his pocket. While walking down the street, he takes a big bite of the apple. He feels the apple's juicy crispness in his mouth and savors its tart-sweet flavor. Life is wonderful. All these sensations, actions, and emotions are made possible by the nervous system, which consists of the brain, spinal cord, nerves, and sensory receptors.

## FUNCTIONS OF THE NERVOUS SYSTEM

The nervous system is involved in some way in nearly every body function. Some major functions of the nervous system are

1. *Receive sensory input.* Sensory receptors monitor numerous external and internal stimuli. We are aware of sensations from some stimuli, such as vision, hearing, taste, smell, touch, pain, body position, and temperature. Other stimuli, such as blood pH, blood gases, and blood pressure, are processed at a subconscious level.
2. *Integrate information.* The brain and spinal cord are the major organs for processing sensory input and initiating responses. The input may produce an immediate response, be stored as memory, or be ignored.
3. *Control muscles and glands.* Skeletal muscles normally contract only when stimulated by the nervous system. Thus, by controlling skeletal muscle, the nervous system controls the major movements of the body. The nervous system also participates in controlling cardiac muscle, smooth muscle, and many glands.
4. *Maintain homeostasis.* The nervous system plays an important role in maintaining homeostasis. This function depends on the nervous system's ability to detect, interpret, and respond to changes in internal and external conditions. In response, the nervous system can stimulate or inhibit the activities of other systems to help maintain a constant internal environment.
5. *Establish and maintain mental activity.* The brain is the center of mental activity, including consciousness, memory, and thinking.

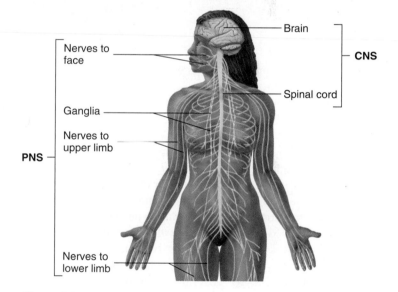

**Figure 8.1** Nervous System

The central nervous system (CNS) consists of the brain and spinal cord. The peripheral nervous system (PNS) consists of nerves and ganglia.

## DIVISIONS OF THE NERVOUS SYSTEM

The nervous system can be divided into the central and the peripheral nervous systems (figure 8.1). The **central nervous system (CNS)** consists of the brain and spinal cord. The **peripheral nervous system (PNS)** consists of all the nervous tissue outside the CNS (nerves and ganglia).

The PNS has two functional subdivisions: The **sensory division,** or *afferent* (toward) *division,* conducts action potentials from sensory receptors to the CNS (figure 8.2). The neurons that transmit

Nervous

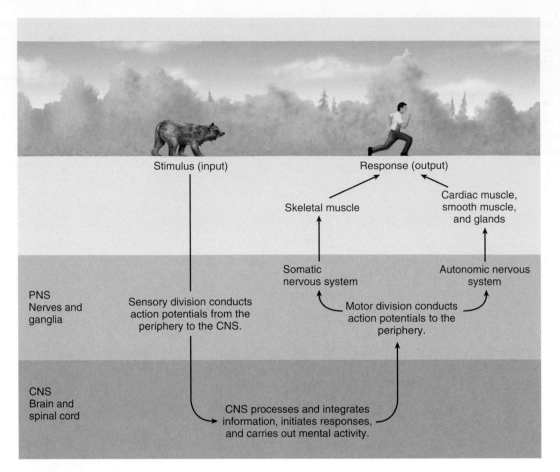

**Figure 8.2** **Organization of the Nervous System**

The sensory division of the peripheral nervous system (PNS; *gold*) detects stimuli and conducts action potentials to the central nervous system (CNS; *pink*). The CNS interprets incoming action potentials and initiates action potentials that are conducted through the motor division to produce a response. The motor division is divided into the *somatic nervous system* and the *autonomic nervous system.*

action potentials from the periphery to the CNS are called **sensory neurons.** The **motor division,** or *efferent* (away) *division,* conducts action potentials from the CNS to effector organs, such as muscles and glands. The neurons that transmit action potentials from the CNS toward the periphery are called **motor neurons.**

The motor division can be further subdivided into the **somatic** (sō-mat′ik; bodily) **nervous system,** which transmits action potentials from the CNS to skeletal muscles, and the **autonomic** (aw-tō-nom′ik; self-governing) **nervous system (ANS),** which transmits action potentials from the CNS to cardiac muscle, smooth muscle, and glands. The autonomic nervous system, in turn, is divided into sympathetic and parasympathetic divisions.

The **enteric nervous system (ENS)** is a unique subdivision of the peripheral nervous system. The ENS has both sensory and motor neurons contained wholly within the digestive tract. The ENS can function without input from the CNS or other parts of the PNS, although it is normally integrated with the CNS by sensory neurons and ANS motor neurons.

## CELLS OF THE NERVOUS SYSTEM

The two types of cells that make up the nervous system are neurons and neuroglia.

## Neurons

**Neurons** (noor′onz; nerve), or *nerve cells* (figure 8.3), receive stimuli, conduct action potentials, and transmit signals to other neurons or effector organs. There are three parts to a neuron: a cell body and two types of processes, called dendrites and axons.

Each neuron **cell body** contains a single nucleus. As with any other cell, the nucleus of the neuron is the source of information for gene expression. Extensive rough endoplasmic reticulum (rough ER), a Golgi apparatus, and mitochondria surround the nucleus. Large numbers of neurofilaments (intermediate filaments) and microtubules organize the cytoplasm into distinct areas.

**Dendrites** (den′drīts; trees) are short, often highly branching cytoplasmic extensions that are tapered from their bases at the neuron cell body to their tips. Most dendrites are extensions of the neuron cell body, but dendritelike structures also project from the peripheral ends of some sensory axons. Dendrites usually receive information from other neurons or from sensory receptors and transmit the information toward the neuron cell body.

An **axon** is a long cell process extending from the neuron cell body. Each neuron has a single axon that extends from the cell body. The area where the axon leaves the neuron cell body is called the **axon hillock.** Each axon has a uniform diameter and may vary in length from a few millimeters to more than a meter. Axons of motor branch to form **collateral** (ko-lat′er-ăl) **axons.** Axons can be surrounded by a highly specialized insulating layer of cells called the myelin sheath (described in more detail later in this chapter).

### Types of Neurons

Three categories of neurons exist, based on the arrangement of their processes (figure 8.4 and table 8.1). **Multipolar neurons** have many dendrites and a single axon. Most of the neurons within the CNS, including nearly all motor neurons, are multipolar. **Bipolar neurons** have two processes: one dendrite and one axon. Bipolar neurons are located in some sensory organs, such as in the retina of the eye and in the nasal cavity. Most other sensory neurons are pseudo-unipolar. **Pseudo-unipolar neurons** have a single process extending from the cell body. This process divides into two processes a short distance from the cell body. One process extends to the periphery, and the other extends to the CNS. The two extensions function as a single axon with small, dendritelike sensory receptors at the periphery. The axon receives sensory information at the periphery and transmits that information in the form of action potentials to the CNS.

## Neuroglia

**Neuroglia** (noo-rog′lē-ă; nerve glue), or *glial* (glī′ăl, glē′ăl) *cells,* are the nonneuronal cells of the CNS and PNS. Neuroglia are far more numerous than neurons. Most neuroglia retain the ability to divide, whereas neurons do not. There are five types of neuroglia. **Astrocytes** (as′trō-sītz) serve as the major supporting tissue in the CNS. In this role, astrocytes can stimulate or inhibit the signaling activity of nearby neurons. In addition, astrocytes participate with the blood vessel endothelium to form a permeability barrier, called the **blood-brain barrier,** between the blood and the CNS. Astrocytes help limit damage to neural tissue; however, the repair process can form a scar that blocks regeneration of damaged axons. **Ependymal** (ep-en′di-măl) **cells** line the fluid-filled cavities (ventricles and canals) within the CNS. Some ependymal cells produce cerebrospinal fluid, and others, with cilia on the surface, help move the cerebrospinal fluid through the CNS. **Microglia** (mī-krog′lē-ă) act as immune cells of the CNS. They help protect the brain by removing bacteria and cell debris. **Oligodendrocytes** (ol′i-gō-den′drō-sītz) in the CNS and **Schwann cells** in the PNS provide an insulating material that surrounds axons (figure 8.5 and table 8.1).

## Myelin Sheaths

Axons are surrounded by the cell processes of oligodendrocytes in the CNS and Schwann cells in the PNS (figure 8.5). **Unmyelinated axons** rest in indentations of the oligodendrocytes in the CNS and the Schwann cells in the PNS (figure 8.6). **Myelinated axons** have specialized sheaths, called **myelin sheaths,** wrapped around them (figure 8.6). Each oligodendrocyte process or Schwann cell repeatedly wraps around a segment of an axon to form a series of tightly wrapped cell membranes. A typical small nerve, which consists of axons of multiple neurons, usually contains more unmyelinated than myelinated axons. Myelin is an excellent insulator that prevents almost all electrical current flow through the cell membrane.

**Figure 8.3** Typical Neuron

Structural features of a typical motor neuron in the PNS include a cell body and two types of cell processes: dendrites and an axon.

neurons conduct action potentials away from the CNS, and axons of sensory neurons conduct action potentials toward the CNS. Axons also conduct action potentials from one part of the brain or spinal cord to another part. An axon may remain unbranched or may

Nervous

(a) A **multipolar neuron** has many dendrites and an axon.

(b) A **bipolar neuron** has a dendrite and an axon.

(c) A **pseudo-unipolar neuron** appears to have an axon and no dendrites.

**Figure 8.4** Types of Neurons

| TABLE 8.1 | Cells of the Nervous System (see figures 8.4 and 8.5) | |
|---|---|---|
| **Cell Type** | **Description** | **Function** |
| **Neuron** | | |
| Multipolar | Many dendrites and one axon | Most motor neurons and most CNS neurons |
| Bipolar | One dendrite and one axon | Found in special sense organs, such as eye and nose |
| Pseudo-unipolar | Appears to have a single axon | Most sensory neurons |
| **Neuroglia** | | |
| Astrocytes | Star-shaped | Provide structural support; regulate neuronal signaling; contribute to blood-brain barrier; help with neural tissue repair |
| Ependymal cells | Epithelial-like | Line ventricles of brain, circulate cerebrospinal fluid (CSF); some form choroid plexuses, which produce CSF |
| Microglia | Small, mobile cells | Protect CNS from infection; become phagocytic in response to inflammation |
| Oligodendrocytes | Cell with processes that can surround several axons | Cell processes form myelin sheaths around axons or enclose unmyelinated axons in the CNS |
| Schwann cells | Single cells surrounding axons | Form myelin sheaths around axons or enclose unmyelinated axons in the PNS |

Gaps in the myelin sheath, called **nodes of Ranvier** (ron′vē-ā), occur about every millimeter between the oligodendrocyte segments or between individual Schwann cells. At the nodes of Ranvier, current flows easily between the extracellular fluid and the axon, and action potentials can develop, as described in the next section.

## Organization of Nervous Tissue

Both the CNS and the PNS contain areas of gray matter and areas of white matter. **Gray matter** consists of groups of neuron cell bodies and their dendrites, where there is very little myelin. In the CNS, gray matter on the surface of the brain is called the **cortex**, and clusters of gray matter located deeper within the brain are called **nuclei.** In the PNS, a cluster of neuron cell bodies is called a **ganglion** (gang′glē-on; pl. ganglia, a swelling or knot).

**White matter** consists of bundles of parallel axons with their myelin sheaths, which are whitish in color. White matter of the CNS forms **nerve tracts,** or *conduction pathways,* which propagate action potentials from one area of the CNS to another. In the PNS, bundles of axons and their connective tissue sheaths are called **nerves.**

**Nervous**

**Figure 8.5 Types of Neuroglia**

(a) Astrocytes, with foot processes surrounding a blood capillary. (b) Ependymal cells, with cilia extending from the surfaces. (c) Microglial cell. (d) Oligodendrocyte, forming a myelin sheath around parts of three axons within the CNS. (e) Schwann cell forming part of the myelin sheath of an axon in the PNS.

(a) **Myelinated axon**

(b) **Unmyelinated axons**

**Figure 8.6  Comparison of Myelinated and Unmyelinated Axons**

(*a*) Myelinated axon with two Schwann cells forming part of the myelin sheath around a single axon. Each Schwann cell surrounds part of one axon. (*b*) Unmyelinated axons with two Schwann cells surrounding several axons in parallel formation. Each Schwann cell surrounds part of several axons.

# ELECTRICAL SIGNALS AND NEURAL PATHWAYS

## Resting Membrane Potential

All cells exhibit electrical properties. The inside of most cell membranes is negatively charged compared to the outside of the cell membrane, which is positively charged (as discussed in chapter 7). This uneven distribution of charge means the cell is **polarized.** In an unstimulated (or resting) cell, the uneven charge distribution is called the **resting membrane potential.** The outside of the cell membrane can be thought of as the positive pole of a battery and the inside as the negative pole. Thus, a small voltage difference, called a potential, can be measured across the resting cell membrane.

The resting membrane potential is generated by three main factors: (1) a higher concentration of $K^+$ immediately inside the cell membrane, (2) a higher concentration of $Na^+$ immediately outside the cell membrane; and (3) greater permeability of the cell membrane $K^+$ than to $Na^+$ (figure 8.7). Thus, the resting membrane potential results from differences in the concentration of ions across the membrane and the permeability characteristics of the membrane.

The difference in membrane permeability is due to the difference in the number of open ion channels. Recall from chapter 3 that ions cannot move freely across the cell membrane; instead, ions must flow through ion channels, which are proteins embedded in the cell membrane. Ions flow through channels due to the differences in their concentration across the membrane. There are two basic types of ion channels: leak channels and gated ion

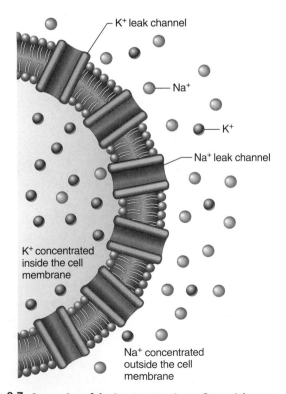

**Figure 8.7  Generation of the Resting Membrane Potential**

A higher concentration of $K^+$ exists inside the cell membrane, and a higher concentration of $Na^+$ exists outside the cell membrane. There are many more $K^+$ leak channels than $Na^+$ leak channels.

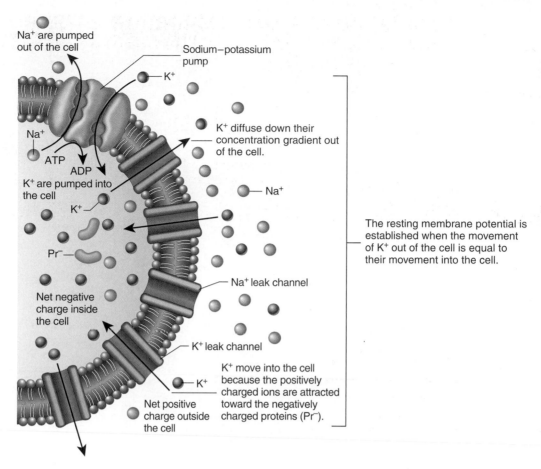

**Figure 8.8**  Control of Ion Concentrations Across a Resting Cell Membrane

In a resting cell, there is a higher concentration of $K^+$ inside the cell membrane and a higher concentration of $Na^+$ outside the cell membrane. In the resting cell, only the leak channels are open; the gated channels are closed. There are many more $K^+$ leak channels than $Na^+$ leak channels. As a result, $K^+$ diffuse out of the cell down their concentration gradient. The membrane is not permeable to the negatively charged proteins inside the cell. The tendency for the $K^+$ to diffuse to the outside of the cell down their concentration gradient is opposed by the tendency for the positively charged $K^+$ to be attracted back into the cell by the negatively charged proteins. A small amount of $Na^+$ diffuses into the cell (not shown). The sodium-potassium pump helps maintain the differential levels of $Na^+$ and $K^+$ by pumping three $Na^+$ out of the cell in exchange for two $K^+$ into the cell. The pump is driven by ATP hydrolysis.

channels (see figure 3.5). **Leak ion channels** are always open. Thus, as the name suggests, ions can "leak" across the membrane down their concentration gradient. When a cell is at rest, the membrane potential is established by diffusion of ions through leak channels. Because there are 50–100 times more $K^+$ leak channels than $Na^+$ leak channels, the resting membrane has much greater permeability to $K^+$ than to $Na^+$; therefore, the $K^+$ leak channels have the greatest contribution to the resting membrane potential. **Gated ion channels** are closed until opened by specific signals. **Chemically gated channels** are opened by neurotransmitters or other chemicals, whereas **voltage-gated channels** are opened by a change in membrane potential. When opened, the gated channels can change the membrane potential and are thus responsible for the action potential, described next.

As mentioned earlier, there is a net negative charge inside and a net positive charge outside a resting cell. The primary source of negative charge inside the cell is a high concentration of negatively charged molecules, such as proteins, that cannot diffuse out of the cell because the cell membrane is impermeable to them (figure 8.8). Consequently, as positively charged $K^+$ leak

out of the cell via the leak channels, the charge inside the cell membrane becomes even more negative. The negatively charged molecules inside the cell tend to attract the positive $K^+$ back into the cell. The resting membrane potential is the point of equilibrium at which the tendency for $K^+$ to move down their concentration gradient out of the cell is balanced by the negative charge within the cell, which tends to attract the $K^+$ back into the cell (figure 8.8).

To compensate for the constant leakage of ions across the membrane, the **sodium-potassium pump** ($Na^+$–$K^+$ pump) is required to maintain the greater concentration of $Na^+$ outside the cell membrane and $K^+$ inside. The pump actively transports $K^+$ into the cell and $Na^+$ out of the cell (figure 8.8). The importance of this pump is indicated by the astounding amount of energy it consumes. It is estimated that the sodium-potassium pump consumes 25% of all the ATP in a typical cell and 70% of the ATP in a neuron.

## Action Potentials

Muscle and nerve cells are **excitable cells,** meaning that the resting membrane potential changes in response to stimuli that activate

Nervous

**1 Resting membrane potential.** Na⁺ channels (*pink*) and most, but not all, K⁺ channels (*purple*) are closed. The outside of the plasma membrane is positively charged compared to the inside.

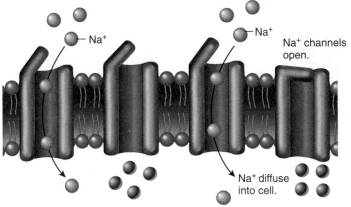

**2 Depolarization.** Na⁺ channels open. K⁺ channels begin to open. Depolarization results because the inward movement of Na⁺ makes the inside of the membrane more positive.

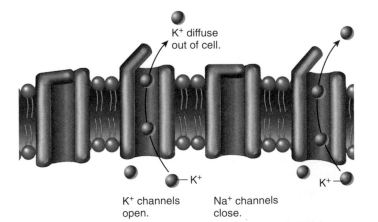

**3 Repolarization.** Na⁺ channels close and additional K⁺ channels open. Na⁺ movement into the cell stops, and K⁺ movement out of the cell increases, causing repolarization.

**PROCESS Figure 8.9  Voltage-Gated Ion Channels and the Action Potential**

Step 1 illustrates the status of voltage-gated Na⁺ and K⁺ channels in a resting cell. Steps 2 and 3 show how the channels open and close to produce an action potential. Next to each step, a graph shows in *red* the membrane potential resulting from the condition of the ion channels.

gated ion channels. The opening and closing of gated channels can change the permeability characteristics of the plasma membrane and hence change the membrane potential.

The channels responsible for the action potential are voltage-gated Na⁺ and K⁺ channels. When the plasma membrane is at rest, the voltage-gated channels are closed (figure 8.9, step 1). When a stimulus is applied to a muscle cell or nerve cell, following neurotransmitter activation of chemically gated channels, Na⁺ channels open very briefly, and Na⁺ diffuse quickly into the cell (figure 8.9, step 2). This movement of Na⁺, which is called a **local**

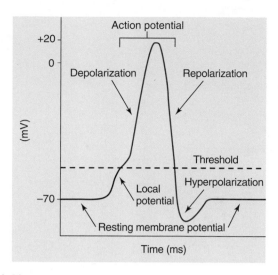

**Figure 8.10** **Action Potential**

Once a local potential reaches threshold, an all-or-none action potential is started. During the depolarization phase, the voltage across the cell membrane changes from approximately −70 mV to approximately +20 mV. During the repolarization phase, the voltage across the cell membrane returns to −70 mV. There is a brief period of hyperpolarization at the end of repolarization before the membrane returns to its resting membrane potential. The entire process lasts 1 or 2 milliseconds (ms).

**current,** causes the inside of the cell membrane to become more positive, a change called **depolarization.** This depolarization results in a **local potential.** If depolarization is not strong enough, the $Na^+$ channels close again, and the local potential disappears without being conducted along the nerve cell membrane. If depolarization is large enough, $Na^+$ enter the cell so that the local potential reaches a **threshold** value. This threshold depolarization causes voltage-gated $Na^+$ channels to open. Threshold is most often reached at the axon hillock, near the cell body. The opening of these channels causes a massive, 600-fold increase in membrane permeability to $Na^+$. Voltage-gated $K^+$ channels also begin to open. As more $Na^+$ enter the cell, depolarization occurs until a brief reversal of charge takes place across the membrane—the inside of the cell membrane becomes positive relative to the outside of the cell membrane. The charge reversal causes $Na^+$ channels to close and more $K^+$ channels to open. $Na^+$ then stop entering the cell, and $K^+$ leave the cell (figure 8.9, step 3). This repolarizes the cell membrane to its resting membrane potential. Depolarization and repolarization constitute an **action potential** (figure 8.10). At the end of repolarization, the charge on the cell membrane briefly becomes more negative than the resting membrane potential; this condition is called **hyperpolarization.** The elevated permeability to $K^+$ lasts only a very short time.

In summary, the resting membrane potential is set by the activity of the leak channels. On stimulation, chemically gated channels are opened and initiate *local* potentials. If sufficiently strong, the local potentials activate voltage-gated channels to initiate an *action* potential.

Action potentials occur in an **all-or-none** fashion. That is, if threshold is reached, an action potential occurs; if the threshold is not reached, no action potential occurs. Action potentials in a cell

are all of the same magnitude—in other words, the amount of charge reversal is always the same. Stronger stimuli produce a greater frequency of action potentials but do not increase the size of each action potential. Thus, neural signaling is based on the number of action potentials.

Action potentials are conducted slowly in unmyelinated axons and more rapidly in myelinated axons. In unmyelinated axons, an action potential in one part of a cell membrane stimulates local currents in adjacent parts of the cell membrane. The local currents in the adjacent membrane produce an action potential. By this means, the action potential is conducted along the entire axon cell membrane. This type of action potential conduction is called **continuous conduction** (figure 8.11).

In myelinated axons, an action potential at one node of Ranvier causes a local current to flow through the surrounding extracellular fluid and through the cytoplasm of the axon to the next node, stimulating an action potential at that node of Ranvier. By this means, action potentials "jump" from one node of Ranvier to the next along the length of the axon. This type of action potential conduction is called **saltatory** (sal′tă-tōr-ē; to leap) **conduction** (figure 8.12). Saltatory conduction greatly increases the conduction velocity because the nodes of Ranvier make it unnecessary for action potentials to travel along the entire cell membrane. Action potential conduction in a myelinated fiber is like a grasshopper jumping, whereas in an unmyelinated axon it is like a grasshopper walking.

Medium-diameter, lightly myelinated axons, characteristic of autonomic neurons, conduct action potentials at the rate of about 3–15 meters per second (m/s), whereas large-diameter, heavily myelinated axons conduct action potentials at the rate of 15–120 m/s. These rapidly conducted action potentials, carried by sensory and motor neurons, allow for rapid responses to changes in the external environment. In addition, several hundred times fewer ions cross the cell membrane during conduction in myelinated cells than in unmyelinated cells. Much less energy is therefore required for the sodium–potassium pump to maintain the ion distribution.

## The Synapse

A **synapse** (sin′aps) is a junction where the axon of one neuron interacts with another neuron or with an effector organ, such as a muscle or gland (figure 8.13). The end of the axon forms a **presynaptic terminal.** The membrane of the dendrite or effector cell is the **postsynaptic membrane,** and the space separating the presynaptic and postsynaptic membranes is the **synaptic cleft.** Chemical substances called **neurotransmitters** (noor′ō-trans-mit′erz; *neuro-,* nerve + *transmitto,* to send across) are stored in **synaptic vesicles** in the presynaptic terminal. When an action potential reaches the presynaptic terminal, voltage-gated $Ca^{2+}$ channels open, and $Ca^{2+}$ move into the cell. This influx of $Ca^{2+}$ causes the release of neurotransmitters by exocytosis from the presynaptic terminal. The neurotransmitters diffuse across the synaptic cleft and bind to specific receptor molecules on the postsynaptic membrane. The binding of neurotransmitters to these membrane receptors causes chemically gated channels for $Na^+$, $K^+$, or $Cl^-$ to open or close in the postsynaptic membrane, depending on the

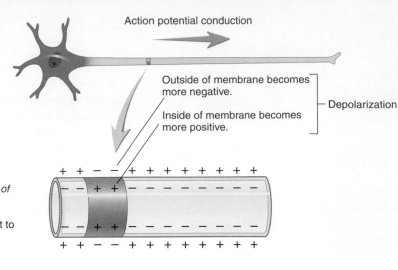

**Action potential conduction**

Outside of membrane becomes more negative.
Inside of membrane becomes more positive.
⎤ Depolarization

**1** An action potential (*orange part of the membrane*) generates local currents that depolarize the membrane immediately adjacent to the action potential.

**2** When depolarization caused by the local currents reaches threshold, a new action potential is produced adjacent to where the original action potential occurred.

**3** The action potential is conducted along the axon cell membrane.

Site of next action potential

**PROCESS Figure 8.11** Continuous Conduction in an Unmyelinated Axon

Node of Ranvier          Schwann cell

**1** An action potential (*orange*) at a node of Ranvier generates local currents. The local currents flow to the next node of Ranvier because the myelin sheath of the Schwann cell insulates the axon between nodes.

**2** When the depolarization caused by the local currents reaches threshold at the next node of Ranvier, a new action potential is produced (*orange*).

**3** Action potential conduction is rapid in myelinated axons because the action potentials are produced at successive nodes of Ranvier (*1–5*) instead of at every part of the membrane along the axon.

Direction of action potential conduction

**PROCESS Figure 8.12** Saltatory Conduction: Action Potential Conduction in a Myelinated Axon

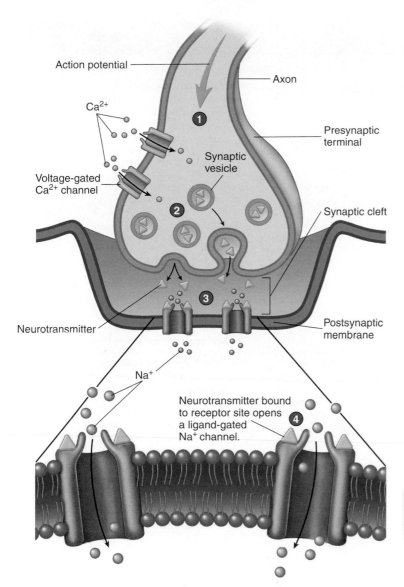

① Action potentials arriving at the presynaptic terminal cause voltage-gated $Ca^{2+}$ channels to open.

② $Ca^{2+}$ diffuse into the cell and cause synaptic vesicles to release neurotransmitter molecules.

③ Neurotransmitter molecules diffuse from the presynaptic terminal across the synaptic cleft.

④ Neurotransmitter molecules combine with their receptor sites and cause ligand-gated $Na^+$ channels to open. $Na^+$ diffuse into the cell (*shown in illustration*) or out of the cell (*not shown*) and cause a change in membrane potential.

**PROCESS Figure 8.13  The Synapse**

A synapse consists of the end of a neuron (presynaptic terminal), a small space (synaptic cleft), and the postsynaptic membrane of another neuron or an effector cell, such as a muscle or gland cell.

type of neurotransmitter in the presynaptic terminal and the type of receptors on the postsynaptic membrane. The response may be either stimulation or inhibition of an action potential in the postsynaptic cell. For example, if $Na^+$ channels open, the postsynaptic cell becomes depolarized, and an action potential will result if threshold is reached. If $K^+$ or $Cl^-$ channels open, the inside of the

postsynaptic cell tends to become more negative, or **hyperpolarized** (hī′per-pō′lăr-ī-zed), and an action potential is inhibited from occurring.

Of the many neurotransmitters or suspected neurotransmitter substances, the best known are **acetylcholine (ACh)** (as′e-til-kō′lēn) and **norepinephrine** (nōr′ep-i-nef′rin). Other neurotransmitters include serotonin (sēr-ō-tō′nin), dopamine (dō′pă-mēn), γ (gamma)-aminobutyric (gam′ă ă-mē′nō-bū-tēr′ik) acid (GABA), glycine, and endorphins (en′dōr-finz; endogenous morphine) (table 8.2). Neurotransmitter substances are rapidly broken down by enzymes within the synaptic cleft or are transported back into the presynaptic terminal. Consequently, they are removed from the synaptic cleft, so their effects on the postsynaptic membrane are very short-term. In synapses where acetylcholine is the neurotransmitter, such as in the neuromuscular junction (see chapter 7), an enzyme called **acetylcholinesterase** (as′e-til-kō-lin-es′-ter-ās) breaks down the acetylcholine. The breakdown products are then returned to the presynaptic terminal for reuse. Norepinephrine is either actively transported back into the presynaptic terminal or broken down by enzymes. The release and breakdown or removal of neurotransmitters occurs so rapidly that a postsynaptic cell can be stimulated many times a second. Drugs can modulate the action of neurotransmitters at the synapse. Cocaine and amphetamines increase the release and block the reuptake of norepinephrine, resulting in overstimulation of postsynaptic neurons and deleterious effects on the body. Drugs that block serotonin reuptake are particularly effective at treating depression and behavioral disorders.

## A CASE IN POINT
### Botulism

Ima Kannar removed a 5-year-old jar of green beans from the shelf in her food storage room. As she dusted off the lid, she noticed that it was a bit bowed out, but the beans inside looked and smelled okay when she opened the jar. She ate some of the beans for dinner, watched some television, and then went to bed. The next afternoon, as Ima was working in her garden, her throat felt dry. Then she started to experience double vision and difficulty breathing, so she dialed 911. By the time the paramedics arrived, 20 minutes later, Ima was not breathing. She was put on a respirator and rushed to the hospital, where physicians diagnosed a case of botulism. Ima recovered very slowly and was in the hospital for several months, much of that time on a respirator. Botulism is caused by the toxin of the bacterium *Clostridium botulinum,* which grows in anaerobic (without oxygen) environments, such as improperly processed canned food. The toxin blocks the release of acetylcholine in synapses. As a result, muscles cannot contract, and the person may die from respiratory failure.

## Reflexes

A **reflex** is an involuntary reaction in response to a stimulus applied to the periphery and transmitted to the CNS. Reflexes allow a person to react to stimuli more quickly than is possible if conscious thought is involved. A **reflex arc** is the neuronal pathway by which a reflex occurs (figure 8.14). The reflex arc is the basic functional unit of the nervous system because it is the smallest,

## TABLE 8.2     Neurotransmitters

| Substance | Site of Release | Effect | Clinical Example |
|---|---|---|---|
| Acetylcholine (ACh) | CNS synapses, ANS synapses, and neuromuscular junctions | Excitatory or inhibitory | Alzheimer disease (a type of senile dementia) is associated with a decrease in acetylcholine-secreting neurons. Myasthenia gravis (weakness of skeletal muscles) results from a reduction in acetylcholine receptors. |
| Norepinephrine (NE) | Selected CNS synapses and some ANS synapses | Excitatory | Cocaine and amphetamines increase the release and block the reuptake of norepinephrine, resulting in overstimulation of postsynaptic neurons. |
| Serotonin | CNS synapses | Generally inhibitory | It is involved with mood, anxiety, and sleep induction. Levels of serotonin are elevated in schizophrenia (delusions, hallucinations, and withdrawal). Drugs that block serotonin transporters, such as prozac, are used to treat depression and anxiety disorders. |
| Dopamine | Selected CNS synapses and some ANS synapses | Excitatory or inhibitory | Parkinson disease (depression of voluntary motor control) results from destruction of dopamine-secreting neurons. |
| Gamma-aminobutyric acid (GABA) | CNS synapses | Inhibitory | Drugs that increase GABA function have been used to treat epilepsy (excessive discharge of neurons). |
| Glycine | CNS synapses | Inhibitory | Glycine receptors are inhibited by the poison strychnine. Strychnine increases the excitability of certain neurons by blocking their inhibition. Strychnine poisoning results in powerful muscle contractions and convulsions. Tetanus of respiratory muscles can cause death. |
| Endorphins | Descending pain pathways | Inhibitory | The opiates morphine and heroin bind to endorphin receptors on presynaptic neurons and reduce pain by blocking the release of a neurotransmitter. |

simplest pathway capable of receiving a stimulus and yielding a response. A reflex arc generally has five basic components: (1) a **sensory receptor;** (2) a **sensory neuron;** (3) in some reflexes, **interneurons,** which are neurons located between and communicating with two other neurons; (4) a **motor neuron;** and (5) an **effector organ.** The simplest reflex arcs do not involve interneurons. Most reflexes occur in the spinal cord or brainstem rather than in the higher brain centers.

One example of a reflex is when a person's finger touches a hot stove. The heat stimulates pain receptors in the skin, and action potentials are produced. Sensory neurons conduct the action potentials to the spinal cord, where they synapse with interneurons. The interneurons, in turn, synapse with motor neurons in the spinal cord that conduct action potentials along their axons to flexor muscles in the upper limb. These muscles contract and pull the finger away from the stove. No conscious thought is required for this reflex, and withdrawal of the finger from the stimulus begins before the person is consciously aware of any pain.

## Neuronal Pathways

Neurons are organized within the CNS to form pathways ranging from relatively simple to extremely complex. The two simplest pathways are converging and diverging pathways. In a **converging pathway,** two or more neurons synapse with (converge on) the same neuron (figure 8.15a). This allows information transmitted in more than one neuronal pathway to converge into a single pathway. In a **diverging pathway,** the axon from one neuron divides (diverges) and synapses with more than one other neuron (figure 8.15b). This allows information transmitted in one neuronal pathway to diverge into two or more pathways.

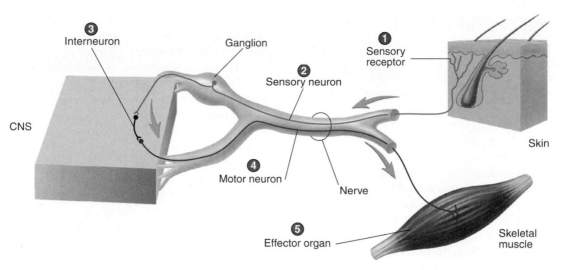

**PROCESS Figure 8.14  Reflex Arc**

The parts of a reflex arc are labeled in the order in which action potentials pass through them. The five components are the (1) sensory receptor, (2) sensory neuron, (3) interneuron, (4) motor neuron, and (5) effector organ. Simple reflex arcs have only four components because the reflex does not involve interneurons.

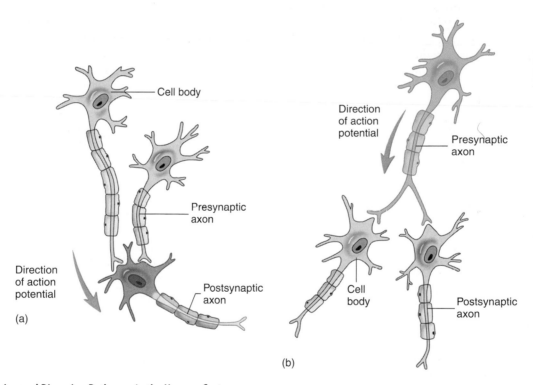

**Figure 8.15  Converging and Diverging Pathways in the Nervous System**

(a) General model of a converging pathway, showing two neurons converging onto one neuron. (b) General model of a diverging pathway, showing one neuron diverging onto two neurons.

Within the CNS and in many PNS synapses, it takes more than a single action potential to have an effect. A single presynaptic action potential usually does not cause a sufficiently large postsynaptic local potential to reach threshold and produce an action potential in the target cell. Instead, many presynaptic action potentials are needed in a process called **summation.** Summation of signals in neuronal pathways allows integration of multiple subthreshold changes in local potentials. Summation of the local potentials can bring the membrane potential to threshold and trigger an action potential. **Spatial summation** occurs when the local potentials originate from different locations on the postsynaptic neuron—for example, from converging pathways. **Temporal summation** occurs when local potentials overlap in time. This can occur from a single input that fires rapidly, which allows the resulting local potentials to overlap briefly. Spatial and temporal summation can lead to stimulation or inhibition, depending on the type of signal. Collectively, this integration of multiple inputs determines whether the postsynaptic neuron will fire an action potential.

# CENTRAL AND PERIPHERAL NERVOUS SYSTEMS

As mentioned previously, the central nervous system (CNS) consists of the brain and spinal cord. The brain is housed within the braincase; the spinal cord is in the vertebral column.

The peripheral nervous system (PNS) consists of all the nerves and ganglia outside the brain and spinal cord. The PNS collects information from numerous sources both inside and on the surface of the body and relays it by way of sensory fibers to the CNS, where one of three results is possible: The information is ignored, triggers a reflex, or is evaluated more extensively. Motor fibers in the PNS relay information from the CNS to muscles and glands in various parts of the body, regulating activity in those structures. The nerves of the PNS can be divided into two groups: 12 pairs of cranial nerves and 31 pairs of spinal nerves.

# SPINAL CORD

The **spinal cord** extends from the foramen magnum at the base of the skull to the second lumbar vertebra (figure 8.16). Spinal nerves communicate between the spinal cord and the body. The inferior end of the spinal cord and the spinal nerves exiting there resemble a horse's tail and are called the **cauda equina** (kaw′dă, tail; ē-kwī′nă, horse).

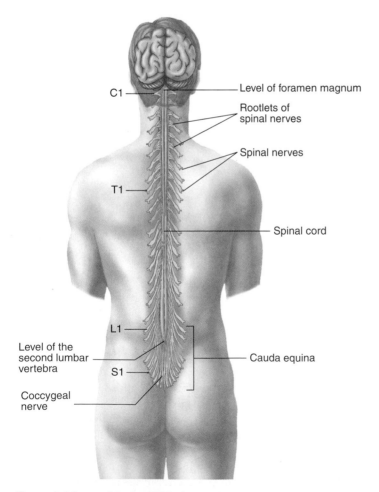

C1 — Level of foramen magnum
Rootlets of spinal nerves
Spinal nerves
T1
Spinal cord
L1
Level of the second lumbar vertebra
S1 — Cauda equina
Coccygeal nerve

**Figure 8.16** Spinal Cord and Spinal Nerve Roots

A cross section reveals that the spinal cord consists of a peripheral white matter portion and a central gray matter portion (figure 8.17a). The white matter consists of myelinated axons, and the gray matter is mainly a collection of neuron cell bodies. The white matter in each half of the spinal cord is organized into three columns, called the **dorsal** (posterior), **ventral** (anterior), and **lateral columns.** Each column of the spinal cord contains ascending and descending tracts, or pathways. **Ascending tracts** consist of axons that conduct action potentials toward the brain, and **descending tracts** consist of axons that conduct action potentials away from the brain. Ascending tracts and descending tracts are discussed more fully in a subsequent section of this chapter.

The gray matter of the spinal cord is shaped like the letter H, with **posterior horns** and **anterior horns.** Small **lateral horns** exist in levels of the cord associated with the autonomic nervous system. The **central canal** is a fluid-filled space in the center of the cord.

Spinal nerves arise from numerous rootlets along the dorsal and ventral surfaces of the spinal cord (figure 8.17a). The ventral rootlets combine to form a **ventral root** on the ventral (anterior) side of the spinal cord, and the dorsal rootlets combine to form a **dorsal root** on the dorsal (posterior) side of the cord at each segment. The ventral and dorsal roots unite just lateral to the spinal cord to form a spinal nerve. The dorsal root contains a ganglion, called the **dorsal root ganglion** (gang′glē-on; a swelling or knot).

The cell bodies of pseudo-unipolar sensory neurons are in the dorsal root ganglia (figure 8.17b). The axons of these neurons originate in the periphery of the body. They pass through spinal nerves and the dorsal roots to the posterior horn of the spinal cord gray matter. In the posterior horn, the axons either synapse with interneurons or pass into the white matter and ascend or descend in the spinal cord.

The cell bodies of motor neurons, which supply muscles and glands, are located in the anterior and lateral horns of the spinal cord gray matter. Somatic motor neurons are in the anterior horn, and autonomic neurons are in the lateral horn. Axons from the motor neurons form the ventral roots and pass into the spinal nerves. Thus, the dorsal root contains sensory axons, the ventral root contains motor axons, and each spinal nerve has both sensory and motor axons.

### Predict 1

*Describe the direction of action potential propagation in the spinal nerves, dorsal roots, and ventral roots.*

An example of a converging pathway involves a motor neuron in the spinal cord that stimulates muscle contraction (figure 8.18a). Sensory fibers from pain receptors carry action potentials to the spinal cord and synapse with interneurons, which in turn synapse with a motor neuron. Neurons in the cerebral cortex, controlling conscious movement, also synapse with the same motor neuron by way of axons in descending tracts. Both the interneurons and the neurons in the cerebral cortex have axons that converge on the motor neuron, which can therefore be stimulated either through the reflex arc or by conscious thought.

An example of a diverging pathway involves sensory neurons within the spinal cord (figure 8.18b). The axon of a sensory neuron carrying action potentials from pain receptors branches

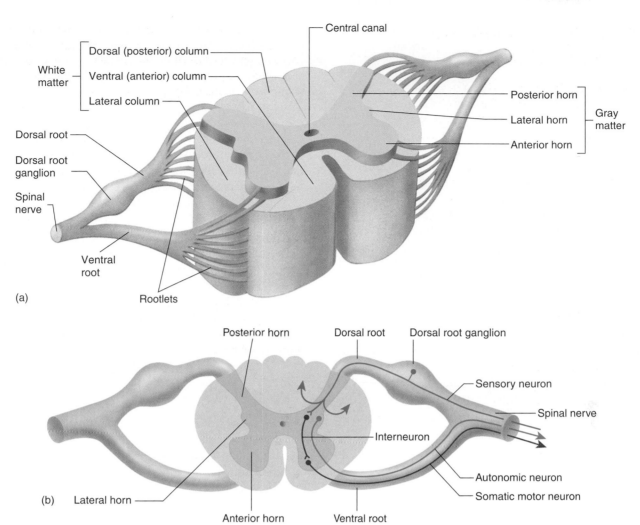

**Figure 8.17** Cross Section of the Spinal Cord

(a) In each segment of the spinal cord, rootlets combine to form a dorsal root on the dorsal (posterior) side and a ventral root on the ventral (anterior) side.
(b) Relationship of sensory and motor neurons to the spinal cord.

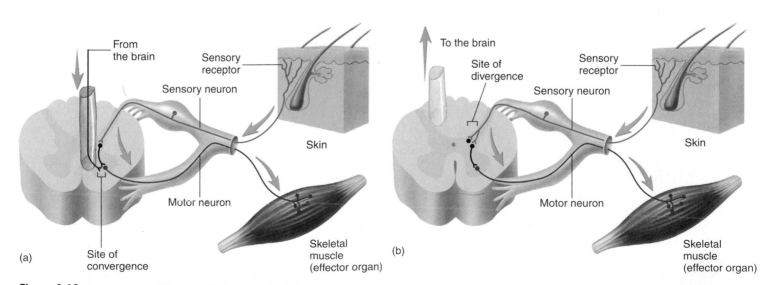

**Figure 8.18** Converging and Diverging Pathways in the Spinal Cord

(a) In a converging circuit, a sensory neuron from the periphery (*green*) and a descending neuron from the brain (*red*) converge on a single motor neuron (*purple*).
(b) In a diverging circuit, a sensory neuron from the periphery (*green*) diverges and sends information to a motor neuron by way of an interneuron (*black*) and sends information to the brain.

**PROCESS Figure 8.19** Knee-Jerk Reflex

1. Sensory receptors in the muscle detect stretch of the muscle.

2. Sensory neurons conduct action potentials to the spinal cord.

3. Sensory neurons synapse with motor neurons. Descending neurons (*black*) within the spinal cord also synapse with the neurons of the stretch reflex and modulate their activity.

4. Stimulation of the motor neurons causes the muscle to contract and resist being stretched.

within the spinal cord. One branch produces a reflex response by synapsing with an interneuron. The interneuron, in turn, synapses with a motor neuron, which stimulates a muscle to withdraw the injured region of the body from the source of the pain. The other branch synapses with an ascending neuron that carries action potentials through a nerve tract to the brain, where the stimulation is interpreted as pain.

## Spinal Cord Reflexes

### Knee-Jerk Reflex

The simplest reflex is the **stretch reflex,** in which muscles contract in response to a stretching force applied to them. The **knee-jerk reflex,** or *patellar reflex* (figure 8.19), is a classic example of the stretch reflex. When the patellar ligament is tapped, the quadriceps femoris muscle tendon and the muscles themselves are stretched. Sensory receptors within these muscles are also stretched, and the stretch reflex is activated. Consequently, contraction of the muscles extends the leg, producing the characteristic knee-jerk response.

Clinicians use the knee-jerk reflex to determine if the higher CNS centers that normally influence this reflex are functional. Descending neurons within the spinal cord synapse with the neurons of the stretch reflex and modulate their activity. This activity is important in maintaining posture and in coordinating muscular activity. Following a severe spinal cord injury, all spinal reflexes are lost below the level of injury. By about 2 weeks after the injury, the knee-jerk reflex returns, but it is often exaggerated. When the stretch reflex is absent or greatly exaggerated, it indicates that the neurons within the brain or spinal cord that modify this reflex have been damaged.

### Withdrawal Reflex

The function of the **withdrawal reflex,** or *flexor reflex,* is to remove a limb or another body part from a painful stimulus. The sensory receptors are pain receptors (see chapter 9). Following painful stimuli, sensory neurons conduct action potentials through the dorsal root to the spinal cord, where the sensory neurons synapse with interneurons, which in turn synapse with motor neurons (figure 8.20). These neurons stimulate muscles, usually flexor muscles, that remove the limb from the source of the painful stimulus.

## SPINAL NERVES

The **spinal nerves** arise along the spinal cord from the union of the dorsal roots and ventral roots (see figure 8.17). All the spinal nerves contain axons of both sensory and somatic motor neurons and thus are called mixed nerves. Some spinal nerves also contain parasympathetic or sympathetic axons. Most of the spinal nerves exit the vertebral column between adjacent vertebrae. Spinal nerves are categorized by the region of the vertebral column from which they emerge—cervical (C), thoracic (T), lumbar (L), sacral (S), and coccygeal (Co). The spinal nerves are also numbered (starting superiorly) according to their order within that region. The 31 pairs of spinal nerves are therefore C1 through C8, T1 through T12, L1 through L5, S1 through S5, and Co (figure 8.21).

Most of the spinal nerves are organized into three major **plexuses** (plek′sŭs-ēz; braids) where nerves come together and then separate: the cervical plexus, the brachial plexus, and the lumbosacral plexus (table 8.3). The major nerves of the neck and limbs are branches of these plexuses. Spinal nerves T2 through T11 do not join a plexus. Instead, these nerves extend around the thorax between the ribs, giving off branches to muscles and skin. Motor nerve fibers derived from plexuses innervate skeletal muscles, and sensory nerve fibers in those plexuses supply sensory innervation to the skin overlying those muscles (table 8.3). In addition to the major plexuses, the small coccygeal plexus supplies motor innervation to the muscles of the pelvic floor and sensory cutaneous innervation to the skin over the coccyx (figure 8.21).

### Cervical Plexus

The **cervical plexus** originates from spinal nerves C1 to C4. Branches from this plexus innervate several of the muscles attached to the hyoid bone, as well as the skin of the neck and posterior portion of the head. One of the most important branches

# CLINICAL IMPACT

## Spinal Cord Injury

About 10,000 new cases of **spinal cord injury** occur each year in the United States. Automobile and motorcycle accidents are the leading causes, followed by gunshot wounds, falls, and swimming accidents. Most spinal cord injuries are acute contusions (kon-too'shŭnz; bruising) of the cervical portion of the cord and do not completely sever the spinal cord.

Spinal cord injuries can interrupt ascending and/or descending tracts. Reflexes can still function below the level of injury, but sensations and/or motor functions and reflex modulation may be disrupted.

At the time of spinal cord injury, two types of tissue damage occur: (1) primary mechanical damage and (2) secondary tissue damage extending into a much larger region of the cord than the primary damage. The only treatment for primary damage is prevention, such as wearing seat belts when riding in automobiles and not diving in shallow water. Once an accident occurs, however, little can be done about primary damage.

Secondary spinal cord damage, which begins within minutes of the primary damage, is caused by ischemia (lack of blood supply), edema (fluid accumulation), ion imbalances, the release of "excitotoxins" (such as glutamate), and inflammatory cell invasion. Unlike primary damage, much secondary damage can be prevented or reversed if treatment is prompt. Giving the patient large doses of methylprednisolone, a synthetic anti-inflammatory steroid, within 8 hours of the injury can dramatically reduce the secondary damage to the cord. Other treatments include anatomical realignment and stabilization of the vertebral column and decompression of the spinal cord. Rehabilitation is based on retraining the patient to use whatever residual connections exist across the site of damage.

Although the spinal cord was formerly considered incapable of regeneration following severe damage, researchers have learned that most neurons of the adult spinal cord survive an injury and begin to regenerate, growing about 1 mm into the site of damage, before regressing to an inactive, atrophic state. The major block to adult spinal cord regeneration is the formation of a scar, consisting mainly of astrocytes, at the site of injury. Myelin in the scar apparently inhibits regeneration. However, implantation of peripheral nerves, Schwann cells, or fetal CNS tissue can bridge the scar and stimulate regeneration. Certain growth factors can also stimulate regeneration. Current research continues to look for the right combination of chemicals and other factors to stimulate regeneration of the spinal cord following injury.

1. Pain receptors detect a painful stimulus.

2. Sensory neurons conduct action potentials to the spinal cord.

3. Sensory neurons synapse with interneurons that synapse with motor neurons.

4. Excitation of the motor neurons results in contraction of the flexor muscles and withdrawal of the limb from the painful stimulus.

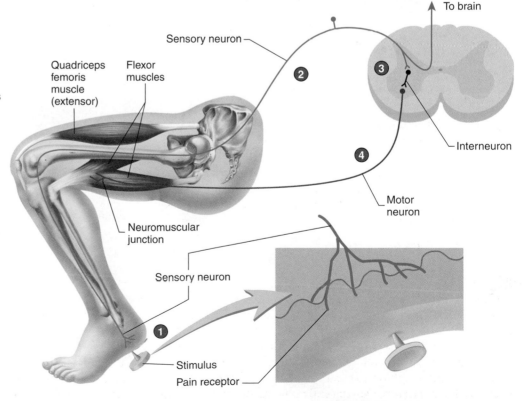

**PROCESS Figure 8.20** Withdrawal Reflex

Withdrawal reflex

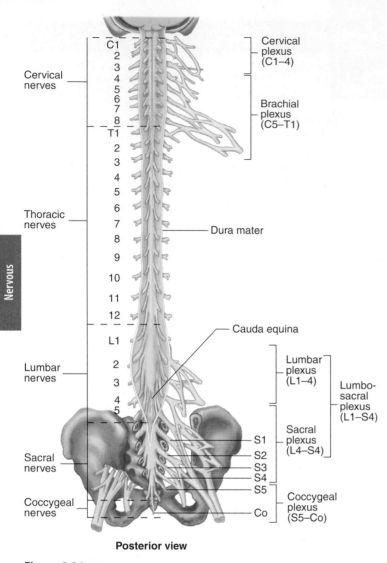

**Posterior view**

**Figure 8.21** Plexuses

of the cervical plexus is the **phrenic nerve,** which innervates the diaphragm. Contraction of the diaphragm is largely responsible for our ability to breathe (see chapter 15).

### Predict 2

*The phrenic nerve may be damaged where it descends along the neck or during open thorax or open heart surgery. Explain how damage to the right phrenic nerve affects the diaphragm. Describe the effect on the diaphragm of completely severing the spinal cord in the thoracic region versus in the upper cervical region.*

## Brachial Plexus

The **brachial plexus** originates from spinal nerves C5 to T1. Five major nerves emerge from the brachial plexus to supply the upper limb and shoulder. The **axillary nerve** innervates two shoulder muscles and the skin over part of the shoulder. The **radial nerve** innervates all the muscles in the posterior arm and forearm as well as the skin over the posterior surface of the arm, forearm, and hand. The **musculocutaneous** (mŭs′kū-lō-kū-tā′nē-ŭs; muscle + skin) **nerve** innervates the anterior muscles of the arm and the skin over the radial surface of the forearm. The **ulnar nerve** innervates two anterior forearm muscles and most of the intrinsic hand muscles. It also innervates the skin over the ulnar side of the hand. The ulnar nerve can be easily damaged where it passes posterior to the medial side of the elbow. The ulnar nerve at this location is called the "funny bone." The **median nerve** innervates most of the anterior forearm muscles and some of the intrinsic hand muscles. It also innervates the skin over the radial side of the hand.

## Lumbosacral Plexus

The **lumbosacral** (lŭm′bō-sā′krăl) **plexus** originates from spinal nerves L1 to S4. Four major nerves exit the plexus to supply the lower limb. The **obturator** (ob′tū-rā-tŏr) **nerve** innervates the muscles of the medial thigh and the skin over the same region. The **femoral**

| TABLE 8.3 | Plexuses of the Spinal Nerves (see figure 8.21) | | | |
|---|---|---|---|---|
| **Plexus** | **Origin** | **Major Nerves** | **Muscles Innervated** | **Skin Innervated** |
| Cervical | C1–C4 | | Several neck muscles | Neck and posterior head |
| | | Phrenic | Diaphragm | |
| Brachial | C5–T1 | Axillary | Two shoulder muscles | Part of shoulder |
| | | Radial | Posterior arm and forearm muscles (extensors) | Posterior arm, forearm, and hand |
| | | Musculocutaneous | Anterior arm muscles (flexors) | Radial surface of forearm |
| | | Ulnar | Two anterior forearm muscles (flexors), most intrinsic hand muscles | Ulnar side of hand |
| | | Median | Most anterior forearm muscles (flexors), some intrinsic hand muscles | Radial side of hand |
| Lumbosacral | L1–S4 | Obturator | Medial thigh muscles (adductors) | Medial thigh |
| | | Femoral | Anterior thigh muscles (extensors) | Anterior thigh, medial leg and foot |
| | | Tibial | Posterior thigh muscles (flexors), anterior and posterior leg muscles, most foot muscles | Posterior leg and sole of foot |
| | | Common fibular | Lateral thigh and leg, some foot muscles | Anterior and lateral leg, dorsal (top) part of foot |

**nerve** innervates the anterior thigh muscles and the skin over the anterior thigh and medial side of the leg. The **tibial nerve** innervates the posterior thigh muscles, the anterior and posterior leg muscles, and most of the intrinsic foot muscles. It also innervates the skin over the sole of the foot. The **common fibular** (fib′ū-lăr) **nerve** innervates the muscles of the lateral thigh and leg and some intrinsic foot muscles. It also innervates the skin over the anterior and lateral leg and the dorsal surface (top) of the foot. The tibial and common fibular nerves are bound together within a connective tissue sheath and together are called the **sciatic** (sī-at′ik) **nerve.**

# BRAIN

The major regions of the brain are the brainstem, the diencephalon, the cerebrum, and the cerebellum (figure 8.22).

## Brainstem

The **brainstem** connects the spinal cord to the remainder of the brain (figure 8.23). It consists of the medulla oblongata, the pons, and the midbrain, and it contains several nuclei involved in vital body functions, such as the control of heart rate, blood pressure, and breathing. Damage to small areas of the brainstem can cause death, whereas damage to relatively large areas of the cerebrum or cerebellum often do not. Nuclei for all but the first two cranial nerves are also located in the brainstem.

### Medulla Oblongata

The **medulla oblongata** (ob′long-gă′tă) is the most inferior portion of the brainstem (figure 8.23) and is continuous with the spinal cord. It extends from the level of the foramen magnum to the pons. In addition to ascending and descending nerve tracts, the medulla

**Anterior**

Diencephalon
- Thalamus
- Hypothalamus

Brainstem
- Midbrain
- Pons
- Medulla oblongata

**Medial view**

Cerebrum

Corpus callosum

**Posterior**

Cerebellum

**Figure 8.22** Regions of the Right Half of the Brain

(a)    **Anterior view**

(b)    **Posterolateral view**

**Figure 8.23** Brainstem and Diencephalon

oblongata contains discrete nuclei with specific functions, such as regulation of heart rate and blood vessel diameter, breathing, swallowing, vomiting, coughing, sneezing, balance, and coordination.

On the anterior surface, two prominent enlargements called **pyramids** extend the length of the medulla oblongata (figure 8.23a). The pyramids consist of descending nerve tracts, which transmit action potentials from the brain to motor neurons of the spinal cord and are involved in the conscious control of skeletal muscles.

**Predict 3**

*A large tumor or* **hematoma** *(hē-mă-tō′mă), a mass of blood that occurs as the result of bleeding into the tissues, can cause increased pressure within the skull. This pressure can force the medulla oblongata downward toward the foramen magnum. The displacement can compress the medulla oblongata and lead to death. Give two likely causes of death, and explain why they would occur.*

## Pons

Immediately superior to the medulla oblongata is the **pons** (ponz; bridge). It contains ascending and descending nerve tracts, as well as several nuclei. Some of the nuclei in the pons relay information between the cerebrum and the cerebellum. Not only is the pons a functional bridge between the cerebrum and cerebellum but on the anterior surface it resembles an arched footbridge (figure 8.23a). Several nuclei of the medulla oblongata, described earlier, extend into the lower part of the pons so functions such as breathing, swallowing, and balance are controlled in the lower pons, as well

as in the medulla oblongata. Other nuclei in the pons control functions such as chewing and salivation.

## Midbrain

The **midbrain,** just superior to the pons, is the smallest region of the brainstem (figure 8.23b). The dorsal part of the midbrain consists of four mounds called the **colliculi** (ko-lik′ū-lī; sing. colliculus, hill). The two inferior colliculi are major relay centers for the auditory nerve pathways in the CNS. The two superior colliculi are involved in visual reflexes and receive touch and auditory input. Turning the head toward a tap on the shoulder, a sudden loud noise, or a bright flash of light is a reflex controlled in the superior colliculi. The midbrain contains nuclei involved in coordinating eye movements and controlling pupil diameter and lens shape. The midbrain also contains a black nuclear mass, called the **substantia nigra** (sŭb-stan′shē-ă nī′gră; black substance), which is part of the basal nuclei (see "Basal Nuclei" later in this chapter) and is involved in regulating general body movements. The rest of the midbrain consists largely of ascending tracts from the spinal cord to the cerebrum and descending tracts from the cerebrum to the spinal cord or cerebellum.

## Reticular Formation

Scattered throughout the brainstem is a group of nuclei collectively called the **reticular formation.** The reticular formation plays important regulatory functions. It is particularly involved in regulating cyclical motor functions, such as respiration, walking, and chewing.

The reticular formation is a major component of the **reticular activating system,** which plays an important role in arousing and maintaining consciousness and in regulating the sleep-wake cycle. Stimuli such as a ringing alarm clock, sudden bright lights, smelling salts, or cold water splashed on the face can arouse consciousness. Conversely, removal of visual or auditory stimuli may lead to drowsiness or sleep. General anesthetics suppress the reticular activating system. Damage to cells of the reticular formation can result in coma.

## Cerebellum

The **cerebellum** (ser-e-bel′ŭm; little brain), shown in figure 8.22, is attached to the brainstem by several large connections called **cerebellar peduncles** (pe-dŭng′klz; *pes,* foot). These connections provide routes of communication between the cerebellum and other parts of the CNS. The structure and function of the cerebellum are discussed in a subsequent section of this chapter.

## Diencephalon

The **diencephalon** (dī′en-sef′ă-lon) is the part of the brain between the brainstem and the cerebrum (figure 8.24). Its main components are the thalamus, the epithalamus, and the hypothalamus.

### Thalamus

The **thalamus** (thal′ a-mŭs) is by far the largest part of the diencephalon. It consists of a cluster of nuclei and is shaped somewhat like a yo-yo, with two large, lateral parts connected in the center by a small **interthalamic adhesion** (figure 8.24). Most sensory input that ascends through the spinal cord and brainstem projects to the thalamus, where ascending neurons synapse with thalamic neurons. Thalamic neurons, in turn, send their axons to the cerebral cortex. The thalamus also influences mood and registers an unlocalized, uncomfortable perception of pain.

### Epithalamus

The **epithalamus** (ep′i-thal′ă-mŭs; *epi,* upon) is a small area superior and posterior to the thalamus (figure 8.24). It consists of a few small nuclei, which are involved in the emotional and visceral response to odors, and the pineal gland. The **pineal gland** (pin′ē-ăl; pinecone-shaped) is an endocrine gland that may influence the onset of puberty and may play a role in controlling some long-term cycles that are influenced by the light-dark cycle. In animals, the pineal gland is known to influence annual behaviors, such as migration in birds, as well as changes in the color and density of fur in some mammals (see chapter 10).

### Hypothalamus

The **hypothalamus** is the most inferior part of the diencephalon; it contains several small nuclei that are very important in maintaining homeostasis. The hypothalamus plays a central role in the control of body temperature, hunger, and thirst. Sensations such as sexual pleasure, rage, fear, and relaxation after a meal are related to hypothalamic functions. Emotional responses that seem inappropriate to the circumstances, such as "nervous perspiration" in response to stress or hunger as a result of depression, also involve the hypothalamus. A funnel-shaped stalk, the **infundibulum** (in-fŭn-dib′ū-lŭm; a funnel), extends from the floor of the hypothalamus to the pituitary gland. The hypothalamus plays a major role in controlling the secretion of hormones from the

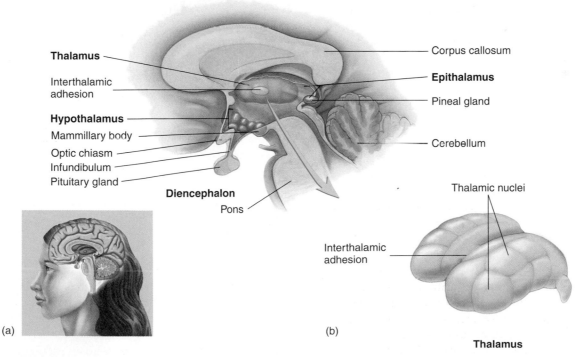

**Figure 8.24  Diencephalon**
(*a*) Median section of the diencephalon showing the thalamus, epithalamus, and hypothalamus. (*b*) Both halves of the thalamus as seen from an anterior, dorsolateral view with the separations between nuclei depicted by indentations on the surface.

Nervous

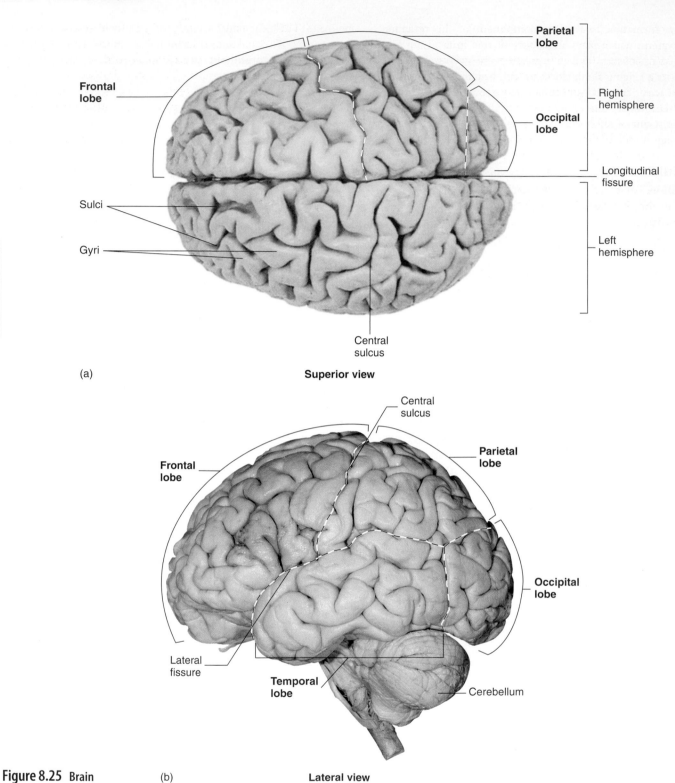

(a)    **Superior view**

**Figure 8.25**  Brain    (b)    **Lateral view**

pituitary gland (see chapter 10). The **mammillary** (mam′i-lār-ē; *mamilla,* nipple) **bodies** form externally visible swellings on the posterior portion of the hypothalamus and are involved in emotional responses to odors and in memory.

## Cerebrum

The **cerebrum** (ser′ĕ-brŭm, sĕ-rē′brŭm; brain) is the largest part of the brain (figure 8.25). It is divided into left and right hemispheres by a **longitudinal fissure.** The most conspicuous features on the surface of each hemisphere are numerous folds called **gyri** (jī′rī; sing. gyrus, *gyros,* circle), which greatly increase the surface area of the cortex, and intervening grooves called **sulci** (sŭl′sī; sing. sulcus, a furrow or ditch).

Each cerebral hemisphere is divided into lobes, which are named for the skull bones overlying them. The **frontal lobe** is important in the control of voluntary motor functions, motivation, aggression, mood, and olfactory (smell) reception. The **parietal lobe** is the principal center for receiving and consciously perceiving most sensory information, such as touch, pain, temperature, balance, and taste. The frontal and parietal lobes are separated by the **central sulcus.** The **occipital lobe** functions in receiving and perceiving visual input and is not distinctly separate from the other lobes. The **temporal lobe** (figure 8.25*b*) is involved in olfactory (smell) and auditory (hearing) sensations and plays an important role in memory. Its anterior and inferior portions, called the "psychic cortex," are associated with functions such as abstract thought and judgment. Most of the temporal lobe is separated from the rest of the cerebrum by the **lateral fissure.**

# SENSORY FUNCTIONS

The CNS constantly receives a variety of stimuli originating both inside and outside the body. We are unaware of a large part of this sensory input, but it is vital to our survival and normal functions. Sensory input to the brainstem and diencephalon helps maintain homeostasis. Input to the cerebrum and cerebellum keeps us informed about our environment and allows the CNS to control motor functions. A small portion of the sensory input results in **sensation,** the conscious awareness of stimuli (see chapter 9).

## Ascending Tracts

The spinal cord and brainstem contain a number of **ascending** (sensory) **tracts,** or pathways, that transmit information via action potentials from the periphery to various parts of the brain (table 8.4

| TABLE 8.4 | Ascending Tracts (see figures 8.26 and 8.27) |
|---|---|
| **Pathway** | **Function** |
| Spinothalamic | Pain, temperature, light touch, pressure, tickle, and itch sensations |
| Dorsal column | Proprioception, touch, deep pressure, and vibration |
| Spinocerebellar (anterior and posterior) | Proprioception to cerebellum |

and figure 8.26). Each tract is involved with a limited type of sensory input, such as pain, temperature, touch, position, or pressure, because each tract contains axons from specific sensory receptors specialized to detect a particular type of stimulus (see chapter 9).

Tracts are usually given composite names that indicate their origin and termination. The names of ascending tracts usually begin with the prefix *spino-*, indicating that they begin in the spinal cord. For example, the spinothalamic tract begins in the spinal cord and terminates in the thalamus.

Most ascending tracts consist of two or three neurons in sequence, from the periphery to the brain. Almost all neurons relaying information to the cerebrum terminate in the thalamus. Another neuron then relays the information from the thalamus to the cerebral cortex. The **spinothalamic tract,** which transmits action potentials dealing with pain and temperature to the thalamus and on to the cerebral cortex, is an example of an ascending tract. The **dorsal column,** which transmits action potentials dealing with touch, position, and pressure, is another example (figure 8.27).

**Figure 8.27** Dorsal Column

The dorsal column transmits action potentials dealing with touch, position, and pressure. Lines on the inset indicate levels of section

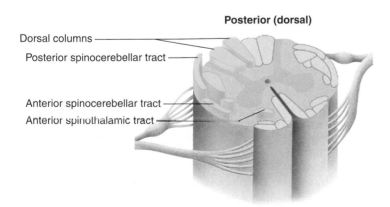

**Figure 8.26** Ascending Tracts of the Spinal Cord

Nervous

**Lateral view**

**Figure 8.28** Sensory and Motor Areas of the Lateral Side of the Left Cerebral Cortex

Sensory tracts typically cross from one side of the body in the spinal cord or brainstem to the other side of the body. Thus, the left side of the brain receives sensory input from the right side of the body, and vice versa.

Ascending tracts also terminate in the brainstem or cerebellum. The anterior and posterior **spinocerebellar tracts,** for example, transmit information about body position to the cerebellum (see "Cerebellum" later in this chapter).

## Sensory Areas of the Cerebral Cortex

Figure 8.28 depicts a lateral view of the left cerebral cortex with some of the sensory and motor areas indicated. The terms *area* and *cortex* are often used interchangeably for these regions of the cerebral cortex. Ascending tracts project to specific regions of the cerebral cortex, called **primary sensory areas,** where sensations are perceived. The **primary somatic sensory cortex,** or *general sensory area,* is located in the parietal lobe posterior to the central sulcus. Sensory fibers carrying general sensory input, such as pain, pressure, and temperature, synapse in the thalamus, and thalamic neurons relay the information to the primary somatic sensory cortex. Sensory fibers from specific parts of the body project to specific regions of the primary somatic sensory cortex so that a topographic map of the body, with the head most inferior, exists in this part of the cerebral cortex (figure 8.28). Other primary sensory areas include the visual cortex in the occipital lobe, the primary auditory cortex in the temporal lobe, and the taste area in the parietal lobe.

Cortical areas immediately adjacent to the primary sensory areas, called **association areas,** are involved in the process of recognition. For example, sensory action potentials originating in the retina of the eye reach the visual cortex, where the image is perceived. Action potentials then pass from the visual cortex to the visual association area, where the present visual information is compared to past visual experience ("Have I seen this before?"). On the basis of this comparison, the visual association area "decides" whether the visual input is recognized and judges whether the input is significant. For example, if you pass a man walking down a street, you usually pay less attention to him if you've never seen him before than if you know him, unless a unique characteristic of the unknown person draws your attention. Other examples of association areas are the auditory association area, adjacent to the primary auditory cortex, and the somatic sensory association area, adjacent to the primary somatic sensory cortex.

## MOTOR FUNCTIONS

The motor system of the brain and spinal cord is responsible for maintaining the body's posture and balance, as well as moving the trunk, head, limbs, tongue, and eyes and communicating through facial expressions and speech. Reflexes mediated through the spinal cord and brainstem are responsible for some body movements. These are called **involuntary movements** because they occur without conscious thought. **Voluntary movements,** on the other hand, are consciously activated to achieve a specific goal, such as walking or typing. Although consciously activated, the details of most voluntary movements occur automatically. For example, once you start walking, you don't have to think about the moment-to-moment control of every muscle because neural circuits in the reticular formation and spinal cord automatically control your limbs. Once learned, complex tasks, such as typing, can be performed relatively automatically.

Voluntary movements result from the stimulation of upper and lower motor neurons. **Upper motor neurons** have cell bodies in the cerebral cortex. The axons of upper motor neurons form descending tracts that connect to lower motor neurons. **Lower motor neurons** have cell bodies in the anterior horn of the spinal cord gray matter or in cranial nerve nuclei. Their axons leave the central nervous system and extend through spinal or cranial nerves to skeletal muscles. Lower motor neurons form the motor units described in chapter 7.

## Motor Areas of the Cerebral Cortex

The **primary motor cortex** is located in the posterior portion of the frontal lobe, directly anterior to the central sulcus (figure 8.28). Action potentials initiated in this region control voluntary movements of skeletal muscles. Upper motor neuron axons project from specific regions of this cortex to specific parts of the body so that a topographic map of the body exists in the primary motor cortex, with the head inferior, analogous to the topographic map of the primary somatic sensory cortex (figure 8.28). The **premotor area** of the frontal lobe is where motor functions are organized before they are actually initiated in the primary motor cortex. For example, if a person decides to take a step, the neurons of the premotor area are first stimulated, and the determination is made there as to which muscles must contract, in what order, and to what degree. Action potentials are then passed to the upper motor neurons of the primary motor cortex, which initiate each planned movement.

The motivation and foresight to plan and initiate movements occur in the anterior portion of the frontal lobes, called the **prefrontal area.** This region of association cortex is well developed only in primates, especially humans. It is involved in motivation and regulation of emotional behavior and mood. The large size of this area in humans may account for our emotional complexity and our relatively well-developed capacity to think ahead and feel motivated.

## Descending Tracts

The names of the descending tracts are based on their origin and termination (table 8.5 and figure 8.29). For example, the corticospinal tracts are so named because they begin in the cerebral cortex and terminate in the spinal cord. The corticospinal tracts are considered **direct** because they extend directly from upper motor neurons in the cerebral cortex to lower motor neurons in the spinal cord (a similar direct tract extends to lower motor neurons in the brainstem). Other tracts are named after the part of the brainstem from which they originate. Although they originate in the brainstem, these tracts are indirectly controlled by the cerebral cortex, basal nuclei, and cerebellum. These tracts are called **indirect** because no direct connection exists between the cortical and spinal neurons.

The descending tracts control different types of movements (table 8.5). Tracts in the lateral columns (see figure 8.17; figure 8.29) are most important in controlling goal-directed limb movements, such as reaching and manipulating. The **lateral corticospinal tracts** are especially important in controlling the speed and precision of skilled movements of the hands. Tracts in the ventral columns, such as the reticulospinal tract, are most important for maintaining posture, balance, and limb position through their control of neck, trunk, and proximal limb muscles.

## A CASE IN POINT
### Recovery from Spinal Cord Injury

Harley Chopper was taking a curve on his motorcycle when he hit a patch of loose gravel, and his motorcycle slid onto its side. Harley fell off and was spun, head-first, into a retaining wall. Even though Harley was wearing a helmet, the compression of hitting the wall with great force injured his neck. Examination at the hospital revealed that Harley had suffered a fracture of the fifth and sixth cervical vertebrae. Physicians administered methylprednisone to reduce inflammation around the spinal cord, and Harley's neck vertebrae were realigned and stabilized. Initially, Harley had no feeling, or movement, in his lower limbs and some feeling, but little movement, in his upper limbs. He was hospitalized for several weeks. Sensation and motor function gradually returned in all four limbs, but he required many months of physical therapy. Three years later, Harley had still not fully regained his ability to walk.

Immediately following Harley's injury, both ascending and descending tracts experienced a lack of function. This is not uncommon with a spinal cord injury, as the entire cord experiences the "shock" of the trauma. Without rapid treatment, that initial shock can become permanent loss. Sometimes, even with treatment, the damage is so severe that permanent loss of function results. In Harley's case, all ascending tracts regained normal function, but some functional deficit still remained in the lateral corticospinal tract to the lower limbs after 3 years.

| TABLE 8.5 | Descending Tracts (see figures 8.29 and 8.30) |
|---|---|
| **Pathway** | **Function** |
| Direct | |
| Lateral corticospinal | Muscle tone and skilled movements, especially of hands |
| Anterior corticospinal | Muscle tone and movement of trunk muscles |
| Indirect | |
| Rubrospinal | Movement coordination |
| Reticulospinal | Posture adjustment, especially during movement |
| Vestibulospinal | Posture and balance |
| Tectospinal | Movement in response to visual reflexes |

**Posterior (dorsal)**

Lateral corticospinal

Rubrospinal

Anterior corticospinal

Reticulospinal

Vestibulospinal

Tectospinal

**Anterior (ventral)**

**Figure 8.29 Descending Tracts of the Spinal Cord**

**Figure 8.30** **Example of a Direct Tract**

The lateral corticospinal tract is responsible for movement below the head. Lines on the inset indicate levels of section.

The lateral corticospinal tract serves as an example of how descending pathways function. It begins in the cerebral cortex and descends into the brainstem (figure 8.30). At the inferior end of the pyramids of the medulla oblongata, the axons cross over to the opposite side of the body and continue into the spinal cord. Crossover of axons in the brainstem or spinal cord to the opposite side of the body is typical of descending pathways. Thus, the left side of the brain controls skeletal muscles on the right side of the body, and vice versa.

## Basal Nuclei

The **basal nuclei** are a group of functionally related nuclei (figure 8.31). Two primary nuclei are the **corpus striatum**

(kōr′-pŭs strī-ā′tŭm), located deep within the cerebrum, and the **substantia nigra,** a group of darkly pigmented cells in the midbrain.

The basal nuclei are important in planning, organizing, and coordinating motor movements and posture. Complex neural circuits link the basal nuclei with each other, with the thalamus, and with the cerebral cortex. These connections form several feedback loops, some of which are stimulatory and others inhibitory. The stimulatory circuits facilitate muscle activity, especially at the beginning of a voluntary movement, such as rising from a sitting position or beginning to walk. The inhibitory circuits facilitate the actions of the stimulatory circuits by inhibiting muscle activity in antagonist muscles. In addition, inhibitory circuits inhibit random movements of the trunk and limbs. Inhibitory circuits also decrease muscle tone when the body, limbs, and head are at rest. Disorders of the basal nuclei result in difficulty rising from a sitting position and difficulty initiating walking. People with basal nuclei disorders exhibit increased muscle tone and exaggerated, uncontrolled movements when they are at rest. A specific feature of some basal nuclei disorders is "resting tremor," a slight shaking of the hands when a person is not performing a task. Parkinson disease, Huntington disease, and cerebral palsy are basal nuclei disorders (see the Diseases and Disorders table at the end of this chapter).

## Cerebellum

The **cerebellum** is attached by cerebellar peduncles to the brainstem (see figures 8.23 and 8.25b). The cerebellar cortex is composed of gray matter and has gyri and sulci, but the gyri are much smaller than those of the cerebrum. Internally, the cerebellum consists of gray nuclei and white nerve tracts. The cerebellum is involved in maintaining balance and muscle tone and in coordinating fine motor movement. If the cerebellum is damaged, muscle tone decreases, and fine motor movements become very clumsy.

A major function of the cerebellum is that of a **comparator** (figure 8.32). A comparator is a sensing device that compares the data from two sources—in this case, the motor cortex and peripheral structures. Action potentials from the cerebral motor cortex descend into the spinal cord to initiate voluntary movements. Collateral branches are also sent from the motor cortex to the cerebellum, giving information representing the intended movement. In addition, simultaneously, reaching the cerebellum are action potentials from **proprioceptive** (prō-prē-ō-sep′tiv) **neurons,** which innervate joints, tendons, and muscles and provide information about the position of body parts. The cerebellum compares information about the intended movement from the motor cortex to sensory information from the moving structures. If a difference is detected, the cerebellum sends action potentials to motor neurons in the motor cortex and the spinal cord to correct the discrepancy. The result is smooth and coordinated movements. For example, if you close your eyes, the cerebellar comparator function allows you to touch your nose smoothly and easily with your finger. If the cerebellum is not functioning, your finger tends to overshoot the target. One effect of alcohol is to inhibit the function of the cerebellum.

Another function of the cerebellum involves participating with the cerebrum in learning motor skills, such as playing the piano. Once the cerebrum and cerebellum "learn" these skills, the specialized movements can be accomplished smoothly and automatically.

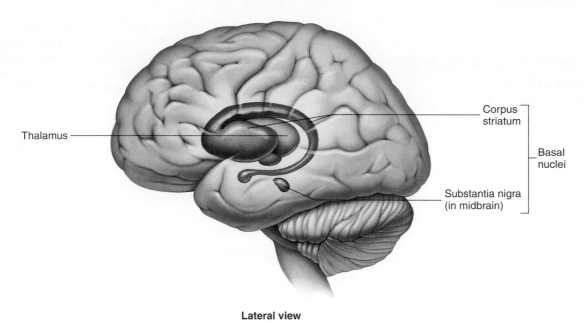

Thalamus

Corpus striatum

Basal nuclei

Substantia nigra (in midbrain)

**Lateral view**

**Figure 8.31**  Basal Nuclei

① The motor area of the cerebral cortex sends action potentials to lower motor neurons in the spinal cord.

② Action potentials from the motor cortex inform the cerebellum of the intended movement.

③ Lower motor neurons in the spinal cord send action potentials to skeletal muscles, causing them to contract.

④ Proprioceptive signals from the skeletal muscles and joints to the cerebellum convey information concerning the status of the muscles and the structure being moved during contraction.

⑤ The cerebellum compares the information from the motor cortex to the proprioceptive information from the skeletal muscles and joints.

⑥ Action potentials from the cerebellum modify the stimulation from the motor cortex to the lower motor neurons in the spinal cord.

⑦ Action potentials from the cerebellum are sent to the motor cortex to modify motor activity.

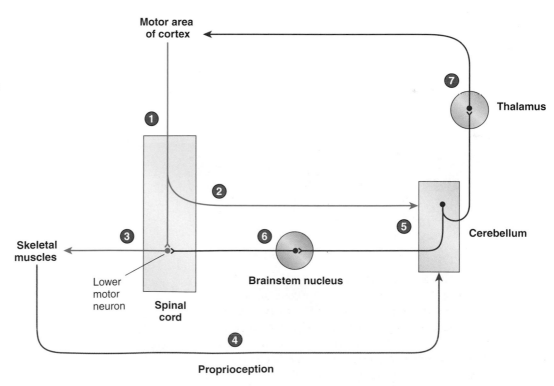

**PROCESS Figure 8.32**  Cerebellar Comparator Function

# OTHER BRAIN FUNCTIONS

## Communication Between the Right and Left Hemispheres

The right cerebral hemisphere receives sensory input from and controls muscular activity in the left half of the body. The left cerebral hemisphere receives input from and controls muscles in the right half of the body. Sensory information received by one hemisphere is shared with the other through connections between the two hemispheres called **commissures** (kom'ĭ-shūrz; joining together). The largest of these commissures is the corpus callosum (kōr'pus kă-lō'sŭm), a broad band of nerve tracts at the base of the longitudinal fissure (see figures 8.22 and 8.24).

Language and perhaps other functions, such as artistic activities, are not shared equally between the two hemispheres. Researchers believe the left hemisphere is the more analytical hemisphere, emphasizing such skills as mathematics and speech, whereas the right hemisphere is more involved in functions such as three-dimensional or spatial perception and musical ability.

## Speech

In most people, the speech area is in the left cerebral cortex. Two major cortical areas are involved in speech: the **sensory speech area** (Wernicke area), a portion of the parietal lobe, and the **motor speech area** (Broca area) in the inferior portion of the frontal lobe (see figure 8.28). Damage to these parts of the brain or to associated brain regions may result in **aphasia** (ă-fā'zē-ă; *a-*, without + *phasis,* speech), absent or defective speech or language comprehension. The most common cause is a stroke. It is estimated that 25–40% of stroke survivors exhibit aphasia.

Speech-related functions require similar primary pathways. For example, to repeat a word that you *hear* involves the following pathway: Action potentials from the ear reach the primary auditory cortex, where the word is perceived; the word is recognized in the auditory association area and comprehended in portions of the sensory speech area. Action potentials representing the word are then conducted through nerve tracts that connect the sensory and motor speech areas. In the motor speech area, the word is formulated as it is to be repeated; action potentials then go to the premotor area, where the movements are programmed, and finally to the primary motor cortex, where specific movements are triggered.

Speaking a *written* word involves a slightly different pathway: The information enters the visual cortex; passes to the visual association area, where it is recognized; and continues to the sensory speech area, where it is understood and formulated as it is to be spoken. From the sensory speech area, it follows the same route for repeating words that you hear: through nerve tracts to the motor speech area, to the premotor area, and then to the primary motor cortex.

### Predict 4

*Describe a neural pathway that allows a blindfolded person to name an object placed in her hand.*

## Brain Waves and Consciousness

Different levels of consciousness can be revealed by different patterns of electrical activity in the brain. Electrodes placed on a person's scalp and attached to a recording device can record the brain's electrical activity, producing an **electroencephalogram** (ē-lek'trō-en-sef'ă-lō-gram) **(EEG)** (figure 8.33). These electrodes are not positioned so that they can detect individual action potentials, but they can detect the simultaneous action potentials in large numbers of neurons. Most of the time, EEG patterns are irregular, with no particular pattern, because the brain's electrical activity is normally not synchronous. At other times, however, EEG patterns can be detected as wavelike patterns known as **brain waves.** Their intensity and frequency differ, based on the state of brain activity. The different levels of consciousness in an awake and a sleeping person are marked by different types of brain waves. **Alpha waves** are observed in a normal person who is awake

(a)

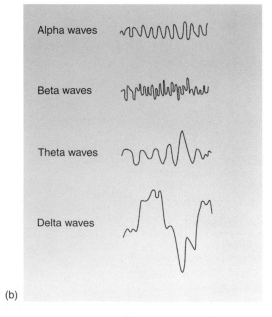

(b)

**Figure 8.33 Electroencephalogram**

(*a*) Patient with electrodes attached to his head. (*b*) Four electroencephalographic tracings.

but in a quiet, resting state with the eyes closed. **Beta waves** have a higher frequency than alpha waves and occur during intense mental activity. During the beginning of sleep, a rapid transition takes place from a beta rhythm to an alpha rhythm. As sleep deepens, progressively more delta waves occur. **Delta waves** occur during deep sleep, in infants, and in patients with severe brain disorders. A fourth type of brain waves, **theta waves,** are usually observed in children, but they can also occur in adults who are experiencing frustration or who have certain brain disorders. Neurologists use these patterns to diagnose and determine the treatment for the disorders.

## Memory

Memory consists of four processes: encoding, consolidation, storage, and retrieval. *Encoding* consists of the brief retention of sensory input received by the brain while something is scanned, evaluated, and acted on. This process, called **sensory memory,** occurs in the temporal lobe and lasts less than a second.

Once a piece of data is encoded, it can be *consolidated* within the temporal lobe into **short-term memory,** or *working memory,* where information is retained for a few seconds to a few minutes. This memory is limited primarily by the number of bits of information (about seven) that can be stored at any one time. When new information is presented, old information, previously stored in short-term memory, is eliminated. What happens to a telephone number you just looked up if you are distracted? If the temporal lobe is damaged, the transition from encoding to short-term

memory may not occur; from that point on, the person always lives only in the present and in the more remote past, before the injury. This person is unable to add new memory.

Certain pieces of information are transferred for *storage* from short-term to **long-term memory,** where it may be stored for only a few minutes or become permanent. The length of time memory is stored may depend on how often it is *retrieved* and used. **Declarative memory,** or *explicit memory,* involves the retention of facts, such as names, dates, and places, as well as related emotional undertones. Emotion and mood apparently serve as gates in the brain and determine what is stored in long-term declarative memory. **Procedural memory,** or *implicit memory,* involves the development of motor skills, such as riding a bicycle. Only a small amount of procedural memory is lost over time.

Long-term memory involves structural and functional changes in neurons that lead to long-term enhancement of synaptic transmission. A whole series of neurons, called **memory engrams,** or *memory traces,* are probably involved in the long-term retention of a given piece of information, a thought, or an idea. Repeating the information and associating it with existing memories help us transfer information from short-term to long-term memory.

## Limbic System

The olfactory cortex and certain deep cortical regions and nuclei of the cerebrum and the diencephalon are grouped together under the title **limbic** (lim′bik; a boundary) **system** (figure 8.34). The

**Medial view**

### Figure 8.34  Limbic System

The limbic system includes the olfactory cortex, the cingulate gyrus (an area of the cerebral cortex on the medial side of each hemisphere), nuclei such as those of the hypothalamus and thalamus, the hippocampus (a mass of neuron cell bodies in the temporal lobe), and connecting nerve tracts, such as the fornix.

limbic system influences long-term declarative memory, emotions, visceral responses to emotions, motivation, and mood. A major source of sensory input to the limbic system is the olfactory nerves. The limbic system responds to olfactory stimulation by initiating responses necessary for survival, such as hunger and thirst. The limbic system is connected to, and functionally associated with, the hypothalamus. Lesions in the limbic system can result in voracious appetite, increased (often perverse) sexual activity, and docility (including loss of normal fear and anger responses).

# MENINGES, VENTRICLES, AND CEREBROSPINAL FLUID

## Meninges

Three connective tissue membranes, the **meninges** (mě-nin'jēz; *meninx,* membrane) (figure 8.35), surround and protect the brain and spinal cord. The most superficial and thickest of the meninges is the **dura mater** (doo'ră mā'ter; tough mother).

The dura mater around the brain consists of two layers, which function as a single layer but are physically separated into several regions to form dural folds and dural venous sinuses. Folds of dura mater extend into the longitudinal fissure between the two cerebral hemispheres and between the cerebrum and the cerebellum. The folds help hold the brain in place within the skull. The dural venous sinuses collect blood from the small veins of the brain and empty into the internal jugular veins, which exit the skull.

Within the skull, the dura mater adheres tightly to the cranial bones. In contrast, within the vertebral canal is an **epidural space** between the dura mater and the vertebrae (figure 8.35c). The epidural space is clinically important as the injection site for **epidural anesthesia** of the spinal nerves, which is often given to women during childbirth.

The second meningeal membrane is the very thin, wispy **arachnoid** (ă-rak'noyd; spiderlike, as in cobwebs) **mater.** The space between the dura mater and the arachnoid mater is the **subdural space,** which is normally only a potential space containing a very small amount of serous fluid.

## A CASE IN POINT

### Subdural Hematoma

May Fall is a 70-year-old woman who fell in her living room and struck her head on the coffee table. Some days later, she complained to her son, Earl E., that she was experiencing headaches and that her left hand seemed weak. Earl also noticed that his mother, who was normally very happy and positive, seemed irritable and short-tempered. Earl convinced his mother to see a physician. Based on May's age and history, her physician ordered a CT scan. The CT scan revealed an image consistent with a subdural hematoma over the right hemisphere, and May was scheduled for surgery. After making a large surgical flap, the surgeon opened the dura mater and removed the hematoma by suction. After the surgery, May required several months of physical therapy, but she recovered.

Damage to the veins crossing between the cerebral cortex and the dural venous sinuses can cause bleeding into the subdural space, resulting in a subdural hematoma, which can put pressure on the brain. The pressure can result in decreased brain function in the affected area. For example, the primary motor cortex is located in the posterior portion of the frontal lobe. Therefore, pressure on the portion of the right primary motor cortex involved in hand movements can cause decreased function in the left hand. The frontal lobe is also involved with mood; thus, pressure in that area can result in mood changes—for example, a normally happy person may become irritable. Subdural hematoma is more common in people over age 60 because their veins are less resilient and more easily damaged.

The spinal cord extends only to approximately the level of the second lumbar vertebra. Spinal nerves surrounded by meninges extend to the end of the vertebral column. Because there is no spinal cord in the inferior portion of the vertebral canal, a needle can be introduced into the subarachnoid space at that level without damaging the spinal cord. Health professionals use such a needle to inject anesthetic into the area as a **spinal block** or to take a sample of cerebrospinal fluid in a **spinal tap.** The cerebrospinal fluid can then be examined for infectious agents (meningitis) or for blood (hemorrhage).

The third meningeal membrane, the **pia mater** (pī'ă, pē'ă; affectionate mother), is very tightly bound to the surface of the brain and spinal cord. Between the arachnoid mater and the pia mater is the **subarachnoid space,** which is filled with cerebrospinal fluid and contains blood vessels.

## Ventricles

The CNS contains fluid-filled cavities, called **ventricles,** which are quite small in some areas and large in others (figure 8.36). Each cerebral hemisphere contains a relatively large cavity called the **lateral ventricle.** The **third ventricle** is a smaller, midline cavity located in the center of the diencephalon between the two halves of the thalamus and connected by foramina (holes) to the lateral ventricles. The **fourth ventricle** is located at the base of the cerebellum and connected to the third ventricle by a narrow canal, called the **cerebral aqueduct.** The fourth ventricle is continuous with the **central canal** of the spinal cord. The fourth ventricle also opens into the subarachnoid space through foramina in its walls and roof.

## Cerebrospinal Fluid

**Cerebrospinal fluid (CSF)** bathes the brain and spinal cord, providing a protective cushion around the CNS. It is produced by the **choroid** (kō'royd; lacy) **plexuses,** specialized structures made of ependymal cells, which are located in the ventricles (figure 8.37). CSF fills the brain ventricles, the central canal of the spinal cord, and the subarachnoid space. The CSF flows from the lateral ventricles into the third ventricle and then through the cerebral aqueduct into the fourth ventricle. A small amount of CSF enters the central canal of the spinal cord. The CSF exits the fourth ventricle through small openings in its walls and roof and enters the subarachnoid space. Masses of arachnoid tissue, called **arachnoid granulations,** penetrate the

Nervous

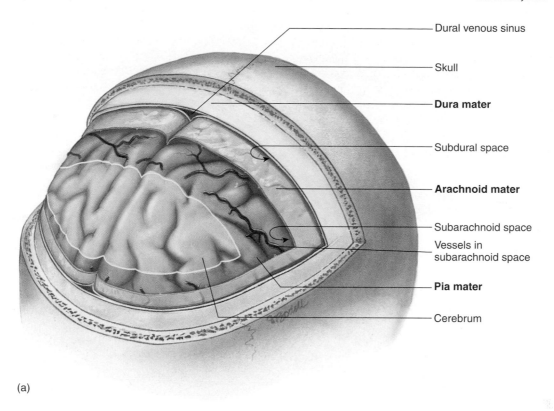

Dural venous sinus

Skull

**Dura mater**

Subdural space

**Arachnoid mater**

Subarachnoid space

Vessels in subarachnoid space

**Pia mater**

Cerebrum

(a)

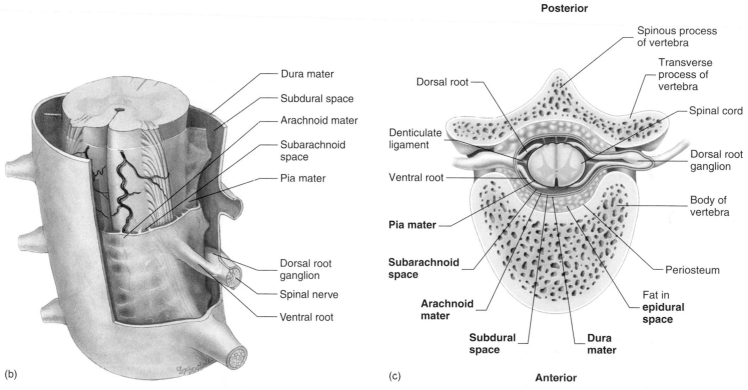

Dura mater

Subdural space

Arachnoid mater

Subarachnoid space

Pia mater

Dorsal root ganglion

Spinal nerve

Ventral root

(b)

**Posterior**

Spinous process of vertebra

Transverse process of vertebra

Dorsal root

Spinal cord

Denticulate ligament

Ventral root

Dorsal root ganglion

**Pia mater**

Body of vertebra

**Subarachnoid space**

Periosteum

**Arachnoid mater**

Fat in **epidural space**

**Subdural space**

**Dura mater**

(c)

**Anterior**

**Figure 8.35  Meninges**

(a) Anterior superior view of the head to show the meninges. (b) Meningeal membranes surrounding the spinal cord. (c) Cross section of a vertebra and the spinal cord.

Nervous

Lateral ventricle
(anterior horn)

Third ventricle

Lateral ventricle
(inferior horn)

Lateral ventricle
(posterior horn)

Cerebral aqueduct

Fourth ventricle

Central canal
of spinal cord

**Lateral view**

**Figure 8.36  Ventricles of the Brain Viewed from the Left**

superior sagittal sinus, a dural venous sinus in the longitudinal fissure, and CSF passes from the subarachnoid space into the blood through these granulations.

Blockage of the openings in the fourth ventricle or the cerebral aqueduct can cause CSF to accumulate in the ventricles, a condition known as **hydrocephalus.** The accumulation of fluid creates increased pressure that dilates the ventricles and compresses the brain tissue, which usually results in irreversible brain damage. If the skull bones are not completely ossified when the hydrocephalus occurs, as in a fetus or newborn, the pressure can also cause severe enlargement of the head. Hydrocephalus is treated by placing a drainage tube (shunt) from the ventricles to the abdominal cavity to eliminate the high internal pressures.

## CRANIAL NERVES

The 12 pairs of **cranial nerves** are listed in table 8.6 and pictured in figure 8.38. They are designated by Roman numerals from I to XII. There are two general categories of cranial nerve function: sensory and motor. Sensory functions can be divided into the special senses, such as vision, and the more general senses, such as touch and pain in the face. Motor functions are further subdivided into somatic motor and parasympathetic. Somatic cranial nerves innervate skeletal muscles in the head and neck. Parasympathetic cranial nerves innervate glands, smooth muscle, and cardiac muscle.

Some cranial nerves are only sensory, and some are only somatic whereas other cranial nerves are mixed nerves with

sensory, somatic and parasympathetic functions. Three cranial nerves—the olfactory (I), optic (II), and vestibulocochlear (VIII) nerves—are sensory only. Four other cranial nerves—the trochlear (IV), abducens (VI), accessory (XI), and hypoglossal (XII) nerves—are considered somatic only (although these motor nerves do provide proprioceptive information).

The trigeminal nerve (V) has sensory and somatic functions. It has the greatest general sensory distribution of all the cranial nerves and is the only cranial nerve supplying sensory information to the brain from the skin of the face. Sensory information from the skin over the rest of the body is carried to the CNS by spinal nerves. Injections of anesthetic by a dentist are designed to block sensory transmission through branches of the trigeminal nerve from the teeth. These dental branches of the trigeminal nerve are probably anesthetized more often than any other nerves in the body.

The oculomotor nerve (III) is somatic and parasympathetic. The facial (VII), glossopharyngeal (IX), and vagus (X) nerves have all three functions: sensory, somatic and parasympathetic (table 8.6). The vagus nerve is probably the most important parasympathetic nerve in the body. It helps regulate the functions of the thoracic and abdominal organs, such as heart rate, respiration rate, and digestion.

As with the spinal nerves, the sensory and motor functions of many cranial nerves cross between the face and the cerebral cortex. For example, sensory input from the right side of the face via the trigeminal nerve (V) crosses in the brainstem and projects to the left cerebral cortex. Motor output from the left cerebral cortex crosses in the brainstem to the right side of the face via the facial nerve (VII).

1. Cerebrospinal fluid (CSF) is produced by the choroid plexuses of each of the four ventricles (*inset, lower left*).

2. CSF from the lateral ventricles flows to the third ventricle.

3. CSF flows from the third ventricle through the cerebral aqueduct to the fourth ventricle.

4. CSF exits the fourth ventricle through openings in the wall of the fourth ventricle and enters the subarachnoid space. Some CSF enters the central canal of the spinal cord.

5. CSF flows through the subarachnoid space to the arachnoid granulations in the superior sagittal sinus, where it enters the venous circulation (*inset, upper right*).

Skull

Superior sagittal sinus (dural venous sinus)

Dura mater

Arachnoid granulation

Arachnoid mater

Subarachnoid space

Subarachnoid space

Pia mater

Cerebrum

Superior sagittal sinus

Arachnoid granulation

Subarachnoid space

Choroid plexus of lateral ventricle

Opening connecting lateral and third ventricles

Choroid plexus of third ventricle

Cerebral aqueduct

Opening in wall of fourth ventricle

Choroid plexus of fourth ventricle

Opening in roof of fourth ventricle

Subarachnoid space

Central canal of spinal cord

Dura mater

Ependymal cells

Connective tissue

Section of choroid plexus

Capillary containing blood

CSF enters the ventricle.

**Sagittal section, medial view**

**PROCESS Figure 8.37   Flow of CSF**

*White arrows* indicate the direction of CSF flow, and *black arrows* indicate the direction of blood flow in the sinuses.

| TABLE 8.6 | Cranial Nerves and Their Functions (see figure 8.38) | | |
|---|---|---|---|
| **Number** | **Name** | **General Function*** | **Specific Function** |
| I | Olfactory | S | Smell |
| II | Optic | S | Vision |
| III | Oculomotor | M, P | Motor to four of six extrinsic eye muscles and upper eyelid; parasympathetic: constricts pupil, thickens lens |
| IV | Trochlear | M | Motor to one extrinsic eye muscle |
| V | Trigeminal | S, M | Sensory to face and teeth; motor to muscles of mastication (chewing) |
| VI | Abducens | M | Motor to one extrinsic eye muscle |
| VII | Facial | S, M, P | Sensory: taste; motor to muscles of facial expression; parasympathetic to salivary and tear glands |
| VIII | Vestibulocochlear | S | Hearing and balance |
| IX | Glossopharyngeal | S, M, P | Sensory: taste and touch to back of tongue; motor to pharyngeal muscles; parasympathetic to salivary glands |
| X | Vagus | S, M, P | Sensory to pharynx, larynx, and viscera; motor to palate, pharynx, and larynx; parasympathetic to viscera of thorax and abdomen |
| XI | Accessory | M | Motor to two neck and upper back muscles |
| XII | Hypoglossal | M | Motor to tongue muscles |

*S, sensory; M, somatic motor; P, parasympathetic

**Figure 8.38** Inferior Surface of the Brain Showing the Origin of the Cranial Nerves

# AUTONOMIC NERVOUS SYSTEM

The autonomic nervous system (ANS) comprises motor neurons that carry action potentials from the CNS to the periphery. The autonomic neurons innervate smooth muscle, cardiac muscle, and glands. Autonomic functions are largely controlled unconsciously.

Axons from autonomic neurons do not extend all the way from the CNS to target tissues. This is in contrast to somatic neurons, which extend axons from the CNS to skeletal muscle. In the autonomic nervous system, two neurons in series extend from the CNS to the effector organs. The first is called the **preganglionic neuron;** the second is the **postganglionic neuron** (figure 8.39). The neurons are so named because preganglionic neurons synapse with postganglionic neurons in **autonomic ganglia** outside the CNS. An exception is the preganglionic neuron that extends to the adrenal gland. There, the postganglionic neurons are actually the hormone-secreting cells of the adrenal medulla.

The autonomic nervous system is composed of the **sympathetic** (sim-pă -thet'ik; *sympatheo,* to feel with + *pathos,* suffering) **division** and the **parasympathetic** (par-ă-sim-pa-thet'ik; *para,* alongside of) **division** (table 8.7 and figure 8.40). Increased activity in sympathetic neurons generally prepares the individual for physical activity, whereas parasympathetic stimulation generally activates involuntary functions, such as digestion, that are normally associated with the body at rest.

## Anatomy of the Sympathetic Division

Cell bodies of sympathetic preganglionic neurons are in the lateral horn of the spinal cord gray matter (see figure 8.17) between the first thoracic (T1) and the second lumbar (L2) segments. The axons of the preganglionic neurons exit through ventral roots and project to either sympathetic chain ganglia or collateral ganglia (figure 8.40). The **sympathetic chain ganglia** are connected to one another and are so named because they form a chain along both sides of the spinal cord. The thoracic cavity is innervated by postganglionic sympathetic nerves. **Collateral ganglia** are located nearer target organs and consist of the celiac, superior mesenteric, and inferior mesenteric ganglia. The **splanchnic nerves** are sympathetic nerves that extend to collateral ganglia in the abdominal and pelvic regions. Some preganglionic neurons synapse in the sympathetic chain ganglia, but others pass through these ganglia and synapse in the collateral ganglia. Postganglionic neurons arise in the sympathetic chain ganglia or in the collateral ganglia and project to target tissues.

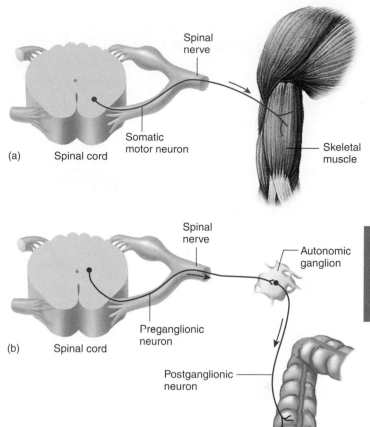

**Figure 8.39** **Organization of Somatic and Autonomic Nervous System Neurons**

(*a*) The cell body of the somatic neuron is in the CNS, and its axon extends to the skeletal muscle. (*b*) The cell body of the preganglionic neuron is in the CNS, and its axon extends to the autonomic ganglion and synapses with the postganglionic neuron. The postganglionic neuron extends to and synapses with its effector organ.

## Anatomy of the Parasympathetic Division

Preganglionic cell bodies of the parasympathetic division are located either within brainstem nuclei of the oculomotor nerve (III), facial nerve (VII), glossopharyngeal nerve (IX), or vagus nerve (X) or within the lateral part of the central gray matter of

| TABLE 8.7 | Sympathetic and Parasympathetic Divisions of the Autonomic Nervous System (see figure 8.40) | | |
|---|---|---|---|
| **Division** | **Location of Preganglionic Cell Body** | **Location of Postganglionic Cell Body** | **General Function** |
| Sympathetic | T1–L2 | Sympathetic chain ganglia or collateral ganglia | "Fight-or-flight"; prepares the body for physical activity |
| Parasympathetic | Cranial nerves III, VII, IX, X; S2–S4 spinal nerves | Terminal ganglia near or embedded in the walls of target organs | Stimulates involuntary activities of the body at rest |

Nervous

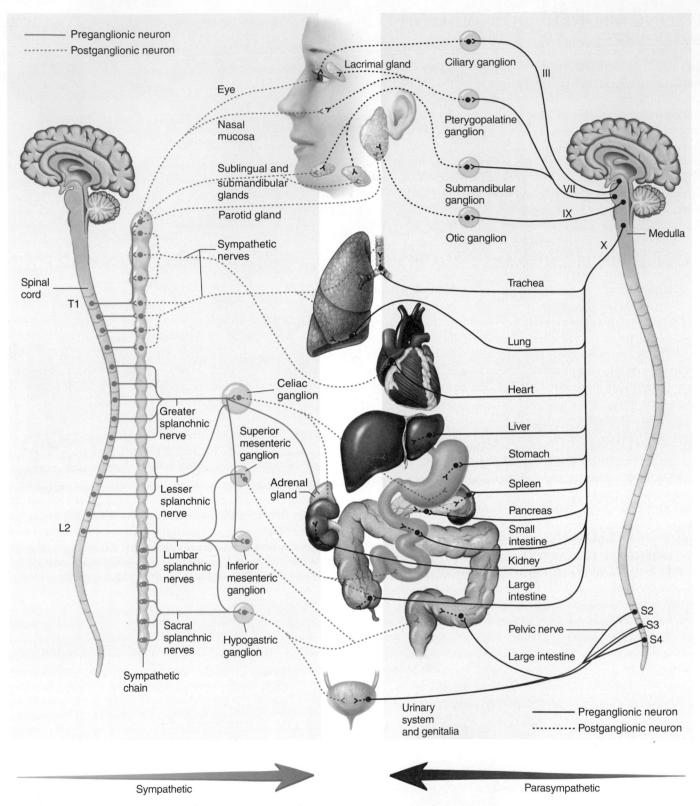

**Figure 8.40  Innervation of Organs by the ANS**

Preganglionic fibers are indicated by *solid lines,* and postganglionic fibers are indicated by *dashed lines.*

## Biofeedback and Meditation

**Biofeedback** is a technique that uses electronic instruments or other methods to monitor subconscious activities, many of which are regulated by the autonomic nervous system. For example, skin temperature, heart rate, and brain waves may be monitored electronically. Then, by watching the monitor and using biofeedback techniques, a person can learn to consciously reduce his or her heart rate and blood pressure or change the pattern of brain waves. The severity of peptic (stomach) ulcers, migraine headaches, high blood pressure, anxiety, or depression can be reduced by using biofeedback techniques.

**Meditation** is another technique that influences autonomic functions. Its practitioners also claim that meditation can improve their spiritual well-being, consciousness, and holistic view of the universe. Some people find meditation techniques useful in reducing heart rate, blood pressure, the severity of ulcers, and other symptoms frequently associated with stress.

the spinal cord in the regions that give rise to spinal nerves S2 through S4.

Axons of the preganglionic neurons extend through spinal nerves or cranial nerves to **terminal ganglia** either located near effector organs in the head (figure 8.40) or embedded in the walls of effector organs in the thorax, abdomen, and pelvis. The axons of the postganglionic neurons extend a relatively short distance from the terminal ganglia to the target organ. Most of the thoracic and abdominal organs are supplied by preganglionic neurons of the **vagus nerve** extending from the brainstem. The vagus nerve branches to provide parasympathetic innervation to the heart, the lungs, the liver, and the stomach and other digestive organs.

## Autonomic Neurotransmitters

All preganglionic neurons of both the sympathetic and the parasympathetic divisions and all the postganglionic neurons of the parasympathetic division secrete the neurotransmitter **acetylcholine.** Most postganglionic neurons of the sympathetic division secrete **norepinephrine.** Many body functions can be stimulated or inhibited by drugs that either mimic these neurotransmitters or prevent the neurotransmitters from activating their target tissues.

## Functions of the Autonomic Nervous System

The sympathetic nervous system prepares a person for physical activity (table 8.8). These actions include increasing heart rate and blood pressure, dilating respiratory passageways to increase airflow, and stimulating the release of glucose from the liver for energy. At the same time, it inhibits digestive activities. In this way, the sympathetic division decreases the activity of organs not essential for the maintenance of physical activity and shunts blood and nutrients to structures that are active during exercise. In addition, excess heat is removed by vasodilation of the vessels near the skin and increased perspiration. The sympathetic division is sometimes referred to as the **fight-or-flight system** because it prepares the person either to stand and face a threat or to leave the situation as quickly as possible.

The parasympathetic division of the autonomic nervous system is generally consistent with resting conditions (table 8.8). Increased activity of the parasympathetic division stimulates involuntary activities, such as digestion, defecation, and urination. The actions of the parasympathetic division in the digestive system illustrate how the ANS can coordinate the activities of multiple organs. Parasympathetic activity causes the release of digestive enzymes from the pancreas and contractions to mix the enzymes with food in the small intestine and move the material through the digestive tract. This cooperativity enhances the digestion and absorption of food. At the same time, parasympathetic stimulation lowers heart rate, which lowers blood pressure and constricts air passageways, which decreases airflow.

The sympathetic and parasympathetic divisions can each produce both stimulatory and inhibitory effects, depending on the target tissue. For example, the sympathetic division stimulates smooth muscle contraction in blood vessel walls and inhibits smooth muscle contraction in lung airway walls. Likewise, the parasympathetic division stimulates contraction of the urinary bladder and inhibits contraction of the heart.

Most organs that receive autonomic neurons are innervated by both the parasympathetic and the sympathetic division. In most cases, the influences of the two autonomic divisions are opposite. For example, sympathetic stimulation of the heart causes an increase in heart rate, whereas parasympathetic stimulation causes a decrease in heart rate. When both divisions innervate a single organ, the sympathetic division tends to play a major role during physical activity or stress, whereas the parasympathetic division has more influence during resting conditions. Despite the general opposing actions of the two divisions, in some situations, both divisions can act together to coordinate the activity of multiple targets. For example, in males, the parasympathetic division initiates erection of the penis, and the sympathetic division stimulates the release of secretions and helps initiate ejaculation. Not all organs receive dual innervation. Sweat glands and blood vessels are innervated by sympathetic neurons almost exclusively, whereas the smooth muscles associated with the lens of the eye are innervated primarily by parasympathetic neurons.

# CLINICAL IMPACT

## Autonomic Dysfunctions

Raynaud (rā-nō') **disease** involves the spasmodic contraction of blood vessels in the periphery, especially in the digits, and results in pale, cold hands that are prone to ulcerations and gangrene as a result of poor circulation. This condition may be caused by exaggerated sensitivity of the blood vessels to sympathetic stimulation. Occasionally, a surgeon may cut the preganglionic neurons to alleviate the condition. **Dysautonomia** (dis'aw-tō-nō'mē-ă) is an inherited condition characterized by reduced tear secretion, poor vasomotor control, dry mouth and throat, and other symptoms. It results from poorly controlled autonomic reflexes.

**Predict 5**

*List some of the responses stimulated by the autonomic nervous system in (a) a person who is extremely angry and (b) a person who has just finished eating and is now relaxing.*

## ENTERIC NERVOUS SYSTEM

The **enteric nervous system (ENS)** consists of plexuses within the wall of the digestive tract (see figure 16.2). The plexuses include (1) sensory neurons that connect the digestive tract to the CNS; (2) sympathetic and parasympathetic neurons that connect the CNS to the digestive tract; and (3) enteric neurons, located entirely within the enteric plexuses. A unique feature of enteric neurons is that they are capable of monitoring and controlling the digestive tract independently of the CNS through local reflexes. For example, stretching of the digestive tract is detected by enteric sensory neurons, which stimulate enteric interneurons. The enteric interneurons stimulate enteric motor neurons, which stimulate glands to secrete. Although the ENS is capable of controlling the activities of the digestive tract completely independently of the CNS, normally the two systems work together. CNS control of parasympathetic branches of the vagus nerve and sympathetic nerves (primarily, the splanchnic nerves) can override the actions of enteric neurons. Hence, the ENS is an independent subdivision of the PNS that is integrated with the ANS.

## EFFECTS OF AGING ON THE NERVOUS SYSTEM

As a person ages, sensory function gradually declines because of decreases in the number of sensory neurons, the function of the remaining neurons, and CNS processing. As a result of decreases

| TABLE 8.8 | Effects of the ANS on Various Tissues | |
|---|---|---|
| **Target** | **Sympathetic Effects** | **Parasympathetic Effects** |
| Heart | Increases rate and force of contraction | Decreases rate |
| Blood vessels | Constricts and dilates | None |
| Lungs | Dilates bronchioles | Constricts bronchioles |
| Eyes | Dilates pupil, relaxes ciliary muscle to adjust lens for far vision | Constricts pupil, contracts ciliary muscle to adjust lens for near vision |
| Intestinal and stomach walls | Decreases motility, contracts sphincters | Increases motility, relaxes sphincters |
| Liver | Breaks down glycogen, releases glucose | Synthesizes glycogen |
| Adipose tissue | Breaks down fat | None |
| Adrenal gland | Secretes epinephrine, norepinephrine | None |
| Sweat glands | Secretes sweat | None |
| Salivary glands | Secretes thick saliva | Secretes watery saliva |
| Urinary bladder | Relaxes muscle, constricts sphincter | Contracts muscle, relaxes sphincter |
| Pancreas | Decreases secretion of digestive enzymes and insulin | Increases secretion of digestive enzymes and insulin |

Nervous

# DISEASES AND DISORDERS  Nervous System

| CONDITION | DESCRIPTION |
|---|---|
| | **CENTRAL NERVOUS SYSTEM DISORDERS** |
| **Infections of the CNS** | |
| Encephalitis | Inflammation of the brain caused by a virus and less often by bacteria or other agents; symptoms include fever, coma, and convulsions |
| Meningitis | Inflammation of meninges caused by viral or bacterial infection; symptoms include stiffness in the neck, headache, and fever; severe cases can cause paralysis, coma, or death |
| Rabies | Viral disease transmitted by infected animal; brain infection results in abnormal excitability, aggression, paralysis, and death |
| Tetanus | Caused by bacterial neurotoxin; affects lower motor neurons in spinal cord and brainstem, leading to muscle contraction; prevents muscle relaxation; body becomes rigid, including "lockjaw"; death results from spasms in respiratory muscles |
| Multiple sclerosis | Autoimmune condition; may be initiated by viral infection; inflammation in brain and spinal cord with demyelination and sclerotic (hard) sheaths result in poor conduction of action potentials; symptoms include exaggerated reflexes, tremor, and speech defects |
| **CNS Movement Disorders** | |
| Dyskinesias | Group of disorders, involving basal nuclei, that result in a resting tremor and brisk, jerky movements |
| Huntington disease | Dominant hereditary disorder; causes progressive degeneration of basal nuclei; characterized by involuntary movements |
| Cerebral palsy | General term for defects in motor functions or coordination due to brain damage caused by abnormal development or injury; symptoms include increased muscle tone, resting tremors, difficulty speaking and swallowing, and slow, writhing, aimless movements |
| Parkinson disease | Caused by a lesion in basal nuclei; characterized by muscular rigidity, resting tremor, general lack of movement, and a slow, shuffling gait |
| **Other CNS Disorders** | |
| Stroke | Caused by bleeding in the brain or a clot or spasm blocking cerebral blood vessels that results in a local area of cell death; symptoms include loss of speech, numbness, or paralysis |
| Alzheimer disease | Mental deterioration, or dementia; usually affects older people; involves loss of neurons in cerebral cortex; symptoms include general intellectual deficiency, memory loss, short attention span, moodiness, disorientation, and irritability |
| Tay-Sachs disease | Hereditary lipid-storage disorder of infants; primarily affects CNS; symptoms include paralysis, blindness, and death |
| Epilepsy | Seizures involving a sudden massive neuronal discharge, which may result in involuntary muscle contractions (convulsions) |
| Headaches | May be due to inflammation, dental irritations, eye disorders, tension in head and neck muscles, or unknown causes |
| | **PERIPHERAL NERVOUS SYSTEM DISORDERS** |
| **General PNS Disorders** | |
| Anesthesia | Loss of sensation; may be a pathological condition or may be induced temporarily to facilitate medical action |
| Neuritis | Inflammation of a nerve from a number of causes; in motor nerves, can result in loss of motor function; in sensory nerves, can result in anesthesia or neuralgia |
| Neuralgia | Nerve pain; involves severe spasms of throbbing or stabbing pain along the pathway of a nerve; can result from inflammation, nerve damage, or unknown cause |
| **Infections of the PNS** | |
| Leprosy | Bacterial disease that kills skin and PNS cells; characterized by disfiguring nodules and tissue necrosis |
| Herpes | Family of diseases characterized by skin lesions due to herpes viruses in sensory ganglia; different viruses cause oral lesions (cold sores), sexually transmitted disease with lesions on genitalia, or chicken pox in children (shingles in adults) |
| Poliomyelitis | Viral infection of the CNS; damages somatic motor neurons, leaving muscles without innervation and leads to paralysis and atrophy |
| **Other PNS Disorders** | |
| Myotonic dystrophy | Dominant hereditary disease characterized by muscle weakness, dysfunction, and atrophy |
| Neurofibromatosis | Genetic disorder; neurofibromas (benign tumors along peripheral nerve tract) occur in early childhood and result in skin growths |
| Myasthenia gravis | Autoimmune disorder affecting acetylcholine receptors; makes the neuromuscular junction less functional; muscle weakness and increased fatigue lead to paralysis |

Go to **www.mhhe.com/seeleyess7** for additional information on these pathologies.

# SYSTEMS PATHOLOGY

## Stroke

Sam, who was approaching middle age, was somewhat overweight and had high blood pressure. He spent most of his free time seated on the edge of his couch, watching TV, surrounded by pizza boxes, bowls of chips and salsa, empty beer cans, and full ash trays. One afternoon, as he cheered on his favorite team in a hotly contested big game, which they would be winning easily if it weren't for the lousy officiating, Sam noticed that he felt drowsy and that the television screen seemed blurry. He began to feel dizzy. As he tried to stand up, he suddenly vomited and collapsed to the floor, unconscious.

Sam was rushed to the local hospital, where the following signs and symptoms were observed: weakness in his limbs, especially on the right side; ataxia (inability to walk); loss of pain and temperature sensation in his right lower limb and the left side of his face; persistent dizziness; disorientation and lack of attentiveness; hoarseness and dysphagia (inability to swallow); and a nystagmus (rhythmic oscillation of the eyes). In addition, his pupils were slightly dilated, his respiration was short and shallow, and his pulse rate and blood pressure were elevated.

## Background Information

Sam had suffered a "stroke," also referred to as a **cerebrovascular accident (CVA)**. The term **stroke** describes a heterogeneous group of conditions involving the death of brain tissue due to disruption of its vascular supply. There are two types of stroke: **Hemorrhagic stroke** results from bleeding of arteries supplying brain tissue, and **ischemic** (is-kē'mik) **stroke** results when arteries supplying brain tissue are blocked (figure 8A). The blockage in ischemic stroke can result from a thrombus, which is a clot that develops in place within an artery, or an embolism, which is a plug composed of a detached thrombus or other foreign body, such as a fat globule or gas bubble, that becomes lodged in an artery, blocking it. Sam was at high risk of developing a stroke. He was approaching middle age, was overweight, did not exercise enough, smoked, was under stress, and had a poor diet.

The combination of motor loss (as exhibited by weakness in the limbs) and sensory loss (evidenced by loss of pain and temperature sensation in the left lower limb and loss of all sensation in the right side of the face), along with the ataxia, dizziness, nystagmus, and hoarseness, suggests that the stroke affected the brainstem and cerebellum.

### Figure 8A

(*a*) MRI (magnetic resonance imaging) of a massive stroke in the left side (the viewer's right) of the brain. (*b*) Colorized NMR (nuclear magnetic resonance) showing disruption of blood flow to the left side of the brain (*yellow*). This disruption could cause a stroke.

(a)                                                        (b)

in the number of skin receptors, elderly people are less conscious of anything touching or pressing on the skin and have more difficulty identifying objects by touch. These changes leave elderly people more prone to skin injuries.

A decreased sense of position of the limbs and in the joints can affect balance and coordination. Information about the position,

tension, and length of tendons and muscles also decreases, further reducing the senses of movement, posture, and position, as well as the control and coordination of movement.

Other sensory neurons with reduced function include those that monitor blood pressure, thirst, objects in the throat, the amount of urine in the urinary bladder, and the amount of feces in

| System Interactions | Effects of Stroke on Other Systems |
|---|---|
| **System** | **Interactions** |
| Integumentary | Decubitus (dē-kŭ′bi-tŭs) ulcers (bedsores) result from immobility because of the loss of motor function. Immobility causes compression of blood vessels in areas of the skin, which leads to death of the skin in areas supplied by those vessels. |
| Skeletal | Loss of bone mass occurs if muscles are dysfunctional for a prolonged time; in the absence of muscular activity, the bones to which those muscles are attached begin to be resorbed by osteoclasts. |
| Muscular | Absence of innervation because of damaged pathways or neurons leads to decreased motor function and may result in muscle atrophy. |
| Endocrine | Strokes in other parts of the brain could involve the functions of the hypothalamus, pineal gland, or pituitary gland. |
| Cardiovascular | Risks: Phlebothrombosis (fleb′-ō-throm-bō′sis; blood clot in a vein) can occur from inactivity. Edema around the brain can increase the pressure inside the skull, activating reflexes that increase the blood pressure. If the cardioregulatory center in the brain is damaged, death may occur rapidly because of dramatically decreased blood pressure. Bleeding may result from the use of anticoagulants, and hypotension may result from the use of antihypertensives. |
| Respiratory | Pneumonia may result from aspiration (fluid from the digestive tract entering the lungs when the patient vomits). Breathing may slow or stop completely if the respiratory center is damaged. |
| Digestive | Vomiting or dysphagia (dis-fā′jē-ă; difficulty swallowing) may occur if the brain centers controlling these functions are damaged. Hypovolemia (decreased blood volume) results from decreased fluid intake because of dysphagia; bowel control may be lost. |
| Urinary | Motor innervation of the bladder is often affected. Urinary tract infection may result from catheter implantation or from urine retention. |
| Reproductive | There may be loss of libido (sex drive); innervation of the reproductive organs is often affected. |

Blockage of the vertebral artery, a major artery supplying the brain or its branches can result in an area of dead tissue called a lateral medullary infarction. Damage to the descending motor tracts in that area, above the medullary crossover point, causes muscle weakness. Damage to ascending tracts can result in loss of pain and temperature sensation or other sensory modalities, depending on the affected tract. Damage to cranial nerve nuclei causes the loss of pain and temperature sensation in the face, dizziness, blurred vision, nystagmus, vomiting, and hoarseness. These signs and symptoms are not observed unless the lesion is in the brainstem, where these nuclei are located. Some damage to the cerebellum, also supplied by branches of the vertebral artery, can account for the ataxia.

General responses to neurological damage include drowsiness, disorientation, inattentiveness, loss of consciousness, and even seizures. Depression, due to either neurological damage or discouragement, is also common. Slight dilation of the pupils, short and shallow respiration, and increased pulse rate and blood pressure are all signs of Sam's anxiety, not about the outcome of the game but about his current condition and his immediate future. Because he lost consciousness, Sam would not remember the last few minutes of what he saw in the game he was watching. People in these circumstances are often worried about how they are going to deal with work tomorrow. They often have no idea that the motor and sensory losses may be permanent or that a long stretch of therapy and rehabilitation lies ahead.

### Predict 6

*Given that Sam exhibited weakness in his right limbs and loss of pain and the temperature sensation in his right lower limb and the left side of his face, which side of the brainstem was most severely affected by the stroke? Explain your answer.*

the rectum. As a result, elderly people are more prone to high blood pressure, dehydration, swallowing and choking problems, urinary incontinence, and constipation or bowel incontinence.

A general decline in the number of motor neurons also occurs. As many as 50% of the lower motor neurons in the lumbar region of the spinal cord may be lost by age 60. Muscle fibers innervated by the lost motor neurons are also lost, resulting in a general decline in muscle mass. Loss of motor units leads to more rapid fatigue as the remaining units must perform compensatory work. However, regular physical exercise can forestall and reduce these declines.

Reflexes slow as people age because both the generation and the conduction of action potentials and synaptic functions slow. The number of neurotransmitters and receptors declines. As reflexes slow, older people are less able to react automatically, quickly, and accurately to changes in internal and external conditions.

The size and weight of the brain decrease as a person ages. At least part of these changes results from the loss of neurons within the cerebrum. The remaining neurons can apparently compensate for much of this loss. Structural changes also occur in neurons. Neuron plasma membranes become more rigid, the endoplasmic reticulum becomes more irregular in structure, neurofibrillar tangles develop in the cells, and amyloid plaques form in synapses. All these changes decrease the ability of neurons to function.

Short-term memory is decreased in most older people. This change varies greatly among individuals, but in general, such changes are slow until about age 60 and then become more rapid, especially after age 70. However, the total amount of memory loss is normally not great for most people. Older people have the most

difficulty assimilating information that is unfamiliar and presented verbally and rapidly. Long-term memory appears to be unaffected or even improved in older people.

As with short-term memory, thinking, which includes problem solving, planning, and intelligence, generally declines slowly to age 60 but more rapidly thereafter. However, these changes are slight and quite variable. Many older people show no change, and about 10% show an increase in thinking ability. Many of these changes are affected by a person's background, education, health, motivation, and experience. It appears that continued mental activity can decrease the loss of mental skills with aging.

Older people tend to require more time to fall asleep and experience more periods of waking during the night, which are also of greater duration. Factors that can affect sleep include pain, indigestion, rhythmic leg movements, sleep apnea, decreased urinary bladder capacity, and poor peripheral circulation.

# SUMMARY

## Functions of the Nervous System   (p. 193)

The functions of the nervous system include receiving sensory input, integrating information, controlling muscles and glands, maintaining homeostasis, and serving as the center of mental activity.

## Divisions of the Nervous System   (p. 193)

1. The central nervous system (CNS) consists of the brain and spinal cord, whereas the peripheral nervous system (PNS) consists of all the nervous tissue outside the CNS.
2. The sensory division of the PNS transmits action potentials to the CNS; the motor division carries action potentials away from the CNS.
3. The motor division is divided into somatic and autonomic systems. The somatic motor nervous system innervates skeletal muscle and is mostly under voluntary control. The autonomic nervous system innervates cardiac muscle, smooth muscle, and glands and is mostly under involuntary control.
4. The autonomic nervous system is divided into sympathetic and parasympathetic divisions.
5. The enteric nervous system contains both sensory and motor neurons, which can function independently of the CNS.

## Cells of the Nervous System   (p. 194)

### Neurons

1. Neurons receive stimuli and conduct action potentials. A neuron consists of a cell body, dendrites, and an axon.
2. Neurons are multipolar, bipolar, or pseudo-unipolar.

### Neuroglia

Neuroglia are the support cells of the nervous system. They include astrocytes, ependymal cells, microglia, oligodendrocytes, and Schwann cells.

### Myelin Sheaths

Axons are either unmyelinated or myelinated.

### Organization of Nervous Tissue

Nervous tissue consists of white matter and gray matter. Gray matter forms the cortex and nuclei in the brain and ganglia in the PNS. White matter forms nerve tracts in the CNS and nerves in the PNS.

## Electrical Signals and Neural Pathways   (p. 198)

### Resting Membrane Potential

1. A resting membrane potential results from the charge difference across the membrane of cells.
2. The resting membrane potential is set by leak ion channels and the sodium-potassium pump.

### Action Potentials

1. An action potential occurs when the charge across the cell membrane is briefly reversed.
2. Ligand-gated and voltage-gated ion channels generate the action potential.

### The Synapse

1. A synapse is a point of contact between two neurons or between a neuron and another cell, such as a muscle or gland cell.
2. An action potential arriving at the synapse causes the release of a neurotransmitter from the presynaptic terminal, which diffuses across the synaptic cleft and binds to the receptors of the postsynaptic membrane.

### Reflexes

1. Reflex arcs are the functional units of the nervous system.
2. A complex reflex arc consists of a sensory receptor, a sensory neuron, interneurons, a motor neuron, and an effector organ; the simplest reflex arcs do not involve interneurons.

### Neuronal Pathways

1. Neuronal pathways are either converging or diverging.
2. Spatial and temporal summation occur in neuronal pathways.

## Central and Peripheral Nervous Systems   (p. 206)

The CNS consists of the brain and spinal cord. The PNS consists of nerves and ganglia outside the CNS.

## Spinal Cord   (p. 206)

1. The spinal cord extends from the foramen magnum to the second lumbar vertebra; below that is the cauda equina.
2. The spinal cord has a central gray part organized into horns and a peripheral white part forming nerve tracts.

3. Roots of spinal nerves extend out of the cord.
4. The knee-jerk reflex occurs when the quadriceps femoris muscle is stretched.
5. The withdrawal reflex removes a body part from a painful stimulus.

## Spinal Nerves    (p. 208)

1. The spinal nerves exit the vertebral column at the cervical, thoracic, lumbar, and sacral regions.
2. The nerves are grouped into plexuses.
3. The phrenic nerve, which supplies the diaphragm, is the most important branch of the cervical plexus.
4. The brachial plexus supplies nerves to the upper limb.
5. The lumbosacral plexus supplies nerves to the lower limb.

## Brain    (p. 211)

### Brainstem

1. The brainstem contains several nuclei.
2. The medulla oblongata contains nuclei that control activities such as heart rate, breathing, swallowing, and balance.
3. The pons contains relay nuclei between the cerebrum and cerebellum.
4. The midbrain is involved in hearing and in visual reflexes.
5. The reticular formation is scattered throughout the brainstem and is important in regulating cyclical motor functions. It is also involved in maintaining consciousness and in the sleep-wake cycle.

### Cerebellum

The cerebellum is attached to the brainstem.

### Diencephalon

The diencephalon consists of the thalamus (main sensory relay center), the epithalamus (the pineal gland may play a role in sexual maturation), and the hypothalamus (important in maintaining homeostasis).

### Cerebrum

The cerebrum has two hemispheres, each divided into lobes: the frontal, parietal, occipital, and temporal lobes.

## Sensory Functions    (p. 215)

1. The CNS constantly receives sensory input.
2. We are unaware of much of the input, but it is vital to our survival.
3. Some sensory input results in sensation.

### Ascending Tracts

1. Ascending tracts transmit action potentials from the periphery to the brain.
2. Each tract carries a specific type of sensory information.

### Sensory Areas of the Cerebral Cortex

1. Ascending tracts project to primary sensory areas of the cerebral cortex.
2. Association areas are involved in recognizing the sensory input.

## Motor Functions    (p. 216)

1. Motor functions include involuntary and voluntary movements.
2. Upper motor neurons in the cerebral cortex connect to lower motor neurons in the spinal cord or cranial nerve nuclei.

### Motor Areas of the Cerebral Cortex

1. Upper motor neurons are located in the primary motor cortex.
2. The premotor and prefrontal areas regulate movements.

### Descending Tracts

Descending tracts project directly from upper motor neurons in the cerebral cortex to lower motor neurons in the spinal cord and brainstem; indirectly, they project from basal nuclei, the cerebellum or the cerebral cortex through the brainstem to lower motor neurons in the spinal cord.

### Basal Nuclei

1. Basal nuclei help plan, organize, and coordinate motor movements and posture.
2. People with basal nuclei disorders exhibit increased muscle tone and exaggerated, uncontrolled movements when at rest.

### Cerebellum

1. The cerebellum is involved in balance, muscle tone, and muscle coordination.
2. Through its comparator function, the cerebellum compares the intended action to what is occurring and modifies the action to eliminate differences.
3. If the cerebellum is damaged, muscle tone decreases and fine motor movements become very clumsy.

## Other Brain Functions    (p. 220)

### Communication Between the Right and Left Hemispheres

1. Each hemisphere controls the opposite half of the body.
2. Commissures connect the two hemispheres.
3. The left hemisphere is thought to be the dominant analytical hemisphere, and the right hemisphere is thought to be dominant for spatial perception and musical ability.

### Speech

Speech involves the sensory speech area, the motor speech area, and the interactions between them and other cortical areas.

### Brain Waves and Consciousness

An EEG monitors brain waves, which are a summation of the electrical activity of the brain.

### Memory

1. The types of memory are sensory (less than 1 second), short-term (lasting a few minutes), and long-term (permanent) memory.
2. Long-term memory includes declarative and procedural memories.

### Limbic System

1. The limbic system includes the olfactory cortex, deep cortical regions, and nuclei.
2. The limbic system is involved with memory, motivation, mood, and other visceral functions. Olfactory stimulation is a major influence.

## Meninges, Ventricles, and Cerebrospinal Fluid    (p. 222)

### Meninges

Three connective tissue meninges cover the CNS: the dura mater, arachnoid mater, and pia mater.

### Ventricles

1. The brain and spinal cord contain fluid-filled cavities: the lateral ventricles in the cerebral hemispheres, a third ventricle in the diencephalon, a cerebral aqueduct in the midbrain, a fourth ventricle at the base of the cerebellum, and a central canal in the spinal cord.
2. The fourth ventricle has openings into the subarachnoid space.

### Cerebrospinal Fluid

Cerebrospinal fluid is formed in the choroid plexuses in the ventricles. It exits through the fourth ventricle and reenters the blood through arachnoid granulations in the superior sagittal sinus.

## Cranial Nerves    (p. 224)

1. There are 12 pairs of cranial nerves: 3 with only sensory function (S), 4 with only somatic motor function (M), 1 with motor and

sensory function, 1 with somatic motor and parasympathetic (P) function, and 3 with all three functions. Four of the cranial nerves have parasympathetic function.

2. The cranial nerves are olfactory (I; S), optic (II; S), oculomotor (III; M, P), trochlear (IV; M), trigeminal (V; S, M), abducens (VI; M), facial (VII; S, M, P), vestibulocochlear (VIII; S), glossopharyngeal (IX; S, M, P), vagus (X; S, M, P), accessory (XI; M), and hypoglossal (XII; M).

## Autonomic Nervous System    (p. 227)

1. The autonomic nervous system contains preganglionic and postganglionic neurons.
2. The autonomic nervous system has sympathetic and parasympathetic divisions.

### Anatomy of the Sympathetic Division

1. Preganglionic cell bodies of the sympathetic division lie in the thoracic and upper lumbar regions of the spinal cord.
2. Postganglionic cell bodies are located in the sympathetic chain ganglia or in collateral ganglia.

### Anatomy of the Parasympathetic Division

1. Preganglionic cell bodies of the parasympathetic division are associated with some of the cranial and sacral nerves.

2. Postganglionic cell bodies are located in terminal ganglia, either near or within target organs.

### Autonomic Neurotransmitters

1. All autonomic preganglionic and parasympathetic postganglionic neurons secrete acetylcholine.
2. Most sympathetic postganglionic neurons secrete norepinephrine.

### Functions of the Autonomic Nervous System

1. The sympathetic division prepares a person for action by increasing heart rate, blood pressure, respiration and release of glucose for energy.
2. The parasympathetic division is involved in involuntary activities at rest, such as the digestion of food, defecation, and urination.

## Enteric Nervous System    (p. 230)

1. The enteric nervous system forms plexuses in the digestive tract wall.
2. Enteric neurons are sensory, motor, or interneurons and receive CNS input but can also function independently.

## Effects of Aging on the Nervous System    (p. 230)

1. In general, sensory and motor functions decline with age.
2. Mental functions, including memory, may decline with age, but this varies from person to person.

# REVIEW AND COMPREHENSION

1. List the functions of the nervous system.
2. Compare and contrast the CNS and PNS.
3. Define the sensory and motor divisions of the PNS and the somatic motor and autonomic nervous systems.
4. What are the functions of neurons? Name the three parts of a neuron.
5. List the three types of neurons based on the organization of their processes.
6. Define neuroglia. Name and describe the functions of the different neuroglia.
7. What are the differences between unmyelinated and myelinated axons? Which conducts action potentials more rapidly? Why?
8. For nuclei, ganglia, nerve tracts, and nerves, name the cells or parts of cells found in each, state if they are white or gray matter, and name the part of the nervous system (CNS or PNS) in which they are found.
9. Explain the resting membrane potential and the roles of the $K^+$ leak channels and the sodium-potassium pump.
10. List the series of events at the membrane that generate an action potential, including the contributions of ligand-gated and voltage-gated ion channels.
11. Describe the sequence of events at a synapse, starting with an action potential in the presynaptic neuron and ending with the generation of an action potential in the postsynaptic neuron.
12. Describe the primary function of neurotransmitters at the synapse.
13. Define a reflex. Name the five components of a reflex arc, and explain its operation.
14. Compare and contrast the two major types of neuronal pathways. Give an example of each.
15. Describe the importance of spatial and temporal summation.
16. Describe the spinal cord gray matter. Where are sensory and motor neurons located in the gray matter?

17. Differentiate between the ventral root and the dorsal root in relation to a spinal nerve. Which contains sensory fibers, and which contains motor fibers?
18. Describe the knee-jerk and withdrawal reflexes.
19. List the spinal nerves by name and number.
20. Name the three main plexuses and the major nerves derived from each.
21. Name the four parts of the brainstem, and describe the general functions of each.
22. Name the three main components of the diencephalon, describing their functions.
23. Name the four lobes of the cerebrum, and describe the location and function of each.
24. List the ascending tracts, and state their functions.
25. Describe the locations in the cerebral cortex of the primary sensory areas and their association areas. How do the association areas interact with the primary areas?
26. Distinguish between upper and lower motor neurons.
27. Distinguish among the functions of the primary motor cortex, premotor area, and prefrontal area.
28. List the descending tracts, and state their functions.
29. Describe the functions of the basal nuclei.
30. Describe the comparator activities of the cerebellum.
31. What are the differences in function between the right and left cerebral hemispheres?
32. Describe the process required to speak a word that is seen compared to one that is heard.
33. Name the three types of memory, and describe the processes that result in long-term memory.
34. What is the function of the limbic system?
35. Name and describe the three meninges that surround the CNS.

36. Describe the production and circulation of the cerebrospinal fluid. Through what structures does the cerebrospinal fluid return to the blood?

37. What are the three principal functional categories of the cranial nerves? List a specific function for each cranial nerve.

38. Define preganglionic neuron and postganglionic neuron.

39. Compare the structure of the somatic nervous system and the autonomic nervous system in terms of the number of neurons between the CNS and the effector organs and the types of effector organs.

40. Contrast the functions of the sympathetic and parasympathetic divisions of the autonomic nervous system.

41. What kinds of neurons (sympathetic or parasympathetic, preganglionic or postganglionic) are found in the following?
    a. cranial nerve nuclei
    b. lateral horn of the thoracic spinal cord gray matter
    c. lateral portion of the sacral spinal cord gray matter
    d. chain ganglia
    e. ganglia in the wall of an organ

42. List the three parts of the enteric nervous system.

43. List all the reasons that an elderly person may have difficulty walking from the living room to the bedroom.

## CRITICAL THINKING

1. Given two series of neurons, explain why action potentials could be propagated along one series more rapidly than along the other series.

2. The left lung of a cancer patient was removed. To reduce the empty space left in the thorax after the lung was removed, the diaphragm on the left side was paralyzed so that the abdominal viscera would push the diaphragm upward into the space. What nerve should be cut to paralyze the left half of the diaphragm?

3. Name the nerve that, if damaged, produces the following symptoms:
   a. The elbow and wrist on one side are held in a flexed position and cannot be extended.
   b. The patient is unable to flex the right hip and extend the knee (as in kicking a ball).

4. A patient suffered brain damage in an automobile accident. Physicians suspected that the cerebellum was affected. On the basis of what you know about cerebellar function, how could you determine that the cerebellum was involved? What symptoms would you expect to see?

5. Louis Ville was accidentally struck in the head with a baseball bat. He fell to the ground, unconscious. Later, when he had regained consciousness, he was unable to remember any of the events that happened during the 10 minutes before the accident. Explain.

6. Name the cranial nerve that, if damaged, produces the following symptoms:
   a. The patient is unable to move the tongue.
   b. The patient is unable to see out of one eye.
   c. The patient is unable to feel one side of the face.
   d. The patient is unable to move the facial muscles on one side.
   e. The pupil of one eye is dilated and does not constrict.

7. Why doesn't injury to the spinal cord at the level of C6 significantly interfere with nervous system control of the digestive system?

*Answers in Appendix D*

Visit this textbook's website at **www.mhhe.com/seeleyess7** for practice quizzes, animations, interactive learning exercises, and other study tools.

McGraw-Hill offers a study CD that features interactive cadaver dissection. ***Anatomy & Physiology Revealed*** includes cadaver photos that allow you to peel away layers of the human body to reveal structures beneath the surface. This program also includes animations, radiologic imaging, audio pronunciations, and practice quizzing.

Many songs and poems have been written about our senses, especially our eyes. Our senses of sight, taste, and touch help define our personalities, protect our bodies, and maintain homeostasis.

# Chapter 9

## Senses

## Chapter Outline and Objectives

### General Senses (p. 239)

1. Define sensation; list the general senses and the receptor type associated with each.

2. Define and describe pain and referred pain.

### Special Senses (p. 241)

3. List the special senses.

### Olfaction (p. 241)

4. Describe olfactory neurons, and explain how airborne molecules can stimulate action potentials in the olfactory nerves.

### Taste (p. 242)

5. Outline the structure and function of a taste bud.

### Vision (p. 242)

6. List the accessory structures of the eye, and explain their functions.

7. Name the tunics of the eye, list the parts of each tunic, and describe the functions of each part.

8. Explain the differences in function between rods and cones.

9. Describe the chambers of the eye and the fluids they contain.

10. Explain how images are focused on the retina.

### Hearing and Balance (p. 251)

11. Describe the structures of the outer and middle ears, and state the function of each.

12. Describe the anatomy of the cochlea, and explain how sounds are detected.

13. Explain how the structures of the vestibule and semicircular canals function in static and dynamic equilibrium.

### Effects of Aging on the Senses (p. 258)

14. Describe changes that occur in the senses with aging.

**S**ense is the ability to perceive stimuli. The senses are the means by which the brain receives information about the environment and the body. **Sensation,** or *perception,* is the conscious awareness of stimuli received by sensory receptors. The brain constantly receives a wide variety of stimuli from both inside and outside the body, but stimulation of sensory receptors does not immediately result in sensation. Sensory receptors respond to stimuli by generating action potentials that are propagated to the spinal cord and brain. Sensations result when action potentials reach the cerebral cortex. Some other parts of the brain are also involved in sensation. For example, the thalamus plays a role in the sensation of pain.

Historically, five senses were recognized: smell, taste, sight, hearing, and touch. Today we recognize many more senses and divide them into two basic groups: general and special senses. The **general senses** have receptors distributed over a large part of the body. They are divided into two groups: the somatic senses and the visceral senses. The **somatic senses** provide sensory information about the body and the environment. The **visceral senses** provide information about various internal organs, primarily involving pain and pressure.

**Special senses** are more specialized in structure and are localized to specific parts of the body. The special senses are smell, taste, sight, hearing, and balance.

# GENERAL SENSES

The general senses are widely distributed throughout the body and include the senses of touch, pressure, pain, temperature, vibration, itch, and proprioception (prō-prē-ō-sep′shun), which is the sense of movement and position of the body and limbs.

**Sensory receptors** are sensory nerve endings or specialized cells capable of responding to stimuli by developing action potentials. Several types of receptors are associated with both the special and the general senses, and each responds to a different type of stimulus:

**Mechanoreceptors** (mek′ă-nō-rē-sep′tŏrz) respond to mechanical stimuli, such as the bending or stretching of receptors.

**Chemoreceptors** (kem′ō-rē-sep′tŏrz) respond to chemicals, such as odor molecules.

**Photoreceptors** (fō′tō-rē-sep′tŏrz) respond to light.

**Thermoreceptors** (ther′mō-rē-sep′tŏrz) respond to temperature changes.

**Nociceptors** (nō′si-sep′tŏrs; *noceo,* to injure) respond to stimuli that result in the sensation of pain.

Many of the receptors for the general senses are associated with the skin (figure 9.1); others are associated with deeper structures, such as tendons, ligaments, and muscles. Structurally, the simplest and most common receptor nerve endings are **free nerve endings,** which are relatively unspecialized neuronal branches similar to dendrites. Free nerve endings are distributed throughout almost all parts of the body. Some free nerve endings respond

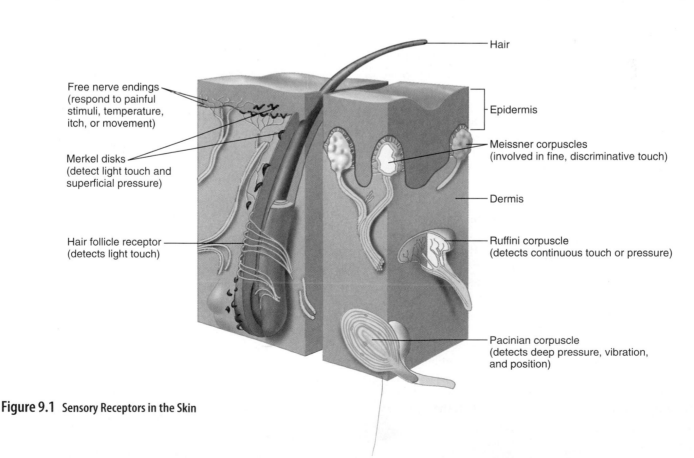

**Figure 9.1  Sensory Receptors in the Skin**

to painful stimuli, some to temperature, some to itch, and some to movement. Receptors for temperature are either **cold receptors** or **warm receptors.** Cold receptors respond to decreasing temperatures but stop responding at temperatures below 12°C (54°F). Warm receptors respond to increasing temperatures but stop responding at temperatures above 47°C (117°F). It is sometimes difficult to distinguish very cold from very warm objects touching the skin because only pain receptors are stimulated at temperatures below 12°C or above 47°C.

Touch receptors are structurally more complex than free nerve endings, and many are enclosed by capsules. There are several types of touch receptors (figure 9.1). **Merkel disks** are small, superficial nerve endings involved in detecting light touch and superficial pressure. **Hair follicle receptors,** associated with hairs, are also involved in detecting light touch. Light touch receptors are very sensitive but not very discriminative, meaning that the point being touched cannot be precisely located. Receptors for fine, discriminative touch, called **Meissner corpuscles,** are located just deep to the epidermis. These receptors are very specific in localizing tactile sensations. Deeper tactile receptors, called **Ruffini corpuscles,** play an important role in detecting continuous pressure in the skin. The deepest receptors are associated with tendons and joints and are called **pacinian corpuscles.** These receptors relay information concerning deep pressure, vibration, and position (proprioception).

## Pain

**Pain** is characterized by a group of unpleasant perceptual and emotional experiences. There are two types of pain sensation: (1) localized, sharp, pricking, or cutting pain resulting from rapidly conducted action potentials, and (2) diffuse, burning, or aching pain resulting from action potentials that are propagated more slowly.

Superficial pain sensations in the skin are highly localized as a result of the simultaneous stimulation of pain receptors and tactile receptors. Deep or visceral pain sensations are not highly localized because of the absence of tactile receptors in the deeper structures. Visceral pain stimuli are normally perceived as diffuse pain.

Action potentials from pain receptors in local areas of the body can be suppressed by chemical anesthetics injected near a sensory receptor or nerve, resulting in reduced pain sensation. This treatment is called **local anesthesia.** Pain sensations can also be suppressed if loss of consciousness is produced. This is usually accomplished by administering chemical anesthetics that affect the reticular formation, a treatment called **general anesthesia.**

Pain sensations can also be influenced by inherent control systems. Sensory axons from tactile receptors in the skin have collateral branches that synapse with neurons in the dorsal horn of the spinal cord. Those neurons, in turn, synapse with and inhibit neurons in the dorsal horn that give rise to the lateral spinothalamic tract. For example, rubbing the skin in the area of an injury stimulates the tactile receptors, which send action potentials along the sensory axons to the spinal cord. According to the **gate control theory,** these action potentials "close the gate" and inhibit action potentials carried to the brain by the lateral spinothalamic tract.

The gate control theory may explain the physiological basis for several techniques that have been used to reduce the intensity of pain. Action potentials carried by the lateral spinothalamic tract can be inhibited by action potentials carried by descending neurons of the dorsal column system (see chapter 8). These neurons are stimulated by mental or physical activity, especially involving movement of the limbs. The descending neurons synapse with and inhibit neurons in the dorsal horn that give rise to the lateral spinothalamic tract. Vigorous mental or physical activity increases the rate of action potentials in neurons of the dorsal column and can reduce the sensation of pain. Exercise programs are important components in the clinical management of chronic pain. Acupuncture and acupressure procedures may also decrease the sensation of pain by stimulating descending dorsal column neurons, which inhibit action potentials in the lateral spinothalamic tract neurons. The gate control theory also explains why the intensity of pain is decreased by diverting a person's attention.

### Referred Pain

**Referred pain** is perceived to originate in a region of the body that is not the source of the pain stimulus. Most commonly, we sense referred pain when deeper structures, such as internal organs, are damaged or inflamed (figure 9.2). This occurs because sensory neurons from the superficial area to which the pain is referred and the neurons from the deeper, visceral area where the pain stimulation originates converge onto the same ascending neurons in the spinal cord. The brain cannot distinguish between the two sources

**Figure 9.2  Areas of Referred Pain on the Body Surface**
Pain from the indicated internal organs is referred to the surface areas shown.

of pain stimuli, and the painful sensation is referred to the most superficial structures innervated, such as the skin.

Referred pain is clinically useful in diagnosing the actual cause of the painful stimulus. For example, during a heart attack, pain receptors in the heart are stimulated when blood flow is blocked to some of the heart muscle. Heart attack victims, however, often do not feel the pain in the heart but instead perceive cutaneous pain radiating from the left shoulder down the arm (figure 9.2).

> *A man has constipation that causes distention and painful cramping in the colon (part of the large intestine). What kind of pain does he experience (localized or diffuse), and where does he perceive it? Explain.*

# SPECIAL SENSES

The senses of smell, taste, sight, hearing, and balance are associated with very specialized, localized sensory receptors. The sensations of smell and taste are closely related, both structurally and functionally, and both are initiated by the interaction of chemicals with sensory receptors. The sense of vision is initiated by the interaction of light with sensory receptors. Both hearing and balance function in response to the interaction of mechanical stimuli with sensory receptors. Hearing occurs in response to sound waves, and balance occurs in response to gravity or motion.

# OLFACTION

The sense of smell, called **olfaction** (ol-fak′shŭn), occurs in response to airborne molecules, called **odorants,** that enter the nasal cavity. **Olfactory neurons** are bipolar neurons within the **olfactory epithelium,** which lines the superior part of the nasal cavity (figure 9.3*a*). The dendrites of the olfactory neurons extend to the epithelial surface, and their ends are modified with long, specialized cilia that lie in a thin mucous film on the epithelial surface. The mucus keeps the nasal epithelium moist, traps and dissolves airborne molecules, and facilitates the removal of molecules and particles from the nasal epithelium.

Airborne odorants become dissolved in the mucus on the surface of the epithelium and bind to receptor molecules on the membranes of the specialized cilia. There are at least 400 functional **olfactory receptors** in humans. Unlike most other receptors in the body, each olfactory receptor can bind multiple types of odorants; conversely, each type of odorant can bind to multiple olfactory receptors. These multiple combinations of odorants and receptors allow us to detect an estimated 10,000 different smells. Once an odorant has bound to its receptor, that receptor is desensitized and does not respond to another odor molecule for some time, which helps with adaptation (described next) to a particular odor. The threshold for the detection of odors is extremely low, so very few odorants bound to an olfactory neuron can initiate an action potential. The olfactory range and sensitivity is even greater in some animals than in humans, due to a larger number and more types of olfactory receptors. For example, dogs are often used to

(a)

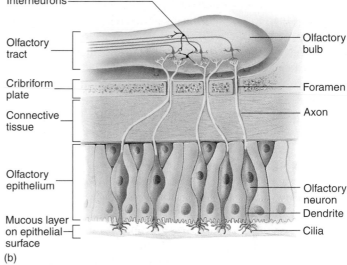

(b)

**Figure 9.3   Olfactory Epithelium and Olfactory Bulb**

(*a*) A sagittal section through the lateral wall of the nasal cavity shows the olfactory nerves, olfactory bulb, and olfactory tract. (*b*) The olfactory neurons lie within the olfactory epithelium. The axons of olfactory neurons pass through the cribriform plate to the olfactory bulb.

detect small traces of explosives and other chemicals that people cannot detect.

## Neuronal Pathways for Olfaction

Axons from olfactory neurons form the olfactory nerves (cranial nerve I), which pass through foramina of the cribriform plate and enter the **olfactory bulb** (figure 9.3*b*). There they synapse with interneurons that relay action potentials to the brain through the **olfactory tracts.** Each olfactory tract terminates in an area of the brain called the **olfactory cortex,** located within the temporal and frontal lobes. Olfaction is the only major sensation that is relayed directly to the cerebral cortex without first passing through the

Nervous

thalamus. This is a reflection of the older, more primitive origin of the olfactory cortex. The olfactory cortex is involved with both the conscious perception of smell and the visceral and emotional reactions that are often linked to odors.

Within the olfactory bulb and olfactory cortex are feedback loops that tend to inhibit transmission of action potentials resulting from prolonged exposure to a given odorant. This feedback, plus the temporary decreased sensitivity at the level of the receptors, results in **adaptation** to a given odor. For example, if you enter a room that has an odor, you adapt to the odor and cannot smell it as well after the first few minutes. If you leave the room for some time and then reenter the room, the odor again seems more intense.

## TASTE

The sensory structures that detect taste stimuli are the **taste buds.** Taste buds are oval structures located on the surface of certain **papillae** (pă-pil′ē; nipples), which are enlargements on the surface of the tongue (figure 9.4). Taste buds are also distributed throughout other areas of the mouth and pharynx, such as on the palate, the root of the tongue, and the epiglottis. Each taste bud consists of two types of cells. Specialized epithelial cells form the exterior supporting capsule of each taste bud, and the interior consists of about 40 **taste cells.** Each taste cell contains hairlike processes, called **taste hairs,** that extend into a tiny opening in the surrounding stratified epithelium, called a **taste pore.** Dissolved molecules or ions bind to receptors on the taste hairs and initiate action potentials, which sensory neurons carry to the parietal lobe of the cerebral cortex.

Taste sensations are divided into five basic types: sour, salty, bitter, sweet, and umami (ū-ma′mē; savory). Although all taste buds are able to detect all five of the basic taste sensations, each taste bud is usually most sensitive to one class of taste stimuli. Presumably, our ability to perceive many different tastes is achieved through various combinations of these five types.

Many taste sensations are strongly influenced by olfactory sensations. This influence can be demonstrated by comparing the taste of some food before and after pinching your nose. It is easy to detect that the sense of taste is reduced while the nose is pinched.

## Neuronal Pathways for Taste

Taste sensations are carried by three cranial nerves. The facial nerve (VII) transmits taste sensations from the anterior two-thirds of the tongue, and the glossopharyngeal nerve (IX) carries taste sensations from the posterior one-third. In addition, the vagus nerve (X) carries some taste sensations from the root of the tongue. Axons from these three cranial nerves synapse in the gustatory (taste) portion of brainstem nuclei. Axons of neurons in these brainstem nuclei synapse in the thalamus, and axons from neurons in the thalamus project to the taste area in the parietal lobe of the cerebral cortex (see figure 8.28).

**Predict 2**

*Why doesn't food taste as good when you have a cold?*

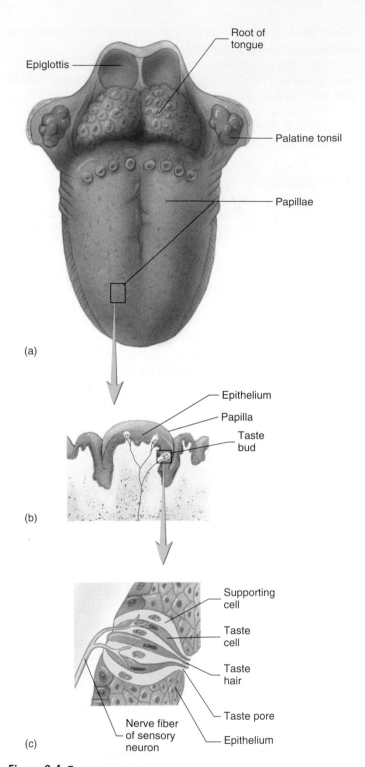

(a)

(b)

(c)

**Figure 9.4  Tongue**

(*a*) Dorsal surface of the tongue. (*b*) Section through a papilla showing the location of taste buds. (*c*) Enlarged view of a section through a taste bud.

## VISION

The visual system includes the eyes, the accessory structures, and sensory neurons. The eyes are housed within bony cavities called orbits. Action potentials convey visual information from the eyes to the brain. We obtain much of our information about the world through the visual system. For example, education is largely based on visual input and depends on our ability to read words and numbers. Visual input includes information about light and dark, movement and color.

## A CASE IN POINT
### Loss of Taste

T. Burnes boiled a pot of water to make some tea. After pouring the water over the tea bags in her tea cup, she absent-mindedly took a large gulp of hot tea, severely burning her tongue. For the next several days, sensation in her tongue, including her sense of touch and sense of taste, was greatly reduced. Heat damage to the tongue epithelial tissue can cause injury or even death to epithelial cells, including taste cells in the taste buds. If the epithelial cells are only damaged, taste sensation returns within a few hours to a few days. If the cells die, it takes about 2 weeks for the epithelial cells to be replaced.

## Accessory Structures of the Eye

Accessory structures protect, lubricate, and move the eye. They include the eyebrows, eyelids, conjunctiva, lacrimal apparatus, and extrinsic eye muscles (figures 9.5 and 9.6).

### Eyebrows

The **eyebrows** protect the eyes by preventing perspiration from running down the forehead and into the eyes, causing irritation. They also help shade the eyes from direct sunlight.

### Eyelids

The **eyelids,** with their associated lashes, protect the eyes from foreign objects. If an object suddenly approaches the eye, the eyelids protect the eye by closing and then opening quite rapidly (blink reflex). Blinking, which normally occurs about 20 times per minute, also helps keep the eyes lubricated by spreading tears over the surface.

### Conjunctiva

The **conjunctiva** (kon-jŭnk-tī′vă) is a thin, transparent mucous membrane covering the inner surface of the eyelids and the anterior surface of the eye. Conjunctivitis is an inflammation of the conjunctiva (see the Diseases and Disorders table at the end of this chapter).

**Figure 9.5  Left Eye and Some of Its Accessory Structures**

**Figure 9.6  Sagittal Section Through the Eye, Showing Its Accessory Structures**

## Lacrimal Apparatus

The **lacrimal** (lak′ri-măl; tear) **apparatus** consists of a lacrimal gland situated in the superior lateral corner of the orbit and a nasolacrimal duct and associated structures in the inferior medial corner of the orbit (figure 9.7). The **lacrimal gland** produces tears, which pass over the anterior surface of the eye. Most of the fluid produced by the lacrimal glands evaporates from the surface of the eye, but excess tears are collected in the medial angle of the eyes by small ducts called **lacrimal canaliculi** (kan-ă-lik′ū-lī; little canals). These canaliculi open into a **lacrimal sac,** an enlargement of the **nasolacrimal** (nā-zō-lak′ri-măl) **duct,** which opens into the nasal cavity. Tears lubricate and cleanse the eye. They also contain an enzyme that helps combat eye infections.

**Predict 3**

*Explain why it is often possible to smell (or "taste") medications, such as eyedrops, that have been placed into the eyes.*

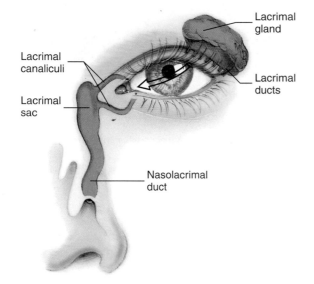

**Figure 9.7  Lacrimal Apparatus**

## Extrinsic Eye Muscles

Movement of each eyeball is accomplished by six skeletal muscles called the **extrinsic eye muscles** (figure 9.8). Four of these muscles run more or less straight from their origins in the posterior portion of the orbit to the eye, to attach to the four quadrants of the eyeball. They are the superior, inferior, medial, and lateral **rectus muscles.** Two muscles, the superior and inferior **oblique muscles,** are located at an angle to the long axis of the eyeball.

# Anatomy of the Eye

The eyeball is a hollow, fluid-filled sphere. The sphere is divided into chambers and contains different types of fluids (figure 9.9), but before discussing those, we will examine the wall of the eye, which is composed of three layers, or **tunics.** The outer **fibrous tunic** consists of the sclera and cornea. The middle **vascular tunic** consists of the choroid, ciliary body, and iris. The inner, **nervous tunic** consists of the retina.

## Fibrous Tunic

The **sclera** (sklēr′ă; hard) is the firm, white, outer connective tissue layer of the posterior five-sixths of the fibrous tunic. The sclera helps maintain the shape of the eye, protects the internal structures and provides attachment sites for the extrinsic eye muscles. A small portion of the sclera can be seen as the "white of the eye."

The **cornea** (kōr′nē-ă) is the transparent anterior sixth of the eye, which permits light to enter. As part of the focusing system of the fibrous tunic, it also bends, or refracts, the entering light.

## Vascular Tunic

The middle tunic of the eye is called the **vascular tunic** because it contains most of the blood vessels of the eye. The posterior portion of the vascular tunic, associated with the sclera, is the **choroid** (kō′royd). This very thin structure consists of a vascular network and many melanin-containing pigment cells, so that it appears black. The black color absorbs light, so that it is not reflected inside the eye. If light were reflected inside the eye, the reflection would interfere with vision. The interiors of cameras are black for the same reason.

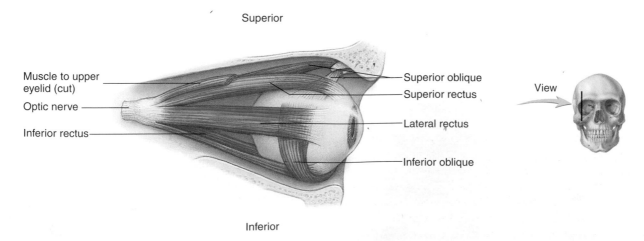

**Figure 9.8  Extrinsic Eye Muscles**

Extrinsic muscles of the right eye as seen from a lateral view with the lateral wall of the orbit removed. The medial rectus muscle cannot be seen from this view.

# CLINICAL IMPACT

## Corneal Transplants

The cornea was one of the first organs to be successfully transplanted. Several characteristics make it relatively easy to transplant: It is easily accessible and relatively easily removed from the donor and grafted to the recipient; it does not have blood vessels and therefore does not require the growth of an extensive circulation into the tissue after grafting; and it is less likely to stimulate the immune system and therefore less likely to be rejected than are other tissues.

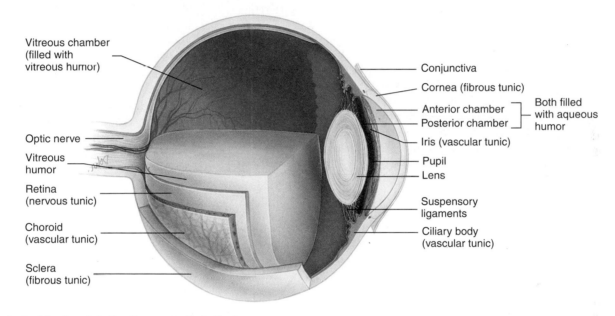

**Figure 9.9** Sagittal Section of the Eye, Demonstrating Its Tunics

Anteriorly, the vascular tunic consists of the ciliary body and the iris. The **ciliary** (sil′ē-ar-ē) **body** is continuous with the anterior margin of the choroid. The ciliary body contains smooth muscles called **ciliary muscles,** which attach to the perimeter of the lens by **suspensory ligaments** (figure 9.10). The **lens** is a flexible, biconvex, transparent disc (see figure 9.9).

The **iris** is the colored part of the eye. It is attached to the anterior margin of the ciliary body, anterior to the lens. The iris is a contractile structure consisting mainly of smooth muscle surrounding an opening called the **pupil.** Light passes through the pupil, and the iris regulates the diameter of the pupil, which controls the amount of light entering the eye. Parasympathetic stimulation from the oculomotor nerve (III) causes the circular smooth muscles of the iris to contract, constricting the pupil, whereas sympathetic stimulation causes radial smooth muscles of the iris to contract, dilating the pupil (figure 9.11). As light intensity increases, the pupil constricts; as light intensity decreases, the pupil dilates.

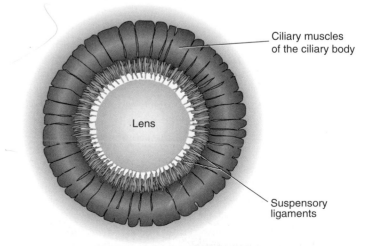

**Figure 9.10** Lens and Ciliary Body

245

**Figure 9.11** Iris

(*a*) Circular smooth muscles of the iris constrict the pupil. (*b*) Radial smooth muscles of the iris dilate the pupil.

## Nervous Tunic

The **nervous tunic** is the innermost tunic and consists of the **retina**. The retina covers the posterior five-sixths of the eye and is composed of two layers: an outer **pigmented retina** and an inner **sensory retina** (figure 9.12*a*). The pigmented retina, with the choroid, keeps light from reflecting back into the eye. The sensory retina contains photoreceptor cells, called **rods** and **cones,** which respond to light. The sensory retina also contains numerous interneurons, some of which are named in figure 9.12. Over most of the retina, rods are 20 times more common than cones. Rods are very sensitive to light and can function in dim light, but they do not provide color vision. Cones require much more light, and they do provide color vision. There are three types of cones, each sensitive to a different color: blue, green, or red.

### Predict 4

*In dim light, colors seem to fade and objects appear in shades of gray. Explain this phenomenon.*

The outer segments of rod and cone cells are modified by numerous foldings of the cell membrane to form discs (figure 9.12*b–d*). Rod cells contain a photosensitive pigment called **rhodopsin** (rō-dop′sin; purple pigment), which is made up of the colorless protein **opsin** (op′sin) in loose chemical combination with a yellow pigment called **retinal** (ret′i-năl) (figure 9.12*e*). When light strikes a rod cell, retinal changes shape (figure 9.13). This causes opsin to change shape, and retinal dissociates from the opsin molecule. The change in rhodopsin's shape stimulates a response in the rod cell, resulting in vision. Retinal then completely detaches from opsin. Energy (ATP) is required to reattach retinal to opsin and, at the same time, to return rhodopsin to the shape it had before being stimulated by light.

The manufacture of retinal in rod cells takes time and requires vitamin A. In bright light, much of the rhodopsin in rod cells is dissociated. For example, suppose you go into a dark building on a bright day. It will take several seconds for your eyes to adjust to the dark as opsin and retinal reassociate to form rhodopsin in the rod cells, which can then react to the dim light. A person with a vitamin A deficiency may have a condition called **night blindness,** characterized by difficulty seeing in dim light. Night blindness can also result from **retinal detachment,** which is the separation of the sensory retina from the pigmented retina. Retinal detachment affects the periphery of the retina, where the rods are located, more than the center of the retina, where the cones are located. Because the rods are more sensitive than the cones to light, retinal detachment affects vision in low light to a greater extent than vision in bright light.

The photosensitive pigments in cone cells are slightly different from those in rod cells. The pigments in cone cells are sensitive to colors. Each color results from stimulation by a certain wavelength of light. Three major types of color-sensitive opsin exist; they are sensitive to blue, red, or green. The many colors that we can see result from the stimulation of combinations of these three types of cones.

The rod and cone cells synapse with bipolar cells of the sensory retina (see figure 9.12). These and the horizontal cells of the retina modify the output of the rod and cone cells. For example, this modification helps us perceive the borders between objects of contrasting brightness. The bipolar and horizontal cells synapse with ganglion cells, whose axons converge at the posterior of the eye to form the **optic nerve** (II; see figures 9.9 and 9.12*a*).

When the posterior region of the retina is examined with an ophthalmoscope (of-thal′mō-skōp), two major features can be observed: the macula and the optic disc (figure 9.14*a*). The **macula** (mak′ū-lā) is a small spot near the center of the posterior retina. In the center of the macula is a small pit, the **fovea** (fō′vē-ă; pit) **centralis.** The fovea centralis is the part of the retina where light is most focused when the eye is looking directly at an object. The fovea centralis contains only cone cells, and the cells are more tightly packed there than anywhere else in the retina. Hence, the fovea centralis is the region with the greatest ability to discriminate fine images, which explains why objects are best seen straight ahead.

The **optic disc** is a white spot just medial to the macula, through which a number of blood vessels enter the eye and spread over the surface of the retina. This is also the spot at which axons from the retina meet, pass through the two outer tunics, and exit the eye as the optic nerve. The optic disc contains no photoreceptor cells and does not respond to light; it is therefore called the **blind spot** of the eye. A small image projected onto the blind spot cannot be seen (figure 9.14*b*).

## Chambers of the Eye

The interior of the eye is divided into the **anterior chamber,** the **posterior chamber,** and the **vitreous** (vit′rē-ŭs; glassy), **chamber** (see figure 9.9). The anterior and posterior chambers are located between the cornea and the lens. The iris separates the anterior chamber from the posterior chamber, which are continuous with each other through the pupil. The much larger vitreous chamber is posterior to the lens.

The anterior and posterior chambers are filled with **aqueous humor** (watery fluid), which helps maintain pressure within the eye, refracts light, and provides nutrients to the inner surface of the eye. Aqueous humor is produced by the ciliary body as a blood filtrate and is returned to the circulation through a venous

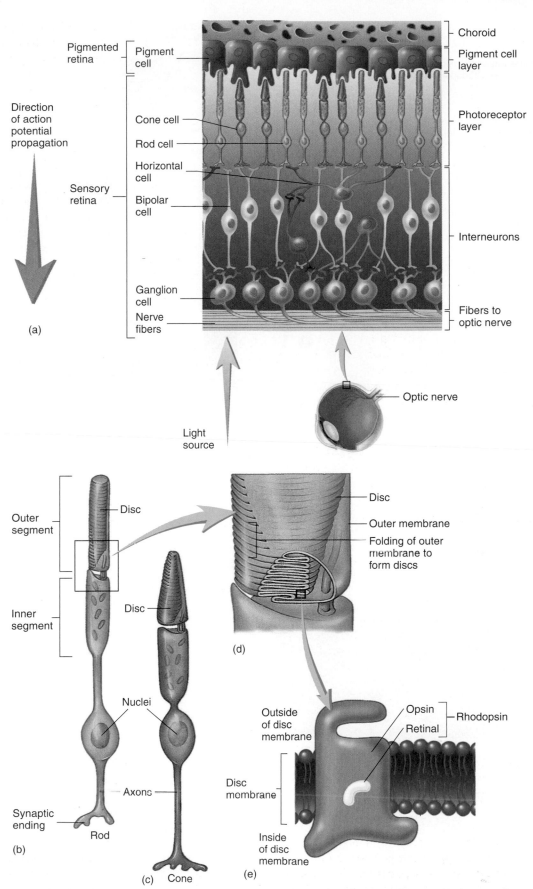

## Figure 9.12  Retina

(a) Enlarged section through the retina, with its major layers labeled. (b) Rod cell. (c) Cone cell. (d) Enlargement of the discs in the outer segment. (e) Enlargement of one of the discs, showing the relationship of rhodopsin to the membrane.

**(a)** labels:
Pigmented retina — Pigment cell — Choroid — Pigment cell layer
Direction of action potential propagation
Cone cell — Rod cell — Photoreceptor layer
Horizontal cell — Bipolar cell
Sensory retina
Interneurons
Ganglion cell — Nerve fibers — Fibers to optic nerve
Light source
Optic nerve

**(b)** Outer segment — Disc — Inner segment — Nuclei — Axons — Synaptic ending — Rod

**(c)** Disc — Cone

**(d)** Disc — Outer membrane — Folding of outer membrane to form discs

**(e)** Outside of disc membrane — Opsin — Rhodopsin — Retinal — Disc membrane — Inside of disc membrane

Nervous

1. Retinal is attached inside opsin to make rhodopsin.

2. Light activates rhodopsin by causing retinal to change shape, which causes opsin to change shape.

3. Activated rhodopsin stimulates cell changes that result in vision.

4. Following rhodopsin activation, retinal detaches from opsin.

5. Energy from ATP is required to bring retinal back to its original form.

6. Retinal attaches to opsin to form rhodopsin (return to step 1).

PROCESS Figure 9.13   Effect of Light on Rhodopsin

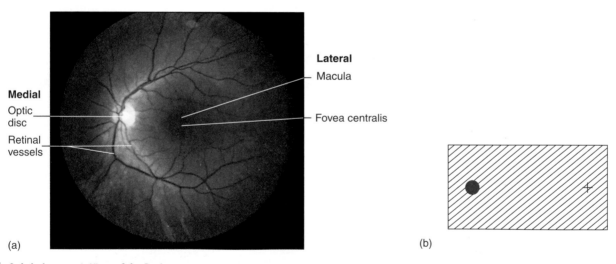

(a)                                    (b)

Figure 9.14   Ophthalmoscopic View of the Retina

(a) This view shows the posterior wall of the left eye as seen through the pupil. Notice the vessels entering the eye through the optic disc. The macula, with the fovea centralis in the center, is located lateral to the optic disc. (b) Demonstration of the blind spot. Close your right eye. Hold the drawing in front of your left eye and stare at the +. Move the drawing toward your eye. At a certain point, when the image of the spot is over the optic disc, the red spot seems to disappear.

ring that surrounds the cornea. The presence of aqueous humor keeps the eye inflated, much like the air in a basketball. If flow of the aqueous humor from the eye through the venous ring is blocked, the pressure in the eye increases, resulting in a condition called **glaucoma** (see the Diseases and Disorders table at the end of this chapter). Glaucoma can eventually lead to blindness because the fluid compresses the retina, thereby restricting blood flow through it.

The vitreous chamber of the eye is filled with a transparent, jellylike substance called **vitreous humor.** The vitreous humor helps maintain pressure within the eye and holds the lens and the retina in place. It also refracts light. Unlike the aqueous humor, the vitreous humor does not circulate.

## Functions of the Eye

The eye functions much as does a camera. The iris allows light into the eye, which is focused by the cornea, lens, and humors onto the retina. The light striking the retina produces action potentials that are relayed to the brain.

### Light Refraction

An important characteristic of light is that it can be refracted (bent). As light passes from air to some other, denser transparent substance, the light rays are refracted. If the surface of a lens is concave, the light rays are bent, so that they diverge as they pass through the lens; if the surface is convex, they converge. As the light rays converge, they finally reach a point at which they cross. The crossing point is called the **focal point (FP)** (figure 9.15), and causing light to converge is called **focusing.** The focal point in the eye occurs just anterior to the retina, and the tiny image that is focused on the retina is inverted compared to the actual object.

### Focusing Images on the Retina

The cornea is a convex structure, and as light rays pass from the air through the cornea, they converge (figure 9.15). Additional convergence occurs as light passes through the aqueous humor, lens, and vitreous humor. The greatest contrast in media density is between the air and the cornea. The greatest amount of convergence therefore occurs at that point. However, the shape of the cornea and its distance from the retina are fixed, so the cornea cannot make any adjustment in focus. Fine adjustments in focus are accomplished by changing the shape of the lens.

When the ciliary muscles are relaxed, the suspensory ligaments of the ciliary body maintain elastic pressure on the perimeter of the lens, keeping it relatively flat and allowing for distant vision (figure 9.15a). When an object is brought closer than 20 feet (about 6½ m) to the eye, the ciliary muscles contract as a result of parasympathetic stimulation, pulling the ciliary body toward the lens. This reduces the tension on the suspensory ligaments of the lens and allows the lens to assume a more spherical form because of its own internal elastic nature (figure 9.15b). The spherical lens then has a more convex surface, causing greater refraction of light. This process is called **accommodation** (ă-kom′ō-dā′shŭn), and it enables the eye to focus on images closer than 20 feet on the retina.

*As you are driving a car, what changes occur when you look down at the speedometer and then back up at the road?*

When a person's vision is tested, a chart is placed 20 feet from the eye, and the person is asked to read a line that has been standardized for normal vision. If the person can read the line, he or she has 20/20 vision, which means that the person can see at 20 feet what people with normal vision see at 20 feet. On the other

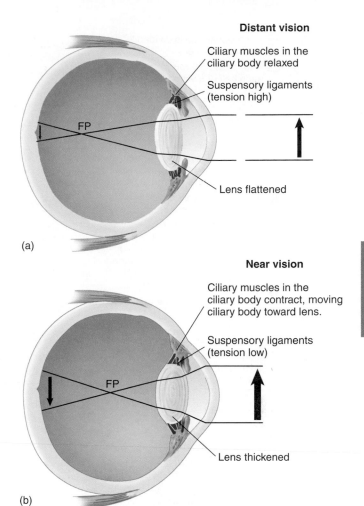

**Figure 9.15  Focus by the Eye**
The focal point (FP) is where light rays cross. (a) When viewing a distant image, the lens is flattened, and the image is focused on the retina. (b) In accommodation for near vision, the lens becomes more rounded, allowing the image to be focused on the retina.

hand, if the person can read only letters at 20 feet that people with normal vision see at 40 feet, the person's eyesight is 20/40, and corrective lenses are probably needed.

## Neuronal Pathways for Vision

Figure 9.16 shows the neuronal pathways that transmit signals generated by light from the time light enters the eye until it reaches the area of the brain where vision is perceived. The **optic nerve** leaves the eye and exits the orbit through the optic foramen to enter the cranial cavity. Just inside the cranial cavity, the two optic nerves connect to each other at the **optic chiasm** (kī′azm; crossing). Axons from the nasal (medial) part of each retina cross through the optic chiasm and project to the opposite side of the brain. Axons from the temporal (lateral) part of each retina pass through the optic nerves and project to the brain on the same side of the body without crossing.

Beyond the optic chiasm, the route of the ganglionic axons is through the two **optic tracts** (figure 9.16). Most of the optic tract

1 Each visual field is divided into temporal and nasal halves.

2 After passing through the lens, light from each half of a visual field projects to the opposite side of the retina.

3 An optic nerve consists of axons extending from the retina to the optic chiasm.

4 In the optic chiasm, axons from the nasal part of the retina cross and project to the opposite side of the brain. Axons from the temporal part of the retina do not cross.

5 An optic tract consists of axons that have passed through the optic chiasm (with or without crossing) to the thalamus.

6 The axons synapse in the thalamus. Collateral branches of the axons in the optic tracts synapse in the superior colliculi.

7 An optic radiation consists of axons from thalamic neurons that project to the visual cortex.

8 The right part of each visual field (*dark green* and *light blue*) projects to the left side of the brain, and the left part of each visual field (*light green* and *dark blue*) projects to the right side of the brain.

(a)

(b)

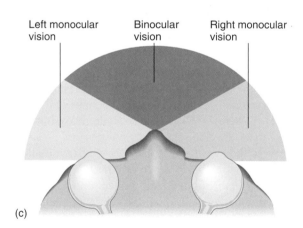

(c)

**PROCESS Figure 9.16** Visual Pathways

(a) Pathways for both eyes (*superior view*). (b) Photograph of the visual nerves, tracts, and pathways (*inferior view*). (c) Overlap of the fields of vision (*superior view*).

axons terminate in the thalamus. Some axons do not terminate in the thalamus but separate from the optic tracts to terminate in the superior colliculi, the center for visual reflexes. An example of a visual reflex is turning the head and eyes toward a stimulus, such as a sudden noise or flash of light. Neurons from the thalamus form the fibers of the **optic radiations,** which project to the **visual cortex** in the occipital lobe of the brain (figure 9.16). The visual cortex is the area of the brain where vision is perceived.

The image seen by each eye is the **visual field** of that eye (figure 9.16*a*). Depth perception (three-dimensional, or binocular, vision) requires both eyes and occurs where the two visual fields overlap (figure 9.16*c*). Each eye sees a slightly different (monocular) view of the same object. The brain then processes the two images into a three-dimensional view of the object. If only one eye is functioning, the view of the object is flat, much like viewing a photograph.

# CLINICAL IMPACT

## Color Blindness

**Color blindness** is the absence of perception of one or more colors (figure 9A). Color perception may be decreased or completely lost. The loss may involve perception of all three colors or of one or two colors. Most forms of color blindness occur more frequently in males and are X-linked genetic traits (see chapter 20). In Western Europe, about 8% of all males have some form of color blindness, whereas only about 1% of the females are color blind.

(a)                    (b)

### Figure 9A

These color blindness charts demonstrate the differences in color perception associated with some forms of color blindness. (*a*) A person with normal vision can see the number 74, whereas a person with red-green color blindness sees the number 21. (*b*) A person with normal vision can see the number 42. A person with red color blindness sees the number 2, and a person with green color blindness sees the number 4.

Reproduced from Ishihara's Tests for Colour Deficiency published by Kanehara Trading, Inc., Tokyo, Japan. Tests for color deficiency cannot be conducted with this material. For accurate testing, the original plates should be used.

## A CASE IN POINT
### Double Vision

I.C. Double awoke one morning and discovered that her vision was blurred. However, when she closed one eye, her vision was clear. She tried to ignore the problem and went about her work, but the condition persisted. Eventually, she consulted an optometrist, who found that her blurred vision resulted from double vision, or **diplopia.** The most common cause of diplopia is misalignment of the two eyes (binocular diplopia). This often results from weakness of the muscles moving the eyes. Most adults cannot easily ignore the double image and may need to wear a patch over one eye to eliminate one image. In children with diplopia, the brain may compensate for the two discordant images by ignoring one of the images; thus, the problem appears to go away. Double vision should not be ignored because it can be a symptom of a serious neurological problem, such as an expanding brain tumor compressing the nerves to the eye muscles, and decreasing their function. A physician should be consulted as soon as possible.

## HEARING AND BALANCE

The organs of hearing and balance are apportioned into three areas: the external, middle, and inner ears (figure 9.17). The external ear is the part extending from the outside of the head to the tympanic membrane, which is also called the eardrum.

The middle ear is an air-filled chamber medial to the tympanic membrane. The inner ear is a set of fluid-filled chambers medial to the middle ear. The external and middle ears are involved in hearing only, whereas the inner ear functions in both hearing and balance.

## Anatomy and Function of the Ear

### External Ear

The **auricle** (aw'ri-kl; ear) is the fleshy part of the external ear on the outside of the head. The auricle opens into the **external auditory canal,** a passageway that leads to the eardrum. The auricle collects sound waves and directs them toward the external acoustic canal, which transmits them to the tympanic membrane. The auditory canal is lined with hairs and **ceruminous** (sĕ-roo'mi-nŭs; *cera,* wax) **glands,** which produce **cerumen** (sĕ-roo'men), a modified sebum commonly called earwax. The hairs and cerumen help prevent foreign objects from reaching the delicate tympanic membrane.

The **tympanic** (tim-pan'ik; drumlike) **membrane,** or *eardrum,* is a thin membrane that separates the external ear from the middle ear. It consists of a thin layer of connective tissue sandwiched between two epithelial layers. Sound waves reaching the tympanic membrane cause it to vibrate.

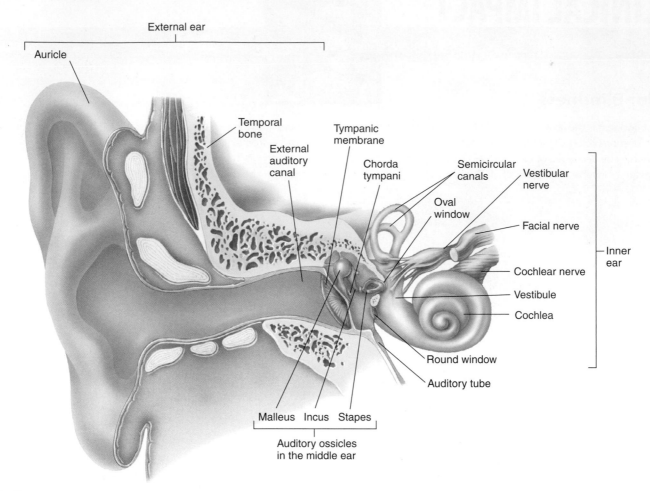

**Figure 9.17** Structure of the Ear

## Middle Ear

Medial to the tympanic membrane is the air-filled cavity of the middle ear. Two covered openings on the medial side of the middle ear, the **oval window** and the **round window,** connect the middle ear with the inner ear. The middle ear contains three **auditory ossicles** (os′i-klz; ear bones): the **malleus** (mal′ē-ŭs; hammer), the **incus** (ing′kŭs; anvil), and the **stapes** (stā′pēz; stirrup). These bones transmit vibrations from the tympanic membrane to the oval window. The malleus is attached to the medial surface of the tympanic membrane. The incus connects the malleus to the stapes. The base of the stapes is seated in the oval window, surrounded by a flexible ligament. As vibrations are transmitted from the malleus to the stapes, the force of the vibrations is amplified about 20-fold because the area of the tympanic membrane is about 20 times that of the oval window. Two small muscles in the middle ear, one attached to the malleus and the other to the stapes, help dampen vibrations caused by loud noises, thus protecting the delicate inner ear structures.

There are two unblocked openings into the middle ear. One opens into the mastoid air cells in the mastoid process of the temporal bone. The other, called the **auditory tube,** or *eustachian* (ū-stā′shŭn) *tube,* opens into the pharynx and enables air pressure to be equalized between the outside air and the middle ear cavity. Unequal pressure between the middle ear and the outside environment can distort the tympanic membrane, dampen its vibrations, and make hearing difficult. Distortion of the tympanic membrane also stimulates pain fibers associated with that structure. That distortion is why, as a person changes altitude, sounds seem muffled and the tympanic membrane may become painful. These symptoms can be relieved by opening the auditory tube to allow air to enter or exit the middle ear, such as by swallowing, yawning, chewing, or holding the nose and mouth shut while gently forcing air out of the lungs.

## Inner Ear

The inner ear consists of interconnecting tunnels and chambers within the temporal bone, called the **bony labyrinth** (lab′i-rinth; maze) (figure 9.18a). Inside the bony labyrinth is a smaller set of membranous tunnels and chambers called the **membranous labyrinth** (figure 9.18b). The membranous labyrinth is filled with a clear fluid called **endolymph** (en′dō-limf), and the space between the membranous and bony labyrinths is filled with a fluid called **perilymph** (per′i-limf). The bony labyrinth can be divided into three regions: the cochlea, the vestibule, and the semicircular canals. The cochlea is involved in hearing. The vestibule and semicircular canals are involved primarily in balance.

The **cochlea** (kok′lē-ă; snail shell) is shaped like a snail shell (figure 9.18a) and contains a bony core shaped like a screw. The

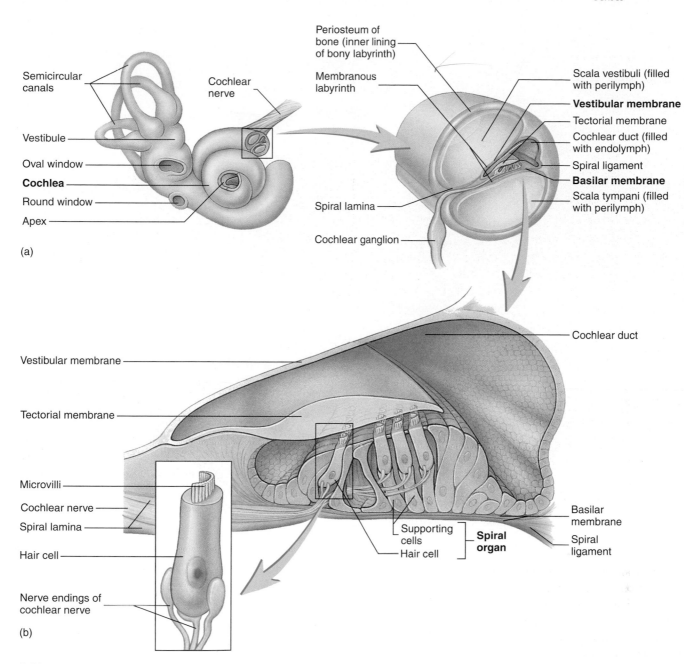

**Figure 9.18** Structure of the Inner Ear

(a) Bony labyrinth. The outer surface (gray) is the periosteum lining the inner surface of the bony labyrinth. (b) In this cross section of the cochlea, the outer layer is the periosteum lining the inner surface of the bony labyrinth. The membranous labyrinth is very small in the cochlea and consists of the vestibular and basilar membranes. The space between the membranous and bony labyrinths consists of two parallel tunnels: the scala vestibuli and the scala tympani. (c) An enlarged section of the cochlear duct (membranous labyrinth). (d) A greatly enlarged individual sensory hair cell.

threads of this screw are called the **spiral lamina.** The cochlea is divided into three channels: the scala vestibuli, the scala tympani, and the cochlear duct (figure 9.18b). The **scala vestibuli** (skā′lă ves-tib′ū-lī; *scala,* stairway) extends from the oval window to the apex of the cochlea. The **scala tympani** (tim-pa′nē) extends in parallel with the scala vestibuli from the apex back to the round window. These two channels are perilymph-filled spaces between the walls of the bony and membranous labyrinths. The wall of the membranous labyrinth that lines the scala vestibuli is called the **vestibular** (ves-tib′ū-lăr) **membrane;** the wall of the membranous

labyrinth that lines the scala tympani is the **basilar membrane.** The **cochlear duct** is formed by the space between the vestibular membrane and the basilar membrane and is filled with endolymph.

Inside the cochlear duct is a specialized structure called the **spiral organ,** or *organ of Corti* (figure 9.18c). The spiral organ contains specialized sensory cells called **hair cells,** which have hairlike microvilli on their surfaces (figure 9.18c,d). The microvilli are stiffened by actin filaments. The hair tips are embedded within an acellular gelatinous shelf called the **tectorial** (tek-tōr′ē-ăl;

1. Sound waves strike the tympanic membrane and cause it to vibrate.

2. Vibration of the tympanic membrane causes the malleus, the incus, and the stapes to vibrate.

3. The foot plate of the stapes vibrates in the oval window.

4. Vibration of the foot plate causes the perilymph in the scala vestibuli to vibrate.

5. Vibration of the perilymph causes the vestibular membrane to vibrate, which causes vibrations in the endolymph.

6. Vibration of the endolymph causes displacement of the basilar membrane. Short waves (high pitch), cause displacement of the basilar membrane near the oval window, and longer waves (low pitch) cause displacement of the basilar membrane some distance from the oval window. Movement of the basilar membrane is detected in the hair cells of the spiral organ, which are attached to the basilar membrane. Vibrations of the perilymph in the scala vestibuli and of the basilar membrane are transferred to the perilymph of the scala tympani.

7. Vibrations in the perilymph of the scala tympani are transferred to the round window, where they are dampended.

**PROCESS Figure 9.19** Effect of Sound Waves on Middle and Inner Ear Structures

a covering) **membrane,** which is attached to the spiral lamina (figure 9.18b,c).

Hair cells have no axons of their own, but each hair cell is associated with axon terminals of sensory neurons, the cell bodies of which are located within the **cochlear ganglion,** or *spiral ganglion.* Axons of the sensory neurons join to form the cochlear nerve. This nerve joins the vestibular nerve to become the **vestibulocochlear nerve** (VIII), which carries action potentials to the brain.

## Hearing

Vibrations create sound waves. Sound waves are collected by the auricle and conducted through the external auditory canal toward the tympanic membrane. Sound waves strike the tympanic membrane and cause it to vibrate. This vibration causes vibration of the three ossicles of the middle ear, and by this mechanical linkage, the force of vibration is amplified and transferred to the oval window (figure 9.19).

*Predict 6*

*When you hear a faint sound, why do you turn your head toward it?*

Vibrations of the base of the stapes, seated in the oval window, produce waves in the perilymph of the cochlea. The two scalae can be thought of as a continuous, U-shaped tube, with the oval window at one end of the scala vestibuli and the round window at the other end of the scala tympani (figure 9.19). The vibrations of the stapes in the oval window cause movement of the perilymph, which pushes against the membrane covering the round window. This phenomenon is similar to pushing against a rubber diaphragm on one end of a fluid-filled glass tube. If the tube has a rubber diaphragm on each end, the fluid can move. If one end of the glass tube or of the cochlear tubes were solid, no fluid movement would occur.

The waves produced in the perilymph pass through the vestibular membrane and cause vibrations of the endolymph. Waves in the endolymph, within the cochlear duct, cause displacement of the basilar membrane. As the basilar membrane is displaced, the hair cells, seated on the basilar membrane, move with the movements of the membrane. The microvilli of the hair cells are embedded in the tectorial membrane, which is a rigid shelf that does not move. Because one end of the microvilli moves with the hair cells and their other ends are embedded in the nonmoving tectorial membrane, the microvilli bend. The bending of the microvilli stimulates the hair cells, which induces action potentials in the cochlear nerves.

The basilar membrane is not uniform throughout its length. The membrane is narrower and denser near the oval window and wider and less dense near the tip of the cochlea. The various regions of the membrane can be compared to the strings in a piano (i.e., some are short and thick, and others are longer and thinner). As a result of this organization, sounds with higher **pitches** cause maximum distortion of the basilar membrane nearer the oval window, whereas sounds with lower pitches cause maximum distortion nearer the apex of the cochlea. In each case, different hair cells are stimulated, and because of the differences in which hair cells are maximally stimulated, a person is able to detect variations in pitch. Sound **volume** is a function of sound wave amplitude, which causes the basilar membrane to distort more intensely and the hair cells to be stimulated more strongly.

Hearing impairment can have many causes. In general, there are two categories of hearing impairment: conduction deafness and sensorineural hearing loss (see Diseases and Disorders table at the end of this chapter). **Conduction deafness** results from mechanical deficiencies—for example, destruction of the ligament that holds the malleus and incus together. **Sensorineural hearing loss** is caused by deficiencies in the spiral organ or nerves; for example, loud sounds can damage the delicate microvilli of the hair cells, leading to destruction of the spiral organ.

## Neuronal Pathways for Hearing

The senses of hearing and balance are both transmitted by the vestibulocochlear nerve (VIII). This nerve functions as two separate nerves, carrying information from two separate but closely related structures. The cochlear nerve is the portion of the vestibulocochlear nerve involved in hearing; the vestibular nerve is involved in balance. The cochlear nerve sends axons to the **cochlear nucleus** in the brainstem. Neurons in the cochlear nucleus project to other areas of the brainstem and to the **inferior colliculus** in the midbrain. Neurons from the inferior colliculus also project to superior colliculus, where reflexes that turn the head and eyes in response to loud sounds are initiated. From the inferior colliculus, fibers project to the thalamus and from there to the auditory cortex of the cerebrum.

## Balance

The sense of balance, or equilibrium, has two components: static equilibrium and dynamic equilibrium. **Static equilibrium** is associated with the vestibule and is involved in evaluating the position of the head relative to gravity. **Dynamic equilibrium** is associated with the semicircular canals and is involved in evaluating changes in the direction and rate of head movements.

The **vestibule** (ves′ti-bool) of the inner ear can be divided into two chambers: the **utricle** (ū′tri-kl) and the **saccule** (sak′ūl) (figure 9.20a). Each chamber contains specialized patches of epithelium called the **maculae** (mak′ū-lē), which are surrounded by endolymph. The maculae, like the spiral organ, contain hair cells. The tips of the microvilli of these cells are embedded in a gelatinous mass weighted by **otoliths** (ō′tō-liths; ear stones), particles composed of protein and calcium carbonate. The weighted gelatinous mass moves in response to gravity, bending the hair cell microvilli (figure 9.20c) and initiating action potentials in the associated neurons. The action potentials from these neurons are carried by axons of the vestibular portion of the vestibulocochlear nerve (VIII) to the brain, where they are interpreted as a change in the position of the head. For example, when a person bends over, the maculae are displaced by gravity, and the resultant action potentials provide information to the brain concerning the position of the head (figure 9.21).

Three **semicircular canals** are involved in dynamic equilibrium. The canals are placed at nearly right angles to one another, enabling a person to detect movements in essentially any direction. The base of each semicircular canal is expanded into an **ampulla** (am-pul′ă) (figure 9.22a). Within each ampulla, the epithelium is

**Figure 9.20  Location and Structure of the Macula**

(a) Location of the utricular and saccular maculae within the vestibule. (b) Enlargement of the utricular macula, showing hair cells and otoliths. (c) Enlarged hair cell, showing the microvilli.

**Figure 9.21  Function of the Vestibule in Maintaining Balance**

(*a*) In an upright position, the maculae don't move. (*b*) When the position of the head changes, as when a person bends over, the maculae respond by moving in the direction of gravity.

specialized to form a **crista ampullaris** (kris′tă am-pūl′ar′is) (figure 9.22*b*). Each crista consists of a ridge of epithelium with a curved, gelatinous mass, the **cupula** (koo′poo-lă; a tub), suspended over the crest. The cupula is structurally and functionally very similar to the maculae, except that it contains no otoliths. The hair-like microvilli of the crista hair cells (figure 9.22*c*) are embedded in the cupula. The cupula functions as a float that is displaced by endolymph movement within the semicircular canals (figure 9.23). As the head begins to move in one direction, the endolymph tends to remain stationary, while the cupula moves with the head. This difference displaces the cupula in a direction opposite that of the movement of the head. As movement continues, the fluid "catches up." When movement of the head and the cupula stops, the fluid continues to move, displacing the cupula in the direction of the movement. Movement of the cupula causes the hair cell microvilli to bend, which initiates depolarization in the hair cells. This depolarization initiates action potentials in the vestibular nerves, which join the cochlear nerves to form the vestibulocochlear nerves.

Continuous stimulation of the semicircular canals—as occurs due to the rocking motion of a boat—can cause **motion sickness,** characterized by nausea and weakness. The brain compares sensory input from the semicircular canals, eyes, and position receptors (proprioceptors) in the back and lower limbs. Conflicting input from these sources can lead to motion sickness.

**Predict 7**

*Explain why closing your eyes or looking at the horizon can help decrease motion sickness.*

## Neuronal Pathways for Balance

Axons forming the vestibular portion of the vestibulocochlear nerve (VIII) project to the vestibular nucleus in the brainstem. Axons run from this nucleus to numerous areas of the CNS, such as the cerebellum and cerebral cortex. Balance is a complex sensation involving sensory input to the vestibular nucleus not only from the inner ear but also from the limbs (proprioception) and

**Figure 9.22 Semicircular Canals**

(a) Location of the ampullae of the semicircular canals. (b) Enlargement of the crista ampullaris, showing the cupula and hair cells. (c) Enlargement of a hair cell.

(a)

(b)    Movement of semicircular canal with body movement

**Figure 9.23 Function of the Crista Ampullaris**

(a) As a person begins to tumble, the semicircular canals (b) move in the same direction as the body (*blue arrow*). The endolymph in the semicircular canals tends to stay in place as the body and the crista ampullaris begin to move. As a result, the cupula is displaced by the moving endolymph (*red arrow*) in a direction opposite the direction of movement.

visual system as well. In sobriety tests, people are asked to close their eyes while their balance is evaluated because alcohol affects the proprioceptive and vestibular components of balance to a greater extent than the visual component of balance.

## A CASE IN POINT

### Seasickness

Earl E. Fisher booked his first trip on a charter fishing boat. After boarding the boat, Earl was surprised that all those warnings about passengers becoming seasick did not seem to apply to him. At last, the boat arrived at the fishing site, the engine was cut, and the sea anchor was set. As the boat drifted, it began to roll and pitch. Earl noticed, for the first time, that the smell of the bait mixed unpleasantly with the diesel fumes. Earl felt a little light-headed and a bit drowsy. He then noticed that he was definitely nauseated. "I'm

seasick," he realized. Seasickness is a form of motion sickness, which is caused by conflicting information reaching the brain from different sensory sources, such as the eyes and the semicircular canals of the inner ear. The brain reacts with a feeling of vertigo (a feeling of spinning) and nausea. In Earl's case, trying to fish seemed to worsen his condition. Eventually, his nausea intensified, and he leaned over the boat rail and vomited into the ocean. "Improves the fishing," the ship owner shouted to him cheerily. Earl noticed that he felt somewhat better, and he found that looking at the horizon rather than at the water helped his condition even more. He enjoyed the rest of the trip and even caught a couple of very nice fish.

## DISEASES AND DISORDERS  Vision, Hearing, and Balance

| CONDITION | DESCRIPTION |
|---|---|
| | **EYE DISORDERS** |
| **Infections** | |
| Conjunctivitis (kon-jŭnk-ti-vi′tis) | Inflammation of conjunctiva, usually from bacterial infection; one form, pinkeye, occurs primarily in children |
| Trachoma (trǎ-kō′mǎ) | Type of conjunctivitis caused by *Chlamydia;* leading cause of infectious blindness in the world; transmitted by contact or flies |
| Stye (stī) | Infection of eyelash hair follicle |
| **Defects of Focus, Alignment, or Color Vision** | |
| Myopia (mī-ō′pē-ǎ) | Nearsightedness—ability to see close but not distant objects; caused when refractive power of cornea and lens is too great relative to length of eye |
| Hyperopia (hī-per-ō′pē-ǎ) | Farsightedness—ability to see distant but not close objects; caused when cornea is too flat or lens has too little refractive power relative to length of eye |
| Presbyopia (prez-bē-ō′pē-ǎ) | Decrease in near vision, a normal part of aging |
| Astigmatism (ǎ-stig′ mǎ-tizm) | Cornea or lens is not uniformly curved, so image is not sharply focused |
| Strabismus (stra-biz′mŭs) | One or both eyes are misdirected; can result from weak eye muscles |
| Diplopia (di-plō′pē-ǎ) | Double vision |
| Color blindness | Complete or partial absence of perception of one or more colors (see figure 9A); most forms are more frequent in males |
| **Blindness** | |
| Cataract (kat′ǎ-rakt) | Clouding of lens as a result of advancing age, infection, or trauma; most common cause of blindness in the world |
| Macular degeneration | Loss of sharp central vision, peripheral vision maintained; leading cause of legal blindness in older Americans; most causes not known |
| Glaucoma (glaw-kō′mǎ) | Excessive pressure buildup in aqueous humor; may destroy retina or optic nerve, resulting in blindness |
| Diabetic retinopathy | Involves optic nerve degeneration, cataracts, retinal detachment; often caused by blood vessel degeneration and hemorrhage |
| Retinal detachment | Separation of sensory retina from pigmented retina; relatively common problem; may result in vision loss |
| | **EAR DISORDERS** |
| Conduction deafness | Mechanical deficiency in transmission of sound waves from outer ear to spiral organ |
| Sensorineural hearing loss | Deficiencies of spiral organ or nerve pathways |
| Tinnitus (ti-nī′tus) | Phantom sound sensations, such as ringing in ears; common problem |
| Middle ear infection | Symptoms are low-grade fever, lethargy, irritability, and pulling at ear; in extreme cases, can damage or rupture tympanic membrane; common in young children |
| Inner ear infection | Can decrease detection of sound and maintenance of balance, may be caused by chronic middle ear infections |
| Motion sickness | Nausea and weakness caused when information to brain from semicircular canals conflicts with information from eyes and position sensors in back and lower limbs |
| Meniere disease | Vertigo, hearing loss, tinnitus, and a feeling of fullness in the affected ear; most common disease involving dizziness from inner ear; cause unknown but may involve a fluid abnormality in ears |

Go to **www.mhhe.com/seeleyess7** for additional information on these pathologies.

# EFFECTS OF AGING ON THE SENSES

As a person ages, both the general and the special sensory functions gradually decline. Among the general senses, free nerve endings and hair follicle receptors in the skin remain largely unchanged. However, the numbers of Meissner corpuscles and pacinian corpuscles decrease with age, and those that remain are often structurally distorted and less functional. As a result, elderly people are less conscious of something touching or pressing the skin, which increases the risk of skin injuries. The sense of two-point discrimination decreases, and the elderly have a more difficult time identifying objects by touch. A loss of pacinian corpuscles also decreases their awareness of limb and joint positions, which can affect balance and coordination. The functions of Golgi tendon organs and muscle spindles also decline with age, which decreases information on the position, tension, and length of tendons and muscles. This can further reduce coordination and control of movements.

Among the special senses, elderly people experience only a slight loss in the ability to detect odors. However, their ability to correctly identify specific odors decreases, especially in men over age 70.

In general, the sense of taste decreases as people age. The number of sensory receptors decreases, and the brain's ability to interpret taste sensations declines.

The lenses of the eyes lose flexibility as a person ages because the connective tissue of the lenses becomes more rigid. Consequently, the lenses' ability to change shape initially declines and eventually is lost. This condition, called **presbyopia,** is the most common age-related change in the eyes. In addition, the number of cones decreases, especially in the fovea centralis, resulting in a gradual decline in visual acuity and color perception.

The most common visual problem in older people requiring medical treatment, such as surgery, is the development of cataracts. Following cataracts in frequency are macular degeneration, glaucoma, and diabetic retinopathy, in that order.

As people age, the number of hair cells in the cochlea decreases, resulting in age-related sensorineural hearing loss, called **presbyacusis.** This decline doesn't occur equally in both ears. Therefore, because direction is determined by comparing sounds coming into each ear, elderly people may experience a decreased ability to localize the origin of certain sounds. This may lead to a general sense of disorientation. In addition, CNS defects in the auditory pathways can lead to difficulty understanding sounds when echoes or background noise is present. This deficit makes it difficult for elderly people to understand rapid or broken speech.

With age, the number of hair cells in the saccule, utricle, and ampullae decreases. The number of otoliths also declines. As a result, elderly people experience a decreased sensitivity to gravity, acceleration, and rotation, which may lead to disequilibrium (instability) and vertigo (a feeling of spinning). Some elderly people feel that they can't maintain posture and are prone to fall.

 # SUMMARY

Sensations result only from those stimuli that reach the cerebral cortex and are consciously perceived. Senses can be defined as general or special.

## General Senses (p. 239)

Receptors for general senses, such as pain, temperature, touch, pressure, and proprioception, are scattered throughout the body.

### Pain

1. Pain is an unpleasant sensation that may be either localized or diffuse.
2. Pain can be reduced or controlled by "gating mechanisms" involving the gate control theory.
3. Pain from deeper structures may be referred to more superficial structures, such as the skin.

## Special Senses (p. 241)

Smell and taste respond to chemical stimulation, vision to light stimulation, and hearing and balance to mechanical stimulation.

## Olfaction (p. 241)

1. Olfactory neurons have enlarged distal ends with long cilia. The cilia have receptors that respond to dissolved substances in the nasal mucus.
2. The wide range of detectable odors results from combinations of receptor responses.

### Neuronal Pathways for Olfaction

Axons of the olfactory neurons form the olfactory nerves, which enter the olfactory bulb. Olfactory tracts carry action potentials from the olfactory bulbs to the olfactory cortex of the brain.

## Taste (p. 242)

1. Taste buds contain taste cells with hairs that extend into taste pores. Receptors on the hairs detect dissolved substances.
2. There are five basic types of taste: sour, salty, bitter, sweet, and umami.

### Neuronal Pathways for Taste

The facial nerve carries taste from the anterior two-thirds of the tongue; the glossopharyngeal nerve carries taste from the posterior one-third of the tongue; and the vagus nerve carries taste from the root of the tongue.

## Vision (p. 242)

### Accessory Structures of the Eye

1. The eyebrows prevent perspiration from entering the eyes.
2. The eyelids protect the eyes from foreign objects.
3. The conjunctiva covers the inner eyelids and the anterior surface of the eye.
4. Lacrimal glands produce tears that flow across the eye surface to lubricate and protect the eye. Excess tears pass through the nasolacrimal duct into the nasal cavity.
5. The extrinsic eye muscles move the eyeball.

### Anatomy of the Eye

1. The fibrous tunic is the outer layer of the eye. It consists of the sclera and cornea.
2. The vascular tunic is the middle layer of the eye. It consists of the choroid, ciliary body, and iris.
3. The lens is held in place by the suspensory ligaments, which are attached to the smooth muscles of the ciliary body.
4. The nervous tunic (retina) is the inner layer of the eye and contains neurons sensitive to light.
5. Rods are responsible for vision in low illumination (night vision).
6. Cones are responsible for color vision.
7. Light causes retinal to change shape, which causes opsin to change shape, leading eventually to cellular changes that result in vision.
8. The fovea centralis in the center of the macula has the highest concentration of cones and is the area where images are detected most clearly.
9. The optic disc, or blind spot, is where the optic nerve exits the eye and blood vessels enter.
10. The anterior and posterior chambers of the eye are anterior to the lens and are filled with aqueous humor. The vitreous chamber is filled with vitreous humor. The humors keep the eye inflated, refract light, and provide nutrients to the inner surface of the eye.

### Functions of the Eye

1. Light passing through a concave surface diverges. Light passing through a convex surface converges.
2. Converging light rays cross at the focal point and are said to be focused.
3. The cornea, aqueous humor, lens, and vitreous humor all refract light. The cornea is responsible for most of the convergence, whereas the lens can adjust the focus by changing shape (accommodation).

### Neuronal Pathways for Vision

1. Axons pass through the optic nerves to the optic chiasm, where some cross. Axons from the nasal retina cross, and those from the temporal retina do not.
2. Optic tracts from the chiasm lead to the thalamus.
3. Optic radiations extend from the thalamus to the visual cortex in the occipital lobe.

## Hearing and Balance    (p. 251)

### Anatomy and Function of the Ear

1. The external ear consists of the auricle and the external auditory canal.
2. The middle ear contains the three auditory ossicles.
3. The tympanic membrane (eardrum) is stretched across the external auditory canal.
4. The malleus, incus, and stapes connect the tympanic membrane to the oval window of the inner ear.
5. The auditory, or eustachian, tube connects the middle ear to the pharynx and equalizes pressure. The middle ear is also connected to the mastoid air cells.
6. The inner ear has three parts: the semicircular canals, the vestibule, and the cochlea.
7. The cochlea is a canal shaped like a snail's shell.
8. The cochlea is divided into three compartments by the vestibular and basilar membranes.
9. The spiral organ consists of hair cells that attach to the basilar and tectorial membranes.

### Hearing

1. Sound waves are funneled through the auricle down the external auditory canal, causing the tympanic membrane to vibrate.

2. The tympanic membrane vibrations are passed along the ossicles to the oval window of the inner ear.
3. Movement of the stapes in the oval window causes the perilymph to move the vestibular membrane, which causes the endolymph to move the basilar membrane. Movement of the basilar membrane causes the hair cells in the spiral organ to move and generate action potentials, which travel along the vestibulocochlear nerve.

### Neuronal Pathways for Hearing

From the vestibulocochlear nerve, action potentials travel to the cochlear nucleus and on to the cerebral cortex.

### Balance

1. Static equilibrium evaluates the position of the head relative to gravity.
2. Maculae, located in the vestibule, consist of hair cells with the microvilli embedded in a gelatinous mass that contains otoliths. The gelatinous mass moves in response to gravity.
3. Dynamic equilibrium evaluates movements of the head.
4. The inner ear contains three semicircular canals, arranged perpendicular to each other. The ampulla of each semicircular canal contains a crista ampullaris, which has hair cells with microvilli embedded in a gelatinous mass, the cupula.

### Neuronal Pathways for Balance

Axons in the vestibular portion of the vestibulocochlear nerve project to the vestibular nucleus and on to the cerebral cortex.

## Effects of Aging on the Senses    (p. 258)

Elderly people experience a general decline in some general senses and in taste, vision, hearing, and balance.

 ## REVIEW AND COMPREHENSION

1. Describe how a stimulus becomes a sensation.
2. Contrast the features of the receptors associated with the general senses.
3. Explain how pain is reduced by analgesics and how it can be modified, according to the gate control theory.
4. Explain referred pain, and give an example.
5. Describe the process by which airborne molecules produce the sensation of smell.
6. How is the sense of taste related to the sense of smell?
7. What are the five primary tastes? How do they produce many different kinds of taste sensations?
8. Describe the following structures and state their functions: eyebrows, eyelids, conjunctiva, lacrimal apparatus, and extrinsic eye muscles.
9. Name the three tunics of the eye. Describe the structures composing each layer, and explain the functions of these structures.
10. Describe the three chambers of the eye, the substances that fill each, and the function of these substances.
11. Describe the lens of the eye and how it is held in place.

12. Describe the arrangement of rods and cones in the fovea centralis and in the periphery of the eye.
13. What is the blind spot of the eye, and what causes it?
14. What causes the pupil to constrict and dilate?
15. What causes light to refract? What is a focal point?
16. Define accommodation. What does accommodation accomplish?
17. Name the three regions of the ear, name the structures in each region, and state the functions of each structure.
18. Describe the relationship among the tympanic membrane, the ear ossicles, and the oval window of the inner ear.
19. Describe the structure of the cochlea.
20. Starting with the auricle, trace sound into the inner ear to the point at which action potentials are generated in the vestibulocochlear nerve.
21. Describe the maculae in the vestibule of the ear and their function in balance.
22. What is the function of the semicircular canals? Describe the crista ampullaris and its mode of operation.

# CRITICAL THINKING

1. An elderly male with normal vision developed cataracts. A surgeon treated his condition by removing the lenses of his eyes. What kind of glasses do you recommend to compensate for the removal of his lenses?

2. On a camping trip, Starr Gazer was admiring the stars in the night sky. She noticed a little cluster of dim stars at the edge of her vision, but when she looked directly at that part of the sky, she could not see the cluster. On the other hand, when she looked toward the stars but not directly at them, she could see them. Explain what was happening.

3. Skin divers are subject to increased pressure as they descend toward the bottom of the ocean. Sometimes this pressure can lead to damage to the ear and loss of hearing. Describe the normal mechanisms that adjust for changes in pressure. Explain how increased pressure might cause reduced hearing, and suggest at least one other common condition that might interfere with this pressure adjustment.

4. If a vibrating tuning fork were placed against the mastoid process of your temporal bone, you would perceive the vibrations as sound, even if the external auditory canal were plugged. Explain how this happens.

5. The main way that people "catch" colds is through their hands. After touching an object contaminated with the cold virus, the person transfers the virus to the nasal cavity, where it causes an infection. Other than the obvious entry of the virus through the nose, how could the virus get into the nasal cavity?

*Answers in Appendix D*

**Nervous**

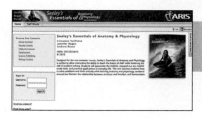

Visit this textbook's website at **www.mhhe.com/seeleyess7** for practice quizzes, animations, interactive learning exercises, and other study tools.

McGraw-Hill offers a study CD that features interactive cadaver dissection. ***Anatomy & Physiology Revealed*** includes cadaver photos that allow you to peel away layers of the human body to reveal structures beneath the surface. This program also includes animations, radiologic imaging, audio pronunciations, and practice quizzing.

The endocrine system produces many different chemical messengers, or hormones, that help control virtually every process our bodies perform. If our bodies lose the ability to produce a particular hormone, modern medicine has been able to find ways to replace it, as shown in this picture. This young girl's father is helping her inject herself with insulin, an important hormone.

# Chapter 10

# Endocrine System

## Chapter Outline and Objectives

### Functions of the Endocrine System (p. 263)
1. Compare the means by which the nervous and endocrine systems regulate body functions.

### Chemical Signals (p. 263)
2. Describe the relationships among chemical signals, receptor molecules, and receptor sites.

### Hormones (p. 265)
3. List the major categories of hormones on the basis of their chemical structure.
4. Describe how hormones interact with tissues to produce a response.

### Receptors (p. 266)
5. Describe how membrane-bound and intracellular receptor molecules mediate responses to intercellular chemical signals.
6. List the mechanisms by which intercellular chemical signals produce responses in their target tissues.

### Regulation of Hormone Secretion (p. 268)
7. Describe three methods of regulating the release of hormones.

### Endocrine Glands and Their Hormones (p. 269)
8. State the location of each of the endocrine glands in the body.
9. Describe how the hypothalamus regulates hormone secretion from the pituitary.
10. Describe how the pituitary gland regulates the secretion of hormones from other endocrine glands.
11. Choose a hormone and use it to explain how negative feedback results in homeostasis.
12. For each of the major hormones in the chapter, describe the endocrine gland from which it is secreted, its target tissue, the response of the target tissue, and the means by which its secretion is regulated.
13. List the effects of over- and undersecretion of the major hormones.

### Other Hormones (p. 286)
14. Describe the functions of hormones secreted by the stomach and small intestine, the functions of prostaglandins, and the functions of erythropoietin.

### Effects of Aging on the Endocrine System (p. 288)
15. Describe the major age-related changes that occur in the endocrine system.

Homeostasis depends on the precise regulation of the body's organs and organ systems. The nervous and endocrine systems are the two major systems responsible for that regulation. Together they regulate and coordinate the activity of nearly all other body structures. When these systems fail to function properly, homeostasis is disturbed. Failure of some component of the endocrine system to function can result in conditions such as diabetes mellitus or Addison disease. Early in the 1900s, there were no effective treatments for most endocrine diseases, and patients often died. But now we know more about the endocrine system, and treatments for many endocrine diseases are available.

The regulatory functions of the nervous and endocrine systems are similar in some respects, but they differ in other important ways. The nervous system controls the activity of tissues by sending action potentials along axons, which release chemical signals at their ends, near the cells they control. The endocrine system releases chemical signals into the blood, which carries them to all parts of the body. The cells that can detect those chemical signals produce responses.

The nervous system usually acts quickly and has short-term effects, whereas the endocrine system usually responds more slowly and has longer-lasting effects. In general, each nervous stimulus controls a specific tissue or organ, whereas each endocrine stimulus controls several tissues or organs.

## FUNCTIONS OF THE ENDOCRINE SYSTEM

The main regulatory functions of the endocrine system are:

1. *Growth, metabolism, and tissue maturation.* The endocrine system regulates the growth of many tissues, such as bone and muscle, and the rate of metabolism of many tissues, which helps maintain a normal body temperature and normal mental functions. Maturation of tissues, which results in the development of adult features and adult behavior, is also influenced by the endocrine system.
2. *Blood glucose.* The endocrine system regulates the levels of blood glucose and other nutrients in the blood.
3. *Ion concentration.* The endocrine system regulates $Na^+$, $K^+$, and $Ca^{2+}$ concentrations in the blood.
4. *Water balance.* The endocrine system regulates water balance by controlling the solute concentration of the blood.
5. *Heart rate and blood pressure.* The endocrine system helps regulate the heart rate and blood pressure and helps prepare the body for physical activity.
6. *Immune function.* The endocrine system helps control the production and functions of immune cells.
7. *Reproductive functions.* The endocrine system controls the development and functions of the reproductive systems in males and females.
8. *Uterine contractions and milk release.* The endocrine system regulates uterine contractions during delivery of the newborn and stimulates milk release from the breasts in lactating females.

## CHEMICAL SIGNALS

The body uses many different types of chemical signals for communication. **Chemical signals,** or *ligands* (lī′gandz; to bind), are molecules released from one location that move to another location to produce a response. **Intracellular chemical signals** are produced in one part of a cell, such as the cell membrane, and then travel to another part of the same cell, where they bind to receptors in either the cytoplasm or the nucleus. **Intercellular chemical signals** are released from one cell and carried in the intercellular fluid; they bind to receptors that are found in some cells, but usually not in all cells, of the body. Intercellular chemical signals can be categorized into six types, depending on the source of the chemical signals or their targets (table 10.1).

**Autocrine** (aw′tō-krin; *autos,* self) **chemical signals** are released by cells and have a local effect on the same cell type from which they are released. Examples include chemicals such as the eicosanoids (*eicosa,* refers to 20-carbon atoms) that are released from smooth muscle cells and from platelets in response to inflammation. These chemicals cause the relaxation of blood vessel smooth muscle cells and the aggregation of platelets. As a result, the blood vessels dilate, and blood clots.

**Paracrine** (par′ă-krin; *para,* alongside) **chemical signals** are released by cells that produce effects on other cell types near the cells from which they are released, without being transported in blood. For example, cells in the pancreas release a peptide called somatostatin, which functions locally to inhibit the secretion of insulin from other cells of the pancreas.

**Hormones** (hōr′mōnz) and **neurohormones** are intercellular chemical signals secreted into the circulatory system (see "Hormones" later in this chapter). They are carried in the blood to the organs they control, where they bind to receptor molecules and produce a response.

**Neurotransmitters** and **neuromodulators** are intercellular chemical signals, secreted by nerve cells, that play important roles in nervous system function (see chapter 8).

**Pheromones** (fer′ō-mōnz) are chemical signals secreted into the environment that modify the behavior and physiology of other individuals of the same species. For example, pheromones released in the urine of cats and dogs at certain times are olfactory signals that indicate fertility. Humans also appear to produce pheromones. For example, pheromones produced by women can influence the timing of the menstrual cycles of other women. More research is needed concerning the role of pheromones in humans.

Endocrine

**TABLE 10.1     Functional Classification of Intercellular Chemical Signals**

| Intercellular Chemical Signal | Description | Example | |
|---|---|---|---|
| Autocrine chemical signal | Secreted by cells in a local area; influences the activity of the same cell type from which it was secreted | Eicosanoids (prostaglandins, thromboxanes, prostacylins, leukotrienes) | Autocrine chemical signal |
| Paracrine chemical signal | Produced by a wide variety of tissues and secreted into tissue spaces; usually has a localized effect on other tissues | Somatostatin histamine, eicosanoids | Paracrine chemical signal |
| Hormone | Secreted into the blood by specialized cells; travels some distance to target tissues; influences specific activities | Thyroid hormones, growth hormone, insulin, epinephrine, estrogen, progesterone, testosterone | Hormone |
| Neurohormone (often simply referred to as a hormone) | Produced by neurons and functions as a hormone | Oxytocin, antidiuretic hormone, hypothalamic-releasing and -inhibiting hormones | Neuron / Neurohormone |
| Neurotransmitter or neuromodulator | Produced by neurons and secreted into extracellular spaces by presynaptic nerve terminals; travels short distances; influences postsynaptic cells | Acetylcholine, epinephrine | Neurotransmitter / Neuron |
| Pheromone | Secreted into the environment; modifies physiology and behavior of other individuals of the same species | Released by humans and many other animals; released in the urine of animals, such as dogs and cats; pheromones produced by women can influence the timing of the menstrual cycles of other women | Pheromone |

Endocrine

Many intercellular chemical signals consistently fit one specific definition, although others do not. For example, norepinephrine functions as both a neurotransmitter and a hormone, whereas prostaglandins function as neurotransmitters, neuromodulators, paracrine chemical signals, and autocrine chemical signals. The schemes that classify chemicals on the basis of their functions are useful, but they do not indicate that a specific molecule always acts as the same type of chemical signal. For that reason, the study of endocrinology often includes the study of autocrine and paracrine chemical signals in addition to hormones and neurohormones.

## HORMONES

The term **endocrine** (en′dō-krin) is derived from the Greek words *endo* and *krino,* meaning "within" and "to secrete." The word implies that intercellular chemical signals are produced within and secreted from **endocrine glands.** The chemical signals are then transported in the blood to tissues some distance from the glands, where they produce their effects. Examples of endocrine glands are the thyroid gland and the adrenal glands. In contrast, **exocrine** (ek′sō-krin; *exō,* outside) **glands** secrete their products into ducts, which exit the glands and carry the secretory products to an external or internal surface, such as the skin or digestive tract. Examples of exocrine glands are the sweat glands and the salivary glands.

The **endocrine system** consists of the endocrine glands of the body (see figure 10.7). The intercellular chemical signals secreted by endocrine glands are called hormones, a term derived from the Greek word *hormon,* meaning to "set into motion" because hormones set responses by cells into motion. Traditionally, a hormone is defined as an intercellular chemical signal, that (1) is produced in minute amounts by a collection of cells; (2) is secreted into the interstitial fluid; (3) enters the circulatory system to be transported some distance; and (4) influences specific tissues at another site in the body to act in a specific way. All hormones exhibit most of these characteristics. Neurohormones, which are secreted from cells of the nervous system, are commonly referred to as hormones.

Hormones are distributed in the blood to all parts of the body, but only certain tissues, called **target tissues,** respond to each type of hormone. A target tissue for a hormone is made up of cells that have receptor molecules for the hormone. Each hormone can bind only to its receptor molecules and cannot influence the function of cells that do not have receptor molecules for the hormone (figure 10.1).

## Chemistry of Hormones

Hormones fall into two major chemical categories: water-soluble hormones and lipid-soluble hormones. The chemical nature of hormones is an important determinant of their interactions with receptors at target tissues.

1. *Water-soluble hormones include proteins, peptides, and amino acid derivatives.* Hormones in this category bind to membrane-bound receptors, with the exception of peptide hormones secreted by the thyroid gland, which diffuse through membranes and bind to intracellular receptors.

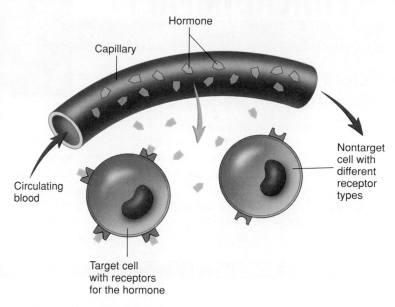

**Figure 10.1  Target Cell Response to Hormones**

Hormones are secreted into the blood and distributed throughout the body, where they diffuse from the blood into the interstitial fluid (*yellow arrow*). Only target cells have receptors to which hormones can bind; therefore, even though a hormone is distributed throughout the body, only target cells for that hormone can respond to it.

 a. Some hormones are proteins, consisting of many amino acids bound together by peptide bonds. Carbohydrate molecules are bound to some of the protein hormones. Most hormones of the anterior pituitary gland, including those that control growth, metabolism, and reproductive functions, are protein hormones.

 b. Peptide hormones consist of short chains of amino acids. The hormones of the posterior pituitary gland that control functions such as milk letdown in the breast and urine volume in the kidneys are peptide hormones.

 c. Some hormones consist of single amino acids that have been chemically modified. Examples include hormones secreted by the adrenal medulla, which help a person prepare for physical activity or respond to emergency conditions.

2. *Lipid-soluble hormones include steroids, thyroid hormones, and eicosanoids.*

 a. Steroids are lipids, which are derived from cholesterol. Steroids have a structure that varies only slightly among the different types. The small differences, however, give each type of steroid unique functional characteristics. Steroid hormones are produced mostly by the adrenal cortex and the gonads (testes and ovaries). Hormones in this category diffuse across the cell membrane and bind to intracellular receptor molecules.

 b. Thyroid hormones are amino acid derivatives, produced by the thyroid gland, that are lipid-soluble. Because these hormones are lipid-soluble, they can diffuse through the cell membrane and bind to intracellular receptor molecules.

# CLINICAL IMPACT

## Lipid- and Water-Soluble Hormones in Medicine

Specific hormones are administered to treat certain illnesses. Hormones that are soluble in lipids, such as steroids, may be taken orally because they can diffuse across the wall of the stomach and intestine into the circulatory system. Examples include the synthetic estrogen and progesterone-like hormones in birth control pills and steroids that reduce the severity of inflammation, such as prednisone (pred′ni-sōn). In contrast to lipid-soluble hormones, protein hormones cannot diffuse across the wall of the intestine. Furthermore, protein hormones are broken down to individual amino acids before they are transported across the wall of the digestive system. The normal structure of a protein hormone is therefore destroyed, and its physiological activity is lost. Consequently, protein hormones must be injected rather than taken orally. The most commonly administered protein hormone is insulin, which is prescribed to treat diabetes mellitus.

---

c. The **eicosanoids** (ī′kō-să-noydz) make up a class of chemicals derived from the fatty acid arachidonic acid. The eicosanoids include the prostaglandins, thromboxanes, prostacyclins, and leukotrienes. A major group of eicosanoids is composed of the **prostaglandins** (pros′tă-glan′dinz), which are named after the prostate gland and other accessory glands, where they were first found. Prostaglandins are produced by many tissues and generally have local effects. They play important roles in regulating smooth muscle contractions and inflammation. They bind to membrane-bound receptors that are associated with G proteins to produce a response.

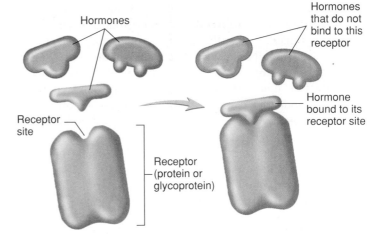

**Figure 10.2** **Receptors and Specificity of Receptor Sites**
Hormones bind to receptor molecules. The shape and chemical characteristics of each receptor site allow certain hormones that have a compatible shape and compatible chemical characteristics to bind to it, not others. This relationship is called specificity.

## RECEPTORS

Hormones bind to proteins or glycoproteins, called **receptor molecules,** to produce a response. The portion of each receptor molecule where a hormone binds is a **receptor site.** The shape and other characteristics of each receptor site allow only a specific hormone to bind to it (figure 10.2). The tendency for each receptor site to bind to a specific hormone and not to others is called **specificity.**

## Receptor Types for Hormones

Two major types of receptor molecules respond to hormones:

1. **Membrane-bound receptors** extend through the cell membrane, with their receptor sites on its outer surface (figure 10.3a). These receptors respond to hormones composed of large, water-soluble molecules that do not diffuse through the cell membrane. When a hormone binds to the receptor site of a membrane-bound receptor, the part of the receptor that extends to the inside of the cell produces a response.

2. **Intracellular receptors** are located in either the cytoplasm or the nucleus of the cell. These receptors respond to hormones that are lipid-soluble and can diffuse through the cell membrane and into the cytoplasm of the cell. They then bind to receptor sites on intracellular receptors (figure 10.3b) in either the cytoplasm or the nucleus of the cell.

## Receptor Responses

Once hormones bind to their receptors, the combinations produce responses.

### Membrane-Bound Receptor Responses

There are many membrane-bound receptors. When hormones bind to their receptors, they can produce responses by two major mechanisms:

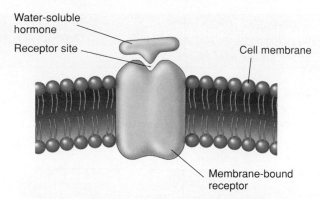

Water-soluble hormone

Receptor site

Cell membrane

Membrane-bound receptor

(a) Hormones that are large and water-soluble interact with membrane-bound receptors that extend across the cell membrane and have receptor sites exposed to the outer surface of the cell membrane. The portion of the receptor inside the cell produces the response.

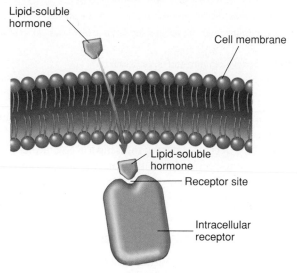

Lipid-soluble hormone

Cell membrane

Lipid-soluble hormone

Receptor site

Intracellular receptor

(b) Small, lipid-soluble hormones diffuse through the cell membrane and combine with the receptor sites of intracellular receptors. The combination of hormones and receptors produces a response.

**Figure 10.3** Membrane-Bound and Intracellular Receptors

1. *Receptors that activate G proteins.* Some hormones bind to membrane-bound receptors and activate a complex of proteins at the inner surface of the cell membrane called G proteins (figure 10.4 and table 10.2a). The inactive G protein complex has alpha ($\alpha$), beta ($\beta$), and gamma ($\gamma$) subunits bound together. Guanosine diphosphate (GDP) is bound to the $\alpha$ subunit. The $\alpha$, $\beta$, and $\gamma$ subunit complex can bind to the intercellular portion of the receptor molecule. When a hormone binds to the membrane-bound receptor, the $\alpha$ subunit separates from the $\beta$ and $\gamma$ subunits. Guanosine triphosphate (GTP) then replaces guanosine diphosphate (GDP) on the $\alpha$ subunit. This is the activated form of the G protein complex. The $\alpha$ subunit with GTP bound to it can (1) open or close membrane channels to produce a response, (2) activate enzymes that produce hormones or (3) affect gene expression.

   Several types of intracellular chemical signals, called **second messengers,** are produced in response to hormonal (the first messenger) activation of G proteins. The type depends on the receptor molecule, the specific function of the G proteins, and the specific enzyme activated in the target tissue. One example of a second messenger is **cyclic adenosine monophosphate (cyclic AMP, or cAMP).** The steps involved in the production of cAMP are shown in figure 10.4. Other second messengers include **diacylglycerol** (dī′as-il-glis′er-ol) (**DAG**) and **inosotol** (in-ō′si-tō1) **triphosphate (IP3).** Second messengers interact with regulatory proteins in the cell and alter their activity to produce responses.

2. *Receptors that directly alter the activity of enzymes.* Some hormones bind to receptor sites and directly increase or decrease the activity of enzymes (figure 10.5 and table 10.2b). In some cases, the enzyme adds phosphate groups to certain proteins inside the cells. The proteins with phosphates attached to them produce the cell's response to the hormone.

1 After a water-soluble hormone binds to a membrane-bound receptor, the G protein is activated.

2 The activated $\alpha$ subunit, with GTP bound to it, binds to and activates an adenylate cyclase enzyme so that it converts ATP to cAMP.

3 The cAMP can activate protein kinase enzymes, which phosphorylate specific enzymes.

4 Phosphodiesterase enzymes inactivate cAMP by converting cAMP to AMP.

Water-soluble hormone bound to membrane-bound receptor.

$\alpha$ subunit of G protein with GTP bound to it

GTP

Adenylate cyclase catalyzes the formation of cAMP.

ATP

cAMP

Phosphodiesterase inactivates cAMP.

cAMP is an intracellular mediator that activates protein kinases.

Protein kinase

AMP (inactive)

**PROCESS Figure 10.4** Membrane-Bound Receptors That Activate G Proteins

| TABLE 10.2 | Some Major Hormones and Their Sources, Target Tissues, and Effects | |
| --- | --- | --- |
| **Hormone** | **Source** | **Target Tissue and Effect** |
| **a. Hormones That Bind to Membrane-Bound Receptors and Activate G Proteins** | | |
| Luteinizing hormone | Anterior pituitary | Ovaries, estrogen secretion and ovulation; testes, testosterone secretion |
| Thyroid-stimulating hormone | Anterior pituitary | Thyroid gland; secretion of thyroid hormones |
| Oxytocin | Posterior pituitary | Uterus contraction and expulsion of milk from mammary glands |
| Vasopressin or antidiuretic hormone | Posterior pituitary | Kidney; water conservation |
| **b. Hormones That Bind to Membrane-Bound Receptors and Alter the Activity of Enzymes** | | |
| Insulin | Pancreatic islets | Most cells; increases glucose and amino acid uptake |
| Growth hormone | Anterior pituitary gland | Most cells; increases gene expression and resists protein breakdown |
| **c. Hormones That Combine with Intracellular Receptors** | | |
| Testosterone | Testis | Responsible for development of the reproductive structures and male secondary sexual characteristics |
| Estrogen | Ovary | Causes increased cell division in the lining of the uterus |
| Cortisol | Adrenal cortex | Increases breakdown of proteins and fats and increases blood levels of glucose |
| Triiodothyronine ($T_3$) | Thyroid gland | Regulates development and metabolism |

## Intracellular Receptor Responses

Hormones that diffuse across cell membranes and bind to intracellular receptors are relatively small and are soluble in lipids. Some intracellular receptors are enzymes. Hormones bind to these kinds of receptors and change the activity of the enzymes. Other intracellular receptors bind to DNA in the nucleus. Hormones bind to these kinds of receptors, which then bind to DNA in the nucleus and increase or decrease specific messenger RNA in the nucleus of the cell. The messenger RNA then moves to the ribosomes in the cytoplasm, where new proteins are produced (figure 10.6, table 10.2c; see chapter 3).

## Comparison of Receptor Responses

There are some important differences in the responses produced by membrane-bound and intracellular receptors. Because of the time required for the synthesis of new mRNA and proteins, several hours can pass between the time the hormones bind to their intracellular receptors and the response. In contrast, hormones that bind to membrane-bound receptors often produce rapid responses. A few hormones can bind to their membrane-bound receptors, and each activated receptor can produce many second messengers. The second messengers, in turn, rapidly activate many specific enzymes inside the cell. This pattern of response is called a **cascade effect** because a few hormones can produce many second messengers, and the second messengers can, in turn, quickly activate many proteins inside the cell.

### Predict 1

*A drug binds to a receptor and prevents target tissue from responding to a hormone. It is known that the drug is lipid-soluble and that it prevents the synthesis of messenger RNA. Explain how the hormone produces a response in its target tissue.*

# REGULATION OF HORMONE SECRETION

The secretion of hormones is controlled by negative-feedback mechanisms (see chapter 1). Negative-feedback mechanisms

① The membrane-bound receptor is associated with an enzyme at the inner surface of the cell membrane. When the receptor site at the outside of the cell membrane is not occupied by a hormone, the enzyme remains inactive.

② The hormone binds to its receptor site on the outside of the cell membrane and causes the enzyme at the inner surface of the cell membrane to become active.

**PROCESS Figure 10.5** Membrane-Bound Receptors That Control Enzyme Activity

1. The lipid-soluble hormone diffuses through the cell membrane and enters the cell's cytoplasm.

2. The hormone combines with a receptor in the cytoplasm (or in the nucleus).

3. The receptor with the hormone bound to it interacts with DNA and increases or decreases the synthesis of specific messenger RNA (mRNA) molecules.

4. The mRNA molecule passes from the nucleus to the cytoplasm.

5. In the cell's cytoplasm, the mRNA molecule combines with ribosomes, and new protein molecules are synthesized.

6. The new proteins produce the cell's response to the hormone.

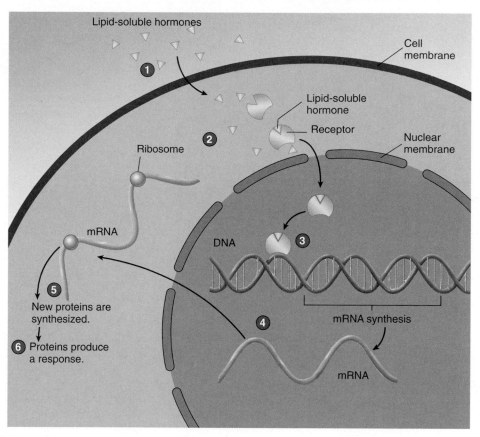

**PROCESS Figure 10.6** Lipid-Soluble Hormones That Diffuse into Cells and Bind to Intracellular Receptors

keep the body functioning within a narrow range of values consistent with life. For example, insulin is a hormone that regulates the concentration of blood glucose, or blood sugar. When blood glucose levels increase after a meal, insulin is secreted. Insulin acts on several target tissues and causes them to take up glucose, causing blood glucose levels to decline. However, as blood glucose levels begin to fall, the rate at which insulin is secreted falls also. As insulin levels fall, the rate at which glucose is taken up by the tissues decreases, keeping the blood glucose levels from declining too much. This negative-feedback mechanism counteracts increases and decreases in blood glucose levels to maintain homeostasis.

Hormone secretion is regulated in three ways. The secretion of some hormones is regulated by one of these methods, whereas the secretion of other hormones can be regulated by two or even all of these methods:

1. *Blood composition.* The secretion of some hormones is directly controlled by the blood levels of certain chemicals. For example, blood glucose levels control insulin secretion, and blood calcium levels control the secretion of parathyroid hormone.
2. *Hormonal control.* The secretion of some hormones is controlled by other hormones. For example, hormones from the pituitary gland act on the ovaries and the testes, causing those organs to secrete sex hormones.
3. *Neural control.* The secretion of some hormones is controlled by the nervous system. An example is epinephrine, which is released from the adrenal medulla as a result of nervous system stimulation.

# ENDOCRINE GLANDS AND THEIR HORMONES

The endocrine system consists of ductless glands that secrete hormones into the interstitial fluid (figure 10.7). The hormones then enter the blood (table 10.3; see figure 10.1). For example, the organs in the body with the richest blood supply are endocrine glands, such as the adrenal gland and the thyroid gland.

Some glands of the endocrine system perform functions in addition to hormone secretion. For example, the endocrine part of the pancreas has cells that secrete hormones, whereas the much larger exocrine portion of the pancreas secretes digestive enzymes. Portions of the ovaries and testes secrete hormones, but other parts of the ovaries and testes produce oocytes (female sex cells) or sperm cells (male sex cells), respectively.

## Pituitary and Hypothalamus

The **pituitary** (pi-too′i-tār-rē; *pituita,* phlegm or thick mucous secretion) **gland** is also called the *hypophysis* (hī-pof′i-sis; *hypo,*

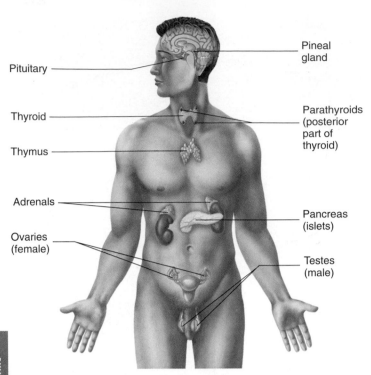

Pituitary

Thyroid

Thymus

Adrenals

Ovaries
(female)

Pineal
gland

Parathyroids
(posterior
part of
thyroid)

Pancreas
(islets)

Testes
(male)

**Figure 10.7** **Major Endocrine Glands and Their Locations**

under + *physis*, growth). It is a small gland about the size of a pea (figure 10.8, *top*). It rests in a depression of the sphenoid bone inferior to the hypothalamus of the brain. The **hypothalamus** (hī′pō-thal′ă-mŭs; *hypo*, under + *thalamos*) is an important autonomic nervous system and endocrine control center of the brain located inferior to the thalamus. The pituitary gland lies posterior to the optic chiasm and is connected to the hypothalamus by a stalk called the **infundibulum** (in-fŭn-dib′ŭ-lŭm; a funnel). The pituitary gland is divided into two parts: The **anterior pituitary** is made up of epithelial cells derived from the embryonic oral cavity; the **posterior pituitary** is an extension of the brain and is composed of nerve cells. The hormones secreted from each lobe of the pituitary gland are listed in table 10.3.

Hormones from the pituitary gland control the functions of many other glands in the body, such as the ovaries, the testes, the thyroid gland, and the adrenal cortex (figure 10.8, *bottom*). The pituitary gland also secretes hormones that influence growth, kidney function, birth, and milk production by the mammary glands. Historically, the pituitary gland was known as the body's *master gland* because it controls the function of so many other glands. However, we now know that the hypothalamus controls the pituitary gland through two ways: hormonal control and direct innervation.

| TABLE 10.3 | Endocrine Glands, Hormones, and Their Target Tissues | | |
|---|---|---|---|
| **Gland** | **Hormone** | **Target Tissue** | **Response** |
| Pituitary gland | | | |
| Anterior | Growth hormone | Most tissues | Increases gene expression, breakdown of lipids, and release of fatty acids from cells; increases blood glucose levels |
| | Thyroid-stimulating hormone (TSH) | Thyroid gland | Increases thyroid hormone secretion (thyroxine and triiodothyronine) |
| | Adrenocorticotropic hormone (ACTH) | Adrenal cortex | Increases secretion of glucocorticoid hormones, such as cortisol; increases skin pigmentation at high concentrations |
| | Melanocyte-stimulating hormone (MSH) | Melanocytes in skin | Increases melanin production in melanocytes to make skin darker in color |
| | Luteinizing hormone (LH) or interstitial cell-stimulating hormone (ICSH) | Ovary in females, testis in males | Promotes ovulation and progesterone production in ovary; promotes testosterone synthesis and support for sperm cell production in testis |
| | Follicle-stimulating hormone (FSH) | Follicles in ovary in females, seminiferous tubules in males | Promotes follicle maturation and estrogen secretion in ovary; promotes sperm cell production in testis |
| | Prolactin | Ovary and mammary gland in females, testis in males | Stimulates milk production and prolongs progesterone secretion following ovulation and during pregnancy in women; increases sensitivity to LH in males |
| Posterior | Antidiuretic hormone (ADH) | Kidney | Conserves water |
| | Oxytocin | Uterus | Increases uterine contractions |
| | | Mammary gland | Increases milk letdown from mammary glands |

*continued*

Endocrine

| TABLE 10.3 | *continued* | | |
|---|---|---|---|
| **Gland** | **Hormone** | **Target Tissue** | **Response** |
| Thyroid gland | Thyroid hormones (thyroxine, triiodothyronine) | Most cells of the body | Increase metabolic rates, essential for normal process of growth and maturation |
| | Calcitonin | Primarily bone | Decreases rate of bone breakdown; prevents large increase in blood $Ca^{2+}$ levels following a meal |
| Parathyroid glands | Parathyroid hormone | Bone, kidney | Increases rate of bone breakdown by osteoclasts; increases vitamin D synthesis, essential for maintenance of normal blood calcium levels |
| Adrenal medulla | Epinephrine mostly, some norepinephrine | Heart, blood vessels, liver, fat cells | Increases cardiac output; increases blood flow to skeletal muscles and heart; increases release of glucose and fatty acids into blood; in general, prepares body for physical activity |
| Adrenal cortex | Mineralocorticoids (aldosterone) | Kidneys; to lesser degree, intestine and sweat glands | Increase rate of sodium transport into body; increase rate of potassium excretion; secondarily favor water retention |
| | Glucocorticoids (cortisol) | Most tissues (e.g., liver, fat, skeletal muscle, immune tissues) | Increase fat and protein breakdown; increase glucose synthesis from amino acids; increase blood nutrient levels; inhibit inflammation and immune response |
| | Adrenal androgens | Most tissues | Insignificant in males; increase female sexual drive, growth of pubic and axillary hair |
| Pancreas | Insulin | Especially liver, skeletal muscle, adipose tissue | Increases uptake and use of glucose and amino acids |
| | Glucagon | Primarily liver | Increases breakdown of glycogen and release of glucose into the circulatory system |
| Reproductive organs Testes | Testosterone | Most tissues | Aids in sperm cell production, maintenance of functional reproductive organs, secondary sexual characteristics, sexual behavior |
| Ovaries | Estrogens, progesterone | Most tissues | Aid in uterine and mammary gland development and function, external genitalia structure, secondary sexual characteristics, sexual behavior, menstrual cycle |
| Uterus, ovaries, inflamed tissues | Prostaglandins | Most tissues | Mediate inflammatory responses; increase uterine contractions and ovulation |
| Thymus | Thymosin | Immune tissues | Promotes immune system development and function |
| Pineal gland | Melatonin | At the least, hypothalamus | Inhibits secretion of gonadotropin-releasing hormone, thereby inhibiting reproduction |

Endocrine

## Hormonal Control of the Anterior Pituitary

Neurons of the hypothalamus produce and secrete neurohormones that act on cells of the anterior pituitary gland (figure 10.9). They act as either releasing hormones or inhibiting hormones. Each releasing hormone stimulates the production and secretion of a specific hormone by the anterior pituitary, whereas each inhibiting hormone decreases the secretion of a specific anterior pituitary hormone. Releasing and inhibiting hormones enter a capillary bed in the hypothalamus and are transported through veins to a second capillary bed in the anterior pituitary. There they leave the blood and bind to membrane-bound receptors involved with regulating anterior pituitary hormone secretion. The capillary beds and veins that transport the releasing and inhibiting hormones are called the **hypothalamic-pituitary portal system.**

## Direct Innervation of the Posterior Pituitary

Stimulation of neurons within the hypothalamus controls the secretion of hormones from the posterior pituitary (figure 10.10). The cell bodies of these neurons are in the hypothalamus, and their axons extend through the infundibulum to the posterior pituitary. Hormones are produced in the nerve cell bodies and transported through the axons to the posterior pituitary, where they are stored in the axon endings. When these nerve cells are stimulated, action potentials from the hypothalamus travel along the axons to the posterior pituitary and cause the release of hormones from the axon endings.

Within the hypothalamus and pituitary, the nervous and endocrine systems are closely interrelated. Emotions such as joy and anger, as well as chronic stress, influence the endocrine system through the hypothalamus. Conversely, hormones of the endocrine

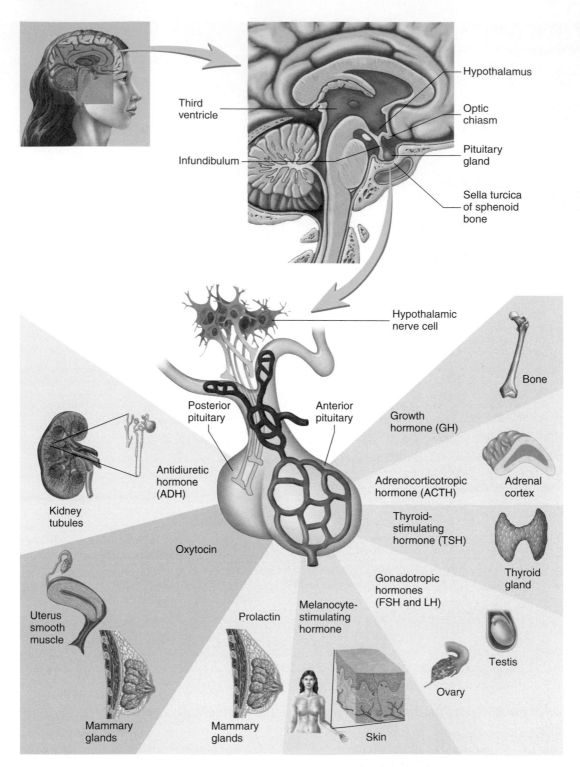

**Figure 10.8** Pituitary Gland, Its Hormones, and Their Target Tissues

(*Top*) The pea-sized pituitary gland lies inferior to the hypothalamus of the brain. The infundibulum connects the hypothalamus and the pituitary gland. (*Bottom*) Hormones secreted by the posterior pituitary are indicated by *blue shading*, and hormones secreted by the anterior pituitary by *peach shading*.

system can influence the functions of the hypothalamus and other parts of the brain.

### Hormones of the Anterior Pituitary

**Growth hormone (GH)** stimulates the growth of bones, muscles, and other organs by increasing gene expression. It also resists protein breakdown during periods of food deprivation and favors fat breakdown. Too little growth hormone secretion can result from abnormal development of the pituitary gland. A young person suffering from a deficiency of growth hormone remains small, although normally proportioned, and is called a **pituitary dwarf** (dwŏrf). This condition can be treated by administering growth hormone.

Stimuli from the
nervous system

Stimulatory

Inhibitory

Hypothalamic
nerve cells
secrete
releasing
and inhibiting
hormones.

**1** Stimuli within the nervous system
cause releasing and inhibiting hormones
(*blue circles*) to be secreted from nerve
cells of the hypothalamus.

**2** Releasing and inhibiting hormones pass
through the hypothalamic-pituitary portal
system to the anterior pituitary.

**Hypothalamic-
pituitary
portal system**

Artery

**3** Releasing and inhibiting hormones leave
capillaries, bind to membrane-bound receptors,
and influence the secretion of hormones
from anterior pituitary cells.

Releasing
and inhibiting
hormones
stimulate
or inhibit
pituitary
hormone
secretions.

Anterior
pituitary
endocrine
cell

**4** Anterior pituitary hormones (*yellow squares*)
are carried in the blood to their target
tissues (*green arrow*), which in some
cases are other endocrine glands.

Posterior
pituitary

Vein

Target tissue
or another endocrine
gland

Endocrine

**PROCESS Figure 10.9** **Hypothalamus and Anterior Pituitary**

Because GH is a protein, it is difficult to produce artificially using conventional techniques. However, human genes for GH have been successfully introduced into bacteria using genetic engineering techniques. The gene in the bacteria causes GH synthesis, and the GH can be extracted from the medium in which the bacteria are grown. Thus, modern genetic engineering has provided a source of human GH for people who produce inadequate quantities.

Excess growth hormone secretion can result from hormone-secreting tumors of the pituitary gland. If excess growth hormone is present before bones finish growing in length, exaggerated bone growth occurs. The person becomes abnormally tall, a condition called **giantism** (jī'an-tizm). If excess hormone is secreted after growth in bone length is complete, growth continues in bone diameter only. As a result, the facial features and hands become abnormally large, a condition called **acromegaly** (ak-rō-meg'ă-lē).

The secretion of growth hormone is controlled by two hormones from the hypothalamus. A releasing hormone stimulates growth hormone secretion, and an inhibiting hormone inhibits its secretion. Most people have a rhythm of growth hormone secretion, with daily peak levels occurring during deep sleep. Growth hormone secretion also increases during periods of fasting and exercise. Blood growth hormone levels do not become greatly elevated during periods of rapid growth, although children tend to have somewhat higher blood levels of growth hormone than do adults. In addition to growth hormone, genetics, nutrition, and sex hormones influence growth.

Part of the effect of growth hormone is influenced by a group of protein hormones called **insulin-like growth factors (IGFs),** or *somatomedins* (sō'mă-tō-mē'dinz). Growth hormone increases IGF secretion from tissues such as the liver, and the IGF molecules bind to receptors on the cells of tissues such as bone and cartilage, where they stimulate growth. The IGFs are similar in structure to insulin and can bind, to some degree, to insulin receptors. Also, insulin, at high concentrations, can bind to IGF receptors.

*Predict 2*

*Mr. Hoops has a son who wants to be a basketball player. Mr. Hoops knows something about GH. He asked his son's doctor if he would prescribe some GH for his son, so that he can grow taller. What do you think the doctor tells Mr. Hoops?*

**Thyroid-stimulating hormone(TSH)** binds to membrane-bound receptors on cells of the thyroid gland and causes the cells to secrete thyroid hormone. When too much TSH is secreted, the thyroid gland enlarges and secretes too much thyroid hormone. When too little TSH is secreted, the thyroid gland decreases in size and secretes too little thyroid hormone. The rate of TSH secretion is increased by a releasing hormone from the hypothalamus.

**Adrenocorticotropic** (a-drē'nō-kōr'ti-kō-trō'pik) **hormone (ACTH)** binds to membrane-bound receptors on cells in the cortex of the adrenal glands. ACTH increases the secretion of a hormone from the adrenal cortex called **cortisol** (kōr'ti-sol), also called

Stimulatory

Inhibitory

**1** Stimuli within the nervous system stimulate hypothalamic nerve cells to produce action potentials.

**2** Action potentials are carried by axons of nerve cells to the posterior pituitary. The axons of nerve cells store hormones in the posterior pituitary.

**3** In the posterior pituitary gland, action potentials cause the release of hormones (*red circles*) from the axons into the circulatory system.

**4** The hormones pass through the circulatory system and influence the activity of their target tissues (*green arrow*).

Stimuli from the nervous system

Hypothalamic nerve cells

Optic chiasm

Posterior pituitary

Hormone

Anterior pituitary

Vein

Target tissue

**PROCESS Figure 10.10**  Hypothalamus and Posterior Pituitary

hydrocortisone. ACTH is required to keep the adrenal cortex from degenerating. ACTH molecules also bind to melanocytes in the skin and increase skin pigmentation (see chapter 5). One symptom of too much ACTH secretion is darkening of the skin. The rate of ACTH secretion is increased by a releasing hormone from the hypothalamus.

**Gonadotropins** (gō′nad-ō-trō′pinz) are hormones that bind to membrane-bound receptors on the cells of the gonads (ovaries and testes). They regulate the growth, development, and functions of the gonads. In females, **luteinizing** (loo′tē-ĭ-nīz-ing) **hormone (LH)** causes the ovulation of oocytes and the secretion of the sex hormones estrogen and progesterone from the ovaries. In males, LH stimulates interstitial cells of the testes to secrete the sex hormone testosterone and thus is sometimes referred to as **interstitial** (in-ter-stish′ăl) **cell–stimulating hormone (ICSH). Follicle-stimulating hormone (FSH)** stimulates the development of follicles in the ovaries and sperm cells in the testes. Without LH and FSH, the ovaries and testes decrease in size, no longer produce oocytes or sperm cells, and no longer secrete hormones. A single releasing hormone from the hypothalamus increases the secretion of both LH and FSH.

**Prolactin** (prō-lak′tin; *pro*, precursor + *lact*, milk) binds to membrane-bound receptors in cells of the breast, where it helps promote development of the breast during pregnancy and stimulates the production of milk following pregnancy. The regulation of prolactin secretion is complex and may involve several substances

released from the hypothalamus. There are two main regulatory hormones; one increases prolactin secretion and one decreases it.

**Melanocyte-stimulating** (mel′ă-nō-sīt) **hormone (MSH)** binds to membrane-bound receptors on melanocytes and causes them to synthesize melanin. The structure of MSH is similar to that of ACTH, and oversecretion of either hormone causes the skin to darken. Regulation of MSH is not well understood, but there appear to be two regulatory hormones from the hypothalamus— one that increases MSH secretion and one that decreases it.

## A CASE IN POINT

### Hypopituitarism

Shortly after Howie R. Mone was delivered, his pediatricians discovered that he had low blood levels of several hormones, including growth hormone, thyroid hormones, and hormones from the adrenal cortex. This condition was due to abnormal development of the anterior pituitary gland. Fortunately, essential hormones can be administered to Howie to either replace the missing anterior pituitary hormones or replace the hormones that are normally secreted in response to the missing anterior pituitary hormones. For example, GH can substitute for the GH from the anterior pituitary gland, which regulates growth. Thyroid hormones can replace those normally produced by the thyroid gland. Cortisol and aldosterone (or related substances) can replace cortisol and aldosterone produced by the adrenal cortex. Howie's thyroid gland and adrenal glands do not secrete hormones because the anterior pituitary does not secrete TSH or ACTH. Reproductive hormones are not needed until Howie reaches puberty. Without GH, growth would be dramatically reduced. Without thyroid hormones, metabolism, growth, and mental development would not be normal. Without cortisol and aldosterone, Howie could not survive because these hormones are essential for the maintenance of blood pressure.

### Hormones of the Posterior Pituitary

**Antidiuretic** (an'tē-dī-ŭ-ret'ik; *anti,* against + *uresis,* urine volume) **hormone (ADH)** binds to membrane-bound receptors and increases water reabsorption by kidney tubules. This results in less water lost as urine. ADH can also cause blood vessels to constrict when released in large amounts. Consequently, it is sometimes also called *vasopressin* (vā-sō-pres'in). Reduced ADH release from the posterior pituitary results in large amounts of dilute urine.

A lack of ADH secretion causes diabetes insipidus, which is the production of a large amount of dilute urine. The consequences of diabetes insipidus are not obvious until the condition becomes severe, producing many liters of urine each day. The large urine volume causes an increase in the concentration of the body fluids and the loss of important electrolytes, such as $Ca^{2+}$, $Na^+$, and $K^+$.

**Oxytocin** (ok'sĭ-tō'sin; swift birth) binds to membrane-bound receptors, and causes contraction of the smooth muscle cells of the uterus as well as milk letdown from the breasts in lactating women. Commercial preparations of oxytocin, known as Pitocin, are given under certain conditions to assist in childbirth and to constrict uterine blood vessels afterwards.

### A CASE IN POINT
#### Syndrome of Inappropriate ADH secretion

Willy Leek was recovering from an emergency appendectomy. Three days after the surgery, he developed a reduced sense of taste, poor appetite, fatigue, weakness, abdominal cramps, and vomiting. Willy's kidneys were retaining a large volume of water. His urine volume was low, his urine was very concentrated, and his blood $Na^+$ levels were reduced (hyponatremia). This condition sometimes occurs following surgery as the result of abnormally increased ADH secretion. In Willy's case, many of the symptoms were caused by the effect of hyponatremia on the nervous system. As is typical, the condition was temporary and lasted only a few days, during which Willy's water intake was restricted and a hypertonic saline solution was administered to keep his blood $Na^+$ levels within the normal range.

## Thyroid Gland

The **thyroid** (thī'royd; shield-shaped) **gland** is made up of two lobes connected by a narrow band called the **isthmus** (is'mŭs; a constriction). The lobes are located on each side of the trachea, just inferior to the larynx (figure 10.11*a,b*). The thyroid gland is one of the largest endocrine glands. It appears more red than the surrounding tissues because it is highly vascular. It is surrounded by a connective tissue capsule. The main function of the thyroid gland is to secrete **thyroid hormones,** which bind to intracellular receptors in cells and regulate

the rate of metabolism in the body (table 10.3). Thyroid hormones are synthesized and stored within the gland in numerous **thyroid follicles,** which are small spheres with walls composed of simple cuboidal epithelium (figure 10.11*c,d*). Each thyroid follicle is filled with the protein thyroglobulin (thī-rō-glob'ū-lin), to which thyroid hormones are attached. Between the follicles is a network of loose connective tissue that contains capillaries and scattered **C cells,** or *parafollicular cells,* which secrete the hormone calcitonin.

Without a normal rate of thyroid hormone secretion, growth and development cannot proceed normally. A lack of thyroid hormones is called **hypothyroidism** (hī'pō-thī'royd-izm). In infants, hypothyroidism can result in **cretinism** (krē'tin-izm), characterized by mental retardation, short stature, and abnormally formed skeletal structures. In adults, the lack of thyroid hormones results in a decreased metabolic rate, sluggishness, a reduced ability to perform routine tasks, and **myxedema** (mik-se-dē'mă), which is the accumulation of fluid and other molecules in the subcutaneous tissue. An elevated rate of thyroid hormone secretion, known as **hyperthyroidism** (hī-per-thī'royd-izm), causes an increased metabolic rate, extreme nervousness, and chronic fatigue. **Graves disease** is a type of hyperthyroidism that results when the immune system produces abnormal proteins that are similar in structure and function to TSH. Graves disease is often accompanied by bulging of the eyes, a condition called **exophthalmia** (ek-sof-thal'mē-ă) (see Systems Pathology, "Graves Disease").

The thyroid gland requires iodine to synthesize thyroid hormones. Iodine is taken up by the thyroid follicles. One thyroid hormone, called **thyroxine** (thī-rok'sin) or *tetraiodothyronine* (tet'ră-ī'ō-dō-thī'rō-nēn), contains four iodine atoms and is abbreviated $T_4$. The other thyroid hormone, **triiodothyronine** (trī-ī'ō-dō-thī'rō-nēn), contains three iodine atoms and is abbreviated $T_3$. If insufficient iodine is present, production and secretion of the thyroid hormones decrease.

A lack of iodine in the diet results in reduced $T_3$ and $T_4$ synthesis. A deficiency of iodine is not as common in the United States as it once was. Table salt with iodine added to it (iodized salt) is available in grocery stores, and vegetables grown in soil rich in iodine can be shipped to most places.

Thyroid hormone secretion is regulated by hormones from the hypothalamus and pituitary. The hypothalamus secretes TSH-releasing hormone, which travels to the anterior pituitary and stimulates the secretion of thyroid-stimulating hormone (TSH) (figure 10.12). In turn, TSH stimulates the secretion of thyroid hormones from the thyroid gland. Small fluctuations in blood TSH levels occur on a daily basis, with a small increase at night. Increasing blood levels of TSH increase the synthesis and release of thyroid hormones from thyroglobulin. Decreasing blood levels of TSH decrease the synthesis and release of thyroid hormones.

The thyroid hormones have a negative-feedback effect on the hypothalamus and pituitary, so that increasing levels of thyroid hormones inhibit the secretion of TSH-releasing hormone from the hypothalamus and inhibit TSH secretion from the anterior pituitary gland. Decreasing thyroid hormone levels allow additional TSH releasing hormone and TSH to be secreted. Because of the negative-feedback effect, the thyroid hormones fluctuate within a narrow concentration range in the blood. However, a loss of negative feedback will result in excess TSH.

Endocrine

(a)   Anterior view

(b)   Posterior view

(c)

(d)

## Figure 10.11  Thyroid and Parathyroid Glands

(a) Anterior view of the thyroid gland. (b) The four small parathyroid glands are embedded in the posterior surface of the thyroid gland. (c) Three-dimensional interpretive drawing of thyroid follicles and parafollicular cells. (d) Light micrograph of thyroid and parathyroid tissue.

This causes the thyroid gland to enlarge, a condition called a **goiter** (goy'ter). One type of goiter develops if iodine in the diet is too low. As less $T_3$ and $T_4$ is synthesized and secreted, TSH-releasing hormone and TSH secretion increase above normal levels and cause dramatic enlargement of the thyroid gland.

**Predict 3**

*In people with Graves disease (hyperthyroidism), the immune system produces a large amount of a protein that is so much like TSH that it binds to cells of the thyroid gland and acts like TSH. Unlike TSH, however, the secretion of this protein does not respond to negative feedback. Predict the effect of this abnormal protein on the structure and function of the thyroid gland and the release of hormones from the hypothalamus and anterior pituitary gland.*

In addition to secreting thyroid hormones, the thyroid gland secretes a hormone called **calcitonin** (kal-si-tō'nin) from the C cells

(see figure 10.11c). Calcitonin is secreted if the blood concentration of $Ca^{2+}$ becomes too high, and it causes $Ca^{2+}$ levels to decrease to their normal range (figure 10.13). Calcitonin binds to membrane-bound receptors of osteoclasts and reduces the rate of $Ca^{2+}$ resorption (breakdown) from bone by inhibiting the osteoclasts. Calcitonin may prevent blood $Ca^{2+}$ levels from becoming overly elevated following a meal that contains a high concentration of $Ca^{2+}$.

Calcitonin helps prevent elevated blood $Ca^{2+}$ levels, but a lack of calcitonin secretion does not result in a prolonged increase in those levels. Other mechanisms controlling blood $Ca^{2+}$ levels compensate for the lack of calcitonin secretion.

## Parathyroid Glands

Four tiny **parathyroid** (par-ă-thī'royd) **glands** are embedded in the posterior wall of the thyroid gland (see figure 10.11b,d). The

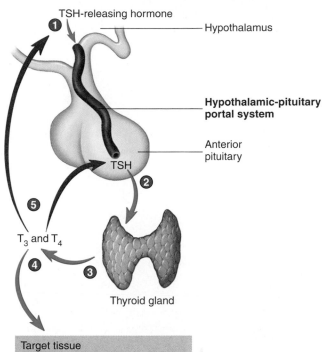

**1** Neurons within the hypothalamus release TSH-releasing hormone into the blood. It passes through the hypothalamic-pituitary portal system to the anterior pituitary.

**2** TSH-releasing hormone causes cells of the anterior pituitary to secrete TSH, which passes through the general circulation to the thyroid gland.

**3** TSH causes increased release of thyroid hormones ($T_3$ and $T_4$) into the general circulation.

**4** $T_3$ and $T_4$ act on target tissues to produce a response.

**5** $T_3$ and $T_4$ also have an inhibitory effect on the secretion of TSH-releasing hormone from the hypothalamus and TSH from the anterior pituitary.

**PROCESS Figure 10.12** Regulation of Thyroid Hormone ($T_3$ and $T_4$) Secretion

parathyroid glands secrete a hormone called **parathyroid hormone (PTH),** which is essential for the regulation of blood calcium levels (table 10.3). In fact, PTH is more important than calcitonin in regulating blood $Ca^{2+}$ levels. PTH binds to membrane-bound receptors of renal tubule cells, which increases active vitamin D formation. Vitamin D causes the epithelial cells of the intestine to increase $Ca^{2+}$ absorption. PTH also binds to receptors on osteoblasts. Substances released by the osteoblasts increase osteoclast activity and cause resorption of bone tissue to release $Ca^{2+}$ into the circulatory system. PTH binds to receptors on cells of the renal tubules and decreases the rate at which $Ca^{2+}$ are lost in the urine. PTH acts on its target tissues to raise blood $Ca^{2+}$ levels to normal.

Vitamin D is produced from precursors in the skin that are modified by the liver and kidneys. Ultraviolet light acting on the skin is required for the first stage of vitamin D synthesis, and the final stage of synthesis in the kidney is stimulated by PTH. Vitamin D can also be supplied in the diet.

### Predict 4

*Explain why a lack of vitamin D results in bones that are softer than normal.*

Decreasing blood $Ca^{2+}$ levels stimulate an increase in PTH secretion (figure 10.13). For example, if too few $Ca^{2+}$ are consumed

in the diet or if a person suffers from a prolonged lack of vitamin D, blood $Ca^{2+}$ levels decrease, and PTH secretion increases. The increased PTH increases the rate of bone resorption. Blood $Ca^{2+}$ levels can be maintained within a normal range, but prolonged resorption of bone results in reduced bone density, as manifested by soft, flexible bones that are easily deformed in young people and porous, fragile bones in older people.

Increasing blood $Ca^{2+}$ levels cause a decrease in PTH secretion (figure 10.13). The decreased PTH secretion leads to reduced blood $Ca^{2+}$ levels. In addition, increasing blood $Ca^{2+}$ levels stimulate calcitonin secretion, which also causes blood $Ca^{2+}$ levels to decline.

An abnormally high rate of PTH secretion is called **hyperparathyroidism.** One cause is a tumor in a parathyroid gland. The elevated blood levels of PTH increase bone resorption and elevate blood $Ca^{2+}$ levels. As a result, bones can become soft, deformed, and easily fractured. In addition, the elevated blood $Ca^{2+}$ levels make nerve and muscle cells less excitable, resulting in fatigue and muscle weakness. The excess $Ca^{2+}$ can be deposited in soft tissues of the body, causing inflammation. In addition, kidney stones can result.

An abnormally low rate of PTH secretion, called **hypoparathyroidism,** can result from injury to or the surgical removal of the thyroid and parathyroid glands. The low blood levels of PTH lead to reductions in the rate of bone resorption and the formation of vitamin D. As a result, blood $Ca^{2+}$ levels decrease. In response

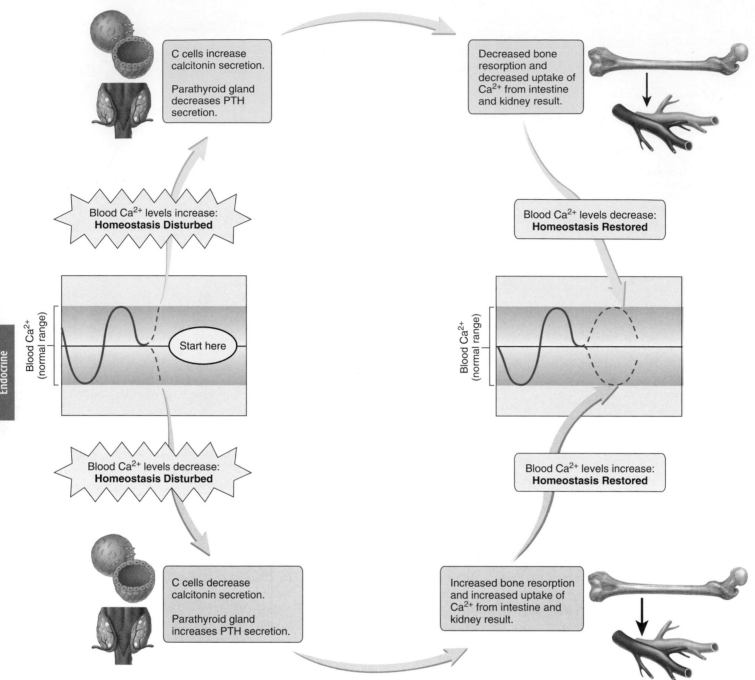

C cells increase calcitonin secretion.

Parathyroid gland decreases PTH secretion.

Decreased bone resorption and decreased uptake of Ca²⁺ from intestine and kidney result.

Blood Ca²⁺ levels increase:
**Homeostasis Disturbed**

Blood Ca²⁺ levels decrease:
**Homeostasis Restored**

Blood Ca²⁺ (normal range)

Blood Ca²⁺ (normal range)

Start here

Blood Ca²⁺ levels decrease:
**Homeostasis Disturbed**

Blood Ca²⁺ levels increase:
**Homeostasis Restored**

C cells decrease calcitonin secretion.

Parathyroid gland increases PTH secretion.

Increased bone resorption and increased uptake of Ca²⁺ from intestine and kidney result.

**Homeostasis Figure 10.13**  Regulation of Blood Calcium Ion Levels

to low blood Ca²⁺ levels, nerves and muscles become excitable and produce spontaneous action potentials that cause frequent muscle cramps or tetanus. Severe tetanus can affect the respiratory muscles; breathing stops, resulting in death.

## Adrenal Glands

The **adrenal** (ă-drē′năl; near or on the kidneys) **glands** are two small glands located superior to each kidney (figure 10.14a; table 10.3). Each adrenal gland has an inner part, called the **adrenal medulla** (marrow, or middle), and an outer part, called the **adrenal cortex**

(bark, or outer). The adrenal medulla and the adrenal cortex function as separate endocrine glands.

### Adrenal Medulla

The principal hormone released from the adrenal medulla is **epinephrine** (ep′i-nef′rin; *epi*, upon + *nephros*, kidney), or *adrenaline* (ă-dren′ă-lin; from the adrenal gland). Small amounts of **norepinephrine** (nōr′ep-i-nef′rin) are also released. The adrenal medulla releases epinephrine and norepinephrine in response to stimulation by the sympathetic nervous system, which becomes most

**Figure 10.14**  Anatomy and Histology of the Adrenal Gland

(a) An adrenal gland lies at the superior pole of each kidney. (b) The adrenal glands have an outer cortex and an inner medulla. (c) The cortex is surrounded by a connective tissue capsule.

active when a person is excited or physically active (figure 10.15). These hormones bind to membrane-bound receptors in their target tissues. Stress and low blood glucose levels can also cause increased sympathetic stimulation of the adrenal medulla. Epinephrine and norepinephrine are referred to as the **fight-or-flight** hormones because of their role in preparing the body for vigorous physical activity. The major effects of the hormones released from the adrenal medulla are:

1. Increases in the breakdown of glycogen to glucose in the liver, the release of the glucose into the blood, and the release of fatty acids from fat cells. The glucose and fatty acids serve as energy sources to maintain the body's increased rate of metabolism
2. Increased heart rate, which causes blood pressure to rise

3. Stimulation of smooth muscle in the walls of arteries supplying the internal organs and the skin, but not those supplying skeletal muscle. Blood flow to internal organs and the skin decreases, as do the functions of the internal organs. Blood flow through skeletal muscles increases
4. Increased blood pressure due to smooth muscle contraction in the walls of blood vessels in the internal organs and the skin
5. Increased metabolic rate of several tissues, especially skeletal muscle, cardiac muscle, and nervous tissue

Responses to hormones from the adrenal medulla reinforce the effect of the sympathetic division of the autonomic nervous system. Thus, the adrenal medulla and the sympathetic division function together to prepare the body for physical activity and to produce the fight-or-flight response and many other responses to stress.

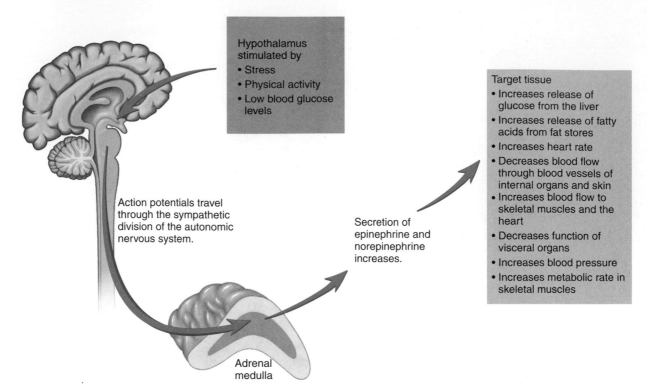

**Figure 10.15  Regulation of Adrenal Medullary Secretions**

Stimulation of the hypothalamus by stress, physical activity, or low blood glucose levels causes action potentials to travel through the sympathetic nervous system to the adrenal medulla. In response, the adrenal medulla releases epinephrine and smaller amounts of norepinephrine into the general circulation. These hormones have several effects that prepare the body for physical activity.

## Adrenal Cortex

The adrenal cortex secretes three classes of steroid hormones: mineralocorticoids, glucocorticoids, and androgens. The molecules of all three classes of steroid hormones enter their target cells and bind to intracellular receptor molecules. However, the hormones and the receptors of each class have unique structural and functional characteristics.

The first class of hormones secreted by the adrenal cortex, the **mineralocorticoids** (min′er-al-ō-kōr′ti-koydz), helps regulate blood volume and blood levels of $K^+$ and $Na^+$. **Aldosterone** (al-dos′ter-ōn) is the major hormone of this class (figure 10.16). Aldosterone primarily binds to receptor molecules in the kidney, but it also affects the intestine, sweat glands, and salivary glands. Aldosterone causes $Na^+$ and $H_2O$ to be retained in the body and increases the rate at which $K^+$ are eliminated.

Blood levels of $K^+$ and $Na^+$ directly affect the adrenal cortex to influence aldosterone secretion. The adrenal gland is much more sensitive to changes in blood $K^+$ levels than to changes in blood $Na^+$ levels. The rate of aldosterone secretion increases when blood $K^+$ levels increase or when blood $Na^+$ levels decrease.

Changes in blood pressure indirectly affect the rate of aldosterone secretion. Low blood pressure causes the release of a protein molecule called **renin** (rē′nin) from the kidney. Renin, which acts as an enzyme, causes a blood protein called **angiotensinogen** (an′jē-ō-ten-sin′ō-jen) to be converted to **angiotensin I** (an-jē-ō-ten′sin). Then, a protein called **angiotensin-converting enzyme** causes angiotensin I to be converted to **angiotensin II.**

Angiotensin II causes smooth muscle in blood vessels to constrict, and angiotensin II acts on the adrenal cortex to increase aldosterone secretion. Aldosterone causes retention of sodium and water, which leads to an increase in blood volume (figure 10.16). Both blood vessel constriction and increased blood volume help raise blood pressure.

**Predict 5**

*Predict the effects of reduced aldosterone secretion on blood levels of $Na^+$ and $K^+$, as well as on blood pressure.*

The second class of hormones secreted by the adrenal cortex, the **glucocorticoids** (gloo-kō-kōr′ti-koydz), helps regulate blood nutrient levels. The major glucocorticoid hormone is **cortisol** (kōr′ti-sol), which increases the breakdown of protein and fat and increases their conversion to forms of energy the body can use. For example, cortisol causes the liver to convert amino acids to glucose, and it acts on adipose tissue, causing fat to be broken down to fatty acids. The glucose and fatty acids are released into the blood, taken up by tissues, and used as a source of energy. Cortisol also causes proteins to be broken down to amino acids, which are then released into the blood.

Cortisol reduces the inflammatory and immune responses. A closely related steroid, **cortisone** (kōr′ti-sōn), or other similar drugs are often given to reduce inflammation caused by injuries. Cortisone can also reduce the immune and inflammatory responses that result from allergic reactions or abnormal immune responses, such as rheumatoid arthritis or asthma.

Stimulatory

① Increased blood K⁺ levels or decreased blood Na⁺ levels cause the adrenal cortex to increase the secretion of aldosterone into the general circulation.

② The kidneys detect a decrease in blood pressure. In response, they increase the secretion of renin into the general circulation. Renin converts angiotensinogen to angiotensin I. A converting enzyme changes angiotensin I to angiotensin II, which causes constriction of blood vessels, resulting in increased blood pressure.

③ Angiotensin II causes increased secretion of aldosterone, which primarily affects the kidneys.

④ Aldosterone stimulation of the kidneys causes Na⁺ retention, K⁺ excretion, and decreased water loss.

**PROCESS Figure 10.16**  Regulation of Aldosterone Secretion from the Adrenal Cortex

In response to stressful conditions, cortisol is secreted in larger than normal amounts, thus, it aids the body by providing energy sources for tissues. However, if stressful conditions are prolonged, the immune system can be suppressed enough to make the body susceptible to stress-related conditions (see Clinical Impact, "Hormones and Stress").

Adrenocorticotropic hormone (ACTH) molecules from the anterior pituitary bind to membrane-bound receptors and regulate the secretion of cortisol from the adrenal cortex in the process depicted in figure 10.17. When blood glucose levels decline, cortisol secretion increases. The low blood glucose acts on the hypothalamus to increase the secretion of the ACTH-releasing hormone, which in turn, stimulates ACTH secretion from the anterior pituitary. ACTH then stimulates cortisol secretion. Without ACTH, the adrenal cortex atrophies and loses its ability to secrete cortisol.

*Predict 6*

*Cortisone is sometimes given to people who have severe allergies or extensive inflammation or to people who suffer from autoimmune diseases. Taking this substance for long periods of time can damage the adrenal cortex. Explain how this damage can occur.*

The third class of hormones secreted by the adrenal cortex is composed of the **androgens** (an'drō-jenz), which stimulate the

development of male sexual characteristics. Small amounts of androgens are secreted from the adrenal cortex in both males and females. In adult males, most androgens are secreted by the testes. In adult females, the adrenal androgens influence the female sex drive. If the secretion of sex hormones from the adrenal cortex is abnormally high, exaggerated male characteristics develop in both males and females. This condition is most apparent in females and in males before puberty, when the effects are not masked by the secretion of androgens by the testes.

## Pancreas, Insulin, and Diabetes

The endocrine part of the **pancreas** (pan'krē-as) consists of **pancreatic islets** (islets of Langerhans), which are dispersed among the exocrine portion of the pancreas (figure 10.18). The islets secrete two hormones—insulin and glucagon—which help regulate the blood levels of nutrients, especially glucose (table 10.4). **Alpha cells** of the pancreatic islets secrete glucagon, and **beta cells** of the pancreatic islets secrete insulin.

It is very important to maintain blood glucose levels within a normal range (figure 10.19). A below-normal blood glucose level causes the nervous system to malfunction because glucose is the nervous system's main source of energy. When blood glucose decreases, other tissues rapidly break down fats and proteins to

# CLINICAL IMPACT

## Hormones and Stress

Stress, in the form of disease, physical injury, or emotional anxiety, initiates a specific response that involves the nervous and endocrine systems. The stressful condition influences the hypothalamus, and through the hypothalamus, the sympathetic division of the autonomic nervous system is activated. The sympathetic division prepares the body for physical activity. It increases heart rate and blood pressure, shunts blood from the intestine and other visceral structures to skeletal muscles, and increases the metabolic rate in several tissues, especially skeletal muscle. Part of the response of the sympathetic division is due to the release of epinephrine from the adrenal medulla.

In addition to sympathetic responses, stress causes the release of ACTH from the pituitary. ACTH acts on the adrenal cortex to cause the release of glucocorticoids. These hormones increase blood glucose levels and break down protein and fat, making nutrients more readily available to tissues.

Although the ability to respond to stress is adaptive for short periods of time, responses triggered by stressful conditions are harmful if they occur for long periods. Prolonged stress can lead to hypertension (elevated blood pressure), heart disease, ulcers, inhibited immune system function, changes in mood, and other conditions. Humans are frequently exposed to prolonged psychological stress from high-

pressure jobs, the inability to meet monetary obligations, or social expectations. Although responses to stress prepare a person for physical activity, increased physical activity is often not an appropriate response to the situation causing the stress. Long-term exposure to stress under conditions in which physical activity and emotions must be constrained may be harmful. Techniques that effectively reduce responses to stressful conditions, such as biofeedback, meditation, or other relaxation exercises can help people who are exposed to chronic stress. Adequate rest, relaxation, and regular physical exercise are also important in maintaining good health and reducing unhealthy responses to stressful situations.

① In response to stress or low blood glucose, ACTH-releasing hormone passes from the hypothalamus through the hypothalamic-pituitary portal system to the anterior pituitary. The releasing hormone binds to and stimulates cells that secrete ACTH into the general circulation.

② ACTH acts on the adrenal cortex and stimulates the secretion of cortisol into the general circulation.

③ Cortisol acts on its target tissues to increase protein breakdown and blood glucose.

④ Cortisol acts on the hypothalamus and anterior pituitary to decrease ACTH secretion.

**PROCESS Figure 10.17**  Regulation of Cortisol Secretion from the Adrenal Cortex

### Figure 10.18  Structure of the Pancreas

The endocrine portion of the pancreas is made up of scattered pancreatic islets. Alpha cells secrete glucagon, and beta cells secrete insulin. The exocrine portion of the pancreas surrounds the pancreatic islets and produces digestive enzymes that are carried through a system of ducts to the small intestine. The stain used for the light micrograph does not distinguish between alpha and beta cells.

LM 400x

Pancreatic islet

Common bile duct from liver

Duodenum (first part of small intestine)

Pancreatic duct

Pancreas

Pancreas

Exocrine portions of pancreas (secrete enzymes that move through the ducts to the small intestine)

Alpha cell (secretes glucagon)

Beta cell (secretes insulin)

To pancreatic duct

To bloodstream

Endocrine

| TABLE 10.4 | Effects of Insulin and Glucagon on Target Tissues | |
|---|---|---|
| **Target Tissue** | **Insulin Responses** | **Glucagon Responses** |
| Skeletal muscle, cardiac muscle, cartilage, bone fibroblasts, blood cells, mammary glands | Increases glucose uptake and glycogen synthesis; increases uptake of amino acids | Has little effect |
| Liver | Increases glycogen synthesis; increases use of glucose for energy | Causes rapid increase in the breakdown of glycogen to glucose and release of glucose into the blood; increases the formation of glucose from amino acids and, to some degree, from fats; increases metabolism of fatty acids |
| Adipose cells | Increases glucose uptake, glycogen synthesis, fat synthesis | High concentrations cause breakdown of fats; probably unimportant under most conditions |
| Nervous system | Has little effect except to increase glucose uptake in the satiety center | Has no effect |

provide an alternative energy source. As fats are broken down, the liver converts some of the fatty acids to acidic **ketones** (kē'tōnz), which are released into the blood. When blood glucose levels are very low, the breakdown of fats can cause the release of enough fatty acids and ketones to reduce the pH of the body fluids below normal, a condition called **acidosis** (as-i-dō'sis). The amino acids of proteins are broken down and used by the liver to synthesize glucose.

If blood glucose levels are too high, the kidneys produce large volumes of urine containing substantial amounts of glucose. Because of the rapid loss of water in the form of urine, dehydration can result.

The hormone **insulin** (in'sŭ-lin) is released from the beta cells primarily in response to the elevated blood glucose levels and increased parasympathetic stimulation associated with digestion of a meal. Increased blood levels of certain amino acids also stimulate insulin secretion. Decreased insulin secretion results from decreasing blood glucose levels and from stimulation of the pancreas by the sympathetic division of the nervous system, which occurs during physical activity. Decreased insulin levels allow blood glucose to be conserved to provide the brain with adequate glucose and to allow other tissues to metabolize fatty acids and glycogen stored in the cells.

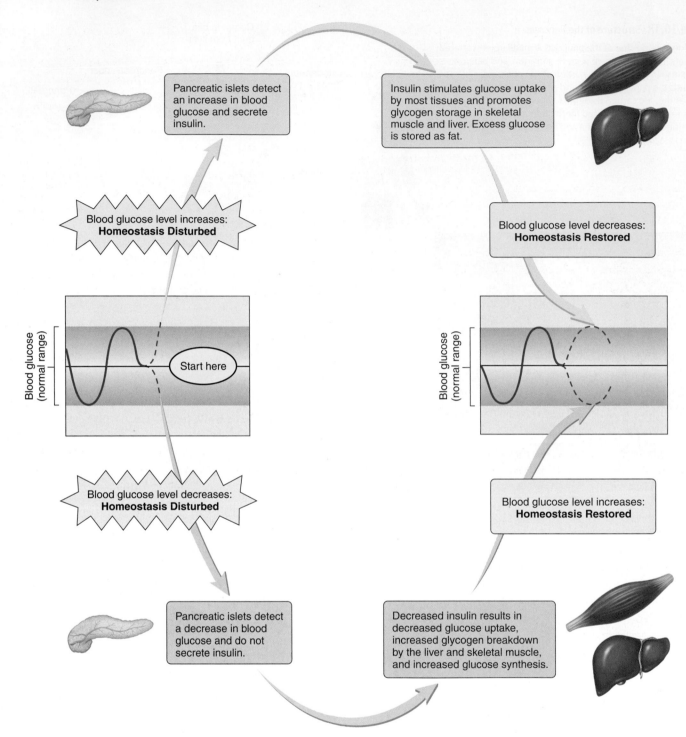

**Homeostasis Figure 10.19** Regulation of Blood Glucose Levels

The major target tissues for insulin are the liver, adipose tissue, muscles, and the area of the hypothalamus that controls appetite, called the **satiety** (sa-tī′-ĕ-tē; fulfillment of hunger) **center.** Insulin binds to membrane-bound receptors and, either directly or indirectly, increases the rate of glucose and amino acid uptake in these tissues. Glucose is converted to glycogen or fat, and the amino acids are used to synthesize protein. The effects of insulin on target tissues are summarized in table 10.4.

**Diabetes mellitus** (dī-ă-bē′tēz′ me-lī′tŭs; much urine + honey or sweetened) has several causes. Type 1 diabetes mellitus

occurs when too little insulin is secreted from the pancreas, and type 2 diabetes mellitus is caused by insufficient numbers of insulin receptors on target cells or by defective receptors that do not respond normally to insulin.

In type 1 diabetes mellitus, tissues cannot take up glucose effectively, causing blood glucose levels to become very high, a condition called **hyperglycemia** (hī′per-glī-sē′mē-ă; *hyper,* above + *glycemia,* blood glucose). Because glucose cannot enter the cells of the satiety center in the brain without insulin, the satiety center responds as if there were very little blood glucose, resulting in an exaggerated appetite. The excess glucose in the blood is excreted in the urine, making the urine volume much greater than normal. Because of excessive urine production, the person has a tendency to become dehydrated and thirsty. Even though blood glucose levels are high, fats and proteins are broken down to provide an energy source for metabolism, resulting in the wasting away of body tissues, acidosis, and ketosis. People with this condition also exhibit a lack of energy. Insulin must be injected regularly to adequately control blood glucose levels. When too much insulin is present, as occurs when a diabetic is injected with too much insulin or has not eaten after an insulin injection, blood glucose levels become very low. The brain, which depends primarily on glucose for an energy source, malfunctions. This condition, called insulin shock, can cause disorientation and convulsions and may result in loss of consciousness. Fortunately genetic engineering has allowed synthetic insulin to become widely available to diabetics.

## A CASE IN POINT
### Type 2 Diabetes Mellitus

Kandy Barr is 60 years old. She is overweight, has been feeling lethargic and weak, and has had two urinary tract infections in the past 6 months. She visited her optometrist because she could not read the handwriting on checks at the bank where she works. Her optometrist recommended that Kandy see her physician, who ordered a blood test. The results indicated high blood glucose and blood lipid levels. After additional tests, Kandy's physician told her that she has type 2 diabetes mellitus, which results from reduced sensitivity of tissues to the effects of insulin because of abnormal insulin receptors or abnormal responses to the insulin receptors. Consequently, insulin is less able to facilitate the entry of glucose into the liver, skeletal muscles, and adipose tissues. Kandy's physician recommended that she try to control her diabetes by restricting food intake, especially carbohydrates and fats, and increasing exercise. Kandy's symptoms were reduced as a result of faithful adherence to her physician's recommendations. People who are obese are approximately 10 times more likely to develop type 2 diabetes than people of normal weight. Frequent infections, changes in vision, and fatigue are common signs of high blood glucose levels.

**Glucagon** (gloo′kă-gon) is released from the alpha cells when blood glucose levels are low. Glucagon binds to membrane-bound receptors primarily in the liver, causing the glycogen stored in the liver to be converted to glucose. The glucose is then released into the blood to increase blood glucose levels. After a meal, when blood glucose levels are elevated, glucagon secretion is reduced.

### Predict 7

*How are the rates of insulin and glucagon secretion affected immediately following a large meal rich in carbohydrates? How are they affected after 12 hours without eating?*

Insulin and glucagon together regulate blood glucose levels (see figure 10.19). When blood glucose levels increase, insulin secretion increases, and glucagon secretion decreases. When blood glucose levels decrease, the rate of insulin secretion declines, and the rate of glucagon secretion increases. Other

hormones, such as epinephrine, cortisol, and growth hormone, also maintain blood levels of nutrients. When blood glucose levels decrease, these hormones are secreted at a greater rate. Epinephrine and cortisol cause the breakdown of protein and fat and the synthesis of glucose to help increase blood levels of nutrients. Growth hormone slows protein breakdown and favors fat breakdown.

## Testes and Ovaries

The testes of the male and the ovaries of the female secrete sex hormones, in addition to producing sperm cells or oocytes. The hormones produced by these organs play important roles in the development of sexual characteristics. Structural and functional differences between males and females, as well as the ability to reproduce, depend on the sex hormones (see table 10.3).

The main sex hormone in the male is **testosterone** (tes′tos′tĕ-rōn), which is secreted by the testes. It is responsible for the growth and development of the male reproductive structures, muscle enlargement, the growth of body hair, voice changes, and the male sexual drive.

In the female, two main classes of sex hormones, secreted by the ovaries, affect sexual characteristics: **estrogen** (es′trō-jen) and **progesterone** (prō-jes′ter-ōn). Together, these hormones contribute to the development and function of female reproductive structures and other female sexual characteristics. Two such characteristics are enlargement of the breasts and the distribution of fat, which influences the shape of the hips, breasts, and thighs. In addition, the female menstrual cycle is controlled by the cyclical release of estrogen and progesterone from the ovaries.

LH and FSH stimulate the secretion of hormones from the ovaries and testes. Releasing hormone from the hypothalamus controls the rate of LH and FSH secretion in males and females. LH and FSH, in turn, control the secretion of hormones from the ovaries and testes. Hormones secreted by the ovaries and testes also have a negative-feedback effect on the hypothalamus and anterior pituitary. The control of hormones that regulate reproductive functions is discussed in greater detail in chapter 19.

## Thymus

The **thymus** lies in the upper part of the thoracic cavity (see figure 10.7 and table 10.3). It is important in the function of the immune system. The thymus secretes a hormone called **thymosin** (thī′mō-sin), which aids the development of white blood cells called T cells. T cells help protect the body against infection by foreign organisms. The thymus is most important early in life; if an infant is born without a thymus, the immune system does not develop normally, and the body is less capable of fighting infections (see chapter 14).

## Pineal Gland

The **pineal** (pin′ē-ăl; pinecone) **gland** is a small, pinecone-shaped structure located superior and posterior to the thalamus of the brain (see chapter 8). The pineal gland produces a hormone called **melatonin** (mel-ă-tōn′in), which is thought to decrease the secretion of LH and FSH by decreasing the release of hypothalamic-releasing hormones (see table 10.3). Thus, melatonin inhibits the functions of

## Graves Disease (Hyperthyroidism)

Grace manages a business, has several employees, and works hard to make time for her husband and two children besides. Over several months, she slowly noticed feeling warm when others did not, sweating excessively, and appearing flushed. In addition, she often felt her heart pounding, was much more nervous than usual, and found it difficult to concentrate. Then Grace began to feel weak and lose weight, even though her appetite was greater than normal. Her family recognized some of these changes and noticed that her eyes seemed larger than usual. They encouraged her to see her physician. After an examination and some blood tests, Grace was diagnosed with Graves disease, a type of hyperthyroidism.

### Background Information

Graves disease is caused by altered regulation of hormone secretion—specifically, the elevated secretion of thyroid hormones from the thyroid gland. In approximately 95% of Graves disease cases, the immune system produces an unusual antibody type, which binds to receptors on the cells of the thyroid follicle and stimulates them to secrete increased amounts of thyroid hormone. The secretion of the releasing hormone and thyroid-stimulating hormone is inhibited by elevated thyroid hormones. However, the antibody is produced in large amounts, and is not inhibited by thyroid hormones. A very elevated rate of thyroid hormone secretion is therefore maintained. In addition, the size of the thyroid gland increases, and connective tissue components are deposited behind the eyes, causing them to bulge (figure 10A). Enlargement of the thyroid gland is called a goiter.

Grace was treated with radioactive iodine ($^{131}$I) atoms that were actively transported into thyroid cells, where they destroyed a substantial portion of the thyroid gland. Data indicate that this treatment has few side effects and is effective in treating most cases of Graves disease. Other options include drugs that inhibit the synthesis and secretion of thyroid hormones and surgery to remove part of the thyroid gland.

**Figure 10A**

(a) A goiter and (b) protruding eyes are symptoms of hyperthyroidism.

**Predict 9**

Explain why removal of part of the thyroid gland is an effective treatment for Graves disease.

the reproductive system. Animal studies have demonstrated that the amount of available light controls the rate of melatonin secretion. In many animals, short day length causes an increase in melatonin secretion, whereas longer day length causes a decrease in melatonin secretion. Some evidence suggests that melatonin plays an important role in the onset of puberty in humans. Tumors may develop in the pineal gland, which increase pineal secretions in some cases but decrease them in others.

**Predict 8**

A tumor can destroy the pineal gland's ability to secrete melatonin. How would that affect the reproductive system of a young person?

## OTHER HORMONES

Cells in the lining of the stomach and small intestine secrete hormones that stimulate the production of digestive juices from the stomach, pancreas, and liver. This secretion occurs when food is present in the digestive system, but not at other times. Hormones secreted from the small intestine also help regulate the rate at which food passes from the stomach into the small intestine (see chapter 16).

Prostaglandins are widely distributed in tissues of the body, where they function as intercellular signals. Unlike most hormones,

| System Interactions | Effects of Graves Disease (Hyperthyroidism) on Other Systems |
|---|---|
| **System** | **Interactions** |
| Integumentary | Excessive sweating, flushing, and warm skin result from the elevated body temperature caused by the increased rate of metabolism. Sweating and dilation of blood vessels in the skin are mechanisms by which the skin increases heat loss. The elevated metabolic rate makes amino acids unavailable for protein synthesis, resulting in fine, soft, straight hair, along with hair loss. |
| Skeletal | Some increased bone resorption occurs, which can decrease bone density; increased blood $Ca^{2+}$ levels can occur in severe cases. |
| Muscular | Muscle atrophy and muscle weakness result from increased metabolism, which causes the breakdown of muscle and the increased use of muscle proteins as energy sources. |
| Nervous | Enlargement of the extrinsic eye muscles, edema in the area of the orbits, and the accumulation of fibrous connective tissue cause protrusion of the eyes in 50–70% of individuals with Graves disease. Damage to the retina and optic nerve and paralysis of the extraocular muscles can occur. The effects on the eyes may be influenced directly by the abnormal antibodies, and they may progress even after the hyperthyroid condition is treated successfully. Restlessness, short attention span, compulsive movement, tremor, insomnia, and increased emotional responses are consistent with hyperactivity of the nervous system. |
| Cardiovascular | Cardiovascular responses are consistent with an increased rate of metabolism and increased stimulation by the sympathetic division of the autonomic nervous system. An increased amount of blood pumped by the heart leads to increased blood flow through the tissues, including the skin. The heart rate is greater than normal, heart sounds are louder than normal, and the heartbeats may be out of rhythm periodically. |
| Lymphatic and Immune | Antibodies that bind to receptors for thyroid-stimulating hormone on the cells of the thyroid gland have been found in nearly all people with Graves disease. The condition, therefore, is classified as an autoimmune disease in which antibodies produced by the lymphatic system result in abnormal functions. |
| Respiratory | Breathing may be labored, and the volume of air taken in with each breath may be decreased. Weak contractions of muscles of inspiration contribute to respiratory difficulties. |
| Digestive | Weight loss occurs, with an associated increase in appetite. Increased peristalsis in the intestines leads to frequent stools or diarrhea. Nausea, vomiting, and abdominal pain also result. Hepatic glycogen stores and adipose and protein stores are increasingly used for energy, and serum lipid levels (including triglycerides, phospholipids, and cholesterol) decrease. The tendency to develop vitamin deficiencies increases. The body's to ability absorb nutrients from the intestine becomes reduced, and metabolism increases, resulting in more use of body stores of fat and more breakdown of proteins for energy. |
| Reproductive | Reduced regularity of menstruation or lack of menstruation may occur in women because of the elevated metabolism. In men, the primary effect is loss of sex drive. |

they are usually not transported long distances in the blood but function mainly as autocrine or paracrine chemical signals (see table 10.1). Thus, their effects occur in the tissues where they are produced. Some prostaglandins cause relaxation of smooth muscle, such as dilation of blood vessels. Others cause contraction of smooth muscle, such as contraction of the uterus during the delivery of a baby. Because of their action on the uterus, prostaglandins have been used medically to initiate abortion. Prostaglandins also play a role in inflammation. They are released by damaged tissues and cause blood vessel dilation, localized swelling, and pain. Prostaglandins produced by platelets appear to be necessary for normal blood clotting. The ability of

aspirin and related substances to reduce pain and inflammation, to help prevent the painful cramping of uterine smooth muscle, and to treat headache is a result of their inhibitory effect on prostaglandin synthesis.

The kidneys secrete the hormone **erythropoietin** (ĕ-rith′rō-poy′ĕ-tin) in response to reduced oxygen levels in the kidney. Erythropoietin acts on bone marrow to increase the production of red blood cells (see chapter 11).

In pregnant women, the placenta is an important source of hormones that maintain pregnancy and stimulate breast development. These hormones are estrogen, progesterone, and **human** chorionic gonadotropin (gō′nad-ō-trō′pin), which is similar in structure and function to LH. Chapter 20 describes the function of these hormones in more detail.

# EFFECTS OF AGING ON THE ENDOCRINE SYSTEM

Age-related changes to the endocrine system include a gradual decrease in the secretion of some, but not all, endocrine glands. Some of the decreases in secretion may be due to the fact that older people commonly engage in less physical activity.

GH secretion decreases as people age, but the decrease is greatest in those who do not exercise, and it may not occur at all in older people who exercise regularly. Decreasing GH levels may explain the gradual decrease in bone and muscle mass and the increase in adipose tissue seen in many elderly people. So far, administering GH to slow or prevent the consequences of aging has not been found to be effective, and unwanted side effects are possible.

A decrease in melatonin secretion may influence age-related changes in sleep patterns, as well as the decreased secretion of some hormones, such as GH and testosterone.

The secretion of thyroid hormones decreases slightly with age. Age-related damage to the thyroid gland by the immune system can occur. Approximately 10% of elderly women experience some reduction in thyroid hormone secretion; this tendency is less common in men.

The kidneys of the elderly secrete less renin, reducing the ability to respond to decreases in blood pressure.

Reproductive hormone secretion gradually declines in elderly men, and women experience menopause (see chapter 19).

Secretion of thymosin from the thymus decreases with age. Fewer functional lymphocytes are produced, and the immune system becomes less effective in protecting the body against infections and cancer.

Parathyroid hormone secretion increases to maintain blood calcium levels if dietary $Ca^{2+}$ and vitamin D levels decrease, as they often do in the elderly. Consequently, a substantial decrease in bone matrix may occur.

In most people, the ability to regulate blood glucose does not decrease with age. However, there is an age-related tendency to develop type 2 diabetes mellitus for those who have a familial tendency, and it is correlated with age-related increases in body weight.

 SUMMARY

The nervous and endocrine systems are the two major regulatory systems in the body.

1. The nervous system controls structures by sending action potentials along axons, which release chemical signals at their ends near the cells they control. The endocrine system, by contrast, releases chemicals into the blood, which carries the chemicals to the cells they control.
2. The endocrine system has a more general and longer-lasting effect and acts more slowly than the nervous system.

## Functions of the Endocrine System (p. 263)

The main functions of the endocrine system are regulation of growth, metabolism, and tissue maturation; blood glucose concentration; ion concentration of blood; water balance; heart rate and blood pressure; immune cell production; reproductive functions in males and females; and uterine contractions during delivery as well as milk release from the breasts.

## Chemical Signals (p. 263)

Chemical signals bind to receptor sites on receptor molecules.
1. Intracellular chemical signals are produced in one part of a cell and travel to another part of the same cell, where they bind to receptors.
2. Intercellular chemical signals are released from one cell, are carried in the intercellular fluid, and bind to receptors in other cells.
3. An intercellular chemical signal can be classified as an autocrine signal, a paracrine signal, a hormone, a neurohormone, a neuromodulator, a neurotransmitter, or a pheromone.

## Hormones (p. 265)

1. Endocrine glands produce hormones that are released into the blood and travel some distance, where they act on target tissues to produce a response.
2. A target tissue for a given hormone has receptor molecules for that hormone.

### Chemistry of Hormones

1. Hormones are either water-soluble or lipid-soluble.
2. Protein hormones and most peptide hormones bind to receptors on the cell membrane and cause permeability changes or the production of second messengers inside the cell receptors on the cell membrane. Lipid-soluble hormones, such as the steroids and thyroid hormones, enter the cell and bind to receptors inside the cell. Eicosanoids bind to receptors on the cell membrane.
3. The combining of hormones with their receptors results in a response.

## Receptors (p. 266)

Hormones bind to receptor sites on receptor molecules to produce a response.

### Receptor Types for Hormones

Membrane-bound receptors extend through the cell membrane, and intracellular receptors are located in the cytoplasm or nucleus of the cell.

### Receptor Responses

1. Membrane-bound receptors can produce a response by
   a. Activating G proteins, which in turn can open or close membrane channels or activate enzymes that produce second messengers.
   b. Altering the activity of enzymes directly. The enzymes increase or decrease the synthesis of second messengers or add phosphate groups to proteins in the cell.
2. Intracellular receptors are located in the cytoplasm or nuclei and can regulate enzyme activity or the synthesis of specific messenger RNA.

## Regulation of Hormone Secretion (p. 268)

1. The secretion of hormones is controlled by negative-feedback mechanisms.

2. The secretion of hormones from a specific gland is controlled by blood levels of some chemical, another hormone, or the nervous system.

## Endocrine Glands and Their Hormones    (p. 269)

1. The endocrine system consists of ductless glands.
2. Some glands of the endocrine system perform more than one function.

### Pituitary and Hypothalamus

1. The pituitary is connected to the hypothalamus in the brain by the infundibulum. It is divided into anterior and posterior portions.
2. Secretions from the anterior pituitary are controlled by hormones that pass through the hypothalamic-pituitary portal system from the hypothalamus.
3. Hormones secreted from the posterior pituitary are controlled by action potentials carried by axons that pass from the hypothalamus to the posterior pituitary.
4. The hormones released from the anterior pituitary are growth hormone (GH), thyroid-stimulating hormone (TSH), adrenocorticotropic hormone (ACTH), luteinizing hormone (LH), follicle-stimulating hormone (FSH), prolactin, and melanocyte-stimulating hormone (MSH).
5. Hormones released from the posterior pituitary include antidiuretic hormone (ADH) and oxytocin.

### Thyroid Gland

The thyroid gland secretes thyroid hormones, which control the metabolic rate of tissues, and calcitonin, which helps regulate blood $Ca^{2+}$ levels.

### Parathyroid Glands

The parathyroid glands secrete parathyroid hormone, which helps regulate blood $Ca^{2+}$ levels. Active vitamin D also helps regulate blood $Ca^{2+}$ levels.

### Adrenal Glands

1. The adrenal medulla secretes primarily epinephrine and some norepinephrine. These hormones help prepare the body for physical activity.
2. The adrenal cortex secretes three classes of hormones.
   a. Glucocorticoids (cortisol) reduce inflammation and break down fat and protein, making them available as energy sources to other tissues.
   b. Mineralocorticoids (aldosterone) help regulate blood $Na^+$ and $K^+$ levels and water volume. Renin, secreted by the kidneys, helps regulate blood pressure by increasing angiotensin II and aldosterone production. These hormones cause blood vessels to constrict and enhance $Na^+$ and water retention by the kidney.
   c. Adrenal androgens increase female sexual drive but normally have little effect in males.

### Pancreas, Insulin, and Diabetes

1. The pancreas secretes insulin in response to elevated levels of blood glucose and amino acids. Insulin increases the rate at which many tissues, including adipose tissue, the liver, and skeletal muscles, take up glucose and amino acids.
2. The pancreas secretes glucagon in response to reduced blood glucose and increases the rate at which the liver releases glucose into the blood.

### Testes and Ovaries

1. The testes secrete testosterone, and the ovaries secrete estrogen and progesterone. These hormones help control reproductive processes.
2. LH and FSH from the pituitary gland control hormone secretion from the ovaries and testes.

### Thymus

The thymus secretes thymosin, which enhances the function of the immune system.

### Pineal Gland

The pineal gland secretes melatonin, which may help regulate the onset of puberty by acting on the hypothalamus.

## Other Hormones    (p. 286)

1. Hormones secreted by cells in the stomach and intestine help regulate stomach, pancreatic, and liver secretions.
2. The prostaglandins are hormones that have a local effect, produce numerous effects on the body, and play a role in inflammation.
3. Erythropoietin from the kidney stimulates red blood cell production.
4. The placenta secretes human chorionic gonadotropin, estrogen, and progesterone, which are essential to the maintenance of pregnancy.

## Effects of Aging on the Endocrine System    (p. 288)

1. Age-related changes include a gradual decrease in
   a. GH in people who do not exercise
   b. melatonin
   c. thyroid hormones (slight decrease)
   d. reproductive hormones
   e. thymosin
2. Parathyroid hormones increase if vitamin D and $Ca^{2+}$ levels decrease.
3. There is an increase in type 2 diabetes in people with a familial tendency.

 REVIEW AND COMPREHENSION

1. What are the major functional differences between the endocrine system and the nervous system?
2. List the functions of the endocrine system.
3. List the major differences between intracellular and intercellular chemical signals.
4. List the intercellular chemical signals that are classified on the basis of the cells from which they are secreted and their target cells.
5. Explain the relationship between a hormone and its receptor.
6. Describe the mechanisms by which membrane-bound receptors produce responses in their target tissues.
7. Describe the mechanisms by which intracellular receptors produce responses in their target tissues.
8. Compare the means by which hormones that can and cannot cross the cell membrane produce a response.
9. Define endocrine gland and hormone.
10. What makes one tissue a target tissue for a hormone and another not a target tissue?
11. Into what chemical categories can hormones be classified?
12. Name three ways that hormone secretion is regulated.

13. Describe how secretions of the anterior and posterior pituitary hormones are controlled.

14. What are the functions of growth hormone? What happens when too little or too much growth hormone is secreted?

15. Describe the effect of gonadotropins on the ovary and testis.

16. What are the functions of the thyroid hormones, and how is their secretion controlled? What happens when too little or too much of the thyroid hormones are secreted?

17. Explain how calcitonin, parathyroid hormone, and vitamin D are involved in maintaining blood $Ca^{2+}$ levels. What happens when too little or too much parathyroid hormone is secreted?

18. List the hormones secreted from the adrenal gland, give their functions, and compare the means by which the secretion rate of each is controlled.

19. What are the major functions of insulin and glucagon? How is their secretion regulated? What is the effect if too little insulin is secreted or the target tissues are not responsive to insulin?

20. List the effects of testosterone, estrogen, and progesterone.

21. What hormones are produced by the thymus and by the pineal gland? Name the effects of these hormones.

22. List the effects of prostaglandins. How is aspirin able to reduce the severity of the inflammatory response?

23. List the hormones secreted by the placenta.

24. List the major age-related changes that affect the endocrine system.

## CRITICAL THINKING

1. A hormone is known to bind to a membrane-bound receptor. A drug that inhibits the breakdown of cyclic AMP causes an increased response to the hormone. A drug that inhibits the binding of GTP to proteins reduces the response to this drug. Based on these observations, describe the mechanism by which the membrane-bound receptor is most likely to produce a response to the hormone.

2. Aldosterone and antidiuretic hormone play important roles in regulating blood volume and concentration of blood. The response to one of these hormones is evident within minutes, and the response to the other requires several hours. Explain the difference in response time for these two hormones.

3. Biceps Benny figured that, if a small amount of a vitamin was good, a lot should be better, so he began to take supplements that included a large amount of vitamin D. Predict the effect of vitamin D on his blood $Ca^{2+}$ levels and on the secretion of hormones that regulate blood $Ca^{2+}$ levels.

4. If a person's adrenal cortex degenerated and no longer secreted hormones, what would be the consequences?

5. Predict the effect of elevated aldosterone secretion from the adrenal cortex.

6. Explain how the blood levels of glucocorticoids, epinephrine, insulin, and glucagon change after a person has gone without food for 24 hours.

7. Stetha Scope wanted to go to medical school to become a physician. While attending college, she knew her grades had to be excellent. Stetha worked very hard and worried constantly. By the end of each school year, she had a cold and suffered from stomach pains. Explain why she might be susceptible to these symptoms.

*Answers in Appendix D*

Visit this textbook's website at **www.mhhe.com/seeleyess7** for practice quizzes, animations, interactive learning exercises, and other study tools.

McGraw-Hill offers a study CD that features interactive cadaver dissection. **Anatomy & Physiology Revealed** includes cadaver photos that allow you to peel away layers of the human body to reveal structures beneath the surface. This program also includes animations, radiologic imaging, audio pronunciations, and practice quizzing.

Blood is so vital to survival of the human body that it has become a metaphor for life and all sorts of passions. For example, many sports fans display bumper stickers proclaiming they bleed the team colors. On a more serious note, blood banking, as shown in the photo, has saved millions of lives.

# 11 Chapter

# Blood

## Chapter Outline and Objectives

### Functions of Blood (p. 292)
1. State the functions of blood.

### Composition of Blood (p. 292)
2. List the components of blood.

### Plasma (p. 292)
3. Name the components of plasma, and give their functions.

### Formed Elements (p. 294)
4. Describe the origin and production of the formed elements.
5. Describe the structure, function, and life history of red blood cells.
6. Compare the structures and functions of the five types of white blood cells.
7. Describe the origin and structure of platelets.

### Preventing Blood Loss (p. 299)
8. Explain the formation and function of platelet plugs and blood clots.
9. Describe the regulation of clot formation and how clots are removed.

### Blood Grouping (p. 303)
10. Explain the basis of ABO and Rh incompatibilities.

### Diagnostic Blood Tests (p. 306)
11. Describe diagnostic blood tests and the normal values for the tests, and give examples of disorders that produce abnormal test values.

Blood has always fascinated humans, and throughout history they have speculated about its function. Some societies consider blood the "essence of life" because the uncontrolled loss of it can result in death. Many cultures around the world, both ancient and modern, believe blood has magical qualities. Blood has also been thought to define our character and emotions. For example, people of a noble bloodline are sometimes described as "blue bloods," whereas criminals are said to have "bad" blood. People commonly say that anger causes their blood to "boil," whereas fear makes it "curdle." The scientific study of blood reveals characteristics as fascinating as any of these fantasies. Blood performs many functions essential to life and can reveal much about our health.

# FUNCTIONS OF BLOOD

The heart pumps blood through blood vessels that extend throughout the body. Blood helps maintain homeostasis in several ways:

1. *Transport of gases, nutrients, and waste products.* Oxygen enters the blood in the lungs and is carried to cells. Carbon dioxide, produced by cells, is carried in the blood to the lungs, from which it is expelled. The blood transports ingested nutrients, ions, and water from the digestive tract to cells, and the blood transports the waste products of the cells to the kidneys for elimination.

2. *Transport of processed molecules.* Many substances are produced in one part of the body and transported in the blood to another part, where they are modified. For example, the precursor to vitamin D is produced in the skin (see chapter 5) and transported by the blood to the liver and then to the kidneys for processing into active vitamin D. Then the blood transports active vitamin D to the small intestine, where it promotes the uptake of calcium. Another example is lactic acid produced by skeletal muscles during anaerobic respiration (see chapter 7). The blood carries lactic acid to the liver, where it is converted into glucose.

3. *Transport of regulatory molecules.* The blood carries many of the hormones and enzymes that regulate body processes from one part of the body to another.

4. *Regulation of pH and osmosis.* Buffers (see chapter 2), which help keep the blood's pH within its normal limits of 7.35–7.45, are found in the blood. The osmotic composition of blood is also critical for maintaining normal fluid and ion balance.

5. *Maintenance of body temperature.* Warm blood is transported from the interior of the body to the surface, where heat is released from the blood. This is one of the mechanisms that help regulate body temperature.

6. *Protection against foreign substances.* Certain cells and chemicals in the blood constitute an important part of the immune system, protecting against foreign substances, such as microorganisms and toxins.

7. *Clot formation.* When blood vessels are damaged, blood clotting protects against excessive blood loss. When tissues are damaged, the blood clot that forms is also the first step in tissue repair and the restoration of normal function (see chapter 4).

# COMPOSITION OF BLOOD

Blood is a type of connective tissue that consists of a liquid matrix containing cells and cell fragments. The liquid matrix is the **plasma** (plaz'mă), and the cells and cell fragments are the **formed elements** (figure 11.1). The plasma accounts for slightly more than half of the total blood volume, and the formed elements account for slightly less than half. The total blood volume in the average adult is about 4–5 liters (L) in females and 5–6 L in males. Blood makes up about 8% of total body weight.

# PLASMA

Plasma is a pale yellow fluid that consists of about 91% water, 7% proteins, and 2% other components, such as ions, nutrients, gases, waste products, and regulatory substances (figure 11.1 and table 11.1). Unlike the fibrous proteins found in other connective tissues, such as loose connective tissue, plasma contains dissolved proteins. Plasma proteins include albumin, globulins, and fibrinogen. **Albumin** (al-bū'min) makes up 58% of the plasma proteins. Although the osmotic pressure (see chapter 3) of blood results primarily from sodium chloride, albumin makes an important contribution. The water balance between the blood and the tissues is determined by the movement of water into and out of the blood by osmosis. **Globulins** (glob'ū-linz; globule) account for 38% of the plasma proteins. Some globulins, such as antibodies and complement, are part of the immune system (see chapter 14). Other globulins and albumin function as transport molecules because they bind to molecules, such as hormones (see chapter 10), and carry them in the blood throughout the body. Some globulins are clotting factors, which are necessary for the formation of blood clots. **Fibrinogen** (fī-brin'ō-jen) is a clotting factor that constitutes 4% of plasma proteins. Activation of clotting factors results in the conversion of fibrinogen to **fibrin** (fī'brin), a threadlike protein that forms blood clots (see "Blood Clotting" later in this chapter). **Serum** (ser'um) is plasma without the clotting factors.

Plasma volume remains relatively constant. Normally, water intake through the digestive tract closely matches water loss through the kidneys, lungs, digestive tract, and skin. Oxygen enters the blood in the lungs, and carbon dioxide enters the blood from tissues. Other suspended or dissolved substances in the blood come from the liver, kidneys, intestines, endocrine glands, and immune tissues, such as the lymph nodes and spleen. The

**Figure 11.1** **Composition of Blood**
Approximate values for the components of blood in a normal adult.

| TABLE 11.1 | Composition of Plasma |
|---|---|
| **Plasma Components** | **Functions and Examples** |
| Water | Acts as a solvent and suspending medium for blood components |
| Proteins | Maintain osmotic pressure (albumin), destroy foreign substances (antibodies and complement), transport molecules (albumin and globulins), and form clots (fibrinogen) |
| Ions | Involved in osmotic pressure (sodium and chloride ions), membrane potentials (sodium and potassium ions), and acid-base balance (hydrogen, hydroxide, and bicarbonate ions) |
| Nutrients | Source of energy and "building blocks" of more complex molecules (glucose, amino acids, triglycerides) |
| Gases | Involved in aerobic respiration (oxygen and carbon dioxide) |
| Waste products | Breakdown products of protein metabolism (urea and ammonia salts), red blood cells (bilirubin), and anaerobic respiration (lactic acid) |
| Regulatory substances | Catalyze chemical reactions (enzymes) and stimulate or inhibit many body functions (hormones) |

concentration of these substances in the blood is also regulated and maintained within narrow limits.

# FORMED ELEMENTS

About 95% of the volume of the formed elements consists of **red blood cells (RBCs),** or *erythrocytes* (ě-rith′rō-sītz; *erythro-,* red + *kytos,* cell). The remaining 5% of the volume of the formed elements consists of **white blood cells (WBCs),** or *leukocytes* (loo′kō-sītz; *leuko-,* white), and cell fragments called **platelets** (plăt′letz), or *thrombocytes* (throm′bō-sītz; *thrombo-,* clot). Red blood cells are 700 times more numerous than white blood cells and 17 times more numerous than platelets. Table 11.2 illustrates the formed elements of the blood.

## Production of Formed Elements

The process of blood cell production is called **hematopoiesis** (hē′mă-tō-poy-ē′sis; *hemato-,* blood + *poiesis,* making). In the fetus, hematopoiesis occurs in several tissues, including the liver, thymus, spleen, lymph nodes, and red bone marrow. After birth, hematopoiesis is confined primarily to red bone marrow, but some white blood cells are produced in lymphatic tissues (see chapter 14).

All the formed elements of blood are derived from a single population of cells called **stem cells,** or *hemocytoblasts.* These stem cells differentiate to give rise to different cell lines, each of which ends with the formation of a particular type of formed element (figure 11.2). The development of each cell line is regulated by specific growth factors. That is, growth factors determine the types of formed elements derived from the stem cells and how many formed elements are produced.

## Red Blood Cells

Normal red blood cells are disk-shaped, with edges that are thicker than the center of the cell (figure 11.3). The biconcave shape increases the cell's surface area compared to a flat disk of the same

| **TABLE 11.2** | **Formed Elements of the Blood** | | |
|---|---|---|---|
| **Cell Type** | **Illustration** | **Description** | **Function** |
| **Red Blood Cell** | | Biconcave disk; no nucleus; contains hemoglobin, which colors the cell red; 6.5–8.5 μm in diameter | Transports oxygen and carbon dioxide |
| **White Blood Cells** | | Spherical cells with a nucleus | Five types of white blood cells, each with specific functions |
| *Granulocytes* | | | |
| Neutrophil | | Nucleus with two to four lobes connected by thin filaments; cytoplasmic granules stain a light pink or reddish purple; 10–12 μm in diameter | Phagocytizes microorganisms and other substances |
| Basophil | | Nucleus with two indistinct lobes; cytoplasmic granules stain blue-purple; 10–12 μm in diameter | Releases histamine, which promotes inflammation, and heparin, which prevents clot formation |
| Eosinophil | | Nucleus often bilobed; cytoplasmic granules stain orange-red or bright red; 11–14 μm in diameter | Releases chemicals that reduce inflammation; attacks certain worm parasites |
| *Agranulocytes* | | | |
| Lymphocyte | | Round nucleus; cytoplasm forms a thin ring around the nucleus; 6–14 μm in diameter | Produces antibodies and other chemicals responsible for destroying microorganisms; contributes to allergic reactions, graft rejection, tumor control, and regulation of immune system |
| Monocyte | | Nucleus round, kidney-shaped, or horseshoe-shaped; contains more cytoplasm than does lymphocyte; 12–20 μm in diameter | Phagocytic cell in the blood; leaves the blood and becomes a macrophage, which phagocytizes bacteria, dead cells, cell fragments, and other debris within tissues |
| **Platelet** | | Cell fragment surrounded by a plasma membrane and containing granules; 2–4 μm in diameter | Forms platelet plugs; releases chemicals necessary for blood clotting |

Circulatory

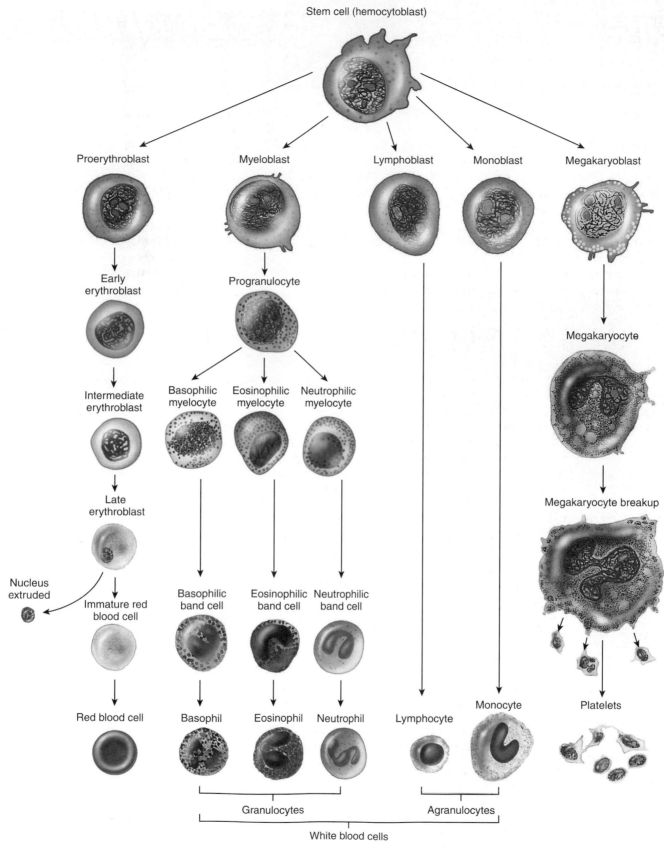

**Figure 11.2  Hematopoiesis**

Stem cells give rise to the cell lines that produce the formed elements.

## Stem Cells and Cancer Therapy

Many cancer therapies attack the types of rapidly dividing cells found in tumors. An undesirable side effect, however, can be the destruction of nontumor cells that divide rapidly, such as the stem cells and their derivatives in red bone marrow. After treatment for cancer, some patients are pre-scribed growth factors to stimulate the rapid regeneration of the red bone marrow. Although not a treatment for the cancer itself, the growth factors can speed recovery from the side effects of cancer therapy.

Some types of leukemia and genetic immune deficiency diseases can be treated with a bone marrow transplant containing blood stem cells. To avoid tissue rejection, families with a history of these disorders can freeze the umbilical cord blood of their newborn children. The cord blood contains many stem cells and can be used instead of a bone marrow transplant.

(a)

(b)  Top view          Side view

**Figure 11.3**  Red Blood Cells and White Blood Cells

(*a*) Scanning electron micrograph of formed elements: red blood cells (*red doughnut shapes*) and white blood cells (*yellow*). (*b*) Shape and dimensions of a red blood cell.

size. The greater surface area makes it easier for gases to move into and out of the red blood cell. In addition, the red blood cell can bend or fold around its thin center, decreasing its size and enabling it to pass more easily through smaller blood vessels.

During their development, red blood cells lose their nuclei and most of their organelles. Consequently, they are unable to divide. Red blood cells live for about 120 days in males and 110 days in females. The main component of a red blood cell is the pigmented protein **hemoglobin** (hē-mō-glō′bin), which accounts for about one-third of the cell's volume and is responsible for its red color.

### Function

The primary functions of red blood cells are to transport oxygen from the lungs to the various tissues of the body and to help transport carbon dioxide from the tissues to the lungs. Oxygen transport is accomplished by hemoglobin, which consists of four protein chains and four heme groups. Each protein, called a **globin** (glō′bin), is bound to one **heme** (hēm), a red-pigmented molecule. Each heme contains one iron atom, which is necessary for the normal function of hemoglobin. Each iron in a heme molecule can reversibly bind to an oxygen molecule. Hemoglobin picks up oxygen in the lungs and releases oxygen in other tissues (see chapter 15). Hemoglobin that is bound to oxygen is bright red, whereas hemoglobin without bound oxygen is a darker red. Hemoglobin is responsible for 98.5% of the oxygen transported in blood. The remaining 1.5% is transported dissolved in plasma.

Because iron is necessary for oxygen transport, it is not surprising that two-thirds of the body's iron is found in hemoglobin. Small amounts of iron are required in the diet to replace the small amounts lost in the urine and feces, but otherwise the existing iron is recycled, as described later in this section. Women need more dietary iron than men do because women lose iron as a result of menstruation.

Although oxygen is the primary molecule that binds to hemoglobin, other molecules can also bind to hemoglobin. For example, carbon monoxide, a gas produced by the incomplete combustion of hydrocarbons, such as gasoline, binds to the iron in hemoglobin about 210 times more readily than does oxygen and does not tend to unbind. As a result, the hemoglobin bound to carbon monoxide no longer transports oxygen. Nausea, headache, unconsciousness, and death are possible consequences of prolonged exposure to carbon monoxide.

Carbon dioxide is produced in tissues and transported in the blood to the lungs, where it is removed from the blood (see chapter 15). Carbon dioxide transport involves bicarbonate ions, hemoglobin, and plasma. Approximately 70% of the carbon dioxide

Circulatory

in blood is transported in the form of bicarbonate ions. The enzyme **carbonic anhydrase** (kar-bon′ik an-hī′drās), found primarily inside red blood cells, catalyzes a reaction that converts carbon dioxide ($CO_2$) and water ($H_2O$) into a hydrogen ion ($H^+$) and a bicarbonate ion ($HCO_3^-$):

$$CO_2 + H_2O \rightleftarrows H^+ + HCO_3^-$$

Carbon dioxide can bind reversibly to the globin part of hemoglobin. About 23% of the $CO_2$ in blood is transported bound to hemoglobin or other blood proteins. The remaining 7% of $CO_2$ is transported dissolved in plasma.

### Life History of Red Blood Cells

Under normal conditions, about 2.5 million red blood cells are destroyed every second. Fortunately, new red blood cells are produced just as rapidly. Stem cells form **proerythroblasts** (prō-ĕ-rith′rō-blastz; pro-, before + erythro-, red + blastos, germ), which give rise to the red blood cell line (see figure 11.2). Red blood cells are the final cells produced through a series of cell divisions. After each cell division, each formed cell changes and becomes more like a mature red blood cell. For example, following one of these cell divisions, the newly formed cells manufacture large amounts of hemoglobin. After the final cell division, the cell loses its nucleus and becomes a completely mature red blood cell.

The process of cell division requires the B vitamins folate and $B_{12}$, which are necessary for the synthesis of DNA (see chapter 3). Iron is required for the production of hemoglobin. Consequently, a lack of folate, vitamin $B_{12}$, or iron can interfere with normal red blood cell production.

Red blood cell production is stimulated by low blood oxygen levels. Typical causes of low blood oxygen are decreased numbers of red blood cells, decreased or defective hemoglobin, diseases of the lungs, high altitude, inability of the cardiovascular system to deliver blood to tissues, and increased tissue demand for oxygen, as occurs during endurance exercises.

Low blood oxygen levels stimulate red blood cell production by increasing the formation and release of the glycoprotein **erythropoietin** (ĕ-rith-rō-poy′ĕ-tin), primarily by the kidneys (figure 11.4). Erythropoietin stimulates red bone marrow to produce more red blood cells. Thus, when oxygen levels in the blood decrease, the production of erythropoietin increases, which increases red blood cell production. The greater number of red blood cells increases the blood's ability to transport oxygen. This mechanism returns blood oxygen levels to normal and maintains homeostasis by increasing the delivery of oxygen to tissues. Conversely, if blood oxygen levels rise, less erythropoietin is released, and red blood cell production decreases.

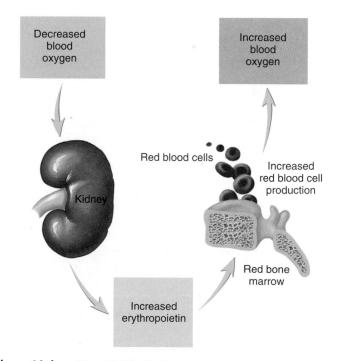

**Figure 11.4  Red Blood Cell Production**

In response to decreased blood oxygen, the kidneys release erythropoietin into the general circulation. The increased erythropoietin stimulates red blood cell production in the red bone marrow. This process increases blood oxygen levels.

1. In macrophages, the globin part of hemoglobin is broken down to individual amino acids (*pink arrow*) and metabolized or used to build new proteins.

2. The heme of hemoglobin releases iron. The heme is converted into bilirubin.

3. Blood transports iron to the red bone marrow, where it is used to produce new hemoglobin (*green arrows*).

4. Blood transports bilirubin (*blue arrows*) to the liver.

5. Bilirubin is excreted as part of the bile into the small intestine.

6. Bilirubin derivatives contribute to the color of feces or are reabsorbed from the intestine into the blood and excreted from the kidneys in the urine.

**PROCESS Figure 11.5  Hemoglobin Breakdown**

Macrophages break down hemoglobin, and the breakdown products are used or excreted.

**Predict 1**

*Cigarette smoke produces carbon monoxide. If a nonsmoker smoked a pack of cigarettes a day for a few weeks, what would happen to the number of red blood cells in the person's blood? Explain.*

When red blood cells become old, abnormal, or damaged, they are removed from the blood by macrophages located in the spleen and liver (figure 11.5). Within the macrophage, the globin part of the hemoglobin molecule is broken down into amino acids that are reused to produce other proteins. The iron released from heme is transported in the blood to the red bone marrow and used to produce new hemoglobin. Thus, the iron is recycled. The heme molecules are converted to **bilirubin** (bil-i-roo′bin), a yellow pigment molecule. Bilirubin is normally taken up by the liver and released into the small intestine as part of the bile (see chapter 16). If the liver is not functioning normally, or if the flow of bile from the liver to the small intestine is hindered, bilirubin builds up in the circulation and produces **jaundice** (jawn′dis; *jaune,* yellow), a yellowish color to the skin. After it enters the intestine, bilirubin is converted by bacteria into other pigments. Some of these pigments give feces their brown color, whereas others are absorbed from the intestine into the blood, modified by the kidneys, and excreted in the urine, contributing to the characteristic yellow color of urine.

## White Blood Cells

White blood cells are spherical cells that lack hemoglobin. White blood cells form a thin, white layer of cells between plasma and red blood cells when the components of blood are separated from each other (see figure 11.1). They are larger than red blood cells, and each has a nucleus (table 11.2). Although white blood cells are components of the blood, the blood serves primarily as a means of transporting these cells to other body tissues. White blood cells can leave the blood and travel by **ameboid** (ă-mē′boyd; like an ameba) **movement** through the tissues. In this process, the cell projects a cytoplasmic extension that attaches to an object. Then the rest of the cell's cytoplasm flows into the extension. Two functions of white blood cells are (1) to protect the body against invading microorganisms and (2) to remove dead cells and debris from the tissues by phagocytosis.

Each white blood cell type is named according to its appearance in stained preparations. Those containing large cytoplasmic granules are *granulocytes* (gran′ū-lō-sītz; *granulo-,* granular), and those with very small granules that cannot be seen easily with the light microscope are **agranulocytes** (ă-gran′ū-lō-sītz; *a-,* without).

There are three kinds of granulocytes: neutrophils, basophils, and eosinophils. **Neutrophils** (noo′trō-filz; *neutro-,* neutral + *philos,* loving), the most common type of white blood cells, have small cytoplasmic granules that stain with both acidic and basic

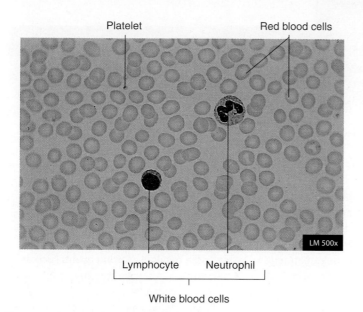

Platelet    Red blood cells

LM 500x

Lymphocyte    Neutrophil

White blood cells

**Figure 11.6  Standard Blood Smear**

A thin film of blood is spread on a microscope slide and stained. The white blood cells have pink cytoplasm and purple nuclei. The red blood cells do not have nuclei. The center of a red blood cell appears whitish because light shines more readily through the thin center of the disk than through the thicker edges. The platelets are purple cell fragments.

dyes (figure 11.6). Their nuclei are commonly lobed, with the number of lobes varying from two to four. Neutrophils usually remain in the blood for a short time (10–12 hours), move into other tissues, and phagocytize microorganisms and other foreign substances. Dead neutrophils, cell debris, and fluid can accumulate as **pus** at sites of infections.

**Basophils** (bā′sō-filz; *baso-*, base), the least common of all white blood cells, contain large cytoplasmic granules that stain blue or purple with basic dyes (table 11.2). Basophils release histamine and other chemicals that promote inflammation (see chapters 4 and 14). They also release heparin, which prevents the formation of clots.

**Eosinophils** (ē-ō-sin′o-filz) contain cytoplasmic granules that stain bright red with eosin, an acidic stain. They often have a two-lobed nucleus (table 11.2). Eosinophils release chemicals that reduce inflammation. In addition, chemicals from eosinophils are involved in destroying certain worm parasites.

There are two kinds of agranulocytes: lymphocytes and monocytes. **Lymphocytes** (lim′fō-sītz; *lympho-*, lymph) are the

smallest of the white blood cells (figure 11.6 and table 11.2). The lymphocytic cytoplasm consists of only a thin, sometimes imperceptible ring around the nucleus. There are several types of lymphocytes, and they play an important role in the body's immune response. Their diverse activities involve the production of antibodies and other chemicals that destroy microorganisms, contribute to allergic reactions, reject grafts, control tumors, and regulate the immune system. Chapter 14 considers these cells in more detail.

**Monocytes** (mon′ō-sītz) are the largest of the white blood cells (table 11.2). After they leave the blood and enter tissues, monocytes enlarge and become **macrophages** (mak′rō-fā-jez; *macro-*, large + *phagō*, to eat), which phagocytize bacteria, dead cells, cell fragments, and any other debris within the tissues. In addition, macrophages can break down phagocytized foreign substances and present the processed substances to lymphocytes, causing activation of the lymphocytes (see chapter 14).

**Predict 2**

*Based on their morphology, identify each of the white blood cells shown in figure 11.7*

## Platelets

Platelets are minute fragments of cells, each consisting of a small amount of cytoplasm surrounded by a cell membrane (see figure 11.6). They are produced in the red bone marrow from **megakaryocytes** (meg-ă-kar′ē-ō-sītz; *mega-*, large + *karyon*, nucleus), which are large cells (see figure 11.2). Small fragments of these cells break off and enter the blood as platelets, which play an important role in preventing blood loss.

## PREVENTING BLOOD LOSS

When a blood vessel is damaged, blood can leak into other tissues and interfere with normal tissue function, or blood can be lost from the body. The body can tolerate a small amount of blood loss and can produce new blood to replace it. But a large amount of blood loss can lead to death. Fortunately, when a blood vessel is damaged, loss of blood is minimized in three ways: vascular spasm, platelet plug formation, and blood clotting.

## Vascular Spasm

**Vascular spasm** is an immediate but temporary constriction of a blood vessel that results when smooth muscle within the wall of the vessel contracts. This constriction can close small vessels completely

(a)

(b)

(c)

(d)

LM 1200x

(e)

**Figure 11.7  Identification of White Blood Cells**

See Predict 2.

## Clinical Importance of Activating Platelets

Platelet activation results in platelet plug formation and the production of chemicals, such as phospholipids, that are important for blood clotting. Thus, inhibition of platelet activation reduces the formation of blood clots.

Thromboxanes, which activate platelets, are derived from certain prostaglandins. Aspirin inhibits prostaglandin synthesis and, therefore, thromboxane synthesis, which results in reduced platelet activation. If an expectant mother ingests aspirin near the end of pregnancy, thromboxane synthesis is inhibited. As a result, the mother experiences excessive bleeding after delivery, and the baby can exhibit numerous localized hemorrhages over the surface of

its body, both due to decreased platelet function. If the quantity of ingested aspirin is large, the infant, the mother, or both may die as a result of bleeding.

Platelet plugs and blood clots can block blood vessels, producing heart attacks and strokes. Suspected heart attack victims are routinely given aspirin en route to the emergency room as part of their treatment. The United States Preventive Services Task Force (USPSTF) and the American Heart Association recommend low-dose aspirin therapy (75–160 mg/day) for all men and women at high risk for cardiovascular disease. Determining high risk involves analyzing many factors and should be done in consultation with a physician. The decreased

risk of cardiovascular disease from aspirin therapy must be weighed against the increased risk of hemorrhagic stroke and gastrointestinal bleeding. Risk factors include age (men over 40 and postmenopausal women), high cholesterol, high blood pressure, a history of smoking, diabetes, a family history of cardiovascular disease and a previous clotting event, such as a heart attack, a transient ischemic attack, or an occlusive stroke.

Plavix (clopidogrel bisulfate) is a drug that reduces the activation of platelets by blocking the ADP receptors on the surface of platelets. It is administered to prevent clotting and, along with other anticlotting drugs, to treat heart attacks.

and stop the flow of blood through them. Damage to blood vessels can activate nervous system reflexes that cause vascular spasm. Chemicals also produce vascular spasm. For example, platelets release **thromboxanes** (throm′bok-zānz), which are derived from certain prostaglandins, and endothelial (epithelial) cells lining blood vessels release the peptide **endothelin** (en-dō-thē′lin).

## Platelet Plug Formation

A **platelet plug** is an accumulation of platelets that can seal up a small break in a blood vessel. Platelet plug formation is very important in maintaining the integrity of the circulatory system because small tears occur in the smaller vessels and capillaries many times each day. People who lack the normal number of platelets tend to develop numerous small hemorrhages in their skin and internal organs.

The formation of a platelet plug can be described as a series of steps, but in actuality many of these steps occur at the same time (figure 11.8). First, platelets stick to the collagen exposed by blood vessel damage; this phenomenon is called **platelet adhesion.** Most platelet adhesion is mediated through **von Willebrand factor,** a protein produced and secreted by blood vessel endothelial cells. Von Willebrand factor forms a bridge between collagen and platelets by binding to platelet surface receptors and collagen. After platelets adhere to collagen, they become activated, change shape, and release chemicals.

In the **platelet release reaction,** platelets release chemicals, such as ADP and thromboxane, which bind to their respective receptors on the surfaces of platelets, activating the platelets. These activated platelets also release ADP and thromboxane, which

activates more platelets. Thus, a cascade of chemical release activates many platelets. This is an example of positive feedback. As platelets become activated, they express surface receptors called **fibrinogen receptors,** which can bind to fibrinogen, a plasma protein. In **platelet aggregation,** fibrinogen forms bridges between the fibrinogen receptors of numerous platelets, resulting in a platelet plug.

## A CASE IN POINT
### Idiopathic Thrombocytopenic Purpura

Following a weekend touch football game, Bruz Moore noticed that he had developed a number of small bruises. Although the bruising was unusual, he dismissed it. A few days later, Bruz suddenly developed a rash on his legs, which he also ignored. But when he noticed bleeding from his gums while brushing his teeth, he went to see his doctor. There, Bruz learned that the rash was caused by many pinhead-sized hemorrhages in the skin. This sign, together with the bruising and the bleeding from the gums, indicated a below-normal ability to stop bleeding. A blood test confirmed that Bruz's platelet count was lower than normal. In the absence of any evidence of toxic exposure to chemicals or diseases associated with a low platelet count, the doctor diagnosed idiopathic thrombocytopenic purpura (id′ē-ō-path′ik throm′bō-sī-tō-pē′nik pŭr′poo-ra) (ITP). In this disorder, the immune system makes antibodies (see chapter 14) that bind to platelets. As a result, the "tagged" platelets are removed from the blood by phagocytic cells in the spleen faster than they are produced in the red bone marrow. Consequently, platelet numbers decrease, causing bleeding problems. For more information on thrombocytopenia, see "Platelet Count" later in this chapter.

Circulatory

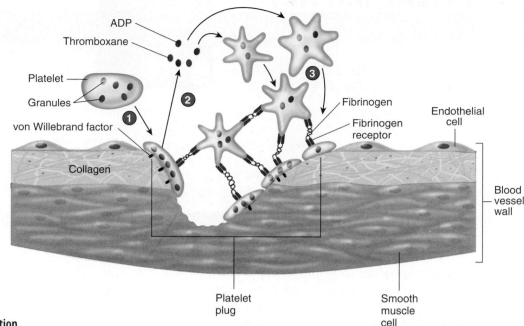

1. Platelet adhesion occurs when von Willebrand factor connects collagen and platelets.

2. During the platelet release reaction, ADP, thromboxanes, and other chemicals are released and activate other platelets.

3. Platelet aggregation occurs when fibrinogen receptors on activated platelets bind to fibrinogen, connecting the platelets to one another. The accumulating mass of platelets forms a platelet plug.

**PROCESS Figure 11.8** Platelet Plug Formation

## Blood Clotting

Blood vessel constriction and platelet plugs alone are not sufficient to close large tears or cuts in blood vessels. When a blood vessel is severely damaged, **blood clotting,** or *coagulation* (kō-ag-ū-lā′shŭn), results in the formation of a clot. A **clot** is a network of threadlike protein fibers, called **fibrin** (fī′brin), that traps blood cells, platelets, and fluid.

The formation of a blood clot depends on a number of proteins found within plasma, called **clotting factors.** Normally, the clotting factors are inactive and do not cause clotting. Following injury, however, the clotting factors are activated. Clot formation is a complex process involving many chemical reactions, but it can be summarized in three stages (figure 11.9).

1. The chemical reactions can be started in two ways:
   (a) Inactive clotting factors come in contact with exposed connective tissue, resulting in their activation, or
   (b) chemicals, such as **thromboplastin,** are released from injured tissues, causing activation of clotting factors. After

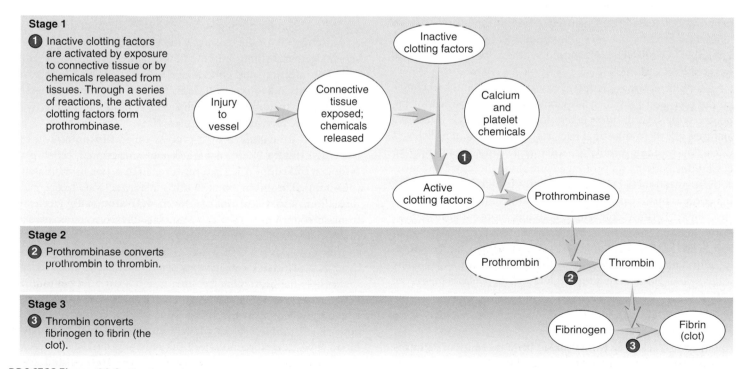

**Stage 1**

1. Inactive clotting factors are activated by exposure to connective tissue or by chemicals released from tissues. Through a series of reactions, the activated clotting factors form prothrombinase.

**Stage 2**

2. Prothrombinase converts prothrombin to thrombin.

**Stage 3**

3. Thrombin converts fibrinogen to fibrin (the clot).

**PROCESS Figure 11.9** Clot Formation

Clot formation has three stages.

## Danger of Unwanted Clots

When platelets encounter damaged or diseased areas of blood vessels or heart walls, an attached clot, called a **thrombus** (throm'bus), can form. A thrombus that breaks loose and begins to float through the circulation is called an **embolus** (em'bō-lŭs). Both thrombi and emboli can cause death if they block vessels that supply blood to essential organs, such as the heart, brain, or lungs. Abnormal coagulation can be prevented or hindered by administering an anticoagulant, such as heparin, which acts rapidly. Warfarin (war'fă-rin), brand name Coumadin (koo'mă-din), acts more slowly than heparin. Warfarin prevents clot formation by suppressing the liver's production of vitamin K–dependent clotting factors.

the initial clotting factors are activated, they in turn activate other clotting factors. A series of reactions results in which each clotting factor activates the next until the clotting factor **prothrombinase** (prō-throm'bi-nās), or *prothrombin activator*, is formed.

2. Prothrombinase converts an inactive clotting factor called **prothrombin** (prō-throm'bin) to its active form, **thrombin** (throm'bin).

3. Thrombin converts the inactive clotting factor **fibrinogen** (fī-brin'ō-jen) to its active form, fibrin.

At each step of the clotting process, each clotting factor activates many additional clotting factors, resulting in the formation of a clot.

Most clotting factors are manufactured in the liver, and many of them require vitamin K for their synthesis. In addition, many of the chemical reactions of clot formation require $Ca^{2+}$ and the chemicals released from platelets. The clotting process can be severely impaired by low levels of vitamin K, low levels of $Ca^{2+}$, low numbers of platelets, or reduced synthesis of clotting factors because of liver dysfunction.

Humans rely on two sources of vitamin K. About half comes from the diet, and the other half comes from bacteria within the large intestine. Antibiotics taken to fight bacterial infections sometimes kill these intestinal bacteria, reducing vitamin K levels and causing bleeding problems. Vitamin K supplements may be necessary for patients on prolonged antibiotic therapy. Newborns lack these intestinal bacteria and thus routinely receive a vitamin K injection at birth. Infants can also obtain vitamin K from food, such as milk. Because cow's milk contains more vitamin K than does human milk, breast-fed infants are more susceptible to bleeding than are bottle-fed infants. However, maternal supplementation with vitamin K, such as with oral vitamins, adequately elevates breast-fed infant vitamin K levels and decreases the risk of bleeding.

### Control of Clot Formation

Without control, clotting would spread from the point of its initiation throughout the entire circulatory system. Fortunately, the blood contains several **anticoagulants** (an'tē-kō-ag'ū-lantz), which prevent clotting factors from forming clots under normal conditions.

For example, **antithrombin** (an-tē-throm'bin) and **heparin** (hep'ă-rin) inactivate thrombin. Without thrombin, fibrinogen is not converted to fibrin, and no clot forms. At an injury site, however, the activation of clotting factors is very rapid. Enough clotting factors are activated that the anticoagulants can no longer prevent a clot from forming. Away from the injury site, there are enough anticoagulants to prevent clot formation from spreading.

### Clot Retraction and Fibrinolysis

After a clot has formed, it begins to condense into a more compact structure through a process known as **clot retraction.** Platelets contain the contractile proteins actin and myosin, which operate in a fashion similar to that of the actin and myosin in muscle (see chapter 7). Platelets form small extensions that attach to fibrin through surface receptors. Contraction of the extensions pulls on the fibrin and leads to clot retraction. During clot retraction, serum, which is plasma without the clotting factors, is squeezed out of the clot.

Retraction of the clot pulls the edges of the damaged blood vessel together, helping stop the flow of blood, reducing the probability of infection, and enhancing healing. The vessel is repaired as fibroblasts move into the damaged area and new connective tissue forms. In addition, epithelial cells around the wound divide and fill in the torn area (see chapter 4).

Clots are dissolved by a process called **fibrinolysis** (fī-bri-nol'-i-sis) (figure 11.10). An inactive plasma protein called **plasminogen** (plaz-min'ō-jen) is converted to its active form, **plasmin** (plaz'min). Thrombin, other clotting factors activated during clot formation, and **tissue plasminogen activator (t-PA)** released from surrounding tissues can stimulate the conversion of plasminogen to plasmin. Over a few days, plasmin slowly breaks down the fibrin.

A heart attack can result when a clot blocks blood vessels that supply the heart. One treatment for a heart attack is to inject certain chemicals into the blood that activate plasmin. Unlike aspirin and anticoagulant therapies, which are used to prevent heart attacks, plasmin activators quickly dissolve the clot and restore blood flow to cardiac muscle, thus reducing damage to tissues. **Streptokinase** (strep-tō-kīn'ās), a bacterial enzyme, and t-PA, produced through genetic engineering, have been used successfully to dissolve clots.

Circulatory

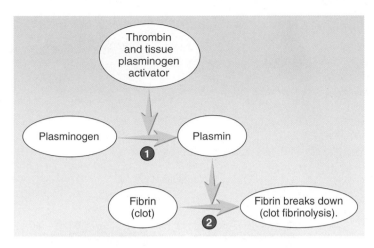

① Thrombin and tissue plasminogen activator convert inactive plasminogen into plasmin.

② Plasmin breaks down the fibrin in a blood clot, resulting in clot fibrinolysis.

**PROCESS Figure 11.10** Fibrinolysis

# BLOOD GROUPING

If large quantities of blood are lost during surgery or due to injury, a patient can go into shock and die unless red blood cells are replaced to restore the blood's oxygen-carrying capacity. In this event, either a transfusion or an infusion is called for. A **transfusion** is the transfer of blood or blood components from one individual to another. An **infusion** is the introduction of a fluid other than blood, such as a saline or glucose solution, into the blood. In many cases, the return of blood volume to normal levels

is all that is necessary to prevent shock. Eventually, the body produces enough red blood cells to replace those that were lost.

Early attempts to transfuse blood were often unsuccessful because they resulted in **transfusion reactions,** characterized by clumping or rupture of blood cells and clotting within blood vessels. We now know that transfusion reactions are caused by interactions between antigens and antibodies (see chapter 14). In brief, the surfaces of red blood cells have molecules called **antigens** (an'ti-jenz), and the plasma includes proteins called **antibodies** (an'te-bod-ēz; *anti-,* against). Antibodies are very specific, meaning that each antibody can bind only to a certain antigen. When the antibodies in the plasma bind to the antigens on the surface of the red blood cells, they form molecular bridges that connect the red blood cells together. As a result, **agglutination** (ă-gloo-ti-nā'shŭn; *ad,* to + *gluten,* glue), or clumping of the cells, occurs. The combination of the antibodies with the antigens can also initiate reactions that cause **hemolysis** (hē-mol'i-sis; *hemo-,* blood + *lysis,* destruction), or rupture of the red blood cells. The debris formed from the ruptured red blood cells can trigger clotting within small blood vessels. As a result, tissue damage and death may occur.

The antigens on the surface of red blood cells have been categorized into **blood groups.** Although many blood groups are recognized, the ABO and Rh blood groups are the most important when discussing transfusion reactions.

## ABO Blood Group

The **ABO blood group** system is used to categorize human blood. ABO antigens appear on the surface of the red blood cells. Type A blood has type A antigens, type B blood has type B antigens, type AB blood has both types of antigens, and type O blood has neither A nor B antigens (figure 11.11). In addition, plasma from type A blood contains anti-B antibodies, which act against type B antigens; plasma

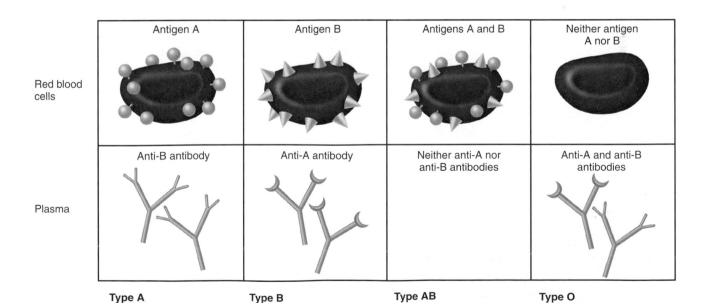

**Figure 11.11** ABO Blood Groups

(a) **No agglutination reaction.** Type A blood donated to a type A recipient does not cause an agglutination reaction because the anti-B antibodies in the recipient do not combine with the type A antigens on the red blood cells in the donated blood.

Type A blood of donor

Anti-B antibody in type A blood of recipient

Antigen and antibody do not match.

No agglutination

(b) **Agglutination reaction.** Type A blood donated to a type B recipient causes an agglutination reaction because the anti-A antibodies in the recipient combine with the type A antigens on the red blood cells in the donated blood.

Type A blood of donor

Anti-A antibody in type B blood of recipient

Antigen and antibody match.

Agglutination

**Figure 11.12** Agglutination Reaction

from type B blood contains anti-A antibodies, which act against type A antigens. Type AB blood plasma has neither type of antibody, and type O blood plasma has both anti-A and anti-B antibodies.

The ABO blood types do not exist in equal numbers. In caucasians in the United States, the distribution is type O, 47%; type A, 41%; type B, 9%; and type AB, 3%. Among African-Americans, the distribution is type O, 46%; type A, 27%; type B, 20%; and type AB, 7%.

Normally, antibodies do not develop against an antigen unless the body is exposed to that antigen; however, the anti-A and/or anti-B antibodies are present in the blood even without exposure to antigens on foreign red blood cells. One possible explanation for the production of anti-A and/or anti-B antibodies is that type A or B antigens on bacteria or food in the digestive tract stimulate the formation of antibodies against antigens that are different from the body's own antigens. In support of this explanation, anti-A and anti-B antibodies are not found in the blood until about 2 months after birth. It is possible that an infant with type A blood would produce anti-B antibodies against the B antigens on bacteria or food. Meanwhile, an infant with A antigens would not produce antibodies against the A antigens on bacteria or food because mechanisms exist in the body to prevent the production of antibodies that would react with the body's own antigens (see chapter 14).

When a blood transfusion is performed, the **donor** is the person who gives blood, and the **recipient** is the person who receives it. Usually, a recipient can successfully receive blood from a donor as long as they both have the same blood type. For example, a person with type A blood can receive blood from a person with type A blood. No ABO transfusion reaction occurs because the recipient has no anti-A antibodies against the type A antigen. On the other hand, if type A blood were donated to a person with

type B blood, a transfusion reaction would occur because the person with type B blood has anti-A antibodies against the type A antigen, causing agglutination (figure 11.12).

Historically, people with type O blood have been called universal donors because they can usually give blood to the other ABO blood types without causing an ABO transfusion reaction. Their red blood cells have no ABO surface antigens and therefore do not react with the recipient's anti-A or anti-B antibodies. For example, if a person with type A blood receives type O blood, the type O red blood cells do not react with the anti-B antibodies in the recipient's blood.

However, the term *universal* donor is misleading. Transfusion of type O blood can still produce a transfusion reaction in one of two ways: First, mismatching blood groups other than the ABO blood group can cause a transfusion reaction. To reduce the likelihood of a transfusion reaction, all the blood groups must be correctly matched (see "Type and Crossmatch" in the next section of this chapter). Second, antibodies in the donor's blood can react with antigens on the recipients's red blood cells. For example, type O blood has anti-A and anti-B antibodies. If type O blood is transfused into a person with type A blood, the anti-A antibodies (in the type O blood) react against the A antigens (on the red blood cells in the type A blood). Usually, such reactions are not serious because the antibodies in the donor's blood are diluted in the large volume of the recipient's blood. Even though such transfusion reactions seldom occur, type O blood is given to a person with another blood type only in life-or-death situations.

*Predict 3*

*Historically, people with type AB blood were called* universal recipients. *What is the rationale for this term? Explain why the term is misleading.*

# Rh Blood Group

Another important blood group is the **Rh blood group,** so named because it was first studied in the rhesus monkey. People are Rh-positive if they have certain Rh antigens on the surface of their red blood cells, and they are Rh-negative if they do not have these Rh antigens. About 85% of caucasians and 95% of African-Americans are Rh-positive. The ABO blood type and the Rh blood type are usually expressed together. For example, a person designated as type A in the ABO blood group and Rh-positive is said to be A-positive. The rarest combination in the United States is AB-negative, which occurs in less than 1% of the population.

Antibodies against the Rh antigens do not develop unless an Rh-negative person is exposed to Rh-positive red blood cells. This can occur through a transfusion or by the transfer of blood across the placenta to a mother from her fetus. When an Rh-negative person receives a transfusion of Rh-positive blood, the recipient becomes sensitized to the Rh antigens and produces anti-Rh antibodies. If the Rh-negative person is unfortunate enough to receive a second transfusion of Rh-positive blood after becoming sensitized, a transfusion reaction results.

Rh incompatibility can pose a major problem in a pregnancy when the mother is Rh-negative and the fetus is Rh-positive. If fetal blood leaks through the placenta and mixes with the mother's blood, the mother becomes sensitized to the Rh antigen. The mother produces anti-Rh antibodies that cross the placenta and cause agglutination and hemolysis of fetal red blood cells. This disorder is called **hemolytic** (hē-mō-lit'ik) **disease of the newborn (HDN),** or *erythroblastosis fetalis* (ĕ-rith'rō-blas-tō'sis fē-ta'lis), (figure 11.13). In the mother's first pregnancy, there is often no problem. The leakage of fetal blood is usually the result of a tear in the placenta that takes place either late in the pregnancy or during delivery. Thus, there is not sufficient time for the mother to produce enough anti-Rh antibodies to harm the fetus. In later pregnancies, however, a problem can arise because the mother has been sensitized to the Rh antigen. Consequently, if the fetus is Rh-positive and if any fetal blood leaks into the mother's blood, she

**1** Maternal circulation

Maternal Rh-negative red blood cell

Fetal Rh-positive red blood cell in the maternal circulation

Fetal Rh-positive red blood cell

Maternal circulation

Maternal Rh-negative red blood cell

Anti-Rh antibodies

**2**

**1** Before or during delivery, Rh-positive red blood cells from the fetus enter the blood of an Rh-negative woman through a tear in the placenta.

**2** The mother is sensitized to the Rh antigen and produces anti-Rh antibodies. Because this usually happens after delivery, the fetus is not affected in the first pregnancy.

**3** During a subsequent pregnancy with an Rh-positive fetus, Rh-positive red blood cells cross the placenta, enter the maternal circulation, and stimulate the mother to produce antibodies against the Rh antigen. Antibody production is rapid because the mother has been sensitized to the Rh antigen.

**4** The anti-Rh antibodies from the mother cross the placenta, causing agglutination and hemolysis of fetal red blood cells, and hemolytic disease of the newborn (HDN) develops.

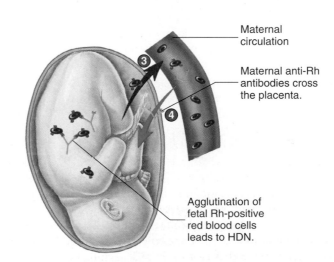

Maternal circulation

Maternal anti-Rh antibodies cross the placenta.

Agglutination of fetal Rh-positive red blood cells leads to HDN.

**PROCESS Figure 11.13   Hemolytic Disease of the Newborn (HDN)**

Circulatory

rapidly produces large amounts of anti-Rh antibodies, which can cross the placenta to the fetus, resulting in HDN. Because HDN can be fatal to the fetus, the levels of anti-Rh antibodies in the mother's blood should be monitored. If they increase to unacceptable levels, the fetus should be tested to determine the severity of the HDN. In severe cases, a transfusion to replace lost red blood cells can be performed through the umbilical cord, or the baby can be delivered if mature enough.

Prevention of HDN is often possible if the Rh-negative mother is injected with a specific preparation called Rho(D) immune globulin (RhoGAM), which contains antibodies against Rh antigens. The injection can be given during the pregnancy, before delivery, or immediately after each delivery, miscarriage, or abortion. The injected antibodies bind to the Rh antigens of any fetal red blood cells that may have entered the mother's blood. This treatment inactivates the fetal Rh antigens and prevents sensitization of the mother.

### A CASE IN POINT
#### Treatment of Hemolytic Disease of the Newborn

Billy Rubin was born with HDN. He was treated with phototherapy, exchange transfusion, and erythropoietin, all effective treatments.

During fetal development, the increased rate of red blood cell destruction caused by the mother's anti-Rh antibodies leads to increased production of bilirubin. Although high levels of bilirubin can damage the brain by killing nerve cells, this is not usually a problem in the fetus because the placenta removes the bilirubin. Following birth, bilirubin levels can increase because red blood cells continue to lyse and the newborn's liver is unable to handle the large bilirubin load. However, using phototherapy, blood passing through the skin is exposed to blue or white lights, which break down bilirubin to less toxic compounds, which the newborn's liver can then remove.

An exchange transfusion replaces Billy Rubin's blood with donor blood, thus decreasing the bilirubin and anti-Rh antibody levels. The presence of fewer anti-Rh antibodies decreases the agglutination and lysis of red blood cells. The exchange transfusion also helps alleviate the low number of red blood cells, a condition called anemia. The exchange transfusion replaces the newborn's blood with blood that has more red blood cells. Finally, administration of erythropoietin also treats the anemia by stimulating the newborn's body to produce more red blood cells.

**Predict 4**

*When treating HDN with an exchange transfusion, should the donor's blood be Rh-positive or Rh-negative? Explain.*

## DIAGNOSTIC BLOOD TESTS

### Type and Crossmatch

To prevent transfusion reactions, the blood must be typed. **Blood typing** determines the ABO and Rh blood groups of a blood sample. Typically, the cells are separated from the serum and then tested with known antibodies to determine the type of antigen on the cell surface. For example, if a patient's blood cells agglutinate when mixed with type A antibodies but do not agglutinate when mixed with type B antibodies, the cells have type A antigen. In a similar fashion, the serum is mixed with known cell types (antigens) to determine the type of antibodies in the serum.

Normally, donor blood must match the ABO and Rh type of the recipient. However, because other blood groups can cause a transfusion reaction, a crossmatch is performed. In a **crossmatch,** the donor's blood cells are mixed with the recipient's serum, and the donor's serum is mixed with the recipient's cells. The donor's blood is considered safe for transfusion only if no agglutination occurs in either match.

## Complete Blood Count

A **complete blood count (CBC)** is an analysis of blood that provides much useful information. A CBC consists of a red blood cell count, hemoglobin and hematocrit measurements, and a white blood cell count.

### Red Blood Count

Blood cell counts are usually performed electronically with a machine, but they can also be done manually with a microscope. A normal **red blood count (RBC)** for a male is 4.6–6.2 million red blood cells per microliter (μL) of blood; for a female, a normal RBC count is 4.2–5.4 million per μL of blood. (A microliter is equivalent to 1 cubic millimeter [$mm^3$] or $10^{-6}$ L, and one drop of blood is approximately 50 μL). The condition called **erythrocytosis** (ĕ-rith′rō-sī-tō′sis) is an overabundance of red blood cells.

### Hemoglobin Measurement

The amount of hemoglobin in a given volume of blood is usually expressed in terms of grams of hemoglobin per 100 mL of blood. The normal hemoglobin measurement for a male is 14–18 grams (g) per 100 mL of blood, and for a female 12 to 16 g per 100 mL of blood. An abnormally low hemoglobin measurement is an indication of **anemia** (ă-nē′mē-ă), which is either a reduced number of red blood cells or a reduced amount of hemoglobin in each red blood cell (see the Diseases and Disorders table at the end of this chapter).

### Hematocrit Measurement

The percentage of the total blood volume that is composed of red blood cells is the **hematocrit** (hē′mă-tō-krit, hem′a-tō-krit). One way to determine hematocrit is to place blood in a capillary tube and spin it in a centrifuge. The formed elements which are heavier than the plasma, are forced to one end of the tube. Of these, the white blood cells and platelets form a thin, whitish layer, called the **buffy coat,** between the plasma and the red blood cells (figure 11.14). The red blood cells account for 40–52% of the total blood volume in males and 38–48% in females. The hematocrit measurement is affected by the number and size of red blood cells because it is based on volume. For example, a decreased hematocrit can result from a decreased number of normal-size red blood cells or a normal number of small red blood cells. The average size of a red blood cell is calculated by dividing the hematocrit by the red blood cell count. A number of disorders cause red blood cells to be smaller or larger than normal. For example, inadequate iron in the diet can impair hemoglobin production. Consequently, red blood

## Erythrocytosis

Erythrocytosis is an over-abundance of red blood cells, leading to increased blood viscosity, reduced flow rates, and, if severe, plugged capillaries. **Relative erythrocytosis** results from decreased plasma volume, such as that caused by dehydration, diuretics, and burns. **Primary erythrocytosis,** often called **polycythemia vera** (pol′ē-sī-thē′mē -ă ve′ra), is a stem cell defect of unknown cause that results in the overproduction of red blood cells, granulocytes, and platelets. Erythropoietin levels are low, and the spleen may be enlarged.

**Secondary erythrocytosis** results from a decreased oxygen supply, as occurs at high altitudes, in chronic obstructive pulmonary disease, or in congestive heart failure. The resulting decrease in oxygen delivery to the kidneys stimulates erythropoietin secretion and red blood cell production. In primary and secondary erythrocytosis, the greater number of red blood cells increases blood viscosity and blood volume. Capillaries may become clogged, and hypertension can develop.

---

cells do not fill up with hemoglobin during their formation, and they remain smaller than normal.

### White Blood Count

A **white blood count (WBC)** measures the total number of white blood cells in the blood. There are normally 5000–9000 white blood cells per microliter of blood. **Leukopenia** (loo-kō-pē′nē-ă) is a lower than normal WBC resulting from decreased production or destruction of the red marrow. Radiation, drugs, tumors, viral infections, or a deficiency of the vitamins folate or $B_{12}$ can cause leukopenia. **Leukocytosis** (loo′kō-sī-tō′sis) is an abnormally high WBC. Bacterial infections often cause leukocytosis by stimulating neutrophils to increase in number. **Leukemia** (loo-kē′mē-ă), cancer of the red marrow characterized by abnormal production of one or more of the white blood cell types, can cause leukocytosis. However, the white blood cells do not function normally. Because these cells are usually immature or abnormal and lack normal immunological functions, people with leukemia are very susceptible to infections. The excess production of white blood cells in the red marrow can also interfere with the formation of red blood cells and platelets and thus lead to anemia and bleeding.

### Differential White Blood Count

A **differential white blood count** determines the percentage of each of the five kinds of white blood cells. Normally, neutrophils account for 60–70%, lymphocytes 20–25%, monocytes 3–8%, eosinophils 2–4%, and basophils 0.5–1% of all white blood cells. Much insight into a patient's condition can be obtained from a differential white blood count. For example, if a bacterial infection is present, the neutrophil count is often greatly increased, whereas in allergic reactions, the eosinophil and basophil counts are elevated.

### Clotting

The blood's ability to clot can be assessed by the platelet count and the prothrombin time measurement.

### Platelet Count

A normal **platelet count** is 250,000–400,000 platelets per microliter of blood. In the condition called **thrombocytopenia** (throm′bō-sī-tō-pē′nē-ă), the platelet count is greatly reduced, resulting in chronic bleeding through small vessels and capillaries. It can be caused by decreased platelet production as a result of hereditary

Centrifuge blood in the hematocrit tube

Hematocrit scale

100
90
80
70
60
50
40
30
20
10
0

Hematocrit tube

Plasma

White blood cells and platelets form the buffy coat.

Red blood cells

Withdraw blood into hematocrit tube.

Male    Female

#### Figure 11.14 Hematocrit

Blood is withdrawn into a capillary tube and spun in a centrifuge. The blood is separated into plasma and red blood cells, with a narrow layer of white blood cells and platelets forming in between. The hematocrit is the percentage of the blood volume that is red blood cells. It does not measure the white blood cells and platelets. As shown here, the normal hematocrit for a male lies in the range of 40%–52% of the total blood volume and for a female within the range of 38%–48%.

Circulatory

# CLINICAL IMPACT

## Anemia

Anemia is a deficiency of normal hemoglobin in the blood, resulting from a decreased number of red blood cells, a decreased amount of hemoglobin in each red blood cell, or both. Anemia can also be caused by abnormal hemoglobin production.

Anemia reduces the blood's ability to transport oxygen. People with anemia lack energy and feel excessively tired and listless. They may appear pale and become short of breath after only slight exertion.

One general cause of anemia is insufficient production of red blood cells. **Aplastic** (ā-plas'tik) **anemia** is caused by an inability of the red bone marrow to produce red blood cells and, often, white blood cells and platelets. It is usually acquired as a result of damage to the stem cells in red marrow by chemicals; such as benzene, certain antibiotics and sedatives; or radiation.

Red blood cell production can also be lower than normal due to nutritional deficiencies. **Iron-deficiency anemia** results from insufficient intake or absorption of iron or from excessive iron loss. Consequently, not enough hemoglobin is produced, the number of red blood cells decreases, and the red blood cells that are manufactured are smaller than normal.

**Folate deficiency** can also cause anemia. The usual cause of folate deficiency is inadequate folate in the diet, with the disorder developing most often in the poor, in pregnant women, and in chronic alcoholics. Because folate helps in the synthesis of

DNA, a folate deficiency results in fewer cell divisions and therefore decreased red blood cell production. A deficiency in folate during pregnancy is also associated with birth disorders called neural tube defects, such as spina bifida.

Another type of nutritional anemia is **pernicious** (per-nish'ŭs) **anemia,** which is caused by inadequate vitamin $B_{12}$. Because vitamin $B_{12}$ is important for folate synthesis, inadequate amounts of vitamin $B_{12}$ can also result in decreased red blood cell production. Although inadequate levels of vitamin $B_{12}$ in the diet can cause pernicious anemia, the usual cause is insufficient absorption of the vitamin. Normally, the stomach produces **intrinsic factor,** a protein that binds to vitamin $B_{12}$. The combined molecules pass into the lower intestine, where intrinsic factor facilitates the absorption of the vitamin. Without adequate levels of intrinsic factor, insufficient vitamin $B_{12}$ is absorbed, and pernicious anemia develops. Most cases of pernicious anemia probably result from an autoimmune disease in which the body's immune system damages the cells in the stomach that produce intrinsic factor.

Another general cause of anemia is loss or destruction of red blood cells. **Hemorrhagic** (hem-ŏ-raj'ik) **anemia** results from a loss of blood due to trauma, ulcers, or excessive menstrual bleeding. Chronic blood loss, in which small amounts of blood are lost over a period of time, can cause

iron-deficiency anemia. **Hemolytic** (hē-mō-lit'ik) **anemia** occurs when red blood cells rupture or are destroyed at an excessive rate. It can be caused by inherited defects in the red blood cells. For example, one kind of inherited hemolytic anemia results from a defect in the cell membrane that causes red blood cells to rupture easily. Many kinds of hemolytic anemia result from unusual damage to the red blood cells by drugs, snake venom, artificial heart valves, autoimmune disease, or hemolytic disease of the newborn.

Anemia can also stem from a reduced rate of synthesis of the globin chains in hemoglobin. **Thalassemia** (thal-ă-sē'mē-ă) is a hereditary disease found in people of Mediterranean, Asian, and African ancestry. If hemoglobin production is severely depressed, death usually occurs before age 20. In less severe cases, thalassemia produces a mild anemia.

Some anemias are caused by defective hemoglobin production. **Sickle-cell anemia** is a hereditary disease found mostly in people of African descent. The red blood cells assume a rigid sickle shape and plug up small blood vessels. They are also more fragile than normal. In severe cases, so much abnormal hemoglobin is produced that the disease is usually fatal before age 30. In many cases, however, enough normal hemoglobin is produced to compensate for the abnormal hemoglobin, and the person exhibits no symptoms.

disorders, lack of vitamin $B_{12}$ (pernicious anemia), drug therapy, or radiation therapy.

### Prothrombin Time Measurement

**Prothrombin time measurement** calculates how long it takes for the blood to start clotting, which is normally 9–12 seconds. Prothrombin time is determined by adding thromboplastin to whole plasma. Thromboplastin is a chemical released from injured tissues that starts the process of clotting (see figure 11.9). Prothrombin time is officially reported as the International Normalized Ratio (INR), which standardizes the time it takes to

clot on the basis of the slightly different thromboplastins used by different labs. Because many clotting factors have to be activated to form fibrin, a deficiency of any one of them can cause the prothrombin time to be abnormal. Vitamin K deficiency, certain liver diseases, and drug therapy can increase prothrombin time.

## Blood Chemistry

The composition of materials dissolved or suspended in the plasma can be used to assess the functioning of many of the body's systems. For example, high blood glucose levels can indicate that the pancreas is not producing enough insulin; high blood urea

## DISEASES AND DISORDERS  Blood

| CONDITION | DESCRIPTION |
|---|---|
| **Erythrocytosis** | |
| Relative erythrocytosis | Overabundance of red blood cells due to decreased blood volume, as may result from dehydration, diuretics, or burns |
| Primary erythrocytosis (polycythemia vera) | Stem cell defect of unknown cause; results in overproduction of red blood cells, granulocytes, and platelets; signs include low erythropoietin levels and enlarged spleen; increased blood viscosity and blood volume can cause clogging of the capillaries and hypertension |
| Secondary erythrocytosis | Overabundance of red blood cells resulting from decreased oxygen supply, as occurs at high altitudes, in chronic obstructive pulmonary disease, or in congestive heart failure; decreased oxygen delivery to the kidney stimulates the secretion of erythropoietin, resulting in increased blood viscosity and blood volume that can cause clogging of the capillaries and hypertension |
| **Clotting Disorders** | |
| Disseminated intravascular coagulation (DIC) | Clotting throughout the vascular system, followed by bleeding; may develop when normal regulation of clotting by anticoagulants is overwhelmed, as occurs due to massive tissue damage; also caused by alteration of the lining of the blood vessels resulting from infections or snake bites |
| von Willebrand disease | Most common inherited bleeding disorder; platelet plug formation and the contribution of activated platelets to blood clotting are impaired; treatments are injection of von Willebrand factor or administration of drugs that increase von Willebrand factor levels in blood, which helps platelets adhere to collagen and become activated |
| Hemophilia | Genetic disorder in which clotting is abnormal or absent; each of the several types results from deficiency or dysfunction of a clotting factor; most often a sex-linked trait that occurs almost exclusively in males (see chapter 20) |
| **Infectious Diseases of the Blood** | |
| Septicemia (blood poisoning) | Spread of microorganisms and their toxins by the blood; often the result of a medical procedure, such as insertion of an intravenous tube; release of toxins by bacteria can cause septic shock, producing decreased blood pressure and possibly death |
| Malaria | Caused by a protozoan introduced into blood by *Anopheles* mosquito; symptoms include chills and fever produced by toxins released when the protozoan causes red blood cells to rupture |
| Infectious mononucleosis | Caused by Epstein-Barr virus, which infects salivary glands and lymphocytes; symptoms include fever, sore throat, and swollen lymph nodes, all probably produced by the immune system response to infected lymphocytes |
| Acquired immunodeficiency syndrome (AIDS) | Caused by human immunodeficiency virus (HIV), which infects lymphocytes and suppresses immune system |

Go to **www.mhhe.com/seeleyess7** for additional information on these pathologies.

nitrogen (BUN) is a sign of reduced kidney function; increased bilirubin can indicate liver dysfunction; and high cholesterol levels can signify an increased risk of cardiovascular disease. A number of blood chemistry tests are routinely done when a blood sample is taken, and additional tests are available.

**Predict 5**

*When a patient complains of acute pain in the abdomen, the physician suspects appendicitis, a bacterial infection of the appendix. What blood test can provide supporting evidence for that diagnosis?*

 ## SUMMARY

### Functions of Blood  (p. 292)

1. Blood transports gases, nutrients, waste products, processed molecules, and regulatory molecules.
2. Blood regulates pH as well as fluid and ion balance.
3. Blood is involved with temperature regulation and protects against foreign substances, such as microorganisms and toxins.
4. Blood clotting prevents fluid and cell loss and is part of tissue repair.

### Composition of Blood  (p. 292)

1. Blood is a connective tissue consisting of plasma and formed elements.
2. Total blood volume in an average adult is approximately 5 L.

### Plasma  (p. 292)

1. Plasma is 91% water and 9% suspended or dissolved substances.
2. Plasma maintains osmotic pressure, is involved in immunity, prevents blood loss, and transports molecules.

### Formed Elements  (p. 294)

The formed elements are cells (red blood cells and white blood cells) and cell fragments (platelets).

#### Production of Formed Elements

Formed elements arise (hematopoiesis) in red bone marrow from stem cells.

Circulatory

### Red Blood Cells

1. Red blood cells are disk-shaped cells containing hemoglobin, which transports oxygen and carbon dioxide. Red blood cells also contain carbonic anhydrase, which is involved with carbon dioxide transport.
2. In response to low blood oxygen levels, the kidneys produce erythropoietin, which stimulates red blood cell production in red bone marrow.
3. Worn-out red blood cells are phagocytized by macrophages in the spleen or liver. Hemoglobin is broken down, iron and amino acids are reused, and heme becomes bilirubin that is secreted in bile.

### White Blood Cells

1. White blood cells protect the body against microorganisms and remove dead cells and debris.
2. Granulocytes contain cytoplasmic granules. The three types of granulocytes are neutrophils, small phagocytic cells; basophils, which promote inflammation; and eosinophils, which reduce inflammation.
3. Agranulocytes have very small granules and are of two types: Lymphocytes are involved in antibody production and other immune system responses; monocytes become macrophages that ingest microorganisms and cellular debris.

### Platelets

Platelets are cell fragments involved with preventing blood loss.

## Preventing Blood Loss   (p. 299)

### Vascular Spasm

Blood vessels constrict in response to injury, resulting in decreased blood flow.

### Platelet Plug Formation

1. Platelet plugs repair minor damage to blood vessels.
2. Platelet plugs form when platelets adhere to collagen, release chemicals (ADP and thromboxanes) that activate other platelets, and connect to one another with fibrinogen.

### Blood Clotting

1. Blood clotting, or coagulation, is the formation of a clot (a network of protein fibers called fibrin).
2. There are three steps in the clotting process: activation of clotting factors by connective tissue and chemicals, resulting in the formation of prothrombinase; conversion of prothrombin to thrombin by prothrombinase; and conversion of fibrinogen to fibrin by thrombin.
3. Anticoagulants in the blood, such as antithrombin and heparin, prevent clot formation.
4. Clot retraction condenses the clot, pulling the edges of damaged tissue closer together.
5. Serum is plasma without clotting factors.
6. Fibrinolysis (clot breakdown) is accomplished by plasmin.

## Blood Grouping   (p. 303)

1. Blood groups are determined by antigens on the surface of red blood cells.
2. In transfusion reactions, antibodies can bind to red blood cell antigens, resulting in agglutination or hemolysis of red blood cells.

### ABO Blood Group

1. Type A blood has A antigens, type B blood has B antigens, type AB blood has A and B antigens, and type O blood has neither A nor B antigens.
2. Type A blood has anti-B antibodies, type B blood has anti-A antibodies, type AB blood has neither anti-A nor anti-B antibodies, and type O blood has both anti-A and anti-B antibodies.
3. Mismatching the ABO blood group can result in transfusion reactions.

### Rh Blood Group

1. Rh-positive blood has Rh antigens, whereas Rh-negative blood does not.
2. Antibodies against the Rh antigen are produced when an Rh-negative person is exposed to Rh-positive blood.
3. The Rh blood group is responsible for hemolytic disease of the newborn, which can occur when the fetus is Rh-positive and the mother is Rh-negative.

## Diagnostic Blood Tests   (p. 306)

### Type and Crossmatch

1. Blood typing determines the ABO and Rh blood groups of a blood sample.
2. A crossmatch tests for agglutination reactions between donor and recipient blood.

### Complete Blood Count

The complete blood count consists of the red blood count (million/μL), the hemoglobin measurement (grams of hemoglobin per 100 mL of blood), the hematocrit measurement (percent volume of red blood cells), and the white blood count (million/μL).

### Differential White Blood Count

The differential white blood count determines the percentage of each type of white blood cell.

### Clotting

Platelet count and prothrombin time measurement determine the blood's ability to clot.

### Blood Chemistry

The composition of materials dissolved or suspended in plasma (e.g., glucose, urea nitrogen, bilirubin, and cholesterol) can be used to assess the functioning and status of the body's systems.

 REVIEW AND COMPREHENSION

1. Describe the functions of blood.
2. Define plasma, and list its functions.
3. Define formed elements, and name the different types of formed elements. Explain how and where the formed elements arise through hematopoiesis.
4. Describe the two basic parts of a hemoglobin molecule. Which part is associated with iron? What gases are transported by each part?
5. What is the role of carbonic anhydrase in gas transport?
6. Why are the vitamins folate and $B_{12}$ important in red blood cell production?
7. Explain how low blood oxygen levels result in increased red blood cell production.
8. Where are red blood cells broken down? What happens to the breakdown products?

9. Give two functions of white blood cells.

10. Name the five types of white blood cells, and state a function for each type.

11. What are platelets, and how are they formed?

12. Describe the role of blood vessel constriction and platelet plugs in preventing bleeding. Describe the three steps of platelet plug formation.

13. What are clotting factors? Describe the three steps of activation that result in the formation of a clot.

14. Explain the function of anticoagulants in the blood, and give an example of an anticoagulant.

15. What is clot retraction, and what does it accomplish?

16. Define fibrinolysis, and name the chemicals responsible for this process.

17. What are blood groups, and how do they cause transfusion reactions? List the four ABO blood types. Why is a person with type O blood considered a universal donor?

18. What is meant by the term *Rh-positive*? How can Rh incompatibility affect a pregnancy?

19. For each of the following tests, define the test and give an example of a disorder that would cause an abnormal test result:
    a. type and crossmatch
    b. red blood count
    c. hemoglobin measurement
    d. hematocrit measurement
    e. white blood count
    f. differential white blood count
    g. platelet count
    h. prothrombin time
    i. blood chemistry

# CRITICAL THINKING

1. Red Packer, a physical education major, wanted to improve his performance in an upcoming marathon race. About 6 weeks before the race, he had 1 L of blood removed from his body and the formed elements separated from the plasma. The formed elements were frozen, and the plasma was reinfused into his body. Just before the race, the formed elements were thawed and injected into his circulatory system. Explain why this procedure, called blood doping or blood boosting, would help Red's performance. Can you suggest any possible bad effects?

2. Chemicals such as benzene can destroy red bone marrow, causing aplastic anemia. What symptoms would you expect as a result of the lack of (a) red blood cells, (b) platelets, and (c) white blood cells?

3. E. Z. Goen habitually used barbiturates to ease feelings of anxiety. Because barbiturates depress the respiratory centers in the brain, they cause hypoventilation (i.e., a slower than normal breathing rate). What happens to the red blood cell count of a habitual user of barbiturates? Explain.

4. What blood problems would you expect in a patient after a total gastrectomy (removal of the stomach)?

5. According to an old saying, "Good food makes good blood." Name three substances in the diet that are essential for "good blood." What blood disorders develop if these substances are absent from the diet?

6. Why do many anemic patients have gray feces? (*Hint:* The feces are lacking their normal coloration.)

7. Reddie Popper has a cell membrane defect in her red blood cells that makes them more susceptible to rupture. Her red blood cells are destroyed faster than they can be replaced. Would her RBC, hemoglobin, hematocrit, and bilirubin levels be below normal, normal, or above normal? Explain.

*Answers in Appendix D*

Visit this textbook's website at **www.mhhe.com/seeleyess7** for practice quizzes, animations, interactive learning exercises, and other study tools.

McGraw-Hill offers a study CD that features interactive cadaver dissection. **Anatomy & Physiology Revealed** includes cadaver photos that allow you to peel away layers of the human body to reveal structures beneath the surface. This program also includes animations, radiologic imaging, audio pronunciations, and practice quizzing.

The heart is essential to human life. Checking the heartbeat is one of the first tests to measure someone's health.

Chapter 12

# Heart

## Chapter Outline and Objectives

### Functions of the Heart (p. 313)
1. List the major functions of the heart.

### Size, Form, and Location of the Heart (p. 313)
2. Describe the size, shape, and location of the heart, and explain why knowing its location is important.

### Anatomy of the Heart (p. 314)
3. Describe the structure of the pericardium.
4. Give the location and function of the coronary arteries.
5. Describe the chambers of the heart.
6. Name the valves of the heart, and state their locations and functions.
7. Describe the flow of blood through the heart, and name each of the chambers and structures through which the blood passes.

### Histology of the Heart (p. 322)
8. List the components of the heart wall, and describe the structure and function of each.
9. Describe the structural and functional characteristics of cardiac muscle cells.

### Electrical Activity of the Heart (p. 324)
10. Describe the characteristics of action potentials in cardiac muscle.

11. Explain the structure and function of the conduction system of the heart.
12. Describe the waves of an electrocardiogram, and relate each of them to contractions of the heart.

### Cardiac Cycle (p. 328)
13. Describe the cardiac cycle and the relationship among the contraction of each of the chambers, the pressure in each of the chambers, the phases of the electrocardiogram, and the heart sounds.

### Heart Sounds (p. 331)
14. Describe the heart sounds and their significance.

### Regulation of Heart Function (p. 332)
15. Describe intrinsic and extrinsic regulation of the heart.
16. Give the conditions for which the major heart medications and treatments are administered.

### Effects of Aging on the Heart (p. 340)
17. List the major age-related changes that affect the heart.

People often refer to the heart as the seat of strong emotions. For example, we may describe a very determined person as having "a lot of heart" or say that a person who has been disappointed romantically has a "broken heart." Emotions, however, are a product of brain function, not heart function. The heart is a muscular organ that is essential for life because it pumps blood through the body.

Fluids flow through a pipe only if they are forced to do so. The force is commonly produced by a pump, which increases the pressure of the liquid at the pump above the pressure in the pipe. Thus, the liquid flows from the pump through the pipe from an area of higher pressure to an area of lower pressure. If the pressure produced by the pump increases, flow of liquid through the pipe increases. If the pressure produced by the pump decreases, flow of liquid through the pipe decreases.

Like a pump that forces water through a pipe, the heart contracts forcefully to pump blood through the blood vessels of the body. Together, the heart, the blood vessels, and the blood make up the **cardiovascular system** (figure 12.1). The heart of a healthy adult, at rest, pumps approximately 5 liters (L) of blood per minute. For most people, the heart continues to pump at approximately that rate for more than 75 years. During short periods of vigorous exercise, the amount of blood pumped per minute increases several-fold. But if the heart loses its pumping ability for even a few minutes, blood flow through the blood vessels stops, and the person's life is in danger.

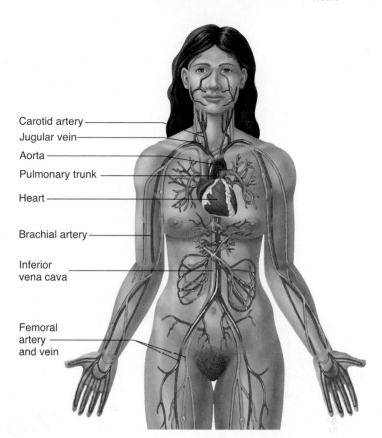

Carotid artery
Jugular vein
Aorta
Pulmonary trunk
Heart
Brachial artery
Inferior vena cava
Femoral artery and vein

**Figure 12.1  Cardiovascular System**
The heart, the blood vessels, and the blood are the major components of the cardiovascular system.

The heart is actually two pumps in one. The right side of the heart pumps blood to the lungs and back to the left side of the heart through vessels of the **pulmonary circulation** (figure 12.2). The left side of the heart pumps blood to all other tissues of the body and back to the right side of the heart through vessels of the **systemic circulation.**

# FUNCTIONS OF THE HEART

The functions of the heart are:

1. *Generating blood pressure.* Contractions of the heart generate blood pressure, which is required to force blood through the blood vessels.
2. *Routing blood.* The heart separates the pulmonary and systemic circulations, which ensures the flow of oxygen-rich blood to tissues.
3. *Ensuring one-way blood flow.* The valves of the heart ensure a one-way flow of blood through the heart and blood vessels.
4. *Regulating blood supply.* Changes in the rate and force of heart contraction match blood flow to the changing metabolic needs of the tissues during rest, exercise, and changes in body position.

# SIZE, FORM, AND LOCATION OF THE HEART

The adult heart is shaped like a blunt cone and is approximately the size of a closed fist. It is larger in physically active adults than in less active but otherwise healthy adults. The heart generally decreases in size after approximately age 65, especially in people who are not physically active. The blunt, rounded point of the heart is the **apex** and the larger, flat part at the opposite end of the heart is the **base.**

The heart is located in the thoracic cavity between the two pleural cavities that surround the lungs. The heart, trachea, esophagus, and associated structures form a midline partition, the **mediastinum** (me′dē-as-tī′nŭm; see figure 1.12). The heart is surrounded by its own cavity, the **pericardial cavity** (*peri*, around + *cardio*, heart) (see chapter 1).

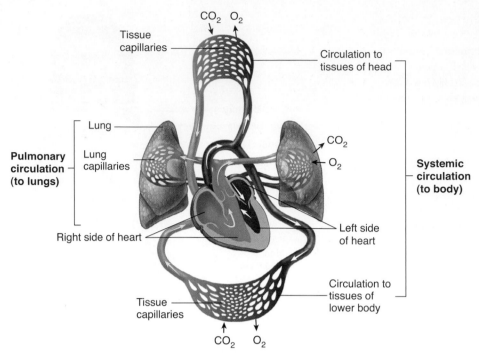

**Figure 12.2  Overview of the Circulatory System**

The circulatory system consists of the pulmonary and systemic circulations. The right side of the heart pumps blood through vessels to the lungs and back to the left side of the heart through the pulmonary circulation. The left side of the heart pumps blood through vessels to the body tissues and back to the right side of the heart through the systemic circulation.

It is important for health professionals to know the location and shape of the heart in the thoracic cavity. This knowledge enables them to accurately place a stethoscope to hear the heart sounds, to place chest leads for an electrocardiogram (described later in this chapter), or to administer **cardiopulmonary resuscitation** (kar′dē-ō-pǔl′mo-nār-ē rē-sǔs′i-tā-shun; **CPR).**

The heart lies obliquely in the mediastinum, with its base directed posteriorly and slightly superiorly and its apex directed anteriorly and slightly inferiorly. The apex is also directed to the left, so that approximately two-thirds of the heart's mass lies to the left of the midline of the sternum (figure 12.3). The base of the heart is located deep to the sternum and extends to the level of the second intercostal space. The apex is deep to the left fifth intercostal space, approximately 7–9 centimeters (cm) to the left of the sternum near the midclavicular line, which is a perpendicular line that extends down from the middle of the clavicle.

# ANATOMY OF THE HEART

## Pericardium

The heart lies in the **pericardial cavity.** The pericardial cavity is formed by the **pericardium** (per-i-kar′dē-ŭm), or *pericardial sac,* tissues that surround the heart and anchor it within the mediastinum (figure 12.4). The pericardium consists of two layers. The tough, fibrous connective tissue outer layer is called the **fibrous pericardium,** and the inner layer of flat epithelial cells, with a thin layer of connective tissue, is called the **serous pericardium.** The portion of the serous pericardium lining the fibrous pericardium is the **parietal pericardium,** whereas the

portion covering the heart surface is the **visceral pericardium,** or *epicardium* (ep-i-kar′dē-ŭm; upon the heart). The parietal and visceral pericardia are continuous with each other where the great vessels enter or leave the heart. The pericardial cavity, located between the visceral and parietal pericardia, is filled with a thin layer of **pericardial fluid** produced by the serous pericardium. The pericardial fluid helps reduce friction as the heart moves within the pericardium.

## External Anatomy

The right and left **atria** (a′trē-ă; sing. atrium, entrance chamber) are located at the base of the heart, and the right and left **ventricles** (ven′tri-klz; cavities) extend from the base of the heart toward the apex (figure 12.5). A **coronary** (kōr′o-nār-ē) **sulcus** (sool′kus) extends around the heart, separating the atria from the ventricles. In addition, two grooves, or sulci, which indicate the division between the right and left ventricles, extend inferiorly from the coronary sulcus. The **anterior interventricular sulcus** extends inferiorly from the coronary sulcus on the anterior surface of the heart, and the **posterior interventricular sulcus** extends inferiorly from the coronary sulcus on the posterior surface of the heart.

Six large veins carry blood to the heart (figure 12.5a,c): The **superior vena cava** and **inferior vena cava** carry blood from the body to the right atrium, and four **pulmonary** (pǔl′mō-nār-ē; lung) **veins** carry blood from the lungs to the left atrium. Two arteries, the **pulmonary trunk** and the **aorta** (ā-ōr′tă), exit the heart. The pulmonary trunk, arising from the right ventricle, splits into the right and left **pulmonary arteries,** which carry blood to

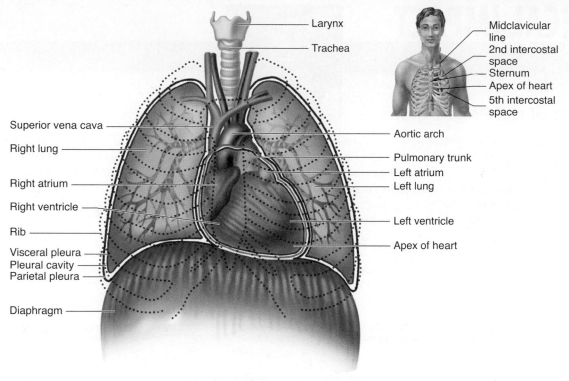

Larynx

Trachea

Midclavicular line
2nd intercostal space
Sternum
Apex of heart
5th intercostal space

Superior vena cava

Right lung

Right atrium

Right ventricle

Rib

Visceral pleura
Pleural cavity
Parietal pleura

Diaphragm

Aortic arch

Pulmonary trunk
Left atrium
Left lung

Left ventricle

Apex of heart

**Anterior view**

## Figure 12.3 Location of the Heart

The heart is in the thoracic cavity between the lungs, deep and slightly to the left of the sternum. The base of the heart, located deep to the sternum, extends superiorly to the second intercostal space, and the apex of the heart is located deep to the fifth intercostal space, approximately 7–9 cm to the left of the sternum where the midclavicular line intersects with the fifth intercostal space (see inset).

Pericardium

Fibrous pericardium

Serous pericardium

Parietal pericardium
Visceral pericardium (or epicardium)

Pericardial cavity filled with pericardial fluid

**Anterior view**

## Figure 12.4 Heart in the Pericardium

The pericardium consists of an outer fibrous pericardium and an inner serous pericardium. The serous pericardium has two parts: The parietal pericardium lines the fibrous pericardium, and the visceral pericardium (epicardium) covers the surface of the heart. The pericardial cavity, between the parietal pericardium and visceral pericardium, is filled with a small amount of pericardial fluid.

Circulatory

## Disorders of the Pericardium

**Pericarditis** (per'i-kar-dī'tis) is an inflammation of the serous pericardium. The cause is frequently unknown, but it can result from infection, diseases of connective tissue, or damage due to radiation treatment for cancer. The condition can cause extremely painful sensations that are referred to the back and to the chest and can be confused with a myocardial infarction (heart attack). Pericarditis can lead to a small amount of fluid accumulation within the pericardial sac.

**Cardiac tamponade** (tam-pŏ-nād'; a pack or plug) is a potentially fatal condition in which fluid or blood accumulates in the pericardial cavity and compresses the heart from the outside. The heart is a powerful muscle, but it relaxes passively. When it is compressed by fluid within the pericardial cavity, it cannot expand when the cardiac muscle relaxes. Consequently, the heart cannot fill with blood during relaxation, which makes pumping impossible.

Cardiac tamponade can cause a person to die quickly unless the fluid is removed. Causes of cardiac tamponade include rupture of the heart wall following a myocardial infarction, rupture of blood vessels in the pericardium after a malignant tumor invades the area, damage to the pericardium due to radiation therapy, and trauma, such as that resulting from a traffic accident.

---

the lungs. The aorta, arising from the left ventricle, carries blood to the rest of the body.

## Heart Chambers and Internal Anatomy

The heart is a muscular pump consisting of four chambers: the right and left atria and the right and left ventricles (figure 12.6).

### Right and Left Atria

The atria of the heart receive blood from veins. The atria function primarily as reservoirs, where blood returning from veins collects before it enters the ventricles. Contraction of the atria forces blood into the ventricles to complete ventricular filling. The right atrium receives blood through three major openings.

Superior vena cava

Branches of right pulmonary artery

Right pulmonary veins

**Right atrium**

Coronary sulcus

Right coronary artery

**Right ventricle**

Inferior vena cava

Aortic arch

Left pulmonary artery

Branches of left pulmonary artery

Pulmonary trunk

Left pulmonary veins

**Left atrium**

Great cardiac vein (in anterior interventricular sulcus)

Anterior interventricular artery (in anterior interventricular sulcus)

**Left ventricle**

(a)

**Anterior view**

**Figure 12.5  Surface of the Heart**

(a) In this anterior view of the heart, the two atria (right and left) are located superiorly, and the two ventricles (right and left) are located inferiorly. The superior and inferior venae cavae enter the right atrium. The pulmonary veins enter the left atrium. The pulmonary trunk exits the right ventricle, and the aorta exits the left ventricle.

Circulatory

(b) **Anterior view**

(c) **Posterior view**

**Figure 12.5** *(continued)*

(b) Photograph of the anterior surface of the heart. (c) In this posterior view of the heart, the two atria (right and left) are located superiorly, and the two ventricles (right and left) are located inferiorly. The superior and inferior venae cavae enter the right atrium, and the four pulmonary veins enter the left atrium.

The superior vena cava and the inferior vena cava drain blood from most of the body (figure 12.6), and the smaller coronary sinus drains blood from most of the heart muscle. The left atrium receives blood through the four pulmonary veins, which drain blood from the lungs. The two atria are separated from each other by a partition called the **interatrial** (between the atria) **septum.**

### Right and Left Ventricles

The ventricles of the heart are its major pumping chambers. They eject blood into the arteries and force it to flow through the circulatory system. The atria open into the ventricles, and each ventricle has one large outflow route located superiorly near the midline of the heart. The right ventricle pumps blood into the pulmonary trunk, and the left ventricle pumps blood into the aorta. The two

## Figure 12.6  Internal Anatomy of the Heart
The heart is cut in a frontal plane to show its internal anatomy.

ventricles are separated from each other by the muscular **interventricular** (between the ventricles) **septum** (figure 12.6).

The wall of the left ventricle is thicker than the wall of the right ventricle, and the wall of the left ventricle contracts more forcefully and generates a greater blood pressure than the wall of the right ventricle. When the left ventricle contracts, the pressure increases to approximately 120 mm Hg. When the right ventricle contracts, the pressure increases to approximately one-fifth of the pressure in the left ventricle. However, the left and right ventricles pump nearly the same volume of blood. The higher pressure generated by the left ventricle moves blood through the larger systemic circulation, whereas the lower pressure generated by the right ventricle moves blood through the smaller pulmonary circulation (see figure 12.2).

## Heart Valves

The **atrioventricular (AV) valves** are located between the right atrium and the right ventricle and between the left atrium and the left ventricle. The AV valve between the right atrium and the right ventricle has three cusps and is called the **tricuspid valve** (figure 12.7a).

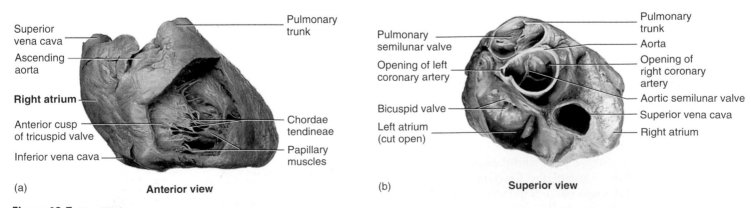

## Figure 12.7  Heart Valves
(a) Anterior view of the tricuspid valve, the chordae tendineae, and the papillary muscles. (b) In superior view, note that the three cusps of each semilunar valve meet to prevent the backflow of blood.

(a) **Anterior view**    (b) **Anterior view**

**Figure 12.8** Function of the Heart Valves

(*a*) As the ventricle relaxes, the pressure in the ventricle becomes lower than the pressure in the atrium. Blood flowing into the left atrium opens the bicuspid valve, and blood flows into the left ventricle. At the same time, in the aorta, blood flowing back toward the relaxed ventricle causes the aortic semilunar valve to close, and no blood can reenter the ventricle from the aorta. (*b*) When the ventricle contracts, blood flowing toward the atrium causes the bicuspid valve to close. The increased pressure in the left ventricle forces the aortic semilunar valve open.

The AV valve between the left atrium and the left ventricle has two cusps and is called the **bicuspid valve** or *mitral* (resembling a bishop's miter, a two-pointed hat) *valve* (figure 12.7*b*). These valves allow blood to flow from the atria into the ventricles but prevent it from flowing back into the atria. When the ventricles relax, the higher pressure in the atria forces the AV valves to open, and blood flows from the atria into the ventricles (figure 12.8*a*). In contrast, when the ventricles contract, blood flows toward the atria and causes the AV valves to close (figure 12.8*b*).

Each ventricle contains cone-shaped, muscular pillars called **papillary** (pap′ĭ-lār-ē) **muscles.** These muscles are attached by thin, strong, connective tissue strings called **chordae tendineae** (kōr′dē ten′di-nē-ē; heart strings) to the free margins of the cusps of the atrioventricular valves. When the ventricles contract, the papillary muscles contract and prevent the valves from opening into the atria by pulling on the chordae tendineae attached to the valve cusps (see figure 12.7*a*; figure 12.8).

The aorta and pulmonary trunk possess **aortic** and **pulmonary semilunar valves,** respectively (see figure 12.6). Each valve consists of three pocketlike semilunar (half-moon-shaped) cusps (see figure 12.7*b*; figure 12.8). When the ventricles contract, the increasing pressure within the ventricles forces the semilunar valves to open (figure 12.8*b*). When the ventricles relax, the pressures in the aorta and pulmonary trunk are higher than in the ventricles. Blood flows back from the aorta or pulmonary trunk toward the ventricles and enters the pockets of the cusps, causing them to bulge toward and meet in the center of the aorta or pulmonary trunk, thus closing the vessels and blocking blood flow back into the ventricles (figure 12.8*a*).

A plate of connective tissue, sometimes called the **cardiac skeleton,** consists mainly of fibrous rings that surround the atrioventricular and semilunar valves and give them solid support (figure 12.9). This connective tissue plate also serves as electrical

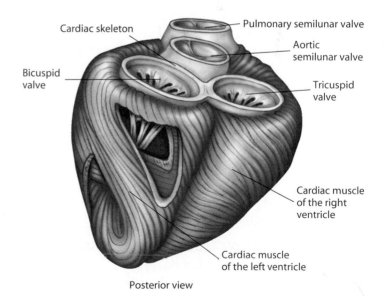

Posterior view

**Figure 12.9** Cardiac Skeleton

The cardiac skeleton consists of fibrous connective tissue rings that surround the heart valves and separate the atria from the ventricles. Cardiac muscle attaches to the fibrous connective tissue. The muscle fibers are arranged so that, when the ventricles contract, a wringing motion is produced, and the distance between the apex and the base of the heart shortens.

Superior vena cava

Branches of right pulmonary arteries

Aortic semilunar valve

Pulmonary veins

Pulmonary semilunar valve

**Right atrium**

Tricuspid valve

Papillary muscles

**Right ventricle**

Inferior vena cava

Aortic arch

Branches of left pulmonary arteries

Pulmonary trunk

Pulmonary veins

**Left atrium**

Bicuspid valve

**Left ventricle**

Interventricular septum

(a)

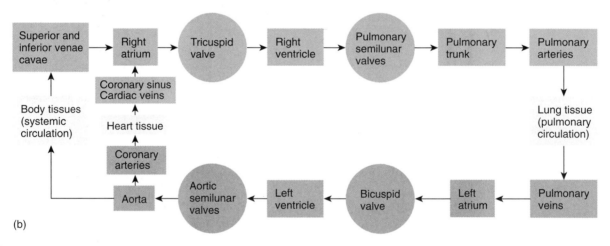

(b)

## Figure 12.10  Blood Flow Through the Heart

(a) Frontal section of the heart, revealing the four chambers and the direction of blood flow. (b) Diagram listing in order the structures through which blood flows in the systemic and pulmonary circulations. The heart valves are indicated by *circles,* deoxygenated blood by *blue,* and oxygenated blood by *red.*

insulation between the atria and the ventricles and provides a rigid attachment site for cardiac muscle.

## Route of Blood Flow Through the Heart

The route of blood flow through the heart is depicted in figure 12.10. Even though blood flow is described for the right and then the left side of the heart, it is important to understand that both atria contract at the same time, and both ventricles contract at the same time. This concept is most important when considering the electrical activity, pressure changes, and heart sounds.

Blood enters the right atrium from the systemic circulation through the superior and inferior venae cavae, and from heart muscle through the coronary sinus. Most of the blood flowing into the right atrium flows into the right ventricle while the right ventricle relaxes following the previous contraction. Before the end of ventricular relaxation, the right atrium contracts, and enough blood is pushed from the right atrium into the right ventricle to complete right ventricular filling.

Following right atrial contraction, the right ventricle begins to contract. This contraction pushes blood against the tricuspid valve, forcing it closed. After pressure within the right ventricle increases, the pulmonary semilunar valve is forced open, and blood flows into the pulmonary trunk. As the right ventricle relaxes, its pressure falls rapidly, and pressure in the pulmonary trunk becomes greater than in the right ventricle. The backflow of blood forces the pulmonary semilunar valve to close.

The pulmonary trunk branches to form the right and left pulmonary arteries, which carry blood to the lungs, where carbon dioxide is released and oxygen is picked up. Blood returning from the lungs enters the left atrium through the four pulmonary veins (figure 12.10). Most of the blood flowing into the left atrium passes into the left ventricle while the left ventricle relaxes following the previous contraction. Before the end of ventricular relaxation, the left atrium contracts, and enough blood is pushed from the left atrium into the left ventricle to complete left ventricular filling.

Following left atrial contraction, the left ventricle begins to contract. This contraction pushes blood against the bicuspid valve, forcing it closed. After pressure within the left ventricle increases, the aortic semilunar valve is forced open, and blood flows into the aorta. Blood flowing through the aorta is distributed to all parts of the body, except to that part of the lung supplied by the pulmonary blood vessels. As the left ventricle relaxes, its pressure falls rapidly, and pressure in the aorta becomes greater than in the left ventricle. The backflow of blood forces the aortic semilunar valve to close.

## Blood Supply to the Heart

### Coronary Arteries

The cardiac muscle in the wall of the heart is thick and metabolically very active. Two **coronary arteries** supply blood to the wall of the heart (figure 12.11*a*). The coronary arteries originate from the base of the aorta, just above the aortic semilunar valves. The **left coronary artery** originates on the left side of the aorta. It has three major branches: The **anterior interventricular artery** lies in the anterior interventricular sulcus; the **circumflex artery** extends around the coronary sulcus on the left to the posterior surface of the heart; and the **left marginal artery** extends inferiorly along the lateral wall of the left ventricle from the circumflex artery. The branches of the left coronary artery supply much of the anterior wall of the heart and most of the left ventricle. The **right coronary artery** originates on the right side of the aorta. It extends around the coronary sulcus on the right to the posterior surface of the heart and gives rise to the **posterior interventricular artery,** which lies in the posterior interventricular sulcus. The **right marginal artery** extends inferiorly along the lateral wall of the right ventricle. The right coronary artery and its branches supply most of the wall of the right ventricle.

In a resting person, blood flowing through the coronary arteries gives up approximately 70% of its oxygen. In comparison, blood flowing through arteries to skeletal muscle gives up only about 25% of its oxygen. The percentage of oxygen the blood releases to skeletal muscle increases to 70% or more during exercise, but the percentage of oxygen the blood releases to cardiac muscle cannot increase substantially during exercise. Therefore, the rate of blood flow through the coronary arteries must increase above its resting level to provide cardiac muscle with adequate oxygen during exercise. Blood flow into the coronary circulation is greatest while the ventricles of the heart are relaxed and contraction of the cardiac muscle does not compress the coronary

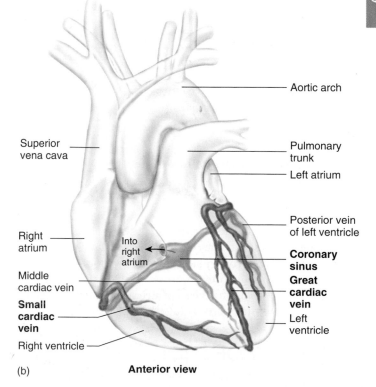

**(a)** **Anterior view**

**(b)** **Anterior view**

**Figure 12.11** **Blood Supply to the Heart**

The vessels of the anterior surface of the heart depicted here are darker in color, whereas the vessels of the posterior surface are seen through the heart and are lighter in color. (*a*) Coronary arteries supply blood to the wall of the heart. (*b*) Cardiac veins carry blood from the wall of the heart back to the right atrium.

# CLINICAL IMPACT

## Heart Attack

When a blood clot, or **thrombus** (throm′bŭs; a clot), suddenly blocks a coronary blood vessel, a **heart attack,** or *coronary thrombosis* (throm-bō′sis), occurs. The area that has been cut off from its blood supply suffers from lack of oxygen and nutrients and dies if the blood supply is not quickly reestablished. The region of dead heart tissue is called an **infarct** (in′farkt), or *myocardial infarction.* If the infarct is large, the heart may be unable to pump enough blood to keep the person alive. People can reduce their risk of having a heart attack by taking small amounts of aspirin daily, which inhibits thrombus formation (see chapter 11).

Aspirin is also administered to many people who are exhibiting clear symptoms of a heart attack. In some cases, it is possible to treat heart attacks with enzymes, such as **streptokinase** (strep-tō-kī′nās) or **tissue plasminogen** (plaz-min′o-jen) **activator (t-PA),** which break down blood clots. One of the enzymes is injected into the circulatory system of a heart attack patient, where it reduces or removes the blockage in the

coronary artery. If the clot is broken down quickly, the blood supply to cardiac muscle is reestablished, and the heart may suffer little permanent damage.

Coronary arteries can become blocked more gradually by **atherosclerotic** (ath′er-ō-skler-ot′ik; *athero,* pasty material + *sclerosis,* hardness) **plaque.** These thickenings in the walls of arteries can contain deposits that are high in cholesterol and other lipids. The lesions narrow the lumen (opening) of the arteries, thus restricting blood flow. The ability of cardiac muscle to function is reduced when it is deprived of an adequate blood supply. The person suffers from fatigue and often experiences pain in the area of the chest and usually in the left arm with the slightest exertion. The pain is called **angina pectoris** (an-jī′nă, pain; pek′tō-ris, in the chest).

**Angioplasty** (an′jē-ō-plas-tē) is a procedure in which a surgeon threads a small balloon through the aorta and into a coronary artery. After entering a partially blocked coronary artery, the balloon is inflated, flattening the atherosclerotic deposits against

the vessel wall and opening the blocked blood vessel. This technique improves the function of cardiac muscle in patients experiencing inadequate blood flow to the cardiac muscle through the coronary arteries. Some controversy exists about its effectiveness, at least in some patients, because dilation of the coronary arteries can be reversed within a few weeks or months and because blood clots can form in coronary arteries following angioplasty. Small rotating blades and lasers are also used to remove lesions from coronary vessels, or a small coil device, called a **stent,** may be placed in a vessel to hold it open following angioplasty.

A **coronary bypass** is a surgical procedure that relieves the effects of obstructions in the coronary arteries. The technique involves taking healthy segments of blood vessels from other parts of the patient's body and using them to bypass, or create an alternative path around, obstructions in the coronary arteries. The technique is common in cases of severe blockage of parts of the coronary arteries.

---

arteries. Blood flow into other arteries of the body is highest during contraction of the ventricles.

### Predict 1

*Predict the effect on the heart if blood flow through the anterior interventricular artery is restricted or completely blocked (Hint: See figure 12.11a).*

### Cardiac Veins

The **cardiac veins** drain blood from the cardiac muscle. Their pathways are nearly parallel to the coronary arteries, and most of them drain blood into the **coronary sinus,** a large vein located within the coronary sulcus on the posterior aspect of the heart. Blood flows from the coronary sinus into the right atrium (figure 12.11*b*). Some small cardiac veins drain directly into the right atrium.

# HISTOLOGY OF THE HEART

## Heart Wall

The heart wall is composed of three layers of tissue: the epicardium, the myocardium, and the endocardium (figure 12.12). The

**epicardium** (ep-i-kar′dē-ŭm), also called the *visceral pericardium,* is a thin, serous membrane forming the smooth outer surface of the heart. It consists of simple squamous epithelium overlying a layer of loose connective tissue and adipose tissue. The thick, middle layer of the heart, the **myocardium** (mī-ō-kar′dē-ŭm), is composed of cardiac muscle cells and is responsible for contraction of the heart chambers. The smooth inner surface of the heart chambers is the **endocardium** (en-dō-kar′dē-ŭm), which consists of simple squamous epithelium over a layer of connective tissue. The endocardium allows blood to move easily through the heart. The heart valves are formed by folds of endocardium that include a thick layer of connective tissue.

The surfaces of the interior walls of the ventricles are modified by ridges and columns of cardiac muscle called **trabeculae carneae.** Smaller muscular ridges are also present in portions of the atria.

## Cardiac Muscle

Cardiac muscle cells are elongated, branching cells that contain one, or occasionally two, centrally located nuclei (figure 12.13). Cardiac muscle cells contain actin and myosin myofilaments organized to

**Figure 12.12 Heart Wall**

This enlarged section of the heart wall illustrates the epicardium (visceral pericardium), the myocardium, and the endocardium.

form sarcomeres, which are joined end-to-end to form myofibrils (see chapter 7). The actin and myosin myofilaments are responsible for muscle contraction, and their organization gives cardiac muscle a striated (banded) appearance much like that of skeletal muscle. However, the striations are less regularly arranged and less numerous than in skeletal muscle.

Like skeletal muscle, cardiac muscle relies on $Ca^{2+}$ and ATP for contraction. $Ca^{2+}$ enter cardiac muscle cells in response to action potentials and activate the process of contraction much as they do in skeletal muscle. ATP production depends on oxygen availability. Cardiac muscle cells are rich in mitochondria, which produce ATP at a rate rapid enough to sustain the normal energy requirements of cardiac muscle. An extensive capillary network provides adequate oxygen to the cardiac muscle cells. Unlike skeletal muscle, cardiac muscle cannot develop a significant oxygen deficit. Development of a large oxygen deficit could result in muscular fatigue and cessation of cardiac muscle contraction.

Cardiac muscle cells are organized into spiral bundles or sheets (see figure 12.9). When cardiac muscle fibers contract, not only do the muscle fibers shorten but the spiral bundles twist to compress the contents of the heart chambers. Cardiac muscle cells are bound end-to-end and laterally to adjacent cells by specialized cell-to-cell contacts called **intercalated** (in-ter′kă-lā-ted) **disks** (figure 12.13). The membranes of the intercalated disks are highly folded, and the adjacent cells fit together, greatly increasing contact between them and preventing cells from pulling apart. Specialized cell membrane structures in the intercalated disks called **gap junctions** (see chapter 4) allow cytoplasm to flow freely between cells. This enables action potentials to pass easily from one cell to the next. The cardiac muscle cells of the atria or ventricles, therefore,

(a)                                    (b)

**Figure 12.13 Cardiac Muscle Cells**

(*a*) Cardiac muscle cells are branching cells with centrally located nuclei. As in skeletal muscle, sarcomeres join end-to-end to form myofibrils, and mitochondria provide ATP for contraction. The cells are joined to one another by intercalated disks, which allow action potentials to pass from one cardiac muscle cell to the next. Sarcoplasmic reticulum and T tubules are visible but are not as numerous as they are in skeletal muscle. (*b*) In a light micrograph of cardiac muscle tissue, the muscle fibers appear striated because of the arrangement of the individual myofilaments.

contract at nearly the same time. The heart's highly coordinated pumping action depends on this characteristic.

# ELECTRICAL ACTIVITY OF THE HEART

## Action Potentials in Cardiac Muscle

Like action potentials in skeletal muscle and neurons, those in cardiac muscle exhibit depolarization followed by repolarization. In cardiac muscle, however, a period of slow repolarization greatly prolongs the action potential (figure 12.14). In contrast to action potentials in skeletal muscle, which take less than 2 milliseconds (ms) to complete, action potentials in cardiac muscle take approximately 200 to 500 ms to complete. In addition, unlike in skeletal muscle, action potentials in cardiac muscle are conducted from cell to cell.

In cardiac muscle, each action potential consists of a **depolarization phase** followed by a period of slow repolarization called the **plateau phase.** At the end of the plateau phase, a rapid **repo-**

**larization phase** takes place. During the final repolarization phase, the membrane potential achieves its maximum degree of repolarization (figure 12.14) and returns to the resting membrane potential.

The opening and closing of membrane channels is responsible for the changes in the permeability of the cell membrane that produce action potentials (see chapter 8). The initial, rapid depolarization phase of the action potential results from the opening of voltage-gated $Na^+$ channels, which increases the permeability of the cell membrane to $Na^+$. Sodium ions then diffuse into the cell, causing depolarization. This depolarization stimulates the opening of voltage-gated $Ca^{2+}$ channels, and $Ca^{2+}$ begin diffusing into the cell, contributing to the overall depolarization. At the peak of depolarization, the $Na^+$ channels close, and a small number of $K^+$ channels open. However, the $Ca^{2+}$ channels remain open. Thus, the exit of $K^+$ from the cell is counteracted by the continued movement of $Ca^{2+}$ into the cell. Consequently, the plateau phase is primarily the result of the opening of voltage-regulated $Ca^{2+}$ channels. The slow diffusion of $Ca^{2+}$ into the cell

**Skeletal Muscle**

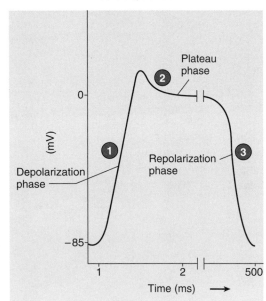

**Cardiac Muscle**

(a)  **1** **Depolarization phase**
  • $Na^+$ channels open.
  • $K^+$ channels begin to open.

**2** **Repolarization phase**
  • $Na^+$ channels close.
  • $K^+$ channels continue to open, causing repolarization.
  • $K^+$ channels close at the end of repolarization and return the membrane potential to its resting value.

(b)  **1** **Depolarization phase**
  • $Na^+$ channels open.
  • $Ca^{2+}$ channels open.

**2** **Plateau phase**
  • $Na^+$ channels close.
  • Some $K^+$ channels open, causing repolarization.
  • $Ca^{2+}$ channels are open, producing the plateau by slowing further repolarization.

**3** **Repolarization phase**
  • $Ca^{2+}$ channels close.
  • Many $K^+$ channels open.

**PROCESS Figure 12.14  Comparison of Action Potentials in Skeletal and Cardiac Muscle**

(*a*) An action potential in skeletal muscle consists of depolarization and repolarization phases. (*b*) An action potential in cardiac muscle consists of depolarization, plateau, and repolarization phases. Cardiac muscle does not repolarize as rapidly as skeletal muscle (*indicated by the break in the curve*) because of the plateau phase.

is the reason the cardiac muscle fiber action potential lasts longer than the action potentials in skeletal muscle fibers. The plateau phase ends, and the repolarization phase begins as the $Ca^{2+}$ channels close and many $K^+$ channels open, allowing $K^+$ to move out of the cell.

Action potentials in cardiac muscle exhibit a **refractory period,** like that of action potentials in skeletal muscle and in neurons. The refractory period lasts about as long as the plateau phase of the action potential in cardiac muscle. The prolonged action potential and refractory period allow cardiac muscle to contract and relax almost completely before another action potential can be produced. Also, the long refractory period in cardiac muscle prevents tetanic contractions from occurring, thus ensuring a rhythm of contraction and relaxation for cardiac muscle. Therefore, action potentials in cardiac muscle are different from those in skeletal muscle because the plateau phase makes the action potential and its refractory period last longer.

**Predict 2**

*Why is it important to prevent tetanic contractions in cardiac muscle but not in skeletal muscle?*

## Conduction System of the Heart

Unlike skeletal muscle that requires neural stimulation to contract, cardiac muscle can contract without neural stimulations. Contraction of the atria and ventricles is coordinated by specialized cardiac muscle cells in the heart wall that form the **conduction system of the heart** (figure 12.15).

All the cells of the conduction system can produce spontaneous action potentials. The sinoatrial node, atrioventricular node, atrioventricular bundle, right and left bundle branches, and Purkinje fibers constitute the conduction system of the heart. The **sinoatrial (SA) node,** which functions as the heart's pacemaker, is located in the superior wall of the right atrium and initiates the contraction of the heart. Action potentials originate in the SA node and spread over the right and left atria, causing them to contract. The SA node produces action potentials at a faster rate than other areas of the heart and has a larger number of $Ca^{2+}$ channels than other cells in the heart. In addition, the $Na^+$ and $Ca^{2+}$ channels in the SA node spontaneously open and close at a rhythmic rate. $Ca^{2+}$ channel blocking agents are drugs that slow the heart by decreasing the rate of action potential production in the SA node. $Ca^{2+}$ channel blockers decrease the rate at which $Ca^{2+}$ move through $Ca^{2+}$ channels. As a result, it takes longer for depolarization to reach threshold, and the interval between action potential production increases.

A second area of the heart, the **atrioventricular (AV)** (ā-trē-ō-ven′trik′-ū′lăr) **node,** is located in the lower portion of the right atrium. When action potentials reach the AV node, they spread slowly through it and then into a bundle of specialized cardiac muscle called the **atrioventricular (AV) bundle.** The slow rate of action potential conduction in the AV node allows the atria to complete their contraction before action potentials are delivered to the ventricles.

After action potentials pass through the AV node, they are rapidly transmitted through the AV bundle, which projects through the fibrous connective tissue plate that separates the atria from the ventricles (see figure 12.9). The AV bundle then divides into two branches of conducting tissue, called the **left** and **right bundle branches** (figure 12.15). At the tips of the left and right bundle branches, the conducting tissue forms many small bundles of **Purkinje** (pŭr-kĭn′jē) **fibers.** The Purkinje fibers pass to the apex of the heart and then extend to the cardiac muscle of the ventricle walls. The AV bundle, the bundle branches, and the Purkinje fibers are composed of specialized cardiac muscle fibers that conduct action potentials more rapidly than do other cardiac muscle fibers. Consequently, action potentials are rapidly delivered to all the cardiac muscle of the ventricles. The coordinated contraction of the ventricles depends on the conduction of action potentials by the conduction system.

1 Action potentials originate in the sinoatrial (SA) node and travel across the wall of the atrium (*arrows*) from the SA node to the atrioventricular (AV) node.

2 Action potentials pass through the AV node and along the atrioventricular (AV) bundle, which extends from the AV node, through the fibrous skeleton, into the interventricular septum.

3 The AV bundle divides into right and left bundle branches, and action potentials descend to the apex of each ventricle along the bundle branches.

4 Action potentials are carried by the Purkinje fibers from the bundle branches to the ventricular walls.

**PROCESS Figure 12.15** **Conduction System of the Heart**

# CLINICAL IMPACT

## Fibrillation of the Heart

In some people, cardiac muscle can malfunction—it is as if thousands of pacemakers were each making a very small portion of the heart contract rapidly and independently of all other areas. This condition, called **fibrillation** (fi-bri-lā'shŭn), reduces the output of the heart to only a few milliliters of blood per minute when it occurs in the ventricles. Unless ventricular fibrillation is stopped, the person dies in just a few minutes.

To stop the process of fibrillation, health professionals use a technique called defibrillation, in which they apply a strong electrical shock to the chest region. The shock causes simultaneous depolarization of all cardiac muscle fibers. Following depolarization, the SA node can recover and produce action potentials before any other area of the heart. Consequently, the normal pattern of action potential generation and the normal rhythm of contraction are reestablished.

Defibrillator machines have changed considerably over the years. Now available are portable models that non-professionals can learn to use for emergencies in workplaces and even in homes. Consequently, the time required to respond to an emergency requiring a defibrillator can be greatly shortened.

Fibrillation is more likely to occur when action potentials originate at ectopic sites in the heart. For example, people who have ectopic beats that originate from one of their ventricles are more likely to develop fibrillation of the heart than are people who have normal heartbeats.

### Predict 3

*If blood supply is reduced in a small area of the heart through which the left bundle branch passes, predict the effect on ventricular contraction.*

Following their contraction, the ventricles begin to relax. After the ventricles have completely relaxed, another action potential originates in the SA node to begin the next cycle of contractions.

The SA node is the pacemaker of the heart, but other cells of the conduction system are also capable of producing action potentials spontaneously. For example, if the SA node is unable to function, another area, such as the AV node, becomes the pacemaker. The resulting heart rate is much slower than normal. When action potentials originate in an area of the heart other than the SA node, the result is called an **ectopic** (ek-top'ik) **beat.**

## Electrocardiogram

Action potentials conducted through the heart during the cardiac cycle produce electrical currents that can be measured at the surface of the body. Electrodes placed on the body surface and attached to a recording device can detect the small electrical changes resulting from the action potentials in all of the cardiac muscle cells. The record of these electrical events is an **electrocardiogram** (ē-lek-trō-kar'dē-ō-gram; **ECG** or **EKG**) (figure 12.16).

The normal ECG consists of a P wave, a QRS complex, and a T wave. The **P wave** results from depolarization of the atrial myocardium, and the beginning of the P wave precedes the onset of atrial contraction. The **QRS complex** consists of three individual waves: the Q, R, and S waves. The QRS complex results from depolarization of the ventricles, and the beginning of the QRS complex precedes ventricular contraction. The **T wave** represents repolarization of the ventricles, and the beginning of the T wave precedes

ventricular relaxation. A wave representing repolarization of the atria cannot be seen because it occurs during the QRS complex.

The time between the beginning of the P wave and the beginning of the QRS complex is the **PQ interval,** commonly called the **PR interval** because the Q wave is very small. During

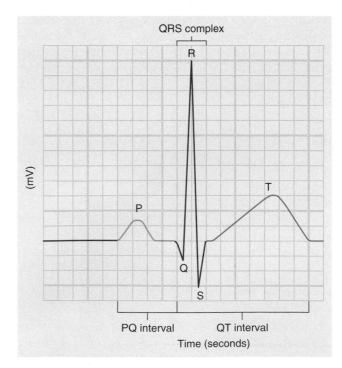

### Figure 12.16 Electrocardiogram

The major waves and intervals of an electrocardiogram are labeled. Each thin horizontal line on the ECG recording represents 1 millivolt (mV), and each thin vertical line represents 0.04 second.

Circulatory

| TABLE 12.1 | Major Cardiac Arrhythmias | |
|---|---|---|
| **Condition** | **Symptoms** | **Possible Causes** |
| **Abnormal Heart Rhythms** | | |
| Tachycardia | Heart rate in excess of 100 beats per minute (bpm) | Elevated body temperature, excessive sympathetic stimulation, toxic conditions |
| Bradycardia | Heart rate less than 60 bpm | Increased stroke volume in athletes, excessive vagus nerve stimulation, nonfunctional SA node, carotid sinus syndrome |
| Sinus arrhythmia | Heart rate varies as much as 5% during respiratory cycle and up to 30% during deep respiration. | Cause not always known; occasionally caused by ischemia, inflammation, or cardiac failure |
| Paroxysmal atrial tachycardia | Sudden increase in heart rate to 95–150 bpm for a few seconds or even for several hours; P waves precede every QRS complex; P wave is inverted and superimposed on T wave | Excessive sympathetic stimulation, abnormally elevated permeability of cardiac muscle to $Ca^{2+}$ |
| Atrial flutter | As many as 300 P waves/min and 125 QRS complexes/min; resulting in two or three P waves (atrial contractions) for every QRS complex (ventricular contraction) | Ectopic beats in atria |
| Atrial fibrillation | No P waves, normal QRS and T waves, irregular timing; ventricles are constantly stimulated by atria; reduced ventricle filling; increased chance of fibrillation | Ectopic beats in atria |
| Ventricular tachycardia | Frequently causes fibrillation | Often associated with damage to AV node or ventricular muscle |
| **Heart Blocks** | | |
| SA node block | No P waves; low heart rate resulting from AV node acting as the pacemaker; normal QRS complexes and T waves | Ischemia, tissue damage resulting from infarction; cause sometimes unknown |
| AV node blocks | | |
| First-degree | PQ interval greater than 0.2 s | Inflammation of AV bundle |
| Second-degree | PQ interval 0.25–0.45 s; some P waves trigger QRS complexes and others do not; examples of 2:1, 3:1, and 3:2 P wave/QRS complex ratios | Excessive vagus nerve stimulation, AV node damage |
| Third-degree (complete heart block) | P wave dissociated from QRS complex; atrial rhythm about 100 bpm; ventricular rhythm less than 40 bpm | Ischemia of AV node or compression of AV bundle |
| **Premature Contractions** | | |
| Premature atrial contractions | Occasional shortened intervals between one contraction and the succeeding contraction; frequently occurs in healthy people | Excessive smoking, lack of sleep, too much caffeine |
| Premature ventricular contractions (PVCs) | Prolonged QRS complex; exaggerated voltage because only one ventricle may depolarize; possible inverted T wave; increased probability of fibrillation | Ectopic beat in ventricles, lack of sleep, too much caffeine, irritability; occasionally occurs with coronary thrombosis |

the PQ interval, the atria contract and begin to relax. At the end of the PQ interval, the ventricles begin to depolarize.

The **QT interval** extends from the beginning of the QRS complex to the end of the T wave and represents the length of time required for ventricular depolarization and repolarization. Table 12.1 describes several conditions associated with abnormal heart rhythms.

The ECG is not a direct measurement of mechanical events in the heart, and neither the force of contraction nor the blood pressure can be determined from it. However, each deflection in the ECG record indicates an electrical event within the heart and correlates with a subsequent mechanical event. Consequently, the ECG is an extremely valuable tool for diagnosing a number of cardiac abnormalities, particularly because it is painless, easy to record, and non surgical. Analysis of an ECG can reveal abnormal heart rates or rhythms; problems in conduction pathways, such as blockages; hypertrophy or atrophy of portions of the heart; and the approximate location of damaged cardiac muscle.

**Predict 4**

Explain how the ECGs would appear for a person who has a damaged left bundle branch (see Predict 3) and for a person who has many ectopic beats originating from her atria.

## A CASE IN POINT
### Cardiopulmonary Resuscitation (CPR)

Willie May Kitt is a 65-year-old bank manager. While walking up a short flight of stairs to his office, he experienced a crushing pain in his chest and exhibited substantial pallor. Willie fell to the floor, lost consciousness, and then stopped breathing. A coworker noticed the pallor, saw Willie fall, and ran to his aid. He could detect no pulse and decided to administer cardiopulmonary resuscitation (CPR). Another coworker called 911 and then assisted the first. One coworker pushed down firmly on Willie's sternum at a rate of approximately 100 compressions per minute. After every 30 compressions, he paused. The other coworker forced air into Willie's lungs by tipping Willie's head back slightly, placing her mouth over Willie's, and blowing air forcefully into his mouth two times while holding his nasal passages closed. Pushing down on the sternum compresses the ventricles of the heart and forces blood into the aorta and pulmonary trunk. Between compressions, blood flows into the ventricles from the atria.

Fortunately, it took only about 5 minutes for emergency medical technicians to arrive. They confirmed the lack of a pulse and used portable equipment to record an electrocardiogram, which indicated that the heart was fibrillating (see Clinical Impact, Fibrillation of the Heart). They quickly used a portable defibrillator to apply a strong electrical shock to Willie's chest. Fortunately, Willie's heart responded by beginning to beat rhythmically.

Willie's heart may have first developed arrhythmia and then ventricular fibrillation. Willie was very fortunate. Most people whose heart suddenly stops pumping do not survive. In Willie's case, CPR was administered quickly and effectively, and emergency help arrived in a very short time.

Willie was transported to a hospital. His condition could be due to a myocardial infarction (see Systems Pathology, "Myocardial Infarction") or to some other cause that needs to be identified and treated.

## CARDIAC CYCLE

The right and left halves of the heart can be viewed as two separate pumps. Each pump consists of a primer pump—the atrium—and a power pump—the ventricle. The atria act as primer pumps because they complete the filling of the ventricles with blood, and the ventricles act as power pumps because they produce the major force that causes blood to flow through the pulmonary and systemic circulations. The term **cardiac cycle** refers to the repetitive pumping process that begins with the onset of cardiac muscle contraction and ends with the beginning of the next contraction (figure 12.17). Pressure changes produced within the heart chambers as a result of cardiac muscle contraction move blood from areas of higher pressure to areas of lower pressure.

**Atrial systole** (sis'tō-lē; a contracting) refers to contraction of the two atria. **Ventricular systole** refers to contraction of the two ventricles. **Atrial diastole** (dī-as'tō-lē; dilation) refers to relaxation of the two atria, and **ventricular diastole** refers to relaxation of the two ventricles. When the terms *systole* and *diastole* are used alone, they refer to ventricular contraction or relaxation. The ventricles contain more cardiac muscle than the atria and produce far greater pressures, which force blood to circulate throughout the vessels of the body.

The major events of the cardiac cycle are:

- *Systole*—At the beginning of systole, contraction of the ventricles pushes blood toward the atria, causing the AV valves to close as the pressure begins to increase (figure 12.17, step 1).
- As systole continues, the increasing pressure in the ventricles exceeds the pressure in the pulmonary trunk and aorta, the semilunar valves are forced open, and blood is ejected into the pulmonary trunk and aorta (figure 12.17, step 2).
- *Diastole*—At the beginning of ventricular diastole, the pressure in the ventricles decreases below the pressure in the aorta and pulmonary trunk. The semilunar valves close and prevent blood from flowing back into the ventricles (figure 12.17, step 3).
- As diastole continues, the pressure continues to decline in the ventricles until atrial pressures are greater than ventricular pressures. Then the AV valves open, and blood flows directly from the atria into the relaxed ventricles. During the previous ventricular systole, the atria were relaxed and blood collected in them. When the ventricles relax and the AV valves open, blood flows into the ventricles (figure 12.17, step 4) and fills them to approximately 70% of their volume.
- At the end of ventricular diastole, the atria contract and then relax. Atrial systole forces additional blood to flow into the ventricles to complete their filling (figure 12.17, step 5). The semilunar valves remain closed.

Figure 12.18 displays the main events of the cardiac cycle and should be examined from top to bottom for each period. The ECG indicates the electrical events that cause the atria and ventricles to contract and relax. The pressure graph shows the pressure changes within the left atrium, left ventricle, and aorta resulting from atrial and ventricular contraction and relaxation. The pressure changes in the right side of the heart are not shown here, but they are similar to those in the left side, only lower. The volume graph presents the changes in left ventricular volume as blood flows into and out of the left ventricle as a result of the pressure changes. The sound graph records the closing of valves caused by blood flow. See figure 12.17 for illustrations of the valves and blood flow.

**Predict 5**

Predict the effect of a leaky (incompetent) aortic semilunar valve on the volume of blood in the left ventricle just before ventricular contraction. Predict the effect of a severely narrowed opening through the aortic semilunar valves on the amount of work the heart must do to pump the normal volume of blood into the aorta during each beat of the heart.

**1** Contraction of the ventricles causes pressure in the ventricles to increase. Almost immediately, the AV valves close (the first heart sound). The pressure in the ventricles continues to increase.

**2** Continued ventricular contraction causes the pressure in the ventricles to exceed the pressure in the pulmonary trunk and aorta. As a result, the semilunar valves are forced open, and blood is ejected into the pulmonary trunk and aorta.

**5** The atria contract and complete ventricular filling.

**3** At the beginning of ventricular diastole, the ventricles relax, and the semilunar valves close (the second heart sound).

**4** The AV valves open, and blood flows into the ventricles. The ventricles fill to approximately 70% of their volume.

Circulatory

**PROCESS Figure 12.17  Cardiac Cycle**

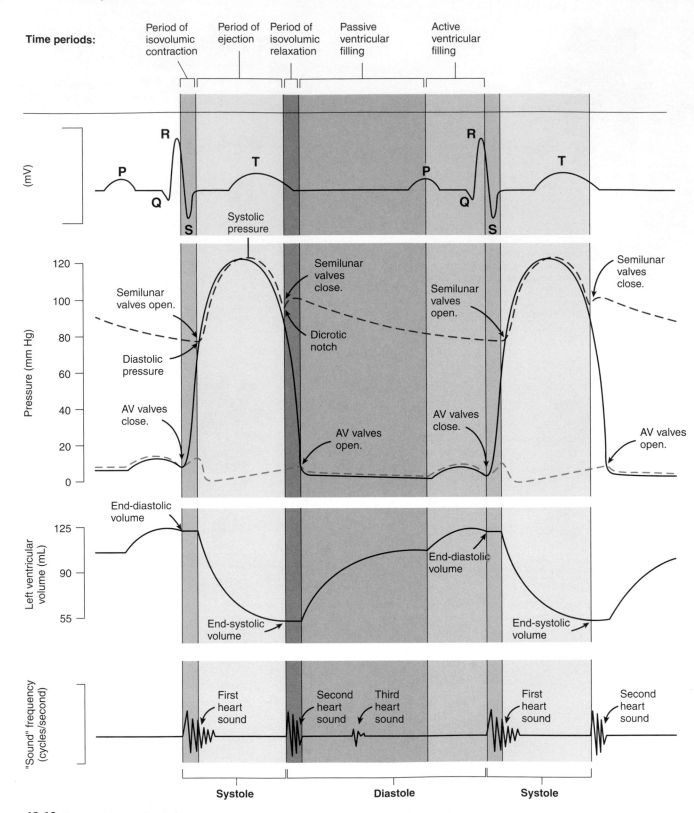

**Time periods:**

Period of isovolumic contraction

Period of ejection

Period of isovolumic relaxation

Passive ventricular filling

Active ventricular filling

(mV)

R    T    P    Q    S    R    T

P    Q    S

Systolic pressure

Semilunar valves close.

Semilunar valves open.

Dicrotic notch

Semilunar valves open.

Semilunar valves close.

Diastolic pressure

AV valves close.

AV valves open.

AV valves close.

AV valves open.

Pressure (mm Hg)

120    100    80    60    40    20    0

End-diastolic volume

Left ventricular volume (mL)

125    90    55

End-systolic volume

End-diastolic volume

End-systolic volume

"Sound" frequency (cycles/second)

First heart sound

Second heart sound

Third heart sound

First heart sound

Second heart sound

Systole    Diastole    Systole

**Figure 12.18  Events of the Cardiac Cycle**

(Top) The cardiac cycle is divided into five periods. *From top to bottom,* these graphs show an electrocardiogram; pressure changes for the left atrium (*blue dashed line*), the left ventricle (*black line*), and the aorta (*red dashed line*); left ventricular volume curve; and the sequence of the heart sounds.

Circulatory

# CLINICAL IMPACT

## Consequences of an Incompetent Bicuspid Valve

In some individuals, a heart valve does not close completely and thus is called an **incompetent valve.** Such valves leak when they are supposed to be closed and allow blood to flow in the reverse direction. For example, an incompetent bicuspid valve allows blood to flow from the left ventricle to the left atrium during ventricular systole.

This reduces the amount of blood pumped into the aorta. It also dramatically increases the blood pressure in the left atrium and in the pulmonary veins during ventricular systole. During diastole, the excess blood pumped into the atrium once again flows into the ventricle, along with the blood that normally flows from the lungs to the left

atrium. Therefore, the volume of blood entering the left ventricle is greater than normal. The increased filling of the left ventricle gradually causes it to hypertrophy and can lead to heart failure. The increased pressure in the pulmonary veins can cause edema in the lungs.

## HEART SOUNDS

The **stethoscope** (steth′ō-skōp; *stetho,* the chest) was originally developed to listen to the sounds of the lungs and heart and is now used to listen to other sounds of the body as well. Figure 12.19 shows the sites in the thorax where the heart sounds can best be heard with a stethoscope. There are two main heart sounds. The first heart sound can be represented by the syllable *lubb,* and the second heart sound can be represented by *dupp.* The first heart

sound has a lower pitch than the second. The first heart sound occurs at the beginning of ventricular systole and results from closure of the AV valves (see figure 12.17, step 1, and figure 12.18). The second heart sound occurs at the beginning of ventricular diastole and results from closure of the semilunar valves (see figure 12.17, step 3, and figure 12.18). The valves usually do not make sounds when they open.

Clinically, ventricular systole occurs between the first and second heart sounds. Ventricular diastole occurs between the second heart sound and the first heart sound of the next beat. Because ventricular diastole lasts longer than ventricular systole, there is less time between the first and second heart sounds than between the second heart sound and the first heart sound of the next beat.

### Predict 6

*Compare the rate of blood flow out of the ventricles between the first and second heart sounds of the same beat to the rate of blood flow out of the ventricles between the second heart sound of one beat and the first heart sound of the next beat.*

Abnormal heart sounds called **murmurs** are usually a result of faulty valves. For example, an incompetent valve fails to close tightly, so that blood leaks through the valve when it is closed (see Clinical Impact). A murmur caused by an incompetent valve makes a swishing sound immediately after the valve closes. For example, an incompetent bicuspid valve produces a swishing sound immediately after the first heart sound.

When the opening of a valve is narrowed, or **stenosed** (sten′ozd; a narrowing), a swishing sound precedes closure of the stenosed valve. For example, when the bicuspid valve is stenosed, a swishing sound precedes the first heart sound.

### Predict 7

*If normal heart sounds are represented by lubb-dupp, lubb-dupp, what does a heart sound represented by lubb-duppshhh, lubb-duppshhh represent? What does lubb-shhhdupp, lubb-shhhdupp represent (assuming that shhh represents a swishing sound)?*

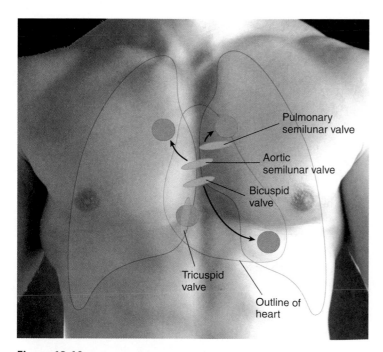

Pulmonary semilunar valve

Aortic semilunar valve

Bicuspid valve

Tricuspid valve

Outline of heart

**Figure 12.19   Location of the Heart Valves in the Thorax**

Surface markings of the heart in an adult male. The positions of the four heart valves are indicated by *blue ellipses,* and the sites where the sounds of the valves are best heard with a stethoscope are indicated by *pink circles.*

Circulatory

## A CASE IN POINT
### Paroxysmal Atrial Tachycardia

Speedy Beat is a 70-year-old man. He and his daughter, Normal, were getting out of the car at a restaurant where they planned to have dinner. Before Speedy could get completely out of the car, he became dizzy. He exhibited substantial pallor and experienced chest pains. Normal checked her father's pulse, which was close to 180–200 bpm and irregular. She helped him back into the car and drove to the emergency room of a nearby hospital. There, physicians determined that Speedy's blood pressure was low, even though his heart rate was rapid. Having been previously diagnosed with paroxysmal atrial tachycardia, Speedy regularly takes a calcium channel blocking agent.

Tachycardia is characterized by rapid and ectopic beats that originate in the atria. The term *paroxysmal* means that the cause of the tachycardia is not known. The heart rate is irregular because the rate at which action potentials responsible for atrial contractions occur is greater than the rate at which the ventricles can contract. Consequently, not every atrial contraction is followed by a ventricular contraction. Speedy's blood pressure is low because his heart rate is so fast that there is little time for blood to fill the rapidly contracting chambers of the heart between the contractions. Consequently, Speedy's stroke volume and cardiac output are low. Chest pains result because the heart muscle is working hard but blood flow to cardiac muscle through the coronary vessels is reduced, so that the heart muscle suffers from an inadequate supply of oxygen (ischemia).

Once Speedy's heart rate and rhythm were stabilized, the amount of calcium channel blocking agent prescribed for him was adjusted, and he was released from the hospital the next day. Along with the calcium channel blocking agent, another drug used to treat this condition is digoxin, which has the overall effect of increasing the force and slowing the rate of cardiac muscle contraction.

# REGULATION OF HEART FUNCTION

Various measurements can be taken to assess the heart's function. **Cardiac output (CO)** is the volume of blood pumped by either ventricle of the heart each minute. **Stroke volume (SV)** is the volume of blood pumped per ventricle each time the heart contracts, and the **heart rate (HR)** is the number of times the heart contracts each minute. Cardiac output can be calculated by multiplying the stroke volume times the heart rate:

$$CO = SV \times HR$$
$$(mL/min) \quad (mL/beat) \quad (beats/min)$$

Under resting conditions, the heart rate is approximately 72 beats/min, and the stroke volume is approximately 70 mL/beat. Consequently, the cardiac output is slightly more than 5 L/min:

$$
\begin{aligned}
CO &= SV \times HR \\
&= 70 \text{ mL/beat} \times 72 \text{ bpm} \\
&= 5040 \text{ mL/min (approximately 5 L/min)}
\end{aligned}
$$

The heart rate and the stroke volume vary considerably among people. Athletes tend to have a higher stroke volume and lower heart rate at rest because exercise has increased the size of their hearts. Nonathletes are more likely to have a higher heart rate and lower stroke volume. During exercise, the heart rate in a nonathlete can increase to 190 bpm, and the stroke volume can increase to 115 mL/beat. Therefore, the cardiac output increases to approximately 22 L/min:

$$
\begin{aligned}
CO &= SV \times HR \\
&= 115 \text{ mL/beat} \times 190 \text{ bpm} \\
&= 21{,}850 \text{ mL/min (approximately 22 L/min)}
\end{aligned}
$$

This cardiac output is several times greater than the cardiac output under resting conditions. Athletes can increase their cardiac output to a greater degree than nonathletes. The control mechanisms that modify the stroke volume and the heart rate are classified as intrinsic and extrinsic.

## Intrinsic Regulation of the Heart

**Intrinsic regulation** refers to mechanisms contained within the heart itself. The force of contraction produced by cardiac muscle is related to the degree of stretch of cardiac muscle fibers. The amount of blood in the ventricles at the end of ventricular diastole determines the degree to which cardiac muscle fibers are stretched. **Venous return** is the amount of blood that returns to the heart, and the degree to which the ventricular walls are stretched at the end of diastole is called **preload.** If venous return increases, the heart fills to a greater volume and stretches the cardiac muscle fibers, producing an increased preload. In response, cardiac muscle fibers contract with a greater force. The greater force of contraction causes an increased volume of blood to be ejected from the heart, resulting in an increased stroke volume. As venous return increases, resulting in an increased preload, cardiac output increases. Conversely, if venous return decreases, resulting in a decreased preload, the cardiac output decreases. The relationship between preload and stroke volume is called **Starling's law of the heart.**

Because venous return is influenced by many conditions, Starling's law of the heart has a major influence on cardiac output. For example, muscular activity during exercise causes increased venous return, resulting in increased preload, stroke volume, and cardiac output. This is beneficial because increased cardiac output is needed during exercise to supply oxygen to exercising skeletal muscles.

**Afterload** refers to the pressure against which the ventricles must pump blood. People suffering from hypertension have an increased afterload because their aortic pressure is elevated during contraction of the ventricles. The heart must do more work to pump blood from the left ventricle into the aorta, which increases the workload on the heart and can eventually lead to heart failure. A reduced afterload decreases the work the heart must do. People who have lower blood pressure have a reduced afterload and develop heart failure less often than people who have hypertension. However, the afterload influences cardiac output less than preload influences it. The afterload must increase substantially before it decreases the volume of blood pumped by a healthy heart.

## Extrinsic Regulation of the Heart

**Extrinsic regulation** refers to mechanisms external to the heart, such as either nervous or chemical regulation.

# CLINICAL IMPACT

## Consequences of Heart Failure

Although heart failure can occur in young people, it usually results from progressive weakening of the heart muscle in elderly people. In heart failure, the heart is not capable of pumping all the blood that is returned to it because further stretching of the cardiac muscle fibers does not increase the stroke volume. Consequently, blood backs up in the veins. For example, heart failure that affects the right ventricle, called **right heart failure,** causes blood to back up in the veins that return blood from systemic vessels to the heart. Filling of the veins with blood causes edema, especially in the legs and feet, due to the accumulation of fluid in the tissues outside blood vessels. Heart failure that affects the left ventricle, called **left heart failure,** causes blood to back up in the veins that return blood from the lungs to the heart. Filling of these veins causes edema in the lungs, which makes breathing difficult.

## Nervous Regulation: Baroreceptor Reflex

Nervous influences are carried through the autonomic nervous system. Both sympathetic and parasympathetic nerve fibers innervate the heart and have a major effect on the SA node. Stimulation by sympathetic nerve fibers causes the heart rate and the stroke volume to increase, whereas stimulation by parasympathetic nerve fibers causes the heart rate to decrease.

The **baroreceptor** (bar′ō-rē-sep′ter; *baro,* pressure) **reflex** is a mechanism of the nervous system that plays an important role in regulating heart function. **Baroreceptors** are stretch receptors that monitor blood pressure in the aorta and in the wall of the internal carotid arteries, which carry blood to the brain. Changes in blood pressure result in changes in the stretch of the walls of these blood vessels—and changes in the frequency of action potentials produced by the baroreceptors. The action potentials are transmitted along nerve fibers from the stretch receptors to the medulla oblongata of the brain.

Within the medulla oblongata is a **cardioregulatory center,** which receives and integrates action potentials from the baroreceptors. The cardioregulatory center controls the action potential frequency in sympathetic and parasympathetic nerve fibers that extend from the brain and spinal cord to the heart. The cardioregulatory center also influences sympathetic stimulation of the adrenal gland. Epinephrine and norepinephrine, released from the adrenal gland, increase the stroke volume and heart rate.

Figure 12.20 shows how the baroreceptor reflex keeps the heart rate and stroke volume within normal ranges. When the blood pressure increases, the baroreceptors are stimulated. Action potentials are sent along the nerve fibers to the medulla oblongata at increased frequency. This prompts the cardioregulatory center to increase parasympathetic stimulation and decrease sympathetic stimulation of the heart. As a result, the heart rate and stroke volume decrease, causing blood pressure to decline.

When the blood pressure decreases, less stimulation of the baroreceptors occurs. A lower frequency of action potentials is sent to the medulla oblongata of the brain, and this triggers a response in the cardioregulatory center. The cardioregulatory center responds by increasing sympathetic stimulation of the heart and decreasing parasympathetic stimulation. Consequently, the heart rate and stroke volume increase. If the decrease in blood pressure is large, sympathetic stimulation of the adrenal medulla also increases. The epinephrine and norepinephrine secreted by the adrenal medulla increase the heart rate and stroke volume, also causing the blood pressure to increase toward its normal value (figure 12.20).

### Predict 8

*In response to a severe hemorrhage, blood pressure lowers, the heart rate increases dramatically, and the stroke volume lowers. If low blood pressure activates a reflex that increases sympathetic stimulation of the heart, why is the stroke volume low?*

## Chemical Regulation: Chemoreceptor Reflex

Epinephrine and small amounts of norepinephrine released from the adrenal medulla in response to exercise, emotional excitement, or stress also influence the heart's function. Epinephrine and norepinephrine bind to receptor proteins on cardiac muscle and cause increased heart rate and stroke volume. Excitement, anxiety, or anger can affect the cardioregulatory center, resulting in increased sympathetic stimulation of the heart and increased cardiac output. Depression, on the other hand, can increase parasympathetic stimulation of the heart, causing a slight reduction in cardiac output.

The medulla oblongata of the brain also contains chemoreceptors that are sensitive to changes in pH and carbon dioxide levels. A decrease in pH, often caused by an increase in carbon dioxide, results in sympathetic stimulation of the heart (figure 12.21).

Changes in the extracellular concentration of $K^+$, $Ca^{2+}$, and $Na^+$, which influence other electrically excitable tissues, also affect cardiac muscle function. An excess of extracellular $K^+$ causes the heart rate and stroke volume to decrease. If the extracellular $K^+$ concentration increases further, normal conduction of action potentials through cardiac muscle is blocked, and death can result. An excess of extracellular $Ca^{2+}$ causes the heart to contract arrhythmically. Reduced extracellular $Ca^{2+}$ cause both the heart rate and stroke volume to decrease.

Figure 12.22 summarizes how nervous and chemical factors interact to regulate the heart rate and stroke volume.

Circulatory

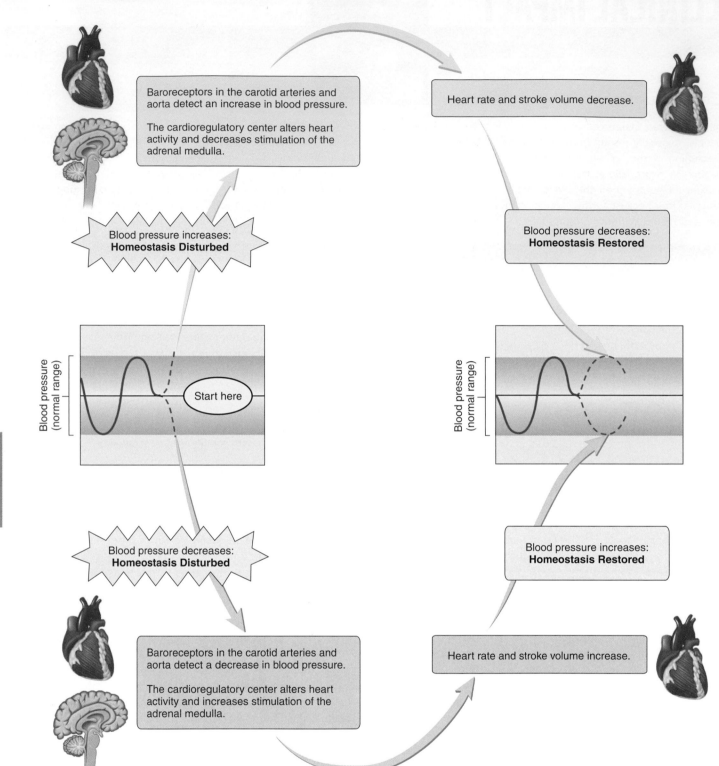

Baroreceptors in the carotid arteries and aorta detect an increase in blood pressure.

The cardioregulatory center alters heart activity and decreases stimulation of the adrenal medulla.

Heart rate and stroke volume decrease.

Blood pressure increases:
**Homeostasis Disturbed**

Blood pressure decreases:
**Homeostasis Restored**

Blood pressure (normal range)

Start here

Blood pressure (normal range)

Blood pressure decreases:
**Homeostasis Disturbed**

Blood pressure increases:
**Homeostasis Restored**

Baroreceptors in the carotid arteries and aorta detect a decrease in blood pressure.

The cardioregulatory center alters heart activity and increases stimulation of the adrenal medulla.

Heart rate and stroke volume increase.

**Homeostasis Figure 12.20  Baroreceptor Reflex**

The baroreceptor reflex maintains homeostasis in response to changes in blood pressure.

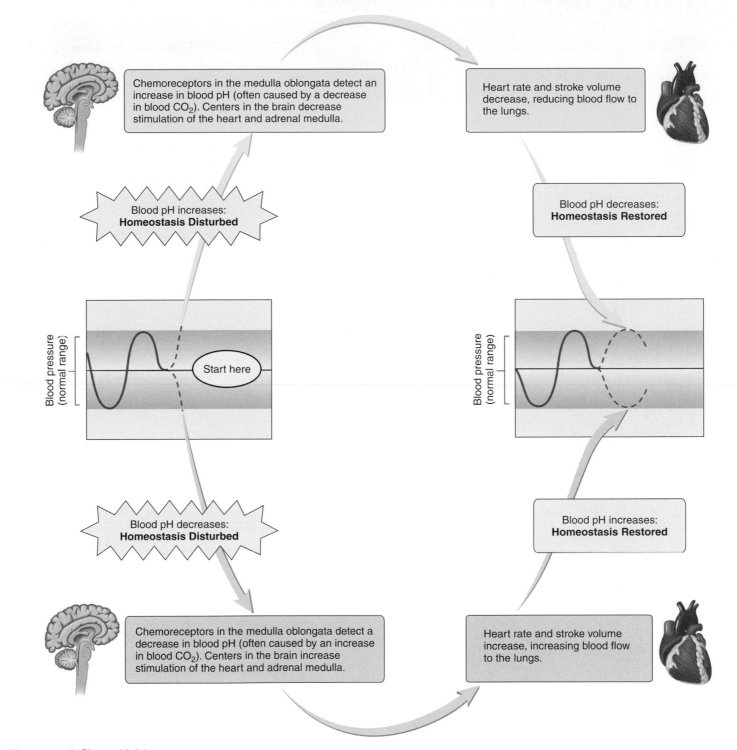

**Chemoreceptors in the medulla oblongata detect an increase in blood pH (often caused by a decrease in blood $CO_2$). Centers in the brain decrease stimulation of the heart and adrenal medulla.**

**Heart rate and stroke volume decrease, reducing blood flow to the lungs.**

Blood pH increases:
**Homeostasis Disturbed**

Blood pH decreases:
**Homeostasis Restored**

Blood pressure (normal range)

Start here

Blood pressure (normal range)

Blood pH decreases:
**Homeostasis Disturbed**

Blood pH increases:
**Homeostasis Restored**

**Chemoreceptors in the medulla oblongata detect a decrease in blood pH (often caused by an increase in blood $CO_2$). Centers in the brain increase stimulation of the heart and adrenal medulla.**

**Heart rate and stroke volume increase, increasing blood flow to the lungs.**

Circulatory

**Homeostasis Figure 12.21  Chemoreceptor Reflex—pH**

The chemoreceptor reflex maintains homeostasis in response to changes in blood concentrations of $CO_2$ and $H^+$ (or pH).

# CLINICAL IMPACT

## Treatment and Prevention of Heart Disease

Heart disease is a leading cause of death in the United States, but fortunately a number of medical and surgical treatments are available.

### Heart Medications

#### Digitalis

**Digitalis** (dij′i-tal′is), an extract of the foxglove plant, slows and strengthens contractions of the heart muscle. This drug is frequently given to heart failure patients, although it can also be used to treat atrial tachycardia.

#### Nitroglycerin

During exercise, when the heart rate and stroke volume rise, dilation of blood vessels in the exercising skeletal muscles and constriction in most other blood vessels result in increased venous return to the heart and increased preload. **Nitroglycerin** (nī-trō-glis′er-in) causes dilation of all the veins and arteries without increasing heart rate or stroke volume. When all blood vessels dilate, a greater volume of blood pools in the dilated blood vessels, decreasing the venous return to the heart. The reduced preload causes cardiac output to decrease, reducing the amount of work the heart must perform. Nitroglycerin is frequently given to people who have coronary artery disease, which restricts coronary blood flow. The decreased work performed by the heart reduces the amount of oxygen the cardiac muscle requires. In addition, dilation of coronary arteries can increase blood flow to cardiac muscle. Consequently, the heart does not suffer from a lack of oxygen, and angina pectoris does not develop.

#### Beta-Adrenergic Blocking Agents

**Beta-adrenergic** (bā-tă ad-rĕ-ner′jik) **blocking agents** decrease the rate and strength of cardiac muscle contractions, thus reducing the heart's oxygen demand. They bind to receptors for norepinephrine and epinephrine and prevent these substances from having their normal effects. Beta-adrenergic blocking agents are often used to treat people who have rapid heart rates, certain types of arrhythmias, and hypertension.

#### Calcium Channel Blockers

**Calcium channel blockers** reduce the rate at which $Ca^{2+}$ diffuse into cardiac and smooth muscle cells. Because the action potentials that produce cardiac muscle contractions depend in part on the flow of $Ca^{2+}$ into the cardiac muscle cells, $Ca^{2+}$ channel blockers can be used to control the force of heart contractions and reduce arrhythmia, tachycardia, and hypertension. Because entry of $Ca^{2+}$ into smooth muscle cells causes contraction, $Ca^{2+}$ channel blockers can dilate coronary blood vessels and increase blood flow to cardiac muscle. Consequently, they are used to treat angina pectoris.

#### Antihypertensive Agents

Several drugs are used specifically to treat hypertension. Because these drugs reduce blood pressure, they also decrease the heart's workload. In addition, the lowered blood pressure reduces the risk of heart attacks and strokes. Medications to treat hypertension include those that reduce the activity of the sympathetic division, dilate arteries and veins, increase urine production (diuretics), and block the conversion of angiotensin I to angiotensin II (see chapter 13).

#### Anticoagulants

**Anticoagulants** (an′tē-kō-ag′ū-lants) prevent clot formation in people whose heart valves or blood vessels have been damaged or in those who have had a myocardial infarction. Aspirin functions as a weak anticoagulant by inhibiting the synthesis of prostaglandins in platelets, which in turn reduces clot formation. Some data suggest that taking a small dose of aspirin regularly reduces the chance of a heart attack. Approximately one baby aspirin each day may benefit people who are likely to experience a coronary thrombosis.

### Instruments

#### Artificial Pacemaker

An **artificial pacemaker** is an instrument placed beneath the skin that is equipped with an electrode extending to the heart. The pacemaker provides an electrical stimulus to the heart at a set frequency. Artificial pacemakers are used in patients whose natural pacemakers do not produce a heart rate high enough to sustain normal physical activity. Modern electronics has made it possible to design artificial pacemakers capable of increasing the heart rate as physical activity increases. In addition, special artificial pacemakers can defibrillate the heart if it becomes arrhythmic. It is likely that continuing developments in electronics will further increase the ability of artificial pacemakers to regulate the heart.

#### Heart-Lung Machine

A **heart-lung machine** serves as a temporary substitute for the patient's heart and lungs. It pumps blood throughout the body and oxygenates and removes carbon dioxide from the blood. Use of a heart-lung machine has made possible many surgeries on the heart and lungs.

### Surgical Procedures

#### Heart Valve Replacement

**Heart valve replacement** is a surgical procedure performed on patients whose valves are deformed and scarred from a condition such as endocarditis. Valves that are severely incompetent or stenosed are replaced with substitute valves made of synthetic materials, such as plastic or Dacron; valves transplanted from pigs are also used.

#### Heart Transplants

A **heart transplant** is possible when the immune characteristics of a donor and a recipient are closely matched. The heart of a recently deceased donor is transplanted into the recipient after the recipient's diseased

# CLINICAL IMPACT (continued)

heart has been removed. People who have received heart transplants must remain on drugs that suppress their immune responses for the rest of their lives. Otherwise, their immune system would reject the transplanted heart.

### Artificial Hearts

**Artificial hearts** have been used on an experimental basis to extend the life of an individual until an acceptable transplant can be found or to replace the heart permanently. However, the technology currently available has not yet produced an artificial heart capable of permanently sustaining a high quality of life.

## Prevention of Heart Disease

Proper nutrition is important in reducing the risk of heart disease. A recommended diet is low in fats, especially saturated fats and cholesterol, and low in refined sugar. The diet should also be high in fiber, whole grains, fruits, and vegetables. Nutritionists advise people to reduce their intake of sodium chloride (salt) and to limit their total food intake to avoid obesity.

In addition, tobacco and excessive use of alcohol should be avoided. Smoking increases the risk of heart disease by at least 10-fold, and excessive use of alcohol substantially increases the risk of heart disease.

Chronic stress, frequent emotional upsets, and lack of physical exercise can increase the risk of cardiovascular disease. Remedies include relaxation techniques and aerobic activities, such as swimming, walking, jogging, or dancing, with gradual increases in duration and difficulty.

Hypertension, an abnormally high systemic blood pressure, affects approximately one-fifth of the population. Regular blood pressure monitoring is important because hypertension does not produce obvious symptoms. If hypertension cannot be controlled by diet and exercise, drugs should be prescribed. The cause of hypertension is unknown in the majority of cases.

---

1. Sensory neurons (*green*) carry action potentials from baroreceptors to the cardioregulatory center. Chemoreceptors in the medulla oblongata influence the cardioregulatory center.

2. The cardioregulatory center controls the frequency of action potentials in the parasympathetic neurons (*red*) extending to the heart. The parasympathetic neurons decrease the heart rate.

3. The cardioregulatory center controls the frequency of action potentials in the sympathetic neurons (*blue*) extending to the heart. The sympathetic neurons increase the heart rate and the stroke volume.

4. The cardioregulatory center influences the frequency of action potentials in the sympathetic neurons (*blue*) extending to the adrenal medulla. The sympathetic neurons increase the secretion of epinephrine and some norepinephrine into the general circulation. Epinephrine and norepinephrine increase the heart rate and stroke volume.

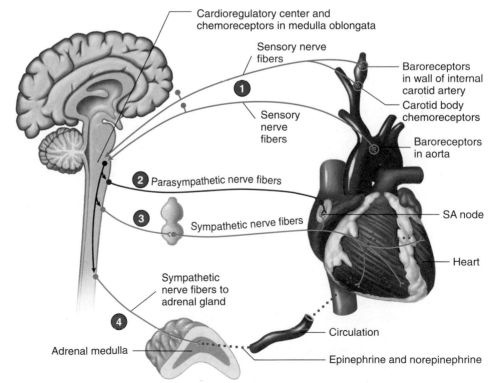

**PROCESS Figure 12.22   Summary of Extrinsic Regulation**

Sensory nerves (*green*) carry action potentials from sensory receptors to the medulla oblongata. Sympathetic nerves (*blue*) and parasympathetic nerves (*red*) exit the spinal cord or medulla oblongata and extend to the heart to regulate its function. Epinephrine and norepinephrine from the adrenal gland also help regulate the heart's action. (SA = sinoatrial)

# SYSTEMS PATHOLOGY

## Myocardial Infarction

Paul was an overweight, out-of-shape executive who regularly smoked and indulged in high-fat foods. He found his job frustrating because he was frequently confronted with stressful deadlines. He had not had a physical examination for several years, so he was not aware that his blood pressure was high. One evening, Paul was walking to his car after work when he began to feel pain in his chest that also radiated down his left arm. Shortly after the onset of pain, he became out of breath, developed marked pallor, felt dizzy, and had to lie down on the sidewalk. The pain in his chest and arm was poorly localized but intense, and he became anxious and then disoriented. Paul lost consciousness, although he did not stop breathing. After a short delay, one of his coworkers noticed him and called for help.

When paramedics arrived, they determined that Paul's blood pressure was low, and he exhibited arrhythmia and tachycardia. The paramedics took an electrocardiogram and transmitted it to a physician by way of their electronic communication system. Then they discussed Paul's symptoms with a physician at the hospital, who directed them to administer oxygen and medication to control the arrhythmia and to transport the patient to the hospital. At the hospital, Paul was given tissue plasminogen activator (t-PA) to improve blood flow to the damaged area of the heart by activating plasminogen, which dissolves blood clots. Over the next few days, blood levels of enzymes, such as creatine phosphokinase, increased in Paul's blood, which confirmed that cardiac muscle had been damaged by an infarction.

In the hospital, Paul began to experience shortness of breath because of pulmonary edema, and after a few days he developed pneumonia. He was treated for the pneumonia, and his condition gradually improved over the next few weeks. An **angiogram** (an'jē-ō-gram; *angeion,* a vessel + *gramma,* a writing) (figure 12A), which is an imaging technique used to visualize the coronary arteries, was performed several days after Paul's infarction. The angiogram indicated a permanent reduction in a significant part of the lateral wall of Paul's left ventricle, but physicians believed neither angioplasty nor bypass surgery was necessary. The angiogram also showed some restricted blood flow in Paul's coronary arteries.

## Background Information

Paul experienced a myocardial infarction in which a thrombosis in one of the branches of the left coronary artery reduced blood supply to the lateral wall of the left ventricle, resulting in ischemia of the left ventricle wall. The fact that t-PA was an effective treatment supports the conclusion that the infarction was due to a thrombosis. An ischemic area of the heart wall was not able to contract normally, therefore, the heart's pumping effectiveness was dramatically reduced. The reduced pumping capacity was responsible for the low blood pressure, which decreased

Occluded coronary artery

### Figure 12A

An angiogram is a picture of a blood vessel that is usually obtained by injecting the blood vessel with dye that can be detected by x-rays. Note the partially occluded (blocked) coronary blood vessel in this angiogram, which has been computer-enhanced to show colors.

the blood flow to Paul's brain, resulting in confusion, disorientation, and unconsciousness.

Low blood pressure, increasing blood carbon dioxide levels, pain, and anxiousness increased sympathetic stimulation of the heart and adrenal glands. Increased sympathetic stimulation of the adrenal medulla caused the release of epinephrine. Increased parasympathetic stimulation of the heart resulted from pain sensations. The heart rate was periodically arrhythmic due to the combined effects of parasympathetic stimulation, sympathetic stimulation, and the release of epinephrine and norepinephrine from the adrenal gland. In addition, the ischemic areas of the left ventricle produced ectopic beats.

Pulmonary edema resulted from the increased pressure in the pulmonary veins because of the left ventricle's reduced ability to pump blood. The edema allowed bacteria to infect the lungs and cause pneumonia.

The heart began to beat rhythmically in response to medication because the infarction had not damaged the heart's conducting

| System Interactions | Effects of Myocardial Infarctions on Other Systems |
|---|---|
| **System** | **Interactions** |
| Integumentary | Pallor of the skin results from intense vasoconstriction of peripheral blood vessels, including those in the skin. |
| Muscular | Skeletal muscle activity is reduced because of lack of blood flow to the brain and because blood is shunted from blood vessels that supply skeletal muscles to those that supply the heart and brain. |
| Nervous | Decreased blood flow to the brain, decreased blood pressure, and pain due to ischemia of heart muscle result in increased sympathetic and parasympathetic stimulation of the heart. Loss of consciousness occurs when the blood flow to the brain decreases so that not enough oxygen is available to maintain normal brain function, especially in the reticular activating system. |
| Endocrine | When blood pressure becomes low, antidiuretic hormone (ADH) is released from the posterior pituitary gland, and renin, released from the kidney, activates the renin-angiotensin-aldosterone mechanism. ADH, secreted in large amounts, and angiotensin II cause vasoconstriction of peripheral blood vessels. ADH and aldosterone act on the kidneys to retain water and ions. Increased blood volume increases venous return, which results in increased stroke volume and blood pressure unless damage to the heart is very severe. |
| Lymphatic or Immune | White blood cells, including macrophages, move to the area of cardiac muscle damage and phagocytize any dead cardiac muscle cells. |
| Respiratory | Decreased blood pressure results in decreased blood flow to the lungs. The decrease in gas exchange leads to increased blood carbon dioxide levels, acidosis, and decreased blood oxygen levels. Initially, respiration becomes deep and labored because of the elevated carbon dioxide levels, decreased blood pH, and depressed oxygen levels. If the blood oxygen levels decrease too much, the person loses consciousness. Pulmonary edema can result when the pumping effectiveness of the left ventricle is substantially reduced. |
| Digestive | Intense sympathetic stimulation reduces blood flow to the digestive system to very low levels, which often results in nausea and vomiting. |
| Urinary | Blood flow to the kidney decreases dramatically in response to sympathetic stimulation. If the kidney becomes ischemic, the kidney tubules can be damaged, resulting in acute renal failure and reduced urine production. Increased blood urea nitrogen, increased blood levels of $K^+$, and edema are indications that the kidneys cannot eliminate waste products and excess water. If damage is not too great, the period of reduced urine production may last up to 3 weeks; then the rate of urine production slowly returns to normal as the kidney tubules heal. |

system, which is an indication that there were no permanent arrhythmias. (Permanent arrhythmias are indications of damage done to cardiac muscle cells specialized to conduct action potentials in the heart.)

Analysis of the electrocardiogram, blood pressure measurements, and the angiogram indicated that the infarction was located on the left side of Paul's heart.

Paul's physician made it very clear that he was lucky to have survived a myocardial infarction. The physician recommended a weight-loss program and a low-sodium and low-fat diet. He also advised Paul to take a small amount of aspirin regularly and to stop smoking. He explained that Paul would have to take medication for high blood pressure if his blood pressure did not decrease in response to the recommended changes. The physician recommended an aerobic exercise program after a period of recovery and suggested that Paul seek ways to reduce the stress associated with his job. Paul followed the doctor's recommendations, and after several months, his blood pressure was normal, and he felt better than he had in years.

### Predict 9

*Severe ischemia in the wall of a ventricle can cause the death of cardiac muscle cells. Inflammation develops around the necrotic (dead) tissue, and macrophages invade and phagocytize dead cells. At the same time, blood vessels and connective tissue grow into the necrotic area and begin to deposit connective tissue to replace the necrotic tissue. While Paul was still recovering in the hospital, another patient was admitted with a similar condition. After about a week, that person's blood pressure suddenly decreased to very low levels, and he died within a short time. An autopsy revealed a large amount of blood in the pericardial sac and a rupture in the wall of the left ventricle. Explain these effects.*

## DISEASES AND DISORDERS  Heart

| CONDITION | DESCRIPTION |
|---|---|
| **Inflammation of Heart Tissue** | |
| Endocarditis | Inflammation of the endocardium; affects the valves more severely than other areas of the endocardium; may lead to scarring, causing stenosed or incompetent valves |
| Pericarditis | Inflammation of the pericardium; see Clinical Impact, "Disorders of the Pericardium" |
| Cardiomyopathy | Disease of the myocardium of unknown cause or occurring secondarily to other disease; results in weakened cardiac muscle, causing all chambers of the heart to enlarge; may eventually lead to congestive heart failure |
| Rheumatic heart disease | Results from a streptococcal infection in young people; toxin produced by the bacteria can cause rheumatic fever several weeks after the infection that can result in rheumatic endocarditis |
| **Reduced Blood Flow to Cardiac Muscle** | |
| Coronary heart disease | Reduces the amount of blood the coronary arteries can deliver to the myocardium; see Clinical Impact "Heart Attack" |
| Coronary thrombosis | Formation of blood clot in a coronary artery |
| Myocardial infarction | Damaged cardiac muscle tissue resulting from lack of blood flow to the myocardium; often referred to as a heart attack; see Systems Pathology |
| **Congenital Heart Diseases (Occur at Birth)** | |
| Septal defect | Hole in the septum between the left and right sides of the heart, allowing blood to flow from one side of the heart to the other and greatly reducing the heart's pumping effectiveness |
| Patent ductus arteriosus | Ductus arteriosus fails to close after birth, allowing blood to flow from the aorta to the pulmonary trunk under a higher pressure, which damages the lungs; also, the left ventricle must work harder to maintain adequate systemic pressure |
| Stenosis of the heart valves | Narrowed opening through one or more of the heart valves; aortic or pulmonary semilunar stenosis increases the heart's workload; bicuspid valve stenosis causes blood to back up in the left atria and lungs, resulting in edema of the lungs; tricuspid valve stenosis results in similar blood flow problems and edema in the peripheral tissues |
| Cyanosis (sī-ă-nō′sis; *cyan,* blue + *osis,* condition of) | Symptom of inadequate heart function in babies with congenital heart disease; the infant's skin appears blue because of low oxygen levels in the blood in peripheral blood vessels |

Go to **www.mhhe.com/seeleyess7** for additional information on these pathologies.

Body temperature affects the metabolism in the heart just as it affects other tissues. Elevated body temperature increases the heart rate, and reduced body temperature slows the heart rate. For example, the heart rate is usually elevated when a person has a fever. During heart surgery, the body temperature is sometimes intentionally lowered to slow the heart rate and metabolism.

# EFFECTS OF AGING ON THE HEART

Gradual changes in heart function are associated with aging. These changes are minor under resting conditions but become more obvious during exercise and in response to age-related diseases.

By age 70, a person's cardiac output has probably decreased by approximately one-third. Because of the decrease in the reserve strength of the heart, many elderly people have a limited ability to respond to emergencies, infections, blood loss, or stress.

Hypertrophy (enlargement) of the left ventricle is a common age-related change. This appears to result from a gradual increase in the pressure in the aorta (afterload), against which the left ventricle must pump. The increased aortic pressure results from a gradual decrease in the aorta's elasticity. Because the left ventricle is enlarged, its ability to pump out blood is reduced, which can cause an increase in left atrial pressure and lead to increased pulmonary edema. Consequently, older people tend to feel out of breath when they exercise strenuously.

Aging cardiac muscle requires a greater amount of time to contract and relax. Thus, the maximum heart rate decreases. Both the resting and maximum cardiac output slowly lower as people grow older, and by age 85, the cardiac output has decreased 30–60%.

Age-related changes also affect the connective tissue of the heart valves. The connective tissue becomes less flexible, and calcium deposits develop in the valves. As a result, the aortic semilunar valve may become stenosed or incompetent.

An age-related increase in cardiac arrhythmias occurs as a consequence of a decreased number of cardiac cells in the SA node and because of cell replacement in the AV bundle.

Coronary artery disease and heart failure are also age-related. Approximately 10% of elderly people over age 80 have heart failure, and a major contributing factor is coronary heart disease. Advanced age, malnutrition, chronic infections, toxins, severe anemias, hyperthyroidism, and hereditary factors can also lead to heart failure.

Exercise has many beneficial effects on the heart. Regular aerobic exercise improves the heart's functional capacity at all ages, providing the person has no other conditions that cause the increased workload on the heart to be harmful.

Circulatory

# ⏤⋀⏤ SUMMARY

## Functions of the Heart   (p. 313)

1. The heart generates blood pressure.
2. The heart routes blood through the systemic and pulmonary circulations.
3. The heart's pumping action and its valves ensure a one-way flow of blood through the heart and blood vessels.
4. The heart helps regulate blood supply to tissues.

## Size, Form, and Location of the Heart   (p. 313)

The heart is approximately the size of a closed fist and is located in the pericardial cavity.

## Anatomy of the Heart   (p. 314)

### Pericardium

1. The pericardium is a sac consisting of fibrous and serous pericardia. The fibrous pericardium is lined by the parietal pericardium.
2. The outer surface of the heart is lined by the visceral pericardium (epicardium).
3. Between the visceral and parietal pericardia is the pericardial cavity, which is filled with pericardial fluid.

### External Anatomy

1. The atria are separated externally from the ventricles by the coronary sulcus. The right and left ventricles are separated externally by the interventricular sulci.
2. The inferior and superior venae cavae enter the right atrium. The four pulmonary veins enter the left atrium.
3. The pulmonary trunk exits the right ventricle, and the aorta exits the left ventricle.

### Heart Chambers and Internal Anatomy

1. There are four chambers in the heart. The left and right atria receive blood from veins and function mainly as reservoirs. Contraction of the atria completes ventricular filling.
2. The atria are separated internally from each other by the interatrial septum.
3. The ventricles are the main pumping chambers of the heart. The right ventricle pumps blood into the pulmonary trunk, and the left ventricle, which has a thicker wall, pumps blood into the aorta.
4. The ventricles are separated internally by the interventricular septum.

### Heart Valves

1. The heart valves ensure one-way flow of blood.
2. The tricuspid valve (three cusps) separates the right atrium and the right ventricle, and the bicuspid valve (two cusps) separates the left atrium and the left ventricle.
3. The papillary muscles attach by the chordae tendineae to the cusps of the tricuspid and bicuspid valves and adjust tension on the valves.
4. The aorta and pulmonary trunk are separated from the ventricles by the semilunar valves.
5. The skeleton of the heart is a plate of fibrous connective tissue that separates the atria from the ventricles, acts as an electrical barrier between the atria and ventricles, and supports the heart valves.

### Route of Blood Flow Through the Heart

1. The left and right sides of the heart can be considered separate pumps.
2. Blood flows from the systemic vessels to the right atrium and from the right atrium to the right ventricle. From the right ventricle, blood flows to the pulmonary trunk and from the pulmonary trunk to the lungs. From the lungs, blood flows through the pulmonary veins to the left atrium, and from the left atrium, blood flows to the left ventricle. From the left ventricle, blood flows into the aorta and then through the systemic vessels.

### Blood Supply to the Heart

1. The left and right coronary arteries originate from the base of the aorta and supply the heart.
2. The left coronary artery has three major branches: the anterior interventricular, the circumflex, and the left marginal arteries.
3. The right coronary artery has two major branches: the posterior interventricular and the right marginal arteries.
4. Blood returns from heart tissue through cardiac veins to the coronary sinus and into the right atrium. Small cardiac veins also return blood directly to the right atrium.

## Histology of the Heart   (p. 322)

### Heart Wall

The heart wall consists of the outer epicardium, the middle myocardium, and the inner endocardium.

### Cardiac Muscle

1. Cardiac muscle is striated; it depends on ATP for energy and on aerobic metabolism.
2. Cardiac muscle cells are joined by intercalated disks that allow action potentials to be propagated throughout the heart.

## Electrical Activity of the Heart   (p. 324)

### Action Potentials in Cardiac Muscle

1. Action potentials in cardiac muscle are prolonged compared to those in skeletal muscle and have a depolarization phase, a plateau phase, and a repolarization phase.
2. The depolarization is due mainly to opening of the voltage-gated $Na^+$ channels, and the plateau phase is due to opened voltage-gated $Ca^{2+}$ channels. Repolarization at the end of the plateau phase is due to the opening of $K^+$ channels for a brief period.
3. The prolonged action potential in cardiac muscle ensures that contraction and relaxation occur and prevents tetany.
4. The SA node located in the upper wall of the right atrium is the normal pacemaker of the heart, and cells of the SA node have more voltage-gated $Ca^{2+}$ channels than do other areas of the heart.

### Conduction System of the Heart

1. The conduction system of the heart is made up of specialized cardiac muscle cells.
2. The SA node produces action potentials that are propagated over the atria to the AV node.
3. The AV node and the atrioventricular bundle conduct action potentials to the ventricles.
4. The right and left bundle branches conduct action potentials from the atrioventricular bundle through Purkinje fibers to the ventricular muscle.
5. An ectopic beat results from an action potential that originates in an area of the heart other than the SA node.

### Electrocardiogram

1. An ECG is a record of electrical events within the heart.
2. An ECG can be used to detect abnormal heart rates or rhythms, abnormal conduction pathways, hypertrophy or atrophy of the heart, and the approximate location of damaged cardiac muscle.
3. A normal ECG consists of a P wave (atrial depolarization), a QRS complex (ventricular depolarization), and a T wave (ventricular repolarization).
4. Atrial contraction occurs during the PQ interval, and the ventricles contract and relax during the QT interval.

Circulatory

## Cardiac Cycle   (p. 328)

1. Atrial systole is contraction of the atria, and ventricular systole is contraction of the ventricles. Atrial diastole is relaxation of the atria, and ventricular diastole is relaxation of the ventricles.
2. During ventricular systole, the AV valves close, pressure increases in the ventricles, the semilunar valves are forced open, and blood flows into the aorta and pulmonary trunk.
3. At the beginning of ventricular diastole, pressure in the ventricles decreases. The semilunar valves close to prevent backflow of blood from the aorta and pulmonary trunk into the ventricles.
4. When the pressure in the ventricles is low enough, the AV valves open, and blood flows from the atria into the ventricles.
5. During atrial systole, the atria contract and complete filling of the ventricles.

## Heart Sounds   (p. 331)

1. The first heart sound results from closure of the AV valves. The second heart sound results from closure of the semilunar valves.
2. Abnormal heart sounds, called murmurs, can result from incompetent (leaky) valves or stenosed (narrowed) valves.

## Regulation of Heart Function   (p. 332)

Cardiac output (volume of blood pumped per ventricle per minute) is equal to the stroke volume (volume of blood ejected per beat) times the heart rate (beats per minute).

### Intrinsic Regulation of the Heart

1. *Intrinsic regulation* refers to regulation mechanisms contained within the heart.
2. As venous return to the heart increases, the heart wall is stretched, and the increased stretch of the ventricular walls is called preload.
3. A greater preload causes the cardiac output to increase because stroke volume increases (Starling's law of the heart).
4. Afterload is the pressure against which the ventricles must pump blood.

### Extrinsic Regulation of the Heart

1. *Extrinsic regulation* refers to nervous and chemical mechanisms.
2. Sympathetic stimulation increases stroke volume and heart rate; parasympathetic stimulation decreases heart rate.
3. The baroreceptor reflex detects changes in blood pressure. If blood pressure increases suddenly, the reflex causes a decrease in heart rate and stroke volume; if blood pressure decreases suddenly, the reflex causes an increase in heart rate and stroke volume.
4. Emotions influence heart function by increasing sympathetic stimulation of the heart in response to exercise, excitement, anxiety, or anger and by increasing parasympathetic stimulation in response to depression.
5. Alterations in body fluid levels of carbon dioxide, pH, and ion concentrations, as well as changes in body temperature, influence heart function.

## Effects of Aging on the Heart   (p. 340)

The following age-related changes are common:
1. By age 70, cardiac output has often decreased by one-third.
2. Hypertrophy of the left ventricle can cause pulmonary edema.
3. Decrease in the maximum heart rate by 30%–60% by age 85 leads to decreased cardiac output.
4. The aortic semilunar valve can become stenotic or incompetent.
5. Coronary artery disease and congestive heart failure can develop.
6. Aerobic exercise improves the functional capacity of the heart at all ages.

 # REVIEW AND COMPREHENSION

1. Describe the size and location of the heart, including its base and apex.
2. Describe the structure and function of the pericardium.
3. What chambers make up the left and right sides of the heart? What are their functions?
4. Describe the structure and location of the tricuspid, bicuspid, and semilunar valves. What is the function of these valves?
5. What are the functions of the atria and ventricles?
6. Starting in the right atrium, describe the flow of blood through the heart.
7. Describe the vessels that supply blood to the cardiac muscle.
8. Define heart attack and infarct. How does atherosclerotic plague affect the heart?
9. Describe the three layers of the heart. Which of the three layers is most important in causing contractions of the heart?
10. Describe the structure of cardiac muscle cells, including the structure and function of intercalated disks.
11. Describe the events that result in an action potential in cardiac muscle.
12. Explain how cardiac muscle cells in the SA node produce action potentials spontaneously and why the SA node is the heart's pacemaker.
13. What is the function of the conduction system of the heart? Starting with the SA node, describe the route of an action potential as it goes through the conduction system of the heart.
14. Explain the electrical events that generate each portion of the electrocardiogram. How do they relate to contraction events?
15. What contraction and relaxation events occur during the P-Q interval and the Q-T interval of the electrocardiogram?
16. Define cardiac cycle, systole, and diastole.
17. Describe blood flow and the opening and closing of heart valves during the cardiac cycle.
18. Describe the pressure changes that occur in the left atrium, left ventricle, and aorta during ventricular systole and diastole (see figure 12.18).
19. What events cause the first and second heart sounds?
20. Define murmur. Describe how either an incompetent or a stenosed valve can cause a murmur.
21. Define cardiac output, stroke volume, and heart rate.
22. What is Starling's law of the heart? What effect does an increase or a decrease in venous return have on cardiac output?
23. Describe the effect of parasympathetic and sympathetic stimulation on heart rate and stroke volume.
24. How does the nervous system detect and respond to the following?
    a. a decrease in blood pressure
    b. an increase in blood pressure
25. What is the effect of epinephrine on the heart rate and stroke volume?
26. Explain how emotions affect heart function.
27. What effects do the following have on cardiac output?
    a. a decrease in blood pH
    b. an increase in blood carbon dioxide
28. How do changes in body temperature influence the heart rate?
29. List the common age-related heart diseases that develop in elderly people.

# CRITICAL THINKING

1. A friend tells you that an ECG revealed that her son has a slight heart murmur. Should you be convinced that he has a heart murmur? Explain.

2. Predict the effect on Starling's law of the heart if the parasympathetic (vagus) nerves to the heart are cut.

3. Predict the effect on heart rate if the sensory nerve fibers from the baroreceptors are cut.

4. An experiment is performed on a dog in which the arterial blood pressure in the aorta is monitored before and after the common carotid arteries are clamped. Explain the change in arterial blood pressure that would occur. (*Hint:* Baroreceptors are located in the internal carotid arteries, which are superior to the site of clamping of the common carotid arteries.)

5. Predict the consequences on the heart if a person took a large dose of a drug that blocks $Ca^{2+}$ channels.

6. What happens to cardiac output following the ingestion of a large amount of fluid?

7. At rest, the cardiac output of athletes and nonathletes can be equal, but the heart rate of athletes is lower than that of nonathletes. At maximum exertion, the maximum heart rate of athletes and nonathletes can be equal, but the cardiac output of athletes is greater than that of nonathletes. Explain these differences.

8. Explain why it is useful that the walls of the ventricles are thicker than those of the atria.

9. Predict the effect of an incompetent aortic semilunar valve on ventricular and aortic pressure during ventricular systole and diastole.

*Answers in Appendix D*

Visit this textbook's website at **www.mhhe.com/seeleyess7** for practice quizzes, animations, interactive learning exercises, and other study tools.

McGraw-Hill offers a study CD that features interactive cadaver dissection. ***Anatomy & Physiology Revealed*** includes cadaver photos that allow you to peel away layers of the human body to reveal structures beneath the surface. This program also includes animations, radiologic imaging, audio pronunciations, and practice quizzing.

Circulatory

In this photo, a nurse is measuring a patient's blood pressure. The heart pumps blood through the rest of the body by forcing it through small tubes, or blood vessels. The force of blood moving through the blood vessels is what creates blood pressure. If blood pressure is not tightly regulated, the body cannot maintain homeostasis.

# Chapter 13

# Blood Vessels and Circulation

## Chapter Outline and Objectives

**Functions of the Circulatory System (p. 345)**

1. List the functions of the circulatory system.

**General Features of Blood Vessel Structure (p. 345)**

2. Describe the structure and function of arteries, capillaries, and veins.

**Blood Vessels of the Pulmonary Circulation (p. 348)**

3. Describe the blood vessels of the pulmonary circulation.

**Blood Vessels of the Systemic Circulation: Arteries (p. 348)**

4. List the major arteries that supply each of the body areas, and describe their functions.

**Blood Vessels of the Systemic Circulation: Veins (p. 356)**

5. List the major veins that carry blood from each of the body areas, and describe their functions.

**Physiology of Circulation (p. 361)**

6. Explain how blood pressure and resistance to flow change as blood flows through the blood vessels.

7. Describe the exchange of material across a capillary wall.

**Control of Blood Flow in Tissues (p. 366)**

8. Explain how local control mechanisms regulate blood flow.

9. Explain how nervous mechanisms control blood flow.

**Regulation of Arterial Pressure (p. 368)**

10. Describe the short-term and long-term mechanisms that regulate arterial pressure.

**Effects of Aging on the Blood Vessels (p. 373)**

11. Describe the effects of aging on blood vessels.

The blood vessels of the body form a network more complex than an interstate highway system. The blood vessels carry blood to within two or three cell diameters of nearly all the trillions of cells that make up the body. Blood flow through them is regulated, so that cells receive adequate nutrients and so that waste products are removed. Blood vessels remain functional, in most cases, in excess of 70 years, and when they are damaged, they repair themselves.

Blood vessels outside the heart are divided into two classes: (1) the **pulmonary vessels,** which transport blood from the right ventricle through the lungs and back to the left atrium, and (2) the **systemic vessels,** which transport blood through all parts of the body, from the left ventricle and back to the right atrium (see chapter 12 and figure 12.2). Together, the pulmonary vessels and the systemic vessels constitute the **circulatory system.**

# FUNCTIONS OF THE CIRCULATORY SYSTEM

The heart provides the major force that causes blood to circulate, and the circulatory system has five functions:

1. *Carries blood.* Blood vessels carry blood from the heart to all the tissues of the body and back to the heart.
2. *Exchanges nutrients, waste products, and gases with tissues.* Nutrients and oxygen diffuse from blood vessels to cells in essentially all areas of the body. Waste products and carbon dioxide diffuse from the cells, where they are produced, to blood vessels.
3. *Transports substances.* Blood transports hormones, components of the immune system, molecules required for coagulation, enzymes, nutrients, gases, waste products, and other substances to all areas of the body.
4. *Helps regulate blood pressure.* The circulatory system and the heart work together to regulate blood pressure within a normal range.
5. *Directs blood flow to the tissues.* The circulatory system directs blood to tissues when increased blood flow is required to maintain homeostasis.

# GENERAL FEATURES OF BLOOD VESSEL STRUCTURE

The three main types of blood vessels are arteries, capillaries, and veins. **Arteries** (ar′ter-ēz) carry blood away from the heart; usually, the blood is oxygen-rich. Blood is pumped from the ventricles of the heart into large, elastic arteries, which branch repeatedly to form progressively smaller arteries. As they become smaller, the artery walls undergo a gradual transition from having more elastic tissue than smooth muscle to having more smooth muscle than elastic tissue (figure 13.1*a,b*). The arteries are normally classified as elastic arteries, muscular arteries, or arterioles, although they form a continuum from the largest to the smallest branches.

Blood flows from arterioles into **capillaries** (kap′i-lār-ēz), where exchange occurs between the blood and the tissue fluid. Capillaries have thinner walls than do arteries. Blood flows through them more slowly, and there are far more of them than any other blood vessel type.

From the capillaries, blood flows into veins. **Veins** (vānz) carry blood toward the heart; usually, the blood is oxygen-poor.

Compared to arteries, the walls of veins are thinner and contain less elastic tissue and fewer smooth muscle cells (figure 13.1*c*). Starting at capillaries and proceeding toward the heart, small-diameter veins come together to form larger-diameter veins, which are fewer in number. Veins increase in diameter and decrease in number as they progress toward the heart, and their walls increase in thickness. Veins may be classified as venules, small veins, medium-sized veins, or large veins.

Except in capillaries and venules, blood vessel walls consist of three layers, or **tunics** (too′niks). From the inner to the outer wall, the tunics are (1) the tunica intima, (2) the tunica media, and (3) the tunica adventitia, or tunica externa (figure 13.2; see figure 13.1).

The **tunica intima** (too′ni-kă in′ti-mă), or innermost layer, consists of an endothelium composed of simple squamous epithelial cells, a basement membrane, and a small amount of connective tissue. In muscular arteries, the tunica intima also contains a layer of thin elastic connective tissue. The **tunica media,** or middle layer, consists of smooth muscle cells arranged circularly around the blood vessel. It also contains variable amounts of elastic and collagen fibers, depending on the size and type of the vessel. In muscular arteries, a layer of elastic connective tissue forms the outer margin of the tunica media. The **tunica adventitia** (ad-ven-tish′ă) is composed of connective tissue that is dense adjacent to the tunica media but becomes loose connective tissue toward the outer portion of the blood vessel wall.

## Arteries

**Elastic arteries** are the largest-diameter arteries and have the thickest walls (see figure 13.1*a*). Compared to other arteries, a greater proportion of their walls is composed of elastic tissue, and a smaller proportion is smooth muscle. The aorta and pulmonary trunk are examples of elastic arteries. Elastic arteries stretch when the ventricles of the heart pump blood into them. The elastic recoil of these arteries prevents blood pressure from falling rapidly and maintains blood flow while the ventricles are relaxed.

The **muscular arteries** include medium-sized and small arteries. The walls of medium-sized arteries are relatively thick compared to their diameter. Most of the wall's thickness results from smooth muscle cells of the tunica media (see figure 13.1*b*). Medium-sized arteries are frequently called **distributing arteries** because the smooth muscle tissue enables these vessels to control blood flow to different body regions. Contraction of the smooth muscle in blood vessels, called **vasoconstriction** (vā′sō-kon-strik′shŭn), decreases blood vessel diameter and blood flow.

Circulatory

(a) **Elastic arteries.** The tunica media is mostly elastic connective tissue. Elastic arteries recoil when stretched, which prevents blood pressure from falling rapidly.

(b) **Muscular arteries.** The tunica media is a thick layer of smooth muscle. Muscular arteries regulate blood flow to different regions of the body.

(c) **Medium-sized and large veins.** All three tunics are present. The tunica media is thin but can regulate vessel diameter because blood pressure in the venous system is low. The predominant layer is the tunica adventitia.

(d) Folds in the tunica intima form the valves of veins, which allow blood to flow toward the heart but not in the opposite direction.

Tunica adventitia

Tunica media
(elastic tissue and smooth muscle)

Connective tissue

Endothelium and basement membrane

Tunica intima

Tunica adventitia

Elastic connective tissue

Smooth muscle

Tunica media

Elastic connective tissue

Connective tissue

Endothelium and basement membrane

Tunica intima

Tunica adventitia

Tunica media

Connective tissue

Endothelium and basement membrane

Tunica intima

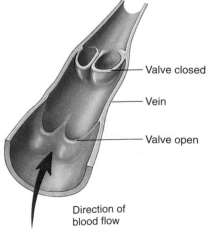

Valve closed

Vein

Valve open

Direction of blood flow

**Figure 13.1** Blood Vessel Structure

**Figure 13.2** **Photomicrograph of an Artery and a Vein**

The typical structure of a medium-sized artery (A) and vein (V). Note that the artery has a thicker wall than the vein. The predominant layer in the wall of the artery is the tunica media with its circular layers of smooth muscle. The predominant layer in the wall of the vein is the tunica adventitia, and the tunica media is thinner.

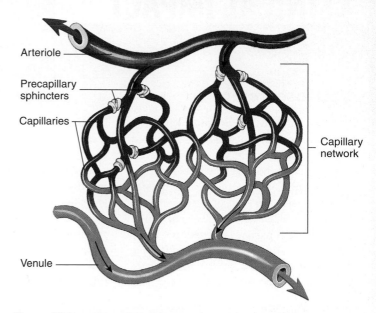

**Figure 13.3** **Capillary Network**

A capillary network stems from an arteriole. The network forms numerous branches. Blood flows through capillaries into venules. Smooth muscle cells, called precapillary sphincters, regulate blood flow through the capillaries. Blood flow decreases when the precapillary sphincters constrict and increases when they dilate.

Relaxation of the smooth muscle in blood vessels, called **vasodilation** (vā′sō-dĭ-lā′shŭn), increases blood vessel diameter and blood flow.

Medium-sized arteries supply blood to small arteries. Small arteries have about the same structure as the medium-sized arteries, except for a smaller diameter and thinner walls. The smallest of the small arteries have only three or four layers of smooth muscle in their walls.

**Arterioles** (ar-tēr′ē-ōlz) transport blood from small arteries to capillaries. Arterioles are the smallest arteries in which the three tunics can be identified; the tunica media consists of only one or two layers of circular smooth muscle cells. Small arteries and arterioles are adapted for vasodilation and vasoconstriction.

## Capillaries

Blood flows from arterioles into capillaries, which branch to form networks (figure 13.3). Blood flow through capillaries is regulated by smooth muscle cells called **precapillary sphincters** located at the origin of the branches. Capillary walls consist of **endothelium** (en-dō-thē′lē-ŭm), which is a layer of simple squamous epithelium surrounded by delicate loose connective tissue. The thin walls of capillaries facilitate diffusion between the capillaries and surrounding cells. Each capillary is 0.5–1 millimeter (mm) long. Capillaries branch without changing their diameter, which is approximately the same as the diameter of a red blood cell (7.5 μm) (figure 13.4).

Red blood cells flow through most capillaries in single file and are frequently folded as they pass through the smaller-diameter capillaries. As blood flows through capillaries, blood gives up oxygen and nutrients to the tissue spaces and takes up carbon dioxide and

other by-products of metabolism. Capillary networks are more numerous and more extensive in the lungs and in highly metabolic tissues, such as the liver, kidneys, skeletal muscle, and cardiac muscle, than in other tissue types.

## Veins

Blood flows from capillaries into venules and from venules into small veins. **Venules** (ven′oolz) have a diameter slightly larger than that of capillaries and are composed of endothelium resting on a delicate connective tissue layer. The structure of venules, except for their diameter, is very similar to that of capillaries. **Small veins** are slightly larger in diameter than venules. All three

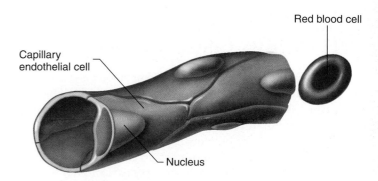

**Figure 13.4** **Capillary**

The capillary wall is a thin layer of simple squamous epithelium, which facilitates exchange of gases, nutrients, and waste products between the blood and the tissues.

Circulatory

## Varicose Veins

**Varicose** (văr'ĭ-kōs) **veins** result when the veins of the lower limbs become so dilated that the cusps of the valves no longer overlap to prevent the backflow of blood. As a consequence, venous pressure is greater than normal in the veins of the lower limbs and can result in edema (swelling) of the limb. Some people have a genetic tendency to develop varicose veins. The condition is further encouraged by activities that increase the pressure in veins. For example, standing in place for prolonged periods allows the pressure of the blood to stretch the veins, and the enlarged uterus in pregnancy compresses the veins in the pelvis, increasing pressure in the veins that drain the lower limbs.

The blood in varicose veins can become so stagnant that it clots. The clots, called **thromboses** (throm-bō'sēz), can cause inflammation of the veins, a condition called **phlebitis** (fle-bī'tis). If the condition becomes severe enough, the blocked veins can prevent blood from flowing through the capillaries that are drained by the veins. The lack of blood flow can lead to tissue death and infection with anaerobic bacteria, a condition called **gangrene** (gang'grēn). In addition, fragments of the clots may dislodge and travel through the veins to the lungs, where they can cause severe damage. Fragments of thromboses that dislodge and float in the blood are called **emboli** (em'bō-lī).

---

tunics are present in small veins. The tunica media contains a continuous layer of smooth muscle cells, and the connective tissue of the tunica adventitia surrounds the tunica media.

**Medium-sized veins** collect blood from small veins and deliver it to large veins. The three thin but distinctive tunics make up the wall of the medium-sized and large veins. The tunica media contains some circular smooth muscle and sparsely scattered elastic fibers. The predominant layer is the outer tunica adventitia, which consists primarily of dense collagen fibers (see figures 13.1c and 13.2). Consequently, veins are more distensible than arteries. The connective tissue of the tunica adventitia determines the degree to which they can distend.

**Large veins,** having diameters greater than 2 mm, contain **valves,** which allow blood to flow toward the heart but not in the opposite direction (see figure 13.1d). Each valve consists of folds in the tunica intima that form two flaps, which are similar in shape and function to the semilunar valves of the heart. There are many valves in medium-sized veins and more valves in veins of the lower limbs than in veins of the upper limbs. This prevents blood from flowing toward the feet in response to the pull of gravity.

## BLOOD VESSELS OF THE PULMONARY CIRCULATION

The **pulmonary circulation** is the system of blood vessels that carries blood from the right ventricle of the heart to the lungs and back to the left atrium of the heart. Blood from the right ventricle is pumped into a short vessel called the **pulmonary** (pŭl'mō-nār-ē) **trunk** (figure 13.5). The pulmonary trunk then branches into the **right** and **left pulmonary arteries,** which extend to the right and left lungs, respectively. These arteries carry oxygen-poor blood to the pulmonary capillaries in the lungs, where the blood takes up oxygen and releases carbon dioxide. Blood rich in oxygen flows from the lungs to the left atrium. Four **pulmonary veins** (two from each lung) exit the lungs and carry the oxygen-rich blood to

the left atrium. Figure 13.5 traces the path of the pulmonary circulation and shows it in relation to the systemic circulation, described next.

## BLOOD VESSELS OF THE SYSTEMIC CIRCULATION: ARTERIES

The **systemic circulation** is the system of blood vessels that carries blood from the left ventricle of the heart to the tissues of the body and back to the right atrium. Oxygen-rich blood from the pulmonary veins passes from the left atrium into the left ventricle and from the left ventricle into the aorta. Arteries distribute blood from the aorta to all portions of the body (figure 13.6).

### Aorta

All arteries of the systemic circulation branch directly or indirectly from the **aorta** (ā-ōr'tă). The aorta is usually considered in three parts—the ascending aorta, the aortic arch, and the descending aorta—the last segment is further divided into the thoracic aorta and the abdominal aorta (figure 13.7a).

The **ascending aorta** is the part of the aorta that passes superiorly from the left ventricle. The right and left **coronary arteries** arise from the base of the ascending aorta and supply blood to the heart (see chapter 12).

The aorta arches posteriorly and to the left as the **aortic arch.** Three major arteries, which carry blood to the head and upper limbs, originate from the aortic arch: the brachiocephalic artery, the left common carotid artery, and the left subclavian artery (figure 13.7b).

The **descending aorta** is the longest part of the aorta. It extends through the thorax and abdomen to the upper margin of the pelvis. The part of the descending aorta that extends through the thorax to the diaphragm is called the **thoracic** (thō-ras'ik) **aorta.** The part of the descending aorta that extends from the diaphragm to the point at which it divides into the two **common iliac**

Circulatory

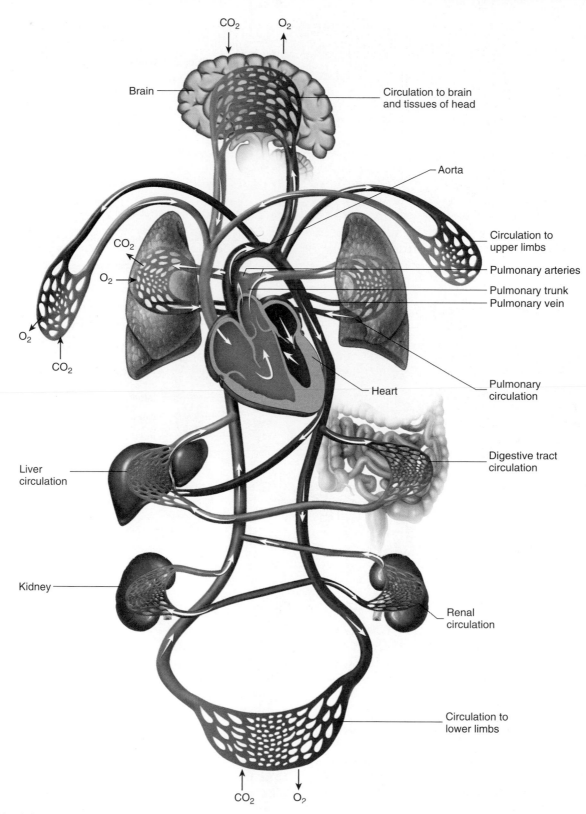

**Figure 13.5 Blood Flow Through the Circulatory System**

Veins and venules return blood from the body to the right atrium. After passing from the right atrium to the right ventricle, blood is pumped into the pulmonary trunk. The pulmonary trunk divides into the right and left pulmonary arteries, which carry oxygen-poor blood to the lungs. In the lung capillaries, carbon dioxide is given off, and the blood picks up oxygen. Blood, now rich in oxygen, flows from each lung to the left atrium. Blood passes from the left atrium to the left ventricle. The left ventricle then pumps the blood into the aorta, which distributes the blood through its branches to all of the body. Blood returns to the heart through the venous system, and the cycle continues.

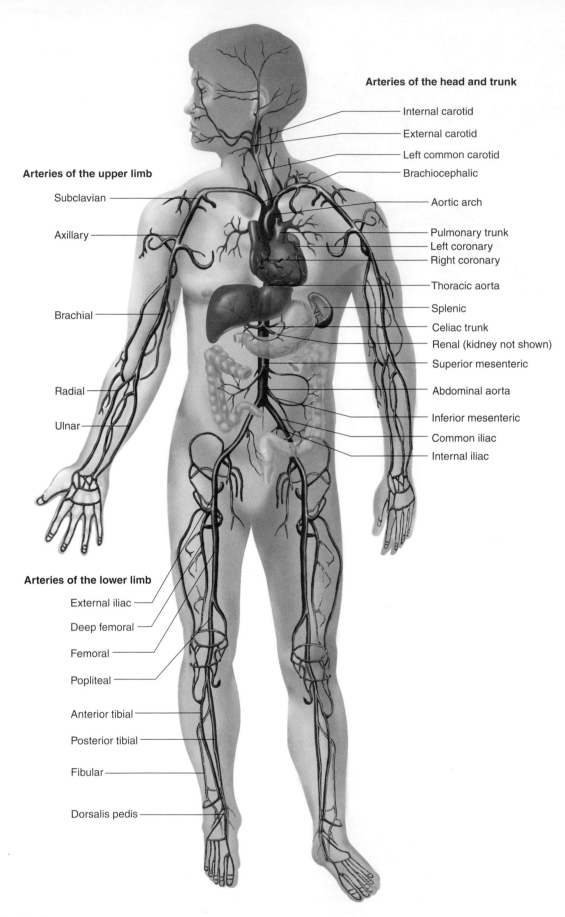

**Arteries of the head and trunk**

Internal carotid
External carotid
Left common carotid
Brachiocephalic
Aortic arch
Pulmonary trunk
Left coronary
Right coronary
Thoracic aorta
Splenic
Celiac trunk
Renal (kidney not shown)
Superior mesenteric
Abdominal aorta
Inferior mesenteric
Common iliac
Internal iliac

**Arteries of the upper limb**

Subclavian
Axillary
Brachial
Radial
Ulnar

**Arteries of the lower limb**

External iliac
Deep femoral
Femoral
Popliteal
Anterior tibial
Posterior tibial
Fibular
Dorsalis pedis

**Figure 13.6** Major Arteries

These major arteries carry blood from the left ventricle of the heart to the body tissues.

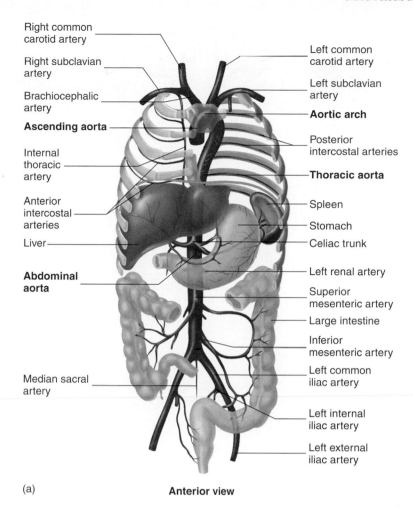

Right common carotid artery
Right subclavian artery
Brachiocephalic artery
**Ascending aorta**
Internal thoracic artery
Anterior intercostal arteries
Liver
**Abdominal aorta**
Median sacral artery

Left common carotid artery
Left subclavian artery
**Aortic arch**
Posterior intercostal arteries
**Thoracic aorta**
Spleen
Stomach
Celiac trunk
Left renal artery
Superior mesenteric artery
Large intestine
Inferior mesenteric artery
Left common iliac artery
Left internal iliac artery
Left external iliac artery

(a)          **Anterior view**

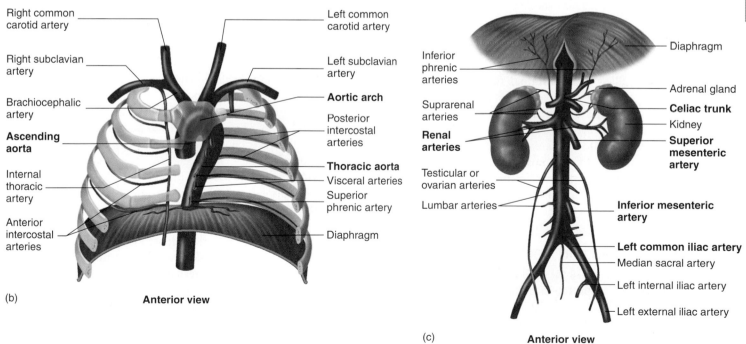

Right common carotid artery
Right subclavian artery
Brachiocephalic artery
**Ascending aorta**
Internal thoracic artery
Anterior intercostal arteries

Left common carotid artery
Left subclavian artery
**Aortic arch**
Posterior intercostal arteries
**Thoracic aorta**
Visceral arteries
Superior phrenic artery
Diaphragm

(b)          **Anterior view**

Inferior phrenic arteries
Suprarenal arteries
**Renal arteries**
Testicular or ovarian arteries
Lumbar arteries

Diaphragm
Adrenal gland
**Celiac trunk**
Kidney
**Superior mesenteric artery**
**Inferior mesenteric artery**
**Left common iliac artery**
Median sacral artery
Left internal iliac artery
Left external iliac artery

(c)          **Anterior view**

**Figure 13.7  Branches of the Aorta**

(a) The aorta is considered in three portions: the ascending aorta, the aortic arch, and the descending aorta. The descending aorta consists of the thoracic and abdominal aortas. (b) The thoracic aorta. (c) The abdominal aorta.

Circulatory

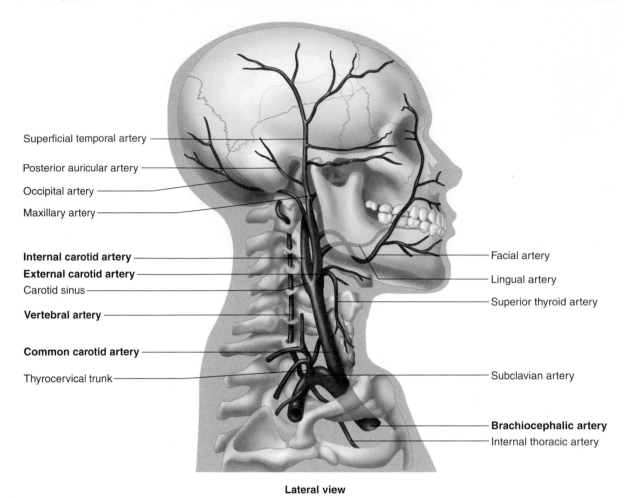

**Lateral view**

**Figure 13.8** Arteries of the Head and Neck

The brachiocephalic artery, the right common carotid artery, and the right vertebral artery supply the head and neck. The right common carotid artery branches from the brachiocephalic artery, and the vertebral artery branches from the subclavian artery.

(il′ē-ak) **arteries** is called the **abdominal** (ab-dom′i-năl) **aorta** (figure 13.7a,c).

An **arterial aneurysm** (an′ū-rizm; a dilation) is a localized dilation of an artery that usually develops in response to trauma or a congenital (existing at birth) weakness of the artery wall. Rupture of a large aneurysm in the aorta is almost always fatal, and rupture of an aneurysm in an artery of the brain causes massive damage to brain tissue and even death. If aneurysms are discovered, they can sometimes be surgically corrected. For example, large aortic aneurysms that leak blood slowly can often be repaired.

## Arteries of the Head and Neck

Figure 13.8 depicts the arteries of the head and neck, and figure 13.9 shows how these arteries connect to each other and to the arteries of the thorax. It may help you to refer to figure 13.9 as you read the following discussion. The first vessel to branch from the aortic arch is the **brachiocephalic** (brā′kē-ō-se-fal′ik; vessel to the arm and head) **artery.** This short artery branches at the level of the clavicle to form the **right common carotid** (ka-rot′id) **artery,** which transports blood to the right side of the head and neck, and the **right subclavian** (sŭb-klā′vē-an; beneath the clavicle) **artery,** which transports blood to the right upper limb (see figures 13.7b and 13.8).

There is no brachiocephalic artery on the left side of the body. Instead, both the left common carotid and the left subclavian arteries branch directly off the aortic arch (see figures 13.6, 13.7a,b, and 13.9). They are the second and third branches of the aortic arch. The **left common carotid artery** transports blood to the left side of the head and neck, and the **left subclavian artery** transports blood to the left upper limb.

The common carotid arteries extend superiorly along each side of the neck to the angle of the mandible, where they branch into **internal** and **external carotid arteries.** The base of each internal carotid artery is slightly dilated to form a **carotid sinus,** which contains structures important in monitoring blood pressure (baroreceptors). The external carotid arteries have several branches that supply the structures of the neck, face, nose, and mouth. The internal carotid arteries pass through the carotid canals and contribute to the **cerebral arterial circle** (circle of Willis) at the base of the brain (figure 13.10). The vessels that supply blood to most of the brain branch from the cerebral arterial circle.

Some of the blood to the brain is supplied by the **vertebral** (ver′tĕ-brăl) **arteries,** which branch from the subclavian arteries and pass to the head through the transverse foramina of the cervical

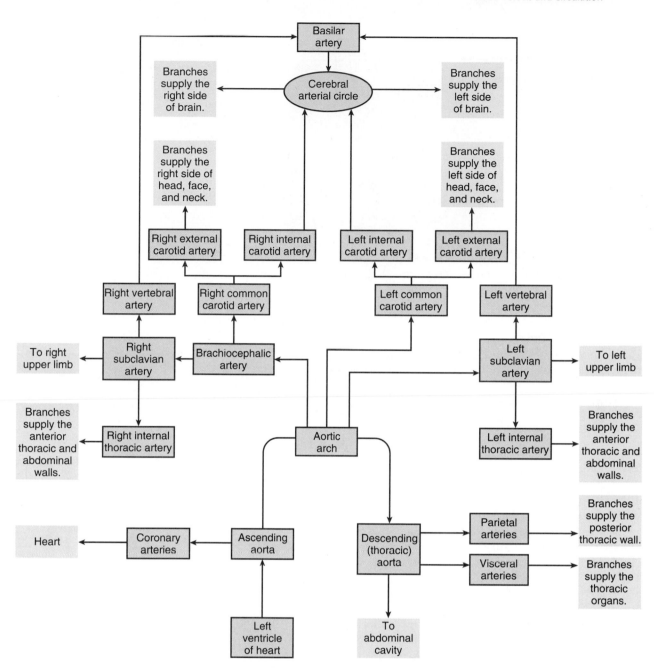

**Figure 13.9** Major Arteries of the Head and Thorax

vertebrae. The vertebral arteries then pass into the cranial cavity through the foramen magnum. Branches of the vertebral arteries supply blood to the spinal cord, as well as to the vertebrae, muscles, and ligaments in the neck.

Within the cranial cavity, the vertebral arteries unite to form a single **basilar** (bas′i-lăr; relating to the base of the brain) **artery** located along the anterior, inferior surface of the brainstem (figure 13.10). The basilar artery gives off branches that supply blood to the pons, cerebellum, and midbrain. It also forms right and left branches that contribute to the cerebral arterial circle. Most of the blood supply to the brain is through the internal carotid arteries; however, not enough blood is supplied to the brain to maintain life if either the vertebral arteries or the carotid arteries are blocked.

**Predict 1**

*The term carotid means to put to sleep, reflecting the fact that, if the carotid arteries are blocked for several seconds, the patient can lose consciousness. Interruption of the blood supply for even a few minutes can result in permanent brain damage. What is the physiological significance of atherosclerosis (lipid deposits that block the vessels) in the carotid arteries?*

## Arteries of the Upper Limbs

The arteries of the upper limbs are named differently as they pass into different body regions, even though no major branching occurs. The subclavian artery, located deep to the clavicle, becomes the **axillary** (ak′sil-ār-ē) **artery** in the axilla (armpit). The **brachial** (brā′kē-ăl) **artery,** located in the arm, is a continuation of the

Circulatory

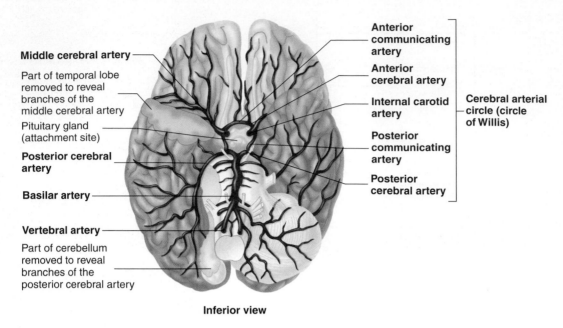

**Figure 13.10  Cerebral Arterial Circle**

The internal carotid and vertebral arteries carry blood to the brain. The vertebral arteries join to form the basilar artery. Branches of the internal carotid arteries and the basilar artery supply blood to the brain and complete a circle of arteries around the pituitary gland and the base of the brain called the cerebral arterial circle (circle of Willis).

axillary artery (figure 13.11). Blood pressure measurements are normally taken from the brachial artery. The brachial artery branches at the elbow to form the **ulnar** (ul′năr) **artery** and the **radial** (rā′dē-ăl) **artery,** which supply blood to the forearm and hand. The radial artery is the one most commonly used for taking a pulse. The pulse can be detected easily on the thumb (radial) side of the anterior surface of the wrist.

## Thoracic Aorta and Its Branches

The branches of the thoracic aorta can be divided into two groups: The **visceral** (vis′er-ăl) **arteries** supply the thoracic organs, and the **parietal** (pă-rī′ĕ-tăl) **arteries** supply the thoracic wall. The visceral branches of the thoracic aorta supply the esophagus, the trachea, the parietal pericardium, and part of the lung. The major parietal arteries are the **posterior intercostal** (in-ter-kos′tăl) **arteries,** which arise from the thoracic aorta and extend between the ribs (see figure 13.7). They supply the intercostal muscles, the vertebrae, the spinal cord, and the deep muscles of the back. The **superior phrenic** (fren′ik; diaphragm) **arteries** supply the diaphragm.

The **internal thoracic arteries** are branches of the subclavian arteries. They descend along the internal surface of the anterior thoracic wall and give rise to branches called the **anterior intercostal arteries,** which extend between the ribs to supply the anterior chest wall (see figure 13.7a,b).

## Abdominal Aorta and Its Branches

The branches of the abdominal aorta, like those of the thoracic aorta, can be divided into visceral and parietal groups. The visceral arteries are divided into paired and unpaired branches. There are three major unpaired branches: the **celiac** (sē′lē-ak; belly) **trunk,** the

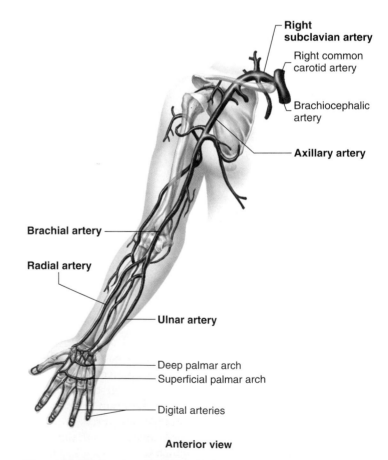

**Figure 13.11  Arteries of the Upper Limb**

Anterior view of the arteries of the right upper limb and their branches.

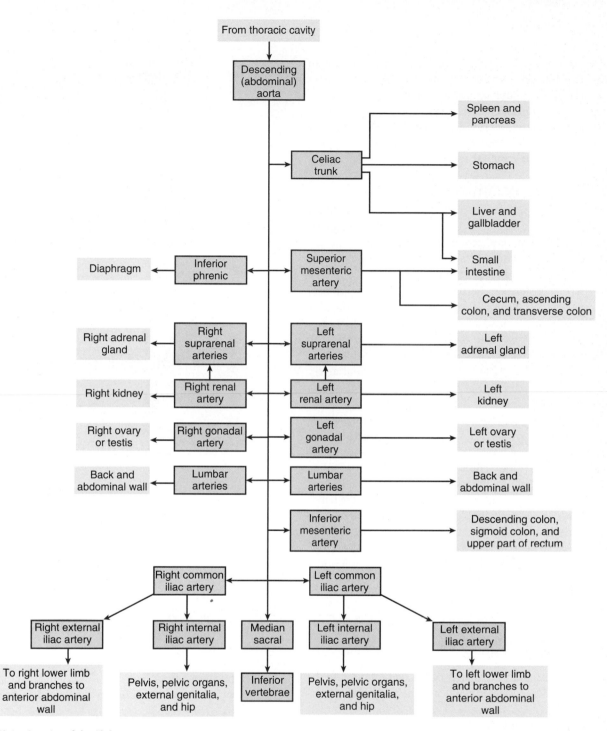

**Figure 13.12** Major Arteries of the Abdomen and Pelvis

Visceral branches include those that are unpaired (celiac trunk, superior mesenteric, inferior mesenteric) and those that are paired (renal, suprarenal, testicular or ovarian). Parietal branches include inferior phrenic, lumbar, and median sacral.

superior mesenteric (mez-en-ter'ik) **artery,** and the **inferior mesenteric artery.** The celiac trunk supplies blood to the stomach, pancreas, spleen, upper duodenum, and liver. The superior mesenteric artery supplies blood to the small intestine and the upper portion of the large intestine, and the inferior mesenteric artery supplies blood to the remainder of the large intestine (figure 13.12).

There are three paired visceral branches of the abdominal aorta. The **renal** (rē'nal; kidney) **arteries** supply the kidneys, and

the **suprarenal** (sū'prǎ-rē'nǎl; superior to the kidney) **arteries** supply the adrenal glands. The **testicular arteries** supply the testes in males; the **ovarian arteries** supply the ovaries in females.

The parietal branches of the abdominal aorta supply the diaphragm and abdominal wall. The **inferior phrenic arteries** supply the diaphragm; the **lumbar arteries** supply the lumbar vertebrae and back muscles; and the **median sacral artery** supplies the inferior vertebrae.

## Arteries of the Pelvis

The abdominal aorta divides at the level of the fifth lumbar vertebra into two **common iliac arteries.** Each common iliac artery divides to form an **external iliac artery,** which enters a lower limb, and an **internal iliac artery,** which supplies the pelvic area (see figure 13.7c). Visceral branches of the internal iliac artery supply organs such as the urinary bladder, rectum, uterus, and vagina. Parietal branches supply blood to the walls and floor of the pelvis; the lumbar, gluteal, and proximal thigh muscles; and the external genitalia.

## Arteries of the Lower Limbs

Like the arteries of the upper limbs, the arteries of the lower limbs are named differently as they pass into different body regions, even though there are no major branches. The external iliac artery in the pelvis becomes the **femoral** (fem′ŏ-răl) **artery** in the thigh, and it becomes the **popliteal** (pop-lit′ē-ăl) **artery** in the popliteal space, which is the posterior region of the knee. The popliteal artery branches slightly inferior to the knee to give off the **anterior tibial artery** and the **posterior tibial artery,** both of which give rise to arteries that supply blood to the leg and foot (figure 13.13). The anterior tibial artery becomes the **dorsalis pedis** (dōr-săl′lis pē′dis; *pes,* foot) **artery** at the ankle. The posterior tibial artery gives rise to the **fibular artery,** or *peroneal artery,* which supplies the lateral leg and foot.

   The femoral triangle is located in the superior and medial area of the thigh. Its margins are formed by the inguinal ligament, the medial margin of the sartorius muscle, and the lateral margin of the adductor longus muscle (see figures 7.13a and 7.25a). Passing through the femoral triangle are the femoral artery, vein, and nerve. A pulse in the femoral artery can be detected in the area of the femoral triangle. This area is also susceptible to serious traumatic injuries that result in hemorrhage and nerve damage. In addition, pressure applied to this area can help prevent bleeding from wounds in more inferior areas of the lower limb. The femoral triangle is an important access point for certain medical procedures as well.

# BLOOD VESSELS OF THE SYSTEMIC CIRCULATION: VEINS

The oxygen-poor blood from the tissues of the body returns to the heart through veins. The **superior vena cava** (vē′nă kā′vă) returns blood from the head, neck, thorax, and upper limbs to the right atrium of the heart, and the **inferior vena cava** returns blood from the abdomen, pelvis, and lower limbs to the right atrium (figure 13.14).

## Veins of the Head and Neck

The two pairs of major veins that drain blood from the head and neck are the **external** and **internal jugular** (jŭg′ū-lar) **veins** (figure 13.15). The external jugular veins are the more superficial of the two sets. They drain blood from the posterior head and neck, emptying primarily into the subclavian veins. The internal jugular veins are much larger and deeper. They drain blood from the brain and the anterior head, face, and neck. The internal jugular veins join the **subclavian veins** on each side of the body to form the **brachiocephalic veins.** The brachiocephalic veins join to form the superior vena cava.

Abdominal aorta
Common iliac artery
Internal iliac artery
**External iliac artery**

**Femoral artery**

**Popliteal artery**

**Anterior tibial artery**

**Posterior tibial artery**

**Fibular artery**

**Dorsalis pedis artery**

**Anterior view**

**Figure 13.13   Arteries of the Lower Limb**
Anterior view of the arteries of the right lower limb and their branches.

## Veins of the Upper Limbs

The veins of the upper limbs (figure 13.16) can be divided into deep and superficial groups. The deep veins, which drain the deep structures of the upper limbs, follow the same course as the arteries and are named for their respective arteries. The only noteworthy deep veins are the **brachial veins,** which accompany the brachial artery and empty into the axillary vein. The superficial veins drain the superficial structures of the upper limbs and then empty into the deep veins. The **cephalic** (sĕ′fal′ik; toward the head) **vein,** which empties into the **axillary vein,** and the **basilic** (ba-sil′ik) **vein,** which becomes the axillary vein, are the major superficial veins. Many of their tributaries in the forearm and hand can be seen through the skin. The **median cubital** (kū′bi-tal) **vein** usually connects the cephalic vein or its tributaries with the basilic vein. Although this vein varies in size among people, it is usually quite prominent on the anterior surface of the upper limb at the level of the elbow, an area called the **cubital fossa,** and is often used as a site for drawing blood.

## Veins of the Thorax

Three major veins return blood from the thorax to the superior vena cava: the **right** and **left brachiocephalic veins** and the **azygos** (az-ī′gos, az′i-gos) **vein** (figure 13.17). Blood drains from the

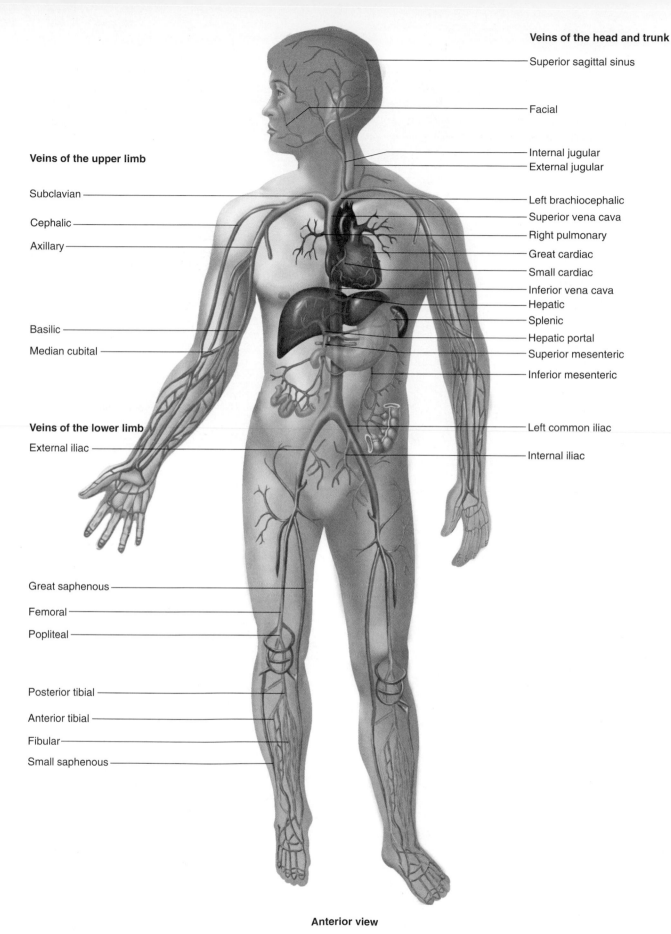

Veins of the head and trunk

Superior sagittal sinus

Facial

Internal jugular
External jugular

Left brachiocephalic
Superior vena cava
Right pulmonary
Great cardiac
Small cardiac
Inferior vena cava
Hepatic
Splenic
Hepatic portal
Superior mesenteric
Inferior mesenteric

Left common iliac

Internal iliac

Veins of the upper limb

Subclavian

Cephalic

Axillary

Basilic

Median cubital

Veins of the lower limb

External iliac

Great saphenous

Femoral

Popliteal

Posterior tibial

Anterior tibial

Fibular

Small saphenous

Anterior view

Circulatory

**Figure 13.14** Major Veins

These major veins return blood from the body tissues to the right atrium.

Superficial temporal vein

Retromandibular vein

**External jugular vein**

Facial vein
Lingual vein
Superior thyroid vein

**Internal jugular vein**

**Subclavian vein**

**Right brachiocephalic vein**

Cephalic vein

Axillary vein

Azygos vein

Basilic vein

Brachial veins

**Left brachiocephalic vein**

**Superior vena cava**

**Inferior vena cava**

**Anterior view**

**Figure 13.15  Veins of the Head and Neck**

The external and internal jugular veins drain blood from the head and neck. The internal jugular veins join the subclavian veins on each side of the body to form the brachiocephalic veins. The external jugular veins drain into the subclavian veins.

**Subclavian vein**

Internal jugular vein

Brachiocephalic vein

**Cephalic vein**

Clavicle

**Axillary vein**

**Brachial veins**

Cephalic vein

**Basilic vein**

**Median cubital vein**

**Anterior view**

**Figure 13.16  Veins of the Upper Limb**

Anterior view of the major veins of the right upper limb and their branches.

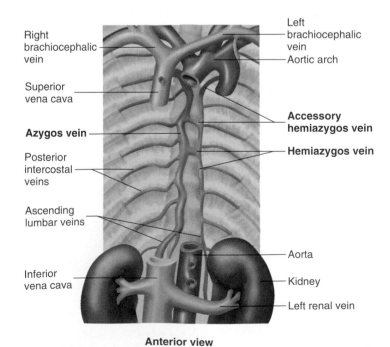

Right brachiocephalic vein

Superior vena cava

**Azygos vein**

Posterior intercostal veins

Ascending lumbar veins

Inferior vena cava

Left brachiocephalic vein

Aortic arch

**Accessory hemiazygos vein**

**Hemiazygos vein**

Aorta

Kidney

Left renal vein

**Anterior view**

**Figure 13.17  Veins of the Thorax**

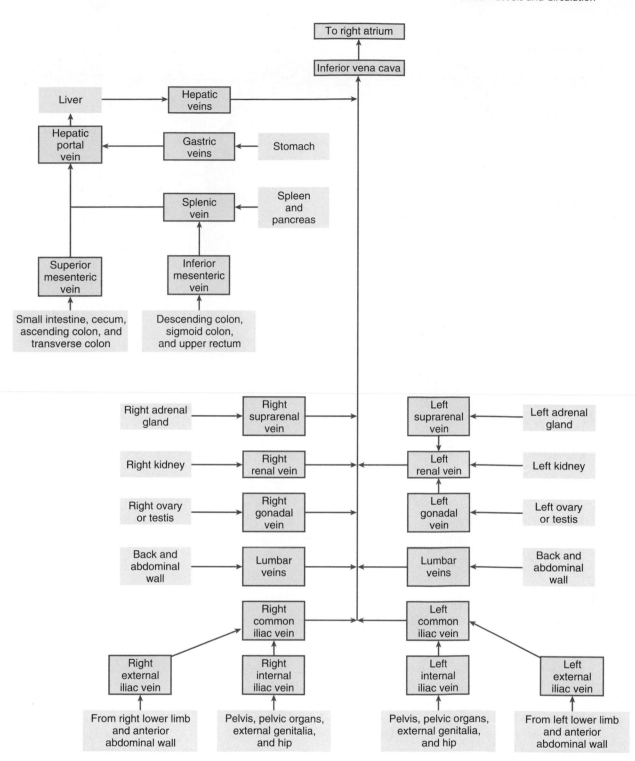

**Figure 13.18  Major Veins of the Abdomen and Pelvis**

anterior thoracic wall by way of the **anterior intercostal veins.** These veins empty into the **internal thoracic veins,** which empty into the brachiocephalic veins. Blood from the posterior thoracic wall is collected by **posterior intercostal veins** that drain into the azygos vein on the right and the **hemiazygos vein** or the **accessory hemiazygos vein** on the left. The hemiazygos and accessory hemiazygos veins empty into the azygos vein, which drains into the superior vena cava (see figure 13.15).

## Veins of the Abdomen and Pelvis

Figure 13.18 provides an overview of the major veins of the abdomen and pelvis. It may help to refer to this figure as you read the following discussion. Blood from the posterior abdominal wall drains through **ascending lumbar veins** into the azygos vein. Blood from the rest of the abdomen and from the pelvis and lower limbs returns to the heart through the inferior vena cava. The gonads (testes or ovaries), kidneys, adrenal glands, and liver are

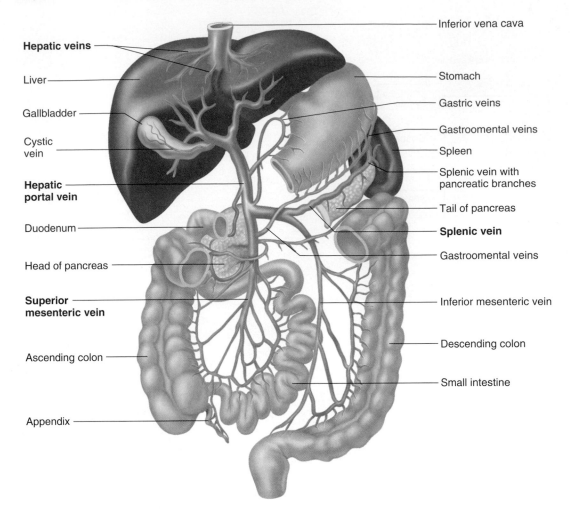

**Figure 13.19  Veins of the Hepatic Portal System**

The hepatic portal system begins as capillary beds in the stomach, pancreas, spleen, small intestine, and large intestine. The veins of the hepatic portal system converge on the hepatic portal vein, which carries blood to a series of capillaries in the liver. Hepatic veins carry blood from capillaries in the liver to the hepatic veins, which then enter the inferior vena cava.

the only abdominal organs outside the pelvis from which blood drains directly into the inferior vena cava. The **internal iliac veins** drain the pelvis and join the **external iliac veins** from the lower limbs to form the **common iliac veins.** The common iliac veins combine to form the inferior vena cava (see figure 13.14).

Blood from the capillaries within most of the abdominal viscera, such as the stomach, intestines, pancreas, and spleen, drains through a specialized system of blood vessels to the liver. The liver is a major processing center for substances absorbed by the intestinal tract. A **portal** (pōr′tăl) **system** is a vascular system that begins and ends with capillary beds and has no pumping mechanism, such as the heart, in between. The **hepatic** (he-pa′tik) **portal system** (figure 13.19) begins with capillaries in the viscera and ends with capillaries in the liver. The major tributaries of the hepatic portal system are the **splenic** (splen′ik) **vein** and the **superior mesenteric vein.** The **inferior mesenteric vein** empties into the splenic vein. The splenic vein carries blood from the spleen and pancreas. The superior and inferior mesenteric veins carry blood from the intestines. The splenic vein and the superior mesenteric vein join to form the **hepatic portal vein,** which enters the liver.

Blood from the liver flows into **hepatic veins,** which join the inferior vena cava. Blood entering the liver through the hepatic portal vein is rich in nutrients collected from the intestines, but it may also contain a number of toxic substances that are potentially harmful to body tissues. Within the liver, nutrients are taken up and stored or modified, so that they can be used by other cells of the body. Also within the liver, toxic substances are converted to nontoxic substances. These substances can be removed from the blood or carried by the blood to the kidneys for excretion. The liver is discussed more fully in chapter 16.

Other veins of the abdomen and pelvis include the renal veins, the suprarenal veins, and the gonadal veins. The **renal veins** drain the kidneys, and the **suprarenal veins** drain the adrenal glands. The **testicular veins** drain the testes in males; the **ovarian veins** drain the ovaries in females.

## Veins of the Lower Limbs

The veins of the lower limbs (figure 13.20), like those of the upper limbs, consist of deep and superficial groups. The deep veins follow the same path as the arteries and are named for the arteries they accompany. The superficial veins consist of the great and small

## Blood Vessels Used for Coronary Bypass Surgery

The great saphenous vein is often surgically removed and used in coronary bypass surgery. Portions of the saphenous vein are grafted to create a route for blood flow that bypasses blocked portions of the coronary arteries. The circulation interrupted by the removal of the saphenous vein flows through other veins of the lower limb instead. The internal thoracic artery is also used for coronary bypasses. The distal end of the artery is freed and attached to a coronary artery at a point that bypasses the blocked portion. This technique appears to be better because the internal thoracic artery does not become blocked as fast as the saphenous vein. However, due to the length of the saphenous vein and the fact that more than one segment can be removed for use in bypass surgery, surgeons use the saphenous vein.

Anterior view

**Figure 13.20  Veins of the Lower Limb**
Anterior view of the major veins of the right lower limb and their branches.

saphenous (să-fē′nŭs, sa′fĕ-nŭs) **veins.** The **great saphenous vein** originates over the dorsal and medial side of the foot and ascends along the medial side of the leg and thigh to empty into the femoral vein. The **small saphenous vein** begins over the lateral side of the foot and joins the **popliteal vein,** which becomes the femoral vein. The femoral vein empties into the external iliac vein.

## A CASE IN POINT
### Venous Thrombosis

Harry Leggs is a 55-year-old college professor who teaches a night class in a small town about 50 miles from his home. One evening, as he walked to his car after teaching the class, Harry became aware of a slight pain in his right leg. By the time he reached home about 90 minutes later, the calf of his right leg was swollen. When he extended his knee and plantar flexed his foot, the pain increased. Harry knew this was a serious condition, so he drove to the emergency room. There, technicians performed a Doppler test to monitor the flow of blood through the blood vessels in Harry's right leg. The test confirmed that a thrombus had formed in one of the deep veins of the leg.

Harry's symptoms stemmed from several events. Small proteins released from fibrinogen during clot formation caused the activation of kinins that stimulate pain receptors (see chapter 14). Also, the blocked vein caused edema and led to tissue ischemia, a decrease in blood flow to tissues, which may resulted in tissue damage and pain. Edema occurred because of the increase in venous pressure inferior to the blocked vein (see figure 13.24).

A major danger associated with a deep venous thrombosis is that portions of the thrombus may form emboli. The emboli are likely to be transported in the blood to the lungs, where they form pulmonary emboli. Pulmonary emboli are life-threatening because they may be large or numerous enough to block blood flow to substantial areas of the lungs.

Harry was admitted to the hospital for a few days. His movements were restricted, and he was given an anticoagulant. Even after his release, he continued taking the anticoagulant for a few weeks until the clot had dissolved.

### Predict 2

*If a thrombus in the posterior tibial vein gave rise to an embolus, name in order the parts of the circulatory system the embolus would pass through before lodging in a blood vessel in the lungs. Why are the lungs the most likely place for the embolus to lodge?*

## PHYSIOLOGY OF CIRCULATION

The function of the circulatory system is to maintain adequate blood flow to all body tissues. Adequate blood flow is required to provide nutrients and oxygen to the tissues and to remove the

1 When the cuff pressure is high enough to keep the brachial artery closed, no blood flows through it, and no sound is heard.

2 When cuff pressure decreases and is no longer able to keep the brachial artery closed, blood is pushed through the partially opened brachial artery, producing turbulent blood flow and a sound. **Systolic pressure** is the pressure at which a sound is first heard.

3 As cuff pressure continues to decrease, the brachial artery opens even more during systole. At first, the artery is closed during diastole, but as cuff pressure continues to decrease, the brachial artery partially opens during diastole. Turbulent blood flow during systole produces Korotkoff sounds, although the pitch of the sounds changes as the artery becomes more open.

4 Eventually, cuff pressure decreases below the pressure in the brachial artery, and it remains open during systole and diastole. Nonturbulent flow is reestablished, and no sounds are heard. **Diastolic pressure** is the pressure at which the sound disappears.

**PROCESS Figure 13.21** Measuring Blood Pressure

waste products of metabolism from the tissues. Blood flows through the arterial system primarily as a result of the pressure produced by contractions of the heart ventricles.

## Blood Pressure

**Blood pressure** is a measure of the force blood exerts against the blood vessel walls. In arteries, blood pressure values go through a cycle that depends on the rhythmic contractions of the heart. When the ventricles contract, blood is forced into the arteries, and the pressure reaches a maximum value called the **systolic pressure.**

When the ventricles relax, blood pressure in the arteries falls to a minimum value called the **diastolic pressure.** The standard unit for measuring blood pressure is millimeters of mercury (mm Hg). For example, if the blood pressure is 100 mm Hg, the pressure is great enough to lift a column of mercury 100 mm.

Health professionals most often use the **auscultatory** (aws-kŭl′tă-tō-rē; to listen) **method** to determine blood pressure (figure 13.21). A blood pressure cuff connected to a **sphygmomanometer** (sfig′mō-mă-nom′ĕ-ter) is wrapped around the patient's arm, and a **stethoscope** (steth′ō-skōp) is placed over the brachial artery. The blood pressure cuff is then inflated until the brachial

artery is completely blocked. Because no blood flows through the constricted area at this point, no sounds can be heard through the stethoscope. The pressure in the cuff is then gradually lowered. As soon as the pressure in the cuff declines below the systolic pressure, blood flows through the constricted area each time the left ventricle contracts. The blood flow is turbulent immediately downstream from the constricted area. This turbulence produces vibrations in the blood and surrounding tissues that can be heard through the stethoscope. These sounds are called **Korotkoff** (Kō-rot′kof) **sounds,** and the pressure at which the first Korotkoff sound is heard is the systolic pressure.

As the pressure in the blood pressure cuff is lowered still more, the Korotkoff sounds change tone and loudness. When the pressure has dropped until the brachial artery is no longer constricted and blood flow is no longer turbulent, the sound disappears completely. The pressure at which the Korotkoff sounds disappear is the diastolic pressure. The brachial artery remains open during systole and diastole, and continuous blood flow is reestablished.

The systolic pressure is the maximum pressure produced in the large arteries. It is also a good measure of the maximum pressure

# CLINICAL IMPACT

## Hypertension

**Hypertension,** or *high blood pressure,* affects at least 20% of all people at sometime in their lives. The following guidelines* categorize blood pressure for adults:

- *Normal:* less than 120 mm Hg systolic and 80 mm Hg diastolic
- *Prehypertension:* from 120 mm Hg systolic and 80 mm Hg diastolic to 139 mm Hg systolic and 89 mm Hg diastolic
- *Stage 1 hypertension:* from 140 mm Hg systolic and 90 mm Hg diastolic to 159 mm Hg systolic and 99 mm Hg diastolic
- *Stage 2 hypertension:* greater than 160 mm Hg systolic and 100 mm Hg diastolic

Individuals with prehypertension pressure should monitor their blood pressure regularly and consider lifestyle changes that can reduce blood pressure.

Hypertension requires the heart to perform a greater than normal amount of work because of the increased afterload (see chapter 12). The extra work leads to hypertrophy of cardiac muscle, especially in the left ventricle, and can result in heart failure. Hypertension also increases the rate of arteriosclerosis development. Arteriosclerosis, in turn, increases the chance that blood clots will form and that blood vessels

will rupture. Common conditions associated with hypertension are cerebral hemorrhage, heart attack, hemorrhage of renal blood vessels, and poor vision resulting from burst blood vessels in the retina.

The most common treatments for hypertension are those that dilate blood vessels, increase the rate of urine production, or decrease cardiac output. A low-salt diet is also recommended to reduce the amount of sodium chloride and water absorbed from the intestine into the bloodstream.

---

*Joint National Committee on Prevention, Detection, Evaluation, and Treatment of High Blood Pressure.

within the left ventricle. The diastolic pressure is close to the lowest pressure within the large arteries. During relaxation of the left ventricle, the aortic semilunar valve closes, trapping the blood that was ejected during ventricular contraction in the aorta. The pressure in the ventricles falls to 0 mm Hg during ventricular relaxation. However, the blood trapped in the elastic arteries is compressed by the recoil of the elastic arteries, and the pressure falls more slowly, reaching the diastolic pressure (see figure 12.18).

## Pressure and Resistance

The values for systolic and diastolic pressure vary among healthy people, making the range of normal values quite broad. In addition, other factors, such as physical activity and emotions, affect blood pressure values in a normal person. A standard blood pressure for a resting young adult male is 120 mm Hg for the systolic pressure and 80 mm Hg for the diastolic pressure, commonly expressed as 120/80.

As blood flows from arteries through the capillaries and veins, blood pressure falls progressively to about 0 mm Hg or even slightly lower by the time blood is returned to the right atrium. In addition, the fluctuations in blood pressure are damped, meaning that the difference between the systolic and diastolic pressures is decreased in the small-diameter vessels. By the time blood reaches the capillaries, there is no variation in blood pressure, and only a steady pressure of about 30 mm Hg remains (figure 13.22).

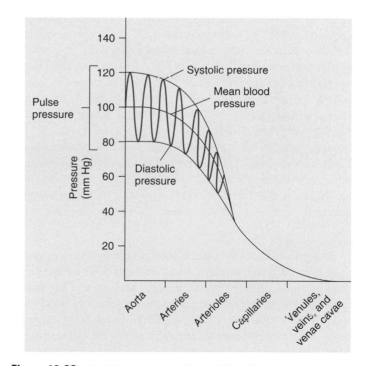

**Figure 13.22  Blood Pressure in the Various Blood Vessel Types**

In small arteries and arterioles, blood pressure fluctuations between systole and diastole are damped, so the fluctuations become smaller. No fluctuations in blood pressure occur in capillaries and veins.

## A CASE IN POINT
### Atherosclerosis

Mori Payne and Leslie Payne were preparing for their fortieth wedding anniversary celebration. While they were at the shopping mall, Mori's legs began to ache, especially his left leg. He had noticed before that, when he walked briskly or very far, his legs hurt. Mori sat down to rest while Leslie shopped at a nearby store. After about 5 minutes, Mori experienced excruciating pain in his left leg that was not relieved by massage or rest. Leslie asked a passerby to help her get Mori to their car, and she drove him to the hospital.

In the emergency room, the examining physician observed that Mori's left leg was mottled and cyanotic (bluish) distal to the knee and was cool to the touch. His right leg was pink and warm. A Doppler test performed on the left leg revealed decreased pulses with faint, abnormal sounds called bruits (broo-ez′) in the distal left popliteal artery. Normal pulses and no obvious bruits were observed in other arteries of Mori's thigh and leg, including the right popliteal artery. His temperature was slightly elevated, and his blood pressure was 165/95 mm Hg. His pulse was 96 beats per minute and regular. His respiratory rate was 20 respirations per minute.

The physician explained to Leslie that reduced pulses indicate reduced blood flow and that bruits indicate turbulent blood flow. Both of these observations and the sudden onset of the pain are consistent with the formation of an arterial thrombus partially blocking Mori's distal left popliteal artery. He was quickly treated with an enzyme that breaks down the fibrin of blood clots, and this treatment successfully dissolved the thrombus. As blood flow to the leg increased, the pain and cyanosis decreased.

An angiogram on Mori's left leg revealed severe stenosis (narrowing) in his left distal popliteal artery due to an atherosclerotic plaque. When an artery is stenosed, blood delivery may be adequate to maintain tissue homeostasis as long as the person is at rest, but not during exercise. In Mori's case, other, smaller atherosclerotic lesions were likely in both legs, because both of his legs hurt in response to brisk walking. The lack of sufficient oxygen resulted in anaerobic respiration and acidic by-products, such as lactic acid.

The sudden, dramatic increase in pain that Mori experienced resulted from the formation of a thrombus on a large atherosclerotic plaque that nearly blocked the blood flow through the left popliteal artery. After the thrombus was dissolved, angioplasty was performed to further increase blood flow through the left popliteal artery.

It was later determined that Mori has high blood cholesterol and blood glucose levels, conditions that are associated with the increased development of atherosclerotic plaque in arteries. The high blood glucose may be a sign of type 2 diabetes mellitus (see chapter 10). From now on, it is important for Mori to reduce his blood pressure, blood cholesterol, and blood glucose levels.

The greater the resistance in a blood vessel, the more rapidly the pressure decreases as blood flows through it. The most rapid decline in blood pressure occurs in the arterioles and capillaries because their small diameters increase the resistance to blood flow. Blood pressure declines slowly as blood flows from large to medium-sized arteries because their diameters are larger and the resistance to blood flow is not great.

Resistance to blood flow in veins is low because of their larger diameters. Also, the valves that prevent the backflow of blood in the veins, as well as skeletal muscle movements that periodically compress veins, force blood to flow toward the heart. Consequently, blood flows through veins, even though the pressure in them is low.

The muscular arteries, arterioles, and precapillary sphincters are capable of constricting (vasoconstriction) and dilating (vasodilation). If vessels constrict, resistance to blood flow increases, and the volume of blood flowing through the vessels declines. Because muscular arteries are able to constrict and dilate, they help control the amount of blood flowing to each region of the body. In contrast, arterioles and precapillary sphincters regulate blood flow through local tissues.

## Pulse Pressure

The difference between the systolic and diastolic pressures is called the **pulse pressure.** For example, if a person has a systolic pressure of 120 mm Hg and a diastolic pressure of 80 mm Hg, the pulse pressure is 40 mm Hg. When the stroke volume increases, the systolic pressure increases more than the diastolic pressure, causing the pulse pressure to increase. During periods of exercise, the stroke volume and pulse pressure increase substantially.

In people who have **arteriosclerosis** (ar-tēr′ē-ō-skler-ō′sis; hardening of the arteries), the arteries are less elastic than normal. Arterial pressure increases rapidly and falls rapidly. The systolic pressure increases substantially, and the diastolic pressure may be somewhat lower than normal or slightly increased. The same amount of blood ejected into a less elastic artery results in a higher systolic pressure than that in a more elastic artery. Therefore, the pulse pressure is greater than normal, even though the same amount of blood is ejected into the aorta as in a normal person. Arteriosclerosis increases the amount of work the heart performs because the left ventricle must produce a greater pressure to eject the same amount of blood into a less elastic artery. In severe cases, the heart's increased workload leads to heart failure.

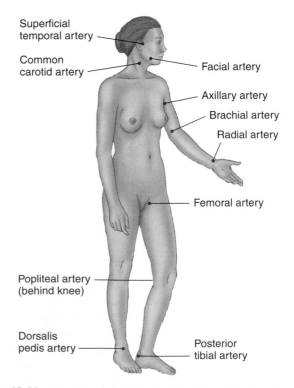

**Figure 13.23  Major Points at Which the Pulse Can Be Monitored**
Each pulse point is named after the artery on which it occurs.

## Edema

**Edema,** or swelling, results from a disruption in the normal inwardly and outwardly directed pressures across the capillary walls. For example, inflammation increases the permeability of capillaries. Proteins, mainly albumin, leak out of the capillaries into the interstitial spaces. The proteins increase the osmotic pressure in the interstitial fluid. Consequently, fluid passes from the arterial end of capillaries into the interstitial spaces at a greater rate, and fluid passes from the interstitial spaces into the venous ends of capillaries at a slower rate. The lymphatic capillaries cannot carry all the fluid away. Thus, fluid accumulates in the interstitial spaces, resulting in edema.

Ejection of blood from the left ventricle into the aorta produces a pressure wave, or **pulse,** which travels rapidly along the arteries. A pulse can be felt at locations where large arteries are close to the surface of the body (figure 13.23). Health professionals need to know the major locations of pulse detection because monitoring the pulse can yield important information about the heart rate, the heart rhythm, and other characteristics. For example, a weak pulse usually indicates a decreased stroke volume or increased constriction of the arteries.

### Predict 3

*A weak pulse occurs in response to premature heartbeats and during cardiovascular shock due to hemorrhage. Stronger than normal pulses occur in a healthy person during exercise. Explain the causes for the changes in the pulse under these conditions.*

## Capillary Exchange

There are about 10 billion capillaries in the body. Nutrients diffuse across the capillary walls into the interstitial spaces, and waste products diffuse in the opposite direction. In addition, a small amount of fluid is forced out of the capillaries into the interstitial space at their arterial ends. Most of that fluid, but not all, reenters the capillaries at their venous ends.

The major forces responsible for moving fluid through the capillary wall are blood pressure and osmosis (figure 13.24). Blood pressure forces fluid out of the capillary, and osmosis moves fluid into it. Fluid moves by osmosis from the interstitial space into the capillary because blood has a greater osmotic pressure than does interstitial fluid. The greater the concentration of molecules dissolved in a fluid, the greater the osmotic pressure of the fluid

Circulatory

1. At the arterial end of the capillary, the movement of fluid out of the capillary due to blood pressure is greater than the movement of fluid into the capillary due to osmosis.

2. At the venous end of the capillary, the movement of fluid into the capillary due to osmosis is greater than the movement of fluid out of the capillary due to blood pressure.

3. Approximately nine-tenths of the fluid that leaves the capillary at its arterial end reenters the capillary at its venous end. About one-tenth of the fluid passes into the lymphatic capillaries.

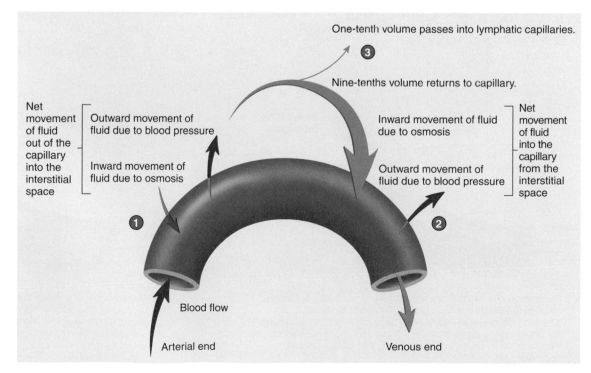

**PROCESS Figure 13.24 Capillary Exchange**

(see chapter 3). The greater osmotic pressure of blood is caused by the large concentration of blood proteins (see chapter 11) that are unable to cross the capillary wall. The concentration of proteins in the interstitial space is much lower than that in the blood. The capillary wall acts as a selectively permeable membrane, which prevents proteins from moving from the capillary into the interstitial space but allows fluid to move across the capillary wall.

At the arterial end of the capillary, the movement of fluid out of the capillary due to blood pressure is greater than the movement of fluid into the capillary due to osmosis. Consequently, there is a net movement of fluid out of the capillary into the interstitial space (figure 13.24).

At the venous end of the capillary, blood pressure is lower than at the arterial end because of the resistance to blood flow through the capillary. Consequently, the movement of fluid out of the capillary due to blood pressure is less than the movement of fluid into the capillary due to osmosis, and there is a net movement of fluid from the interstitial space into the capillary (figure 13.24).

Approximately nine-tenths of the fluid that leaves the capillary at the arterial end reenters the capillary at its venous end. The remaining one-tenth of the fluid enters the lymphatic capillaries and is eventually returned to the general circulation (see chapter 14).

**Predict 4**

*Explain edema (a) in response to a decrease in plasma protein concentration and (b) as a result of increased blood pressure within a capillary.*

# CONTROL OF BLOOD FLOW IN TISSUES

Blood flow provided to the tissues by the circulatory system is highly controlled and matched closely to the metabolic needs of tissues. Mechanisms that control blood flow through tissues are classified as (1) local control or (2) nervous and hormonal control.

## Local Control of Blood Flow

Local control of blood flow is achieved by the periodic relaxation and contraction of the precapillary sphincters. When the sphincters relax, blood flow through the capillaries increases. When the sphincters contract, blood flow through the capillaries decreases. The precapillary sphincters are controlled by the metabolic needs of the tissues. For example, blood flow increases when by-products of metabolism build up in tissue spaces. During exercise, the metabolic needs of skeletal muscle increase dramatically, and the by-products of metabolism are produced more rapidly. The precapillary sphincters relax, increasing blood flow through the capillaries.

Other factors that control blood flow through the capillaries are the tissue concentrations of nutrients, such as oxygen, glucose, amino acids, and fatty acids (figure 13.25 and table 13.1). Blood flow increases when oxygen levels decrease or, to a lesser degree, when glucose, amino acids, fatty acids, and other nutrients decrease. An increase in carbon dioxide or a decrease in pH also causes the precapillary sphincters to relax, thereby increasing blood flow.

**Predict 5**

*A student has been sitting for a short time with her legs crossed. After getting up to walk out of class, she notices a red blotch on the back of one of her legs. On the basis of what you know about local control of blood flow, explain why this happens.*

In addition to the control of blood flow through existing capillaries, if the metabolic activity of a tissue increases often, additional capillaries gradually grow into the area. The additional capillaries allow local blood flow to increase to a level that matches the metabolic needs of the tissue. For example, the density of capillaries in the well-trained skeletal muscles of athletes is greater than that in poorly trained skeletal muscles (table 13.1).

**Blood flow increases.**

Smooth muscle of precapillary sphincter relaxes.

Blood flow

**Blood flow decreases.**

Smooth muscle of precapillary sphincter contracts.

Blood flow

1 **Relaxation of precapillary sphincters.** Precapillary sphincters relax as the tissue concentration of nutrients, such as $O_2$, glucose, amino acids, and fatty acids, decreases. The sphincters also relax as the concentration of tissue metabolic by-products increases as a result of increased $CO_2$ and lactic acid and decreased pH.

2 **Contraction of precapillary sphincters.** Precapillary sphincters contract as the tissue concentration of nutrients, such as $O_2$, glucose, amino acids, and fatty acids, increases. The sphincters also contract as the tissue concentration of metabolic by-products decreases as a result of decreased $CO_2$ and lactic acid and increased pH.

**PROCESS Figure 13.25  Local Control of Blood Flow Through Capillary Beds**

| TABLE 13.1 | Homeostasis: Control of Blood Flow* |
|---|---|
| **Stimulus** | **Response** |
| **Regulation by Metabolic Needs of Tissues** | |
| Increased carbon dioxide and decreased pH or decreased nutrients, such as oxygen, glucose, amino acids, and fatty acids, due to increased metabolism | Relaxation of precapillary sphincters and subsequent increase in blood flow through capillaries |
| Decreased carbon dioxide and increased pH or increased nutrients, such as oxygen, glucose, amino acids, and fatty acids | Contraction of precapillary sphincters and subsequent decrease in blood flow through capillaries |
| **Regulation by Nervous Mechanisms** | |
| Increased physical activity or increased sympathetic activity | Constriction of blood vessels in skin and viscera |
| Increased body temperature detected by neurons of hypothalamus | Dilation of blood vessels in skin (see chapter 5) |
| Decreased body temperature detected by neurons of hypothalamus | Constriction of blood vessels in skin (see chapter 5) |
| Decrease in skin temperature below a critical value | Dilation of blood vessels in skin (protects skin from extreme cold) |
| Anger or embarrassment | Dilation of blood vessels in skin of face and upper thorax |
| **Regulation by Hormonal Mechanisms (Reinforces Increased Activity of Sympathetic Division)** | |
| Increased physical activity and increased sympathetic activity, causing release of epinephrine and small amounts of norepinephrine from adrenal medulla | Constriction of blood vessels in skin and viscera; dilation of blood vessels in skeletal and cardiac muscle |
| **Long-Term Local Blood Flow** | |
| Increased metabolic activity of tissues over a long period, as occurs in athletes who train regularly | Increased number of capillaries |
| Decreased metabolic activity of tissues over a long period, as occurs during periods of reduced physical activity | Decreased number of capillaries |

*The mechanisms operate when the systemic blood pressure is maintained within a normal range of values.

## Nervous and Hormonal Control of Blood Flow

Nervous control of blood flow is carried out primarily through the sympathetic division of the autonomic nervous system. Sympathetic nerve fibers innervate most blood vessels of the body, except the capillaries and precapillary sphincters, which have no nerve supply (figure 13.26).

An area of the lower pons and upper medulla oblongata, called the **vasomotor center,** continually transmits a low frequency of action potentials to the sympathetic nerve fibers. As a consequence, the peripheral blood vessels are continually in a partially constricted state, a condition called **vasomotor** (vă-sō-mō′ter) **tone.** An increase in vasomotor tone causes blood vessels to constrict further and blood pressure to increase. A decrease in vasomotor tone causes blood vessels to dilate and blood pressure to decrease. Nervous control of blood vessel diameter is an important way that blood pressure is regulated.

Nervous control of blood vessels also causes blood to be shunted from one large area of the body to another. For example, nervous control of blood vessels during exercise increases vasomotor tone in the viscera and skin and reduces vasomotor tone in exercising skeletal muscles. As a result, blood flow to the viscera and skin decreases, and blood flow to skeletal muscle increases. Nervous control of blood vessels during exercise and dilation of precapillary sphincters as muscle activity increases together increase blood flow through exercising skeletal muscle several-fold (table 13.1).

The sympathetic division also regulates hormonal control of blood flow through the release of epinephrine and norepinephrine from the adrenal medulla. These hormones are transported in the

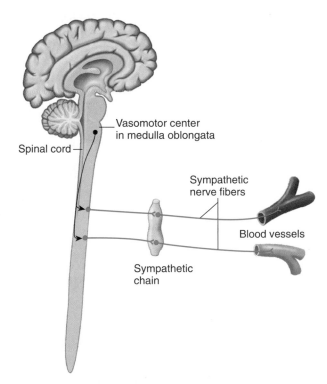

**Figure 13.26  Nervous Regulation of Blood Vessels**

Most arteries and veins are innervated by sympathetic nerve fibers. The vasomotor center within the medulla oblongata regulates the frequency of action potentials in nerve fibers that innervate blood vessels. In most blood vessels, increased action potential frequencies cause vasoconstriction, and decreased action potential frequencies cause vasodilation.

blood to all parts of the body. In most blood vessels, these hormones cause constriction, which reduces blood flow. But in some tissues, such as skeletal muscle and cardiac muscle, these hormones cause the blood vessels to dilate, increasing blood flow.

*Raynaud syndrome is a treatable condition in which blood vessels, primarily in the fingers and toes undergo exaggerated vasoconstriction in response to emotions or exposure to cold. Predict the effects on the fingers and toes of a person who has severe, untreated Raynaud syndrome. Explain why these consequences occur.*

# REGULATION OF ARTERIAL PRESSURE

Adequate blood pressure is required to maintain blood flow through the blood vessels of the body, and several regulatory mechanisms ensure that blood pressure remains adequate for this task. The **mean arterial blood pressure (MAP)** is slightly less than the average of the systolic and diastolic pressures in the aorta because diastole lasts longer than systole. The mean arterial pressure is about 70 mm Hg at birth, is maintained at about 95 mm Hg from adolescence to middle age, and may reach 110 mm Hg in a healthy older person.

The body's MAP is equal to the **cardiac output (CO)** times the **peripheral resistance (PR),** which is the resistance to blood flow in all the blood vessels:

$$MAP = CO \times PR$$

Because the cardiac output is equal to the **heart rate (HR)** times the **stroke volume (SV),** the mean arterial pressure is equal to the heart rate times the stroke volume times the peripheral resistance (PR):

$$MAP = HR \times SV \times PR$$

Thus, the MAP increases in response to increases in HR, SV, or PR, and the MAP decreases in response to decreases in HR, SV, or PR. The MAP is controlled on a minute-to-minute basis by changes in these variables. For example, when blood pressure suddenly drops because of hemorrhage or some other cause, control systems attempt to reestablish blood pressure by increasing HR, SV, and PR, so that blood pressure is maintained at a value consistent with life. Mechanisms are also activated to increase the blood volume to its normal value.

## Baroreceptor Reflexes

Baroreceptor reflexes activate responses that keep the blood pressure within its normal range. **Baroreceptors** respond to stretch in arteries caused by increased pressure. They are scattered along the walls of most of the large arteries of the neck and thorax, and many are located in the carotid sinus at the base of the internal carotid artery and in the walls of the aortic arch. Action potentials travel from the baroreceptors to the medulla oblongata along sensory nerve fibers (figure 13.27).

A sudden increase in blood pressure stretches the artery walls and increases action potential frequency in the baroreceptors. The increased action potential frequency delivered to the vasomotor and cardioregulatory centers in the medulla oblongata causes responses that lower the blood pressure. One major response is a decrease in vasomotor tone, resulting in dilation of blood vessels and decreased peripheral resistance. Other responses, controlled by the cardioregulatory center, are an increase in the parasympathetic stimulation of

1. Baroreceptors in the carotid sinus and aortic arch monitor blood pressure.

2. Sensory nerves conduct action potentials to the cardioregulatory and vasomotor centers in the medulla oblongata.

3. Increased parasympathetic stimulation of the heart decreases the heart rate.

4. Increased sympathetic stimulation of the heart increases the heart rate and stroke volume.

5. Increased sympathetic stimulation of blood vessels increases vasoconstriction.

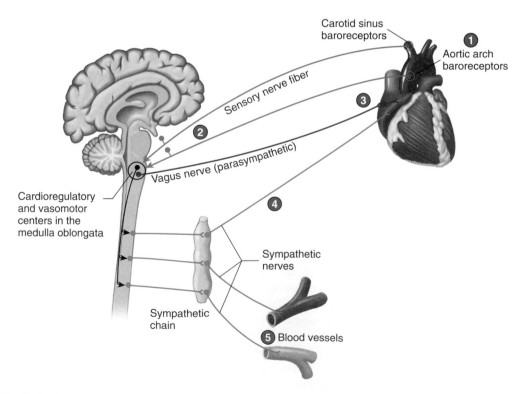

**PROCESS Figure 13.27  Baroreceptor Reflex Mechanisms**

The baroreceptor reflex helps control blood pressure.

# CLINICAL IMPACT

## Circulatory Shock

**Circulatory shock** is inadequate blood flow throughout the body that causes tissue damage due to lack of oxygen. Severe shock may damage vital body tissues and lead to death.

There are several causes of circulatory shock. One cause is excessive blood loss, which leads to the type of shock known as hemorrhagic shock. Here, we use this condition to illustrate the general characteristics of shock. If hemorrhagic shock is not severe, blood pressure decreases only a moderate amount, and the mechanisms that normally regulate blood pressure are able to reestablish normal pressure and blood flow. The baroreceptor reflexes produce strong sympathetic responses, resulting in intense vasoconstriction and increased heart rate.

As a result of the reduced blood flow through the kidneys, increased amounts of renin are released. The elevated renin level results in a greater rate of angiotensin II formation, causing vasoconstriction and increased aldosterone release from the adrenal cortex. Aldosterone, in turn, promotes water and salt retention by the kidneys. In response to reduced blood pressure, antidiuretic hormone (ADH) is released from the posterior pituitary gland; ADH also enhances the kidneys' retention of water.

An intense sensation of thirst leads to increased water intake, which helps restore the normal blood volume.

In mild cases of shock, the baroreceptor reflexes can be adequate to compensate for blood loss until the blood volume is restored. In more severe cases of shock, all of the regulatory mechanisms are needed to sustain life. But in the most severe cases, the regulatory mechanisms are not adequate to compensate for the effects of shock. As a consequence, a positive-feedback cycle begins to develop: The blood pressure regulatory mechanisms lose their ability to control the blood pressure, and shock worsens. As shock becomes worse, the effectiveness of the regulatory mechanisms deteriorates even further. The positive-feedback cycle proceeds until death occurs or until treatment, such as a transfusion, terminates the cycle. Five types of shock are classified based on their cause:

1. **Hypovolemic shock** is the result of reduced blood volume. **Hemorrhagic shock,** caused by internal or external bleeding, is one type of hypovolemic shock. **Plasma loss shock** results from loss of plasma, as may occur in severely burned areas of the body. **Interstitial**

**fluid loss shock** is reduced blood volume resulting from the loss of interstitial fluid, as may occur as a result of diarrhea, vomiting, or dehydration.

2. **Neurogenic shock** is caused by vasodilation in response to emotional upset or anesthesia.

3. **Anaphylactic shock** is caused by an allergic response that results in the release of inflammatory substances that cause vasodilation and increased capillary permeability. Large amounts of fluid then move from capillaries into the interstitial spaces.

4. **Septic shock,** or *blood poisoning,* is caused by infections that release toxic substances into the circulatory system, depressing the heart's activity and leading to vasodilation and increased capillary permeability.

5. **Cardiogenic shock** results from a decrease in cardiac output caused by events that decrease the heart's ability to function. Heart attack (myocardial infarction) is a common cause of cardiogenic shock. Fibrillation of the heart, which can be initiated by stimuli such as cardiac arrhythmias or exposure to electrical shocks, also results in cardiogenic shock.

---

the heart, which decreases the heart rate, and a decrease in the sympathetic stimulation of the heart, which reduces the stroke volume. The decreased heart rate, stroke volume, and peripheral resistance lower the blood pressure toward its normal value (figure 13.28).

A sudden decrease in blood pressure results in a decreased action potential frequency in the baroreceptors. The decreased frequency of action potentials delivered to the vasomotor and cardioregulatory centers in the medulla oblongata produces responses that raise blood pressure. Sympathetic stimulation of the heart increases, which increases the heart rate and stroke volume. In addition, vasomotor tone increases, which increases peripheral resistance. The increased heart rate, stroke volume, and peripheral resistance raise the blood pressure toward its normal value (figure 13.28).

These **baroreceptor reflexes** regulate blood pressure on a moment-to-moment basis. When a person rises rapidly from a sitting or lying position, blood pressure in the neck and thoracic regions

drops dramatically due to the pull of gravity on the blood. This reduction in blood pressure can be so great that it reduces blood flow to the brain enough to cause dizziness or even loss of consciousness. The falling blood pressure activates the baroreceptor reflexes, which reestablish normal blood pressure within a few seconds. A healthy person usually experiences only a temporary sensation of dizziness.

## Chemoreceptor Reflexes

**Carotid bodies** are small structures that lie near the carotid sinuses, and **aortic bodies** lie near the aortic arch. These structures contain sensory receptors that respond to changes in blood oxygen concentration, carbon dioxide concentration, and pH. Because they are sensitive to chemical changes in the blood, they are called **chemoreceptors.** They send action potentials along sensory nerve fibers to the medulla oblongata. There are also chemoreceptors in the medulla oblongata.

Baroreceptors in the carotid arteries and aorta detect an increase in blood pressure.

The cardioregulatory center and the vasomotor center in the brain alter activity of the heart and blood vessels (baroreceptor reflex), and the adrenal medulla decreases secretion of epinephrine.

Heart rate and stroke volume decrease; blood vessels dilate.

Blood pressure increases:
**Homeostasis Disturbed**

Blood pressure decreases:
**Homeostasis Restored**

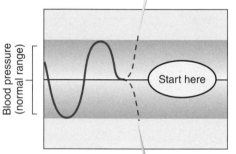

Start here

Blood pressure decreases:
**Homeostasis Disturbed**

Blood pressure increases:
**Homeostasis Restored**

Baroreceptors in the carotid arteries and aorta detect a decrease in blood pressure.

The cardioregulatory center and the vasomotor center in the brain alter activity of the heart and blood vessels (baroreceptor reflex), and the adrenal medulla increases secretion of epinephrine (adrenal medullary mechanism).

Heart rate and stroke volume increase; blood vessels constrict.

**Homeostasis Figure 13.28**  Baroreceptor Effects on Blood Pressure

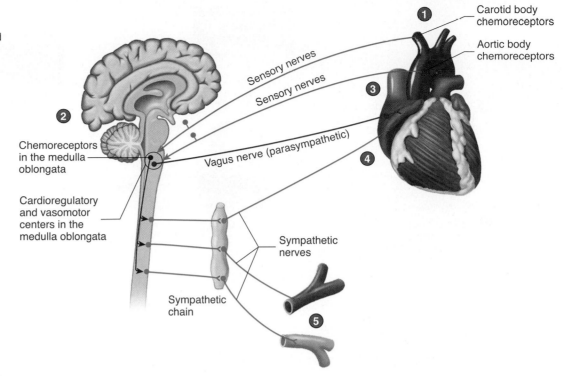

1. Chemoreceptors in the carotid and aortic bodies monitor blood $O_2$, $CO_2$, and pH.

2. Chemoreceptors in the medulla oblongata monitor blood $CO_2$ and pH.

3. Decreased blood $O_2$, increased $CO_2$, and decreased pH decrease parasympathetic stimulation of the heart, which increases the heart rate.

4. Decreased blood $O_2$, increased $CO_2$, and decreased pH increase sympathetic stimulation of the heart, which increases the heart rate and stroke volume.

5. Decreased blood $O_2$, increased $CO_2$, and decreased pH increase sympathetic stimulation of blood vessels, which increases vasoconstriction.

**PROCESS Figure 13.29  Chemoreceptor Reflex Mechanisms**

The chemoreceptor reflex helps control blood pressure.

When oxygen or pH levels decrease or when carbon dioxide levels increase, the chemoreceptors respond with an increased frequency of action potentials and activate the **chemoreceptor reflexes** (figure 13.29). In response, the vasomotor and cardiovascular centers decrease parasympathetic stimulation of the heart, which increases the heart rate. The vasomotor and cardioregulatory centers also increase sympathetic stimulation of the heart, which further increases heart rate, stroke volume, and vasomotor tone. All these changes result in increased blood pressure. This increased blood pressure causes a greater rate of blood flow to the lungs, which helps raise blood oxygen levels and reduce blood carbon dioxide levels. The chemoreceptor reflexes function under emergency conditions and usually do not play an important role in regulating the cardiovascular system. They respond strongly only when the oxygen levels in the blood fall to very low levels or when carbon dioxide levels become substantially elevated.

## Hormonal Mechanisms

In addition to the rapidly acting baroreceptor and chemoreceptor reflexes, four important hormonal mechanisms help control blood pressure.

### Adrenal Medullary Mechanism

Stimuli that lead to increased sympathetic stimulation of the heart and blood vessels also cause increased stimulation of the adrenal medulla. The adrenal medulla responds by releasing epinephrine and small amounts of norepinephrine into the blood (figure 13.30). Epinephrine increases heart rate and stroke volume and causes vasoconstriction, especially of blood vessels in the skin

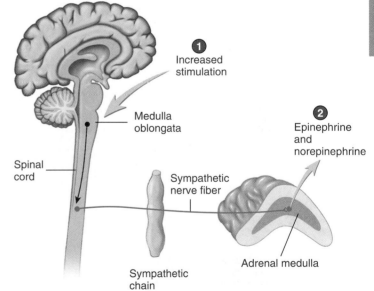

1. The same stimuli that increase sympathetic stimulation of the heart and blood vessels cause action potentials to be carried to the medulla oblongata.

2. Descending pathways from the medulla oblongata to the spinal cord increase sympathetic stimulation of the adrenal medulla, resulting in secretion of epinephrine and some norepinephrine.

**PROCESS Figure 13.30  Hormonal Regulation: Adrenal Medullary Mechanism**

Stimuli that increase sympathetic stimulation of the heart and blood vessels also increase sympathetic stimulation of the adrenal medulla and lead to secretion of epinephrine and some norepinephrine.

Circulatory

Circulatory

1 The kidneys detect decreased blood pressure, resulting in increased renin secretion.

2 Renin converts angiotensinogen, a protein secreted from the liver, to angiotensin I.

3 Angiotensin-converting enzyme in the lungs converts angiotensin I to angiotensin II.

4 Angiotensin II is a potent vasoconstrictor, resulting in increased blood pressure.

5 Angiotensin II stimulates the adrenal cortex to secrete aldosterone.

6 Aldosterone acts on the kidneys to increase Na⁺ reabsorption. As a result, urine volume decreases and blood volume increases, resulting in increased blood pressure.

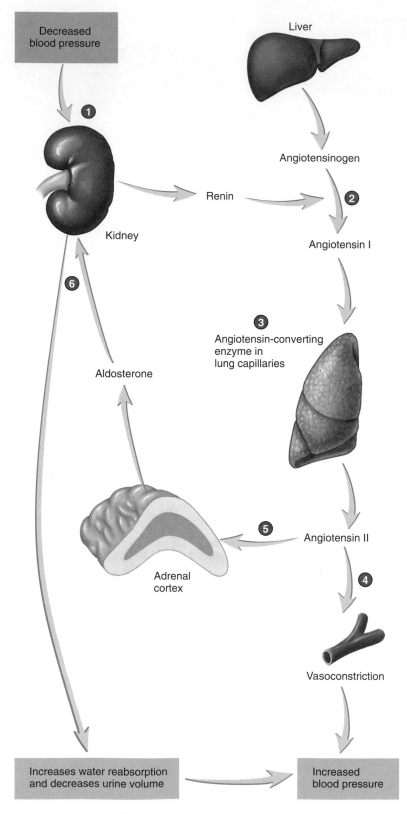

**PROCESS Figure 13.31** Hormonal Regulation: Renin-Angiotensin-Aldosterone Mechanism and viscera. Epinephrine also causes vasodilation of blood vessels in skeletal muscle and cardiac muscle, thereby increasing the supply of blood flowing to those muscles and preparing the body for physical activity.

### Renin-Angiotensin-Aldosterone Mechanism

In response to reduced blood flow, the kidneys release an enzyme called **renin** (rē′nin) into the circulatory system (figure 13.31). Renin acts on the blood protein **angiotensinogen** (an′jē-ō-ten-sin′ō-jen) to produce **angiotensin I** (an-jē-ō-ten′sin). Another enzyme, called **angiotensin-converting enzyme,** found in large amounts in organs, such as the lungs, acts on angiotensin I to convert it to its most active form, **angiotensin II.** Angiotensin II is a potent vasoconstrictor. Thus, in response to reduced blood pressure, the kidneys' release of renin increases the blood pressure toward its normal value.

Angiotensin II also acts on the adrenal cortex to increase the secretion of **aldosterone** (al-dos′ter-ōn). Aldosterone acts on the kidneys, causing them to conserve Na⁺ and water. As a result, the volume of water lost from the blood into the urine is reduced. The decrease in urine volume results in less fluid loss from the body, which maintains blood volume. Adequate blood volume is essential to maintain normal venous return to the heart and thereby maintain blood pressure (see chapter 12).

### Antidiuretic Hormone Mechanism

When the concentration of solutes in the plasma increases or when blood pressure decreases substantially, nerve cells in the hypothalamus respond by causing the release of **antidiuretic** (an′tē-dī-ū-ret′ik; to decrease urine production) **hormone (ADH),** also called *vasopressin* (vā-sō-pres′in; to cause vasoconstriction), from the posterior pituitary gland (figure 13.32). ADH acts on the kidneys and causes them to absorb more water, thereby decreasing urine volume. This response helps maintain blood volume and blood pressure. The release of large amounts of ADH causes vasoconstriction of blood vessels, which causes blood pressure to increase.

### Atrial Natriuretic Mechanism

A peptide hormone called **atrial natriuretic** (ā′trē-ăl nā′trē-ū-ret′ik) **hormone** is released primarily from specialized cells of the right atrium in response to elevated blood pressure. Atrial natriuretic hormone causes the kidneys to promote the loss of Na⁺ in the urine and to increase urine volume. Loss of water in the urine causes blood volume to decrease, thus decreasing the blood pressure.

## Summary of Regulatory Mechanisms

Baroreceptor mechanisms are most important in controlling blood pressure on a short-term basis (see figures 13.27 and 13.28). They are sensitive to sudden changes in blood pressure, and they respond quickly. The chemoreceptor and adrenal medullary reflexes are also sensitive to sudden changes in blood pressure and respond quickly, but they respond to large changes in blood pressure. The renin-angiotensin-aldosterone, antidiuretic hormone, and atrial natriuretic mechanisms are more important in maintaining blood pressure on a long-term basis. They are influenced by small changes in blood pressure or concentration and respond by gradually bringing the blood pressure back into its normal range (figure 13.33).

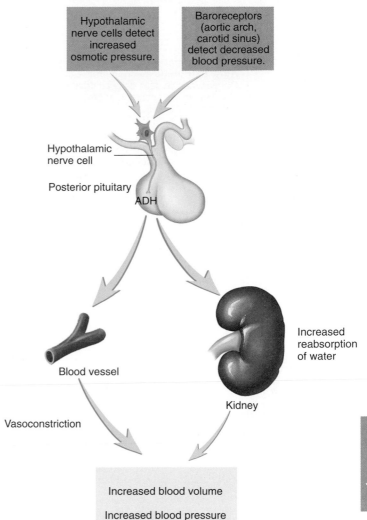

**Figure 13.32  Hormonal Regulation: Antidiuretic Hormone Mechanism**

Increases in the osmolality of blood or decreases in blood pressure result in antidiuretic hormone (ADH) secretion. ADH increases water reabsorption by the kidneys, and large amounts of ADH result in vasoconstriction. These changes maintain blood pressure.

**Predict 7**

*Suppose that a hemorrhage results in rapid loss of a large volume of blood. What blood pressure regulation mechanisms will come into play? If the hemorrhage results in the loss of the same volume of blood over a period of several hours, will the same mechanisms respond? Explain.*

## EFFECTS OF AGING ON THE BLOOD VESSELS

The walls of all arteries undergo changes as people age. Some arteries change more rapidly than others, and some individuals are more susceptible to change than others. The most significant effects of aging occur in the large, elastic arteries, such as the aorta; in large arteries carrying blood to the brain; and in the coronary arteries.

**Atrial natriuretic mechanism:**
Cardiac muscle cells detect increased atrial blood pressure; secretion of atrial natriuretic hormone increases.

**Renin-angiotensin-aldosterone mechanism:**
The kidneys detect increased blood pressure; production of angiotensin II and secretion of aldosterone from the adrenal cortex decrease.

Vasodilation decreases peripheral resistance to blood flow. More Na$^+$ and water are lost in the urine, decreasing blood volume.

Blood pressure increases:
**Homeostasis Disturbed**

Blood pressure decreases:
**Homeostasis Restored**

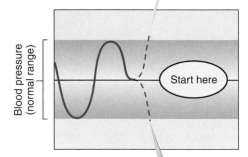

Blood pressure (normal range)

Start here

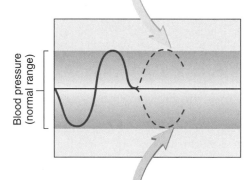

Blood pressure (normal range)

Blood pressure decreases:
**Homeostasis Disturbed**

Blood pressure increases:
**Homeostasis Restored**

**Renin-angiotensin-aldosterone mechanism:**
The kidneys detect decreased blood pressure; production of angiotensin II and secretion of aldosterone from the adrenal cortex increase.

**ADH (vasopressin) mechanism:**
Baroreceptors detect decreased blood pressure, resulting in decreased stimulation of the hypothalamus and increased ADH secretion by the posterior pituitary.

Vasoconstriction increases peripheral resistance to blood flow. Less Na$^+$ and water are lost in the urine, increasing blood volume.

**Homeostasis Figure 13.33**  Long-Term Control of Blood Pressure

Circulatory

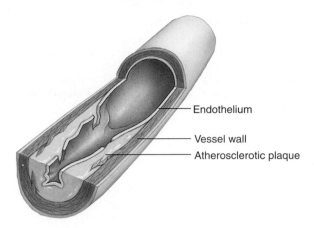

Endothelium

Vessel wall

Atherosclerotic plaque

**Figure 13.34  Atherosclerotic Plaque**
Atherosclerotic plaque develops within the tissue of the artery wall.

Changes in arteries that make them less elastic, referred to as arteriosclerosis, occur in nearly every individual, and become more severe with advancing age. A type of arteriosclerosis called **atherosclerosis** (ath′er-ō-skler-ō′sis) results from the deposition of material in the walls of arteries that forms plaques (figure 13.34). The material is composed of a fatlike substance containing cholesterol. The fatty material can eventually be dominated by the deposition of dense connective tissue and calcium salts.

Several factors influence the development of atherosclerosis. Lack of exercise, smoking, obesity, and a diet high in cholesterol and fats appear to increase the severity of atherosclerosis and the rate at which it develops. Severe atherosclerosis is more prevalent in some families than in others, which suggests a genetic influence. Some evidence suggests that a low-fat diet, mild exercise, and relaxation activities slow the progression of atherosclerosis and may even reverse it to some degree.

Atherosclerosis greatly increases resistance to blood flow because the deposits narrow the inside diameter of the arteries. The added resistance hampers normal circulation to tissues and greatly increases the work the heart must perform. Furthermore, the rough atherosclerotic plaques attract platelets, which adhere to them and increase the chance of thrombus formation.

Capillaries narrow and become more irregular in shape with age. Their walls become thicker, and consequently the efficiency of capillary exchange decreases.

Veins tend to develop patchy thickenings in their walls, resulting in narrowing in these areas. The tendency to develop varicose veins increases with age (see Clinical Impact earlier in this chapter), as does the tendency to develop hemorrhoids (varicose veins of the rectum or anus) because some veins increase in diameter due to weakening of the connective tissue in their walls. There is a related increase in the development of thrombi and emboli, especially in veins that are dilated or those in which blood flow is sluggish.

Circulatory

 **SUMMARY**

The circulatory system can be divided into the pulmonary vessels and the systemic vessels. The circulatory system and the heart maintain sufficient blood flow to tissues.

## Functions of the Circulatory System  (p. 345)

The circulatory system carries blood; exchanges nutrients, waste products, and gases; transports hormones; regulates blood pressure; and directs blood flow.

## General Features of Blood Vessel Structure  (p. 345)

1. The heart pumps blood through elastic arteries, muscular arteries, and arterioles to the capillaries.
2. Blood returns to the heart from the capillaries through venules, small veins, medium-sized veins, and large veins.
3. Except for capillaries and venules, blood vessels have three layers:
   a. The tunica intima, the innermost layer, consists of endothelium, a basement membrane, and connective tissue.
   b. The tunica media, the middle layer, contains circular smooth muscle and elastic fibers.
   c. The outer tunica adventitia is composed of connective tissue.

### Arteries

1. Large, elastic arteries have many elastic fibers but little smooth muscle in their walls. They carry blood from the heart to smaller arteries with little decrease in pressure.

2. The walls of muscular arteries have much smooth muscle and some elastic fibers. They undergo vasodilation and vasoconstriction to control blood flow to different regions of the body.
3. Arterioles, the smallest arteries, have smooth muscle cells and a few elastic fibers. They undergo vasodilation and vasoconstriction to control blood flow to local areas.

### Capillaries

1. Capillaries consist of only endothelium and are surrounded by a basement membrane and loose connective tissue.
2. Nutrient and waste exchange is the principal function of capillaries.
3. Blood is supplied to capillaries by arterioles. Precapillary sphincters regulate blood flow through capillary networks.

### Veins

1. Venules are composed of endothelium surrounded by a basement membrane.
2. Small veins are venules covered with a layer of smooth muscle and a layer of connective tissue.
3. Medium-sized and large veins contain less smooth muscle and fewer elastic fibers than arteries of the same size.
4. Valves prevent the backflow of blood in the veins.

## Blood Vessels of the Pulmonary Circulation   (p. 348)

The pulmonary circulation moves blood to and from the lungs. The pulmonary trunk carries oxygen-poor blood from the heart to the lungs, and pulmonary veins carry oxygen-rich blood from the lungs to the left atrium of the heart.

## Blood Vessels of the Systemic Circulation: Arteries   (p. 348)

### Aorta

The aorta leaves the left ventricle to form the ascending aorta, the aortic arch, and the descending aorta, which consists of the thoracic aorta and the abdominal aorta.

### Arteries of the Head and Neck

1. The brachiocephalic, left common carotid, and left subclavian arteries branch from the aortic arch to supply the head and the upper limbs.
2. The common carotid arteries and the vertebral arteries supply the head. The common carotid arteries divide to form the external carotids (which supply the face and mouth) and the internal carotids (which supply the brain).

### Arteries of the Upper Limbs

The subclavian artery continues as the axillary artery and then as the brachial artery, which branches to form the radial and ulnar arteries.

### Thoracic Aorta and Its Branches

The thoracic aorta has visceral branches, which supply the thoracic organs, and parietal branches, which supply the thoracic wall.

### Abdominal Aorta and Its Branches

The abdominal aorta has visceral branches, which supply the abdominal organs, and parietal branches, which supply the abdominal wall.

### Arteries of the Pelvis

Branches of the internal iliac arteries supply the pelvis.

### Arteries of the Lower Limbs

The common iliac arteries give rise to the external iliac arteries, and the external iliac artery continues as the femoral artery and then as the popliteal artery in the leg. The popliteal artery divides to form the anterior and posterior tibial arteries.

## Blood Vessels of the Systemic Circulation: Veins   (p. 356)

The superior vena cava drains the head, neck, thorax, and upper limbs. The inferior vena cava drains the abdomen, pelvis, and lower limbs.

### Veins of the Head and Neck

1. The internal jugular veins drain the brain, anterior head, and anterior neck.
2. The external jugular veins drain the posterior head and posterior neck.

### Veins of the Upper Limbs

The deep veins are the brachial, axillary, and subclavian; the superficial veins are the cephalic, basilic, and median cubital.

### Veins of the Thorax

The left and right brachiocephalic veins and the azygos veins return blood to the superior vena cava.

### Veins of the Abdomen and Pelvis

1. Posterior abdominal wall veins join the azygos veins.
2. Veins from the kidneys, adrenal glands, and gonads directly enter the inferior vena cava.
3. Veins from the stomach, intestines, spleen, and pancreas connect with the hepatic portal vein, which transports blood to the liver for processing. The hepatic veins from the liver join the inferior vena cava.

### Veins of the Lower Limbs

1. The deep veins course with the deep arteries and have similar names.
2. The superficial veins are the great and small saphenous veins.

## Physiology of Circulation   (p. 361)

### Blood Pressure

1. Blood pressure is a measure of the force exerted by blood against the blood vessel walls.
2. Blood pressure moves blood through vessels.
3. Blood pressure can be measured by listening for Korotkoff sounds produced as blood flows through arteries partially constricted by a blood pressure cuff.

### Pressure and Resistance

In a normal adult, blood pressure fluctuates between 120 mm Hg (systolic) and 80 mm Hg (diastolic) in the aorta. If blood vessels constrict, resistance to blood flow increases, and blood flow decreases.

### Pulse Pressure

1. Pulse pressure is the difference between systolic and diastolic pressures. Pulse pressure increases when stroke volume increases.
2. A pulse can be detected when large arteries are near the body surface.

### Capillary Exchange

1. Most exchange across the wall of the capillary occurs by diffusion.
2. Blood pressure, capillary permeability, and osmosis affect movement of fluid across the wall of the capillaries. There is a net movement of fluid from the blood into the tissues. The fluid gained by the tissues is removed by the lymphatic system.

## Control of Blood Flow in Tissues   (p. 366)

### Local Control of Blood Flow

Blood flow through a tissue is usually proportional to the metabolic needs of the tissue and is controlled by the precapillary sphincters.

### Nervous and Hormonal Control of Blood Flow

1. The vasomotor center (sympathetic division) controls blood vessel diameter. Other brain areas can excite or inhibit the vasomotor center.
2. Vasomotor tone is the state of partial constriction of blood vessels.
3. The nervous system is responsible for routing the flow of blood, except in the capillaries and for maintaining blood pressure.
4. Epinephrine and norepinephrine released by the adrenal medulla alter blood vessel diameter.

## Regulation of Arterial Pressure   (p. 368)

Mean arterial pressure (MAP) is proportional to cardiac output times the peripheral resistance.

### Baroreceptor Reflexes

1. Baroreceptors are sensitive to stretch.
2. Baroreceptors are located in the carotid sinuses and the aortic arch.
3. The baroreceptor reflex changes peripheral resistance, heart rate, and stroke volume in response to changes in blood pressure.

Circulatory

### Chemoreceptor Reflexes

1. Chemoreceptors are sensitive to changes in blood oxygen, carbon dioxide, and pH.
2. Chemoreceptors are located in the carotid bodies and the aortic bodies.
3. The chemoreceptor reflex increases peripheral resistance in response to low oxygen levels, high carbon dioxide levels, and reduced blood pH.

### Hormonal Mechanisms

1. Epinephrine released from the adrenal medulla as a result of sympathetic stimulation increases heart rate, stroke volume, and vasoconstriction.
2. The kidneys release renin in response to low blood pressure. Renin promotes the production of angiotensin II, which causes vasoconstriction and increased secretion of aldosterone. Aldosterone reduces urine output. Angiotensin II can also cause vasoconstriction.
3. ADH released from the posterior pituitary causes vasoconstriction and reduces urine output.

4. The heart releases atrial natriuretic hormone when atrial blood pressure increases. Atrial natriuretic hormone stimulates an increase in urine production, causing a decrease in blood volume and blood pressure.

### Summary of Regulatory Mechanisms

1. The baroreceptor, chemoreceptor, and adrenal medullary reflex mechanisms are most important in short-term regulation of blood pressure.
2. Hormonal mechanisms, such as the renin-angiotensin-aldosterone system, antidiuretic hormone, and atrial natriuretic hormone, are more important in long-term regulation of blood pressure.

## Effects of Aging on the Blood Vessels  (p. 373)

1. Reduced elasticity and thickening of arterial walls result in hypertension and decreased ability to respond to changes in blood pressure.
2. Atherosclerosis is an age-related condition.
3. The efficiency of capillary exchange decreases with age.
4. Walls of veins thicken in some areas and dilate in others. Thrombi, emboli, varicose veins, and hemorrhoids are age-related conditions.

---

 ## REVIEW AND COMPREHENSION

1. Name, in order, all the types of blood vessels, starting at the heart, going to the tissues and those returning to the heart.
2. Name the three layers of a blood vessel. What kinds of tissue are in each layer?
3. Relate the structures of the different types of arteries to their functions.
4. Describe a capillary network. Name the structure that regulates blood flow through the capillary network.
5. Describe the structure of capillaries, and explain their major function.
6. Describe the structure of veins.
7. What is the function of valves in blood vessels, and which blood vessels have valves?
8. List the parts of the aorta. Name the major arteries that branch from the aorta and deliver blood to the vessels that supply the heart, the head and upper limbs, and the lower limbs.
9. Name the arteries that supply the major areas of the head, upper limbs, thorax, abdomen, and lower limbs. Describe the area each artery supplies.
10. Name the major vessels that return blood to the heart. What area of the body does each drain?
11. List the veins that drain blood from the thorax, abdomen, and pelvis. What specific area of the body does each drain? Describe the hepatic portal system.
12. List the major veins that drain the upper and lower limbs.

13. Define blood pressure, and describe how it is normally measured.
14. Describe the changes in blood pressure, starting in the aorta, moving through the vascular system, and returning to the right atrium.
15. Define pulse pressure, and explain what information can be determined by monitoring the pulse.
16. Explain how blood pressure and osmosis affect the movement of fluid between capillaries and tissues. What happens to excess fluid that enters the tissues?
17. Explain what is meant by local control of blood flow through tissues, and describe what carries out local control.
18. Describe nervous control of blood vessels. Define vasomotor tone.
19. Define mean arterial pressure. How is it related to heart rate, stroke volume, and peripheral resistance?
20. Where are baroreceptors located? Describe the baroreceptor reflex when blood pressure increases and when it decreases.
21. Where are the chemoreceptors for carbon dioxide and pH located? Describe what happens when oxygen levels in the blood decrease.
22. For each of the following hormones—epinephrine, renin, angiotensin II, aldosterone, ADH, and atrial natriuretic hormone—state where each is produced, what stimulus causes increased hormone production, and the hormone's effect on the circulatory system.
23. Describe the changes that occur in arteries as people age.

---

 ## CRITICAL THINKING

1. For each of the following destinations, name all the arteries a red blood cell encounters if it starts its journey in the left ventricle:
   a. brain
   b. external part of the skull
   c. left hand
   d. anterior portion of the right leg

2. For each of the following starting places, name all the veins a red blood cell encounters on its way back to the right atrium:
   a. left side of the brain
   b. external part of the right side of the skull
   c. left hand
   d. medial portion of the right leg
   e. kidney
   f. small intestine

3.  In angioplasty, a surgeon threads a catheter through blood vessels to a blocked coronary artery. The tip of the catheter can expand, stretching the coronary artery and unblocking it, or the tip of the catheter can be equipped with tiny blades capable of removing the blockage. Typically, the catheter is first inserted into a large blood vessel in the superior, medial part of the thigh. Starting with this blood vessel, name all the blood vessels the catheter passes through to reach the anterior interventricular artery.

4.  A 55-year-old man has a colonoscopy that reveals a large tumor in his colon. His doctor orders a liver scan to determine if the cancer has spread from the colon to the liver. Based on your knowledge of blood vessels, explain how cancer cells from the colon can end up in the liver.

5.  High blood pressure can be caused by advanced atherosclerosis of the renal arteries, even though blood flow appears sufficient to allow a normal volume of urine to be produced. Explain how atherosclerosis of the renal arteries can result in high blood pressure.

6.  Hugo Faster ran a race during which his stroke volume and heart rate increased. Vasoconstriction occurred in his viscera, and his blood pressure rose, but not dramatically. Explain these changes in his circulatory system.

7.  Nitroglycerin is often given to people who experience angina pains. This drug causes vasodilation of arteries and veins, which reduces the amount of work the heart performs and increases blood flow through the coronary arteries. Explain why dilation of arteries and veins reduces the heart's amount of work.

*Answers in Appendix D*

Visit this textbook's website at **www.mhhe.com/seeleyess7** for practice quizzes, animations, interactive learning exercises, and other study tools.

McGraw-Hill offers a study CD that features interactive cadaver dissection. ***Anatomy & Physiology Revealed*** includes cadaver photos that allow you to peel away layers of the human body to reveal structures beneath the surface. This program also includes animations, radiologic imaging, audio pronunciations, and practice quizzing.

Circulatory

Ouch! But It's well worth the pain. This photo shows a young boy receiving an immunization. The nurse injects a weakened form of a pathogen into the boy's body. Once inside, the weakened pathogen will elicit white blood cells to mount an attack to kill the pathogen. In the process, some white blood cells will retain memory of that pathogen, resulting in immunity to it. The immune system does all this to help the body maintain homeostasis in the face of a foreign invader.

# 14 Chapter

# Lymphatic System and Immunity

## Chapter Outline and Objectives

### Functions of the Lymphatic System (p. 380)
1. Describe the functions of the lymphatic system.
2. Explain how lymph is formed.

### Anatomy of the Lymphatic System (p. 380)
3. Describe how lymph is transported.
4. Describe the structure and function of tonsils, lymph nodes, the spleen, and the thymus.

### Immunity (p. 386)
5. Define the concepts of specificity and memory as they apply to immunity.

### Innate Immunity (p. 386)
6. Define innate immunity, and describe the cells and chemical mediators involved.
7. List the events of an inflammatory response, and explain their significance.

### Adaptive Immunity (p. 388)
8. Define antigen.
9. Describe the origin, development, activation, and proliferation of lymphocytes.

10. Define antibody-mediated immunity, and cell-mediated immunity, and name the cells responsible for each.
11. Diagram the structure of an antibody, and describe the effects produced by antibodies.
12. Discuss the primary and secondary responses to an antigen. Explain the basis for long-lasting immunity.
13. Describe the functions of T cells.

### Acquired Immunity (p. 397)
14. Explain the four ways that adaptive immunity can be acquired.

### Overview of Immune Interactions (p. 398)
15. Explain how innate, antibody-mediated, and cell-mediated immunity can function together to eliminate an antigen.

### Immunotherapy (p. 398)
16. Define and give examples of immunotherapy.

### Effects of Aging on the Lymphatic System and Immunity (p. 400)
17. Describe how aging affects the lymphatic system and immunity.

379

One of the basic tenets of life is that many organisms consume or use other organisms in order to survive. Some microorganisms, such as certain bacteria or viruses, use humans as a source of nutrients and as an environment where they can survive and reproduce. As a result, some of these microorganisms can damage the body, causing disease or even death. Not surprisingly, the body has ways to resist or destroy harmful microorganisms. This chapter considers how the lymphatic system and the components of other systems, such as white blood cells and phagocytes, continually provide protection against invading microorganisms.

# FUNCTIONS OF THE LYMPHATIC SYSTEM

The **lymphatic** (lim-fat′ik) **system** functions are:

1. *Fluid balance.* About 30 liters (L) of fluid pass from the blood capillaries into the interstitial spaces each day, whereas only 27 L pass from the interstitial spaces back into the blood capillaries (see chapter 13). If the extra 3 L of interstitial fluid remained in the interstitial spaces, edema would result, causing tissue damage and eventually death. Instead, the 3 L of fluid enter the lymphatic capillaries, where it is called **lymph** (limf), and it passes through the lymphatic vessels to return to the blood. In addition to water, lymph contains solutes derived from two sources: (a) Substances in plasma, such as ions, nutrients, gases, and some proteins, pass from blood capillaries into the interstitial spaces and become part of the lymph; (b) substances such as hormones, enzymes, and waste products, derived from cells within the tissues, are also part of the lymph.

2. *Fat absorption.* The lymphatic system absorbs fats and other substances from the digestive tract (see figure 16.14) through lymphatic vessels called **lacteals** (lak′tē-ălz) located in the lining of the small intestine. Fats enter the lacteals and pass through the lymphatic vessels to the venous circulation. The lymph passing through these lymphatic vessels appears white because of its fat content and is called **chyle** (kīl).

3. *Defense.* Microorganisms and other foreign substances are filtered from lymph by lymph nodes and from blood by the spleen. In addition, lymphocytes and other cells are capable of destroying microorganisms and foreign substances. Because the lymphatic system is involved with fighting infections, as well as filtering blood and lymph to remove microorganisms, many infectious diseases produce symptoms associated with the lymphatic system (see the Diseases and Disorders table at the end of this chapter).

# ANATOMY OF THE LYMPHATIC SYSTEM

## Lymphatic Capillaries and Vessels

The lymphatic system includes lymph, lymphocytes, lymphatic vessels, lymph nodes, the tonsils, the spleen, and the thymus (figure 14.1a). The lymphatic system, unlike the circulatory system, does not circulate fluid to and from tissues. Instead, the lymphatic system carries fluid in one direction, from tissues to the circulatory system. Fluid moves from blood capillaries into tissue spaces (see figure 13.24). Most of the fluid returns to the blood, but some of the fluid moves from the tissue spaces into lymphatic capillaries to become lymph (figure 14.2a). The **lymphatic capillaries** are tiny, closed-ended vessels consisting of simple squamous epithelium. The lymphatic capillaries are more permeable than blood capillaries because they lack a basement membrane, and fluid moves easily into them. Overlapping squamous cells of the lymphatic capillary walls act as valves that prevent the backflow of fluid (figure 14.2b). After fluid enters lymphatic capillaries, it flows through them.

Lymphatic capillaries are present in most tissues of the body. Exceptions are the central nervous system, bone marrow, and tissues lacking blood vessels, such as the epidermis and cartilage. A superficial group of lymphatic capillaries drains the dermis and hypodermis, and a deep group drains muscle, the viscera, and other deep structures.

The lymphatic capillaries join to form larger **lymphatic vessels,** which resemble small veins (figure 14.2b). Small lymphatic vessels have a beaded appearance because they have one-way valves that are similar to the valves of veins (see chapter 13). When a lymphatic vessel is compressed, the valves prevent backward movement of lymph. Consequently, compression of the lymphatic vessels causes lymph to move forward through them. Three factors cause compression of the lymphatic vessels: (1) contraction of surrounding skeletal muscle during activity, (2) periodic contraction of smooth muscle in the lymphatic vessel wall, and (3) pressure changes in the thorax during breathing.

The lymphatic vessels converge and eventually empty into the blood at two locations in the body. Lymphatic vessels from the right upper limb and the right half of the head, neck, and chest form the **right lymphatic duct,** which empties into the right subclavian vein. Lymphatic vessels from the rest of the body enter the **thoracic duct,** which empties into the left subclavian vein (see figure 14.1b).

## Lymphatic Organs

The **lymphatic organs** include the tonsils, the lymph nodes, the spleen, and the thymus. **Lymphatic tissue,** which consists of many lymphocytes and other cells, such as macrophages, is found within lymphatic organs. The lymphocytes originate from red bone marrow (see chapter 11) and are carried by the blood to lymphatic organs. These lymphocytes divide and increase in number when the body is exposed to microorganisms or foreign substances. The increased number of lymphocytes is part of the immune response that causes the destruction of microorganisms

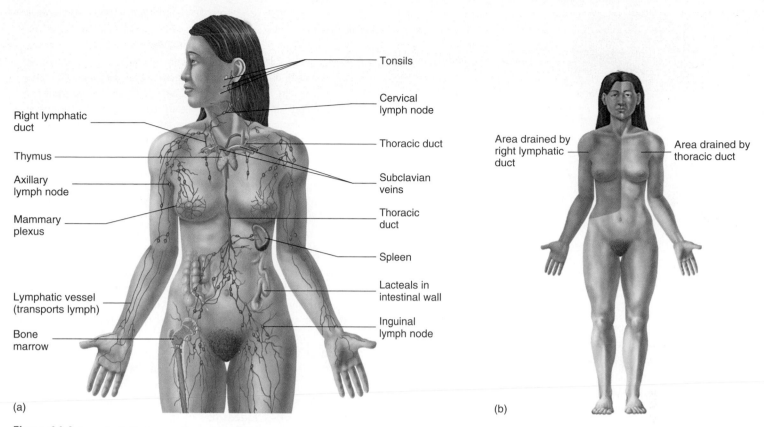

**Figure 14.1  Lymphatic System and Lymph Drainage**

(a) The major lymphatic organs are the tonsils, the lymph nodes, the spleen, and the thymus. Lymph nodes are located along lymphatic vessels throughout the body, but aggregations of them are found in the cervical, axillary, and inguinal areas. (b) Lymph from the uncolored areas drains through the thoracic duct. Lymph from the darkened area drains through the right lymphatic duct.

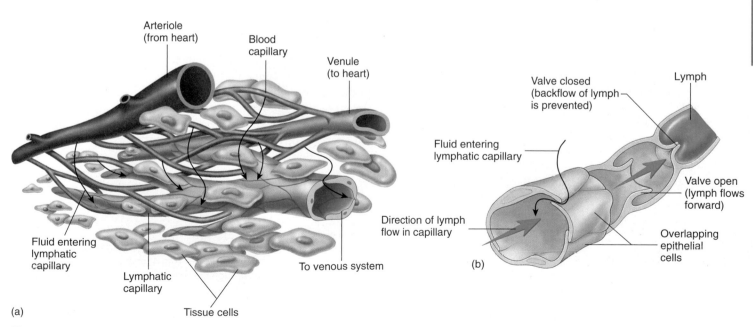

**Figure 14.2  Lymph Formation and Movement**

(a) Fluid moves from blood capillaries into tissues and from tissues into lymphatic capillaries to form lymph. (b) The overlap of epithelial cells of the lymphatic capillary allows fluid to enter easily but prevents it from moving back into the tissue. Valves, located farther along in lymphatic vessels, also ensure one-way flow of lymph.

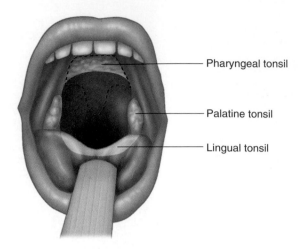

Pharyngeal tonsil

Palatine tonsil

Lingual tonsil

## Figure 14.3

Anterior view of the oral cavity, showing the tonsils. Part of the palate is removed (*dotted line*) to show the pharyngeal tonsil.

and foreign substances. In addition to cells, lymphatic tissue has very fine reticular fibers (see chapter 4). These fibers form an interlaced network that holds the lymphocytes and other cells in place. When lymph or blood filters through lymphatic organs, the fiber network also traps microorganisms and other items in the fluid.

### Tonsils

There are three groups of **tonsils** (figure 14.3; see figure 15.2). The **palatine** (pal´ă-tīn; palate) **tonsils** are located on each side of the posterior opening of the oral cavity; these are the ones usually referred to as "the tonsils." The **pharyngeal** (fă-rin´jē-ăl) **tonsil** is located near the internal opening of the nasal cavity. When the pharyngeal tonsil is enlarged, it is commonly called the **adenoid** (ad´ĕ-noid), or adenoids. An enlarged pharyngeal tonsil can interfere with normal breathing. The **lingual** (ling´gwăl; tongue) **tonsil** is on the posterior surface of the tongue.

The tonsils form a protective ring of lymphatic tissue around the openings between the nasal and oral cavities and the pharynx. They protect against pathogens and other potentially harmful material entering from the nose and mouth. Sometimes the palatine or pharyngeal tonsils become chronically infected and must be removed. The lingual tonsil becomes infected less often than the other tonsils and is more difficult to remove. In adults, the tonsils decrease in size and may eventually disappear.

## A CASE IN POINT
### Tonsillectomy and Adenoidectomy

Audie Tory is a 4-year-old girl who has some hearing loss and an associated delay in speech development. She has a history of frequent sore throats and middle ear infections, which have been treated with antibiotics. Recently, she has experienced difficulty in swallowing; she also snores and sleeps with her mouth open. Audie's physician tells her parents that her palatine tonsils and adenoid are enlarged and chronically infected. Her enlarged tonsils make swallowing difficult, and the infection stimulates inflammation that causes her throat to hurt. Her enlarged adenoid is restricting airflow, causing her to snore and sleep with her mouth open—and it is also the probable cause of her middle ear infections because the openings to the auditory tubes are located next to the adenoid. Chronic middle ear infections are associated with loss of hearing, which affects speech development. The doctor recommends a procedure called a **T&A**—that is, a **tonsillectomy** (ton´si-lek´tō-mē), which is removal of the palatine tonsils, and an **adenoidectomy** (ad´ĕ-noy-dek´tō-mē), which is removal of the adenoid.

### Lymph Nodes

**Lymph nodes** are rounded structures, varying from the size of a small seed to that of a shelled almond. Lymph nodes are distributed along the various lymphatic vessels (see figure 14.1a), and most lymph passes through at least one lymph node before entering the blood. Although lymph nodes are found throughout the body, there are three superficial aggregations of lymph nodes on each side of the body: inguinal nodes in the groin, axillary nodes in the axilla (armpit), and cervical nodes in the neck.

A dense connective tissue **capsule** surrounds each lymph node (figure 14.4). Extensions of the capsule, called **trabeculae,** subdivide a lymph node into compartments containing lymphatic tissue and lymphatic sinuses. The lymphatic tissue consists of lymphocytes and other cells that can form dense aggregations of tissue called **lymphatic nodules. Lymphatic sinuses** are spaces between the lymphatic tissue that contain macrophages on a network of fibers. Lymph enters the lymphatic node through afferent vessels, passes through the lymphatic tissue and sinuses, and exits through efferent vessels.

As lymph moves through the lymph nodes, two functions are performed. One function is to activate the immune system. Microorganisms or other foreign substances in the lymph can stimulate lymphocytes in the lymphatic tissue to divide. The lymphatic nodules containing the rapidly dividing lymphocytes are called **germinal centers.** The newly produced lymphocytes are released into the lymph and eventually reach the blood, where they circulate and enter other lymphatic tissues. The lymphocytes are part of the adaptive immune response (described later in this chapter) that destroys microorganisms and foreign substances. The second function of the lymph nodes is to remove microorganisms and foreign substances from the lymph through the action of macrophages.

### Predict 1

*Cancer cells can spread from a tumor site to other areas of the body through the lymphatic system. At first, however, as the cancer cells pass through the lymphatic system, they are trapped in the lymph nodes, which filter the lymph. During radical cancer surgery, malignant (cancerous) lymph nodes are removed, and their vessels are cut and tied off to prevent the cancer from spreading. Predict the consequences of tying off the lymphatic vessels.*

Lymphatic

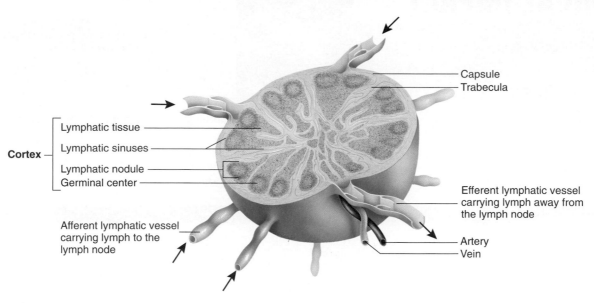

**Figure 14.4  Lymph Node**

*Arrows* indicate direction of lymph flow. As lymph moves through the sinuses, phagocytic cells remove foreign substances. The germinal centers are sites of lymphocyte production.

## Spleen

The **spleen** (splēn) is roughly the size of a clenched fist and is located in the left, superior corner of the abdominal cavity (figure 14.5). The spleen has an outer **capsule** of dense connective tissue and a small amount of smooth muscle. **Trabeculae** from the capsule divide the spleen into small, interconnected compartments containing two specialized types of lymphatic tissue. **White pulp** is lymphatic tissue surrounding the arteries within the spleen. **Red pulp** is associated with the veins. It consists of a fibrous network, filled with macrophages and red blood cells, and enlarged capillaries that connect to the veins.

The spleen filters blood instead of lymph. Cells within the spleen detect and respond to foreign substances in the blood and destroy worn-out red blood cells. Lymphocytes in the white pulp can be stimulated in the same manner as in lymph nodes. Before blood leaves the spleen through veins, it passes through the red pulp. Macrophages in the red pulp remove foreign substances and worn-out red blood cells through phagocytosis.

The spleen also functions as a blood reservoir, holding a small volume of blood. In emergency situations, such as hemorrhage, smooth muscle in splenic blood vessels and in the splenic capsule can contract, allowing a small amount of blood to move out of the spleen into the general circulation.

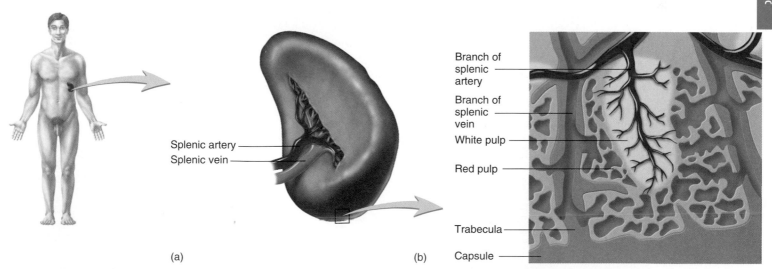

**Figure 14.5  Spleen**

(*a*) Inferior view of the spleen. (*b*) Section showing the arrangement of arteries, veins, white pulp, and red pulp. White pulp is associated with arteries, and red pulp is associated with veins.

### Thymus

The **thymus** (thi'mŭs) is a bilobed gland roughly triangular in shape (figure 14.6a). It is located in the superior mediastinum, the partition dividing the thoracic cavity into left and right parts. Each lobe of the thymus is surrounded by a thin connective tissue **capsule. Trabeculae** from the capsule divide each lobe into **lobules** (figure 14.6b). Near the capsule and trabeculae, the lymphocytes are numerous and form dark-staining areas called the **cortex.** A lighter-staining, central portion of the lobules, called the **medulla,** has fewer lymphocytes.

The thymus is the site for the maturation of lymphocytes (described in "Adaptive Immunity" later in this chapter). Large numbers of lymphocytes are produced in the thymus, but most degenerate. The lymphocytes that survive the maturation process are capable of reacting to foreign substances. The mature lymphocytes migrate to the medulla, enter the blood, and travel to other lymphatic tissues, where they help protect against microorganisms and other foreign substances.

## Overview of the Lymphatic System

Figure 14.7 summarizes the processes performed by the lymphatic system. Lymphatic capillaries and vessels remove fluid from tissues and absorb fats from the small intestine. Lymph nodes filter lymph, and the spleen filters blood.

Figure 14.7 also illustrates two types of lymphocytes, called B cells and T cells. B cells originate and mature in red bone marrow. Pre-T cells are produced in red bone marrow and migrate to the thymus, where they mature to become T cells. B cells from red bone marrow and T cells from the thymus circulate to, and populate, other lymphatic tissues.

B cells and T cells are responsible for much of immunity. In response to infections, B cells and T cells increase in number and circulate to lymphatic and other tissues. How B cells and T cells protect the body is discussed later in this chapter (see "Adaptive Immunity").

(a)

(b)

**Figure 14.6**

(a) Location and shape of the thymus. (b) Histology of thymic lobules, showing the outer cortex and the inner medulla.

Lymphatic

**1** Lymphatic capillaries remove fluid from tissues. The fluid becomes lymph (see figure 14.2*a*).

**2** Lymph flows through lymphatic vessels, which have valves that prevent the backflow of lymph (see figure 14.2*b*).

**3** Lymph nodes filter lymph (see figure 14.4) and are sites where lymphocytes respond to infections.

**4** Lymph enters the thoracic duct or the right lymphatic duct.

**5** Lymph enters the blood.

**6** Lacteals in the small intestine (see figure 16.14) absorb fats, which enter the thoracic duct.

**7** Chyle, which is lymph containing fats, enters the blood.

**8** The spleen (see figure 14.5) filters blood and is a site where lymphocytes respond to infections.

**9** Lymphocytes (pre-B and pre-T cells) originate from stem cells in the red bone marrow (see figure 14.9). The pre-B cells become mature B cells in the red bone marrow and are released into the blood. The pre-T cells enter the blood and migrate to the thymus.

**10** The thymus (see figure 14.6) is where pre-T cells derived from red bone marrow increase in number and become mature T cells that are released into the blood (see figure 14.9).

**11** B cells and T cells from the blood enter and populate all lymphatic tissues. These lymphocytes can remain in tissues or pass through them and return to the blood. B cells and T cells can also respond to infections by dividing and increasing in number. Some of the newly formed cells enter the blood and circulate to other tissues.

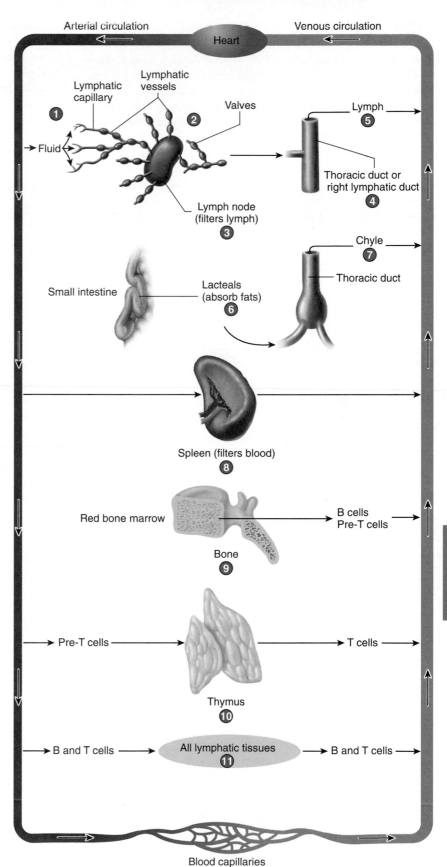

**PROCESS Figure 14.7  Overview of the Lymphatic System**

Lymphatic

# IMMUNITY

**Immunity** (i-mū′ni-tē) is the ability to resist damage from foreign substances—such as microorganisms; harmful chemicals, such as toxins released by microorganisms; and internal threats, such as cancer cells. Immunity is categorized as **innate** (i′nāt, i-nāt′) **immunity** (also called *nonspecific resistance*) and **adaptive immunity** (also called *specific immunity*), although the two systems are fully integrated in the body. In innate immunity, the body recognizes and destroys certain foreign substances, but the response to them is the same each time the body is exposed. In adaptive immunity, the body recognizes and destroys foreign substances, but the response to them improves each time the foreign substance is encountered.

Specificity and memory are characteristics of adaptive immunity, but not innate immunity. **Specificity** is the ability of adaptive immunity to recognize a particular substance. For example, innate immunity can act against bacteria in general, whereas adaptive immunity can distinguish among various kinds of bacteria. **Memory** is the ability of adaptive immunity to "remember" previous encounters with a particular substance. As a result, the response is faster, stronger, and longer-lasting.

In innate immunity, each time the body is exposed to a substance, the response is the same because specificity and memory of previous encounters are not present. For example, each time a bacterial cell is introduced into the body, it is phagocytized with the same speed and efficiency. In adaptive immunity, the response during the second exposure is faster and stronger than the response to the first exposure because the immune system exhibits memory for the bacteria from the first exposure. For example, following the first exposure to the bacteria, the body can take many days to destroy them. During this time, the bacteria damage tissues, producing the symptoms of disease. Following the second exposure to the same bacteria, the response is rapid and effective. Bacteria are destroyed before any symptoms develop, and the person is said to be **immune.**

Innate and adaptive immunity are intimately linked. Most importantly, mediators of innate immunity are required for the initiation and regulation of the adaptive response.

# INNATE IMMUNITY

Innate immunity is accomplished by physical barriers, chemical mediators, white blood cells, and the inflammatory response.

## Physical Barriers

**Physical barriers** prevent microorganisms and chemicals from entering the body in two ways: (1) The skin and mucous membranes form barriers that prevent their entry, and (2) tears, saliva, and urine wash these substances from body surfaces. Microorganisms cannot cause a disease if they cannot get into the body.

## Chemical Mediators

**Chemical mediators** are molecules responsible for many aspects of innate immunity. Some chemicals on the surface of cells kill microorganisms or prevent their entry into the cells. For example, lysozyme in tears and saliva kills certain bacteria, and mucus on the mucous membranes prevents the entry of some microorganisms. Other chemical mediators, such as histamine (his′tă-mēn), complement, prostaglandins (pros-tă-glan′dinz), and leukotrienes (loo-kō-trī′ēnz), promote inflammation by causing vasodilation, increasing vascular permeability, and stimulating phagocytosis. In addition, interferons protect cells against viral infections.

### Complement

**Complement** (kom′plĕ-ment) is a group of approximately 20 proteins found in plasma. The operation of complement proteins is similar to that of clotting proteins (see chapter 11). Normally, complement proteins circulate in the blood in an inactive form. Certain complement proteins can be activated by combining with foreign substances, such as parts of a bacterial cell, or by combining with antibodies (see "Effects of Antibodies" later in this chapter). Once activation begins, a series of reactions results, in which each complement protein activates the next. Once activated, certain complement proteins promote inflammation and phagocytosis and can directly lyse (rupture) bacterial cells.

### Interferons

**Interferons** (in-ter-fēr′onz) are proteins that protect the body against viral infections. When a virus infects a cell, the infected cell produces viral nucleic acids and proteins, which are assembled into new viruses. The new viruses are then released to infect other cells. Because infected cells usually stop their normal functions or die during viral replication, viral infections are clearly harmful to the body. Fortunately, viruses often stimulate infected cells to produce interferons, which do not protect the cell that produces them. Instead, interferons bind to the surface of neighboring cells, where they stimulate those cells to produce antiviral proteins. These antiviral proteins inhibit viral reproduction by preventing the production of new viral nucleic acids and proteins.

Some interferons play a role in activating immune cells, such as macrophages and natural killer cells (see next section).

## White Blood Cells

**White blood cells** (see chapter 11) and the cells derived from them are the most important cellular components of immunity. White blood cells are produced in red bone marrow and lymphatic tissue and released into the blood. Chemicals released from microorganisms or damaged tissues attract the white blood cells, and they leave the blood and enter affected tissues. Important chemicals known to attract white blood cells include complement, leukotrienes, kinins (kī′ninz), and histamine. The movement of white blood cells toward these chemicals is called **chemotaxis** (kem-ō-tak′sis, kē-mō-tak′sis).

### Phagocytic Cells

**Phagocytosis** (fag′ō-sī-tō′sis) is the ingestion and destruction of particles by cells called **phagocytes** (fag′ō-sītz) (see chapter 3). The particles can be microorganisms or their parts, foreign substances, or dead cells from the body. The most important phagocytes are neutrophils and macrophages, although other white blood cells also have limited phagocytic ability.

**Neutrophils** (noo′trō-filz) are small phagocytic cells that are usually the first cells to enter infected tissues from the blood in large numbers. They release chemical signals that increase the inflammatory response by recruiting and activating other immune cells. Neutrophils often die after phagocytizing a single microorganism. **Pus** is an accumulation of fluid, dead neutrophils, and other cells at a site of infection.

**Macrophages** (mak′rō-fā′jes) are monocytes that leave the blood, enter tissues, and enlarge about fivefold. Monocytes and macrophages form the **mononuclear phagocytic system** because they are phagocytes with a single (mono), unlobed nucleus. Sometimes macrophages are given specific names, such as *dust cells* in the lungs, *Kupffer cells* in the liver, and *microglia* in the central nervous system. Macrophages can ingest more and larger items than can neutrophils. Macrophages usually appear in infected tissues after neutrophils do, and they are responsible for most of the phagocytic activity in the late stages of an infection, including cleaning up dead neutrophils and other cellular debris.

In addition to leaving the blood in response to an infection, macrophages are also found in uninfected tissues. If microorganisms enter uninfected tissue, the macrophages may phagocytize the microorganisms before they can replicate or cause damage. For example, macrophages are located at potential points where microorganisms may enter the body, such as beneath the skin and mucous membranes, and around blood and lymphatic vessels. They also protect lymph in lymph nodes and blood in the spleen and liver.

### Cells of Inflammation

**Basophils,** which are derived from red bone marrow, are motile white blood cells that can leave the blood and enter infected tissues. **Mast cells,** which are also derived from red bone marrow, are nonmotile cells in connective tissue, especially near capillaries. Like macrophages, mast cells are located at points where microorganisms may enter the body, such as the skin, lungs, gastrointestinal tract, and urogenital tract.

Basophils and mast cells can be activated through innate immunity (e.g., by complement) or through adaptive immunity (see "Adaptive Immunity" later in this chapter). When activated, they release chemicals, such as histamine and leukotrienes, that produce an inflammatory response or activate other mechanisms, such as smooth muscle contraction in the lungs.

**Eosinophils** are produced in red bone marrow, enter the blood, and within a few minutes enter tissues. Enzymes released by eosinophils break down chemicals released by basophils and mast cells. Thus, while inflammation is initiated, mechanisms are activated that contain and reduce the inflammatory response.

Inflammation is beneficial in the fight against microorganisms, but too much inflammation can be harmful, destroying healthy tissues as well as the microorganisms.

### Natural Killer Cells

**Natural killer (NK) cells,** a type of lymphocyte produced in red bone marrow, account for up to 15% of lymphocytes. NK cells recognize classes of cells, such as tumor cells or virus-infected cells in general, rather than specific tumor cells or cells infected by a specific virus. For this reason, and because NK cells do not exhibit a memory response, they are classified as part of innate immunity. NK cells use a variety of methods to kill their target cells, including releasing chemicals that damage cell membranes and cause the cells to lyse.

## Inflammatory Response

The **inflammatory response** to injury involves many of the chemicals and cells previously discussed. Most inflammatory responses are very similar, although some details vary, depending on the intensity of the response and the type of injury. Here, we use a bacterial infection to illustrate an inflammatory response (figure 14.8). Bacteria enter the tissue, causing damage that stimulates the release or activation of chemical mediators, such as histamine, prostaglandins, leukotrienes, complement, and kinins. The chemicals produce several effects: (1) Vasodilation increases blood flow and brings phagocytes and other white blood cells to the area; (2) phagocytes leave the blood and enter the tissue; and (3) increased vascular permeability allows fibrinogen and complement to enter the tissue from the blood. Fibrinogen is converted to fibrin (see chapter 11), which isolates the infection by walling off the infected area. Complement further enhances the inflammatory response and attracts additional

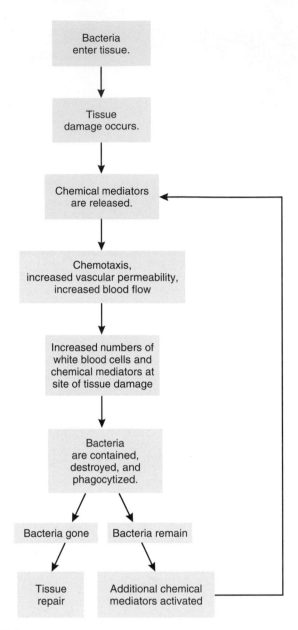

**Figure 14.8 Inflammatory Response**

Bacteria cause tissue damage and the release of chemical mediators that initiate inflammation and phagocytosis, resulting in the destruction of the bacteria. If any bacteria remain, additional chemical mediators are activated. After all the bacteria are destroyed, the tissue is repaired.

phagocytes. This process of releasing chemical mediators and attracting phagocytes and other white blood cells continues until the bacteria are destroyed. Phagocytes remove microorganisms and dead tissue, and the damaged tissues are repaired.

Inflammation can be local or systemic. **Local inflammation** is an inflammatory response confined to a specific area of the body. Symptoms of local inflammation include redness, heat, and swelling due to increased blood flow and increased vascular permeability, as well as pain caused by swelling and by chemical mediators acting on pain receptors. The tissue destruction, swelling, and pain lead to loss of function (see chapter 4).

**Systemic inflammation** is an inflammatory response that is generally distributed throughout the body. In addition to the local symptoms at the sites of inflammation, three additional features can be present:

1. Red bone marrow produces and releases large numbers of neutrophils, which promote phagocytosis.
2. **Pyrogens** (pī′rō-jenz; fever-producing), chemicals released by microorganisms, neutrophils, and other cells, stimulate fever production. Pyrogens affect the body's temperature-regulating mechanism in the hypothalamus in the brain. As a consequence, heat production and conservation increase, raising body temperature. Fever promotes the activities of the immune system, such as phagocytosis, and inhibits the growth of some microorganisms.
3. In severe cases of systemic inflammation, vascular permeability can increase so much that large amounts of fluid are lost from the blood into the tissues. The decreased blood volume can cause shock and death.

## ADAPTIVE IMMUNITY

Adaptive immunity exhibits specificity and memory. As explained earlier in this chapter, specificity is the ability to recognize a particular substance, and memory is the ability to respond with increasing effectiveness to successive exposures to the antigen. Substances that stimulate adaptive immune responses are called **antigens** (an′ti-jenz; *anti*(body) + -*gen*, producing). Antigens can be divided into two groups: foreign antigens and self-antigens.

**Foreign antigens** are introduced from outside the body. Microorganisms, such as bacteria and viruses, and chemicals released

| TABLE 14.1 | Comparison of Innate and Adaptive Immunity | |
|---|---|---|
| **Primary Cells** | **Origin of Cells** | **Site of Maturation** |
| **Innate Immunity** | | |
| Neutrophils, eosinophils, basophils, mast cells, monocytes, and macrophages | Red bone marrow | Red bone marrow (neutrophils, eosinophils, basophils, and monocytes) and tissues (mast cells and macrophages) |
| **Adaptive Immunity** | | |
| Antibody-mediated immunity (B cells) | Red bone marrow | Red bone marrow |
| Cell-mediated immunity (T cells) | Red bone marrow | Thymus |

by microorganisms are examples of foreign antigens. Pollen, animal hairs, foods, and drugs can cause an **allergic reaction** because they are foreign antigens that produce an overreaction of the immune system (see the Diseases and Disorders table at the end of this chapter). Transplanted tissues and organs contain foreign antigens, and the response to these antigens can cause rejection of the transplant.

**Self-antigens** are molecules the body produces to stimulate an immune system response. The response to self-antigens can be beneficial. For example, the recognition of tumor antigens can result in destruction of the tumor. But the response to self-antigens can also be harmful. **Autoimmune disease** results when self-antigens stimulate unwanted destruction of normal tissue. An example is rheumatoid arthritis, which destroys the tissue within joints.

Adaptive immunity can be divided into antibody-mediated immunity and cell-mediated immunity. **Antibody-mediated immunity** involves proteins called **antibodies** (an′tĕ-bod-ēz), which are found in the plasma. Antibodies are produced by a type of lymphocyte called **B cells. Cell-mediated immunity** involves the actions of a second type of lymphocyte, called **T cells.** Several subpopulations of T cells exist. For example, **cytotoxic** (sī-tō-tok′sik; destructive to cells) **T cells** produce the effects of cell-mediated immunity, and **helper T cells** can promote or inhibit the activities of both antibody-mediated immunity and cell-mediated immunity.

Table 14.1 summarizes and contrasts the main features of innate immunity and the two categories of adaptive immunity (i.e., antibody-mediated immunity and cell-mediated immunity).

## Origin and Development of Lymphocytes

To understand how lymphocytes are responsible for antibody-mediated and cell-mediated immunity, it is important to know how lymphocytes originate and become specialized immune cells. **Stem cells** in red bone marrow are capable of giving rise to all the blood cells (see figure 11.2). Some stem cells give rise to pre-T cells, which migrate through the blood to the thymus, where they divide and are processed into T cells (figure 14.9). Other stem cells produce pre-B cells, which are processed in the red bone marrow into B cells.

B cells are released from red bone marrow, and T cells are released from the thymus. Both types of cells move through the blood to lymphatic tissues (see figure 14.7). These lymphocytes

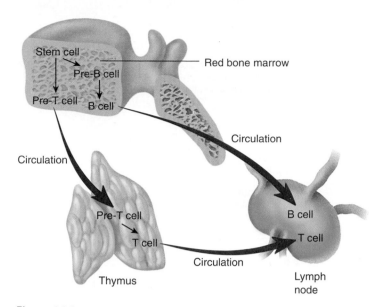

**Figure 14.9  Origin and Processing of B Cells and T Cells**

Both B cells and T cells originate from stem cells in red bone marrow. B cells are processed from pre-B cells in the red marrow, whereas T cells are processed from pre-T cells in the thymus. Both B cells and T cells circulate to other lymphatic tissues, such as lymph nodes.

live for a few months to many years and continually circulate between the blood and the lymphatic tissues. Normally, there are about five T cells for every B cell in the blood. When stimulated by an antigen, B cells and T cells divide, producing cells that are responsible for the destruction of antigens.

Evidence suggests that small groups of identical B cells or T cells, called **clones,** form during embryonic development. Each clone is derived from a single, unique B cell or T cell. Each clone can respond only to a particular antigen. However, there is such a large variety of clones that the immune system can react to most antigens. Among the antigens to which the clones can respond are self-antigens. Because this response could destroy the body's own cells, clones acting against self-antigens are normally eliminated or suppressed. Most of this process occurs during prenatal development, but it also continues after birth and throughout a person's lifetime.

| Location of Mature Cells | Primary Secretory Product | Primary Actions | Allergic Reactions |
|---|---|---|---|
| Blood, connective tissue, and lymphatic tissue | Histamine, complement, prostaglandins, leukotrienes, kinins, and interferons | Inflammatory response and phagocytosis | None |
| Blood and lymphatic tissue | Antibodies | Protection against extracellular antigens (bacteria, toxins, and viruses outside of cells) | Immediate hypersensitivity |
| Blood and lymphatic tissue | Cytokines | Protection against intracellular antigens (viruses and intracellular bacteria) and tumors; responsible for graft rejection | Delayed hypersensitivity |

Lymphatic

1 Antigen-presenting cells, such as macrophages, phagocytize, process, and display antigens on the cell's surface.

2 The antigens are bound to MHC molecules, which present the processed antigen to the T-cell receptor of the helper T cell.

3 Costimulation results from interleukin-1, secreted by the macrophage, and the CD4 glycoprotein of the helper T cell.

4 Interleukin-1 stimulates the helper T cell to secrete interleukin-2 and to produce interleukin-2 receptors.

5 The helper T cell stimulates itself to divide when interleukin-2 binds to interleukin-2 receptors.

6 The "daughter" helper T cells resulting from this division can be stimulated to divide again if they are exposed to the same antigen that stimulated the "parent" helper T cell. This greatly increases the number of helper T cells.

7 The increased number of helper T cells can facilitate the activation of B cells or effector T cells.

**PROCESS Figure 14.10  Proliferation of Helper T Cells**

An antigen-presenting cell (macrophage) stimulates helper T cells to divide.

## Activation and Multiplication of Lymphocytes

The specialized B-cell or T-cell clones can respond to antigens and produce an adaptive immune response. For the adaptive immune response to be effective, two events must occur: (1) antigen recognition by lymphocytes and (2) proliferation of the lymphocytes recognizing the antigen.

### Antigen Recognition

Lymphocytes have proteins, called **antigen receptors,** on their surfaces. The antigen receptors on B cells are called **B-cell receptors,** and those on T cells are called **T-cell receptors.** Each receptor binds with only a specific antigen. Each clone consists of lymphocytes that have identical antigen receptors on their surfaces. When antigens combine with the antigen receptors of a clone, the lymphocytes in that clone can be activated, and the adaptive immune response begins.

B cells and T cells typically recognize antigens after large molecules have been processed or broken down into smaller fragments. Antigen-presenting cells, such as macrophages, present antigens to B and T cells. The antigens are taken into macrophages by phagocytosis and are broken down into smaller antigen fragments. The processed antigen fragments are bound to major histocompatibility complex molecules, transported to the surface of the macrophages, and presented to B cells and T cells (figure 14.10, step 1).

**Major histocompatibility complex (MHC) molecules** are glycoproteins that have binding sites for antigens. Different MHC molecules have different binding sites—that is, they are specific for certain antigens. The MHC molecules function as "serving trays" that hold and present a processed antigen on the outer surface of the cell membrane. The combined MHC molecule and processed antigen can then bind to the antigen receptor on a B cell or T cell and stimulate it. For example, figure 14.10, step 2, illustrates how helper T cells are stimulated when combined with MHC class II molecules.

The MHC molecule/antigen combination is usually only the first signal necessary to produce a response from a B cell or T cell. In many cases, **costimulation** by a second signal is also required. Costimulation can be achieved by **cytokines** (sī'tō-kīnz), which are proteins or peptides secreted by one cell as a regulator of neighboring cells. For example, **interleukin-1** (in-ter-loo'kin) is a cytokine released by macrophages that can stimulate helper T cells (figure 14.10, step 3).

Lymphocytes have other surface molecules besides MHC molecules that help bind cells together and stimulate a response. For example, helper T cells have a glycoprotein called CD4, which helps connect helper T cells to the macrophage by binding to MHC molecules. The CD4 protein is also bound by the virus that causes AIDS (see the Diseases and Disorders table at the end of this chapter). As a result, the virus preferentially infects helper T cells. Cytotoxic T cells have a glycoprotein called CD8, which helps connect cytotoxic T cells to cells displaying MHC molecules.

## Lymphocyte Proliferation

Before exposure to a particular antigen, the number of helper T cells that can respond to that antigen is too small to produce an effective response against it. After the antigen is processed and presented to a helper T cell by a macrophage, the helper T cell responds by producing **interleukin-2** and **interleukin-2 receptors** (figure 14.10, step 4). Interleukin-2 binds to the receptors and stimulates the helper T cell to divide (figure 14.10, step 5). The "daughter" helper T cells produced by this division can again be presented with the antigen by macrophages and again be stimulated to divide. Thus, the number of helper T cells is greatly increased (figure 14.10, step 6)

It is important for the number of helper T cells to increase because helper T cells are necessary for the activation of most B cells or T cells (figure 14.10, step 7). For example, B cells have receptors that can recognize antigens. Most B cells, however, do not respond to antigens without stimulation from helper T cells.

B-cell proliferation begins when a B cell takes in the same kind of antigen that stimulated the helper T cell (figure 14.11, step 1). The antigen is processed by the B cell and presented on

1 Before a B cell can be activated by a helper T cell, the B cell must phagocytize and process the same antigen that activated the helper T cell. The antigen binds to a B-cell receptor, and both the receptor and the antigen are taken into the cell by endocytosis.

2 The B cell uses an MHC class II molecule to present the processed antigen to the helper T cell.

3 The T-cell receptor binds to the MHC class II/antigen complex.

4 There is costimulation of the B cell by CD4 and other surface molecules.

5 There is costimulation by interleukins (cytokines) released from the helper T cell.

6 The B cell divides, the resulting daughter cells divide, and so on, eventually producing many cells (only two are shown here) that recognize the same antigen.

7 Many of the daughter cells differentiate to become plasma cells, which produce antibodies. Antibodies are part of the immune response that eliminates the antigen.

**PROCESS Figure 14.11 Proliferation of B Cells**

A helper T cell stimulates a B cell to divide and produce antibodies.

the B-cell surface by an MHC class II molecule (figure 14.11, step 2). A helper T cell is stimulated when it binds to the MHC class II/antigen complex (figure 14.11, step 3). There is also costimulation involving CD4 and interleukins (figure 14.11, steps 4 and 5). As a result, the B cell divides into two "daughter" cells (figure 14.11, step 6). The division process continues, eventually producing many cells capable of producing antibodies (figure 14.11, step 7) and resulting in sufficient antibodies to destroy all the antigen.

### Predict 2

*How does elimination of the antigen stop the production of antibodies?*

## Antibody-Mediated Immunity

Exposure of the body to an antigen can lead to the activation of B cells and the production of antibodies. The antibodies bind to the antigens, which can be destroyed through several different mechanisms. Because antibodies are in body fluids, antibody-mediated immunity is effective against extracellular antigens, such as bacteria, viruses (when they are outside cells), and toxins. Antibody-mediated immunity is also involved in certain allergic reactions.

### Structure of Antibodies

Antibodies are proteins produced in response to an antigen. They are Y-shaped molecules consisting of four polypeptide chains: two identical heavy chains and two identical light chains (figure 14.12). The end of each "arm" of the antibody is the **variable region,** the part of the antibody that combines with the antigen. The variable region of a particular antibody can join only with a particular antigen; this is similar to the lock-and-key model of enzymes (see chapter 2). The rest of the antibody is the **constant region,** and it has several functions. For example, the constant region can activate complement, or it can attach the antibody to cells, such as macrophages, basophils, and mast cells.

Antibodies make up a large portion of the proteins in plasma. Most plasma proteins can be separated into albumin and alpha, beta, and gamma globulin portions (see chapter 11). Antibodies are sometimes called **gamma globulins** (glob'ū-linz), because they are found mostly in the gamma globulin

**Figure 14.12 Structure of an Antibody**

The Y-shaped antibody has two "arms." Each arm has a variable region that functions as an antigen-binding site. The constant region can activate complement or bind to other immune system cells, such as macrophages, basophils, or mast cells.

part of plasma, or **immunoglobulins (Ig),** because they are globulin proteins involved in immunity. The five general classes of antibodies are denoted IgG, IgM, IgA, IgE, and IgD (table 14.2).

### Effects of Antibodies

Antibodies can affect antigens either directly or indirectly. Direct effects occur when a single antibody binds to an antigen and inactivates the antigen, or when many antigens are bound together and are inactivated by many antibodies (figure 14.13a,b). The ability of antibodies to join antigens together is the basis for many clinical tests, such as blood typing, because when enough antigens are bound together, they form visible clumps.

Most of the effectiveness of antibodies results from indirect effects (figure 14.13c–e). After an antibody has attached by

Lymphatic

# CLINICAL IMPACT

## Use of Monoclonal Antibodies

A **monoclonal antibody** is a pure antibody preparation that is specific for only one antigen. Monoclonal antibodies are grown in laboratories and have many clinical uses. Monoclonal antibodies are produced by injecting an antigen into a laboratory animal to activate a B-cell clone against the antigen. These B cells are removed from the animal and fused with tumor cells. The resulting cells have two ideal characteristics: (1) They produce only one (mono) specific antibody because they are derived from one B-cell clone, and (2) they divide rapidly because they are derived from tumor cells. The result is many cells producing a specific antibody.

Monoclonal antibodies are used for determining pregnancy and for diagnosing diseases such as gonorrhea, syphilis, hepatitis, rabies, and cancer. These tests are specific and rapid because the monoclonal antibodies bind only to the antigen being tested. Monoclonal antibodies were initially hailed as "magic bullets" because of their potential for delivering drugs to cancer cells (see "Immunotherapy" later in this chapter). This potential is now being realized, as over 20 monoclonal antibodies are being used to treat diseases, including some types of cancer, and hundreds more monoclonal antibodies are in clinical trials.

| TABLE 14.2 | Classes of Antibodies and Their Functions | | |
|---|---|---|---|
| **Antibody** | **Total Serum Antibody (%)** | **Structure** | **Description** |
| IgG | 80–85 | IgG | Activates complement and increases phagocytosis; can cross the placenta and provide immune protection to the fetus and newborn; responsible for Rh reactions, such as hemolytic disease of the newborn |
| IgM | 5–10 | IgM / IgA | Activates complement and acts as an antigen-binding receptor on the surface of B cells; responsible for transfusion reactions in the ABO blood system; often the first antibody produced in response to an antigen |
| IgA | 15 | | Secreted into saliva, into tears, and onto mucous membranes to protect body surfaces; found in colostrum and milk to provide immune protection to the newborn |
| IgE | 0.002 | IgE / IgD — Heavy chain / Light chain | Binds to mast cells and basophils and stimulates the inflammatory response |
| IgD | 0.2 | | Functions as an antigen-binding receptor on B cells |

Lymphatic

its variable region to an antigen, the constant region of the antibody can activate other mechanisms that destroy the antigen. For example, the constant region of antibodies can activate complement, which stimulates inflammation, attracts white blood cells through chemotaxis, and lyses bacteria. When an antigen combines with the antibody, the constant region triggers the release of inflammatory chemicals from mast cells and basophils. For example, people who have hay fever inhale the antigens (usually plant pollens), which are then absorbed through the respiratory mucous membrane. The combination of the antigen with antibodies stimulates mast cells to release inflammatory chemicals, such as histamine. The resulting localized inflammatory response produces swelling and excess mucus production in the respiratory tract. Finally, macrophages can attach to the constant region of the antibody and phagocytize both the antibody and the antigen.

## Antibody Production

The production of antibodies after the first exposure to an antigen is different from that following a second or subsequent exposure. The **primary response** results from the first exposure of a B cell to

Antigen

(a) **Inactivate the antigen.** An antibody binds to an antigen and inactivates it.

Antibody

(b) **Bind antigens together.** Antibodies bind several antigens together.

(c) **Activate the complement cascade.** An antigen binds to an antibody. As a result, the antibody can activate complement proteins, which can produce inflammation, chemotaxis, and lysis.

Complement cascade activated → Inflammation, chemotaxis, lysis

(d) **Initiate the release of inflammatory chemicals.** An antibody binds to a mast cell or a basophil. When an antigen binds to the antibody, it triggers the release of chemicals that cause inflammation.

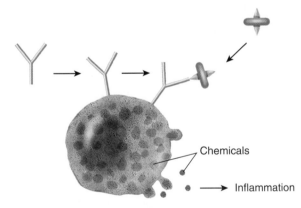

Chemicals

Inflammation

Mast cell or basophil

(e) **Facilitate phagocytosis.** An antibody binds to an antigen and then to a macrophage, which phagocytizes the antibody and antigen.

Macrophage

**Figure 14.13 Effects of Antibodies**

Antibodies directly affect antigens by inactivating the antigens or by binding the antigens together. Antibodies indirectly affect antigens by activating other mechanisms through the constant region of the antibody. Indirect mechanisms include activation of complement, increased inflammation resulting from the release of inflammatory chemicals from mast cells or basophils, and increased phagocytosis resulting from antibody attachment to macrophages.

Lymphatic

**1 Primary response.** The primary response occurs when a B cell is first activated by an antigen. The B cell proliferates to form plasma cells and memory cells. The plasma cells produce antibodies.

**2 Secondary response.** The secondary response occurs when another exposure to the same antigen causes the memory cells to rapidly form plasma cells and additional memory cells. The secondary response is faster and produces more antibodies than the primary response.

**PROCESS Figure 14.14** Antibody Production

an antigen (figure 14.14, step 1). When the antigen binds to the antigen-binding receptor on the B cell, the B cell undergoes several divisions to form plasma cells and memory B cells. **Plasma cells** produce antibodies. The primary response normally takes 3–14 days to produce enough antibodies to be effective against the antigen. In the meantime, the individual usually develops disease symptoms because the antigen has had time to cause tissue damage.

**Memory B cells** are responsible for the **secondary response,** or *memory response,* which occurs when the immune system is exposed to an antigen against which it has already produced a primary response (figure 14.14, step 2). When exposed to the antigen, the memory B cells quickly divide to form plasma cells, which rapidly produce antibodies. The secondary response provides better protection than the primary response for two reasons: (1) The time required to start producing antibodies is less (hours to a few days), and (2) more plasma cells and antibodies are produced. As a consequence, the antigen is quickly destroyed, no disease symptoms develop, and the person is immune.

The secondary response also includes the formation of new memory cells, which provide protection against additional exposures to a specific antigen. Memory cells are the basis of adaptive immunity. After destruction of the antigen, plasma cells die, the antibodies they released are degraded, and antibody levels decline to the point where they can no longer provide adequate protection. However, memory cells persist for many years—for life, in some cases. If memory cell production is not stimulated, or if the memory cells produced are short-lived, it is possible to have repeated infections of the same disease. For example, the same cold virus can cause the common cold more than once in the same person.

**Predict 3**

*One theory for long-lasting immunity assumes that humans are continually exposed to disease-causing agents. Explain how this exposure can produce lifelong immunity.*

## Cell-Mediated Immunity

Cell-mediated immunity is a function of cytotoxic T cells and is most effective against microorganisms that live inside body cells. Viruses and some bacteria are examples of intracellular microorganisms. Cell-mediated immunity is also involved with some allergic reactions, the control of tumors, and graft rejection.

① An MHC class I molecule displays an antigen, such as a viral protein, on the surface of a target cell.

② The activation of a cytotoxic T cell begins when the T-cell receptor binds to the MHC class I/antigen complex.

③ There is costimulation of the cytotoxic T cell by CD8 and other surface molecules.

④ There is costimulation by cytokines, such as interleukin-2, released from helper T cells.

⑤ The activated cytotoxic T cell divides, the resulting daughter cells divide, and so on, eventually producing many cytotoxic T cells (only two are shown here).

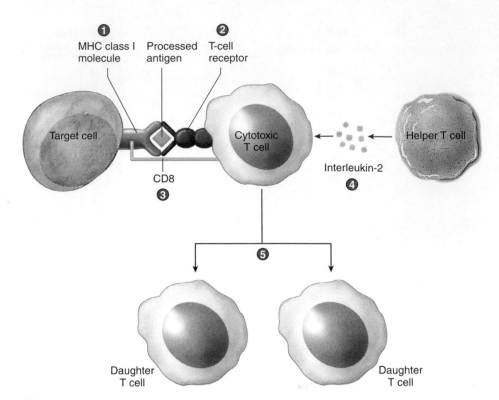

**PROCESS Figure 14.15** Proliferation of Cytotoxic T Cells

Cell-mediated immunity is essential for fighting viral infections. When viruses infect cells, they direct the cells to make new viruses, which are then released to infect other cells. Thus, cells are turned into virus-manufacturing plants. While inside the cell, viruses have a safe haven from antibody-mediated immunity because antibodies cannot cross the cell membrane. Cell-mediated immunity fights viral infections by destroying virally infected cells. When viruses infect cells, some viral proteins are broken down and become processed antigens that are combined with MHC class I molecules and displayed on the surface of the infected cell (figure 14.15, step 1). Cytotoxic T cells can distinguish between virally infected cells and noninfected cells because the T-cell receptor can bind to the MHC class I/viral antigen complex, which is not present on uninfected cells.

The T-cell receptor binding with the MHC class I/antigen complex is a signal for activating cytotoxic T cells (figure 14.15, step 2). Costimulation by other surface molecules, such as CD8, also occurs (figure 14.15, step 3). Helper T cells provide costimulation by releasing cytokines, such as interleukin-2, which stimulate activation and cell division of cytotoxic T cells (figure 14.15, step 4).

Increasing the number of "daughter" helper T cells (figure 14.15, step 5) results in greater stimulation of cytotoxic T cells. In cell-mediated responses, helper T cells are activated and stimulated to divide in the same fashion as in antibody-mediated responses (see figure 14.10).

After cytotoxic T cells are activated by an antigen on the surface of a target cell, they undergo a series of divisions to produce additional cytotoxic T cells and memory T cells (figure 14.16). The cytotoxic T cells are responsible for the cell-mediated immune response, and the **memory T cells** provide a secondary response and long-lasting immunity in the same fashion as memory B cells.

Cytotoxic T cells have two main effects:

1. They release cytokines that activate additional components of the immune system. For example, some cytokines attract innate immune cells, especially macrophages. These cells are then responsible for phagocytosis of the antigen and the production of an inflammatory response. Cytokines also activate additional cytotoxic T cells, which increases the effectiveness of the cell-mediated response.

2. Cytotoxic T cells can come in contact with other cells and kill them. Virus-infected cells have viral antigens, tumor cells have tumor antigens, and tissue transplants have foreign antigens that can stimulate cytotoxic T-cell activity. The cytotoxic T cells bind to the antigens on the surfaces of these cells and cause the cells to lyse.

## A CASE IN POINT
### Sjogren Syndrome

Ima Akin has a toothache, so she goes to her dentist, Dr. Dekay, for help. Dr. Dekay finds an abscessed tooth, several dental cavities, and enlarged parotid salivary glands, but little saliva production. On questioning, Ima tells the dentist she has been experiencing a dry mouth and dry eyes and she feels fatigued all the time. Dr. Dekay refers Ima to her family physician, Dr. Hurtt, for further testing. Eventually, the physician determines that Ima Akin has **Sjogren** (show′grin) **syndrome,** a systemic, autoimmune inflammatory disorder affecting glands and mucous membranes.

**Figure 14.16  Stimulation and Effects of T Cells**

When activated, cytotoxic T cells form many additional cytotoxic T cells, as well as memory T cells. The cytotoxic T cells release cytokines that promote the destruction of the antigen or cause the lysis of target cells, such as virus-infected cells, tumor cells, or transplanted cells. The memory T cells are responsible for the secondary response.

Innate, antibody-mediated, and cell-mediated immunity normally work together to protect us against foreign antigens. In autoimmune disorders, self-antigens activate immune responses, resulting in the destruction of healthy tissues. In Sjogren syndrome, damage to the salivary glands leads to decreased saliva production and a dry mouth, which increases the likelihood of developing cavities. Damage to the lacrimal glands results in decreased tear production and dry eyes, which damage the conjunctiva. Sjogren syndrome is one of the most common autoimmune disorders. About 50% of the time, it occurs alone, but the remainder of cases occur in conjunction with other autoimmune diseases, such as rheumatoid arthritis, systemic lupus erythematosus, and scleroderma. Nine out of 10 people with Sjogren syndrome are women.

# ACQUIRED IMMUNITY

There are four ways to acquire adaptive immunity: active natural, active artificial, passive natural, and passive artificial (figure 14.17). **Active immunity** results when an individual is exposed to an antigen (either naturally or artificially) and the response of the individual's own immune system is the cause of the immunity. **Passive immunity** occurs when another person or an animal develops immunity and the immunity is transferred to a nonimmune individual. *Natural* and *artificial* refer to the method of exposure. *Natural exposure* implies that contact with the antigen occurs as part of everyday living and is not deliberate. Artificial exposure is the deliberate introduction of an antigen or antibody into the body.

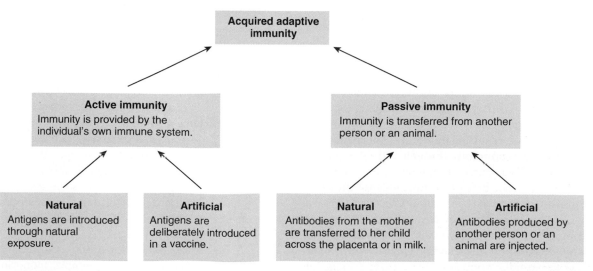

**Figure 14.17  Ways to Acquire Adaptive Immunity**

Lymphatic

## Active Natural Immunity

**Active natural immunity** results from natural exposure to an antigen, such as a disease-causing microorganism, that stimulates the immune system to respond against the antigen. Because the individual is not immune during the first exposure, he or she usually develops the symptoms of the disease.

## Active Artificial Immunity

In **active artificial immunity,** an antigen is deliberately introduced into an individual to stimulate the immune system. This process is called **vaccination** (vak′si-nā′shŭn), and the introduced antigen is a **vaccine** (vak′sēn, vak-sēn′). A vaccine is usually administered by injection. Examples of vaccinations are the DTP injection against diphtheria, tetanus, and pertussis (whooping cough) and the MMR injection against mumps, measles, and rubella (German measles).

The vaccine usually consists of part of a microorganism—either a dead microorganism or a live, altered one. The antigen has been changed so that it will stimulate an immune response but will not cause the disease symptoms. Because active artificial immunity produces long-lasting immunity without disease symptoms, it is the preferred method of acquiring adaptive immunity.

> **Predict 4**
>
> *Some vaccination procedures require a "booster" shot, another dose of the original vaccine given some time after the original dose. Why are booster shots administered?*

## Passive Natural Immunity

**Passive natural immunity** results when antibodies are transferred from a mother to her child across the placenta before birth. During her life, the mother has been exposed to many antigens, either naturally or artificially, and she has antibodies against many of these antigens, which protect her and the developing fetus against disease. Some of the antibodies (IgG) can cross the placenta and enter the fetal blood. Following birth, the antibodies protect the baby for the first few months. Eventually, the antibodies break down, and the baby must rely on its own immune system. If the mother nurses her baby, antibodies (IgA) in the mother's milk may also provide some protection for the baby.

## Passive Artificial Immunity

Achieving **passive artificial immunity** begins with vaccinating an animal, such as a horse. After the animal's immune system responds to the antigen, antibodies are removed from the animal and injected into the human requiring immunity. Alternatively, a human who has developed immunity through natural exposure or vaccination can serve as a source of antibodies. Passive artificial immunity provides immediate protection because the antibodies either directly or indirectly destroy the antigen. Passive artificial immunity is therefore the preferred treatment when not enough time is available for the individual to develop his or her own active immunity. However, the technique provides only temporary immunity because the antibodies are used or eliminated by the recipient.

Antibodies that provide passive artificial immunity are referred to by the general term **antiserum** because the antibodies are found in serum, which is plasma minus the clotting factors. Antisera are available against microorganisms that cause disease, such as rabies, hepatitis, and measles; bacterial toxins, such as those that cause tetanus, diphtheria, and botulism; and venoms from poisonous snakes and spiders.

# OVERVIEW OF IMMUNE INTERACTIONS

Although the immune system can be described in terms of innate, antibody-mediated, and cell-mediated immunity, these categories are artificial divisions used to emphasize particular aspects of immunity. In actuality, there is only one immune system, but its responses often involve components of more than one type of immunity. For example, although adaptive immunity can recognize and remember specific antigens, once recognition has occurred the antigen is destroyed with the help of many innate immunity activities, including inflammation and phagocytosis (figure 14.18; see table 14.1).

# IMMUNOTHERAPY

Knowledge of how the immune system operates has produced two fundamental benefits: (1) an understanding of the cause and progression of many diseases and (2) the development or proposed development of methods to prevent, stop, or even reverse diseases.

**Immunotherapy** treats disease by altering immune system function or by directly attacking harmful cells. Some approaches attempt to boost immune system function in general. For example, administering cytokines or other agents can promote inflammation and activate immune cells, which can help destroy tumor cells. On the other hand, sometimes inhibiting the immune system is helpful. For example, multiple sclerosis is an autoimmune disease in which the immune system treats self-antigens as foreign antigens, destroying the myelin that covers axons. The cytokine interferon beta blocks the expression of MHC molecules that display self-antigens and is used to treat multiple sclerosis.

Some immunotherapy takes a more specific approach. For example, vaccination can prevent many diseases (see "Acquired Immunity"). The ability to produce monoclonal antibodies can result in effective therapies for treating tumors. If an antigen unique to tumor cells can be found, then monoclonal antibodies can be used to deliver radioactive isotopes, drugs, toxins, enzymes, or cytokines that kill the tumor cell directly or activate the immune system to kill the cell. Unfortunately, so far researchers have found no antigen on tumor cells that is not also present on normal cells. Nonetheless, this approach can be useful if damage to normal cells is minimal.

Some uses of monoclonal antibodies to treat tumors are yielding promising results. For example, monoclonal antibodies with radioactive iodine ($^{131}$I) have been found to cause the regression

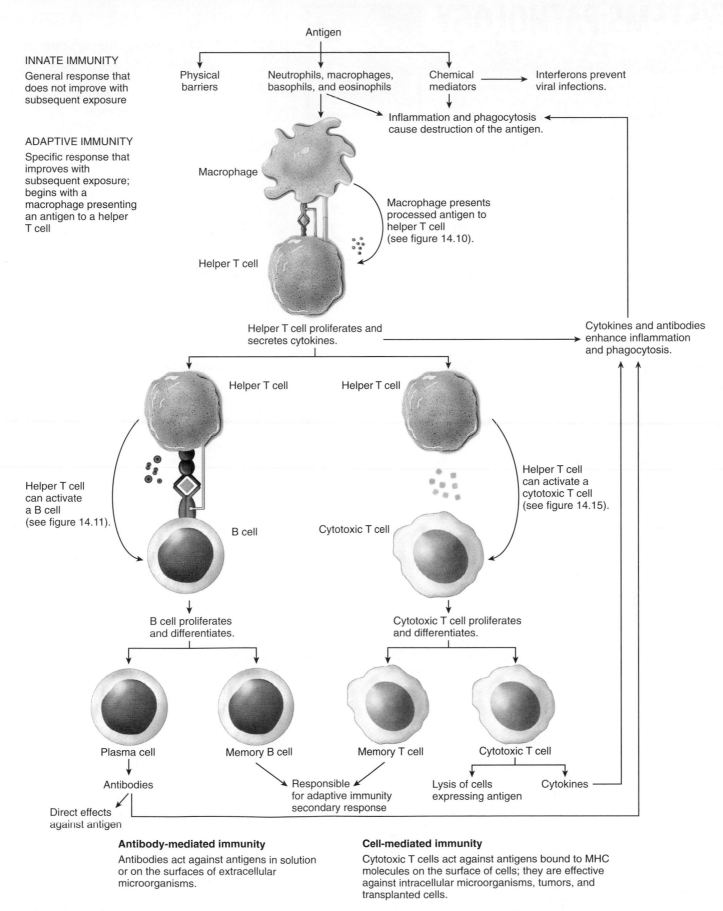

Antigen

**INNATE IMMUNITY**
General response that does not improve with subsequent exposure

Physical barriers

Neutrophils, macrophages, basophils, and eosinophils

Chemical mediators → Interferons prevent viral infections.

Inflammation and phagocytosis cause destruction of the antigen.

**ADAPTIVE IMMUNITY**
Specific response that improves with subsequent exposure; begins with a macrophage presenting an antigen to a helper T cell

Macrophage

Macrophage presents processed antigen to helper T cell (see figure 14.10).

Helper T cell

Helper T cell proliferates and secretes cytokines.

Cytokines and antibodies enhance inflammation and phagocytosis.

Helper T cell

Helper T cell

Helper T cell can activate a B cell (see figure 14.11).

Helper T cell can activate a cytotoxic T cell (see figure 14.15).

B cell

Cytotoxic T cell

B cell proliferates and differentiates.

Cytotoxic T cell proliferates and differentiates.

Plasma cell

Memory B cell

Memory T cell

Cytotoxic T cell

Antibodies

Responsible for adaptive immunity secondary response

Lysis of cells expressing antigen

Cytokines

Direct effects against antigen

**Antibody-mediated immunity**
Antibodies act against antigens in solution or on the surfaces of extracellular microorganisms.

**Cell-mediated immunity**
Cytotoxic T cells act against antigens bound to MHC molecules on the surface of cells; they are effective against intracellular microorganisms, tumors, and transplanted cells.

**Figure 14.18 Immune Interactions**

The major interactions and responses of innate and adaptive immunity to an antigen.

Lymphatic

## Systemic Lupus Erythematosus

Lucy is a 30-year-old divorced woman with two children. Even though she had to work to support herself and the children, she entered college, determined to become a nurse and provide a better life for her family. Lucy was an excellent student, but her class attendance and her performance on tests were somewhat erratic. Sometimes she was very energetic and earned high grades, but other times she seemed depressed and did not do as well. Toward the end of the course, she developed a rash on her face and a large red lesion on her arm, and she was obviously not feeling well.

Lucy asked the instructor if she could receive an incomplete grade and take the last exam at a later time. She explained that she had been diagnosed with lupus at the age of 25. Normally, medication controlled her symptoms, but the stress of being a single parent combined with the challenges of school seemed to be making her condition worse. She further explained that the symptoms of lupus come and go, and bedrest is often helpful. Lucy finished the course requirements later that summer. She went on to complete her education and now works full-time as a nurse at a local hospital.

### Background Information

**Systemic lupus erythematosus (SLE)** (lū′pŭs er-i-thē-mǎ-tō′sŭs) is a disease in which tissues and cells are damaged by the immune system. The name describes the skin rash that is characteristic of the disease (figure 14A). The term *lupus* literally means "wolf" and originally referred to eroded (as if gnawed by a wolf) lesions of the skin. *Erythematosus* refers to redness of the skin resulting from inflammation. Unfortunately, as the term *systemic* implies, the disorder is not confined to the skin but can affect tissues and cells throughout the body. Another systemic effect is the presence of low-grade fever in most cases of active SLE.

SLE is an autoimmune disorder in which a large variety of antibodies are produced that recognize self-antigens, such as nucleic acids, phospholipids, coagulation factors, red blood cells, and platelets. The combination of the antibodies with self-antigens forms immune complexes that circulate throughout the body and are deposited in various tissues, where they stimulate inflammation and tissue destruction.

**Figure 14A**

Inflammation in the skin caused by systemic lupus erythematosus results in the characteristic "butterfly rash" on the face.

Thus, SLE can affect many body systems. For example, the most common antibodies act against DNA released from damaged cells. Normally, the liver removes the DNA, but when DNA and antibodies form immune complexes, they tend to be deposited in the kidneys and other tissues. Approximately 40–50% of individuals with SLE develop renal disease. In some cases, the antibodies can bind to antigens on cells, causing the cells to lyse. For example, antibodies binding to red blood cells cause hemolysis and anemia.

The cause of SLE is unknown. The most popular hypothesis suggests that a viral infection disrupts the function of the T cells that normally prevent an immune response to self-antigens. Genetic factors probably contribute to the development of the disease. The likelihood of developing SLE is much higher if a family member also has it.

Approximately 1 out of 2000 individuals in the United States has SLE. The first symptoms usually appear between 15 and 25 years of age, affecting women approximately nine times as often as men. The progress of the disease is unpredictable, with flare-ups followed by periods of remission. The survival after diagnosis is greater than 90% after 10 years. The most frequent causes of death are kidney failure, CNS dysfunction, infections, and cardiovascular disease.

---

of B-cell lymphomas while producing few side effects. Herceptin, a monoclonal antibody, binds to a growth factor receptor that is overexpressed in 25–30% of primary breast cancers. The antibody "tags" cancer cells, which are then lysed by natural killer cells. Herceptin slows disease progression and increases survival time, but it is not a cure for breast cancer.

Many other immunotherapy approaches are being studied, and more treatments that use the immune system are sure to develop in the future.

## EFFECTS OF AGING ON THE LYMPHATIC SYSTEM AND IMMUNITY

Aging appears to have little effect on the lymphatic system's ability to remove fluid from tissues, absorb fats from the digestive tract, or remove defective red blood cells from the blood. However, aging has a severe impact on the immune system.

| System Interactions | Effects of Systemic Lupus Erythematosus on Other Systems |
|---|---|
| **System** | **Interactions** |
| Integumentary | Skin lesions frequently occur and are made worse by exposure to the sun. Skin lesions take three forms: (1) an inflammatory redness that can be in the shape of a butterfly, extending from the bridge of the nose to the cheeks; (2) small, localized, pimplelike eruptions accompanied by scaling of the skin; (3) areas of atrophied, depigmented skin with borders of increased pigmentation. Hair loss results in diffuse thinning of the hair. |
| Skeletal | Arthritis, tendinitis, and death of bone tissue can develop. |
| Muscular | Destruction of muscle tissue and muscular weakness occur. |
| Nervous | Memory loss, intellectual deterioration, disorientation, psychosis, reactive depression, headache, seizures, nausea, and loss of appetite can occur. Stroke is a major cause of dysfunction and death. Cranial nerve involvement results in facial muscle weakness, drooping of the eyelid, and double vision. Central nervous system lesions can cause paralysis. |
| Endocrine | Sex hormones may play a role in SLE because 90% of the cases occur in females, and females with SLE have reduced levels of androgens. |
| Cardiovascular | Inflammation of the pericardium (pericarditis) with chest pain can develop. Damage to heart valves, inflammation of cardiac tissue, tachycardia, arrhythmias, angina, and myocardial infarction also occur. Hemolytic anemia and leukopenia can be present (see chapter 11). Antiphospholipid antibody syndrome, through an unknown mechanism, increases coagulation and thrombus formation, which increases the risk of stroke and heart attack. |
| Respiratory | Chest pain may be caused by inflammation of the pleural membranes; fever, shortness of breath, and hypoxemia may occur due to inflammation of the lungs; alveolar hemorrhage can develop. |
| Digestive | Ulcers develop in the oral cavity and pharynx. Abdominal pain and vomiting are common, but no cause can be found. Inflammation of the pancreas and occasionally enlargement of the liver and minor abnormalities in liver function tests occur. |
| Urinary | Renal lesions and glomerulonephritis can result in progressive failure of kidney functions. Excess proteins are lost in the urine, resulting in lower than normal blood proteins, which can produce edema. |

There is no cure for SLE, nor is there one standard of treatment because the course of the disease is highly variable, and patient histories differ widely. Treatment usually begins with mild medications and proceeds to increasingly potent therapies as conditions warrant. Aspirin and nonsteroidal anti-inflammatory drugs are prescribed to suppress inflammation. Antimalarial drugs are used to treat skin rash and arthritis in SLE, but the mechanism of action is unknown. Patients who do not respond to these drugs or who have severe SLE are helped by glucocorticoids. Although these steroids effectively suppress inflammation, they can produce undesirable side effects, including suppression of normal adrenal gland functions. In patients with life-threatening SLE, very high doses of glucocorticoids are used.

### Predict 5

*The red lesion Lucy developed on her arm is called purpura (pŭr′pū-rӑ), and it is caused by bleeding into the skin. The lesions gradually change color and disappear in 2 or 3 weeks. Explain how SLE produces purpura.*

With age, people eventually lose the ability to produce new, mature T cells in the thymus. By age 40, much of the thymus has been replaced with adipose tissue, and after age 60, the thymus decreases in size to the point that it can be difficult to detect. While the number of T cells remains stable in most individuals due to the replication (not maturation) of T cells in lymphatic tissues, the T cells are less functional. In many individuals, the ability of helper T cells to proliferate in response to antigens decreases. Thus, antigen exposure produces fewer helper T cells, which results in less stimulation of B cells and cytotoxic T cells. Consequently, both antibody-mediated immunity and cell-mediated immunity responses to antigens decrease.

Both primary and secondary antibody responses decrease with age. More antigen is required to produce a response, the response is slower, less antibody is produced, and fewer memory cells result. Thus, a person's ability to resist infections and develop immunity decreases.

The ability of cell-mediated immunity to resist intracellular pathogens also decreases with age. For example, the elderly are

# DISEASES AND DISORDERS Lymphatic System

| CONDITION | DESCRIPTION |
|---|---|
| | **LYMPHATIC SYSTEM** |
| **Infections** | |
| Lymphadenitis (lim-fad′ē-nī′tis) | Inflammation of the lymph nodes; nodes become enlarged and tender as microorganisms are trapped and destroyed |
| Lymphangitis (lim-fan-jī′tis) | Inflammation of the lymphatic vessels; often results in visible red streaks in the skin that extend from the site of infection |
| Bubonic (bū-bon′ik) plague | Enlarged lymph nodes caused by bacterial infection (transferred by flea bites from rats); without treatment, bacteria enter blood and death occurs rapidly due to septicemia (see chapter 11); known as the Black Death in the Middle Ages |
| **Lymphedema (limf′e-de′ma)** | |
| Abnormal accumulation of lymph in tissues, often the limbs; 70%–90% cases in women; can be caused by developmental defects, disease, or damage to the lymphatic system | |
| Elephantiasis (el-ĕ-fan-tī′ă-sis) | Caused by long, slender roundworms transferred to humans by mosquito bites; adult worms lodge in lymphatic vessels and block lymph flow, so that a limb can become permanently swollen and enlarged; major cause of lymphedema worldwide |
| Lymphedema following cancer treatment | Caused by removal of lymph nodes near a tumor by surgery or radiation therapy; sentinel lymph nodes (those nearest the tumor) are first examined for cancer cells |
| **Lymphoma (lim-fō′mă)** | Cancer of lymphocytes that often begins in lymph nodes; immune system becomes depressed, with increased susceptibility to infections |
| | **IMMUNE SYSTEM** |
| **Immediate Allergic Reactions** | |
| Symptoms occur within a few minutes of exposure to an antigen because antibodies are already present from prior exposure. | |
| Hay fever | Often caused by inhalation of plant pollen antigens |
| Asthma (az′mă) | Antigen combines with antibodies on mast cells or basophils in the lungs, which then release inflammatory chemicals that cause constriction of the air tubes, so that the patient has trouble breathing |
| Urticaria (er′ti-kar′i-á) | Skin rash or localized swelling; can be caused by an ingested antigen; also called hives |
| Anaphylaxis (an′ă -fĭ-lak′sis, an′ă-fī-lak′sis) | Systemic allergic reaction, often resulting from insect stings or drugs such as penicillin; chemicals released from mast cells and basophils cause systemic vasodilation, increased vascular permeability, drop in blood pressure, and possibly death |
| **Delayed Allergic Reactions** | |
| Symptoms occur in hours to days following exposure to the antigen because these types of reactions involve migration of T cells to the antigen, followed by release of cytokines. | |
| Poison ivy and poison oak | Antigen absorbed by epithelial cells, which are then destroyed by T cells, causing inflammation and tissue destruction; itching can be intense |
| **Autoimmune Diseases** | |
| Similar to allergic reactions, except that the immune system incorrectly treats self-antigens as foreign antigens. Many types of autoimmune diseases exist, including type 1 diabetes, celiac disease, rheumatoid arthritis, multiple sclerosis, systemic lupus erythematosus, and Graves disease. | |
| **Congenital Immunodeficiencies** | |
| Usually involve failure of the fetus to form adequate numbers of B cells, T cells, or both | |
| Severe combined immunodeficiency (SCID) | Both B cells and T cells fail to form; unless patient kept in a sterile environment or provided with a compatible bone marrow transplant, death from infection results |
| **Acquired Immunodeficiencies** | |
| Many causes—for example, diseases, stress, or drugs | |
| Acquired immunodeficiency syndrome (AIDS) | Life-threatening disease caused by the human immunodeficiency virus (HIV); HIV is transmitted in body fluids; infection begins when the virus binds to the CD4 protein found primarily on helper T cells; without helper T cells, cytotoxic T-cell and B-cell activation is impaired, and adaptive immunity is suppressed; course of HIV infection varies, with most people surviving 10 or more years |
| Transplanted tissue rejection | Caused by a normal immune response to foreign antigens encoded by the major histocompatibility complex genes, also celled human leukocyte antigen (HLA) genes; drugs that suppress the immune system must be administered for life to prevent graft rejection |

Go to **www.mhhe.com/seeleyess7** for additional information on these pathologies.

more susceptible to influenza (flu) and should be vaccinated every year. Some pathogens cause disease but are not eliminated from the body; with age, decreased immunity can lead to reactivation of the pathogen. An example is the virus that causes chicken pox in children, which can remain latent within nerve cells, even though the disease seems to have disappeared. Later in life, the virus can leave the nerve cells and infect skin cells, causing painful lesions known as herpes zoster, or shingles.

Autoimmune disease occurs when immune responses destroy otherwise healthy tissue. There is very little increase in the number of new-onset autoimmune diseases in the elderly. However, the chronic inflammation and immune responses that began earlier in life have a cumulative, damaging effect. Likewise, the increased incidence of cancer in the elderly is likely to be caused by a combination of repeated exposure and damage from cancer-causing agents and decreased immunity.

 # SUMMARY

## Functions of the Lymphatic System    (p. 380)

The lymphatic system maintains fluid balance in tissues, absorbs fats from the small intestine, and defends against microorganisms and foreign substances.

## Anatomy of the Lymphatic System    (p. 380)

The lymphatic system consists of lymph, lymphocytes, lymphatic vessels, lymph nodes, tonsils, the spleen, and the thymus.

### Lymphatic Capillaries and Vessels

1. Lymphatic vessels carry lymph away from tissues. Valves in the vessels ensure the one-way flow of lymph.
2. Skeletal muscle contraction, contraction of lymphatic vessel smooth muscle, and thoracic pressure changes move the lymph through the vessels.
3. The thoracic duct and right lymphatic duct empty lymph into the blood.

### Lymphatic Organs

1. Lymphatic tissue produces lymphocytes when exposed to foreign substances, and it filters lymph and blood.
2. The tonsils protect the openings between the nasal and oral cavities and the pharynx.
3. Lymph nodes, located along lymphatic vessels, filter lymph.
4. The white pulp of the spleen responds to foreign substances in the blood, whereas the red pulp phagocytizes foreign substances and worn-out red blood cells. The spleen also functions as a reservoir for blood.
5. The thymus processes lymphocytes that move to other lymphatic tissue to respond to foreign substances.

### Overview of the Lymphatic System

The lymphatic system removes fluid from tissues, absorbs fats from the small intestine, and produces B cells and T cells, which are responsible for much of immunity.

## Immunity    (p. 386)

1. Immunity is the ability to resist the harmful effects of microorganisms and other foreign substances.
2. Immunity is classified as innate or adaptive.

## Innate Immunity    (p. 386)

### Physical Barriers

1. The skin and mucous membranes are barriers that prevent microorganisms from entering the body.
2. Tears, saliva, and urine wash away microorganisms.

### Chemical Mediators

1. Chemical mediators kill microorganisms, promote phagocytosis, and increase inflammation.
2. Lysozyme in tears and complement in plasma are examples of chemicals involved in innate immunity.
3. Interferons prevent the replication of viruses.

### White Blood Cells

1. Chemotaxis is the ability of cells to move toward microorganisms or sites of tissue damage.
2. Neutrophils are the first phagocytic cells to respond to microorganisms.
3. Macrophages are large phagocytic cells that are active in the latter part of an infection. Macrophages are positioned at sites where microorganisms may enter tissues.
4. Basophils and mast cells promote inflammation, whereas eosinophils inhibit inflammation.
5. Natural killer cells lyse tumor cells and virus-infected cells.

### Inflammatory Response

1. Chemical mediators cause vasodilation and increase vascular permeability, allowing chemicals to enter damaged tissues. Chemicals also attract phagocytes.
2. The amount of chemical mediators and phagocytes increases until the cause of the inflammation is destroyed. Then the tissues undergo repair.
3. Local inflammation produces redness, heat, swelling, pain, and loss of function. Symptoms of systemic inflammation include an increase in neutrophil numbers, fever, and shock.

## Adaptive Immunity    (p. 388)

1. Antigens are molecules that stimulate adaptive immunity.
2. B cells are responsible for antibody-mediated immunity; T cells are involved with cell-mediated immunity.

### Origin and Development of Lymphocytes

1. B cells and T cells originate in red bone marrow. T cells are processed in the thymus, and B cells are processed in red bone marrow.
2. B cells and T cells move to lymphatic tissue from their processing sites. They continually circulate from one lymphatic tissue to another.

### Activation and Multiplication of Lymphocytes

1. B cells and T cells have antigen receptors on their surfaces. Clones are lymphocytes with the same antigen receptor.
2. Major histocompatibility complex (MHC) molecules present processed antigens to B or T cells.

Lymphatic

3. Costimulation by cytokines, such as interleukins, and surface molecules, such as CD4, are required in addition to MHC molecules.
4. Macrophages present processed antigens to helper T cells, which divide and increase in number.
5. Helper T cells stimulate B cells to divide and differentiate into cells that produce antibodies.

### Antibody-Mediated Immunity

1. Antibodies are proteins. The variable region combines with antigens and is responsible for antibody specificity. The constant region activates complement or attaches the antibody to cells. The five classes of antibodies are IgG, IgM, IgA, IgE, and IgD.
2. Antibodies directly inactivate antigens or cause them to clump together. Antibodies indirectly destroy antigens by promoting phagocytosis and inflammation.
3. The primary response results from the first exposure to an antigen. B cells form plasma cells, which produce antibodies, and memory B cells.
4. The secondary (memory) response results from exposure to an antigen after a primary response. Memory B cells quickly form plasma cells and new memory B cells.

### Cell-Mediated Immunity

1. Exposure to an antigen activates cytotoxic T cells and produces memory T cells.
2. Cytotoxic T cells lyse virus-infected cells, tumor cells, and tissue transplants. Cytotoxic T cells produce cytokines, which promote inflammation and phagocytosis.

## Acquired Immunity   (p. 397)

1. Active natural immunity results from everyday exposure to an antigen against which the person's own immune system mounts a response.
2. Active artificial immunity results from deliberate exposure to an antigen (vaccine) to which the person's own immune system responds.
3. Passive natural immunity is the transfer of antibodies from a mother to her fetus during gestation or to her baby during breastfeeding.
4. Passive artificial immunity is the transfer of antibodies from an animal or another person to a person requiring immunity.

## Overview of Immune Interactions   (p. 398)

Innate immunity, antibody-mediated immunity, and cell-mediated immunity can function together to eliminate an antigen.

## Immunotherapy   (p. 398)

Immunotherapy stimulates or inhibits the immune system to treat diseases.

## Effects of Aging on the Lymphatic System and Immunity   (p. 400)

1. Aging has little effect on the lymphatic system's ability to remove fluid from tissues, absorb fats from the digestive tract, or remove defective red blood cells from the blood.
2. Decreased helper T-cell proliferation results in decreased antibody-mediated and cell-mediated immune responses.
3. The primary and secondary antibody responses decrease with age.
4. The ability to resist intracellular pathogens decreases with age.

 ## REVIEW AND COMPREHENSION

1. List the parts of the lymphatic system, and describe the three main functions of the lymphatic system.
2. What is the function of the valves in lymphatic vessels? What causes lymph to move through lymphatic vessels?
3. Which parts of the body are drained by the right lymphatic duct and which by the thoracic duct?
4. Describe the cells and fibers of lymphatic tissue, and explain the functions of lymphatic tissue.
5. Name the three groups of tonsils. What is their function?
6. Where are lymph nodes found? What is the function of the germinal centers within lymph nodes?
7. Where is the spleen located? What is the function of white pulp and red pulp within the spleen? What other function does the spleen perform?
8. Where is the thymus located, and what function does it perform?
9. What is the difference between innate immunity and adaptive immunity?
10. How do physical barriers and chemical mediators provide protection against microorganisms? Describe the effects of complement and interferons.
11. Describe the functions of the two major phagocytic cell types of the body. What is the mononuclear phagocytic system?
12. Name the cells involved in promoting and inhibiting inflammation.
13. What protective function do natural killer cells perform?
14. Describe the effects that take place during an inflammatory response. What are the symptoms of local and systemic inflammation?
15. Define antigen. What is the difference between a self-antigen and a foreign antigen?

16. Which cells are responsible for antibody-mediated immunity and for cell-mediated immunity?
17. Describe the origin and development of B cells and T cells.
18. What is the function of antigen receptors and major histocompatibility proteins?
19. What is costimulation? Give an example.
20. Describe the process by which an antigen can cause an increase in helper T-cell numbers.
21. Describe the process by which helper T cells can stimulate B cells to divide, differentiate, and produce antibodies.
22. What are the functions of the variable and constant regions of an antibody?
23. Describe the direct and indirect ways that antibodies destroy antigens.
24. What are the functions of plasma cells and memory B cells?
25. Define primary and secondary responses. How do they differ from each other in regard to speed of response and amount of antibody produced?
26. Explain how cytotoxic T cells are activated.
27. What are the functions of cytotoxic T cells and memory T cells?
28. Define active natural, active artificial, passive natural, and passive artificial immunity. Give an example of each.
29. What effect does aging have on the major functions of the lymphatic system?
30. Describe the effects of aging on B cells and T cells. Give examples of how this affects antibody-mediated and cell-mediated immunity responses.

Lymphatic

# CRITICAL THINKING

1.  A patient is suffering from edema in the right lower limb. Explain why elevation and massage of the limb help remove the excess fluid.

2.  If the thymus of an experimental animal is removed immediately following birth, the animal exhibits the following characteristics:
    a.  increased susceptibility to infections
    b.  decreased numbers of lymphocytes
    c.  greatly decreased ability to reject grafts
    Explain these observations.

3.  Adjuvants are substances that slow, but do not stop, the release of an antigen from an injection site into the blood. Suppose injection A of a given amount of antigen is given without an adjuvant and injection B of the same amount of antigen is given with an adjuvant that caused the release of antigen over a period of 2 to 3 weeks. Does injection A or injection B result in the greater amount of antibody production? Explain.

4.  Compare how long active immunity and passive immunity last. Explain the difference between the two types of immunity. In what situations is one type preferred over the other?

5.  Tetanus is caused by bacteria (*Clostridium tetani*) that enter the body through wounds in the skin. The bacteria produce a toxin that causes spastic muscle contractions. Death often results from failure of the respiratory muscles. A patient goes to the emergency room after stepping on a nail. If the patient has been vaccinated against tetanus, he is given a tetanus booster shot, which consists of the toxin altered so that it is harmless. If the patient has never been vaccinated against tetanus, he is given an antiserum shot against tetanus. Explain the rationale for this treatment strategy. Sometimes both a booster and an antiserum shot are given, but at different locations on the body. Explain why this is done and why the shots are given in different locations.

6.  A child appears healthy until approximately 9 months of age. Then the child develops severe bacterial infections, one after another. Fortunately, the infections are successfully treated with antibiotics. When infected with measles and other viral diseases, the child recovers without difficulty. Explain.

7.  A patient had many allergic reactions. As part of the treatment scheme, physicians decided to try to identify the allergens that stimulated the allergic reactions. A series of solutions, each containing a common allergen, was composed. Each solution was then injected into the skin at different locations on the patient's back. The following results were obtained:
    a.  Within a few minutes, one injection site became red and swollen.
    b.  At another injection site, swelling and redness did not appear until 2 days later.
    c.  No redness or swelling developed at the other sites.

    Explain what happened for each observation and what caused the redness and swelling.

8.  Ivy Hurtt developed a poison ivy rash after a camping trip. Her doctor prescribed a cortisone ointment to relieve the inflammation. A few weeks later, Ivy scraped her elbow, which became inflamed. Because she had some leftover cortisone ointment, she applied it to the scrape. Was the ointment a good idea for the poison ivy? Was it an appropriate treatment for the scrape?

9.  Suzy Withitt has just had her ears pierced. To her dismay, she finds that wearing inexpensive (but tasteful) jewelry causes an inflammatory (allergic) reaction to the metal in the jewelry by the end of the day. Is this because of antibodies or cytokines?

*Answers in Appendix D*

Visit this textbook's website at **www.mhhe.com/seeleyess7** for practice quizzes, animations, interactive learning exercises, and other study tools.

McGraw-Hill offers a study CD that features interactive cadaver dissection. *Anatomy & Physiology Revealed* includes cadaver photos that allow you to peel away layers of the human body to reveal structures beneath the surface. This program also includes animations, radiologic imaging, audio pronunciations, and practice quizzing.

Lymphatic

The man in this photo is putting a heavy load on his respiratory system. The increased demand from jogging is met by an increased breathing rate. But notice the mountains in the background. Simply being at high altitude can cause an increased breathing rate; add jogging to that and the respiratory system is doing double-duty. No wonder many elite athletes train in high-altitude locales.

# Chapter 15

# Respiratory System

## Chapter Outline and Objectives

### Functions of the Respiratory System (p. 407)

1. Describe the functions of the respiratory system.

### Anatomy of the Respiratory System (p. 407)

2. Describe the anatomy of the respiratory passages, beginning at the nose and ending with the alveoli.
3. Describe the anatomy of the lungs, and define respiratory membrane.

### Ventilation and Respiratory Volumes (p. 416)

4. Explain how contraction of the muscles of respiration causes changes in thoracic volume during quiet breathing and during labored breathing.
5. Describe the changes in alveolar pressure that are responsible for moving air into and out of the lungs.
6. Explain how surfactant and pleural pressure prevent the lungs from collapsing and how changes in pleural pressure cause alveolar volume to change.
7. List the respiratory volumes and capacities, and define each.

### Gas Exchange (p. 421)

8. Explain the factors that affect gas movement through the respiratory membrane.

9. Describe the partial pressure gradients for oxygen and carbon dioxide.

### Gas Transport in the Blood (p. 424)

10. Explain how the blood transports oxygen and carbon dioxide.

### Rhythmic Breathing (p. 424)

11. Describe the respiratory areas of the brainstem and how they produce a rhythmic pattern of ventilation.
12. Name the neural mechanisms that can modify the normal rhythmic pattern of ventilation.
13. Explain how alterations in blood pH, carbon dioxide, and oxygen levels affect ventilation.

### Respiratory Adaptations to Exercise (p. 430)

14. Describe the regulation of ventilation during exercise and the changes in the respiratory system that result from exercise training.

### Effects of Aging on the Respiratory System (p. 433)

15. Describe the effects of aging on the respiratory system.

Studying, sleeping, talking, eating, and exercising all involve breathing. From our first breath at birth, the rate and depth of our breathing are unconsciously matched to our activities. Although we can voluntarily stop breathing, within a few minutes we must breathe again. Breathing is so characteristic of life that, along with the pulse, it is one of the first things health professionals check to determine if an unconscious person is alive.

**Respiration** includes the following processes: (1) ventilation, or breathing, which is the movement of air into and out of the lungs; (2) the exchange of oxygen and carbon dioxide between the air in the lungs and the blood; (3) the transport of oxygen and carbon dioxide in the blood; and (4) the exchange of oxygen and carbon dioxide between the blood and the tissues. The term *respiration* is also used in reference to cell metabolism. In aerobic respiration, for example, cells use oxygen and produce carbon dioxide. We discuss cellular respiration in chapter 17.

# FUNCTIONS OF THE RESPIRATORY SYSTEM

Respiration is necessary because all living cells of the body require oxygen and produce carbon dioxide. The respiratory system performs the following functions:

1. *Gas exchange.* The respiratory system allows oxygen from the air to enter the blood and carbon dioxide to leave the blood and enter the air. The cardiovascular system transports oxygen from the lungs to the body cells and carbon dioxide from the body cells to the lungs. Thus, the respiratory and cardiovascular systems work together to supply oxygen to all cells and to remove carbon dioxide. Without healthy respiratory and cardiovascular systems, the body's capacity to carry out normal activity is reduced, and without adequate respiratory and cardiovascular system functions, life itself is impossible.
2. *Regulation of blood pH.* The respiratory system can alter blood pH by changing blood carbon dioxide levels.
3. *Voice production.* Air movement past the vocal cords makes sound and speech possible.
4. *Olfaction.* The sensation of smell occurs when airborne molecules are drawn into the nasal cavity.
5. *Innate immunity* (see chapter 14). The respiratory system protects against some microorganisms by preventing them from entering the body and by removing them from respiratory surfaces.

# ANATOMY OF THE RESPIRATORY SYSTEM

The **respiratory system** is divided into the upper respiratory tract and the lower respiratory tract (figure 15.1). The **upper respiratory tract** includes the external nose, the nasal cavity, the pharynx, and associated structures; the **lower respiratory tract** includes the larynx, the trachea, the bronchi, and the lungs. Keep in mind, however, that *upper* and *lower respiratory tract* are not official anatomical terms. Rather, they are arbitrary divisions for the purposes of discussion, and some anatomists define them differently. For example, one alternative places the larynx in the

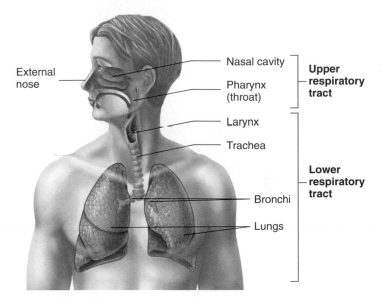

**Figure 15.1** **Respiratory System**

The upper respiratory tract consists of the external nose, nasal cavity, and pharynx (throat). The lower respiratory tract consists of the larynx, trachea, bronchi, and lungs.

upper respiratory tract. Also, the oral cavity is considered part of the digestive system, not the respiratory system, even though air frequently passes through the oral cavity.

## Nose

The **nose** consists of the external nose and the nasal cavity. The **external nose** is the visible structure that forms a prominent feature of the face. Most of the external nose is composed of hyaline cartilage, although the bridge of the external nose consists of bone (see figure 6.12). The bone and cartilage are covered by connective tissue and skin.

The **nares** (nā′rēs; sing. nā′ris), or *nostrils,* are the external openings of the nose, and the **choanae** (kō′an-ē; funnels) are the openings into the pharynx. The **nasal** (nā′zăl) **cavity** extends from the nares to the choanae (figure 15.2). The **nasal septum** is a partition dividing the nasal cavity into right and left parts. A **deviated nasal septum** occurs when the septum bulges to one side. The **hard palate** (pal′ăt) forms the floor of the nasal cavity, separating the nasal cavity from the oral cavity. Air can flow

Respiratory

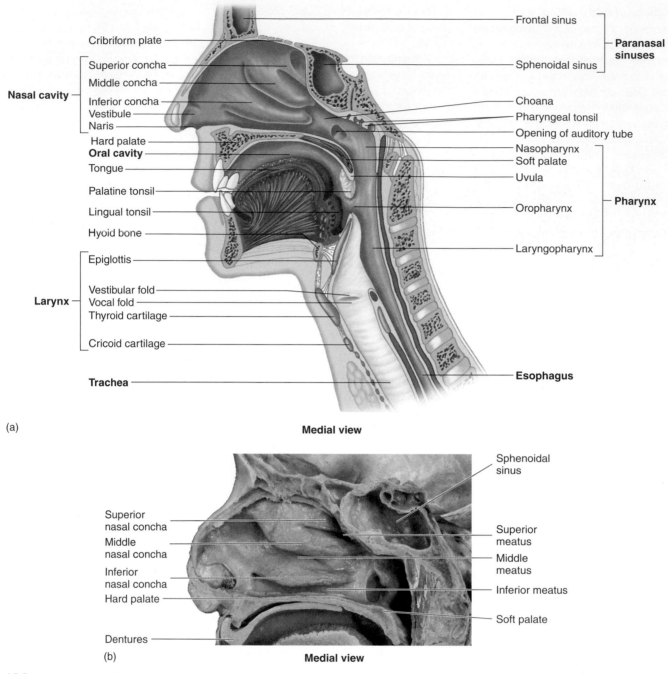

**Figure 15.2** Nasal Cavity and Pharynx

(*a*) Sagittal section through the nasal cavity and pharynx. (*b*) Photograph of sagittal section of the nasal cavity.

through the nasal cavity when the oral cavity is closed or full of food.

Three prominent bony ridges called **conchae** (kon′kē) are present on the lateral walls on each side of the nasal cavity. The conchae increase the surface area of the nasal cavity and cause air to churn, so that it can be cleansed, humidified, and warmed.

The **paranasal** (par-ă-nā′săl) **sinuses** are air-filled spaces within bone. They include the maxillary, frontal, ethmoidal, and sphenoidal sinuses, each named for the bones in which they are located. The paranasal sinuses open into the nasal cavity and are lined with a mucous membrane. They reduce the weight of the skull, produce mucus, and influence the quality of the voice by acting as resonating chambers.

The **nasolacrimal** (nā-zō-lak′ri-măl) **ducts,** which carry tears from the eyes, also open into the nasal cavity. Sensory receptors for the sense of smell are in the superior part of the nasal cavity (see chapter 9).

Air enters the nasal cavity through the nares. Just inside the nares, the lining of the cavity is composed of stratified squamous epithelium containing coarse hairs. The hairs trap some of the large particles of dust suspended in the air. The rest of the nasal cavity is lined with pseudostratified columnar epithelial cells containing cilia and many mucus-producing goblet cells (see chapter 4). Mucus produced by the goblet cells also traps debris in the air. The cilia sweep the mucus posteriorly to the pharynx, where it is swallowed. As air flows through the nasal cavities, it is humidified by moisture from the mucous epithelium and warmed by blood flowing through the superficial capillary networks underlying the mucous epithelium.

The **sneeze reflex** dislodges foreign substances from the nasal cavity. Sensory receptors detect the foreign substances, and action potentials are conducted along the trigeminal nerves to the medulla oblongata, where the reflex is triggered. During the sneeze reflex, the uvula and the soft palate are depressed, so that rapidly flowing air from the lungs is directed primarily through the nasal passages, although a considerable amount passes through the oral cavity.

About 17–25% of people have a photic sneeze reflex, which is stimulated by exposure to bright light, such as the sun. Another reflex, the pupillary reflex, causes the pupils to constrict in response to bright light. Physiologists speculate that the complicated "wiring" of the pupillary and sneeze reflexes are intermixed in some people, so that, when bright light activates a pupillary reflex, it also activates a sneeze reflex. Sometimes the photic sneeze reflex is fancifully called ACHOO, which stands for *a*utosomal-dominant-*c*ompelling-*h*elio-*o*phthalmic-*o*utburst.

## A CASE IN POINT
### Sinusitis

Cy Ness has a cold; his symptoms include sneezing, achiness, and fatigue. Although Cy recovers from the cold, a stuffy nose and recurrent headaches linger for several weeks. Cy's doctor listens to his history of symptoms, examines his nose and throat, and orders a sinus x-ray. He concludes that Cy Ness has **sinusitis** (sī-nŭ-sī′tis), inflammation of the mucous membrane of a sinus, especially one or more of the paranasal sinuses. Viral infections, such as the common cold, can cause mucous membranes to become inflamed and swollen and to produce excess mucus. As a result, the sinus opening into the nasal cavity can be partially or completely blocked. In addition, mucus accumulation within the sinus can promote the development of a bacterial infection. The combination of built-up mucus and inflamed and infected mucous membranes produces pain.

Treatment of sinusitis consists of taking antibiotics to kill the bacteria and decongestants to promote sinus drainage; drinking fluids to maintain hydration; and inhaling steam. Decongestants reduce tissue swelling, or edema. When mucous membranes become less swollen, breathing is easier, and more mucus can move out of the paranasal sinuses and nasal cavity. Decongestants, such as pseudoephedrine hydrochloride (e.g., Sudafed), reduce swelling by causing the release of norepinephrine from sympathetic neurons supplying blood vessels. Increased vasoconstriction of blood vessels in mucous membranes reduces blood flow and the movement of fluid from the blood into tissues. Sinusitis can also result from inflammation caused by allergies or from benign growths, called polyps, that obstruct a sinus opening into the nasal cavity.

**Predict 1**

*Explain what happens to your throat when you sleep with your mouth open, especially when your nasal passages are plugged as a result of a cold. Explain what may happen to your lungs when you run a long way in very cold weather while breathing rapidly through your mouth.*

## Pharynx

The **pharynx** (far′ingks; throat) is the common passageway for both the respiratory and the digestive systems. Air from the nasal cavity and air, food, and water from the mouth pass through the pharynx. Inferiorly, the pharynx leads to the rest of the respiratory system through the opening into the larynx and to the digestive system through the esophagus. The pharynx can be divided into three regions: the nasopharynx, the oropharynx, and the laryngopharynx (figure 15.2a).

The **nasopharynx** (nā′zō-far′ingks) is the superior part of the pharynx. It is located posterior to the choanae and superior to the **soft palate,** which is an incomplete muscle and connective tissue partition separating the nasopharynx from the oropharynx. The **uvula** (ū′vū-lă; a little grape) is the posterior extension of the soft palate. The soft palate forms the floor of the nasopharynx. The nasopharynx is lined with pseudostratified ciliated columnar epithelium that is continuous with the nasal cavity. The auditory tubes extend from the middle ears and open into the nasopharynx. The posterior part of the nasopharynx contains the **pharyngeal** (fă-rin′jē-ăl) **tonsil,** which helps defend the body against infection (see chapter 14). The soft palate is elevated during swallowing; this movement closes the nasopharynx and prevents food from passing from the oral cavity into the nasopharynx.

The **oropharynx** (ōr′ō-far′ingks) extends from the uvula to the epiglottis, and the oral cavity opens into the oropharynx. Thus, food, drink, and air all pass through the oropharynx. The oropharynx is lined with stratified squamous epithelium, which protects against abrasion. Two sets of tonsils, the palatine tonsils and the lingual tonsil, are located near the opening between the mouth and the oropharynx. The **palatine** (pal′ă-tīn) **tonsils** are located in the lateral walls near the border of the oral cavity and the oropharynx. The **lingual tonsil** is located on the surface of the posterior part of the tongue.

The **laryngopharynx** (lă-ring′gō-far-ingks) passes posterior to the larynx and extends from the tip of the epiglottis to the esophagus. Food and drink pass through the laryngopharynx to the esophagus. A small amount of air is usually swallowed with the food and drink. Swallowing too much air can cause excess gas in the stomach and may result in belching. The laryngopharynx is lined with stratified squamous epithelium and ciliated columnar epithelium.

## Larynx

The **larynx** (lar′ingks) is located in the anterior throat and extends from the base of the tongue to the trachea (figure 15.2a). It is a passageway for air between the pharynx and the trachea. The larynx consists of an outer casing of nine cartilages connected to one another by muscles and ligaments (figure 15.3). Three of the nine cartilages are unpaired, and six of them form three pairs. The largest

(a) **Anterior view**    (b) **Posterior view**    (c) **Medial view of sagittal section**

**Figure 15.3**  Anatomy of the Larynx

cartilage is the unpaired **thyroid** (thī′royd; shield-shaped) **cartilage,** or *Adam's apple.* The thyroid cartilage is attached superiorly to the hyoid bone. The most inferior cartilage of the larynx is the unpaired **cricoid** (krī′koyd; ring-shaped) **cartilage,** which forms the base of the larynx on which the other cartilages rest. The thyroid and cricoid cartilages maintain an open passageway for air movement.

The third unpaired cartilage is the **epiglottis** (ep-i-glot′is; on the glottis). It differs from the other cartilages in that it consists of elastic cartilage rather than hyaline cartilage. Its inferior margin is attached to the thyroid cartilage anteriorly, and the superior part of the epiglottis projects superiorly as a free flap toward the tongue. The epiglottis helps prevent swallowed materials from entering the larynx. As the larynx elevates during swallowing, the epiglottis tips posteriorly to cover the opening of the larynx.

The six paired cartilages consist of three cartilages on each side of the posterior part of the larynx (figure 15.3*b*). The top cartilage on each side is the **cuneiform** (kū′nē-i-fōrm; wedge-shaped) **cartilage,** the middle cartilage is the **corniculate** (kōr-nik′ū-lāt; horn-shaped) **cartilage,** and the bottom cartilage is the **arytenoid** (ar-i-tē′noyd; ladle-shaped) **cartilage.** The arytenoid cartilages articulate with the cricoid cartilage inferiorly. The paired cartilages form an attachment site for the vocal folds.

Two pairs of ligaments extend from the posterior surface of the thyroid cartilage to the paired cartilages. The superior pair forms the **vestibular** (ves-tib′ū-lăr) **folds,** or *false vocal cords,* and the inferior pair composes the **vocal folds,** or *true vocal cords* (figure 15.4). When the vestibular folds come together, they prevent air from leaving the lungs, as when a person holds his or her breath. Along with the epiglottis, the vestibular folds also prevent food and liquids from entering the larynx.

The vocal folds are the primary source of voice production. Air moving past the vocal folds causes them to vibrate, producing sound. Muscles control the length and tension of the vocal folds. The force of air moving past the vocal folds controls the loudness, and the tension of the vocal folds controls the pitch of the voice. An inflammation of the mucous epithelium of the vocal folds is called **laryngitis** (lar-in-jī′tis). Swelling of the vocal folds during laryngitis inhibits voice production.

## Trachea

The **trachea** (trā′kē-ă), or *windpipe,* is a membranous tube attached to the larynx. It consists of connective tissue and smooth muscle, reinforced with 16–20 C-shaped pieces of hyaline cartilage (see figure 15.3). The adult trachea is about 1.4–1.6 centimeters (cm)

## Establishing Airflow

In cases of extreme emergency, when the upper air passageway is blocked by a foreign object to the extent that the victim cannot breathe, quick action is required to save the person's life. In the **Heimlich maneuver,** the sudden application of pressure to the abdomen can force air up the trachea to dislodge the obstruction. The person who performs the maneuver stands behind the victim with his or her arms under the victim's arms and his or her hands over the victim's abdomen between the navel and the rib cage. With one hand formed into a fist, the other hand suddenly pulls the fist toward the abdomen with an accompanying upward motion. The pressure pushes up on the diaphragm and therefore increases air pressure in the lungs. If this maneuver is done properly, it causes air to flow from the lungs with sufficient force to dislodge most foreign objects.

There are other ways to establish airflow, but they should be performed only by trained medical personnel. **Intubation** is the insertion of a tube into an opening, a canal, or a hollow organ. A tube can be passed through the mouth or nose into the pharynx and then through the larynx to the trachea. Sometimes it is necessary to make an opening for the tube. The preferred point of entry in an emergency is through the membrane between the cricoid and thyroid cartilages (see figure 15.3a), a procedure referred to as a **cricothyrotomy** (krī′kō-thī-rot′ō-mē).

A tube is then inserted into the opening to facilitate the passage of air.

A **tracheostomy** (trā′kē-os′tō-mē; tracheo- + stoma, mouth) is an operation to make an opening into the trachea. Usually, the opening is intended to be permanent, and a tube is inserted into the trachea to allow airflow and provide a way to remove secretions. The term **tracheotomy** (trā-kē-ot′ō-mē; tracheo- + tome, incision) refers to the actual cutting into the trachea (although sometimes the terms tracheostomy and tracheotomy are used interchangeably). It is not advisable to enter the air passageway through the trachea in emergency cases because arteries, nerves, and the thyroid gland overlie the anterior surface of the trachea.

in diameter and about 10–11 cm long. It begins immediately inferior to the cricoid cartilage, which is the most inferior cartilage of the larynx. The trachea projects through the mediastinum and divides into the right and left primary bronchi at the level of the fifth thoracic vertebra (figure 15.5). The esophagus lies immediately posterior to the trachea (see figure 15.2a).

C-shaped cartilages form the anterior and lateral sides of the trachea. The cartilages protect the trachea and maintain an open passageway for air. The posterior wall of the trachea has no cartilage and consists of a ligamentous membrane and smooth muscle (see figure 15.3b). The smooth muscle can alter the diameter of the trachea. This occurs, for example, when the

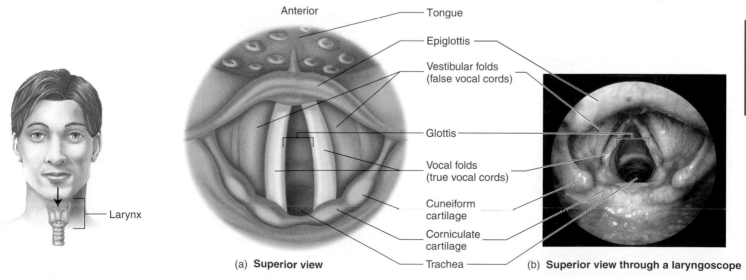

Anterior — Tongue
Epiglottis
Vestibular folds (false vocal cords)
Glottis
Vocal folds (true vocal cords)
Cuneiform cartilage
Corniculate cartilage
Larynx
Trachea

(a) **Superior view**  (b) **Superior view through a laryngoscope**

**Figure 15.4  Vestibular and Vocal Folds**

(*Far left*) The arrow shows the direction of viewing the vestibular and vocal folds. (*a*) The relationship of the vestibular folds to the vocal folds and the laryngeal cartilages. (*b*) Superior view of the vestibular and vocal folds as seen through a laryngoscope.

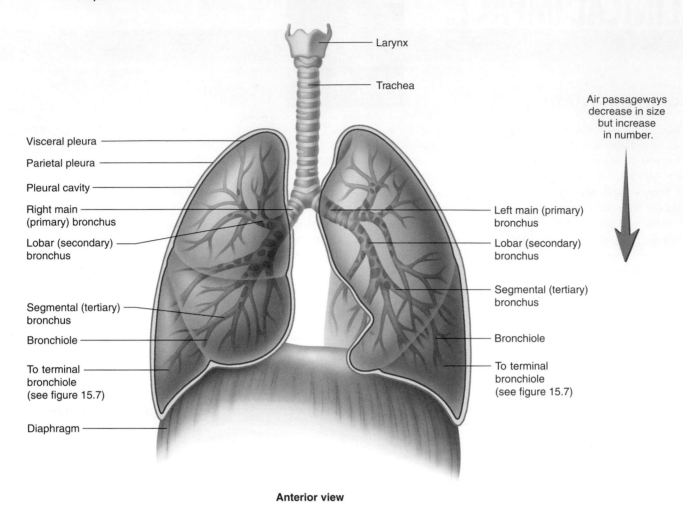

Air passageways decrease in size but increase in number.

Larynx

Trachea

Visceral pleura

Parietal pleura

Pleural cavity

Right main (primary) bronchus

Lobar (secondary) bronchus

Left main (primary) bronchus

Lobar (secondary) bronchus

Segmental (tertiary) bronchus

Segmental (tertiary) bronchus

Bronchiole

Bronchiole

To terminal bronchiole (see figure 15.7)

To terminal bronchiole (see figure 15.7)

Diaphragm

**Anterior view**

**Figure 15.5  Anatomy of the Trachea and Lungs**

The trachea and lungs and the branching of the bronchi are shown. Each lung is surrounded by a pleural cavity, formed by the visceral and parietal pleurae.

cough reflex dislodges foreign substances from the trachea. Sensory receptors detect the foreign substance, and action potentials travel along the vagus nerves to the medulla oblongata, where the cough reflex is triggered. During coughing, the smooth muscle of the trachea contracts, decreasing the trachea's diameter. As a result, air moves rapidly through the trachea, which helps expel mucus and foreign substances. Also, the uvula and soft palate are elevated, so that air passes primarily through the oral cavity.

Predict 2

*Explain what happens to the shape of the trachea when a person swallows a large mouthful of food. Why is this process advantageous?*

The trachea is lined with pseudostratified columnar epithelium, which contains numerous cilia and goblet cells. The cilia propel mucus produced by the goblet cells, as well as foreign particles embedded in the mucus, out of the trachea, through the larynx, and into the pharynx, from which they are swallowed.

Constant, long-term irritation of the trachea by cigarette smoke can cause the tracheal epithelium to change to stratified squamous epithelium. The stratified squamous epithelium has no cilia and therefore cannot clear the airway of mucus and debris. The accumulations of mucus provide a place for microorganisms to grow, resulting in respiratory infections. Constant irritation and inflammation of the respiratory passages stimulate the cough reflex, resulting in "smoker's cough."

## Bronchi

The trachea divides into the left and right **main bronchi** (brong′kī; sing. bronchus, brong′kŭs; windpipe), or *primary bronchi,* each of which connects to a lung. The left main bronchus is more horizontal than the right main bronchus because it is displaced by the heart (figure 15.5). Foreign objects that enter the trachea usually lodge in the right main bronchus, because it is more vertical than the left main bronchus and therefore more in direct line with the trachea. The main bronchi extend from the trachea to the lungs. Like the trachea, the main bronchi are lined with pseudostratified ciliated columnar epithelium and are supported by C-shaped pieces of cartilage.

Respiratory

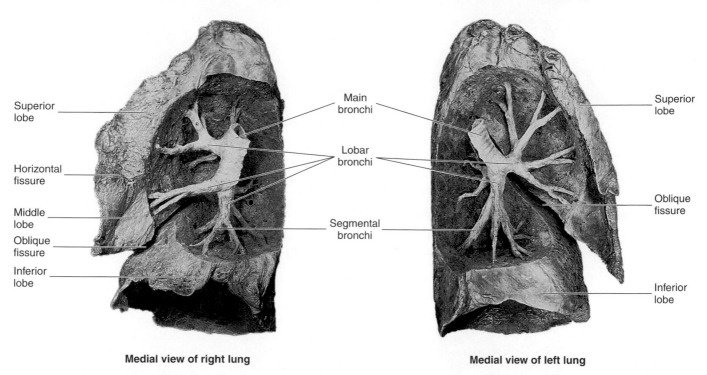

Superior lobe

Horizontal fissure

Middle lobe

Oblique fissure

Inferior lobe

Main bronchi

Lobar bronchi

Segmental bronchi

Superior lobe

Oblique fissure

Inferior lobe

**Medial view of right lung**    **Medial view of left lung**

**Figure 15.6  Lungs, Lung Lobes, and Bronchi**

The right lung is divided into three lobes by the horizontal and oblique fissures. The left lung is divided into two lobes by the oblique fissure. A main bronchus supplies each lung, a lobar bronchus supplies each lung lobe, and segmental bronchi supply the bronchopulmonary segments (not visible).

## Lungs

The **lungs** are the principal organs of respiration. Each lung is cone-shaped, with its base resting on the diaphragm and its apex extending superiorly to a point about 2.5 cm above the clavicle (figure 15.5). The right lung has three **lobes,** called the superior, middle, and inferior lobes. The left lung has two lobes, called the superior and inferior lobes (figure 15.6). The lobes of the lungs are separated by deep, prominent fissures on the lung surface. Each lobe is divided into **bronchopulmonary segments** separated from one another by connective tissue septa, but these separations are not visible as surface fissures. Individual diseased bronchopulmonary segments can be surgically removed, leaving the rest of the lung relatively intact, because major blood vessels and bronchi do not cross the septa. There are 9 bronchopulmonary segments in the left lung and 10 in the right lung.

The main bronchi branch many times to form the **tracheobronchial tree** (see figure 15.5). Each main bronchus divides into lobar bronchi as they enter their respective lungs (figure 15.6). The **lobar bronchi** (or *secondary bronchi*), two in the left lung and three in the right lung, conduct air to each lobe. The lobar bronchi in turn give rise to **segmental bronchi** (or *tertiary bronchi*), which extend to the bronchopulmonary segments of the lungs. The bronchi continue to branch many times, finally giving rise to **bronchioles** (brong′kē-ōlz). The bronchioles also subdivide numerous times to give rise to **terminal bronchioles,** which then subdivide into **respiratory bronchioles** (figure 15.7). Each respiratory bronchiole subdivides to form **alveolar** (al-vē′ō-lăr) **ducts,** which are like long, branching hallways with many open doorways. The doorways open into **alveoli** (al-vē′ō-lī; hollow sacs),

which are small air sacs. The alveoli become so numerous that the alveolar duct wall is little more than a succession of alveoli. The alveolar ducts end as two or three **alveolar sacs,** which are chambers connected to two or more alveoli. There are about 300 million alveoli in the lungs.

As the air passageways of the lungs become smaller, the structure of their walls changes. The amount of cartilage decreases and the amount of smooth muscle increases until, at the terminal bronchioles, the walls have a prominent smooth muscle layer but no cartilage. Relaxation and contraction of the smooth muscle within the bronchi and bronchioles can change the diameter of the air passageways. For example, during exercise the diameter can increase, thus increasing the volume of air moved. During an **asthma attack,** however, contraction of the smooth muscle in the terminal bronchioles can result in greatly reduced airflow (see Systems Pathology at the end of this chapter). In severe cases, air movement can be so restricted that death results. Fortunately, medications, such as albuterol (al-bū′ter-ol), help counteract the effects of an asthma attack by promoting smooth muscle relaxation in the walls of terminal bronchioles, so that air can flow more freely.

As the air passageways of the lungs become smaller, the lining of their walls also changes. The trachea and bronchi have pseudostratified ciliated columnar epithelium, the bronchioles have ciliated simple columnar epithelium, and the terminal bronchioles have ciliated simple cuboidal epithelium. The ciliated epithelium of the air passageways functions as a mucus-cilia escalator, which traps debris in the air and removes it from the respiratory system.

Respiratory

Smooth muscle

Bronchial vein, artery, and nerve

Branch of pulmonary artery

Deep lymphatic vessel

Terminal bronchiole

Respiratory bronchioles

Alveolus

Alveolar ducts

Superficial lymphatic vessel

Alveoli

Lymph nodes

Alveolar sac

Connective tissue

Visceral pleura

Pulmonary capillaries

Pleural cavity

Branch of pulmonary vein

Parietal pleura

Elastic fibers

**Figure 15.7  Bronchioles and Alveoli**

A terminal bronchiole branches to form respiratory bronchioles, which give rise to alveolar ducts. Alveoli connect to the alveolar ducts and respiratory bronchioles. The alveolar ducts end as two or three alveolar sacs.

As the air passageways beyond the terminal bronchioles become smaller, their walls become thinner. The walls of the respiratory bronchioles are composed of cuboidal epithelium, and those of the alveolar ducts and alveoli are simple squamous epithelium. The **respiratory membrane** of the lungs is where gas exchange between the air and blood takes place. It is mainly formed by the walls of the alveoli and the surrounding capillaries (figure 15.8), but the alveolar ducts and respiratory bronchioles also contribute. The respiratory membrane is very thin to facilitate the diffusion of gases. It consists of six layers:

1. a thin layer of fluid lining the alveolus
2. the alveolar epithelium, composed of simple squamous epithelium
3. the basement membrane of the alveolar epithelium
4. a thin interstitial space
5. the basement membrane of the capillary endothelium
6. the capillary endothelium, composed of simple squamous epithelium

The elastic fibers surrounding the alveoli (figure 15.8) allow them to expand during inspiration and recoil during expiration. The lungs are very elastic and when inflated are capable of expelling the air and returning to their original, uninflated state. Specialized secretory cells within the walls of the alveoli (figure 15.8) secrete a chemical, called surfactant, that reduces the tendency of alveoli to recoil (see "Lung Recoil" later in this chapter).

## Pleural Cavities

The lungs are contained within the thoracic cavity. In addition, each lung is surrounded by a separate **pleural** (ploor′ăl; relating to the ribs) **cavity.** Each pleural cavity is lined with a serous membrane called the **pleura.** The pleura consists of a parietal and a visceral part. The **parietal pleura,** which lines the walls of the thorax, diaphragm, and mediastinum, is continuous with the **visceral pleura,** which covers the surface of the lung (figure 15.9; see figure 15.5).

The pleural cavity, between the parietal and visceral pleurae, is filled with a small volume of pleural fluid produced by the pleural membranes. The pleural fluid performs two functions: (1) It acts as a lubricant, allowing the visceral and parietal pleurae to slide past each other as the lungs and thorax change shape

**Figure 15.8   Alveolus and the Respiratory Membrane**

(a) Section of an alveolus, showing the air-filled interior and thin walls composed of simple squamous epithelium. The alveolus is surrounded by elastic connective tissue and blood capillaries. (b) $O_2$ and $CO_2$ diffuse across the six thin layers of the respiratory membrane.

during respiration, and (2) it helps hold the pleural membranes together. The pleural fluid acts as a thin film of water between two sheets of glass (the visceral and parietal pleurae); the glass sheets can slide over each other easily, but it is difficult to separate them.

**Predict 3**

*Pleurisy is an inflammation of the pleural membranes. Explain why this condition is so painful, especially when a person takes deep breaths.*

## Lymphatic Supply

The lungs have two lymphatic supplies (see figure 15.7). The **superficial lymphatic vessels** are deep to the visceral pleura; they drain lymph from the superficial lung tissue and the visceral pleura. The **deep lymphatic vessels** follow the bronchi; they drain lymph from the bronchi and associated connective tissues. No lymphatic vessels are located in the walls of the alveoli. Both the superficial and deep lymphatic vessels exit the lungs at the main bronchi.

**Figure 15.9  Pleural Cavities and Membranes**

Transverse section of the thorax, showing the relationship of the pleural cavities to the thoracic organs. Each lung is surrounded by a pleural cavity. The parietal pleura lines the wall of each pleural cavity, and the visceral pleura covers the surface of the lungs. The space between the parietal and visceral pleurae is small and filled with pleural fluid.

Phagocytic cells within the lungs phagocytize most carbon particles and other debris from inspired air and move them to the lymphatic vessels. In older people, the surface of the lungs can appear gray to black because of the accumulation of these particles, especially if the person smoked or lived primarily in a city with air pollution. Cancer cells from the lungs can also spread to other parts of the body through the lymphatic vessels.

## VENTILATION AND RESPIRATORY VOLUMES

**Ventilation,** or *breathing,* is the process of moving air into and out of the lungs. There are two phases of ventilation: (1) **Inspiration,** or *inhalation,* is the movement of air into the lungs; (2) **expiration,** or *exhalation,* is the movement of air out of the lungs. Ventilation is regulated by changes in thoracic volume, which produce changes in air pressure within the lungs.

### Changing Thoracic Volume

The muscles associated with the ribs are responsible for ventilation (figure 15.10). The **muscles of inspiration** include the diaphragm and the muscles that elevate the ribs and sternum, such as the external intercostals. The **diaphragm** (dī′a-fram; partition) is a large dome of skeletal muscle that separates the thoracic cavity from the abdominal cavity (see figure 7.18). The **muscles of** expiration, such as the internal intercostals, depress the ribs and sternum.

At the end of a normal, quiet expiration, the respiratory muscles are relaxed (figure 15.10a). During quiet inspiration, contraction of the diaphragm causes the top of the dome to move inferiorly, which increases the volume of the thoracic cavity. The largest change in thoracic volume results from movement of the diaphragm. Contraction of the external intercostals also elevates the ribs and sternum (figure 15.10b), which increases thoracic volume by increasing the diameter of the thoracic cage.

**Predict 4**

*During inspiration, the abdominal muscles relax. How is this advantageous?*

Expiration during quiet breathing occurs when the diaphragm and external intercostals relax and the elastic properties of the thorax and lungs cause a passive decrease in thoracic volume.

There are several differences between normal, quiet breathing and labored breathing. During labored breathing, all the inspiratory muscles are active, and they contract more forcefully than during quiet breathing, causing a greater increase in thoracic volume (figure 15.10b). Also during labored breathing, forceful contraction of the internal intercostals and the abdominal muscles produces a faster and greater decrease in thoracic volume than would be produced by the passive recoil of the thorax and lungs.

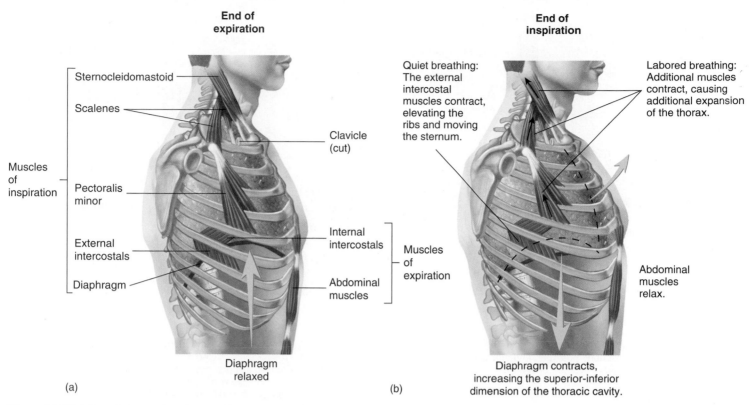

**End of expiration**

- Sternocleidomastoid
- Scalenes
- Clavicle (cut)
- Muscles of inspiration
- Pectoralis minor
- External intercostals
- Diaphragm
- Internal intercostals
- Muscles of expiration
- Abdominal muscles

Diaphragm relaxed

(a)

**End of inspiration**

Quiet breathing: The external intercostal muscles contract, elevating the ribs and moving the sternum.

Labored breathing: Additional muscles contract, causing additional expansion of the thorax.

Abdominal muscles relax.

Diaphragm contracts, increasing the superior-inferior dimension of the thoracic cavity.

(b)

**Figure 15.10    Effect of the Muscles of Respiration on Thoracic Volume**
(a) Muscles of respiration at the end of expiration. (b) Muscles of respiration at the end of inspiration.

## Pressure Changes and Airflow

Two physical principles govern the flow of air into and out of the lungs:

1. *Changes in volume result in changes in pressure.* As the volume of a container increases, the pressure within the container decreases. As the volume of a container decreases, the pressure within the container increases. In the same way, the muscles of respiration change the volume of the thorax and therefore the pressure within the thoracic cavity.
2. *Air flows from an area of higher pressure to an area of lower pressure.* If the pressure is higher at one end of a tube than at the other, air or fluid (see chapter 13) flows from the area of higher pressure toward the area of lower pressure. The greater the pressure difference, the greater the rate of airflow. Air flows through the respiratory passages because of pressure differences between the outside of the body and the alveoli inside the body. These pressure differences are produced by changes in thoracic volume.

The volume and pressure changes responsible for one cycle of inspiration and expiration can be described as follows:

1. At the end of expiration, **alveolar pressure,** which is the air pressure within the alveoli, is equal to **atmospheric pressure,** which is the air pressure outside the body. No air moves into or out of the lungs because alveolar pressure and atmospheric pressure are equal (figure 15.11, step 1).
2. During inspiration, contraction of the muscles of inspiration increases the volume of the thoracic cavity. The increased thoracic volume causes the lungs to expand, resulting in an increase in alveolar volume (see "Changing Alveolar Volume" later in this section). As the alveolar volume increases, alveolar pressure becomes less than atmospheric pressure, and air flows from outside the body through the respiratory passages to the alveoli (figure 15.11, step 2).
3. At the end of inspiration, the thorax and alveoli stop expanding. When the alveolar pressure and atmospheric pressure become equal, airflow stops (figure 15.11, step 3).
4. During expiration, the thoracic volume decreases, producing a corresponding decrease in alveolar volume. Consequently, alveolar pressure increases above atmospheric pressure, and air flows from the alveoli through the respiratory passages to the outside (figure 15.11, step 4).

As expiration ends, the decrease in thoracic volume stops, and the process repeats, beginning at step 1.

Respiratory

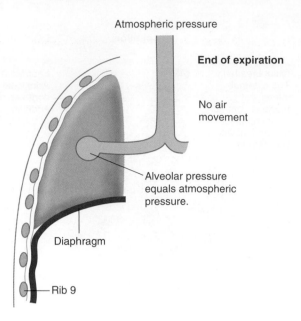

Atmospheric pressure

**End of expiration**

No air movement

Alveolar pressure equals atmospheric pressure.

Diaphragm

Rib 9

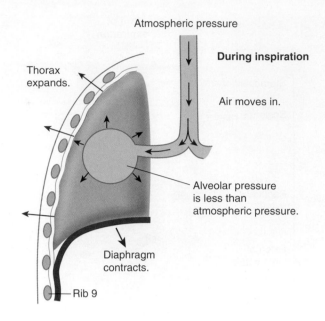

Atmospheric pressure

**During inspiration**

Thorax expands.

Air moves in.

Alveolar pressure is less than atmospheric pressure.

Diaphragm contracts.

Rib 9

**1** At the end of expiration, alveolar pressure is equal to atmospheric pressure, and there is no air movement.

**2** During inspiration, increased thoracic volume results in increased alveolar volume and decreased alveolar pressure. Atmospheric pressure is greater than alveolar pressure, and air moves into the lungs.

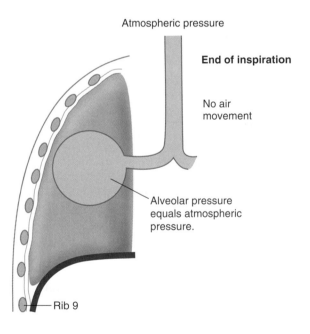

Atmospheric pressure

**End of inspiration**

No air movement

Alveolar pressure equals atmospheric pressure.

Rib 9

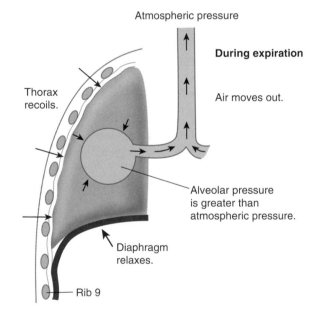

Atmospheric pressure

**During expiration**

Thorax recoils.

Air moves out.

Alveolar pressure is greater than atmospheric pressure.

Diaphragm relaxes.

Rib 9

**3** At the end of inspiration, alveolar pressure is equal to atmospheric pressure, and there is no air movement.

**4** During expiration, decreased thoracic volume results in decreased alveolar volume and increased alveolar pressure. Alveolar pressure is greater than atmospheric pressure, and air moves out of the lungs.

**PROCESS Figure 15.11   Alveolar Pressure Changes During Inspiration and Expiration**

The combined space of all the alveoli is represented by a large "bubble" (*blue*). The alveoli are actually microscopic and cannot be seen in the illustration.

Respiratory

## A CASE IN POINT
### Infant Respiratory Distress Syndrome

Tu Soun was born 3 months prematurely. She presented with a respiration rate of 68 breaths per minute; blue lips, tongue, and nail beds; nasal flaring during inspiration; inward movement of the thoracic cage and outward movement of the abdomen during inspiration; an expiratory grunt; and a negative shake test. Tu Soun has **infant respiratory distress syndrome (IRDS),** caused by too little surfactant, a substance that covers the lining of the lung alveoli, where it helps reduce surface tension. IRDS, also called *hyaline membrane disease,* is common in premature infants because surfactant is not produced in adequate quantities until about the seventh month of gestation. Thereafter, the amount produced increases as the fetus matures. Pregnant women who are likely to deliver prematurely can be given cortisol, which crosses the placenta into the fetus and stimulates surfactant synthesis.

The normal respiration rate for a newborn is around 40 breaths per minute. Tu's high respiratory rate is stimulated by high blood carbon dioxide levels and low blood oxygen levels (see "Chemical Control of Breathing" later in this chapter). Her blue lips, tongue, and nail beds are signs of cyanosis caused by deoxygenated blood. The nasal flaring occurs to maximize air intake.

If too little surfactant has been produced by the time of birth, the lungs tend to collapse, and the muscles of respiration must exert a great deal of energy to keep the lungs inflated; even then, ventilation is inadequate. The thoracic cage in newborns is very pliable. During labored inspiration, the increased inferior movement of the diaphragm causes such a decrease in thoracic cavity pressure that the thoracic cage is pulled inward as the abdomen expands. The more labored the breathing, the more exaggerated the expansion of the abdomen and the inward movement of the thoracic cage.

An expiratory grunt is a gruff, throaty sound made during expiration. It is caused by partial closure of the vestibular and vocal folds. Expiratory grunting increases airway pressure and helps prevent alveolar collapse.

The shake test determines the presence of surfactant in lung fluid. Fetal lung fluid is either swallowed by the fetus or passed through the mouth into the amniotic fluid, so 30 minutes after delivery the fetus's gastric fluid still contains swallowed lung fluid and amniotic fluid. To perform the shake test, a sample of gastric fluid is collected and then mixed with saline and alcohol, placed in a tube, and shaken. A positive shake test produces bubbles, but a negative shake test does not, which indicates that very little surfactant is present because the bubbles collapse.

Without specialized treatment, most babies with IRDS die soon after birth due to inadequate ventilation of the lungs and respiratory muscle fatigue.

Treatment strategies include forcing enough oxygen-rich air into the lungs to inflate them and administering purified, natural surfactant via an intubation tube directly into the lungs.

## Lung Recoil

During quiet expiration, thoracic volume and lung volume decrease because of **lung recoil,** the tendency for an expanded lung to decrease in size. The thoracic wall also recoils due to the elastic properties of its tissues. Lung recoil is able to occur because the connective tissue of the lungs contains elastic fibers and because the film of fluid lining the alveoli has surface tension. **Surface tension** exists because the oppositely charged ends of water molecules attract each other (see chapter 2). As the water molecules pull together, they also pull on the alveolar walls, causing the alveoli to recoil and become smaller.

Two factors keep the lungs from collapsing: (1) surfactant and (2) pressure in the pleural cavity.

### Surfactant

**Surfactant** (ser-fak'tănt; *surface acting agent*) is a mixture of lipoprotein molecules produced by secretory cells of the alveolar epithelium. The surfactant molecules form a single layer on the surface of the thin fluid layer lining the alveoli, reducing surface tension. Without surfactant, the surface tension causing the alveoli to recoil can be 10 times greater than when surfactant is present. Thus, surfactant greatly reduces the tendency of the lungs to collapse.

### Pleural Pressure

When **pleural pressure,** the pressure in the pleural cavity, is less than alveolar pressure, the alveoli tend to expand. This principle can be understood by considering a balloon. The balloon expands when the pressure outside it is less than the pressure inside. This pressure difference is normally achieved by increasing the pressure inside the balloon by blowing into it. This pressure difference, however, can also be achieved by decreasing the pressure outside the balloon. For example, if the balloon is placed in a chamber from which air is removed, the pressure around the balloon becomes lower than atmospheric pressure, and the balloon expands. The lower the pressure outside the balloon, the greater the tendency for the higher pressure inside the balloon to cause it to expand. In a similar fashion, decreasing pleural pressure can result in expansion of the alveoli.

Normally, the alveoli are in the expanded state because pleural pressure is lower than alveolar pressure. Pleural pressure is lower than alveolar pressure because of a suction effect caused by fluid removal by the lymphatic system (see chapter 14) and by lung recoil. As the lungs recoil, the visceral and parietal pleurae tend to be pulled apart. Normally, the lungs do not pull away from the thoracic wall because pleural fluid holds the visceral and parietal pleurae together. Nonetheless, this pull decreases pressure in the pleural cavity. You can appreciate this effect by putting water on the palms of your hands and then placing them together. As you gently pull your hands apart, you will feel a sensation of negative pressure.

When pleural pressure is lower than alveolar pressure, the alveoli tend to expand. This expansion is opposed by the tendency of the lungs to recoil. Therefore, the alveoli expand when the pleural pressure is low enough that lung recoil is overcome. If the pleural pressure is not low enough to overcome lung recoil, the alveoli collapse.

## Changing Alveolar Volume

Changes in alveolar volume cause the changes in alveolar pressure that are responsible for moving air into and out of the lungs (see figure 15.11). Alveolar volume changes result from changes in pleural pressure. For example, during inspiration, pleural pressure

# CLINICAL IMPACT

## Pneumothorax

A **pneumothorax** (noo-mō-thōr′aks) is the introduction of air into the pleural cavity, the space between the parietal and visceral pleurae that normally contains only pleural fluid. Air can enter by an external route, as when a sharp object, such as a bullet or a broken rib, penetrates the thoracic wall, or air can enter the pleural cavity by an internal route if alveoli at the lung surface rupture, as can occur in a patient with emphysema. When the pleural cavity is connected to the outside by such openings, the pressure in the pleural cavity increases and becomes equal to the air pressure outside the body. Thus, pleural pressure is also equal to alveolar pressure because pressure in the alveoli at the end of expiration is equal to air pressure outside the body. When pleural pressure and alveolar pressure are equal, there is no tendency for the alveoli to expand, lung recoil is unopposed, and the lungs collapse. A pneumothorax can occur in one lung while the other remains inflated because the two pleural cavities are separated by the mediastinum.

---

**Predict 5**

*Treatment of a pneumothorax involves closing the opening into the pleural cavity that caused the pneumothorax, then placing a tube into the pleural cavity. In order to inflate the lung, should this tube pump in air under pressure (as in blowing up a balloon), or should the tube apply suction? Explain.*

decreases, and the alveoli expand. The decrease in pleural pressure occurs for two reasons:

1. Increasing the volume of the thoracic cavity results in a decrease in pleural pressure because a change in volume affects pressure.
2. As the lungs expand, lung recoil increases, increasing the suction effect and lowering the pleural pressure. The increased lung recoil of the stretched lung is similar to the increased force generated in a stretched rubber band.

The events of inspiration and expiration can be summarized as follows:

1. During inspiration, pleural pressure decreases because of increased thoracic volume and increased lung recoil. As pleural pressure decreases, alveolar volume increases, alveolar pressure decreases, and air flows into the lungs.
2. During expiration, pleural pressure increases because of decreased thoracic volume and decreased lung recoil. As pleural pressure increases, alveolar volume decreases, alveolar pressure increases, and air flows out of the lungs.

## Respiratory Volumes and Capacities

**Spirometry** (spī-rom′ĕ-trē) is the process of measuring volumes of air that move into and out of the respiratory system, and the **spirometer** (spī-rom′ĕ-ter) is the device that measures these respiratory volumes. Measurements of the respiratory volumes can provide information about the health of the lungs. **Respiratory volumes** are measures of the amount of air movement during different portions of ventilation, whereas **respiratory capacities** are sums of two or more respiratory volumes. The four respiratory volumes and their normal values for a young adult male are shown in figure 15.12:

1. **Tidal volume** is the volume of air inspired or expired with each breath. At rest, quiet breathing results in a tidal volume of about 500 milliliters (mL).
2. **Inspiratory reserve volume** is the amount of air that can be inspired forcefully beyond the resting tidal volume (about 3000 mL).
3. **Expiratory reserve volume** is the amount of air that can be expired forcefully beyond the resting tidal volume (about 1100 mL).
4. **Residual volume** is the volume of air still remaining in the respiratory passages and lungs after maximum expiration (about 1200 mL).

The tidal volume increases when a person is more active. Because the maximum volume of the respiratory system does not change from moment to moment, an increase in the tidal volume causes a decrease in the inspiratory and expiratory reserve volumes.

---

**Predict 6**

*The **minute ventilation** is the total amount of air moved into and out of the respiratory system each minute, and it is equal to the tidal volume times the respiratory rate. The **respiratory rate** is the number of breaths taken per minute. Calculate the minute ventilation of a resting person who has a tidal volume of 500 mL and a respiratory rate of 12 respirations/min and the minute ventilation of an exercising person who has a tidal volume of 4000 mL and a respiratory rate of 24 respirations/min.*

Values of respiratory capacities, the sum of two or more pulmonary volumes, are shown in figure 15.12:

1. **Functional residual capacity** is the expiratory reserve volume plus the residual volume. This is the amount of air remaining in the lungs at the end of a normal expiration (about 2300 mL at rest).

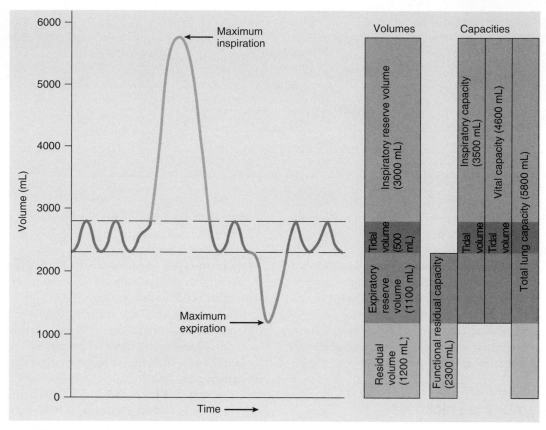

**Figure 15.12  Respiratory Volumes and Respiratory Capacities**

The tidal volume shown here is during resting conditions. Respiratory volumes are measurements of the volume of air moved into and out of the lungs during breathing. Respiratory capacities are the sum of two or more respiratory volumes.

2. **Inspiratory capacity** is the tidal volume plus the inspiratory reserve volume. This is the amount of air a person can inspire maximally after a normal expiration (about 3500 mL at rest).

3. **Vital capacity** is the sum of the inspiratory reserve volume, the tidal volume, and the expiratory reserve volume. It is the maximum volume of air that a person can expel from the respiratory tract after a maximum inspiration (about 4600 mL).

4. **Total lung capacity** is the sum of the inspiratory and expiratory reserves and the tidal and residual volumes (about 5800 mL). The total lung capacity is also equal to the vital capacity plus the residual volume.

Factors such as sex, age, and body size influence the respiratory volumes and capacities. For example, the vital capacity of adult females is usually 20–25% less than that of adult males. The vital capacity reaches its maximum amount in young adults and gradually decreases in the elderly. Tall people usually have a greater vital capacity than short people, and thin people have a greater vital capacity than obese people. Well-trained athletes can have a vital capacity 30–40% above that of untrained people. In patients whose respiratory muscles are paralyzed by spinal cord injury or diseases such as poliomyelitis or muscular dystrophy, the vital capacity can be reduced to values not consistent with survival (less than 500–1000 mL).

The **forced expiratory vital capacity** is the rate at which lung volume changes during direct measurement of the vital capacity. It is a simple and clinically important pulmonary test. The individual inspires maximally and then exhales maximally and as rapidly as possible into a spirometer. The spirometer records the volume of air expired per second. This test can help identify conditions in which the vital capacity might not be affected but the expiratory flow rate is reduced. Abnormalities that increase the resistance to airflow slow the rate at which air can be forced out of the lungs. For example, in people who have asthma, contraction of the smooth muscle in the bronchioles increases the resistance to airflow. In people who have emphysema, changes in the lung tissue result in the destruction of the alveolar walls, collapse of the bronchioles, and decreased elasticity of the lung tissue. The collapsed bronchioles increase the resistance to airflow. In people who have chronic bronchitis, the air passages are inflamed. The swelling, increased mucus secretion, and gradual loss of cilia result in narrowed bronchioles and increased resistance to airflow.

## GAS EXCHANGE

Ventilation supplies atmospheric air to the alveoli. The next step in the process of respiration is the diffusion of gases between the alveoli and the blood in the pulmonary capillaries. As previously

Respiratory

**TABLE 15.1    Partial Pressures of Gases at Sea Level**

| Gases | Dry Air | | Humidified Air | | Alveolar Air | | Expired Air | |
|---|---|---|---|---|---|---|---|---|
| | mm HG | % | mm HG | % | mm HG | % | mm HG | % |
| Nitrogen | 600.2 | 78.98 | 563.4 | 74.09 | 569.0 | 74.9 | 566.0 | 74.5 |
| Oxygen | 159.5 | 20.98 | 149.3 | 19.67 | 104.0 | 13.6 | 120.0 | 15.7 |
| Carbon dioxide | 0.3 | 0.04 | 0.3 | 0.04 | 40.0 | 5.3 | 27.0 | 3.6 |
| Water vapor | 0.0 | 0.0 | 47.0 | 6.20 | 47.0 | 6.2 | 47.0 | 6.2 |

stated, gas exchange between air and blood occurs in the respiratory membrane of the lungs (see figure 15.8). The major area of gas exchange is in the alveoli, although some takes place in the respiratory bronchioles and alveolar ducts. Gas exchange between blood and air does not occur in other areas of the respiratory passageways, such as the bronchioles, bronchi, and trachea. The volume of these passageways is therefore called **dead space.**

The exchange of gases across the respiratory membrane is influenced by the thickness of the membrane, the total surface area of the respiratory membrane, and the partial pressure of gases across the membrane.

## Respiratory Membrane Thickness

The thickness of the respiratory membrane increases during certain respiratory diseases. For example, in patients with pulmonary edema, fluid accumulates in the alveoli, and gases must diffuse through a thicker than normal layer of fluid. If the thickness of the respiratory membrane is doubled or tripled, the rate of gas exchange is markedly decreased. Oxygen exchange is affected before carbon dioxide exchange because oxygen diffuses through the respiratory membrane about 20 times less easily than does carbon dioxide.

## Surface Area

The total surface area of the respiratory membrane is about 70 square meters ($m^2$) in the normal adult, which is approximately the floor area of a 25- $\times$ 30-ft room, or roughly half the size of a basketball court. Under resting conditions, a decrease in the surface area of the respiratory membrane to one-third or one-fourth of normal can significantly restrict gas exchange. During strenuous exercise, even small decreases in the surface area of the respiratory membrane can adversely affect gas exchange. Possible reasons for having a decreased surface area include the surgical removal of lung tissue, the destruction of lung tissue by cancer, and the degeneration of the alveolar walls by emphysema. Collapse of the lung—as occurs in pneumothorax—dramatically reduces the volume of the alveoli, as well as the surface area for gas exchange.

## Partial Pressure

Gas molecules move randomly from higher concentration to lower concentration until an equilibrium is achieved. One measurement of the concentration of gases is partial pressure. The **partial pressure** of a gas is the pressure exerted by a specific gas in

a mixture of gases, such as air. For example, if the total pressure of all the gases in a mixture of gases is 760 millimeters of mercury (mm Hg), which is the atmospheric pressure at sea level, and 21% of the mixture is made up of oxygen, the partial pressure for oxygen is 160 mm Hg ($0.21 \times 760$ mm Hg = 160 mm Hg). If the composition of air is 0.04% carbon dioxide at sea level, the partial pressure for carbon dioxide is 0.3 mm Hg ($0.0004 \times 760 = 0.3$ mm Hg) (table 15.1). It is traditional to designate the partial pressure of individual gases in a mixture with a capital $P$ followed by the symbol for the gas. Thus, the partial pressure of oxygen is $P_{O_2}$, and that of carbon dioxide is $P_{CO_2}$.

When air is in contact with a liquid, such as water, gases in the air, such as carbon dioxide and oxygen, dissolve in the liquid. The gases dissolve until the partial pressure of each gas in the liquid is equal to the partial pressure of that gas in the air. Gases in a liquid, like gases in air, diffuse from areas of higher partial pressure toward areas of lower partial pressure, until the partial pressures of the gases are equal throughout the liquid.

### Diffusion of Gases in the Lungs

The cells of the body use oxygen and produce carbon dioxide. Thus, blood returning from tissues and entering the lungs has a decreased $P_{O_2}$ and an increased $P_{CO_2}$ compared to alveolar air (figure 15.13). Oxygen diffuses from the alveoli into the pulmonary capillaries because the $P_{O_2}$ in the alveoli is greater than that in the pulmonary capillaries. In contrast, carbon dioxide diffuses from the pulmonary capillaries into the alveoli because the $P_{CO_2}$ is greater in the pulmonary capillaries than in the alveoli.

When blood enters a pulmonary capillary, the $P_{O_2}$ and $P_{CO_2}$ in the capillary are different from the $P_{O_2}$ and $P_{CO_2}$ in the alveolus. By the time blood flows through the first third of the pulmonary capillary, an equilibrium is achieved, and the $P_{O_2}$ and $P_{CO_2}$ in the capillary are the same as in the alveolus. Thus, in the lungs, the blood gains oxygen and loses carbon dioxide.

During breathing, atmospheric air mixes with alveolar air. The air entering and leaving the alveoli keeps the $P_{O_2}$ higher in the alveoli than in the pulmonary capillaries. Increasing the breathing rate makes the $P_{O_2}$ even higher in the alveoli than it is during slow breathing. During labored breathing, the rate of oxygen diffusion into the pulmonary capillaries increases because the difference in partial pressure between the alveoli and the pulmonary capillaries has increased.

Increasing the rate of breathing also makes the $P_{CO_2}$ lower in the alveoli than it is during normal, quiet breathing. Because the

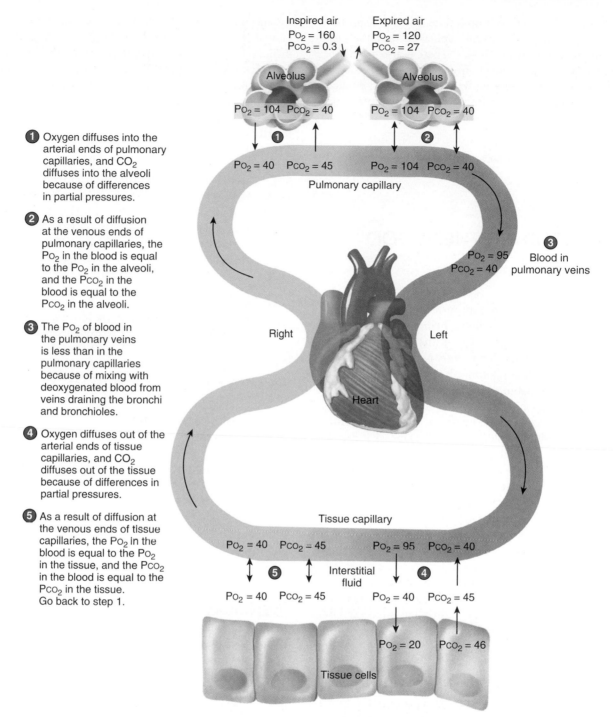

Inspired air
$Po_2 = 160$
$Pco_2 = 0.3$

Expired air
$Po_2 = 120$
$Pco_2 = 27$

Alveolus

Alveolus

$Po_2 = 104$  $Pco_2 = 40$

$Po_2 = 104$  $Pco_2 = 40$

1. Oxygen diffuses into the arterial ends of pulmonary capillaries, and $CO_2$ diffuses into the alveoli because of differences in partial pressures.

$Po_2 = 40$   $Pco_2 = 45$

$Po_2 = 104$  $Pco_2 = 40$

Pulmonary capillary

2. As a result of diffusion at the venous ends of pulmonary capillaries, the $Po_2$ in the blood is equal to the $Po_2$ in the alveoli, and the $Pco_2$ in the blood is equal to the $Pco_2$ in the alveoli.

$Po_2 = 95$
$Pco_2 = 40$

Blood in pulmonary veins

3. The $Po_2$ of blood in the pulmonary veins is less than in the pulmonary capillaries because of mixing with deoxygenated blood from veins draining the bronchi and bronchioles.

Right

Left

Heart

4. Oxygen diffuses out of the arterial ends of tissue capillaries, and $CO_2$ diffuses out of the tissue because of differences in partial pressures.

Tissue capillary

5. As a result of diffusion at the venous ends of tissue capillaries, the $Po_2$ in the blood is equal to the $Po_2$ in the tissue, and the $Pco_2$ in the blood is equal to the $Pco_2$ in the tissue. Go back to step 1.

$Po_2 = 40$   $Pco_2 = 45$

$Po_2 = 95$   $Pco_2 = 40$

Interstitial fluid

$Po_2 = 40$   $Pco_2 = 45$

$Po_2 = 40$   $Pco_2 = 45$

$Po_2 = 20$   $Pco_2 = 46$

Tissue cells

**PROCESS Figure 15.13  Gas Exchange**

Differences in partial pressure are responsible for the exchange of oxygen and carbon dioxide that occurs between the alveoli and the pulmonary capillaries and between the tissues and the tissue capillaries.

alveolar $Pco_2$ decreases, the difference in partial pressure between the alveoli and the pulmonary capillaries increases, which increases the rate of carbon dioxide diffusion from the pulmonary capillaries into the alveoli.

**Predict 7**

*How would inadequate ventilation affect the difference in $Po_2$ and $Pco_2$ across the respiratory membrane? How would the rate of oxygen and carbon dioxide diffusion across the membrane change?*

### Diffusion of Gases in the Tissues

Blood flows from the lungs through the left side of the heart to the tissue capillaries. Figure 15.13 illustrates the partial pressure differences for oxygen and carbon dioxide across the wall of a tissue capillary. Oxygen diffuses from the capillary into the interstitial fluid because the $Po_2$ in the interstitial fluid is lower than in the capillary. Oxygen diffuses from the interstitial fluid into cells, in which the $Po_2$ is less than in the interstitial fluid. Within the cells, oxygen is used in aerobic respiration. There is a constant difference in $P_{O_2}$ between the tissue capillaries and the cells because the cells continuously use oxygen. There is also a constant diffusion gradient for carbon dioxide from the cells to the tissue capillaries because the cells continuously produce carbon dioxide. Carbon dioxide therefore diffuses from cells into the interstitial fluid and from the interstitial fluid into the tissue capillaries.

**Predict 8**

*During exercise, more oxygen moves into skeletal muscle cells, and more carbon dioxide moves out of skeletal muscle cells. Explain how this happens.*

# GAS TRANSPORT IN THE BLOOD

## Oxygen Transport

After oxygen diffuses through the respiratory membrane into the blood, about 98.5% of the oxygen transported in the blood combines reversibly with the iron-containing heme groups of hemoglobin (see chapter 11). About 1.5% of the oxygen remains dissolved in the plasma. Hemoglobin with oxygen bound to its heme groups is called **oxyhemoglobin** (ok′sē-hē-mō-glō′bin).

The ability of hemoglobin to bind to oxygen depends on the $P_{O_2}$. At high $P_{O_2}$, hemoglobin binds to oxygen, and at low $P_{O_2}$, hemoglobin releases oxygen. In the lungs, $P_{O_2}$ normally is sufficiently high that hemoglobin holds as much oxygen as it can. In the tissues, $P_{O_2}$ is lower because the tissues are using oxygen. Consequently, hemoglobin releases oxygen in the tissues. Oxygen then diffuses into the cells, which use it in aerobic respiration. At rest, approximately 23% of the oxygen picked up by hemoglobin in the lungs is released to the tissues.

The amount of oxygen released from oxyhemoglobin is influenced by four factors. More oxygen is released from hemoglobin if (1) the $P_{O_2}$ is low, (2) the $P_{CO_2}$ is high, (3) the pH is low, and (4) the temperature is high. Increased muscular activity results in a decreased $P_{O_2}$, an increased $P_{CO_2}$, a reduced pH, and an increased temperature. Consequently, during physical exercise, as much as 73% of the oxygen picked up by hemoglobin in the lungs is released into skeletal muscles.

## Carbon Dioxide Transport and Blood pH

Carbon dioxide diffuses from cells, where it is produced, into the tissue capillaries. After carbon dioxide enters the blood, it is transported in three ways: (1) About 7% is transported as carbon dioxide dissolved in the plasma; (2) 23% is transported in combination with blood proteins, primarily hemoglobin; and (3) 70% is transported in the form of bicarbonate ions.

Carbon dioxide ($CO_2$) reacts with water to form carbonic acid ($H_2CO_3$), which then dissociates to form $H^+$ and bicarbonate ions ($HCO_3^-$):

$$CO_2 + H_2O \overset{\text{Carbonic anhydrase}}{\rightleftharpoons} H_2CO_3 \rightleftharpoons H^+ + HCO_3^-$$

Carbon | Water | Carbonic | Hydrogen | Bicarbonate
dioxide | | acid | ion | ion

An enzyme called **carbonic anhydrase** (kar-bon′ik an-hī′drās) is located inside red blood cells and on the surface of capillary epithelial cells. Carbonic anhydrase increases the rate at which $CO_2$ reacts with water to form $H^+$ and $HCO_3^-$ in the tissue capillaries (figure 15.14a). Thus, carbonic anhydrase promotes the uptake of $CO_2$ by red blood cells.

In the capillaries of the lungs, the process is reversed, so that the $HCO_3^-$ and $H^+$ combine to produce $H_2CO_3$, which then forms $CO_2$ and $H_2O$ (figure 15.14b). The $CO_2$ diffuses into the alveoli and is expired.

Carbon dioxide has an important effect on the pH of blood. As $CO_2$ levels increase, the blood pH decreases (becomes more acidic) because $CO_2$ reacts with $H_2O$ to form $H_2CO_3$. The $H^+$ that result from the dissociation of $H_2CO_3$ are responsible for the decrease in pH. Conversely, as blood levels of $CO_2$ decline, the blood pH increases (becomes less acidic, or more basic).

**Predict 9**

*What effect does rapid breathing have on blood pH? What effect does holding your breath have on blood pH? Explain.*

# RHYTHMIC BREATHING

The normal rate of breathing in adults is between 12 and 20 breaths per minute. In children, the rates are higher and may vary from 20 to 40 per minute. The rate of breathing is determined by the number of times respiratory muscles are stimulated. The basic rhythm of breathing is controlled by neurons within the medulla oblongata that stimulate the muscles of respiration. An increased depth of breathing results from stronger contractions of the respiratory muscles caused by recruitment of muscle fibers and increased frequency of stimulation of muscle fibers.

## Respiratory Areas in the Brainstem

Neurons involved with respiration are located in the brainstem. The neurons that are active during inspiration and those active during expiration are intermingled in these areas.

The **medullary respiratory center** consists of two **dorsal respiratory groups,** each forming a longitudinal column of cells located bilaterally in the dorsal part of the medulla oblongata, and two **ventral respiratory groups,** each forming a longitudinal column of cells located bilaterally in the ventral part of the medulla oblongata (figure 15.15). The dorsal respiratory groups are primarily responsible for stimulating contraction of the diaphragm. The ventral respiratory groups are primarily responsible for stimulating the external intercostal, internal intercostal, and abdominal muscles. A part of the ventral respiratory group, the **pre-Bötzinger complex,** is believed to establish the basic rhythm of breathing.

The **pontine respiratory group** is a collection of neurons in the pons (figure 15.15). It has connections with the medullary respiratory center and appears to play a role in switching between inspiration and expiration.

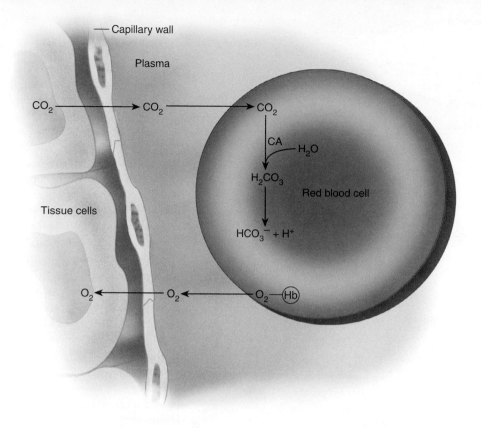

(a) **Gas exchange in the tissues**

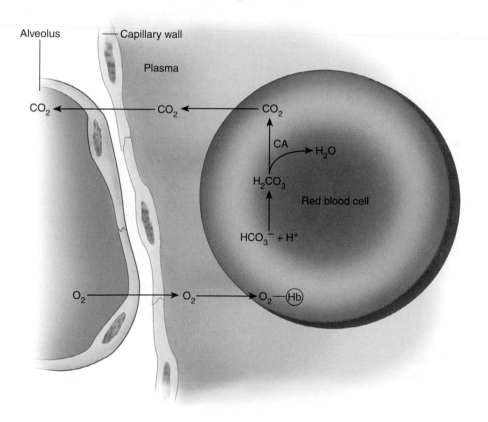

(b) **Gas exchange in the lungs**

**Figure 15.14** **Gas Exchange in the Tissues and in the Lungs**

(a) In the tissues, $CO_2$ diffuses into red blood cells, where the enzyme carbonic anhydrase (CA) is located. CA catalyzes the reaction of $CO_2$ with $H_2O$ to form carbonic acid ($H_2CO_3$). $H_2CO_3$ dissociates to form bicarbonate ions ($HCO_3^-$) and hydrogen ions ($H^+$). Oxygen is released from hemoglobin (Hb) and diffuses into tissue cells. (b) In the lungs, $CO_2$ diffuses from red blood cells into the alveoli. CA catalyzes the formation of $CO_2$ and $H_2O$ from $H_2CO_3$. $H^+$ and $HCO_3^-$ combine to replace $H_2CO_3$. Oxygen diffuses into red blood cells and binds to hemoglobin.

**Medial view**

**Anterior view**

**Figure 15.15  Respiratory Structures in the Brainstem**
Specific structures in the brainstem correlate with the nerves that innervate the muscles of respiration.

## Generation of Rhythmic Breathing

The medullary respiratory center generates the basic pattern of spontaneous, rhythmic breathing. Although the precise mechanism is not well understood, the generation of rhythmic breathing involves the integration of stimuli that start and stop inspiration.

1. *Starting inspiration.* The neurons in the medullary respiratory center that promote inspiration are continuously active. The medullary respiratory center constantly receives stimulation from many sources, such as receptors that monitor blood gas levels and the movements of muscles and joints. In addition, stimulation can come from parts of the brain concerned with voluntary respiratory movements and emotions. When the inputs from all these sources reach a threshold level, the neurons that stimulate respiratory muscles produce action potentials, and inspiration starts.
2. *Increasing inspiration.* Once inspiration begins, more and more neurons are activated. The result is progressively stronger stimulation of the respiratory muscles, which lasts for approximately 2 seconds (s).
3. *Stopping inspiration.* The neurons stimulating the muscles of respiration also stimulate the neurons in the medullary respiratory center that are responsible for stopping inspiration. The neurons responsible for stopping inspiration also receive input from the pontine respiratory neurons, stretch receptors in the lungs, and probably other sources. When the inputs to these neurons exceed a threshold level, they cause the neurons stimulating respiratory muscles to be inhibited. Relaxation of respiratory muscles results in expiration, which lasts approximately 3 s. The next inspiration begins with step 1.

Although the medullary neurons establish the basic rate and depth of breathing, their activities can be influenced by input from other parts of the brain and from peripherally located receptors.

## Nervous Control of Breathing

Higher brain centers can modify the activity of the respiratory center (figure 15.16*a*). For example, controlling air movements out of the lungs makes speech possible, and emotions can make us sob or gasp. In addition, breathing can be consciously controlled—that is, it is possible to breathe or to stop breathing voluntarily. Some people can hold their breath until they lose consciousness due to lack of oxygen in the brain. Children have used this strategy to encourage parents to give them what they want. However, as soon as conscious control of respiration is lost, automatic control resumes, and the person starts to breathe again.

Several reflexes, such as sneeze and cough reflexes, can modify breathing. The **Hering-Breuer** (her′ing broy′er) **reflex** supports rhythmic respiratory movements by limiting the extent of inspiration (figure 15.16*d*). As the muscles of inspiration contract, the lungs fill with air. Sensory receptors that respond to stretch are located in the lungs, and as the lungs fill with air, the stretch receptors are stimulated. Action potentials from the lung stretch receptors are then sent to the medulla oblongata, where they inhibit the respiratory center neurons and cause expiration. In infants, the Hering-Breuer reflex plays an important role in regulating the basic rhythm of breathing and in preventing overinflation of the lungs. In adults, however, the reflex is important only when the tidal volume is large, as occurs during heavy exercise.

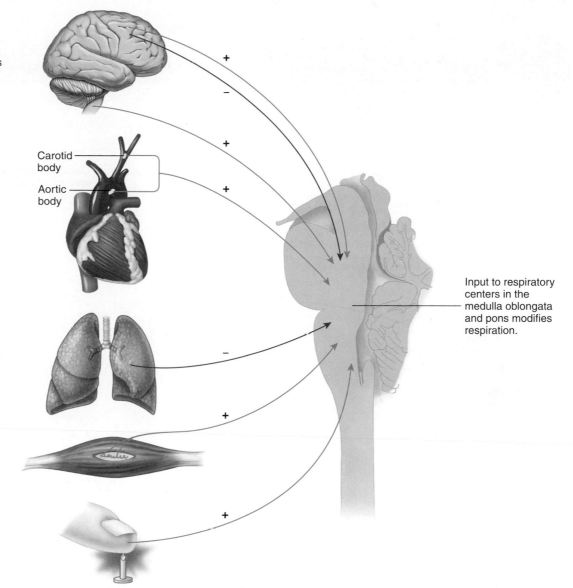

(a) Higher centers of
the brain (speech,
emotions, voluntary
control of breathing,
and action potentials
in motor pathways)

(b) Medullary
chemoreceptors
↓pH, ↑$CO_2$

(c) Carotid and
aortic body
chemoreceptors
↓$O_2$

Carotid body

Aortic body

(d) Hering-Breuer reflex
(stretch receptors
in lungs)

(e) Proprioceptors
in muscles
and joints

(f) Receptors for
touch, temperature,
and pain stimuli

Input to respiratory
centers in the
medulla oblongata
and pons modifies
respiration.

**Figure 15.16** **Nervous and Chemical Mechanisms of Breathing**

Several regulatory mechanisms affect the rate and depth of breathing. A plus sign indicates that the mechanism increases breathing and a minus sign indicates that it results in a decrease in breathing.

Touch, thermal, and pain receptors in the skin also stimulate the respiratory center, which explains why we gasp in response to being splashed with cold water or being pinched (figure 15.16e,f).

## Chemical Control of Breathing

During aerobic respiration, the body cells consume oxygen and produce carbon dioxide (see chapter 17). The respiratory system adds oxygen and removes carbon dioxide from the blood. The respiratory system is able to maintain these gases at homeostatic levels by monitoring gas levels and responding appropriately.

Carbon dioxide levels in the blood are the major driving force for regulating breathing. Even a small increase in the carbon dioxide level can cause an increase in ventilation. A large increase, as may occur while holding your breath, can result in a powerful

urge to breathe. A greater than normal amount of carbon dioxide in the blood is called **hypercapnia** (hī-per-kap′nē-ă).

Oxygen is not normally as important as carbon dioxide in regulating breathing because hemoglobin is very effective at picking up oxygen in the lungs. Normally, there is little variation in the oxygen content of blood leaving the lungs, despite changes in the body's oxygen demand. As long as blood carbon dioxide levels are normal, blood oxygen levels are usually normal as well.

Changes in blood carbon dioxide levels are not directly detected, however. Instead, changes in blood pH are monitored. This can occur because changes in carbon dioxide cause changes in pH (see "Carbon Dioxide Transport and Blood pH" in the previous section). Thus, the chemical regulation of ventilation accomplishes two things: (1) Homeostatic levels of carbon dioxide and oxygen are maintained, and (2) pH homeostasis is maintained.

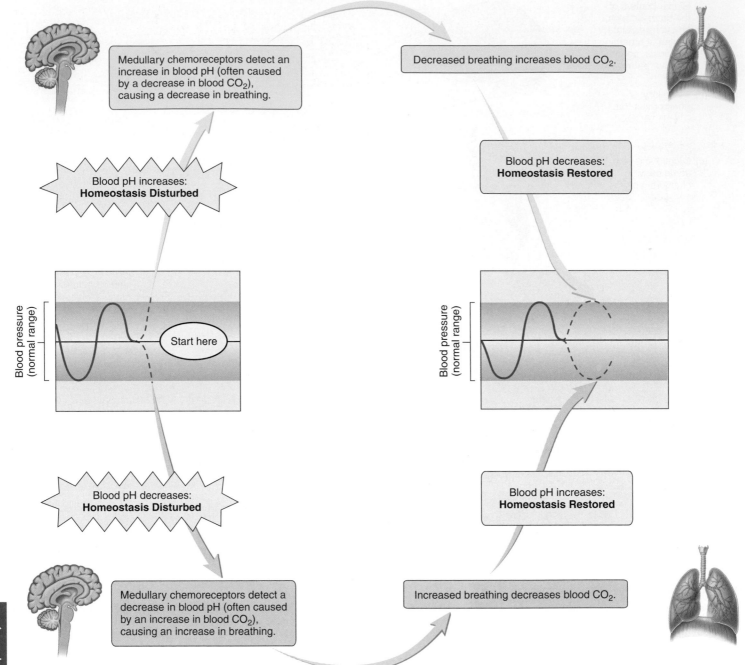

**Homeostasis Figure 15.17**   Regulation of Blood pH

---

*Predict 10*

*Explain why a person who breathes rapidly and deeply (hyperventilates) for several seconds experiences a short period of time during which breathing does not occur (apnea) before normal breathing resumes.*

**Chemoreceptors** in the medulla oblongata are sensitive to small changes in blood pH (figure 15.16b). Chemoreceptors can detect an increase in blood pH, typically caused by a decrease in blood carbon dioxide. As a result, the respiratory center decreases breathing and therefore less carbon dioxide is removed from the

blood. Because carbon dioxide is continually produced, carbon dioxide levels now increase, causing pH to decrease; thus, homeostasis is maintained (figure 15.17).

The medullary chemoreceptors also respond to a decrease in blood pH, typically caused by an increase in blood carbon dioxide. As a result, the respiratory center increases breathing and therefore carbon dioxide is removed from the blood. As blood carbon dioxide levels decrease, blood pH increases; again, homeostasis is maintained (figure 15.17).

Chemoreceptors in the carotid and aortic bodies also provide input to the respiratory center (see figure 15.16c). These

## Effects of High Altitude and Emphysema

Air is composed of 21% oxygen at both low and high altitudes. At sea level, the atmospheric pressure is 760 mm Hg, and $P_{O_2}$ is about 160 mm Hg (760 mm Hg $\times$ 0.21 = 160 mm Hg). At higher altitudes, the atmospheric pressure is lower, and $P_{O_2}$ is decreased. For example, at 10,000 ft above sea level, the atmospheric pressure is 523 mm Hg. Consequently, $P_{O_2}$ is 110 mm Hg (523 mm Hg $\times$ 0.21 = 110 mm Hg). Because $P_{O_2}$ is lower at high altitudes, the blood levels of oxygen can decline enough to stimulate the carotid and aortic bodies. Oxygen then becomes an important stimulus for elevating the rate and depth of breathing. At high altitudes, the respiratory system's ability to eliminate carbon dioxide is not adversely affected by the low atmospheric pressure. Thus, the blood carbon dioxide levels become lower than normal because of the increased rate and depth of breathing stimulated by the low blood oxygen levels. The decreased blood carbon dioxide levels cause the blood pH to become abnormally high.

A similar situation can exist in people who have emphysema. Carbon dioxide diffuses across the respiratory membrane more readily than does oxygen. Because emphysema is characterized by a decrease in the surface area of the respiratory membrane, levels of oxygen and carbon dioxide are both low. Just as occurs at high altitude, the low blood oxygen stimulates the carotid and aortic bodies to elevate the rate and depth of breathing. More severe emphysema, in which the surface area of the respiratory membrane is reduced to a minimum, can result in elevated blood levels of carbon dioxide as well.

chemoreceptors primarily respond to changes in blood oxygen. When blood oxygen declines to a low level, a condition called **hypoxia** (hī-pok′sē-ă; *hypo,* under + oxygen) results. Causes include exposure to high altitudes, emphysema, shock, and asphyxiation due to reduced oxygen levels but normal carbon dioxide levels. In these situations, the chemoreceptors of the carotid and aortic bodies are strongly stimulated. They send action potentials to the respiratory center and produce an increase in the rate and depth of breathing (figure 15.17). The increased breathing increases oxygen diffusion from the alveoli into the blood, resulting in more oxygen in the blood.

## Effect of Exercise on Breathing

The mechanisms by which breathing is regulated during exercise are controversial, and no single factor can account for all the observed responses. Breathing during exercise can be divided into two phases:

1. *Breathing increases abruptly.* At the onset of exercise, the rate of breathing immediately increases. This initial increase can be as much as 50% of the total increase that will occur. The immediate increase occurs too quickly to be explained by changes in metabolism or blood gases. As axons pass from the motor cortex of the cerebrum through the motor pathways, numerous collateral fibers project to the respiratory center. During exercise, action potentials in the motor pathways stimulate skeletal muscle contractions, and action potentials in the collateral fibers stimulate the respiratory center (see figure 15.16).

    Furthermore, during exercise, body movements stimulate proprioceptors in the joints of the limbs. Nerve fibers from these proprioceptors extend to the spinal cord to connect with sensory nerve tracts ascending to the brain. Collateral fibers from these nerve tracts connect to the respiratory center; therefore, movement of the limbs has a strong stimulatory influence on the respiratory center (see figure 15.16).

    There may also be a learned component in the breathing response during exercise. After a period of training, the brain "learns" to match breathing with the intensity of the exercise. Well-trained athletes match their respiratory movements more efficiently with their level of physical activity than do untrained individuals. Thus, centers in the brain involved in learning have an indirect influence on the respiratory center, but the exact mechanism is unclear.

2. *Breathing increases gradually.* After the immediate increase in breathing, breathing continues to increase gradually and then levels off within 4–6 minutes after the onset of exercise. Factors responsible for the immediate increase in breathing may play a role in the gradual increase as well.

Despite large changes in oxygen consumption and carbon dioxide production during exercise, the *average* arterial oxygen, carbon dioxide, and pH levels remain constant and close to resting levels as long as the exercise is aerobic (see chapter 7). This suggests that changes in blood gases and pH do not play an important role in regulating breathing during aerobic exercise. However, during exercise, the values of arterial oxygen, carbon dioxide, and pH levels rise and fall more than they do at rest. Thus, even though their average values do not change their oscillations may be a signal for helping control breathing.

The highest level of exercise that can be performed without causing a significant change in blood pH is the **anaerobic threshold.** If the exercise intensity becomes high enough to exceed the anaerobic threshold, skeletal muscles produce lactic

Respiratory

## Asthma

Will is an 18-year-old track athlete in seemingly good health. Despite suffering from a slight cold, Will went jogging one morning with his running buddy, Al. After a few minutes of exercise, Will felt that he could hardly get enough air. Even though he stopped jogging, he continued to breathe rapidly and wheeze forcefully. Because his condition was not improving, Al took him to the emergency room of a nearby hospital.

The emergency room doctor used a stethoscope to listen to Will's lungs and noted that air movement was poor. Although Will had no history of asthma, the emergency room doctor was convinced that he was having an asthma attack. Will inhaled a bronchodilator drug, which rapidly improved his condition. He was released from the emergency room and referred to his personal physician for further treatment and education about asthma.

### Background Information

**Asthma** (az'mă; difficult breathing) is characterized by abnormally increased constriction of the trachea and bronchi in response to various stimuli, which results in narrowed air passageways and decreased ventilation efficiency. Symptoms include rapid and shallow breathing, wheezing, coughing, and shortness of breath. In contrast to many other respiratory disorders, the symptoms of asthma typically reverse either spontaneously or with therapy.

In the United States, asthma occurs in 3–6% of the general population. In approximately half the cases, the symptoms appear before age 10, and twice as many boys as girls develop asthma. The exact cause of asthma is unknown, but it is more common in some families, where it often occurs in conjunction with allergies. There is no definitive pathological feature or diagnostic test for asthma, but three important characteristics of the disease are chronic airway inflammation, airway hyperreactivity, and airflow obstruction. The inflammation results in tissue damage, edema, and mucus buildup, which can block airflow through the bronchi. Airway hyperreactivity means that the smooth muscle in the trachea and bronchi contracts greatly in response to a stimulus, thus decreasing the diameter of the airway and increasing resistance to airflow. The effects of inflammation and airway hyperreactivity combine to cause airflow obstruction.

Many cases of asthma appear to be associated with a chronic inflammatory response by the immune system. The number of immune cells in the bronchi, including mast cells, eosinophils, neutrophils, macrophages,

**Figure 15A**

Strenuous exercise is one of the many factors that can bring on an asthma attack.

and lymphocytes, increases. Inflammation appears to be linked to airway hyperreactivity by some chemical mediators released by immune cells (e.g., leukotrienes, prostaglandins, and interleukins), which increase the airway's sensitivity to stimulation and cause smooth muscle contraction.

The stimuli that prompt airflow obstruction in asthma vary from one individual to another. Some asthmatics react to particular allergens, which are foreign substances that evoke an inappropriate immune system response (see chapter 14). Examples include inhaled pollen, animal dander, and dust mites. Many cases of asthma are caused by an allergic reaction to substances in the droppings and carcasses of cockroaches, which may explain the higher rate of asthma in poor, urban areas. However, other inhaled substances, such as chemicals in the workplace or cigarette smoke, can provoke an asthma attack without stimulating an allergic reaction. Over 200 substances have been associated with occupational asthma. An asthma attack can also

acid through anaerobic respiration (see figure 17.5). Lactic acid released into the blood decreases blood pH, which stimulates the carotid bodies, resulting in increased breathing. In fact, ventilation can increase so much that arterial carbon dioxide levels fall below resting levels, and arterial oxygen levels rise above resting levels.

## RESPIRATORY ADAPTATIONS TO EXERCISE

In response to training, athletic performance increases because the cardiovascular and respiratory systems become more efficient at delivering oxygen and picking up carbon dioxide. In

| System Interactions | Effects of Asthma on Other Systems |
|---|---|
| **System** | **Interactions** |
| Integumentary | Cyanosis, a bluish skin color, results from decreased blood oxygen content. |
| Muscular | Skeletal muscles are necessary for respiratory movements and the cough reflex. Increased muscular work during a severe asthma attack can cause metabolic acidosis because of anaerobic respiration and excessive lactic acid production. |
| Skeletal | Many of the immune cells responsible for the inflammatory response of asthma are produced in the red bone marrow. The thoracic cage is necessary for respiration. |
| Nervous | Emotional upset or stress can provoke an asthma attack. Peripheral and central chemoreceptor reflexes affect ventilation. The cough reflex helps remove mucus from respiratory passages. Pain, anxiety, and death from asphyxiation can result from the altered gas exchange caused by asthma. One theory of the cause of asthma is an imbalance in the autonomic nervous system (ANS) control of bronchiolar smooth muscle, and drugs that enhance the sympathetic effects or block the parasympathetic effects are used to treat asthma. |
| Endocrine | Steroids from the adrenal gland help regulate inflammation and are used in asthma therapy. |
| Cardiovascular | Increased vascular permeability of lung blood vessels results in edema. Blood carries ingested substances that provoke an asthma attack to the lungs. Blood also carries immune cells from red bone marrow to the lungs. Tachycardia commonly occurs during an asthma attack, and the normal effects of respiration on venous return are exaggerated, resulting in large fluctuations in blood pressure. |
| Lymphatic and Immune | Immune cells release chemical mediators that promote inflammation, increase mucus production, and cause bronchiolar constriction, which is believed to be a major factor in asthma. Ingested allergens, such as aspirin or sulfites in food, can provoke an asthma attack. |
| Digestive | Ingested substances, such as aspirin, sulfiting agents (preservatives), tartrazine, certain foods, and reflux of stomach acid into the esophagus can provoke an asthma attack. |
| Urinary | Modifying hydrogen ion secretion into the urine helps compensate for acid-base imbalances caused by asthma. |

be stimulated by ingested substances, such as aspirin; nonsteroidal anti-inflammatory compounds, such as ibuprofen (ī-bū′prō-fen); sulfites in food preservatives; and tartrazine (tar′tră-zēn) in food colorings. Asthmatics can substitute acetaminophen (as-et-ă-mē′nō-fen, a-set-ă-min′ō-fen) (e.g., Tylenol) for aspirin.

Other stimuli, such as strenuous exercise (especially in cold weather) can precipitate an asthma attack (figure 15A). Such episodes can often be avoided by using a bronchodilator prior to exercise. Viral infections, emotional upset, stress, air pollution, and even reflux of stomach acid into the esophagus are known to elicit an asthma attack.

Treatment of asthma involves avoiding the causative stimulus and taking medications. Steroids and mast cell–stabilizing agents, which prevent the release of chemical mediators from mast cells, can reduce airway inflammation. Theophylline (thē-of′i-lēn) and other drugs are commonly used to dilate the bronchi. Treatment is generally effective in controlling asthma. In rare cases, death by asphyxiation occurs, but it can be prevented through intensive therapy.

### Predict 11

*It is not usually necessary to assess arterial blood gases when diagnosing and treating asthma. However, this information can sometimes be useful in severe asthma attacks. Suppose that Will had a $P_{O_2}$ of 60 mm Hg and a $P_{CO_2}$ of 30 mm Hg when he first went to the emergency room. Explain how that could happen.*

most individuals, breathing does not limit performance because breathing can increase to a greater extent than can cardiovascular function.

After training, vital capacity increases slightly, and residual volume decreases slightly. Tidal volume at rest and during standardized, submaximal exercise (activities normally encountered in everyday life) does not change. At maximal exercise, however, the tidal volume increases. After training, the respiratory rate at rest or during standardized submaximal exercise is slightly lower, but at maximal exercise, the respiratory rate is generally increased.

Minute ventilation is affected by changes in tidal volume and breathing rate. After training, minute ventilation is essentially

# DISEASES AND DISORDERS  Respiratory System

| CONDITION | DESCRIPTION |
|---|---|
| | **RESPIRATORY DISORDERS** |
| **Bronchi and Lungs** | |
| Bronchitis (brong-kī'tis) | Inflammation of the bronchi caused by irritants, such as cigarette smoke or infections; swelling impairs breathing; bronchitis can progress to emphysema |
| Emphysema (em-fi-sē'mă) | Destruction of alveolar walls; increased coughing increases pressure on the alveoli, causing rupture and destruction; loss of alveoli decreases surface area for gas exchange and decreases the lungs' ability to expel air; progression can be slowed, but there is no cure; in combination with bronchitis, the condition is known as **chronic obstructive pulmonary disease (COPD)** |
| Adult respiratory distress syndrome (ARDS) | Caused by damage to the respiratory membrane, which promotes inflammation; amount of surfactant is reduced, and fluid fills the alveoli, lessening gas exchange; ARDS usually develops after an injurious event, such as inhaling smoke from a fire or breathing toxic fumes |
| Cystic fibrosis (fi-brō'sis) | Genetic disorder that affects mucus secretions throughout the body due to an abnormal transport protein; mucus is much more viscous and accumulates in ducts and tubes, such as the bronchioles; airflow is restricted, and infections are more likely |
| Pulmonary fibrosis | Replacement of lung tissue with fibrous connective tissue, making the lungs less elastic; exposure to asbestos or coal dust are common causes |
| Lung cancer | Occurs in the epithelium of the respiratory tract; can easily spread to other parts of the body because of the rich blood and lymphatic supply to the lungs |
| **Circulatory System** | Inadequate blood flow through the pulmonary capillaries, affecting respiratory function |
| Thrombosis of the pulmonary arteries | Blood clot in lung blood vessels |
| Anemia | Reduced hemoglobin lowers oxygen-carrying capacity of blood |
| Carbon monoxide | Carbon monoxide binds more strongly to hemoglobin than does oxygen and prevents already-bound oxygen from entering tissues |
| **Nervous System** | |
| Sudden infant death syndrome (SIDS) | Most frequent cause of death of infants between 2 weeks and 1 year of age; cause is still unknown, but at-risk babies can be placed on monitors that warn if breathing stops |
| Paralysis of the respiratory muscles | Damage to the spinal cord in the cervical or thoracic region interrupts nervous signals to the muscles of respiration |
| **Thoracic Wall** | Decreased elasticity of the thoracic wall prevents it from expanding to full capacity and reduces air movement; two spine curvature conditions that reduce elasticity of the thoracic wall are scoliosis (skō-lē-ō'sis) and kyphosis (kī-fō'sis) |
| | **INFECTIOUS DISEASES OF THE RESPIRATORY SYSTEM** |
| **Upper Respiratory Tract** | |
| Strep throat | Caused by streptococcal bacteria (*Streptococcus pyogenes*); characterized by inflammation of the pharynx and fever |
| Diphtheria (dif-thēr'ē-ă) | Caused by the bacterium *Corynebacterium diphtheriae*; a grayish membrane forms in the throat and can completely block respiratory passages; DPT immunization for children partially targets diphtheria |
| Common cold | Results from a viral infection |
| **Lower Respiratory Tract** | |
| Whooping cough (pertussis; per-tŭs'is) | Caused by the bacterium *Bordetella pertussis,* which destroys cilia lining the respiratory epithelium, allowing mucus to accumulate; leads to a very severe cough; DPT immunization for children partially targets pertussis |
| Tuberculosis (tū-ber'kyū-lō'sis) | Caused by the bacterium *Clostridium tuberculosis,* which forms small, lumplike lesions called tubercles; immune system targets tubercles and causes larger lesions; certain strains of tuberculosis are resistant to antibiotics |
| Pneumonia (noo-mō'nē-ă) | Many bacterial or viral infections of the lungs that cause fever, difficulty in breathing, and chest pain; edema in the lungs reduces their inflation ability and reduces gas exchange |
| Flu (influenza; in-flū-en'ză) | Viral infection of the respiratory system; does not affect the digestive system, as is commonly misunderstood; causes chills, fever, headache, and muscle aches |
| Fungal diseases | Fungal spores enter the respiratory tract attached to dust particles, usually resulting in minor respiratory infections that in some cases can spread to other parts of the body |

Go to **www.mhhe.com/seeleyess7** for additional information on these pathologies.

Respiratory

unchanged or slightly reduced at rest, slightly reduced during standardized submaximal exercise, and greatly increased at maximal exercise. For example, an untrained person with a minute ventilation of 120 liters per minute (L/min) can increase his or her minute ventilation to 150 L/min after training. Increases to 180 L/min are typical of highly trained athletes.

# EFFECTS OF AGING ON THE RESPIRATORY SYSTEM

Aging affects most aspects of the respiratory system. But even though vital capacity, maximum ventilation rates, and gas exchange decrease with age, the elderly can engage in light to moderate exercise because the respiratory system has a large reserve capacity.

With age, mucus accumulates within the respiratory passageways. The mucus-cilia escalator is less efficient because the mucus becomes more viscous and the number of cilia and their rate of movement decrease. As a consequence, the elderly are more susceptible to respiratory infections and bronchitis.

Vital capacity decreases with age because of reduced ability to fill the lungs (decreased inspiratory reserve volume) and to empty the lungs (decreased expiratory reserve volume). As a result, maximum minute ventilation rates decrease, which in turn decreases the ability to perform intense exercise. These changes are related to the weakening of respiratory muscles and the stiffening of cartilage and ribs.

Residual volume increases with age as the alveolar ducts and many of the larger bronchioles increase in diameter. This increases the dead space, which decreases the amount of air available for gas exchange. In addition, gas exchange across the respiratory membrane declines because parts of the alveolar walls are lost, which decreases the surface area available for gas exchange, and the remaining walls thicken, which decreases the diffusion of gases. A gradual increase in resting tidal volume with age compensates for these changes.

 SUMMARY

Respiration includes ventilation, the movement of air into and out of the lungs; the exchange of gases between the air and the blood; the transport of gases in the blood; and the exchange of gases between the blood and the tissues.

## Functions of the Respiratory System  (p. 407)

The respiratory system exchanges oxygen and carbon dioxide between the air and the blood, regulates blood pH, helps produce sounds, moves air over the sensory receptors that detect smell, and protects against some microorganisms.

## Anatomy of the Respiratory System  (p. 407)

### Nose

1. The nose consists of the external nose and the nasal cavity.
2. The bridge of the nose is bone, and most of the external nose is cartilage.
3. The nasal cavity warms, humidifies, and cleans the air. The nares open to the outside, and the choanae lead to the pharynx. The nasal cavity is divided by the nasal septum into right and left parts. The paranasal sinuses and the nasolacrimal duct open into the nasal cavity. Hairs just inside the nares trap debris. The nasal cavity is lined with pseudostratified epithelium containing cilia that traps debris and moves it to the pharynx.

### Pharynx

1. The nasopharynx joins the nasal cavity through the choanae and contains the opening to the auditory tube and the pharyngeal tonsils.
2. The oropharynx joins the oral cavity and contains the palatine and lingual tonsils.
3. The laryngopharynx opens into the larynx and the esophagus.

### Larynx

1. The larynx consists of three unpaired cartilages and six paired ones. The thyroid cartilage and cricoid cartilage form most of the larynx. The epiglottis covers the opening of the larynx during swallowing.
2. The vestibular folds can prevent air, food, and liquids from passing into the larynx.

3. The vocal folds (true vocal cords) vibrate and produce sounds when air passes through the larynx. The force of air movement controls loudness, and changes in the length and tension of the vocal folds determine pitch.

### Trachea

The trachea connects the larynx to the main bronchi.

### Bronchi

The main bronchi extend from the trachea to each lung.

### Lungs

1. There are two lungs.
2. The airway passages of the lungs branch and decrease in size. The main bronchi form the lobar bronchi, which go to each lobe of the lungs. The lobar bronchi form the segmental bronchi, which extend to each bronchopulmonary segment of the lungs. The segmental bronchi branch many times to form the bronchioles. The bronchioles branch to form the terminal bronchioles, which give rise to the respiratory bronchioles, from which alveolar ducts branch. Alveoli are air sacs connected to the alveolar ducts and respiratory bronchioles.
3. The epithelium from the trachea to the terminal bronchioles is ciliated to facilitate removal of debris. Cartilage helps hold the tube system open (from the trachea to the bronchioles). Smooth muscle controls the diameter of the tubes (especially the bronchioles). The alveoli are formed by simple squamous epithelium, and they facilitate diffusion of gases.
4. The respiratory membrane has six layers, including a film of water, the walls of the alveolus and the capillary, and an interstitial space. The respiratory membrane is thin and has a large surface area that facilitates gas exchange.

### Pleural Cavities

The pleural membranes surround the lungs and provide protection against friction.

## Lymphatic Supply

Superficial and deep lymphatic vessels drain the lungs.

## Ventilation and Respiratory Volumes    (p. 416)

### Changing Thoracic Volume

1. Inspiration occurs when the diaphragm contracts and the external intercostal muscles lift the rib cage, thus increasing the volume of the thoracic cavity. During labored breathing, additional muscles of inspiration increase rib movement.
2. Expiration can be passive or active. Passive expiration during quiet breathing occurs when the muscles of inspiration relax. Active expiration during labored breathing occurs when the diaphragm relaxes and the internal intercostal and abdominal muscles depress the rib cage to forcefully decrease the volume of the thoracic cavity.

### Pressure Changes and Airflow

1. Respiratory muscles cause changes in thoracic volume, which in turn cause changes in alveolar volume and pressure.
2. During inspiration, air flows into the alveoli because atmospheric pressure is greater than alveolar pressure.
3. During expiration, air flows out of the alveoli because alveolar pressure is greater than atmospheric pressure.

### Lung Recoil

1. The lungs tend to collapse because of the elastic recoil of the connective tissue and the surface tension of the fluid lining the alveoli.
2. The lungs normally do not collapse because surfactant reduces the surface tension of the fluid lining the alveoli, and pleural pressure is lower than alveolar pressure.

### Changing Alveolar Volume

1. Increasing thoracic volume results in decreased pleural pressure, increased alveolar volume, decreased alveolar pressure, and air movement into the lungs.
2. Decreasing thoracic volume results in increased pleural pressure, decreased alveolar volume, increased alveolar pressure, and air movement out of the lungs.

### Respiratory Volumes and Capacities

1. There are four measurements of respiratory volume: tidal, inspiratory reserve, expiratory reserve, and residual.
2. Respiratory capacities are the sum of two or more respiratory volumes; they include vital capacity and total lung capacity.
3. The forced expiratory vital capacity measures the rate at which air can be expelled from the lungs.

## Gas Exchange    (p. 421)

1. Gas exchange between air and blood occurs in the respiratory membrane.
2. The parts of the respiratory passageways where gas exchange between air and blood does not occur constitute the dead space.

### Respiratory Membrane Thickness

Increases in the thickness of the respiratory membrane result in decreased gas exchange.

### Surface Area

Small decreases in surface area adversely affect gas exchange during strenuous exercise. When the surface area is decreased to one-third to one-fourth of normal, gas exchange is inadequate under resting conditions.

### Partial Pressure

1. The pressure exerted by a specific gas in a mixture of gases is reported as the partial pressure of that gas.

2. Oxygen diffuses from a higher partial pressure in the alveoli to a lower partial pressure in the pulmonary capillaries. Oxygen diffuses from a higher partial pressure in the tissue capillaries to a lower partial pressure in the tissue spaces.
3. Carbon dioxide diffuses from a higher partial pressure in the tissues to a lower partial pressure in the tissue capillaries. Carbon dioxide diffuses from a higher partial pressure in the pulmonary capillaries to a lower partial pressure in the alveoli.

## Gas Transport in the Blood    (p. 424)

### Oxygen Transport

1. Most (98.5%) oxygen is transported bound to hemoglobin. Some (1.5%) oxygen is transported dissolved in plasma.
2. Oxygen is released from hemoglobin in tissues when the partial pressure for oxygen is low, the partial pressure for carbon dioxide is high, pH is low, and temperature is high.

### Carbon Dioxide Transport and Blood pH

1. Carbon dioxide is transported in solution as plasma (7%), in combination with blood proteins (23%), and as bicarbonate ions (70%).
2. In tissue capillaries, carbon dioxide combines with water inside the red blood cells to form carbonic acid that dissociates to form $HCO_3^-$ and $H^+$. This reaction promotes the transport of carbon dioxide.
3. In lung capillaries, $HCO_3^-$ combine with $H^+$ to form carbonic acid. The carbonic acid dissociates to form carbon dioxide, which diffuses out of the red blood cells.
4. As blood carbon dioxide levels increase, blood pH decreases; as blood carbon dioxide levels decrease, blood pH increases. Changes in breathing change blood carbon dioxide levels and pH.

## Rhythmic Breathing    (p. 424)

### Respiratory Areas in the Brainstem

1. The medullary respiratory center (dorsal respiratory and ventral respiratory groups) establishes rhythmic breathing.
2. The pontine respiratory group appears to be involved with the switch between inspiration and expiration.

### Generation of Rhythmic Breathing

1. Inspiration begins when stimuli from many sources, such as receptors that monitor blood gases, reach a threshold.
2. Expiration begins when the neurons causing inspiration are inhibited.

### Nervous Control of Breathing

1. Higher brain centers allow voluntary control of breathing. Emotions and speech production affect breathing.
2. The Hering-Breuer reflex inhibits the respiratory center when the lungs are stretched during inspiration.
3. Touch, thermal, and pain receptors can stimulate breathing.

### Chemical Control of Breathing

1. Chemoreceptors in the medulla oblongata respond to changes in blood pH. Usually, changes in blood pH are produced by changes in blood carbon dioxide.
2. Carbon dioxide is the major chemical regulator of breathing. An increase in blood carbon dioxide causes a decrease in blood pH, resulting in increased breathing.
3. Low blood levels of oxygen can stimulate chemoreceptors in the carotid and aortic bodies, also resulting in increased breathing.

### Effect of Exercise on Breathing

Input from higher brain centers and from proprioceptors stimulates the respiratory center during exercise.

## Respiratory Adaptations to Exercise  (p. 430)

Training results in increased minute volume at maximal exercise because of increased tidal volume and respiratory rate.

## Effects of Aging on the Respiratory System  (p. 433)

1. Vital capacity and maximum minute ventilation decrease with age because of weakening of the respiratory muscles and stiffening of the thoracic cage.

2. Residual volume and dead space increase because the diameter of respiratory passageways increases.

3. An increase in resting tidal volume compensates for increased dead space, loss of alveolar walls (surface area), and thickening of alveolar walls.

4. The ability to remove mucus from the respiratory passageways decreases with age.

 REVIEW AND COMPREHENSION

1. Define respiration.

2. What are the functions of the respiratory system?

3. Describe the structures of the nasal cavity and their functions.

4. Name the three parts of the pharynx. With what structures does each part communicate?

5. Name and give the functions of the three unpaired cartilages of the larynx.

6. What are the functions of the vestibular and vocal folds? How are sounds of different loudness and pitch produced?

7. Starting at the larynx, name in order all the tubes air passes through to reach an alveolus.

8. What is the function of the C-shaped cartilages in the trachea? What happens to the amount of cartilage in the tube system of the respiratory system as the tubes become smaller? Explain why breathing becomes more difficult during an asthma attack.

9. What is the function of the ciliated epithelium in the tracheobronchial tree?

10. Distinguish among the lungs, a lobe of the lung, and a bronchopulmonary segment.

11. List the components of the respiratory membrane.

12. Describe the pleurae of the lungs. What is their function?

13. Describe the lymphatic supply of the lungs. What is its function?

14. Explain how the muscles of inspiration and expiration change thoracic volume.

15. Describe the pressure changes that cause air to move into and out of the lungs. What causes these pressure changes?

16. Give two reasons the lungs tend to recoil. What two factors keep the lungs from collapsing?

17. Explain how changes in thoracic volume result in changes in pleural pressure, alveolar volume, alveolar pressure, and airflow during inspiration and expiration.

18. Define tidal volume, inspiratory reserve volume, expiratory reserve volume, and residual volume. Define vital capacity, total lung capacity, and forced expiratory vital capacity.

19. Describe the factors that affect the diffusion of gases across the respiratory membrane. Give some examples of diseases that decrease diffusion by altering these factors.

20. What is the partial pressure of a gas? Describe the diffusion of oxygen and carbon dioxide between the alveoli and the pulmonary capillaries and between the tissue capillaries and the tissues in terms of partial pressures.

21. List the ways in which blood transports oxygen. What factors promote the release of oxygen in tissues?

22. List the ways in which blood transports carbon dioxide.

23. How does the level of blood carbon dioxide affect blood pH? How can changes in ventilation affect blood pH?

24. Name the respiratory areas of the brainstem, and explain how rhythmic breathing is generated.

25. Describe how higher brain centers and the Hering-Breuer reflex can modify breathing.

26. Explain the role of blood pH, carbon dioxide, and oxygen in modifying breathing.

27. How is breathing regulated during exercise?

28. What effect does exercise training have on the respiratory system?

29. Why do vital capacity, alveolar ventilation, and diffusion of gases across the respiratory membrane decrease with age? Why are the elderly more likely to develop respiratory infections and bronchitis?

 CRITICAL THINKING

1. Cardiopulmonary resuscitation (CPR) has replaced former methods of sustaining ventilation. The back pressure/arm lift method is one such technique that is no longer used. This procedure must be performed with the victim lying face down. The rescuer presses firmly on the base of the scapulae for several seconds, then grasps the arms and lifts them. The sequence is then repeated. Although this procedure is less efficient than CPR, it does result in ventilation of the lungs. Explain why.

2. Another technique for artificial respiration is mouth-to-mouth resuscitation. The rescuer takes a deep breath, blows air into the victim's mouth, and then lets air flow out of the victim's lungs. The process is repeated. Explain the following:
   a. Why do the victim's lungs expand?
   b. Why does air move out of the victim's lungs?

3. A person's vital capacity was measured while she was standing and while she was lying down. What difference, if any, in the measurement would you predict, and why?

4. If water vapor forms 10% of the gases in air at sea level, what is the partial pressure of water?

5. A patient has pneumonia, and fluids accumulate within the alveoli. Explain why this results in an increased rate of respiration. How can oxygen therapy return this rate to normal?.

6. A patient has severe emphysema that has extensively damaged the alveoli and reduced the surface area of the respiratory membrane. Although the patient is receiving oxygen therapy, he still has a tremendous urge to take a breath (i.e., he does not feel as if he is getting enough air). Why does this occur?

7.  Patients with diabetes mellitus who are not being treated with insulin therapy rapidly metabolize lipids, and acidic by-products of lipid metabolism accumulate in the bloodstream. How does this affect ventilation? Why is the change in ventilation beneficial?

8.  Ima Anxious was hysterical and hyperventilating. The doctor made her breathe into a paper bag. Because you are an especially astute student, you say to the doctor, "When Ima was hyperventilating, she was reducing blood carbon dioxide levels; when she breathed into the paper bag, carbon dioxide was trapped in the bag, and she was rebreathing it, thus causing her blood carbon dioxide levels to increase. As blood carbon dioxide levels increase, the urge to breathe should have increased. Instead, she began to breathe more slowly. Please explain." How do you think the doctor would respond? (*Hint:* Recall that the effect of decreased blood carbon dioxide on the vasomotor center results in vasodilation and a sudden decrease in blood pressure.)

9.  Hyperventilating before swimming underwater can increase the time spent under water. Explain how that can happen. Sometimes a person who has hyperventilated before swimming under water passes out while still under water and drowns. Explain.

10. The blood pH of a runner was monitored during a race. Shortly after the beginning of the race, her blood pH increased for a short time. Propose an explanation to account for the increased pH values following the start of the race.

*Answers in Appendix D*

Visit this textbook's website at **www.mhhe.com/seeleyess7** for practice quizzes, animations, interactive learning exercises, and other study tools.

McGraw-Hill offers a study CD that features interactive cadaver dissection. ***Anatomy & Physiology Revealed*** includes cadaver photos that allow you to peel away layers of the human body to reveal structures beneath the surface. This program also includes animations, radiologic imaging, audio pronunciations, and practice quizzing.

The pizza this boy is eating will activate all aspects of his digestive system. Pizza contains most of the major nutrients: carbohydrates, proteins, and fat. The body's ability to break down, separate, and appropriately distribute these nutrients is amazing.

# 16 Chapter

# Digestive System

## Chapter Outline and Objectives

**Functions of the Digestive System (p. 438)**

1. List the major functions of the digestive system.

**Anatomy and Histology of the Digestive System (p. 438)**

2. Describe the general histology of the digestive tract.

**Oral Cavity, Pharynx, and Esophagus (p. 440)**

3. Describe the structure of a tooth.
4. Describe the major salivary glands. Compare their structures and functions.
5. Describe mastication and swallowing.

**Stomach (p. 446)**

6. Outline the anatomical and physiological characteristics of the stomach.
7. Describe the stomach secretions, their functions, and their regulation.
8. Describe gastric movements and stomach emptying and how they are regulated.

**Small Intestine (p. 451)**

9. List the characteristics of the small intestine that account for its large surface area.

10. Describe the secretions and movements that occur in the small intestine.

**Liver and Pancreas (p. 453)**

11. Describe the anatomy, histology, and ducts of the liver and pancreas.
12. Describe the major functions of the liver and pancreas, and explain how they are regulated.

**Large Intestine (p. 459)**

13. List the parts of the large intestine, and describe its anatomy and histology.
14. Describe the major functions of the large intestine, and explain how movement is regulated.

**Digestion, Absorption, and Transport (p. 460)**

15. Describe the digestion, absorption, and transport of carbohydrates, lipids, and proteins.
16. Discuss water movement into and out of the digestive tract.

**Effects of Aging on the Digestive System (p. 465)**

17. Describe the effects of aging on the digestive system.

Every cell of the body needs nourishment, yet most cells cannot leave their position in the body and travel to a food source; therefore, the food must be delivered. The **digestive system** (figure 16.1), with the help of the circulatory system, is like a gigantic "meals on wheels," serving over 100 trillion customers the nutrients they need. It also has its own quality control and waste disposal methods. Food is taken into the digestive system, where it is broken down into smaller and smaller particles. Enzymes in the digestive system break the particles down into very small molecules, which are absorbed into the circulation and transported all over the body. There, those molecules are broken down by other enzymes to release energy or are assembled into new molecules to build tissues and organs. This chapter describes the structure and function of the digestive organs and their accessory glands.

## FUNCTIONS OF THE DIGESTIVE SYSTEM

The functions of the digestive system are to

1. *Take in food.* Food and water enter the body through the mouth.
2. *Break down food.* During the process of digestion, food is broken down from complex molecules to smaller molecules that can be absorbed.
3. *Absorb digested molecules.* The small molecules that result from digestion are absorbed through the walls of the intestine for use in the body.
4. *Provide nutrients.* The process of digestion and absorption provides the body with water, electrolytes, and other nutrients, such as vitamins and minerals.
5. *Eliminate wastes.* Undigested material, such as fiber from food, plus waste products excreted into the digestive tract are eliminated in the feces.

## ANATOMY AND HISTOLOGY OF THE DIGESTIVE SYSTEM

The digestive system consists of the **digestive tract,** or *gastrointestinal (GI;* gas′trō-in-tes′tin-ăl) *tract,* a tube extending from the mouth to the anus, plus the associated organs, which secrete fluids into the digestive tract. The inside of the digestive tract is continuous with the outside environment, where it opens at the mouth and anus. Nutrients cross the wall of the digestive tract to enter the circulation.

The digestive tract consists of the oral cavity, pharynx, esophagus, stomach, small intestine, large intestine, and anus. Accessory glands are associated with the digestive tract (figure 16.1). The salivary glands empty into the oral cavity, and the liver and pancreas are connected to the small intestine.

Various parts of the digestive tract are specialized for different functions, but nearly all parts consist of four layers, or tunics: the mucosa, the submucosa, the muscularis, and a serosa or an adventitia, shown in figure 16.2:

1. The innermost tunic, the **mucosa** (mū-kō′să), consists of **mucous epithelium,** a loose connective tissue called the **lamina propria,** and a thin smooth muscle layer, the **muscularis mucosae.** The epithelium in the mouth, esophagus, and anus resists abrasion, and the epithelium in the stomach and intestine absorbs and secretes.

2. The **submucosa** lies just outside the mucosa. It is a thick layer of loose connective tissue containing nerves, blood vessels, and small glands. An extensive network of nerve cell processes forms a **plexus** (network). The plexus is innervated by autonomic nerves.

3. The next tunic is the **muscularis,** which in most parts of the digestive tract consists of an inner layer of **circular smooth muscle** and an outer layer of **longitudinal smooth muscle.** Another nerve plexus, also innervated by autonomic nerves, lies between the two muscle layers. Together, the nerve plexuses of the submucosa and muscularis compose the **enteric** (en-tĕr′ik) **nervous system.** This nervous system is extremely important in controlling movement and secretion within the tract.

4. The fourth, or outermost, layer of the digestive tract is either a serosa or an adventitia. The **serosa** consists of the peritoneum, which is a smooth epithelial layer, and its underlying connective tissue. Regions of the digestive tract not covered by peritoneum are covered by a connective tissue layer called the **adventitia** (ad′ven-tish′ă; foreign, coming from outside), which is continuous with the surrounding connective tissue.

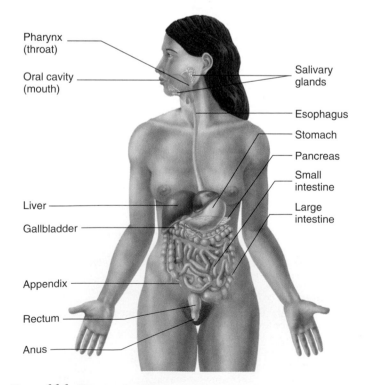

**Figure 16.1** **Digestive System**

Pharynx (throat)
Oral cavity (mouth)
Salivary glands
Esophagus
Stomach
Pancreas
Small intestine
Liver
Large intestine
Gallbladder
Appendix
Rectum
Anus

## Peritonitis

Peritonitis (per'i-tō-nī'tis) is a potentially life-threatening inflammation of the peritoneal membranes. The inflammation can result from chemical irritation by substances, such as bile, that have escaped from the digestive tract. Or it can result from infection originating in the digestive tract, as may occur when an infected appendix ruptures. The main symptoms of peritonitis are acute abdominal pain and tenderness that are worsened by movement.

## Peritoneum

The body wall of the abdominal cavity and the abdominal organs are covered with serous membranes (figure 16.3). The serous membrane that covers the organs is the serosa, or **visceral peritoneum** (per'i-tō-nē'ŭm; to stretch over). The serous membrane that lines the wall of the abdominal cavity is the **parietal peritoneum.**

Many of the organs of the abdominal cavity are held in place by connective tissue sheets called **mesenteries** (mes'en-ter-ēz). *Mesentery* is a general term referring to the serous membranes attached to the abdominal organs. The mesenteries consist of two layers of serous membranes with a thin layer of loose connective tissue between them. The mesentery connecting the lesser curvature of the stomach to the liver and diaphragm is called the **lesser omentum** (ō-men'tŭm), and the mesentery connecting the greater curvature of the stomach to the transverse colon and posterior body wall is called the **greater omentum.** The greater omentum is unusual in that it is a long, double fold of mesentery that extends inferiorly from the stomach before looping back to the transverse colon to create a cavity, or pocket, called the **omental bursa** (ber'să). Adipose tissue accumulates in the greater omentum, giving it the appearance of a fat-filled apron that covers the anterior surface of the abdominal viscera. The mesentery that attaches the small intestine to the posterior abdominal wall is called the **mesentery proper.**

### Predict 1

*If you placed a pin completely through both folds of the greater omentum, through how many layers of simple squamous epithelium would the pin pass?*

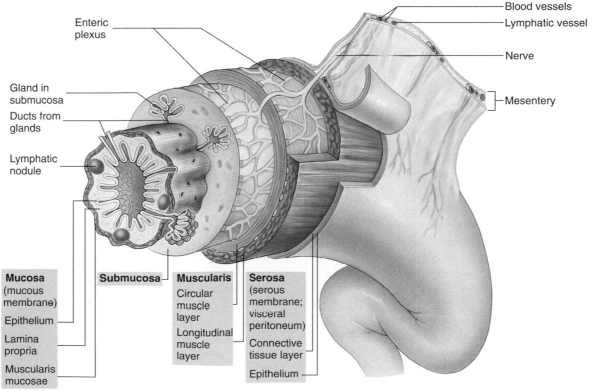

**Figure 16.2 Digestive Tract Histology**

The four tunics are the mucosa, the submucosa, the muscularis, and a serosa or an adventitia. Glands may exist along the digestive tract as part of the epithelium, within the submucosa, or as large glands outside the digestive tract.

Labels:
- Enteric plexus
- Gland in submucosa
- Ducts from glands
- Lymphatic nodule
- Blood vessels
- Lymphatic vessel
- Nerve
- Mesentery
- **Mucosa** (mucous membrane)
  - Epithelium
  - Lamina propria
  - Muscularis mucosae
- **Submucosa**
- **Muscularis**
  - Circular muscle layer
  - Longitudinal muscle layer
- **Serosa** (serous membrane; visceral peritoneum)
  - Connective tissue layer
  - Epithelium

**Figure 16.3 Peritoneum and Mesenteries**
                                                            **Medial view**

The parietal peritoneum lines the abdominal cavity (*blue*), and the visceral peritoneum covers abdominal organs (*red*). Retroperitoneal organs are behind the parietal peritoneum. The mesenteries are membranes that connect abdominal organs to each other and to the body wall.

Other abdominal organs lie against the abdominal wall, have no mesenteries, and are described as **retroperitoneal** (re′trō-per′i-tō-nē′ăl; behind the peritoneum). The retroperitoneal organs include the duodenum, pancreas, ascending colon, descending colon, rectum, kidneys, adrenal glands, and urinary bladder.

# ORAL CAVITY, PHARYNX, AND ESOPHAGUS

## Anatomy of the Oral Cavity

The **oral cavity** (figure 16.4), or mouth, is the first part of the digestive tract. It is bounded by the lips and cheeks and contains the teeth and tongue. The **lips** are muscular structures, formed mostly by the **orbicularis oris** (ōr-bik′ū-lā′ris ōr′is) **muscle** (see figure 7.15). The outer surfaces of the lips are covered by skin. The keratinized stratified epithelium of the skin becomes thin at the margin of the lips. The color from the underlying blood vessels can be seen through the thin, transparent epithelium, giving the lips a reddish-pink appearance. At the internal margin of the lips, the epithelium is continuous with the moist stratified squamous epithelium of the mucosa in the oral cavity.

The cheeks form the lateral walls of the oral cavity. The **buccinator** (bŭk′si-nā-tōr) **muscles** (see figure 7.15), located within the cheeks, flatten the cheeks against the teeth. The lips and cheeks are important in the process of **mastication** (mas-ti-kā′shŭn), or chewing. They help manipulate the food within the oral cavity and hold the food in place while the teeth crush or tear it. Mastication begins the process of mechanical digestion, which breaks down large food particles into smaller ones. The cheeks also help form words during the speech process.

The **tongue** is a large, muscular organ that occupies most of the oral cavity. The major attachment of the tongue is in the posterior part of the oral cavity. The anterior part of the tongue is relatively free, except for an anterior attachment to the floor of the mouth by a thin fold of tissue called the **frenulum** (fren′ū-lŭm) (figure 16.4). The muscles associated with the tongue are described in chapter 7.

The tongue moves food in the mouth and, in cooperation with the lips and cheeks, holds the food in place during mastication. It also plays a major role in the process of swallowing. In addition, the tongue is a major sensory organ for taste, as well as one of the major organs of speech.

### Teeth

There are 32 **teeth** in the normal adult mouth, located in the mandible and maxillae. The teeth can be divided into quadrants: right upper, left upper, right lower, and left lower. In adults, each quadrant contains one central and one lateral **incisor** (in-sī′zŏr; to cut); one **canine** (kā′nīn; dog); first and second **premolars** (prē-mō′lărz; *molaris,* a millstone); and first, second, and third **molars** (mō′lărz). The third molars are called **wisdom teeth** because they usually appear in the late teens or early twenties, when the person is old enough to have acquired some degree of wisdom.

The teeth of adults are called **permanent teeth,** or *secondary teeth* (figure 16.5a). Most of them are replacements for the 20 **primary teeth,** or *deciduous* (dē-sid′ū-ŭs) *teeth,* also called milk or baby teeth, which are lost during childhood (figure 16.5b).

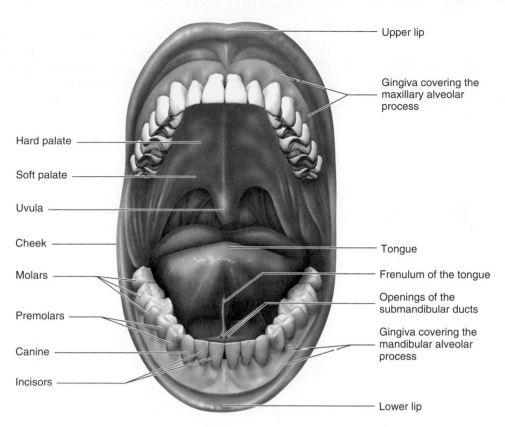

- Upper lip
- Gingiva covering the maxillary alveolar process
- Hard palate
- Soft palate
- Uvula
- Cheek
- Molars
- Premolars
- Canine
- Incisors
- Tongue
- Frenulum of the tongue
- Openings of the submandibular ducts
- Gingiva covering the mandibular alveolar process
- Lower lip

**Figure 16.4  Oral Cavity**

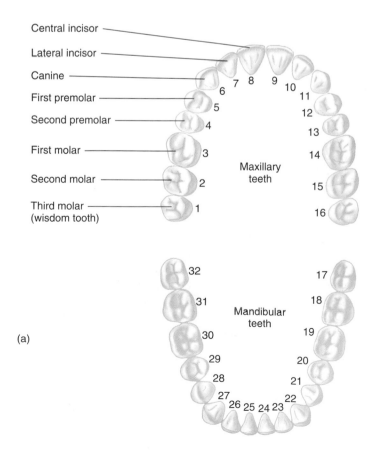

Central incisor
Lateral incisor
Canine
First premolar
Second premolar
First molar
Second molar
Third molar (wisdom tooth)

Maxillary teeth

Mandibular teeth

(a)

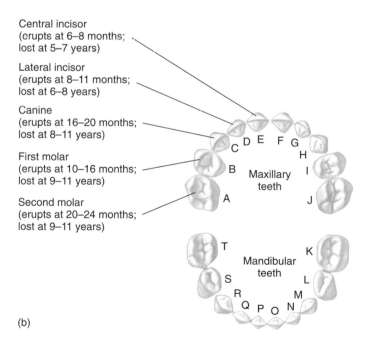

Central incisor (erupts at 6–8 months; lost at 5–7 years)

Lateral incisor (erupts at 8–11 months; lost at 6–8 years)

Canine (erupts at 16–20 months; lost at 8–11 years)

First molar (erupts at 10–16 months; lost at 9–11 years)

Second molar (erupts at 20–24 months; lost at 9–11 years)

Maxillary teeth

Mandibular teeth

(b)

**Figure 16.5  Teeth**

(a) Permanent teeth. (b) Deciduous teeth. Dental professionals have developed a "universal" numbering and lettering system for convenience in identifying individual teeth.

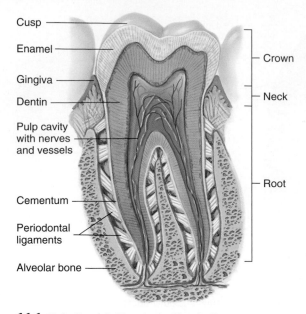

**Figure 16.6** Molar Tooth in Place in the Alveolar Bone

A tooth consists of a crown, a neck, and a root. The root is covered with cementum, and the tooth is held in the socket by periodontal ligaments. Nerves and vessels enter and exit the tooth through a foramen in the part of the root deepest in the alveolus.

Each tooth consists of a **crown** with one or more **cusps** (points), a **neck,** and a **root** (figure 16.6). The center of the tooth is a **pulp cavity,** which is filled with blood vessels, nerves, and connective tissue, called **pulp.** The pulp cavity is surrounded by a living, cellular, bonelike tissue called **dentin** (den′tin; *dens,* tooth). The dentin of the tooth crown is covered by an extremely hard, acellular substance called **enamel,** which protects the tooth against abrasion and acids produced by bacteria in the mouth. The surface of the dentin in the root is covered with **cementum** (se-men′tŭm), which helps anchor the tooth in the jaw.

The teeth are rooted within **alveoli** (al-vē′ō-lī; sockets) along the alveolar processes of the mandible and maxillae. The alveolar processes are covered by dense fibrous connective tissue and moist stratified squamous epithelium, referred to as the **gingiva** (jin′ji-vă), or gums. The teeth are held in place by **periodontal** (per′ē-ō-don′tăl; around the teeth) **ligaments,** which are connective tissue fibers that extend from the alveolar walls and are embedded into the cementum.

Formation of **dental caries** (kār′ēz), or tooth decay, is the result of the breakdown of enamel by acids produced by bacteria on the tooth surface. Enamel is nonliving and cannot repair itself. Consequently, a dental filling is necessary to prevent further damage. **Periodontal disease** is inflammation and degeneration of the periodontal ligaments, gingiva, and alveolar bone. This disease is the most common cause of tooth loss in adults.

**A CASE IN POINT**

**Lymph Nodes and a Toothache**

Tu Thake went to his dentist for a checkup because he had experienced pain in his right mandible for several days. During the routine exam, the dentist felt along the anterior and posterior edges of his sternocleidomastoid muscle on

each side and along the inferior edges of the mandible on each side. Just inferior to the angle of the right mandible, the dentist noted a lump, suggesting an enlargement of the superior cervical lymph nodes. Enlargement of the superior cervical lymph nodes indicates a problem in the face, because all lymphatic drainage from the face goes through these nodes. Additional swellings along the inferior edge of the anterior or central part of the body of the mandible suggest a problem in the mandible. The "problem" could be an infection or a cancerous growth, or it could be idiopathic (of unknown origin). An infection may occur in a tooth, in the bone, or in the soft tissues of the area. In Tu's case, further examination revealed a small abscess near his right, first mandibular molar. The dentist opened the abscess and treated it with topical and systemic antibiotics. The infection disappeared, as did the swelling in the superior cervical lymph nodes.

### Palate and Tonsils

The **palate** (pal′ăt), or roof of the oral cavity, separates the oral cavity from the nasal cavity and prevents food from passing into the nasal cavity during chewing and swallowing. The palate consists of two parts. The anterior part contains bone and is called the **hard palate,** whereas the posterior portion consists of skeletal muscle and connective tissue and is called the **soft palate** (see figure 16.4). The **uvula** (ū′vū-lă; a grape) is a posterior extension of the soft palate.

The **tonsils** (ton′silz) are located in the lateral posterior walls of the oral cavity, in the nasopharynx, and in the posterior surface of the tongue. The tonsils are described in chapter 14.

### Salivary Glands

There are three pairs of **salivary** (sal′i-văr-ē) **glands:** the parotid, submandibular, and sublingual glands (figure 16.7). They produce

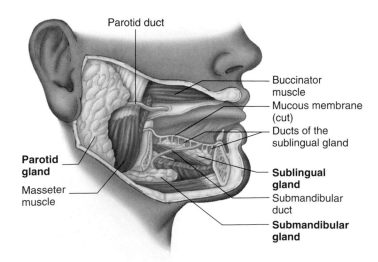

**Figure 16.7** Salivary Glands

The large salivary glands are the parotid glands, the submandibular glands, and the sublingual glands.

| TABLE 16.1 | Functions of Digestive Secretions | |
|---|---|---|
| **Fluid or Enzyme** | **Source** | **Function** |
| **Mouth** | | |
| Saliva (water, bicarbonate ions, mucus) | Salivary glands | Moistens and lubricates food, neutralizes bacterial acids, flushes bacteria from oral cavity |
| Salivary amylase | Salivary glands | Digests starch |
| Lysozyme | Salivary glands | Has weak antibacterial action |
| **Stomach** | | |
| Hydrochloric acid | Gastric glands | Kills bacteria, converts pepsinogen to pepsin |
| Pepsin* | Gastric glands | Digests protein |
| Mucus | Mucous cells | Protects stomach lining |
| Intrinsic factor | Gastric glands | Binds to vitamin $B_{12}$, aids in its absorption |
| **Small Intestine and Associated Glands** | | |
| Bile salts | Liver | Emulsify fats |
| Bicarbonate ions | Pancreas | Neutralize stomach acid |
| Trypsin*, chymotrypsin*, carboxypeptidase* | Pancreas | Digest protein |
| Pancreatic amylase | Pancreas | Digests starch |
| Lipase | Pancreas | Digests lipid (triglycerides) |
| Nucleases | Pancreas | Digest nucleic acid |
| Mucus | Duodenal glands and goblet cells | Protects duodenum from stomach acid and digestive enzymes |
| Peptidases** | Small intestine | Digest polypeptide |
| Sucrase** | Small intestine | Digests sucrose |
| Lactase** | Small intestine | Digests lactose |
| Maltase** | Small intestine | Digests maltose |

*These enzymes are secreted as inactive forms, then activated.

**These enzymes remain in the microvilli.

saliva, which is a mixture of **serous** (watery) and **mucous** fluids. The salivary glands are compound alveolar glands. They have branching ducts with clusters of alveoli, resembling grapes, at the ends of the ducts (see chapter 4).

The largest of the salivary glands, the **parotid** (pă-rot′id; beside the ear) **glands,** are serous glands located just anterior to each ear. Parotid ducts enter the oral cavity adjacent to the second upper molars.

**Mumps** (mŭmpz) is an inflammation of the parotid gland caused by a viral infection. The inflamed parotid glands become swollen, often making the cheeks quite large. The virus causing mumps can also infect other structures. Mumps in an adult male may involve the testes and can result in sterility.

The **submandibular** (sŭb-man-dib′ū-lăr; below the mandible) **glands** produce more serous than mucous secretions. Each gland can be felt as a soft lump along the inferior border of the mandible. The submandibular ducts open into the oral cavity on each side of the frenulum of the tongue (see figure 16.4).

The **sublingual** (sŭb-ling′gwăl; below the tongue) **glands,** the smallest of the three paired salivary glands, produce primarily mucous secretions. They lie immediately below the mucous membrane in the floor of the oral cavity. Each

sublingual gland has 10–12 small ducts opening onto the floor of the oral cavity.

## Saliva

**Saliva** (să-lī′vă) helps keep the oral cavity moist and contains enzymes that begin the process of digestion. Saliva is secreted at the rate of approximately 1 liter (L) per day. The serous part of saliva, produced mainly by the parotid and submandibular glands, contains a digestive enzyme called **salivary amylase** (am′il-ās) (table 16.1), which breaks the covalent bonds between glucose molecules in starch and other polysaccharides to produce the disaccharides maltose and isomaltose. Maltose and isomaltose have a sweet taste; thus, the digestion of polysaccharides by salivary amylase enhances the sweet taste of food.

Food spends very little time in the mouth. Consequently, only about 5% of the total carbohydrates humans absorb are digested in the mouth. Also, most starches are contained in plant cells, which are surrounded by cell walls composed primarily of the polysaccharide **cellulose** (sel′ū-lōs). Humans lack the necessary enzymes to digest cellulose. Cooking and thorough chewing of food disrupt the cellulose covering and increase the efficiency of the digestive process.

Digestive

Saliva prevents bacterial infection in the mouth by washing the oral cavity, and it contains **lysozyme** (lī'sō-zīm), an enzyme that has a weak antibacterial action. The bicarbonate ions in saliva act as a buffer to neutralize the acids produced by oral bacteria. This reduces the harmful effects of bacterial acids on tooth enamel. Lack of salivary gland secretion (which can result from radiation therapy) increases the chance of ulceration and infection of the oral mucosa and caries (cavities) formation in the teeth.

The serous part of saliva dissolves molecules, which must be in solution to stimulate taste receptors. The mucous secretions of the submandibular and sublingual glands contain a large amount of **mucin** (mū'sin), a proteoglycan that gives a lubricating quality to the secretions of the salivary glands.

Salivary gland secretion is regulated primarily by the autonomic nervous system, with parasympathetic stimulation the most important. Salivary secretions increase in response to a variety of stimuli, such as tactile stimulation in the oral cavity and certain tastes, especially sour. Higher brain centers can stimulate parasympathetic activity and thus increase the activity of the salivary glands in response to the thought of food, to odors, or to the sensation of hunger. Sympathetic stimulation increases the mucous content of saliva. When a person becomes frightened and the sympathetic division of the autonomic nervous system is stimulated, the person may have a dry mouth with thick mucus.

## Mastication

Food taken into the mouth is chewed, or masticated, by the teeth. The incisors and canines primarily cut and tear food, whereas the premolars and molars primarily crush and grind it. Mastication breaks large food particles into many small ones, which have a much larger total surface area than a few large particles would have. Because digestive enzymes act on molecules only at the surface of the food particles, mastication increases the efficiency of digestion.

## Pharynx

The **pharynx** (far'ingks), or throat, which connects the mouth with the esophagus, consists of three parts: the nasopharynx, the oropharynx, and the laryngopharynx (see chapter 15). Normally, only the oropharynx and laryngopharynx transmit food. The posterior walls of the oropharynx and laryngopharynx are formed by the superior, middle, and inferior **pharyngeal constrictor muscles.**

## Esophagus

The **esophagus** (ē-sof'ă-gŭs; gullet) is a muscular tube, lined with moist stratified squamous epithelium, that extends from the pharynx to the stomach. It is about 25 centimeters (cm) long and lies anterior to the vertebrae and posterior to the trachea within the mediastinum. It passes through the diaphragm and ends at the stomach. The esophagus transports food from the pharynx to the stomach. Upper and lower **esophageal sphincters,** located at the upper and lower ends of the esophagus, respectively, regulate the movement of food into and out of the esophagus. The lower esophageal sphincter is sometimes called the **cardiac sphincter.** Numerous mucous glands produce a thick, lubricating mucus that coats the inner surface of the esophagus.

## Swallowing

**Swallowing,** or *deglutition* (dē-gloo-tish'ŭn), can be divided into three phases: the voluntary phase, the pharyngeal phase, and the esophageal phase (figure 16.8). During the **voluntary phase,** a bolus, or mass of food, is formed in the mouth. The bolus is pushed by the tongue against the hard palate, forcing the bolus toward the posterior part of the mouth and into the oropharynx.

The **pharyngeal phase** of swallowing is a reflex that is initiated when a bolus of food stimulates receptors in the oropharynx. This phase of swallowing begins with the elevation of the soft palate, which closes the passage between the nasopharynx and oropharynx. The pharynx elevates to receive the bolus of food from the mouth. The three **pharyngeal constrictor muscles** then contract in succession, forcing the food through the pharynx. At the same time, the upper esophageal sphincter relaxes, and food is pushed into the esophagus. As food passes through the pharynx, the vestibular and vocal folds close, and the **epiglottis** (ep-i-glot'is; upon the glottis, opening of the larynx) is tipped posteriorly, so that the opening into the larynx is covered. These movements prevent food from passing into the larynx.

### Predict 2

*What would happen if a person had a cleft of the soft palate, so that the soft palate did not completely close the passage between the nasopharynx and the oropharynx during swallowing? What happens if a person has an explosive burst of laughter while trying to swallow a liquid? What happens if a person tries to swallow and speak at the same time?*

Digestive

1 During the **voluntary phase**, a bolus of food (*yellow*) is pushed by the tongue against the hard and soft palates and posteriorly toward the oropharynx (*blue arrow* indicates tongue movement; *black arrow* indicates movement of the bolus). *Tan:* bone; *purple:* cartilage; *red:* muscle.

2 During the **pharyngeal phase**, the soft palate is elevated, closing off the nasopharynx. The pharynx and larynx are elevated (*blue arrows* indicate muscle movement).

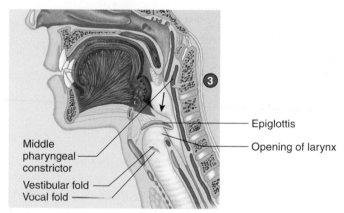

3 Successive constriction of the pharyngeal constrictors from superior to inferior (*blue arrows*) forces the bolus through the pharynx and into the esophagus. As this occurs, the vestibular and vocal folds expand medially to close the passage of the larynx. The epiglottis is bent down over the opening of the larynx largely by the force of the bolus pressing against it.

4 As the inferior pharyngeal constrictor contracts, the upper esophageal sphincter relaxes (outwardly directed *blue arrows*), allowing the bolus to enter the esophagus.

5 During the **esophageal phase**, the bolus is moved by peristaltic contractions of the esophagus toward the stomach (inwardly directed *blue arrows*).

**PROCESS Figure 16.8  Events During the Three Phases of Swallowing**

Digestive

## Hypertrophic Pyloric Stenosis

**Hypertrophic pyloric stenosis** is a common defect of the stomach in infants. It occurs in 1 in 150 males and 1 in 750 females. The pyloric sphincter is greatly thickened and thus interferes with normal stomach emptying. Infants with this defect develop symptoms 3 to 10 weeks after birth. The infants exhibit projectile (forceful) vomiting. Because the pyloric opening is blocked, little food enters the intestine, and the infant fails to gain weight. The defect can be corrected with surgery to enlarge the pyloric opening.

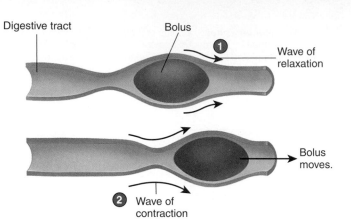

① A wave of smooth muscle relaxation moves ahead of the bolus, allowing the digestive tract to expand.

② A wave of contraction of the smooth muscle behind the bolus propels it through the digestive tract.

**PROCESS Figure 16.9** Peristalsis

The **esophageal phase** of swallowing is responsible for moving food from the pharynx to the stomach. Muscular contractions of the esophagus occur in **peristaltic** (per-i-stal′tik; *peri,* around + *stalsis,* constriction) **waves** (figure 16.9). A wave of relaxation of the esophageal muscles precedes the bolus of food down the esophagus, and a wave of strong contraction of the circular muscles follows and propels the bolus through the esophagus. Gravity assists the movement of material, especially liquids, through the esophagus. However, the peristaltic contractions that move material through the esophagus are sufficiently forceful to allow a person to swallow even while doing a headstand or floating in the zero-gravity environment of space. The peristaltic contractions cause relaxation of the lower esophageal sphincter in the esophagus as the peristaltic waves approach the stomach.

## STOMACH

### Anatomy of the Stomach

The **stomach** (figure 16.10) is an enlarged segment of the digestive tract in the left superior part of the abdomen. The opening from the esophagus into the stomach is called the **cardiac opening** because it is near the heart. The region of the stomach around the cardiac opening is called the **cardiac region.** The most superior part of the stomach is the **fundus** (fŭn′dŭs; the bottom of a round-bottomed leather bottle). The largest part of the stomach is the **body,** which turns to the right, forming a **greater curvature** on the left and a **lesser curvature** on the right. The opening from the stomach into the small intestine is the **pyloric** (pī-lōr′ik; gate-keeper) **opening,** which is surrounded by a relatively thick ring of smooth muscle called the **pyloric sphincter.** The region of the stomach near the pyloric opening is the **pyloric region.**

The muscular layer of the stomach is different from other regions of the digestive tract in that it consists of three layers: an outer longitudinal layer, a middle circular layer, and an inner oblique layer. These muscular layers produce a churning action in the stomach, important in the digestive process. The submucosa and mucosa of the stomach are thrown into large folds called **rugae** (roo′gē; wrinkles) (figure 16.10*a*) when the stomach is empty. These folds allow the mucosa and submucosa to stretch, and the folds disappear as the stomach is filled.

The stomach is lined with simple columnar epithelium. The mucosal surface forms numerous tubelike **gastric pits** (figure 16.10*b*), which are the openings for the **gastric glands.** The epithelial cells of the stomach can be divided into five groups. The first group consists of **surface mucous cells** on the inner surface of the stomach and lining the gastric pits. Those cells produce mucus, which coats and protects the stomach lining. The remaining four cell types are in the gastric glands. They are **mucous neck cells,** which produce mucus; **parietal cells,** which produce hydrochloric acid and intrinsic factor; **endocrine cells,** which produce regulatory chemicals and **chief cells,** which produce **pepsinogen** (pep-sin′ō-jen), a precursor of the protein-digesting enzyme **pepsin** (pep′sin; *pepsis,* digestion).

Digestive

**Figure 16.10  Anatomy and Histology of the Stomach**

(*a*) Cutaway section reveals muscular layers and internal anatomy. (*b*) A section of the stomach wall illustrates its histology, including several gastric pits and glands. (*c*) Photomicrograph of gastric glands.

Digestive

## Secretions of the Stomach

The stomach functions primarily as a storage and mixing chamber for ingested food. As food enters the stomach, it is mixed with stomach secretions to become a semifluid mixture called **chyme** (kīm; juice). Although some digestion occurs in the stomach, that is not its principal function.

Stomach secretions from the gastric glands include hydrochloric acid, pepsin, mucus, and intrinsic factor (table 16.1). **Hydrochloric acid** produces a pH of about 2.0 in the stomach. The acid kills microorganisms and activates **pepsin** from its inactive form called pepsinogen. Pepsin breaks covalent bonds of proteins to form smaller peptide chains. Pepsin exhibits optimum enzymatic activity at a pH of about 2.0. A thick layer of **mucus** lubricates and protects the epithelial cells of the stomach wall from the damaging effect of the acidic chyme and pepsin. Irritation of the stomach mucosa stimulates the secretion of a greater volume of mucus. **Intrinsic** (in-trin′sik) **factor** binds with vitamin B$_{12}$ and makes it more readily absorbed in the small intestine. Vitamin B$_{12}$ is important in deoxyribonucleic acid (DNA) synthesis and in red blood cell production.

## Regulation of Stomach Secretions

Approximately 2 L of gastric secretions (gastric juice) are produced each day. Both nervous and hormonal mechanisms regulate gastric secretions. The neural mechanisms involve central nervous system (CNS) reflexes integrated within the medulla oblongata. Higher brain centers can influence these reflexes. Local reflexes are integrated within the enteric plexus in the wall of the digestive tract and do not involve the CNS. Hormones produced by the stomach and intestine help regulate stomach secretions.

Regulation of stomach secretions can be divided into three phases: the cephalic, gastric, and intestinal phases. The cephalic phase can be viewed as the "get started" phase, when the stomach secretions are increased in anticipation of incoming food. This is followed by the gastric, "go for it," phase, when most of the stimulation of secretion occurs. Finally, the intestinal phase is the "slow-down" phase, during which stomach secretion decreases.

In the **cephalic** (se-fal′ik; *kephale,* head) **phase** (figure 16.11*a*), sensations of taste, the smell of food, stimulation of tactile receptors during the process of chewing and swallowing, and pleasant thoughts of food stimulate centers within the medulla oblongata that influence gastric secretions. Action potentials are sent from the medulla oblongata along parasympathetic axons within the vagus nerves to the stomach. Within the stomach wall, the preganglionic neurons stimulate postganglionic neurons in the enteric plexus. The postganglionic neurons stimulate secretory activity in the cells of the stomach mucosa, causing the release of hydrochloric acid, pepsin, mucus, and intrinsic factor. The neurons also stimulate the release of gastrin and histamine from endocrine cells. **Gastrin** (gas′trin) is a hormone that enters the circulation and is carried back to the stomach, where it stimulates additional secretory activity (table 16.2). Histamine is a paracrine chemical signal that acts locally and enters the blood to stimulate gastric gland secretory activity. Histamine is the most potent stimulator of hydrochloric acid secretion. Drugs that block the actions of histamine can lower acid levels.

The **gastric phase** is the period during which the greatest volume of gastric secretion occurs (figure 16.11*b*). The gastric phase is activated by the presence of food in the stomach. During the gastric phase, the food in the stomach is mixed with gastric secretions. Distention of the stomach stimulates stretch receptors. Action potentials generated by these receptors activate CNS reflexes and local reflexes, resulting in the cascade of events that increase secretion, as in the cephalic phase. Peptides, produced by the action of pepsin on proteins, stimulate the secretion of gastrin, which in turn stimulates additional hydrochloric acid secretion.

The **intestinal phase** of gastric secretion primarily inhibits gastric secretions (figure 16.11*c*). It is controlled by the entrance of acidic chyme into the duodenum, which initiates both neural and hormonal mechanisms. When the pH of the chyme entering the duodenum drops to 2.0 or below, the inhibitory influence of the intestinal phase is greatest. The hormone **secretin** (se-krē′tin), which inhibits gastric secretions, is released from the duodenum in response to low pH (table 16.2). Fatty acids and peptides in the duodenum initiate the release of the hormone **cholecystokinin** (kō′lē-sis-tō-kī′nin; *chole,* bile + *kystis,* bladder + *kineo,* to move), which also inhibits gastric secretions (table 16.2). Acidic chyme (pH < 2.0) in the duodenum also inhibits CNS stimulation and initiates local reflexes that inhibit gastric secretion.

To summarize, once gastric acid secretion begins, further secretion is controlled by negative-feedback loops involving nerves and hormones. First, during the gastric phase, high acid levels in the

| TABLE 16.2 | Functions of the Major Digestive System Hormones | |
|---|---|---|
| **Hormone** | **Source** | **Functions** |
| Gastrin | Gastric glands | • Increases gastric secretions |
| Secretin | Duodenum | • Decreases gastric secretions<br>• Increases pancreatic and bile secretions high in bicarbonate ions<br>• Decreases gastric motility |
| Cholecystokinin | Duodenum | • Decreases gastric secretions<br>• Strongly decreases gastric motility<br>• Increases gallbladder contraction<br>• Increases pancreatic enzyme secretion |

### Cephalic Phase

**1** The taste, smell, or thought of food or tactile sensations of food in the mouth stimulate the medulla oblongata (*green arrows*).

**2** Vagus nerves carry parasympathetic action potentials to the stomach (*pink arrow*), where enteric plexus neurons are activated.

**3** Postganglionic neurons stimulate secretion by parietal and chief cells and stimulate gastrin and histamine secretion by endocrine cells.

**4** Gastrin is carried through the circulation back to the stomach (*purple arrow*), where, along with histamine, it stimulates secretion.

(a)

### Gastric Phase

**1** Distention of the stomach stimulates mechanoreceptors (stretch receptors) and activates a parasympathetic reflex. Action potentials generated by the mechanoreceptors are carried by the vagus nerves to the medulla oblongata (*green arrow*).

**2** The medulla oblongata increases action potentials in the vagus nerves that stimulate secretions by parietal and chief cells and stimulate gastrin and histamine secretion by endocrine cells (*pink arrow*).

**3** Distention of the stomach also activates local reflexes that increase stomach secretions (*orange arrow*).

**4** Gastrin is carried through the circulation back to the stomach (*purple arrow*), where, along with histamine, it stimulates secretion.

(b)

### Intestinal Phase

**1** Chyme in the duodenum with a pH less than 2 or containing fat digestion products (lipids) inhibits gastric secretions by three mechanisms (2–4).

**2** Chemoreceptors in the duodenum are stimulated by H⁺ (low pH) or lipids. Action potentials generated by the chemoreceptors are carried by the vagus nerves to the medulla oblongata (*green arrow*), where they inhibit parasympathetic action potentials (*pink arrow*), thereby decreasing gastric secretions.

**3** Local reflexes activated by H⁺ or lipids also inhibit gastric secretion (*orange arrows*).

**4** Secretin and cholecystokinin produced by the duodenum (*brown arrows*) decrease gastric secretions in the stomach.

(c)

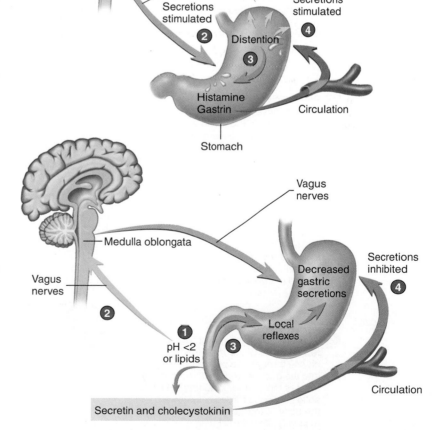

**PROCESS Figure 16.11** **Regulation of Stomach Secretions**
(*a*) Cephalic phase. (*b*) Gastric phase. (*c*) Intestinal phase.

Digestive

stomach trigger a decrease in additional acid secretion. Second, during the intestinal phase, acidic chyme entering the duodenum triggers a decrease in gastric acid secretion. These negative-feedback loops ensure that the acidic chyme entering the duodenum is neutralized, which is required for the digestion of food by pancreatic enzymes and for the prevention of peptic ulcer formation.

## A CASE IN POINT

### Heartburn

Hart Burne had a large meat-lover's pizza delivered to his apartment. He had been eating chips and drinking beer before the pizza came. Hart consumed the whole pizza and two more bottles of beer as he watched the last half of a game on TV. By the end of the game, he was feeling uncomfortably full, so he lay down on the couch. Within half an hour, Hart felt a severe pain in his chest and headed for the bathroom to find an antacid.

**Heartburn,** or *gastritis,* is a painful or burning sensation in the chest usually associated with an increase in gastric acid secretion and/or a backflush of acidic chyme into the esophagus. Overeating, eating fatty foods, lying down immediately after a meal, consuming too much alcohol or caffeine, smoking, and wearing extremely tight clothing can all cause heartburn. Medications can relieve the symptoms of heartburn by blocking gastric acid secretion. Cimetidine (si-met'i-dēn; Tagamet) and ranitidine (ră-nī'ti-dēn; Zantac) block histamine stimulation of acid release from parietal cells, and esomeprazole (eh-sō-meh'pra-zōl; Nexium) blocks the proton pump in parietal cells that generates gastric acid. Antacids (e.g., Tums) neutralize acids already secreted into the stomach.

## Movement in the Stomach

Two types of stomach movement aid digestion and help move chyme through the digestive tract: mixing waves and peristaltic waves (figure 16.12). Both types of movements result from smooth

**1** A mixing wave initiated in the body of the stomach progresses toward the pyloric sphincter (*pink arrows directed inward*).

**2** The more fluid part of the chyme is pushed toward the pyloric sphincter (*blue arrows*), whereas the more solid center of the chyme squeezes past the peristaltic constriction back toward the body of the stomach (*orange arrow*).

**3** Peristaltic waves (*purple arrows*) move in the same direction and in the same way as the mixing waves but are stronger.

**4** Again, the more fluid part of the chyme is pushed toward the pyloric region (*blue arrows*), whereas the more solid center of the chyme squeezes past the peristaltic constriction back toward the body of the stomach (*orange arrow*).

**5** Peristaltic contractions force a few milliliters of the most fluid chyme through the pyloric opening into the duodenum (*small red arrows*). Most of the chyme, including the more solid portion, is forced back toward the body of the stomach for further mixing (*yellow arrow*).

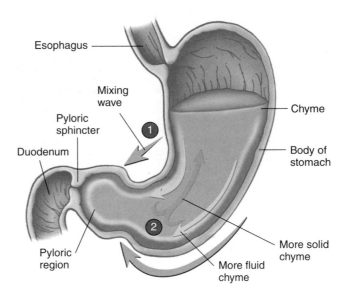

**PROCESS Figure 16.12   Movements in the Stomach**

muscle contractions in the stomach wall. The contractions occur about every 20 seconds and proceed from the body of the stomach toward the pyloric sphincter. Relatively weak contractions result in **mixing waves,** which thoroughly mix ingested food with stomach secretions to form chyme. The more fluid part of the chyme is pushed toward the pyloric sphincter, whereas the more solid center moves back toward the body of the stomach. Stronger contractions result in **peristaltic waves,** which force the chyme toward and through the pyloric sphincter. The pyloric sphincter usually remains closed because of mild tonic contraction. Each peristaltic contraction is sufficiently strong to cause partial relaxation of the pyloric sphincter and to pump a few milliliters of chyme through the pyloric opening and into the duodenum. Increased motility leads to increased emptying.

If the stomach empties too fast, the efficiency of digestion and absorption in the small intestine is reduced. However, if the rate of emptying is too slow, the highly acidic contents of the stomach may damage the stomach wall. To prevent these two extremes, stomach emptying is regulated. The hormonal and neural mechanisms that stimulate stomach secretions also increase stomach motility, so that the increased secretions are effectively mixed with the stomach contents. The major stimulus of gastric motility and emptying is distension of the stomach wall. Inhibition of gastric motility and emptying is accomplished by the same negative-feedback loops associated with the intestinal phase of gastric secretion. In particular, cholecystokinin is a major inhibitor of motility and emptying (table 16.2) Hence, stomach emptying is slower after a fatty meal due to the release of cholecystokinin.

In an empty stomach, peristaltic contractions that approach tetanic contractions can occur for about 2 to 3 minutes. The contractions are increased by low blood glucose levels and are sufficiently strong to create an uncomfortable sensation called a "hunger pang." Hunger pangs usually begin 12 to 24 hours after the previous meal, in less time for some people. If nothing is ingested, hunger pangs reach their maximum intensity within 3 or 4 days and then become progressively weaker.

# SMALL INTESTINE

## Anatomy of the Small Intestine

The **small intestine** is about 6 meters (m) long and consists of three parts: the duodenum, the jejunum, and the ileum (figure 16.13). The **duodenum** (doo-od′ĕ-nŭm, doo-ō-dĕ′nŭm) is about 25 cm long (the term *duodenum* means 12, suggesting that it is 12 in. long). The **jejunum** (jĕ-joo′nŭm) is about 2.5 m long and makes up two-fifths of the total length of the small intestine. The **ileum** (il′ē-ŭm) is about 3.5 m long and makes up three-fifths of the small intestine.

The duodenum nearly completes a 180-degree arc as it curves within the abdominal cavity. Part of the pancreas lies within this arc. The **common bile duct** from the liver and the **pancreatic duct** from the pancreas join and empty into the duodenum (see figure 16.17).

The small intestine is the major site of digestion and absorption of food, which are accomplished due to the presence of a large surface area. The small intestine has three modifications that increase its surface area about 600-fold: circular folds, villi, and

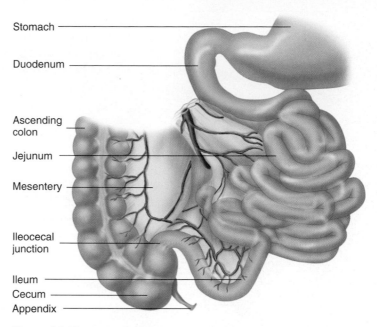

Stomach

Duodenum

Ascending colon

Jejunum

Mesentery

Ileocecal junction

Ileum

Cecum

Appendix

**Figure 16.13**  Small Intestine

microvilli. The mucosa and submucosa form a series of **circular folds** that run perpendicular to the long axis of the digestive tract (figure 16.14a). Tiny, fingerlike projections of the mucosa form numerous **villi** (vil′i; sing. villus), which are 0.5–1.5 mm long (figure 16.14b). Most of the cells composing the surface of the villi have numerous cytoplasmic extensions, called **microvilli** (mī′krō-vil′ī) (figure 16.14c,d). Each villus is covered by simple columnar epithelium. Within the loose connective tissue core of each villus are a blood capillary network and a lymphatic capillary called a **lacteal** (lak′tē-ăl; resembling milk) (figure 16.14c). The blood capillary network and the lacteal are very important in transporting absorbed nutrients.

The mucosa of the small intestine is simple columnar epithelium with four major cell types: (1) **absorptive cells,** which have microvilli, produce digestive enzymes, and absorb digested food; (2) **goblet cells,** which produce a protective mucus; (3) **granular cells,** which may help protect the intestinal epithelium from bacteria; and (4) **endocrine cells,** which produce regulatory hormones.

The epithelial cells are produced within tubular glands of the mucosa, called **intestinal glands,** at the base of the villi. Granular and endocrine cells are located in the bottom of the glands. The submucosa of the duodenum contains mucous glands, called **duodenal glands,** which open into the base of the intestinal glands.

The duodenum, jejunum, and ileum are similar in structure. However, progressing from the duodenum through the ileum, there are gradual decreases in the diameter of the small intestine, in the thickness of the intestinal wall, in the number of circular folds, and in the number of villi. Lymphatic nodules are common along the entire length of the digestive tract, and clusters of lymphatic nodules, called **Peyer patches,** are numerous in the ileum. These lymphatic tissues help protect the intestinal tract from harmful microorganisms.

**Figure 16.14** Anatomy and Histology of the Duodenum

(a) Wall of the duodenum, showing the circular folds. (b) Villi on a circular fold. (c) A single villus, showing the lacteal and capillary network. (d) Transmission electron micrograph of microvilli on the surface of a villus.

The site where the ileum connects to the large intestine is called the **ileocecal** (il'ē-ō-sē'kăl) **junction.** It has a ring of smooth muscle, the **ileocecal sphincter,** and an **ileocecal valve** (see figure 16.21a), which allow the intestinal contents to move from the ileum to the large intestine, but not in the opposite direction.

## Secretions of the Small Intestine

Secretions from the mucosa of the small intestine contain mainly mucus, ions, and water. Intestinal secretions lubricate and protect the intestinal wall from the acidic chyme and the action of digestive enzymes. They also keep the chyme in the small intestine in a liquid form to facilitate the digestive process. Most of the secretions entering the small intestine are produced by the intestinal mucosa, but the secretions of the liver and the pancreas also enter the small intestine and play important roles in digestion.

The epithelial cells in the walls of the small intestine have enzymes, bound to their free surfaces, that are significant in the final steps of digestion. **Peptidases** (pep'ti-dās-ez) break the peptide bonds in proteins to form amino acids. **Disaccharidases** (dī-sak'ă-rid-ās-ez) break down disaccharides, such as maltose, into monosaccharides, such as glucose. The amino acids and monosaccharides can be absorbed by the intestinal epithelium (see table 16.1).

## Lactose Intolerance

**Lactase** (lak'tās) is a digestive enzyme, bound to the epithelium of the small intestine, that digests lactose (milk sugar). A deficiency in this enzyme leads to **lactose intolerance.** Lactase is made at birth; however, in 5–15% of the European-American population and 80–90% of the African-American and Asian-American populations, lactase is no longer synthesized after about age 6, so lactose can no longer be digested. Symptoms include cramps, bloating, and diarrhea after ingesting milk or milk products.

Lactose intolerance is one of the most common gastrointestinal disorders seen by primary care physicians. An increasing selection of foods for lactose intolerant people can be found in the supermarket.

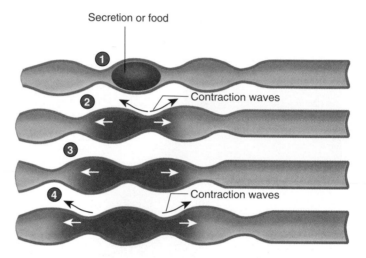

1. A secretion introduced into the digestive tract or into food within the tract begins in one location.

2. Segments of the digestive tract alternate between contraction and relaxation.

3. Material (*brown*) in the intestine is spread out in both directions from the site of introduction.

4. The secretion or food is spread out in the digestive tract and becomes more diffuse (*lighter color*) through time.

Secretion or food

Contraction waves

Contraction waves

**PROCESS Figure 16.15** Segmental Contractions in the Small Intestine

---

Mucus is produced by duodenal glands and by goblet cells, which are dispersed throughout the epithelial lining of the entire small intestine and within intestinal glands. Hormones released from the intestinal mucosa stimulate liver and pancreatic secretions. Secretion by duodenal glands is stimulated by the vagus nerve, secretin release, and chemical or tactile irritation of the duodenal mucosa.

## Movement in the Small Intestine

Mixing and propulsion of chyme are the primary mechanical events that occur in the small intestine. **Peristaltic contractions** proceed along the length of the intestine for variable distances and cause the chyme to move along the small intestine (see figure 16.9). **Segmental contractions** are propagated for only short distances and mix intestinal contents (figure 16.15).

The ileocecal sphincter at the juncture of the ileum and the large intestine remains mildly contracted most of the time, but peristaltic contractions reaching the ileocecal sphincter from the small intestine cause the sphincter to relax and allow chyme to move from the small intestine into the cecum. The ileocecal valve tends to prevent movement from the large intestine back into the ileum.

## Absorption in the Small Intestine

A major function of the small intestine is the **absorption** of nutrients. Most absorption occurs in the duodenum and jejunum, although some absorption also occurs in the ileum (See "Digestion, Absorption, and Transport" later in this chapter).

# LIVER AND PANCREAS

Two large accessory glands, the liver and the pancreas, produce secretions that empty into the duodenum.

## Anatomy of the Liver

The **liver** (figure 16.16*a,b*; see figure 16.1) weighs about 1.36 kilograms (kg) (3 lb) and is located in the right upper quadrant of the abdomen, tucked against the inferior surface of the diaphragm. The posterior surface of the liver is in contact with the right ribs 5–12. It is divided into two major lobes, the **right lobe** and the **left lobe,** which are separated by a connective tissue septum, the **falciform** (fal'si-fōrm) **ligament.** Two smaller lobes, the **caudate** (kaw'dāt; having a tail) **lobe** and the **quadrate** (kwah'drāt; square) **lobe,** can be seen from an inferior view. Also seen from the inferior

Digestive

Liver

Inferior vena cava

**Right lobe**

**Left lobe**

Falciform
ligament

Round ligament

Gallbladder

(a)

Gallbladder

**Quadrate lobe**

**Right lobe**

Hepatic ducts
Hepatic portal vein ⎤ Porta
Hepatic artery

**Caudate lobe**

**Left lobe**

Lesser
omentum

Inferior vena cava

(b)

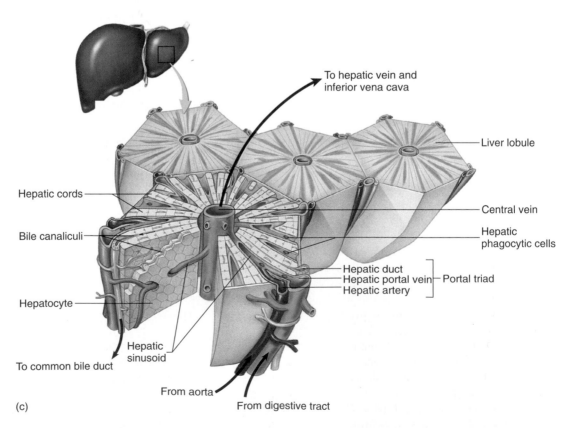

To hepatic vein and
inferior vena cava

Liver lobule

Hepatic cords

Central vein

Bile canaliculi

Hepatic
phagocytic cells

Hepatic duct
Hepatic portal vein ⎤ Portal triad
Hepatic artery

Hepatocyte

To common bile duct

Hepatic
sinusoid

From aorta

From digestive tract

(c)

**Figure 16.16  Liver**

(a) Anterior view. (b) Inferior view. (c) Histology.

Digestive

① The hepatic ducts from the liver lobes combine to form the common hepatic duct.

② The common hepatic duct combines with the cystic duct from the gallbladder to form the common bile duct.

③ The common bile duct joins the pancreatic duct.

④ The combined duct empties into the duodenum at the duodenal papilla.

⑤ Pancreatic secretions may also enter the duodenum through an accessory pancreatic duct, which also empties into the duodenum.

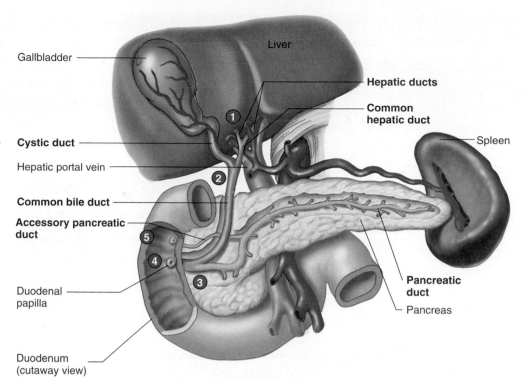

**PROCESS Figure 16.17**  Flow of Bile and Pancreatic Secretions Through the Duct System of the Liver, Gallbladder, and Pancreas

view is the **porta,** which is the "gate" through which blood vessels, ducts, and nerves enter or exit the liver.

The liver receives blood from two sources (see chapter 13). The **hepatic** (he-pa′tik; associated with the liver) **artery** takes oxygen-rich blood to the liver, which supplies liver cells with oxygen. The **hepatic portal vein** carries blood that is oxygen-poor but rich in absorbed nutrients and other substances from the digestive tract to the liver. Liver cells process nutrients and detoxify harmful substances from the blood. Blood exits the liver through **hepatic veins,** which empty into the inferior vena cava.

Many delicate connective tissue septa divide the liver into **lobules** with portal triads at their corners. The **portal triads** contain three structures: the hepatic artery, the hepatic portal vein, and the hepatic duct (figure 16.16c). **Hepatic** (he-pa′tik) **cords,** formed by platelike groups of cells called **hepatocytes** (hep′ă-tō-sīts), are located between the center and the margins of each lobule. The hepatic cords are separated from one another by blood channels called **hepatic sinusoids** (si′nŭ-soydz, sī′nū-soydz; resembling cavities). The sinusoid epithelium contains phagocytic cells that help remove foreign particles from the blood. Blood from the hepatic portal vein and the hepatic artery flows into the sinusoids and mixes together. The mixed blood flows toward the center of each lobule into a **central vein.** The central veins from all the lobes unite to form the hepatic veins, which carry blood out of the liver to the inferior vena cava.

A system of ducts from the liver to the duodenum serves as a pathway for bile and other secretions (figure 16.17). A cleftlike lumen, the **bile canaliculus** (kan′ă-lik′ū-lŭs; pl. kan′ă lik′ū-lī, little canals), is between the cells of each hepatic cord. Bile, produced by the hepatocytes, flows through the bile canaliculi to the hepatic ducts in the portal triads. The hepatic ducts converge and empty into the right and left **hepatic ducts,** which transport bile out of the liver. The right and left hepatic ducts unite to form a single **common hepatic duct.** The common hepatic duct is joined by the **cystic** (sis′tik; *kystis,* bladder) **duct** from the gallbladder to form the **common bile duct.** The **gallbladder** is a small sac on the inferior surface of the liver that stores and concentrates bile (see figure 16.16a,b). The common bile duct joins the pancreatic duct and opens into the duodenum at the **duodenal papilla** (pă-pil′ă) (figure 16.17). The opening into the duodenum is regulated by a sphincter.

## Functions of the Liver

The liver performs important digestive and excretory functions, stores and processes nutrients, detoxifies harmful chemicals, and synthesizes new molecules (table 16.3).

The liver secretes about 700 mL of bile each day. **Bile** (bīl) contains no digestive enzymes, but it plays an important role in digestion by diluting and neutralizing stomach acid and by dramatically increasing the efficiency of fat digestion and absorption. Digestive enzymes cannot act efficiently on large fat globules. **Bile salts** emulsify fats, breaking the fat globules into smaller droplets, much like the action of detergents in dishwater (see tables 16.1 and 16.3). The small droplets are more easily digested by digestive enzymes. Bile also contains excretory products, such as cholesterol, fats and bile pigments, including **bilirubin** (bil-i-roo′bin), a bile pigment that results from the breakdown of hemoglobin.

Digestive

| TABLE 16.3 | Functions of the Liver |
|---|---|
| **Function** | **Explanation** |
| Digestion | Bile neutralizes stomach acid and emulsifies fats, which facilitates fat digestion. |
| Excretion | Bile contains excretory products, such as cholesterol, fats, and bile pigments (e.g., bilirubin), that result from hemoglobin breakdown. |
| Nutrient storage | Liver cells remove sugar from the blood and store it in the form of glycogen; they also store fat, vitamins (A, $B_{12}$, D, E, and K), copper, and iron. |
| Nutrient conversion | Liver cells convert some nutrients into others; for example, amino acids can be converted to lipids or glucose, fats can be converted to phospholipids, and vitamin D is converted to its active form. |
| Detoxification of harmful chemicals | Liver cells remove ammonia from the circulation and convert it to urea, which is eliminated in the urine; other substances are detoxified and secreted in the bile or excreted in the urine. |
| Synthesis of new molecules | The liver synthesizes blood proteins, such as albumin, fibrinogen, globulins, and clotting factors. |

Neural and hormonal stimuli regulate the secretion and release of bile (figure 16.18). Bile secretion by the liver is stimulated by parasympathetic stimulation through the vagus nerve. Secretin, which is released from the duodenum, also stimulates bile secretion and release. Cholecystokinin stimulates the gallbladder to contract and release bile into the duodenum. In addition, most (90%) bile salts are reabsorbed in the ileum. The blood carries the bile salts back to the liver, where they stimulate additional bile salt secretion and are once again secreted into the bile. The loss of bile salts in the feces is reduced by this recycling process.

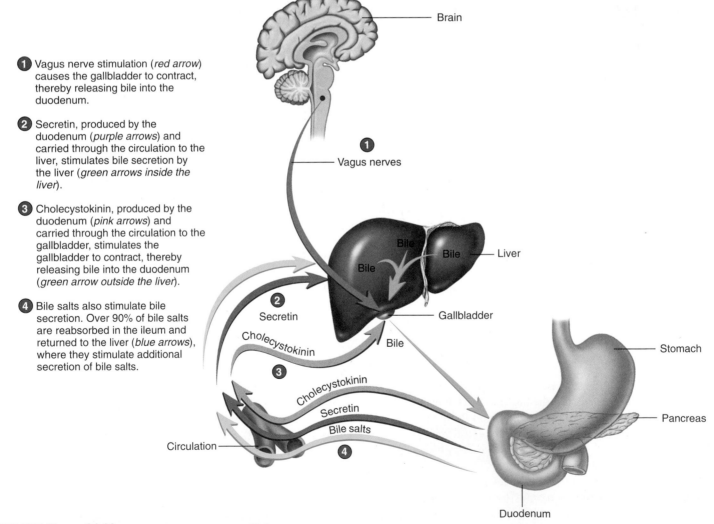

1. Vagus nerve stimulation (*red arrow*) causes the gallbladder to contract, thereby releasing bile into the duodenum.

2. Secretin, produced by the duodenum (*purple arrows*) and carried through the circulation to the liver, stimulates bile secretion by the liver (*green arrows inside the liver*).

3. Cholecystokinin, produced by the duodenum (*pink arrows*) and carried through the circulation to the gallbladder, stimulates the gallbladder to contract, thereby releasing bile into the duodenum (*green arrow outside the liver*).

4. Bile salts also stimulate bile secretion. Over 90% of bile salts are reabsorbed in the ileum and returned to the liver (*blue arrows*), where they stimulate additional secretion of bile salts.

**PROCESS Figure 16.18** Control of Bile Secretion and Release

The liver can remove sugar from the blood and store it in the form of glycogen (table 16.3). It can also store fat, vitamins, copper, and iron. This storage function is usually short-term.

Foods are not always ingested in the proportion needed by the tissues. If this is the case, the liver can convert some nutrients into others (table 16.3). For example, if a person eats a meal that is very high in protein, a large amount of amino acids and only a small amount of lipids and carbohydrates are delivered to the liver. The liver can break down the amino acids and cycle many of them through metabolic pathways to produce ATP and to synthesize lipids and glucose (see chapter 17).

The liver also transforms some nutrients into more readily usable substances. Ingested fats, for example, can be combined with choline and phosphorus in the liver to produce phospholipids, which are essential components of cell membranes.

Many ingested substances are harmful to body cells. In addition, the body itself produces many by-products of metabolism that, if accumulated, are toxic. The liver is an important line of defense against many of those harmful substances. It detoxifies them by altering their structure, making their excretion easier (table 16.3). For example, the liver removes ammonia, a toxic by-product of amino acid metabolism, from the circulation and converts it to urea, which is then secreted into the circulation and eliminated by the kidneys in the urine. The liver also removes other substances from the circulation and excretes them into the bile.

The liver can also produce unique new compounds (table 16.3). Many of the blood proteins, such as albumins, fibrinogen, globulins, and clotting factors, are synthesized in the liver and released into the circulation.

## Anatomy of the Pancreas

The **pancreas** is located retroperitoneal, posterior to the stomach in the inferior part of the left upper quadrant (see figure 16.1). It has a **head** near the midline of the body and a **tail** that extends to the left, where it touches the spleen (figure 16.19; see figure 16.17). It is a complex organ composed of both endocrine and exocrine tissues that perform several functions. The endocrine part of the pancreas consists of **pancreatic islets** (islets of Langerhans). The islet cells produce the hormones insulin and glucagon, which

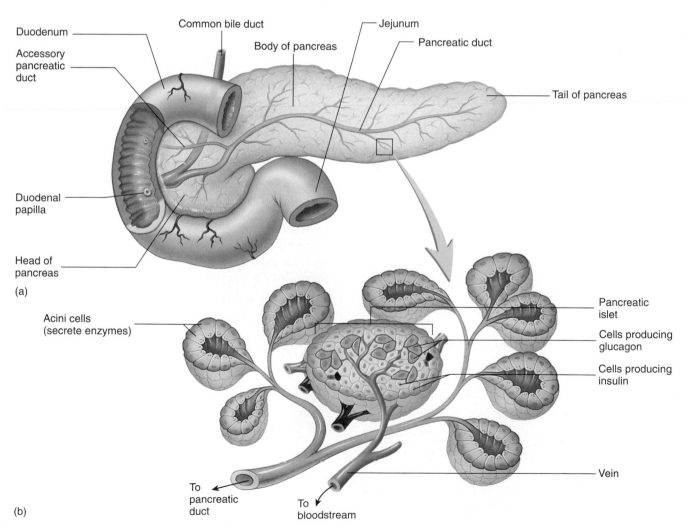

**Figure 16.19** **Anatomy and Histology of the Duodenum and Pancreas**

(a) The head of the pancreas lies within the duodenal curvature, with the pancreatic duct emptying into the duodenum. (b) Histology of the pancreas, showing both the acini and the pancreatic duct system.

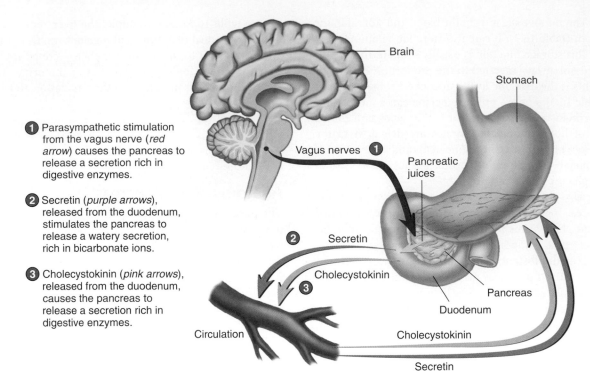

① Parasympathetic stimulation from the vagus nerve (*red arrow*) causes the pancreas to release a secretion rich in digestive enzymes.

② Secretin (*purple arrows*), released from the duodenum, stimulates the pancreas to release a watery secretion, rich in bicarbonate ions.

③ Cholecystokinin (*pink arrows*), released from the duodenum, causes the pancreas to release a secretion rich in digestive enzymes.

**PROCESS Figure 16.20  Control of Pancreatic Secretion**

enter the blood. These hormones are very important in controlling blood levels of nutrients, such as glucose and amino acids (see chapter 10).

The exocrine part of the pancreas is a compound acinar gland (see chapter 4). The **acini** (as′i-nī; grapes) produce digestive enzymes. Clusters of acini are connected by small ducts, which join to form larger ducts, and the larger ducts join to form the **pancreatic duct.** The pancreatic duct joins the common bile duct and empties into the duodenum.

## Functions of the Pancreas

The exocrine secretions of the pancreas include bicarbonate ions ($HCO_3^-$), which neutralize the acidic chyme that enters the small intestine from the stomach. The increased pH resulting from the secretion of $HCO_3^-$ stops pepsin digestion but provides the proper environment for the function of pancreatic enzymes. Pancreatic enzymes are also present in the exocrine secretions and are important in digesting all major classes of food (see table 16.1). Without the enzymes produced by the pancreas, lipids, proteins, and carbohydrates cannot be adequately digested.

The major proteolytic (protein-digesting) enzymes are **trypsin** (trip′sin), **chymotrypsin** (kī-mō-trip′sin), and **carboxypeptidase** (kar-box′ē-pep′ti-dās). These enzymes continue the protein digestion that started in the stomach, and **pancreatic amylase** (am′il-ās) continues the polysaccharide digestion that began in the oral cavity. The pancreatic enzymes also include **lipase** (lip′ās), a lipid-digesting enzyme, and **nucleases** (noo′klē-ās-ez), which are enzymes that degrade DNA and ribonucleic acid to their component nucleotides.

The exocrine secretory activity of the pancreas is controlled by both hormonal and neural mechanisms (figure 16.20; see table 16.2). Secretin initiates the release of a watery pancreatic solution that contains a large amount of $HCO_3^-$. The primary stimulus for secretin release is the presence of acidic chyme in the duodenum. Cholecystokinin stimulates the pancreas to release an enzyme-rich solution. The primary stimulus for cholecystokinin release is the presence of fatty acids and amino acids in the duodenum, and the enzymes secreted by the pancreas digest fatty acids and amino acids. Parasympathetic stimulation through the vagus nerves also stimulates the secretion of pancreatic juices rich in pancreatic enzymes. Sympathetic action potentials inhibit pancreatic secretion.

**Predict 3**

*Explain how secretin production in response to acidic chyme and bicarbonate ion secretion in response to secretin constitute a negative-feedback mechanism.*

## A CASE IN POINT
### Appendicitis

Lowe Payne had been feeling nauseous for a couple of days and had lost his appetite. Suddenly, he felt a sharp pain in his lower right abdomen. The pain was so intense that his mother, Lotta, took Lowe to the hospital. There, he was diagnosed as having **appendicitis.** Part of that diagnosis includes applying slight pressure, such as by pushing with the fingertips, to a specific point in the right lower quadrant of the abdomen. That point, called the McBurney point, is midway between the umbilicus and the right anterior superior iliac spine of the coxal bone.

Appendicitis is an inflammation of the appendix that usually occurs because of obstruction. Secretions from the appendix cannot pass the obstruction; therefore, they accumulate, causing enlargement and pain. Bacteria in the area can cause infection. Symptoms include sudden abdominal pain, particularly in the right lower quadrant, along with a slight fever, loss of appetite, constipation or diarrhea, nausea, and vomiting. If the appendix bursts, the infection can spread throughout the peritoneal cavity, causing **peritonitis** (see Clinical Impact earlier in this chapter), with life-threatening results. Each year, 500,000 people in the United States suffer from appendicitis. The usual treatment is an **appendectomy,** surgical removal of the appendix.

# LARGE INTESTINE

## Anatomy of the Large Intestine

The **large intestine** consists of the cecum, colon, rectum, and anal canal (figure 16.21; see figure 16.1).

### Cecum

The **cecum** (sē′kŭm) is the proximal end of the large intestine where it joins with the small intestine at the ileocecal junction. The cecum is located in the right lower quadrant of the abdomen near the iliac fossa. The cecum is a sac that extends inferiorly about 6 cm past the ileocecal junction. Attached to the cecum is a tube about 9 cm long called the **appendix.**

### Colon

The **colon** (kō′lon) is about 1.5–1.8 m long and consists of four parts: the ascending colon, the transverse colon, the descending colon, and the sigmoid colon (figure 16.21). The **ascending colon** extends superiorly from the cecum to the right colic flexure, near the liver, where it turns to the left. The **transverse colon** extends from the right colic flexure to the left colic flexure near the spleen, where the colon turns inferiorly; and the **descending colon** extends from the left colic flexure to the pelvis, where it becomes the **sigmoid colon.** The sigmoid colon forms an S-shaped tube that extends medially and then inferiorly into the pelvic cavity and ends at the rectum.

The mucosal lining of the colon contains numerous straight, tubular glands called **crypts,** which contain many mucus-producing goblet cells. The longitudinal smooth muscle layer of the colon does not completely envelop the intestinal wall but forms three bands called **teniae coli** (tē′nē-ē kō′lī).

### Rectum

The **rectum** is a straight, muscular tube that begins at the termination of the sigmoid colon and ends at the anal canal (figure 16.21). The muscular tunic is composed of smooth muscle and is relatively thick in the rectum compared to the rest of the digestive tract.

### Anal Canal

The last 2–3 cm of the digestive tract is the **anal canal.** It begins at the inferior end of the rectum and ends at the **anus** (external digestive tract opening). The smooth muscle layer of the anal canal

**Figure 16.21** Large Intestine

(a) The large intestine consists of the cecum, colon, rectum, and anal canal. The teniae coli are bands of smooth muscle along the length of the colon. (b) Radiograph of the large intestine following a barium enema.

is even thicker than that of the rectum and forms the **internal anal sphincter** at its superior end. The **external anal sphincter** at the inferior end of the anal canal is formed by skeletal muscle.

Hemorrhoids are enlarged or inflamed rectal, or hemorrhoidal, veins, that supply the anal canal. Hemorrhoids may cause pain, itching, and/or bleeding around the anus. Treatments include increasing bulk (indigestible fiber) in the diet, taking sitz baths, and using hydrocortisone suppositories. Surgery may be necessary if the condition is extreme and does not respond to other treatments.

## Functions of the Large Intestine

Normally, 18–24 hours are required for material to pass through the large intestine in contrast to the 3–5 hours required for chyme to move through the small intestine. While in the colon, chyme is converted to **feces** (fē′sēz). The formation of feces involves the absorption of water and salts, the secretion of mucus, and extensive action of microorganisms. The colon stores the feces until they are eliminated by the process of **defecation** (def-ĕ-kā′shŭn).

Numerous microorganisms inhabit the colon. They reproduce rapidly and ultimately constitute about 30% of the dry weight of the feces. Some bacteria in the intestine synthesize vitamin K and other vitamins, which are passively absorbed in the colon.

Every 8–12 hours, large parts of the colon undergo several strong contractions, called **mass movements,** which propel the colon contents a considerable distance toward the anus. Each mass movement contraction extends over 20 or more centimeters of the large intestine, which is a much longer part of the digestive tract than that covered by a peristaltic contraction. These mass movements are very common following some meals, especially breakfast.

Faces distend the rectal wall and stimulate the **defecation reflex,** which involves local and parasympathetic reflexes. Local reflexes cause weak contractions, whereas parasympathetic reflexes cause strong contractions and are normally responsible for most of the defecation reflex. Action potentials produced in response to the distention travel along sensory nerve fibers to the sacral region of the spinal cord, where motor action potentials are initiated that reinforce peristaltic contractions in the lower colon and the rectum. Action potentials from the spinal cord also cause the internal anal sphincter to relax. The external anal sphincter, which is composed of skeletal muscle and is under conscious cerebral control, prevents feces from moving out of the rectum and through the anal opening. If this sphincter is relaxed voluntarily, feces are expelled. The defecation reflex persists for only a few minutes and quickly subsides. Generally, the reflex is reinitiated after a period that may be as long as several hours. Mass movements in the colon are usually the reason for the reinitiation of the defecation reflex.

Defecation can be initiated by voluntary actions that stimulate a defecation reflex. These actions include a large inspiration of air, followed by closure of the larynx and forceful contraction of the abdominal muscles. As a consequence, the pressure in the abdominal cavity increases and forces feces into the rectum. Stretch of the rectum initiates a defecation reflex. The increased abdominal pressure also helps push feces through the rectum.

**Predict 4**

*Explain how an enema stimulates defecation.*

# DIGESTION, ABSORPTION, AND TRANSPORT

**Digestion** is the breakdown of food to molecules that are small enough to be absorbed into the circulation. **Mechanical digestion** breaks large food particles into smaller ones. **Chemical digestion** involves the breaking of covalent chemical bonds in organic molecules by digestive enzymes. Carbohydrates break down into monosaccharides, lipids break down into fatty acids and monoglycerides, and proteins break down into amino acids (figure 16.22).

**Absorption** begins in the stomach, where some small, lipid-soluble molecules, such as alcohol and aspirin, can diffuse through the stomach epithelium into the circulation. Most absorption occurs

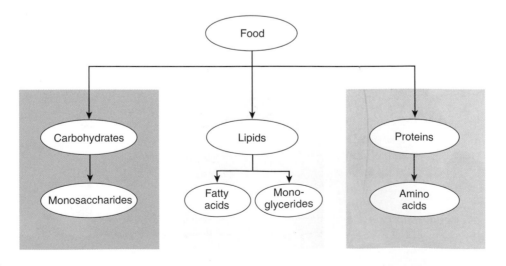

**Figure 16.22 Digestion**
Food consists primarily of carbohydrates, lipids, and proteins. Carbohydrates are broken down into monosaccharides, lipids into fatty acids and monoglycerides, and proteins into amino acids.

in the duodenum and jejunum, although some occurs in the ileum. Some molecules can diffuse through the intestinal wall, whereas others must be transported across the intestinal wall. **Transport** requires carrier molecules and includes facilitated diffusion, cotransport, and active transport. Cotransport and active transport require energy to move the transported molecules across the intestinal wall.

## Carbohydrates

Ingested **carbohydrates** (kar-bō-hī′drāts) consist primarily of starches, cellulose, sucrose (table sugar), and small amounts of fructose (fruit sugar) and lactose (milk sugar). Starches, cellulose, sucrose, and fructose are derived from plants, and lactose is derived from animals. **Polysaccharides** (pol-ē-sak′ă-rīdz) are large carbohydrates,

such as starches, cellulose, and glycogen, that consist of many sugars linked by chemical bonds. Starch is an energy-storage molecule in plants. Cellulose forms the walls of plant cells. Glycogen is an energy-storage molecule in animals and is contained in muscle and in the liver. When uncooked meats are processed or stored, the glycogen is broken down to glucose, which is further broken down, so that little, if any, glycogen remains. Therefore, almost all dietary carbohydrates come from plants. Starch is broken down by enzymes. Cellulose is a polysaccharide that is not digested but is important for providing fiber in the diet.

**Salivary amylase** begins the digestion of carbohydrates in the mouth (figure 16.23). The carbohydrates then pass to the stomach, where digestion continues until the food is well mixed

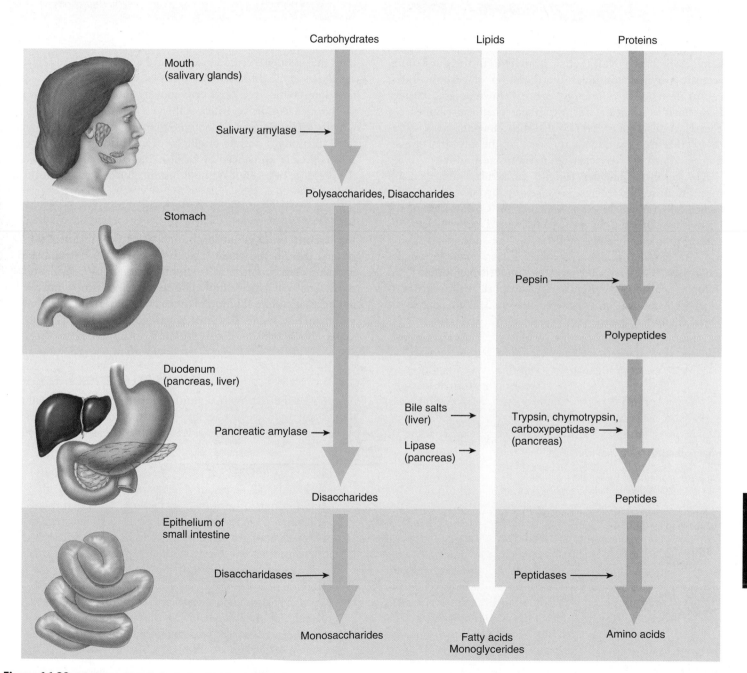

**Figure 16.23  Digestion of Carbohydrates, Lipids, and Proteins**

The enzymes involved in digesting carbohydrates, lipids, and proteins are depicted in relation to the region of the digestive tract where each functions.

Digestive

with acid, which inactivates salivary amylase. In the duodenum, **pancreatic amylase** continues the digestion of carbohydrates, and absorption begins. The amylases break down polysaccharides to **disaccharides** (dī-sak′ă-rīdz; two sugars chemically linked; see chapter 2). A group of enzymes called **disaccharidases** (dī-sak′ă-rid-ăs-ez) that are bound to the microvilli of the intestinal epithelium break down the disaccharides to monosaccharides.

The **monosaccharides** (mon-ō-sak′ă-rīdz; single sugars) glucose, galactose, and fructose are taken up through the intestinal epithelial cells (figure 16.24; see chapter 3). Cotransport of the major monosaccharide, glucose, and Na$^+$ is driven by a Na$^+$ concentration gradient that is established by the sodium-potassium pump. Diffusion of Na$^+$ down its concentration gradient provides the energy to transport glucose across the cell membrane. This mechanism is also used for galactose transport, while fructose is taken up by facilitated diffusion (see chapter 3). Once inside the intestinal epithelial cell, monosaccharides are transported into the capillaries of the intestinal villi and are carried by the hepatic portal system to the liver. Liver cells convert different types of monosaccharides to **glucose,** which is then carried from the liver by the circulation throughout the body. Glucose enters the cells by facilitated diffusion. The rate of glucose transport into most types of cells is greatly influenced by **insulin** and can increase 10-fold in the presence of insulin. Without insulin, glucose enters most cells very slowly.

## Lipids

**Lipid** molecules are insoluble or only slightly soluble in water (see chapter 2). They include triglycerides, phospholipids, steroids, and fat-soluble vitamins. **Triglycerides** (trī-glis′er-īdz) are the most common type of lipid. They consist of three fatty acids bound to glycerol. Triglycerides are often referred to as fats. Fats

**Monosaccharide (glucose) transport**

1. Glucose is absorbed by symport with Na$^+$ into intestinal epithelial cells.

2. Symport is driven by a sodium gradient established by a Na$^+$–K$^+$ pump.

3. Glucose moves out of the intestinal epithelial cells by facilitated diffusion.

4. Glucose enters the capillaries of the intestinal villi and is carried through the hepatic portal vein to the liver.

**PROCESS Figure 16.24** Transport of Glucose Across the Intestinal Epithelium are **saturated** if their fatty acids have only single bonds between carbons and **unsaturated** if they have one (monounsaturated) or more (polyunsaturated) double bonds between carbons (see chapter 2). Saturated fats are solid at room temperature, whereas polyunsaturated fats are liquid at room temperature. Saturated fats are found in meat, dairy products, eggs, nuts, coconut oil, and palm oil. Unsaturated fats are found in fish and most plant oils.

The first step in lipid digestion is **emulsification** (ē-mŭl′si-fi-kā′shŭn), by which large lipid droplets are transformed into much smaller droplets. Emulsification is accomplished by bile salts secreted by the liver. The enzymes that digest lipids are soluble in water and can digest the lipids only by acting at the surface of the droplets. The emulsification process increases the surface area of the lipid droplets exposed to the digestive enzymes by increasing the number of lipid droplets and by decreasing the size of each droplet.

**Lipase,** secreted by the pancreas, digests lipid molecules (see figure 16.23). The primary products of this digestive process are fatty acids and monoglycerides.

In the intestine, bile salts aggregate around small droplets of digested lipids to form **micelles** (mi-selz′, mī-selz′; small morsels) (figure 16.25). The hydrophobic (water-fearing) ends of the bile salts are directed toward the lipid particles, and the hydrophilic (water-loving) ends are directed outward, toward the water environment. When a micelle comes in contact with the epithelial cells of the small intestine, the lipids, fatty acids, and monoglyceride molecules pass, by simple diffusion, from the micelles through the cell membranes of the epithelial cells.

**Lipid transport**

1. Bile salts surround fatty acids and monoglycerides to form micelles.

2. Micelles attach to the cell membranes of intestinal epithelial cells, and the fatty acids and monoglycerides pass by simple diffusion into the intestinal epithelial cells.

3. Within the intestinal epithelial cell, the fatty acids and monoglycerides are converted to triglycerides; proteins coat the triglycerides to form chylomicrons, which move out of the intestinal epithelial cells by exocytosis.

4. The chylomicrons enter the lacteals of the intestinal villi and are carried through the lymphatic system to the general circulation.

Digestive

**PROCESS Figure 16.25** Transport of Lipids Across the Intestinal Epithelium

## High- and Low-Density Lipids

**Cholesterol** is a steroid molecule that plays important roles in the body (see chapter 2). However, cholesterol levels in the blood are of great concern to many adults because people with high blood cholesterol run a much greater risk of heart disease and stroke than do people with low cholesterol. A total cholesterol level of less than 180 milligrams (mg)/dL is considered low, which is usually good, although an extremely low cholesterol level can be harmful. A cholesterol level above 200 mg/dL is considered too high. People with high levels should seek advice

from their physician, reduce their intake of foods rich in cholesterol and other fats, and increase their level of exercise. Some people with very high cholesterol levels may have to take medication to lower their cholesterol.

Fats, including cholesterol, are not soluble in water; thus, they are transported in the blood as lipid-protein complexes, or lipoproteins (figure 16.26). Cholesterol is found primarily in two types of lipoproteins. Low-density lipoproteins (LDLs) carry cholesterol to the tissues for use by the cells. High-density lipoproteins (HDLs) transport cholesterol

from the tissues to the liver. In addition to assessing total cholesterol levels, blood tests can reveal the levels of HDLs and LDLs in a person's blood. LDL is commonly considered "bad" because it deposits cholesterol in arterial walls, which leads to atherosclerosis. HDL is considered "good" because it transports cholesterol to the liver for removal from the body by excretion in the bile. A high HDL/LDL ratio in the bloodstream is related to a lower risk of heart disease. Aerobic exercise is one way to elevate HDL and decrease LDL levels.

---

Once inside the intestinal epithelial cells, the fatty acids and monoglycerides are recombined to form triglycerides. These, and other lipids, are packaged inside a protein coat. The packaged lipid-protein complexes, or **lipoproteins,** are called **chylomicrons** (kī-lō-mi′kronz). Chylomicrons leave the epithelial cells and enter the lacteals, lymphatic capillaries within the intestinal villi. Lymph containing large amounts of absorbed lipid is called **chyle** (kīl; milky lymph). The lymphatic system carries the chyle to the bloodstream. Chylomicrons are transported to the liver, where the lipids are stored, converted into other molecules, or used as energy. They are also transported to adipose tissue, where they are stored until an energy source is needed elsewhere in the body. Other lipoproteins, called **low-density lipoproteins (LDLs)** and **high-density lipoproteins (HDLs),** transport cholesterol and fats in the blood (figure 16.26).

## Proteins

**Proteins** are chains of **amino acids.** They are found in most of the plant and animal products we eat. **Pepsin** is an enzyme secreted by the stomach that breaks down proteins, producing shorter amino acid chains called **polypeptides** (see figure 16.23). Only about 10–20% of the total ingested protein is digested by pepsin. After the remaining proteins and polypeptide chains leave the stomach and enter the small intestine, the enzymes **trypsin, chymotrypsin,** and **carboxypeptidase,** produced by the pancreas in their inactive forms and activated in the intestine, continue the digestive process. These enzymes produce small peptides, which are further broken down into tripeptides (three amino acids), dipeptides (two amino acids), or single amino acids by digestive enzymes called **peptidases.** The peptidases are bound to the microvilli of the small intestine.

Absorption of tripeptides, dipeptides, or individual amino acids occurs through the intestinal epithelial cells by various cotransport mechanisms. Many, but not all, amino acids are taken up by

**Chylomicron**
- Phospholipid (4%)
- Triglyceride (90%)
- Cholesterol (5%)
- Protein (1%)

**Low-density lipoprotein (LDL)**
- Phospholipid (20%)
- Triglyceride (10%)
- Cholesterol (45%)
- Protein (25%)

**High-density lipoprotein (HDL)**
- Phospholipid (30%)
- Triglyceride (5%)
- Cholesterol (20%)
- Protein (45%)

**Figure 16.26** Lipoproteins

cotransport with $Na^+$ similar to glucose cotransport (figure 16.27). Within the intestinal epithelial cells, tripeptides and dipeptides are broken down into amino acids. The amino acids then enter blood capillaries in the villi and are carried by the hepatic portal vein to the liver. The amino acids may be modified in the liver, or they may be released into the bloodstream and distributed throughout the body.

Amino acids are actively transported into the various cells of the body. This transport is stimulated by growth hormone and insulin. Most amino acids are used as building blocks to form new proteins, but some may be metabolized, with a portion of the

**Digestive**

**Amino acid transport**

**1** Acidic and most neutral amino acids are absorbed by symport into intestinal epithelial cells.

**2** Symport is driven by a sodium gradient established by a $Na^+$–$K^+$ pump.

**3** Amino acids move out of intestinal epithelial cells.

**4** Amino acids enter the capillaries of the intestinal villi and are carried through the hepatic portal vein to the liver.

**PROCESS Figure 16.27** Transport of Amino Acids Across the Intestinal Epithelium

released energy used to produce ATP. The body cannot store excess amino acids. Instead, they are partially broken down and used to synthesize glycogen or fat, which can be stored. The body can store only small amounts of glycogen, so most of the excess amino acids are converted to fat.

## Water and Minerals

Approximately 9 L of water enter the digestive tract each day (figure 16.28). We ingest about 2 L in food and drink, and the remaining 7 L are from digestive secretions. Approximately 92% of that water is absorbed in the small intestine, about 7% is absorbed in the large intestine, and about 1% leaves the body in the feces. Water can move in either direction by osmosis across the wall of the digestive tract. The direction of its movement is determined by osmotic gradients across the epithelium. When the chyme is diluted, water moves out of the intestine into the blood. If the chyme is concentrated and contains little water, water moves into the lumen of the small intestine.

Sodium, potassium, calcium, magnesium, and phosphate ions are actively transported from the small intestine. Vitamin D is required for the transport of $Ca^{2+}$. Negatively charged $Cl^-$ move passively through the wall of the duodenum and jejunum with the positively charged $Na^+$, but $Cl^-$ are actively transported from the ileum.

## EFFECTS OF AGING ON THE DIGESTIVE SYSTEM

As a person ages, the connective tissue layers of the digestive tract—the submucosa and serosa—tend to thin. The blood supply to the digestive tract decreases. There is also a decrease in the number of smooth muscle cells in the muscularis, resulting in decreased motility in the digestive tract. In addition, goblet cells

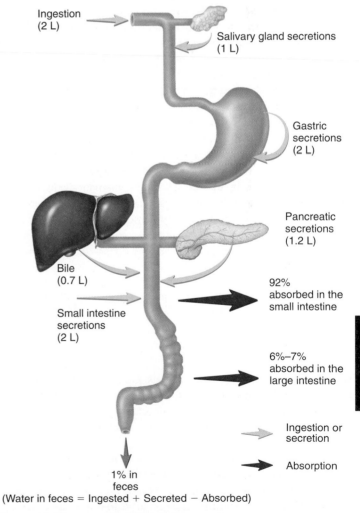

(Water in feces = Ingested + Secreted − Absorbed)

**Figure 16.28** Fluid Volumes in the Digestive Tract

# SYSTEMS PATHOLOGY

## Diarrhea

While on vacation in a foreign country, Tom was shopping with his wife when he started to experience sharp pains in his abdominal region. He also began to feel hot and sweaty and felt an extreme urge to defecate. Tom anxiously inquired about the nearest facility. Once his immediate crisis was taken care of, Tom and his wife went back to their hotel room, where they remained while Tom recovered. During the next 2 days, his stools were frequent and watery. He also vomited a couple of times. Because they were in a foreign country, Tom did not consult a physician. Instead, he rested, drank plenty of fluids, and was feeling much better, although a little weak, in a couple of days.

### Background Information

Diarrhea is one of the most common complaints in clinical medicine. Diarrhea affects more than half the tourists in developing countries (figure 16A), where it may result from eating food to which the digestive tract is not accustomed or from ingesting food or water contaminated with microorganisms.

Diarrhea is any change in bowel habits involving increased stool frequency of volume or increased stool fluidity. It is not a disease in itself, but it can be a symptom of a wide variety of disorders. Diarrhea that lasts less than 2–3 weeks is acute diarrhea; diarrhea lasting longer is considered chronic. Acute diarrhea is usually self-limiting, but some forms of diarrhea can be fatal if not treated. Diarrhea results from either a decrease in fluid absorption in the intestine or an increase in fluid secretion. It can also be caused by increased bowel motor activity that moves chyme rapidly through the small intestine, so that more water enters the colon. Normally, about 600 mL of fluid enter the colon each day, and all but 150 mL are reabsorbed. The loss of more than 200 mL of fluid per day in the stool is considered abnormal.

Mucus secretion by the colon increases dramatically in response to diarrhea. This mucus contains large quantities of bicarbonate ions, which come from the dissociation of carbonic acid into bicarbonate ions and hydrogen ions within the blood supply to the colon. The bicarbonate ions enter the mucus secreted by the colon, whereas the hydrogen ions remain in the circulation; as a result, the blood pH

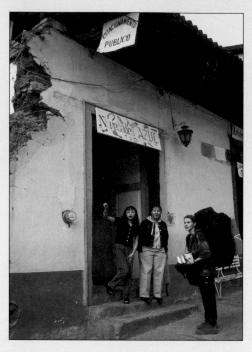

**Figure 16A**

Many tourists develop diarrhea.

decreases. Thus, a condition called metabolic acidosis can develop (see chapter 18).

Diarrhea is usually caused by bacteria, viruses, amoebic parasites, and chemical toxins. Symptoms can begin from as little as 1–2 hours after bacterial toxins are ingested to as long as 24 hours or more for some strains of bacteria. Nearly any bacterial species is capable of causing diarrhea. Some types of bacterial diarrhea are associated with severe vomiting, whereas others are not. Some bacterial toxins also induce fever. Identifying the causal organism usually requires laboratory

within the mucosa secrete less mucus. Glands along the digestive tract, such as the gastric glands, the liver, and the pancreas, also tend to secrete less with age. However, these changes by themselves don't appreciably decrease the function of the digestive system.

Through the years, the digestive tract, like the skin and lungs, is directly exposed to materials from the outside environment. Some of those substances can cause mechanical damage to the digestive tract, and others may be toxic to the tissues. Because the connective tissue of the digestive tract becomes thin with age

and because the protective mucous covering is reduced, the digestive tract of elderly people becomes less and less protected from these outside influences. In addition, the mucosa of elderly people tends to heal more slowly following injury. The liver's ability to detoxify certain chemicals tends to decline; the ability of the hepatic phagocytic cells to remove particulate contaminants decreases; and the liver's ability to store glycogen decreases.

This overall decline in the defenses of the digestive tract leaves elderly people more susceptible to infections and to the

Digestive

| System Interactions | Effects of Diarrhea on Other Systems |
|---|---|
| **System** | **Interactions** |
| Integumentary | Pallor is due to vasoconstriction of blood vessels in the skin, resulting from a decrease in blood volume. Pallor and sweating increase in response to abdominal pain and anxiety. |
| Muscular | Muscular weakness may result due to ion loss, metabolic acidosis, fever, and general malaise. The stimulus to defecate may become so strong that it overcomes the voluntary control mechanisms. |
| Nervous | Local reflexes in the colon respond to increased colon fluid volume by stimulating mass movements and the defecation reflex. Abdominal pain, much of which is felt as referred pain, can occur as the result of inflammation and distention of the colon. Nervous system function decreases due to ion loss. Reduced blood volume stimulates a sensation of thirst in the CNS. |
| Endocrine | A decrease in extracellular fluid volume, due to the loss of fluid in the feces, stimulates the release of hormones (antidiuretic hormone from the posterior pituitary and aldosterone from the adrenal cortex) that increase water retention and sodium reabsorption in the kidney. In addition, decreased extracellular fluid volume and anxiety result in increased release of epinephrine and norepinephrine from the adrenal medulla. |
| Cardiovascular | Movement of extracellular fluid into the colon results in decreased blood volume. The reduced blood volume activates the baroreceptor reflex, which maintains blood volume and blood pressure. |
| Lymphatic and Immune | White blood cells migrate to the colon in response to infection and inflammation. In the case of bacterial diarrhea, the immune response is initiated to begin production of antibodies against bacteria and bacterial toxins. |
| Respiratory | Increased $HCO_3^-$ secretion and $H^+$ absorption reduce blood pH. As the result of reduced blood pH, the rate of respiration increases to eliminate carbon dioxide, which helps eliminate excess $H^+$. |
| Urinary | A decrease in urine volume and an increase in urine concentration result from activation of the baroreceptor reflex, which decreases blood flow to the kidney; antidiuretic hormone secretion, which increases water reabsorption in the kidney; and aldosterone secretion, which increases $Na^+$ and water reabsorption in the kidney. After approximately 24 hours, the kidney is activated to compensate for metabolic acidosis by increasing hydrogen ion secretion and bicarbonate ion reabsorption. |

analysis of the food or stool, but in cases of acute diarrhea, the infectious agent is seldom identified.

Treatment of diarrhea involves replacing lost fluids and ions. The diet should be limited to clear fluids during at least the first day or so. Medicines that may help combat diarrhea include bismuth subsalicylate (sŭb-să-lis′i-lāt), which increases mucus and $HCO_3^-$ secretion and decreases pepsin activity, and loperamide (lō-per′ă-mīd), which slows intestinal motility. Patients should avoid milk and milk products. Breads, rice, and baked fish or chicken can be added to the diet as the person's condition improves. A normal diet can be resumed after 2–3 days.

**Predict 5**

*Predict the effects of prolonged diarrhea on the cardiovascular system.*

effects of toxic agents. Elderly people are more likely to develop ulcerations and cancers of the digestive tract. Colorectal cancers, for example, are the second leading cause of cancer deaths in the United States, with an estimated 135,000 new cases and 57,000 deaths each year.

Gastroesophageal reflux disorder (GERD) increases with advancing age. It is probably the main reason that elderly people take antacids and inhibitors of hydrochloric acid secretion. Disorders that are not necessarily age-induced, such as hiatal hernia and irregular or inadequate esophageal motility, may be worsened by the effects of aging, because of general decreased motility in the digestive tract.

The enamel on the surface of elderly people's teeth becomes thinner with age and may expose the underlying dentin. In addition, the gingiva covering the tooth root recedes, exposing additional dentin. Exposed dentin may become painful and change the person's eating habits. Many elderly people also lose teeth, which can have a marked effect on eating habits unless artificial teeth are provided. The muscles of mastication tend to become weaker; as a result, older people tend to chew their food less before swallowing.

## DISEASES AND DISORDERS  Digestive System

| CONDITION | DESCRIPTION |
|---|---|
| **Stomach** | |
| Vomiting | Contraction of the diaphragm and abdominal muscles and relaxation of the esophageal sphincters to forcefully expel gastric contents; vomiting reflex is initiated by irritation of the stomach or small intestine |
| Peptic ulcer | Lesions in the lining of the stomach or duodenum, usually due to infection by the bacterium *Helicobacter pylori*; stress, diet, smoking, or alcohol may be predisposing factors; antibiotic therapy is the accepted treatment |
| **Liver** | |
| Cirrhosis (sir-ō′sis) | Characterized by damage and death of hepatic cells and replacement by connective tissue; results in loss of normal liver function and interference with blood flow through the liver; a common consequence of alcoholism |
| Hepatitis (hep-ă-tī′tis) | Inflammation of the liver that causes liver cell death and replacement by scar tissue; if not corrected, results in loss of liver function and eventually death; symptoms include nausea, abdominal pain, fever, chills, malaise, and jaundice; caused by any of seven distinct viruses |
|     Hepatitis A | Infectious hepatitis; usually transmitted by poor sanitation practices or from mollusks living in contaminated waters |
|     Hepatitis B | Serum hepatitis; usually transmitted through blood or other body fluids through either sexual contact or contaminated hypodermic needles |
|     Hepatitis C | Often a chronic disease leading to cirrhosis and possibly cancer of the liver |
| Gallstones | Most often due to excess cholesterol in the bile; gallstones can occasionally enter the cystic duct, where they block the release of bile and/or pancreatic enzymes, which interferes with digestion |
| **Intestine** | |
| Inflammatory bowel disease (IBD) | Localized inflammatory degeneration that may occur anywhere along the digestive tract but most commonly involves the distal ileum and proximal colon; the intestinal wall often becomes thickened, constricting the lumen, with ulcers and fissures in the damaged areas; symptoms include diarrhea, abdominal pain, fever, fatigue, and weight loss; cause is unknown; treatments involve anti-inflammatory drugs, avoidance of foods that produce symptoms, and surgery in some cases; also called Crohn's disease or ulcerative colitis |
| Irritable bowel syndrome (IBS) | Disorder of unknown cause marked by alternating bouts of constipation and diarrhea; may be linked to stress or depression; high familial incidence |
| Gluten enteropathy (celiac disease) | Malabsorption in the small intestine due to the effects of gluten, a protein in certain grains, especially wheat; the reaction can destroy newly formed epithelial cells, causing the intestinal villi to become blunted and decreasing the intestinal surface, which reduces absorption of nutrients |
| Constipation (kon-sti-pā′ shŭn) | Slow movement of feces through the large intestine, causing the feces to become dry and hard because of increased fluid absorption while being retained; often results from inhibiting normal defecation reflexes; spasms of the sigmoid colon resulting from irritation can also result in slow feces movement and constipation; high-fiber diet can be preventative |
| | **INFECTIONS OF THE DIGESTIVE TRACT** |
| Food poisoning | Caused by ingesting bacteria or toxins, such as *Staphylococcus aureus, Salmonella,* or *Escherichia coli;* symptoms include nausea, abdominal pain, vomiting, and diarrhea; in severe cases, death can occur |
| Typhoid (tī′foyd) fever | Caused by a virulent strain of the bacterium *Salmonella typhi,* which can cross the intestinal wall and invade other tissues; symptoms include severe fever, headaches, and diarrhea; usually transmitted through poor sanitation practices; leading cause of death in many underdeveloped countries |
| Cholera (kol′er-a) | Caused by a bacterium, *Vibrio cholerae,* in contaminated water; bacteria produce a toxin that stimulates the secretion of chloride, bicarbonate, and water into the intestine, resulting in severe diarrhea; the loss of as much as 12–20 L of fluid per day causes shock, circulatory collapse, and even death; still a major health problem in parts of Asia |
| Giardiasis (jē-ar-dī′a-sis) | Caused by a protozoan, *Giardia lamblia,* that invades the intestine; symptoms include nausea, abdominal cramps, weakness, weight loss, and malaise; bacteria are transmitted in the feces of humans and other animals, often by drinking from contaminated wilderness streams |
| Intestinal parasites | Common under conditions of poor sanitation; parasites include tapeworms, pinworms, hookworms, and roundworms |
| Diarrhea (dī-ă-rē′ă) | Intestinal mucosa secretes large amounts of water and ions due to irritation, inflammation, or infection; diarrhea moves feces out of the intestine more rapidly and speeds recovery |
| Dysentery (dis′ en-tār-ē) | Severe form of diarrhea with blood or mucus in the feces; can be caused by bacteria, protozoa, or amoebae |

Go to **www.mhhe.comseeleyess7** for additional information on these pathologies.

Digestive

# ⎯◦⎯Λ⎯◦⎯ SUMMARY

## Functions of the Digestive System (p. 438)

The functions of the digestive system are to take in food, break down the food, absorb the digested molecules, provide nutrients to the body, and eliminate wastes.

## Anatomy and Histology of the Digestive System (p. 438)

The digestive tract is composed of four tunics: the mucosa, the submucosa, the muscularis, and a serosa or an adventitia.

### Peritoneum

1. The peritoneum is a serous membrane that lines the abdominal cavity and covers the organs.
2. Mesenteries are double layers of peritoneum that extend from the body wall to many of the abdominal organs.
3. Retroperitoneal organs are located behind the parietal peritoneum.

## Oral Cavity, Pharynx, and Esophagus (p. 440)

### Anatomy of the Oral Cavity

1. The lips and cheeks are involved in mastication and speech.
2. The tongue is involved in speech, taste, mastication, and swallowing.
3. There are 32 permanent teeth, including incisors, canines, premolars, and molars. Each tooth consists of a crown, a neck, and a root.
4. The roof of the oral cavity is divided into the hard and soft palates.
5. Salivary glands produce serous and mucous secretions. The three pairs of large salivary glands are the parotid, submandibular, and sublingual glands.

### Saliva

Saliva helps protect the mouth from oral bacteria, starts starch digestion, and provides lubrication.

### Mastication

Mastication is accomplished by the teeth, which cut, tear, and crush the food.

### Pharynx

The pharynx consists of the nasopharynx, oropharynx, and laryngopharynx.

### Esophagus

The esophagus connects the pharynx to the stomach. The upper and lower esophageal sphincters regulate movement.

### Swallowing

1. During the voluntary phase of swallowing, a bolus of food is moved by the tongue from the oral cavity to the pharynx.
2. During the pharyngeal phase of swallowing, the soft palate closes the nasopharynx, and the epiglottis closes the opening into the larynx. Pharyngeal muscles elevate the pharynx and larynx and then move the bolus to the esophagus.
3. During the esophageal phase of swallowing, a wave of constriction (peristalsis) moves the food down the esophagus to the stomach.

## Stomach (p. 446)

### Anatomy of the Stomach

1. The stomach connects to the esophagus at the cardiac opening and to the duodenum at the pyloric opening.
2. The wall of the stomach consists of three muscle layers: longitudinal, circular, and oblique.
3. Gastric glands contain mucous neck cells, parietal cells, endocrine cells, and chief cells.

### Secretions of the Stomach

1. Mucus protects the stomach lining.
2. Hydrochloric acid kills microorganisms and activates pepsin.
3. Pepsin starts protein digestion.
4. Intrinsic factor aids in vitamin $B_{12}$ absorption.

### Regulation of Stomach Secretions

1. Parasympathetic stimulation, gastrin, and histamine increase stomach secretions.
2. During the cephalic phase, the stomach secretions are initiated by the sight, smell, taste, or thought of food.
3. During the gastric phase, partially digested proteins and distention of the stomach promote secretion.
4. During the intestinal phase, acidic chyme in the duodenum stimulates neuronal reflexes and the secretion of hormones that inhibit gastric secretions by negative-feedback loops. Secretin and cholecystokinin inhibit gastric secretion.

### Movement in the Stomach

1. Mixing waves mix the stomach contents with the stomach secretions to form chyme.
2. Peristaltic waves move the chyme into the duodenum. Increased motility increases emptying.
3. Distention of the stomach increases gastric motility. Neural and hormonal feedback loops from the duodenum inhibit gastric motility. Cholecystokinin is a major inhibitor of gastric motility.

## Small Intestine (p. 451)

### Anatomy of the Small Intestine

1. The small intestine is divided into the duodenum, jejunum, and ileum.
2. Circular folds, villi, and microvilli greatly increase the surface area of the intestinal lining.
3. Goblet cells and duodenal glands produce mucus.

### Secretions of the Small Intestine

1. Mucus protects against digestive enzymes and stomach acids.
2. Chemical or tactile irritation, vagal stimulation, and secretion stimulate intestinal secretion.

### Movement in the Small Intestine

1. Peristaltic contractions occur over the length of the intestine and propel chyme through the intestine.
2. Segmental contractions occur over short distances and mix the intestinal contents.

### Absorption in the Small Intestine

Most absorption occurs in the duodenum and jejunum.

## Liver and Pancreas (p. 453)

### Anatomy of the Liver

1. The liver consists of four lobes. It receives blood from the hepatic artery and the hepatic portal vein.
2. Branches of the hepatic artery and hepatic portal vein empty into hepatic sinusoids, which empty into a central vein in the center of each lobe. The central veins empty into hepatic veins, which exit the liver.
3. The liver is divided into lobules with portal triads at the corners. Portal triads contain branches of the hepatic portal vein, hepatic artery, and hepatic duct.

4. Hepatic cords, formed by hepatocytes, form the substance of each lobule. A bile canaliculus, between the cells of each cord, joins the hepatic duct system.
5. Bile leaves the liver through the hepatic duct system. The right and left hepatic ducts join to form the common hepatic duct. The gallbladder stores bile. The cystic duct joins the common hepatic duct to form the common bile duct. The common bile duct joins the pancreatic duct and empties into the duodenum.

### Functions of the Liver

1. The liver produces bile, which contains bile salts that emulsify fats, and excretory products.
2. The liver stores and processes nutrients, produces new molecules, and detoxifies molecules.

### Anatomy of the Pancreas

The pancreas is both an endocrine and an exocrine gland. Its endocrine function is to control blood nutrient levels. Its exocrine function is to produce bicarbonate ions and digestive enzymes.

### Functions of the Pancreas

1. The pancreas produces $HCO_3^-$ and digestive enzymes.
2. Acidic chyme stimulates the release of a watery bicarbonate solution that neutralizes acidic chyme. Fatty acids and amino acids in the duodenum stimulate the release of pancreatic enzymes.

## Large Intestine    (p. 459)

### Anatomy of the Large Intestine

1. The cecum forms a blind sac at the junction of the small and large intestines. The appendix is a blind sac off the cecum.
2. The colon consists of ascending, transverse, descending, and sigmoid portions.
3. The large intestine contains mucus-producing crypts.
4. The rectum is a straight tube that ends at the anal canal.
5. The anal canal is surrounded by an internal anal sphincter (smooth muscle) and an external anal sphincter (skeletal muscle).

### Functions of the Large Intestine

1. The functions of the large intestine are feces production and water absorption.
2. Mass movements occur three or four times a day.
3. Defecation is the elimination of feces. Reflex activity moves feces through the internal anal sphincter. Voluntary activity regulates movement through the external anal sphincter.

## Digestion, Absorption, and Transport    (p. 460)

Digestion is the chemical breakdown of organic molecules into their component parts. After the molecules are digested, some diffuse through the intestinal wall; others must be transported across the intestinal wall.

### Carbohydrates

1. Polysaccharides are split into disaccharides by salivary and pancreatic amylases.
2. Disaccharides are broken down to monosaccharides by disaccharidases on the surface of the intestinal epithelium.
3. The major monosaccharide, glucose, is absorbed by cotransport with $Na^+$ into intestinal epithelial cells. The transport is driven by the $Na^+$ gradient generated by the sodium-potassium pump.
4. Glucose is carried by the hepatic portal vein to the liver and enters most cells in the body by facilitated diffusion. Insulin increases the rate of glucose transport into most cells.

### Lipids

1. Bile salts emulsify lipids.
2. Lipase breaks down lipids. The breakdown products aggregate with bile salts to form micelles.
3. Micelles come in contact with the intestinal epithelium, and their contents diffuse into the cells, where they are packaged and released into the lacteals.
4. Lipids are stored in adipose tissue and in the liver, both of which release the lipids into the blood when energy sources are needed elsewhere in the body.

### Proteins

1. Proteins are split into small polypeptides by enzymes secreted by the stomach and pancreas, and peptidases on the surface of intestinal epithelial cells.
2. Tripeptides, dipeptides, and amino acids are absorbed into intestinal epithelial cells.
3. Amino acids are actively transported into cells under the influence of growth hormone and insulin.
4. Amino acids are used to build new proteins or to serve as a source of energy.

### Water and Minerals

Water can move across the intestinal wall in either direction, depending on osmotic conditions. Approximately 99% of the water entering the intestine is absorbed. Most minerals are actively transported across the wall of the small intestine.

## Effects of Aging on the Digestive System    (p. 465)

1. With advancing age, the layers of the digestive tract thin, and the blood supply decreases.
2. Mucus secretion and motility also decrease in the digestive tract.
3. The defenses of the digestive tract decline, leaving it more sensitive to infection and the effects of toxic agents.
4. Tooth enamel becomes thinner, and the gingiva recede, exposing dentin, which may become painful and affect eating habits.

## REVIEW AND COMPREHENSION

1. What are the functions of the digestive system?
2. What are the major layers, or tunics, of the digestive tract?
3. What are the peritoneum, mesenteries, and retroperitoneal organs?
4. List the functions of the lips, cheeks, and tongue.
5. Distinguish between the deciduous teeth and the permanent teeth. Name the different kinds of teeth.
6. Describe the three parts of a tooth. What are dentin, enamel, and pulp?
7. What is the function of the palate? What are the hard and soft palates?
8. Name and give the locations of the three pairs of salivary glands.
9. What are the functions of saliva?
10. Where is the esophagus located?
11. Describe the three phases of swallowing.
12. Describe the parts of the stomach. How are the muscles in the stomach different from those in the esophagus?

13. What are gastric pits and gastric glands? Name the secretions they produce.

14. List the stomach secretions, and give their functions.

15. Describe the three phases of the regulation of stomach secretion.

16. What are the two kinds of stomach movements? What do they accomplish?

17. Name and describe the three parts of the small intestine.

18. What are circular folds, villi, and microvilli in the small intestine? What are their functions?

19. List the secretions of the small intestine, and give their functions.

20. Describe the kinds of movements in the small intestine, and explain what they accomplish.

21. Describe the anatomy and location of the liver and pancreas, including their duct systems.

22. What are the functions of the liver?

23. Name the exocrine secretions of the pancreas. What are their functions?

24. Describe the parts of the large intestine.

25. How is chyme converted to feces?

26. Describe the defecation reflex.

27. Describe carbohydrate digestion, absorption, and transport.

28. Describe the role of bile salts in lipid digestion and absorption.

29. Describe protein digestion and amino acid absorption. What enzymes are responsible for the digestion?

30. Describe the movement of water into and out of the digestive tract.

31. Describe the effects of aging on the digestive system.

 ## CRITICAL THINKING

1. While anesthetized, patients sometimes vomit. Given that the anesthetic eliminates the swallowing reflex, explain why it is dangerous for an anesthetized patient to vomit.

2. Achlorhydria (ā-klōr-hī′drē-ă) is a condition in which the stomach stops producing hydrochloric acid and other secretions. What effect does achlorhydria have on the digestive process?

3. Victoria Worrystudent developed a duodenal ulcer during final examination week. Describe the possible reasons. Explain what habits could have contributed to the ulcer, and recommend a reasonable remedy.

4. Gallstones sometimes obstruct the common bile duct. What would be the consequences of such a blockage?

5. Many people have a bowel movement shortly after a meal, especially breakfast. Why does this occur?

*Answers in Appendix D*

 Visit this textbook's website at **www.mhhe.com/seeleyess7** for practice quizzes, animations, interactive learning exercises, and other study tools.

 McGraw-Hill offers a study CD that features interactive cadaver dissection. *Anatomy & Physiology Revealed* includes cadaver photos that allow you to peel away layers of the human body to reveal structures beneath the surface. This program also includes animations, radiologic imaging, audio pronunciations, and practice quizzing.

Digestive

Our bodies require many different nutrients to maintain homeostasis. The two young children in this photo are eating part of a balanced diet, which is essential for our bodies to acquire all the appropriate building blocks.

# Chapter 17

# Nutrition, Metabolism, and Body Temperature Regulation

## Chapter Outline and Objectives

### Nutrition (p. 473)

1. Define nutrition, essential nutrient, and kilocalorie.
2. For carbohydrates, lipids, and proteins, describe their dietary sources, their uses in the body, and the daily recommended amounts of each in the diet.
3. List the common vitamins and minerals, and give a function for each.

### Metabolism (p. 481)

4. Define metabolism, anabolism, and catabolism.
5. List three ways in which enzyme activity is controlled.
6. Describe glycolysis, and name its products.
7. Describe the citric acid cycle and its products.

8. Describe the electron-transport chain and how ATP is produced in the process.
9. Explain how the breakdown of glucose yields 2 ATP molecules in anaerobic respiration and 38 ATP molecules in aerobic respiration.
10. Describe the basic steps involved in using lipids and amino acids as energy sources.
11. Differentiate between the absorptive and postabsorptive metabolic states.
12. Define metabolic rate.

### Body Temperature Regulation (p. 491)

13. Describe heat production and regulation in the body.

W hen choosing from a menu or selecting food to prepare, we are often
more concerned with the food's
taste than with its nutritional value. However,
knowing about nutrition is important because
the food we eat provides us with the energy
and the building blocks necessary to synthesize
new molecules. What happens if we don't
obtain enough vitamins, or if we eat too much
sugar and fat? Health claims about foods and
food supplements bombard us every day.
Which ones are ridiculous, and which ones have
merit? A basic understanding of nutrition can
help us answer these and other questions, so
that we can develop a healthy diet. It can also
allow us to know which questions currently do
not have good answers.

# NUTRITION

**Nutrition** (noo-trish′ŭn; to nourish) is the process by which food
is taken into and used by the body; it includes digestion, absorption, transport, and metabolism. Nutrition is also the study of food
and drink requirements for normal body function.

## Nutrients

**Nutrients** are the chemicals taken into the body that provide
energy and building blocks for new molecules. Some substances in
food are not nutrients but provide bulk (fiber) in the diet.
Nutrients can be divided into six major classes: carbohydrates,
lipids, proteins, vitamins, minerals, and water. Carbohydrates,
lipids, and proteins are the major organic nutrients and are broken
down by enzymes into their individual subunits during digestion.
Subsequently, many of these subunits are broken down further to
supply energy, whereas other subunits are used as building blocks
for making new carbohydrates, lipids, and proteins. Vitamins,
minerals, and water are taken into the body without being broken
down. They are essential participants in the chemical reactions
necessary to maintain life. The body requires some nutrients in fairly
substantial quantities and others, called **trace elements,** in only
minute amounts.

   **Essential nutrients** are nutrients that must be ingested
because the body cannot manufacture them—or it cannot manufacture them in adequate amounts. The essential nutrients include
certain amino acids, certain fatty acids, most vitamins, minerals,
water, and some carbohydrates. The term *essential* does not mean
that the body requires only the essential nutrients. Other nutrients
are necessary, but if they are not ingested, they can be synthesized
from the essential nutrients. Most of this synthesis takes place in
the liver, which has a remarkable ability to transform and manufacture molecules. A balanced diet consists of enough nutrients in
the correct proportions to support normal body functions.

   Every 5 years, the U.S. Department of Health and Human
Services (HHS) and the Department of Agriculture (USDA)
jointly make recommendations on what Americans should eat to
be healthy. The latest recommendations, "The Dietary Guidelines
for Americans 2005," were published in January 2005. Unlike the
previous, single food guide pyramid, there are now 12 pyramids
that take into account a person's age, gender, and activity level.
Thus, you can pick the pyramid, called MyPyramid, that best
describes you (www.mypyramid.gov). All of the new pyramids
have the same form (figure 17.1). Six colored bands represent the
approximate recommended proportions of grains (orange), vegetables (green), fruits (red), fats and oils (yellow), milk and milk

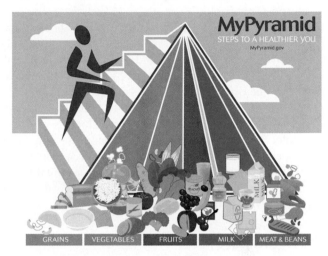

**Figure 17.1  MyPyramid**

The pyramid represents the approaches to a healthy diet: Eat different foods
(different-colored bands), eat different amounts of each food type (band
width), eat in moderation (bands narrow from bottom to top), use fats and
sugars sparingly (the wide base stands for foods with little or no solid fats or
added sugars), and exercise (stick figure climbing stairs).

Source: U.S. Department of Agriculture.

products (blue), and meat and beans (purple). A balanced diet
includes a variety of foods from each of the major food groups.
Variety is necessary because no one food contains all of the nutrients necessary for health. Moderation is indicated by the narrowing of each food group from bottom to top. The wider base
represents foods with little or no solid fats, added sugars, or caloric
sweeteners. The climbing stick figure stresses the importance of
daily exercise.

   The importance of a healthy diet has been proposed as a key
to a healthy life. Two studies completed in 2000 compared the eating habits of 51,529 men and 67,272 women to the government's
Healthy Eating index, a measure of how well a diet conforms to
recommended dietary guidelines. Results from these studies demonstrate that eating a healthy diet does provide certain health
benefits. People who ate the best, according to the index, were
compared to those who ate the worst. Men who ate best had a 28%
reduction in heart disease and an 11% decrease in chronic diseases
compared to men who ate worst. Women who ate best had a 14%
reduction in heart disease but no significant decrease in chronic
diseases compared to women who ate worst. There was no significant difference in cancer rates between the men and women who
ate best compared to those who ate worst.

# Kilocalories

The energy the body uses is stored within the chemical bonds of certain nutrients. A **calorie** (kal′ō-rē; heat) **(cal)** is the amount of energy (heat) necessary to raise the temperature of 1 gram (g) of water 1°C. A **kilocalorie** (kil′ō-kal-ō-rē) **(kcal)** is 1000 cal and is used to express the larger amounts of energy supplied by foods and released through metabolism. For example, one slice of white bread contains about 75 kcal, 1 cup of whole milk contains 150 kcal, a banana contains 100 kcal, a hot dog contains 170 kcal (not counting the bun and dressings), a McDonald's Big Mac has 563 kcal, and a soft drink adds another 145 kcal. For each gram of carbohydrate or protein metabolized by the body, about 4 kcal of energy are released. Fats contain more energy per unit of weight than carbohydrates and proteins, and they yield about 9 kcal/g. Table 17.1 lists the kilocalories supplied by some typical foods. A typical diet in the United States consists of 50–60% carbohydrates, 35–45% fats, and 10–15% proteins. Table 17.1 also lists the carbohydrate, fat, and protein composition of some foods.

A kilocalorie is often called a *Calorie* (with a capital *C*). Unfortunately, this usage has resulted in confusion of the term *calorie* (with a lowercase *c*) with *Calorie* (with a capital *C*). It is common practice on food labels to use the term *calorie* when *Calorie* (*kilocalorie*) is the proper term.

# Carbohydrates

## Sources in the Diet

**Carbohydrates** (kar-bō-hī′drāts) include monosaccharides, disaccharides, and polysaccharides (see chapter 2). Although most of the carbohydrates we ingest are derived from plants, lactose is derived from animals. The most common monosaccharides in the diet are glucose and fructose. Plants capture energy from sunlight and use that energy to produce glucose, which can be found in vegetables, fruits, molasses, honey, and syrup. Fructose is most often derived from fruits and berries.

The disaccharide sucrose (table sugar) is what most people think of when they use the term *sugar*. Sucrose consists of one glucose and one fructose molecule joined together, and its principal sources are sugarcane and sugar beets. Maltose (malt sugar), derived from germinating cereals, is a combination of two glucose molecules, and lactose (milk sugar) consists of one glucose molecule and one galactose molecule.

**Complex carbohydrates** are large polysaccharides, which are composed of long chains of glucose (see figure 2.11). Examples are starch, glycogen, and cellulose, which differ from one another in the arrangement of the glucose molecules and the structure of the chemical bonds holding them together.

Starch is an energy-storage molecule in plants and is found primarily in vegetables, fruits, and grains. Glycogen is an energy-storage molecule in animals and is located primarily in muscle and in the liver. Cellulose forms plant cell walls.

## Uses of Carbohydrates in the Body

During digestion, polysaccharides and disaccharides are split into monosaccharides, which are absorbed into the blood (see chapter 16). Humans have enzymes that can break the bonds between the glucose molecules of starch and glycogen, but they do not have the enzymes necessary to digest cellulose. Therefore, it is important to thoroughly cook or chew plant matter. Cooking and chewing break down the plant cell walls and expose the starches contained inside the cells to digestive enzymes. The undigested cellulose provides fiber, or "roughage," which increases the bulk of feces, making it easier to defecate.

Fructose and other monosaccharides absorbed into the blood are converted into glucose by the liver. Glucose, whether absorbed directly from the digestive tract or synthesized by the liver, is an energy source used to produce ATP. Because the brain relies almost entirely on glucose for its energy, the body carefully regulates blood glucose levels.

If an excess amount of glucose is present in the diet, the glucose is used to make glycogen, which is stored in muscle and in the liver. Glycogen can be rapidly converted back to glucose when energy is needed. Because cells can store only a limited amount of glycogen, additional glucose that is ingested is converted into fat for long-term storage in adipose tissue.

In addition to being used as an energy source, sugars form part of deoxyribonucleic acid (DNA), ribonucleic acid (RNA), and ATP molecules. Sugars also combine with proteins to form glycoproteins, some of which function as receptor molecules on the outer surface of the cell membrane.

## Recommended Consumption of Carbohydrates

According to the Dietary Guidelines Advisory Committee, the **Acceptable Macronutrient Distribution Range (AMDR)** for carbohydrates is 45–65% of total kilocalories. Although a minimum level of carbohydrates is not known, it is assumed that amounts of 100 g or less per day result in overuse of the body's proteins and fats for energy sources. Because muscles are primarily protein, the use of proteins for energy can result in the breakdown of muscle tissue. The extensive use of fats as an energy source can result in acidosis (see chapter 18).

Complex carbohydrates are recommended in the diet because many starchy foods contain other valuable nutrients, such as vitamins and minerals, and because the slower rate of digestion and absorption of complex carbohydrates does not result in large increases and decreases in blood glucose level, as the consumption of large amounts of simple sugars does. Foods containing large amounts of simple sugars, such as soft drinks and candy, are rich in carbohydrates but have few other nutrients. For example, a typical soft drink is mostly sucrose, containing 9 teaspoons of sugar per 12-oz container. In excess, the consumption of these kinds of foods usually results in obesity and tooth decay.

# Lipids

## Sources in the Diet

**Lipids** (lip′idz) include triglycerides, steroids, phospholipids, and fat-soluble vitamins. **Triglycerides** (trī-glis′er-īdz), also called *triacylglycerols* (trī-as′il-glis′er-olz), are the most common type of lipid in the diet, accounting for about 95% of the total lipid intake. Triglyceride molecules consist of three fatty acids bound to one glycerol molecule (see figure 2.12). Triglycerides are often referred to as fats. If the fat is a liquid at room temperature, it is referred to

## TABLE 17.1 Food Composition

| Food | Quantity | Food Energy (kcal) | Carbohydrate (g) | Fat (g) | Protein (g) |
|---|---|---|---|---|---|
| **Dairy Products** | | | | | |
| Whole milk (3.3% fat) | 1 cup | 150 | 11 | 8 | 8 |
| Low-fat milk (2% fat) | 1 cup | 120 | 12 | 5 | 8 |
| Butter | 1 T | 100 | — | 12 | — |
| **Grain** | | | | | |
| Bread, white enriched | 1 slice | 75 | 24 | 1 | 2 |
| Bread, whole-wheat | 1 slice | 65 | 14 | 1 | 3 |
| **Fruit** | | | | | |
| Apple | 1 | 80 | 20 | 1 | — |
| Banana | 1 | 100 | 25 | — | 1 |
| Orange | 1 | 65 | 16 | — | 1 |
| **Vegetables** | | | | | |
| Corn, canned | 1 cup | 140 | 33 | 1 | 4 |
| Peas, canned | 1 cup | 150 | 29 | 1 | 8 |
| Lettuce | 1 cup | 5 | 2 | — | — |
| Celery | 1 cup | 20 | 5 | — | 1 |
| Potato, baked | 1 large | 145 | 33 | — | 4 |
| **Meat, Fish, and Poultry** | | | | | |
| Lean ground beef (10% fat) | 3 oz | 185 | — | 10 | 23 |
| Shrimp, french fried | 3 oz | 190 | 9 | 9 | 17 |
| Tuna, canned | 3 oz | 170 | — | 7 | 24 |
| Chicken breast, fried | 3 oz | 160 | 1 | 5 | 26 |
| Bacon | 2 slices | 85 | — | 8 | 4 |
| Hot dog | 1 | 170 | 1 | 15 | 7 |
| **Fast Foods** | | | | | |
| McDonald's Egg McMuffin | 1 | 327 | 31 | 15 | 19 |
| McDonald's Big Mac | 1 | 563 | 41 | 33 | 26 |
| Taco Bell beef burrito | 1 | 466 | 37 | 21 | 30 |
| Arby's roast beef | 1 | 350 | 32 | 15 | 22 |
| Pizza Hut Super Supreme | 1 slice | 260 | 23 | 13 | 15 |
| Long John Silver's fish | 2 pieces | 366 | 21 | 22 | 22 |
| Dairy Queen malt, large | 1 | 840 | 125 | 28 | 22 |
| **Desserts** | | | | | |
| Chocolate chip cookie | 1 | 50 | 7 | 2 | 1 |
| Apple pie | 1 piece | 135 | 49 | 14 | 3 |
| Soft ice cream | 1 cup | 377 | 38 | 23 | 7 |
| **Beverage** | | | | | |
| Cola soft drink | 12 oz | 145 | 37 | — | — |
| Beer | 12 oz | 144 | 13 | — | — |
| Wine | $3\frac{1}{2}$ oz | 73 | 2 | — | — |
| Hard liquor (86 proof) | $1\frac{1}{2}$ oz | 105 | — | — | — |
| **Miscellaneous** | | | | | |
| Egg | 1 | 80 | 1 | 6 | 6 |
| Mayonnaise | 1 T | 100 | — | 11 | — |
| Sugar | 1 T | 45 | 12 | — | — |

## Fatty Acids and Blood Clotting

The essential fatty acids are used to synthesize molecules that affect blood clotting. Linoleic acid can be converted to **arachidonic** (ă-rak-i-don′ik; *arakis*) **acid,** which is used to produce prostaglandins that *increase* blood clotting. Alpha-linolenic acid can be converted to **eicosapentaenoic** (ī′kō-să-pen-tă-nō′ik) **acid (EPA),** which is used to produce prostaglandins that *decrease* blood clotting. Normally, most prostaglandins are synthesized from linoleic acid because it is more plentiful in the diet. However, individuals who consume foods rich in EPA, such as herring, salmon, tuna, and sardines, increase the synthesis of prostaglandins from EPA. People who eat these fish two or more times per week have a lower risk of heart attack than those who don't, probably because of reduced blood clotting. Although EPA can be obtained using fish oil supplements, this is not currently recommended because fish oil supplements contain high amounts of cholesterol, vitamins A and D, and uncommon fatty acids, all of which can cause health problems when taken in large amounts.

as an oil. Fats are **saturated** if their fatty acids have only single covalent bonds between carbon atoms and **unsaturated** if they have one or more double bonds (see figure 2.13). **Monounsaturated fats** have one double bond, and **polyunsaturated fats** have two or more double bonds. Saturated fats are found in meat, dairy products, eggs, nuts, coconut oil, and palm oil (table 17.1). Monounsaturated fats include olive and peanut oils; polyunsaturated fats are found in fish, safflower, sunflower, and corn oils.

Solid fats, mainly shortening and margarine, work better than liquid oils in preparing some foods, such as pastries. Polyunsaturated vegetable oils can be changed from a liquid to a solid by making them more saturated—that is, by decreasing the number of double covalent bonds in their polyunsaturated fatty acids. To saturate an unsaturated oil, the oil can be *hydrogenated,* which means that hydrogen gas is bubbled through the oil, producing a change in molecular shape that solidifies the oil. The more saturated the product, the harder it becomes at room temperature. These processed fats are usually referred to as *trans* **fats.**

Processed foods and oils account for most of the *trans* fats in the American diet, although some *trans* fats occur naturally in food from animal sources. *Trans* fatty acids raise the concentration of low-density lipoproteins and lower the concentration of high-density lipoproteins in the blood (see chapter 16). These changes are associated with a greater risk of cardiovascular disease. Beginning in 2006, the Food and Drug Administration (FDA) required that food labels include a detailed list of the amounts of saturated and *trans* fats, allowing the consumer to make better dietary choices.

The remaining 5% of ingested lipids include steroids and phospholipids. **Cholesterol** (kō-les′ter-ol) is a steroid (see chapter 2) found in high concentrations in the brain, the liver, and egg yolks, but it is also present in whole milk, cheese, butter, and meats. Cholesterol is not found in plants. Phospholipids, such as **lecithin** (les′i-thin; *lekithos,* egg yolk), are major components of cell membranes and are found in a variety of foods. A good source of lecithin is egg yolks.

### Uses of Lipids in the Body

Triglycerides are an important source of energy that can be used to produce ATP. A gram of triglyceride delivers over twice as many calories as does a gram of carbohydrate or protein. Some cells, such as skeletal muscle cells, derive most of their energy from triglycerides.

Ingested triglyceride molecules not immediately used are stored in adipose tissue or in the liver. When energy is required, the stored triglycerides are broken down, and the fatty acids are released into the blood. The fatty acids can be taken up and used by various tissues. In addition to storing energy, adipose tissue surrounds, pads, and protects organs. Adipose tissue located under the skin is an insulator, which helps reduce heat loss.

Cholesterol is an important molecule with many functions in the body. It is obtained in food, or it can be manufactured by the liver and most other tissues. Cholesterol is a component of the cell membrane, and it can be modified to form other useful molecules, such as bile salts and steroid hormones. Bile salts emulsify fats, which is important for fat digestion and absorption (see chapter 16). Steroid hormones include the sex hormones estrogen, progesterone, and testosterone, which regulate the reproductive system. Eicosanoids, which are derived from fatty acids, are involved in inflammation, tissue repair, smooth muscle contraction, and other functions.

Phospholipids (see chapter 2) are part of the cell membrane and are used to construct myelin sheaths around the axons of nerve cells. Lecithin is found in bile and helps emulsify fats.

### Recommended Consumption of Lipids

The AMDR for fats is 20–35% for adults, 25–35% for children and adolescents 4 to 18 years of age, and 30–35% for children 2 to 3 years of age. Saturated fats should be no more than 10% of total kilocalories, or as low as possible. Most dietary fat should come from sources of polyunsaturated and monounsaturated fats. Cholesterol should be limited to 300 mg (the amount in one egg yolk) or less per day, and *trans* fat consumption should be as low as possible. These guidelines reflect the belief that excess amounts of fats, especially saturated fats, *trans* fats, and cholesterol, contribute to cardiovascular disease. The typical American diet derives 35–45% of its kilocalories from fats, indicating that most Americans need to reduce their fat consumption. See table 17.1 for a sampling of the fat composition in foods.

If a person does not consume enough fats, the body can synthesize fats from carbohydrates and proteins. **Linoleic** (lin-ō-lē'ik; *linum,* flax + *oleum,* oil) **acid** and **alpha linolenic** (lin-ō-len'ik) **acid** are **essential fatty acids** because the body cannot synthesize them, so they must be ingested. They are found in plant oils, such as canola or soybean oil.

## Proteins

### Sources in the Diet

**Proteins** (prō'tēnz) are chains of amino acids (see figure 2.16). They are found in most of the plant and animal products we eat (see table 17.1). Proteins in the body are constructed of 20 different kinds of amino acids, which are divided into two groups: essential and nonessential amino acids. The body cannot synthesize **essential amino acids,** so they must be obtained in the diet. The nine essential amino acids are histidine, isoleucine, leucine, lysine, methionine, phenylalanine, threonine, tryptophan, and valine. **Nonessential amino acids** are necessary to construct our proteins but do not necessarily need to be ingested, since they can be synthesized from the essential amino acids. A **complete protein** food contains all nine essential amino acids in the needed proportions, whereas an incomplete protein food does not. Animal proteins tend to be complete proteins, whereas plant proteins tend to be incomplete. Examples of complete proteins are red meat, fish, poultry, milk, cheese, and eggs. Examples of incomplete proteins are leafy green vegetables, grains, and legumes (peas and beans). If two incomplete proteins, such as rice and beans, are ingested, each can provide amino acids lacking in the other. Thus, a vegetarian diet, if balanced correctly, provides all the essential amino acids.

### Uses of Proteins in the Body

Proteins perform numerous functions in the human body, as the following examples illustrate. Collagen provides structural strength in connective tissue, as does keratin in the skin. The combination of actin and myosin makes muscle contraction possible. Enzymes are responsible for regulating the rate of chemical reactions in the body, and protein hormones regulate many physiological processes (see chapter 10). Proteins in the blood act as clotting factors, transport molecules, and buffers (which prevent changes in pH). Proteins also function as ion channels, carrier molecules, and receptor molecules in the cell membrane. Antibodies, lymphokines, and complement are all proteins that function in the immune system.

Proteins can also be used as a source of energy, yielding approximately the same amount of energy as that derived from carbohydrates. If excess proteins are ingested, the body can use the energy from the proteins to produce glycogen and fat molecules, which can be stored. When protein intake is adequate in a healthy adult, the synthesis and breakdown of proteins occur at the same rate.

### Recommended Consumption of Proteins

The AMDR for protein is 10–35% of total kilocalories. See table 17.1 for a sampling of the protein composition in foods.

## Vitamins

**Vitamins** (vīt'ă-minz; life-giving chemicals) are organic molecules that exist in minute quantities in food and are essential to normal metabolism (table 17.2). **Essential vitamins** cannot be produced by the body and must be obtained through the diet. Because no single food item or nutrient class provides all the essential vitamins, it is necessary to maintain a balanced diet by eating a variety of foods. The absence of an essential vitamin in the diet can result in a specific deficiency disease. A few vitamins, such as vitamin K, are produced by intestinal bacteria, and a few others can be formed by the body from substances called provitamins. A **provitamin** is a part of a vitamin that the body can assemble or modify into a functional vitamin. Beta carotene is an example of a provitamin that can be modified by the body to form vitamin A. The other provitamins are **7-dehydrocholesterol** (dē-hī'drō-kō-les'ter-ol), which can be converted to vitamin D, and **tryptophan** (trip'tō-fan), which can be converted to niacin.

Vitamins are not broken down by catabolism but are used by the body in their original or slightly modified forms. After the chemical structure of a vitamin is destroyed, its function is lost. The chemical structure of many vitamins is destroyed by heat, as when food is overcooked.

Most vitamins function as **coenzymes,** which combine with enzymes to make the enzymes functional (see chapter 2). Vitamins $B_2$ and $B_3$, biotin (bī'ō-tin), and pantothenic (pan-tō-then'ik) acid are critical for some of the chemical reactions involved in the production of ATP. Folate (fō'lāt) and vitamin $B_{12}$ are involved in nucleic acid synthesis. Vitamins A, $B_1$, $B_6$, $B_{12}$, C, and D are necessary for growth. Vitamin K is necessary for the synthesis of proteins involved in blood clotting (table 17.2).

---

**Predict 1**

*Predict what would happen if vitamins were broken down during the process of digestion rather than being absorbed intact into the circulation.*

---

Vitamins are either fat-soluble or water-soluble. **Fat-soluble vitamins,** such as vitamins A, D, E, and K, are absorbed from the intestine along with lipids. Some of them can be stored in the body for a long time. Because they can be stored, these vitamins can accumulate in the body to the point of toxicity. **Water-soluble vitamins,** such as the B vitamins and vitamin C, are absorbed with water from the intestinal tract and typically remain in the body only a short time before being excreted in the urine.

Vitamins were first identified at the beginning of the twentieth century. They were found to be associated with certain foods that were known to protect people from diseases such as rickets and beriberi. In 1941, the first Food and Nutrition Board established the **recommended dietary allowances (RDAs),** which are the nutrient intakes that are sufficient to meet the needs of nearly all people in certain age and gender groups. RDAs were established for different-aged males and females, starting with infants and continuing on to adults. RDAs were

## TABLE 17.2 The Principal Vitamins

| Vitamin | Fat- (F) or Water- (W) Soluble | Source | Function | Symptoms of Deficiency | Reference Daily Intake (RDI)s[a] |
|---|---|---|---|---|---|
| A (retinol) | F | From provitamin carotene found in yellow and green vegetables: preformed in liver, egg yolk, butter, and milk | Necessary for rhodopsin synthesis, normal health of epithelial cells, and bone and tooth growth | Rhodopsin deficiency, night blindness, retarded growth, skin disorders, and increased infection risk | 900 RE[b] |
| B$_1$ (thiamine) | W | Yeast, grains, and milk | Involved in carbohydrate and amino acid metabolism; necessary for growth | Beriberi—muscle weakness (including cardiac muscle), neuritis, and paralysis | 1.2 mg |
| B$_2$ (riboflavin) | W | Green vegetables, liver, wheat germ, milk, and eggs | Component of flavin adenine dinucleotide; involved in citric acid cycle | Eye disorders and skin cracking, especially at corners of the mouth | 1.3 mg |
| B$_3$ (niacin) | W | Fish, liver, red meat, yeast, grains, peas, beans, and nuts | Component of nicotinamide adenine dinucleotide; involved in glycolysis and citric acid cycle | Pellagra—diarrhea, dermatitis, and nervous system disorder | 16 mg |
| Pantothenic acid | W | Liver, yeast, green vegetables, grains, and intestinal bacteria | Constituent of coenzyme-A, glucose production from lipids and amino acids, and steroid hormone synthesis | Neuromuscular dysfunction and fatigue | 5 mg |
| Biotin | W | Liver, yeast, eggs, and intestinal bacteria | Fatty acid and nucleic acid synthesis; movement of pyruvic acid into citric acid cycle | Mental and muscle dysfunction, fatigue, and nausea | 30 μg |
| B$_6$ (pyridoxine) | W | Fish, liver, yeast, tomatoes, and intestinal bacteria | Involved in amino acid metabolism | Dermatitis, retarded growth, and nausea | 1.7 mg |
| Folate | W | Liver, green leafy vegetables, and intestinal bacteria | Nucleic acid synthesis, hematopoiesis; prevents birth defects | Macrocytic anemia (enlarged red blood cells) and spina bifida | 0.4 mg |
| B$_{12}$ (cobalamins) | W | Liver, red meat, milk, and eggs | Necessary for red blood cell production, some nucleic acid and amino acid metabolism | Pernicious anemia and nervous system disorders | 2.4 μg |
| C (ascorbic acid) | W | Citrus fruit, tomatoes, and green vegetables | Collagen synthesis; general protein metabolism | Scurvy—defective bone formation and poor wound healing | 90 mg |
| D (cholecalciferol, ergosterol) | F | Fish liver oil, enriched milk, and eggs; provitamin D converted by sunlight to cholecalciferol in the skin | Promotes calcium and phosphorus use; normal growth and bone and tooth formation | Rickets—poorly developed, weak bones, osteomalacia; bone reabsorption | 10 μg[c] |
| E (tocopherol, tocotrienols) | F | Wheat germ, cottonseed, palm, and rice oils; grain; liver; and lettuce | Prevents oxidation of cell membranes and DNA | Hemolysis of red blood cells | 15 mg |
| K (phylloquinone) | F | Alfalfa, liver, spinach, vegetable oils, cabbage, and intestinal bacteria | Required for synthesis of a number of clotting factors | Excessive bleeding due to retarded blood clotting | 120 μg |

[a] RDIs for people over 4 years of age; IU = international units.
[b] Retinol equivalents (RE). 1 retinol equivalent = 1 μg retinol or 6 μg β-carotene.
[c] As cholecalciferol. 1 μg cholecalciferol = 40 IU (international units) vitamin D.

also set for pregnant and lactating women. The RDAs have been reevaluated every 4–5 years and updated, when necessary, on the basis of new information.

The RDAs establish a minimum intake of vitamins and minerals that should protect almost everyone (97%) in a given group from diseases caused by vitamin or mineral deficiencies. Although personal requirements can vary, the RDAs are a good benchmark. The further below the RDAs an individual's dietary intake is, the more likely that person is to develop a nutritional deficiency. On the other hand, consuming too much of some nutrients can have harmful effects. For example, the long-term ingestion of 3–10 times the RDA for vitamin A can cause bone and muscle pain, skin disorders, hair loss, and an enlarged liver. Consuming 5–10 times the RDA of vitamin D over the long term can result in calcium deposits in the kidneys, heart, and blood vessels, and regularly consuming more than 2 g of vitamin C daily can cause stomach inflammation and diarrhea.

## Minerals

**Minerals** (min′er-ălz) are inorganic nutrients that are essential for normal metabolic functions. People ingest minerals alone or in combination with organic molecules. They constitute about 4–5% of total body weight and are involved in a number of important functions, such as establishing resting membrane potentials and generating action potentials; adding mechanical strength to bones and teeth; combining with organic molecules; and acting as coenzymes, buffers, or regulators of osmotic pressure. A balanced diet can provide all the necessary minerals, with a few possible exceptions. For example, individuals suffering from chronic bleeding or women who experience excessive menstrual bleeding may need an iron supplement. Table 17.3 lists some minerals and their functions.

## Daily Values

**Daily Values** appear on food labels to help consumers plan a healthy diet and to minimize confusion. However, not all possible Daily Values are required to be listed on food labels. Daily Values

are based on two other sets of reference values—Reference Daily Intakes and Daily Reference Values:

- **Reference Daily Intakes (RDIs)** are based on the 1968 RDAs for certain vitamins and minerals. RDIs have been set for four categories of people: infants, toddlers, people over 4 years of age, and pregnant or lactating women. Generally, the RDIs are set to the highest 1968 RDA value of an age category. For example, the highest RDA for iron in males over 4 years of age is 10 mg/day and for females over 4 years of age is 18 mg/day. Thus, the RDI for iron is set at 18 mg/day.
- **Daily Reference Values (DRVs)** are set for total fat, saturated fat, cholesterol, total carbohydrate, dietary fiber, sodium, potassium, and protein.

The Daily Values appearing on food labels are based on a 2000-kcal reference diet, which approximates the weight maintenance requirements of postmenopausal women, women who exercise moderately, teenage girls, and sedentary men (figure 17.2). On large food labels, additional information is listed based on a daily intake of 2500 kcal, which is adequate for young men.

The Daily Values for energy-producing nutrients are determined as a percentage of daily kilocaloric intake: 60% for carbohydrates, 30% for total fats, 10% for saturated fats, and 10% for proteins. The Daily Value for fiber is 14 g for each 1000 kcal of intake. The Daily Values for a nutrient in a 2000 kcal/day diet can be calculated on the basis of the recommended daily percentage of the nutrient and the kilocalories in a gram of the nutrient. For example, carbohydrates should comprise 60% of a 2000 kcal/day diet, or 1200 kcal/day ($0.60 \times 2000$). Because there are 4 kilocalories in a gram of carbohydrate, the Daily Value for carbohydrate is 300 g/day (1200/4).

The Daily Values for some nutrients are the uppermost limit considered desirable because of the link between these nutrients and certain diseases. Thus, the Daily Values for total fats are less than 65 g; saturated fats, less than 20 g; and cholesterol, less than 300 mg because of their association with increased risk of heart

**TABLE 17.3    Important Minerals**

| Mineral | Function | Symptoms of Deficiency | Reference Daily Intake (RDIs)[a] |
|---|---|---|---|
| Calcium | Bone and teeth formation, blood clotting, muscle activity, and nerve function | Spontaneous action potential generation in neurons and tetany | 1300 mg |
| Chlorine | Blood acid-base balance; hydrochloric acid production in stomach | Acid-base imbalance | 2.3 g[b] |
| Chromium | Associated with enzymes in glucose metabolism | Unknown | 35 µg |
| Cobalt | Component of vitamin $B_{12}$; red blood cell production | Anemia | Unknown |
| Copper | Hemoglobin and melanin production; electron-transport system | Anemia and loss of energy | 0.9 mg |
| Fluorine | Provides extra strength in teeth; prevents dental caries | No real pathology | 4 mg |
| Iodine | Thyroid hormone production; maintenance of normal metabolic rate | Goiter and decrease in normal metabolism | 150 µg |
| Iron | Component of hemoglobin; ATP production in electron-transport system | Anemia, decreased oxygen transport, and energy loss | 18 mg |
| Magnesium | Coenzyme constituent; bone formation; muscle and nerve function | Increased nervous system irritability, vasodilation, and arrhythmias | 420 mg |
| Manganese | Hemoglobin synthesis; growth; activation of several enzymes | Tremors and convulsions | 2.3 mg |
| Molybdenum | Enzyme component | Unknown | 45 µg |
| Phosphorus | Bone and teeth formation; important in energy transfer (ATP); component of nucleic acids | Loss of energy and cellular function | 1250 mg |
| Potassium | Muscle and nerve function | Muscle weakness, abnormal electrocardiogram, and alkaline urine | 4.7 g |
| Selenium | Component of many enzymes | Unknown | 55 µg |
| Sodium | Osmotic pressure regulation; nerve and muscle function | Nausea, vomiting, exhaustion, and dizziness | 1.5 g[b] |
| Sulfur | Component of hormones, several vitamins, and proteins | Unknown | Unknown |
| Zinc | Component of several enzymes; carbon dioxide transport and metabolism; necessary for protein metabolisma | Deficient carbon dioxide transport and deficient protein metabolism | 11 mg |

[a] RDIs for people over 4 years of age, except for sodium.
[b] 3.8 g sodium chloride (table salt)

disease. The Daily Value for sodium is less than 2400 mg because of its association with high blood pressure in some people.

For a particular food, the Daily Values are used to calculate the **Percent Daily Value (% Daily Value)** for some of the nutrients in one serving of the food (figure 17.2). For example, if a serving of food has 3 g of fat and the Daily Value for total fat is 65 g, then the % Daily Value is 5% (3/65 5 = 0.05, or 5%). The Food and Drug Administration (FDA) requires % Daily Values to be on food labels, so that the public has useful and accurate dietary information.

The % Daily Values for nutrients related to energy consumption are based on a 2000 kcal/day diet. For people who maintain their weight on a 2000 kcal/day diet, the total of the % Daily Values for each of these nutrients should be no more than 100%. However, for individuals consuming more or fewer than 2000 kcal per day, the total of the % Daily Values can be more than 100%. For example, for a person consuming 2200 kcal/day, the total of the % Daily Values for each of these nutrients should be no more than 110% because 2200/2000 = 1.10, or 110%.

*Predict 2*

One serving of a food has 30 g of carbohydrate. What % Daily Value for carbohydrate is on the food label for this food?

*Predict 3*

Suppose a person consumes 1800 kcal/day. What total % Daily Values for energy-producing nutrients is recommended?

**Nutrition Facts**
Serving Size 1 oz. (28g/About 32 chips)
Servings Per Container 2.5

**Amount Per Serving**

**Calories** 160          Calories from Fat 90

                                           **% Daily Value***

| | |
|---|---|
| **Total Fat** 10g | **16%** |
| Saturated Fat 1.5g | **7%** |
| **Cholesterol** 0mg | **0%** |
| **Sodium** 170mg | **7%** |
| **Total Carbohydrate** 15g | **5%** |
| Dietary Fiber 1g | **4%** |
| Sugars less than 1g | |
| **Protein** 2g | |

| | | | |
|---|---|---|---|
| Vitamin A 0% | ● | Vitamin C 0% | |
| Calcium 2% | ● | Iron 0% | |

* Percent Daily Values are based on a 2,000 calorie diet. Your daily values may be higher or lower depending on your calorie needs:

| | Calories: | 2,000 | 2,500 |
|---|---|---|---|
| Total Fat | Less than | 65g | 80g |
| Sat Fat | Less than | 20g | 25g |
| Cholesterol | Less than | 300mg | 300mg |
| Sodium | Less than | 2,400mg | 2,400mg |
| Total Carbohydrate | | 300g | 375g |
| Dietary Fiber | | 25g | 30g |

Calories per gram:
Fat 9   ●   Carbohydrate 4   ●   Protein 4

**Figure 17.2  Food Label**
Source: U.S. Food and Drug Administration.

When using the % Daily Values of a food to determine how the amounts of certain nutrients in the food fit into the overall diet, the number of servings in a container or package needs to be considered. For example, suppose a small (2.25-oz) bag of corn chips has a % Daily Value of 16% for total fat. A person might suppose that eating the bag of chips accounts for 16% of total fat for the day. The bag, however, contains 2.5 servings. All the chips in the bag account for 40% (16% × 2.5) of the maximum recommended total fat.

# METABOLISM

**Metabolism** (mĕ-tab′ō-lizm; change) is the total of all the chemical reactions that occur in the body. It consists of **catabolism** (kă-tab′ō-lizm), the energy-releasing process by which large molecules are broken down into smaller ones, and **anabolism** (ă-nab′ō-lizm), the energy-requiring process by which small molecules are joined to form larger ones. Catabolism begins during the process of digestion and is concluded within individual cells. Anabolism occurs in all cells of the body as they divide to form new cells, maintain their own intracellular structure, and produce molecules such as hormones, neurotransmitters, or extracellular matrix molecules for export. The energy derived from catabolism is used to drive anabolic reactions.

Metabolism can be divided into the chemical reactions that occur during digestion and the chemical reactions that occur after the products of digestion are taken up by cells. The chemical reactions that occur within cells are often referred to as **cellular metabolism.** The digestive products of carbohydrates, proteins, and lipids can be further broken down inside cells. The energy released during this breakdown can be used to combine **ADP** and an inorganic phosphate group ($P_i$) to form **ATP** (figure 17.3).

ATP is often called the energy currency of the cell. When ATP is broken down to ADP, cells can use the released energy for active transport, muscle contraction, and molecule synthesis. Because the body has high energy demands, it uses ATP rapidly.

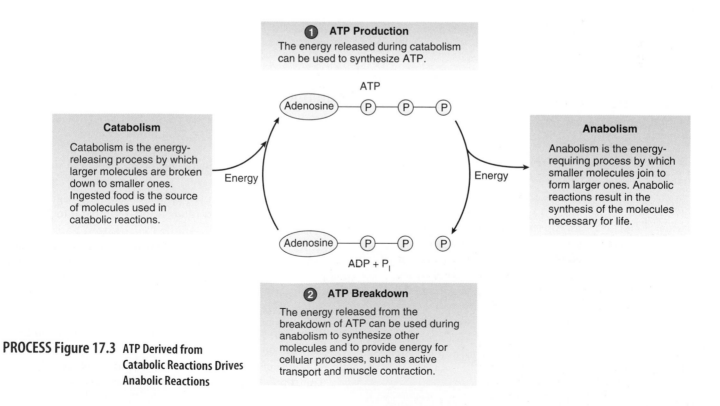

**① ATP Production**
The energy released during catabolism can be used to synthesize ATP.

**Catabolism**
Catabolism is the energy-releasing process by which larger molecules are broken down to smaller ones. Ingested food is the source of molecules used in catabolic reactions.

**Anabolism**
Anabolism is the energy-requiring process by which smaller molecules join to form larger ones. Anabolic reactions result in the synthesis of the molecules necessary for life.

ATP

Adenosine —(P)—(P)—(P)

Energy          Energy

Adenosine —(P)—(P)   (P)

ADP + $P_i$

**② ATP Breakdown**
The energy released from the breakdown of ATP can be used during anabolism to synthesize other molecules and to provide energy for cellular processes, such as active transport and muscle contraction.

**PROCESS Figure 17.3  ATP Derived from Catabolic Reactions Drives Anabolic Reactions**

# CLINICAL IMPACT

## Enzymes and Disease

Many metabolic disorders result from missing or dysfunctional enzymes. For example, in **Tay-Sachs disease,** the breakdown of lipids within lysosomes is impaired. Thus, the intermediate products of lipid metabolism accumulate abnormally, resulting in the destruction of neurons and death by age 3–4 years. **Phenylketonuria** (fen′il-kē′tō-noo′rē-a) **(PKU)** results from the inability to convert the amino acid phenylalanine to tyrosine. Therefore, phenylalanine accumulates in nerve cells and causes brain damage. Unlike Tay-Sachs disease, the extent of damage resulting from PKU can be limited by controlling the diet. Restricting the intake of phenylalanine is an effective treatment for PKU because it reduces the accumulation of phenylalanine in the body. This subsequently reduces the amount of damage to the brain.

**Figure 17.4  Biochemical Pathway**

Each step in the pathway is regulated by a specific enzyme. Substance D can inhibit enzyme 1, thus regulating its own production. Some of the energy released by reactions in the pathway can be used to synthesize ATP.

## A CASE IN POINT
### Vegetarian Diet

Hazel Nutt decides to switch from a typical American diet to a vegetarian diet because she believes a vegetarian diet is better for her **health,** and she no longer wants to eat animals. A strict **vegetarian** (or vegan) diet includes only plant foods.

Plants alone can provide all the protein required for **good health.** But to get adequate amounts of the essential amino acids, a person following a vegan diet should consume a variety of protein sources, such as grains and legumes. In addition, the Dietary Guidelines for Americans recommends that vegan diets be supplemented with vitamin $B_{12}$, vitamin D, calcium, iron, and zinc. This is especially important for children and pregnant and lactating women. Plant sources do not supply vitamin $B_{12}$ or sufficient amounts of vitamin D, although the body can produce vitamin D with adequate exposure to sunlight (see chapter 5). Calcium is found in green leafy vegetables and nuts. Iron and zinc are in whole grains, nuts, and legumes. However, these minerals are either present in low amounts or not easily absorbed.

## Regulation of Metabolism

The products of digestion, such as glucose, fatty acids, and amino acids, are molecules containing energy within their chemical bonds. A series of chemical reactions, called a **biochemical pathway,** controls the energy release from these molecules. At some of the steps, small amounts of energy are released, part of which is used to synthesize ATP (figure 17.4). About 40% of the energy in foods is incorporated into ATP; the rest is lost as heat.

There are several different biochemical pathways inside cells. Which pathways function and how much each pathway is used is determined by enzymes because each step in the pathway requires a specific enzyme (see chapter 2). In turn, enzymes are regulated in several ways:

1. *Enzyme synthesis.* Enzymes are proteins, and their synthesis depends on DNA (see chapter 3). Thus, the types and amounts of enzymes present in cells are under genetic control.

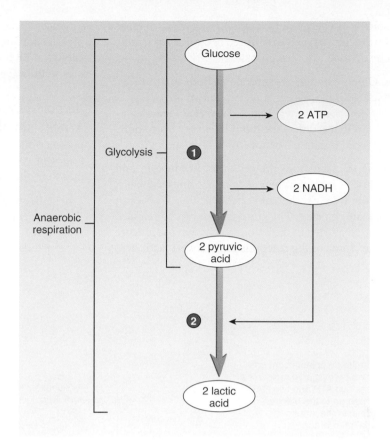

① Glycolysis converts glucose to two pyruvic acid molecules. There is a net gain of two ATP and two NADH from glycolysis.

② Anaerobic respiration, which does not require oxygen, includes glycolysis and converts the two pyruvic acid molecules produced by glycolysis to two lactic acid molecules. This conversion requires energy, which is derived from the NADH generated in glycolysis.

**PROCESS Figure 17.5** Glycolysis and Anaerobic Respiration

2. *Receptor-mediated enzyme activity.* The combination of a chemical signal, such as a neurotransmitter or hormone, with a membrane-bound or intracellular receptor can activate or inhibit enzyme activity (see chapter 10).

3. *Product control of enzyme activity.* The end product of a biochemical pathway can inhibit the enzyme responsible for the first reaction in the pathway. This negative-feedback regulation prevents accumulation of the intermediate products and the end product of the pathway (figure 17.4).

## Carbohydrate Metabolism

Monosaccharides are the breakdown products of carbohydrate digestion. Of these, glucose is the most important in terms of cellular metabolism. Glucose is transported in the circulation to all tissues of the body, where it is used as a source of energy. Any excess glucose in the blood following a meal can be used to form **glycogen** (glī′kō-jen; *glyks,* sweet), or it can be partially broken down and the components used to form fat. Glycogen is a short-term energy-storage molecule that the body can only store in limited amounts, whereas fat is a long-term energy-storage molecule that the body can store in large amounts. Most of the body's glycogen is in skeletal muscle and in the liver.

### Glycolysis

**Glycolysis** (glī-kol′i-sis) is a series of chemical reactions that occurs in the fluid part of cytoplasm surrounding the organelles. It results in the breakdown of glucose to two **pyruvic** (pī-roo′vik)

**acid** molecules (figure 17.5). When glucose is converted to pyruvic acid, two ATP molecules are used and four ATP molecules are produced, for a net gain of two ATP molecules.

Glucose consists of many hydrogen atoms covalently bonded to the carbon atoms of the molecule. During the breakdown of glucose, a hydrogen ion ($H^+$) and two electrons ($e^-$) are released and can attach to a **carrier molecule,** which moves the $H^+$ and electrons to other parts of the cell. A very common carrier molecule in cells is **nicotinamide adenine dinucleotide** (nik-ō-tin′a-mīd ad′ě-nēn dī-noo′klē-ō-tīd) **(NADH):**

$$NAD^+ + 2\,e^- + H^+ \rightarrow NADH$$

The $H^+$ and high-energy electrons in the NADH molecules can be used in other chemical reactions or in the production of ATP molecules in the electron-transport chain (described later in this section).

The pyruvic acid and NADH produced in glycolysis can be used in two different biochemical pathways, depending on the availability of oxygen. When the amounts of oxygen are limited, anaerobic respiration can take place. Anaerobic respiration does not require oxygen and can quickly produce a few ATP molecules for a short time. If the cell has adequate amounts of oxygen, the pyruvic acid and NADH produced in glycolysis are used in aerobic respiration to produce many more ATP.

### Anaerobic Respiration

**Anaerobic** (an-ār-ō′bik) **respiration** is the breakdown of glucose in the absence of oxygen to produce two molecules of **lactic** (lak′tik) **acid** and two molecules of ATP (figure 17.5). The ATP thus produced is a source of energy during activities, such as intense exercise, when insufficient oxygen is delivered to tissues. Anaerobic respiration can be divided into two phases:

1. *Glycolysis.* Glucose undergoes several reactions to produce two pyruvic acid molecules, two ATP, and two NADH.
2. *Lactic acid formation.* Pyruvic acid is converted to lactic acid, a reaction that requires the input of energy from the NADH produced in phase 1 of anaerobic respiration.

Lactic acid is released from the cells that produce it, and blood transports it to the liver. When oxygen becomes available, the lactic acid in the liver can be converted through a series of chemical reactions into glucose. The glucose then can be released from the liver and transported in the blood to cells that use glucose as an energy source. Some of the reactions that convert lactic acid

to glucose require the input of energy derived from ATP that is produced by aerobic respiration. The oxygen necessary to make enough ATP for the synthesis of glucose from lactic acid contributes to the **oxygen deficit** (see chapter 7).

### Aerobic Respiration

**Aerobic** (ār-ō′bik) **respiration** is the breakdown of glucose in the presence of oxygen to produce carbon dioxide, water, and 38 molecules of ATP (figure 17.6). Aerobic respiration can be divided into four phases:

1. *Glycolysis.* The six-carbon glucose molecule is broken down to form two molecules of pyruvic acid, each consisting of three carbon atoms. Two ATP and two NADH molecules are also produced.

**① Glycolysis in the cytoplasm converts** glucose to two pyruvic acid molecules and produces two ATP and two NADH. The NADH can go to the electron-transport chain in the inner mitochondrial membrane.

**② The two pyruvic acid molecules** produced in glycolysis are converted to two acetyl-CoA molecules, producing two $CO_2$ and two NADH. The NADH can go to the electron-transport chain.

**③ The two acetyl-CoA molecules enter the** citric acid cycle, which produces four $CO_2$, six NADH, two $FADH_2$, and two ATP. The NADH and $FADH_2$ can go to the electron-transport chain.

**④ The electron-transport chain uses NADH** and $FADH_2$ to produce 34 ATP. This process requires $O_2$, which combines with $H^+$ to form $H_2O$.

**PROCESS Figure 17.6** Aerobic Respiration

Aerobic respiration involves four phases: (1) glycolysis, (2) acetyl-CoA formation, (3) the citric acid cycle, and (4) the electron-transport chain. The number of carbon atoms in a molecule is indicated after the molecule's name. As glucose is broken down, the carbon atoms from glucose are incorporated into carbon dioxide.

2. *Acetyl-CoA formation.* Each pyruvic acid moves from the cytoplasm into a mitochondrion, where enzymes remove a carbon atom from the three-carbon pyruvic acid molecule to form carbon dioxide and a two-carbon acetyl (as′e-til, a-set′il) group. Hydrogen ions and electrons are released in the reaction and can be used to produce NADH. Each acetyl group combines with coenzyme A (CoA), derived from vitamin $B_2$, to form **acetyl-CoA.** Because two pyruvic acid molecules are produced in phase 1, phase 2 results in two acetyl-CoAs, two carbon dioxide molecules, and two NADH molecules for each glucose molecule.

3. *Citric acid cycle.* Each acetyl-CoA combines with a four-carbon molecule to form a six-carbon citric acid molecule, which enters the citric acid cycle. The **citric** (sit′rik) **acid cycle** is also called the *tricarboxylic* (trī-kar-bok′sil-ik) *acid (TCA) cycle* (citric acid has three carboxylic acid groups) or the *Krebs cycle,* after its discoverer, British biochemist Sir Hans Krebs (1900–1981). The citric acid cycle is a series of reactions wherein the six-carbon citric acid molecule is

converted, in a number of steps, into a four-carbon molecule (figure 17.6). The four-carbon molecule can then combine with another acetyl-CoA molecule to form another citric acid molecule and reinitiate the cycle. During the cycle, two carbon atoms are used to form carbon dioxide, and energy, $H^+$, and electrons are released. Some of the energy can be used to produce ATP. Most of the energy, $H^+$, and electrons are used to form NADH molecules and another carrier molecule, called **flavin** (flā′vin) **adenine dinucleotide (FADH$_2$).** These molecules are used in the electron-transport chain to generate additional ATP. Carbon dioxide diffuses out of the cell and into the blood. It is transported by the circulatory system to the lungs, where it is expired. Thus, the carbon atoms that constitute food molecules, such as glucose, are eventually eliminated from the body as carbon dioxide. We literally breathe out part of the food we eat!

4. *Electron-transport chain.* The **electron-transport chain** is a series of electron-transport molecules attached to the inner mitochondrial membrane (figure 17.7). This membrane

① NADH or FADH$_2$ transfer their electrons to the electron-transport chain.

② As the electrons move through the electron-transport chain, some of their energy is used to pump $H^+$ into the outer compartment, resulting in a higher concentration of $H^+$ in the outer than in the inner compartment.

③ The $H^+$ diffuses back into the inner compartment through special channels (ATP synthase) that couple the $H^+$ movement with the production of ATP. The electrons, $H^+$, and $O_2$ combine to form $H_2O$.

④ ATP is transported out of the inner compartment by a carrier molecule that exchanges ATP for ADP. A different carrier molecule moves phosphate into the inner compartment.

**PROCESS Figure 17.7** Electron-Transport Chain

The electron-transport chain in the inner membrane consists of four protein complexes (*purple;* numbered I to IV) with carrier molecules.

divides the interior of the mitochondrion into an inner compartment and an outer compartment. Electrons are transferred from NADH and $FADH_2$ to the electron-transport carriers, and $H^+$ are released into the inner mitochondrial compartment. Some of the electron-transport carriers are also $H^+$ pumps, which use some of the energy from the transported electrons to pump $H^+$ from the inner to the outer mitochondrial compartment. Because of an increased $H^+$ concentration in the outer compartment, the $H^+$ pass by diffusion back into the inner compartment. The $H^+$ pass through channels in the inner mitochondrial membrane that couple the movement of the $H^+$ to ATP production. In the last step of the electron-transport chain, two $H^+$ and two electrons combine with an $O_2$ atom to form $H_2O$:

$$2\,H^+ + 2\,e^- + \tfrac{1}{2}O_2 \rightarrow H_2O$$

Without $O_2$ to accept the $H^+$ and electrons, the citric acid cycle and the electron-transport chain cannot function. Note that the $O_2$ we breathe in is eventually bound to two hydrogen atoms to become water, which has many uses in the body (see chapter 2).

**Predict 4**

*Many poisons function by blocking certain steps in the metabolic pathways. For example, cyanide blocks the last step in the electron-transport chain. Explain why this blockage causes death.*

### Summary of Anaerobic and Aerobic Respiration

In anaerobic respiration, each glucose molecule ($C_6H_{12}O_6$) yields 2 ATP and 2 lactic acid molecules ($C_3H_6O_3$) through glycolysis:

$$C_6H_{12}O_6 + 2\,ADP + P_i \rightarrow 2\,C_3H_6O_3 + 2\,ATP$$

In contrast, in aerobic respiration, each glucose molecule yields 38 ATP:

$$C_6H_{12}O_6 + 6\,O_2 + 38\,ADP + 38\,P_i \rightarrow 6\,CO_2 + 6\,H_2O + 38\,ATP$$

Of the 38 ATP molecules, 2 result from glycolysis, 2 result from the citric acid cycle, and 34 form through the electron-transport chain. Thus, aerobic respiration is much more efficient at producing ATP than is anaerobic respiration. In addition, many of the chemical reactions of aerobic respiration can be used to produce energy from other food molecules, such as lipids and proteins (see "Lipid Metabolism" and "Protein Metabolism" next).

The number of ATP molecules produced during aerobic respiration can also be reported as 36. The 2 NADH molecules produced by glycolysis cannot cross the inner mitochondrial membrane; thus, their electrons are donated to a shuttle molecule that carries the electrons to the electron-transport chain. Depending on the shuttle molecule, each glycolytic NADH molecule can produce 2 or 3 ATP molecules. In skeletal muscle and in the brain, only 2 ATP molecules are produced for each NADH molecule formed during glycolysis, resulting in a total number of 36 ATP molecules; however, in the liver, kidneys, and heart, 3 ATP molecules are produced for each NADH molecule, and the total number of ATP molecules formed is 38.

## Lipid Metabolism

Triglycerides, or fat, are the body's main energy-storage molecules. In a normal person, fat is responsible for about 99% of the body's energy storage, and glycogen accounts for the remaining 1%.

Between meals, triglycerides in adipose tissue are broken down into fatty acids and glycerol. Some of the fatty acids produced are released into the blood. Other tissues, especially skeletal muscle and the liver, use the fatty acids as a source of energy.

The metabolism of fatty acids takes place in the mitochondria. It occurs by a series of reactions wherein two carbon atoms are removed from the end of a fatty acid chain to form acetyl-CoA (figure 17.8). As the process continues, carbon atoms are removed two at a time until the entire fatty acid chain is converted into acetyl-CoA. Acetyl-CoA can enter the citric acid cycle and be used to generate ATP. In the liver, two acetyl-CoA molecules can also combine to form **ketones** (kē′tōnz). The ketones are released into the blood and travel to other tissues, especially skeletal muscle. In these tissues, the ketones are converted back to acetyl-CoA, which can enter the citric acid cycle to produce ATP.

The presence of small amounts of ketones in the blood is normal and beneficial, but excessive production of ketones is called **ketosis** (ke-to′sis). If the increased number of ketones exceeds the capacity of the body's buffering systems, acidosis, a decrease in blood pH, can occur (see "Regulation of Acid-Base Balance," chapter 18). Conditions that increase fat metabolism can increase the rate of ketone formation. Examples are starvation (see Clinical Impact: Starvation and Obesity later in this chapter), diets consisting of proteins and fats with little carbohydrates, and untreated diabetes mellitus (see chapter 10). Because ketones are excreted by the kidneys and lungs, characteristics of untreated diabetes mellitus include ketones in the urine and "acetone breath."

## Protein Metabolism

Amino acids are the products of protein digestion. Once amino acids are absorbed into the body, they are quickly taken up by cells, especially in the liver. Amino acids are used primarily to synthesize needed proteins and only secondarily as a source of energy. If serving as a source of energy, amino acids can be used in two ways (figure 17.8): (1) The amino acids can be converted into the molecules of carbohydrate metabolism, such as pyruvic acid and acetyl-CoA. These molecules can be metabolized to yield ATP. (2) The amine group ($-NH_2$) can be removed from the amino acid, leaving ammonia and an $\alpha$-keto acid. This process produces NADH, which can enter the electron-transport chain to produce ATP. Ammonia is toxic to cells, so the liver converts it to urea, which the blood carries to the kidneys, where it is eliminated. The $\alpha$-keto acid can enter the citric acid cycle or can be converted into pyruvic acid, acetyl-CoA, or glucose. Although proteins can serve as an energy source, they are not considered major storage molecules.

## Metabolic States

The body experiences two major metabolic states. The first is the **absorptive state,** the period immediately after a meal, when nutrients are being absorbed through the intestinal wall into the circulatory

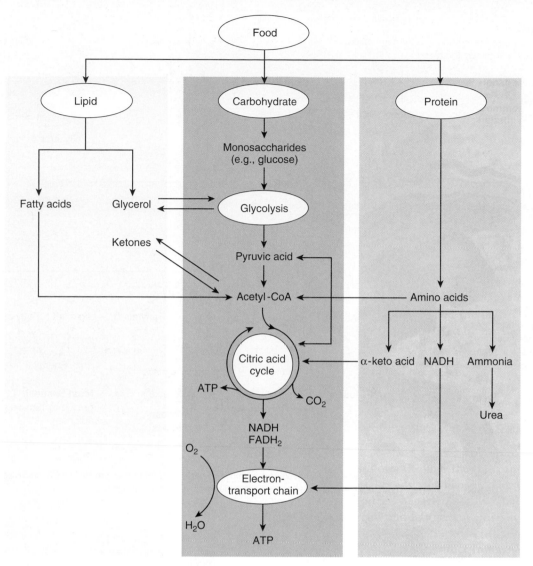

**Figure 17.8  Overall Pathways for the Metabolism of Food**
Carbohydrates, lipids, and proteins enter biochemical pathways to produce energy in the cell.

and lymphatic systems (figure 17.9). The absorptive state usually lasts about 4 hours after each meal, and most of the glucose that enters the circulation is used by cells for the energy they require. The remainder of the glucose is converted into glycogen or fats. Most of the absorbed fats are deposited in adipose tissue. Many of the absorbed amino acids are used by cells in gene expression, and some are used for energy; still others enter the liver and are converted to fats or carbohydrates.

The second state, the **postabsorptive state,** occurs late in the morning, late in the afternoon, or during the night after each absorptive state is concluded (figure 17.10). Normal blood glucose levels range between 70 and 110 mg/dL, and it is vital to the body's homeostasis that this range be maintained. During the postabsorptive state, blood glucose levels are maintained by the conversion of other molecules to glucose. The first source of blood glucose during the postabsorptive state is the glycogen stored in the liver. However, this glycogen supply can provide glucose for only about 4 hours. The glycogen stored in skeletal muscles can also be used during vigorous exercise. As the glycogen stores are

depleted, the body uses fats as an energy source. The glycerol from triglycerides can be converted to glucose. The fatty acids from fat can be converted to acetyl-CoA, moved into the citric acid cycle, and used as a source of energy to produce ATP. In the liver, acetyl-CoA can be used to produce ketone bodies that other tissues can use for energy. The use of fatty acids as an energy source can partly eliminate the need to use glucose for energy, so that less glucose is removed from the blood and homeostasis is maintained. The amino acids of proteins can be converted to glucose or can be used to produce energy, again sparing blood glucose.

## Metabolic Rate

**Metabolic rate** is the total amount of energy produced and used by the body per unit of time. Metabolic rate is usually estimated by measuring the amount of oxygen used per minute. One liter of oxygen consumed by the body is estimated to produce 4.825 kcal of energy.

Metabolic energy can be used in three ways: for basal metabolism, for muscle contraction, and for the assimilation of

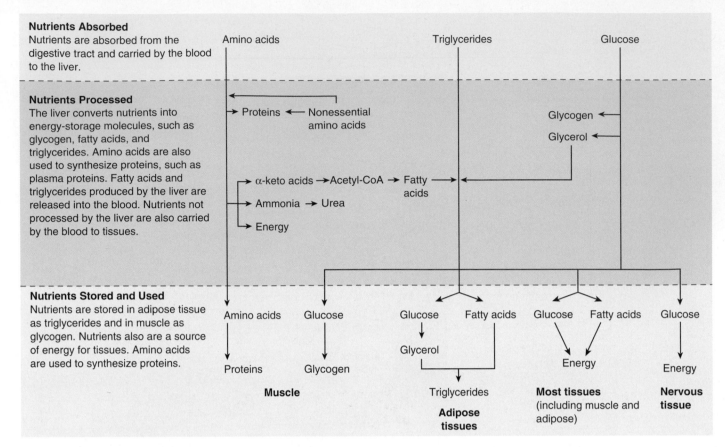

**Nutrients Absorbed**
Nutrients are absorbed from the digestive tract and carried by the blood to the liver.

**Nutrients Processed**
The liver converts nutrients into energy-storage molecules, such as glycogen, fatty acids, and triglycerides. Amino acids are also used to synthesize proteins, such as plasma proteins. Fatty acids and triglycerides produced by the liver are released into the blood. Nutrients not processed by the liver are also carried by the blood to tissues.

**Nutrients Stored and Used**
Nutrients are stored in adipose tissue as triglycerides and in muscle as glycogen. Nutrients also are a source of energy for tissues. Amino acids are used to synthesize proteins.

**Figure 17.9 Events of the Absorptive State**
Absorbed molecules, especially glucose, are used as sources of energy. Molecules not immediately needed for energy are stored: Glucose is converted to glycogen or triglycerides, triglycerides are deposited in adipose tissue, and amino acids are converted to triglycerides or carbohydrates.

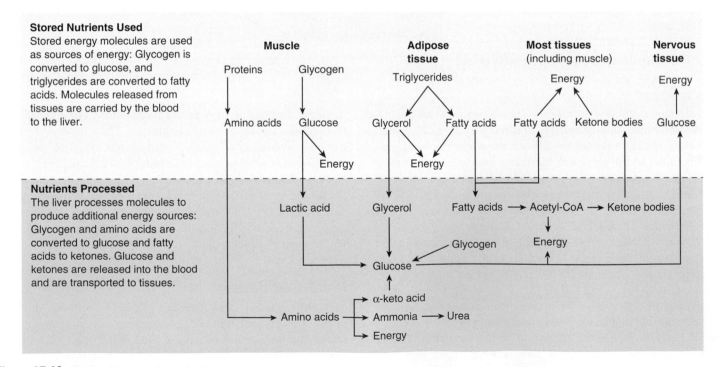

**Stored Nutrients Used**
Stored energy molecules are used as sources of energy: Glycogen is converted to glucose, and triglycerides are converted to fatty acids. Molecules released from tissues are carried by the blood to the liver.

**Nutrients Processed**
The liver processes molecules to produce additional energy sources: Glycogen and amino acids are converted to glucose and fatty acids to ketones. Glucose and ketones are released into the blood and are transported to tissues.

**Figure 17.10 Events of the Postabsorptive State**
Stored energy molecules are used as sources of energy: Glycogen is converted to glucose; triglycerides are broken down to fatty acids, some of which are converted to ketones; and proteins are converted to glucose.

## Starvation and Obesity

### Starvation

**Starvation** is the inadequate intake of nutrients or the inability to metabolize or absorb nutrients. Starvation can be caused by a number of factors, such as prolonged fasting, anorexia, deprivation, or disease. No matter what the cause, starvation takes approximately the same course and consists of three phases. The events of the first two phases occur even during relatively short periods of fasting or dieting. The third phase occurs in prolonged starvation and ends in death.

During the first phase of starvation, blood glucose levels are maintained through the production of glucose from glycogen, proteins, and fats. At first, glycogen is broken down into glucose. However, enough glycogen is stored in the liver to last only a few hours. Thereafter, blood glucose levels are maintained by the breakdown of proteins and fats. Fats are decomposed into fatty acids and glycerol. Fatty acids can be used as a source of energy, especially by skeletal muscle, thus decreasing the use of glucose by tissues other than the brain. The brain cannot use fatty acids as an energy source, so the conservation of glucose is critical to normal brain function. Glycerol can be used to make a small amount of glucose, but most of the glucose is formed from the amino acids of proteins. In addition, some amino acids can be used directly for energy.

In the second stage, which can last for several weeks, fats are the primary energy source. The liver metabolizes fatty acids into ketone bodies that can serve as a source of energy. After about a week of fasting, the brain begins to use ketone bodies, as well as glucose, for energy. This usage decreases the demand for glucose, and the rate of protein breakdown diminishes but does not stop. In addition, the use of proteins is selective; that is, proteins not essential for survival are used first.

The third stage of starvation begins when the fat reserves are nearly depleted and the body switches to proteins as the major energy source. Muscles, the largest source of protein in the body, are rapidly depleted. At the end of this stage, proteins essential for cellular functions are broken down, and cell function degenerates. Death can occur very rapidly.

Symptoms of starvation, in addition to weight loss, include apathy, listlessness, withdrawal, and increased susceptibility to infectious diseases. Few people die directly from starvation because they usually die of an infectious disease first. Other signs of starvation include changes in hair color, flaky skin, and massive edema in the abdomen and lower limbs, causing the abdomen to appear bloated.

During starvation, the body's ability to consume normal volumes of food also decreases. Foods high in bulk but low in protein content often cannot reverse the process of starvation. Intervention involves feeding the starving person low-bulk foods containing ample protein, calories, vitamins, and minerals. Starvation also results in dehydration; thus, rehydration is an important part of intervention. Even with intervention, a victim may be so affected by disease or weakness that he or she cannot recover.

### Obesity

**Obesity** is the storage of excess fat, and it results from the ingestion of more food than is necessary for the body's energy needs. Obesity can be defined on the basis of body weight, body mass index, or body fat. "Desirable body weight" is listed in a table produced by the Metropolitan Life Insurance Company and indicates, for any height, the weight associated with a maximum life span. Overweight is defined as 10% more than the desirable weight, and obesity is 20% more than the desirable weight. **Body mass index (BMI)** can be calculated by dividing a person's weight (Wt) in kilograms by the square of his or her height (Ht) in meters:

$$BMI = Wt/Ht^2$$

A BMI greater than 25–27 is overweight, and a value greater than 30 is defined as obese. About 10% of Americans have a BMI of 30 or greater. In terms of the percent of total body weight contributed by fat, 15% body fat in men and 25% body fat in women is associated with reduced health risks. Obesity is defined as more than 25% body fat in men and 30–35% in women.

The distribution of fat in obese individuals can vary. Fat can accumulate mainly in the upper body, such as in the abdominal region, or it can be associated with the hips and buttocks. These distribution differences can be clinically significant because upper body obesity is associated with an increased likelihood of diabetes mellitus, cardiovascular disease, stroke, and death.

In some cases, obesity is caused by a medical condition. For example, a tumor in the hypothalamus can stimulate overeating. In most cases, however, no specific cause can be recognized. In fact, obesity can occur for many reasons and can have more than one cause in the same individual. There seems to be a genetic component for obesity; if one or both parents are obese, their children are more likely to be obese. Environmental factors, such as eating habits, can also play an important role. For example, adopted children can exhibit obesity similar to that of their adoptive parents. In addition, psychological factors, such as overeating as a way to deal with stress, can contribute to obesity.

In **hypertrophic** (hī-per-trof′ik; *hyper-*, above normal + *trophē*, nourishment) **obesity,** the number of adipose cells is usually normal, but the amount of fat contained in

(continued)

## CLINICAL IMPACT (continued)

each adipose cell is increased. This type of obesity is characteristic of adult-onset obesity. People who were thin or of average weight and quite active when young become less active as they age. Although they no longer use as many kilocalories, they still consume the same amount of food as when they were younger. The excess kilocalories (see "Metabolic Rate" earlier in this section) are used to synthesize fat. In this

type of obesity, the amount of fat in each adipose cell increases, and if the amount of stored fat continues to increase, the total number of adipose cells may also increase. It is estimated that the average U.S. resident gains 1.25–1.5 lb of fat per year after age 25 and, at the same time, loses 0.25–0.5 lb of lean body weight (muscle mass) per year.

In **hyperplastic** (hī-per-plas'tik; *hyper-* + *plasis,* a molding) **obesity,** which is character-

istic of juvenile-onset obesity, the number of adipose cells is increased. This condition may also be accompanied by an increase in cell size (hypertrophic obesity). Hyperplastic obesity has a very strong hereditary component, but family eating habits can also have a great influence. People with hyperplastic obesity are obese as children and become more obese with age. This type of obesity is a major health problem in school-age children.

---

food, which involves processes such as the production of digestive enzymes and the active transport of digested molecules. The **basal metabolic rate (BMR)** is the energy needed to keep the resting body functional. It is the metabolic rate calculated in expended kilocalories per square meter of body surface area per hour. BMR is measured when a person is awake but restful and has not eaten for 12 hours. A typical BMR for a 70-kg (154-lb) male is 38 kcal/m$^2$/hour.

Basal metabolism supports active transport mechanisms, muscle tone, maintenance of body temperature, beating of the heart, and other activities. A number of factors can affect the BMR. Males have a higher BMR than females, younger people have a higher BMR than older people, and fever can increase BMR. Greatly reduced kilocaloric input, as occurs during dieting or fasting, depresses the BMR.

The daily input of energy should equal the energy demands of metabolism; otherwise, a person will gain or lose weight. For a 23-year-old, 70-kg (154-lb) male to maintain his weight, the input should be 2700 kcal/day; for a 58-kg (128-lb) female of the same age, 2000 kcal/day are necessary. A pound of body fat contains about 3500 kcal. Reducing kilocaloric intake by 500 kcal/day can result in the loss of 1 lb of fat per week. Clearly, adjusting kilocaloric input is an important way to control body weight.

The other way to control weight is through energy expenditure. Physical activity through skeletal muscle movement (exercise) greatly increases the metabolic rate. In the average person, basal metabolism accounts for about 60% of energy expenditure, muscular activity 30%, and assimilation of food about 10%. Of these amounts, energy loss through muscular activity is the only component that a person can reasonably control. Comparing the number of kilocalories gained from food and the number of kilocalories burned during exercise reveals why losing weight is difficult. For example, if brisk walking uses 225 kcal/h, it takes 20 minutes of brisk walking to burn off the 75 kcal in one slice of bread (75/225 = 0.33 h). Research suggests that a combination of appropriate physical activity and appropriate kilocaloric intake is

the best approach to maintaining a healthy body composition and weight.

### Predict 5

*If watching TV uses 95 kcal/h, how long does it take to burn off the kilocalories in one cola or beer (see table 17.1)? If jogging at a pace of 6 mph uses 580 kcal/h, how long does it take to use the kilocalories in one cola or beer?*

## A CASE IN POINT
### Gastric Bypass Surgery

Les Moore is a 35-year-old male who weighs 420 pounds. Although at one time he weighed less as a result of dieting, he has not only regained the lost weight but has also gained additional weight. Les has type 2 diabetes, high blood cholesterol, and high blood pressure. Because of his extreme obesity and related health problems, Les elects to have a Roux-en-Y gastric bypass. In this procedure, the digestive tract is surgically rearranged to form a Y-shaped structure. One arm of the Y is formed by separating the superior part of the stomach to form a small pouch that is connected to the jejunum. The small size of the stomach pouch dramatically reduces the amount of food a person can eat. The other arm of the Y consists of the remainder of the stomach, the duodenum, and part of the jejunum. The body of the Y is the distal jejunum. Food from the esophagus enters the small stomach pouch and bypasses the rest of the stomach and the duodenum by entering the jejunum. Bile and pancreatic juices enter the duodenum and pass into the jejunum.

Roux-en-Y is the most commonly performed gastric bypass surgery in the United States. It results in the permanent loss of over 50% of presurgery body weight for over 90% of patients. It also cures most cases of type 2 diabetes, lowers blood pressure, decreases cholesterol levels, reduces the risk of premature death, and improves the quality of life in most patients. But despite the benefits, gastric bypass has some negatives. It is major surgery—complications and even death is possible. Also, the normal digestive functions of the stomach and duodenum are bypassed as food moves quickly through the stomach pouch into the jejunum. As a result, vitamin and mineral deficiencies can develop—particularly, deficiencies of vitamin B$_{12}$, calcium, and iron.

# BODY TEMPERATURE REGULATION

Humans can maintain a relatively constant internal body temperature despite changes in the temperature of the surrounding environment. A constant body temperature is very important to homeostasis. For example, environmental temperatures are too low for normal enzyme function. The heat produced by metabolism and muscle contraction helps maintain the body temperature at a steady, elevated level that is high enough for normal enzyme function. Excessively high temperatures can alter enzyme structure, resulting in loss of the enzyme's function.

**Free energy** is the total amount of energy that can be liberated by the complete catabolism of food. About 40% of the total energy released by catabolism is used to accomplish biological work, such as anabolism, muscular contraction, and other cellular activities. The remaining energy is lost as **heat.**

### Predict 6

*Explain why we become warm during exercise. Why is shivering useful when it is cold?*

Normal body temperature is regulated in the same way as other homeostatic conditions in the body. The average normal temperature is usually considered to be 37°C (98.6°F) when measured orally and 37.6°C (99.7°F) when measured rectally. Rectal temperature comes closer to the true core body temperature, but an oral temperature is more easily obtained in older children and adults and therefore is the preferred measure. The normal oral temperature may vary from person to person, with a range of approximately 36.1–37.2°C (97–99°F).

Body temperature is maintained by balancing heat input with heat loss. The body and the environment exchange heat in a number of ways (figure 17.11). **Radiation** is the gain or loss of heat as infrared energy between two objects that are not in physical contact with each other. For example, heat can be gained by radiation from the sun, a hot coal, or the hot sand of a beach. On the other hand, heat can be lost as radiation to cool vegetation, water in the ocean, or snow on the ground. **Conduction** is the exchange of heat between objects that are in direct contact with each other, such as the bottom of the feet and the ground. **Convection** is a transfer of heat between the body and the air or water. A cool breeze moves air over the body and causes loss of heat from the body. **Evaporation** is the conversion of water from a liquid to a gaseous form. As water evaporates from body surfaces, heat is lost.

The amount of heat exchanged between the environment and the body is determined by the difference in temperature between the two. The greater the temperature difference, the greater the rate of heat exchange. Control of the temperature difference can be used to regulate body temperature. For example, if the environmental temperature is very cold, as occurs on a winter day, there is a large temperature difference between the body and the environment, and therefore a large loss of heat. Behaviorally, we can reduce the loss of heat by selecting a warmer environment, such as by going inside a heated house or by putting on extra clothes. Physiologically, the body controls the

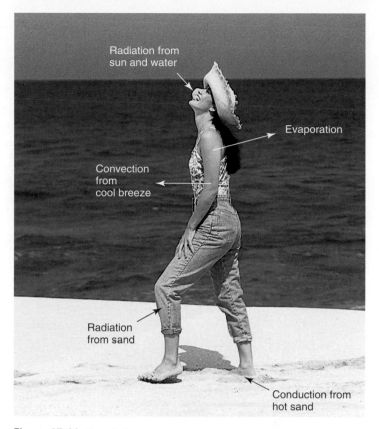

**Figure 17.11  Heat Exchange**

Heat exchange between a person and the environment occurs by convection, radiation, evaporation, and conduction. *Arrows* show the direction of net heat gain or loss in this environment.

temperature difference through dilation and constriction of blood vessels in the skin. When the blood vessels dilate, they bring warm blood to the surface of the body, raising skin temperature; when the blood vessels constrict, blood flow and skin temperature decrease (see figure 5.8).

### Predict 7

*Explain why constriction of skin blood vessels on a cold winter day is beneficial.*

When environmental temperature is greater than body temperature, dilation of blood vessels in the skin brings blood to the skin, causing an increase in skin temperature that decreases the gain of heat from the environment. At the same time, evaporation carries away excess heat to prevent heat gain and overheating.

Body temperature regulation is an example of a negative-feedback system (figure 17.12). Maintenance of a specific body temperature is accomplished by neurons in the hypothalamus, which regulate body temperature around a set point. A small area in the anterior part of the hypothalamus can detect slight increases in body temperature through changes in blood temperature. As a result, mechanisms that cause heat loss, such

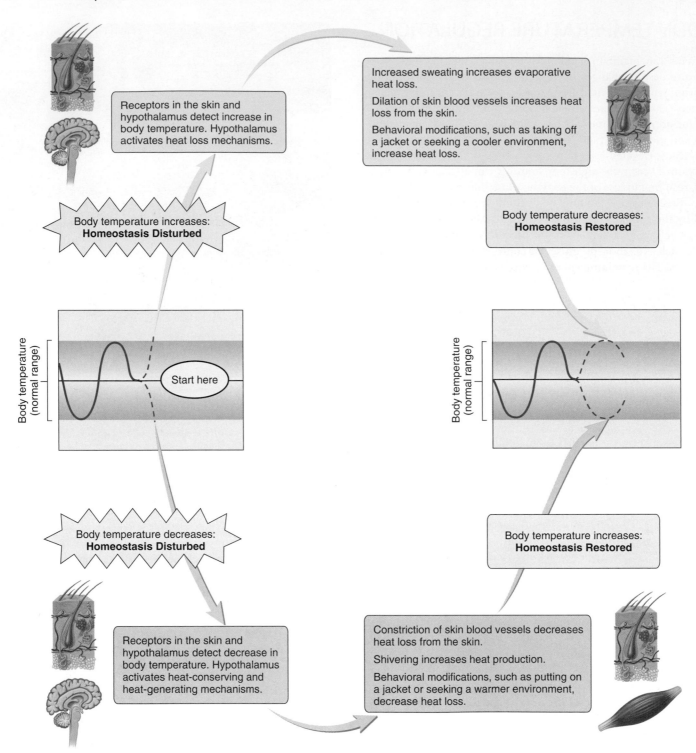

**Homeostasis Figure 17.12**   Temperature Regulation

as dilation of blood vessels in the skin and sweating, are activated, and body temperature decreases. A small area in the posterior hypothalamus can detect slight decreases in body temperature and can initiate heat gain by increasing muscular activity (shivering) and by initiating constriction of blood vessels in the skin.

Under some conditions, the hypothalamus set point is actually changed. For example, during a fever, the set point is raised. Heat-conserving and heat-producing mechanisms are stimulated, and body temperature increases. To recover from a fever, the set point is reduced to normal, heat loss mechanisms are initiated, and body temperature decreases.

# CLINICAL IMPACT

## Too Hot or Too Cold

**Hyperthermia** is a condition in which heat gain in the body exceeds heat loss. Hyperthermia can result from exposure to a hot environment, exercise, or fever. Prolonged exposure to a hot environment can lead to **heat exhaustion,** in which normal temperature reduction mechanisms cannot keep pace with the excessive environmental heat, thus allowing the body temperature to rise. Heat exhaustion is characterized by cool, wet skin due to heavy sweating. Weakness, dizziness, and nausea usually occur as well. The heavy sweating can lead to dehydration, decreased blood volume, decreased blood pressure, and increased heart rate.

Treatment involves increasing heat loss by moving the person to a cooler environment, decreasing heat production by decreasing muscular activity, and replacing lost body fluids. **Heat stroke** results from an increase in the hypothalamic set point and is characterized by dry, flushed skin because sweating is inhibited. The person becomes confused and irritable, and can even become comatose. Treatment is the same as for heat exhaustion but also involves increasing evaporation from the skin by applying water to the skin or by placing the person in cool water.

**Hypothermia** is a condition in which heat loss exceeds heat gain. The body's normal temperature increase mechanisms are working, but they cannot keep pace with heat loss, and the body temperature decreases. Hypothermia usually results from prolonged exposure to a cold environment or even to a cool, damp environment because the moisture draws heat away from the body. Treatment for hypothermia calls for rewarming the body at a rate of a few degrees per hour. **Frostbite** is local damage to the skin or deeper tissues resulting from prolonged exposure to a cold environment. The best treatment for frostbite is immersion in a warm-water bath. Rubbing the affected area and applying local, dry heat should be avoided.

---

 SUMMARY

## Nutrition (p. 473)

Nutrition is the ingestion and use of food.

### Nutrients

1. Nutrients, the chemicals used by the body, consist of carbohydrates, lipids, proteins, vitamins, minerals, and water.
2. Essential nutrients either cannot be produced by the body or cannot be produced in adequate amounts.
3. The MyPyramid guide recommends the proper amounts of different food types and fiber necessary for good health, based on a person's age, gender, and physical activity.

### Kilocalories

A kilocalorie is the energy required to raise the temperature of 1000 g of water 1°C. A kilocalorie (Calorie) is the unit of measurement used to express the energy content of food.

### Carbohydrates

1. Carbohydrates include monosaccharides, disaccharides, and polysaccharides.
2. Most of the carbohydrates we ingest are from plants.
3. Carbohydrates are used as an energy source and for making DNA, RNA, and ATP.
4. The Acceptable Macronutrient Distribution Range (AMDR) for carbohydrates is 45–65% of total kilocalories.

### Lipids

1. Lipids include triglycerides, steroids, phospholipids, and fat-soluble vitamins. Triglycerides are a major source of energy. Eicosanoids are involved in inflammation, tissue repair, and smooth muscle contraction. Cholesterol and phospholipids are part of the cell membrane. Some steroid hormones regulate the reproductive system.
2. The AMDR for lipids is 20–35%.

### Proteins

1. Proteins are chains of amino acids.
2. Animal proteins tend to be complete proteins, whereas plant proteins tend to be incomplete.
3. Proteins are involved in structural strength, muscle contraction, regulation, buffering, clotting, transport, ion channels, receptors, and the immune system.
4. The AMDR for protein is 10–35% of total kilocalories.

### Vitamins

1. Most vitamins are not produced by the body and must be obtained in the diet. Some vitamins can be formed from provitamins.
2. Vitamins are important in energy production, nucleic acid synthesis, growth, and blood clotting.
3. Vitamins are classified as either fat-soluble or water-soluble.
4. Recommended dietary allowances (RDAs) are a guide for estimating the nutritional needs of groups of people on the basis of their age, gender, and other factors.

### Minerals

Minerals are essential for normal metabolic functions. They are involved with establishing the resting membrane potential; generating action potentials; adding mechanical strength to bones and teeth; combining with organic molecules; and acting as coenzymes, buffers, or regulators of osmotic pressure.

## Daily Values

1. Daily Values are dietary references that can be used to plan a healthy diet.
2. Daily Values for vitamins and minerals are based on Reference Daily Intakes (RDIs), which are generally the highest 1968 RDA value of an age category.
3. Daily Values are based on Daily Reference Values. The Daily Reference Values for energy-producing nutrients (carbohydrates, total fat, saturated fat, and proteins) and dietary fiber are recommended percentages of the total kilocalories ingested daily for each nutrient. The Daily Reference Values for total fats, saturated fats, cholesterol, and sodium are the uppermost limits considered desirable because of their link to diseases.
4. The % Daily Value is the percent of the recommended Daily Value of a nutrient found in one serving of a particular food.

## Metabolism (p. 481)

1. Metabolism consists of catabolism and anabolism. Catabolism, the breakdown of molecules, gives off energy. Anabolism, the synthesis of molecules, requires energy.
2. The energy in carbohydrates, lipids, and proteins is used to produce ATP.
3. The energy from ATP can be used for active transport, muscle contraction, and the synthesis of molecules.

### Regulation of Metabolism

1. A biochemical pathway is a series of chemical reactions, some of which release energy that can be used to synthesize ATP.
2. Each step in a biochemical pathway requires enzymes.
3. Enzyme synthesis is determined by DNA. Enzyme activity is modified by receptor-mediated and end-product processes.

### Carbohydrate Metabolism

1. Glycolysis is the breakdown of glucose to two pyruvic acid molecules. Two ATP molecules are also produced.
2. Anaerobic respiration is the breakdown of glucose in the absence of oxygen to two lactic acid molecules and two ATP molecules.
3. Lactic acid can be converted to glucose using aerobically produced ATP; the necessary oxygen contributes to the oxygen deficit.
4. Aerobic respiration is the breakdown of glucose in the presence of oxygen to produce carbon dioxide, water, and 38 molecules of ATP. The first phase of aerobic respiration is glycolysis; the second

phase is the conversion of pyruvic acid to acetyl-CoA; the third phase is the citric acid cycle; and the fourth phase is the electron-transport chain, which uses carrier molecules, such as NADH, to synthesize ATP.

### Lipid Metabolism

1. Lipids are broken down in adipose tissue, and fatty acids are released into the blood.
2. Fatty acids are taken up by cells and broken down into acetyl-CoA, which can enter the citric acid cycle. Acetyl-CoA can also be converted into ketones by the liver. Ketones released from the liver into the blood are used as energy sources by other cells.

### Protein Metabolism

1. Amino acids are used to synthesize proteins.
2. Amino acids can be used for energy, yielding ammonia as a by-product. Ammonia is converted to urea and excreted by the kidneys.

### Metabolic States

1. In the absorptive state, nutrients are used as energy, with the remainder being stored.
2. In the postabsorptive state, stored nutrients are used for energy.

### Metabolic Rate

1. The metabolic rate is the total energy expenditure per unit of time.
2. Metabolic energy is used for basal metabolism, muscular activity, and the assimilation of food.

## Body Temperature Regulation (p. 491)

1. Body temperature is a balance between heat gain and heat loss.
2. Heat is produced through metabolism.
3. Heat is exchanged through radiation, conduction, convection, and evaporation.
4. The greater the temperature difference, the greater the rate of heat exchange.
5. Body temperature is maintained around a set point by neural circuits in the hypothalamus.
6. Dilation of blood vessels in the skin and sweating increase heat loss from the body.
7. Constriction of blood vessels in the skin and shivering promote heat gain by the body.

## REVIEW AND COMPREHENSION

1. Define a nutrient, and list the six major classes of nutrients. What is an essential nutrient?
2. What is a kilocalorie (Calorie)? Distinguish between a Calorie and a calorie.
3. List some sources of carbohydrates, fats, and proteins in the diet.
4. List the recommended consumption amounts of carbohydrates, fats, and proteins.
5. What are vitamins and provitamins? Name the water-soluble vitamins and the fat-soluble vitamins. List some of the functions of vitamins.
6. What are the Recommended Dietary Allowances (RDAs)?
7. List some of the minerals, and give their functions.
8. What are the Daily Values? How are the Daily Values related to total daily kilocaloric intake? Why are some Daily Values considered the uppermost amounts that should be consumed?
9. Define a % Daily Value.
10. Define a biochemical pathway. How are the steps in a biochemical pathway controlled? What are three ways in which enzymes are regulated?
11. Describe glycolysis. What molecule is the end product of glycolysis? How many ATP and NADH molecules are produced?
12. What determines whether the pyruvic acid produced in glycolysis becomes lactic acid or acetyl-CoA?
13. Describe the two phases of anaerobic respiration. How many ATP molecules are produced? What happens to the lactic acid produced when oxygen becomes available?
14. Define aerobic respiration, and state how many ATP molecules are produced.
15. Describe the citric acid cycle.
16. What is the function of the electron-transport chain?

17. What happens to the carbon atoms in ingested food during metabolism? What happens to the oxygen we breathe in during metabolism?

18. Describe the events occurring during the absorptive and postabsorptive metabolic states.

19. What is meant by metabolic rate? Name three ways that the body uses metabolic energy.

20. Describe how heat is produced by and lost from the body. How is body temperature regulated?

 ## CRITICAL THINKING

1. One serving of a food contains 2 g of saturated fat. What % Daily Value for saturated fat would appear on the food label? (See bottom of figure 17.2 for information needed to answer this question.)

2. An active teenage boy consumes 3000 kcal/day. What is the maximum amount (weight) of total fats he should consume, according to the Daily Values?

3. If the teenager in question 2 eats a food that has a total fat content of 10 g/serving, what is his total fat % Daily Value?

4. Suppose the food in question 3 is in a package that lists a serving size of $\frac{1}{2}$ cup, with four servings in the package. If the teenager eats half the contents of the package (1 cup), how much of his % Daily Value does he consume?

5. Why can some people lose weight on a 1200 kcal/day diet, whereas other people cannot?

6. Lotta Bulk, a bodybuilder, wanted to increase her muscle mass. Knowing that proteins are the main components of muscle, she consumed large amounts of protein daily (high-protein diet), along with small amounts of lipids and carbohydrates. Explain why this strategy will or will not work.

7. After consuming a high-protein diet for several days, does Lotta Bulk's urine contain less, the same, or more urea than before she consumed the proteins? Explain.

8. Thyroid hormone is known to increase the activity of the sodium-potassium exchange pump, which is an active-transport mechanism, thereby increasing the breakdown of ATP. If a person produced excess amounts of thyroid hormone, what effect would this have on basal metabolic rate, body weight, and body temperature? How might the body attempt to compensate for the changes in body weight and temperature?

9. On learning that sweat evaporation results in the loss of calories, an anatomy and physiology student decides that sweating is an easier way to lose weight than dieting. He knows that a liter (about a quart) of water weighs 1000 g, which is equivalent to 580,000 cal (or 580 kcal) of heat when lost as sweat. He believes that, instead of reducing his caloric intake by 580 kcal/day, he can lose about a pound of fat a week by losing a liter of sweat every day in the sauna. Will this approach work? Explain.

10. It is recommended that a person on a diet drink six to eight glasses of cool water per day. How could this practice aid weight loss?

11. In some diseases, an infection results in a high fever. The patient is on the way to recovery when the crisis is over and body temperature begins to return to normal. If you were looking for symptoms in a patient who had just passed through the crisis state, would you look for a dry, pale skin or a wet, flushed skin? Explain.

*Answers in Appendix D*

 Visit this textbook's website at **www.mhhe.com/seeleyess7** for practice quizzes, animations, interactive learning exercises, and other study tools.

 McGraw-Hill offers a study CD that features interactive cadaver dissection. **Anatomy & Physiology Revealed** includes cadaver photos that allow you to peel away layers of the human body to reveal structures beneath the surface. This program also includes animations, radiologic imaging, audio pronunciations, and practice quizzing.

The urinary system is a central player in the maintenance of homeostasis. Removal of toxins, regulation of body fluid pH, and regulation of body fluid concentration are only a few of the tasks the urinary system performs. When the urinary system malfunctions, medical interventions—such as urinary dialysis, as shown in the photo—are required for survival.

# Chapter 18

# Urinary System and Fluid Balance

## Chapter Outline and Objectives

**Functions of the Urinary System (p. 497)**
1. List the structures that make up the urinary system.
2. List the major functions of the urinary system.

**Anatomy of the Kidneys (p. 497)**
3. Describe the location and anatomy of the kidneys.
4. Describe the structure of the nephron and the location of the parts of the nephron in the kidney.

**Urine Formation (p. 502)**
5. Identify the principal factors that influence filtration pressure, and explain how they affect the rate of filtrate formation.
6. Give the function of the proximal convoluted tubule, descending and ascending limbs of the loop of Henle, distal convoluted tubule, and collecting duct.
7. Explain how the movement of substances across the wall of the nephron and collecting duct influences the composition of the filtrate.

**Regulation of Urine Concentration and Volume (p. 509)**
8. Explain how antidiuretic hormone, aldosterone, and atrial natriuretic hormone influence the volume and concentration of urine.

**Urine Movement (p. 514)**
9. Describe the ureters, urinary bladder, and urethra.
10. Describe the micturition reflex.

**Body Fluid Compartments (p. 514)**
11. List the major body fluid compartments.

**Regulation of Extracellular Fluid Composition (p. 516)**
12. Describe the mechanisms by which $Na^+$, $K^+$, and $Ca^{2+}$ are regulated in the extracellular fluid.

**Regulation of Acid-Base Balance (p. 518)**
13. Illustrate how the mechanisms that regulate the body fluid pH function by explaining how they respond to decreasing and increasing pH in the body fluids.

Although you probably know that each person has two kidneys and you most likely can identify their general location, you may not be aware of the many functions the kidneys perform. On the other hand, you probably have a much better understanding of the function of the urinary bladder and a great appreciation for the attention required when it is filled with the urine produced by the kidneys.

The **urinary** (ūr′i-nār-ē) **system** consists of two kidneys, two ureters, the urinary bladder, and the urethra (figure 18.1). A large volume of blood flows through the kidneys, which remove substances from the blood to form urine. The urine contains excess water and ions, metabolic wastes (such as urea), and toxic substances consumed with food. The urine produced by the kidneys flows through the ureters to the urinary bladder, where it is stored until it is eliminated through the urethra.

The kidneys can suffer extensive damage and still maintain their extremely important role in the maintenance of homeostasis. As long as about one-third of one kidney remains functional, survival is possible. However, if the functional ability of the kidneys fails completely, death will result unless the person receives medical treatment.

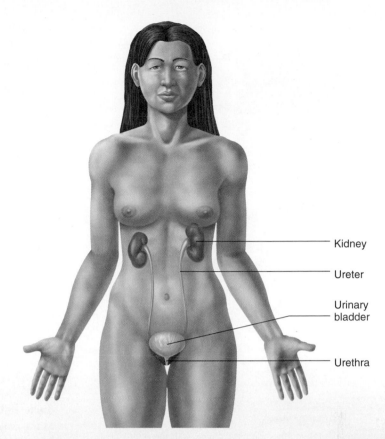

Kidney

Ureter

Urinary bladder

Urethra

**Figure 18.1** Urinary System

The urinary system consists of two kidneys, two ureters, the urinary bladder, and the urethra.

# FUNCTIONS OF THE URINARY SYSTEM

The major function of the urinary system is to control the composition and volume of body fluids. The kidneys perform this function through multiple processes:

1. *Excretion.* The kidneys are the major excretory organs of the body. They remove waste products, many of which are toxic, from the blood. Most waste products are metabolic by-products of cells and substances absorbed from the intestine. The skin, liver, lungs, and intestines eliminate some of these waste products, but they cannot compensate if the kidneys fail to function.

2. *Regulation of blood volume and pressure.* The kidneys play a major role in controlling the extracellular fluid volume in the body by producing either a large volume of dilute urine or a small volume of concentrated urine. Consequently, the kidneys regulate blood volume and blood pressure.

3. *Regulation of the concentration of solutes in the blood.* The kidneys help regulate the concentration of the major molecules and ions, such as glucose, $Na^+$, $Cl^-$, $K^+$, $Ca^{2+}$, $HCO_3^-$, and $HPO_4^{2-}$.

4. *Regulation of extracellular fluid pH.* The kidneys secrete variable amounts of $H^+$ to help regulate extracellular fluid pH.

5. *Regulation of red blood cell synthesis.* The kidneys secrete a hormone, erythropoietin, which regulates the synthesis of red blood cells in bone marrow (see chapter 11).

6. *Regulation of vitamin D synthesis.* The kidneys play an important role in controlling blood levels of $Ca^{2+}$ by regulating the synthesis of vitamin D (see chapter 6).

# ANATOMY OF THE KIDNEYS

The **kidneys** are bean-shaped organs, each about the size of a tightly clenched fist. They lie on the posterior abdominal wall, behind the peritoneum, with one kidney on each side of the vertebral column (figure 18.2). Structures that are behind the peritoneum are said to be **retroperitoneal** (re′trō-per′i-tō-nē′ăl). A layer of connective tissue called the **renal** (derived from the Latin word for kidney) **capsule** surrounds each kidney. Around the renal capsule is a thick layer of fat, which protects the kidney from mechanical shock. On the medial side of each kidney is the **hilum** (hī′lŭm), where the renal artery and nerves enter and where the

Urinary

**Anterior view**

(a) The kidneys are located in the abdominal cavity, with the right kidney just below the liver and the left kidney below the spleen. The ureters extend from the kidneys to the urinary bladder within the pelvic cavity. An adrenal gland is located at the superior pole of each kidney.

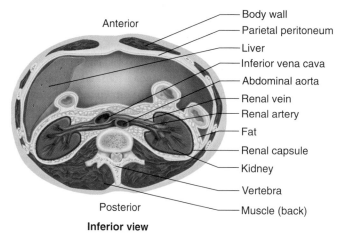

**Inferior view**

(b) The kidneys are located behind the parietal peritoneum. Fat surrounds each kidney. The renal arteries extend from the abdominal aorta to each kidney, and the renal veins extend from the kidneys to the inferior vena cava.

**Figure 18.2   Anatomy of the Urinary System**

renal vein and ureter exit the kidney (figure 18.3). The hilum opens into a cavity called the **renal sinus,** which contains blood vessels, part of the system for collecting urine, and fat.

The kidney is divided into an outer **cortex** and an inner **medulla,** which surround the renal sinus. The bases of several cone-shaped **renal pyramids** are located at the boundary between the

cortex and the medulla, and the tips of the renal pyramids project toward the center of the kidney. A funnel-shaped structure called a **calyx** (kā′liks; pl. calyces kal′i-sēz) surrounds the tip of each renal pyramid. The calyces from all the renal pyramids join to form a larger funnel called the **renal pelvis.** The renal pelvis then narrows to form a small tube, the **ureter** (ū-rē′ter, ū′re-ter), which exits the kidney and connects to the urinary bladder. Urine passes from the tips of the renal pyramids into the calyces. From the calyces, urine collects in the renal pelvis and exits the kidney through the ureter (figure 18.3).

The functional unit of the kidney is the **nephron** (nef′ron; Greek for kidney), and there are approximately 1.3 million of them in each kidney. Each nephron consists of a **renal corpuscle,** a **proximal convoluted tubule,** a **loop of Henle,** and a **distal convoluted tubule** (figure 18.4). Fluid enters the renal corpuscle and then flows into the proximal convoluted tubule. From there, it flows into the loop of Henle. Each loop of Henle consists of a descending limb and an ascending limb. The limbs are further categorized into segments: the thin segment of the descending limb, the thin segment of the ascending limb, and the thick segment of the ascending limb. The descending limb extends toward the renal sinus, where it makes a hairpin turn, and the ascending limb extends back toward the cortex. The fluid flows through the ascending limb of the loop of Henle to the distal convoluted tubule. Many distal convoluted tubules empty into a **collecting duct,** which carries the fluid from the cortex, through the medulla. Many collecting ducts empty into a **papillary duct,** and the papillary ducts empty their contents into a calyx.

The renal corpuscle and both convoluted tubules are in the renal cortex (figure 18.4). The collecting duct and loop of Henle enter the medulla. Approximately 15% of the nephrons, called **juxtamedullary** (next to the medulla) **nephrons,** have loops of Henle that extend deep into the medulla of the kidney. The other nephrons (85%), called **cortical nephrons,** have loops of Henle that do not extend deep into the medulla.

The renal corpuscle of the nephron consists of Bowman's capsule and the glomerulus (figure 18.5; see figure 18.4). **Bowman's capsule** consists of the enlarged end of the nephron, which is indented to form a double-walled chamber. The **glomerulus** (glō-mār′ū-lŭs) is a tuft of capillaries that resembles a ball of yarn and lies within the indentation of Bowman's capsule. The cavity of Bowman's capsule opens into the proximal convoluted tubule, which carries fluid away from the capsule. The inner layer of Bowman's capsule consists of specialized cells called **podocytes** (pod′ō-sīts), which wrap around the glomerular capillaries. The outer layer of Bowman's capsule consists of simple squamous epithelial cells.

The glomerular capillaries have pores in their walls, and the podocytes have numerous cell processes with gaps between them. The endothelium of the glomerular capillaries, the podocytes, and the basement membrane between them form a **filtration membrane** (figure 18.5d). In the first step of urine formation, fluid, consisting of water and solutes smaller than proteins, passes from the blood in the glomerular capillaries through the filtration membrane into Bowman's capsule. The fluid that passes across the filtration membrane is called **filtrate.**

The proximal convoluted tubules, the thick segment of the loops of Henle, the distal convoluted tubules, and the collecting ducts consist of simple cuboidal epithelium. The cuboidal epithelial cells

(a)

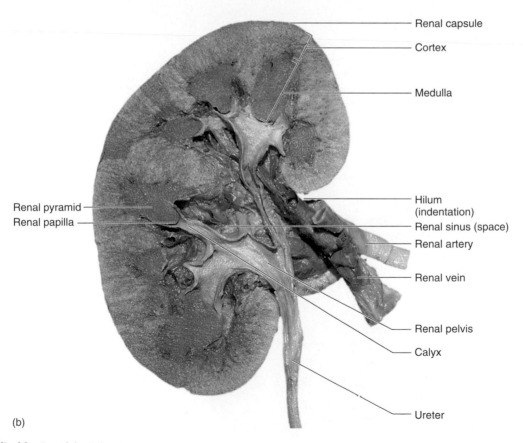

(b)

## Figure 18.3 Longitudinal Section of the Kidney

(a) The cortex and the medulla of the kidney surround the renal sinus. The renal sinus is a space containing the renal pelvis, calyces, blood vessels, fat, and other connective tissues. The renal pyramids extend from the cortex of the kidney to the renal sinus. The tip of each renal pyramid is surrounded by a calyx. The calyces connect to the renal pelvis. Urine flows from the tip of the renal pyramid through the calyx and renal pelvis into the ureter. (b) Photograph of a longitudinal section through the kidney.

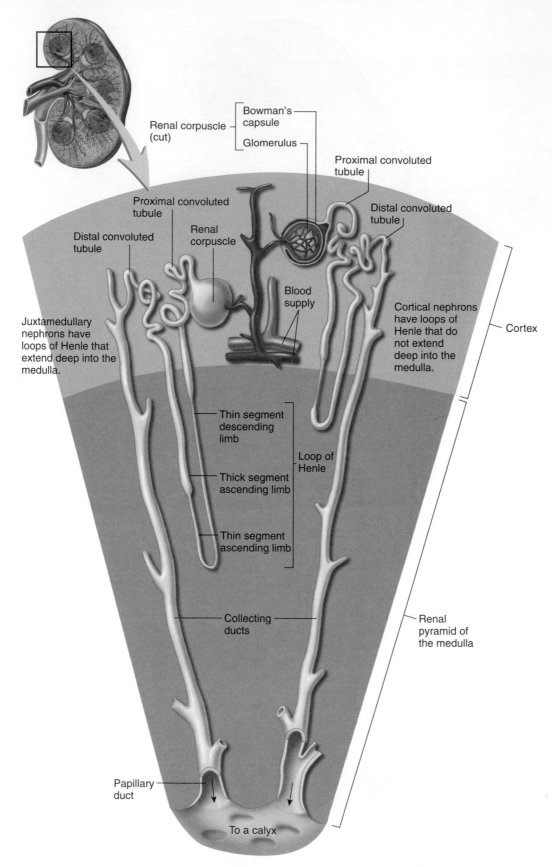

**Figure 18.4  Functional Unit of the Kidney—the Nephron**

A nephron consists of a renal corpuscle, a proximal convoluted tubule, a loop of Henle, and a distal convoluted tubule. The distal convoluted tubule empties into a collecting duct. Juxtamedullary nephrons (those near the medulla of the kidney) have loops of Henle that extend deep into the medulla of the kidney, whereas cortical nephrons do not. Collecting ducts merge into larger papillary ducts, which empty into a calyx.

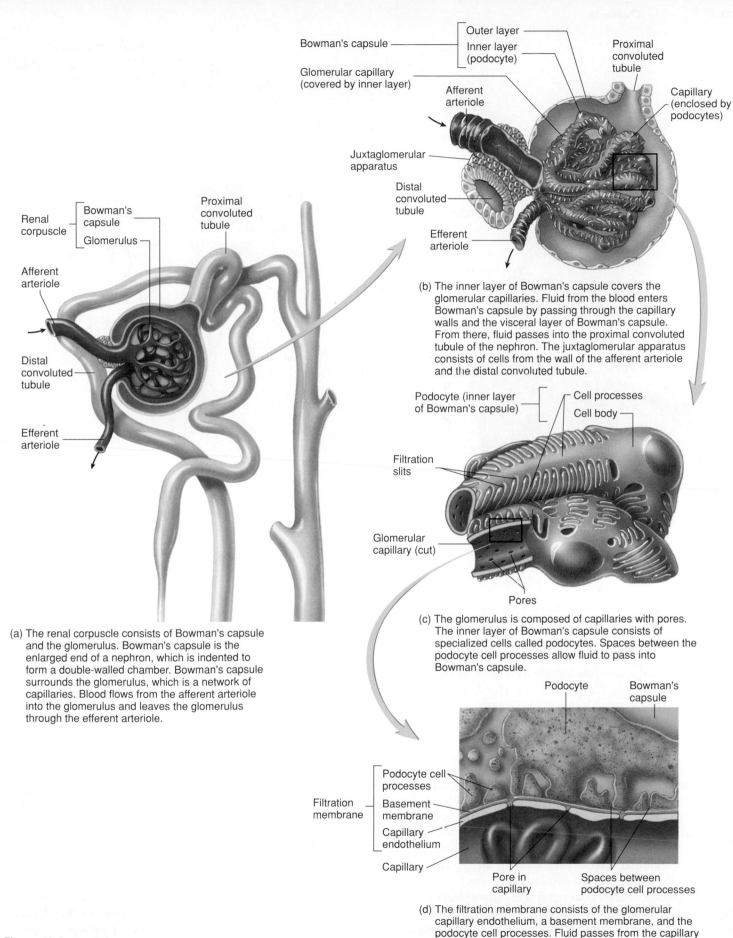

**Bowman's capsule**
— Outer layer
— Inner layer (podocyte)

**Glomerular capillary** (covered by inner layer)

**Proximal convoluted tubule**

**Afferent arteriole**

**Capillary** (enclosed by podocytes)

**Juxtaglomerular apparatus**

**Distal convoluted tubule**

**Efferent arteriole**

(b) The inner layer of Bowman's capsule covers the glomerular capillaries. Fluid from the blood enters Bowman's capsule by passing through the capillary walls and the visceral layer of Bowman's capsule. From there, fluid passes into the proximal convoluted tubule of the nephron. The juxtaglomerular apparatus consists of cells from the wall of the afferent arteriole and the distal convoluted tubule.

**Renal corpuscle**
— Bowman's capsule
— Glomerulus

**Proximal convoluted tubule**

**Afferent arteriole**

**Distal convoluted tubule**

**Efferent arteriole**

(a) The renal corpuscle consists of Bowman's capsule and the glomerulus. Bowman's capsule is the enlarged end of a nephron, which is indented to form a double-walled chamber. Bowman's capsule surrounds the glomerulus, which is a network of capillaries. Blood flows from the afferent arteriole into the glomerulus and leaves the glomerulus through the efferent arteriole.

**Podocyte (inner layer of Bowman's capsule)**
— Cell processes
— Cell body

**Filtration slits**

**Glomerular capillary (cut)**

**Pores**

(c) The glomerulus is composed of capillaries with pores. The inner layer of Bowman's capsule consists of specialized cells called podocytes. Spaces between the podocyte cell processes allow fluid to pass into Bowman's capsule.

**Podocyte**      **Bowman's capsule**

**Podocyte cell processes**

**Filtration membrane**
— Basement membrane
— Capillary endothelium

**Capillary**

**Pore in capillary**

**Spaces between podocyte cell processes**

(d) The filtration membrane consists of the glomerular capillary endothelium, a basement membrane, and the podocyte cell processes. Fluid passes from the capillary through the filtration membrane into Bowman's capsule.

**Figure 18.5** Renal Corpuscle and Filtration Membrane

Urinary

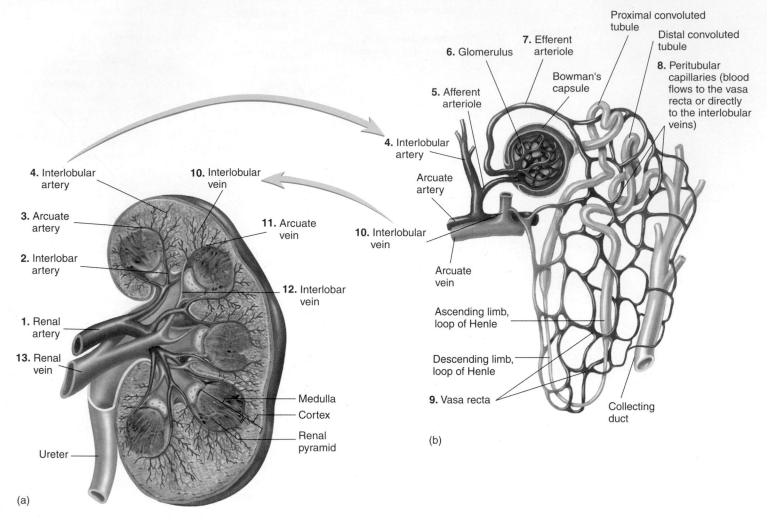

**Figure 18.6  Blood Flow Through the Kidney**

(a) Blood flow through the larger arteries and veins of the kidney. (b) Blood flow through the arteries, capillaries, and veins that provide circulation to the nephrons.

have microvilli and many mitochondria. These portions of the nephron actively transport molecules and ions across the wall of the nephron. The thin segments of the descending and ascending limbs of the loops of Henle have very thin walls made up of simple squamous epithelium. Water and solutes pass through the walls of these portions of the nephron by diffusion. The thin segment of the descending limb of the loop of Henle is permeable to water and solutes, and the thin segment of the ascending limb is permeable to solutes, but not to water.

## Arteries and Veins

A system of blood vessels allows the exchange of materials that occurs in the kidneys. The **renal arteries** branch off the abdominal aorta and enter the kidneys (figure 18.6a). They give rise to several branches. The **interlobar** (in-ter-lō′bar; between the lobes) **arteries** pass between the renal pyramids and give rise to the **arcuate** (ar′kū-āt; arched) **arteries,** which arch between the cortex and the medulla. **Interlobular arteries** branch off the arcuate arteries and project into the cortex. The **afferent arterioles** arise from branches of the interlobular arteries and extend to the glomerular capillaries (figure 18.6b). **Efferent arterioles** extend from the glomerular capillaries to the **peritubular** (around the tubes) **capillaries,** which surround the proximal convoluted and distal convoluted

tubules and the loops of Henle. The **vasa recta** (vā′să rek′tă; straight vessels) are specialized portions of the peritubular capillaries that extend deep into the medulla of the kidney and surround the loops of Henle and collecting ducts. Blood from the peritubular capillaries, including the vasa recta, enters the interlobular veins. The veins of the kidney run parallel to the arteries and have similar names (figure 18.6).

A structure called the **juxtaglomerular** (jŭks′-tă-glŏ-mer′ū-lăr) **apparatus** (pl. apparati) is formed where the distal convoluted tubule comes in contact with the afferent arteriole next to Bowman's capsule (see figure 18.5b). The juxtaglomerular apparatus consists of specialized cells of the walls of the distal convoluted tubules and collecting ducts. These cells secrete renin and play an important role in blood pressure regulation (see "Hormonal Mechanism" later in the chapter).

## URINE FORMATION

Three major processes are essential for urine formation: filtration, tubular reabsorption, and tubular secretion (figure 18.7). Urine is mostly water and contains organic waste products, as well as excess ions (table 18.1).

Urine forms through the following three processes:

**1 Filtration.**    Filtration (*blue arrow*) is the movement of materials across the filtration membrane into Bowman's capsule to form filtrate.

**2 Tubular reabsorption.** Solutes are reabsorbed (*purple arrow*) across the wall of the nephron into the interstitial fluid by transport processes, such as active transport and cotransport. Water is reabsorbed (*green arrow*) across the wall of the nephron by osmosis. Water and solutes pass from the interstitial fluid into the peritubular capillaries.

**3 Tubular secretion.**    Solutes are secreted (*orange arrow*) across the wall of the nephron into the filtrate.

**PROCESS Figure 18.7**  Urine Formation

**Filtration** is the movement of water, ions, and small molecules through the filtration membrane into Bowman's capsule. The portion of the plasma entering the nephron is called the filtrate.

**Tubular reabsorption** is the movement of substances from the filtrate across the wall of the nephron back into the blood of the peritubular capillaries. Certain solute molecules and ions are reabsorbed by processes such as active transport and cotransport into the cells of the nephron wall and then from there into the interstitial fluid. Water reabsorption occurs by osmosis across the nephron wall. The molecules and ions that enter the interstitial fluid surrounding the nephron pass into the peritubular capillaries. In general, the useful substances that enter the filtrate are reabsorbed, and metabolic waste products remain in the filtrate and are eliminated. For example, when proteins are metabolized, ammonia is a by-product. Ammonia, which is toxic to humans, is converted into urea by the liver. Urea forms part of the filtrate; although some of it is reabsorbed, much of it is eliminated in the urine.

**Tubular secretion** is the active transport of solutes across the nephron walls into the filtrate. Consequently, urine consists of substances that are filtered across the filtration membrane and those that are secreted from the peritubular capillaries into the nephron, minus the substances that are reabsorbed.

## Filtration

An average of 21% of the blood pumped by the heart each minute flows through the kidneys. Of the total volume of blood plasma that flows through the glomerular capillaries, about 19% passes through the filtration membrane into Bowman's capsule to become filtrate. In all the nephrons of both kidneys, about 180 liters (L) of filtrate are produced each day, but only about 1% or less of the filtrate becomes urine because most of the filtrate is reabsorbed.

The filtration membrane allows some substances, but not others, to pass from the blood into Bowman's capsule. Water and small solutes readily pass through the openings of the filtration membrane, but blood cells and most proteins are too large to pass through the filtration membrane and do not enter Bowman's capsule. However, albumin, a small blood protein, enters the filtrate in very small amounts. Consequently, the filtrate contains no cells and little protein.

The formation of filtrate depends on a pressure gradient, called the **filtration pressure,** which forces fluid from the glomerular capillary across the filtration membrane into Bowman's capsule (figure 18.8). The filtration pressure results from forces that move fluid out of the glomerular capillary into Bowman's capsule minus the forces that move fluid out of Bowman's capsule into the glomerular capillary. The **glomerular capillary pressure** is the blood pressure in the glomerular capillary. It is the major force causing fluid to move from the glomerular capillary across the filtration membrane into Bowman's capsule. Opposing the movement of fluid into the lumen of Bowman's capsule is the **capsular pressure,** caused by the pressure of filtrate already inside Bowman's capsule, and the **colloid osmotic pressure** within the glomerular capillary. Because most plasma proteins do not pass through the filtration

| TABLE 18.1 | The Average Concentration of Major Urine Substances | |
|---|---|---|
| **Substance** | **Plasma** | **Urine** |
| Water (L/day) | | 1.4 |
| Organic molecules (mg/dL) | | |
| Protein | 3900–5000 | 0* |
| Glucose | 100 | 0 |
| Urea | 26 | 1820 |
| Uric acid | 3 | 42 |
| Creatinine | 1 | 196 |
| Ions (mEq/L) | | |
| $Na^+$ | 142 | 128 |
| $K^+$ | 5 | 60 |
| $Cl^-$ | 103 | 134 |
| $HCO_3^-$ | 28 | 14 |
| Specific gravity (g/ml)† | | 1.005–1.030 |
| pH | | 4.5–8.0 |

*Trace amounts of protein can be found in the urine.
†The specific gravity increases as the concentration of solutes in urine increases.

**1** Glomerular capillary pressure, the blood pressure within the glomerulus, moves fluid from the blood into Bowman's capsule.

**2** Capsular pressure, the pressure inside Bowman's capsule, moves fluid from the capsule into the blood.

**3** Colloid osmotic pressure, produced by the concentration of blood proteins, moves fluid from Bowman's capsule into the blood by osmosis.

**4** Filtration pressure is equal to the glomerular capillary pressure minus the capsular and colloid osmotic pressures.

**1** Glomerular capillary pressure

**2** Capsular pressure

Glomerular capillary pressure
− Capsular pressure
− Colloid osmotic pressure

**4** Filtration pressure

**3** Colloid osmotic pressure

**PROCESS Figure 18.8** Filtration Pressure

Filtration pressure across the filtration membrane is equal to the glomerular capillary pressure minus the capsular pressure and minus the colloid osmotic pressure.

membrane, they produce an osmotic pressure that favors fluid movement into the glomerular capillary from Bowman's capsule. Therefore, the filtration pressure is calculated as follows:

$$
\begin{array}{r}
\text{Glomerular capillary pressure} \\
- \text{ Capsular pressure} \\
\underline{- \text{ Colloid osmotic pressure}} \\
\text{Filtration pressure}
\end{array}
$$

The filtration pressure forces fluid from the glomerulus into Bowman's capsule because the glomerular capillary pressure is greater than both the capsular and the colloid osmotic pressures. Under most conditions, the filtration pressure remains within a narrow range of values. However, when the filtration pressure increases, both the filtrate volume and the urine volume increase, and when the filtration pressure decreases, both the filtrate volume and the urine volume decrease.

The filtration pressure is influenced by the blood pressure in the glomerular capillaries, the blood protein concentration, and the pressure in Bowman's capsule. The blood pressure is normally higher in the glomerular capillaries than it is in most capillaries. The filtration pressure increases if the blood pressure in the glomerular capillaries increases further. The filtration pressure decreases if the blood pressure in the glomerular capillary decreases.

The concentration of proteins in the blood opposes the effect of blood pressure on the filtration pressure because of osmosis (see chapter 3). An increase in blood protein concentration increases the movement of water by osmosis into the blood and therefore reduces the filtration pressure. A decrease in blood protein concentration decreases the movement of water by osmosis into the blood, which increases the filtration pressure.

### Regulation of Filtration

The blood pressure within the glomerular capillaries is fairly constant because the afferent and efferent arterioles either dilate or constrict to regulate the blood pressure there, even though the systemic blood pressure may fluctuate substantially. Also, the concentration of blood proteins and the pressure inside Bowman's

capsule are fairly constant. As a consequence, the filtration pressure and the rate of filtrate formation are maintained within a narrow range of values most of the time.

However, the filtration pressure does change dramatically under some conditions. Sympathetic neurons innervate the blood vessels of the kidney. Sympathetic stimulation constricts the arteries, causing a decrease in renal blood flow and filtrate formation in proportion to the intensity of the stimulation. It is possible to decrease filtrate formation to only a few milliliters per minute. Consequently, only a small volume of urine is produced (see figure 18.17). For instance, during **cardiovascular shock,** the filtration pressure and filtrate formation fall dramatically. One of the dangers of cardiovascular shock is that the renal blood flow can be so low that the kidneys suffer from lack of oxygen. If the oxygen level remains too low for a long enough time, permanent kidney damage or complete kidney failure results. One important reason for treating cardiovascular shock quickly is to avoid damage to the kidneys. Other conditions, such as intense physical activity or trauma, also increase sympathetic stimulation of renal arteries and decrease urine production to very low levels. On the other hand, increased blood pressure decreases sympathetic stimulation of renal blood arteries, and urine volume increases (see figure 18.17).

In addition to the effects of blood pressure on filtration pressure, decreases in the concentration of plasma proteins, caused by conditions such as inflammation of the liver, where most of the blood proteins are produced, increase the filtration pressure. The increased filtration pressure causes the filtrate and urine volume to increase.

## Tubular Reabsorption

As the filtrate flows from Bowman's capsule through the proximal convoluted tubule, loop of Henle, distal convoluted tubule, and collecting duct, many of the solutes in the filtrate are reabsorbed. About 99% of the original filtrate volume is reabsorbed and enters the peritubular capillaries. The reabsorbed filtrate flows through the renal veins to enter the general circulation. Only 1% of the original filtrate volume becomes urine (figure 18.9). Because excess ions and metabolic waste products are not readily reabsorbed, the

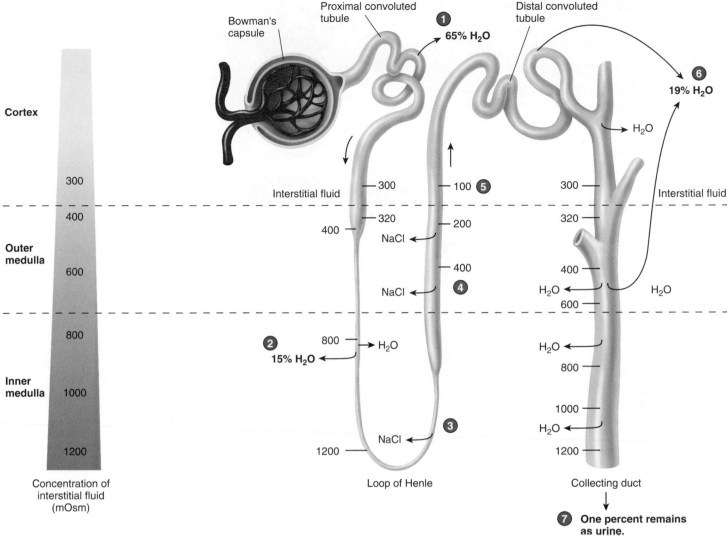

**Cortex**

Concentration of
interstitial fluid
(mOsm)

**Loop of Henle**

**Collecting duct**

⑦ **One percent remains
as urine.**

**PROCESS Figure 18.9** Urine-Concentrating Mechanism

The concentration gradient from the cortex to the inner medulla is shown on the left. Interstitial fluid increases in concentration from 300 mOsm/L in the cortex to 1200 mOsm/L in the medulla. The concentrations of the filtrate in different parts of the nephron are also shown.

① Approximately 180 L of filtrate enters the nephrons each day; of that volume, 65% is reabsorbed in the proximal convoluted tubule. In the proximal convoluted tubule, solute molecules move by active transport and cotransport from the lumen of the tubule into the interstitial fluid. Water moves by osmosis because the cells of the tubule wall are permeable to water (see figure 18.10).

② Approximately 15% of the filtrate volume is reabsorbed in this segment of the descending limb of the loop of Henle. The descending limb passes through the concentrated interstitial fluid of the medulla. Because the wall of the descending limb is permeable to water, water moves by osmosis from the tubule into the more concentrated interstitial fluid (see figure 18.11, step 1). By the time the filtrate reaches the tip of the renal pyramid, the concentration of the filtrate is equal to the concentration of the interstitial fluid.

③ The ascending limb of the loop of Henle is not permeable to water. Solutes diffuse out of the thin segment (see figure 18.11, step 2).

④ $Na^+$ are actively transported, and $K^+$ and $Cl^-$ are cotransported from the filtrate of the thick segment into the interstitial fluid (see figure 18.12).

⑤ The volume of the filtrate doesn't change as it passes through the ascending limb, but the concentration is greatly reduced (see figure 18.12). By the time the filtrate reaches the cortex of the kidney, the concentration is approximately 100 mOsm/L, which is less concentrated than the interstitial fluid of the cortex (300 mOsm/L).

⑥ The distal convoluted tubule and collecting duct are permeable to water if ADH is present. If ADH is present, water moves by osmosis from the less concentrated filtrate into the more concentrated interstitial fluid (see figure 18.13). By the time the filtrate reaches the tip of the renal pyramid, an additional 19% of the filtrate is reabsorbed.

⑦ One percent or less remains as urine, when ADH is present.

1 Solute molecules, such as amino acids, glucose, and fructose, as well as $Na^+$, $K^+$, $Ca^{2+}$, $HCO_3^-$, and $Cl^-$, are reabsorbed by active transport and cotransport out of the proximal convoluted tubule by the epithelial cells.

2 Solutes plus 65% of the filtrate volume are reabsorbed from the proximal convoluted tubule and enter the peritubular capillary.

3 Water moves by osmosis out of the proximal convoluted tubule.

Blood flow

Solutes  $H_2O$

Peritubular capillary

Interstitial fluid

1  2  3

Filtrate flow

Proximal convoluted tubule

■ Active transport
■ Cotransport
■ Osmosis

**PROCESS Figure 18.10   Reabsorption in the Proximal Convoluted Tubule**

small volume of urine produced contains a high concentration of ions and metabolic waste products.

The proximal convoluted tubule is the primary site for the reabsorption of solutes and water. The cuboidal cells of the proximal convoluted tubule have numerous microvilli and mitochondria, and they are well adapted to transport molecules and ions across the nephron wall by active transport and cotransport. Substances transported from the proximal convoluted tubule include proteins, amino acids, glucose, and fructose molecules, as well as $Na^+$, $K^+$, $Ca^{2+}$, $HCO_3^-$, and $Cl^-$. The proximal convoluted tubule is permeable to water. As solute molecules are transported out of the proximal convoluted tubule into the interstitial fluid, water moves by osmosis in the same direction. The solutes and water then enter the peritubular capillaries. Consequently, 65% of the filtrate volume is reabsorbed from the proximal convoluted tubule (figures 18.9, step 1, and 18.10).

The descending limb of the loop of Henle further concentrates the filtrate. The renal medulla contains very concentrated interstitial fluid that has large amounts of $Na^+$, $Cl^-$, and urea. The wall of the thin segment of the descending limb is permeable to water and moderately permeable to solutes. As the filtrate passes through the descending limb of the loop of Henle into the medulla of the kidney, water moves out of the nephron by osmosis, and some solutes move into the nephron by diffusion. By the time the filtrate has passed through the descending limb, another 15% of the filtrate volume has been reabsorbed, and the filtrate is as concentrated as the interstitial fluid of the medulla. The reabsorbed filtrate enters the vasa recta (figure 18.11; see figure 18.9, step 2).

The ascending limb of the loop of Henle dilutes the filtrate by removing solutes. The thin segment of the ascending limb is not permeable to water, but it is permeable to solutes. Consequently, solutes diffuse out of the nephron (figure 18.11; see figure 18.9, step 3).

The cuboidal epithelial cells of the thick segment of the ascending limb actively transport $Na^+$ out of the nephron, and $K^+$ and $Cl^-$ are cotransported with $Na^+$. The thick segment of the ascending limb is not permeable to water. As a result, $Na^+$, $K^+$, and $Cl^-$, but little water, are removed from the filtrate (see figure 18.9, step 4). Because of the efficient removal of these solutes, the highly concentrated filtrate that enters the ascending limb of the loop of Henle is converted to a dilute solution by the time it reaches the distal convoluted tubule (figure 18.12; see figure 18.9, step 5). As the filtrate enters the distal convoluted tubule, it is more dilute than the interstitial fluid of the renal cortex. Also, because of the volume of filtrate reabsorbed in the proximal convoluted tubule and the descending limb of the loop of Henle, only about 20% of the original filtrate volume remains. The solutes transported from the ascending limb of the loop of Henle enter the interstitial fluid of the medulla and help keep the concentration of solutes in the medulla high. Excess solutes enter the vasa recta.

The cuboidal cells of the distal convoluted tubule and collecting duct remove water and additional solutes. $Na^+$ and $Cl^-$ are reabsorbed. Sodium ions are actively transported, and $Cl^-$ are cotransported. Also, 19% of the original filtrate volume is reabsorbed by osmosis, leaving about 1% of the original filtrate as urine (figure 18.13; see figure 18.9, steps 6 and 7). The reabsorbed water

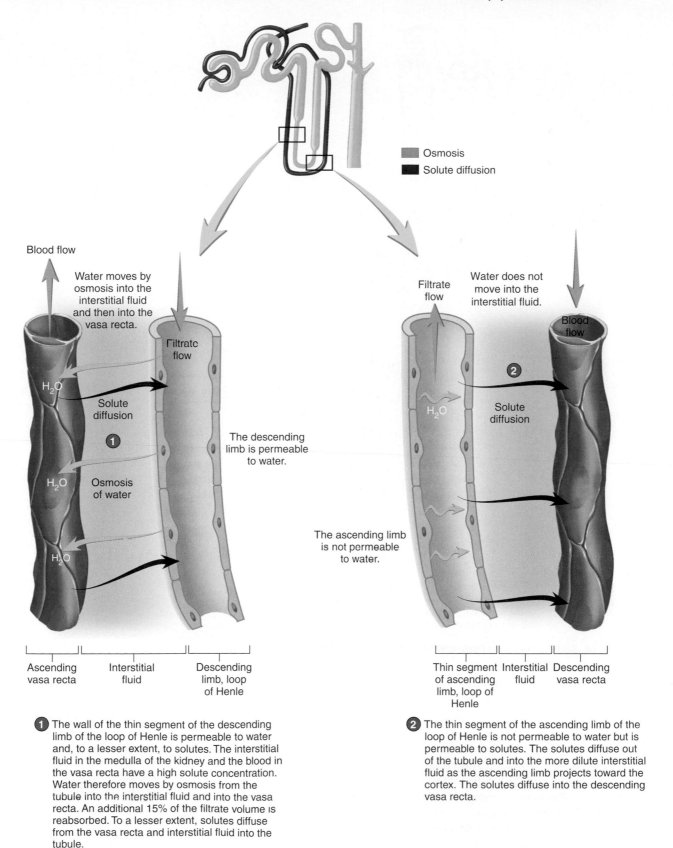

Osmosis

Solute diffusion

Blood flow

Water moves by osmosis into the interstitial fluid and then into the vasa recta.

Filtrate flow

$H_2O$

Solute diffusion

1

$H_2O$

Osmosis of water

$H_2O$

The descending limb is permeable to water.

Ascending vasa recta | Interstitial fluid | Descending limb, loop of Henle

Filtrate flow

Water does not move into the interstitial fluid.

Blood flow

2

$H_2O$

Solute diffusion

The ascending limb is not permeable to water.

Thin segment of ascending limb, loop of Henle | Interstitial fluid | Descending vasa recta

1 The wall of the thin segment of the descending limb of the loop of Henle is permeable to water and, to a lesser extent, to solutes. The interstitial fluid in the medulla of the kidney and the blood in the vasa recta have a high solute concentration. Water therefore moves by osmosis from the tubule into the interstitial fluid and into the vasa recta. An additional 15% of the filtrate volume is reabsorbed. To a lesser extent, solutes diffuse from the vasa recta and interstitial fluid into the tubule.

2 The thin segment of the ascending limb of the loop of Henle is not permeable to water but is permeable to solutes. The solutes diffuse out of the tubule and into the more dilute interstitial fluid as the ascending limb projects toward the cortex. The solutes diffuse into the descending vasa recta.

**PROCESS Figure 18.11   Reabsorption in the Loop of Henle: The Descending Limb and the Thin Segment of the Ascending Limb**

Urinary

Blood flow

Filtrate flow

■ Active transport
■ Cotransport

**1** The thick segment of the ascending limb of the loop of Henle is not permeable to water.

**2** Na⁺ move by active transport, and K⁺ and Cl⁻ move by cotransport out of the ascending limb of the loop of Henle.

Thick segment of the ascending limb, loop of Henle | Interstitial fluid | Descending vasa recta

**PROCESS Figure 18.12   Reabsorption in the Loop of Henle: The Thick Segment of the Ascending Limb**

and solutes from the distal convoluted tubule enter the peritubular capillaries and the vasa recta from the collecting ducts.

The reabsorption of water and solutes from the distal convoluted tubule and collecting duct is controlled by hormones, which have a great influence on urine concentration and volume (see "Regulation of Urine Concentration and Volume").

In summary, most of the useful solutes that pass through the filtration membrane into Bowman's capsule are reabsorbed in the proximal convoluted tubule. Filtrate volume is reduced by 65% in the proximal convoluted tubule and by 15% in the descending limb of the loop of Henle. In the ascending limb of the loop of Henle, Na⁺, K⁺, and Cl⁻, but little water, are removed from the filtrate. Consequently, the filtrate becomes dilute. In the distal convoluted tubule and the collecting duct, additional Na⁺ and Cl⁻ are removed, water moves out by osmosis, and the filtrate volume is reduced by another 19%, leaving 1% of the original filtrate volume as urine.

**Predict 1**

*People who suffer from untreated diabetes mellitus can experience very high levels of glucose in the blood. The glucose can easily cross the filtration membrane into Bowman's capsule. Normally, all the glucose is reabsorbed from the nephron. However, if the concentration of glucose in the nephron becomes too high, not all the glucose can be reabsorbed because the number of transport molecules in the cells of the proximal convoluted tubule is limited. How does the volume of urine produced by a person with untreated diabetes mellitus differ from that of a normal person, and how does the concentration of the urine differ from that of a normal person?*

## Tubular Secretion

Some substances, including by-products of metabolism that become toxic in high concentrations and drugs or molecules not

Active transport
Cotransport
Osmosis

Filtrate flow

**1** Water moves by osmosis from the distal convoluted tubule and collecting duct.

**2** $Na^+$ move by active transport and $Cl^-$ move by cotransport out of the distal convoluted tubule and collecting duct.

**3** The reabsorbed water and solutes enter the peritubular capillaries and vasa recta.

① $H_2O$
② $Na^+$
③ $Cl^-$

Peritubular capillary or vasa recta | Interstitial fluid | Distal convoluted tubule or collecting duct

**PROCESS Figure 18.13   Reabsorption in the Distal Convoluted Tubule and Collecting Duct**

Approximately 19% of the filtrate is reabsorbed from the distal convoluted tubule and collecting duct.

normally produced by the body, are secreted into the nephron from the peritubular capillaries. As with tubular reabsorption, tubular secretion can be either active or passive. For example, ammonia diffuses into the lumen of the nephron, whereas $H^+$, $K^+$, creatinine, histamine, and penicillin are actively transported into the nephron.

Hydrogen ions are actively transported into the proximal convoluted tubule. The epithelial cells actively transport large quantities of $H^+$ across the nephron wall into the filtrate. The secretion of $H^+$ plays an important role in regulating the body fluid pH.

In the proximal convoluted tubule, $K^+$ are reabsorbed. However, in the distal convoluted tubule and collecting duct, $K^+$ are secreted, resulting in a net loss of $K^+$ in the urine.

# REGULATION OF URINE CONCENTRATION AND VOLUME

Given a solution in a container, such as a pan on a stove, it is possible to change its concentration by adding water to it or by boiling it. Similarly, the kidneys maintain the concentration of the body fluids by increasing water reabsorption from the filtrate when the body fluid concentration increases and by reducing water reabsorption from the filtrate when the body fluid concentration decreases. The volume and composition of urine therefore change, depending on conditions in the body. If body fluid concentration increases above normal levels, the kidneys produce a small volume of concentrated urine. This eliminates solutes and conserves water, both of which help lower the body fluid concentration back to normal. On the other hand, if the body fluid concentration decreases, the kidneys produce a large volume of dilute urine. As a result, water is lost, solutes are conserved, and the body fluid concentration increases.

Urine production also maintains blood volume and therefore blood pressure. An increase in blood volume can increase blood pressure, and a decrease in blood volume can decrease blood pressure. When blood volume increases above normal, the kidneys produce a large volume of urine. The loss of water in the urine lowers blood volume. Conversely, if blood volume decreases below normal, the kidneys produce a small volume of urine to conserve water and maintain blood volume.

## Hormonal Mechanisms

Three major hormonal mechanisms are involved in regulating urine concentration and volume: the renin-angiotensin-aldosterone mechanism, the antidiuretic hormone (ADH) mechanism and the atrial natriuretic hormone (ANH) mechanism. Each mechanism is activated by different stimuli, but they work together to achieve homeostasis. Both the renin-angiotensin-aldosterone mechanism and the ANH mechanism are more sensitive to changes in blood pressure, whereas the ADH mechanism is more sensitive to changes in blood concentration.

### Renin-Angiotensin-Aldosterone Mechanism

**Renin** (rē'nin, ren'in) and **angiotensin** (an'jē-ō-ten'sin) help regulate **aldosterone** (al-dos'ter-ōn) secretion. Renin, an enzyme, is secreted by cells of the juxtaglomerular apparati in the kidneys (see figure 18.5b). Renin acts on **angiotensinogen** (an'jē-ō-tensin'ō-jen), a plasma protein produced by the liver, and converts it to **angiotensin I.** Angiotensin I is rapidly converted to a smaller peptide called **angiotensin II** by **angiotensin-converting enzyme (ACE).** Angiotensin II acts on the adrenal cortex, causing it to secrete aldosterone (see chapter 13).

Aldosterone increases the rate of active transport of $Na^+$ in the distal convoluted tubules and collecting ducts. In the absence of aldosterone, large amounts of $Na^+$ remain in the nephron and become part of the urine. A high $Na^+$ concentration in the filtrate causes water to remain in the nephrons and increases urine volume. Therefore, when the rate of active transport of $Na^+$ is slow, urine

volume increases, and the urine contains a high concentration of $Na^+$. Because $Cl^-$ are attracted by the positive charge on $Na^+$, $Cl^-$ are cotransported with $Na^+$.

When blood pressure suddenly decreases (figure 18.14; see figure 18.17) or when the concentration of $Na^+$ in the blood becomes too low, the kidney releases renin. The resultant increase in aldosterone causes an increase in $Na^+$ and $Cl^-$ reabsorption from the nephrons. Water follows the $Na^+$ and $Cl^-$. Thus, the volume of water lost in the form of urine declines. This method of conserving water helps prevent a further decline in blood pressure (figure 18.14; see figure 18.17).

**Predict 2**

*Drugs that increase the urine volume are called **diuretics** (dī-ū-ret' iks). Some diuretics inhibit the active transport of $Na^+$ in the nephron. Explain how these diuretic drugs can cause an increase in urine volume.*

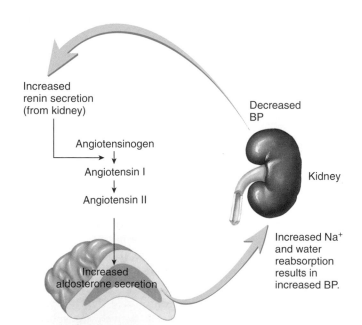

**Figure 18.14 Aldosterone and the Regulation of $Na^+$ and Water in Extracellular Fluid**

Low blood pressure (BP) stimulates renin secretion from the kidney. Renin stimulates the production of angiotensin I, which is converted to angiotensin II, which in turn stimulates aldosterone secretion from the adrenal cortex. Aldosterone increases $Na^+$ and water reabsorption in the kidney.

## Diabetes Insipidus

**Diabetes insipidus** is a pathological condition involving antidiuretic hormone (ADH) abnormality. There are two main causes of diabetes insipidus: The posterior pituitary fails to secrete ADH, or the kidney tubules have abnormal receptors for ADH and do not respond to the presence of ADH. In people suffering from diabetes insipidus, much of the filtrate entering both the proximal convoluted and the distal convoluted tubules becomes urine. People with this condition can produce as much as 20–30 L of urine each day. Because they lose so much water, they are continually in danger of severe dehydration. Even though their urine is dilute, producing such a large volume of urine leads to the loss of $Na^+$, $Ca^{2+}$, and other ions. The resulting ionic imbalances cause the nervous system and cardiac muscle to function abnormally. People suffering from diabetes insipidus due to reduced ADH secretion can be successfully treated by taking ADH in the form of either a nasal spray or an injection.

## Antidiuretic Hormone Mechanism

**Antidiuretic** (an'tē dī ū-ret'ik) **hormone (ADH),** secreted by the posterior pituitary gland, passes through the circulatory system to the kidneys. ADH regulates the amount of water reabsorbed by the distal convoluted tubules and collecting ducts. When ADH levels increase, the permeability of the distal convoluted tubules and collecting ducts to water increases, and more water is reabsorbed from the filtrate. Consequently, an increase in ADH results in the production of a small volume of concentrated urine. On the other hand, when ADH levels decrease, the distal convoluted tubules and collecting ducts become less permeable to water. As a result, less water is reabsorbed, and a large volume of dilute urine is produced (figure 18.15; see figure 18.17).

The release of ADH from the posterior pituitary is regulated by the hypothalamus. Certain cells of the hypothalamus are sensitive to changes in solute concentration. When the solute concentration in the blood increases, action potentials are sent along the axons of the ADH-secreting neurons of the hypothalamus to the posterior pituitary, and ADH is released from the ends of the axons (figure 18.15; see chapter 10). A reduced solute concentration in the blood causes inhibition of ADH release.

Baroreceptors that monitor blood pressure also influence ADH secretion. A large decrease in blood pressure causes an increase in ADH secretion (figure 18.15; see figure 18.17), and a large increase in blood pressure decreases ADH secretion.

## Atrial Natriuretic Hormone

**Atrial natriuretic** (nā'trē-ū-ret'ik) **hormone (ANH)** is secreted from cardiac muscle cells in the right atrium of the heart when blood pressure in the right atrium increases above normal (figures 18.16 and 18.17). ANH acts on the kidney to decrease $Na^+$ reabsorption. Therefore, sodium ions and water remain in the nephron to become urine. The increased loss of $Na^+$ and water as urine reduces the blood volume and the blood pressure.

Increased blood solute concentration or large decrease in BP

Increased ADH release

Kidney

Increased water reabsorption results in decreased solute concentration and increased BP.

**Figure 18.15 ADH and the Regulation of Extracellular Fluid Concentration**

Increased blood solute concentration affects hypothalamic neurons, and decreased blood pressure affects baroreceptors. As a result of these stimuli, the posterior pituitary secretes ADH, which increases water reabsorption by the kidney.

Increased blood pressure in right atrium

ANH

ANH secretion

Kidney

Increased $Na^+$ excretion and increased water loss result in decreased BP.

**Figure 18.16 ANH and the Regulation of $Na^+$ and Water in Extracellular Fluid**

Increased blood pressure in the right atrium of the heart causes increased secretion of ANH, which increases $Na^+$ excretion and water loss in the form of urine.

Urinary

**Pituitary:**
Baroreceptors inhibit posterior pituitary ADH secretion when blood volume increases.

**Kidney:**
Juxtaglomerular apparati inhibit renin release when blood volume increases, which decreases aldosterone secretion.

**Heart:**
Atrial cardiac muscle cells secrete ANH when blood volume increases.

**Blood vessels:**
Sympathetic division baroreceptors detect increased blood volume, which causes vasodilation of renal arteries.

High blood volume induces elevated blood pressure:
**Homeostasis Disturbed**

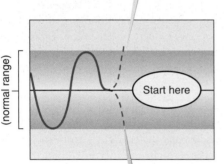
Start here

Low blood volume induces lowered blood pressure:
**Homeostasis Disturbed**

**Blood vessels:**
Sympathetic division baroreceptors detect decreased blood volume, which causes vasoconstriction of renal arteries.

**Heart:**
Atrial cardiac muscle cells do not secrete ANH when blood volume decreases.

**Kidney:**
Juxtaglomerular apparati stimulate renin release when blood volume decreases, which increases aldosterone secretion.

**PItuitary:**
Baroreceptors stimulate posterior pituitary ADH secretion when blood volume decreases.

Decreased ADH decreases water reabsorption by the distal convoluted tubules and collecting ducts. Less water returns to the blood and more water is lost in the urine, which decreases blood volume.

Decreased aldosterone and increased ANH decrease Na$^+$ reabsorption into the distal convoluted tubule and collecting duct. More Na$^+$ and water are lost in the urine, which decreases blood volume.

Increased renal blood flow increases the rate of filtrate formation, and more water is lost in the urine.

Reduced blood volume due to loss of water and Na$^+$ in the urine lowers blood pressure:
**Homeostasis Restored**

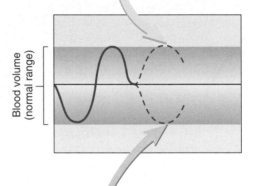

Increased blood volume due to decreased Na$^+$ and water loss in the urine raises blood pressure:
**Homeostasis Restored**

Decreased renal blood flow decreases filtrate formation, and less water is lost in urine, which increases blood volume.

Increased aldosterone and decreased ANH increase Na$^+$ reabsorption in the distal convoluted tubule and the collecting duct. Less Na$^+$ and water are lost in the urine, which increases blood volume.

Increased ADH increases the permeability of the distal convoluted tubule and the collecting duct to water. Increased ADH also increases the sensation of thirst. Less water is lost in the urine.

Urinary

**Homeostasis Figure 18.17**   Hormonal Regulation of Blood Volume and Its Effect on Urine Volume and Concentration

**Figure 18.18  Ureters and Urinary Bladder**

(*a*) Ureters extend from the renal pelvis to the urinary bladder. (*b*) The walls of the ureters and the urinary bladder are lined with transitional epithelium, which is surrounded by a connective tissue layer (lamina propria), smooth muscle layers, and a fibrous adventitia. (*c*) Section through the wall of the urinary bladder.

**Predict 3**

*Ivy Saline worked as a nurse in a hospital. Because she was very observant, she recognized that one of her patients had received a much larger volume of an intravenous (IV) saline solution than had been ordered. A saline solution consists of NaCl but also contains other solutes, such as small amounts of KCl. Saline solutions have the same concentration as body fluids. Predict the effect of the large volume of IV saline solution on the patient's rate of urine production, and describe the role of ADH, ANH, and aldosterone in controlling the change in urine production.*

## Anatomy and Histology of the Ureters, Urinary Bladder, and Urethra

The **ureters** are small tubes that carry urine from the renal pelvis of the kidney to the posterior inferior portion of the urinary bladder (figure 18.18). The **urinary bladder** is a hollow, muscular container that lies in the pelvic cavity just posterior to the pubic symphysis. It stores urine; thus, its size depends on the quantity of urine present. The urinary bladder can hold from a few milliliters (mL) to a maximum of about 1000 mL of urine. When the urinary bladder reaches a volume of a few hundred mL, its wall is stretched enough to activate a reflex that causes the smooth muscle of the urinary bladder to contract, and most of the urine flows out of the urinary bladder through the urethra.

The **urethra** is the tube that carries urine from the urinary bladder to the outside of the body. The triangle-shaped portion of the urinary bladder located between the opening of the ureters and the opening of the urethra is called the **trigone** (trī′gōn; triangle).

The ureters and the urinary bladder are lined with transitional epithelium, which is specialized to stretch (see chapter 4). As the volume of the urinary bladder increases, the epithelial cells change in shape from columnar to flat, and the number of epithelial cell layers decreases. As the volume of the urinary bladder decreases, transitional epithelial cells assume their columnar shape and form a greater number of cell layers.

The walls of the ureter and urinary bladder are composed of layers of smooth muscle and connective tissue. Regular waves of smooth muscle contraction in the ureters produce the force that causes urine to flow from the kidneys to the urinary bladder. Contractions of smooth muscle in the urinary bladder force urine to flow from the bladder through the urethra.

At the junction of the urinary bladder and the urethra, the smooth muscle of the bladder wall forms the **internal urinary sphincter** in males. (There is no functional urinary sphincter in females.) In males, the internal urinary sphincter contracts to keep semen from entering the urinary bladder during sexual intercourse (see chapter 19). In both males and females, the **external urinary sphincter** is formed of skeletal muscle that surrounds the urethra as the urethra extends through the pelvic floor. The external urinary sphincter is under voluntary control, allowing a person to start or stop the flow of urine through the urethra.

In males, the urethra extends to the end of the penis, where it opens to the outside. The female urethra is much shorter (approximately 4 cm) than the male urethra (approximately 20 cm) and opens into the vestibule anterior to the vaginal opening.

### Predict 4

*Cystitis (sis-tī'tis) is an inflammation of the urinary bladder often caused by a bacterial infection. Typically, bacteria from outside the body enter the bladder. Are males or females more prone to cystitis? Explain.*

## A CASE IN POINT
### Cystitis

Ima Burning was attending a 3-day business meeting out of town. On the last day of the meeting, she noticed a frequent urge to urinate, even though her urine volume was small. She also felt a burning sensation when urinating. By the time Ima returned home, she was experiencing lower abdominal pain. Her urine appeared cloudy and had an unpleasant odor. Ima made an appointment with her physician, Dr. Blatter, who requested a urine sample. Urine is normally sterile, but Ima's urine contained abundant bacteria. Her physician diagnosed cystitis, an inflammation of the urinary bladder usually resulting from a bacterial infection. Infection by the bacterium *E. coli* is the most common cause of cystitis. In Ima's case, Dr. Blatter was unable to identify a specific cause of the infection, and she explained to Ima that 30% of women experience cystitis during their lifetime. She prescribed an antibiotic. Within 3 days, Ima was feeling normal again. A urine sample taken several days later showed no sign of infection. It is important to recognize cystitis early and treat it, because the infection can ascend along the ureters and affect the kidneys.

## URINE MOVEMENT

The **micturition** (mik-choo-rish'ŭn) **reflex** is activated by stretch of the urinary bladder wall. As the urinary bladder fills with urine, pressure increases, stimulating stretch receptors in the wall of the urinary bladder. Action potentials are conducted from the urinary bladder to the spinal cord through the pelvic nerves. Integration of the reflex occurs in the spinal cord, and action potentials are conducted along parasympathetic nerve fibers to the urinary bladder. Parasympathetic action potentials cause the urinary bladder to contract (figure 18.19).

The external urinary sphincter is normally contracted as a result of stimulation from the somatic motor nervous system. Because of the micturition reflex, action potentials conducted along somatic motor nerves to the external urinary sphincter decrease, which causes the sphincter to relax. The micturition reflex is an automatic reflex, but it can be inhibited or stimulated by higher centers in the brain. The higher brain centers prevent micturition by sending action potentials through the spinal cord to decrease the intensity of the autonomic reflex that stimulates urinary bladder contractions and to stimulate nerve fibers that

keep the external urinary sphincter contracted. The ability to voluntarily inhibit micturition develops at the age of 2–3 years.

When a person feels the urge to urinate, the higher brain centers alter action potentials sent to the spinal cord to facilitate the micturition reflex and relax the external urinary sphincter. Awareness of the need to urinate occurs because stretch of the urinary bladder stimulates sensory nerve fibers that increase action potentials carried to the brain by ascending tracts in the spinal cord. Irritation of the urinary bladder or the urethra by a bacterial infection or some other condition can also initiate the urge to urinate, even though the urinary bladder is nearly empty.

## A CASE IN POINT
### Kidney Stones (Renal Calculi)

After a long day at work, Harry Payne had just sat down to watch the news when he noticed rapidly developing discomfort in his left lateral abdominal region. Before long, the pain had radiated to the inguinal region on his left side and had become excruciating and debilitating. Harry suspected a kidney stone, or renal calculus, because he had passed a kidney stone about 2 years ago. His wife helped him to the car and drove him to the emergency room. X-rays confirmed the presence of a kidney stone in the upper portion of Harry's left ureter. Peristaltic contractions of the ureter were forcing the irregularly shaped kidney stone through the narrow ureter, causing inflammation and pain. Harry's physician, Dr. Stone, prescribed strong analgesics to reduce the intense pain. Over the next 24 hours, urine was collected, and finally a kidney stone was recovered. Once the kidney stone had passed through the ureter, the pain decreased dramatically.

Like the majority of kidney stones, this one consisted mainly of calcium oxylate. Dr. Stone explained that Harry needed to increase his fluid intake to produce at least 2 liters of urine each day. Keeping the urine dilute reduces the likelihood of calcium salts precipitating and forming additional kidney stones. Also, as a precaution, Dr. Stone recommended that Harry reduce his intake of animal protein.

In addition, Dr. Stone ordered a CT scan, which revealed a large kidney stone in Harry's left renal pelvis. This large kidney stone could give rise to additional kidney stones, obstruct the flow of urine from the left kidney through the left ureter, or become a chronic source of irritation, which could lead to kidney infections. Therefore, Dr. Stone recommended lithotripsy (lith'ō-trip-sē), an ultrasound technique that pulverizes kidney stones into small particles that can pass easily through the ureter.

## BODY FLUID COMPARTMENTS

Approximately 60% of the total body weight of an adult male consists of water. Approximately 50% of the total body weight of an adult female is water. Because the water content of adipose tissue is relatively low, the fraction of the body's weight composed of water decreases as the amount of adipose tissue increases. A smaller percentage of the body weight of an adult female consists of water because females generally have a greater percentage of body fat than do males. Water and the ions dissolved in it are distributed in two major compartments: the intracellular fluid compartment and the extracellular fluid compartment (table 18.2). Water and ions move between these compartments, but their movement is regulated.

The **intracellular fluid compartment** includes the fluid inside all the cells of the body. The cell membranes of the individual cells enclose the intracellular compartment, which actually consists of

### Control of the micturition reflex by higher brain centers

**A** Ascending pathways carry an increased frequency of action potentials up the spinal cord to the pons and cerebrum when the urinary bladder becomes stretched. This increases the conscious urge to urinate.

**B** Descending pathways carry action potentials to the sacral region of the spinal cord to tonically inhibit the micturition reflex, preventing automatic urination when the urinary bladder is full. Descending pathways carry action potentials from the cerebrum to the sacral region of the spinal cord to facilitate the reflex when stretch of the urinary bladder produces the conscious urge to urinate and when a person voluntarily chooses to urinate. This reinforces the micturition reflex.

### Micturition reflex

**1** Urine in the urinary bladder stretches the urinary bladder wall.

**2** Action potentials produced by stretch receptors are carried along pelvic nerves (*green line*) to the sacral region of the spinal cord.

**3** Action potentials are carried by parasympathetic nerves (*red line*) to contract the smooth muscles of the urinary bladder. Decreased action potentials carried by somatic motor nerves (*purple line*) cause the external urinary sphincter to relax.

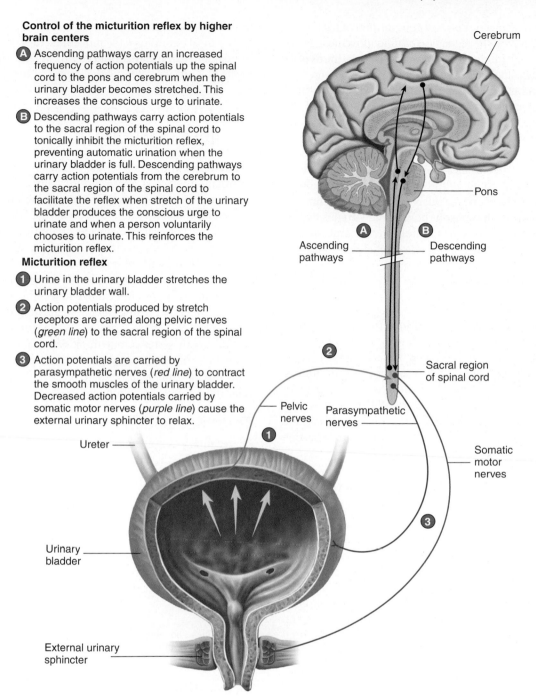

**PROCESS Figure 18.19** Micturition Reflex

| TABLE 18.2 | Approximate Volumes of Body Fluid Compartments* | | | | |
|---|---|---|---|---|---|
| Age of Person | Total Body Water | Intracellular Fluid | Plasma | Extracellular Fluid Interstitial | Total |
| Infant | 75 | 45 | 4 | 26 | 30 |
| Adult male | 60 | 40 | 5 | 15 | 20 |
| Adult female | 50 | 35 | 5 | 10 | 15 |

*Expressed as percentages of body weight.

Urinary

trillions of small compartments. Both the composition of the fluid in all these compartments and the regulation of fluid movement across all these cell membranes are similar. Approximately two-thirds of all the water in the body is in the intracellular fluid compartment.

The **extracellular fluid compartment** includes all the fluid outside the cells. It constitutes approximately one-third of the total body water. The extracellular fluid compartment includes the interstitial fluid, the plasma within blood vessels, and the fluid in the lymphatic vessels. A small portion of the extracellular fluid volume is separated by membranes into subcompartments. These special subcompartments contain fluid with a composition different from that of the other extracellular fluid. Fluids within the subcompartments include the aqueous humor and vitreous humor of the eye, cerebrospinal fluid, synovial fluid in the joint cavities, serous fluid in the body cavities, and fluid secreted by glands.

## Composition of the Fluid in the Body Fluid Compartments

Intracellular fluid has a similar composition from cell to cell. It contains a relatively high concentration of ions, such as $K^+$, magnesium ($Mg^{2+}$), phosphate ($PO_4^{3-}$), and sulfate ($SO_4^{2-}$), compared to the extracellular fluid. It has a lower concentration of $Na^+$, $Ca^{2+}$, $Cl^-$, and $HCO_3^-$ than that of the extracellular fluid. The concentration of protein in the intracellular fluid is also greater than in the extracellular fluid. Like intracellular fluid, the extracellular fluid has a fairly consistent composition from one area of the body to another.

## Exchange Between Body Fluid Compartments

The cell membranes that separate the body fluid compartments are selectively permeable. Water continually passes through them, but they are much less permeable to ions dissolved in the water. Water movement is regulated mainly by hydrostatic pressure differences and osmotic differences between the compartments. For example, water moves across the wall of the capillary at the arterial end of the capillary because the blood pressure there is great enough to force fluid into the interstitial space. At the venous end of the capillary, the blood pressure is much lower, and fluid returns to the capillary because the osmotic pressure is higher inside the capillary than outside it (see chapter 13).

The major influence controlling the movement of water between the intracellular and extracellular spaces is osmosis. For example, if the extracellular concentration of ions increases, water moves by osmosis from cells into the extracellular fluid.

The intracellular fluid can help maintain the extracellular fluid volume if it is depleted. When a person becomes dehydrated, the concentration of ions in the extracellular fluid increases. As a consequence, water moves from the intracellular fluid to the extracellular fluid, thus maintaining the extracellular fluid volume. Because blood is an important component of the extracellular fluid volume, this process helps maintain blood volume. Movement of water from the intracellular fluid compartment to the extracellular fluid compartment can help prolong the time a person can survive a condition such as dehydration or cardiovascular shock.

If the concentration of ions in the extracellular fluid decreases, water moves by osmosis from the extracellular fluid

into the cells. This water movement can cause the cells to swell. Under most conditions, the movement of water between the intracellular and extracellular fluid compartments is maintained within limits that are consistent with survival of the individual.

# REGULATION OF EXTRACELLULAR FLUID COMPOSITION

Homeostasis requires that the intake of substances equals their elimination. Needed water and ions enter the body by ingestion; excess water and ions exit the body by excretion. The amounts of water and ions entering and leaving the body can vary over the short term. For example, greater quantities of water and ions are lost in the form of perspiration on warm days than on cool days, and varying amounts of water and ions may be lost in the form of feces. However, over a long period, the total amount of water and ions in the body does not change unless the individual is growing, gaining weight, or losing weight. Regulating the amounts of water and ions in the body involves the coordinated participation of several organ systems, but the kidneys are the most important, with the skin, liver, and digestive tract playing supporting roles. Two mechanisms help regulate the levels of ions in the extracellular fluid: thirst regulation and ion concentration regulation.

## Thirst Regulation

Water intake is controlled by neurons in the hypothalamus, collectively called the **thirst center.** When the concentration of blood increases, the thirst center responds by initiating the sensation of thirst (figure 18.20). When water or another dilute solution is consumed, the concentration of the blood decreases, and the sensation of thirst decreases. Similarly, when blood pressure decreases, as occurs during shock, the thirst center is activated, and the sensation of thirst is triggered. Consumption of water increases the blood volume and allows the blood pressure to increase toward its normal value. Other stimuli can also trigger the sensation of thirst.

Increased concentration or large decrease in blood pressure

Increased thirst

Hypothalamus

**Figure 18.20  Thirst and the Regulation of Extracellular Fluid Concentration**

Increased blood concentration affects hypothalamic neurons, and large decreases in blood pressure affect baroreceptors in the aortic arch, carotid sinuses, and atrium. These stimuli cause an increase in thirst, which increases water intake. Increased water intake reduces blood concentration and increases blood volume.

For example, if the mucosa of the mouth becomes dry, the thirst center is activated. Thirst is one of the important means of regulating extracellular fluid volume and concentration.

## Ion Concentration Regulation

If the water content or concentration of ions in the extracellular fluid deviates from its normal range, cells cannot control the movement of substances across their cell membranes or the composition of their intracellular fluid. The consequence is abnormal cell function or even cell death. Keeping the extracellular fluid composition within a normal range is therefore required to sustain life.

Regulating the concentrations of positively charged ions, such as $Na^+$, $K^+$, and $Ca^{2+}$, in the body fluids is particularly important. Action potentials, muscle contraction, and normal cell membrane permeability depend on the maintenance of a narrow range of concentrations for these ions. Important mechanisms control the concentrations of these ions in the body. Negatively charged ions, such as $Cl^-$, are secondarily regulated by the mechanisms that control the positively charged ions. The negatively charged ions are attracted to the positively charged ions; when the positively charged ions are transported, the negatively charged ions move with them.

### Sodium Ions

**Sodium ions** ($Na^+$) are the dominant ions in the extracellular fluid. About 90–95% of the osmotic pressure of the extracellular fluid results from $Na^+$ and from the negative ions associated with them.

The recommended intake of $Na^+$ is 2.4 grams per day (g/day), because of their association with high blood pressure in some people. Most people in the United States consume two to three times the recommended amount of $Na^+$. The kidneys provide the major route by which the excess $Na^+$ are excreted.

Stimuli that control aldosterone secretion influence the reabsorption of $Na^+$ from nephrons of the kidneys and the total amount of $Na^+$ in the body fluids. Reabsorption of $Na^+$ from the distal convoluted tubules and collecting ducts is very efficient, and little $Na^+$ are lost in the urine when aldosterone is present. When aldosterone is absent, reabsorption of $Na^+$ in the nephron is greatly reduced, and the amount of $Na^+$ lost in the urine increases. Aldosterone also plays an essential role in regulating the extracellular $K^+$ concentration (figure 18.21).

Sodium ions are also excreted from the body in **sweat.** Normally, only a small quantity of $Na^+$ is lost each day in the form of sweat, but the amount increases during heavy exercise in a warm environment.

Because $Na^+$ have such a large effect on the osmotic pressure of the extracellular fluid, mechanisms that influence $Na^+$ concentrations in the extracellular fluid also influence the extracellular fluid volume. The mechanisms that play important roles in controlling these levels are the renin-angiotensin-aldosterone mechanism, the atrial natriuretic (ANH) mechanism, and antidiuretic hormone (ADH) (see figures 18.14, 18.15, 18.16, and 18.17). For example, low blood pressure increases renin and ADH secretion. The result is an increase in $Na^+$ and water reabsorption in the kidney to bring blood pressure and the $Na^+$ concentration back to their normal ranges. Increased blood pressure inhibits renin and ADH secretion, and stimulates ANH secretion. The result is a decrease in $Na^+$ reabsorption and an increase in urine production to bring blood pressure and the blood's $Na^+$ concentration into their normal ranges.

### Potassium Ions

Electrically excitable tissues, such as muscles and nerves, are highly sensitive to slight changes in the extracellular potassium ion ($K^+$) concentration. The extracellular concentration of $K^+$ must be maintained within a narrow range for these tissues to function normally.

Aldosterone plays a major role in regulating the concentration of $K^+$ in the extracellular fluid. Dehydration, circulatory system shock resulting from plasma loss, and tissue damage due to injuries such as severe burns, all cause extracellular $K^+$ to become more concentrated than normal. In response, aldosterone secretion from the adrenal cortex increases and causes $K^+$ secretion to increase (figure 18.21).

If the $K^+$ concentration in the extracellular fluid decreases, aldosterone secretion from the adrenal cortex decreases. In response, the rate of $K^+$ secretion by the kidney is reduced (figure 18.21).

### Calcium Ions

The extracellular concentration of calcium ions ($Ca^{2+}$), like that of other ions, is maintained within a narrow range. Increases and decreases in the extracellular concentration of $Ca^{2+}$ have dramatic effects on the electrical properties of excitable tissues. For example, decreased extracellular $Ca^{2+}$ concentrations make cell membranes more permeable to $Na^+$, thus making them more electrically excitable. Decreased extracellular concentrations of $Ca^{2+}$ cause spontaneous action potentials in nerve and muscle cells, resulting in hyperexcitability and muscle tetany. Increased extracellular $Ca^{2+}$ concentrations make cell membranes less permeable to $Na^+$, thus making them less electrically excitable. Increased extracellular concentrations of $Ca^{2+}$ inhibit action potentials in nerve and muscle cells, resulting in reduced excitability and either muscle weakness or paralysis.

**Parathyroid hormone (PTH),** secreted by the parathyroid glands, increases extracellular $Ca^{2+}$ concentrations. The rate of PTH secretion is regulated by the extracellular $Ca^{2+}$ concentration (see figure 10.13). An elevated $Ca^{2+}$ concentration inhibits, and a reduced $Ca^{2+}$ concentration stimulates, the secretion of PTH. PTH causes osteoclasts to degrade bone and release $Ca^{2+}$ into the body fluids. PTH also increases the rate of $Ca^{2+}$ reabsorption from kidney nephrons.

Vitamin D increases $Ca^{2+}$ concentration in the blood by increasing the rate of $Ca^{2+}$ absorption by the intestine. Some vitamin D is consumed in food, and the body produces the rest (see chapter 5). PTH affects the intestinal uptake of $Ca^{2+}$ because PTH increases the rate of vitamin D production in the body.

**Calcitonin** (kal-si-tō′nin) is secreted by the thyroid gland. Calcitonin reduces the blood $Ca^{2+}$ concentration when it is too high. An elevated blood $Ca^{2+}$ concentration causes the thyroid gland to secrete calcitonin, and a low blood $Ca^{2+}$ concentration inhibits calcitonin secretion. Calcitonin reduces the rate at which bone is broken down and decreases the release of $Ca^{2+}$ from bone (see figure 10.13).

### Phosphate and Sulfate Ions

Some ions, such as **phosphate ions** ($PO_4^{3-}$) and **sulfate ions** ($SO_4^{2-}$), are reabsorbed by active transport in the kidneys. The rate of reabsorption is slow, so that, if the concentration of these

Urinary

**Homeostasis Figure 18.21  Regulation of Blood K$^+$ Levels**

Boxed labels in figure (clockwise from "Start here"):

- Increased blood levels of K$^+$ act on the adrenal cortex to increase aldosterone secretion.
- Blood K$^+$ levels increase: **Homeostasis Disturbed**
- Start here
- Increased aldosterone increases the rate of K$^+$ secretion from the distal convoluted tubules and collecting ducts of the kidney into the urine.
- Blood K$^+$ levels decrease: **Homeostasis Restored**
- Blood K+ levels increase: **Homeostasis Restored**
- Decreased aldosterone reduces the rate of K$^+$ secretion from the distal convoluted tubules and collecting ducts of the kidneys into the urine.
- Blood K$^+$ levels decrease: **Homeostasis Disturbed**
- Decreased blood levels of K$^+$ act on the adrenal cortex to decrease aldosterone secretion.

Axis labels: Blood K$^+$ (normal range)

ions in the filtrate exceeds the nephron's ability to reabsorb them, the excess is excreted into the urine. As long as the concentration of these ions is low, nearly all of them are reabsorbed by active transport. This mechanism plays a major role in regulating the concentration of $PO_4^{3-}$ and $SO_4^{2-}$ in the body fluids.

# REGULATION OF ACID-BASE BALANCE

The concentration of H$^+$ in the body fluids is reported as the pH. The body fluid pH is maintained between 7.35 and 7.45; any deviation from that range is life-threatening. Consequently, the mechanisms that regulate body fluid pH are critical for survival. The pH of body fluids is controlled by three factors: buffers, the respiratory system, and the kidneys. When the pH of body fluids is not properly maintained, the result is acidosis or alkalosis.

## Buffers

**Buffers** are chemicals that resist a change in the pH of a solution when either acids or bases are added to the solution. The buffers in the body fluids contain salts of either weak acids or weak bases that combine with H$^+$ when H$^+$ increase in those fluids or release H$^+$ when H$^+$ decrease in those fluids. Buffers tend to keep the H$^+$ concentration, and thus the pH, within a narrow range of values (figure 18.22) because of these characteristics. The three major buffers in the body fluids are the proteins, the $PO_4^{3-}$ buffer system, and the $HCO_3^-$ buffer system.

Proteins and $PO_4^{3-}$ in the body fluids combine with a large number of $H^+$. When the $H^+$ concentration increases, proteins and $PO_4^{3-}$ prevent a decrease in pH by combining with the $H^+$. Conversely, when the $H^+$ concentration decreases, proteins and $PO_4^{3-}$ release $H^+$, preventing an increase in pH.

The following reaction illustrates how $PO_4^{3-}$ buffers work:

$$HPO_4^{2-} + H^+ \rightleftarrows H_2PO_4^-$$

Monohydrogen phosphate ion     Hydrogen ion     Dihydrogen phosphate ion

Monohydrogen phosphate ions ($HPO_4^{2-}$) combine with $H^+$ to form dihydrogen phosphate ions ($H_2PO_4^-$) when excess $H^+$ are present. When $H^+$ concentration declines, some of the $H^+$ separate from the $H_2PO_4^-$.

Proteins are able to function as buffers because their amino acids have side chains that function as weak acids and weak bases. Many side chains contain carboxyl groups (–COOH) or amine groups (–NH₂). Both of these groups are able to function as buffers because of the following reactions:

$$-COO^- + H^+ \rightleftarrows -COOH$$

Carboxyl group (ionized)     Hydrogen ion     Carboxyl group

$$-NH_2 + H^+ \rightleftarrows -NH_3$$

Amine group     Hydrogen ion     Ammonium group

The bicarbonate ($HCO_3^-$) buffer system is unable to combine with as many $H^+$ as can proteins and $PO_4^{3-}$ buffers, but the $HCO_3^-$ buffer system is critical because it can be regulated by the respiratory and urinary systems. Carbon dioxide ($CO_2$) combines with water ($H_2O$) to form carbonic acid ($H_2CO_3$), which in turn forms $H^+$ and $HCO_3^-$ as follows:

$$H_2O + CO_2 \rightleftarrows H_2CO_3 \rightleftarrows H^+ + HCO_3^-$$

Water     Carbon dioxide     Carbonic acid     Hydrogen ion     Bicarbonate ion

The reaction between $CO_2$ and $H_2O$ is catalyzed by an enzyme, called carbonic anhydrase, which is found in red blood cells and on the surface of capillary epithelial cells (see chapter 15). The enzyme accelerates the rate at which the reaction proceeds in either direction.

The higher the concentration of $CO_2$, the greater the amount of $H_2CO_3$ formed, and the greater the number of $H^+$ and $HCO_3^-$ formed. This results in a decreased pH. The reaction is reversible, however. If $CO_2$ levels decline, the equilibrium shifts in the opposite direction. That is, $H^+$ and $HCO_3^-$ combine to form $H_2CO_3$, which then forms $CO_2$ and $H_2O$, and the pH increases.

## Respiratory System

The **respiratory system** responds rapidly to a change in pH and helps bring the pH of body fluids back toward normal. Increasing carbon dioxide levels and decreasing body fluid pH stimulate neurons in the respiratory center of the brain and cause the rate and depth of ventilation to increase. As a result, carbon dioxide is eliminated from the body through the lungs at a greater rate, and the concentration of carbon dioxide in the body fluids decreases. As carbon dioxide levels decline, the concentration of $H^+$ also declines. The pH therefore rises back toward its normal range (figure 18.22).

**Predict 5**

*Under stressful conditions, some people hyperventilate. Predict the effect of the rapid rate of ventilation on the pH of body fluids. In addition, explain why a person who is hyperventilating may benefit from breathing into a paper bag.*

If carbon dioxide levels become too low or the pH of the body fluids is elevated, the rate and depth of respiration decline. As a consequence, the rate at which carbon dioxide is eliminated from the body is reduced. Carbon dioxide then accumulates in the body fluids because it is continually produced as a by-product of metabolism. As carbon dioxide accumulates in the body fluids, so do $H^+$, resulting in a decreased pH.

## Kidneys

The nephrons of the kidneys secrete $H^+$ into the urine and therefore can directly regulate the pH of the body fluids. The kidney is a powerful regulator of pH, but it responds more slowly than does the respiratory system. Cells in the walls of the distal convoluted tubule are primarily responsible for the secretion of $H^+$. As the pH of the body fluids decreases below normal, the rate at which the distal convoluted tubules secrete $H^+$ increases (figure 18.22). At the same time, reabsorption of $HCO_3^-$ increases. The increased rate of $H^+$ secretion and the increased rate of $HCO_3^-$ reabsorption both cause the blood pH to increase toward its normal value. On the other hand, as the body fluid pH increases above normal, the rate of $H^+$ secretion by the distal convoluted tubules declines, and the amount of $HCO_3^-$ lost in the urine increases. Consequently, the blood pH decreases toward its normal value.

## Acidosis and Alkalosis

Failure of the buffer systems, the respiratory system, or the urinary system to maintain normal pH levels can result in acidosis or alkalosis.

### Acidosis

**Acidosis** (as-i-dō′sis) occurs when the blood pH falls below 7.35. The central nervous system malfunctions, and the individual becomes disoriented and, as the condition worsens, may become comatose. Acidosis is separated into two categories. **Respiratory acidosis** results when the respiratory system is unable to eliminate adequate amounts of carbon dioxide. Carbon dioxide accumulates in the circulatory system, causing the pH of the body fluids to decline. **Metabolic acidosis** results from excess production of acidic substances, such as lactic acid and ketone bodies, because of increased metabolism or decreased ability of the kidneys to eliminate $H^+$ in the urine.

**Predict 6**

*Adam suffers from severe emphysema. Gas exchange in his lungs is not adequate, and he requires supplementary oxygen. But even though his blood carbon dioxide levels are elevated, his blood pH is close to normal. Explain.*

## Acute Renal Failure

A large piece of machinery overturned at the construction site where Roger worked, trapping him beneath it. His legs were severely crushed, although they healed after several months. However, Roger nearly lost his life because of the acute renal failure that developed because of his injury. Roger was trapped in a very difficult place to reach for several hours. During that time, his blood pressure decreased to very low levels because of the blood loss, the edema in the inflamed tissues, and emotional shock. After he was rescued, fluid replacement in the form of both intravenous saline solutions and blood transfusions were given, and his blood pressure returned to its normal range. Twenty-four hours after the accident, however, Roger's urine volume began to decrease. His urinary $Na^+$ concentration increased, but his urine osmolality decreased. Cellular debris was evident in his urine.

For approximately 7 days, Roger exhibited reduced urine production and required renal dialysis to maintain his blood volume and ion concentrations within normal ranges. After that week, his kidneys gradually began to produce large quantities of urine. Roger's medical team carefully monitored his blood pressure and ion concentrations to make sure they remained within normal ranges and had to administer substantial quantities of water, $Na^+$, and $K^+$ to him. After about 3 weeks, his kidney functions slowly began to improve, although many months passed before they returned to normal.

### Background Information

In Roger's case, the events after 24 hours are consistent with acute renal failure caused by prolonged low blood pressure and lack of blood flow to the kidneys. The reduced blood flow to the kidneys was severe enough to cause damage to the epithelial lining of the kidney tubules. The period of reduced urine volume resulted from tubular damage. Dead and damaged tubular cells sloughed off into the tubules and blocked them, so that filtrate could not flow through. In addition, the filtrate leaked from the blocked or partially blocked tubules back into the interstitial spaces and therefore back into the circulatory system. As a result, the amount of filtrate that became urine was markedly reduced.

Blood levels of urea and of creatine increased because of the reduced filtrate formation and the reduced function of the tubular epithelium. The kidney's ability to eliminate metabolic waste products was therefore reduced. The small amount of urine produced had a high $Na^+$ concentration, although the osmolality was close to the concentration of the body fluids because the kidney was not able to reabsorb $Na^+$ and because the kidney's urine-concentrating ability was severely damaged.

After 7 days, the nephrons were partially healed and could produce urine, but the nephrons' ability to concentrate urine was not yet normal. Therefore, large volumes of urine that contained significant amounts of $Na^+$ and $K^+$ were produced. The kidneys produced urine that was more concentrated than the body fluids, but the concentrating ability of the kidneys was still below normal. As time passed, the kidneys' concentrating ability improved and eventually became normal once again.

### Treatments for Renal Failure

**Hemodialysis** (hē′mō-dī-al′i-sis) is used when a person is suffering from severe acute or chronic kidney failure. The procedure substitutes

From an artery

Blood pump

Bubble trap

To a vein

Dialysis membrane

Diffusion of waste products, such as urea

Compressed $CO_2$ and air

Fresh dialysis fluid

Constant temperature bath

Used dialysis fluid

Blood

Diffusion of waste products across the dialysis membrane

Dialysis fluid

### Figure 18A

During hemodialysis, blood flows through a system of tubes composed of a selectively permeable membrane. Dialysis fluid, which has a composition similar to that of normal blood (except that the concentration of waste products is very low), flows in the opposite direction on the outside of the dialysis tubes. Waste products, such as urea, diffuse from the blood into the dialysis fluid. Other substances, such as $Na^+$, $K^+$, and glucose, can diffuse from the blood into the dialysis fluid if they are present in higher than normal concentrations, because these substances are present in the dialysis fluid at the same concentrations found in normal blood.

Urinary

| System Interactions | Effects of Renal Failure on Other Systems |
|---|---|
| **System** | **Interactions** |
| Integumentary | Anemia causes pallor, and bruising results from clotting proteins lacking in the blood because they are lost in the urine. A waxy yellow coloring develops in the skin of light-skinned people, an ashen gray cast in black-skinned people, or a yellowish brown cast in brown-skinned people due to accumulation of urinary pigments. When the urea concentration in the blood is very high, it can also give a yellow cast to light-skinned people, and white crystals of urea, called uremic frost, may appear on areas of the skin where there is heavy perspiration. |
| Skeletal | Changes in the skeletal system are not significant unless kidney damage results in chronic kidney failure. Bone resorption can result because of excessive and chronic loss of $Ca^{2+}$ in the urine when the kidneys produce large volumes of urine. Also, vitamin D levels may be reduced. |
| Muscular | Neuromuscular irritability results from the toxic effect of metabolic wastes on the central nervous system and ionic imbalances, such as elevated blood $K^+$ levels. Involuntary jerking and twitching can occur as neuromuscular irritability develops. Tremor of the hands is an indication of the toxic effects of metabolic wastes on the cerebrum. |
| Nervous | Elevated blood $K^+$ levels and the toxic effects of metabolic wastes result in depolarization of neurons. Slowing of action potential conduction, burning sensations, pain, numbness, or tingling results. Also, decreased mental acuity, reduced ability to concentrate, apathy, and lethargy occur. Periods of lethargy can alternate with restlessness and insomnia. In severe cases, the patient can become confused and comatose. |
| Endocrine | Major hormone deficiencies include vitamin D deficiency. In addition, the secretion of reproductive hormones decreases due to the effects of metabolic wastes and ionic imbalances on the hypothalamus. |
| Cardiovascular | Water and $Na^+$ retention can result in edema in peripheral tissues and in the lungs. Also, increased blood pressure and congestive heart failure may result. Elevated blood $K^+$ levels result in dysrhythmias and can cause cardiac arrest. Anemia due to decreased erythropoietin production by the damaged kidney and decreased half-life of red blood cells can result. Nosebleeds and bruising occur due to reduced concentration of clotting factors that have been lost in the urine. |
| Lymphatic | There are no major direct effects on the lymphatic system, except that increased lymph flow occurs as a result of edema. |
| Respiratory | Early during acute renal failure, the depth of breathing increases and becomes labored as acidosis develops because the kidneys are not able to secrete $H^+$. Pulmonary edema often develops because of water and $Na^+$ retention as a result of reduced urine production. The likelihood of infection increases as a result of pulmonary edema. |
| Digestive | Decreased appetite, nausea, and vomiting result from altered digestive tract functions due to the effects of ionic imbalances on the nervous system. The breath can have the odor of ammonia, and there can be a metallic taste in the mouth. These effects result from the accumulation of metabolic waste products in the digestive tract and the action of the normal digestive tract organisms on the waste products, which convert urea to ammonia. The ammonia and other metabolic waste products predispose the mouth to inflammation and infection. |

for the excretory functions of the kidney. Hemodialysis is based on blood flow through tubes composed of a selectively permeable membrane. Blood is usually taken from an artery, passed through the tubes of the dialysis machine, and then returned to a vein (figure 18A). On the outside of the dialysis tubes is a fluid, called dialysis fluid, which contains the same concentration of solutes as normal plasma, except for the metabolic waste products. As a consequence, the metabolic wastes diffuse from the blood to the dialysis fluid. The dialysis membrane has pores that are too small to allow plasma proteins to pass through them, and because the dialysis fluid contains the same beneficial solutes as the plasma, the net movement of these substances is zero.

Peritoneal (per'i-tō-nē'ăl) dialysis is sometimes used to treat kidney failure. The principles by which peritoneal dialysis works are the same as for hemodialysis, but the dialysis fluid flows through a tube inserted into the peritoneal cavity. The visceral and parietal peritonea act as the dialysis membrane. Waste products diffuse from the blood vessels beneath the peritoneum, across the peritoneum, and into the dialysis fluid.

Kidney transplants are sometimes performed on people who have severe renal failure. Usually, the donor suffered an accidental death and had granted permission to have his or her kidneys used for transplantation. The major cause of kidney transplant failure is rejection by the recipient's immune system. Physicians therefore attempt to match the immune characteristics of the donor and recipient to reduce the tendency for rejection. Even with careful matching, recipients have to take medication for the rest of their lives to suppress their immune reactions. But in most cases, the transplanted kidney functions well, and the tendency for the recipient's immune system to reject the transplanted kidney can be controlled.

**Predict 7**

*Nine days after the accident, Roger began to appear pale, became dizzy on standing, and was very weak and lethargic. His hematocrit was elevated, and his heart was arrhythmic. Explain these manifestations.*

**Kidney:**
The distal convoluted tubules decrease $H^+$ secretion into the urine and decrease $HCO_3^-$ reabsorption into the blood.

**Lungs:**
The respiratory center in the brain decreases the rate and depth of respiration which increases blood $CO_2$.

**Buffers:**
Buffers release $H^+$.
$$H_2O + CO_2 \rightarrow H_2CO_3 \rightarrow H^+ + HCO_3^-$$

Fewer $H^+$ are removed from the blood, and fewer $HCO_3^-$ are available to bind $H^+$.

Increased blood $CO_2$ reacts with water to produce carbonic acid, which dissociates to increase $H^+$.
$$H_2O + CO_2 \rightarrow H_2CO_3 \rightarrow H^+ + HCO_3^-$$

The number of $H^+$ in the blood increases.

Blood pH increases ($H^+$ decreases):
**Homeostasis Disturbed**

Blood pH decreases ($H^+$ increases):
**Homeostasis Restored**

Start here

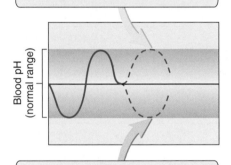

Blood pH decreases ($H^+$ increases):
**Homeostasis Disturbed**

Blood pH increases ($H^+$ decreases):
**Homeostasis Restored**

**Buffers:**
Buffers bind $H^+$.
$$H_2O + CO_2 \leftarrow H_2CO_3 \leftarrow H^+ + HCO_3^-$$

**Lungs:**
The respiratory center in the brain increases the rate and depth of respiration, which decreases blood $CO_2$.

**Kidney:**
The distal convoluted tubules increase $H^+$ secretion into the urine, and increase $HCO_3^-$ reabsorption into the blood.

The number of $H^+$ in the blood decreases.

Decreased blood $CO_2$ causes $H^+$ to react with $HCO_3^-$ to form carbonic acid, which decreases $H^+$ in the blood.
$$H_2O + CO_2 \leftarrow H_2CO_3 \leftarrow H^+ + HCO_3^-$$

More $H^+$ are removed from the blood, and more $HCO_3^-$ are available to bind $H^+$.

**Homeostasis Figure 18.22**   Regulation of Acid-Base Balance

## DISEASES AND DISORDERS  Urinary System

| CONDITION | DESCRIPTION |
|---|---|
| **Inflammation of the Kidneys** | |
| Glomerulonephritis (glō-măr′ū-lō-nef-rī′tis) | Inflammation of the filtration membrane within the renal corpuscle, causing increased membrane permeability; plasma proteins and blood cells enter the filtrate, causing increased urine volume due to increased osmotic concentration of the filtrate |
| Acute glomerulonephritis | Often occurs 1–3 weeks after a severe bacterial infection, such as strep throat; normally subsides after several days |
| Chronic glomerulonephritis | Long-term and progressive process whereby the filtration membrane thickens and is eventually replaced by connective tissue and the kidneys become nonfunctional |
| **Renal Failure** | Can result from any condition that interferes with kidney function |
| Acute renal failure | Occurs when damage to the kidney is rapid and extensive; leads to accumulation of wastes in the blood; can lead to death in 1–2 weeks if renal failure is complete |
| Chronic renal failure | Results from permanent damage to so many nephrons that the remaining nephrons are inadequate for normal kidney function; can be caused by chronic glomerulonephritis, trauma to the kidneys, tumors, or kidney stones |

Go to **www.mhhe.com/seeleyess7** for additional information on these pathologies.

### Alkalosis

**Alkalosis** (al-kă-lō′sis) occurs when the blood pH increases above 7.45. A major effect of alkalosis is hyperexcitability of the nervous system. Peripheral nerves are affected first, resulting in spontaneous nervous stimulation of muscles. Spasms and tetanic contractions result, as can extreme nervousness or convulsions.

Tetany of respiratory muscles can cause death. **Respiratory alkalosis** results from hyperventilation, as can occur in response to stress. **Metabolic alkalosis** usually results from the rapid elimination of $H^+$ from the body, as occurs during severe vomiting or when excess aldosterone is secreted by the adrenal cortex.

## SUMMARY

The urinary system consists of two kidneys, two ureters, the urinary bladder, and the urethra.

### Functions of the Urinary System    (p. 497)

1. The kidneys excrete waste products.
2. The kidneys control blood volume and blood pressure by regulating the volume of urine produced.
3. The kidneys help regulate the concentration of major ions in the body fluids.
4. The kidneys help regulate the pH of the extracellular fluid.
5. The kidneys regulate the concentration of red blood cells in the blood.
6. The kidneys participate, with the skin and liver, in regulating vitamin D synthesis.

### Anatomy of the Kidneys    (p. 497)

1. Each kidney is behind the peritoneum and surrounded by a renal capsule and adipose tissue.
2. The kidney is divided into an outer cortex and an inner medulla.
3. Each renal pyramid has a base located at the boundary between the cortex and the medulla. The tip of the renal pyramid extends toward the center of the kidney and is surrounded by a calyx.
4. Calyces are extensions of the renal pelvis, which is the expanded end of the ureter within the renal sinus.
5. The functional unit of the kidney is the nephron. The parts of a nephron are the renal corpuscle, the proximal convoluted tubule, the loop of Henle, and the distal convoluted tubule.
6. The filtration membrane is formed by the glomerular capillaries, the basement membrane, and the podocytes of Bowman's capsule.

### Arteries and Veins

1. Renal arteries give rise to branches that lead to afferent arterioles.
2. Afferent arterioles supply the glomeruli.
3. Efferent arterioles carry blood from the glomeruli to the peritubular capillaries.
4. Blood from the peritubular capillaries flows to the renal veins.

### Urine Formation    (p. 502)

Urine is produced by filtration, tubular reabsorption, and tubular secretion.

### Filtration

1. The renal filtrate passes from the glomerulus into Bowman's capsule and contains no blood cells and few blood proteins.
2. Filtration pressure is responsible for filtrate formation.
3. Increased sympathetic activity decreases blood flow to the kidney, decreases filtrate formation, and decreases urine formation. Decreased sympathetic activity has the opposite effect.

### Tubular Reabsorption

1. About 99% of the filtrate volume is reabsorbed; 1% becomes urine.
2. Among the substances reabsorbed are proteins, amino acids, glucose, fructose, $Na^+$, $K^+$, $Ca^{2+}$, $HCO_3^-$, and $Cl^-$.
3. About 65% of the filtrate volume is reabsorbed in the proximal convoluted tubule, 15% is reabsorbed in the descending limb of the loop of Henle, and another 19% is reabsorbed in the distal convoluted tubule and collecting duct.

### Tubular Secretion

Hydrogen ions, some by-products of metabolism, and some drugs are actively secreted into the nephron.

## Regulation of Urine Concentration and Volume    (p. 509)

### Hormonal Mechanisms

1. Renin is secreted from the kidney when the blood pressure decreases. Renin converts angiotensinogen to angiotensin I, which is then converted to angiotensin II by angiotensin-converting enzyme. Angiotensin II stimulates aldosterone secretion, and aldosterone increases the rate of $Na^+$ and $Cl^-$ reabsorption from the nephron.
2. ADH is secreted from the posterior pituitary when the concentration of blood increases or when blood pressure decreases. ADH increases the permeability to water of the distal convoluted tubule and collecting duct. It increases water reabsorption by the kidney.
3. Atrial natriuretic hormone, secreted from the right atrium in response to increases in blood pressure, acts on the kidney to increase $Na^+$ and water loss in the urine.

### Anatomy and Histology of the Ureters, Urinary Bladder, and Urethra

1. Each ureter carries urine from a renal pelvis to the urinary bladder.
2. The urethra carries urine from the urinary bladder to the outside of the body.
3. The ureters and urinary bladder are lined with transitional epithelium and have smooth muscle in their walls.
4. The external urinary sphincter regulates the flow of urine through the urethra.

## Urine Movement    (p. 514)

1. Increased volume in the urinary bladder stretches its wall and activates the micturition reflex.
2. Parasympathetic action potentials cause the urinary bladder to contract. Reduced somatic motor action potentials cause the external urinary sphincter to relax.
3. Higher brain centers control the micturition reflex. Stretching of the urinary bladder stimulates sensory neurons that carry impulses to the brain and inform the brain of the need to urinate.

## Body Fluid Compartments    (p. 514)

1. Water and the ions dissolved in it are distributed in the intracellular and extracellular fluid compartments.
2. Approximately two-thirds of the total body water is found within cells.
3. Approximately one-third of the total body water is found outside cells, mainly in interstitial fluid, blood plasma, and lymph.

### Composition of the Fluid in the Body Fluid Compartments

1. Intracellular fluid contains more $K^+$, $Mg^{2+}$, $PO_4^{3-}$, $SO_4^{2-}$, and protein than does extracellular fluid.
2. Extracellular fluid contains more $Na^+$, $Ca^{2+}$, $Cl^-$, and $HCO_3^-$ than does intracellular fluid.

### Exchange Between Body Fluid Compartments

Water moves continually between compartments in response to hydrostatic pressure differences and osmotic differences between the compartments.

## Regulation of Extracellular Fluid Composition    (p. 516)

The total amount of water and electrolytes in the body does not change unless the person is growing, gaining weight, or losing weight.

### Thirst Regulation

The sensation of thirst increases if extracellular fluid becomes more concentrated or if blood pressure decreases.

### Ion Concentration Regulation

1. Sodium ions are the dominant extracellular ions. Aldosterone increases $Na^+$ reabsorption from the filtrate, ADH increases water reabsorption from the nephron, and ANH increases $Na^+$ loss in the urine.
2. Aldosterone increases $K^+$ secretion in the urine. Increased blood levels of $K^+$ stimulate, and decreased blood levels of $K^+$ inhibit, aldosterone secretion.
3. Parathyroid hormone secreted from the parathyroid glands increases extracellular $Ca^{2+}$ levels by causing bone resorption and increased $Ca^{2+}$ uptake in the kidney. Parathyroid hormone increases vitamin D synthesis. Calcitonin, secreted by the thyroid gland, inhibits bone resorption and lowers blood $Ca^{2+}$ levels when they are too high.
4. When $PO_4^{3-}$ and $SO_4^{2-}$ levels in the filtrate are low, nearly all $PO_4^{3-}$ and $SO_4^{2-}$ are reabsorbed. When levels are high, excess is lost in the urine.

## Regulation of Acid-Base Balance    (p. 518)

### Buffers

Three principal classes of buffers in the circulatory system resist changes in the pH: proteins, the phosphate buffer system, and the bicarbonate buffer system.

### Respiratory System

The respiratory system rapidly regulates pH. An increased respiratory rate raises the pH because the rate of carbon dioxide elimination is increased, and a reduced respiratory rate reduces the pH because the rate of carbon dioxide elimination is reduced.

### Kidneys

The kidneys excrete $H^+$ in response to a decreasing blood pH, and they reabsorb $H^+$ in response to an increasing blood pH.

### Acidosis and Alkalosis

1. Acidosis occurs when the pH of the blood falls below 7.35. The two major types are respiratory acidosis and metabolic acidosis.
2. Alkalosis occurs when the pH of the blood increases above 7.45. The two major types are respiratory alkalosis and metabolic alkalosis.

 REVIEW AND COMPREHENSION

1. Name the structures that make up the urinary system. List the functions of the urinary system.
2. What structures surround the kidney?
3. Describe the relationships of the renal pyramids, calyces, renal pelvis, and ureter.
4. What is the functional unit of the kidney? Name its parts.
5. Describe the blood supply to the kidney.
6. Name the three general processes involved in the production of urine.
7. Describe the filtration membrane. What substances do not pass through it?

8. How do changes in blood pressure in the glomerulus affect the volume of filtrate produced?

9. What effect does sympathetic stimulation have on the kidneys?

10. What substances are reabsorbed in the nephron? What happens to most of the filtrate volume that enters the nephron?

11. In what parts of the nephron are large volumes of filtrate reabsorbed? In what part of the nephron is no filtrate reabsorbed?

12. In general, what substances are secreted into the nephron?

13. What effect does ADH have on urine volume? Name the factors that cause an increase in ADH secretion.

14. Where is renin produced, and what stimulates its secretion?

15. Explain how renin controls the synthesis of angiotensin I. What enzyme regulates the conversion of angiotensin I to angiotensin II?

16. Describe the effect of angiotensin II on aldosterone secretion.

17. Where is aldosterone produced, and what effect does it have on urine volume? What factors stimulate aldosterone secretion?

18. Where is atrial natriuretic hormone produced, and what effect does it have on urine production?

19. What are the functions of the ureters, urinary bladder, and urethra? Describe their structure.

20. Describe the micturition reflex. How is voluntary control of micturition accomplished?

21. What stimuli result in an increased sensation of thirst?

22. Describe how $Na^+$ levels are regulated in the body fluids.

23. Describe how $K^+$ levels are regulated in the body fluids.

24. Describe how $Ca^{2+}$ levels are regulated in the body fluids.

25. Explain how buffers respond to changes in the pH of body fluids.

26. Explain how the respiratory system and the kidneys respond to changes in the pH of body fluids.

27. Define respiratory acidosis, metabolic acidosis, respiratory alkalosis, and metabolic alkalosis.

#  CRITICAL THINKING QUESTIONS

1. Mucho McPhee decided to do an experiment after reading the urinary system chapter in his favorite anatomy and physiology textbook. He drank 2 L of water in 15 min and then monitored his rate of urine production and urine concentration over the next 2 hours. What did he observe? Explain the major mechanism involved.

2. A student ate a full bag of salty (NaCl) potato chips but drank no liquids. What effect did this have on urine concentration and the rate of urine production? Explain the mechanisms involved.

3. During severe exertion in a hot environment, a person can lose up to 4 L of sweat per hour (sweat is less concentrated than extracellular fluid in the body). What effect would this loss have on urine concentration and rate of production? Explain the mechanisms involved.

4. Which of the following symptoms are consistent with reduced secretion of aldosterone: excessive urine production, low blood pressure, high plasma potassium levels, high plasma sodium levels? Explain.

5. Propose as many ways as you can to decrease the rate at which filtrate enters Bowman's capsule.

6. Swifty Trotts has an enteropathogenic *Escherichia coli* infection that causes severe diarrhea. Diarrhea produces a large volume of mucus that contains high concentrations of $HCO_3^-$. How would this diarrhea affect this patient's blood pH, urine pH, and respiration rate?

7. Spanky and his mother went to a grocery store, where Spanky eyed some candy he wanted. His mother refused to buy it, so Spanky became angry. He held his breath for 2 min. What effect did this have on his body fluid pH? After the 2 min, what mechanisms were most important in reestablishing the normal body fluid pH?

8. Martha suffered from severe nausea for 2 days. She vomited frequently and was so nauseated that she could not stand to eat or drink anything. Explain how each of the following levels had changed in her body by the second day: blood pH, blood ADH, blood aldosterone, and urine pH.

*Answers in Appendix D*

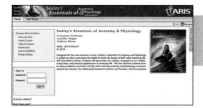 Visit this textbook's website at **www.mhhe.com/seeleyess7** for practice quizzes, animations, interactive learning exercises, and other study tools.

 McGraw-Hill offers a study CD that features interactive cadaver dissection. *Anatomy & Physiology Revealed* includes cadaver photos that allow you to peel away layers of the human body to reveal structures beneath the surface. This program also includes animations, radiologic imaging, audio pronunciations, and practice quizzing.

In order for the human population to continue, sexual reproduction is necessary to produce the next generation. Human sexual reproduction requires a cell from a male to unite with a cell from a female. The sexual reproductive cycle in both males and females involves a set of exquisitely timed events.

# Chapter 19

# Reproductive System

## Chapter Outline and Objectives

### Functions of the Reproductive System (p. 527)
1. List the functions of the male and female reproductive systems.

### Formation of Gametes (p. 527)
2. Describe the function of meiosis in the formation of sperm cells and oocytes.

### Male Reproductive System (p. 528)
3. Describe the scrotum and its role in regulating the temperature of the testes.
4. Describe the structure of the testes, the specialized cells of the testes, and the process of spermatogenesis.
5. Describe the ducts of the male reproductive system and their functions.
6. Describe the structure of the penis, seminal vesicles, bulbourethral glands, and prostate gland, and explain their functions.

### Physiology of Male Reproduction (p. 535)
7. List the hormones that influence the male reproductive system, and describe their functions.
8. Describe the changes that occur in males during puberty.
9. Explain the events that occur during the male sexual act.

### Female Reproductive System (p. 539)
10. Name the organs of the female reproductive system, and describe their structure.
11. Describe the anatomy and histology of the ovaries.
12. Discuss the development of the oocyte and the follicle, and describe ovulation and fertilization.
13. Describe the structure of the uterine tubes, uterus, vagina, external genitalia, and mammary glands.

### Physiology of Female Reproduction (p. 545)
14. Describe the changes that occur in females during puberty and the changes in the ovary and uterus that occur during the menstrual cycle.
15. List the hormones of the female reproductive system, and explain how their secretion is regulated.
16. Explain the events that occur during the female sexual act.
17. Define menopause, and describe the changes that result from it.

### Effects of Aging on the Reproductive System (p. 552)
18. Describe the major age-related changes in the reproductive system.

The human species could not survive without functional male and female reproductive systems. The reproductive systems play essential roles in the development of the structural and functional differences between males and females, influence human behavior, and produce offspring. However, a reproductive system, unlike other organ systems, is not necessary for the survival of an individual human.

Most of the body's organ systems show little difference between males and females, but the male and female reproductive systems exhibit striking differences. On the other hand, they have a number of similarities. For example, many reproductive organs of males and females are derived from the same embryological structures (see chapter 20), and some hormones are the same in males and females, even though they produce different responses.

# FUNCTIONS OF THE REPRODUCTIVE SYSTEM

The reproductive system performs the following functions:

1. *Production of gametes.* The reproductive system produces gametes: sperm cells by the testes in males and oocytes (eggs) by the ovaries in females.
2. *Fertilization.* The reproductive system enhances fertilization of the oocyte by the sperm. The duct system in males nourishes sperm until they are mature and are deposited in the female reproductive tract by the penis. The female reproductive system receives the male's sperm and transports them to the fertilization site.
3. *Development and nourishment of a new individual.* The female reproductive system nurtures the development of a new individual in the uterus until birth and provides nourishment (milk) after birth.
4. *Production of sex hormones.* Hormones produced by the reproductive system control its development and the development of the gender-specific body form. These hormones are also essential for the normal function of the reproductive system and for reproductive behavior.

# FORMATION OF GAMETES

The testes in males and the ovaries in females (figure 19.1) produce **gametes** (gam′ētz), or *sex cells*. The formation of gametes in males and females occurs by a type of cell division called **meiosis** (mī-ō′sis; a lessening) (see Chapter 3).

Meiosis occurs only in the testis and ovary. During meiosis, one cell undergoes two consecutive cell divisions to produce four

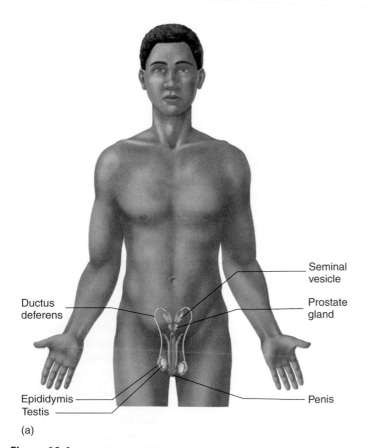

(a)    (b)

**Figure 19.1  Major Organs of the Reproductive System**

(a) The male reproductive system: testes, epididymis, ductus deferens, seminal vesicles, prostate gland, and penis. (b) The female reproductive system: ovaries, uterine tubes, uterus, vagina, and mammary glands.

Labels (a): Ductus deferens, Seminal vesicle, Prostate gland, Epididymis, Testis, Penis

Labels (b): Mammary gland (in breast), Uterine tube, Ovary, Uterus, Vagina daughter cells, each having half as many chromosomes as the parent cell.

The two divisions of meiosis are called meiosis I and meiosis II. Like mitosis, each division of meiosis has prophase, metaphase, anaphase, and telophase. However, there are distinct differences between meiosis and mitosis.

Before meiosis begins, all the chromosomes are duplicated. At the beginning of meiosis, each of the 46 chromosomes consists of 2 chromatids connected by a centromere (figure 19.2, step 1). The chromosomes align as pairs in a process called **synapsis** (si-nap′sis; a connection) (figure 19.2, step 2). Because each chromosome consists of 2 chromatids, the pairing of the chromosomes brings 2 chromatids of each chromosome close together. Occasionally, part of a chromatid of 1 chromosome breaks off and is exchanged with part of another chromatid from the other chromosome. This event, called **crossing over,** allows the exchange of genetic material between chromosomes.

The chromosomes align along the center of the cell (figure 19.2, step 3), and then the pairs of chromosomes are separated to each side of the cell (figure 19.2, step 4). As a consequence, when meiosis I is complete, each daughter cell has 1 chromosome from each of the pairs (figure 19.2, step 5), or 23 chromosomes. Each of the 23 chromosomes in each daughter cell consists of 2 chromatids joined by a centromere.

It is during the first meiotic division that the chromosome number is reduced from 46 (23 pairs) to 23 chromosomes. The first meiotic division is therefore called a reduction division.

The second meiotic division is similar to mitosis. The chromosomes, each consisting of 2 chromatids (figure 19.2, steps 6 and 7) align along the center of the cell. Then the chromatids separate at the centromere, and each daughter cell receives 1 of the chromatids from each chromosome (figure 19.2, steps 8 and 9). When the centromere separates, each of the chromatids is called a chromosome. Consequently, each of the 4 daughter cells produced by meiosis contains 23 chromosomes.

During fertilization, the zygote receives 1 set of chromosomes (23) from each parent. Although half the genetic material of a zygote comes from each parent, the genetic makeup of each zygote is unique.

# MALE REPRODUCTIVE SYSTEM

The male reproductive system consists of the testes (sing. testis), a series of ducts, accessory glands, and supporting structures. The ducts include the epididymides (sing. epididymis), the ducta deferentia (sing. ductus deferens; also, vas deferens), and the urethra. Accessory glands include the seminal vesicles, the prostate gland, and the bulbourethral glands. Supporting structures include the scrotum and the penis (figure 19.3). The sperm cells are very heat-sensitive and must develop at a temperature slightly less than normal body temperature. The testes, in which the sperm cells develop, are located outside the body cavity in the scrotum, where the temperature is lower. Sperm cells travel from each testis to the prostate gland and then empty into the urethra within the prostate gland. The urethra exits the pelvis, passes through the penis, and opens to the outside of the body.

## Scrotum

The **scrotum** (skrō′tum) is a saclike structure containing the testes. It is divided into right and left internal compartments by an incomplete connective tissue septum. Externally, the scrotum consists of skin. Beneath the skin are a layer of loose connective tissue and a layer of smooth muscle called the **dartos** (dar′tōs) **muscle.**

In cold temperatures, the dartos muscle contracts, causing the skin of the scrotum to become firm and wrinkled and reducing the overall size of the scrotum. At the same time, extensions of abdominal muscles into the scrotum, called **cremaster** (krē-mas′ter) **muscles,** contract (see figure 19.6). Consequently, the testes are pulled nearer the body, and their temperature is elevated. During

**First Meiotic Division (Meiosis I)**

**1 Early prophase I**
The duplicated chromosomes become visible chromatids.

Centromere

Chromosome

Nucleus

Chromatids

Centrioles

**2 Middle prophase I**
Pairs of chromosomes synapse. Crossing over may occur at this stage.

Pair of chromosomes

Spindle fibers

**3 Metaphase I**
Pairs of chromosomes align along the center of the cell. Random assortment of chromosomes occurs.

Equatorial plane

**4 Anaphase I**
Chromosomes move apart to opposite sides of the cell.

**5 Telophase I**
New nuclei form, and the cell divides. Each cell now has two sets of half the chromosomes.

Cleavage furrow

Prophase II (top of next column)

**Second Meiotic Division (Meiosis II)**
(continued from the bottom of previous column)

**6 Prophase II**
Each chromosome consists of two chromatids.

**7 Metaphase II**
Chromosomes align along the center of the cell.

**8 Anaphase II**
Chromatids separate, and each is now called a chromosome.

**9 Telophase II**
New nuclei form around the chromosomes. The cells divide to form four daughter cells with half as many chromosomes as the parent cell.

**Figure 19.2** Meiosis

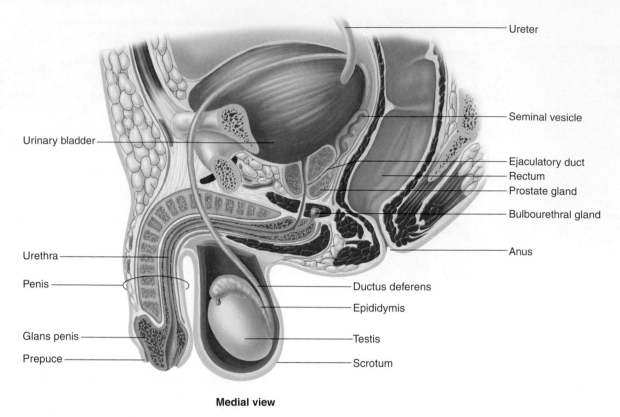

**Figure 19.3** Male Reproductive Structures

Medial view of the male pelvis, showing the male reproductive structures.

## Testes

The **testes** (tes'tēz), or male gonads (gō'nădz; *gonē*, seed), are oval organs, each about 4–5 cm long, within the scrotum (see figure 19.3). The outer part of each testis consists of a thick, white connective tissue capsule. Extensions of the capsule project into the interior of the testis and divide each testis into about 250 cone-shaped lobules (figure 19.4a). The lobules contain **seminiferous** (sem'ĭ-nif'er-ŭs) **tubules,** in which sperm cells develop. Delicate connective tissue surrounding the seminiferous tubules contains clusters of endocrine cells called **interstitial** (in-ter-stish'ăl) **cells,** or *Leydig* (lī'dig) *cells,* which secrete testosterone. The seminiferous tubules contain **germ cells** and **Sustentacular cells,** or *Sertoli* (ser-tō'lē) *cells* (figure 19.4b). Sustentacular cells are large and extend from the periphery to the lumen of the seminiferous tubule. They nourish the germ cells and produce a number of hormones.

## Spermatogenesis

**Spermatogenesis** (sper'mă-tō-jen'ĕ-sis) is the formation of sperm cells. Before puberty, the testes remain relatively simple and unchanged from the time of their initial development. The

Before the discussion continues, warm weather or exercise, the dartos and cremaster muscles relax, the skin of the scrotum becomes loose and thin, and the testes descend away from the body, which lowers their temperature. The response of the dartos and cremaster muscles is important in regulating the temperature in the testes. If the testes become too warm or too cold, normal sperm cell development does not occur.

interstitial cells are not prominent, and the seminiferous tubules are small and not yet functional. At the time of puberty, the interstitial cells increase in number and size, the seminiferous tubules enlarge, and spermatogenesis begins.

Germ cells are partially embedded in the sustentacular cells. The most peripheral germ cells are **spermatogonia** (sper'mă-tō-gō'nē-ă; undeveloped sperm cells), which divide through mitosis (figure 19.5). Some daughter cells produced from these mitotic divisions remain as spermatogonia and continue to divide by mitosis. Other daughter cells form **primary spermatocytes** (sper'mă-tō-sītz; sperm cells), which divide by meiosis and become sperm cells.

A primary spermatocyte contains 46 chromosomes, each consisting of 2 chromatids. Each primary spermatocyte passes through the first meiotic division to produce 2 **secondary spermatocytes.** Each secondary spermatocyte undergoes a second meiotic division to produce 2 smaller cells called **spermatids** (sper'mă-tidz), each having 23 chromosomes. After the second meiotic division, the spermatids undergo major structural changes to form sperm cells (figure 19.5; see figure 19.4b,c). Much of the cytoplasm of the spermatids is eliminated, and each spermatid develops a head, midpiece, and flagellum (tail) to become a **sperm cell,** or *spermatozoon* (figure 19.5; see figure 19.4b–d). The nucleus of the sperm cell is located in the head of the sperm cell. Just anterior to the nucleus is a vesicle called the **acrosome** (ak'rō-sōm), which contains enzymes that are released during the process of fertilization and are necessary for the sperm cell to penetrate the oocyte, or egg cell.

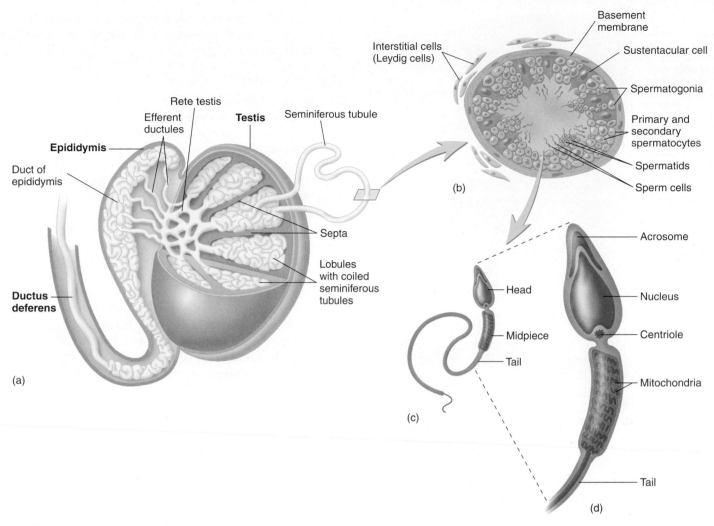

**Figure 19.4  Structure of the Testis and Sperm Cell**

(*a*) Gross anatomy of the testis with a section cut away to reveal internal structures. The epididymis is also shown. (*b*) Cross section of a seminiferous tubule. Spermatogonia are near the periphery, and mature sperm cells are near the lumen of the seminiferous tubule. (*c*) The head, midpiece, and tail of a sperm cell. (*d*) Enlargement of the head and midpiece of a sperm cell shows the nucleus and acrosome in the head and the mitochondria in the midpiece.

At the end of spermatogenesis, the developing sperm cells are located around the lumen of the seminiferous tubules, with their heads directed toward the surrounding sustentacular cells and their tails directed toward the center of the lumen (figure 19.5; see figure 19.4*b*). Finally, sperm cells are released into the lumen of the seminiferous tubules.

**Infertility** (in-fer-til′i-tē) is reduced or diminished fertility. The most common cause of infertility in males is a low sperm cell count. If the sperm cell count drops to below 20 million sperm cells per milliliter, the male is usually sterile.

The sperm cell count can decrease because of damage to the testes as a result of trauma, radiation, cryptorchidism, or infections, such as mumps, which block the ducts in the epididymis. Reduced sperm cell counts can also result from inadequate secretion of luteinizing hormone and follicle-stimulating hormone, which can be caused by hypothyroidism, trauma to the hypothalamus, infarctions of the hypothalamus or anterior pituitary gland, or tumors. Decreased testosterone secretion reduces the sperm cell count as well.

Even when the sperm cell count is normal, fertility can be reduced if sperm cell structure is abnormal, as occurs due to chromosomal abnormalities caused by genetic factors. Reduced sperm cell motility also results in infertility. A major cause of reduced sperm cell motility is the presence of antisperm antibodies, which are produced by the immune system and bind to sperm cells.

In cases of infertility due to low sperm cell count or reduced motility, fertility can sometimes be achieved by collecting several ejaculations, concentrating the sperm cells, and inserting them into the female's reproductive tract, a process called **artificial insemination** (in-sem-i-nā′shŭn).

## Ducts

After their production, sperm cells are transported through the seminiferous tubules and a series of ducts to the exterior of the body.

### Epididymis

The seminiferous tubules of each testis empty into a tubular network called the **rete** (rē′tē; net) **testis** (see figure 19.4*a*). The rete

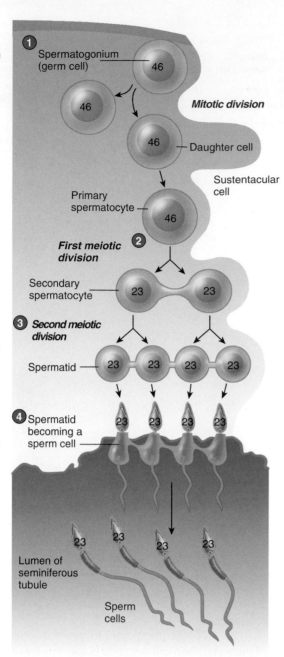

**1** Spermatogonia are the cells from which sperm cells arise. The spermatogonia divide by mitosis. One daughter cell remains a spermatogonium that can divide again by mitosis. One daughter cell becomes a primary spermatocyte.

**2** The primary spermatocyte divides by meiosis to form secondary spermatocytes.

**3** The secondary spermatocytes divide by meiosis to form spermatids.

**4** The spermatids differentiate to form sperm cells.

**PROCESS Figure 19.5 Spermatogenesis**

A section of the seminiferous tubule illustrates the process of meiosis and sperm cell formation. The number in each cell indicates the total chromosome number.

testis empties into 15–20 tubules called the **efferent ductules** (efʹer-ent dŭkʹtoolz). The efferent ductules carry sperm cells from the testis to a tightly coiled series of threadlike tubules that form a comma-shaped structure on the posterior side of the testis called the **epididymis** (ep-i-didʹi-mis) (figure 19.6; see figures 19.3 and 19.4). The sperm cells continue to mature within the epididymis, developing the capacity to swim and the ability to bind to the oocyte. Sperm cells taken directly from the testes are not capable of fertilizing oocytes, but after maturing for several days in the epididymis, the sperm cells develop the capacity to function as

gametes. Final changes in sperm cells, called **capacitation** (kă-pasʹi-tăʹshun), occur after ejaculation of semen into the vagina and prior to fertilization.

### Ductus Deferens

The **ductus deferens** (dŭkʹtŭs defʹer-enz), or *vas deferens,* emerges from the epididymis and ascends along the posterior side of the testis to become associated with the blood vessels and nerves that supply the testis. These structures form the **spermatic cord** (figure 19.6a). Each spermatic cord consists of the ductus deferens, testicular artery and veins, lymphatic vessels, and testicular nerve. It is surrounded by the cremaster muscle and two connective tissue sheaths.

Each ductus deferens extends, in the spermatic cord, through the abdominal wall by way of the inguinal canal. Each ductus deferens then crosses the lateral wall of the pelvic cavity and loops behind the posterior surface of the urinary bladder to approach the prostate gland (figure 19.6a; see figure 19.3). The total length of the ductus deferens is about 45 cm. Just before reaching the prostate gland, the ductus deferens increases in diameter to become the **ampulla of the ductus deferens** (figure 19.6a). The wall of the ductus deferens contains smooth muscle, which contracts in peristaltic waves to propel the sperm cells from the epididymis through the ductus deferens.

### Seminal Vesicle and Ejaculatory Duct

Near the ampulla of each ductus deferens is a sac-shaped gland called the **seminal vesicle** (semʹi-năl vesʹi-kl). A short duct extends from the seminal vesicle to the ampulla of the ductus deferens. The ducts from the seminal vesicle and the ampulla of the ductus deferens join at the prostate gland to form the **ejaculatory** (ē-jakʹū-lă-tōr-ē) **duct.** Each ejaculatory duct extends into the prostate gland and ends by joining the urethra within the prostate gland (figure 19.6a).

### Urethra

The male **urethra** (ū-rēʹthră) extends from the urinary bladder to the distal end of the penis (figure 19.6a; see figure 19.3). The urethra can be divided into three parts: the **prostatic urethra,** which passes through the prostate gland; the **membranous urethra,** which passes through the floor of the pelvis and is surrounded by the external urinary sphincter; and the **spongy urethra,** which extends the length of the penis and opens at its end. The urethra is a passageway for both urine and male reproductive fluids. However, urine and the reproductive fluids do not exit the urethra at the same time. While male reproductive fluids are passing through the urethra, a sympathetic reflex causes the internal urinary sphincter to contract, which keeps semen from passing into the urinary bladder and prevents urine from entering the urethra.

## Penis

The **penis** (peʹnis) is the male organ of copulation and functions in the transfer of sperm cells from the male to the female. The penis contains three columns of erectile tissue (figure 19.6). Engorgement of this erectile tissue with blood causes the penis to enlarge and become firm, a process called **erection** (ē-rekʹshŭn).

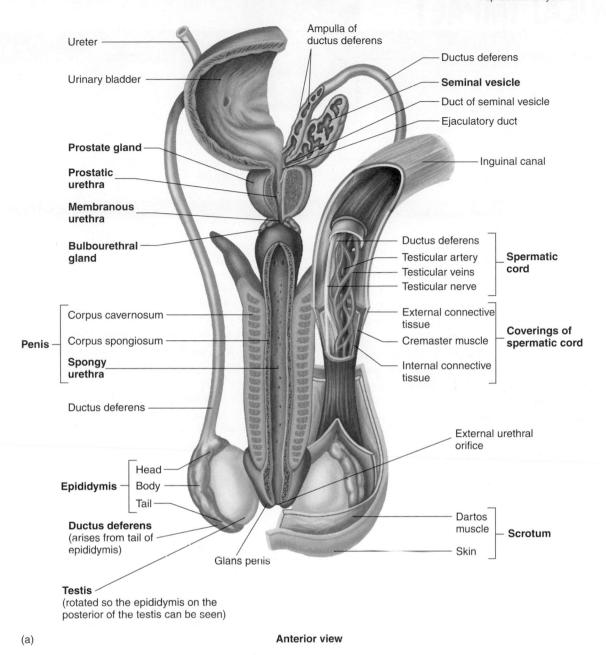

Ureter

Ampulla of ductus deferens

Ductus deferens

Urinary bladder

**Seminal vesicle**

Duct of seminal vesicle

Ejaculatory duct

**Prostate gland**

Inguinal canal

**Prostatic urethra**

**Membranous urethra**

Ductus deferens

**Spermatic cord**

Testicular artery

Testicular veins

Testicular nerve

**Bulbourethral gland**

External connective tissue

**Coverings of spermatic cord**

Cremaster muscle

Internal connective tissue

**Penis**
Corpus cavernosum

Corpus spongiosum

**Spongy urethra**

Ductus deferens

External urethral orifice

**Epididymis** — Head

Body

Tail

**Ductus deferens** (arises from tail of epididymis)

Dartos muscle

**Scrotum**

Skin

Glans penis

**Testis** (rotated so the epididymis on the posterior of the testis can be seen)

(a)    **Anterior view**

**Dorsal surface**

Dorsal vein

Dorsal artery

Dorsal nerve

Connective tissue

**Corpora cavernosa**

Corpus spongiosum

Spongy urethra

(b)    **Ventral surface**

## Figure 19.6  Male Reproductive Organs

(a) Frontal view of a testis, an epididymis, and a ductus deferens, along with the penis and glands of the male reproductive system. The testis is viewed in the scrotal sac with the smooth muscle and cremaster muscles on one side. The ductus deferens extends from the epididymis in the scrotal sac and passes through the inguinal canal and pelvic cavity to the prostate gland. Note that the ductus deferens and the testicular artery, vein, and nerves are surrounded by the cremaster muscle and connective tissue to form the spermatic cord. (b) A cross section of the penis illustrates the two dorsal corpora cavernosa and the ventral corpus spongiosum. Connective tissue sheaths and skin cover the three erectile bodies. Blood vessels, including the dorsal artery and vein and the dorsal nerve of the penis, are visible.

**Reproductive**

# CLINICAL IMPACT

## Circumcision

**Circumcision** (ser-kŭm-sizh'ŭn) is the surgical removal of the prepuce, usually shortly after birth. There are few compelling medical reasons for circumcision. Uncircumcised males have a higher incidence of penile cancer, but the underlying causes seem related to chronic infections and poor hygiene. In those few cases in which the prepuce is "too tight" to be moved over the glans penis, circumcision can be necessary to avoid chronic infections and maintain normal circulation.

---

Two columns of erectile tissue form the dorsal portion and the sides of the penis and are called the **corpora cavernosa** (kōr′pōr-ă kav-er-nōs′ă). The third, smaller erectile column occupies the ventral portion of the penis and is called the **corpus spongiosum** (kōr′pŭs spŭn-gē-ō′sŭm). It expands over the distal end of the penis to form a cap, the **glans** (glanz) **penis.** The spongy urethra passes through the corpus spongiosum, including the glans penis, and opens to the exterior as the **external urethral orifice.**

The shaft of the penis is covered by skin that is loosely attached to the connective tissue surrounding the penis. The skin is firmly attached at the base of the glans penis, and a thinner layer of skin tightly covers the glans penis. The skin of the penis, especially the glans penis, is well supplied with sensory receptors. A loose fold of skin, called the **prepuce** (prē′pūs), or *foreskin,* covers the glans penis (see figure 19.3).

## Glands

The seminal vesicles are glands consisting of many saclike structures located next to the ampulla of the ductus deferens (figure 19.6a; see figure 19.3). There are two seminal vesicles. Each is about 5 cm long and tapers into a short duct that joins the ampulla of the ductus deferens to form the ejaculatory duct, as previously mentioned.

The **prostate** (pros′tāt) **gland** consists of both glandular and muscular tissue and is about the size and shape of a walnut (figure 19.6a; see figure 19.3). The prostate gland surrounds the urethra and the two ejaculatory ducts. It consists of a capsule and numerous partitions. The cells lining the partitions secrete prostatic fluid. There are 10–20 short ducts (not seen in figure 19.6a) that carry secretions of the prostate gland to the prostatic urethra.

## A CASE IN POINT
### Prostate Cancer

Sixty-five-year-old Chance O'Prostancer has a wellness checkup every year. Ten years ago, a prostate-specific antigen (PSA) test indicated that Chance's PSA levels were higher than the results from his previous tests. His physician reported moderate enlargement of the prostate gland but detected no obvious tumorlike structures in a physical examination. Because of the increasing PSA levels, Chance's physician recommended a needle biopsy of the prostate

gland through the rectum. The pathology report described suspicious cells consistent with prostate cancer in one of the tissue samples. Chance's physician had the biopsy samples examined by another pathology laboratory, which did not confirm the first pathology report. As a consequence, Chance's physician explained that one option was to do nothing and continue having regular checkups because prostate cancer typically develops slowly. This is what Chance chose to do.

Eight years later, a PSA test showed another substantial increase in Chance's PSA levels, although no tumor could be detected by a physical exam, and Chance had no complaints, such as difficulty in urinating. This time, a needle biopsy of the prostate gland revealed cancer cells in two of the six biopsy samples. Chance's physician explained that so far the cancer had not metastasized (spread) to areas outside the prostate gland. Therefore, Chance could choose to do nothing, have his prostate gland surgically removed, or treat the cancer with radiation therapy, hormonal therapy, or chemotherapy. Chance elected to have radiation therapy, which focuses radiation on the prostate gland to kill the cancer cells. Statistics indicate that surgery and radiation therapy have similar success rates for small, localized tumors like Chance's. The trauma of surgery and the higher probability of erectile dysfunction following surgery convinced him that radiation therapy was preferable. Chance's physician indicated that doing nothing is a reasonable option for men who are significantly older than Chance because older men diagnosed with prostate cancer often die of other conditions before they succumb to prostate cancer. Chance's physician explained that, for patients like him, approximately 85% are cancer free after 5 years. Chance was grateful that he had had annual checkups. Prostate cancer represents 29% of cancers in males in the United States and 14% of the deaths due to cancer. Only lung cancer results in more cancer deaths in men.

### Predict 1

*Changes in the size and texture of the prostate gland can be an indication of developing prostate cancer. Suggest a way that the size and texture of the prostate gland can be examined by palpation without surgical techniques (see figure 19.3).*

The **bulbourethral** (bul′bō-ū-rē′thrăl) **glands,** or *Cowper glands,* are a pair of small, mucus-secreting glands located near the base of the penis (figure 19.6a; see figure 19.3). In young adults, each is about the size of a pea but decreases in size with age. A single duct from each gland enters the urethra.

## Secretions

**Semen** (sē'men) is a mixture of sperm cells and secretions from the male reproductive glands. The seminal vesicles produce about 60% of the fluid, the prostate gland contributes approximately 30%, the testes contribute 5%, and the bulbourethral glands contribute 5%.

The bulbourethral glands and the mucous glands in the urethra produce a mucous secretion, which lubricates the urethra, helps neutralize the contents of the normally acidic urethra, provides a small amount of lubrication during intercourse, and helps reduce acidity in the vagina.

Testicular secretions include sperm cells and a small amount of fluid. The thick, mucuslike secretion of the seminal vesicles contains the sugar fructose and other nutrients that nourish sperm cells. The seminal vesicle secretions also contain proteins that weakly coagulate after ejaculation and enzymes that are thought to help destroy abnormal sperm cells. Prostaglandins, which stimulate smooth muscle contractions, are present in high concentrations in the secretions of the seminal vesicles and can cause contractions of the female reproductive tract, which help transport sperm cells through the tract.

The thin, milky secretions of the prostate have an alkaline pH and help neutralize the acidic urethra, as well as the acidic secretions of the testes, the seminal vesicles, and the vagina. The increased pH is important for normal sperm cell function. The movement of sperm cells is not optimal until the pH is increased to between 6.0 and 6.5. In contrast, the pH of the secretions of the vagina is between 3.5 and 4.0. Prostatic secretions also contain proteolytic enzymes that break down the coagulated proteins of the seminal vesicles and make the semen more liquid. The normal volume of semen is 2–5 milliliters (mL), with each milliliter of semen typically containing about 100 million sperm cells.

# PHYSIOLOGY OF MALE REPRODUCTION

The male reproductive system depends on both hormonal and neural mechanisms to function normally. Hormones control the development of reproductive structures, the development of secondary sexual characteristics, spermatogenesis, and some aspects of sexual behavior. The mature neural mechanisms are primarily involved in controlling the sexual act and in the expression of sexual behavior.

## Regulation of Sex Hormone Secretion

The hypothalamus of the brain, the anterior pituitary gland, and the testes (figure 19.7) produce hormones that influence the male reproductive system. **Gonadotropin-releasing** (gō'nad-ō-trō'pin) **hormone (GnRH)** is released from neurons in the hypothalamus and passes to the anterior pituitary gland (table 19.1). GnRH

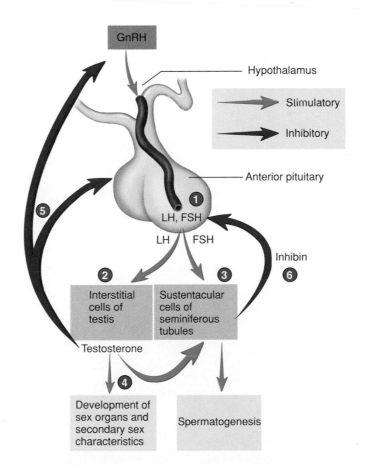

1. Gonadotropin-releasing hormone (GnRH) from the hypothalamus stimulates the secretion of luteinizing hormone (LH) and follicle-stimulating hormone (FSH) from the anterior pituitary.

2. LH stimulates testosterone secretion from the interstitial cells.

3. FSH stimulates sustentacular cells of the seminiferous tubules to increase spermatogenesis and to secrete inhibin.

4. Testosterone has a stimulatory effect on the sustentacular cells of the seminiferous tubules, as well as on the development of sex organs and secondary sex characteristics.

5. Testosterone has a negative-feedback effect on the hypothalamus and pituitary to reduce GnRH, LH, and FSH secretion.

6. Inhibin has a negative-feedback effect on the anterior pituitary to reduce FSH secretion.

**PROCESS Figure 19.7** Regulation of Reproductive Hormone Secretion in Males

## TABLE 19.1  Major Reproductive Hormones in Males and Females

| Hormone | Source | Target Tissue | Response |
|---|---|---|---|
| **Male Reproductive System** | | | |
| Gonadotropin-releasing hormone (GnRH) | Hypothalamus | Anterior pituitary | Stimulates secretion of LH and FSH |
| Luteinizing hormone (LH) | Anterior pituitary | Interstitial cells of the testes | Stimulates synthesis and secretion of testosterone |
| Follicle-stimulating hormone (FSH) | Anterior pituitary | Seminiferous tubules (Sustentacular cells) | Supports spermatogenesis and inhibin secretion |
| Testosterone | Interstitial cells of testes | Testes; body tissues | Development and maintenance of reproductive organs; supports spermatogenesis and causes the development and maintenance of secondary sexual characteristics |
| | | Anterior pituitary and hypothalamus | Inhibits GnRH, LH, and FSH secretion through negative feedback |
| Inhibin | Sustentacular cells | Anterior pituitary | Inhibits FSH secretion through negative feedback |
| **Female Reproductive System** | | | |
| Gonadotropin-releasing hormone (GnRH) | Hypothalamus | Anterior pituitary | Stimulates secretion of LH and FSH |
| Luteinizing hormone (LH) | Anterior pituitary | Ovaries | Causes follicles to complete maturation and undergo ovulation; causes ovulation; causes the ovulated follicle to become the corpus luteum |
| Follicle-stimulating hormone (FSH) | Anterior pituitary | Ovaries | Causes follicles to begin development |
| Estrogen | Follicles of ovaries and corpus luteum | Uterus | Proliferation of endometrial cells |
| | | Breasts | Development of mammary glands (especially duct systems) |
| | | Anterior pituitary and hypothalamus | Positive feedback before ovulation, resulting in increased LH and FSH secretion; negative feedback with progesterone on the hypothalamus and anterior pituitary after ovulation, resulting in decreased LH and FSH secretion |
| | | Other tissues | Development and maintenance of secondary sexual characteristics |
| Progesterone | Corpus luteum of ovaries | Uterus | Enlargement of endometrial cells and secretion of fluid from uterine glands; maintenance of pregnant state |
| | | Breasts | Development of mammary glands (especially alveoli) |
| | | Anterior pituitary | Negative feedback, with estrogen, on the hypothalamus and anterior pituitary after ovulation, resulting in decreased LH and FSH secretion |
| | | Other tissues | Secondary sexual characteristics |
| Oxytocin | Posterior pituitary | Uterus and mammary glands | Contraction of uterine smooth muscle and contraction of cells in the breast, resulting in milk letdown in lactating women |
| Human chorionic gonadotropin | Placenta | Corpus luteum of ovaries | Maintains corpus luteum and increases its rate of progesterone secretion during the first one-third (first trimester) of pregnancy; increases testosterone production in testes of male fetuses |

Reproductive

# CLINICAL IMPACT

## Anabolic Steroids

Some athletes, especially those who depend on muscle strength, may either ingest or inject synthetic **androgens** (an'drō-jenz; *andros,* male), which are hormones that have testosterone-like effects, such as stimulating the development of male sexual characteristics. The synthetic androgens are commonly called **anabolic steroids,** or simply **steroids,** and they are used in an attempt to increase muscle mass. Many of the synthetic androgens are structurally different from testosterone. Their effect on muscle is greater than their effect on the reproductive organs. However, when taken in large amounts, they can influence the reproductive system. Large doses of synthetic androgens have a negative-feedback effect on the hypothalamus and pituitary, reducing GnRH, LH, and FSH levels. As a result, the testes can atrophy, and sterility can develop.

Other side effects of large doses of synthetic androgens include kidney and liver damage, heart attack, and stroke. In addition, synthetic androgens cause abrupt mood swings, usually toward intense anger and rage. Taking synthetic androgens is highly discouraged by the medical profession, violates the rules of most athletic organizations, and is illegal without a prescription.

---

causes cells in the anterior pituitary gland to secrete two hormones, **luteinizing** (loo'tē-ĭ-nīz-ing) **hormone (LH)** and **follicle-stimulating hormone (FSH),** into the blood. LH and FSH are named for their functions in females, but they are also essential reproductive hormones in males.

LH binds to the interstitial cells in the testes and causes them to secrete testosterone. FSH binds primarily to sustentacular cells in the seminiferous tubules and promotes sperm cell development. It also increases the secretion of a hormone called **inhibin** (in-hib'in; to inhibit).

The blood levels of reproductive hormones are under negative-feedback control. Testosterone has a negative-feedback effect on the secretion of GnRH from the hypothalamus and the secretion of LH and FSH from the anterior pituitary gland. Inhibin has a negative-feedback effect on the secretion of FSH from the anterior pituitary gland.

For GnRH to stimulate LH and FSH release, the pituitary gland must be exposed to a series of brief increases and decreases in GnRH. If GnRH is maintained at a high level in the blood for days or weeks, the anterior pituitary cells become insensitive to it. GnRH can be produced synthetically and is useful in treating some people who are infertile. Synthetic GnRH must be administered in small amounts in frequent pulses or surges. GnRH can also inhibit reproduction because long-term administration of GnRH can sufficiently reduce LH and FSH levels to prevent sperm cell production in males or ovulation in females.

## Puberty in Males

**Puberty** (pū'ber-tē) is the sequence of events by which a child is transformed into a young adult. The reproductive system matures and assumes its adult functions, and the structural differences between adult males and females become more apparent. In boys, puberty commonly begins at ages 12–14 and is largely completed by age 18. Before puberty, small amounts of testosterone, secreted by the testes and the adrenal cortex, inhibit GnRH, LH, and FSH secretion. Beginning just before puberty and continuing throughout puberty, developmental changes in the hypothalamus cause the hypothalamus and the anterior pituitary gland to become much less sensitive to the inhibitory effect of testosterone, and the rate of GnRH, LH, and FSH secretion increases. Consequently, elevated FSH levels promote spermatogenesis, and elevated LH levels cause the interstitial cells to secrete larger amounts of testosterone. Testosterone still has a negative-feedback effect on the hypothalamus and anterior pituitary gland, but GnRH, LH, and FSH secretion occurs at substantially higher levels.

## Effects of Testosterone

**Testosterone** (tes'tos'tě-rōn) is the major male hormone secreted by the testes. Testosterone influences reproductive organs and nonreproductive structures (table 19.2). During puberty, testosterone causes the enlargement and differentiation of the male genitals and the reproductive duct system. It is necessary for spermatogenesis and for the development of male secondary sexual characteristics. The **secondary sexual characteristics** are those structural and behavioral changes, other than in the reproductive organs, that develop at puberty and distinguish males from females. Secondary sexual characteristics in males include hair distribution and growth, skin texture, fat distribution, skeletal muscle growth, and changes in the larynx. After puberty, testosterone maintains the adult structure of the male genitals, reproductive ducts, and secondary sexual characteristics.

## Male Sexual Behavior and the Male Sex Act

Testosterone is required for normal sexual behavior. Testosterone enters certain cells within the brain, especially within the hypothalamus, and influences their functions. The blood levels of testosterone remain relatively constant throughout the lifetime of a male, from puberty until about 40 years of age. Thereafter, the levels slowly decline to approximately 20% of this value by 80 years of age, causing a slow decrease in sex drive and fertility.

# CLINICAL IMPACT

## Male Pattern Baldness

Some men have a genetic tendency called **male pattern baldness,** which develops in response to testosterone and other androgens. When testosterone levels increase at puberty, the density of hair on top of the head begins to decrease. Baldness usually reaches its maximum rate of development when the individual is in the third or fourth decade of life. Minoxidil (mi-noks′si-dil; Rogaine) is a drug that effectively prevents a decrease in hair growth in many men who exhibit male pattern baldness. It is most effective in those who are young and just starting to show evidence of baldness. Minoxidil causes blood vessels to dilate, including those close to hair follicles, which may explain how it works. However, the mechanism has not been confirmed.

| TABLE 19.2 | Effects of Testosterone on Target Tissues |
|---|---|
| **Target Tissue** | **Response** |
| Penis and scrotum | Enlargement and differentiation |
| Hair follicles | Hair growth and coarser hair in the pubic area, legs, chest, axillary region, face, and occasionally back; male pattern baldness on the head if the person has the appropriate genetic makeup |
| Skin | Coarser texture of skin; increased rate of secretion of sebaceous glands, frequently resulting in acne at the time of puberty; increased secretion of sweat glands in axillary regions |
| Larynx | Enlargement of larynx and deeper masculine voice |
| Most tissues | Increased rate of metabolism |
| Red blood cells | Increased rate of red blood cell production; a red blood cell count increase by about 20% as a result of increased erythropoietin secretion |
| Kidneys | Retention of sodium and water to a small degree, resulting in increased extracellular fluid volume |
| Skeletal muscle | A skeletal muscle mass increase at puberty; average increase is greater in males than in females |
| Bone | Rapid bone growth, resulting in increased rate of growth and early cessation of bone growth; males who mature sexually at a later age do not exhibit a rapid period of growth, but they grow for a longer time and can become taller than men who mature earlier |

### Predict 2

*Predict the effect on secondary sexual characteristics, external genitalia, and sexual behavior if the testes fail to produce normal amounts of testosterone at puberty.*

The male sex act is a complex series of reflexes that result in erection of the penis, secretion of mucus into the urethra, emission, and ejaculation. **Emission** (ē-mish′ŭn) is the movement of sperm cells, mucus, prostatic secretions, and seminal vesicle secretions into the prostatic, membranous, and spongy urethra. **Ejaculation** (ē-jak′ū-lā′shŭn) is the forceful expulsion of the secretions that have accumulated in the urethra to the exterior. Sensations, normally interpreted as pleasurable, occur during the male sex act and result in an intense sensation called an **orgasm** (ōr′gazm), or *climax.* In males, orgasm is closely associated with ejaculation, although they are separate functions and do not always occur simultaneously. A phase called **resolution** occurs after ejaculation. During resolution, the penis becomes flaccid, an overall feeling of satisfaction exists, and the male is unable to achieve erection and a second ejaculation.

### Sensory Impulses and Integration

Sensory action potentials from the genitals are carried to the sacral region of the spinal cord, where reflexes that result in the male sex act are integrated. Action potentials also travel from the spinal cord to the cerebrum to produce conscious sexual sensations.

Rhythmic massage of the penis, especially the glans, and surrounding tissues, such as the scrotal, anal, and pubic regions, provide important sources of sensory action potentials. Engorgement of the prostate gland and seminal vesicles with secretions or irritation of the urethra, urinary bladder, ducta deferentia, and testes can also cause sexual sensations.

Psychic stimuli, such as sight, sound, odor, or thoughts, have a major effect on male sexual reflexes. Ejaculation while sleeping (nocturnal emission) is a relatively common event in young males and is thought to be triggered by psychic stimuli associated with dreaming.

## Erection, Emission, and Ejaculation

**Erection** is the first major component of the male sex act. Parasympathetic action potentials from the sacral region of the spinal cord cause the arteries that supply blood to the erectile tissues to dilate. Blood then fills small venous sinuses called **sinusoids** in the erectile tissue and compresses the veins, which reduces blood flow from the penis. The increased blood pressure in the sinusoids causes the erectile tissue to become inflated and rigid. Parasympathetic action potentials also cause the mucous glands within the urethra and the bulbourethral glands to secrete mucus.

## Infertility in Males

Failure to achieve erections, or **erectile dysfunction (ED),** sometimes called *impotence,* can be a major source of frustration. The inability to achieve erections can be due to reduced testosterone secretion resulting from hypothalamic, pituitary, or testicular complications. In other cases, ED can be due to defective stimulation of the erectile tissue by nerve fibers or reduced response of the blood vessels to neural stimulation. Some men can achieve erections by taking oral medications, such as sildenafil (Viagra), tadalafil (Cialis), or verdenafil (Livitra), or by having specific drugs injected into the base of the penis. These drugs increase blood flow into the erectile tissue of the penis, resulting in erection for many minutes.

Before ejaculation, the ductus deferens begins to contract rhythmically, propelling sperm cells and testicular fluid from the epididymis through the ductus deferens. Contractions of the ductus deferens, seminal vesicles, and ejaculatory ducts cause the sperm cells, testicular secretions, and seminal fluid to move into the urethra, where they mix with prostatic secretions released by contraction of the prostate.

Emission is stimulated by sympathetic action potentials that originate in the lumbar region of the spinal cord. Action potentials cause the reproductive ducts to contract and stimulate the seminal vesicles and the prostate gland to release secretions. Consequently, semen accumulates in the urethra.

Ejaculation results from the contraction of smooth muscle in the wall of the urethra and skeletal muscles surrounding the base of the penis. Just before ejaculation, action potentials are sent to the skeletal muscles that surround the base of the penis. Rhythmic contractions are produced that force the semen out of the urethra, resulting in ejaculation. In addition, muscle tension increases throughout the body.

# FEMALE REPRODUCTIVE SYSTEM

The female reproductive organs consist of the ovaries, the uterine tubes (or fallopian tubes), the uterus, the vagina, the external genitalia, and the mammary glands (see figure 19.1*b*). The internal reproductive organs of the female are located within the pelvis, between the urinary bladder and the rectum (figure 19.8). The

**Medial view**

**Figure 19.8  Female Pelvis**

The female reproductive tract, including the uterus, the vagina, and the surrounding pelvic structures, is shown in a medial view of the female pelvis. Note that the female reproductive and urinary tracts open separately to the exterior in the vestibule.

**Anterior view**

**Figure 19.9  Female Reproductive Organs**

Anterior view of the uterus, uterine tubes, and associated ligaments. The uterus and uterine tubes are cut in section (on the left side), and the vagina is cut to show the internal anatomy. The inset shows the relationships among the ovary, the uterine tube, and the ligaments that suspend them in the pelvic cavity.

uterus and the vagina are in the midline, with an ovary to each side of the uterus (figure 19.9). The internal reproductive organs are held in place within the pelvis by a group of ligaments. The most conspicuous is the **broad ligament,** which spreads out on both sides of the uterus and attaches to the ovaries and uterine tubes.

## Ovaries

The two **ovaries** (ō′vă-rēz; *ovum,* egg) are small organs suspended in the pelvic cavity by ligaments. The **suspensory ligament** extends from each ovary to the lateral body wall, and the **ovarian ligament** attaches the ovary to the superior margin of the uterus (figure 19.9). In addition, the ovaries are attached to the posterior surface of the broad ligament by folds of peritoneum called the **mesovarium** (mez′ō-vā′rē-ŭm). The ovarian arteries, veins, and nerves traverse the suspensory ligament and enter the ovary through the mesovarium.

A layer of visceral peritoneum covers the surface of the ovary. The outer part of the ovary is composed of dense connective tissue and contains **ovarian follicles** (figure 19.10). Each of the ovarian follicles contains an **oocyte** (ō′ō-sīt; *ōon,* egg), the female sex cell. Loose connective tissue makes up the inner part of the ovary, where blood vessels, lymphatic vessels, and nerves are located.

## Oogenesis and Fertilization

The formation of female gametes begins in the fetus. By the fourth month of development, the ovaries contain 5 million **oogonia**

(ō-ō-gō′nē-ă), the cells from which oocytes develop (figure 19.11). By the time of birth, many of the oogonia have degenerated, and the remaining ones have begun meiosis. Also, some data indicate that oogonia can form after birth from stem cells, but the extent to which this occurs, and how long it occurs, is not clear. As in meiosis in males, the genetic material is duplicated, and two cell divisions occur (see "Meiosis" earlier in this chapter). Meiosis stops, however, during the first meiotic division at prophase I. The cell at this stage is called a **primary oocyte,** and at birth there are about 2 million of them. From birth to puberty, many primary oocytes degenerate. The number of primary oocytes decreases to around 300,000 to 400,000; of these, only about 400 will complete development and be released from the ovaries. Nearly all others degenerate after partial development.

**Ovulation** is the release of an oocyte from an ovary (figure 19.11, step 8). Just before ovulation, the primary oocyte completes the first meiotic division to produce a **secondary oocyte** and a **polar body.** Unlike meiosis in males, cytoplasm is not split evenly between the two cells. Most of the cytoplasm of the primary oocyte remains with the secondary oocyte. The polar body either degenerates or divides to form two polar bodies. The secondary oocyte begins the second meiotic division but stops in metaphase II.

After ovulation, the secondary oocyte may be fertilized by a sperm cell (figure 19.11, step 9). **Fertilization** (fer′til-i-zā′shŭn) begins when a sperm cell penetrates the cytoplasm of a secondary oocyte. Subsequently, the secondary oocyte completes the second

Degenerated corpus luteum

Visceral peritoneum

Outer part of ovary

Inner part of ovary

Corpus luteum

Mature, or
graafian, follicle

Mesovarium

Blood vessels

Primordial follicles

Primary oocyte

Primary follicles

Granulosa cells

Zona pellucida

Degenerating follicle

Secondary follicle

Vesicles

Zona pellucida

Theca

Zona pellucida

Primary oocyte

Cumulus cells

Antrum

Theca

**Figure 19.10  Structure of the Ovary and Ovarian Follicles**

The ovary is sectioned to illustrate its internal structure (inset shows plane of section). Represented are ovarian follicles from each major stage of development, as well as a corpus luteum.

meiotic division to form 2 cells, each containing 23 chromosomes. One of these cells has very little cytoplasm and is another polar body that degenerates. In the other, larger cell, the 23 chromosomes from the male join with the 23 from the female to form a **zygote** (zī′gōt) and complete fertilization. The zygote has 23 pairs of chromosomes (a total of 46 chromosomes). All cells of the human body contain 23 pairs of chromosomes, except for the male and female gametes. The zygote divides by mitosis to form 2 cells, which divide to form 4 cells, and so on. The mass of cells formed may eventually implant in, or attach to, the uterine wall and develop into a new individual (see chapter 20).

## Follicle Development

As we discussed, when a female is in her mother's uterus, her ovaries have already begun oocyte formation. The primary oocytes present at birth are surrounded by a primordial follicle. A **primordial follicle** is a primary oocyte surrounded by a single layer of flat cells, called **granulosa cells** (figure 19.11). Once puberty begins, some of the primordial follicles are converted to **primary follicles** when the oocyte enlarges and the single layer of granulosa cells

becomes enlarged and cuboidal. Subsequently, several layers of granulosa cells form, and a layer of clear material called the **zona pellucida** (zō′nă pel-lū′sid-dă) is deposited around the primary oocyte.

Approximately every 28 days, hormonal changes stimulate some of the primary follicles to continue to develop (figure 19.11). The primary follicle becomes a **secondary follicle** as fluid-filled spaces called **vesicles** form among the granulosa cells, and a capsule called the **theca** (thē′kă; a box) forms around the follicle.

The secondary follicle continues to enlarge, and when the fluid-filled vesicles fuse to form a single, fluid-filled chamber called the **antrum** (an′trŭm), the follicle is called the **mature follicle,** or *graafian* (graf′e-ăn) *follicle.* The primary oocyte is pushed off to one side and lies in a mass of granulosa cells called the **cumulus cells.**

The mature follicle forms a lump on the surface of the ovary. During ovulation, the mature follicle ruptures, forcing a small amount of blood, follicular fluid, and the primary oocyte, surrounded by the cumulus cells, into the peritoneal cavity. In most cases, only one of the follicles that begin to develop forms a mature

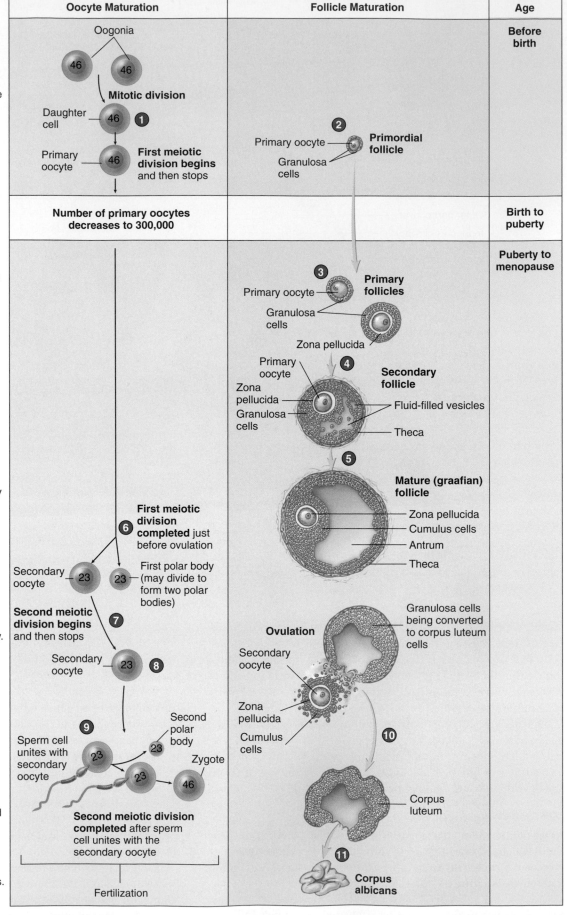

| Oocyte Maturation | Follicle Maturation | Age |
|---|---|---|

**Before birth**

1. Oogonia give rise to oocytes. Before birth, oogonia multiply by mitosis. During development of the fetus, many oogonia begin meiosis, but stop in prophase I and are now called primary oocytes. They remain in this state until puberty.

Oogonia

46    46

**Mitotic division**

Daughter cell    46    ①

Primary oocyte    46    **First meiotic division begins** and then stops

2. Before birth, the primary oocytes become surrounded by a single layer of granulosa cells, creating a primordial follicle. These are present until puberty.

② Primary oocyte — **Primordial follicle**
Granulosa cells

**Number of primary oocytes decreases to 300,000**

**Birth to puberty**

3. After puberty, primordial follicles develop into primary follicles when the granulosa cells enlarge and increase in number.

**Puberty to menopause**

③ Primary oocyte — **Primary follicles**
Granulosa cells
Zona pellucida

4. Secondary follicles from when fluid-filled vesicles develop and thecal cells arise on the outside of the follicle.

④ **Secondary follicle**
Primary oocyte
Zona pellucida
Granulosa cells
Fluid-filled vesicles
Theca

5. Mature follicles form when the vesicles create a single antrum.

⑤ **Mature (graafian) follicle**
Zona pellucida
Cumulus cells
Antrum
Theca

6. Just before ovulation, the primary oocyte completes meiosis I, creating a secondary oocyte and a nonviable polar body.

**First meiotic division completed** just before ovulation
⑥
Secondary oocyte    23    23    First polar body (may divide to form two polar bodies)
**Second meiotic division begins** and then stops

7. The secondary oocyte begins meiosis II, but stops at metaphase II.

⑦
Secondary oocyte    23    ⑧

8. During ovulation, the secondary oocyte is released from the ovary.

**Ovulation**
Secondary oocyte
Zona pellucida
Cumulus cells
Granulosa cells being converted to corpus luteum cells
⑩

9. The secondary oocyte only completes meiosis II if it is fertilized by a sperm cell. The completion of meiosis II forms a secondary oocyte and a second polar body. Fertilization is complete when the secondary oocyte nucleus and the sperm cell nucleus unite, creating a zygote.

⑨
Sperm cell unites with secondary oocyte    23
Second polar body    23
Zygote    23 → 46
**Second meiotic division completed** after sperm cell unites with the secondary oocyte

10. Following ovulation, the granulosa cells divide rapidly and enlarge to form the corpus

Corpus luteum
⑪

11. The corpus luteum degenerates to form a scar, or corpus albicans.

**Corpus albicans**

Fertilization

## PROCESS Figure 19.11  Maturation of the Oocyte and Follicle

The events leading to oocyte and follicle maturation are closely linked. Once these structures are mature, fertilization can result. The numbers written in the cells are total numbers of chromosomes.

follicle and undergoes ovulation. The other follicles degenerate. After ovulation, the remaining cells of the ruptured follicle are transformed into a glandular structure called the **corpus luteum** (kōr′pŭs, body; loo′tē-ŭm, yellow). If pregnancy occurs, the corpus luteum enlarges in response to a hormone secreted by the placenta called **human chorionic gonadotropin hormone (hCG)** (kō-rē-on′ik gō′nad-o-trō′pin) (see table 19.1). If pregnancy does not occur, the corpus luteum lasts for 10–12 days and then begins to degenerate.

## Uterine Tubes

A **uterine tube,** also called a *fallopian* (fa-lō′pē-an) *tube* or *oviduct* (ō′vi-dŭct), is associated with each ovary. The uterine tubes extend from the area of the ovaries to the uterus. They open directly into the peritoneal cavity near each ovary and receive the secondary oocyte. The opening of each uterine tube is surrounded by long, thin processes called **fimbriae** (fim′brē-ē; fringes) (see figure 19.9).

The fimbriae nearly surround the surface of the ovary. As a result, as soon as the secondary oocyte is ovulated, it comes into contact with the surface of the fimbriae. Cilia on the fimbriae surface sweep the oocyte into the uterine tube. Fertilization usually occurs in the part of the uterine tube near the ovary, called the **ampulla** (am-pul′lă). The fertilized oocyte then travels to the uterus, where it embeds in the uterine wall in a process called **implantation.**

## Uterus

The **uterus** (ū′ter-ŭs; womb) is as big as a medium-sized pear (see figures 19.8 and 19.9). It is oriented in the pelvic cavity with the larger, rounded part directed superiorly. The part of the uterus superior to the entrance of the uterine tubes is called the **fundus** (fŭn′dŭs). The main part of the uterus is called the **body,** and the narrower part, the **cervix** (ser′viks; neck), is directed inferiorly. Internally, the **uterine cavity** in the fundus and uterine body continues through the cervix as the **cervical canal,** which opens into the vagina. The cervical canal is lined by mucous glands.

The uterine wall is composed of three layers: a serous layer, a muscular layer, and a layer of endometrium (see figure 19.9).

The outer layer, called the **perimetrium** (per-i-mē′trē-ŭm), or *serous layer,* of the uterus is formed from visceral peritoneum. The middle layer, called the **myometrium** (mī′ō-mē′trē-ŭm), or *muscular layer,* consists of smooth muscle, is quite thick, and accounts for the bulk of the uterine wall. The innermost layer of the uterus is the **endometrium** (en′dō-mē′trē-ŭm), which consists of simple columnar epithelial cells with an underlying connective tissue layer. Simple tubular glands, called endometrial glands, are formed by folds of the endometrium. The superficial part of the endometrium is sloughed off during menstruation.

The uterus is supported by the broad ligament and the **round ligament.** In addition to these ligaments, much support is provided inferiorly to the uterus by skeletal muscles of the pelvic floor. If ligaments that support the uterus or muscles of the pelvic floor are weakened, as may occur due to childbirth, the uterus can extend inferiorly into the vagina, a condition called a **prolapsed uterus.** Severe cases require surgical correction.

## Vagina

The **vagina** (vă-jī′nă) is the female organ of copulation; it receives the penis during intercourse. It also allows menstrual flow and childbirth. The vagina extends from the uterus to the outside of the body (see figures 19.8 and 19.9). The superior portion of the vagina is attached to the sides of the cervix, so that a part of the cervix extends into the vagina.

The wall of the vagina consists of an outer muscular layer and an inner mucous membrane. The muscular layer is smooth muscle and contains many elastic fibers. Thus, the vagina can increase in size to accommodate the penis during intercourse, and it can stretch greatly during childbirth. The mucous membrane is moist stratified squamous epithelium that forms a protective surface layer. Lubricating fluid passes through the vaginal epithelium into the vagina.

In young females, the vaginal opening is covered by a thin mucous membrane called the **hymen** (hī′men; membrane). The hymen may completely close the vaginal orifice, in which case it must be removed to allow menstrual flow. More commonly, the hymen is perforated by one or several holes. The openings in the hymen are usually greatly enlarged during the first sexual intercourse. The hymen can also be perforated or torn earlier in a young female's life during a variety of activities, including strenuous exercise. The condition of the hymen is therefore not a reliable indicator of virginity.

## External Genitalia

The external female genitalia, also called the **vulva** (vŭl′vă) or *pudendum* (pū-den′dŭm), consist of the vestibule and its surrounding structures (figure 19.12). The **vestibule** (ves′ti-bool) is the space into which the vagina and urethra open. The urethra opens just anterior to the vagina. The vestibule is bordered by a pair of thin, longitudinal skin folds called the **labia minora** (lā′bē-ă, lips; mī-nō′ră, small). A small, erectile structure called the **clitoris** (klit′ŏ-ris, klī′tŏ-ris) is located in the anterior margin of the vestibule. The two labia minora unite over the clitoris to form a fold of skin called the **prepuce.**

The clitoris (see figure 19.8) consists of a shaft and a distal glans. Like the glans penis, the clitoris is well supplied with sensory receptors, and it is made up of erectile tissue. Additional erectile tissue is located on each side of the vaginal opening.

On each side of the vestibule, between the vaginal opening and the labia minora, are openings of the **greater vestibular glands.** These glands produce a lubricating fluid that helps maintain the moistness of the vestibule.

Lateral to the labia minora are two prominent, rounded folds of skin called the **labia majora** (mă-jō′ră; large). The two labia majora unite anteriorly at an elevation of tissue over the pubic symphysis called the **mons pubis** (monz pū′bis) (figure 19.12). The lateral surfaces of the labia majora and the surface of the mons pubis are covered with coarse hair. The medial surfaces of the labia majora are covered with numerous sebaceous and sweat glands. The space between the labia majora is called the **pudendal cleft.** Most of the time, the labia majora are in contact with each other across the midline, closing the pudendal cleft and covering the deeper structures within the vestibule.

The region between the vagina and the anus is the **clinical perineum** (per′i-nē′um; area between the thighs). The skin and muscle of this region can tear during childbirth. To prevent such tearing, an incision called an **episiotomy** (e-piz-ē-ot′ō-mē) is sometimes made in the clinical perineum. Traditionally, this clean, straight incision has been thought to result in less injury, less trouble in healing, and less pain. However, many studies report less injury and pain when no episiotomy is performed.

## Mammary Glands

The **mammary** (mam′ă-rē; relating to breasts) **glands** are the organs of milk production and are located in the **breasts** (figure 19.13). The mammary glands are modified sweat glands. Externally, each of the breasts of both males and females has a raised **nipple** surrounded by a circular, pigmented area called the **areola** (ă-rē′ō-lă).

In prepubescent children, the general structure of the male and female breasts is similar, and both males and females possess a rudimentary duct system. The female breasts begin to enlarge during puberty, under the influence of estrogen and progesterone. Some males also experience a minor and temporary enlargement of the breasts at puberty. Occasionally, the breasts of a male can become permanently enlarged, a condition called **gynecomastia** (gī′nĕ-kō-mas′tē-ă). Causes of gynecomastia include hormonal imbalances and the abuse of anabolic steroids.

Each adult female breast contains mammary glands consisting of usually 15–20 glandular **lobes** covered by adipose tissue (figure 19.13*a,b*). It is primarily this superficial adipose tissue that gives the breast its form. Each lobe possesses a single **lactiferous**

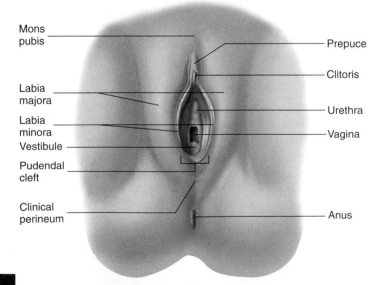

Mons pubis
Labia majora
Labia minora
Vestibule
Pudendal cleft
Clinical perineum

Prepuce
Clitoris
Urethra
Vagina
Anus

**Figure 19.12  Female External Genitalia**

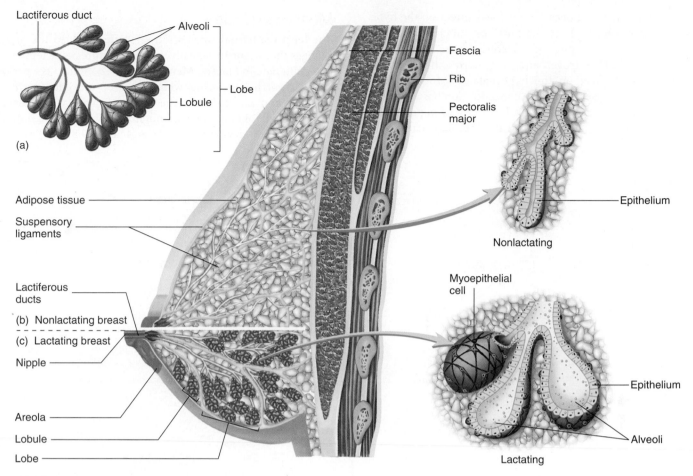

**Figure 19.13** **Anatomy of the Breast**

(a) Each lactiferous duct of the mammary gland branches. At the end of each branch is one or more alveoli. (b) The nonlactating breast has a duct system that is not extensively developed. The branches of the lactiferous ducts end as small, tubelike structures. (c) The lactating breast has a well-developed duct system with many branches. The branches of the lactiferous duct end with well-developed alveoli. Adipose tissue is abundant in both the nonlactating and the lactating breast.

**duct** that opens independently to the surface of the nipple. The duct of each lobe is formed as several smaller ducts, which originate from **lobules,** converge. Within a lobule, the ducts branch and become even smaller. In the milk-producing, or lactating, mammary gland, the ends of these small ducts expand to form secretory sacs called **alveoli. Myoepithelial cells** surround the alveoli and contract to expel milk from the alveoli (figure 19.13c).

The breasts are supported by suspensory ligaments that extend from the fascia over the pectoralis major muscles to the skin over the breasts (figure 19.13b).

The nipples are very sensitive to tactile stimulation and contain smooth muscle. When the smooth muscle contracts in response to stimuli, such as touch, cold, and sexual arousal, the nipple becomes erect.

# PHYSIOLOGY OF FEMALE REPRODUCTION

As in the male, female reproduction is controlled by hormonal and nervous system mechanisms.

## Puberty in Females

The initial change that results in puberty is most likely maturation of the hypothalamus. In girls, puberty, which typically begins between ages 11 and 13 and is largely completed by age 16, is marked by the first episode of menstrual bleeding, which is called **menarche** (me-nar′kē; *mēn,* month + *archē,* beginning). During puberty, the vagina, uterus, uterine tubes, and external genitalia begin to enlarge. Adipose tissue is deposited in the breasts and around the hips, causing them to enlarge and assume an adult form. In addition, pubic and axillary hair grows. The development of sexual drive is also associated with puberty.

The changes associated with puberty primarily result from the increasing rate of estrogen and progesterone secretion by the ovaries. Before puberty, estrogen and progesterone are secreted in very small amounts. At puberty, the cyclical adult pattern of hormone secretion is gradually established.

Before puberty, the rate of GnRH secretion from the hypothalamus and the rate of LH and FSH secretion from the anterior pituitary are very low. Estrogen and progesterone from the ovaries have a strong negative-feedback effect on the hypothalamus and pituitary. After the onset of puberty, the hypothalamus and anterior pituitary secrete larger amounts of GnRH, LH, and FSH. Estrogen and progesterone have less of a negative-feedback effect on the hypothalamus and pituitary, and a sustained increase in estrogen concentration has a positive-feedback effect. The normal cyclical pattern of reproductive hormone secretion that occurs during the menstrual cycle becomes established.

## Menstrual Cycle

The term **menstrual** (men′stroo-ăl) **cycle** refers to the series of changes that occur in sexually mature, nonpregnant females and that culminate in menses. **Menses** (men′sēz; month) is a period of mild hemorrhage, during which part of the endometrium is sloughed and expelled from the uterus. Typically, the menstrual cycle is about 28 days long, although it can be as short as 18 days or as long as 40 days (figure 19.14 and table 19.3). The menstrual

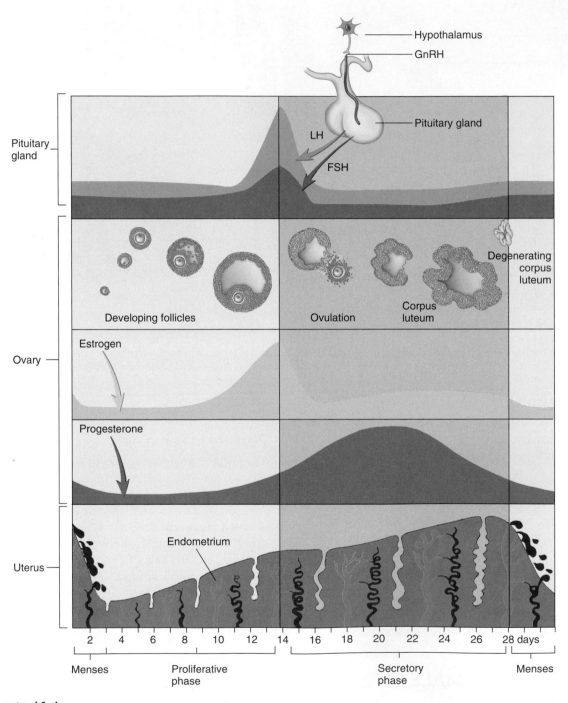

### Figure 19.14  Menstrual Cycle

Over approximately 30 days, fluctuations occur in the levels of follicle-stimulating hormone (FSH) and luteinizing hormone (LH) secretion from the anterior pituitary gland and in the levels of estrogen and progesterone secretion from the ovary. In addition, changes in the ovary and changes in the endometrium of the uterus are correlated with the changes in hormone secretion throughout the menstrual cycle. Ovulation occurs on about day 14.

## TABLE 19.3    Events During the Menstrual Cycle

**Menses (Day 1 to Day 4 or 5 of the Menstrual Cycle)**

| | |
|---|---|
| Pituitary gland | The rate of FSH and LH secretion is low, but the rate of FSH secretion increases as progesterone levels decline. |
| Ovary | The rate of estrogen and progesterone secretion is low after degeneration of the corpus luteum produced during the previous menstrual cycle. |
| Uterus | In response to declining progesterone levels, the endometrial lining of the uterus sloughs off, resulting in menses followed by repair of the endometrium. |

**Proliferative Phase (from Day 4 or 5 Until Ovulation on About Day 14)**

| | |
|---|---|
| Pituitary gland | The rate of FSH and LH secretion is only slightly elevated during most of the proliferative phase; FSH and LH secretions increase near the end of the proliferative phase in response to increasing estrogen secretion from the ovaries. |
| Ovary | Developing follicles secrete increasing amounts of estrogen, especially near the end of the proliferative phase; increasing FSH and LH cause additional estrogen secretion from the ovaries near the end of the proliferative phase. |
| Uterus | Estrogen causes endometrial cells of the uterus to divide. The endometrium of the uterus thickens, and tubelike glands form. Estrogen causes the cells of the uterus to be more sensitive to progesterone by increasing the number of progesterone receptors in uterine tissues. |

**Ovulation (About Day 14)**

| | |
|---|---|
| Pituitary gland | The rate of FSH and LH secretion increases rapidly just before ovulation in response to increasing estrogen levels. Increasing FSH and LH levels stimulate estrogen secretion, resulting in a positive-feedback cycle. |
| Ovary | LH causes final maturation of a mature follicle and initiates the process of ovulation. FSH acts on immature follicles and causes several of them to begin to enlarge. |
| Uterus | The endometrium continues to divide in response to estrogen. |

**Secretory Phase (from About Day 14 to Day 28)**

| | |
|---|---|
| Pituitary gland | Estrogen and progesterone reach levels high enough to inhibit FSH and LH secretion from the pituitary gland. |
| Ovary | After ovulation, the follicle is converted to the corpus luteum; the corpus luteum secretes large amounts of progesterone and smaller amounts of estrogen from shortly after ovulation until about day 24 or 25. If fertilization does not occur, the corpus luteum degenerates after about day 25, and the rate of progesterone secretion rapidly declines to low levels. |
| Uterus | In response to progesterone, the endometrial cells enlarge, the endometrial layer thickens, and the glands of the endometrium reach their greatest degree of development; the endometrial cells secrete a small amount of fluid. After progesterone levels decline, the endometrium begins to degenerate. |

**Menses (Day 1 to Day 4 or 5 of the Next Menstrual Cycle)**

| | |
|---|---|
| Pituitary gland | The rate of LH remains low, and the rate of FSH secretion increases as progesterone levels decline. |
| Ovary | The rate of estrogen and progesterone secretion is low. |
| Uterus | In response to declining progesterone levels, the endometrial lining of the uterus sloughs off, resulting in menses followed by repair of the endometrium. |

cycle results from the cyclical changes that occur in the endometrium of the uterus. These changes, in turn, result from the cyclical changes that occur in the ovary and are controlled by the secretions of FSH and LH from the anterior pituitary gland.

The first day of menstrual bleeding (menses) is considered day 1 of the menstrual cycle. Menses typically lasts 4 or 5 days and can be accompanied by strong uterine contractions, called menstrual cramps, that are extremely uncomfortable in some women. Menstrual cramps are the result of forceful myometrial contractions that occur before and during menstruation. The cramps can result from excessive secretion of prostaglandins. As the endometrium of the uterus sloughs off, it becomes inflamed, and prostaglandins are produced as part of the inflammation. Sloughing of the endometrium is inhibited by progesterone but stimulated by estrogen. Many women can alleviate painful cramps by taking medications, such as aspirin-like drugs, that inhibit prostaglandin biosynthesis just before the onset of menstruation. These medications,

however, are not effective in treating all painful menstruation, especially when the cause of the pain, such as that experienced by women who have tumors of the myometrium, is not due to the inflammatory response.

Ovulation occurs on about day 14 of the menstrual cycle, although the timing of ovulation varies from individual to individual and can vary within an individual from one menstrual cycle to the next.

The time between the end of menses and ovulation is called the **proliferative phase,** referring to proliferation of the endometrium. During the proliferative phase, the secondary follicles in the ovary mature; as they do so, they secrete increasing amounts of estrogen. Estrogen acts on the uterus and causes the epithelial cells of the endometrium to divide rapidly. The endometrium thickens, and endometrial glands form.

The sustained increase of estrogen secreted by the developing follicles stimulates GnRH secretion from the hypothalamus.

# CLINICAL IMPACT

## Amenorrhea

The absence of a menstrual cycle is called **amenorrhea** (ă-men-ō-rē'ă; without menses). If the pituitary gland does not function properly because of abnormal development, a woman does not begin to menstruate at puberty. This condition is called **primary amenorrhea.** In contrast, if a woman has had normal menstrual cycles and later stops menstruating, the condition is called **secondary amenorrhea.** One cause of secondary amenorrhea is anorexia, in which a lack of food causes the hypothalamus of the brain to decrease GnRH secretion to levels so low that the menstrual cycle cannot occur.

Many female athletes and ballet dancers who have rigorous training schedules have secondary amenorrhea. The physical stress that can be coupled with inadequate food intake also results in very low GnRH secretion. Increased food intake for anorexic women or reduced training for dancers and athletes generally restores normal hormone secretion and normal menstrual cycles.

Secondary amenorrhea can also be the result of a pituitary tumor that decreases FSH and LH secretion or a lack of GnRH secretion from the hypothalamus due to head trauma or a tumor.

In addition, secondary amenorrhea can occur due to a lack of normal hormone secretion from the ovaries, which can result from autoimmune diseases that attack the ovary or occur due to **polycystic ovarian disease,** in which cysts in the ovary produce large amounts of androgen, which is converted to estrogen by other body tissues. The increased estrogen prevents the normal cycle of FSH and LH secretion required for ovulation to occur. Other hormone-secreting tumors of the ovary can also disrupt the normal menstrual cycle and result in amenorrhea.

---

GnRH, in turn, triggers FSH and LH secretion from the anterior pituitary gland. FSH stimulates estrogen secretion at an increasing rate from the developing follicles. This positive-feedback loop produces a series of larger and larger surges of FSH and LH secretion. Ovulation occurs in response to the large increases in LH levels that normally occur on about day 14 of the menstrual cycle. This large increase in LH is also responsible for the development of the corpus luteum.

Following ovulation, the corpus luteum begins to secrete progesterone and smaller amounts of estrogen. Progesterone acts on the uterus, causing the cells of the endometrium to become larger and to secrete a small amount of fluid. Together, progesterone and estrogen act on the hypothalamus and anterior pituitary gland to inhibit GnRH, LH, and FSH secretion. Thus, LH and FSH levels decline after ovulation.

The time between ovulation and the next menses is called the **secretory phase** of the menstrual cycle because of the small amount of fluid secreted by the cells of the endometrium. During the secretory phase, the lining of the uterus reaches its greatest degree of development.

If fertilization occurs, the zygote undergoes several cell divisions to produce a collection of cells called the **blastocyst** (blas'tō-sist). The blastocyst passes through the uterine tube and arrives in the uterus by 7 or 8 days after ovulation. The endometrium is prepared to receive the blastocyst, which becomes implanted in the endometrium, where it continues to develop. If the secondary oocyte is not fertilized, the endometrium sloughs away as a result of declining blood progesterone levels. Unless the secondary oocyte is fertilized, the corpus luteum begins to produce less progesterone by day 24 or 25 of the menstrual cycle. By day 28, the declining progesterone causes the endometrium to

slough away to begin menses and the next menstrual cycle. The declining progesterone secretion results in a small increase in FSH secretion at the beginning of the next menses, which triggers more follicles to mature.

### Predict 3

*Predict the effect of administering a relatively large amount of progesterone and estrogen just before the increase in LH that precedes ovulation.*

An **ectopic pregnancy** results if implantation occurs anywhere other than in the uterine cavity. The most common site of ectopic pregnancy is the uterine tube. Implantation in the uterine tube eventually is fatal to the fetus and can cause the tube to rupture. In some cases, implantation occurs in the mesenteries of the abdominal cavity; the fetus can develop normally but must be delivered by cesarean section.

## Menopause

When a woman is 40–50 years old, the menstrual cycles become less regular, and ovulation does not occur consistently during each cycle. Eventually, the cycles stop completely. The cessation of menstrual cycles is called **menopause** (men'ō-pawz; *mēn,* month + *pausis,* cessation), and the whole time period from the onset of irregular cycles to their complete cessation is called the female **climacteric** (klī-mak'ter-ik).

The major cause of menopause is age-related changes in the ovaries. The number of follicles remaining in the ovaries of menopausal women is small. In addition, the follicles that remain become less sensitive to stimulation by FSH and LH, and therefore fewer mature follicles and corpora lutea are produced.

| TABLE 19.4 | Possible Changes in Postmenopausal Women Caused by Decreased Ovarian Hormone Secretion |
|---|---|
| | **Changes** |
| Menstrual Cycle | Five to seven years before menopause, the cycle becomes irregular; the number of cycles in which ovulation does not occur and in which corpora lutea do not develop increases. |
| Uterus | Gradual increase in irregular menstruations is followed by no menstruation; the endometrium finally atrophies, and the uterus becomes smaller. |
| Vagina and External Genitalia | The epithelial lining becomes thinner; the external genitalia become thinner and less elastic; the labia majora become smaller; the pubic hair decreases; reduced secretion leads to dryness; the vagina is more easily inflamed and infected. |
| Skin | The epidermis becomes thinner. |
| Cardiovascular System | Hypertension and atherosclerosis occur more frequently. |
| Vasomotor Instability | Hot flashes and increased sweating are correlated with vasodilation of cutaneous blood vessels; the hot flashes are related to decreased estrogen levels. |
| Libido | Temporary changes, usually a decrease in libido, are associated with the onset of menopause. |
| Fertility | Fertility begins to decline about 10 years before the onset of menopause; by age 50, almost all the oocytes and follicles have degenerated. |
| Pituitary Function | Low levels of estrogen and progesterone produced by the ovaries cause the pituitary gland to secrete larger than normal amounts of LH and FSH; increased levels of these hormones have little effect on the postmenopausal ovaries. |

Older women experience gradual changes in response to the reduced amount of estrogen and progesterone produced by the ovaries (table 19.4). For example, during the climacteric, some women experience sudden episodes of uncomfortable sweating (hot flashes), fatigue, anxiety, temporary decrease in libido, and occasionally emotional disturbances. Many of these symptoms can be treated successfully with hormone replacement therapy (HRT), which usually consists of small amounts of estrogen and progesterone. HRT has been linked to a slightly increased risk of developing breast cancer, uterine cancer, heart attack, stroke, or blood clots. On the positive side, HRT slows the decrease in bone density that can become severe in some women after menopause, and it decreases the risk of developing colorectal cancer.

## Female Sexual Behavior and the Female Sex Act

Sexual drive in females, like sexual drive in males, is dependent on hormones. Testosterone-like hormones, and possibly estrogen, affect brain cells (especially in the area of the hypothalamus) and influence sexual behavior. Testosterone-like hormones are produced primarily in the adrenal cortex. Psychological factors also play a role in sexual behavior. The sensory and motor neural pathways involved in controlling female sexual responses are similar to those found in the male.

During sexual excitement, erectile tissue within the clitoris and around the vaginal opening becomes engorged with blood. The mucous glands within the vestibule, especially the greater vestibular glands, secrete small amounts of mucus. Larger amounts of mucuslike fluid are also extruded into the vagina through its wall. These secretions provide lubrication to allow easy entry and movement of the penis in the vagina during intercourse. Tactile stimulation of the female's genitals during sexual intercourse and psychological stimuli normally trigger an **orgasm,** or *climax.* The vaginal and uterine smooth muscle, as well as the surrounding skeletal muscles, contract rhythmically, and muscle tension increases throughout much of the body. After the sex act, there is a period of **resolution,** which is characterized by an overall sense of satisfaction and relaxation. Females are sometimes receptive to further immediate stimulation, however, and can experience successive orgasms. Orgasm is not necessary for fertilization to occur. Ovulation results from hormonal stimuli and is not dependent on the female sex act.

## Infertility in Females

Causes of infertility in females include malfunctions of the uterine tubes, reduced hormone secretion from the pituitary gland or the ovaries, and interruption of implantation. Uterine tube malfunction can occur when infections result in pelvic inflammatory disease (PID), which causes adhesions to form in one or both of the uterine tubes. Inadequate secretion of LH and FSH can occur for a variety of reasons, including hypothyroidism or a tumor in or trauma to the hypothalamus or anterior pituitary. Decreased secretion of LH and FSH interrupts ovulation.

Interruption of implantation can result from uterine tumors or conditions causing abnormal ovarian hormone secretion. For example, premature degeneration of the corpus luteum causes progesterone levels to decline and menses to occur. If the corpus luteum degenerates before the placenta begins to secrete progesterone, the endometrium and the developing embryonic mass will degenerate and be eliminated from the uterus. The conditions that result in secondary amenorrhea also reduce fertility.

# CLINICAL IMPACT

## Control of Pregnancy

Many methods are used to prevent pregnancy, either by preventing fertilization (contraception) or by preventing implantation of the developing embryo. Many of these techniques are quite effective when used *perfectly* and consistently. But most of these methods also have disadvantages, and the use of some of them is controversial.

### Behavioral Methods

**Abstinence,** or refraining from sexual intercourse, is 100% effective in preventing pregnancy when it is practiced consistently. It is not an effective method when used only occasionally.

**Coitus interruptus** (kō′i-tŭs int-ĕ-rŭp′tŭs), or **withdrawal,** is removal of the penis from the vagina just before ejaculation. This is a very unreliable method of preventing pregnancy because it requires perfect awareness and willingness to withdraw the penis at the correct time. Statistically, about 23 women out of 100 become pregnant while relying on this method. The withdrawal method also ignores the fact that some sperm cells are present in preejaculatory emissions.

The **calendar method** requires abstaining from sexual intercourse near the time of ovulation. A major factor in the success of this method is the ability to predict accurately the time of ovulation. Although the calendar method provides some protection against becoming pregnant, it has a relatively high failure rate because of both the inability to predict the time of ovulation and the failure to abstain from intercourse around that time. About 9 women out of 100 become pregnant while using the calendar method.

Continuous breastfeeding, or **lactation** (lak-tā′shŭn) (also known as lactation amenorrhea, or LAM) often stops the menstrual cycle for up to the first 6 months after childbirth, as long as the baby is exclusively breastfed and the mother does not resume menstruation while lactating. This method is 99% effective. Continuous breastfeeding works because action potentials sent to the hypothalamus in response to infant suckling inhibit GnRH release from the hypothalamus. Reduced GnRH reduces LH, which prevents ovulation. Eventually, the menstrual cycle resumes. Because ovulation normally precedes menstruation, relying on lactation to prevent pregnancy after the first 6 months postdelivery is not effective.

### Barrier Methods

A male **condom** (kon′dom) is a sheath made of animal membrane, rubber, or plastic (figure 19A*a*). When placed over the erect penis, a condom is a barrier device because it collects the semen instead of allowing it to be released into the vagina. Condoms also provide protection against sexually transmitted diseases. Condoms alone are 98% effective, 99% effective when used with spermicide.

A **vaginal condom** (or female condom) also acts as a barrier. A woman can place the vaginal condom into the vagina before sexual intercourse. Female condoms are 95% effective. Using spermicide further increases their effectiveness.

Methods to prevent sperm cells from reaching the oocyte once they are in the vagina include a diaphragm, spermicidal agents, and a vaginal sponge. The **diaphragm** and the **cervical cap** (figure 19A*b*) are flexible latex domes that are placed over the cervix within the vagina, where they prevent sperm cells from passing from the vagina through the cervical canal of the uterus. The diaphragm is a larger, shallow latex cup, and the cervical cap is a smaller, thimble-shaped latex cup. Diaphragms are 94% effective, whereas cervical cap effectiveness ranges from 71% in a woman who has previously been pregnant to 86% in a woman who has never been pregnant. The most commonly used **spermicidal agents** are foams or creams that kill sperm cells (figure 19A*c*). They are inserted into the vagina before sexual intercourse, often in conjunction with diaphragm or condom use. Alone, spermicidal agents are only about 85% effective.

**Intrauterine** (in′tră-yū′ter-in) **devices (IUDs)** (figure 19A*d*) are inserted into the uterus through the cervix. The two types of IUDs now available in the Untied States are the copper-containing ParaGard and the progestin hormone–coated Mirena. The ParaGard may be left in place for 12 years, whereas the Mirena may be left in place for 5 years. Both types of IUDs thicken cervical mucus, which bars sperm from entering the uterus. Some women stop ovulating when they have an IUD implanted. IUDs also alter the endometrium, which in theory may prevent implantation of an embryo. IUDs are 99.99% effective in preventing pregnancy.

### Chemical Methods

Synthetic estrogen and progesterone in **oral contraceptives** (birth control pills) (figure 19A*e*) are among the most effective contraceptives, providing 99.9% effectiveness. The synthetic steroids can have more than one action, but they reduce LH and FSH release from the anterior pituitary. Estrogen and progesterone are present in high enough concentrations to have a negative-feedback effect on the pituitary, which prevents the large increase in LH and FSH secretion that triggers ovulation. Over the years, the dose of estrogen and progesterone in birth control pills has been reduced. The current lower-dose of birth control pills has fewer side effects than earlier dosages. However, the risk of heart attack or stroke increases in females using oral contraceptives who smoke or who have a history of hypertension or coagulation disorders. For most females, the pill is effective and has a minimum frequency of complications, until at least age 35. The **mini-pill** is an oral contraceptive that contains only synthetic progesterone. It reduces and thickens the mucus of the cervix, which prevents sperm cells from reaching the oocyte. It also prevents blastocysts from implanting in the uterus.

Progesterone-like chemicals, such as medroxyprogesterone (med-rok′-sē-prō-jes′ter-ōn) (Depo-Provera), which are

(a)                                    (b)

(c)

(d)

(e)

Ductus deferens within
spermatic cord

Ductus deferens
(vas deferens)
cut and tied

(f)

Ovary

Uterus

Uterine tube
cut and tied

(g)

## Figure 19A

Contraceptive devices and techniques include (*a*) condom, (*b*) cervical cap and diaphragm used with spermicidal jelly, (*c*) spermicidal foam, (*d*) intrauterine device (IUD), (*e*) oral contraceptives, (*f*) vasectomy, and (*g*) tubal ligation.

(continued)

## CLINICAL IMPACT (continued)

injected intramuscularly and slowly released into the circulatory system, can act as effective contraceptives. Injected progesterone-like chemicals can protect against pregnancy for approximately 1 month, depending on the amount injected and are 99.9% effective. The **patch** (Ortho Evra) is an adhesive skin patch containing synthetic estrogen and progesterone. It is worn on the lower abdomen, buttocks, or upper body and is 99.9% effective. The **vaginal contraceptive ring** (Nuva Ring) is inserted into the vagina, where it releases synthetic estrogen and progesterone; it is 99.9% effective.

A drug called **RU486**, or *mifepristone* (mif′pris-tōn), blocks the action of progesterone, causing the endometrium of the uterus to slough off as it does at the time of menstruation. It can therefore be used to induce menstruation and reduce the possibility of implantation when sexual intercourse has occurred near the time of ovulation. It can also be used to terminate pregnancies.

**Morning-after pills,** similar in composition to birth control pills, are available. Alternatively, doubling the number of birth control pills after sexual intercourse within 3 days and again after 12 more hours is sometimes recommended. These techniques can be used after intercourse, but they are only about 75% effective. The elevated blood levels of estrogen and progesterone may inhibit the increase in LH that causes ovulation, alter the rate at which the fertilized oocyte is transported through the uterine tube to the uterus, or inhibit implantation.

### Surgical Methods

**Vasectomy** (va-sek′tō-mē) is a common method used to render males permanently infertile without affecting the performance of the sex act. Vasectomy is a surgical procedure in which the ductus deferens from each testis is cut and tied off within the scrotal sac (figure 19A*f*). This procedure prevents sperm cells from passing through the ductus deferens and becoming part of the ejaculate. Because such a small volume of ejaculate comes from the testis and epididymis, vasectomy has little effect on the volume of the ejaculated semen. The sequestered sperm cells are reabsorbed in the epididymis. Only 1–4 in 1000 surgeries of this type fail.

A common method of permanent birth control in females is **tubal ligation** (lī-gā′shŭn), in which the uterine tubes are tied and cut or clamped by means of an incision through the wall of the abdomen (figure 19A*g*). This procedure closes off the path between the sperm cells and the oocyte. Commonly, a technique called **laparoscopy** (lap-ă-ros′kŏ-pē) is used, in which an instrument is inserted into the abdomen through a small incision, so that only small openings need to be made to perform the operation.

---

## A CASE IN POINT
### Endometriosis

Helen Hurtz is in her mid-twenties and has been married for 4 years. She has become very frustrated because she experiences pain during and after sexual intercourse, and the pain has become worse over the past 2 years. She also has an increasing, persistent pain in her pelvic region that is especially uncomfortable before and during menstruation and becomes more intense during urination and bowel movements. In addition, she recently developed periodic bouts of diarrhea. She reported all these symptoms to her physician.

Helen's physician suspects **endometriosis** (en′dō-mē-trē-ō′sis), a condition in which endometrial tissue migrates from the lining of the uterus into the peritoneal cavity, where it attaches to the surface of organs. Common sites of attachment are the ovaries and the pelvic peritoneum. Other possible sites are the intestines, uterus, urinary bladder, and vagina. If the attached endometrial tissue has an adequate blood supply, it proliferates, breaks down, and bleeds in response to the hormones produced during the menstrual cycle. Unlike the normal endometrium, which is shed each month during menstruation, endometrial tissue attached outside of the uterus causes lesions or tumors to develop, resulting in internal bleeding, scar tissue formation, inflammation, and pain. Other major complications of endometriosis are infertility and ovarian cyst formation. Between 40% and 50% of infertile women have endometriosis.

Helen's physician explained that the most accurate method of confirming the diagnosis is by laparoscopy, a procedure that allows the physician to visually observe the abdominopelvic cavity. After a few weeks of thinking about the procedure, Helen agreed to the laparoscopic examination. During the procedure, the physician observed several lesions characteristic of endometriosis and removed them with a laser instrument that vaporizes them.

Helen's physician explained to her that there is no cure for endometriosis, but the condition can be managed by administering medication and removing the endometrial lesions periodically.

## EFFECTS OF AGING ON THE REPRODUCTIVE SYSTEM

Aging affects the reproductive system in both men and women in several ways. Sexual activity is often maintained in men and women as they age, but the frequency of sexual intercourse usually decreases gradually.

In men, benign prostatic enlargement is common after 50 years of age. A major consequence of prostatic enlargement is blockage of the prostatic urethra. Although benign prostatic enlargement is not preventable, treatments are available to reverse its negative effects. The frequency of prostate cancer also increases as men age and is a significant cause of death in men. In addition, the tendency for erectile dysfunction increases as men age. However, less than 15% of men age 60 or under experience abnormal erectile

# SYSTEMS PATHOLOGY

## Benign Uterine Tumors

Molly had four children and was 43. She noticed that menstruation was becoming gradually more severe and lasting up to several days longer each time. After menstruating almost continuously for 2 months, Molly made an appointment with her physician. The physician performed a pelvic examination, including tests for conditions such as cervical and uterine cancer. Palpation of the uterus indicated the presence of enlarged masses in Molly's uterus. The doctor performed a D&C, which is dilation of the cervix and scraping (curettage) of the endometrium to remove growths or other abnormal tissues. The results of the D&C indicated that Molly had developed leiomyomas.

### Background Information

**Leiomyomas** (lī′ō-mī-ō′măz), also called uterine fibroids, are fibroid tumors of the uterus (figure 19B). They are one of the most common disorders of the uterus, and the most frequent tumor in women, affecting one of every four. However, three-fourths of the women with this condition experience no symptoms. The enlarged masses that originate from smooth muscle tissue compress the uterine lining (endometrium), resulting in ischemia and inflammation. The increased inflammation, which shares some characteristics with menstruation, results in frequent and severe menses, with associated abdominal cramping due to strong uterine contractions. Constant menstruation is a frequent manifestation of these tumors, and it is one of the most common reasons women elect to have the uterus removed, a procedure called a **hysterectomy** (his-ter-ek′tō-mē).

Interstitial leiomyoma

Uterus

Submucous leiomyoma

Subserous leiomyoma

Vagina

### Figure 19B

Leiomyomas, or fibroid tumors, are enlarged masses of smooth muscle. They are located near the mucosa (submucous), within the myometrium (interstitial), or near the serosa (subserous).

**Predict 4**

*When discussing her condition with her mother, Molly discovered that her mother had experienced frequent menses that were irregular and prolonged when she was in her late 40s. Molly's mother did not have a hysterectomy, and in a few years, the frequency gradually began to subside. Explain.*

| System Interactions | Effects of Benign Uterine Tumors on Other Systems |
|---|---|
| **System** | **Interactions** |
| Integumentary | If anemia develops, the skin can appear pale because of the reduced hemoglobin in red blood cells. |
| Muscular | If anemia develops and is severe, muscle weakness results because of the reduced ability of the cardiovascular system to deliver adequate oxygen to muscles. |
| Skeletal | The rate of red blood cell synthesis in the red bone marrow increases. |
| Cardiovascular | A chronic loss of blood, as occurs in prolonged menstruation over many months to years, frequently results in iron-deficiency anemia. Manifestations of anemia include pale skin, reduced hematocrit, reduced hemoglobin concentration, smaller than normal red blood cells (microcytic anemia), and increased heart rate. |
| Respiratory | Because of anemia, the oxygen-carrying capacity of the blood is reduced. Increased respiration during physical exertion and rapid fatigue are likely to occur if anemia develops. |
| Urinary | The kidney increases erythropoietin secretion in response to the reduced oxygen-carrying capacity of the blood. The erythropoietin increases red blood cell synthesis in red bone marrow. An enlarged tumor can put pressure on the urinary bladder, resulting in frequent and painful urination. |

## DISEASES AND DISORDERS  Reproductive System

| CONDITION | DESCRIPTION |
| --- | --- |
| **Infectious Diseases** | |
| Pelvic inflammatory disease (PID) | Bacterial infection of the female pelvic organs; commonly caused by a vaginal or uterine infection with the bacteria gonorrhea or chlamydia; early symptoms include increased vaginal discharge and pelvic pain; antibiotics are effective; if untreated, can lead to sterility or be life-threatening |
| **Sexually Transmitted Diseases** | Commonly known as STDs; spread by intimate sexual contact |
| Nongonococcal urethritis (non-gon′ō-kok′ăl u-rē-thrī′tis) | Inflammation of the urethra that is not caused by gonorrhea; can be caused by trauma, insertion of a nonsterile catheter, or sexual contact; usually due to infection with the bacterium *Chlamydia trachomatis* (kla-mid′ē-ă tra-kō′mă-tis); may go unnoticed and result in pelvic inflammatory disease or sterility; antibiotics are effective treatment |
| Trichomoniasis (trik-ō-mō-nī′ă-sis) | Caused by *Trichomonas* (trik′ō-mō′nas), a protozoan commonly found in the vagina of women and in the urethra of men; results in a greenish-yellow discharge with a foul odor; more common in women than in men |
| Gonorrhea (gon-ō-rē′ă) | Caused by the bacterium *Neisseria gonorrhoeae* (nī-sē′rē-ă gon-ō-rē′ă), which attaches to the epithelial cells of the vagina or male urethra and causes pus to form; pain and discharge from the penis occur in men; asymptomatic in women in the early stages; can lead to sterility in men and pelvic inflammatory disease in women |
| Genital herpes (her′pēz) | Caused by herpes simplex 2 virus; characterized by lesions on the genitals that progress into blisterlike areas, making urination, sitting, and walking painful; antiviral drugs can be effective |
| Genital warts | Caused by a viral infection; very contagious; warts vary from separate, small growths to large, cauliflower-like clusters; lesions are not painful, but sexual intercourse with lesions is; treatments include topical medicines and surgery to remove the lesions |
| Syphilis (sif′i-lis) | Caused by the bacterium *Treponema pallidum* (trep-ō-nē′mă pal′i-dŭm); can be spread by sexual contact; multiple disease stages occur; children born to infected mothers may be developmentally delayed; antibiotics are effective |
| Acquired immunodeficiency syndrome (AIDS) | Caused by the human immunodeficiency virus (HIV), which ultimately destroys the immune system (see chapter 14); transmitted through intimate sexual contact or by allowing infected body fluids into the interior of another person |

Go to **www.mhhe.com/seeleyess7** for additional information on these pathologies.

dysfunction. Keeping physically healthy can minimize some factors leading to abnormal erectile dysfunction, and medical treatments are available.

In women, the most significant age-related change is menopause. By age 50, the amount of estrogen and progesterone produced by the ovaries has decreased. The uterus decreases in size, and the endometrium decreases in thickness. The times between menses become irregular and longer until menstruation stops. The vaginal wall becomes thinner and less elastic, and there is less lubricant in the vagina, resulting in an increased tendency for vaginal yeast infections. However, wearing cotton underwear and loose clothing reduces this possibility.

In the event of infections, very effective medical treatments are available.

Approximately 10% of all women will develop breast cancer. The incidence of breast cancer is greatest between 45 and 65 years of age and is greater for women who have a family history of breast cancer. The single most important measure to guard against death from breast cancer is early detection through breast self-exams and yearly mammograms after age 40. The incidences of uterine cancer, ovarian cancer, and cervical cancer all increase between 50 and 65 years of age. Annual medical checkups, including Pap smears for cervical cancer, are important in order to detect cancer at early stages, when it can be easily treated.

# SUMMARY

## Functions of the Reproductive System    (p. 527)

The reproductive system produces male and female gametes, enhances fertilization of an oocyte by a sperm, nurtures the new individual until birth (in the female), and produces sex hormones.

## Formation of Gametes    (p. 527)

The reproductive organs in males and females produce gametes by meiosis.

### Meiosis

1. Two consecutive cell divisions halve the chromosome number from 46 total chromosomes to 23 total chromosomes.
2. Meiosis forms male and female gametes.

## Male Reproductive System    (p. 528)

### Scrotum

1. The scrotum is a sac containing the testes.
2. The dartos and cremaster muscles help regulate testes temperature.

### Testes

The testes are divided into lobules containing the seminiferous tubules and interstitial cells.

### Spermatogenesis

1. Spermatogenesis begins in the seminiferous tubules at the time of puberty.

2. Sustentacular cells nourish the sperm cells and produce small amounts of hormones.
3. Spermatogonia divide (mitosis) to form primary spermatocytes.
4. Primary spermatocytes divide by meiosis to produce first secondary spermatocytes and then spermatids. The spermatids then mature to form sperm cells.
5. Spermatids develop a head, midpiece, and flagellum to become a sperm cell. The head contains the acrosome and the nucleus.

### Ducts

1. The epididymis, a coiled tube system, is located on the testis and is the site of sperm maturation. Final changes, called capacitation of sperm cells, occur after ejaculation.
2. The seminiferous tubules lead to the rete testis, which opens into the efferent ductules that extend to the epididymis.
3. The ductus deferens passes from the epididymis into the abdominal cavity.
4. The ejaculatory duct is formed by the joining of the ductus deferens and the duct from the seminal vesicle.
5. The ejaculatory ducts join the prostatic urethra within the prostate gland.
6. The urethra extends from the urinary bladder through the penis to the outside of the body.

### Penis

1. The penis consists of erectile tissue.
2. The two corpora cavernosa form the dorsum and the sides.
3. The corpus spongiosum forms the ventral portion and the glans penis, and it encloses the spongy urethra. The prepuce covers the glans penis.

### Glands

1. The seminal vesicles empty into the ejaculatory duct.
2. The prostate gland consists of glandular and muscular tissue and empties into the urethra.
3. The bulbourethral glands empty into the urethra.

### Secretions

1. Semen is a mixture of sperm cells and gland secretions.
2. The bulbourethral glands and the urethral mucous glands produce mucus that neutralizes the acidic pH of the urethra.
3. The testicular secretions contain sperm cells.
4. The seminal vesicle fluid contains nutrients, prostaglandins, and proteins that coagulate.
5. The prostate fluid contains nutrients and proteolytic enzymes, and it neutralizes the pH of the vagina.

## Physiology of Male Reproduction    (p. 535)

### Regulation of Sex Hormone Secretion

1. GnRH is produced in the hypothalamus and released in surges.
2. GnRH stimulates release of LH and FSH from the anterior pituitary.
3. LH stimulates the interstitial cells to produce testosterone.
4. FSH binds to sustentacular cells and stimulates spermatogenesis and secretion of inhibin.
5. Testosterone has a negative-feedback effect on GnRH, LH, and FSH secretion.
6. Inhibin has a negative-feedback effect on FSH secretion.

### Puberty in Males

1. Before puberty, small amounts of testosterone inhibit GnRH release.
2. During puberty, testosterone does not completely suppress GnRH release, resulting in increased production of FSH, LH, and testosterone.

### Effects of Testosterone

1. Testosterone causes enlargement of the genitals and is necessary for spermatogenesis.
2. Testosterone is responsible for the development of secondary sexual characteristics.

### Male Sexual Behavior and the Male Sex Act

1. Testosterone is required for normal sex drive.
2. Stimulation of the sex act can be tactile or psychological.
3. Sensory impulses pass to the sacral region of the spinal cord.
4. Motor stimulation causes erection, mucus production, emission, and ejaculation.

### Infertility in Males

The most common cause of male infertility is a low sperm cell count.

## Female Reproductive System    (p. 539)

### Ovaries

1. By the fourth month of development, the ovaries contain 5 million oogonia.
2. By birth, many oogonia have degenerated, and for the remaining oogonia meiosis has stopped in prophase I, causing them to become primary oocytes.
3. By puberty, 300,000 to 400,000 primary oocytes remain, of which about 400 will be released from the ovaries.
4. Ovulation is the release of an oocyte from an ovary. The first meiotic division is completed, and a secondary oocyte is released.
5. A sperm cell penetrates the secondary oocyte, the second meiotic division is completed, and the nuclei of the oocyte and sperm cell are united to complete fertilization.
6. A primordial follicle is a primary oocyte surrounded by a single layer of flat granulosa cells.
7. In primary follicles, the oocyte enlarges, and granulosa cells become cuboidal and form more than one layer. A zona pellucida is present.
8. In a secondary follicle, fluid-filled vesicles appear, and a theca forms around the follicle.
9. In a mature follicle, vesicles fuse to form an antrum, and the primary oocyte is surrounded by cumulus cells.
10. During ovulation, the mature follicle ruptures, releasing the secondary oocyte, surrounded by cumulus cells, into the peritoneal cavity.
11. The remaining granulosa cells in the follicle develop into the corpus luteum.
12. If fertilization occurs, the corpus luteum persists. If there is no fertilization, it degenerates.

### Uterine Tubes

1. The ovarian end of the uterine tube is surrounded by fimbriae.
2. Cilia on the fimbriae move the oocyte into the uterine tube.
3. Fertilization usually occurs in the ampulla of the uterine tube, which is near the ovary.

### Uterus

1. The uterus is a pear-shaped organ. The uterine cavity and the cervical canal are the spaces formed by the uterus.
2. The wall of the uterus consists of the perimetrium, or serous layer; the myometrium (smooth muscle); and the endometrium.

### Vagina

1. The vagina connects the uterus (cervix) to the vestibule.
2. The vagina consists of a layer of smooth muscle and an inner lining of moist stratified squamous epithelium.
3. The wall of the vagina produces lubricating fluid.
4. The hymen covers the vestibular opening of the vagina in young females.

### External Genitalia

1. The vestibule is a space into which the vagina and the urethra open.
2. The clitoris is composed of erectile tissue and contains many sensory receptors important in detecting sexual stimuli.
3. The labia minora are folds that cover the vestibule and form the prepuce.
4. The greater vestibular glands produce a mucous fluid.
5. The labia majora cover the labia minora, and the pudendal cleft is a space between the labia majora.
6. The mons pubis is an elevated area superior to the labia majora.

### Mammary Glands

1. Mammary glands are the organs of milk production.
2. The mammary glands are modified sweat glands that consist of glandular lobes and adipose tissue.
3. The lobes connect to the nipple through ducts. The nipple is surrounded by the areola.
4. The female breast enlarges during puberty under the influence of estrogen and progesterone.

## Physiology of Female Reproduction    (p. 545)

### Puberty in Females

1. Puberty begins with the first menstrual bleeding (menarche).
2. Puberty begins when GnRH, LH, and FSH levels increase.

### Menstrual Cycle

1. The cyclical changes in the uterus are controlled by estrogen and progesterone produced by the ovary.
2. Menses (from day 1 to day 4 or 5): Menstrual fluid is produced by degeneration of the endometrium.
3. Proliferative phase (from day 5 to day of ovulation): Epithelial cells multiply and form glands.
4. Secretory phase (from day of ovulation to day 28): The endometrium becomes thicker, and endometrial glands secrete. The uterus is prepared for implantation of the developing blastocyst by day 21.
5. Estrogen stimulates proliferation of the endometrium, and progesterone causes thickening of the endometrium. Decreased progesterone causes menses.

6. FSH initiates the development of the follicles.
7. Estrogen produced by the follicles stimulates GnRH, FSH, and LH secretion, and FSH and LH stimulate more estrogen secretion. This positive-feedback mechanism causes FSH and LH levels to increase near the time of ovulation.
8. LH stimulates ovulation and formation of the corpus luteum.
9. Estrogen and progesterone inhibit LH and FSH secretion following ovulation.
10. If fertilization does not occur, progesterone secretion by the corpus luteum decreases and menses begins.
11. If fertilization does occur, the corpus luteum continues to secrete progesterone and menses does not occur.

### Menopause

The cessation of the menstrual cycle is called menopause.

### Female Sexual Behavior and the Female Sex Act

1. Female sex drive is partially influenced by testosterone-like hormones produced by the adrenal cortex and estrogen produced by the ovary.
2. Autonomic nerves cause erectile tissue to become engorged with blood, the vestibular glands to secrete mucus, and the vagina to produce a lubricating fluid.

### Infertility in Females

Causes of infertility in females include malfunctions of the uterine tubes, reduced hormone secretion from the pituitary or ovary, and interruption of implantation.

## Effects of Aging on the Reproductive System    (p. 552)

1. Benign prostatic enlargement affects men as they age, and it blocks urine flow through the prostatic urethra.
2. Prostate cancer is more common in elderly men.
3. Menopause is the most common age-related change in females.
4. Cancers of the breast, the cervix, and the ovaries increase in elderly women.
5. Early detection is key to the prevention of most cancers.

# REVIEW AND COMPREHENSION

1. List the functions of the male and female reproductive systems.
2. What is the scrotum? Explain the function of the dartos and cremaster muscles.
3. Where, specifically, are sperm cells produced in the testes? Describe the process of spermatogenesis.
4. Name the ducts the sperm cells traverse from their site of production to the outside of the body.
5. In which duct do sperm cells develop their ability to fertilize?
6. Name the erectile tissues of the penis, and describe how erectile tissue becomes erect.
7. State where the seminal vesicles, prostate gland, and bulbourethral glands empty into the male reproductive duct system.
8. Define emission and ejaculation.
9. Define semen. What structures give rise to secretions that make up the semen? Describe the composition of the secretions of each gland.
10. Describe where GnRH, FSH, LH, and testosterone are produced, and explain how their secretion is regulated.
11. Describe the effects of testosterone on males during puberty and in adulthood.
12. Describe the regulation of the male sex act.

13. Describe the process of follicle development and ovulation.
14. What is the corpus luteum? What happens to the corpus luteum if fertilization occurs? If fertilization does not occur?
15. Describe the normal pathway the oocyte follows after ovulation. Where does fertilization usually take place?
16. Describe the relationship among the uterus, vagina, vestibule, and external genitalia.
17. Describe the labia minora, the prepuce, the labia majora, the pudendal cleft, and the mons pubis.
18. What are the effects of estrogen and progesterone on the uterus?
19. Describe the hormonal changes that result in ovulation. Explain the sequence of events during each phase of the menstrual cycle.
20. Define menopause and female climacteric. What causes these changes?
21. List the major age-related diseases and conditions that occur in the male reproductive system.
22. List the major age-related changes that occur in the female reproductive system.
23. List the major age-related types of cancer that occur in the female reproductive system.

## ⎯⎯√⎮⎯⎯ CRITICAL THINKING

1. If an adult male were castrated by having his testes removed, what would happen to the levels of GnRH, FSH, LH, and testosterone in his blood?

2. Birth control pills for women contain estrogen and progesterone compounds. Explain how these hormones can prevent pregnancy.

3. During the secretory phase of the menstrual cycle, you would normally expect
   a. the highest levels of progesterone that occur during the menstrual cycle.
   b. a follicle present in the ovary that is ready to undergo ovulation.
   c. the endometrium to reach its greatest degree of development.
   d. a and b
   e. a and c

4. During approximately days 12–14 of the menstrual cycle, you would normally expect
   a. increasing blood levels of estrogen.
   b. increasing blood levels of LH.

   c. blood levels of progesterone are near their maximum.
   d. a and b
   e. a, b, and c

5. On day 15 of the menstrual cycle, you would normally expect
   a. decreasing blood levels of LH.
   b. decreasing blood levels of estrogen.
   c. increasing blood levels of progesterone.
   d. a and b
   e. a, b, and c

6. Predict the consequences if a drug that blocks the effect of progesterone is taken by a woman 2 or 3 days following ovulation or by a woman who is pregnant.

7. During menopause, which reproductive hormones are reduced in the blood and which are increased?

*Answers in Appendix D*

Visit this textbook's website at **www.mhhe.com/seeleyess7** for practice quizzes, animations, interactive learning exercises, and other study tools.

McGraw-Hill offers a study CD that features interactive cadaver dissection. ***Anatomy & Physiology Revealed*** includes cadaver photos that allow you to peel away layers of the human body to reveal structures beneath the surface. This program also includes animations, radiologic imaging, audio pronunciations, and practice quizzing.

Human development begins when a sperm cell fertilizes an egg cell. The fertilized egg undergoes a series of cell divisions, which lead to the eventual formation of a human body. Development continues into adulthood.

Chapter 20

# Development, Heredity, and Aging

## Chapter Outline and Objectives

### Prenatal Development (p. 559)

1. List the prenatal periods, and state the major developmental events associated with each.

2. Describe the process of fertilization.

3. Describe the blastocyst, the process of implantation, and placental formation.

4. Describe the maternal hormonal changes that occur during pregnancy.

5. List the three germ layers, describe their formation, and list the adult derivatives of each layer.

6. Describe the formation of the neural tube and the neural crest cells.

7. Describe the formation of the limbs, the face, and the digestive tract.

8. Explain how the single heart tube is divided into four chambers.

### Parturition (p. 572)

9. Explain the events that occur during parturition.

### The Newborn (p. 574)

10. Discuss the respiratory, circulatory, and digestive changes that occur in the newborn at the time of birth.

### Lactation (p. 576)

11. Describe the events of lactation.

### First Year Following Birth (p. 577)

12. Describe the changes that occur during the first year after birth.

### Life Stages (p. 577)

13. List the stages of life, and describe the major events associated with each stage.

14. Describe the process of aging.

15. Describe the events that occur at the time of death.

### Genetics (p. 579)

16. Define genetics, and explain how chromosomes are related to inheritance.

17. Describe the major types of inheritance.

The stages of life and the activities associated with those stages are issues of great interest in today's society. We view life stages very differently today than we did in the past. For example, in 1960, 20% of males and 12% of females graduating from high school attended college. Today, over half of all people 25 and older have attended college for some period of time. In addition, there are many more nontraditional college students than there were just a few years ago.

In 1900, only 5% of the U.S. population was over age 65. Today, about 16% of the population is over age 65, and by 2030 more than 20% will be older than 65. The average life expectancy in 1900 was about 47 years, in 1940 it was about 63 years, and today it is about 78 years. In 1900, nearly 70% of all males over age 65 were still working; today, only about 20% are still working past age 65. Older people are healthier and more active than they have ever been.

The human life span is usually considered the period between birth and death; however, the 9 months before birth are a critical part of existence. What happens in these 9 months profoundly affects the rest of a person's life. Although most people develop normally and are born without defects, approximately 3 of every 100 people are born with a birth defect so severe that it requires medical attention during the first year of life. Later in life, many more people discover previously unknown problems, such as the tendency to develop asthma, certain brain disorders, or cancer.

# PRENATAL DEVELOPMENT

The **prenatal** (prē-nā′tăl; before birth) **period,** the period from conception to birth, can be divided into three parts: (1) During approximately the first 2 weeks of development, the primitive germ layers are formed; (2) from about the second to the eighth week of development, the major organ systems come into existence; and (3) during the last 7 months of the prenatal period, the organ systems grow and become more mature. Between the time of fertilization and 8 weeks of development, the developing human is called an **embryo** (em′brē-ō). From 8 weeks to birth, the developing human is called a **fetus** (fē′tus; offspring).

To calculate the **clinical age** of an unborn child, the medical community uses the mother's **last menstrual** (men′stroo-ăl) **period (LMP).** An embryo or a fetus is therefore considered to be a certain number of days post-LMP. Most embryologists, on the other hand, use **developmental age,** which begins with fertilization, to describe the timing of developmental events. Because fertilization is assumed to occur approximately 14 days after LMP, the developmental age is 14 days less than the clinical age. The times presented in this chapter are based on developmental age.

## Fertilization

**Fertilization** is the union of a sperm cell and an oocyte, along with their genetic material (chromosomes) to produce a new individual. After sperm cells are ejaculated into the vagina, they are transported through the cervix and the body of the uterus to the uterine tubes, where fertilization occurs. The swimming ability of the sperm cells and the muscular contractions of the uterus and uterine tubes are responsible for the movement of sperm cells through the female reproductive tract. Both oxytocin released by the female posterior pituitary and prostaglandins within the semen stimulate contractions in the uterus and uterine tubes.

While passing through the uterus and the uterine tubes, the sperm cells undergo capacitation. **Capacitation** (kă-pas′i-tā′shŭn) makes the sperm cells capable of releasing the concentrated enzymes contained in the acrosome, a region of the sperm cell head. The enzymes digest a pathway through the cumulus cells and the zona pellucida of the **secondary oocyte** (ō′ō-sīt; egg cell). One sperm cell attaches to the oocyte cell membrane and enters the oocyte (figure 20.1, step 1).

The secondary oocyte is capable of being fertilized for up to about 1 day after ovulation, and some sperm cells remain viable in the female reproductive tract for up to 6 days, although most of them degenerate before that time.

**Predict 1**

*During what days of the menstrual cycle is sexual intercourse most likely to result in pregnancy?*

Hundreds of sperm cells reach the secondary oocyte, but normally a change in the oocyte cell membrane prevents more than one sperm cell from entering the secondary oocyte. The secondary oocyte undergoes the second meiotic division only after a sperm cell enters it (figure 20.1, step 2). After the second meiotic division, the oocyte nucleus moves to the center of the cell, where it meets the nucleus of the sperm cell. Each of these nuclei has 23 chromosomes. Their fusion, which completes the process of fertilization, produces a **zygote** (zī′gōt; having a yoke) that has 46 chromosomes (figure 20.1, step 3). The zygote develops into the embryo.

## Early Cell Division

About 18–36 hours after fertilization, the zygote divides to form two cells (figure 20.2). Those two cells divide to form four cells, which divide to form eight, and so on. Even though the number of cells increases, the size of each cell decreases, so that the total mass of cells remains about the same size as the zygote. These cells have the ability to develop into a wide range of tissues. As

1. Many sperm cells come in contact with the cumulus cells of the secondary oocyte, but only one sperm cell will penetrate the oocyte's cell membrane.

Cumulus cells

Sperm cell contacting oocyte cell membrane

Oocyte nucleus

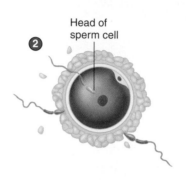

2. The head of the sperm, carrying the genetic material, enters the oocyte, which completes the second meiotic division.

Head of sperm cell

3. The two nuclei fuse to form a single nucleus. Fertilization is complete, and a zygote results.

Single nucleus

**PROCESS Figure 20.1** Fertilization

a result, the total number of cells can be decreased, increased, or reorganized during this period without affecting normal development.

## A CASE IN POINT

### Twins

The parents of twin boys named the babies Juan and Hamall because, at birth, they looked so much alike that, if you've seen Juan, you've seen Hamall. As the twins grew older, they continued to look very much alike but not exactly the same. Their parents wanted to know if they were identical or fraternal twins, so they had an expensive genetic analysis performed. The results indicated that the twins were identical, despite their slight differences in appearance.

In rare cases, following early cell divisions, the cells separate and form two individuals, called "identical twins," or **monozygotic** (mon-ō-zī-got′ik) **twins.** Identical twins therefore have identical genetic information in their cells. Identical twins can also occur by other mechanisms a little later in development.

Occasionally, a woman can ovulate two or more secondary oocytes at the same time. Fertilization of multiple oocytes by different sperm cells results in "fraternal twins," or **dizygotic** (dī′zī-got′ik) **twins.**

Even though who we are and what we look like are determined in large part by genes, identical twins—although they have the same genes—may not look exactly alike. In fact, identical twins often look more like mirror images of each other. Genes interact with minute environmental cues in the embryo and in the child to determine the final form of the individual.

**Figure 20.2  Development of the Blastocyst and Implantation**

Successive cell divisions produce a multicellular morula by day 5, which becomes a hollow blastocyst on day 6. In the figure of the blastocyst, *green* cells are trophoblastic and *orange* cells are embryonic.

## Blastocyst

After fertilization, multiple cell divisions have produced an embryonic mass of about 12–16 cells, called a **morula** (mōr′ū-lă; mulberry) (figure 20.2). Most of the cells of the morula will not form the embryo proper but will form support structures, such as the placenta.

When a cavity begins to appear within the mass of cells, the whole structure is called a **blastocyst** (blas′tō-sist) (figure 20.2). The fluid-filled cavity is called the **blastocele** (blas′tō-sēl). Most of the blastocele is surrounded by a single layer of cells, but at one end of the blastocyst, the cells are several layers thick. This thickened area is called the **inner cell mass.** Not all cells of the blastocyst give rise to the new individual. The embryo proper, which will become the new individual, will develop from only a few cells of the inner cell mass. These cells are commonly referred to as **stem cells** because they give rise to all the cell types within the body. The remaining cells of the blastocyst are called the **trophoblast** (trof′ō-blast, trō′fō-blast), which forms the embryonic part of the placenta and the membranes (chorion and amnion) surrounding the embryo.

## Implantation of the Blastocyst and Development of the Placenta

All these early events, from the first cell division to formation of the blastocele, occur as the embryonic mass moves from the site of

fertilization in the uterine tube to the site of **implantation** (im-plan-tā′shun) in the uterus. By 7 or 8 days after ovulation, the endometrium of the uterus is prepared for implantation. About 7 days after fertilization, the blastocyst attaches itself to the uterine wall and begins the process of implantation. The trophoblast cells of the blastocyst digest the uterine tissues as the blastocyst burrows into the uterine wall. Before implantation and for a short time afterward, the embryo is insensitive to environmental toxins. During the first few days of development, each cell has enough yolk to supply its own energy needs and requires few external nutrients. Furthermore, during the first couple of weeks of development, large numbers of cells can die, yet the embryo can fully recover.

As the blastocyst burrows into the uterine wall, trophoblast cells, called the **chorion** (kō′rē-on), form the embryonic portion of the **placenta** (plă-sen′tă), the organ of nutrient and waste product exchange between the embryo and the mother. Fingerlike projections, called **chorionic villi,** protrude into cavities formed within the maternal endometrium. Those cavities, called **lacunae** (lă-koo′nē), are filled with maternal blood (figure 20.3). In the mature placenta, the embryonic blood supply is separated from the maternal blood supply by the embryonic capillary wall, a basement membrane, and a thin layer of chorion. As a result, the embryonic blood and maternal blood do not mix. Nutrients and waste products must cross this semipermeable barrier between the two circulations.

# CLINICAL IMPACT

## In Vitro Fertilization and Embryo Transfer

In a small number of women, normal pregnancy is not possible because of an anatomical or physiological condition. In 87% of these cases, the uterine tubes are incapable of transporting the zygote to the uterus or allowing sperm cells to reach the oocyte. Since 1978, in vitro fertilization and embryo transfer have made pregnancy possible in hundreds of such women. **In vitro fertilization (IVF)** involves removing secondary oocytes from a woman, placing the oocytes into a petri dish, and adding sperm cells to the dish, where fertilization and early development occur in vitro, which means "in glass." **Embryo transfer** involves removing the developing embryo from the petri dish and introducing the embryo into the uterus of a recipient woman.

For in vitro fertilization and embryo transfer to be accomplished, a woman is first injected with a substance similar to luteinizing hormone (LH), which causes more than one follicle to ovulate at a time. Just before the follicles rupture, the secondary oocytes are surgically removed from the ovary. The oocytes are then incubated in a dish and maintained at body temperature for 6 hours. Then, sperm cells are added to the dish.

After 24–48 hours, several of the embryos are transferred to the uterus. Several embryos are introduced into the uterus to increase the success rate as much as possible, because only a few of them survive. The woman is usually required to lie perfectly still for several hours after the embryos have been introduced into the uterus to prevent possible expulsion before implantation can occur. It is not fully understood why such expulsion does not occur in natural fertilization and implantation. Implantation and subsequent development then proceed in the uterus as they would for natural implantation.

The success rate of embryo transfer varies from clinic to clinic, with the age of the embryo at the time of transfer, and with the age of the patient. The current success rate for achieving pregnancy following IVF is 31%. Of these pregnancies, 83% result in live births. Multiple births have occurred frequently following embryo transfer because of the practice of introducing more than one embryo into the uterus.

**Figure 20.3 Interface Between Maternal and Fetal Circulation**

(a) Location of the placenta and umbilical cord in relation to the fetus. (b) As the chorion encounters maternal blood vessels, lacunae (cavities) form and become filled with maternal blood. In the mature placenta, the embryonic blood vessels and other tissue form chorionic villi, and nutrients are exchanged between embryonic and maternal blood. (c) Under normal conditions, the maternal and fetal blood are separated by the chorion and a basement membrane and do not mix.

Initially, the embryo is attached to the placenta by a connecting stalk. As the embryo matures, the connecting stalk elongates and becomes the **umbilical** (ŭm-bil′i-kăl; navel) **cord** (figure 20.3b). Within the umbilical cord, blood vessels carry blood from the embryo to the placenta and from the placenta to the embryo.

## Maternal Hormonal Changes

The chorion secretes **human chorionic gonadotropin** (gō′nad-ō-trō′pin) **(hCG),** which travels in the blood to the maternal ovary and causes the corpus luteum to remain functional. The secretion of hCG begins shortly after implantation, increases rapidly, and reaches a peak about 8 or 9 weeks after fertilization.

Subsequently, hCG levels decline to a lower level, where they are maintained throughout the remainder of the pregnancy (figure 20.4). Most pregnancy tests are designed to detect hCG in either urine or blood.

The estrogen and progesterone secreted by the corpus luteum (see chapter 19) are essential for maintaining the endometrium for the first 3 months of pregnancy. After the placenta forms, it also begins to secrete estrogen and progesterone. By the third month of pregnancy, the placenta has become an endocrine gland that secretes sufficient quantities of estrogen and progesterone to maintain pregnancy, and the corpus luteum is no longer needed. Estrogen and progesterone levels increase in the mother's blood throughout pregnancy.

**1** Human chorionic gonadotropin (*red line*) (hCG) increases until it reaches a maximum concentration near the end of the first 3 months of pregnancy and then decreases to a low level thereafter.

**2** Progesterone (*blue line*) continues to increase until it levels off near the end of pregnancy. Early in pregnancy, progesterone is produced by the corpus luteum in the ovary; by the second trimester, production shifts to the placenta.

**3** Estrogen (*green line*) increases slowly throughout pregnancy but increases more rapidly as the end of pregnancy approaches. Early in pregnancy, estrogen is produced only in the ovary; by the second trimester, production shifts to the placenta.

**PROCESS Figure 20.4   Changes in Hormone Concentration During Pregnancy**

During pregnancy, hCG, progesterone, and estrogen are secreted. The placenta secretes hCG. Early in pregnancy, the ovary secretes estrogen and progesterone. During midpregnancy, there is a shift toward estrogen and progesterone secretion by the placenta.

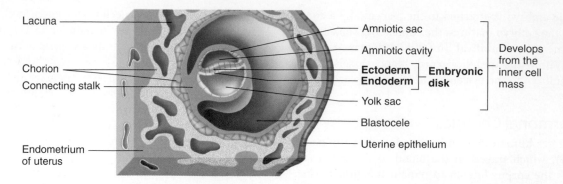

**Figure 20.5** Early Embryo and Surrounding Structures in the Placenta

The embryonic disk consists of ectoderm and endoderm surrounded by the amniotic cavity and yolk sac. The connecting stalk, which attaches the embryo to the uterus, becomes part of the umbilical cord.

## Formation of the Germ Layers

After implantation, a new cavity, called the **amniotic** (am-nē-ot′ik) **cavity,** forms inside the inner cell mass and causes the part of the inner cell mass nearest the blastocele to separate as a flat disk of tissue called the **embryonic disk** (figure 20.5). The amniotic cavity is bounded by a membrane called the **amniotic sac** and is filled with **amniotic fluid.** The embryo will grow in the amniotic cavity, where the amniotic fluid forms a protective cushion. The embryonic disk at first is composed of two layers of cells: an **ectoderm** (ek′tō-derm; outside layer) adjacent to the amniotic cavity and an **endoderm** (en′dō-derm; inside layer) on the side of the disk opposite the amniotic cavity. A third cavity, the **yolk sac,** forms inside the blastocele from the endoderm.

At about 14 days after fertilization, the embryonic disk has become a slightly elongated, oval structure. Some of the ectoderm cells migrate toward the center of the disk, forming a thickened line called the **primitive streak** (figure 20.6). The formation of the primitive streak establishes the central axis of the embryo. Some of the ectoderm cells migrate through the primitive streak and emerge between the ectoderm and endoderm as a new germ layer, called the **mesoderm** (mez′ō-derm; middle layer). This process of cell migration and the formation of mesoderm is called **gastrulation.** The embryo is now three-layered, having ectoderm, mesoderm, and endoderm (figure 20.6). All the tissues of an adult can be traced to these three germ layers (table 20.1).

From about day 14 until about day 35, the embryo is at maximum risk from environmental toxins and drugs that can cause birth defects. The causes of birth defects are a major

1. Cells in the surface ectoderm move toward the primitive streak and migrate through the streak (*blue arrow tails*).

2. Cells of the ectoderm that migrate through the primitive streak become mesodermal cells (*red arrows*).

3. The mesoderm (*pink*) lies between the ectoderm (*blue*) and the endoderm (*yellow*).

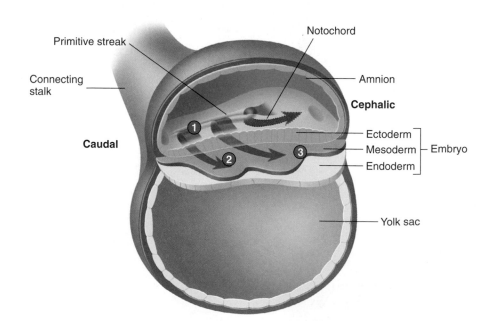

**PROCESS Figure 20.6** Primitive Streak and Germ Layers

The head of the embryo develops over the notochord.

## TABLE 20.1    Tissues Derived from Each Germ Layer

| Endoderm | Ectoderm | Mesoderm |
|---|---|---|
| Lining of digestive tract | Epidermis of skin | Dermis of skin |
| Lining of lungs | Tooth enamel | Heart and blood vessels |
| Lining of hepatic, pancreatic, and other exocrine ducts | Lens and cornea of eye | Parenchyma (substance) of glands |
| Kidney ducts and bladder | Outer ear | Kidneys |
| Thymus | Nasal cavity | Gonads |
| Thyroid gland | Anterior pituitary | Muscle |
| Parathyroid glands | Neuroectoderm | Bones (except facial) |
| Tonsils |    Brain and spinal cord | Microglia |
| |    Somatic motor neurons | |
| |    Preganglionic autonomic neurons | |
| |    Posterior pituitary | |
| |    Neuroglial cells (except microglia) | |
| | Neural crest cells | |
| |    Melanocytes | |
| |    Sensory neurons | |
| |    Postganglionic autonomic neurons | |
| |    Adrenal medulla | |
| |    Facial bones | |
| |    Teeth: dentin and pulp | |
| |    A few skeletal muscles in head | |

unsolved issue in biology at present. However, it appears that oxidative damage to key molecules in certain developing cells and/or changes in the DNA function within those cells may be involved.

A specialized group of cells at the cephalic end of the primitive streak moves from one end of the primitive streak to the other and, in some as yet unknown way, organizes the embryo. A cordlike structure called the **notochord** (nō′tō-kōrd) is formed by these cells as they move down the primitive streak. The notochord marks the central axis of the developing embryo (figure 20.6).

### Predict 2

*Predict the result if two primitive streaks form in one embryonic disk. What if the two primitive streaks are touching each other?*

## Neural Tube and Neural Crest Formation

At about 18 days after fertilization, the ectoderm overlying the notochord thickens to form the **neural plate**. The lateral edges of the plate begin to rise like two ocean waves coming together. These edges are called the **neural folds,** and between them lies a **neural groove.** The neural folds begin to meet in the midline and fuse into a **neural tube** (figure 20.7). The cells of the neural tube are called

**neuroectoderm** (noor-ō-ek′tō-derm) (table 20.1). Neuroectoderm becomes the brain, the spinal cord, and parts of the peripheral nervous system. The neural tube becomes completely closed by day 26. If the neural tube fails to close, major defects of the central nervous system can result.

**Anencephaly** (an′en-sef′ă-lē; no brain) is a birth defect wherein much of the brain fails to form because the neural tube did not close in the region of the head. A baby born with anencephaly cannot survive. **Spina bifida** (spī′nă bif′i-dă; split spine) is a general term describing defects of the spinal cord or vertebral column. Spina bifida can range from a simple defect with one or more vertebral spinous processes split or missing but no clinical manifestation to a more severe defect that can result in paralysis of the limbs or the bowels and bladder, depending on where the defect occurs. More severe forms of spina bifida result from failure of the neural tube in the area of the spinal cord to close. It has been demonstrated that adequate amounts of folic acid, the B vitamin folate, in the diet during pregnancy can reduce the risk of such defects.

As the neural folds come together and fuse, a population of cells breaks away from the neuroectoderm all along the crests of the folds. Most of these **neural crest cells** become part of the peripheral nervous system or become melanocytes in the skin. In the head, neural crest cells also contribute to the skull, the dentin of teeth, blood vessels, and general connective tissue.

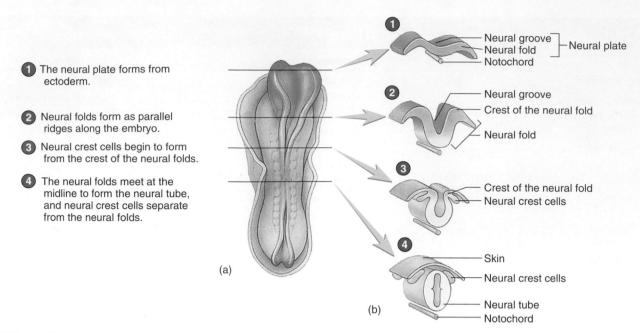

**1** The neural plate forms from ectoderm.

**2** Neural folds form as parallel ridges along the embryo.

**3** Neural crest cells begin to form from the crest of the neural folds.

**4** The neural folds meet at the midline to form the neural tube, and neural crest cells separate from the neural folds.

(a)

**1** Neural groove
Neural fold ⎤ Neural plate
Notochord

**2** Neural groove
Crest of the neural fold
Neural fold

**3** Crest of the neural fold
Neural crest cells

**4** Skin
Neural crest cells
Neural tube
Notochord

(b)

**PROCESS Figure 20.7  Formation of the Neural Tube**

The neural folds, which consist of neuroectoderm, come together at the midline and fuse to form a neural tube. This fusion begins in the center and moves both cranially and caudally. (*a*) The embryo shown is about 21 days after fertilization. (*b*) These cross sections represent progressive closure of the neural tube.

## Formation of the General Body Structure

Arms and legs first appear at about 28 days after fertilization as **limb buds** (figure 20.8) and quickly begin to elongate. At about 35 days, expansions called hand and foot plates form at the ends of the limb buds. Zones of cell death between the future fingers and toes of the hand and foot plates help sculpt the fingers and toes.

Eye

Umbilical cord

Lower limb bud

Heart

Upper limb bud

**Figure 20.8  Human Embryo 35 Days After Fertilization**

The face develops by fusion of five growing masses of tissue, called **processes** (figure 20.9). One, the **frontonasal process,** forms the forehead, nose, and center of the upper jaw and lip. Two **maxillary processes** form the maxillae (upper lip and jaw), and two **mandibular processes** form the mandible (lower lip and jaw).

The nose begins as two structures, one on each side of the forehead mass. As the brain enlarges and the face matures, the two parts of the nose approach each other in the midline and fuse (figure 20.9). The two masses forming the upper jaw expand toward the midline and fuse with part of the nose to form the upper jaw and lip. A **cleft lip** results from failure of these structures to fuse. Cleft lips usually do not occur in the midline, but to one side (or both sides). The cleft can vary in severity from a slight indentation in the lip to a fissure that extends from the mouth to the naris (nostril).

The roof of the mouth, or palate, begins to form as vertical shelves of tissue that grow on the inside of the maxillary masses. These shelves swing to a horizontal position and begin to fuse with each other at about 56 days of development. If the palate does not fuse, a midline cleft in the roof of the mouth called a **cleft palate** results. A cleft palate can range in severity from a slight cleft of the uvula to a fissure extending the entire length of the palate. A cleft lip and cleft palate can occur together, forming a continuous fissure.

## Development of the Organ Systems

The major organ systems appear and begin to develop during the embryonic period (second to eighth week of development). This

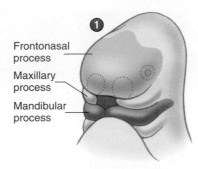

**1  28 days after fertilization**
The face develops from five
processes: frontonasal (*blue*),
two maxillary (*yellow*), and
two mandibular (*orange;*
*already fused*).

Frontonasal
process

Maxillary
process

Mandibular
process

**2  33 days after fertilization**
Nasal placodes appear in
the frontonasal process.

Eye

Frontonasal
process

Nose
beginning
to form

Maxillary
process

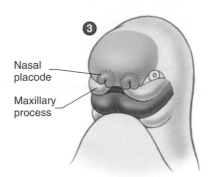

**3  40 days after fertilization**
Maxillary processes extend
toward the midline. The nasal
placodes also move toward
the midline and fuse with the
maxillary processes to form
the jaw and lip.

Nasal
placode

Maxillary
process

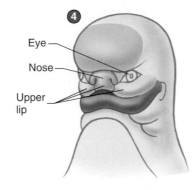

**4  48 days after fertilization**
Continued growth brings
structures more toward the
midline.

Eye

Nose

Upper
lip

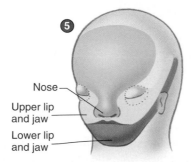

**5  14 weeks after fertilization**
*Colors* show the contributions
of each process to the adult
face.

Nose

Upper lip
and jaw

Lower lip
and jaw

**PROCESS Figure 20.9**  Development of the Face

period is therefore also called the period of **organogenesis**
(ōr′gă-nō-jen′ĕ-sis). The individual organ systems are listed in
table 20.2; only general comments about a few select systems are
presented in the text.

While the neural tube is forming (18–26 days), the remainder of the embryo is folding to form a tube along the upper part of
the yolk sac (figure 20.10). The developing digestive tract pinches
off from the yolk sac as a tube but remains attached in the center
to the yolk sac by a yolk stalk.

A considerable number of outpocketings appear at about
28 days after fertilization along the entire length of the digestive
tract (figure 20.11). A surprisingly large number of important
internal organs develop from those outpocketings, including the
auditory tubes, tonsils, thymus, anterior pituitary gland, thyroid
gland, parathyroid glands, lungs, liver, pancreas, and urinary
bladder.

The heart develops from two blood vessels, which lie side by
side in the early embryo and fuse about 21 days after fertilization
into a single, midline heart (figure 20.12, steps 1 and 2). At about
this time, the primitive heart begins to beat. Blood vessels form
from "blood islands" on the surface of the yolk sac and inside the
embryo. These islands expand and fuse to form the circulatory
system.

The major chambers of the heart, the atrium and ventricle,
expand rapidly. The single ventricle is subdivided into two chambers by the development of an **interventricular** (in-ter-ven-trik′ŭ-
lăr) **septum** (figure 20.12, steps 3 and 4). If the interventricular
septum does not grow enough to completely separate the ventricles, a ventricular septal defect results.

An **interatrial** (in-ter-ā′trē-ăl) **septum** forms to separate
the two atria (figure 20.12, steps 3–5). An opening in the interatrial septum called the **foramen ovale** (ō-val′ē) connects the
two atria and allows blood to flow from the right to the left
atrium in the fetus. Because of the foramen ovale, some blood in
the fetus passes from the right atrium to the left atrium and
bypasses the right ventricle and the lungs. The foramen ovale
normally closes off at the time of birth, and blood then circulates through the right ventricle and the lungs. If this does not
occur, an interatrial septal defect occurs. An interatrial septal
defect or a ventricular septal defect usually results in a heart
murmur.

The kidneys develop from mesoderm located along the lateral wall of the body cavity (see figure 20.11). The embryonic
kidneys are much more extensive than the adult kidneys, extending the entire length of the body cavity. They are closely associated
with internal reproductive organs, such as the ovaries or testes,
and reproductive ducts, such as the uterine tubes or ductus deferens. Most of the embryonic kidneys degenerate, with only a very
small part forming the adult kidney.

## Growth of the Fetus

The embryo becomes a fetus about 8 weeks after fertilization
(figure 20.13). The beginning of the fetal period is marked
by the beginning of bone ossification. In the embryo, most of
the organ systems are developing, whereas in the fetus the
organs are present. During the fetal period, the organ systems

## TABLE 20.2  Development of the Organ Systems

| | Age (Days Since Fertilization) | | | | | |
|---|---|---|---|---|---|---|
| | **1–5** | **6–10** | **11–15** | **16–20** | **21–25** | **26–30** |
| **General Features** | Fertilization, blastocyst | Blastocyst implants. | Primitive streak, three germ layers | Neural plate | Neural tube closed | Limb buds and other "buds" appear. |
| **Integumentary System** | | | Ectoderm, mesoderm | | Melanocytes form from neural crest. | |
| **Skeletal System** | | | Mesoderm | | Neural crest cells (form facial bones) | Limb buds |
| **Muscular System** | | | Mesoderm | Somites (body segments) begin to form. | | Somites are all formed. |
| **Nervous System** | | | Ectoderm | Neural plate | Neural tube complete; neural crest forms; eyes and ears begin to form. | Lens begins to form. |
| **Endocrine System** | | | Ectoderm, mesoderm, endoderm | Thyroid begins to develop. | | Parathyroid glands and pancreas appear. |
| **Cardiovascular System** | | | Mesoderm | Blood islands form; two-tubed heart forms. | Single-tubed heart begins to beat. | Interatrial septum forms. |
| **Lymphatic System** | | | Mesoderm | | | Thymus appears. |
| **Respiratory System** | | | Mesoderm, endoderm | | Diaphragm begins to form. | Trachea, lung buds |
| **Digestive System** | | | Mesoderm, endoderm | | Tooth dentin forms; foregut and hindgut form. | Liver and pancreas appear as buds; tongue bud appears. |
| **Urinary System** | | | Mesoderm, endoderm | | Embryonic kidneys appear. | Embryonic kidneys elongate. |
| **Reproductive System** | | | Mesoderm, endoderm | | Primordial germ cells form on yolk sac. | Male reproductive ducts appear; external genital structures begin to form. |

enlarge and mature. The fetus grows from about 3 cm and 2.5 g (0.09 oz) at 8 weeks to 50 cm and 3300 g (7 lb, 4 oz) at the end of pregnancy. The growth during the fetal period represents more than a 15-fold increase in length and a 1400-fold increase in weight.

Fine, soft hair called **lanugo** (lă-noo′gō) covers the fetus, and a waxy coat of loose epithelial cells called **vernix caseosa** (ver′niks kā′sē-ō′să) forms a protective layer between the fetus and the amniotic fluid. The amniotic fluid contains toxic waste products from the fetus's digestive tract and kidneys.

| Age (Days Since Fertilization) | | | | | |
|---|---|---|---|---|---|
| 31–35 | 36–40 | 41–45 | 46–50 | 51–55 | 56–60 |
| Hand and foot plates on limbs | Fingers and toes appear; lips form; embryo 15 mm. | External ear forming; embryo 20 mm | Embryo 25 mm | Limbs elongate to adult proportions; embryo 35 mm | Face is distinctly human in appearance. |
| Sensory receptors appear in skin. | | Collagen fibers are clearly present in skin. | | Extensive sensory nerve endings in skin | |
| Mesoderm condensation in areas of future bone | Cartilage in site of future humerus | Cartilage in site of future ulna and radius | Cartilage in site of future hand and fingers | | Ossification begins in clavicle and then in other bones. |
| Muscle precursor cells enter limb buds. | | | Functional muscle | | Nearly all muscles appearing in an adult form |
| Nerve processes enter limb buds. | | External ear forming; olfactory nerves begin to form | | Semicircular canals in inner ear are complete. | Eyelids form; cochlea in inner ear is complete. |
| Pituitary appears as evaginations from brain and mouth. | Gonads begin to form; adrenal glands form. | | Pineal body appears. | Thyroid gland in adult position | Anterior pituitary loses its connection to mouth. |
| Interventricular septum begins to form. | | Interventricular septum is complete. | Interatrial septum is complete but foramen ovale remains until birth. | | |
| Large lymphatic vessels form in neck. | Spleen appears. | | | Adult lymph pattern is formed. | |
| Secondary bronchi to lobes form. | Tertiary bronchi to lobules form. | | Tracheal cartilage begins to form. | | |
| Mouth opens to outside. | | Palate begins to form; tooth buds begin to form. | | | Palate begins to fuse (fusion complete by 90 days); anus opens. |
| Adult kidneys begin to develop. | | | | Embryonic kidneys degenerate. | |
| | Gonads begin to form. | Primordial germ cells enter gonads. | Female reproductive ducts appear. | | Uterus is forming; external genitalia begin to differentiate in male and female. |

Subcutaneous fat that accumulates in the fetus provides a nutrient reserve, helps insulate, and aids the newborn in sucking by strengthening and supporting the cheeks, so that a small vacuum can be developed in the oral cavity.

Peak body growth occurs late in gestation, but as the placenta reaches its maximum size, the oxygen and nutrient supply to the fetus becomes limited. Growth of the placenta essentially stops at about 35 weeks, limiting fetal growth.

At approximately 38 weeks of development, the fetus is ready to be delivered. The average weight at this point is 3250 g (7 lb, 2 oz) for a female fetus and 3300 g (7 lb, 4 oz) for a male fetus.

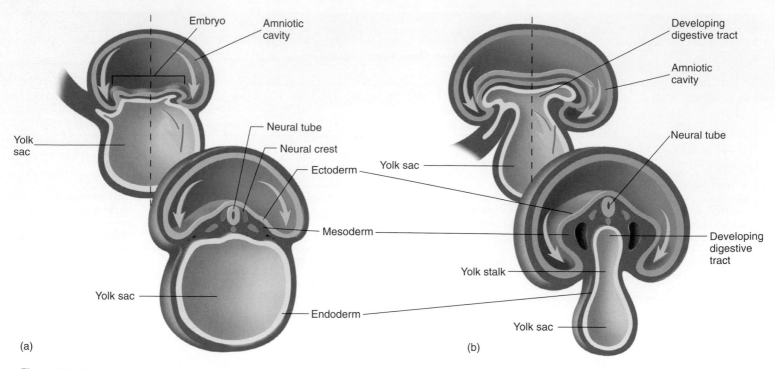

**Figure 20.10  Development of the Digestive Tract**

The digestive tract develops along the dorsal side of the yolk sac (*yellow*) as the body folds into a tube (*blue arrows*). The figures in back are shown in sagittal section. The figures in front are shown in cross section. The *dashed line* on the figures in back shows the plane of section in the figures in front. (*a*) An early embryo (about 24 days). (*b*) A slightly older embryo (about 28 days).

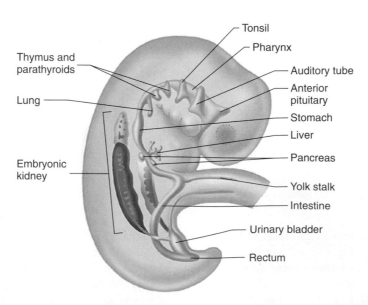

**Figure 20.11  Embryonic Digestive and Urinary Systems**

Outpocketings of the digestive tract (*yellow*), form many adult structures, such as the lungs and glands. The embryonic kidney is also shown (*purple*).

**1** **20 days after fertilization**
At this age, the heart consists of
two parallel tubes that will fuse into
a single, midline heart.

**2** **22 days after fertilization**
The two parallel tubes have fused to
form one tube. This tube bends as it
elongates (*blue arrows* suggest the
direction of bending) within the
confined space of the pericardium.

**3** **31 days after fertilization**
The interatrial septum (*green*) and the
interventricular septum grow toward
the center of the heart.

**4** **35 days after fertilization**
The interventricular septum is nearly
complete. A foramen, which will
become the left side of the foramen
ovale, opens in the left side of the
interatrial septum (*green*) as the right
side of the interatrial septum begins to
form (*blue*).

**5** **Final embryonic condition
of the interatrial septum**
A foramen remains in the right side
of the interatrial septum (*blue*),
which forms the right part of the
foramen ovale. Blood from the right
atrium can flow through the foramen
ovale into the left atrium. After birth,
as blood begins to flow in the other
direction, the left side of the
interatrial septum is forced against
the right side, closing the foramen
ovale.

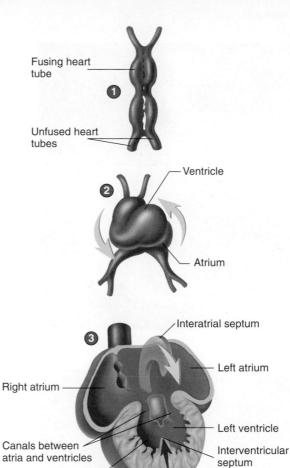

Fusing heart
tube

Unfused heart
tubes

Ventricle

Atrium

Interatrial septum

Left atrium

Right atrium

Left ventricle

Canals between
atria and ventricles

Interventricular
septum

Right ventricle

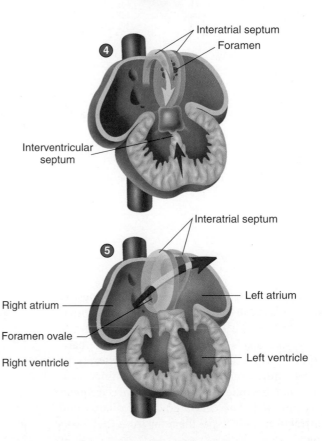

Interatrial septum

Foramen

Interventricular
septum

Interatrial septum

Right atrium

Left atrium

Foramen ovale

Right ventricle

Left ventricle

**PROCESS Figure 20.12** **Formation of the Heart**

(a)

(b)

**Figure 20.13  Late Embryo and Fetus**

(a) Embryo at 50 days of development. (b) Fetus at 3 months of development.

Ruptured amniotic sac

**1 First stage.** The cervix dilates, and the amniotic sac ruptures.

Placenta

**2 Second stage.** The fetus is expelled from the uterus.

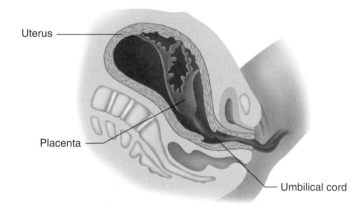

Uterus

Placenta

Umbilical cord

**3 Third stage.** The placenta is expelled.

**PROCESS Figure 20.14  Parturition**

# PARTURITION

Physicians usually calculate the **gestation** (jes-tā′shŭn) **period** (length of pregnancy) as 280 days (40 weeks, or 10 lunar months) from the LMP to the date of delivery of the fetus. **Parturition** (par-toor-ish′ŭn) is the process by which the baby is born (figure 20.14). Near the end of pregnancy, the uterus becomes progressively more excitable and usually exhibits occasional contractions that become stronger and more frequent until parturition is initiated. The cervix gradually dilates, and strong uterine contractions help expel the fetus from the uterus through the vagina.

**Labor** is the period during which uterine contractions occur that result in expulsion of the fetus. Although labor may differ greatly from woman to woman and from one pregnancy to another for the same woman, it can usually be divided into three stages.

1. The **first stage** of labor, often called the *dilation phase,* begins with the onset of regular uterine contractions and extends until the cervix dilates to a diameter about the size of the fetus's head (10 cm) (figure 20.14). This stage takes approximately 24 hours, but it may be as short as a few minutes in some women who have had more than one child. During this phase, the amniotic sac surrounding the fetus ruptures, and amniotic fluid flows through the vagina to the exterior. This event is commonly referred to as the "water breaking."

2. The **second stage** of labor, often called the *expulsion phase,* lasts from the time of maximum cervical dilation until the time the baby exits the vagina. This stage may last from 1 minute to 1 hour or more. During this stage, contraction of the woman's abdominal muscles assists the uterine contractions.

3. The **third stage** of labor, often called the *placental stage,* involves the expulsion of the placenta from the uterus. Contractions of the uterus cause the placenta to tear away from the wall of the uterus. Some bleeding from the uterine wall occurs because of the intimate contact between the placenta and the uterus. However, bleeding is normally limited because uterine smooth muscle contractions compress the blood vessels.

**Predict 3**

*Compare and contrast clinical age and developmental age for fertilization, implantation, the beginning of the fetal period, and parturition.*

During the 4 or 5 weeks following parturition, the uterus becomes much smaller, but it remains somewhat larger than it was before pregnancy. A vaginal discharge composed of small amounts of blood and degenerating endometrium can persist for several weeks after parturition.

The precise signal that triggers parturition is unknown, but many factors that support it have been identified (figure 20.15). Before parturition, the progesterone concentration in the mother's blood has reached its highest level. Progesterone has an inhibitory effect on uterine smooth muscle cells. However, estrogen levels continually increase in the maternal circulation, exciting uterine smooth muscle. Thus, the inhibitory influence of progesterone on smooth muscle is overcome by the stimulatory effect of estrogen near the end of pregnancy.

The fetus also plays a role in stimulating parturition. Stress on the fetus triggers the secretion of a releasing hormone from the fetal hypothalamus, which in turn causes adrenocorticotropic hormone (ACTH) to be released from the fetal anterior pituitary. ACTH stimulates the fetal adrenal gland to secrete hormones from the adrenal cortex that reduce progesterone secretion, increase estrogen secretion, and increase prostaglandin production by the placenta. Prostaglandins strongly stimulate uterine contractions.

During parturition, stretching of the uterine cervix initiates nervous reflexes that cause the mother's posterior pituitary gland to release oxytocin. Oxytocin stimulates uterine contractions, which move the fetus further into the cervix, causing further stretch.

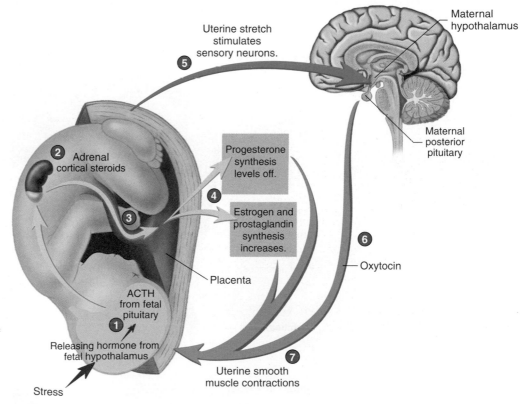

1. The fetal hypothalamus secretes a releasing hormone that stimulates adrenocorticotropic hormone (ACTH) secretion from the pituitary. The fetal pituitary secretes ACTH in greater amounts near parturition.

2. ACTH causes the fetal adrenal gland to secrete greater quantities of adrenal cortical steroids.

3. Adrenal cortical steroids travel in the umbilical blood to the placenta.

4. In the placenta, the adrenal cortical steroids cause progesterone synthesis to level off and estrogen and prostaglandin synthesis to increase, making the uterus more excitable.

5. The stretching of the uterus produces action potentials that are transmitted to the brain through ascending pathways.

6. Action potentials stimulate the secretion of oxytocin from the mother's posterior pituitary.

7. Oxytocin causes the uterine smooth muscle to contract.

**PROCESS Figure 20.15  Factors That Influence Parturition**

Although the precise control of parturition in humans is unknown, these changes appear to play a role.

A positive-feedback mechanism is established. This positive-feedback mechanism stops after delivery, when the cervix is no longer stretched.

## THE NEWBORN

The newborn, or **neonate** (nē'ō-nāt; newborn), experiences several dramatic changes at the time of birth. The major and earliest changes are the separation of the infant from the maternal circulation and the transfer from a fluid to a gaseous environment.

### Respiratory and Circulatory Changes

The large, forced gasps of air that occur when an infant cries at the time of delivery help inflate the lungs. The fetal lungs produce a substance called **surfactant** (ser-fak'tănt), which coats the inner surface of the alveoli, reduces surface tension in the lungs, and allows the newborn lungs to inflate (see chapter 15).

Surfactant is not manufactured in the fetal lungs before about 6 months after fertilization. If a fetus is born before the lungs can produce surfactant, the surface tension inside the lungs is too great for the lungs to inflate. Under these conditions, the newborn may die of respiratory distress. Therefore, premature newborns are treated with bovine or synthetic surfactant.

The initial inflation of the lungs causes important changes in the cardiovascular system (figure 20.16). Expansion of the lungs reduces the resistance to blood flow through the lungs, resulting in increased blood flow from the right ventricle of the heart through

1. Blood bypasses the lungs by flowing from the pulmonary trunk through the ductus arteriosus to the aorta.

2. Blood also bypasses the lungs by flowing from the right to the left atrium through the foramen ovale.

3. Blood bypasses the liver sinusoids by flowing through the ductus venosus.

4. Oxygen-rich blood from the placenta is passed to the fetus by the umbilical vein.

5. Oxygen-poor blood is carried from the fetus to the placenta through the umbilical arteries.

Superior vena cava
Aortic arch
**Ductus arteriosus**
Pulmonary trunk
**Foramen ovale**
Inferior vena cava
Liver
**Ductus venosus**
Hepatic portal vein
**Umbilical vein**
Fetal umbilicus
**Umbilical cord**
**Umbilical arteries**

Abdominal aorta
Kidney
Common iliac artery
Internal iliac arteries

(a)

**PROCESS Figure 20.16   Circulation in the Fetus and Newborn**
(a) Circulatory conditions in the fetus.

the pulmonary arteries. Consequently, an increased amount of blood flows from the right atrium to the right ventricle and into the pulmonary arteries, and less blood flows from the right atrium through the foramen ovale to the left atrium. The reduced resistance to blood flow through the lungs and the increasing volume of blood returning from the lungs through the pulmonary veins to the left atrium make the pressure in the left atrium greater than that in the right atrium. This pressure difference forces blood against the inter-atrial septum, closing a flap of tissue that develops in that region over the foramen ovale. This action completes the separation of the heart into two pumps: the right side and the left side of the heart.

A short artery, the **ductus arteriosus** (ar-tēr′ē-ō-sŭs), con-nects the pulmonary trunk to the aorta. Before birth, the ductus

arteriosus carries blood from the pulmonary trunk to the aorta, bypassing the fetal lungs. This artery closes off shortly after birth, forcing blood to flow through the lungs.

Also before birth, the fetal blood supply passes to the placenta through **umbilical** (ŭm-bil′i-kăl) **arteries,** which originate in the internal iliac arteries. The fetal blood returns through an **umbilical vein.** The umbilical vein passes through the liver but bypasses the sinusoids of the liver by way of the **ductus venosus** (vē-nō′sŭs) and joins the inferior vena cava. When the umbilical cord is tied and cut, no more blood flows through the umbilical vein and arteries, and they degenerate. The remnant of the umbilical vein becomes the round ligament of the liver.

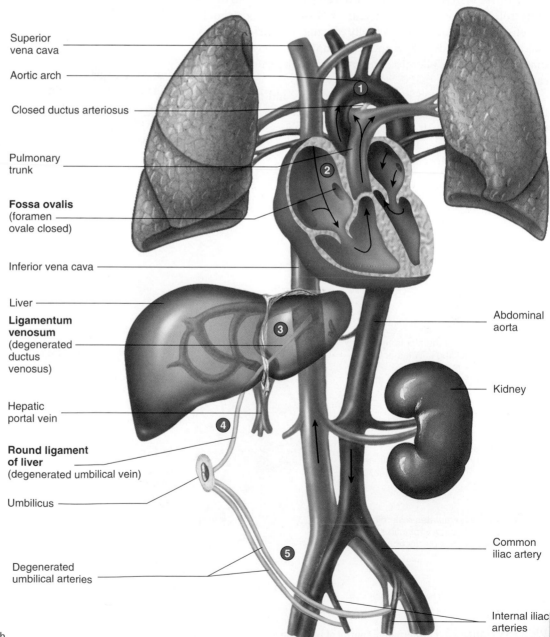

**1** When air enters the lungs, blood is forced through the pulmonary arteries to the lungs. The ductus arteriosus closes (*gray*).

**2** The foramen ovale closes and becomes the fossa ovalis. Blood can no longer flow from the right to the left atrium.

**3** The ductus venosus degenerates and becomes the ligamentum venosum (*gray*).

**4** The umbilical arteries and vein are cut. The umbilical vein becomes the round ligament of the liver (*gray*).

**5** The umbilical arteries also degenerate (*gray*).

(b)

**PROCESS Figure 20.16** (*continued*)
(b) Circulatory changes that occur after birth.

## Digestive Changes

Late in gestation, the fetus swallows amniotic fluid from time to time. Shortly after birth, this swallowed fluid plus cells sloughed from the mucosal lining of the digestive tract, mucus produced by intestinal mucous glands, and bile from the liver are eliminated as a greenish anal discharge called **meconium** (mē-kō′nē-ŭm).

After birth, the neonate is suddenly separated from its source of nutrients, the maternal circulation. Because of this separation and the shock of birth, the neonate usually loses 5–10% of its total body weight during the first few days of life. Although the digestive system of the fetus becomes somewhat functional late in development, it is still very immature in comparison to that of the adult and can digest only a limited number of food types. The newborn digestive system is capable of digesting lactose (milk sugar) from the time of birth. The pancreatic secretions are sufficiently mature for a milk diet, but the digestive system only gradually develops the ability to digest more solid foods over the first year or two. New foods should therefore be introduced gradually during the first 2 years. Parents are also advised to introduce only one new food at a time, so that, if an allergic reaction occurs, the cause is more easily determined.

Amylase secretion by the salivary glands and the pancreas remains low until after the first year. Lactase activity in the small intestine is high at birth but declines during infancy, although the levels still exceed those in adults. In many adults, lactase activity is lost, and an intolerance to milk develops.

## LACTATION

**Lactation** (lak-tā′shŭn) is the production of milk by the mammary glands (figure 20.17). It normally occurs in women following parturition and may continue for up to 2 or 3 years.

During pregnancy, the high concentration and continuous presence of estrogen and progesterone cause expansion of the duct system and the secretory units within the breast. Other hormones, including a prolactin-like hormone produced by the placenta, help support the development of the breasts. Also, additional adipose tissue is deposited; thus, the size of the breasts increases throughout pregnancy. Estrogen and progesterone prevent the secretory part of the breast from producing milk during pregnancy.

Blood levels of estrogen and progesterone fall dramatically after parturition. Once the placenta has been dislodged from the uterus, the source of these hormones is gone. After parturition, in the absence of estrogen and progesterone, prolactin produced by the anterior pituitary stimulates milk production. During suckling, sensory action potentials are sent from the nipple to the brain, stimulating the release of prolactin from the anterior pituitary (figure 20.17). For the first few days following parturition, the mammary glands secrete **colostrum** (kō-los′trŭm), a high-protein material that contains many antibodies. Although colostrum is high in proteins, it contains little fat and less lactose than milk. Eventually, milk with a higher fat and lactose content is produced. Colostrum and milk provide nutrition and antibodies that help protect the nursing baby from infections.

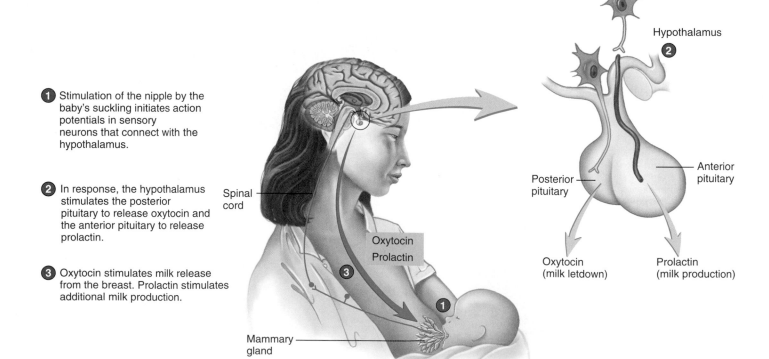

1. Stimulation of the nipple by the baby's suckling initiates action potentials in sensory neurons that connect with the hypothalamus.

2. In response, the hypothalamus stimulates the posterior pituitary to release oxytocin and the anterior pituitary to release prolactin.

3. Oxytocin stimulates milk release from the breast. Prolactin stimulates additional milk production.

Hypothalamus

Spinal cord

Oxytocin
Prolactin

Mammary gland

Posterior pituitary

Anterior pituitary

Oxytocin (milk letdown)

Prolactin (milk production)

**PROCESS Figure 20.17   Hormonal Control of Lactation**

At the time of nursing, milk contained in the alveoli and ducts of the breast is forced out of the breast by contractions of cells surrounding the alveoli. Suckling produces action potentials that are carried to the hypothalamus, where they cause the release of oxytocin from the posterior pituitary (figure 20.17). Oxytocin stimulates cells surrounding the alveoli to contract. As a result, milk flows from the breasts, a process called **milk letdown.** Higher brain centers can also cause the release of oxytocin as a result of a conditioned reflex in response to such stimuli as hearing an infant cry or thinking about breastfeeding.

Repeated stimulation of prolactin release makes breastfeeding possible for several years. If nursing is stopped, prolactin release stops, and within a few days the breasts' ability to respond to prolactin is lost, and milk production ceases.

### Predict 4

*While breastfeeding her baby, a woman felt cramps in her abdomen. Explain what was happening.*

## FIRST YEAR FOLLOWING BIRTH

A great number of changes occur in the infant from the time of birth until 1 year of age. The time when these changes occur may vary considerably from child to child, and the dates given are only rough estimates. The brain is still developing at this time, and much of what the infant can accomplish depends on the amount of brain development achieved. It is estimated that the newborn's central nervous system contains nearly all the adult number of neurons, but subsequent growth and maturation of the brain add new neuroglial cells, myelin sheaths, and new connections between neurons, which may continue throughout life.

By 6 weeks, the baby is usually able to hold up his or her head when placed in a prone position and begins to smile in response to people or objects. At 3 months of age, the infant exercises his limbs aimlessly. However, he can control his arms and hands enough that voluntary thumb-sucking can occur. The infant can follow a moving person with his eyes. At 4 months, the infant begins to do push-ups—that is, raise himself by the arms. The infant can begin to grasp objects placed in his hands, coo and gurgle, roll from back to side, listen quietly when hearing a person's voice or music, hold the head erect, and play with his hands. At 5 months, the infant can usually laugh out loud, reach for objects, turn her head to follow an object, lift her head and shoulders, sit with support, and roll over. At 8 months, the infant can recognize familiar people, sit up without support, and reach for specific objects. At 12 months, the infant may pull herself to a standing position and may be able to walk without support. The child can pick up objects in her hands and examine them carefully. A 12-month-old child can understand much of what is said and may say several words.

## LIFE STAGES

The stages of life from fertilization to death are divided into three prenatal stages and five postnatal stages as follows:

1. **Germinal** (jer′mi-năl) **period,** fertilization to 14 days
2. **Embryo,** 14–56 days after fertilization
3. **Fetus,** 56 days after fertilization to birth
4. **Neonate,** birth to 1 month after birth
5. **Infant** (in′fănt), 1 month to 1 or 2 years after birth (the end of infancy is sometimes set at the time that the child begins to walk)
6. **Child,** 1 or 2 years old to puberty (about 11–14 years)
7. **Adolescent** (ad-ō-les′ent), teenage years, from puberty to 20 years old
8. **Adult,** 20 years old to death

Adulthood is sometimes divided into three periods: **young adult,** 20–40 years old; **middle age,** 40–65 years old; and **older adult,** or *senior citizen,* 65 years old to death. Much of this designation is associated more with social norms than with physiology.

During childhood, the individual grows in size and develops considerably. Many of the emotional characteristics a person possesses throughout life are formed during early childhood.

Major physical and physiological changes occur during adolescence, and many of these changes also affect the emotions and behavior of the individual. Other emotional changes occur as the adolescent attempts to fit into an adult world. **Puberty** (pū′ber-tē) is the time when reproductive cells begin to mature and when gonadal hormones are first secreted in substantial amounts. These hormones stimulate the development and maturation of secondary sexual characteristics, such as enlargement of the female breasts and growth of body hair in both sexes. Puberty usually begins in females at about 11–13 years of age and in males at about 12–14 years of age. The onset of puberty is usually accompanied by a growth spurt, followed by a period of slower growth. Full adult stature is usually achieved before age 17 or 18 in females and before age 19 or 20 in males.

### The Aging Process

Development of a new human being begins at fertilization, and so does the process of aging. Cell division occurs at an extremely rapid rate during early development and then begins to slow as various cells become committed to specific functions within the body.

Many cells of the body, such as skin cells, continue to divide throughout life, replacing dead or damaged tissue. But other cells, such as the neurons in the brain, cease to divide once they have reached a certain number. Dead neurons tend not to be replaced. After the number of neurons reaches a peak (at approximately the time of birth), the number begins to decline. Neuronal loss is most rapid early in life and then decreases to a slower, steady rate.

Young embryonic tissue has relatively small amounts of collagen, and the collagen that is present is not highly cross-linked, making it very flexible and elastic. However, many of the collagen fibers produced during development are permanent components of the individual. As the individual ages, more and more cross-links form between the collagen molecules, rendering the tissues more rigid and less flexible.

The tissues with the highest collagen content and the greatest dependency on collagen for their function are the most severely affected by the collagen cross-linking and tissue rigidity associated with aging. One of the first structures to exhibit age-related changes as a result of this increased rigidity is the lens of the eye. Seeing close objects becomes more difficult with advancing age until most middle-aged people require reading glasses (see chapter 9). Loss of elasticity also affects other tissues, including the joints, kidneys, lungs, and heart, and greatly reduces their functional ability.

As with nervous tissue, the number of skeletal muscle fibers declines with age. The strength of skeletal muscle reaches a peak between 20 and 35 years of life and usually declines steadily thereafter. Skeletal muscle strength depends primarily on the size of the muscle fibers, but the total number of fibers is probably also important to muscle strength. As most people age, both the number of fibers and the size of each tend to decline. The decline in muscle fiber size may be more related to a general decrease in activity than to any specific age-related changes. Like the collagen of connective tissue, however, the macromolecules of skeletal muscle cells undergo biochemical changes during aging, rendering the muscle tissue less functional. A good exercise program can slow or even partly reverse the process of muscular aging.

Cardiac muscle cells also do not normally divide after birth. Age-related changes in cardiac muscle cell function probably contribute to a decline in cardiac function with advancing age. The heart loses elastic recoil ability and muscular contractility. As a result, total cardiac output declines, and less oxygen and fewer nutrients reach the body cells supplied by the cardiovascular system. This decline in blood flow can be particularly harmful to cells that require high oxygen levels, such as neurons of the brain, and cells that are already compromised, such as cartilage cells of the joints, contributing to the general decline in these tissues.

Reduced cardiac function also can result in decreased blood flow to the kidneys, contributing to decreases in the kidneys' filtration ability. Degeneration of the connective tissues as a result of collagen cross-linking and other factors also decreases the filtration efficiency of the glomerular basement membrane.

**Arteriosclerosis** (ar-tēr′ē-ō-skler-ō′sis) is general hardening of the arteries affecting mainly arterioles. **Atherosclerosis** (ath′er-ō-skler-ō′sis) is the gradual formation of lipid-containing plaques in the arterial wall of large and medium-sized arteries (see chapter 13). These plaques then may become fibrotic and calcified, resulting in arteriosclerosis. Atherosclerosis interferes with normal blood flow and may result in a **thrombosis** (throm-bō′sis), the formation of a clot or plaque inside a vessel. An **embolus** (em′bō-lŭs; a patch) is a piece of a clot that has broken loose and floats through the circulation. An embolus can lodge in smaller arteries and cause heart attacks or strokes. Although atherosclerosis affects all middle-aged and elderly people to some extent and can even occur in young people, some people appear more at risk because of high blood cholesterol levels. This condition seems to have a hereditary component, and blood tests are available to screen people for high blood cholesterol levels.

Many other organs, including the liver, pancreas, stomach, and colon, undergo degenerative changes with age. The ingestion of harmful agents can accelerate such changes. Examples are the degenerative changes induced in the lungs (aside from lung cancer) by cigarette smoke and sclerotic changes in the liver resulting from excessive alcohol consumption.

In addition to the previously described changes, cellular wear and tear, or cellular aging, contributes to aging. Progressive damage from many sources, such as radiation and toxic substances, can result in irreversible cellular insults and may be one of the major factors leading to aging. Although the data are mixed and their interpretation controversial, some studies suggest that ingesting moderate amounts of vitamins E and C in combination may help slow aging by stimulating cell repair. Vitamin C also stimulates collagen production and may slow the loss of tissue elasticity associated with aging collagen.

According to the **free radical theory of aging,** free radicals, which are atoms or molecules with an unpaired electron, can react with and alter the structure of molecules that are critical for normal cell function. Alteration of these molecules can result in cell dysfunction, cancer, or other types of cellular damage. Free radicals are produced as a normal part of metabolism and are introduced into the body from the environment through the air we breathe and the food we eat. The damage caused by free radicals may accumulate with age. Antioxidants, such as beta carotene (provitamin A), vitamin C, and vitamin E, can donate electrons to free radicals without themselves becoming harmful. Thus, antioxidants may prevent the damage caused by the free radicals and may ward off age-related disorders ranging from wrinkles to cancer. Again, the data are mixed and their interpretation controversial.

One characteristic of aging is an overall decrease in ATP production. This decline is associated with a decrease in oxidative phosphorylation, which has been shown in many cases to be associated with **mitochondrial DNA mutations.**

Immune system changes may also be a major contributing factor to aging. The aging immune system loses its ability to respond to outside antigens and begins to be more sensitive to the body's own antigens. Immune responses to one's own tissues can result in the degeneration of the tissues and may be responsible for such conditions as arthritic joint disorders, chronic glomerular nephritis, and hyperthyroidism. In addition, T lymphocytes tend to lose their functional capacity with aging and cannot destroy abnormal cells as efficiently. This change may be one reason that certain types of cancer occur more frequently in older people.

One of the greatest disadvantages of aging is the increasing lack of ability to adjust to stress. Homeostasis is far more precarious in elderly people, and eventually the body encounters some stressor so great that the body's ability to recover is surpassed and death results.

## Death

Death is usually not attributed to old age. Some other problem, such as heart failure, renal failure, or stroke, is usually listed as the cause of death.

Death was once defined as the loss of heartbeat and respiration. In recent years, however, more precise definitions of death have been developed because both the heart and the lungs can be kept working artificially, as occurs during cardiopulmonary resuscitation, and the heart can even be temporarily replaced by an artificial device. Modern definitions of death are based on the permanent cessation of life functions and the cessation of integrated tissue and organ function. The most widely accepted indication of death in humans is whole brain death, which is manifested clinically by the absence of (1) response to stimulation, (2) natural respiration and heart function, and (3) brainstem reflexes, in addition to an electroencephalogram that remains isoelectric ("flat") for at least 30 minutes.

When determining death, certain conditions also need to be ruled out. For example, some central nervous system poisons can cause a flat electroencephalogram, but the patient can be revived if the effects of the poison are eliminated. Hypothermia slows all chemical reactions, including those involved in degenerative changes that begin at the time of death. As a result, a person suffering from hypothermia can exhibit no response to stimulation, exhibit no respiration or heartbeat, and have a flat electroencephalogram for more than 30 minutes and still be revived.

**Neocortical** (nē-ō -kōr′ti-kăl) **death** is a condition in which major portions of the cerebrum are no longer functioning. The patient is comatose and incapable of responding to stimuli. However, heartbeat and respiration still continue because of some relatively unimpaired brainstem functions. Also, because some brainstem function remains, the electroencephalogram is not flat but exhibits some level of activity. Under these conditions, some state laws require that the patient be kept alive by intravenous feeding and by other support equipment. The patient may have previously stated in a living will that, if neocortical death occurs and he or she cannot be returned to a reasonably normal level of function, no artificial support should be applied in an attempt to keep the body alive.

**Figure 20.18 Human Karyotype**

The 23 pairs of chromosomes in humans consist of 22 pairs of autosomal chromosomes (numbered 1–22) and 1 pair of sex chromosomes. The autosomal chromosome pairs are numbered in order from largest to smallest. This is the karyotype of a male, as evidenced by the presence of an X and a Y sex chromosome. A female karyotype would have 2 X chromosomes.

# GENETICS

**Genetics** is the study of heredity—that is, the characteristics children inherit from their parents. Many of a person's characteristics, including specific abilities, susceptibility to certain diseases, and even life span, are influenced by heredity. The functional units of heredity are **genes,** which are carried on chromosomes.

## Chromosomes

**Deoxyribonucleic** (dē-oks′ē-rī′bō-noo-klē′ic) **acid (DNA)** molecules and their associated proteins become visible as densely stained bodies, called **mitotic chromosomes** (krō′mō-sōmz) during cell division (see chapter 3). **Somatic** (sō-mat′ik) **cells,** all the cells of the body except the sex cells, contain 23 pairs of chromosomes, or 46 total chromosomes. The sex cells, or **gametes** (gam′ētz), contain 23 unpaired chromosomes.

A **karyotype** (kar′ē-ō-tīp) is a display of the chromosomes in a somatic cell (figure 20.18; see chapter 3). There are 22 pairs of **autosomal** (aw-tō-sō′măl) **chromosomes,** which are all the chromosomes except the sex chromosomes, and there is 1 pair of **sex chromosomes.** A normal female has 2 **X chromosomes** (XX) in each somatic cell, whereas a normal male has 1 X and 1 **Y chromosome** (XY).

Gametes are produced by **meiosis** (mī-ō′sis) (see chapter 19). Meiosis is called a reduction division because it produces gametes that have half the number of chromosomes that somatic cells have. When a sperm cell and a secondary oocyte fuse during fertilization, each contributes one-half of the chromosomes necessary to produce new somatic cells. Therefore, half of an individual's genetic makeup comes from the father, and half comes from the mother.

During meiosis, the chromosomes are distributed in such a way that each gamete receives only one chromosome from each **homologous** (hŏ-mol′ō-gŭs) pair of chromosomes (see chapter 19). Homologous chromosomes contain the same complement of genetic information. The determination of sex illustrates, in part, how chromosomes are distributed during gamete formation and fertilization. During meiosis and gamete formation, the pair of sex chromosomes separates, so that each secondary oocyte receives one of a homologous pair of X chromosomes, whereas each sperm cell receives either an X chromosome or a Y chromosome (figure 20.19). When a sperm cell fertilizes an oocyte to form a single cell, the sex of the individual is determined randomly. If the oocyte is fertilized by a sperm cell with a Y chromosome, a male results; if the oocyte is fertilized by a sperm cell with an X chromosome, a female results. Estimating the probability of any given zygote being male or female is much like flipping a coin. When all the possible combinations of sperm cells with oocytes are considered, half the individuals should be female and the other half should be male.

## Genes

Each chromosome contains thousands of genes. Each **gene** consists of a certain portion of a DNA molecule, but not necessarily a continuous stretch of DNA. Genes determine the proteins in a cell. Both chromosomes of a given pair contain similar but not

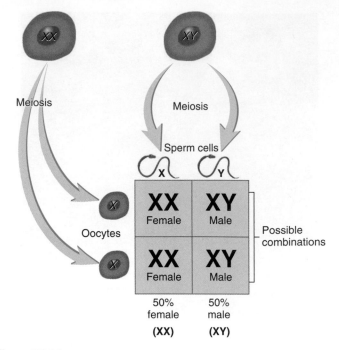

**Figure 20.19  Inheritance of Sex**

The female produces oocytes containing one X chromosome, whereas the male produces sperm cells with either an X or a Y chromosome. There are four possible combinations of an oocyte with a sperm cell, half of which produce females and half of which produce males.

necessarily identical genes. Similar genes on homologous chromosomes are called **alleles** (ă-lēlz′). If the two allelic genes are identical, the person is **homozygous** (hō-mō-zī′gŭs) for the trait specified by that gene. If the two alleles are slightly different, the person is **heterozygous** (het′er-ō-zī′gŭs) for the trait. All the genes in one homologous set of 23 chromosomes in one individual constitute that person's **genome** (jē′nōm).

Through the processes of meiosis, gamete formation, and fertilization, the distribution of genes received from each parent is essentially random. This random distribution is influenced by several factors, however. First, all the genes on a given chromosome are **linked;** that is, they tend to be inherited as a set rather than as individual genes because chromosomes, not individual genes, segregate during meiosis. Second, homologous chromosomes may exchange genetic information during meiosis by **crossing over.**

Furthermore, segregation errors can occur during meiosis. As the chromosomes separate, the 2 members of a homologous pair may not segregate. As a result, 1 of the daughter cells receives both chromosome pairs, and the other daughter cell receives none. When the gametes are fertilized, the resulting zygote has either 47 chromosomes or 45 chromosomes, rather than the normal 46. When this condition results in an abnormal autosomal chromosome number, it is usually, but not always, lethal and is one reason for a high rate of early embryo loss. **Down syndrome,** or *trisomy 21,* which results when there are 3 #21 chromosomes, is one of the few autosomal trisomies that is not lethal. In contrast, sex chromosome abnormalities are not usually lethal. For example, **Turner Syndrome** or *monosomy X,* in which only 1 X chromosome is

present, results in sterility and abnormal sexual development but does not affect the mental development of the female.

## Dominant and Recessive Genes

Most human genetic traits are recognized because defective alleles for those traits exist in the population. For example, on chromosome 11 is a gene that produces an enzyme necessary for the synthesis of melanin, the pigment responsible for skin, hair, and eye color (see chapter 5). An abnormal allele, however, produces a defective enzyme not capable of catalyzing one of the steps in melanin synthesis. If a given person inherits two defective alleles, a homozygous condition, the person is unable to produce melanin and therefore lacks normal pigment. This condition is referred to as **albinism** (al′bi-nizm).

For many genetic traits, the effects of one allele for that trait can mask the effect of another allele for the same trait. For example, a person who is heterozygous for the melanin-producing enzyme gene has one normal gene for melanin production and one defective gene for melanin production. One copy of the gene and its resulting enzymes are enough to make normal melanin. As a result, the person who is heterozygous produces melanin and appears normal. In this case, the allele that produces the normal enzyme is said to be **dominant,** whereas the allele producing the abnormal enzyme is **recessive.** By convention, dominant traits are indicated by uppercase letters, and recessive traits are indicated by lowercase letters. In this example, the letter *A* designates the dominant normal, pigmented allele, the letter *a* the recessive albino. Not all dominant traits are the normal condition, and not all recessive traits are abnormal. Many examples exist of abnormal dominant traits. For example, **Huntington disease,** a degenerative neurological disease, is caused by an abnormal dominant gene.

The possible combinations of dominant and recessive alleles for normal melanin production versus albinism are *AA* (homozygous dominant, normal), *Aa* (heterozygous, normal), and *aa* (homozygous recessive, albino). The alleles a person has for a given trait are called the **genotype** (jen′ō-tīp). The person's appearance is called the **phenotype** (fē′nō-tīp). A person with the genotype *AA* or *Aa* has the phenotype of normal pigmentation, whereas a person with the genotype *aa* has the phenotype of albinism. The recessive trait is expressed only when no allele for the dominant trait is present. A **carrier** is a heterozygous person with an abnormal recessive gene but a normal phenotype because the normal dominant allele for that gene is also present.

**Predict 5**

*Polydactyly (pol-ē-dak′ti-lē) is the condition of having extra fingers or toes. Given that polydactyly is a dominant trait, list all the possible genotypes and phenotypes for polydactyly. Use the letters D and d for the alleles.*

The inheritance of dominant and recessive traits can be determined if the genotypes of the parents are known. For example, if an albino person (*aa*) mates with a heterozygous normal person (*Aa*), the probability that the child will be albino (*aa*) is one-half, and the probability that the child will be normal (*Aa*) is one-half. If two carriers (*Aa*) mate, the probability that the child will be homozygous dominant (*AA*) or homozygous recessive (*aa*)

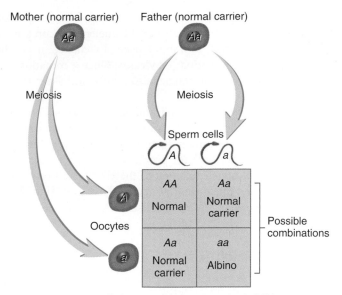

Genotype and phenotype probabilities:
$\frac{1}{4}$ AA (normal) : $\frac{1}{2}$ Aa (normal carrier) : $\frac{1}{4}$ aa (albino)

**Figure 20.20  Inheritance of a Recessive Trait: Albinism**

*A* represents the normal pigmented condition, and *a* represents the recessive unpigmented condition. This Punnett square represents a mating between two normal carriers.

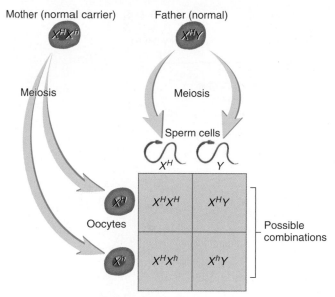

Genotype and phenotype probabilities:
$\frac{1}{4}$ $X^H X^H$ (normal female) : $\frac{1}{4}X^H X^h$ (carrier female) :

$\frac{1}{4}$ $X^H Y$ (normal male) : $\frac{1}{4}X^h Y$ (male with hemophilia)

**Figure 20.21  Inheritance of an X-Linked Trait: Hemophilia**

$X^H$ represents the normal X chromosome condition with all clotting factors, and $X^h$ represents the X chromosome lacking an allele for one clotting factor. This Punnett square represents a mating between a normal male and a normal carrier female.

is one-fourth for either genotype. The probability that the child will be heterozygous (*Aa*) is one-half. Such a probability can be determined easily by using a table called a **Punnett square** (figure 20.20).

## Sex-Linked Traits

Traits affected by genes on the sex chromosomes are called **sex-linked traits.** Most sex-linked traits are **X-linked;** that is, they are on the X chromosome. Only a few **Y-linked** traits exist, largely because the Y chromosome is very small. An example of an X-linked trait is **hemophilia A** (classic hemophilia), in which the person is unable to produce one of the clotting factors (see chapter 11). Consequently, clotting is impaired, and persistent bleeding can occur, either spontaneously or as a result of an injury. Hemophilia A is a recessive trait, and the allele for the trait is located on the X chromosome. The possible genotypes and phenotypes are therefore

$X^H X^H$ (normal homozygous female)
$X^H X^h$ (normal heterozygous female)
$X^h X^h$ (hemophiliac homozygous female)
$X^H Y$ (normal male)
$X^h Y$ (hemophiliac male)

Note that a female must have both recessive alleles to exhibit hemophilia, whereas a male has hemophilia if he has one recessive allele, because he has only one X chromosome. A Punnett square representing the inheritance of hemophilia is illustrated in figure 20.21. If a woman who is a carrier for hemophilia mates with a man who does not have hemophilia, none of their daughters will have hemophilia; however, the probability that a son will have hemophilia is one-half. On the other hand, the probability that a daughter will be a carrier is one-half.

### Predict 6

*Predict the probability of a girl with Turner syndrome having hemophilia if her mother is a carrier for hemophilia.*

## Other Types of Gene Expression

In some cases, the dominant allele does not completely mask the effects of the recessive allele. This is called **incomplete dominance.** An example of incomplete dominance is **sickle-cell anemia,** in which the hemoglobin produced by the gene is abnormal. The result is sickle-shaped red blood cells, which are likely to stick in capillaries and tend to rupture more easily than normal red blood cells. The normal hemoglobin allele (*S*) is dominant over the sickle-cell allele (*s*). A normal person (*SS*) has normal hemoglobin, and a person with sickle-cell anemia (*ss*) has abnormal hemoglobin. A person who is heterozygous (*Ss*) has half normal hemoglobin and half abnormal hemoglobin and usually has only a few sickle-shaped red blood cells. This condition is called **sickle-cell trait.**

In another type of gene expression, called **codominance** (kō-dom′i-năns), two alleles can combine to produce an effect without either of them being dominant or recessive. For example, a person with type AB blood has A antigens and B antigens on the surface of his or her red blood cells (see chapter 11). The antigens result from a gene that causes the production of the A antigen and a different gene that causes the production of the B antigen, and neither gene is dominant or recessive in relation to the other.

Many traits, called **polygenic** (pol-ē-jen′ik) **traits,** are determined by the expression of multiple genes on different chromosomes. Examples are height, intelligence, eye color, and skin color. Polygenic

**Reproductive**

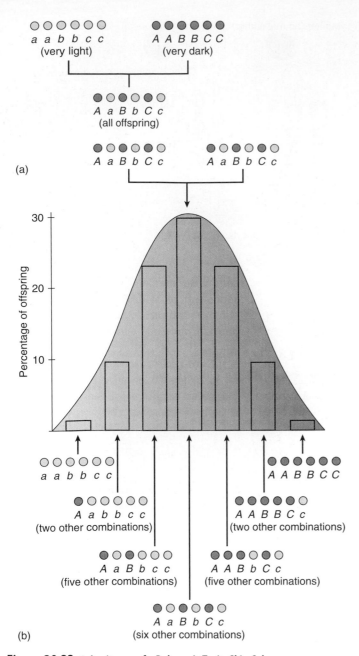

(a)

(b)

**Figure 20.22  Inheritance of a Polygenic Trait: Skin Color**

This example shows three genes for skin color. Each of the dominant alleles (*A, B, C*) contributes one "unit of dark color" to the offspring (indicated by a *dark dot*). Each of the recessive alleles (*a, b, c*) contributes one "unit of light color" to the offspring (indicated by a *light dot*). The dominant alleles are incompletely dominant over the recessive alleles. (*a*) In a mating between a very light-skinned person (*aabbcc*) and a very dark-skinned person (*AABBCC*), all the offspring are of intermediate color. (*b*) In a mating between two people of intermediate skin color (*AaBbCc*), a very low percentage of the offspring (less than 2%) are either very light or very dark, and most are of intermediate color.

traits typically have a great amount of variability. For example, there are many shades of eye color and skin color (figure 20.22).

## Genetic Disorders

**Genetic disorders** are caused by abnormalities in a person's genetic makeup—that is, in his or her DNA. They may involve a single gene or an entire chromosome. Some genetic disorders

result from a **mutation** (mū-tā′shŭn; to change), a change in a gene that usually involves a change in the nucleotides composing the DNA (see chapter 2). Mutations occur by chance or can be caused by chemicals, radiation, or viruses. Once a mutation has occurred, the abnormal trait can be passed from one generation to the next.

The importance of genes is dramatically illustrated by situations in which the alteration of a single gene results in a genetic disorder. For example, in **phenylketonuria** (fen′il-kē′tō-nū′rē-ă) **(PKU),** the gene responsible for producing an enzyme that converts the amino acid phenylalanine to the amino acid tyrosine is defective. Therefore, phenylalanine accumulates in the blood and is eventually converted to harmful substances that can cause mental retardation.

## Genetic Counseling

**Genetic counseling** includes predicting the possible results of matings involving carriers of harmful genes and talking to parents or prospective parents about the possible outcomes and treatments of a genetic disorder. With this knowledge, prospective parents can make informed decisions about having children.

The first step in genetic counseling is to attempt to determine the genotype of the individuals involved. A family tree, or **pedigree,** provides historical information about family members. Figure 20.23 shows the pedigree for a simple dominant trait, such as Huntington disease, a neurological disorder. Sometimes, knowing the phenotypes of relatives makes it possible to determine a person's genotype. As part of the process of collecting information, a karyotype can be prepared. For some genetic disorders, the amount of a given substance, such as an enzyme, produced by a carrier can be tested. For example, carriers for cystic fibrosis produce more salt in their sweat than is normal.

**Figure 20.23  Pedigree of a Simple Dominant Trait**

Males are indicated by *squares,* females by *circles.* People who express the trait are indicated in *purple.* The *horizontal line* between symbols represents a mating. The symbols connected to the mating line by *vertical* and *horizontal lines* represent the children resulting from the mating in order of birth from left to right. Matings not related to the pedigree are not shown.

## Human Genome Project

The **human genome** comprises all the genes in one homologous set of human chromosomes. Researchers estimate that humans have 20,000–25,000 genes. A **genomic** (jĕ-nom'ik, je-nōm'ik) **map** is a description of the DNA nucleotide sequences of the genes and their locations on the chromosomes (figure 20A). The **Human Genome Project,** a large-scale project to characterize the entire human genome, was completed in February 2003.

Chromosome pairs

1.
a. Gaucher disease
b. Prostate cancer
c. Glaucoma
d. Alzheimer disease*

2.
a. Familial colon cancer*
b. Waardenburg syndrome

3.
a. Lung cancer
b. Retinitis pigmentosa*

4.
a. Huntington disease
b. Parkinson disease

5.
a. Cockayne syndrome
b. Familial polyposis of the colon
c. Asthma

6.
a. Spinocerebellar ataxia
b. Diabetes*
c. Epilepsy*

7.
a. Diabetes*
b. Osteogenesis imperfecta
c. Cystic fibrosis
d. Obesity*

8.
a. Werner syndrome
b. Burkitt lymphoma

9.
a. Malignant melanoma
b. Friedreich ataxia
c. Tuberous sclerosis

10.
a. Multiple endocrine neoplasia, type 2
b. Gyrate atrophy

11.
a. Sickle-cell anemia
b. Multiple endocrine neoplasia

12.
a. Zollwogor oyndromc
b. Phenylketonuria (PKU)

13.
a. Breast cancer*
b. Retinoblastoma
c. Wilson disease

14.
a. Alzheimer disease*

15.
a. Marfan syndrome
b. Tay-Sachs disease

16.
a. Polycystic kidney disease
b. Crohn disease*

17.
a. Tumor suppressor protein
b. Breast cancer*
c. Osteogenesis imperfecta

18.
a. Amyloidosis
b. Pancreatic cancer*

19.
a. Familial hypercholesterolemia
b. Myotonic dystrophy

20.
a. Severe combined immunodeficiency

21.
a. Amyotrophic lateral sclerosis*

22.
a. DiGeorge syndrome
b. Neurofibromatosis, type 2

X
a. Duchenne muscular dystrophy
b. Menkes syndrome
c. X-linked severe combined immunodeficiency
d. Factor VIII deficiency (hemophilia A)

*Gene responsible for only some cases.

**Figure 20A**

Representative genetic defects mapped to date. The bars and lines indicate the locations of the genes listed for each chromosome.

## DISEASES AND DISORDERS  Pregnancy

| CONDITION | DESCRIPTION |
|---|---|
| Ectopic pregnancy | Pregnancy that occurs outside the uterus, usually in the uterine tube, as the pregnancy progresses, the tube ruptures, which may lead to life-threatening internal bleeding |
| Miscarriage | Death or early delivery of the fetus; approximately 15% of all pregnancies end in miscarriage |
| Placenta previa | The placenta implants near the opening of the uterus in the cervical canal; as the fetus grows and the uterus stretches, the placenta may tear away from the uterine wall, resulting in death of the fetus; associated hemorrhaging can be life-threatening to the mother as well |
| Toxemia (tok-sē′mē-ă) (pregnancy-induced hypertension) | Increased blood pressure associated with pregnancy; may result in convulsions, kidney failure, and death of both the mother and fetus |
| Teratogens (ter′ă-tō-jenz) | Drugs or other chemicals that can cross the placenta and cause birth defects in the developing embryo; for example, thalidomide (tha-lid′o-mīd) causes abnormal limb development |
| Fetal alcohol syndrome (FAS) | Brain dysfunction, growth retardation, and facial peculiarities in children of women who consumed substantial amounts of alcohol while pregnant |
| **Infections**<br>Rubella (German measles) | If rubella is contracted by a pregnant female, the fetus may be severely affected, including visual and hearing defects as well as mental retardation |
| Neonatal gonorrheal ophthalmia | Severe form of conjunctivitis contracted by an infant passing through the birth canal of a mother infected with gonorrhea; carries a high risk of blindness; can be prevented by treating a newborn's eye with silver nitrate and antibiotics |
| Chlamydial conjunctivitis | Contracted by an infant passing through the birth canal of a mother infected with chlamydia |
| Human immunodeficiency virus (HIV) | Virus that causes acquired immunodeficiency syndrome (AIDS); can infect the fetus in utero, during parturition, or during breastfeeding; the drug azidothymidine (AZT) inhibits HIV replication and can reduce the number of infants who contract AIDS if given to HIV-infected pregnant women and to their newborn infants |

Go to **www.mhhe.com/seeleyess7** for additional information on these pathologies.

# SUMMARY

Prenatal development is an important part of an individual's life. About 3 of every 100 people are born with a birth defect requiring early medical attention.

## Prenatal Development  (p. 559)

1. The developing human is called an embryo during the first 8 weeks of the prenatal period and a fetus from 8 weeks until birth.
2. Developmental age is 14 days less than clinical age.

### Fertilization

Fertilization, the union of the secondary oocyte and sperm cell, results in a zygote.

### Early Cell Division

The zygote undergoes divisions until it becomes a mass of cells.

### Blastocyst

1. The embryonic mass develops a cavity and is known as the blastocyst.
2. The blastocyst consists of a trophoblast and an inner cell mass, where the embryo forms.

### Implantation of the Blastocyst and Development of the Placenta

1. The blastocyst implants in the uterus about 7 days after fertilization.
2. The embryonic portion of the placenta is derived from the trophoblast of the blastocyst.

### Maternal Hormonal Changes

1. Human chorionic gonadotropin levels are high in early pregnancy but then decline to low levels.
2. Progesterone and estrogen levels are low in early pregnancy but increase to high levels late in pregnancy.

### Formation of the Germ Layers

1. The embryo forms around the primitive streak, which forms about 14 days after fertilization.
2. All tissues of the body are derived from the three primary germ layers: ectoderm, mesoderm, and endoderm.

### Neural Tube and Neural Crest Formation

The nervous system develops from a neural tube that forms in the ectodermal surface of the embryo and from neural crest cells derived from the developing neural tube.

### Formation of the General Body Structure

1. The limbs develop as outgrowths called limb buds.
2. The face develops by fusion of five tissue masses.

### Development of the Organ Systems

1. The digestive tract develops as the developing embryo closes off part of the yolk sac.
2. The heart develops as two blood vessels fuse into a single tube that develops septa to form four chambers.

Reproductive

3. The kidneys and the reproductive system are closely related in their development.

### Growth of the Fetus

During the fetal period, the fetus increases 15-fold in length and 1400-fold in weight.

## Parturition    (p. 572)

1. Uterine contractions force the baby out of the uterus during labor.
2. Increased estrogen, decreased progesterone, and secretions from the fetal adrenal cortex initiate parturition.
3. Stretching of the uterus stimulates oxytocin secretion, which stimulates uterine contractions.

## The Newborn    (p. 574)

### Respiratory and Circulatory Changes

1. Inflation of the lungs at birth results in closure of the foramen ovale and the ductus arteriosus.
2. When the umbilical cord is cut, blood no longer flows through the umbilical vessels.

### Digestive Changes

The digestive system only gradually develops the ability to digest a variety of foods.

## Lactation    (p. 576)

1. Estrogen and progesterone help stimulate the growth of the breasts during pregnancy.
2. Suckling stimulates prolactin and oxytocin synthesis. Prolactin stimulates milk production, and oxytocin stimulates milk letdown.

## First Year Following Birth    (p. 577)

Many of the important changes that occur during the first year after birth are linked to continued development of the brain.

## Life Stages    (p. 577)

The eight stages of life are germinal period (fertilization to 14 days); embryo (14–56 days after fertilization); fetus (56 days after fertilization to birth); neonate (birth–1 month); infant (1 month–1 or 2 years); child (1 or 2 years–puberty); adolescent (puberty–20 years); adult (20 years–death).

### The Aging Process

1. Aging occurs as irreplaceable cells wear out and the tissue becomes more brittle and less able to repair damage.
2. Atherosclerosis is the deposit of lipids in the arteries. Arteriosclerosis is hardening of the arteries.

### Death

Death is defined as the absence of brain response to stimulation, the absence of natural respiration and heart function, and a flat electroencephalogram for 30 minutes.

## Genetics    (p. 579)

### Chromosomes

1. Humans have 46 chromosomes in 23 pairs.
2. Males have the sex chromosomes XY, and females have XX.
3. During gamete formation, the chromosomes of each pair of chromosomes separate; therefore, half of a person's genetic makeup comes from the father and half from the mother.

### Genes

1. A gene is a portion of a DNA molecule. Genes determine the proteins in a cell.
2. Genes are paired (located on the paired chromosomes).
3. Dominant genes mask the effects of recessive genes.
4. Sex-linked traits result from genes on the sex chromosomes.
5. In incomplete dominance, the heterozygote expresses a trait that is intermediate between the two homozygous traits.
6. In codominance, neither gene is dominant or recessive, but both are fully expressed.
7. Polygenic traits result from the expression of multiple genes.

### Genetic Disorders

1. A mutation is a change in the DNA.
2. Some genetic disorders result from an abnormal distribution of chromosomes during gamete formation.

### Genetic Counseling

1. A pedigree (family history) can be used to determine the risk of having children with a genetic disorder.
2. Specific chemical tests or an examination of a person's karyotype can be used to determine a person's genotype.

 # REVIEW AND COMPREHENSION

1. Define clinical age and developmental age, and state the difference between the two in number of days. Define embryo and fetus.
2. What are the events during the first week after fertilization? Define zygote, morula, and blastocyst.
3. How does the placenta develop?
4. Describe the formation of the germ layers and the role of the primitive streak.
5. How are the neural tube and neural crest cells formed? What do they become?
6. Describe the formation of the limbs and face.
7. Describe the formation of the digestive tract.
8. How does the single heart tube become four-chambered?
9. What major events distinguish embryonic and fetal development?
10. Describe the hormonal changes that take place before and during parturition.

11. What changes take place in the newborn's circulatory system and digestive system shortly after birth?
12. What hormones are involved in preparing the breasts for lactation? What hormones are involved in milk production and milk letdown?
13. Describe the changes in motor and language skills that take place during the first year of life.
14. List the different life stages.
15. How does the loss of cells that are not replaced affect the aging process? Give examples.
16. How does the loss of tissue elasticity affect the aging process? Give examples.
17. How does aging affect the immune system?
18. Define death.
19. Give the number and type of chromosomes in the karyotype of a human somatic cell. How do the chromosomes of a male and a female differ from each other?

20. How do the chromosomes in somatic cells and gametes differ from each other?

21. What is a gene, and how are genes responsible for the structure and function of cells?

22. Define homozygous dominant, heterozygous, and homozygous recessive.

23. What is a sex-linked trait? Give an example.

24. How do sickle-cell anemia, type AB blood, and a person's height result from the expression of genes?

25. What is a mutation?

26. What is the cause of the genetic disorder Down syndrome?

27. How are pedigrees, karyotypes, and chemical tests used in genetic counseling?

 CRITICAL THINKING

1. A physician tells a woman that her pregnancy has progressed 44 days since her last menstrual period (LMP). How many days has the embryo been developing, and what developmental events are occurring?

2. A high fever can prevent neural tube closure. If a woman has a high fever about 35 to 45 days post-LMP, what kinds of birth defects may be seen in the developing embryo?

3. A drug that stops the production of milk in the breast after a few days probably has which effect?
   a. inhibits prolactin secretion
   b. inhibits oxytocin secretion
   c. increases estrogen secretion
   d. increases progesterone secretion
   e. increases prolactin secretion

4. Dimpled cheeks are inherited as a dominant trait. Is it possible for two parents, each of whom has dimpled cheeks, to have a child without dimpled cheeks? Explain.

5. The ability to roll the tongue to form a "tube" results from a dominant gene. Suppose that a woman and her son can both roll their tongues, but her husband cannot. Is it possible to determine if the husband is the father of her son based on this trait?

6. A woman who does not have hemophilia marries a man who has the disorder. Determine the genotypes of both parents if half of their children have hemophilia.

*Answers in Appendix D*

 Visit this textbook's website at **www.mhhe.com/seeleyess7** for practice quizzes, animations, interactive learning exercises, and other study tools.

 McGraw-Hill offers a study CD that features interactive cadaver dissection. *Anatomy & Physiology Revealed* includes cadaver photos that allow you to peel away layers of the human body to reveal structures beneath the surface. This program also includes animations, radiologic imaging, audio pronunciations, and practice quizzing.

# Appendix A

## Table of Measurements

### TABLE A.1   Table of Measurements

| Unit | Metric Equivalent | Symbol | U.S. Equivalent |
|------|-------------------|--------|-----------------|
| **Measures of Length** | | | |
| 1 kilometer | = 1000 meters | km | 0.62137 mile |
| 1 meter | = 10 decimeters or 100 centimeters | m | 39.37 inches |
| 1 decimeter | = 10 centimeters | dm | 3.937 inches |
| 1 centimeter | = 10 millimeters | cm | 0.3937 inch |
| 1 millimeter | = 1000 micrometers | mm | |
| 1 micrometer[a] | = 1/1000 millimeter or 1000 nanometers | µm | |
| 1 nanometer[a] | = 10 angstroms or 1000 picometers | nm | No U.S. equivalent |
| 1 angstrom | = 1/10,000,000 millimeter | Å | |
| 1 picometer | = 1/1,000,000,000 millimeter | pm | |
| **Measures of Volume** | | | |
| 1 cubic meter | = 1000 cubic decimeters | $m^3$ | 1.308 cubic yards |
| 1 cubic decimeter | = 1000 cubic centimeters | $dm^3$ | 0.03531 cubic foot |
| 1 cubic centimeter | = 1000 cubic millimeters or 1 milliliter | $cm^3$ (cc) | 0.06102 cubic inch |
| **Measures of Capacity** | | | |
| 1 kiloliter | = 1000 liters | kL | 264.18 gallons |
| 1 liter | = 10 deciliters | L | 1.0567 quarts |
| 1 deciliter | = 100 milliliters | dL | 0.4227 cup |
| 1 milliliter | = volume of 1 gram of water at standard temperature and pressure | mL | 0.3381 ounce |
| 1 microliter | = 1 $mm^3$ (cubic millimeter) or $10^{-6}$ L | µL | |
| **Measures of Mass** | | | |
| 1 kilogram | = 1000 grams | kg | 2.2046 pounds |
| 1 gram | = 100 centigrams or 1000 milligrams | g | 0.0353 ounce |
| 1 centigram | = 10 milligrams | cg | 0.1543 grain |
| 1 milligram | = 1/1000 gram | mg | |

[a]A micrometer was formerly called a micron (µm), and a nanometer was formerly called a millimicron (mµ).

### TABLE A.2   Comparative Temperature Scales

# Appendix B

## Some Reference Laboratory Values

| **TABLE B.1** | **Blood, Plasma, or Serum Values** |
|---|---|

| Test | Normal Values | Clinical Significance |
|---|---|---|
| Acetoacetate plus acetone | 0.32–2 mg/100 mL | Values increase in diabetic acidosis, fasting, high-fat diet, and toxemia of pregnancy. |
| Ammonia | 9–33 µmol/L (micromol/L) | Values decrease with proteinuria and as a result of severe burns; values increase in multiple myeloma. |
| Amylase | 4–25 U/mL[a] | Values increase in acute pancreatitis, intestinal obstruction, and mumps; values decrease in cirrhosis of the liver, toxemia of pregnancy, and chronic pancreatitis. |
| Barbiturate | 0 | Coma level: phenobarbital, approximately 10 mg/100 mL; most other drugs, 1–3 mg/100 mL |
| Bilirubin | 0.4 mg/100 mL | Values increase in conditions causing red blood cell destruction, biliary obstruction, or liver inflammation. |
| Blood volume | 8.5–9% of body weight in kilograms | |
| Calcium | 8.5–10.5 mg/100 mL | Values increase in hyperparathyroidism, vitamin D hypervitaminosis; values decrease in hypoparathyroidism, malnutrition, and severe diarrhea. |
| Carbon dioxide content | 24–30 mEq/L[b] 20–26 mEq/L in infants (as $HCO_3^-$) | Values increase in respiratory diseases, vomiting, and intestinal obstruction; values decrease in acidosis, nephritis, and diarrhea. |
| Carbon monoxide | 0 | Symptoms with over 20% saturation |
| Chloride | 100–106 mEq/L | Values increase in Cushing syndrome, nephritis, and hyperventilation; values decrease in diabetic acidosis, Addison disease, diarrhea, and after severe burns. |
| Creatine phosphokinase (CPK) | Female: 5–35 mU/mL Male: 5–55 mU/mL | Values increase in myocardial infarction and skeletal muscle diseases, such as muscular dystrophy. |
| Creatinine | 0.6–1.5 mg/100 mL | Values increase in certain kidney diseases. |
| Ethanol | 0 | 0.3–0.4%, marked intoxication 0.4–0.5%, alcoholic stupor Over 0.5%, alcoholic coma |
| Glucose | Fasting 70–110 mg/100 mL | Values increase in diabetes mellitus, liver diseases, nephritis, hyperthyroidism, and pregnancy; values decrease in hyperinsulinism, hypothyroidism, and Addison disease. |
| Iron | 50–150 µg/100 mL | Values increase in various anemias and liver disease; values decrease in iron-deficiency anemia. |
| Lactic acid | 0.6–1.8 mEq/L | Values increase with muscular activity and in congestive heart failure, severe hemorrhage, shock, and anaerobic exercise. |
| Lactic dehydrogenase | 60–120 U/mL | Values increase in pernicious anemia, myocardial infarction, liver diseases, acute leukemia, and widespread carcinoma. |

[a]A unit (U) is the quantity of a substance that has a physiological effect.
[b]A milliequivalent (mEq) is defined in Appendix C.

*Continued on next page*

## TABLE B.1 Blood, Plasma, or Serum Values *(continued)*

| Test | Normal Values | Clinical Significance |
|---|---|---|
| Lipids | Cholesterol 120–220 mg/100 mL; cholesterol esters 60–75% of cholesterol; phospholipids 9–16 mg/100 mL as lipid phosphorus; total fatty acids 190–420 mg/100 mL<br>Total lipids 450–1000 mg/100 mL<br>Triglycerides 40–150 mg/100 mL | Increased values for cholesterol and triglycerides are connected with increased risk of cardiovascular disease, such as heart attack and stroke. |
| Lithium | Toxic levels 2 mEq/L | |
| Osmolality | 285–295 mOsm/kg water | |
| Oxygen saturation (arterial) (see $P_{O_2}$) | 96–100% | |
| $P_{CO_2}$ | 35–43 mm Hg | Values decrease in acidosis, nephritis, and diarrhea; values increase in respiratory diseases, intestinal obstruction, and vomiting. |
| pH | 7.35–7.45 | Values decrease as a result of hypoventilation, severe diarrhea, Addison disease, and diabetic acidosis; values increase due to hyperventilation, Cushing syndrome, and vomiting. |
| $P_{O_2}$ | 75–100 mm Hg (breathing room air) | Values increase in polycythemia and decrease in anemia and obstructive pulmonary diseases. |
| Phosphatase (acid) | Male: total 0.13–0.63 U/mL<br>Female: total 0.01–0.56 U/mL | Values increase in cancer of the prostate gland, hyperparathyroidism, some liver diseases, myocardial infarction, and pulmonary embolism. |
| Phosphatase (alkaline) | 13–39 IU/L[c] (infants and adolescents up to 104 IU/L) | Values increase in hyperparathyroidism, some liver diseases, and pregnancy. |
| Phosphorus (inorganic) | 3–4.5 mg/100 mL (infants in first year up to 6 mg/100 mL) | Values increase in hypoparathyroidism, acromegaly, vitamin D hypervitaminosis, and kidney diseases; values decrease in hyperparathyroidism. |
| Potassium | 3.5–5 mEq/100 mL | |
| Protein | Total 6–8.4 g/100 mL<br>Albumin 3.5–5 g/100 mL<br>Globulin 2.3–3.5 g/100 mL | Total protein values increase in severe dehydration and shock; values decrease in severe malnutrition and hemorrhage. |
| Salicylate<br>    Therapeutic<br>    Toxic | 0 | 20–25 mg/100 mL<br>Over 30 mg/100 mL<br>Over 20 mg/100 mL after age 60 |
| Sodium | 135–145 mEq/L | Values increase in nephritis and severe dehydration; values decrease in Addison disease, myxedema, kidney disease, and diarrhea. |
| Sulfonamide<br>    Therapeutic | 0 | 5–15 mg/100 mL |
| Urea nitrogen | 8–25 mg/100 mL | Values increase in response to increased dietary protein intake; values decrease in impaired renal function. |
| Uric acid | 3–7 mg/100 mL | Values increase in gout and toxemia of pregnancy and as a result of tissue damage. |

[c]A unit (U) is the quantity of a substance that has a physiological effect.

## TABLE B.2  Blood Count Values

| Test | Normal Values | Clinical Significance |
| --- | --- | --- |
| Clotting (coagulation) time | 5–10 min | Values increase in afibrinogenemia and hyperheparinemia, severe liver damage. |
| Fetal hemoglobin | Newborns: 60–90%<br>Before age 2: 0–4%<br>Adults: 0–2% | Values increase in thalassemia, sickle-cell anemia, and leakage of fetal blood into maternal bloodstream during pregnancy. |
| Hemoglobin | Male: 14–16.5 g/100 mL<br>Female: 12–15 g/100 mL<br>Newborn: 14–20 g/100 mL | Values decrease in anemia, hyperthyroidism, cirrhosis of the liver, and severe hemorrhage; values increase in polycythemia, congestive heart failure, and obstructive pulmonary disease and at high altitude. |
| Hematocrit | Male: 40–54%<br>Female: 38–47% | Values increase in polycythemia, severe dehydration, and shock; values decrease in anemia, leukemia, cirrhosis, and hyperthyroidism. |
| Ketone bodies | 0.3–2 mg/100 mL<br>Toxic level: 20 mg/100 mL | Values increase in ketoacidosis, fever, anorexia, fasting, starvation, and a high-fat diet. |
| Platelet count | 250,000–400,000/µL | Values decrease in anemias and allergic conditions and during cancer chemotherapy; values increase in cancer, trauma, heart disease, and cirrhosis. |
| Prothrombin time | 11–15 s | Values increase in prothrombin and vitamin deficiency, liver disease, and hypervitaminosis A. |
| Red blood cell count | Males: 4.6–6.2 million/µL<br>Females: 4.2–5.4 million/µL | Values decrease in systemic lupus erythematosus, anemias, and Addison disease; values increase in polycythemia and dehydration and following hemorrhage. |
| Reticulocyte count | 1–3% | Values decrease in iron-deficiency and pernicious anemias and radiation therapy; values increase in hemolytic anemia, leukemia, and metastatic carcinoma. |
| White blood cell count, differential | Neutrophils 60–70%<br>Eosinophils 2–4%<br>Basophils 0.5–1%<br>Lymphocytes 20–25%<br>Monocytes 3–8% | Neutrophils increase in acute infections; eosinophils and basophils increase in allergic reactions; monocytes increase in chronic infections; lymphocytes increase during antigen-antibody reactions. |
| White blood cell count, total | 5000–9000/µL | Values decrease in diabetes mellitus, anemias and following cancer chemotherapy; values increase in acute infections, trauma, some malignant diseases, and some cardiovascular diseases. |

## TABLE B.3   Urine Values

| Test | Normal Values | Clinical Significance |
|---|---|---|
| Acetone and acetoacetate | 0 | Values increase in diabetic acidosis and during fasting. |
| Albumin | 0 to trace | Values increase in glomerular nephritis and hypertension. |
| Ammonia | 20–70 mEq/L | Values increase in diabetes mellitus and liver disease. |
| Bacterial count | Under 10,000/mL | Values increase in urinary tract infection. |
| Bile and bilirubin | 0 | Values increase in biliary tract obstruction. |
| Calcium | Under 250 mg/24 h | Values increase in hyperparathyroidism and decrease in hypoparathyroidism. |
| Chloride | 110–254 mEq/24 h | Values decrease in pyloric obstruction and diarrhea; values increase in Addison disease and dehydration. |
| Potassium | 25–100 mEq/L | Values decrease in diarrhea, malabsorption syndrome, and adrenal cortical insufficiency; values increase in chronic renal failure, dehydration, and Cushing syndrome. |
| Sodium | 75–200 mg/24 h | Values decrease in diarrhea, acute renal failure, and Cushing syndrome; values increase in dehydration, starvation, and diabetic acidosis. |
| Creatinine clearance | 100–140 mL/min | Values increase in renal diseases. |
| Creatinine | 1–2 g/24 h | Values increase in infections and decrease in muscular atrophy, anemia, and certain kidney diseases. |
| Glucose | 0 | Values increase in diabetes mellitus and certain pituitary gland disorders. |
| Urea clearance | Over 40 mL of blood cleared of urea per min | Values increase in certain kidney diseases. |
| Urea | 25–35 g/24 h | Values decrease in complete biliary obstruction and severe diarrhea; values increase in liver diseases and hemolytic anemia. |
| Uric acid | 0.6–1 g/24 h | Values increase in gout and decrease in certain kidney diseases. |
| Casts | | |
| Epithelial | Occasional | Values increase in nephrosis and heavy metal poisoning. |
| Granular | Occasional | Values increase in nephritis and pyelonephritis. |
| Hyaline | Occasional | Values increase in glomerular membrane damage and fever. |
| Red blood cell | Occasional | Values increase in pyelonephritis; blood cells appear in urine in response to kidney stones and cystitis. |
| White blood cell | Occasional | Values increase in kidney infections. |
| Color | Amber, straw, transparent yellow | Color varies with hydration, diet, and disease states. |
| Odor | Aromatic | Odor becomes acetone-like in diabetic ketosis. |
| Osmolality | 500–800 mOsm/kg water | Values decrease in aldosteronism and diabetes insipidus; values increase in high-protein diets, heart failure, and dehydration. |
| pH | 4.6–8 | Values decrease in acidosis, emphysema, starvation, and dehydration; values increase in urinary tract infections and severe alkalosis. |

## TABLE B.4  Hormone Levels

| Test | Normal Values |
| --- | --- |
| **Steroid hormones** | |
| Aldosterone | Excretion: 5–19 µg/24 h[a] |
|    Fasting, at rest, 210 mEq sodium diet | Supine: 48 ± 29 pg/mL[b] |
| | Upright: 65 ± 23 pg/mL |
|    Fasting, at rest, 10 mEq sodium diet | Supine: 175 ± 75 pg/mL |
| | Upright: 532 ± 228 pg/mL |
| Cortisol | |
|    Fasting | 8 AM: 5–25 µg/100 mL |
|    At rest | 8 PM: Below 10 µg/100 mL |
| Testosterone | Adult male: 300–1100 ng/100 mL[c] |
| | Adolescent male: over 100 ng/100 mL |
| | Female: 25–90 ng/100 mL |
| **Peptide hormones** | |
| Adrenocorticotropin (ACTH) | 15–170 pg/mL |
| Calcitonin | Undetectable in normal individuals |
| Growth hormone (GH) | |
|    Fasting, at rest | Below 5 ng/mL |
|    After exercise | Children: over 10 ng/mL |
| | Male: below 5 ng/mL |
| | Female: up to 30 ng/mL |
| Insulin | |
|    Fasting | 6–26 µU/mL |
|    During hypoglycemia | Below 20 µU/mL |
|    After glucose | Up to 150 µU/mL |
| Luteinizing hormone (LH) | Male: 6–18 mU/mL |
| | Preovulatory or postovulatory female: 5–22 mU/mL |
| | Midcycle peak 30–250 mU/mL |
| Parathyroid hormone | Less than 10 µEq/L[d] |
| Prolactin | 2–15 ng/mL |
| Renin activity | |
|    Normal diet | |
|       Supine | 1.1 ± 0.8 ng/mL/h |
|       Upright | 1.9 ± 1.7 ng/mL/h |
|    Low-sodium diet | |
|       Supine | 2.7 ± 1.8 ng/mL/h |
|       Upright | 6.6 ± 2.5 ng/mL/h |
| Thyroid-stimulating hormone (TSH) | 0.5–3.5 µU/mL |
| Thyroxine-binding globulin | 15.25 µg $T_4$/100 mL |
| Total thyroxine | 4–12 µg/100 mL |

[a] 1 microgram (1 µg) is equal to $10^{-6}$ g.
[b] 1 picogram (1 pg) is equal to $10^{-12}$ g.
[c] 1 nanogram (1 ng) is equal to $10^{-9}$ g.
[d] µEq microequivalent, 1/1000 mEq, which is defined in Appendix C.

# Appendix C

## Solution Concentrations

Physiologists often express solution concentration in terms of percent, molarity, molality, and equivalents.

### Percent

The weight-volume method of expressing percent concentrations states the weight of a solute in a given volume of solvent. For example, to prepare a 10% solution of sodium chloride, 10 g of sodium chloride is dissolved in a small amount of water (solvent) to form a salt solution. Then additional water is added to the salt solution to form 100 mL of salt solution. Note that the sodium chloride was dissolved in water and then diluted to the required volume. The sodium chloride was not dissolved directly in 100 mL of water.

### Molarity

**Molarity** determines the number of moles of solute dissolved in a given volume of solvent. A **mole** (abbreviated mol) of a substance contains $6.023 \times 10^{23}$ number (Avogadro number) of particles, such as atoms, ions, compounds, or molecules. A 1-molar (1-M) solution is made by dissolving 1 mol of a substance in enough water to make 1 L of solution. For example, 1 mol of sodium chloride solution is made by dissolving 58.44 g of sodium chloride in enough water to make 1 L of solution. One mole of glucose solution is made by dissolving 180.2 g of glucose in enough water to make 1 L of solution. Both solutions have the same number (Avogadro number) of sodium chloride or glucose compounds in solution.

### Molality

Although 1-M solutions have the same number of solute particles, they do not have the same number of solvent (water) molecules. Because 58.5 g of sodium chloride occupies less volume than 180 g of glucose, the sodium chloride solution has more water molecules. **Molality** is a method of calculating concentrations that takes into account the number of solute and solvent molecules. A 1-molal solution (1-*m*) is 1 mol of a substance dissolved in 1 kg of water. Thus, a 1-*m* solution of sodium chloride and a 1-*m* solution of glucose contain the same number of sodium chloride and glucose compounds dissolved in the same amount of water.

When sodium chloride (NaCl) is dissolved in water, it dissociates, or separates, to form two ions: a $Na^+$ and a $Cl^-$. Glucose, however, does not dissociate when dissolved in water. Although 1-*m* solutions of sodium chloride and of glucose have the same number of sodium chloride and glucose compounds, because of dissociation the sodium chloride solution contains twice as many particles as the glucose solution (one $Na^+$ and one $Cl^-$ for each glucose molecule). To report the concentration of these substances in a way that reflects the number of particles in a given mass of solvent, the concept of **osmolality** is used. The osmolality of a solution is the molality of the solution times the number of particles into which the solute dissociates in 1 kg of solvent. Thus, 1 mol of sodium chloride in 1 kg of water is a 2-osmolal (2-osm) solution because sodium chloride dissociates to form two ions.

The osmolality of a solution is a reflection of the number, not the type, of particles in a solution. Thus, a 1-osm solution contains 1 osm of particles per kilogram of solvent, but the particles may be all one type or a complex mixture of different types.

The concentration of particles in body fluids is so low that the measurement milliosmole (mOsm), 1/1000 of an osmole, is used. Most body fluids have an osmotic concentration of approximately 300 mOsm and consist of many different ions and molecules. The osmotic concentration of body fluids is important because it influences the movement of water into or out of cells (see chapter 3).

### Equivalents

**Equivalents** are a measure of the concentrations of ionized substances. One equivalent (Eq) is 1 mol of an ionized substance multiplied by the absolute value of its charge. For example, 1 mol of NaCl dissociates into 1 mol of $Na^+$ and 1 mol of $Cl^-$. Thus, there are 1 Eq of $Na^+$ (1 mol $\times$ 1) and 1 Eq of $Cl^-$ (1 mol $\times$ 1). One mole of $CaCl_2$ dissociates into 1 mol of $Ca^{2+}$ and 2 mol of $Cl^-$. Thus, there are 2 Eq of $Ca^{2+}$ (1 mol $\times$ 2) and 2 Eq of $Cl^-$ (2 mol $\times$ 1). In an electrically neutral solution, the equivalent concentration of positively charged ions is equal to the equivalent concentration of the negatively charged ions. One milliequivalent (mEq) is 1/1000 of an equivalent.

# Appendix D

## Answers to Critical Thinking Questions

### Chapter One

1. D is correct. Positive-feedback mechanisms result in movement away from homeostasis and are usually harmful. The continually decreasing blood pressure is an example. Negative-feedback mechanisms result in a return to homeostasis. The elevated heart rate is a negative-feedback mechanism that attempts to return blood pressure to a normal value. In this case, the negative-feedback mechanism was inadequate to restore homeostasis, and blood pressure continued to decrease. Medical intervention (a transfusion) increased blood volume and blood pressure. With the increase in blood pressure, the positive-feedback mechanism is interrupted and the negative-feedback mechanism is able to maintain blood pressure.
2. Student B is correct. As the muscles become more active, they use more oxygen. The increased rate of respiration maintains blood and muscle tissue oxygen levels (homeostasis) and is a negative-feedback mechanism.
3. a. Inferior
   b. Posterior (dorsal) or deep
   c. Distal or inferior
   d. Lateral
4. The heel is distal, inferior, and posterior to the kneecap. The kneecap is proximal, superior, and anterior to the heel.
5. The wedding band should be worn proximal to the engagement ring.
6. The pancreas is located in the right-upper and left-upper quadrants; it is located in the epigastric and left hypochondriac regions. The top of the urinary bladder is located in the right-lower quadrant and the left-lower quadrant; it is located in the hypogastric region. The rest of the urinary bladder is located in the pelvic cavity.
7. The pelvic cavity contains the uterus, which increases greatly in size during pregnancy as the fetus within the uterus grows. However, the pelvic cavity is surrounded by bones, which are not expandable. Therefore, the uterus moves superiorly into the abdominal cavity, crowding abdominal organs and dramatically increasing the size of the abdominal cavity.

8. After passing through the left thoracic wall, the first membrane to be encountered is the parietal pleura. Continuing through the pleural cavity, the visceral peritoneum and the left lung are pierced. Leaving the lung, the bullet penetrates the visceral pleura, the pleural cavity, and the parietal pleura (remember that the lung is surrounded by a double membrane sac). Next the parietal pericardium, the pericardial cavity, the visceral pericardium, and the heart are encountered.
9. The kidneys are located in the abdominal cavity but are retroperitoneal. When a person is lying prone, it is possible to cut through the posterior abdominal wall and remove a kidney without cutting through parietal peritoneum.

### Chapter Two

1. Because atoms are electrically neutral, the iodine atom has the same number of protons and electrons. The gain of an electron means the iodine ion has one more electron than protons and therefore a charge of minus one. The correct symbol is $I^{-1}$.
2. a. Dissociation
   b. Synthesis
   c. Decomposition
3. Muscle contains proteins and, to increase muscle mass, proteins must be synthesized. The synthesis of molecules in living organisms results from chemical reactions that require the input of energy. The energy needed to drive these synthesis reactions comes from chemical reactions that release energy. The energy-releasing reactions occur during the decomposition of food molecules.
4. The sodium bicarbonate dissociates to form sodium ions ($Na^+$) and bicarbonate ions ($HCO_3^-$). Because this is a reversible reaction, the $HCO_3^-$ added to the solution bind with $H^+$ to form carbon dioxide and water. The decrease in $H^+$ causes the pH to increase (become more basic).
5. The slight amount of heat functioned as activation energy and started a chemical reaction. The reaction released energy, especially heat, which caused the solution to become very hot.

6. There was most likely at least one enzyme present that was required for one of the initial reactions. Boiling denatured any enzymes present in the solution. Without enzymes, the reaction(s) will not occur.

### Chapter Three

1. B is the most logical conclusion. Swollen lung tissue suggests that the tissues had been submerged in a hypotonic solution, such as fresh water. Because the bay contains salt water, which is slightly hypertonic to blood, it is unlikely he drowned in the bay. It is more likely he drowned in fresh water and was later placed in the bay. Although this is the most logical conclusion of those given, there are possibilities other than murder. For example, he may have accidentally drowned in a fresh water stream and then been washed into the bay.
2. B is correct. A solution that is isotonic causes cells to neither shrink nor swell. Therefore, there is no net movement of water between the blood and the dialysis fluid. Also, if the solution has the same composition as blood except for having no urea, only urea diffuses from a higher concentration in the blood to the lower concentration in the dialysis fluid. A solution that is isotonic and contains only protein will cause many different small molecules (not just urea) that are found in blood to diffuse into the dialysis fluid. Distilled water is hypotonic and causes blood cells to swell and undergo lysis. Also, molecules that are able to pass through the dialysis membrane diffuse from the blood into the distilled water. Blood is not a good solution to use because it would have to be blood that has already had urea removed in order to be useful as dialysis fluid.
3. Because the plasma membrane stays the same size, even though small pieces of membrane from secretory vesicles are continually added to the plasma membrane, one must conclude that some process is removing small pieces of membrane at the same rate that they are added. The cell membrane is constantly being recycled.

4. The main function of this cell is to secrete proteins. Well-developed rough endoplasmic reticulum for protein production, well-developed Golgi apparatuses for the packaging of proteins for secretion, and many vesicles for exocytosis are typical of many cells that secrete proteins. Many mitochondria produce large amounts of ATP. The ATP serves as an energy source for forming the bonds between the amino acids, for the active transport of amino acids into the cell, and for exocytosis of the completed proteins. Many microvilli, which greatly increase surface area for transport, can also be found in cells that produce and secrete proteins. However, large numbers of mitochondria or large amounts of microvilli can be found in other types of cells as well.

5. The sickle-cell gene for producing the protein has a different nucleotide than the normal gene for producing the protein. Because the nucleotides are the set of instructions for making the protein, the change means the instructions contain an error. When mRNA copies the instructions from the faulty DNA, the error is also copied. Consequently, when the amino acids are joined together to form the protein at the ribosome, the instructions are wrong and the protein is incorrectly assembled. A substitution of a single nucleotide results in a different amino acid in the chain of amino acids that make up the protein. This change alters the shape of the protein. Just as the function of enzymes depends on their shape, the functions of other proteins depend on their shape. The incorrectly manufactured proteins in the red blood cells have the wrong shape and do not stack correctly. As a result of the abnormal stacking, the red blood cells have an abnormal, sickle shape.

## Chapter Four

1. Pseudostratified columnar ciliated epithelium is found in the trachea. It produces mucus that traps dust and debris in air. The cilia move the mucus with entrapped dust and debris to the throat, where it is swallowed. In heavy smokers, the pseudostratified columnar epithelium is replaced by stratified squamous epithelium, which serves a protective function against the irritating materials in the smoke. Unfortunately, this type of epithelium is not ciliated, so removal of foreign materials from the trachea is more difficult. After 2 years without smoking, the original pseudostratified columnar ciliated epithelium replaces the stratified squamous epithelium and debris removal resumes.

2. Tight junctions prevent the passage of materials between the epithelial cells.

3. A secretory epithelium is generally a simple epithelium. To manufacture large amounts of enzymes, a cuboidal or columnar cell with the appropriate organelles, such as rough endoplasmic reticulum and Golgi aparatuses, would be expected. The pancreas is formed of simple columnar epithelium. The epithelium has microvilli on its free surface, which increase the surface area and facilitate secretion. The tight junctions that connect the epithelial cells of the pancreatic glands and duct system to each other prevent damage to the underlying tissues (by the action of pancreatic digestive enzymes).

4. a. The stratified squamous epithelium that lines the mouth provides protection. Replacement of it with simple columnar epithelium makes the lining of the mouth much more susceptible to damage because the single layer of epithelial cells is easier to damage.
   b. Tendons attach muscles to bone. When muscles contract, muscles pull on the tendons and, because the tendons are attached to bone, the bones move. If the tendons contained elastic tissue, the tendons would be more like elastic bands. The muscles would contract and stretch the tendons. Not all of the force of muscle contraction would be transferred to the bones to cause them to move.
   c. If bones were made of elastic cartilage, they would be much more flexible and could bend and then return to their original shape. They would not be rigid structures, like bone, that support our weight and result in efficient movement.

5. The fibers are organized along the direction of pull on the ligaments—that is, parallel to the length of the ligament. When the back is bent, the ligaments are stretched. The elastic recoil of the stretched ligaments helps straighten the back.

6. When blood is ejected into the aorta, the aortic wall expands and the diameter of the aorta increases. The elastic fibers are arranged in a circular fashion, so that, when the aortic wall expands, the elastic fibers are stretched. Recoil of the fibers causes the aorta to resume its original shape, which helps force blood to flow through the aorta.

7. Chemical mediators of inflammation normally produce beneficial responses, such as dilation of small-diameter blood vessels and increased blood vessel permeability. Blocking these effects could reduce the body's ability to deal with harmful agents, such as bacteria. On the other hand, antihistamines also reduce many of the unpleasant symptoms of inflammation, making the patient more comfortable, and this can be considered beneficial. Antihistamines are commonly taken to prevent allergy symptoms that are often an overreaction of the inflammatory response to foreign substances, such as pollen.

8. Collagen synthesis is required for granulation tissue and scar formation. If collagen synthesis does not occur because of a lack of vitamin C, or if collagen synthesis is slowed, wound healing does not occur or is slower than normal. One might expect that the density of collagen fibers in a scar is reduced and the scar is not as durable as a normal scar.

## Chapter Five

1. Yes, the skin (dermis) can be over stretched as a result of obesity.

2. Alcohol is a solvent that dissolves lipids (see chapter 2). It removes the lipids from the skin, especially in the stratum corneum. The rate of water loss increases after soaking the hand in alcohol because of the removal of the lipids that normally prevent water loss.

3. Carotene, a yellow pigment from ingested plants, accumulates in lipids. The stratum corneum of a callus has more layers of cells than other noncallused parts of the skin and the cells in each layer are surrounded by lipids. The carotene in the lipids make the callus appear yellow.

4. The vermillion border is covered by keratinized epithelium that is a transition between the nonkeratinized stratified epithelium of the mucous membrane and the keratinized stratified epithelium of the facial skin. The mucous membrane has mucous glands, which secrete mucus (see chapter 4). The mucus helps keep the inner surface of the lips moist. In addition, the inner surface of the lips is "sealed off" most of the time and is moistened by saliva. In contrast to facial skin, the skin of the vermillion border does not have sebaceous or sweat glands. Without sebaceous glands, the surface of the vermillion border is not protected against drying by sebum. Also, the vermillion border is not as heavily keratinized as facial skin; that is, the number of cell layers with surrounding lipids is less. Consequently, the vermillion border dries out more easily.

5. The hair follicle, but not the hair, is surrounded with nerve endings that can detect movement or pulling of the hair. The hair is dead, keratinized epithelium, so cutting the hair is not painful.

6. Several methods have some degree of success in treating acne: (1) Kill the bacteria. One effective agent is benzoyl peroxide in some acne medications. Antibiotics have also been used to treat severe cases of acne. (2) Prevent sebum production. Isotretivioin, or Accutane (a derivative of vitamin A), has proven effective in preventing sebum production. However, Accutane produces birth defects and should be used with caution. Estrogen, a female sex hormone, has also been used to treat severe cases of acne. (3) Unplug the follicle. Some sulfur compounds speed up peeling of the skin and unplug the follicle.

7. Rickets is a disease of children resulting from inadequate vitamin D. With inadequate vitamin D, there is insufficient absorption of calcium from the intestine, resulting in soft bones. If adequate vitamin D is ingested, rickets is prevented, whether one is dark- or fair-skinned. However, if dietary vitamin D is inadequate, when the skin is exposed to ultraviolet light, it can produce a precursor molecule that can be converted to vitamin D. Dark-skinned children are more susceptible to rickets because the additional melanin in their skin screens out the ultraviolet light and they produce less vitamin D.

8. When first exposed to the cold temperature just before starting the run, the blood vessels in the skin constrict to conserve heat. This produces a pale skin color. Dilation of skin blood vessels does not occur at this time because the skin has not been exposed to the cold long enough to cause skin temperature to fall below 15°C (27°F). After running awhile, as a result of the excess heat generated by the exercise, the blood vessels in the skin dilate. This results in heat loss and helps prevent overheating. Increased blood flow through the skin causes it to turn red. After the run, the body still has excess heat to eliminate, so for awhile the skin remains red.

## Chapter Six

1. The injury separated the head (epiphysis) from the shaft (diaphysis) at the epiphyseal plate, which is cartilage. Because the epiphyseal plate is the site of bone elongation, damage to the epiphyseal plate can interfere with bone elongation, resulting in a shortened limb.

2. If Justin landed on his heels, he could have broken one or both of his heel bones, the calcaneus. When Hefty Stomper stepped on Justin's foot, the most likely bones broken would be the metatarsal bones, which make up the main structure of the foot.

3. The humerus articulating with the ulna and radius forms the elbow joint. The ulna fits tightly over the end of the humerus. The radius attaches to both the humerus and the ulna, but not as tightly to the humerus as does the ulna. Pulling forcefully on the forearm would, therefore, be more likely to pull the radius away from the humerus than the ulna.

4. Because the female pelvis is relatively wider than the male pelvis, the heads of the femora (pl. of *femur*) are placed wider apart. For the lower limbs to be placed directly below the center of gravity in the body, the femora must be angled in more sharply in females than in males. This angle may bring the knees closer together, resulting in the condition called knock-kneed.

5. All skeletons, including those of humans, have features unique to each species. A human skeleton can be identified based on its unique anatomical features. In addition, female skeletons have several anatomical features that distinguish them from male skeletons. For example, the female pelvis differs in several features from the male pelvis (see figure 6.32).

## Chapter Seven

1. Botulism toxin decreases acetylcholine release in the neuromuscular junction. This prevents action potentials in skeletal muscle cells. Thus, the respiratory muscles (e.g., the diaphragm) relax and do not contract. Other explanations that you could have proposed because they would also lead to respiratory failure are that the toxin prevents acetylcholine from binding to its receptor on skeletal muscle cells and that the toxin inhibits the production of acetylcholine.

2. Harvey's hands became fatigued as a result of ATP depletion. Without ATP, the muscles moving the fingers could not contract and, so, the fingers could not flex or extend. In addition, his inability to release his grip is an indication of physiological contracture, in which too little ATP is present to allow cross-bridge release.

3. Both the 100-m dash and weight lifting (to a lesser extent) involve short bursts of anaerobic activity, so in both cases the researcher should expect to find more fast-twitch muscle in the gastrocnemius of these athletes. The 10,000-m run involves aerobic metabolism, and the outstanding athlete should have more slow-twitch fibers.

4. The exercise that best builds the anterior arm muscles is flexion of the elbow against force, such as in pull-ups. A pull-up with the hand supinated builds both the biceps and the brachialis, but a curl with the hand pronated builds the brachialis to a greater extent. The posterior arm is best built by extension of the elbow, as in push-ups or dips. The anterior forearm muscles are best built up by forcefully flexing the wrist and fingers. The anterior thigh is best built by extending the knee, as in partial knee bends. The posterior leg muscles are best built by forcefully plantar flexing the ankle and toes, as in standing on one's toes. The abdominal muscles are best built by flexing the vertebral column against resistance, as in leg lifts or partial sit-ups. Of course, each exercise has to be repeated many times to increase the size of the muscles.

5. This is a typical hamstring injury (pulled hamstring), in which one or more of the hamstring muscles are either pulled away from their attachment to bone or torn. The slight flexion of the knee and the bulge in the muscle occur because muscle fibers contract in the hamstrings, but the tear does not allow antagonist muscles in the anterior thigh to pull on the muscle fibers to lengthen them. The pain results from the damaged tissue in the muscle (flexion and pain are also due to spasmodic contractions of the muscle in response to the injury). Because of the tissue damage and cramps, the muscles could not respond to voluntary nervous control and could not contract.

## Chapter Eight

1. If one series of neurons had more neurons, it would have more synapses, which should slow down the rate of action potential propagation. Also, if one series were unmyelinated and the other myelinated, the unmyelinated series would be slower.

2. The phrenic nerves are the nerves to the diaphragm. The left phrenic nerve would be cut to paralyze the left side of the diaphragm.

3. a. Radial nerve. This nerve supplies posterior (extensor) muscles. When these muscles are paralyzed, the flexor muscles are unopposed and result in the flexed limb described.
   b. Femoral nerve

4. The cerebellum controls motor functions, such as balance, muscle tone, and fine motor movements. Cerebellar dysfunction is characterized as a loss of balance, a loss of motor tone, and an inability to control fine motor movements, such as touching one's nose with the eyes closed. Such patients appear to be drunk (alcohol apparently most directly affects the cerebellum).

5. The blow to the head erased Louis's short-term memory, which would have extended to about 10 minutes before the blow, and that information was never transferred to long-term memory.

6. a. Hypoglossal nerve (cranial nerve XII)
   b. Optic nerve (cranial nerve II)
   c. Trigeminal nerve (cranial nerve V)
   d. Facial nerve (cranial nerve VII)
   e. Oculomotor nerve (cranial nerve III)

7. Nervous system stimulation of the digestive system occurs primarily through the vagus nerve (cranial nerve X), which arises from the brain. Therefore, an injury to the spinal cord at the level of C6 does not affect the functioning of this nerve.

## Chapter Nine

1. The lenses of the eye are biconvex structures and are most important in close vision as they are thickened (made more convex) by relaxation of the suspensory ligaments (accommodation). Therefore, if they were removed, close vision would be most affected, and convex glasses would do the best job of replacing the lenses for near vision, such as when reading.

2. The fovea centralis, where a person focuses on an object, contains almost entirely cones. Cones do not detect faint light, so dim stars, which can be detected by the rods, are lost from vision when the person looks directly at them. When the person looks slightly to the side, the more peripherally located rods can again detect the faint light from the stars.

3. Under normal conditions, pressure changes in the middle ear occur through the auditory tube. The auditory tube connects the middle ear to the pharynx, which opens to the outside of the body through the oral and nasal cavities. This allows air pressure to equalize on both sides of the eardrum. When a person is under water, the increased water pressure on the outside of the eardrum forces the eardrum toward the middle ear because the external water pressure exceeds the air pressure in the middle ear. As a result, the eardrum cannot vibrate as freely as when the pressure is equalized, and sound transmission is dampened. If the pressure difference becomes too great, it can rupture the eardrum.

4. Sound is normally transmitted through the middle ear by vibration of the auditory ossicles. A vibrating tuning fork touching the mastoid process sets up vibrations in the temporal bone, causing the perilymph and endolymph of the inner ear to vibrate. These vibrations cause the basilar membrane to vibrate and are therefore perceived as sound.

5. When a contaminated hand rubs the eyes, the virus can be introduced into tears on the conjunctiva. From there the virus can spread into the lacrimal canaliculi and pass through the nasolacrimal duct into the nasal cavity.

## Chapter Ten

1. The hormone binds to a membrane-bound receptor and activates a G protein mechanism. Because a drug inhibits the binding of GTP to a protein, GTP cannot bind to the $\alpha$ subunit of the G protein complex. Also, the fact that a drug that inhibits the breakdown of cyclic-AMP causes an increased response suggests that the $\alpha$ subunit of the G protein complex, with GTP bound to it, activates the enzymes that are responsible for the synthesis of cyclic-AMP.

2. The effect of ADH on cells is mediated through membrane-bound receptors, whereas the effects of aldosterone on cells is mediated through intracellular receptors. In addition, ADH levels increase more rapidly than aldosterone levels in response to a decrease in blood volume. ADH is secreted within minutes by the posterior pituitary gland in response to dehydration of the body. The hypothalamus detects a decrease in blood volume and an increase in the concentration of the blood, which stimulates neurons in the hypothalamus to secrete ADH. ADH binds to membrane-bound receptors in the kidney. The cells of the kidney respond within minutes. As a result, the kidney produces a small volume of very concentrated urine. Aldosterone is secreted in response to decreased blood pressure (caused by decreased blood volume), but the process involves other compounds. The kidneys respond to a decrease in blood pressure by secreting renin. Renin acts as an enzyme to break down angiotensinogen into angiotensin I. Angiotensin-converting enzyme converts angiotensin I to angiotensin II. Finally, angiotensin II stimulates the adrenal cortex to increase aldosterone secretion. Aldosterone binds to intracellular receptors in cells of the kidney. The response of these cells is to increase specific mRNA and protein synthesis. The new mRNA and protein synthesis requires a significant amount of time. The response to the synthesis of new proteins is to increase the rate of solute reabsorption in the kidney, including $Na^+$. This enhances water reabsorption by the kidney.

3. Vitamin D increases the rate of transport of $Ca^{2+}$ across the wall of the intestine. Therefore, in response to high concentrations of vitamin D, blood $Ca^{2+}$ levels can become abnormally high. Because the blood $Ca^{2+}$ levels are elevated, the rate of PTH secretion decreases to low levels. Also, the elevated blood $Ca^{2+}$ levels stimulate calcitonin secretion.

4. If the adrenal cortex degenerated, glucocorticoids, mineralocorticoids, and the androgens normally secreted by the adrenal cortex would no longer be secreted. A lack of glucocorticoid hormone secretion results in a reduced ability to maintain blood nutrient levels, such as glucose, between meals or during periods when no food is available. The ability to inhibit inflammation is also lost. Reduced levels of mineralocorticoids result in the loss of $Na^+$ in the urine and the loss of water in the urine. Thus, the ability to regulate blood pressure is reduced. Reduced secretion of androgens also occurs. Because the testes secrete androgens, the reduced androgen secretion in males is not significant. The effect of reduced androgens in females is unclear, although adrenal androgens may influence sexual behavior, to some degree, in females.

5. Aldosterone increases $Na^+$ reabsorption from the distal nephron and it increases $K^+$ secretion into the nephron. Elevated aldosterone secretion results in the increased retention of $Na^+$ and $H_2O$ in the kidney and an increased rate of $K^+$ elimination in the urine. As a result, blood pressure increases and blood $K^+$ levels can get very low.

6. After 24 hours without food, blood glucose levels will be decreasing. This will activate mechanisms that cause blood glucose levels to increase. The decreasing blood glucose levels will result in an increased rate of secretion of cortisol, a glucocorticoid, from the adrenal cortex, epinephrine from the adrenal medulla, and glucagon from pancreatic islets. The decreasing blood glucose levels result in increased ACTH secretion from the anterior pituitary gland; increased sympathetic nervous system stimulation of the adrenal medulla, resulting in increased epinephrine secretion; and

increased secretion of glucagon from the pancreatic islets. Cortisol increases fat and protein breakdown, so that these substances can be used as energy sources. It also causes increased glucose synthesis, primarily from amino acids. Epinephrine and glucagon both bind to receptors on the liver and increase the release of glucose from the liver. The decreasing blood glucose levels inhibit insulin secretion. The decreased insulin secretion slows the uptake of glucose by tissues. Thus, blood glucose levels are maintained within normal levels.

7. Stressful conditions cause increased ACTH secretion. Increased ACTH stimulates the secretion of glucocorticoids, such as cortisol. One of the effects of increased cortisol secretion is inhibition of the immune response. This could lead to an increased likelihood of a variety of infections, including colds and other diseases. Stomach pains may be due to inflammation of the stomach. Many stomach ulcers result from inflammation caused by bacteria.

## Chapter Eleven

1. Blood doping increases the number of red blood cells in the blood, increasing its oxygen-carrying capacity. However, the increased number of red blood cells also makes it more difficult for the blood to flow through the blood vessels, increasing the workload on the heart. The same effect is seen in erythrocytosis (polycythemia). In addition, if the process is not done properly, there is a risk of infection.

2. Symptoms resulting from decreased red blood cells are associated with a decreased ability of the blood to carry oxygen: shortness of breath, weakness, fatigue, and pallor. Symptoms resulting from decreased platelets are associated with a decreased ability to form platelet plugs and clots: small areas of hemorrhage in the skin, bruises, and decreased ability to stop bleeding. A symptom resulting from decreased lymphocytes is an increased susceptibility to infections and diseases.

3. The hypoventilation results in decreased blood oxygen, which stimulates the release of erythropoietin from the kidneys. The erythropoietin stimulates red blood cell production in the red bone marrow. Consequently, red blood cell numbers increase.

4. Removal of the stomach results in removal of intrinsic factor, which is necessary for adequate vitamin $B_{12}$ absorption in the small intestine. Therefore, the patient would develop pernicious anemia. Vitamin $B_{12}$ injections can be used to prevent pernicious anemia.

5. Vitamin $B_{12}$ and folate are necessary for blood cell division. Lack of these vitamins results in anemia. Iron is necessary for the production of hemoglobin. Lack of iron results in iron-deficiency anemia. Vitamin K is necessary for the production of many blood-clotting factors. Lack of vitamin K can greatly increase blood-clotting time.

6. The anemia results from too little hemoglobin. Because there is less hemoglobin, less hemoglobin is broken down into bilirubin. Consequently, less bilirubin is excreted as part of the bile into the small intestine. With decreased bilirubin in the small intestine, bacteria produce fewer of the pigments that normally color the feces.

7. Reddie Popper has hemolytic anemia. The RBC is lower than normal because the red blood cells are being destroyed faster than they are replaced. With fewer red blood cells, hemoglobin and hematocrit are lower than normal. Bilirubin levels are above normal because of the breakdown of the hemoglobin released from the ruptured red blood cells.

## Chapter Twelve

1. A heart murmur is an abnormal heart sound. It is unlikely that an ECG will reveal that he has a heart murmur because an ECG monitors the electrical activity of the heart, not heart sounds. A stethoscope is a more likely instrument for detecting a heart murmur. A condition that causes a heart murmur could also cause an obvious abnormality in the ECG, but that is not always the case.

2. Starling's law of the heart is an intrinsic regulatory mechanism, whereas parasympathetic innervation of the heart is a component of the extrinsic regulation of the heart. Cutting the vagus nerve does not significantly affect the ability of Starling's law of the heart to operate.

3. Cutting sensory nerve fibers from baroreceptors would reduce the frequency of action potentials delivered to the medulla oblongata from the baroreceptors. This results because the normal blood pressure stimulates the baroreceptors. An increase in blood pressure increases the action potential frequency, and a decrease in blood pressure decreases the action potential frequency. Because cutting the sensory nerve fibers decreases the action potential frequency, this acts as a signal to the medulla oblongata that a decrease in blood pressure has occurred, even though the blood pressure has not decreased. The medulla oblongata responds by increasing sympathetic action potentials and reducing parasympathetic action potentials delivered to the heart. Consequently, the heart rate increases.

4. When the internal carotid arteries are clamped, blood flow and blood pressure in the clamped internal carotid arteries decrease dramatically. Thus, blood pressure in the area of the baroreceptors in the internal carotid arteries is low. Sensory neurons carry a lower frequency of action potentials from the baroreceptors of the internal carotid arteries to the medulla oblongata. The cardioregulatory center within the medulla oblongata responds as if the blood pressure has decreased by causing an increase in sympathetic stimulation and a decrease in parasympathetic stimulation of the heart. In addition, epinephrine and norepinephrine are released from the adrenal medulla. Consequently, there is an increased heart rate and stroke volume. Therefore, blood pressure in the aorta increases.

5. A drug that blocks $Ca^{2+}$ channels decreases the heart rate and the force of contraction of the heart. This occurs because $Ca^{2+}$ are involved in the depolarization of the cardiac muscle cells. If fewer $Ca^{2+}$ flow into the cardiac muscle cells, the rate and degree of depolarization decrease. The result is that action potentials develop more slowly and depolarization does not occur to the normal degree. Slower development of action potentials slows the heart rate. The degree of depolarization causes fewer $Ca^{2+}$ to enter the cell; consequently, the force of contraction decreases.

6. Consuming a large amount of fluid increases the total volume of the blood, at least until the mechanisms that regulate blood volume decrease the blood volume to normal values. Increased blood volume causes an increase in venous return to the heart. Because of Starling's law of the heart, the increased venous return results in an increased stroke volume. A slight increase in the heart rate also occurs. Mechanisms that regulate blood pressure, such as the baroreceptor reflex, would prevent a large increase in blood pressure. Therefore, there may be an increased stroke volume but the blood pressure would not increase dramatically.

7. Cardiac output is influenced by the heart rate and stroke volume (CO = SV × HR). An athlete's cardiac output can be equal to

a nonathlete's cardiac output while they are both at rest, even though the athlete's heart rate is lower than the nonathlete's heart rate because the stroke volume of the athlete's heart is greater than the stroke volume of the nonathlete's heart. Athletic training causes a gradual hypertrophy (enlargement) of the heart. Therefore, athletes can maintain a cardiac output that is equal to a nonathlete's because they have an increased stroke volume but a decreased heart rate. During exercise, an athlete's heart rate and stroke volume increase. The stroke volume of the athlete's heart is much greater than the stroke volume of the nonathlete's. Therefore, the cardiac output of the athlete's heart is greater than the cardiac output of the nonathlete's during exercise.

8. The walls of the ventricles are thicker than the walls of the atria because the force that the ventricles must produce is greater than the force the atria must produce. The pressure in the ventricles during ventricular systole is substantially higher than the pressure in the atria during atrial systole. In addition, most of the ventricular filling (approximately 70%) occurs before contraction of the atria. Contraction of the atria is responsible for only about 30% of ventricular filling. In contrast, all of the blood ejected from the ventricles during ventricular systole is ejected because of contraction of the ventricles.

9. An incompetent aortic semilunar valve is leaky. Thus, when it is closed, blood is able to leak through it back into the left ventricle from the aorta. The aortic semilunar valve is normally closed from the beginning of ventricular diastole until just after the beginning of ventricular systole. During ventricular diastole, blood flows from the left atrium into the left ventricle. If the aortic semilunar valve is leaky, blood will also flow through the aortic semilunar valve into the left ventricle. Consequently, the volume of the left ventricle and the pressure in the left ventricle become greater than normal during ventricular diastole. Because blood is leaking from the aorta, the aortic pressure, during ventricular diastole, is lower than normal in the aorta. Because of Starling's law of the heart during ventricular systole, the ventricular muscle contracts with a greater force and forces a greater volume of blood into the aorta. Consequently, the pressure in the left ventricle is greater than normal during ventricular systole. Also, pressure in the aorta is greater than normal during ventricular systole.

# Chapter Thirteen

1. a. For the right side of the brain: the aorta, the brachiocephalic artery, the right common carotid artery, the right internal carotid artery, the cerebral arterial circle, and then the brain tissue. For the left side of the brain: the aorta, the left common carotid artery, the left internal carotid artery, the cerebral arterial circle, and then the brain tissue.
   b. For the left external portion of the skull: the aorta, the brachiocephalic artery, the right common carotid artery, the right external carotid artery, and then the external portion of the skull. For the left side of the external portion of the skull: the aorta, the left common carotid artery, the left external carotid arery, and then the external portion of the skull.
   c. The aorta, the left subclavian artery, the left axillary artery, the left brachial artery, either the left radial or the ulnar artery, and the left hand
   d. The aorta, the right common iliac artery, the external iliac artery, the femoral artery, the popliteal artery, the anterior tibial artery, and the anterior portion of the right leg
2. a. The left internal jugular vein, the left brachiocephalic vein, the superior vena cava, and the right atrium of the heart
   b. The right external jugular vein, the right subclavian vein, the right brachiocephalic vein, the superior vena cava, and the right atrium of the heart
   c. The superficial veins of the left hand and forearm; either the left cephalic or the left basilic vein; the left cephalic vein and the left basilic vein empty into the axillary vein; the left axillary vein, the left subclavian vein, the left brachiocephalic vein, the superior vena cava, and the right atrium of the heart
   d. The right great saphenous vein, the right femoral vein, the right external iliac vein, the right common iliac vein, the inferior vena cava, and the right atrium of the heart
   e. The renal vein, the inferior vena cava, and right atrium of the heart
   f. The superior mesenteric vein, the hepatic portal vein, the liver, the hepatic veins, the inferior vena cava, and right atrium of the heart
3. The femoral artery and vein are close to the surface in the femoral triangle, which is in the superior and medial part of the thigh. The femoral artery is the vessel into which the catheter will be placed. The femoral vein cannot be used because

extending a catheter superiorly in the femoral vein will deliver the catheter to the right side of the heart. The anterior interventricular artery is a branch of the left coronary artery, which originates from the ascending aorta, just superior to the aortic semilunar valve. The catheter is inserted into the femoral artery. From there, it passes through the external iliac, through the common iliac artery, and through the aorta. It passes through the abdominal aorta, the thoracic aorta, and the aortic arch to the beginning of the left coronary artery and then to the anterior interventricular artery.

4. Cells from the tumor can spread from the colon to the liver through the hepatic portal vein. Cells from the tumor can enter the superior mesenteric or the inferior mesenteric vein. The cells can pass from the superior mesenteric vein to the hepatic portal vein. The cells can also pass from the inferior mesenteric vein to the splenic vein and then the hepatic portal vein. The cells can pass through the hepatic portal vein to the liver.

5. Reduced blood flow to the kidney stimulates renin secretion. Renin acts on angiotensinogen to produce angiotensin I. Angiotensin I is converted to angiotensin II by the action of angiotensin-converting enzyme. Angiotensin II causes vasoconstriction, which increases blood pressure. In addition, angiotensin II stimulates the secretion of aldosterone from the adrenal cortex. Aldosterone acts on the kidney, causing the reabsorption of $Na^+$ and water, thus increasing the blood volume. The increased blood volume results in increased blood pressure.

6. During exercise, vasoconstriction occurs in the viscera, but vasodilation occurs in the exercising muscles. Even though the cardiac output is increased because of the increase in stoke volume and heart rate, the blood pressure does not go up as much as it would if vasoconstriction occurred in the viscera without vasodilation of the blood vessels in the exercising skeletal muscles.

7. Dilation of arteries and veins allows blood to accumulate in the dilated blood vessels. Consequently, less blood is returned to the heart (decreased venous return). Because the venous return is reduced, the stroke volume of the heart decreases (see Starling's law of the heart—chapter 12). Consequently, the heart does less work and less oxygen is required to support the contraction of the heart. Therefore, angina pains, which are caused by inadequate oxygen delivery to the heart muscle, are reduced.

# Chapter Fourteen

1. Massage moves lymph through the lymphatic vessels in the same fashion as does contraction of skeletal muscle. The application of pressure periodically to lymphatic vessels forces lymph to flow toward the trunk of the body, but valves prevent the flow of lymph in the reverse direction. The removal of lymph from the tissue helps relieve edema. Elevation of the limb allows gravity to assist the movement of lymph toward the heart.

2. Normally, T cells are processed in the thymus and then migrate to other lymphatic tissues. Without the thymus, this processing is prevented. Because there are normally five T cells for every B cell, the number of lymphocytes is greatly reduced. The loss of T cells results in an increased susceptibility to infection and an inability to reject grafts because of the loss of cell-mediated immunity. In addition, because T cells are involved with the activation of B cells, antibody-mediated immunity is also depressed.

3. Injection B results in the greatest amount of antibody production. At first, the antigen causes a primary response. A few weeks later, the slowly released antigen causes a secondary response, resulting in a greatly increased production of antibodies. Injection A does not cause a secondary response because all of the antigen is eliminated by the primary response.

4. Active immunity varies from a few weeks (e.g., the common cold) to lifelong (e.g., measles). Immunity can be long-lasting if enough memory cells (B or T) are produced and persist to respond to later antigen exposure. Passive immunity is not long-lasting because the individual does not produce his or her own memory cells. Because active immunity can last longer than passive immunity, it is the preferred method in most cases. However, passive immunity is preferred in situations in which immediate protection is needed, because it takes time for active immunity to develop.

5. If the patient has already been vaccinated, the booster shot stimulates a memory (secondary) response and rapid production of antibodies against the toxin. If the patient has never been vaccinated, vaccinating him is not effective because there is not enough time for the patient to develop his own primary response. Therefore, an antiserum is given to provide immediate, but temporary, protection. Sometimes both are given: The antiserum provides short-term protection and the tetanus vaccine stimulates the patient's immune system to provide long-term protection. If the shots are given at the same location in the body, the antiserum (antibodies against the tetanus toxin) could cancel the effects of the tetanus vaccine (tetanus toxin altered to be nonharmful).

6. Resistance to extracellular bacterial infections is accomplished by antibody-mediated immunity, which is not functioning properly here. Maternal antibodies (IGg) that crossed the placenta provided protection following birth but eventually were degraded. Resistance to intracellular viral infections is accomplished by cell-mediated immunity, which appears to be working normally.

7. At the first location, an antibody-mediated response results in an inflammatory response (immediate hypersensitivity). The combination of antibodies with the allergen triggers the release and activation of inflammatory chemicals. At the second location, a cell-mediated response also results in an inflammatory response (delayed hypersensitivity). This probably involves the release of lymphokines and the lysis of cells. At the other locations, there is neither an antibody-mediated nor a cell-mediated response.

8. The ointment was a good idea for the poison ivy, which caused a delayed hypersensitivity reaction—that is, too much inflammation. For the scrape, it is a bad idea because a normal amount of inflammation is beneficial and helps fight infection in the scrape.

9. Because both antibodies and cytokines produce inflammation, the fact that the metal in the jewelry resulted in inflammation is not enough information to answer the question. However, the fact that it took most of the day (many hours) to develop the reaction indicates a delayed hypersensitivity reaction and, therefore, cytokines.

# Chapter Fifteen

1. Pressing firmly on the base of the scapulae decreases the volume of the thoracic cavity, pleural pressure increases, and alveolar pressure increases to a value greater than atmospheric pressure. Thus, air flows from the lungs. When the arms are lifted, gravity causes the thoracic cavity to sag downward and expand. The volume of the thoracic cavity increases, pleural pressure decreases, and alveolar pressure decreases to less than atmospheric pressure, causing air to flow into the lungs.

2. a. The victim's lungs expand because the rescuer is blowing air into the lungs at a pressure higher than alveolar pressure. When alveolar pressure exceeds pleural pressure, the lungs tend to expand.

   b. Air flows out of the lungs because of the natural recoil of the lungs. Elastic fibers in the lungs and surface tension of water in the alveoli cause the recoil of the lungs.

3. The best prediction would be that her vital capacity would be greatest when she is standing up. In the upright position, gravity tends to pull the abdominal organs downward. As a result, movement of the diaphragm is not as restricted and thoracic volume is increased. Lying down allows abdominal organs to exert pressure on the diaphragm, decreasing the thoracic volume.

4. At sea level, all the gases in the atmosphere exert a pressure of 760 mm Hg. If water vapor is 10% of that gas mixture, then water vapor must have a partial pressure of $760 \times 0.10$, or 76 mm Hg.

5. As fluid accumulates in the alveoli, the layer through which oxygen and carbon dioxide must diffuse in the alveoli becomes thicker. As the layer thickens, the rate at which gases diffuse slows. Consequently, the amount of oxygen that diffuses into the pulmonary capillaries is reduced and the amount of carbon dioxide that diffuses out of the pulmonary capillaries decreases. The decreased blood oxygen levels and the increased carbon dioxide levels stimulate the respiratory center and cause the rate and depth of respiration to increase. Oxygen therapy increases the $P_{O_2}$ in the alveoli and reduces the $P_{CO_2}$ in the alveoli. Therefore, oxygen diffuses more rapidly from the alveoli into the pulmonary capillaries and carbon dioxide diffuses more rapidly from the pulmonary capillaries into the alveoli. This establishes more normal blood levels of oxygen and carbon dioxide. Because of increased levels of oxygen and decreased levels of carbon dioxide in the blood, the rate of respiratory movements is decreased.

6. In severe emphysema, the surface area for gas exchange is reduced, so that not enough oxygen can diffuse from the alveoli into the pulmonary capillaries and not enough carbon dioxide can diffuse from the pulmonary capillaries into the alveoli even when oxygen therapy is given. The elevated carbon dioxide and the reduced oxygen levels in the blood both stimulate the respiratory center to produce an urge to take a breath.

7. The pH of the body fluids in patients who suffer from untreated diabetes mellitus decreases (becomes more acidic). A

decreasing pH acts as a strong stimulus to the respiratory center. Consequently, the rate and depth of respiration increase. The increased rate and depth of respiration cause $CO_2$ to be released from the circulation at a greater rate. The lower blood level of $CO_2$ opposes the reduced pH because of the following reaction:

$$CO_2 + H_2O \rightleftharpoons H_2CO_3 \rightleftharpoons H^+ + HCO_3^-$$

As more $CO_2$ is removed from the blood, more $H^+$ and $HCO_3^-$ combine to form $H_2CO_3$. Removal of the $H^+$ from the circulation resists a further reduction in the body fluid pH.

8. When Ima is hyperventilating, the stimulus for the hyperventilation is anxiousness, and the anxiousness is more important than the carbon dioxide in controlling respiratory movements. As the blood levels of carbon dioxide decrease during hyperventilation, vasodilation occurs in the periphery. As a result, the systemic blood pressure decreases. The systemic blood pressure can decrease enough that the blood flow to the brain decreases. Decreased blood flow to the brain results in a reduced oxygen level in the brain tissue, causing dizziness. Breathing into a paper bag increases Ima's blood levels of carbon dioxide toward normal. Because the carbon dioxide does not increase above normal, it does not increase the urge to breathe. The more normal level of carbon dioxide prevents the peripheral vasodilation. As Ima breathes into the paper bag, the anxiousness is likely to subside. As the anxiousness subsides, the normal regulation of respiration resumes.

9. Hyperventilating before swimming decreases the amount of carbon dioxide in the blood. The decreased amount of carbon dioxide allows a longer than normal period of time before the swimmer has a strong urge to breathe. This can be dangerous, however, because the reduced level of carbon dioxide causes vasodilation in the periphery and reduced blood flow to the brain and could cause a person to pass out. Passing out under water could lead to drowning.

10. Immediately after the beginning of a race, runners increase their rate and depth of respiration before blood carbon dioxide levels have a chance to increase. The rate and depth of respiration increase in anticipation of increased muscular activity. As a result, for a short time the increased rate and depth of respiration can actually cause blood levels of carbon dioxide to decrease. The lower levels of blood carbon dioxide result in a slightly increased blood pH. After the race has progressed, the increased metabolic activity of the muscles can produce enough extra carbon dioxide to cause blood carbon dioxide levels to increase above resting levels, even with an increased rate and depth of respiration.

## Chapter Sixteen

1. With the loss of the swallowing reflex, vomit can enter the larynx and trachea, block the respiratory tract, and keep the patient from breathing. Acidic stomach secretions cause severe inflammation and swelling of the respiratory passages.

2. Without adequate amounts of hydrochloric acid, the pH in the stomach is not low enough for the activation of pepsin. This loss of pepsin function results in inadequate protein digestion in the stomach. However, if the food is well chewed, proteolytic enzymes in the small intestine (e.g., trypsin, chymotrypsin) can still digest the protein. If the stomach secretion of intrinsic factor decreases, the absorption of vitamin $B_{12}$ is hindered.

3. Even though bacteria apparently cause most ulcers, overproduction of hydrochloric acid because of stress is a possible contributing factor. Antibiotic therapy is an effective treatment, but reducing hydrochloric acid production is recommended. Possible solutions also include drugs that reduce stomach acid secretion, antacids to neutralize the stomach acid, smaller meals (distention of the stomach stimulates acid production), and proper diet. The person should avoid alcohol, caffeine, and large amounts of protein because they stimulate acid production. He should ingest fatty acids because they inhibit acid production, causing the release of gastric inhibitory polypeptide and cholecystokinin. Stress also stimulates the sympathetic nervous system, which inhibits duodenal gland secretion. As a result, the duodenum has less of a mucous coating and is more susceptible to gastric acid and enzymes. Relaxing after a meal helps decrease sympathetic activities and increase parasympathetic activities.

4. Obstruction of the common bile duct blocks the flow of bile from the gallbladder to the small intestine. As a result, stomach acids are not diluted and neutralized to as great an extent as if bile were present. Fats are not emulsified by the bile, resulting in decreased fat digestion and absorption. Excretory products, such as bile pigments, cholesterol, and fats, are not as readily removed from the body.

5. Introduction of food into the stomach increases the frequency of mass movements. The mass movements move feces into the rectum. Also, during the night, much of the material in the intestine has moved to the lower part of the large intestine. Mass movements following breakfast, therefore, are more likely to move feces into the rectum. Stretch of the rectum triggers the defecation reflex.

## Chapter Seventeen

1. The recommended daily consumption of saturated fats is less than 20 g. Therefore, the % Daily Value for 2 g of fat is 10% (2 / 20 = 0.10, or 10%).

2. It is recommended that no more than 30% of the daily total kilocaloric intake be fats. For a 3000 kcal/day diet, that is 900 kcal (3000 × 0.30). There are 9 kcal/g of fat. Therefore, the maximum amount (weight) of total fat consumed per day is 100 g (900 / 9).

3. From question 2, the teenager's daily maximum amount (weight) of fat consumed per day is 100 g. If he consumes a 10-g serving of fat, his % Daily Value is 10% (10 / 100 = 0.10, or 10%).

4. From question 3, one serving equals 10% Daily Value. If he eats half of the contents of the package, he consumes two servings, or 20% Daily Value.

5. An active person uses many more kilocalories in a day than a person who is not active and, so, can lose weight on a higher-kilocalorie diet. In addition, some people have a higher basal metabolic rate and tend to burn more kilocalories.

6. Ingested proteins are digested to amino acids in the stomach and intestine and are then absorbed into the circulation. The amino acids can be used by cells as building blocks for proteins, which is very important if a person is attempting to build up muscle mass. However, beyond a certain point, the excess amino acids are broken down in the liver to make glucose. These amino acids are, therefore, no more helpful than any other source of energy. If the amount of amino acids is excessive, ammonia or keto acids, the breakdown products of amino acid metabolism, can accumulate to toxic levels.

7. Excess protein is metabolized and used as a source of energy. One of the breakdown products of protein is ammonia. Ammonia is toxic, so it is converted in the liver to urea, which is carried by the blood to the kidneys and eliminated.

8. As ATP breakdown increases, more ATP is what is produced to replace what is used. Over an extended time, the ATP must be

produced through aerobic respiration. Therefore, oxygen consumption and basal metabolic rate increase. The production of ATP requires the metabolism of carbohydrates, lipids, or proteins. As these molecules are used at a faster than normal rate, body weight decreases. Increased appetite and increased food consumption resist the loss in body weight. As ATP is produced and used, heat is released as a by-product. The heat raises body temperature, which is resisted by the dilation of blood vessels in the skin and by sweating.

9. No, this approach does not work because he is not losing stored energy from adipose tissue. In the sauna, he gains heat, primarily by convection from the hot air and by radiation from the hot walls. The evaporating sweat is removing heat gained from the sauna. The loss of water will make him thirsty, and he will regain the lost weight from drinking fluids and eating foods containing water.

10. Drinking cool water could help in two ways. Because the water is cool, raising the water to body temperature requires the expenditure of energy. Also, stretch of the stomach decreases appetite (see chapter 16).

11. During fever production, the body produces heat by shivering. The body also conserves heat by constricting blood vessels in the skin (producing pale skin) and by reducing sweat loss (producing dry skin). When the fever breaks—that is, "the crisis is over"—heat is lost from the body to lower body temperature to normal. This is accomplished by dilation of blood vessels in the skin (producing flushed skin) and increased sweat loss (producing wet skin).

# Chapter Eighteen

1. The rate of urine production increased over the next 2 hours, and the urine was dilute. The water he consumed increased his blood volume and decreased the concentration (osmolality) of his blood. Both of these changes inhibit ADH secretion from the posterior pituitary gland. The reduced ADH levels in the blood cause the kidney to produce a large volume of dilute urine.

2. The sodium chloride increased the concentration (osmolality) of his blood without affecting the volume of blood. The increased concentration (osmolality) of his blood stimulated ADH secretion from the posterior pituitary gland. The ADH caused the kidney to produce a small volume of concentrated urine.

3. The excessive sweating results in a reduced blood volume and the blood has a greater than normal concentration (osmolality). Both of these changes stimulate ADH secretion from the posterior pituitary gland. The ADH acts on the kidney to produce a small volume of concentrated urine.

4. Excessive urine production, low blood pressure, high plasma $K^+$ levels, but not high plasma $Na^+$ levels, result. Low aldosterone levels result in excessive loss of $Na^+$ in the urine. The $Na^+$ attract water molecules from the peritubular capillaries into the nephron, which results in a greater than normal urine volume. The loss of $Na^+$ in the urine reduces the plasma $Na^+$ concentration. The loss of water in the urine reduces the blood volume and therefore reduces the blood pressure. Because aldosterone causes $Na^+$ reabsorption from the nephron and $K^+$ secretion into the nephron, $K^+$ accumulate in the plasma.

5. The following all decrease the blood pressure in the glomerular capillaries and therefore decrease the glomerular filtration rate: decreased blood pressure and increased sympathetic stimulation that constricts the renal arteries and the afferent arterioles. Blocking the nephron causes the capsule pressure in the nephron to increase, so that it is equal to the pressure in the glomerular capillaries, which reduces the filtration rate. An increase in plasma proteins also decreases the filtration rate. The increased plasma protein concentration attracts water by osmosis and therefore reduces the tendency for water to pass from the glomerulus to Bowman's capsule.

6. Removal of a large amount of bicarbonate ion ($HCO_3^-$) from the body fluids causes a decrease in the body fluid pH because there are fewer $HCO_3^-$ to combine with hydrogen ions ($H^+$) according to the following formula:

$$CO_2 + H_2O \rightleftharpoons H_2CO_3 \rightleftharpoons H^+ + HCO_3^-$$

Carbon dioxide  Water  Carbonic acid  Hydrogen ion  Bicarbonate ion

Consequently, the $H^+$ accumulate in the body fluids and decrease the body fluid pH. The rate and depth of respiration increase because a decreased body fluid pH stimulates the respiratory center. The decrease in the body fluid pH causes the kidneys to secrete additional $H^+$ into the nephron. Therefore, the pH of the urine decreases.

7. While Spanky held his breath, carbon dioxide accumulated in his body fluids. The increased carbon dioxide combined

with water to form carbonic acid that, in turn, produced $H^+$ and $HCO_3^-$. The increased concentration of $H^+$ in the body fluids caused the pH of the body fluids to decrease. After 2 minutes, the reduced pH and increased body fluid carbon dioxide levels strongly stimulated the respiratory center to increase the rate and depth of respiration. The increased rate of respiration reduced the body fluid carbon dioxide levels and increased the body fluid pH back toward its normal value. The kidneys did not respond to the change in pH within 2 minutes. The kidneys are more powerful regulators of body fluid pH, but they require several hours to become maximally active.

8. Martha's blood pH increased. Her frequent vomiting resulted in the loss of $H^+$ from her stomach and an increase in $HCO_3^-$ in her blood. Since Martha lost fluid from her body and she consumed no fluid, the concentration of her body fluids increased and her blood volume decreased. Consequently, ADH secretion increased. Because blood volume decreases, the blood flow to the kidneys decrease. This stimulates renin secretion from the kidney. This, in turn, converts angiotensinogen to angiotensin I. Angiotensin I is converted to angiotensin II, which stimulates aldosterone secretion from the adrenal cortex. In response to an increase in blood pH, the kidneys decrease their rate of $H^+$ secretion and $HCO_3^-$ reabsorption. Consequently, this causes an increase in blood $H^+$ concentration and a decrease in blood $HCO_3^-$. The kidney's response to changes in pH takes many hours, but it has a large capacity to maintain pH homeostasis.

# Chapter Nineteen

1. Removing the testes eliminates the source of testosterone. Therefore, blood levels of testosterone would decrease. Because testosterone has a negative-feedback effect on the hypothalamus and pituitary gland, GnRH, FSH, and LH secretion increases and the blood levels of these hormones increase.

2. The estrogen and progesterone in birth control pills inhibit the large increase in LH secretion from the anterior pituitary, which is responsible for ovulation. Without the large increase in LH secretion, ovulation cannot occur.

3. E is the correct answer. The secretory phase of the menstrual cycle occurs after ovulation. It is following ovulation that the corpus luteum forms and produces progesterone. In addition, the

progesterone acts on the endometrium of the uterus to cause its maximum development. Therefore, progesterone secretion reaches its maximum levels and the endometrium reaches its greatest degree of development during the secretory phase of the menstrual cycle.

4. Between approximately 12 and 14 days of the menstrual cycle, you would normally expect increasing blood levels of estrogen and LH. In the average menstrual cycle, ovulation occurs on day 14. Prior to that time, estrogen levels are increasing. The increasing estrogen stimulates LH secretion. The increasing LH, in turn, causes increased estrogen secretion from the developing follicle. Blood progesterone levels are low at this time. Progesterone is not secreted in large amounts until the corpus luteum is formed after ovulation.

5. On day 15 of the menstrual cycle, one would expect decreasing blood levels of LH, decreasing blood levels of estrogen, and increasing blood levels of progesterone. In the average menstrual cycle, ovulation occurs on day 14. After ovulation, the ovulated follicle develops into the corpus luteum. Estrogen levels decrease as the follicle is converted to the corpus luteum. The corpus luteum secretes progesterone in increasing quantities and small amounts of estrogen as the corpus luteum develops. The increasing blood levels of progesterone inhibit LH secretion from the pituitary gland. Consequently, blood levels of LH fall to low levels.

6. A drug that blocks the effect of progesterone on its target tissues will cause the tissue to respond as if no progesterone were present. Progesterone is secreted by the corpus luteum for about 7 days after ovulation. The progesterone affects the endometrium of the uterus and, in response, the endometrium becomes prepared for implantation. A decline in

progesterone causes menstruation to occur. If a drug is taken 3–4 days following ovulation that blocks the effect of progesterone, the endometrium will not become fully prepared for implantation and menstruation will begin early. If a drug that blocks progesterone is taken by a woman who is pregnant, the effect will be to cause the endometrium to slough. This is similar to the events that occur during menstruation. If this occurs, it will terminate the pregnancy. Progesterone is necessary to maintain pregnancy. The corpus luteum of the ovary secretes progesterone until the end of the first 3 months of pregnancy; prior to the third month of pregnancy, the placenta begins to secrete progesterone. The placenta becomes the primary source of progesterone after the third month of pregnancy.

7. After menopause, estrogen and progesterone are produced by the ovary in only small amounts. Consequently, estrogen and progesterone levels in the blood are low. These hormones have a negative-feedback effect on the secretion of FSH and LH from the anterior pituitary gland. Therefore, in the absence of estrogen and progesterone, FSH and LH are secreted in greater amounts and the blood levels of FSH and LH increase. However, there is no large increase in LH, like the increases that occur prior to ovulation. The average concentration of LH and FSH is greater than the levels that occur either before or after ovulation.

## Chapter Twenty

1. Fertilization occurs approximately 14 days after the LMP, so the embryo in this case is 30 days old. It has just completed the "budding" period; limb buds and other "buds" have just formed (see table 20.2).

2. At 35–45 days post-LMP, the embryo would be 21–31 days old, which overlaps the period of neural tube closure (18–26 days).

A high fever during this period could prevent neural tube closure in the embryo, so that the newborn baby may have neural tube defects, such as anencephaly (open neural tube in the area of the brain, resulting in absence of the upper brain) or spina bifida (open spinal cord resulting from failure of the neural tube to close in that area).

3. The correct answer is A. Prolactin is responsible for the production of milk by the breast. Surges of prolactin are stimulated by the process of suckling and are required to maintain milk production. Drugs that inhibit prolactin release will cause the breast to cease milk production after a few days.

4. Yes. Because dimpled cheeks are dominant, a person with dimpled cheeks may have a genotype of $Dd$ ($D$ being the dominant gene for dimples and $d$ being the normal gene). If two $Dd$, dimpled people have children, it may be expected that approximately one-fourth of their children would not have dimples ($dd$).

5. No, not without additional information. To be able to roll their tongue into a tube, both mother and son need to have only one dominant gene. The son could inherit this dominant gene from his mother, so it is possible that his father is a nontongue roller. It is also possible that his father is a tongue roller, but there is no proof for either hypothesis.

6. Hemophilia is an X-linked recessive gene. A female must have the genotype $X^hX^h$ to have hemophilia, whereas because males have only one X chromosome, a single hemophilia gene will cause them to have the disorder ($X^hY$). For half their children to have hemophilia, the mother's genotype must be $XX^h$ (she does not have hemophilia) and the father's genotype must be $X^hY$ (he has hemophilia). If the woman did not carry the hemophilia gene, none of her daughters or sons could have the disorder.

# Appendix E

## Answers to Predict Questions

### Chapter One

1. Donating a pint of blood results in a decrease in blood pressure. Negative-feedback mechanisms, such as an increase in heart rate, return blood pressure toward a normal value. When a negative-feedback mechanism fails to return a value to its normal level, the value can continue to deviate from its normal range. Homeostasis is not maintained in this situation, and the person's health or life can be threatened.

2. The thirst sensation is associated with a decrease in body fluid levels. The thirst mechanism causes the person to drink fluids, which returns the fluid level to normal. Thirst is therefore a sensation involved in negative-feedback control of body fluids.

3. When a boy is standing on his head, his nose is superior to his mouth. Remember that directional terms refer to a person in the anatomical position, not to the body's current position.

4. The spleen is in the left-upper quadrant, the gallbladder is in the right-upper quadrant, the left kidney is in the left-upper quadrant, the right kidney is in the right-upper quadrant, the stomach is mostly in the left-upper quadrant, and the liver is mostly in the right-upper quadrant.

5. There are two ways. First, the visceral peritoneum wraps around organs. Thus, the peritoneal cavity surrounds the organ, but the organ is not inside the peritoneal cavity. The peritoneal cavity contains only peritoneal fluid. Second, retroperitoneal organs are in the abdominopelvic cavity, but they are between the wall of the abdominopelvic cavity and the parietal peritoneal membrane.

### Chapter Two

1. The mass (amount of matter) of the astronaut on the surface of earth and in outer space does not change. In outer space, where the force of gravity from earth is very small, the astronaut is "weightless" compared with his weight on earth's surface.

2. Fluorine has nine protons (the atomic number), ten neutrons (the mass number minus the atomic number), and nine electrons (equal to the number of protons).

3. Because atoms are electrically neutral, the iron (Fe) atom has the same number of protons and electrons. The loss of three electrons results in an iron ion that has three more protons than electrons and therefore a charge of $+3$. The correct symbol is $Fe^{3+}$.

4. Carbon dioxide and water are in equilibrium with $H^+$ and $HCO_3^-$. A decrease in $CO_2$ causes some $H^+$ to react with $HCO_3^-$ to form $CO_2$ and $H_2O$. Consequently, the $H^+$ concentration decreases.

5. During exercise, muscle contractions increase. This requires the release of potential energy during chemical reactions. For example, the potential energy in the phosphate bond of ATP is released when ATP is broken down into ADP and $P_i$. Some of the energy is used to drive muscle contractions, and some of it is released as heat. Because the rate of these reactions increases during exercise, more heat is produced than when at rest, and body temperature increases.

6. A base added to a solution combines with $H^+$, which decreases the $H^+$ concentration. By definition, a decrease in $H^+$ concentration is an increase in pH. Note that buffers take up $H^+$ when an acid is added to a buffered solution. It is reasonable to expect a buffer to release $H^+$ when a base is added to a buffered solution. The released $H^+$ can combine with the base to prevent an increase in pH. For example, $H^+$ can combine with $OH^-$ to form $H_2O$.

### Chapter Three

1. Urea is produced continually by liver cells and diffuses from the cells into the blood. If the kidneys stop eliminating urea, it begins to accumulate in the blood and in the liver cells. The urea finally reaches concentrations high enough to be toxic to cells, causing cell damage, followed by cell death.

2. Glucose transported by facilitated diffusion across the cell membrane moves from a higher to a lower concentration. If glucose molecules are converted quickly to some other molecule as they enter the cell, a large concentration difference is maintained; thus glucose transport into the cell continues proportional to the magnitude of the concentration difference.

3. a. Cells specialized to synthesize and secrete proteins have abundant rough ER because this is an important site of protein synthesis. Well-developed Golgi apparatuses exist to package proteins in secretory vesicles, and numerous secretory vesicles are present.

   b. Cells highly specialized to actively transport substances into the cell have a large surface area, such as microvilli, exposed to the fluid from which substances are actively transported. Numerous mitochondria are present near the membrane across which active transport occurs because the process requires energy.

   c. Cells highly specialized to ingest foreign substances by endocytosis have numerous lysosomes in their cytoplasm and evidence of vesicles containing foreign substances.

4. Changing a single nucleotide within a DNA molecule also changes the nucleotide sequence of the mRNA produced from that segment of DNA. The change in mRNA results in a different codon, and a different amino acid is placed in the amino acid chain for which the mRNA codes. Because a change in the amino acid sequence of a protein can change its structure, one substitution of a nucleotide in a DNA chain can result in altered protein structure and function.

5. Cancer cells generally appear to be undifferentiated. Instead of dividing and then undergoing differentiation, they continue to divide and do not differentiate. One measure of the severity of cancer is related to the degree of differentiation the cancer cells have undergone. Those that are more differentiated divide more slowly and are less dangerous than those that differentiate little.

6. Chloride ions do not move in normal amounts out of the cells of people with cystic fibrosis because $Cl^-$ channels are defective. Instead, the $Cl^-$ tend to accumulate inside the cell. Potassium ions tend to move out of muscle and nerve cells down their concentration gradient. The positively charged $K^+$, however, are attracted by the negatively charged $Cl^-$ accumulated inside the cell. This attraction reduces the movement of $K^+$ out of the cell and causes more $K^+$ to accumulate inside the cell.

## Chapter Four

1. a. If nonkeratinized stratified squamous epithelium lined the digestive tract, it would provide protection against abrasion but would hinder the secretion of digestive enzymes and the absorption of digested food. Nonkeratinized stratified squamous epithelium is not specialized to absorb or secrete because the many layers of

in others. The collagen fibers, which are like microscopic ropes, are therefore all arranged in the same direction to maximize their strength. In the skin, collagen fibers are oriented in many directions because the skin can be pulled in many directions. The collagen fibers can be somewhat randomly oriented, or they can be organized into alternating layers. The fibers within a layer run in the same direction, but the fibers of different layers run in different directions.

3. Collagen synthesis is required for scar formation. If collagen synthesis does not occur because of a lack of vitamin C or if collagen synthesis is slowed, wound healing does not occur or is slower than normal. One might expect that the density of collagen fibers in a scar is reduced and the scar is not as durable as a normal scar.

4. There is more than one way to organize a table that summarizes the characteristics of the major muscle types. See the following table.

lipid-soluble can easily diffuse through the barrier. This fact is used as a basis for administering some medications through the skin. On the other hand, water-soluble substances have difficulty diffusing through the skin. The lipid barrier of the skin prevents water loss from the body.

2. a. The lips are pinker or redder than the palms of the hand. Several explanations for this are possible. There could be more blood vessels in the lips, there could be increased blood flow in the lips, or the blood vessels could be easier to see through the epidermis of the lips. The last possibility actually explains most of the difference in color between the lips and palms. The epidermis of the lips is thinner and not as heavily keratinized as that of the palms. In addition, the dermal papillae containing the blood vessels in the lips are "high" and closer to the surface.

| TABLE E.1 | Major Characteristics of the Three Muscle Types | | | | | |
|---|---|---|---|---|---|---|
| Muscle Type | Nuclei in Each Cell | Location of Nuclei | Control | Cell Shape | Striated | Branching Fibers |
| Skeletal | Many | Peripheral | Voluntary | Long and cylindrical | Yes | No |
| Cardiac | One | Central | Involuntary | Branching cylinders joined by intercalated disks | Yes | Yes |
| Smooth | One | Central | Involuntary | Spindle-shaped cells | No | No |

cells hinder the movement of materials across the epithelium. Squamous cells, as opposed to cuboidal or columnar cells, do not contain the number of organelles necessary to support the production and transport of large numbers of complex molecules.

b. Keratinized stratified epithelium forms a tough layer that is a barrier to the movement of water. Replacing the epithelium of skin with moist stratified squamous epithelium would increase the loss of water across the skin because water can diffuse through nonkeratinized stratified squamous epithelium, and it is more delicate and provides less protection than keratinized stratified squamous epithelium.

2. When a muscle contracts, the pull it exerts is transmitted along the length of its tendons. The tendons need to be very strong in that direction but not as strong

5. In severely damaged tissue in which cells are killed and blood vessels are destroyed, the usual symptoms of inflammation cannot occur. Surrounding these areas of severe tissue damage, however, where blood vessels are still intact and cells are still living, the classic signs of inflammation do develop. The signs of inflammation therefore appear around the periphery of severely injured tissues.

6. Suturing large wounds brings the edges of the wounds close together. Healing is therefore more rapid, there is less danger of infections, less scar tissue is formed, and wound contracture is greatly reduced.

## Chapter Five

1. Because the permeability barrier is composed mainly of lipids surrounding the epidermal cells, substances that are

b. A person who does manual labor has a thicker stratum corneum (and possibly calluses) than a person who does not perform manual labor. The thicker epidermis masks the underlying blood vessels, and the palms do not appear as pink. In addition, carotene accumulating in the lipids of the stratum corneum might give the palms a yellowish cast.

c. The posterior surface of the forearm appears darker because of the tanning effect of ultraviolet light from the sun. The posterior surface of the forearm is usually exposed to more sunlight than the anterior surface of the forearm.

3. The story is impossible. Hair color results from melanin that is added to the hair in the hair bulb as the hair grows. The hair itself is dead. To turn white, the hair must grow out without the addition of melanin. This, of course, takes considerably more time than 1 night.

4. On cold days, skin blood vessels of the ears and nose can dilate, bringing warm blood to the ears and nose, thus preventing tissue damage from the cold. The increased blood flow makes the ears and nose appear red.

5. Reducing water loss is one of the normal functions of the skin. Loss of or damage to the skin can greatly increase water loss, resulting in dehydration and reduced urine production. To counteract the increased loss of fluid from burns, during the first 24 hours following the injury large volumes of fluid are administered. But how much fluid should be given? The amount of fluid given should be sufficient to match that lost, plus enough to allow the kidneys to function. Urine output is therefore monitored. If it is too low, more fluid is administered; if it is too high, less fluid is administered. An adult receiving intravenous fluids should produce 30–50 mL of urine per hour, and children should produce 1 mL/kg of body weight per hour.

## Chapter Six

1. If all the mineral is removed, the bone becomes so flexible that it can be tied into a knot. This can be accomplished by soaking a bone in vinegar (a weak acid) for an extended time. The bone will not be rigid enough to support weight.

   If all the collagen is removed, the bone becomes very brittle and can be easily broken. Because of collagen loss, the bones of many older people break easily.

2. If cartilage growth fails to occur, the bone is normal in diameter (or even greater in diameter than normal), because growth in diameter does not require cartilage growth, but much shorter than normal. This is the condition seen in one type of dwarfism, in which the head and trunk are normal in size but the long bones of the limbs are very short.

3. Tears are produced in lacrimal glands in the superior lateral corner of the orbit. The tears run across the surface of the eye and enter the duct that passes through the nasolacrimal canal to the nasal cavity. The extra moisture in the nasal cavity causes a "runny nose."

4. Just before the swimmer begins the power stroke, the arm is flexed and medially rotated and the forearm is extended and pronated. During the power stroke, the arm is powerfully extended, slightly abducted, and medially rotated. During the recovery stroke, the arm is circumducted, laterally rotated, and flexed in preparation for the next stroke. The forearm is flexed during the first part of the recovery stroke and extended during the last part.

5. Taking in adequate calcium and vitamin D through the digestive system during adulthood increases calcium absorption from the small intestine. The increased calcium is used to increase bone mass. The greater the bone mass before the onset of osteoporosis, the greater the tolerance for bone loss later in life. For this reason, it is important for adults, especially women in their 20s and 30s, to ingest adequate amounts of calcium. Exercising the muscular system places stress on bone, which also increases bone density. The granddaughter should not smoke because this reduces estrogen levels.

## Chapter Seven

1. a. If ATP levels are low in a muscle fiber before stimulation, the force of contraction is reduced because there is insufficient ATP for all motor units to contract.

   b. If action potentials occur at a frequency so great that calcium is not transported back into the sarcoplasmic reticulum between individual action potentials, the muscle continues to contract and does not relax.

2. During a 10-mile run, aerobic metabolism is the primary source of ATP production for muscle contraction. Anaerobic metabolism provides the short (15–20 s) burst of energy for the sprint at the finish. After the race, aerobic metabolism is elevated for a time to repay the oxygen deficit, causing the heavy breathing after the race.

3. Raising eyebrows—occipitofrontalis; winking—orbicularis oculi; whistling—orbicularis oris and buccinator; smiling—zygomaticus; frowning—depressor anguli oris; sneering—levator labii superioris

4. Shortening the right sternocleidomastoid muscle rotates the head to the left and slightly elevates the chin.

5. In the sprinter's stance and the bicyclist's racing posture, the thigh is flexed at a 45-degree angle because at that angle the gluteus maximus functions at its maximum in extending the thigh, thus providing maximum force.

6. DMD affects the muscles of respiration and causes deformity of the thoracic cavity. The reduced capacity of muscle tissue to contract is one factor that reduces the ability to breathe deeply or cough effectively. In addition, the thoracic cavity can become severely deformed because of the replacement of skeletal muscle with connective tissue. The deformity can reduce the ability to breathe deeply. DMD can also affect the muscle of the heart and cause heart failure.

## Chapter Eight

1. Dorsal root ganglia contain neuron cell bodies, which are larger in diameter than the axons of the dorsal roots. Action potentials are propagated in both directions in spinal nerves, toward the spiral cord in the dorsal root and away from the spinal cord in the ventral root.

2. Damage to the right phrenic nerve results in the absence of muscular contraction in the right half of the diaphragm. Because the phrenic nerves originate in the cervical region of the spinal cord, damage to the spinal cord in the thoracic region does not affect the diaphragm. Damage to the upper cervical region, however, cuts the connection between the upper and lower motor neurons. This eliminates phrenic nerve stimulation of the diaphragm and dramatically interferes with breathing. Death is likely to occur.

3. Nuclei within the medulla oblongata regulate heart rate, blood vessel diameter, breathing, swallowing, vomiting, coughing, sneezing, balance, and coordination. Even though all of these functions are important, loss of some of them may not necessarily result in death. Loss of cardiovascular regulation or loss of breathing regulation, however, could result in death. Because both blood flow and respiration are vital functions, interference with either of them may inhibit normal homeostatic functions. If not corrected, the loss of homeostasis results in death. Neuronal control of breathing is more critical than cardiovascular control. Death can occur in minutes if neuronal control of respiration is lost. Neuronal control is not absolutely necessary for the heart to continue beating. Without cardiovascular regulation, however, the blood pressure is reduced, resulting in shock, which can ultimately result in death (see chapters 12 and 13).

4. If a person holds an object in her hand, sensations from the skin of the hand are sent to the primary somatic sensory cortex. The information is then passed to the somatic sensory association area, where the object is recognized. Action potentials then travel to the sensory speech area, where the object is given a name. From there, action potentials travel to the motor speech area, where the spoken word is initiated. Action potentials

from the motor speech area travel to the premotor area and then to the primary motor cortex, where action potentials are initiated that stimulate the muscles necessary to formulate the word.

5. a. In a person who is extremely angry, the sympathetic division of the autonomic nervous system is activated, and the expected responses include increased heart rate, increased blood pressure, dilated pupils, and perspiration.

   b. In a person who has just finished eating and is now relaxing, the parasympathetic division of the autonomic nervous system is the primary autonomic division functioning. The responses include decreased heart rate, decreased blood pressure, and increased digestive activities.

6. The stroke was on the left side of the brainstem. Both the motor and sensory neurons to the right side of the body are located in the left cerebral cortex. At the level of the upper medulla oblongata, neither the motor nor the sensory pathways to the limbs have yet crossed over to the left side of the CNS. Loss of pain and temperature to the left side of the face indicates that the lesion occurred at a level where the nerve fibers from the face had entered the CNS but had not yet crossed (in the brainstem).

## Chapter Nine

1. The pain is diffuse because deep, or visceral pain, such as in the colon, is not highly localized because of the absence of tactile receptors in the deeper structures. The pain is referred to the skin over the lower central abdomen (see figure 9.2). This occurs because sensory neurons from the superficial area to which the pain is referred and the neurons from the colon, where the pain stimulation originated, converge onto the same ascending neurons in the spinal cord.

2. Much of taste is based on olfactory function. A cold may include a stuffy nose, which decreases airflow and increases the thickness of the mucus covering the olfactory epithelium and may interfere with olfaction and, thus, with taste.

3. Medications placed into the eyes can pass through the nasolacrimal duct into the nasal cavity, where their odor can be detected. Because much of our taste sensation is actually smell, the medication is perceived to have a taste.

4. Cones are responsible for color vision, but they are not nearly as sensitive to light as rods. Rods are much more sensitive to light, but they do not detect differences in color. In dim light, when rods rather than cones are responsible for vision, objects therefore appear only in shades of gray.

5. While you are driving, your vision is focused out on the road, some distance in front of the car. As a result, the ciliary muscles are relaxed and the lens is relatively flat, allowing for distant vision. When you look down at the speedometer, the eye accommodates for close vision and the ciliary muscles contract, pulling the ciliary body toward the lens. This reduces the tension on the suspensory ligaments of the lens and allows the lens to assume a more spherical form. When you look back up at the road, the ciliary muscles again relax and the lens flattens, allowing for distant vision. Even though these changes occur very quickly, there is a brief moment, before accommodation occurs, when you first look at the speedometer that it is out of focus. When you look back at the road, it is also briefly out of focus, as the lens changes back for distant vision.

6. When you hear a faint sound, you turn your head toward it because sound waves are collected by the auricle and conducted through the external acoustic meatus toward the tympanic membrane. The auricle is shaped in such a way that sound waves coming from the side and front of the head are most efficiently conducted into the external auditory canal. Because a specific sound reaches each ear at a slightly different time, a person can localize the origin of the sound. Turning the head toward the sound facilitates maximum accumulation of the sound waves by the ear. In addition, reflexes integrated in the superior colliculi cause the head and eyes to turn toward a sound, so that you can see what is making the sound.

7. The brain compares sensory input from the semicircular canals, eyes, and proprioceptors in the lower limbs; perceived differences in the input may result in motion sickness. Closing the eyes eliminates one of these inputs, so that the brain has less input to compare. Less input may decrease the perceived differences and thus reduce the symptoms of motion sickness. Likewise, we can perceive more motion in close objects than in distant objects. Therefore, looking at distant objects where there is less perceived motion, reduces the visual input of perceived motion to the brain. This reduced input may decrease the perceived differences and reduce the symptoms of motion sickness.

## Chapter Ten

1. Because the drug binds to a receptor and prevents the response of a target tissue to a chemical signal and because the drug is lipid-soluble, it is likely that the drug diffuses across the cell membrane and binds to the receptor for the chemical signal, which is inside the cell, and prevents the chemical signal from binding to its receptor site. Thus, the chemical signal functions by diffusing across the cell membrane and binding to an intracellular receptor. The response of the intracellular receptor is to produce new messenger RNA, which leads to the synthesis of new proteins (see chapter 3 for a description of the role of messenger RNA in protein synthesis). The new proteins produce the cell's response to the chemical signal.

2. The doctor might explain that GH administration would cause his son to grow taller. He might also explain that he might develop unwanted changes in his skeletal structure consistent with acromegaly. Other side effects are also possible. Examples include abnormal joint formation and diabetes mellitus.

3. The protein that is similar to TSH causes oversecretion of the thyroid gland (hyperthyroidism). Because the production of the protein cannot be inhibited by thyroid hormones, oversecretion of the thyroid gland is prolonged, and the symptoms associated with hypersecretion of thyroid hormones become obvious and the thyroid gland enlarges. In addition, the increased thyroid hormones have a negative-feedback effect on the hypothalamus and pituitary gland. TSH-releasing hormone secretion from the hypothalamus and TSH secretion from the anterior pituitary gland are inhibited.

4. Insufficient vitamin D results in insufficient $Ca^{2+}$ absorption by the intestine. As a result, blood $Ca^{2+}$ levels begin to fall. In response to the low blood $Ca^{2+}$ levels, PTH is secreted from the parathyroid glands. PTH acts primarily on bone, causing bone to be broken down and $Ca^{2+}$ to be released into the blood to maintain blood $Ca^{2+}$ levels within the normal range. In adults, so much $Ca^{2+}$ can eventually be removed from bones that they become soft, fragile, and easily broken. In adults the condition is called osteomalacia, and in children the condition is called ricketts. The bones of children can become bent and deformed.

5. Reduced aldosterone secretion causes Na$^+$ and water to not be retained. Consequently, blood Na$^+$ levels and blood volume decrease, resulting in low blood pressure. Also, blood K$^+$ levels increase. High blood K$^+$ levels lead to altered nerve and muscle function. These changes can be life-threatening.

6. Large doses of cortisone can damage the adrenal cortex because cortisone inhibits ACTH secretion from the anterior pituitary gland. ACTH is required to keep the adrenal cortex from undergoing atrophy. Prolonged use of large doses of cortisone can cause the adrenal gland to atrophy so much that it cannot recover if ACTH secretion does increase again.

7. After a large meal, glucose enters the blood from the intestine. The increasing blood glucose stimulates insulin secretion and decreases glucagon secretion. Well before 12 hours without eating, blood glucose would have started to decrease. Decreasing blood glucose levels result in a decreased rate of insulin secretion and a stimulation of glucagon secretion.

8. The pineal body secretes melatonin, which inhibits the release of reproductive hormones by acting on the hypothalamus of the brain. If the pineal body secretes less melatonin, it no longer should have an inhibitory effect on the hypothalamus. As a result, reproductive hormones could be secreted in greater amounts, which would result in exaggerated development of the reproductive system in young people with this condition. The evidence for this mechanism is not as clear in humans as it is in other animals.

9. Removal of part of the thyroid gland reduces the amount of thyroid hormone secreted by the gland. Usually, enough thyroid tissue can be removed to cause the amount of thyroid hormone secreted to be reduced to a normal range of values. In addition, the remaining thyroid tissue normally does not hypertrophy enough to cause the thyroid tissue to produce more than enough thyroid hormone, although there are exceptions. The removal of the thyroid tissue does not remove the influence of the abnormal antibodies on the tissues behind the eyes. Thus, in many cases the condition's effect on the eyes is not improved.

## Chapter Eleven

1. Carbon monoxide binds to the iron of hemoglobin and prevents the transport of oxygen. The decreased oxygen stimulates the release of erythropoietin from the kidneys. Erythropoietin increases red blood cell production in red bone marrow, causing the number of red blood cells in the blood to increase.

2. The white blood cells shown in figure 11.7 are (a) lymphocyte, (b) basophil, (c) monocyte, (d) neutrophil, and (e) eosinophil.

3. People with type AB blood were called universal recipients because they could receive type A, B, AB, or O blood with little likelihood of a transfusion reaction. Type AB blood does not have antibodies against type A or B antigens. Transfusion of these antigens in type A, B, or AB blood therefore does not cause a transfusion reaction in a person with type AB blood. The term is misleading, however, for two reasons. First, other blood groups can cause a transfusion reaction. Second, antibodies in the donor's blood can cause a transfusion reaction. For example, type O blood contains anti-A and anti-B antibodies that can react against the A and B antigens in type AB blood.

4. The donor blood used in exchange transfusions for the treatment of HDN should be Rh-negative, even though the newborn is Rh-positive. Rh-negative red blood cells do not have Rh antigens. Therefore, any anti-Rh antibodies in the newborn's blood do not react with the transfused Rh-negative red blood cells. Giving Rh-negative blood to an Rh-positive newborn does not change the blood type of the newborn because blood type is determined genetically. Eventually, all of the Rh-negative red blood cells die, and only Rh-positive red blood cells are produced by the newborn.

5. An increase in the white blood cell count often indicates a bacterial infection. A white blood cell differential count with an abnormally high neutrophil percentage supports the diagnosis.

## Chapter Twelve

1. The anterior interventricular artery supplies blood to the anterior wall of the heart and to much of the left ventricle. A blocked anterior interventricular artery reduces the oxygen supply to the portion of the heart that is supplied by that artery, and the cardiac muscle in that area is not able to contract effectively. Thus, the left ventricle on the anterior surface of the heart does not contract normally.

2. It is important to prevent tetanic contractions in cardiac muscle because the cycle of contraction and relaxation stops during tetanic contractions. This would cause the pumping action of the heart to stop. In skeletal muscle, the cycle of contraction and relaxation is not important as a pump, but it is important to maintain a static contracted state. This is essential to maintaining posture or to holding a limb in a specific position.

3. If the normal blood supply is reduced in a small area of the heart through which the left bundle branch passes, conduction of action potentials through that side of the heart is slowed or blocked. As a consequence, the left side of the heart contracts more slowly. The right side of the heart contracts more normally. The reduced rate of contraction of the left ventricle reduces the pumping effectiveness of the left ventricle.

4. A person who has a damaged left bundle branch will exhibit the consequences outlined for Predict question 3, but the electrocardiogram will also be altered. The QRS complex results from depolarization of the ventricles. Action potentials pass through the right bundle branch normally but conduction of action potentials through the left bundle branch is slowed because of damage. The QRS complex has an abnormal shape and it is prolonged. If many ectopic action potentials arise in the atria, heart rate increases. Each ectopic action potential initiates a new heartbeat. It is possible for some ectopic action potentials arising in the atria to occur while the ventricle is depolarized, but these action potentials do not initiate ventricular contractions. Therefore, there can be more P waves than QRS complexes in the electrocardiogram. If ectopic action potentials do not occur in a regular fashion, they can cause the heart to beat at an irregular rate, or arrhythmically.

5. A leaky aortic semilunar valve results in an increased left ventricular volume just before ventricular contraction. During ventricular relaxation, the aortic semilunar valve closes in a normal person, and blood flows out of the left ventricle into the aorta. When the aortic semilunar valve is incompetent, some blood leaks back into the left ventricle from the aorta during ventricular relaxation. When this blood is added to the blood that normally enters the left ventricle from the left atrium, there is a greater than normal volume of blood in the left ventricle just before ventricular contraction.

   A severely narrowed opening through the aortic semilunar valve increases the amount of work the heart must do to pump the normal volume of blood into the aorta. A greater pressure is required in the ventricle to force the same amount of blood through the narrowed opening during ventricular contraction.

6. Most of the ventricular contraction occurs between the first and second heart sounds of the same beat. Between the first and second heart sounds, blood is ejected from the ventricles into the pulmonary trunk and the aorta. Between the second heart sound of one beat and the first heart sound of the next beat, the ventricles are relaxing and the semilunar valves are closed. No blood passes from the ventricles into the aorta or pulmonary trunk during that period.

7. The shhh sound made after a heart sound is created by the backward flow of blood after closure of a leaky or incompetent valve. A swishing sound immediately after the second heart sound (lubb–duppshhh) represents a leaky aortic semilunar or pulmonary semilunar valve. The shhh sound before a heart sound is created by blood being forced through a narrowed, or stenosed, valve just before the valve closes. The lubb–shhhdupp suggests that there is a swishing sound immediately before the second heart sound, indicating a stenosed aortic or pulmonary semilunar valve.

8. In response to severe hemorrhage, blood pressure decreases, which is detected by baroreceptors. A reduced frequency of action potentials is sent from the baroreceptors to the medulla oblongata. This causes the cardioregulatory center to increase sympathetic stimulation of the heart and increase the heart rate. Sympathetic stimulation of the heart also increases stroke volume, as long as the volume of blood returned to the heart is adequate. Following hemorrhage, however, the blood volume in the body is reduced, and the venous return to the heart from the body is reduced. As a consequence, the volume of blood in the heart is lower than normal. Because of Starling's law, the stroke volume is reduced. The heart rate is increased, but the volume of blood returning to the heart is decreased; thus, the ventricle does not fill with blood. As a consequence, the stroke volume is low and the heart rate is high.

9. Rupture of the left ventricle can occur several days after a myocardial infarction. As the necrotic tissues are being removed by macrophages, the wall of the ventricle becomes thinner and may bulge during systole. If the wall of the ventricle becomes very thin before new connective tissue is deposited, it can rupture. If the left ventricle ruptures, blood flows from the left ventricle into the pericardial sac, resulting in cardiac tamponade. As blood fills the pericardial sac, it compresses the ventricle from the outside. As a consequence, the ventricle is not able to fill with blood and its pumping ability is rapidly eliminated. Rupture of the left ventricular wall quickly results in death.

## Chapter Thirteen

1. Atherosclerosis slowly increases the resistance to blood flow. Blood flow through atherosclerotic carotid arteries to the brain decreases. In advanced stages of arteriosclerosis, the resistance to blood flow increases so much that the blood flow to the brain is reduced significantly. Results include confusion, loss of memory, and a reduced ability to perform other normal brain functions.

2. If a thrombus in the posterior tibial vein gave rise to an embolus, it would travel through the following vessels before lodging in the lungs: posterior tibial vein, popliteal vein, femoral vein, external iliac vein, common iliac vein, inferior vena cava, right atrium, right ventricle, pulmonary trunk, and pulmonary artery. The lung is the most likely place for the embolus to lodge because the pulmonary artery branches and gives rise to smaller and smaller arteries until the arteries deliver blood to the pulmonary capillaries. The embolus will lodge in one of the branches of the pulmonary artery in the lung and block it. All of the other vessels through which the embolus passes are large in diameter.

3. Premature beats of the heart result in contraction of the heart muscle before the heart has time to fill to its normal capacity. The volume of blood ejected by the left ventricle of the heart is therefore reduced (reduced stroke volume) (see chapter 12). The reduced stroke volume is responsible for the weak pulse. In response to cardiovascular shock that is due to hemorrhage, the stroke volume is also reduced because less blood flows into the heart between beats. The reduced stroke volume is responsible for the weak pulse. In a person who is exercising, both the heart rate and the stroke volume increase. The increased stroke volume results in a stronger than normal pulse.

4. a. Decreased plasma protein concentration reduces the osmotic pressure of the blood. Edema results primarily because the movement of fluid into the venous end of the capillary as a result of osmosis is reduced. Consequently, less fluid returns to the capillary at its venous end and fluid accumulates in the interstitial space, resulting in edema.

  b. Increased blood pressure within the capillary forces more fluid to leave the capillary at its arteriolar end. There is no increased tendency for fluid to reenter the capillary at its venous end. The extra fluid that leaves the capillary can accumulate in the tissue space and result in edema.

5. When a blood vessel is blocked (when the legs are crossed, for instance), oxygen and nutrients are depleted and waste products accumulate in tissue supplied by the blocked blood vessel. The reduced supply of oxygen and nutrients and the accumulated waste products all cause relaxation of the precapillary sphincters and greatly increase blood flow through the area when the block is removed.

6. Severe vasoconstriction that results from Raynaud syndrome causes the digits to appear white because of the lack of blood flow through the capillary beds in the digits. The intensity of the vasoconstriction is more severe when the digits are exposed to cold temperatures. If the vasoconstriction is severe enough, there is not enough blood flow to the digits to supply nutrients to the tissues in the digits. Necrotic tissue and gangrene may develop.

7. The rapid loss of a large volume of blood activates the mechanisms responsible for maintaining blood pressure. This includes the baroreceptor reflexes and, if blood oxygen levels decrease substantially and carbon dioxide levels increase (and pH decreases), the chemoreceptor reflexes. In addition, the adrenal medullary mechanism, the reninangiotensin-aldosterone mechanism, and the vasopressin mechanism are all activated. The baroreceptor mechanism, adrenal medullary mechanism, and chemoreceptor mechanism increase the heart rate and result in vasoconstriction of blood vessels, especially of the skin and the viscera. Angiotensin II is produced quickly and it causes vasoconstriction and stimulates aldosterone secretion. Aldosterone requires up to 24 hours to become maximally active. It increases water reabsorption from the kidneys and reduces the loss of water in the form of urine. All of these mechanisms increase blood pressure back to its normal value.

  If blood is lost over several hours, the decrease in blood pressure is not as dramatic as when blood loss occurs quickly. Consequently, mechanisms that respond to a rapid and large decrease in blood pressure are stimulated to a lesser degree. These include the chemoreceptor reflex, the vasopressin mechanism, and the adrenal medullary mechanism. The baroreceptor reflexes are most sensitive to sudden decreases in blood pressure, but the baroreceptor reflex is sensitive to a decrease in blood pressure even if it occurs over a period of several hours. The baroreceptor reflexes trigger vasoconstriction if the blood loss is substantial. The kidneys detect even small decreases in blood volume. Consequently, the renin-angiotensin-aldosterone mechanism is activated and remains active until the blood pressure is returned to its normal range of values. Aldosterone secretion increases and, though it requires several hours to become maximally active, it continues to stimulate water reabsorption by the kidneys until the blood pressure returns to its normal range of values.

## Chapter Fourteen

1. Cutting and tying off the lymphatic vessels prevents the movement of fluid from the affected tissue. The result is edema.

2. When the antigen is eliminated, it is no longer available for processing and combining with MHC molecules. Consequently, there is no signal to cause lymphocytes to proliferate and produce antibodies.

3. The first exposure to the disease-causing agent (antigen) evokes a primary immune system response. Gradually, however, the antibodies degrade and memory cells die. If, before all the memory cells are eliminated, a second exposure to the antigen occurs, a secondary response results. The memory cells produced could provide immunity until the next exposure to the antigen.

4. The booster shot stimulates a memory (secondary) response, resulting in the formation of large amounts of antibodies and memory cells. Consequently, there is better, longer-lasting immunity.

5. SLE is an autoimmune disorder in which self-antigens activate immune responses. Often, this results in the formation of immune complexes and inflammation. Sometimes antibodies bind to antigens on cell membranes, resulting in the rupture of the cell membranes. Purpura results from bleeding into the skin. One cause of purpura is thrombocytopenia, a condition in which the number of platelets is greatly reduced, resulting in decreased platelet plug formation and

blood clotting (see chapter 11). In SLE, antibodies can stimulate the destruction of platelets.

## Chapter Fifteen

1. When you sleep with your mouth open, less air passes through the nasal passages. This is especially true when nasal passages are plugged because you have a cold. As a consequence, air is not humidified and warmed. The dry air dries the throat and the trachea, irritating them. Breathing through the mouth while running in very cold weather results in air that is not humidified and warmed, irritating the respiratory passageways.

2. When a large mouthful of food is swallowed, the esophagus is enlarged in the area through which the food passes. The bulge in the esophagus applies pressure on the trachea, which is immediately anterior to the esophagus. Because the C-shaped cartilages of the trachea have their open portion facing the esophagus, the posterior wall of the trachea collapses momentarily as the food passes. Thus, the passage of food through the esophagus is not hampered by the trachea.

3. During respiratory movements, the parietal and visceral pleurae slide over each other. Normally, the pleural fluid in the pleural cavities lubricates the surfaces of these membranes. When the pleural membranes are inflamed, their surfaces become roughened. The rough surfaces rub against each other and create an intense pain. The pain is increased when a person takes a deep breath because the movement of the membranes is greater than during normal breaths.

4. Relaxation of the abdominal muscles allows the abdominal organs to move inferiorly. Thus, it is easier for the diaphragm to increase the volume of the thoracic cavity.

5. The tube should apply suction. In order for the lung to expand, pressure in the alveoli must be greater than the pressure in the pleural cavity. This can be accomplished by lowering the pressure in the pleural cavity through suction. Applying air under pressure would make the pressure in the pleural cavity greater than the pressure in the alveoli, which would keep the alveoli collapsed.

6. The resting person with a tidal volume of 500 mL and a breathing rate of 12 respirations/min has a minute ventilation of 6000 mL (500 mL × 12 respirations/min). The exercising person with a tidal volume of 4000 mL and a breathing rate of

24 respirations/min has a minute ventilation of 96,000 mL (4000 mL × 24 respirations/min). The difference between the two is 90,000 mL, which means that the exercising person breathed 90,000 mL more air per minute than the person at rest.

7. Inadequate ventilation causes a smaller difference in the $P_{O_2}$ and $P_{CO_2}$ across the respiratory membrane. Therefore, the rate of oxygen and carbon dioxide diffusion across the membrane decreases, causing oxygen levels in the blood to decrease and carbon dioxide levels to increase.

8. During exercise, skeletal muscle cells increase oxygen use in order to produce the ATP molecules required for muscle contraction. Therefore, the $P_{O_2}$ inside the cells decreases, which increases the diffusion gradient for oxygen, resulting in increased movement of oxygen into the cells. The aerobic production of ATP also produces carbon dioxide (see chapter 17). Therefore, the $P_{CO_2}$ inside the cell increases, which increases the diffusion gradient for carbon dioxide, resulting in increased movement of carbon dioxide out of cells.

9. A rapid rate of breathing increases the blood pH because $CO_2$ is eliminated from the blood more rapidly during rapid respiration. As $CO_2$ is lost, $H^+$ and $HCO_3^-$ combine to form $H_2CO_3$, which in turn dissociates to form $CO_2$ and $H_2O$. The decrease in $H^+$ causes an increase in blood pH. Holding one's breath results in a decrease in pH because $CO_2$ accumulates in the blood. The $CO_2$ combines with $H_2O$ to form $H_2CO_3$, which dissociates to form $H^+$ and $HCO_3^-$. The increase in $H^+$ causes a decrease in blood pH.

10. When a person breathes rapidly and deeply for several seconds, the carbon dioxide levels decrease and blood pH increases. Carbon dioxide is an important regulator of respiratory movements. A decrease in blood carbon dioxide and an increase in blood pH result in a reduced stimulus to the respiratory center. As a consequence, respiratory movements stop until blood carbon dioxide levels build up again in the body fluid. This normally takes only a short time.

11. A $P_{O_2}$ of 60 mm Hg and a $P_{CO_2}$ of 30 mm Hg are both below normal. The movement of air into and out of the lungs is restricted because of the asthma, and there is a mismatch between ventilation of the alveoli and blood flow to the alveoli. Consequently, because of the ineffective ventilation, blood oxygen levels decrease. Will hyperventilates, which helps

maintain blood oxygen levels but also results in lower than normal blood carbon dioxide levels. (If there were no hyperventilation, one would expect decreased blood oxygen but increased blood carbon dioxide.)

## Chapter Sixteen

1. Four. Each portion of the mesentery has two layers, with a layer of connective tissue in between. The mesentery is folded back on itself to form the greater omentum.

2. It is important to close off the nasopharynx during swallowing, so that food, and especially liquid, doesn't pass into the nasal cavity. If a person has a cleft of the soft palate, there is an opening between the oral and nasal cavities, and the nasopharynx is not closed off during swallowing. If a person has an explosive burst of laughter while trying to swallow a liquid, the liquid may be explosively expelled from the mouth and even from the nose. Speaking requires that the epiglottis be elevated, so that air can pass out of the larynx. If the epiglottis is elevated while one is swallowing, the food, and especially liquid, may pass into the larynx, causing a person to choke.

3. Secretin production in response to acidic chyme in the small intestine stimulates bicarbonate ion secretion from the pancreas, which neutralizes the acidic chyme. Thus, secretin prevents the acid levels in the chyme from becoming too great. Because this mechanism keeps the pH of the intestinal contents within a normal range, this is an example of a negative-feedback system.

4. Introducing fluid into the rectum by way of an enema causes distention of the rectum. Distention stimulates the defecation reflex.

5. The effects of prolonged diarrhea result from a continued loss of fluid and ions. The major effect is on the cardiovascular system, and the effects are much like massive blood loss. Hypovolemia would continue to increase. Blood pressure would decline in a positive-feedback cycle and, without intervention, could lead to heart failure.

## Chapter Seventeen

1. If vitamins are broken down during the process of digestion, their structures are destroyed and, as a result, their ability to function as vitamins is lost. Because vitamins cannot be synthesized in adequate amounts, vitamin deficiency diseases occur.

2. The Daily Value for carbohydrate is 300 g/day. One serving of food with 30 g of carbohydrate has a % Daily Value of 10% (30 / 300 = 0.10, or 10%).

3. On a 1800 kcal/day diet, the total percentage of Daily Values for energy-producing nutrients should add up to no more than 90% because 1800 / 2000 = 0.9, or 90%.

4. If electrons in the electron-transport chain cannot be donated to oxygen atoms in the last step of the electron-transport chain, the entire chain and the citric acid cycle stop, no ATP is produced aerobically, and death results because too little energy is available for the body to maintain vital functions. Anaerobic metabolism can provide energy for only very short periods of time and cannot sustain life very long.

5. The kilocalories in a beer or cola is about 145 kcal. It takes about 1.5 h to burn off these kilocalories while watching TV (145 / 95 = 1.5 h) and about 15 min while jogging (145 / 580 = 0.25 h). Although it may be difficult to burn off kilocalories through exercise, it is clear that exercise can significantly increase kilocalorie use.

6. When muscles contract, they must use ATP as the energy source for the contractions. As more ATP is produced, heat is also produced. During exercise, the large amounts of heat can raise body temperature, and we feel warm. Shivering consists of small, rapid muscle contractions that produce heat in an effort to prevent a decrease in body temperature in the cold. When the body temperature declines below normal, shivering is initiated involuntarily.

7. Constriction of blood vessels to the skin reduces blood flow, which cools the skin. The benefit is that less heat is lost through the skin to the environment and the internal body temperature is maintained. As the difference in temperature between the skin and the environment decreases, less heat is lost. If the skin temperature decreases too much, however, dilation of blood vessels to the skin occurs, which prevents the skin from becoming so cold that it is damaged.

## Chapter Eighteen

1. If large amounts of glucose enter the nephron and are not reabsorbed, the glucose causes the concentration of solutes in the filtrate to increase. The glucose molecules attract water and, because the glucose molecules are trapped in the nephron, the amount of water that remains in the nephron is increased. A large volume of urine that contains glucose is a symptom of diabetes mellitus.

2. Without the normal active transport of $Na^+$ and $Cl^-$, their concentration within the nephron remains elevated. The normal movement of water out of the nephron cannot occur because of the osmotic effects of the $Na^+$ and $Cl^-$ trapped in the nephron. The result is an increased urine volume.

3. Because the solution was a saline solution, it had the same concentration of solutes as the body fluids. Therefore, the excess IV solution did not change the concentration of the body fluids, but it did increase the volume of the body fluids. An increased volume of saline solution increases the blood volume and blood pressure. Increased blood pressure stimulates baroreceptors, which results in inhibition of ADH secretion. The reduced ADH secretion causes the kidneys to produce a large volume of urine. At the same time, the increased blood volume stretches the walls of the atria, especially the right atrium, and causes the release of atrial natriuretic hormone. Atrial natriuretic hormone acts on the kidneys to reduce $Na^+$ reabsorption. Because $Na^+$ reabsorption is decreased, both $Na^+$ and water are lost in the urine. The increased pressure also results in less renin secretion from the kidney. The reduced renin causes less angiotensinogen to be converted to angiotensin. Consequently, less angiotensin II is formed, which reduces aldosterone secretion from the adrenal cortex. The decreased aldosterone slows $Na^+$ and water reabsorption, causing more $Na^+$ and water to be lost in the urine. Consequently, the urine volume and the amount of NaCl in the urine increase until the excess saline solution is eliminated.

4. The female urinary bladder is more accessible to bacteria from the exterior because the urethra of a female is much shorter than that of a male. For this reason, urinary bladder infections are more common in females than in males.

5. Hyperventilation results in a greater than normal rate of carbon dioxide loss from the circulatory system. Because carbon dioxide is lost from the circulatory system, $H^+$ concentration decreases and the pH of body fluids increases. Breathing into a paper bag corrects for the effects of hyperventilation because the person rebreathes air that has a higher concentration of carbon dioxide. The

result is an increase in carbon dioxide in the body. Consequently, the $H^+$ concentration increases, and pH decreases toward normal levels.

6. Elevated blood carbon dioxide levels cause an increase in $H^+$ and a decrease in blood pH due to the following reaction:

$$CO_2 + H_2O \rightleftarrows H_2CO_3 \rightleftarrows H^+ + HCO_3^-$$

However, the kidney plays an important role in the regulation of blood pH. The kidney's rate of $H^+$ secretion into the urine and reabsorption of $HCO_3^-$ increase. This helps prevent high blood $H^+$ levels and low blood pH in Adam.

7. After 7 days Roger's kidneys began to produce a large volume of urine with larger than normal $Na^+$ and $K^+$ concentrations. As a result, Roger became dehydrated by day 9. Dehydration results in reduced blood volume and blood pressure. His hematocrit was increased because the volume of his blood was decreased, but there was no decrease in the number of red blood cells. The percentage of the blood made of red blood cells therefore increased. The pale skin was the result of vasoconstriction, which was triggered by the reduced blood pressure. Dizziness resulted from reduced blood flow to the brain when Roger tried to stand and walk. He was lethargic in part because of reduced blood volume, but also because of low blood levels of $K^+$ and $Na^+$, caused by the loss of these ions in the urine. Low blood levels of $Na^+$ and $K^+$ alter the electrical activity of nerve and muscle cells and result in muscular weakness. The arrythmia of his heart was due to low blood levels of $K^+$ and increased sympathetic stimulation, which was also triggered by low blood pressure.

## Chapter Nineteen

1. The prostate gland is located just anterior to the rectum. It can be palpated through the wall of the rectum. A physician can insert a finger into the rectum and palpate the prostate through the wall of the rectum. The procedure does not require surgical procedures and involves relatively minor discomfort.

2. Because secondary sexual characteristics, external genitalia, and sexual behavior develop in response to testosterone, if the testes fail to produce normal amounts of testosterone at puberty they do not develop normally. Secondary sexual characteristics and external genitalia remain juvenile, and normal adult sexual behavior does not develop.

3. Administration of a large amount of progesterone and estrogen just before the preovulatory LH surge inhibits the release of GnRH, LH, and FSH. Consequently, ovulation does not occur. A small amount of estrogen administered, without progesterone, before the preovulatory LH surge, however, could stimulate GnRH, LH, and FSH secretion.

4. Molly's mother could have had leiomyomas also, although, without direct data of medical examinations, one cannot be certain. If that was the cause of her irregular menstruations, they may have become less frequent as Molly's mother experienced menopause. During menopause, the uterus gradually becomes smaller, and eventually the cyclical changes in the endometrial lining cease. If the condition was relatively mild, the onset of menopause could explain the gradual disappearance of the irregular and prolonged menstruations. (*Note:* If the tumors are large, constant and severe menstruations are likely even if regular menstrual cycles stop due to menopause.)

## Chapter Twenty

1. Because some sperm cells remain viable in the female reproductive tract for up to 6 days and the secondary oocyte is capable of being fertilized for up to 1 day after ovulation, fertilization could occur if sexual intercourse occurred between 5 days before ovulation and 1 day following ovulation. That will be between 10 days and 15 days post-LMP. Data indicate, however, that the most fertile period during the menstrual cycle is between 2 days just before ovulation and the day of ovulation.

2. Two primitive streaks forming in one embryonic disk can result in identical twins because one embryo is formed in association with each primitive streak. If the two streaks are touching, the twins will be conjoined (Siamese), or attached to each other. This attachment can be fairly simple, and the twins can be separated fairly easily by surgery, or the attachment can be extensive, involving internal organs, and may not be corrected easily.

3. 

| | Clinical Age | Developmental Age |
|---|---|---|
| Fertilization | 14 days | 0 days |
| Implantation | 21 days | 7 days |
| Fetal period | 70 days | 56 days |
| Parturition | 280 days | 266 days |

4. Suckling causes a reflex release of oxytocin from the mother's posterior pituitary. Oxytocin causes expulsion of milk from the breast, but it also causes contraction of the uterus. Contraction of the uterus is responsible for the sensation of cramps.

5. Genotype *DD* (homozygous dominant) has the polydactyly phenotype, genotype *Dd* (heterozygous) has the polydactyly phenotype, and genotype *dd* (homozygous recessive) has the normal phenotype.

6. The probability of a girl with Turner syndrome (XO; that is, with one X chromosome and no other sex chromosome) having hemophilia if her mother is a carrier for hemophilia is the same as for a male because she has only one X chromosome. If her mother is a carrier for hemophilia ($X^H X^h$), the daughter will be either $X^H O$ (normal), or $X^h O$ (hemophiliac). The probability is 1/2, or 50%.

# Glossary

Many of the words in this glossary and throughout the text are followed by a simplified phonetic spelling showing pronunciation. The pronunciation key reflects standard clinical usage, with minor modifications, as presented in *Stedman's Medical Dictionary* (27th edition), which has long been a leading reference volume in the health sciences.

> ā as in day, ate, way
> a as in mat, hat, act
> ă as in alone, abortion, media
> ah as in father
> ar as in far
> aw as in fall (fawl)
> ē as in be, bee, meet
> ě as in taken, genesis
> er as in term, earn, learn
> ī as in pie, pine, side
> i as in pit, tip, fit
> ĭ as in pencil
> ō as in no, note, toe
> o as in not, box, cot
> ŏ as in occult, lemon, son
> oo as in food, to, tool
> ow as in cow, brow, plow, now
> oy as in boy, toy, oil
> u as in wood, foot, took
> ŭ as in but, sun, bud, cup, up
> ū as in pure, unit, union, future

## A

**abdomen** (ab-dō′men, ab′dō-men) Belly, between the thorax and the pelvis.

**abdominal cavity** (ab-dom′i-năl) Space bounded by the diaphragm, the abdominal wall, and the pelvis.

**abdominopelvic cavity** (ab-dom′i-nō-pel′vik) Abdominal and pelvic cavities considered together.

**abduction** (ab-dŭk′shun) [abductio] Movement away from the midline.

**absorption** (ab-sōrp′shŭn) The taking in or reception of gases, liquids, light, heat, or solutes, such as the movement of digested molecules across the intestinal wall and into the bloodstream, the movement of substances through the skin, and the movement of fluid into the lymphatics from the interstitial fluid.

**accommodation** (ă-kom′ŏ-dā′shŭn) The act or state of adjustment or adaptation, such as the increase in the thickness and convexity of the lens of the eye in order to focus an object on the retina as the object moves closer to the eyes; decreasing sensitivity of a nerve cell to a stimulus of constant strength.

**acetabulum** (as-ě-tab′ū-lŭm) [L., shallow vinegar vessel or cup] Cup-shaped depression on the lateral surface of the coxal bone, where the head of the femur articulates.

**acetylcholine** (as-e-til-kō′lēn) Neurotransmitter substance released from motor neurons that innervate skeletal muscle fibers, all autonomic eganglionic neurons, all postganglionic parasympathetic neurons, some postganglionic sympathetic neurons, and some central nervous system neurons.

**acetylcholinesterase** (as′e-til-kō-lin-es′ter-ās) Enzyme that breaks down acetylcholine to acetic acid and choline.

**acetyl-CoA** (as′e-til) Acetyl-coenzyme A; formed by the combination of the two-carbon acetyl group with coenzyme A; the molecule that combines with a four-carbon molecule to enter the citric acid cycle.

**Achilles** (ă-kil′ēz) **tendon** Common tendon of the calf muscles that attaches to the heel (calcaneus); named after a mythical Greek warrior who was vulnerable only in the heel.

**acid** (as′id) Any substance that is a proton donor; or any substance that releases hydrogen ions.

**acidic solution** Solution with more hydrogen ions than hydroxide ions; has a pH of less than 7.

**acidosis** (as-i-dō′sis) Condition characterized by a lower than normal blood pH (pH of 7.35 or lower).

**acinus** (as′i-nŭs), pl. acini (as′i-nī) [L., berry, grape] Grape-shaped secretory portion of a gland.

**acromegaly** (ak-rō-meg′ă-lē) [G. *acro; megas,* large] Disorder marked by progressive enlargement of the bones of the head, face, hands, feet, and thorax as a result of excessive secretion of growth hormone by the anterior pituitary gland.

**acromion** (ă-krō′me-on) [Fr. *akron,* tip + *omos,* shoulder] Lateral end of the spine of the scapula that projects as a broad, flattened process overhanging the glenoid fossa; articulates with the clavicle.

**acrosome** (ak′rō-sōm) [*acro,* tip + G. *soma,* body]. A caplike organelle surrounding the anterior portion of a sperm cell, containing enzymes that facilitate entry of the sperm cell through the zona pellucida.

**actin myofilament** (ak′tin mī-ō-fil′ă-ment) One of the two major kinds of protein fibers that make up a sarcomere; thin filaments; resemble two minute strands of pearls twisted together.

**action potential** All-or-none change in membrane potential in an excitable tissue that is propagated as an electrical signal.

**activation energy** Energy that must be added to atoms or molecules to start a chemical reaction.

**active transport** Carrier-mediated process that requires ATP and can move substances into or out of cells from a lower to a higher concentration.

**adaptive immunity** Immune system response in which there is an ability to recognize, remember, and destroy a specific antigen.

**adduction** (ă-dŭk′shŭn) [L. *adductus,* to bring toward] Movement toward the midline.

**adductor** (a-dŭk′ter, -tōr) [L. *adductus,* to bring toward] Muscle causing movement toward the midline.

**adenoid** (ad′ě-noyd) Enlarged pharyngeal tonsil.

**adenosine triphosphate** (ă-den′ō-sēn trī-fos′făt) (ATP) Adenosine, an organic base, with three phosphate groups attached to it; energy stored in adenosine triphosphate is used in nearly all the energy-requiring reactions in the body.

**ADH** See *antidiuretic hormone.*

**adipose** (ad′i-pōs) [L. *adeps,* fat] Fat; relating to fat tissue.

**adrenal cortex** (ă-drē′năl kōr′teks) The outer part of the adrenal gland, which secretes the following steroid hormones: glucocorticoids, mainly cortisol; mineralocorticoids, mainly aldosterone; and androgens.

**adrenal gland** (ă-drē′năl) [L. *ad,* to; *ren,* kidney, near or on the kidneys] One of two endocrine glands located on the superior pole of each kidney; secretes the hormones epinephrine, norepinephrine, aldosterone, cortisol, and androgens.

**adrenal medulla** (ă-drē′năl me-dool′ă) Inner part of the adrenal gland, which secretes mainly epinephrine but also small amounts of norepinephrine.

**adrenaline** (ă-dren′ă-lin) [from the adrenal gland] Synonym for *epinephrine.*

**adrenocorticotropic hormone** (ă-drē′nō-kōr′ti-kō-trō′pik) [L. *ad,* near + *ren,* kidney + *cortico,* cortex + *trophre,* nurture] (ACTH) Hormone of e anterior pituitary gland that stimulates the adrenal cortex to secrete cortisol.

**adventitia** (ad-ven-tish′ă) [L. *adventicius,* coming from abroad or outside; foreign] Outermost covering of an organ that is continuous with the surrounding connective tissue.

**aerobic respiration** (ār-ō′bik) Breakdown of glucose in the presence of oxygen to produce carbon dioxide, water, and approximately 38 ATP molecules; includes glycolysis, the citric acid cycle, and the electron-transport chain.

**afferent** (af′er-ent) [L. *afferens,* to bring to] Inflowing; conducting toward a center, denoting certain arteries, veins, lymphatics, and sensory nerves. Opposite of *efferent.*

**afferent arteriole** (ar-tēr′ē-ōl) Small artery in the renal cortex that supplies blood to the glomerulus.

**afferent fiber** Sensory nerve fiber going from the peripheral to the central nervous system; sensory or afferent fiber.

**afterload** Resistance against which the ventricles must pump blood; it is increased in people who have hypertension.

**agglutination** (ă-gloo′ti-nā′shŭn) [L. *ad,* to + *gluten,* glue] Process by which cells stick together to form clumps.

**agonist** (ăg′ōn-ist) [G. *agon,* a contest] Denoting a muscle in a state of contraction, with reference to its opposing muscle, or antagonist.

**agranulocyte** (ă-gran′ū-lō-sīt) [G. *a-,* without + *granular* + *kytos,* cell] White blood cell with very small cytoplasmic granules that cannot be easily seen with the light microscope; lymphocytes and monocytes.

**aldosterone** (al-dos′ter-ōn) Steroid hormone produced by the adrenal cortex that facilitates potassium exchange for sodium in the distal convoluted tubule and collecting duct, causing sodium ion reabsorption and potassium and hydrogen ion secretion.

**alkaline solution** (al′kă-līn) See *basic solution.*

**alkalosis** (al-kă-lō′sis) Condition characterized by a higher than normal blood pH (pH of 7.45 or above).

**alveolar duct** (al-vē′ō-lăr) Part of the respiratory passages beyond a respiratory bronchiole; from it arise alveolar sacs and alveoli.

**alveolar sac** Two or more alveoli that share a common opening.

**alveolus** (al-vē′ō-lŭs), pl. alveoli (al-vē′ō-lī) [L., small cavity or hollow sac] Cavity; examples include the sockets into which the teeth fit and the ends of the respiratory system.

**amino acid** (ă-mē′nō) Class of organic acids containing an amine group ($NH_2$) that makes up the building blocks of proteins.

**amniotic cavity** (am-nē-ot′ik) [G. *amnios,* lamb] Fluid-filled cavity surrounding and protecting the developing embryo.

**amylase** (am′il-ās) One of a group of starch-splitting enzymes that cleave starch, glycogen, and related polysaccharides.

**anabolism** (ă-nab′ō-lizm) [G. *anabole,* a raising up] All the synthesis reactions that occur within the body; requires energy.

**anaerobic respiration** (an-ār-ō′bik) Breakdown of glucose in the absence of oxygen to produce lactic acid and two ATP molecules; consists of glycolysis and the reduction of pyruvic acid to lactic acid.

**anaphase** (an′ă-fāz) [G. *ana,* up + *phases,* appearance] Stage of mitosis or meiosis in which the chromosomes move from the center area of the cell, the equatorial plane, toward the poles of the cell.

**anatomical position** (an′ă-tom′i-kăl) Position in which a person is standing erect with the feet facing forward, the arms hanging to the sides, and the palms of the hands facing forward.

**anatomy** (ă-nat′ŏ-mē) [G. *ana,* apart + *tome,* a cutting] Scientific discipline that investigates the structure of the body.

**androgen** (an′drō-jen) [G. *andros,* male] Hormone that stimulates the development of male sexual characteristics; includes testosterone.

**anemia** (ă-nē′mē-ă) [G. *an,* without + *haima,* blood] Condition that results in less than normal hemoglobin in the blood or a lower than normal number of red blood cells.

**anencephaly** (an′en-sef′ă-lē) Defective development of the brain with absence of the cerebral and cerebellar hemispheres and with only a rudimentary brainstem.

**angina pectoris** (an′ji-nă pek′tō-ris, an-jī′nă) Pain resulting from a reduced blood supply to cardiac muscle.

**angioplasty** (an′jē-ō-plas-tē) [G. *angio,* blood vessel] Technique used to dilate the coronary arteries by threading a small, balloonlike device into a partially blocked coronary artery and then inflating the balloon to enlarge the diameter of the vessel.

**angiotensin** (an-jē-ō-ten′sin) [*angio,* blood vessel + *tensus,* to stretch] Angiotensin I is a peptide derived when renin acts on angiotensinogen; angiotensin II is formed from angiotensin I when angiotensin-converting enzyme acts on angiotensin I; angiotensin II is a potent vasoconstrictor, and it stimulates the secretion of aldosterone from the adrenal cortex.

**angiotensinogen** (an′jē-ō-ten-sin′ō-jen) Protein found in the blood that gives rise to angiotensin I after renin, an enzyme secreted from the kidney, acts on it.

**ANH** See *atrial natriuretic hormone.*

**antagonist** (an-tag′ŏ-nist) Muscle that works in opposition to another muscle.

**anterior** (an-tēr′ē-ōr) [L., to go before] That which goes first; in humans, toward the belly or front.

**anterior horn** Part of the spinal cord gray matter containing motor neurons; also called the ventral horn or motor horn.

**anterior pituitary gland** Portion of the pituitary gland derived from the oral epithelium.

**antibody** (an′tē-bod-ē) [G. *anti,* against + *body,* a thing] Protein in the plasma that is responsible for antibody-mediated (humoral) immunity; binds specifically to an antigen.

**antibody-mediated immunity** Immunity resulting from B cells and the production of antibodies.

**anticoagulant** (an′tē-kō-ag′ū-lant) Chemical that prevents coagulation or blood clotting; an example is antithrombin.

**antidiuretic hormone** (an′tē-dī-ū-ret′ik) [G. *anti,* against + *uresis,* urine volume] (ADH) Hormone secreted from the posterior pituitary gland that acts on the kidney to reduce the output of urine; also called vasopressin.

**antigen** (an′ti-jen) [G. *anti* (*body*) + *-gen,* a thing] Substance that induces a state of sensitivity or resistance to microorganisms or toxic substances after a latent period; substance that stimulates the adaptive immune system; self-antigens are produced by the body, and foreign antigens are introduced into the body.

**antigen-receptor** Molecule on the surface of lymphocytes that specifically binds antigens.

**aorta** (ā-ōr′tă) [G. *aorte,* from + *aeiro,* to lift up] Large, elastic artery that is the main trunk of the systemic arterial system, which carries blood from the left ventricle of the heart and passes through the thorax and abdomen.

**aortic semilunar valve** Semilunar valve consisting of three cusps of tissue located at the base of the aorta where it arises from the left ventricle; the cusps overlap during ventricular diastole to prevent leakage of blood from the aorta into the left ventricle.

**apex** (ā′peks) [L., tip] Extremity of a conical or pyramidal structure; the apex of the heart is the rounded tip directed anteriorly and slightly inferiorly.

aphasia (ă-fā′zē-ă) [G. *a-*, without + *phasis*, speech or speechlessness] Impaired or absent communication by speech, writing, or signs because of dysfunction of brain centers in the dominant cerebral hemisphere.

apocrine (ap′ō-krin) [G. *apo*, away from + *krino*, to separate] Type of gland whose cells contribute cytoplasm to its secretion; sweat gland that produces organic secretions traditionally are called apocrine; however, these sweat glands are now known to be merocrine glands; see also *merocrine* and *holocrine*.

aponeurosis (ap′ō-noo-rō′sis) [G. *neuron*, sinew; end of a muscle where it becomes a tendon] Sheet of fibrous connective tissue, or an expanded tendon, serving as the origin or insertion of a flat muscle.

appendicular (ap′en-dik′ū-lăr) [L. *appendo*, to hang something on] Relating to an appendage, such as the limbs and their associated girdles.

appendix (ă-pen′diks), pl. appendices (ă-pen′di-sēs) [L. *appendo*, to hang something on] Smaller structure usually attached by one end to a larger structure; a small blind extension of the colon attached to the cecum.

appositional growth (ap-ō-zish′ŭn-ăl) [L. *ap* + *pono*, to place at or to] To place one layer of bone, cartilage, or other connective tissue against an existing layer; increases the width or diameter of bones.

aqueous humor (ak′wē-ŭs, ā′kwē-ŭs) Watery, clear fluid that fills the anterior chamber and posterior chamber of the eye.

arachnoid mater (ă-rak′noyd ma′ter) [G. *arachne*, spiderlike, cobweb] Thin, cobweblike meningeal layer surrounding the brain and spinal cord; the middle of three layers.

areola (ă-rē′ō-lă), pl. areolae (ă-rē′ō-lē) [small areas] Pigmented area surrounding the nipple of a mammary gland.

areolar (ă-rē′ō-lăr) Relating to connective tissue with small spaces within it; loose connective tissue.

arrector pili (ă-rek′tōr pī′lī) [L., that which raises hair] Smooth muscle attached to the hair follicle and dermis that raises the hair when it contracts.

arteriosclerosis (ar-tēr′ē-ō-skler-ō′sis) [L. *arterio-* + G. *sklerosis*, hardness] Hardness of the arteries.

arteriosclerotic lesion (ar-tēr′ē-ō-skler-ot′ik) Lesion or growth in arteries that narrows the lumen, or passage, and makes the walls of the arteries less elastic.

artery (ar′ter-ē) [G. *arteria*, the windpipe] Blood vessel that carries blood away from the heart.

articulation (ar-tik-ū-lā′shŭn) [L. *articulatio*, a forming of vines] Place where two bones come together; a joint.

artificial heart Mechanical pump used to replace a diseased heart.

artificial pacemaker Electronic device implanted beneath the skin with an electrode that extends to the heart; provides periodic electrical stimuli to the heart and substitutes for a faulty SA node.

astrocyte (as′trō-sīt) [G. *astron*, star + *kytos*, a hollow cell] Star-shaped neuroglial cell that helps regulate the composition of fluid around the neurons of the central nervous system.

atherosclerosis (ath′er-ō-skler-ō′sis) [G. *athere*, gruel or soft, pasty material + *sklerosis*, hardness] Lipid deposits (plaques) in the tunica intima of large and medium-sized arteries.

atom (at′ŏm) [G. *atomos*, indivisible, uncut] Smallest particle into which an element can be divided using chemical methods; composed of neutrons, protons, and electrons.

atomic number (ă-tom′ik) Number of protons in an element.

ATP See *adenosine triphosphate.*

atrial natriuretic (ā′trē-ăl nā′tre-yū-ret′ik) hormone (ANH) Hormone released from cells in the atrial wall of the heart when atrial blood pressure is increased; lowers blood pressure by increasing the rate of urine production.

atrioventricular (AV) bundle (ā-trē-ō-ven-trik′ū-lar) Bundle of modified cardiac muscle fibers that projects from the AV node through the interventricular septum; conducts action potentials from the AV node rapidly through the interventricular septum; also called the bundle of His.

atrioventricular (AV) node Small collection of specialized cardiac muscle fibers located in the inferior part of the right atrium; delays action potential transmission to the atrioventricular bundle.

atrioventricular valve Valve between the atrium and the ventricle of the heart, the tricuspid valve between the right atrium and right ventricle, and the bicuspid (or mitral valve) between the left atrium and left ventricle.

atrium (ā′trē-ŭm), pl. atria (ā′trē-ă) [L., entrance chamber] One of the two chambers of the heart that collect blood during ventricular contraction and pump blood into the ventricles to complete ventricular filling at the end of ventricular relaxation; the right atrium receives blood from the inferior and superior venae cavae and from the coronary sinus and delivers blood to the right ventricle; the left atrium receives blood from the pulmonary veins and delivers blood to the left ventricle.

auditory (aw′di-tōr-ē) Relating to hearing.

auditory ossicles (os′i-klz) Bones of the middle ear; the malleus, incus, and stapes.

auditory tube Air-filled passageway between the middle ear and pharynx.

auricle (aw′ri-kl) [L. *auris*, ear] Fleshy part of the external ear on the outside of the head; a small, conical pouch projecting from the upper anterior part of each atrium of the heart.

auscultatory (aws-kŭl′tă-tō-rē) [L. *ausculto*, to listen] To listen to the sounds made by the various body structures, especially to Korotkoff sounds when determining blood pressure.

autocrine (aw′tō-krin) [G. *autos*, self + *krino*, to separate] Denoting self-stimulation through cellular production of a factor and a specific receptor for it.

autoimmune disease (aw-tō-i-mūn′) Disorder resulting from a specific immune system reaction against self-antigens.

autonomic nervous system (ANS) (aw-tō-nom′ik) Part of the peripheral nervous system composed of efferent fibers that reach from the central nervous system to smooth muscle, cardiac muscle, and glands.

autosome (aw′tō-sōm) [G. *auto-*, self + *soma*, body] Any chromosome other than a sex chromosome; normally occurs in pairs in somatic cells and singly in gametes.

AV See *atrioventricular.*

axial (ak′sē-ăl) [L. *axle*, axis] Head, neck, and trunk as distinguished from the extremities.

axon (ak′son) [G., axis] Main process of a neuron; usually conducts action potentials away from the neuron cell body.

# B

baroreceptor (bar′ō-rē-sep′ter) [G. *baro*, weight or pressure] Sensory nerve endings in the walls of the atria of the heart, aortic arch, and carotid sinuses; sensitive to stretching of the wall caused by increased blood pressure; also called a pressoreceptor.

baroreceptor reflex Process in which baroreceptors detect changes in blood pressure and produce changes in heart rate, force of heart contraction, and blood vessel diameter that return blood pressure toward normal levels.

basal nuclei Nuclei at the base of the cerebrum, diencephalon, and midbrain involved in controlling motor functions.

base Any substance that is a proton acceptor or any substance that binds to hydrogen ions; lower part or bottom of a structure; the base of the heart is the flat portion directed posteriorly and superiorly; veins and arteries project into and out of the base, respectively.

basement membrane Structure that attaches most epithelia (exceptions include lymphatic vessels and the liver sinusoids) to underlying tissue; consists of carbohydrates and proteins secreted by the epithelia and the underlying connective tissue.

basic solution (bā′sik) Solution with fewer hydrogen ions than hydroxide ions; has a pH greater than 7.

basilar membrane (bas′i-lăr) One of two membranes forming the cochlear duct; supports the spiral organ.

basophil (bā′sō-fil) [G. basis, base + phileo, to love] White blood cell with granules that stain purple with basic dyes; promotes inflammation and prevents clot formation.

belly Largest part of a muscle between the origin and insertion.

benign (bē-nīn′) [L. benignus, kind] Mild in character or nonmalignant; does not spread to distant sites.

beta-adrenergic (bā′tă ad-rě-ner′jik) blocking agent Drug that binds to and prevents adrenergic receptors from responding to adrenergic compounds that normally bind to beta-adrenergic receptors and cause them to function; beta-adrenergic blocking agents are used to treat certain arrhythmias in the heart and to treat tachycardia (rapid heart rate).

biceps brachii (bī′seps brā′kē-ī) Muscle in the anterior arm with two heads, or origins, on the scapula and an insertion onto the radius; flexes and supinates the forearm.

bicuspid valve (bī-kŭs′pid) Valve closing the opening between the left atrium and left ventricle of the heart; has two cusps; also called the mitral valve.

bile (bīl) Fluid secreted from the liver, stored in the gallbladder, and released into the duodenum; consists of bile salts, bile pigments, bicarbonate ions, fats, and other materials.

bile salt Organic salt secreted by the liver that emulsifies lipids.

bilirubin (bil-i-roo′bin) [L., bile + ruber, red] Bile pigment formed from the heme in hemoglobin during the destruction of red blood cells by macrophages.

biopsy (bī′op-sē) Process of removing tissue from living patients for diagnostic examination, or a specimen obtained by biopsy.

blastocele (blas′tō-sēl) [G. blastos, germ + koilos, hollow] Cavity in the blastocyst.

blastocyst (blas′tō-sist) [G. blastos, germ + kystis, bladder] Early stage of mammalian embryo development consisting of a hollow ball of cells with an inner cell mass and an outer trophoblast layer.

blood-brain barrier Cellular and matrix barrier made up primarily of blood vessel endothelium, with some help from the surrounding astrocytes; it allows some (usually small) substances to pass from the circulation into the brain but does not allow other (larger) substances to pass.

blood group Category of red blood cells based on the type of antigen on the surface of the red blood cell; for example, the ABO blood group is involved with transfusion reactions.

blood pressure [L. pressus, to press] The force blood exerts against the blood vessel walls; expressed relative to atmospheric pressure and reported in the form of millimeters of mercury (mm Hg) of pressure.

bony labyrinth (lab′i-rinth) Interconnecting tunnels and chambers within the temporal bone in which the inner ear is located.

Bowman's capsule Enlarged end of the nephron; Bowman's capsule and the glomerulus make up the renal corpuscle.

brachialis (brā′kē-ăl-is) Muscle of the anterior arm that originates on the humerus and inserts onto the ulna; flexes the forearm.

brachial plexus (brā′kē-ăl) [L. brachium, arm] Nerve plexus to the upper limb; originates from spinal nerves C5 to T1.

brainstem Portion of the brain consisting of the midbrain, pons, and medulla oblongata.

breathing (brēthing) Movement of air into and out of the lung; see ventilation.

bronchiole (brong′kē-ōl) One of the finer subdivisions of the bronchial tubes, less than 1 mm in diameter, that has no cartilage in its wall but has relatively more smooth muscle and elastic fibers than do larger bronchial tubes.

bronchus (brong′kŭs), pl. bronchi (brong′kī) [G. bronchos, windpipe] Any one of the air ducts conducting air from the trachea to the bronchioles.

buccinator (buk′si-nā′tōr) Muscle making up the lateral sides of the oral cavity; flattens the cheeks.

buffer (bŭf′er) Chemical that resists changes in pH when either an acid or a base is added to a solution containing the buffer.

bundle of His See atrioventricular bundle.

burn Lesion caused by heat, acid, or other agents; a partial-thickness burn of the skin damages only the epidermis (first-degree burn) or the epidermis and part of the dermis (second-degree burn); a full-thickness (third-degree) burn destroys the epidermis and the dermis and sometimes the underlying tissue.

bursa (ber′să) [L., purse or pocket] Closed sac or pocket containing synovial fluid; usually found in areas where friction occurs.

# C

calcaneus (kal-kā′nē-ŭs) [L., the heel] Largest tarsal bone forming the heel.

calcitonin (kal-si-tō′nin) Hormone, released from cells of the thyroid gland, that acts on tissues, especially bone, to cause a decrease in blood levels of calcium ions.

calcium channel blocker (kal′sē-ŭm) Class of drugs that specifically block channels in cell membranes through which calcium ions pass; calcium channel blockers are used to treat some kinds of cardiac arrhythmias.

callus (kal′ŭs) [L., hard skin] Thickening of the stratum corneum of skin in response to friction; the zone of tissue repair between fragments of a broken bone.

calorie (kal′ō-rē) [L. calor, heat] Unit of heat or energy content; the quantity of energy required to raise the temperature of 1 gram of water 1°C. A Calorie (Cal), or kilocalorie (kcal), is the amount of heat or energy required to raise the temperature of 1000 grams of water from 14°C to 15°C.

calyx (kā′liks), pl. calyces (kal′i-sēz) [G., flower petal or cup of a flower] Small container into which urine flows as it leaves the collecting ducts at the tip of the renal pyramids; the calyces come together to form the renal pelvis.

canaliculus (kan-ă-lik′ū-lŭs) Tiny canal in bone between osteocytes containing osteocyte cell processes; a cleftlike lumen between the cells of each hepatic cord, connects medial corner of the eye to the lacrimal sac.

cancer (kan′ser) [L., a crab, suggesting crablike movement] Malignant neoplasm, or tumor.

capacitation (kă-pas′i-tā′shŭn) Process whereby the sperm cells develop the ability to fertilize oocytes.

capillary (kap′i-lăr-ē) [L. capillaris, relating to hair, resembling a fine hair] Minute blood vessel consisting only of simple squamous epithelium and a basement membrane; major site for the exchange of substances between the blood and tissues.

carbohydrate (kar-bō-hī′drāt) Organic molecule made up of one or more monosaccharides chemically bound together; sugars and starches.

carbonic anhydrase (kar-bon′ik an-hī′drās) Enzyme that increases the rate at which carbon dioxide reacts with water to form hydrogen ions and bicarbonate ions.

carcinoma (kar-si-nō′mă) [G. karkinoma, cancer + oma, tumor] Malignant tumor derived from epithelial tissue.

cardiac cycle (kar′dē-ak) One complete sequence of cardiac systole and diastole.

cardiac output  Volume of blood pumped by either ventricle of the heart per minute; about 5 L/min for the heart of a healthy adult at rest.

cardioregulatory center  Specialized area within the medulla oblongata of the brain that receives sensory input and controls parasympathetic and sympathetic stimulation of the heart.

carotene  (kar′ō-tēn)  Yellow pigment in plants such as squash and carrots; accumulates in the lipids of the stratum corneum and in the fat cells of the dermis and hypodermis and is used as a source of vitamin A.

carotid bodies  (ka-rot′id)  Small organ near the carotid sinuses that detects changes in blood oxygen, carbon dioxide, and pH.

carotid sinus  Enlargement of the internal carotid artery near the point where the internal carotid artery branches from the common carotid artery; contains baroreceptors.

carpal  (kar′păl)  [G. karpos, wrist]  Associated with the wrist; bones of the wrist.

carrier molecule  Protein that extends from one side of the cell membrane to the other; binds to molecules to be transported and moves them from one side of the membrane to the other.

cartilage  (kar′ti-lij)  [L. cartilage, gristle]  Firm, smooth, resilient, nonvascular connective tissue.

cascade  (kas-kād′)  [Fr. cascare, to fall]  Series of sequential interactions, which once initiated continues to the final one; each interaction is activated by the preceding one, with cumulative effect.

catabolism  (kă-tab′ō-lizm)  [G. katabole, a casting down]  All the decomposition reactions that occur in the body; releases energy.

catalyst  (kat′ă-list)  Substance that increases the rate of a chemical reaction; in the process, the catalyst is not permanently changed or used up.

cecum  (sē′kŭm)  [L. caecus, blind]  Blind sac forming the beginning of the large intestine.

cell  (sel)  [L. cella, chamber]  Basic living unit of all plants and animals.

cell-mediated immunity  Immunity resulting from the actions of T cells.

cell membrane  Plasma membrane; outermost component of the cell, surrounding and binding the rest of the cell contents.

central canal  Small canal containing blood vessels, nerves, and loose connective tissue and running parallel to the long axis of a bone; also called a haversian canal.

central nervous system (CNS)  Brain and spinal cord.

centriole  (sen′trē-ōl)  Small organelle that divides and migrates to each pole of the nucleus; spindle fibers extend from the centromeres to the centrioles during mitosis.

centromere  (sen′trō-mēr)  [G. kentron, center + meros, part]  Specialized region where chromatids are linked together in a chromosome.

cerebellum  (ser-e-bel′ŭm)  [L., little brain]  Part of the brain attached to the brainstem; important in maintaining muscle tone, balance, and coordination of movements.

cerebral aqueduct  (ser′ĕ-brăl, sĕ-rē′brăl)  Small connecting tube through the midbrain between the third and fourth ventricles.

cerebrospinal fluid  (ser′ĕ-brō-spī-năl, sĕ-rē′brō-spī-năl)  (CSF)  Fluid filling the ventricles and surrounding the brain and spinal cord.

cerebrum  (ser′ĕ-brŭm, sĕ-rē′brŭm)  [L., brain]  Largest part of the brain, consisting of two hemispheres and including the cortex, nerve tracts, and basal nuclei.

cerumen  (sĕ-roo′men)  [L. cera, wax]  Specific type of sebum produced in the external auditory canal; earwax.

cervical  (ser′vĭ-kal)  Neck.

cervical plexus  Nerve plexus of the neck; originates from spinal nerves C1–C4.

cervix  (ser′viks)  [L., neck]  Lower part of the uterus extending to the vagina.

chemical  (kem′i-kăl)  Relating to chemistry, especially to the characteristics of atoms and molecules and to their interactions.

chemical bond  Association between two atoms formed when the outermost electrons are transferred or shared between atoms.

chemical mediator of inflammation  Chemical released or activated by injured tissues and adjacent blood vessels; produces vasodilation, increases vascular permeability, and attracts blood cells; includes histamine, kinins, prostaglandins, and leukotrienes.

chemical reaction  Process by which atoms or molecules interact to form or break chemical bonds.

chemistry  (kem′is-trē)  [G. chemeia, alchemy]  Science dealing with the atomic composition of substances and the reactions they undergo.

chemoreceptor reflex  (kem′ō-rē-sep′tŏr)  Process in which chemoreceptors detect changes in oxygen levels, carbon dioxide levels, and pH in the blood and produce changes in heart rate, force of heart contraction, and blood vessel diameter that return these values toward their normal levels.

cholecystokinin  (kō′lē-sis-tō-kī′nin)  [G. chole, bile + kysis, bladder + kineo, to move]  Hormone released from the duodenum; inhibits gastric acid secretion and stimulates contraction of the gallbladder.

chondrocyte  (kon′drō-sīt)  [G. chondrion, gristle + cyte]  Cartilage cell.

chordae tendineae  (kōr′dē ten′di-nē-ē)  [L., cord]  Tendinous strands running from the papillary muscles to the free margin of the cusps that make up the tricuspid and bicuspid valves; prevent the cusps of these valves from extending up into the atria during ventricular contraction.

choroid  (kō′royd)  [G. chorioeides, membranelike or lacy]  Portion of the vascular tunic associated with the sclera of the eye; prevents scattering of light.

choroid plexus  Specialized group of ependymal cells in the ventricles; secretes cerebrospinal fluid.

chromatid  (krō′mă-tid)  [G. chroma, color]  One of a pair of duplicated chromosomes, joined by the centromere, which separates from its partner during cell division.

chromatin  (krō′ma-tin)  [G. chroma, color]  Genetic material of the nucleus consisting of deoxyribonucleic acid (DNA) associated with proteins.

chromosome  (krō′mō-sōm)  [G. chroma, color + soma, body]  One of the bodies (normally 46 in humans) in the cell nucleus that carry the cell's genetic information.

chyle  (kīl)  [G. chylos, juice]  Milky colored lymph with a high fat content.

chylomicron  (kī-lō-mi′kron)  [chylo- + G. micros, small]  Lipid droplet synthesized in the epithelial cells of the small intestine containing triglycerides, cholesterol, and lipoproteins.

chyme  (kīm)  [G. chymos, juice]  Semifluid mass of partly digested food passed from the stomach into the duodenum.

ciliary body  (sil′ē-ar-ē)  [like an eyelash]  Structure continuous with the choroid layer of the eye at its anterior margin that contains smooth muscle cells and is attached to the lens by suspensory ligaments; regulates the thickness of the lens and produces aqueous humor.

cilium  (sil′ē-ŭm), pl. cilia  (sil′ē-ă)  [L., eyelid]  Mobile extension of a cell surface; varies from one to thousands per cell and contains specialized microtubules enclosed by the cell membrane.

citric acid cycle  (sit′rik)  Series of chemical reactions in which citric acid (six-carbon molecule) is converted into a four-carbon molecule, carbon dioxide is formed, and energy is released; the released energy is used to form ATP; the four-carbon molecule can combine with acetyl-CoA (two-carbon) to form citric acid and start the cycle again.

**clavicle** (klav′i-kl) [L., a small key] Bone between the sternum and shoulder; the collarbone.

**climacteric** (klī-mak′ter-ik, klī-mak-ter′ik) [G., the rung of a ladder] Period of endocrine, somatic, and transitory psychological changes occurring in the transition to menopause.

**clitoris** (klit′ō-ris) Small, erectile structure located in the anterior margin of the vestibule.

**clot** (klot) To coagulate; a soft, insoluble mass formed when blood coagulates.

**clot retraction** Condensation of the clot into a denser, more compact structure.

**clotting factor** One of many proteins found in the blood in an inactive state; activated in a series of chemical reactions that result in the formation of a blood clot.

**coagulation** (kō-ag-ū-lā′shŭn) Process of changing from a liquid to a solid, especially blood.

**cochlea** (kok′lē-ă) Portion of the inner ear involved in hearing; shaped like a snail shell.

**codon** (kō′don) Sequence of three nucleotides in mRNA that codes for a specific amino acid in a protein.

**coenzyme** (kō-en′zīm) Substance that enhances or is necessary for the function of an enzyme.

**collagen** (kol′lă-jen) [G. koila, glue + gen, producing] Ropelike protein of the extracellular matrix.

**collecting duct** Straight tubule that extends from the cortex of the kidney to the tip of the renal pyramid; filtrate from the distal convoluted tubules enters the collecting duct and is carried to the calyces.

**colliculus** (ko-lik′ū-lŭs) [L. collis, hill] One of four small mounds on the dorsal side of the midbrain; the superior two are involved in visual reflexes, and the inferior two are involved in hearing.

**colon** (kō′lon) Division of the large intestine that extends from the cecum to the rectum.

**commissure** (kom′ĭ-shūr) [L., a joining together] Bundle of nerve fibers passing from one side to the other in the brain or spinal cord.

**common bile duct** Duct formed by the union of the common hepatic and cystic ducts; it joins the pancreatic duct and empties into the duodenum.

**common hepatic duct** Duct formed by union of the right and left hepatic ducts; it joins the cystic duct to form the common bile duct.

**compact bone** Bone that is denser and has fewer spaces than spongy bone.

**complement** (kom′plĕ-ment) Group of serum proteins that stimulate phagocytosis, inflammation, and lysis of cells.

**compound** (kom′pound) [to place together] Substance containing two or more different kinds of atoms that are chemically combined.

**concha** (kon′kă) [L., shell] Structure resembling a shell in shape; the three bony ridges on the lateral wall of the nasal cavity.

**condyle** (kon′dīl) [G. kondyles, knuckle] Rounded, articulating surface of a joint.

**cone** Photoreceptor cell in the retina of the eye with cone-shaped photoreceptive process; important in color vision and visual acuity.

**conjunctiva** (kon-jŭnk-tī′vă) [L. conjungo, to bind together] Mucous membrane covering the anterior surface of the eye and the inner lining of the eyelids.

**connective tissue** One of the four major tissue types; consists of cells usually surrounded by large amounts of extracellular material; holds other tissues together and provides a supporting framework for the body.

**constant region** Part of an antibody that does not combine with an antigen and is the same in different antibodies; responsible for activating complement and binding the antibody to cells such as macrophages, basophils, and mast cells.

**corn** [L. cornu, horn] Thickening of the stratum corneum of the skin over a bony projection in response to friction or pressure.

**cornea** (kōr′nē-ă) [hornlike] Transparent, anterior part of the fibrous tunic of the eye, through which light enters the eye.

**corneum** (kōr′nē-ŭm) See stratum corneum.

**coronal plane** (kōr′ō-năl) [G. korone, crown] Plane separating the body into anterior and posterior portions; also called a frontal plane.

**coronary artery** (kōr′o-nār-ē) [circling like a crown] Artery that carries blood to the muscles of the heart; the left and right coronary arteries arise from the aorta.

**coronary bypass** Surgery in which a vein from another part of the body is grafted to a coronary artery in such a way as to allow blood flow past a blockage in the coronary artery.

**coronary vein** Vein that carries blood from the heart muscle, primarily to the right atrium.

**corpus callosum** (kōr′pus kă-lō′sŭm) [L., body; callous] Large, thick nerve fiber tract connecting the two cerebral hemispheres.

**corpus luteum** (loo′tē′ŭm) Yellow endocrine body formed in the ovary in the site of a ruptured follicle immediately after ovulation; secretes progesterone and estrogen.

**cortex** (kōr′teks), pl. **cortices** (kōr′ti-sēz) [L., bark] Outer part of an organ such as the brain, kidney, adrenal gland, or hair.

**cortisol** (kōr′ti-sol) Steroid hormone released by the adrenal cortex; increases blood glucose and inhibits inflammation; it is a glucocorticoid.

**cotransport** (kō-trans′pōrt) Transport of one substance across a cell membrane, coupled with the simultaneous transport of another substance across the same membrane in the same direction.

**covalent bond** (kō-vāl′ent) Chemical bond formed when two atoms share one or more pairs of electrons.

**coxal** (kok′săl), **bone** [L., hip] Bone of the hip.

**cranial nerve** (krā′nē-ăl) Peripheral nerve originating in the brain.

**cranial vault** Eight skull bones that surround and protect the brain; braincase.

**cremaster muscle** (krē-mas′ter) Extension of abdominal muscles; in the male, it raises the testis.

**crenation** (krē-nā′shŭn) [L. crena, a notch] Cell shrinkage that occurs when water moves by osmosis from a cell into a hypertonic solution.

**cretinism** (krē′tin-izm) Hypothyroidism in an infant; appears during the first years of life and results in stunting of body growth and mental development; hypothyroid dwarfism.

**cricoid cartilage** (krī′koyd) Most inferior laryngeal cartilage.

**cricothyrotomy** (krī′kō-thī′rot′ō-mē) Formation of an artificial opening in a victim's air passageway through the membrane between the cricoid and thyroid cartilage.

**crown** Part of the tooth formed of and covered by enamel.

**crypt** (kript) [G. kryptos, hidden] Pitlike depression.

**cryptorchidism** (krip-tōr′ki-dizm) Failure of the testes to descend into the scrotal sac.

**cupula** (koo′poo-lă) [L. cupa, a tub] Gelatinous mass that overlies the hair cells of the cristae ampullaris of the semicircular canals; responds to fluid movement.

**cutaneous** (kū-tā′nē-ŭs) [L. cutis, skin] Relating to the skin.

**cuticle** (kū′ti-kl) [L. cutis, skin] Outer thin layer, usually horny; for example, the outer covering of hair or the growth of the stratum corneum onto the nail.

**cyanosis** (sī-ă-nō′sis) [G., dark blue color] Blue coloration of the skin and mucous membranes caused by insufficient oxygenation of blood.

**cystic duct** (sis′tik) Duct from the gallbladder; it joins the common hepatic duct to form the common bile duct.

**cytoplasm** (sī′tō-plazm) [G. cyto, cell + plasma, a thing formed] Cellular material surrounding the nucleus.

cytoskeleton (sī-tō-skel′ĕ-ton) Collection of microtubules, microfilaments, and intermediate filaments that supports the cytoplasm and organelles; also involved with cell movements.

## D

dartos muscle (dar′tōs) [Fr. derō, to skin] Layer of smooth muscle beneath the skin of the scrotum.

deciduous teeth (dē-sid′ū-ŭs) [L. deciduus, falling off] Primary teeth, which fall out to be replaced by the permanent teeth.

decomposition reaction (dē′kom-pō-zish′ŭn) Breakdown of a larger molecule into smaller molecules, ions, or atoms.

deep [O.E. deop, deep] Away from the surface, internal.

defecation (def-ĕ-kā′shŭn) [L. defaeco, to purify] Discharge of feces from the rectum.

deglutition (dē-gloo-tish′ŭn) [L. de-, from, away + glutio, to swallow] Swallowing.

deltoid (del′toyd) [triangular] Triangular muscle over the shoulder; inserts onto the humerus; abducts the arm.

denaturation (dē-na-tū-rā′shŭn) Change in shape of a protein caused by breaking hydrogen bonds; agents that cause denaturation include heat and changes in pH.

dendrite (den′drīt) [G. dendrite, tree] Short, treelike cell process of a neuron; receives stimuli.

dentin (den′tin) Bonelike material forming the mass of a tooth.

deoxyribonucleic (dē-oks′ē-rī′bō-noo-klē′ic) acid (DNA) Type of nucleic acid containing the sugar deoxyribose; the genetic material of cells.

depolarize Decrease in the difference in potential (charge) between two points, as between the inside and outside of a cell membrane.

dermis (der′mis) [G. derma, skin] Dense connective tissue that forms the deep layer of the skin; responsible for the structural strength of the skin.

desmosome (dez′mō-sōm) [G. desmos, a band + soma, body] Point of adhesion between two cells.

diabetes mellitus (dī-ă-bē′tēz me-lī′tŭs) Condition resulting from too little insulin secreted from the pancreatic islets, insufficient numbers of insulin receptors on target cells, or defective receptors that do not respond to insulin.

diaphragm (dī′ă-fram) [a partition wall] Muscular separation between the thoracic and abdominal cavities; its contraction results in inspiration.

diaphysis (dī-af′i-sis) [G., growing between] Shaft of a long bone.

diastole (dī-as′tō-lē) [G. diastole, dilation] Relaxation of the heart chambers, during which they fill with blood; usually refers to ventricular relaxation.

diastolic pressure Minimum arterial blood pressure achieved during ventricular diastole.

diencephalon (dī-en-sef′ă-lon) [G. dia, through + enkephalos, brain] Part of the brain inferior to and nearly surrounded by the cerebrum and connecting posteriorly and inferiorly to the brainstem.

diffusion (di-fū′zhŭn) [L. diffundo, to pour in different directions] Tendency for solute molecules to move from an area of higher concentration to an area of lower concentration in a solution; the product of the constant random motion of all atoms, ions, or molecules in a solution.

digestion (di-jes′chŭn, dī-jes′chŭn) Breakdown of carbohydrates, lipids, proteins, and other large molecules to their component parts.

digestive tract (di-jes′tiv, dī-jes′tiv) Tract from the mouth to the anus, including the stomach and intestines, where food is taken in, broken down, and absorbed.

digitalis (dij′i-tal′is) [L., relating to fingerlike flowers] Steroid used in the treatment of heart diseases, such as heart failure; increases the force of contraction of the heart; extracted from the foxglove plant (Digitalis purpura).

diploid (dip′loyd) Condition in which there are 2 copies of each autosome and 2 sex chromosomes (46 total chromosomes in humans).

disaccharidase (dī-sak′ă-rid-ās) Enzyme that breaks disaccharides down to monosaccharides; commonly found in the microvilli of the intestinal epithelium.

disaccharide (dī-sak′ă-rīd) [two sugars] Two monosaccharides chemically bound together; glucose and fructose chemically join to form sucrose.

dissociate (di-sō-sē-āt′) [L. dis- + socio, to disjoin, separate] Separation of positive and negative ions when they dissolve in water and are surrounded by water molecules.

distal (dis′tăl) [L. di- + sto, to be distant] Farther from the point of attachment to the body than another structure.

distal convoluted tubule Convoluted tubule of the nephron that extends from the ascending limb of the loop of Henle and ends in a collecting duct.

DNA See deoxyribonucleic acid.

dominant (dom′i-nant) [L. dominus, a master] In genetics, a gene that is expressed phenotypically to the exclusion of a contrasting recessive trait.

dorsal (dōr′săl) [L. dorsum, back] Back surface of the body; in humans, synonymous with posterior.

dorsal root Sensory root of a spinal nerve.

ductus arteriosus (dŭk′tŭs ar-tēr′ē-ō-sŭs) Short artery that extends from the pulmonary trunk to the aorta; in the fetus, blood flows through the ductus arteriosus from the pulmonary trunk into the aorta and bypasses the lungs.

ductus deferens (dŭk′tŭs def′er-enz) Duct of the testis, running from the epididymis to the ejaculatory duct; also called the vas deferens.

duodenum (doo-ō-dē′nŭm, doo-od′ĕ-nŭm) [L. duodeni, 12] First division of the small intestine; connects to the stomach.

dura mater (doo′ră mā′ter) [L., tough mother] Tough, fibrous membrane forming the outermost meningeal covering of the brain and spinal cord.

## E

eardrum See tympanic membrane.

eccrine (ek′rin) [G. ek, out + krino, to separate] Exocrine; refers to water-producing sweat glands; see merocrine.

ECG See electrocardiogram.

ectoderm (ek′tō-derm) Outermost of the three germ layers of the embryo.

ectopic beat (ek-top′ik) Heartbeat that originates from an area of the heart other than the SA node.

edema (e-dē′mă) [G. oidema, a swelling] Excessive accumulation of fluid, usually causing swelling.

efferent (ef′er-ent) [L. efferens, to bring out] Conducting outward from a given organ or part, denoting certain arteries, veins, lymphatics, and motor nerves. Opposite of afferent.

efferent arteriole (ar-tēr′ē-ōl) Vessel that carries blood from the glomerulus to the peritubular capillaries.

efferent ductule (dŭk′tool) Small duct that leads from the testis to the epididymis.

efferent fiber Nerve fiber going from the central nervous system toward the peripheral nervous system; motor fiber.

ejaculation (ē-jak′ū-lā′shŭn) [to shoot out] Reflexive expulsion of semen from the penis.

ejaculatory duct (ē-jak′ū-lă-tōr-ē) Duct formed by the union of the ductus deferens and the excretory duct of the seminal vesicle, which opens into the urethra.

electrocardiogram (ē-lek-trō-kar′dē-ō-gram) (ECG) Graphic record of the heart's electrical currents obtained with an electronic recording instrument.

electrolyte (ē-lek′trō-līt) [G. *electro*, + *lytos*, soluble] Positive and negative ions that conduct electricity in solution.

electron (ē-lek′tron) Negatively charged particle around the nucleus of an atom.

electron-transport chain Series of energy-transfer molecules in the inner mitochondrial membrane; they receive energy and use it in the formation of ATP and water.

element (el′ĕ-ment) [L. *elementum*, a rudiment] Simplest type of matter with unique chemical properties.

embolus (em′bō-lŭs) [G. *embolos*, a plug] Detached clot or other foreign body that occludes a blood vessel.

embryo (em′brē-ō) [Fr. *en*, in + *bryō*, to swell] In prenatal development, the developing human from the time of fertilization to approximately the end of the second month.

emission (ē-mish′ŭn) [L. *emissio*, to send out] Discharge; formation and accumulation of semen prior to ejaculation.

emulsification (ē-mŭl′si-fi-kā-shŭn) Dispersal of one liquid, or very small globules of the liquid, within another liquid.

emulsify (ē-mŭl′si-fī) To form an emulsion, which is one liquid dispersed within another liquid.

enamel (ē-nam′ĕl) Hard substance covering the exposed portion of a tooth.

endocardium (en-dō-kar′dē-ŭm) [G. *endon*, within + *kardia*, heart] Innermost layer of the heart, including endothelium and connective tissue.

endochondral (en-dō-kon′drăl) [*endo* + G. *chondrion*, gristle] Growth of cartilage, which is then replaced by bone.

endochondral ossification (en-dō-kon′drăl os′i-fi-kā′shŭn) Bone formation within cartilage.

endocrine (en′dō-krin) [*endo* + G. *krino*, to separate] Ductless gland that secretes internally, usually into the circulatory system.

endocytosis (en′dō-sī-tō′sis) [*endo* + G. *kytos*, cell + *-osis*, condition] Bulk uptake of material through the cell membrane by taking it into a vesicle.

endoderm (en′dō-derm) [*endo* + G. *derma*, skin] Innermost of the three germ layers of the embryo.

endolymph (en′dō-limf) [*endo* + G. *lympha*, clear fluid or springwater] Fluid inside the membranous labyrinth of the inner ear.

endometrium (en′dō-mē′trē-ŭm) [*endo* + G. *mētra*, uterus] Mucous membrane that constitutes the inner layer of the uterine wall; consists of a simple columnar epithelium and a lamina propria that contains simple tubular uterine glands.

endomysium (en′dō-mis′ē-ŭm, en′dō-miz′ē-ŭm,) [*endo* + G. *mys*, muscle] Fine connective tissue sheath surrounding a muscle fiber.

endoplasmic reticulum (en′dō-plas′mik re-tik′ū-lŭm) [*endo* + G. *plastos*, formed a network] Membranous network inside the cytoplasm; rough endoplasmic reticulum has ribosomes attached to the surface; smooth endoplasmic reticulum does not have ribosomes attached.

endosteum (en-dos′tē-ŭm) [*endo* + G. *osteon*, bone] Membranous lining of the medullary cavity and the cavities of spongy bone.

endothelium, pl. endothelia (en-dō-thē′lē-ŭm) [G. *endo* + *thēlē*, nipple] Layer of flat cells lining especially blood and lymphatic vessels and the heart.

enzyme (en′zīm) [G. *en*, in + *zyme*, leaven] Protein molecule that increases the rate of a chemical reaction without being permanently altered; an organic catalyst.

eosinophil (ē-ō-sin′ō-fil) [eosin, an acidic dye + G. *phileo*, to love] White blood cell with granules that stain red with acidic dyes; inhibits inflammation.

ependymal (ep-en′di-măl) Neuroglial cell layer lining the ventricles of the brain.

epicardium (ep-i-kar′dē-ŭm) [G. *epi*, upon + *kardia*, heart] Serous membrane covering the surface of the heart; also called the visceral pericardium.

epicondyle (ep′i-kon′dīl) [*epi* + G. *kondyle*, knuckle] Projection on (usually to the side of) a condyle.

epidermis (ep-i-derm′is) [*epi* + G. *derma*, skin] Outer portion of the skin formed of epithelial tissue that rests on the dermis; resists abrasion and forms a permeability barrier.

epididymis (ep-i-did′i-mis) [*epi* + G. *didymos*, twin] Elongated structure connected to the posterior surface of the testis; site of storage and maturation of the sperm cells.

epiglottis (ep-i-glot′is) [*epi* + G. *glottis*, the mouth of the windpipe] Plate of elastic cartilage, covered with mucous membrane, that serves as a valve over the opening of the larynx during swallowing to prevent materials from entering the larynx.

epimysium (ep-i-mis′ē-ŭm, ep-i-miz′ē-ŭm) [*epi* + G. *mys*, muscle] Fibrous connective tissue layer surrounding a skeletal muscle.

epinephrine (ep′i-nef′rin) [*epi* + G. *nephros*, kidney] Hormone similar in structure to the neurotransmitter norepinephrine; major hormone released from the adrenal medulla; increases cardiac output and blood glucose levels.

epiphyseal line (ep-i-fiz′ē-ăl) Dense plate of bone in a bone that is no longer growing, indicating the former site of the epiphyseal plate.

epiphyseal plate Site at which bone growth in length occurs; located between the epiphysis and diaphysis of a long bone; area of cartilage where cartilage growth is followed by ossification; also called the growth plate.

epiphysis (e-pif′-i-sis) [*epi*, upon + G. *physis*, growth] End of a bone; separated from the remainder of the bone by the epiphyseal plate or epiphyseal line.

epiploic appendage (ep′i-plō′ik) One of a number of little, fat-filled processes of peritoneum projecting from the serous coat of the large intestine.

episiotomy (e-piz-ē-ot′ō-mē, e-pis-e-ot′ō-mē) [pubic region + G. *tōme*, incision] Incision in the clinical perineum, sometimes performed during childbirth.

epithalamus (ep′i-thal′ă-mŭs) [G. *epi*, upon + thalamus] Small dorsomedial area of the thalamus corresponding to the habenula and its associated structures, the stria medullaris of the thalamus, pineal gland, and habenular commissure.

epithelial tissue (ep-i-thē′lē-ăl) One of the four major tissue types consisting of cells with a basement membrane (exceptions are lymphatic vessels and liver sinusoids), little extracellular material, and no blood vessels; covers the surfaces of the body and forms glands.

epithelium (ep-i-thē′lē-ŭm) [G. *epi*, upon + *thele*, covering or lining] pl. epithelia (ep-i-thē′lē-ă) See *epithelial tissue*.

eponychium (ep-ō-nik′ē-ŭm) [G. *epi* + *onyx*, nail] Thin skin that attaches to the proximal part of the nail.

equilibrium (ē-kwi-lib′rē-ŭm) [G. *aequus*, equal + *libra*, a balance] State created by a chemical reaction proceeding in opposite directions (e.g., from reactants to products and from products to reactants) at equal speed.

erection (ē-rek′shŭn) Engorgement of erectile tissue with blood, such as in the erectile tissues of the penis, causing the penis to enlarge and become firm.

erector spinae (ē-rek′tōr spī′nē) Common name of the muscle group of the back; holds the back erect.

erythroblastosis fetalis (ĕ-rith′rō-blas-tō′sis fē-tă′lis) [*erythroblast* + G. *-osis*, condition] See *hemolytic disease of the newborn*.

erythrocyte (ĕ-rith′rō-sīt) [G. *erythro*, red + *kytos*, cell] See *red blood cell*.

erythropoietin (ĕ-rith′rō-poy′ĕ-tin) [*erythrocyte* + G. *poiesis*, a making] Protein hormone that stimulates red blood cell formation in red bone marrow.

esophagus (ē-sof′ă-gŭs) [G. *oisophagos,* gullet] Part of the digestive tract between the pharynx and stomach.

estrogen (es′trō-jen) Steroid hormone secreted primarily by the ovaries; involved in the maintenance and development of female reproductive organs, secondary sexual characteristics, and the menstrual cycle.

eustachian tube (ū-stā′shŭn) See *auditory tube.*

exchange reaction Combination of a decomposition reaction, in which molecules are broken down, and a synthesis reaction, in which the products of the decomposition reaction are combined to form new molecules.

exocrine (ek′sō-krin) [G. *exo-,* outside + *krino,* to separate] Gland that secretes to a surface or outward through a duct.

exocytosis (ek′sō-sī-tō′sis) Elimination of material from a cell through the formation of vesicles.

exophthalmia (ek-sof-thal′mē-ă) [G. *ex,* cut + *ophthalmos,* eye] Bulging of the eyes that frequently accompanies Graves disease, due to accumulation of a type of connective tissue behind the eye.

expiration (eks-pi-rā′shŭn) To breathe out; to move air out of the lungs.

extension [L. *extensio*] To stretch out, usually to straighten out a joint.

extracellular (eks-tră-sel′ū-lăr) Outside of a cell.

extracellular matrix (mā′triks) Nonliving chemical substances located between cells; often consists of protein fibers, ground substance, and fluid.

extrinsic muscle (eks-trin′sik) Muscle located outside of the structure on which it acts.

extrinsic regulation Regulation of the heart that involves mechanisms outside the heart, including nervous and hormonal regulation.

# F

facet (fas′et) [Fr., little face] Small, smooth articular surface.

facilitated diffusion (fă-sil′i-tā-tĭd di-fū′zhŭn) Carrier-mediated process that does not require ATP and moves substances into or out of cells from a higher to a lower concentration.

fascia (fash′ē-ă) [L., band or fillet] Loose areolar connective tissue beneath the skin (hypodermis), or dense connective tissue that encloses and separates muscles.

fasciculus (fă-sik′ū-lus) [L. *fascis,* bundle] Band or bundle of nerve or muscle fibers bound together by connective tissue.

fat Greasy, soft-solid lipid found in animal tissues and many plants; composed of glycerol and fatty acids.

fatty acid Straight chain of carbon atoms with a carboxyl group (–COOH) attached at one end; a building block of fats.

feces (fē′sēz) Matter discharged from the digestive tract during defecation, consisting of the undigested residue of food, epithelial cells, intestinal mucus, bacteria, and waste material.

fertilization (fer′til-i-zā′shŭn) Union of the sperm cell and oocyte to form a zygote.

fetus (fē′tŭs) In prenatal development, the developing human between approximately 56 days and birth.

fibrillation (fī-bri-lā′shŭn, fib-ri-lā′shŭn) Very rapid contraction of cardiac muscle fibers, but not of the muscle as a whole; results in dramatically reduced pumping action of the heart.

fibrin (fī′brin) [L. *fibra,* fiber] Threadlike protein fiber derived from fibrinogen by the action of thrombin; forms a clot—that is, a network of fibers that traps blood cells, platelets, and fluid—which stops bleeding.

fibrinogen (fī-brin′ō-jen) [L. *fibra,* fiber + *gen,* produce] Protein in plasma that gives rise to fibrin when acted on by thrombin to form a clot.

fibrinolysis (fī-bri-nol′i-sis) [L. *fibra,* fiber + G. *lysis,* dissolution] Breakdown of a clot by plasmin.

fibroblast (fī′brō-blast) Cell in connective tissue responsible for the production of collagen.

filtration (fil-trā′shŭn) Movement, resulting from a pressure difference, of a liquid through a filter, which prevents some or all of the substances in the liquid from passing.

filtration membrane Membrane formed by the glomerular capillary endothelium, the basement membrane, and the podocytes of Bowman's capsule.

fimbria (fim′brē-ă), pl. fimbriae (fim′brē-ē) Long, thin process that surrounds the opening of the uterine tube.

first heart sound Heart sound that results from the simultaneous closure of the tricuspid and bicuspid valves.

flagellum (flă-jel′ŭm), pl. flagella (flă-jel′ă) [L., whip] Whiplike locomotor organelle similar to cilia except longer, and there is usually one per sperm cell.

flexion (flek′shŭn) [L. *flectus*] To bend.

focal point Point at which light rays cross after passing through a concave lens.

follicle-stimulating hormone (fol′i-kl) (FSH) Hormone of the anterior pituitary gland that, in the female, stimulates the follicles of the ovary, assists in maturation of the follicle, and causes secretion of estrogen from the follicle; in the male, it stimulates the epithelium of the seminiferous tubules and is partially responsible for inducing spermatogenesis.

fontanel (fon′tă-nel) [Fr., fountain] One of several membranous gaps between bones of the skull.

foramen (fō-rā′men) Hole; referring to a hole or opening in a bone.

foramen ovale (ō-val′ē) In the fetal heart, the oval opening in the interatrial septum with a valve that allows blood to flow from the right to the left atrium, but not in the opposite direction; becomes the fossa ovalis after birth.

formed element Cell, such as a red blood cell or white blood cell, or cell fragments, such as a platelet, in blood.

fossa (fos′ă) Depression below the level of the surface of a bone; usually longitudinal in shape.

fovea centralis (fō′vē-ă) Depression in the center of the macula of the eye; has the greatest visual acuity and only cones.

free energy Total amount of energy that can be liberated by the complete catabolism of food.

frenulum (fren′ū-lŭm) [L. *frenum,* bridle] Fold extending from the floor of the mouth to the middle of the undersurface of the tongue.

frontal plane Plane separating the body into anterior and posterior portions; also called a coronal plane.

FSH See *follicle-stimulating hormone.*

full-thickness burn Burn that destroys the epidermis and the dermis and sometimes the underlying tissue; sometimes called a third-degree burn.

fundus (fŭn′dŭs) [L., bottom] Bottom, or area farthest from the opening, of a hollow organ, such as the stomach or the uterus.

# G

gamete (gam′ēt) [G. *gametēs,* husband; *gametē,* wife] Germ cell, such as an oocyte or a sperm cell.

gamma globulin (gam′ă glob′ū-lin) Family of proteins found in plasma.

ganglion (gang′glē-on), pl. ganglia (gang′glē-ă) [G., knot] Group of neuron cell bodies in the peripheral nervous system.

gap junction Small channels that allow materials to pass from one cell to an adjacent cell; provides a means of intercellular communication.

gastric gland (gas′trik) Gland within the stomach.

gastric inhibitory polypeptide Hormone released from the duodenum; inhibits gastric acid secretion.

**gastrin** (gas′trin) Hormone secreted in the mucosa of the stomach and duodenum that stimulates secretion of hydrochloric acid by the gastric glands.

**gastrointestinal tract** (gas′trō-in-tes′tin-ăl) Technically, only the stomach and intestines. Often used as a synonym for *digestive tract,* which extends from the mouth to the anus.

**gene** Sequence of nucleotides in DNA that is a chemical set of instructions for making a specific protein.

**genetics** (jĕ-net′iks) Branch of science that deals with heredity.

**genotype** (jen′ō-tīp) Genetic makeup of an individual.

**GH** See *growth hormone.*

**giantism** (jī′an-tizm) Abnormal growth in young people because of hypersecretion of growth hormone by the pituitary gland.

**gingiva** (jin′ji-vă) Dense fibrous tissue, covered by mucous membrane, that covers the alveolar processes of the upper and lower jaws and surrounds the necks of the teeth.

**girdle** (ger′dl) Bony ring or belt that attaches a limb to the body, such as the pectoral (shoulder) and pelvic girdles.

**gland** Single cell or multicellular structure that secretes substances into the blood, into a cavity, or onto a surface.

**glia** (glī′ă) See *neuroglia.*

**glomerulus** (glō-mār′ū-lŭs) [L. *glomus,* ball of yarn] Mass of capillary loops at the beginning of each nephron, nearly surrounded by Bowman's capsule.

**glucagon** (gloo′kă-gon) [*glucose* + *agō,* to lead] Hormone secreted from the pancreatic islets of the pancreas that acts primarily on the liver to release glucose into the circulatory system.

**glucocorticoid** (gloo-kō-kōr′ti-koyd) Hormone from the adrenal cortex capable of increasing the rate at which lipids are broken down to fatty acids and proteins are broken down to amino acids; elevates blood glucose levels and acts as an anti-inflammatory substance.

**glycerol** (glis′er-ol) Three-carbon molecule with a hydroxyl group attached to each carbon; a building block of fats.

**glycogen** (glī′kō-jen) Animal starch; composed of many glucose molecule bound together in chains that are highly branched; functions as a carbohydrate reserve; stores glucose molecules; in animal cells.

**glycolysis** (glī-kol′i-sis) [G. *glykys,* sweet + *lysis,* a loosening] Anaerobic process during which one glucose molecule is converted to two pyruvic acid molecules; a net of two ATP molecules is produced during glycolysis.

**glycoprotein** (glī-kō-prō′tēn) Organic molecule composed of a protein and a carbohydrate.

**GnRH** See *gonadotropin-releasing hormone.*

**goblet cell** Epithelial cell that has the end of the cell at the free surface distended with mucin.

**goiter** (goy′ter) [L. *guttur,* throat] Enlargement of the thyroid gland, not due to a neoplasm, usually caused by a lack of iodine in the diet.

**Golgi apparatus** (gol′jē) Named for Camillo Golgi, Italian histologist and Nobel laureate, 1843–1926; stacks of flattened sacks, formed by membranes, that collect, modify, package, and distribute proteins and lipids.

**gonad** (gō′nad)[L. *gonē,* seed] Organ that produces gametes; a testis or an ovary.

**gonadotropin** (gō′nad-ō-trō′pin) [*gonē,* seed + *trope,* a turning] Hormone capable of promoting gonadal growth and function; two major gonadotropins are luteinizing hormone (LH) and follicle-stimulating hormone (FSH).

**gonadotropin-releasing hormone (GnRH)** Hypothalamic hormone that stimulates the secretion of LH and FSH from the anterior pituitary gland.

**granulation tissue** (gran′ū-lā′shŭn) Vascular connective tissue formed in wounds.

**granulocyte** (gran′ū-lō-sīt) [*granular* + G. *kytos,* cell] White blood cell named according to the appearance, in stained preparations, of large cytoplasmic granules; neutrophils, basophils, and eosinophils.

**Graves disease** Type of hyperthyroidism resulting from abnormal proteins produced by the immune system that are similar in structure and function to thyroid-stimulating hormone; often accompanied by exophthalmia.

**growth hormone (GH)** Protein hormone of the anterior pituitary gland; it promotes body growth, increases fat mobilization, and increases blood glucose levels because it inhibits glucose utilization.

**gynecomastia** (gī′nĕ-kō-mas′tē-ă) Enlarged breasts in males.

**gyrus** (jī′rŭs) [L. *gyros,* circle] Rounded elevation or fold on the surface of the brain.

# H

**hair** Threadlike outgrowth of the skin, consisting of columns of dead, keratinized epithelial cells.

**hair cell** Cell of the inner ear containing hairlike processes (microvilli) that respond to bending of the hairs by depolarization.

**hamstring muscle** One of the three major muscles of the posterior thigh.

**haploid** (hap′loyd) Condition in which a cell has 1 copy of each autosome and 1 sex chromosome (23 total chromosomes in humans); characteristic of gametes.

**haustra** (haw′stră) Sacs of the colon, formed by the teniae coli, which are slightly shorter than the gut, so that the gut forms pouches.

**haversian canal** (ha-ver′shan) Named for seventeenth-century English anatomist Clopton Havers (1650–1702); see *central canal.*

**haversian system** See *osteon.*

**hCG** See *human chorionic gonadotropin.*

**heart-lung machine** Machine that pumps blood and carries out the process of gas exchange; it substitutes for the heart and lungs during heart surgery.

**heart rate** Number of complete cardiac cycles (heartbeats) per minute.

**heart transplant** Process of taking a healthy heart from a recently deceased donor and transplanting it into a recipient who has a diseased heart.

**hematocrit** (hē′mă-tō-krit, hem′a-tō-krit) [G. *hemato,* blood + *krino,* to separate] Percentage of total blood volume composed of red blood cells.

**hematopoiesis** (hē′mă-tō-poy-ē′sis) [G. *hemato,* blood + *poiesis,* a making] Production of blood cells.

**hemidesmosome** (hem-ē-des′mō-sōm) [G. *hemi,* one half] Half desmosome that occurs on the basal surface epithelial cells that rest on the basement membrane.

**hemoglobin** (hē-mō-glō′bin) [G. *hemato,* blood + *glob,* a ball] Red protein of red blood cells consisting of four globin proteins with an iron-containing red pigment, heme, bound to each globin protein; transports oxygen and carbon dioxide.

**hemolysis** (hē-mol′i-sis) [G. *hemo,* blood + *lysis,* destruction] Rupture of red blood cells.

**hemolytic** (hē-mō-lit′ik) **disease of the newborn** Destruction of red blood cells in the fetus or newborn, caused by antibodies produced in the Rh-negative mother acting on the Rh-positive blood of the fetus or newborn.

**hemorrhage** (hem′ŏ-rij) [G. *haima,* blood + *rhegnymi,* to burst forth] Rupture or leaking of blood from vessels.

**hepatic** (he-pat′ik) [G. *hepar,* liver] Associated with the liver.

**hepatic portal system** [L. *porta,* gate] Blood flow through the veins that begin as capillary beds in the small intestine, spleen, pancreas, and stomach and carry blood to the liver, where they end as a capillary bed.

hepatic portal vein  Vein that carries blood from the intestines, stomach, spleen, and pancreas to the liver.

Hering-Breuer reflex  Named for German physiologist Heinrich E. Hering (1866–1948) and Austrian internist Josef Breuer (1842–1925), process in which action potentials from stretch receptors in the lungs arrest inspiration; expiration then occurs.

heterozygous (het′er-ō-zī′gŭs)  Having two different genes for a given trait.

hilum (hī′lŭm) [L., a small amount or trifle]  Part of an organ where the nerves and vessels enter and leave.

histology (his-tol′ō-jē) [G. histo, web (tissue) + logos, study]  Science that deals with the structure of cells, tissues, and organs in relation to their function.

holocrine (hol′ō-krin) [G. holo, whale + krino, to separate]  Gland whose secretion consists of disintegrated cells of the gland; an example is a sebaceous gland.

homeostasis (hō′mē-ō-stā′sis) [G. homoio, like + stasis, a standing]  Existence and maintenance of a relatively constant environment within the body with respect to functions and the composition of fluids and tissues.

homeotherm (hō′mē-ō-therm) [G. homoiois, like + thermos, warm]  Any animal, including mammals and birds, that tends to maintain a constant body temperature; also referred to as warm-blooded.

homozygous (hō-mō-zī′gŭs)  Having two identical genes for a given trait.

hormone (hōr′mōn) [G. hormon, to set into motion]  Substance secreted by endocrine tissues into the blood that acts on a target tissue to produce a specific response.

human chorionic gonadotropin (hCG) (kō-rē-on′ik gō′nad-ō-trō′pin)  Hormone, similar to LH, secreted from the placenta and essential for the maintenance of pregnancy for the first 3 months; prevents the corpus luteum from degenerating.

humerus (hū′mer-ŭs) [L., shoulder]  Bone of the arm.

humoral immunity (hū′-mōr-ăl i-mū′ni-tē)  See antibody-mediated immunity.

hydrogen bond (hī′drō-jen)  Weak attraction between the oppositely charged ends of two polar covalent molecules; the weak attraction between the end of a polar covalent molecule and an ion.

hydrophilic (hī-drō-fil′ik) [G. hydro, water + philos, fond or loving]  Attracting or associating with water molecules; tending to dissolve or associate with water molecules; tending to dissolve in water; polar.

hydrophobic (hī-drō-fōb′ik) [G. hydro, water + phobos, fear]  Lacking affinity for water molecules; tending not to dissolve in water; nonpolar.

hydroxyapatite (hī-drok′sē-ap-ă-tīt)  Complex crystal structure that makes up the mineral portion of bones and teeth.

hymen (hī′men) [G. hymen, membrane]  Thin, membranous fold highly variable in appearance; partly occludes the opening of the vagina prior to its rupture; may occur for a variety of reasons and is frequently absent.

hyoid (hī′oyd) [G., shaped like the letter epsilon, ε]  U-shaped bone in the throat.

hypercapnia (hī-per-kap′nē-ă) [hyper + G. kapnos, smoke, vapor]  Abnormally increased arterial carbon dioxide tension or levels.

hyperpolarize (hī′per-pō′lăr-īz) [G. hyper, above + polaris, polar]  To increase polarization of membranes of nerve or muscle cells.

hypertension (hī′per-ten′shŭn) [G. hyper, above + tensio, tension]  High blood pressure; generally, blood pressure greater than 140/90 is considered too high.

hyperthyroidism (hī-per-thī′royd-izm)  Abnormality of the thyroid gland in which thyroid hormone secretion is increased.

hypertonic (hī-per-ton′ik) [G. hyper, above + tonos, tension]  Solution that causes cells to shrink.

hypodermis (hī-pō-der′mis) [G. hypo, under + dermis, skin]  Loose connective tissue under the dermis that attaches the skin to muscle and bone.

hypophysis (hī-pof′i-sis) [hypo + G., an undergrowth or growth]  Endocrine gland attached to the hypothalamus by the infundibulum; the pituitary gland.

hypothalamic-pituitary portal system (hī′pō-thal′ă-mǐk-pi-too′i-tar-ē)  Series of blood vessels that carry blood from the area of the hypothalamus to the anterior pituitary gland; originates from capillary beds in the hypothalamus and terminates as a capillary bed in the anterior pituitary gland.

hypothalamus (hī′pō-thal′ă-mŭs) [hypo + G. thalamos, bedroom]  Important autonomical and endocrine control center in the brain located beneath the thalamus.

hypothyroidism (hī′pō-thī′royd-izm)  Reduced secretion of thyroid hormones from the thyroid gland, leading to cretinism in infants and symptoms of inadequate thyroid hormone secretion in adults.

hypotonic (hī-pō-ton′ik) [hypo, under + G. tonos, tension or tone]  Solution that causes cells to swell.

hypoxia (hī-pok′sē-ă) [hypo + oxygen]  Below-normal levels of oxygen in arterial blood.

# I

ICSH  See interstitial cell–stimulating hormone.

Ig  See immunoglobulin.

ileocecal junction (il′ē-ō-se′kăl)  Junction of the ileum of the small intestine and the cecum of the large intestine.

ileum (il′ē-ŭm) [L. eileo, to roll up, twist]  Third portion of the small intestine, about 3.5 meters in length; extends from the jejunum to the ileocecal opening.

ilium (il′ē-ŭm)  Broad, flaring portion of the hipbone; becomes fused with the ischium and pubis.

immunity (i-mū′ni-tē)  Ability to resist damage from foreign substances, such as microorganisms, and harmful chemicals, such as toxins released by microorganisms.

immunoglobulin (im′ū-nō-glob-ū-lin) (Ig)  Antibodies.

implantation (im-plan-tā′shŭn)  Attachment of the blastocyst to the endometrium of the uterus; occurring 6 or 7 days after fertilization of the oocyte.

impotence (im′pǒ-tens)  Inability of the male to achieve or maintain an erection and thus engage in sexual intercourse.

incompetent valve  Leaky valve; usually refers to a heart valve that allows blood to flow through it when it is closed.

incus (ing′kus) [L., anvil]  Middle bone of the middle ear; the anvil.

infarct (in′farkt)  Area of necrosis resulting from a sudden insufficiency of arterial blood supply.

inferior (in-fē′rē-ōr) [L., lower]  Down, or lower, with reference to the anatomical position.

inferior vena cava (vē′nă kā′vă)  Receives blood from the lower limbs, pelvis, and abdominal organs and empties into the right atrium of the heart.

inflammatory (in-flam′ă-tōr-ē) response  Complex sequence of events, involving chemicals and immune system cells, that results in the isolation and destruction of foreign substances, such as bacteria; symptoms include redness, heat, swelling, pain, and disturbance of function.

infundibulum (in-fŭn-dib′ū-lŭm) [L., a funnel]  Funnel-shaped structure or passage—for example, the infundibulum that attaches the pituitary gland to the hypothalamus; funnel-like expansion of the uterine tube near the ovary.

**inguinal canal** (ing′gwi-năl) Passageway through which a testis passes as it descends from the abdominopelvic cavity to the scrotum.

**inguinal hernia** (her′nē-ă) Rupture that allows the potential protrusion of abdominal organs, such as the small intestine, through the inguinal canal.

**innate immunity** (i′nāt, i-nāt′) Immune system response that is the same on each exposure to an antigen; there is no ability to remember a previous exposure to a specific antigen.

**inner cell mass** Group of cells at one end of the blastocyst from which the embryo develops.

**inorganic** (in-ōr-gan′ik) Molecules that do not contain carbon atoms; originally defined as molecules that came from nonliving sources; the original definition is no longer valid because carbon dioxide produced by living organisms is considered an inorganic molecule.

**insertion** (in-ser′shŭn) The more movable attachment point of a muscle.

**inspiration** (in-spi-rā′shŭn) To breathe in; to move air into the lungs, or inhale.

**insulin** (in′sŭ-lin) Protein hormone, secreted from the pancreas, that increases the uptake of glucose and amino acids by most tissues.

**interatrial septum** (in-ter-ā′trē-ăl) Cardiac muscle partition separating the right and left atria.

**intercalated disk** (in-ter′kă-lā-ted) [inserted between] Connection between cardiac muscle cells; important in coordinating the contractions of cardiac muscle cells; contains gap junctions that allow action potentials to pass from one cardiac muscle cell to adjacent cardiac muscle cells.

**intercostal muscle** (in-ter-kos′tăl) Muscle located between ribs.

**interferon** (in-ter-fĕr′on) Protein released by virally infected cells that binds to other cells and stimulates them to produce antiviral proteins that inhibit viral replication.

**interkinesis** (in′ter-ki-nē′sis) Short time period between the formation of the daughter cells of the first meiotic division and the second meiotic division.

**interstitial cell** (in-ter-stish′ăl) Cell between the seminiferous tubules of the testes; secretes testosterone; also called Leydig cell.

**interstitial cell–stimulating hormone (ICSH)** Luteinizing hormone in males. Hormone of the anterior pituitary gland that stimulates the secretion of testosterone in the testes. See *luteinizing hormone*.

**interventricular septum** (in-ter-ven-trik′ū-lăr) Cardiac muscle partition separating the right and left ventricles.

**intestinal glands** (in-tes′ti-năl) Tubular glands in the mucous membrane of the small intestine.

**intracellular** (in-tră-sel′ū-lăr) Inside a cell.

**intramembranous ossification** (in′tră-mem′brā-nŭs os′i-fi-kā′shŭn) Bone formation within connective tissue membranes.

**intramural plexus** (in′tră-mū′răl plek′sŭs) [L., within the wall] Nerve plexus within the walls of the digestive tract; involved in local and autonomic control of digestion.

**intrinsic factor** (in-trin′sik) Factor secreted by the gastric glands and required for adequate absorption of vitamin B$_{12}$.

**intrinsic muscle** Muscle located within the structure on which it acts.

**ion** (ī′on) Atom or group of atoms carrying an electrical charge because of a loss or gain of one or more electrons.

**ionic bond** (ī-on′ik) Chemical bond resulting from the attraction between ions of opposite charge.

**iris** (ī′ris) Specialized part of the vascular tunic of the eye; the "colored" part of the eye that can be seen through the cornea; consists of smooth muscles that regulate the amount of light entering the eye.

**isometric contraction** (ī-sō-met′rik) Muscle contraction in which the length of the muscle does not change but the amount of tension increases.

**isotonic** (ī′sō-ton′ik) [G. *iso*, equal + *tonos*, tension] Solution that causes cells to neither shrink nor swell.

**isotonic contraction** Muscle contraction in which the amount of tension is constant and the muscle shortens.

**isotope** (ī′sō-tōp) [G. *isos*, equal + *topos*, part] One of two or more elements that have the same number of protons and electrons but a different number of neutrons.

## J

**jaundice** (jawn′dis) [Fr. *jaune*, yellow] Yellowish staining of the skin, sclerae, and deeper tissues and excretions with bile pigments.

**jejunum** (jĕ-joo′nŭm) [L. *jejunus*, empty] Portion of the small intestine, about 2.5 meters in length, between the duodenum and the ileum.

**juxtaglomerular apparatus** (jŭks′tă-glŏ-mer′ū-lăr) [L. *juxta*, close to + *glomerulus*] Specialized wall of the distal convoluted tubule and afferent arteriole that secretes renin.

## K

**keratin** (ker′ă-tin) Protein that accumulates in cells of nails, hair, and the superficial layers of the epidermis of the skin.

**keratinization** (ker′ă-tin-i-zā′shŭn) Production of keratin and changes in the structure and shape of epithelial cells as they move to the skin surface.

**kinetic** (ki-net′ik) [G. *kinetikos*, of motion] Relating to motion or movement.

**Korotkoff sound** (kō-rot′kof) Named for Russian physician Nikolai Korotkoff (1874–1920). Sound heard over an artery when blood pressure is determined by the auscultatory method.

**kyphosis** (kī-fō′sis) [G., humpback] Abnormal posterior curvature, or flexion, of the spine.

## L

**labia majora** (lā′bē-a) Two rounded folds of skin surrounding the labia minora and vestibule.

**labia minora** Two narrow, longitudinal folds of mucous membrane enclosed by the labia majora; they unite anteriorly to form the prepuce.

**labyrinth** (lab′i-rinth) Series of membranous and bony tunnels in the temporal bone; the part of the inner ear involved in hearing and balance.

**lacrimal** (lak′ri-măl) [L., a tear] Relating to tears or tear production.

**lactation** (lak′tā′shŭn) [L. *lactatio*, suckle] Period following childbirth during which milk is formed in the breasts.

**lacteal** (lak′tē-ăl) [relating to milk] Lymphatic vessel in the wall of the small intestine that carries chyle from the intestine and absorbs fat.

**lactic acid** (lak′tik) Three-carbon molecule derived from pyruvic acid as a product of anaerobic respiration.

**lacuna** (lă-koo′nă), pl. **lacunae** (lă-koo-nē) [L., a pit] Small space, cavity, or depression; a space in cartilage in which a chondrocyte is located; a space in bone matrix in which an osteocyte is located; a cavity containing maternal blood in the placenta.

**lamella** (lă-mel′ă), pl. **lamellae** (lă-mel′ē) [L. *lamina*, plate, leaf] Thin sheet or layer of bone.

**lamina** (lam′i-nă), pl. **laminae** (lam′i-nē) [L. *lamina*, plate, leaf] Layer; a portion of the vertebra that extends from the transverse process to the spinous process.

**lamina propria** (prō′prē-ă) Layer of connective tissue underlying the epithelium of a mucous membrane.

**lanugo** (lă-noo′gō) [L. *lana*, wool] Fine, soft fetal or embryonic hair.

laryngitis (lar-in-jī′tis) Inflammation of the mucous membrane of the larynx.

laryngopharynx (lă-ring′gō-far-ingks) Part of the pharynx lying below the tip of the epiglottis extending to the level of the cricoid cartilage of the larynx.

larynx (lar′ingks) Organ of voice production located between the pharynx and the trachea; it consists of a framework of cartilages and elastic membranes housing the vocal folds (true vocal cords) and the muscles that control the position and tension of these elements.

lateral [L. *latus,* side] Away from the middle or midline of the body.

lateral horn Small, lateral extension of spinal cord gray matter; located only in spinal cord regions T1–L2; containing preganglionic sympathetic neuron cell bodies.

lens Biconvex structure in the anterior part of the eye capable of being flattened or thickened to adjust the focus of light entering the eye.

leukemia (loo-kē′mē-ă) [G. *leukos,* white + *haima,* blood] Tumor of the red bone marrow that results in the production of large numbers of abnormal white blood cells; often accompanied by decreased production of red blood cells and platelets.

leukocyte (loo′kō-sīt) [G. *leukos,* white + *kytos,* cell] See *white blood cell.*

leukocytosis (loo′kō-sī-tō′sis) [*leukocyte* + G. *-osis,* a condition] Higher than normal number of white blood cells.

leukopenia (loo-kō-pē′nē-ă) [*leukocyte* + G. *penia,* poverty] Lower than normal number of white blood cells.

Leydig cell (lī′dig) Named for German anatomist Franz von Leydig (1821–1908). See *interstitial cell.*

LH See *luteinizing hormone.*

ligament (lig′ă-ment) Tough connective tissue band usually connecting bone to bone.

ligand (lig′and, lī′gand) [L. *ligo,* to bind] Molecule that binds to a macromolecule— for example, a ligand binding to a receptor.

limbic system (lim′bik) [L. *limbus,* a border or boundary] Primitive part of the brain involved in visceral and emotional response and in the response to odor.

linea alba (lin′ē-ă al′bă) White line in the center of the abdomen where muscles of the abdominal wall insert.

lipase (lip′ās) Enzyme that breaks down lipids.

lipid (lip′ĭd) [G. *lipos,* fat] Substance composed principally of carbon, oxygen, and hydrogen; generally soluble in nonpolar solvents; fats and cholesterol.

local inflammation Inflammation confined to a specific area of the body; symptoms include redness, heat, swelling, pain, and loss of function.

longitudinal section Cut made through the long axis of an organ.

loop of Henle U-shaped part of the nephron extending from the proximal to the distal convoluted tubule and consisting of descending and ascending limbs; many of the loops of Henle extend into the renal pyramids.

lordosis (lōr-dō′sis) [G., a bending backward; swayback] Abnormal anterior curvature of the spine, usually in the lumbar region; saddle back or swayback.

lower motor neuron Motor neuron located in the brainstem or spinal cord, as opposed to the cerebral cortex.

lumbosacral plexus (lŭm′bō-sā′krăl) [L. *lumbus,* loin + *sacrum,* sacred] Nerve plexus that innervates the lower limbs; originates from spinal nerves L1–S4.

lunula (loo′noo-lă) [L. *luna,* moon] White, crescent-shaped portion of the nail matrix visible through the proximal end of the nail.

luteinizing hormone (LH) (loo′tē-ĭ-nīz-ing) Hormone of the anterior pituitary gland that, in the female, initiates final maturation of the follicles, their rupture to release the oocyte, the conversion of the ruptured follicle into the corpus luteum, and the secretion of progesterone; in the male, it stimulates the secretion of testosterone in the testes and is sometimes referred to as interstitial cell–stimulating hormone (ICSH).

lymph (limf) [L. *lympha,* clear spring water] Clear or yellowish fluid derived from interstitial fluid and found in lymphatic vessels.

lymph node Encapsulated mass of lymphatic tissue found along lymphatic vessels; filters lymph and produces lymphocytes.

lymphocyte (lim′fō-sīt) [L. *lympho,* lymph + G. *kytos,* cell] Nongranulocytic white blood cell involved in the immune system; there are several types of lymphocytes with diverse functions, including antibody production, allergic reactions, graft rejections, tumor control, and regulation of the immune system.

lymphokine (lim′fō-kīn) Class of chemicals, produced by T cells, that activate macrophages and other immune cells; promote phagocytosis and inflammation.

lymphoma (lim-fō′mă) Neoplasm (tumor) of lymphatic tissue that is almost always malignant.

lysis (lī′sis) [G., dissolution or loosening] Rupturing or breaking of a cell membrane.

lysosome (lī′sō-sōm) [G. *lysis,* a loosening + *soma,* body] Membrane-bound vesicle containing intracellular digestive enzymes.

# M

macrophage (mak′rō-fāj) [G. *makros,* large + *phago,* to eat] Large mononuclear, phagocytic cell.

macula (mak′ū-lă) [L., a spot] One of the sensory structures in the vestibule, consisting of hair cells and a gelatinous mass embedded with otoliths; responds to gravity. Also a small yellow spot in the posterior retina of the eye where the cones are concentrated; has no red tint because it is devoid of blood vessels.

malignant (mă-lig′nănt) [L. *maligno,* to do anything malicious, with malice or intent to do harm] In reference to a neoplasm, the property of spreading locally and to distant sites.

malleus (mal′ē-ŭs) [L., hammer] Most lateral of the middle ear bones, attached to the tympanic membrane; the hammer.

mamma (mam′ă) pl. mammae (mam′ē) See *mammary gland.*

mammary gland (mam′ă-rē) Organ of milk secretion, located in the breast, or mamma.

mastication (mas-ti-kā′shŭn) [L. *mastico,* to chew] Chewing.

mastoid (mas′toyd) [*mastos,* breast + *eidos,* resemblance] Resembling a breast—for example, the mastoid process of the temporal bone.

matrix (mā′triks) Substance between the cells of a tissue.

matter Anything that occupies space.

mean arterial blood pressure Average of the arterial blood pressure; it is slightly less than the average of the systolic and diastolic blood pressure because diastole lasts longer than systole.

meatus (mē-ā′tŭs) [L., to go, pass] Passageway or tunnel.

meconium (mē-kō′nē-ŭm) [L. *mēkōn,* poppy] Greenish anal discharge from the fetus; consists of fluid swallowed, epithelial cells from the mucosa of the gut, mucus from the intestinal glands, and bile from the liver.

medial (mē′dē-ăl) [L. *medialis,* middle] Toward the middle or midline of the body.

mediastinum (me′dē-as-tī′nŭm) [L., middle septum or wall] Middle wall of the thorax, consisting of the trachea, esophagus, thymus, heart, and other structures.

medulla (me-dool′ă) [L. *medius,* middle, marrow] Center, or core, of an organ, such as the adrenal gland, kidney, or hair.

medulla oblongata (ob-long-gah′tă) Inferior portion of the brainstem that connects the spinal cord with the brain; contains nuclei of cranial nerves plus autonomic control centers for heart rate, respiration, and so forth.

**medullary cavity** (med′ŭl-er-ē) Large, marrow-filled cavity in the diaphysis of a long bone.

**medullary respiratory center** (res′pi-ră-tōr-ē, rĕ-spīr′ă-tōr-ē) Nerve cells in the medulla oblongata and pons of the brain that control inspiration and expiration.

**megakaryocyte** (meg-ă-kar′ē-ō-sīt) Large cell in red bone marrow that gives rise to platelets.

**meiosis** (mī-ō′sis) [G., a lessening] Process of cell division that results in gametes. Consists of two cell divisions that result in four cells, each of which contains half the number of chromosomes as the parent cell; occur in the testes and ovaries.

**melanin** (mel′ă-nin) [G. melas, black] Brown to black pigment responsible for skin and hair color.

**melanocyte** (mel′ă-nō-sīt) [G. melas, black + kytos, cell] Cells, found mainly in the stratum basale of skin, that produce the brown or black pigment melanin.

**melanocyte-stimulating hormone (MSH)** (mel′ă-nō-sīt) [G. melas, black + kytos, cell] Peptide hormone secreted by the anterior pituitary gland; increases melanin production by melanocytes, making the skin darker in color.

**melanoma** (mel′ă-nō′mă) [G. melas, black + oma, tumor] Malignant tumor derived from melanocytes.

**melanosome** (mel′ă-nō′sōm) [G. melano, black + soma, body] Pigment granule produced by melanocytes.

**melatonin** (mel-ă-tōn′in) Hormone secreted by the pineal gland; may inhibit gonadotropin-releasing hormone secretion from the hypothalamus.

**membranous labyrinth** (mem′bră-nŭs lab′i-rinth) Membrane-bound set of tunnels and chambers of the inner ear.

**memory cell** Lymphocyte derived from a B cell or T cell that has been exposed to an antigen; when exposed to the same antigen a second time, the memory cell rapidly responds to provide immunity.

**memory response** Immune response that occurs when the immune system is exposed to an antigen against which it has already had a primary response; results in the production of large amounts of antibodies and memory cells; also called a secondary response.

**menarche** (me-nar′kē) [G. mēn, month + archē, beginning] First menstrual period or flow.

**meninges** (mĕ-nin′jēz) [G. meninx, membrane] Series of three connective tissue membranes: the dura mater, arachnoid mater, and pia mater; they surround and protect the brain and spinal cord.

**menopause** (men′ō-pawz) [L. mēn, month + pausis, cessation] Permanent cessation of the menstrual cycle.

**menses** (men′sēz) [L. mensis, month] Loss of blood and tissue as the endometrium of the uterus sloughs away at the end of the menstrual cycle; occurring at about 28-day intervals in the nonpregnant female of reproductive age.

**menstrual cycle** (men′stroo-ăl) Series of changes that occur in sexually mature, nonpregnant females and result in menses; specifically includes the cyclical changes that occur in the uterus and ovary.

**merocrine** (mer′ō-krin) [G. meros, part + krino, to separate] Gland that secretes products with no loss of cellular material; an example is water-producing sweat glands; see apocrine and holocrine.

**mesentery** (mes′en-ter-ē) [G. mesos, middle + enteron, intestine] Double layer of peritoneum extending from the abdominal wall to the abdominopelvic organs; conveys blood vessels and nerves to abdominopelvic organs; holds and supports abdominopelvic organs.

**mesoderm** (mez′ō-derm) Middle of the three germ layers of the embryo.

**mesovarium** (mez′ō-vā′rē-ŭm) Mesentery of the ovary; mesentery that attaches the ovary to the posterior surface of the broad ligament.

**metabolic rate** Total amount of energy produced and used by the body per unit of time.

**metabolism** (mĕ-tab′ō-lizm) [G. metabole, change] Sum of the chemical changes that occur in tissues, consisting of the breakdown of molecules (catabolism) to produce energy and the buildup of molecules (anabolism), which requires energy.

**metaphase** (met′ă-fāz) [G. meta, after + phasis, an appearance] Stage of mitosis or meiosis in which the chromosomes become aligned near the center of the cell, at the equatorial plane, separating the centromeres, or chromosome pairs (see also meiosis). The centromeres of each chromosome divide, and the two daughter chromosomes are directed toward opposite poles of the cell.

**metastasis** (mĕ-tas′tă-sis) Shifting of a disease or a neoplasm from one part of the body to another remote from the original location.

**micelle** (mi-sel′, mī-sel′) [L. micella, small morsel] Droplet of digested lipid surrounded by bile salts in the small intestine.

**microglia** (mī-krog′lē-ă) [G. micro, small + glia, glue] Small neuroglial cells that become phagocytic and mobile in response to inflammation; considered to be macrophages of the central nervous system.

**microtubule** (mī-krō-too′būl) Hollow tube composed of tubulin; microtubules help support the cytoplasm of the cell and are components of certain cell organelles, such as cilia and flagella.

**microvillus** (mī′krō-vil′ŭs), pl. microvilli (mī′krō-vil′ī) One of the minute projections of the cell membrane that greatly increase the surface area of the cell membrane.

**micturition reflex** (mik-choo-rish′ŭn) Contraction of the urinary bladder stimulated by stretching of the urinary bladder wall; it results in emptying of the urinary bladder.

**midbrain** Superior end of the brainstem; located between the pons and diencephalon; contains fibers crossing from the brain to the spinal cord and vice versa, as well as nuclei and visual reflex centers.

**midsagittal** (mid′saj′i-tăl) Plane running vertically through the body and dividing it into equal right and left parts.

**mineral** (min′er-ăl) Inorganic nutrient necessary for normal metabolic functions.

**mineralocorticoid** (min′er-al-ō-kōr′ti-koyd) Steroid hormone released from the adrenal cortex; acts on the kidney to increase the rate of sodium ion reabsorption from the nephron and potassium and hydrogen ion secretion into the nephron of the kidney; an example is aldosterone.

**mitochondrion** (mī-tō-kon′drē-on), pl. mitochondria (mī-tō-kon′drē-ă) [G. mitos, thread + chandros, granule] Small, spherical, rod-shaped or thin filamentous structure in the cytoplasm that is a major site of ATP production.

**mitosis** (mī-tō′sis) [G., thread] Division of the nucleus. Process of cell division that results in two daughter cells with exactly the same number and type of chromosomes as the parent cell.

**mitral valve** (mī′trăl) [resembling a bishop's miter, a two-pointed hat] See bicuspid valve.

**molecule** (mol′ĕ-kūl) Two or more atoms chemically combined to form a structure that behaves as an independent unit.

**monocyte** (mon′ō-sīt) [G. mono, one or single + kytos, a cell] Type of white blood cell that transforms to become a macrophage.

**mononuclear phagocytic system** (mon-ō-noo′klē-ăr fag-ō-sit′ik) Phagocytic cells with a single nucleus, derived from monocytes; the cells either enter a tissue by chemotaxis in response to infection or tissue damage or are positioned to intercept microorganisms entering tissues.

monosaccharide (mon-ō-sak′ă-rīd) Basic building block from which more complex carbohydrates are constructed—for example, glucose and fructose.

mons pubis (monz pū′bis) [L., mountain] Prominence formed by a pad of fatty tissue over the pubic symphysis in the female.

morula (mōr′oo-lă) [L. *morus,* mulberry] Solid mass of cells resulting from the early cleavage divisions of the zygote.

motor neuron Neuron in the brain or spinal cord that innervates skeletal, smooth, or cardiac muscle cells or glands. Somatic motor neurons directly innervate skeletal muscle cells. Two autonomic motor neurons in series extend from the central nervous system to smooth or cardiac muscle cells or glands.

motor unit Single motor neuron and all the skeletal muscle fibers it innervates.

MSH See *melanocyte-stimulating hormone.*

mucin (mū′sin) Secretion containing mucopolysaccharides (proteoglycans), produced by mucous gland cells.

mucosa (mū-kō′să) [mucus-producing membrane] Mucous membrane consisting of the epithelium and connective tissue; in the digestive tract, there is also a layer of smooth muscle.

mucous membrane (mū′kŭs) Thin sheet consisting of epithelium and connective tissue that lines cavities opening to the outside of the body; many contain mucous glands, which secrete mucus.

mucus (mū′kŭs) Viscous secretion produced by and covering mucous membranes; lubricates and protects the mucous membrane and traps foreign substances.

murmur (mer′mer) Abnormal sound produced within the heart.

muscle fiber (mŭs′ĕl) Muscle cell.

muscle tissue One of the four major tissue types; consists of cells with the ability to contract; includes skeletal, cardiac, and smooth muscle.

muscle twitch Contraction of an entire muscle in response to a stimulus that causes an action potential in one or more muscle fibers.

muscularis (mŭs-kū-lā′ris) Outermost, smooth muscle coat of a hollow organ.

muscularis mucosa Inner, thin layer of smooth muscle found in most parts of the digestive tract outside the lamina propria.

myelinated (mī′ĕ-li-nāt-ed) [G. *myelos,* marrow] Nerve fibers having a myelin sheath.

myelin sheath (mī′ĕ-lin) Lipoprotein envelope made by wrapping the cell membrane of a Schwann cell or an oligodendrocyte around an axon.

myocardium (mī-ō-kar′dē-ŭm) [*myo-* + G. *kordin,* heart] Middle layer of the heart, consisting of cardiac muscle.

myofibril (mī-ō-fī′bril) Fine, longitudinal fibril within a skeletal muscle fiber; consisting of sarcomeres composed of thick (myosin) and thin (actin) myofilaments, placed end to end.

myofilament (mī-ō-fil′ă-ment) Ultramicroscopic protein thread helping form myofibrils in skeletal muscle; thin myofilaments are composed of actin, and thick myofilaments are composed of myosin.

myometrium (mī′ō-mē′trē-ŭm) Muscular wall of the uterus, composed of smooth muscle.

myosin myofilament (mī′ō-sin) One of the two major kinds of protein fibers of a sarcomere; thick filament resembling a bundle of golf clubs.

myxedema (mik-se-dē′mă) Hypothyroidism characterized by edema beneath the skin due to a change in the structure of the subcutaneous connective tissue.

# N

NADH See *nicotinamide adenine dinucleotide.*

nail (nāl) Thin, horny plate at the ends of the fingers and toes, consisting of several layers of dead epithelial cells containing a hard keratin.

naris (nā′ris), pl. nares (nā′res) Nostril, the opening into the nasal cavity.

nasal cavity (nā′zăl) Cavity divided by the nasal septum and extending from the external nares anteriorly to the nasopharynx posteriorly; bounded inferiorly by the hard palate.

nasolacrimal duct (nā-zō-lak′ri-măl) [L. *nasus,* nose + *lacrima,* tear] Duct that leads from the lacrimal sac to the nasal cavity.

nasopharynx (nā′zō-far′ingks) Part of the pharynx that lies above the soft palate; anteriorly, it opens into the nasal cavity.

negative feedback Mechanism by which any deviation from an ideal normal value or set point is resisted or negated; returns a parameter to its normal range and thereby maintains homeostasis.

neonate (nē′ō-nāt) [G. *neos,* new + L. *natalis,* relating to birth] Newborn, from birth to 1 month.

neoplasm (nē′ō-plazm) [*neo* + G. *plasma,* thing formed] New growth; an abnormal tissue that grows by cellular proliferation; may be benign or malignant.

nephron (nef′ron) [G. *nephros,* kidney] Functional unit of the kidney, consisting of the renal corpuscle, the proximal convoluted tubule, the loop of Henle, and the distal convoluted tubule.

nerve (nerv) Collecti[on]... peripheral nervous sys[tem]... potentials to and from th[e]... system.

nerve cell Cell capable of receiving... and propagating an action potential; neuron.

nerve tract Bundle of axons, their sheaths, and accompanying connective tissues located in the central nervous system.

nervous tissue (ner′vŭs) One of the four major tissue types; consists of neurons, which have the ability to conduct action potentials, and neuroglia, which are support cells.

neural crest cell (noor′ăl) Cell derived during embryonic development from the crests of the neural folds; gives rise to facial structures, pigment cells, and peripheral nerve ganglia.

neural tube Tube formed from the neuroectoderm in the embryo by closure of the neural groove; develops into the brain and spinal cord.

neuroectoderm (noor-ō-ek′tō-derm) Part of the ectoderm that forms the neural tube and neural crest.

neuroglia (noo-rog′lē-ă) [G. *neuro,* nerve + *glia,* glue] Cells of the nervous system other than neurons; play a support role in the nervous system; include astrocytes, ependymal cells, microglia, oligodendrocytes, and Schwann cells; also called glia.

neurolemmocyte (noor-ō-lem′ō-sīt) [G. *neuro,* nerve + *lemma,* husk + *kytos,* cell] See *Schwann cell.*

neuromuscular junction (noor-rō-mŭs′kū-lăr) Synaptic junction between a nerve axon and a muscle fiber.

neuron (noor′on) [G., nerve] Nerve cell.

neurotransmitter (noor′ō-trans-mit′er) [G. *neuro,* nerve + L. *transmitto,* to send across] Chemical that is released by a presynaptic cell into the synaptic cleft and that acts on the postsynaptic cell to cause a response.

neutral solution (noo′trăl) Solution with equal numbers of hydrogen and hydroxide ions; has a pH of 7.0.

neutron (noo′tron) [L. *neuter,* neither] Electrically neutral particle in the nucleus of atoms.

neutrophil (noo′trō-fil) [L. *neuter,* neither + G. *phileo,* to love] White blood cell with granules that stains equally with either basic or acidic dyes; phagocytic white blood cell.

nevus (nē′vŭs), pl. nevi (nē′vī) Benign, localized overgrowth of the melanin-forming cells of the skin present at birth or appearing early in life; a mole.

namide adenine dinucleotide
(NADH) (nik-ō-tin′ă-mīd ad′ĕ-nēn dī-
noo′klē-ō-tīd) Base-containing organic
molecule capable of accepting hydrogen
atoms and transferring energy from
glycolysis and the citric acid cycle to the
electron-transport chain.

nitroglycerin (nī-trō-glis′er-in) Glyceryl
trinitrate used as a vasodilator, especially in
angina pectoris.

nociceptor (nō′si-sep′tŏr) [L. oceo, hurt, pain,
injury + capio, to take] Peripheral sensory
receptor or mechanism for the reception
and transmission of painful or injurious
stimuli.

node of Ranvier (ron′vē-ā) Unmyelinated
area of an axon, every 0.1–1.0 mm, between
adjacent oligodendrocytes of an axon in the
central nervous system and between
individual Schwann cells of the peripheral
nervous system.

norepinephrine (nōr′ep-i-nef′rin)
Neurotransmitter substance released from
most of the postganglionic neurons of the
sympathetic division; hormone released
from the adrenal cortex that increases
cardiac output and blood glucose levels.

notochord (nō′tō-kōrd) [G. notor, back +
chorde, cord or string] Small rod of tissue
lying ventral to the neural tube;
characteristic of all vertebrates; in humans,
it becomes the nucleus pulposus of the
intervertebral disks.

nuclear pore (noo′klē-er) Point where the
inner and outer membranes of the nuclear
envelope come together to form a hole.

nuclease (noo′klē-ās) Enzyme that breaks
down nucleic acids.

nucleic acid (noo-klē′ik, -klā′ik) Molecule
consisting of many nucleotides chemically
bound together; deoxyribonucleic acid and
ribonucleic acid.

nucleolus (noo-klē′ō-lŭs), pl. nucleoli (noo-
klē′ō-lī) Rounded, dense, well-defined
nuclear body with no surrounding
membrane; subunits of ribosomes are
manufactured within the nucleolus.

nucleotide (noo′klē-ō-tīd) Basic building
block of nucleic acids, consisting of a sugar
molecule (either ribose or deoxyribose), one
of several types of organic bases, and a
phosphate group.

nucleus (noo′klē-ŭs), pl. nuclei (noo′klē-ī)
[L., a little nut, stone of a fruit] Cell
organelle containing most of the genetic
material of the cell; center of an atom
consisting of protons and neutrons;
collection of neuron cell bodies in the
central nervous system.

nutrient (noo′trē-ent) [L. nutriens, to
nourish] Chemical taken into the body and
used to produce energy, provide building
blocks for new molecules, or function in
other chemical reactions.

nutrition (noo-trish′ŭn) Process by which
nutrients are obtained and used in the body.

# O

oblique section (ob-lēk′) Cut made at other
than a right angle to the long axis of an organ.

obturator (ob′toor-ā-tōr) [L., to occlude or
stop up] Any occluding structure or a
foramen so occluded, as with the obturator
foramen of the hip.

occipital (ok-sip′i-tăl) Back of the head.

odorant (ō′-dōr-ănt) Substance with an
odor.

olecranon (ō-lek′ră-non) Point of the elbow.

olfaction (ol-fak′shŭn) [L., to smell] Sense of
smell.

olfactory (ol-fak′tŏ-rē) Relating to the sense
of smell.

oligodendrocyte (ol′i-gō-den′drō-sīt)
Neuroglial cell with multiple cell processes
that form myelin sheaths around axons in
the central nervous system.

omental bursa (ō-men′tăl ber′să) Pocketlike
sac inside the fold of the greater omentum.

omentum (ō-men′tŭm) [L., membrane of the
bowels] Fold of peritoneum extending
from the stomach to another organ.

oncology (ong-kol′ō-jē) [G. onco, a tumor +
logos, to study] Study of cancer and its
associated problems.

oocyte (ō′ō-sīt) [G. oon, egg + kytos, cell]
Female gamete, or sex cell; a secondary
oocyte and a polar body result from the first
meiotic division, which occurs prior to
ovulation; a zygote and a polar body result
from the second meiotic division, which
occurs following union of the sperm cell
with the secondary oocyte.

oogonium (ō-ō-gō′nē-ŭm), pl. oogonia (ō-ō-
gō′nē-ă) [G. oon, egg + gone, generation]
Cell that gives rise to oocytes; has a diploid
number of chromosomes.

opsin Protein portion of the rhodopsin
molecule; at least three different opsins are
located in cone cells.

optic (op′tik) Relating to vision.

optic disc Region in the posterior wall of the
eye where the optic nerve exits the eye; the
blind spot.

optic nerve Nerve that leaves the eye and exits
the orbit through the optic foramen to enter
the cranial vault.

oral cavity (ōr′ăl) Mouth; the first portion of
the digestive tract.

orbit (ōr′bit) Seven skull bones that surround
and protect the eye; eye socket.

organ (ōr′găn) [G. organon, tool] Part of the
body composed of two or more tissue types
and performing one or more specific
functions.

organ of Corti Named for Italian anatomist
Alfonso Corti (1822–1888). Specialized
region of the cochlear duct consisting of
hair cells; produces action potentials in
response to sound waves.

organ system Group of organs classified as a
unit because of a common function or set of
functions.

organelle (or′gă-nel) [G. organon, a tool + L.
-elle, small, a little organ] Specialized part
of a cell performing one or more specific
functions.

organic (ōr-gan′ik) Molecules that contain a
carbon atom (carbon dioxide is an
exception); originally defined as molecules
extracted from living organisms; the
original definition became obsolete when it
became possible to manufacture these
molecules in the laboratory.

organism (ōr′gă-nizm) Any living thing
considered as a whole, whether composed
of one cell or many.

organogenesis (ōr′gă-nō-jen′ĕ-sis)
Formation of organs during embryonic
development.

orgasm (ōr′gazm) [G. orgao, to swell, be
excited] Climax of the sexual act, often
associated with a pleasurable sensation.

origin (ōr′i-jin) Less movable attachment
point of a muscle.

oropharynx (ōr′ō-far′ingks) Portion of the
pharynx that lies posterior to the mouth; it
is continuous above with the nasopharynx
and below with the laryngopharynx.

osmosis (os-mō′sis) [G. osmos, thrusting or
an impulsion] Diffusion of solvent (water)
through a selectively permeable membrane
from a region of higher water concentration
to one of lower water concentration.

osmotic pressure (os-mot′ik) Force required
to prevent the movement of water across a
selectively permeable membrane.

ossification (os′i-fi-kā′shŭn) [L. os, bone +
facio, to make] Bone formation.

osteoblast (os′tē-ō-blast) [G. osteo,
bone] Cell that makes bone.

osteoclast (os′tē-ō-klast) [bone eating] Cell
that digests and removes bone.

osteocyte (os′tē-ō-sīt) [G. osteon, bone +
kytos, cell] Mature bone cell surrounded by
bone matrix.

osteon (os′tē-on) Single central canal, with its
contents, and the associated lamellae and
osteocytes surrounding it. Also called a
haversian system.

otolith (ō′tō-lith) [G. ous, ear + lithos, stone]
Small protein and calcium carbonate weight
in the maculae of the vestibule.

ovary (ō′vă-rē) One of two female
reproductive glands located in the pelvic
cavity; produces the oocyte, estrogen, and
progesterone.

**ovulation** (ov′ū-lā′shŭn) Release of an oocyte from the mature follicle.

**oxidative metabolism** (ok-si-dā′tiv mĕ-tab′ō-lizm) Metabolism in which oxygen is required to produce ATP.

**oxygen deficit** (ok′sē-jen) Amount of oxygen required to convert the lactic acid produced during anaerobic respiration to glucose and to replenish creatine phosphate stores.

**oxytocin** (ok′sĭ-tō′sin) [G., swift birth] Peptide hormone, secreted by the posterior pituitary gland, that increases uterine contraction and stimulates milk ejection from the mammary glands.

# P

**palate** (pal′ăt) Roof of the oral cavity; consists of the anterior bony part, the hard palate, and the posterior soft palate that is composed mainly of skeletal muscle and connective tissue.

**pancreas** (pan′krē-as) Elongated gland extending from the duodenum to the spleen; consists of a head, a body, and a tail. There is an exocrine portion, which secretes digestive enzymes, which are carried by the pancreatic duct to the duodenum, and pancreatic islets, which secrete insulin and glucagon.

**pancreatic duct** (pan-krē′at′ik) Duct of the pancreas; it joins the common bile duct to empty into the duodenum.

**pancreatic islet** (i′let) Cellular mass in the tissue of the pancreas; composed of different cell types that constitute the endocrine portion of the pancreas and are the source of insulin and glucagon.

**papilla** (pă-pil′ă), pl. **papillae** (pă-pil′ē) [L., nipple] Small, nipplelike process; projection of the dermis, containing blood vessels and nerves, into the epidermis; projections on the surface of the tongue.

**papillary muscle** (pap′i-l-ār-ē) Raised area of cardiac muscle in the ventricle to which the chordae tendineae attach.

**paracrine** (par′ă-krin) [G. *para*, alongside + *krino*, to separate] Kind of hormone function in which the effects of the hormone are restricted to the local environment.

**parafollicular cell** (par-ă-fo-lik′ū-lăr) Cell type scattered in a network of loose connective tissue between the thyroid follicles of the thyroid gland; secretes calcitonin.

**paranasal sinus** (par-ă-nā′săl) Air-filled cavity within certain skull bones that connects to the nasal cavity; the four sets of paranasal sinuses are the frontal, maxillary, sphenoidal, and ethmoidal.

**parasympathetic** (par-ă-sim-pa-thet′ik) [G. *para*, alongside + *sympathetic*] Subdivision of the autonomic nervous system with preganglionic neurons in the brainstem and sacral part of the spinal cord; involved in involuntary functions, such as digestion, defecation, and urination.

**parathyroid gland** (par-ă-thī′royd) One of four glandular masses embedded in the posterior surface of the thyroid gland; secretes parathyroid hormone.

**parathyroid hormone (PTH)** (hōr′mōn) Hormone produced by the parathyroid gland; increases bone breakdown and blood calcium levels.

**parietal** (pă-rī′ĕ-tăl) [L. *paries,* wall] Relating to the wall of any cavity; parietal serous membranes are in contact with the walls of cavities. The parietal bones form part of the skull.

**parietal peritoneum** (pĕ′rĭ-tō-nē′ŭm) [L., wall] Portion of the serous membranes of the abdominal cavity lining the inner surface of the body wall.

**parotid gland** (pă-rot′id) Largest of the salivary glands; one of a pair of salivary glands located anterior and inferior to each ear.

**partial pressure** Pressure exerted by a single gas in a mixture of gases.

**partial-thickness burn** Burn that damages only the epidermis (first-degree burn) or the epidermis and part of the dermis (also called a second-degree burn).

**parturition** (par-toor-ish′ŭn) [L. *parturio,* to be in labor] Childbirth; the delivery of a baby at the end of pregnancy.

**patella** (pa-tel′ă) [L. *patina,* shallow disk] Kneecap.

**pectoral** (pek′tŏ-răl) [L. *pectoralis,* breastbone] Relating to the chest.

**pedicle** (ped′ĭ-kl) [L. *pedicellus,* foot] Portion of a vertebra that extends from the body to the transverse process.

**pelvic cavity** (pel′vik) Space completely surrounded by the pelvic bones.

**pepsin** (pep′sin) [G. *pepsis,* digestion] Principal digestive enzyme produced by the stomach; digests proteins into smaller peptide chains.

**peptidase** (pep′ti-dās) Enzyme capable of breaking peptide chains into smaller chains and amino acids.

**peptide bond** (pep′tīd) Covalent chemical bond between adjacent amino acids in a polypeptide chain.

**pericardial cavity** (per-i-kar′dē-ăl) [G. *peri-,* around + *kardia,* the heart] Space between the visceral and parietal pericardia, filled with pericardial fluid; a cavity that surrounds the heart.

**pericardial fluid** Serous fluid within the pericardial cavity.

**pericardial membrane** Serous membranes associated with the heart.

**pericardium** (per-i-kar′dē-ŭm) [G. *pericardion,* the membrane around the heart] Membrane consisting of the epicardium and parietal pericardium (of the serous layers) and the outer fibrous pericardium; also called the pericardial sac.

**perilymph** (per′i-limf) [*peri* + G. *lympha,* clear fluid] Fluid between the bony labyrinth and the membranous labyrinth of the inner ear.

**perimetrium** (per-i-mē′trē-ŭm) Outer layer of the uterus; also called the serous layer.

**perimysium** (per′-i-mis′ē-ŭm, per′-i-miz′ē-ŭm) [*peri-* + G. *mys,* muscle] Fibrous sheath enveloping each of the skeletal muscle fascicles.

**perineum** (per′-i-nē′ŭm) Area inferior to the pelvic diaphragm between the thighs; extends from the coccyx to the pubis.

**periodontal** (per′ē-ō-don′tăl) [*peri-* + G. *odous,* tooth] Referring to structures surrounding the tooth, primarily in the alveolus.

**periosteum** (per-ē-os′tē-ŭm) [*peri-* + G. *osteon,* bone] Thick, double-layered connective tissue sheath covering the entire surface of a bone, except the articular surface, which is covered with cartilage.

**peripheral nervous system (PNS)** Part of the nervous system not surrounded by the skull or vertebral column; consists of nerves and ganglia.

**peristaltic waves** (per-i-stal′tik) [*peri-* + G. *stalsis,* constriction] Waves of relaxation followed by waves of contraction moving along a tube; propel food along the digestive tract.

**peritoneal cavity** (per′i-tō-nē′ăl) [to stretch over] Space between the visceral and parietal peritoneum filled with peritoneal fluid; cavity that surrounds many abdominopelvic organs.

**peritoneal membrane** Serous membrane associated with the peritoneal cavity.

**peritubular capillary** (per′ĭ-too′bū-lăr) Capillary network in the cortex of the kidney; associated with the distal and proximal convoluted tubules.

**peroxisome** (per-ok′si-sōm) Membrane-bound body similar to a lysosome in appearance but often smaller and irregular in shape; contains enzymes that either decompose or synthesize hydrogen peroxide.

**Peyer patch** Named for Swiss anatomist Johann Peyer (1653–1712). Collection of lymphatic nodules in the distal half of the small intestine and in the appendix.

**pH scale** Measure of the hydrogen ion concentration of a solution; the scale extends from 0 to 14.0—a pH of 7.0 is neutral, a pH of less than 7 acidic, and a pH of greater than 7 basic.

phagocytosis (fag′ō-sī-tō′sis) [G. *phagein,* to eat + L. *kytos,* cell + *osis,* condition] Ingestion and digestion by cells of substances, such as other cells, bacteria, cell debris, and foreign particles.

pharynx (far′ingks) [G. *pharynx,* throat] Joint openings of the digestive tract and the windpipe. The part of the digestive and respiratory tracts superior to the larynx and esophagus and inferior and posterior to the oral and nasal cavities.

phenotype (fē′nō-tīp) [G. *phaino,* to display + *typos,* model] Characteristic observed in the individual resulting from expression of the genotype.

pheromones (fer′ō-mōnz) [G. *pherō,* to carry + *hormāo,* to excite] Chemical signals secreted by an individual into the environment and perceived by a second individual of the same, or similar, species, producing a change in sexual or social behavior of that individual.

phlebitis (fle-bī′tis) Inflammation of a vein.

phospholipid (fos-fō-lip′id) Lipid with phosphorus resulting in a molecule with a polar and a nonpolar end; main component of cell membranes.

physiology (fiz-ē-ol′ō-jē) [G. *physis,* nature + *logos,* study] Scientific discipline that deals with the processes or functions of living things.

pia mater (pī′ă mā′ter, pē′ă ma′ter) [L., affectionate mother] Innermost meningeal layer; tightly attached to the brain and spinal cord.

pineal gland (pin′ē-ăl) [L. *pineus,* pinecone-shaped] Small endocrine gland attached to the dorsal surface of the diencephalon; may influence the onset of puberty and may play a role in some long-term cycles.

pinocytosis (pī′nō-sī-tō′sis, pī′nō-sī-tō′sis) [G. *pineo,* to drink + *kytos,* cell; *osis,* condition] Cell drinking; uptake of liquid by a cell.

pituitary dwarf (dwōrf) Individual of short stature, of relatively normal proportion, as a result of insufficient growth hormone secreted from the anterior pituitary gland.

pituitary gland (pi-too′i-tār-rē) [L. *pituita,* phlegm or a thick mucous secretion] Endocrine gland attached to the hypothalamus by the infundibulum; secretes hormones that influence the function of several other glands and tissues.

placenta (plă-sen′tă) Structure derived from embryonic and maternal tissues by which the embryo and fetus are attached to the uterus.

plasma (plaz′mă) Fluid portion of blood; blood minus the formed elements.

plasma membrane Cell membrane; outermost component of the cell, surrounding and binding the rest of the cell contents.

plasmin (plaz′min) Enzyme that breaks down the fibrin in blood clots; derived from plasminogen.

platelet (plāt′let) Minute fragment of cells derived from megakaryocytes; plays an important role in preventing blood loss.

platelet plug Accumulation of platelets that stick to connective tissue and to one another and prevent blood loss from damaged blood vessels.

pleural (ploor′ăl) [G., a rib or cavity] cavity Space between the visceral and parietal pleura, filled with pleural fluid; a cavity that surrounds each lung.

pleural membrane Serous membrane associated with the lungs.

plexus (plek′sŭs) [L., a braid] Intertwining of nerves or blood vessels.

PMS See *premenstrual syndrome.*

pneumothorax (noo-mō-thōr′aks) Presence of air in the pleural cavity.

podocyte (pod′ō-sīt) [Fr. *pous, podos,* foot + G. *kytos,* a hollow (cell)] Epithelial cell of Bowman's capsule attached to the outer surface of the glomerular capillary basement membrane; forms part of the filtration membrane.

polar body Oocyte receiving little cytoplasm; results from the first and the second meiotic division.

polar covalent bond Chemical bond in which electrons are shared unequally between two atoms.

polarize To create a difference in potential (charge) between two points, as between the inside and outside of a cell membrane.

polycythemia (pol′ē-sī-thē′mē-ă) [G. *polys,* many + *kytos,* cell] Increase in red blood cell numbers above the normal value.

polysaccharide (pol-ē-sak′ă-rīd) [many sugars] Many monosaccharides chemically bound together, such as glycogen and starch.

pons (ponz) [L., bridge] Part of the brainstem between the medulla oblongata and midbrain; contains nerve tracts between the cerebrum and cerebellum, as well as ascending and descending tracts.

portal system (pōr′tăl) System of vessels in which blood, after passing through one capillary bed, is conveyed through a second capillary network.

positive feedback Mechanism by which any deviation from an ideal normal value or set point is made greater.

posterior (pos-tēr′ē-ŏr) [L. *posterus,* following] That which follows; in humans, toward the back.

posterior horn Posterior extension of spinal cord gray matter; contains neuron cell

bodies that receive input from primary sensory neurons and relay that input to the brain; also called the dorsal horn.

posterior pituitary gland Posterior portion of the pituitary gland, which consists of processes of nerve cells that have their cell bodies located in the hypothalamus; secretes oxytocin and antidiuretic hormone.

postganglionic (pōst′gang-glē-on′ik) Autonomic neurons whose cell bodies are located outside the central nervous system and that receive synaptic stimulation from preganglionic autonomic neurons.

preganglionic (prē′gang-glē-on′ik) Autonomic neurons whose cell bodies are located in the central nervous system and that synapse with postganglionic neurons.

preload (prē′lōd) Degree to which the ventricular wall is stretched at the end of diastole; increases as the venous return increases.

premenstrual syndrome (PMS) (prē-men′stroo-al sin′drōm) In some women of reproductive age, the regular monthly experience of physiological and emotional distress, usually during the few days preceding menses, typically involving fatigue, edema, irritability, tension, anxiety, and depression.

prenatal period (prē-nā′tăl) [L. *prae,* before + *natalis,* relating to birth] Period before birth.

prepuce (prē′poos) In the male, a free fold of skin that almost completely covers the glans penis; the foreskin; in the female, a fold of mucous membrane that covers the clitoris.

primary response Immune response that occurs as a result of the first exposure to an antigen; results in the production of antibodies and memory cells.

prime mover Muscle that plays the principal role in accomplishing a movement.

primitive streak Shallow groove in the ectodermal surface of the embryonic disk; cells migrating through the streak become mesoderm.

process (pros′es, prō′ses) Projection on a bone.

product (prod′ŭkt) Substance produced in a chemical reaction.

progesterone (prō-jes′ter-ōn) Hormone secreted primarily by the corpus luteum and the placenta; aids in growth and development of female reproductive organs and secondary sexual characteristics; causes growth and maturation of the endometrium of the uterus during the menstrual cycle.

prolactin (prō-lak′tin) [L. *pro,* precursor + *lact,* milk] Hormone of the anterior pituitary gland that stimulates the secretion of milk.

pronation (prō-nā′shŭn) [L. *pronare,* to bend forward] Rotation, as of the forearm, starting in the anatomical position, so that the anterior surface faces posteriorly.

prophase (prō′fāzs) [G. *prophasis,* to foreshadow] First stage of mitosis or meiosis, consisting of contraction and increase in thickness of the chromosomes.

proprioceptive neurons (prō′prē-ō-sep′tiv) [L. *proprius,* one's own + *capio,* to take] Nerves that innervate the joints and tendons and provide information about the position of the body and its various parts.

prostaglandin (pros-tă-glan′din) Class of physiologically active substances present in many tissues; effects include vasodilation, stimulation and contraction of uterine smooth muscle, and promotion of inflammation and pain.

prostate gland (pros′tāt) [G. *prostates,* one standing before] Gland that surrounds the beginning of the urethra in the male. The secretion of the gland is a milky fluid that is discharged into the urethra as part of the semen.

protein (prō′tēn) [G. *proteios,* primary] Large molecule consisting of long sequences of amino acids (polypeptides) linked by peptide bonds.

proteoglycan (prō′tē-ō-glī′kan) [G. *proteo,* protein + *glycan,* polysaccharide] Macromolecule consisting of numerous polysaccharides attached to a common protein core; attracts and retains large amounts of water.

proteolytic (prō′tē-ō-lit′ik) Enzyme capable of digesting proteins or polypeptides.

proton (prō′ton) [G. *protos,* first] Positively charged particle in the nuclei of atoms.

provitamin (prō-vīt′ă-min) Substance that can be converted into a vitamin.

proximal (prok′si-măl) [L. *proximus,* nearest] Closer to the point of attachment to the body than another structure.

proximal convoluted tubule Convoluted portion of the nephron that extends from Bowman's capsule to the descending limb of the loop of Henle.

pterygoid (ter′ĭ-goyd) [G. *pteryx,* wing] Wing-shaped structure; two of the muscles of mastication, attached to wing-shaped, bony projections.

PTH See *parathyroid hormone.*

puberty (pū′ber-tē) [L. *pubertas,* grown up] Series of events that transform a child into a sexually mature adult; involves an increase in the secretion of all reproductive hormones.

pudendal cleft (pū-den′dal) [L. *pudeo,* to feel ashamed] Cleft between the labia majora.

pudendum (pū-den′dŭm), pl. pudenda (pū-den′da) External genitals, especially the female genitals. See *vulva.*

pulmonary capacity (pŭl′mō-nār-ē) Sum of two or more pulmonary volumes.

pulmonary circulation Blood flow through the system of blood vessels that carry blood from the right ventricle of the heart to the lungs and back from the lungs to the left atrium.

pulmonary semilunar valve Semilunar valve at the base of the pulmonary trunk where it exits from the right ventricle.

pulmonary trunk Large, elastic artery that carries blood from the right ventricle of the heart to the right and left pulmonary arteries.

pulmonary volume Lung volume, measured by spirometry; deviations from a normal value can be used to diagnose certain lung diseases; the pulmonary volumes are the tidal volume, inspiratory reserve volume, expiratory reserve volume, and residual volume.

pulp (pŭlp) [L. *pulpa,* flesh] Soft tissue inside a tooth, consisting of connective tissue, blood vessels, nerves, and lymphatic vessels.

pulse (pŭls) Pressure wave that travels rapidly along the arteries when blood is ejected from the left ventricle into the aorta.

pulse pressure Difference between systolic and diastolic pressures.

pupil (pū′pĭl) [L. *pupa,* a doll, so called because you can see a little reflection, or doll, in the pupil of another person's eye] Opening in the iris of the eye through which light enters.

Purkinje fiber (pŭr-kĭn′jē) Named for Bohemian anatomist/physiologist Johannes Purkinje (1787–1869). A specialized cardiac muscle fiber that conducts action potentials through cardiac muscle; forms part of the conduction system of the heart.

pus (pŭs) Product of inflammation, consisting of a liquid containing white blood cells, dead cells, and cell fragments.

pyloric sphincter (pī-lōr′ik) [G., gatekeeper] Thickened ring of smooth muscle at the distal end of the stomach.

pyrogen (pī′rō-jen) Chemical released by microorganisms, neutrophils, monocytes, and other cells that stimulates fever production by acting on the hypothalamus.

pyruvic acid (pī-roo′vik) Three-carbon end product of glycolysis; two pyruvic acid molecules are produced from each glucose molecule.

## Q

quadrant (kwăh′drant) [L. *quadrans,* a quarter] One-quarter of a circle; the abdomen is divided into right upper, right lower, left upper, and left lower quadrants by a horizontal and a vertical line intersecting at the umbilicus.

## R

RBC See *red blood cell.*

reactant (rē-ak′tant) Substance taking part in a chemical reaction.

receptor (rē-sĕp′tōr) [L., receiver] Protein molecule on the cell surface or within the cytoplasm that binds to a specific factor, such as a drug, a hormone, an antigen, or a neurotransmitter; one of the sensory nerve endings in the skin, deep tissues, viscera, and special sense organs.

recessive (rē-ses′iv) In genetics, a gene that may not be expressed phenotypically because of the expression of a contrasting dominant gene.

rectum (rek′tŭm) [L. *rectus,* straight] Last, straight part of the large intestine; between the colon and the anal canal.

rectus (rek′tŭs) Straight.

red blood cell (RBC) Biconcave disk that contains hemoglobin, which transports oxygen and carbon dioxide; red blood cells do not have a nucleus.

reflex (rē′fleks) Automatic response to a stimulus; does not require conscious thought.

reflex arc Consists of a sensory receptor, an afferent (sensory) neuron, an association neuron, an efferent (motor) neuron, and an effector organ.

regeneration (rē′jen-er-ā′shŭn) Tissue repair in which the damaged cells are replaced by cells of the same type as those damaged.

releasing hormone Hormone that is released from neurons in the hypothalamus and flows through the hypothalamic-pituitary portal system to the anterior pituitary gland; regulates the secretion of hormones from the cells of the anterior pituitary gland.

renal capsule (rē′năl) Connective tissue capsule that surrounds each kidney.

renal corpuscle (kōr′pŭs-l) Structure composed of a Bowman's capsule and its glomerulus.

renal pyramid (pir′ă-mid) Cone-shaped structure that extends from the renal sinus, where the apex is located, into the cortex of the kidney, where the base is located.

renal sinus (sī′nŭs) Cavity central to the medulla of the kidney that is filled with adipose tissue and contains the renal pelvis.

renin (rē′nin, ren′in) Enzyme secreted by the kidney that converts the plasma protein angiotensinogen to angiotensin I.

replacement Tissue repair in which the damaged cells are replaced by cells of a type different from those damaged.

respiration (res-pi-rā′shŭn) [L. *respiratio,* to breathe] Process in which oxygen is used to oxidize organic fuel molecules, providing a source of energy as well as carbon dioxide and water; includes ventilation, gas exchange, transport of oxygen and carbon dioxide in the blood, gas exchange between the blood and the tissues, and cell metabolism.

respiratory membrane (res′pi-ră-tōr-ē, rĕ-spīr′ă-tōr-ē) Membrane in the lungs across which gas exchange occurs with blood; consists of a thin layer of fluid, the alveolar epithelium, a basement membrane of the alveolar epithelium, interstitial space, the basement membrane of the capillary endothelium, and the capillary endothelium.

respiratory system Nose, nasal cavity, pharynx, larynx, trachea, bronchi, and lungs.

resting membrane potential Charge difference across the membrane of a resting cell (i.e., a cell that has not been stimulated to produce an action potential).

rete testis (rē′tē) Network of canals at the termination of the straight portion of the seminiferous tubules.

reticular formation (rē-tik′ū-lăr) [L. *rete,* net] Loose network of neuron cell bodies scattered throughout the brainstem; involved in the regulation of cycles such as the sleep-wake cycle.

retina (ret′i-nă) [L. *rete,* a net] Inner, light-sensitive tunic of the eye; nervous tunic.

retinaculum (ret-i-nak′ū-lŭm) [L., band, bracelet, halter, to hold back] Dense regular connective tissue sheath holding down the tendons at the wrist, ankle, or other sites.

retinal (rĕt′i-năl) Relating to the retina; retinaldehyde most commonly referring to the all-*trans* form (all-*trans*-retinal).

retroperitoneal (re′trō-per′i-tō-nē′ăl) Located behind the parietal peritoneum; includes the kidneys, the adrenal glands, the pancreas, portions of the intestines, and the urinary bladder.

reversible reaction (rē-ver′si-bl) Chemical reaction in which the reaction can proceed from reactants to products or from products to reactants; the amount of reactants relative to products is constant at equilibrium.

rhodopsin (rō-dop′sin) [G. *rhodon,* rose or red color + *opsin,* protein portion of rhodopsin] Purplish red protein in the external segment of the rods of the retina. Action of light converts it to opsin and all-*trans*-retinal.

ribonucleic acid (rī′bō-noo-klē′ik) (RNA) Type of nucleic acid containing the sugar ribose; involved in protein synthesis.

ribosomal RNA (rRNA) (ri′bō-sōm-ăl) RNA that is associated with certain proteins to form ribosomes.

ribosome (ri′bō-sōm) [ribose, a specific sugar] Small, spherical, cytoplasmic organelle where protein synthesis occurs.

right lymphatic duct Lymphatic duct that empties into the right subclavian vein; drains the right side of the head and neck, the right upper thorax, and the right upper limb.

RNA See *ribonucleic acid.*

rod Photoreceptor cell in the retina of the eye with a rod-shaped photoreceptive process; very light-sensitive cell that is important in dim light.

rotator cuff (rō-tā′tor, rō-tā′tŏr) Four deep muscles that attach the humerus to the scapula.

rRNA See *ribosomal RNA.*

ruga (roo′gă) Ridge or fold in the mucous membrane of the stomach.

# S

SA See *sinoatrial.*

sagittal plane (saj′i-tăl) [L. *sagitta,* the flight of an arrow] Plane running vertically through the body and dividing it into right and left parts.

saliva (să-lī′vă) Fluid containing enzymes and mucus; produced by the salivary glands and released into the oral cavity.

salivary gland (sal′i-vār-rē) Gland opening into the mouth and producing saliva.

salt Molecule consisting of a positively charged ion other than hydrogen and a negatively charged ion other than hydroxide.

sarcolemma (sar′kō-lem′ă) [G. *sarx,* flesh, muscle; *lemma,* husk] Cell membrane of a muscle fiber.

sarcomere (sar′kō-mēr) [sarco- + G. *meros,* part] Part of a myofibril formed of actin and myosin myofilaments, extending from Z disk to Z disk; the structural and functional unit of a muscle.

sarcoplasm (sar′kō-plazm) [sarco- + *plasma,* a thing formed] Cytoplasm of a muscle fiber.

sarcoplasmic reticulum (sar′kō-plaz′mik re-tik′ū-lŭm) Endoplasmic reticulum of a muscle fiber.

scapula (skap′ū-lă) Shoulder blade.

Schwann cell Named for German histologist/physiologist Theodor Schwann (1810–1882). Neuroglial cell forming myelin sheaths around axons in the peripheral nervous system.

sciatic (sī-at′ik) [Fr. *ischion,* the hip joint] Ischiadic or sciatic nerve.

sclera (sklēr′ă) [L. *skleros,* hard] Dense, white, opaque posterior four-fifths of the fibrous tunic of the eye; white of the eye.

scoliosis (skō-lē-ō′sis) [G., a crookedness] Abnormal lateral curvature of the spine.

scrotum (skrō′tŭm) Musculocutaneous sac containing the testes.

sebaceous gland (sē-bā′shŭs) [L. *sebum,* tallow] Gland of the skin that produces sebum; usually associated with a hair follicle.

sebum (sē′bŭm) [L., tallow] Oily, white, fatty substance produced by the sebaceous glands; lubricates hair and the surface of the skin.

secondary response See *memory response.*

secretin (se-krē′tin) Hormone released from the epithelium of the duodenum; inhibits gastric secretion.

sella turcica (sel′ă tŭr′sĭ-kă) [L., saddle, Turkish] Saddle-shaped depression in the inner surface of the skull where the pituitary gland is located.

semen (sē′men) [L., seed] Penile ejaculate; thick, yellowish-white, viscous fluid containing sperm cells and secretions of the testes, seminal vesicles, prostate gland, and bulbourethral glands.

semicircular canal (sem′ē-sir′kū-lăr) One of three canals in each temporal bone; involved in the detection of motion.

semilunar valve (sem-ē-loo′năr) One of two valves in the heart composed of three crescent-shaped cusps that prevent blood flow into the ventricles following ejection; located at the beginning of the aorta and pulmonary trunk.

seminal vesicle (sem′i-năl ves′i-kl) One of two glandular structures that empty into the ejaculatory ducts; its secretion is one of the components of semen.

seminiferous tubule (sem′i-nif′er-ŭs) Tubule in the testis in which sperm cells develop.

sensory neuron Neuron that extends from sensory receptors in the periphery to the central nervous system.

septum (sep′tŭm) [L. *saeptum,* a partition or wall] Thin wall dividing two cavities or masses of softer tissue.

serosa (se-rō′să) Smooth, outermost covering of an organ where it faces a cavity and is not surrounded by connective tissue.

serous membrane (sēr′ŭs) Thin sheet consisting of epithelium and connective tissue that lines cavities not opening to the outside of the body; does not contain glands but does secrete serous fluid.

serum (sēr′ŭm) Fluid portion of blood after the removal of fibrin and formed elements.

sex chromosome Chromosome other than an autosome; responsible for sex determination.

sinoatrial (SA) node (sī′nō-ā′trē-ăl) Mass of specialized cardiac muscle fibers in the right atrium near the opening of the superior vena cava that acts as the "pacemaker" of the cardiac conduction system.

**sliding filament model** Mechanism by which actin and myosin myofilaments slide over one another during muscle contraction.

**solute** (sol′ūt, sō′loot) [L. *solutus*, dissolved] Dissolved substance in a solution.

**solution** (sō-loo′shŭn) Homogeneous mixture formed when a solute dissolves in a solvent (liquid).

**solvent** (sol′vent) [L. *solvens*, dissolve] Liquid that holds another substance in solution.

**somatic motor** (sō-mat′ik) [G. *soma*, body or bodily] Type of motor (efferent) neuron of the peripheral nervous system that innervates skeletal muscle.

**somesthetic** (sō′mes-thet′ik) [G. *soma*, body + *aisthesis*, sensation] Consciously perceived.

**somesthetic cortex** Part of the cerebral cortex involved with the conscious perception and localization of general body sensations.

**spermatid** (sper′mă-tid) Cell in the late stage of the development of the sperm cell (male gamete). It is haploid and is derived from the secondary spermatocyte.

**spermatocyte** (sper′mă-tō′sīt) Cell arising from a spermatogonium and destined to give rise to spermatozoa.

**spermatogenesis** (sper′mă-tō-jen′ĕ-sis) Formation and development of sperm cells.

**spermatogonium** (sper′mă-tō-gō′nē-ŭm), pl. **spermatogonia** (sper′mă-tō-gō′nē-ă) Most peripheral germ cell in the seminiferous tubules scattered between the sustentacular cells; divide by mitosis and some form primary spermatocytes.

**spermatozoon** (sper′mă-tō-zō′on), pl. **spermatazoa** (sper′mă-tō-zō′ă) [G. *sperma*, seed + *zoon*, animal] Male gamete, or sex cell, composed of a head, midpiece, and tail; contains the genetic information transmitted by the male; sperm cell.

**sperm cell** [G. *sperma*, seed] Male reproductive cell; see also *spermatozoon*.

**sphenoid** (sfē′noyd) [G. *sphenoeides*, wedge + *eidos*, resemblance] Sphenoid bone or relating to the sphenoid bone.

**sphygmomanometer** (sfig′mō-mă-nom′ĕ-ter) [G. *sphygmos*, pulse; *manos* + *metron*, measure] Instrument for measuring blood pressure consisting of an arm sleeve and an inflating bulb with a device attached for measuring pressure in the arm sleeve.

**spina bifida** (spī′nă bif′ĭ-dă, bī′fĭ-dă) Defect in the spinal column, consisting in absence of the vertebral arches, through which the spinal membranes, with or without spinal cord tissue, may protrude.

**spinal cord** Portion of the central nervous system extending from the foramen magnum at the base of the skull to the second lumbar vertebra; consists of a central gray portion and a peripheral white portion.

**spinal nerve** Peripheral nerve exiting from the spinal cord.

**spirometer** (spī-rom′ĕ-ter) [L. *spiro*, to breathe + G. *metron*, measure] Meter used for measuring the volume of respiratory gases; usually consists of a counterbalanced, cylindrical bell sealed by dipping into a circular trough of water.

**spirometry** (spī-rom′ĕ-trē) Process of making pulmonary measurements with a spirometer.

**spleen** (splēn) Large lymphatic organ in the left upper part of the abdominal cavity, between the stomach and the diaphragm; composed of white and red pulp; responds to foreign substances in the blood, destroys worn-out red blood cells, and is a reservoir for blood.

**spongy bone** [L., grating or lattice] Bone with a latticelike appearance.

**squamous** (skwā′mŭs) [L. *squama*, a scale] Scalelike, flat.

**stapes** (stā′pēz) [L., stirrup] The third of the three middle ear bones; attached to the oval window; the stirrup.

**Starling's law of the heart** Named for English physiologist Ernest Starling (1866–1927). Force of contraction of cardiac muscle is a function of the length of its muscle fibers at the end of diastole; the greater the degree of filling of the heart (the greater the venous return), the greater the force of contraction of the cardiac muscle.

**stem cell** Single population of cells that differentiate to give rise to the formed elements of blood.

**stenosed valve** (sten′ōzd) Valve that has its opening narrowed or partially closed.

**sternum** (ster′nŭm) [L. *sternon*, chest] Breastbone.

**steroid** (stēr′oyd, ster′oyd) Large family of lipids, including some hormones, vitamins, and cholesterol.

**stethoscope** (steth′ō-skōp) [G. *stetho-*, chest + *skopeo*, to view] Instrument originally devised for aid in hearing the respiratory and cardiac sounds in the chest and now used in hearing other sounds in the body as well.

**strabismus** (stra-biz′mŭs) [G. *strabismos*, a squinting] Lack of parallelism of the visual axes of the eyes.

**stratum** (strat′ŭm), pl. **strata** (strat′tă) [L., bed cover, layer] Layer of tissue.

**stratum basale** (bā-săl′ē) Deepest layer of the epidermis; consists of columnar cells that undergo mitotic divisions.

**stratum corneum** (kōr′nē-ŭm) Most superficial layer of the epidermis; consists of dead squamous cornified cells that have undergone keratinization.

**stria, pl., striae** (stri′e) [L., channel, furrow] Bands of thin, wrinkled skin, becoming red and white, that occur commonly on the abdomen, buttocks, and thighs at puberty and/or during and following pregnancy and result from overextension of the skin.

**stroke volume** Volume of blood ejected from either the right or the left ventricle during each heartbeat.

**styloid** (stī′loyd) [G. *stylos*, a stake or pen] Slender, pencil-shaped process.

**subarachnoid space** (sŭb-ă-rak′noyd) Fluid-filled space below the arachnoid layer covering the brain and spinal cord; contains cerebrospinal fluid.

**subcutaneous** (sŭb-koo-tā′nē-ŭs) [L. *sub*, under + *cutis*, skin] Under the skin; same tissue as the hypodermis.

**sublingual gland** (sŭb-ling′gwăl) One of a pair of salivary glands located below the tongue.

**submandibular gland** (sŭb-man-dib′ū-lăr) One of a pair of salivary glands located below the mandible.

**submucosa** (sŭb-moo-kō′să) Layer of connective tissue deep to the mucous membrane.

**sulcus** (sool′kŭs), pl. **sulci** (sŭl′sī) [L., ditch] Groove on the surface of the brain between gyri.

**superficial** (soo-per-fish′ăl) [L. *superficialis*, surface] Toward or on the surface.

**superior** (soo-pēr′ē-ōr) [L., higher] Up, or higher, with reference to the anatomical position.

**superior vena cava** (vē′nă kā′vă) Blood vessel that receives blood from the head, neck, and upper limbs and empties into the right atrium of the heart.

**supination** (soo′pi-nā′shŭn) [L. *supino*, to place something on its back] Rotation of the forearm so that the anterior surface is anterior; that is, the forearm is in the anatomical position.

**surfactant** (ser-fak′tănt) Mixture of lipoprotein molecules produced by the secretory cells of the alveolar epithelium of the lung; reduces water surface tension.

**sustentacular cell** (sŭs-ten-tak′ū-lăr) Cell in the wall of the seminiferous tubules to which spermatogonia and spermatids are attached; also called a Sertoli cell.

**suture** (soo′choor) [L. *surtura*, a seam] Fibrous joint between flat bones of the skull.

**sweat gland** (swet) Usually, a secretory organ that produces a watery secretion, called sweat, that is released onto the surface of the skin; some sweat glands, however, produce an organic secretion.

**sympathetic** (sim-pă-thet′ik) [G. *sympatheo*, to feel with + *pathos*, suffering] Subdivision of the autonomic nervous system with preganglionic nerve cell bodies located in the thoracic and lumbar regions of the spinal cord; generally involved in preparing the body for immediate physical activity.

**synapse** (sin′aps) [G. *syn*, together + *haptein*, to clasp] Junction between a nerve cell and another nerve cell, muscle cell, or gland cell; in a chemical synapse, chemicals are released from the nerve cell as a result of an action potential in the nerve cell, the chemicals cross the cleft between the cells, and they cause a response in the postsynaptic cell.

**synapsis** (si-nap′sis) Pairing of homologous chromosomes during prophase of the first meiotic division.

**synergist** (sin′er-jist) Muscle that works with another muscle to cause a movement.

**synovial cavity** (si-nō′vē-ăl) [G. *syn*, coming together + *ovia*, resembling egg albumin] Cavity surrounding articulating bones of a freely movable or synovial joint; contains synovial fluid.

**synovial fluid** Somewhat viscous substance serving as a lubricant in movable joints, tendon sheaths, and bursae.

**synovial joint** Freely movable joint.

**synovial membrane** Membrane that lines the inside of a joint cavity; produces synovial fluid.

**synthesis reaction** (sin′thĕ-sis) Combination of atoms, ions, or molecules to form a new, larger molecule.

**systemic circulation** (sis-tem′ik) Blood flow through the system of blood vessels that carry blood from the left ventricle of the heart to the tissues of the body and back from the body to the right atrium.

**systemic inflammation** Inflammation that occurs in many areas of the body; in addition to the symptoms of local inflammation, can include increased neutrophil numbers in the blood, fever, and shock.

**systole** (sis′tō-lē) [G., a contracting] Contraction of the heart chambers, during which blood leaves the chambers; usually refers to ventricular contraction.

**systolic pressure** (sis-tol′ik) Maximum arterial blood pressure reached during ventricular systole.

# T

**target tissue** Tissue on which a hormone acts.

**tarsal bone** (tar′săl) [G. *tarsos*, sole of foot] Bone of the instep of the foot.

**taste bud** Sensory structure mostly on the tongue; functions as a taste receptor.

**tectorial membrane** (tek-tōr′ē-ăl) [L., a covering] Membrane attached to the spiral lamina and extending over the hair cells; hairs of the hair cells have their tips embedded in the membrane.

**telophase** (tel′ō-fāz) [G. *telos*, end + *phasis*, an appearance] Final stage of mitosis or meiosis that begins when migration of chromosomes to the poles of the cells has been completed.

**temporal** (tem′pŏ-răl) [L. *tempus*, time] Indicating the temple; the temple of the head is so named because it is there that the hair first begins turning white, indicating the passage of time.

**tendinous intersection** (ten′di-nŭs) One of the bands of connective tissue crossing the rectus abdominis muscle, subdividing it and attaching it to adjacent connective tissue.

**tendon** (ten′dŏn) Tough connective tissue band connecting a muscle to bone.

**teniae coli** (tē′nē-ē kō′lī) [G. *tainia*, band, tapeworm + *coli*, colon] Segmented, longitudinal smooth muscle layer of the colon.

**testis** (tes′tis), pl. **testes** (test′tēz) One of two male reproductive glands in the scrotum; produces testosterone and sperm cells.

**testosterone** (tes′tos′tĕ-rōn) Steroid hormone secreted primarily by the testes; aids in spermatogenesis, controls maintenance and development of male reproductive organs and secondary sexual characteristics, and influences sexual behavior.

**tetanus** (tet′ă-nŭs) [L. *tetanus*, convulsive tension] Sustained muscular contraction caused by a series of nerve stimuli repeated so rapidly that the individual contractions are fused, producing a sustained tetanic contraction; also a disease marked by painful tonic muscular contractions, caused by the neurotoxin of *Clostridium tetani* action on the central nervous system.

**tetany** (tet′ă-nē) Condition in muscle contraction in which there is no relaxation between muscle twitches.

**tetraiodothyronine** (tet′ră-ī-ō-dō-thī′rō-nēn) ($T_4$) One of the thyroid hormones; contains four iodine atoms; also called thyroxine.

**thalamus** (thal′ă-mŭs) [G., a bedroom] Large mass of gray matter making up the bulk of the diencephalon; involved in the relay of sensory input to the cerebrum.

**thoracic cavity** (thō-ras′ik) Space bounded by the neck, the thoracic wall, and the diaphragm.

**thoracic duct** Largest lymphatic vessel in the body; drains the left side of the head and neck, the left upper thorax, the left upper limb, and the inferior half of the body into the left subclavian vein.

**thorax** (thō′raks) [G., breastplate] Chest; the upper part of the trunk between the neck and the abdomen.

**thrombocyte** (throm′bō-sīt) [*thrombos*-, clot + G. *kytos*, cell] Cell fragment involved in platelet plug and clot formation; also called a platelet.

**thrombosis** (throm′bō′sis) [G. *thrombos*, clot] Formation or presence of a clot (thrombus) inside a blood vessel.

**thrombus** (throm′bŭs) [G. *thrombos*, clot] Clot within the cardiovascular system.

**thymosin** (thī′mō-sin) Hormone secreted from the thymus that helps activate the immune system.

**thymus** (thī′mŭs) [G. *thymos*, sweetbread] Bilobed lymphatic organ located in the inferior neck and superior mediastinum; involved with the maturation of T cells.

**thyroid cartilage** (thī′royd) [G. *thyreoeides*, shield] Largest laryngeal cartilage; forms the laryngeal prominence, or Adam's apple.

**thyroid follicle** One of many small spheres with walls consisting of cuboidal epithelial cells in the thyroid gland; filled with proteins to which thyroid hormones are attached until they are secreted.

**thyroid gland** Endocrine gland located inferior to the larynx and consisting of two lobes connected by a narrow band; secretes the thyroid hormones.

**thyroid hormone** Any hormone secreted by the thyroid gland, especially those, such as thyroxine, that contain iodine and regulate metabolism and the maturation of tissues.

**thyroid-stimulating hormone (TSH)** Hormone released from the hypothalamus that stimulates thyroid hormone secretion from the thyroid gland.

**thyroxine** (thī-rok′sēn, thī-rok′sin) See *tetraiodothyronine*.

**tissue** (tish′ū) [L. *texo*, to weave] Collection of cells with similar structure and function and the substances between the cells.

**tissue repair** Substitution of viable cells for damaged or dead cells by regeneration or replacement.

**tonsil** (ton′sil) Collection of lymphoid tissue; usually refers to large collections of lymphoid tissue beneath mucous membranes of the oral cavity and pharynx; lingual, pharyngeal, and palatine tonsils.

**trabecula** (tră-bek′ū-lă) [L. *trabs*, beam] Beam or plate of spongy bone or other tissue.

**trachea** (tră′kē-ă) [G. *tracheia arteria*, rough artery] Air tube extending from the larynx into the thorax, where it divides to form bronchi; has 16–20 C-shaped pieces of cartilage in its walls.

**tracheostomy** (tră′kē-os′tō-mē) Incision into the trachea.

**tract** (trakt) Nerve tract; a bundle of neuron cell processes (axons) in the central nervous system, usually having a common function.

**transfer RNA (tRNA)** RNA that attaches to individual amino acids and transports them to the ribosomes, where they are connected to form a protein polypeptide chain.

transverse plane (trans-vers′) Plane separating the body into superior and inferior parts.

transverse section Cut made at right angles to the long axis of an organ.

trapezius (tra-pē′zē-ŭs) Back muscle, shaped like a trapezium (a four-sided geometric figure in which no two sides are parallel), that rotates the scapula.

triacylglycerol (trī-as′il-glis′er-ol) See triglyceride.

triceps brachii (trī′seps brā′kē-ī) Three-headed muscle in the posterior arm that extends the forearm.

tricuspid valve (trī-kŭs′pid) Valve closing the opening between the right atrium and right ventricle of the heart.

triglyceride (trī-glis′er-īd) Common type of lipid, or fat, with three fatty acids bound to a glycerol molecule; also called a triacylglycerol.

trigone (trī′gōn) [L. trigonium, triangle] Triangular, smooth area at the base of the urinary bladder between the openings of the two ureters and that of the urethra.

triiodothyronine (trī-ī′ō-dō-thī′rō-nēn) ($T_3$) One of the thyroid hormones; contains three iodine atoms.

tRNA See transfer RNA.

trochanter (trō′kanter) [G., a runner] One of the large tubercles of the proximal femur.

trophoblast (trō′fō-blast) [G. trophe, nourishment + blastos, germ] Outer part of the blastocyst; enters the uterus and becomes the embryonic portion of the placenta.

trypsin (trip′sin) Enzyme released from the pancreas that digests proteins.

TSH See thyroid-stimulating hormone.

tubercle (too′ber-kl) Lump or knob on a bone.

tuberosity (too′ber-os′i-tē) Lump on a bone, usually larger than a tubercle.

tubular reabsorption Movement of materials, by means of diffusion or active transport, from the filtrate within a nephron into the blood.

tubular secretion Movement of materials, by means of active transport, from the blood into the filtrate of a nephron.

tumor (too′mŏr) Swelling, one of the cardinal signs of inflammation, or a new growth of tissue in which the multiplication of cells is uncontrolled and progressive; see also neoplasm.

tunic (too′nik) [L., coat] Layer or coat; one of the three enveloping layers of the wall of the eye; the three tunics are the fibrous, vascular, and nervous tunics; one of the three layers of blood vessels: tunica intima, tunica media, and tunica adventitia.

tunica adventitia (too′ni-kă ad-ven-tish′ă) Outermost fibrous coat of a vessel or an organ that is derived from the surrounding connective tissue.

tunica intima (in′ti-mă) Innermost layer of a blood or lymphatic vessel; consists of endothelium and a small amount of connective tissue.

tunica media Middle, usually muscular, coat of an artery or another tubular structure.

tympanic membrane (tim-pan′ik) [drumlike] Cellular membrane that covers the inner opening of the external auditory canal and separates the middle and external ears; vibrates in response to sound waves; the eardrum.

## U

ulcer (ŭl′ser) [L. ulcus, a sore] Lesion on the surface of the skin or a mucous membrane, such as in the stomach or intestine, caused by a superficial loss of tissue, usually with inflammation.

umbilical cord (ŭm-bil′i-kăl) [L., navel] Cord connecting the fetus to the placenta; contains two umbilical arteries, which originate from the embryo's internal iliac arteries, that carry blood from the embryo to the placenta and one umbilical vein that carries blood back to the fetus.

umbilical vein Vein in the umbilical cord of the fetus by which the fetus receives nourishment from the placenta; becomes the round ligament of the liver in the adult.

upper motor neuron Motor neuron located in the cerebral cortex and synapsing with a lower motor neuron in the brainstem or spinal cord.

ureter (ū-rē′ter, ū′re-ter) [G. oureter, urinary canal] Tube conducting urine from the kidney to the urinary bladder.

urethra (ū-rē′thră) Duct leading from the urinary bladder, discharging the urine externally.

uterus (ū′ter-ŭs) Hollow muscular organ in which the fertilized oocyte develops into a fetus.

utricle (ū′trĭ-kl) Larger of the two membranous sacs in the vestibule of the labyrinth. The semicircular canals arise from it.

uvula (ū′vū-lă) [L. uva, grape] Small, grapelike appendage at the posterior margin of the soft palate.

## V

vaccine (vak′sēn, vak-sēn′) Preparation of killed microorganisms, altered microorganisms, or derivatives of microorganisms intended to produce immunity; usually administered by injection, but sometimes ingestion is preferred.

vagina (vă-jī′nă) [L., sheath] Genital canal in the female, extending from the uterus to the vulva.

variable region Part of an antibody that combines with an antigen; responsible for the specificity of the antibody.

varicose (văr′ĭ-kōs) vein Vein that is so dilated that the cusps of the valves are no longer capable of preventing backflow of blood; usually the veins in the lower legs or the hemorrhoidal veins.

vasoconstriction (vă′sō-kon-strik′shŭn) Decreased diameter of blood vessels.

vasodilation (vă′sō-dī-lā′shŭn) Increased diameter of blood vessels.

vasomotor center (vă-sō-mō′ter) Area of the lower pons and upper medulla oblongata that continually transmits a low frequency of action potentials through sympathetic neurons to smooth muscle in blood vessels; can cause vasoconstriction and vasodilation.

vasomotor tone Partial constriction of blood vessels in the periphery, which results from relatively constant sympathetic stimulation.

vasopressin (vă-sō-pres′in) [L. vaso, blood vessel + pressum, to press down] Peptide hormone, related to oxytocin, secreted from the posterior pituitary gland. In large doses, it causes contraction of blood vessel smooth muscle; see also antidiuretic hormone.

vein (vān) Blood vessel that carries blood toward the heart.

venous return (vē′nŭs) Volume of blood returning to the heart.

ventilation (ven-ti-lā′shŭn) Movement of air in and out of the lungs.

ventral (ven′trăl) [L. venter, belly] In humans, synonymous with anterior.

ventral root Motor (efferent) root of a spinal nerve.

ventricle (ven′tri-kl) [L. venter, belly] Cavity; in the brain, one of four cavities filled with cerebrospinal fluid; one of two chambers of the heart that pump blood into arteries; there is a left and a right ventricle.

vernix caseosa (ver′niks kā′sē-ō′să) Epithelial cells and sebaceous matter that cover the skin of the fetus.

vesicle (ves′i-kl) [L. vesicula, blister or bladder] Small, membrane-bound sac containing material to be transported across the cell membrane.

vestibular fold (ves-tib′ū-lăr) [L., entrance hall] False vocal fold.

vestibule (ves′ti-bool) Small cavity or space at the entrance of a canal; see also vulva.

villus (vil′ŭs), pl. villi (vil′ī) [L., shaggy hair] Projection of the mucous membrane in the small intestine that increases surface area.

visceral (vis′er-ăl) [L. viscus, the soft parts, internal organs] Relating to the internal organs.

visceral peritoneum (per′i-tō-nē′ŭm) [L., organ] Part of the serous membrane in the abdominal cavity covering the surface of some abdominal organs.

vitamin (vīt′ă-min) [L. *vita,* life + *amine,* from ammonia] One of a group of organic substances, present in minute amounts in natural foods, that are essential to normal metabolism; insufficient amounts in the diet may cause deficiency diseases.

vitamin D Fat-soluble vitamin produced from a precursor molecule in skin exposed to ultraviolet light; increases calcium and phosphate uptake in the intestine.

vitreous humor (vit′rē-ŭs) Transparent, jellylike substance that fills the posterior compartment of the eye; helps maintain pressure within the eye and holds the lens and retina in place.

vocal fold (vō′kăl) One of the ligaments that extends from the posterior surface of the thyroid cartilage to the paired cartilages of the larynx; the superior pair are the false vocal folds, and the inferior pair are the true vocal folds.

vulva (vŭl′vă) [L., a wrapper or covering, seed covering, womb] External genitalia of the female; the mons pubis, labia majora and minora, the clitoris, the vestibule and its glands, the opening of the urethra, and the opening of the vagina.

## W

white blood cell (WBC) Round, nucleated blood cell involved in immunity; includes neutrophils, basophils, eosinophils, lymphocytes, and monocytes; also called a leukocyte.

## X

X-linked Trait caused by a gene on the X chromosome.

## Y

yolk sac (yōk, yōlk) Highly vascular endodermal layer surrounding the yolk of an embryo.

## Z

zona pellucida (zō′nă pe-loo′sid-ă) [L. *zone,* girdle + *pellucidus,* passage of light] Extracellular coat surrounding the oocyte; appears translucent.

zygomatic (zī′gō-mat′ik) [G. *zygon,* yoke] Referring to the zygomatic, or cheek, bone; the zygomatic arch is a bony arch created by the junction of the zygomatic and temporal bones.

zygomaticus muscle (zī′gō-mat′i-kŭs) Muscle originating on the zygomatic bone and inserting onto the corner of the mouth; involved in smiling.

zygote (zī′gōt) [G. *zygotos,* yoked] Single-celled, diploid product of fertilization, resulting from the union of a sperm cell and an oocyte.

# Credits

## Photographs

### Chapter 1
Opener: © Digital Vision RF; Figure 1.1b: © Bart Harris/Corbis; 1.7a&b, 1.8a-b, 1.10a: © The McGraw-Hill Companies, Inc./Eric Wise, photographer; 1.10b-d: © R. T Hutchings

### Chapter 2
Opener: © Brand X Pictures/ PunchStock RF; Figure 2.3c: © Trent Stephens; 2A: © ISM/ Sovereign/Phototake.com; 2B: © Simon Fraser/Royal Victoria Infirmary/Photo Researchers, Inc.

### Chapter 3
Opener: © ISM/Phototake.com; Figure 3.2b: © J. David Robertson, from Charles Flickinger, Medical Cell Biology, Philadelphia; 3.12b: © Courtesy of Dr. Birgit H. Satir; 3.13b: © Don Fawcett/Photo Researchers, Inc.; 3.13c: © Bernard Gilula/Photo Researchers, Inc.; 3.16b: © J. David Robertson, from Charles Flickinger, Medical Cell Biology, Philadelphia; 3.17b: © Robert Bollender/Don Fawcett/ Visuals Unlimited; 3.19b, 3.20b: © Don Fawcett/Photo Researchers, Inc.; 3.21b: © Biology Media/Photo Researchers, Inc.; 3.26a-f: © Ed Reschke.

### Chapter 4
Opener: © The McGraw-Hill Companies, Inc./Al Tesler, photographer; Figure 4.1b: © Victor Eroschenko; 4.1c: © Ed Reschke; Table 4.2a (2): © The McGraw-Hill Companies, Inc./Dennis Strete, photographer; Table 4.2b (2), Table 4.2c (2), Table 4.2d (2), Table 4.3a (2), Table 4.3b (2), Table 4.3b (3): © Victor Eroschenko; Table 4.5, Table 4.6: © Ed Reschke; Table 4.7a-b: © Victor Eroschenko; Table 4.8a: © Carolina Biological Supply/Phototake.com; Table 4.8b&c: © Victor Eroschenko; Table 4.9: © Trent Stephens; Table 4.10, Table 4.11 a-b (2): © Ed Reschke; Table 4.11c (2): © Victor Eroschenko; Table 4.12 (2): © Trent Stephens

### Chapter 5
Opener: CO 5: © BananaStock/ PunchStock RF; Figure 5.2b: © John Cunningham/Visuals Unlimited; 5.10a-c: © Thomas B. Habif; 5.A: © James Stevenson/ Science Photo Library/Photo Researchers, Inc.; 5.B: © Stan Levy/ Photo Researchers, Inc.

### Chapter 6
Opener: © Creatas/PunchStock RF; Figure 6.3a: © Trent Stephens; 6.5b: © Visuals Unlimited; 6.7a © Ed Reschke/Peter Arnold, Inc.; 6.7b: © Bio-Photo Assocs/Photo Researchers, Inc.; 6.12b: © The McGraw-Hill Companies, Inc./Eric Wise, photographer; 6.23d: © Trent Stephens; 6.24a: © The McGraw-Hill Companies, Inc./Andy Resek, photographer; 6.24b, 6.28, 6.36, 6.41a-b, 6.41c (1-2), 6.41d, 6.41e (1-2), 6.41f: © The McGraw-Hill Companies, Inc./Eric Wise, photographer; 6.B: © Princess Margaret Rose Orthopaedic Hospital/Science Photo Library/ Photo Researchers, Inc.; 6.C: © Dr. Michael Klein/Peter Arnold, Inc.

### Chapter 7
Opener: © Royalty-Free/Corbis RF; Figure 7.3a: © Richard Rodewald; 7.5b: © Fred Hossler/Visuals Unlimited; 7.14a-b: © Steve Barden/ Action Plus; 7.24a-b, 7.26a,b&d: © The McGraw-Hill Companies, Inc./Eric Wise, photographer; 7.A: © Roberta Seidman.

### Chapter 8
Opener: © Brand X Pictures/ PunchStock RF; Figure 8.22: © Courtesy of Branislav Vidic; 8.25a: © R. T. Hutchings; 8.25b: © The McGraw-Hill Companies, Inc./Rebecca Gray, photographer/ Don Kincaid, dissections; 8.33a: © Deep Light Productions/SPL/ Photo Researchers, Inc.; 8.Aa: © Howard J. Radzyner/ Phototake.com; 8.Ab: © CNRI/ Science Photo Library/Photo Researchers, Inc.

### Chapter 9
Opener: © Royalty-Free/Corbis RF; Figure 9.5: © The McGraw-Hill Companies, Inc./Eric Wise, photographer; 9.14a: © A. L. Blum/ Visuals Unlimited; 9.16b: © The McGraw-Hill Companies, Inc./ Rebecca Gray, photographer/Don Kincaid, dissections; 9.Aa-b: Reproduced from Isihara's Tests for Color Blindness published by Kanehara & Co., Ltd., Tokyo, Japan, but tests for color blindness cannot be conducted with this material. For accurate testing, the original plates should be used.; 9.21a & b2: © Trent Stephens; 9.23a: © Jerry Wachter/ Photo Researchers, Inc.

### Chapter 10
Opener: © Stockdisc/PunchStock RF; Figure 10.11d, 10.14c: © Victor Eroschenko; 10.18: © Bio-Photo Assocs/Photo Researchers, Inc.; 10. Aa-b: © Ken Greer/Visuals Unlimited

### Chapter 11
Opener: © Keith Brofsky/Getty Images RF; Figure 11.1: © The McGraw-Hill Companies, Inc./Eric Wise, photographer; 11.3a: © National Cancer Institute/Science Photo Library/Photo Researchers, Inc.; 11.6: © Ed Reschke; 11.7a-e: © Victor Eroschenko.

### Chapter 12
Opener: © Dynamic Graphics/ JupiterImages RF; Figure 12.5b: © R. T. Hutchings; 12.7a-b: © McMinn & Hutchings, Color Atlas of Human Anatomy/Mosby; 12.13b: © Ed Reschke; 12.19: © Terry Cockerham/Cynthia Alexander/ Synapse Media Productions; 12.A: © Hank Morgan/Science Source/ Photo Researchers, Inc.

### Chapter 13
Opener: © Keith Brofsky/Getty Images RF; Figure13.2: © Visuals Unlimited

### Chapter 14
Opener: © The McGraw-Hill Companies, Inc./Jill Braaten, photographer; Figure 14.6b: © Trent Stephens; 14.A: © Kenneth Greer/Visuals Unlimited

### Chapter 15
Opener: © Photodisc/Getty Images RF; Figure 15.2b: © R. T. Hutchings; 15.4b: © CNRI/ Phototake.com; 15.6a-b: © Branislav Vidic; 15.A: © Deni McIntyre/Photo Researchers, Inc.

### Chapter 16
Opener: © SW Productions/ Photodisc/Getty Images RF; Figure16.10c: © Victor Eroschenko; 16.14d: © David M. Phillips/Visuals Unlimited; 16.21b: © CNRI/Science Photo Library/Photo Researchers, Inc. 16.A: © Maxine Cass, 2003

### Chapter 17
Opener & Figure 17.1: U.S. Department of Agriculture; 17.11: © Kyle Rothenborg/Pacific Stock

### Chapter 18
Opener: © Digital Vision RF; Figure 18.3b: © The McGraw-Hill Companies, Inc./Rebecca Gray, photographer/Don Kincaid, dissections.

### Chapter 19
Opener: © Steve Mason/Getty Images RF; Figure 19.Aa-b: © The McGraw-Hill Companies, Inc./Jill Braaten, photographer 19.Ac: © The McGraw-Hill Companies, Inc.; 19. Ad: © The McGraw-Hill Companies, Inc./Jill Braaten, photographer 19.Ae: © The McGraw-Hill Companies, Inc./Bob Coyle, photographer.

### Chapter 20
Opener: © Creatas/PictureQuest RF; Figure 20.8: © John Giannicchi/Photo Researchers, Inc.; 20.13a-b: © Petit Format/ Nestle/Photo Researchers, Inc.; 20.18: © CNRI/Science Photo Library/Photo Researchers, Inc.

# Index

*Note:* Page references followed by the letters *f* and *t* indicate figures and tables, respectively.

A band, 153, 155*f*, 158, 160*f*
Abdomen, 10
  arteries of, 354–355, 355*f*
  quadrants of, 12, 13*f*
  regions of, 12, 13*f*
  transverse section through, 14*f*
  veins of, 359–360, 359*f*
Abdominal aorta, 348–352, 350*f*, 351*f*, 354–355, 355*f*, 356*f*, 498*f*
Abdominal cavity, 15*f*, 16
Abdominal muscles, 152*f*, 175–176
  actions of, 175–176, 176*t*
  in respiration, 416, 417*f*
Abdominal oblique muscle
  external, 168*f*, 169*f*, 176, 176*f*, 176*t*, 178*f*
  internal, 176, 176*f*, 176*t*, 178*f*
Abdominal region, 12*f*
Abdominal wall, muscles of. *See* Abdominal muscles
Abdominopelvic cavity, 15*f*, 16
  mesenteries of, 16–17, 17*f*
  serous membrane of, 16, 17*f*, 88, 89*f*
Abducens (CN VI) nerve, 224
  functions of, 224, 226*t*
  origin of, 226*f*
Abduction, 144*f*, 145
ABO blood group, 303–304, 303*f*, 581
Absorption
  in digestive system, 438, 453, 460–465
  in epithelial tissue, 72
  in small intestine, 453
Absorptive cells, of small intestine, 451, 452*f*
Absorptive state, metabolic, 486–487, 488*f*
Abstinence, 550
AB type blood, 303–304, 303*f*
Acceptable Macronutrient Distribution Range (AMDR), 474, 476, 477
Accessory hemiazygos vein, 358*f*, 359
Accessory (CN XI) nerve, 224
  functions of, 224, 226*t*
  origin of, 226*f*
Accessory pancreatic duct, 457*f*
Accommodation, 249
ACE. *See* Angiotensin-converting enzyme
Acetabulum, 135, 136*f*, 137*f*
Acetoacetate plus acetone, reference laboratory values for, A-2, A-5
Acetone breath, 486
Acetylcholine (ACh)
  in action potential transmission, 158, 158*f*
  as neurotransmitter, 158, 203, 204*t*
Acetylcholine antagonists, 159
Acetylcholinesterase, 158, 159, 203
Acetyl-CoA
  in citric acid cycle, 484*f*, 485
  in lipid metabolism, 486, 487*f*
  in postabsorptive state, 487, 488*f*
  production in aerobic respiration, 484*f*, 485
ACh. *See* Acetylcholine
Achilles tendon, 169*f*, 184*f*, 185, 186*f*
ACHOO (sneeze reflex), 409
Acid(s), 30–31
Acid-base balance
  buffer systems in, 518–519, 522*f*
  kidneys in, 519, 522*f*
  regulation of, 518–523, 522*f*
  respiratory system in, 519, 522*f*
Acidic solutions, 30, 30*f*
Acidosis, 30, 283, 519
  metabolic, 519
  respiratory, 519
Acinar glands, 78, 79*f*
Acinus (pl., acini)
  of exocrine gland, 78, 79*f*
  of pancreas, 457*f*, 458
Acne, 103
Acquired immunity, 397–398, 397*f*
Acquired immunodeficiency syndrome (AIDS), 309*t*, 391, 402*t*, 554*t*, 584*t*

Acromegaly, 274
Acromial region, 13*f*
Acromioclavicular joint, 142*f*
Acromion process, 132, 132*f*, 133*f*, 135*f*, 143*f*, 179*f*
Acrosome, of sperm, 530, 531*f*
ACTH. *See* Adrenocorticotropic hormone
Actin
  in blood clot retraction, 302
  uses in body, 477
Actin myofilaments
  in cardiac muscle, 322–323
  in muscle contraction, 158–159, 160*f*
  in skeletal muscle, 153, 154*f*
  in smooth muscle, 165–166
  structure of, 153, 154*f*
Action potential(s), 199–201, 200*f*, 201*f*
  all-or-none principle of, 201, 201*f*
  of cardiac muscle, 166, 323–326, 324*f*, 325*f*
  conduction of
    continuous, 201, 202*f*
    saltatory, 201, 202*f*
  in defecation, 460
  depolarization phase of, 200–201, 200*f*, 201*f*
    in cardiac muscle, 324, 324*f*
    in skeletal muscle, 156*f*, 157, 324*f*
    in smooth muscle, 166
  in labor (childbirth), 573*f*
  in male sex act, 538
  in myelinated axons, 201, 202*f*
  in nervous tissue, 87
  in neuromuscular junction, 158, 158*f*
  in pain sensation, 240
  plateau phase of, 324, 324*f*
  refractory period of, 325
  repolarization phase of, 200*f*, 201, 201*f*
    in cardiac muscle, 324, 324*f*
    in skeletal muscle, 156*f*, 157, 324*f*
    in smooth muscle, 166
  of skeletal muscle, 153–159, 156*f*, 324*f*
  of smooth muscle, 166
  spatial summation of, 205
  and stomach secretions, 448, 449*f*
  in synapses, 201–203, 203*f*
  temporal summation of, 205
  threshold for, 200*f*, 201, 201*f*
  in unmyelinated axons, 201, 202*f*
  in urine movement, 514, 515*f*
  in vision, 242
Activation energy, 37, 37*f*
Active immunity, 397, 397*f*
  artificial, 397–398, 397*f*
  natural, 397–398, 397*f*
Active transport, 46*t*, 50–51, 50*f*, 461
  energy required for, 46*t*, 50
  secondary, 46*t*, 51, 51*f*
Adam's apple (thyroid cartilage), 408*f*, 409–410, 410*f*
Adaptation, 242
Adaptive immunity, 386, 388–396, 399*f*
  acquisition of, 397–398, 397*f*
  *vs.* innate immunity, 388*t*–389*t*
  lymphocytes in, 389–392
    activation of, 390–392
    antigen recognition by, 390–391
    clones, 389
    costimulation of, 390*f*, 391, 391*f*
    development of, 389, 389*f*
    proliferation of, 390*f*, 391–392, 391*f*
  memory of, 386, 388
  specificity of, 386, 388
Addison disease, 51, 263
Adduction, 144*f*, 145
Adductor longus muscle, 168*f*, 183*f*
Adductor magnus muscle, 169*f*, 183*f*
Adductor muscle, 183*f*, 184*f*, 185
Adenine
  in DNA, 38, 39*f*, 60
  in RNA, 38, 60
Adenohypophysis. *See* Anterior pituitary gland
Adenoid(s), 382
Adenoidectomy, 382

Adenosine diphosphate (ADP)
  ATP breakdown to, 26, 28, 29*f*, 37*f*, 481, 481*f*
  ATP synthesis from, 26, 29, 481, 481*f*
  in platelet release reaction, 300, 301*f*
  structure of, 38, 39*f*
Adenosine triphosphate (ATP), 38
  in active transport, 46*t*, 50, 50*f*
  aging and, 578
  breakdown to ADP, 26, 28, 29*f*, 37*f*
  as energy currency of cells, 38, 481
  in endocytosis, 46*t*, 52
  in exocytosis, 46*t*, 52
  in muscle contraction, 159, 161*f*, 163–164, 163*f*
  production of, 26
    in aerobic respiration, 163–164, 163*f*, 484–486, 484*f*
    in amino acid metabolism, 486
    in anaerobic respiration, 163–164, 163*f*, 483–484, 483*f*, 486
    in biochemical pathway, 482, 482*f*
    in catabolism and anabolism, 481, 481*f*
    in citric acid cycle, 484*f*, 485, 486
    from creatine phosphate, 163–164, 163*f*
    in electron-transport chain, 484*f*, 485–486, 485*f*
    energy required for, 29
    in glycolysis, 483, 483*f*, 484, 484*f*, 486
    in mitochondria, 2, 56–57
  structure of, 38, 39*f*
ADH. *See* Antidiuretic hormone
Adipocytes (adipose cells), 80, 81*t*
Adipose tissue, 80
  energy storage in, 80
  functions of, 80, 81*t*
  location of, 81*t*
  structure of, 80, 81*t*
Adolescent
  definition of, 577
  development of, 577
ADP. *See* Adenosine diphosphate
Adrenal glands, 6*f*, 270*f*, 278–281
  anatomy of, 278, 279*f*
  cortex of
    anatomy of, 278, 279*f*
    hormones of, 271*t*, 279*f*, 280–281, 281*f*
    fetal, in labor (childbirth), 573, 573*f*
    histology of, 279*f*
  medulla of
    anatomy of, 278, 279*f*
    in blood flow control, 367–368
    in blood pressure regulation, 279, 371–373, 371*f*, 373
    in cardiac regulation, 279
    hormones of, 271*t*, 278–279, 279*f*, 280*f*
    prenatal development of, 565*t*
    regulation of, 278–279, 280*f*
    stress and, 279
Adrenaline. *See* Epinephrine
Adrenocortical insufficiency, chronic, 53, 261
Adrenocorticotropic hormone (ACTH), 272–273, 281
  action of, 270*t*, 273–274, 281
  in labor (childbirth), 573, 573*f*
  reference laboratory values for, A-6
  secretion of, 270*t*, 273–274, 281
  target tissue of, 270*t*
Adult, definition of, 577
Adult respiratory distress syndrome (ARDS), 432*t*
Adventitia, of digestive tract, 438, 439*f*
Aerobic respiration, 56–57, 484–486, 484*f*
  acetyl-CoA formation in, 484*f*, 485
  *vs.* anaerobic, 486
  citric acid cycle in, 484*f*, 485, 486
  electron-transport chain in, 484*f*, 485–486, 485*f*
  in muscle contraction, 163–164, 163*f*
  oxygen in, 483–486
Afferent arteriole, 502, 502*f*
Afferent division, of peripheral nervous system, 193–194, 194*f*

Afferent lymphatic vessels, 382, 383*f*
Afferent neurons. *See* Sensory neurons
African Americans, bone mass in, 145
Afterload, cardiac, 332
Age
  clinical, 559
  developmental, 559
Age spots, 108
Agglutination, 303, 304, 304*f*
Aging, 577–578
  effects of
    on balance, 259
    on blood vessels, 345, 373–375
    on cells, 67
    on circulation, 373–375, 578
    on digestive system, 465–467
    on endocrine system, 288
    on hearing, 259
    on heart, 340, 578
    on immunity, 400–403, 578
    on integumentary system, 2, 108
    on joints, 145–148
    on kidneys, 578
    on lymphatic system, 400–403
    on nervous system, 230–234, 577–578
    on olfaction, 258
    on reproductive system, 552–554
    on respiratory system, 433
    on senses, 258–259, 578
    on skeletal muscle, 187–188, 578
    on skeletal system, 145–148
    on taste, 259
    on tissues, 92
    on vision, 259
  free radical theory of, 578
Agonist, 167
Agranulocytes, 298–299
  function of, 294*t*
  production of, 295*f*
  structure of, 294*t*
AIDS, 309*t*, 391, 402*t*, 554*t*, 584*t*
Airflow
  establishing, 411
  pressure changes and, 417, 418*f*
  principles of, 417
Albinism, 98, 580–581, 581*f*
Albumin
  hepatic synthesis of, 457
  in plasma, 292, 293*f*, 293*t*
  reference laboratory values for, A-5
Alcohol
  as diuretic, 510
  fetal effects of, 584*t*
  and heart disease, 337
  reference laboratory values for, A-2
Aldosterone
  action of, 271*t*, 280
  and blood pressure, 280, 369, 372*f*, 373
  deficiency of, 51
  and potassium concentration, 517, 518*f*
  reference laboratory values for, A-6
  secretion of, 271*t*, 280, 281*f*
  and sodium concentration, 517
  target tissues of, 271*t*, 280
  and urine regulation, 510, 510*f*, 512*f*
Alendronate, 147
Alkaline solutions, 30, 30*f*
Alkalosis, 30, 519, 523
  metabolic, 523
  respiratory, 523
Alleles, 580–582
Allergic reactions, 389
  delayed, 402*t*
  immediate, 402*t*
All-or-none principle, of action potentials, 201, 201*f*
Alpha cells, of pancreatic islets, 281, 283*f*, 285
Alpha linolenic acid, 477
Alpha waves, 220–221, 220*f*
Altitude
  and tympanic membrane, 252
  and ventilation, 429

Alveolar ducts, 413–414, 414f
Alveolar fluid, 415f
Alveolar glands, 78, 79f
Alveolar pressure, 417, 418f
Alveolar sacs, 413, 414f
Alveolar volume, changing, 419–420
Alveoli
    breast, 545, 545f
    dental, 442, 442f
    exocrine gland, 78, 79f
    lung, 413–414, 414f, 415f
        aging and, 433
        elasticity of, 414
        epithelial tissue of, 72–73, 76, 415f
        gas exchange in, 421–424, 423f
        wall of, 415f
Alzheimer disease, 231t
Ameboid movement, of white blood cells, 298
Amenorrhea, 548
    lactation, and birth control, 550
    primary, 548
    secondary, 548, 549
American Heart Association (AHA), 300
Amine group, in amino acids, 35–37, 36f
Amino acid(s), 35–37
    as energy source, 486
    essential, 477
    hormones, 265
    hydrogen bonds of, 36f, 37
    metabolism of, 486, 487f
        in absorptive state, 487, 488f
        in postabsorptive state, 487, 488f
    nonessential, 477
    protein breakdown to, 460, 460f, 461f,
        464–465
    structure of, 36f, 37
    transport of, 464–465, 465f
Ammonia
    conversion to urea, 457, 486, 487f, 503
    reference laboratory values for, A-2, A-5
Ammonium ions, 23t
Amniotic cavity, 564, 564f
Amniotic fluid, 564, 568
Amniotic sac, 564, 564f, 573
Amphetamines, action of, 203
Amphiarthrosis, 140
Ampulla(e)
    of ductus deferens, 532, 533f
    of semicircular canals, 255–256, 257f, 259
    of uterine tube, 540f, 543
Amygdaloid body (amygdala), 221f
Amylase
    pancreatic, 443t, 458, 461f, 462, 576
    reference laboratory values for, A-2
    salivary, 443, 443t, 461–462, 461f, 576
Anabolic steroids, 537
Anabolism, 481, 481f, 491
Anaerobic respiration, 429–430, 483–484,
        483f, 486
    vs. aerobic, 486
    in muscle contraction, 163–164, 163f
Anaerobic threshold, 429–430
Anal canal, 459–460, 459f
Anal sphincter
    external, 177f, 177t, 459f, 460
    internal, 459–460, 459f
Anaphase
    of meiosis, 528, 529f
    of mitosis, 63, 64f
Anaphylactic shock, 369
Anaphylaxis, 402t
Anatomical imaging, 2
Anatomy
    definition of, 2
    regional, 2
    surface, 2
    systemic, 2
    terminology of, 10–17
Androgen(s), 281
    action of, 271t, 281
    secretion of, 271t, 281
    synthetic, 537
    target tissues of, 271t
Anemia, 306, 308, 432t
    aplastic, 308
    hemolytic, 308
    hemorrhagic, 308
    iron-deficiency, 104, 308
    pernicious, 307–308, 308
    sickle-cell, 308, 581
Anencephaly, 565
Anesthesia
    general, 240
    local, 240
    spinal, 222

Anesthesia (disorder), 231t
Aneurysms, arterial, 352
Angina pectoris, 322
Angiogram, 338, 338f
Angioplasty, 322
Angiotensin, 280, 281f
    and blood pressure, 280, 372f, 373
    and urine regulation, 510, 510f
Angiotensin-converting enzyme (ACE), 280,
        281f, 372f, 373
Angiotensin I, 280, 281f, 372f, 373, 510, 510f
Angiotensin II, 280, 281f, 372f, 373, 510, 510f
Angiotensinogen, 280, 281f
ANH. See Atrial natriuretic hormone
Ankle (ankle joint), 12, 12f
    bones of, 138, 139f
    as hinge joint, 142f
    movements of, muscles of, 185, 186f, 187t
Anorexia, and amenorrhea, 548
ANS. See Autonomic nervous system
Antacids, 450
Antagonist muscles, 167
Antebrachial region, 12f
Antecubital region, 12f
Anterior, 10, 11f, 11t
Anterior chamber, of eye, 245f, 246–248
Anterior commissure, 221f
Anterior communicating artery, 354f
Anterior corticospinal tract, 217, 217f, 217t
Anterior fontanel, 140f
Anterior horn, 206, 207f
Anterior interventricular artery, 321, 321f
Anterior interventricular sulcus, 314,
        316f–317f
Anterior nucleus of thalamus, 221f
Anterior pituitary gland, 270
    hormones of, 270t, 272–275, 272f
    hypothalamic regulation of, 271, 272f, 273f
    in labor (childbirth), 573, 573f
    prenatal development of, 565t, 567, 570f
Anterior superior iliac spine, 135, 136f
Antibodies
    in adaptive immunity, 389, 392–395
    aging and, 401
    in blood grouping, 303–306
    classes of, 392, 393t
    constant region of, 392, 392f
    effects of, 392–393, 394f
    monoclonal, 393, 398–400
    primary response of, 393–395, 395f
    production of, 393–395, 395f
    secondary response of, 395, 395f
    structure of, 392, 392f
    variable region of, 392, 392f
Antibody-mediated immunity, 389, 392–395,
        399f
Anticoagulants, 302, 336
Anticodons, 61, 62f
Antidiuretic hormone (ADH), 275
    action of, 268t, 270t, 275
    and blood pressure, 275, 369, 373, 373f, 374f
    diuretics and, 510
    and extracellular fluid regulation, 517
    inappropriate secretion of, 275
    secretion of, 268t, 270t, 275, 511
    target tissue of, 268t, 270t
    and urine regulation, 505f, 511, 511f, 512f
Antigen(s)
    in adaptive immunity, 388–389
    antibody binding to, 392–393, 394f
    in blood grouping, 303–306
    foreign, 388–389
    inactivation of, 394f
    lymphocyte recognition of, 390–391
    self-, 388–389
Antigen receptors, 390
Antihistamines, 89
Antihypertensive agents, 336
Anti-inflammatory agents, 89
Antioxidants, 479, 578
Antiserum, 398
Antithrombin, 302
Antrum, 541, 541f, 542f
Anus, 5f, 438, 438f, 459
    epithelial tissue of, 76, 77t, 438
    female, 544f
    male, 530f
Aorta, 6f, 175f, 314–316, 317f, 318f, 319f,
        348, 349f
    abdominal, 348–352, 350f, 351f, 354–355,
        355f, 356f, 498f
    aging and, 340, 373
    ascending, 318f, 348, 351f, 353f
    blood pressure in, 363f
    branches of, 348–352, 351f

descending, 348–352, 351f, 353f, 355f
    thoracic, 348, 350f, 351f, 353f, 355f
Aortic arch, 4f, 315f, 316f, 318f, 320f, 321f,
        348, 350f, 351f, 353f
Aortic bodies
    and blood pressure, 369
    and respiration, 427f, 428–429
Aortic semilunar valve, 318f, 319, 319f,
        320f, 321f
    incompetent, 340
    location of, 331f
    stenosis of, 340
Apex, of heart, 313–314, 315f, 317f
Aphasia, 220
Aplastic anemia, 308
Apocrine sweat glands, 101, 101f
Aponeurosis, 166
Apoptosis, 66
Appendectomy, 459
Appendicitis, 16, 458–459
Appendicular skeleton, 121t, 122, 123f, 131–140
Appendix, 5f, 438f, 451f, 459f
Appositional growth, 116
Aqueous humor, 246–248
Arachidonic acid, 476
Arachnoid granulations, 222–224, 225f
Arachnoid mater
    of brain, 222, 223f, 225f
    of spinal cord, 222, 223f
Arches, of foot, 140
Arcuate arteries, 502, 502f
Areola, 544, 545f
Arm. See also Upper limb
    bones of, 132–134, 133f
    circumduction of, 144f, 145
    definition of, 12, 132
    movements, muscles of, 177–179, 179f, 180t
    prenatal development of, 566, 566f
    rotation of, 144f, 145
Arrector pili muscle, 97f, 100f, 101, 101f
Arrhythmias, cardiac, 327t
Arterial aneurysm, 352
Arterial circle, cerebral, 352, 353f, 354f
Arteries. See also specific arteries
    of abdomen, 354–355, 355f
    aging of, 92, 373–375, 578
    blood flow in, 345–347
    blood pressure in, 363–364, 363f, 368–373
    distributing, 345
    elastic, 345, 346f, 373
    of head, 350f, 352–354, 352f, 353f, 354f
    of kidneys, 502, 502f
    of lower limb, 350f, 356, 356f
    major, 350f
    medium, 345, 347f
    muscular, 345–347, 346f, 364
    of neck, 352–354, 352f
    of pelvis, 355f, 356
    photomicrograph of, 347f
    in pulmonary circulation, 348, 349f
    small, 345, 347
    structure of, 345–347, 346
    in systemic circulation, 348–356
    of trunk, 350f
    of upper limb, 350f, 353–354, 354f
Arterioles, 345, 347
    afferent, 502, 502f
    blood pressure in, 363f, 364
    efferent, 502, 502f
    maternal and fetal, 562f
Arteriosclerosis, 364, 578
Arthritis, 148, 148t
Articular cartilage, 113, 114f, 141, 141f, 143f
Articular facets, 129
    of cervical vertebrae, 129, 129f
    inferior, 129–130
    of lumbar vertebrae, 129f, 130
    superior, 129–130, 129f
    of thoracic vertebrae, 129, 129f
Articular processes
    inferior, 129
    superior, 128f, 129
Articulation. See Joint(s)
Artificial hearts, 337
Artificial immunity
    active, 397–398, 397f
    passive, 397–398, 397f
Artificial insemination, 531
Artificial pacemaker, 336
Arytenoid cartilage, 410, 410f
Ascending aorta, 318f, 348, 351f, 353f
Ascending colon, 451f, 459, 459f
Ascending lumbar vein, 358f, 359, 359f
Ascending tracts, 206, 215–216, 215f, 215t
Ascorbic acid. See Vitamin C

Aspirin
    as anti-inflammatory, 89
    and asthma, 430–431
    in myocardial infarction, 300, 322
    and platelet activation, 300
Association areas, 216, 216f
Asthma, 402t, 430–431
Asthma attack, 413, 430–431
Astigmatism, 258t
Astrocytes, 195
    functions of, 195, 196t
    structure of, 195, 196t, 197f
Atherosclerosis, 92, 364, 375, 578
Atherosclerotic plaque, 322, 364, 375, 375f, 578
Athletes
    female, amenorrhea in, 548
    pulmonary volumes and capacities in, 421
    respiratory adaptations in, 430–433
Atlantoaxial joint, medial, 142f
Atlantooccipital joint, 142f
Atlas (vertebra), 128f, 129, 129f
Atmospheric pressure, 417
Atom(s), 21–22
    clinical applications of, 27
    definition of, 21
    model of, 22, 22f
Atomic mass, 22
Atomic number, 21t, 22
Atomic structure, 22, 22f
ATP. See Adenosine triphosphate
Atria, 314, 316–317
    blood flow through, 320–321, 320f
    contraction of, 316, 320–321, 328, 329f
    function of, 316, 328
    left, 314, 315f, 316–317, 316f–317f, 318f
    prenatal development of, 567, 571f
    relaxation of, 328, 329f
    right, 314, 315f, 316–317, 316f–317f, 318f
Atrial contractions, premature, 327t
Atrial diastole, 328
Atrial fibrillation, 327t
Atrial flutter, 327t
Atrial natriuretic hormone (ANH)
    and blood pressure, 373, 374f
    and extracellular fluid regulation, 517
    and urine regulation, 511, 511f, 512f
Atrial systole, 328
Atrial tachycardia, paroxysmal, 327t, 332
Atrioventricular (AV) bundle, 325, 325f
Atrioventricular (AV) node, 325, 325f
Atrioventricular (AV) node block, 327t
    complete, 327t
    first-degree, 327t
    second-degree, 327t
    third-degree, 327t
Atrioventricular (AV) valves
    anatomy of, 318–319, 318f
    in cardiac cycle, 328, 329f, 330f
    function of, 319, 319f
    sounds of, 328, 330f, 331
Atrium. See Atria
A type blood, 303–304, 303f
Auditory association area, 216f
Auditory canal
    external, 122, 124f, 127f, 251, 252f, 254f
    internal, 126f
Auditory cortex, primary, 216f
Auditory ossicles, 121t, 122, 252, 252f, 254
Auditory tube, 252, 252f, 254
    elastic cartilage in, 84, 84t
    epithelial tissue of, 75t, 76
    prenatal development of, 567, 570f
Auricle, 251, 252f
Auricular artery, posterior, 352f
Auscultatory method, of blood pressure
        measurement, 362, 362f
Autocrine chemical signals, 263, 264t, 265
Autoimmune disease, 389, 402t, 403
Autonomic ganglion, 227, 227f
Autonomic nervous system (ANS), 194, 194f,
        227–230. See also Parasympathetic
        nervous system; Sympathetic nervous
        system
    anatomy of, 227–229
    divisions of, 227, 227t, 228f
    dysfunctions of, 230
    functions of, 229, 230t
    innervation by, 227, 228f
    neurotransmitters in, 229
    and salivary gland, 444
Autorhythmicity
    of cardiac muscle, 166, 166t
    of smooth muscle, 166, 166t
Autosomal chromosomes, 579, 579f
Autosomes, 62–63

Avastin (bevacizumab), 65
AV node. See Atrioventricular node
Avogadro number, A-7
AV valves. See Atrioventricular valves
Axial skeleton, 121t, 122–131
Axillary artery, 350f, 353, 354f, 364f
Axillary lymph node, 5f, 381f, 382
Axillary nerve, 210
Axillary region, 12f
Axillary vein, 356, 357f, 358f
Axis (vertebra), 128f, 129, 129f
Axon(s), 194–195
    collateral, 195, 195f
    functions of, 87
    myelinated, 195, 198f
        action potentials in, 201, 202f
    structure of, 87, 88t, 194–195, 195f
    unmyelinated, 195, 198f
        action potentials in, 201, 202f
Axon hillock, 194, 195f
Azygos vein, 356, 358f, 359

Baby teeth, 440
Backbone. See Vertebral column
Back muscles, 172–173, 174f, 174t
Back pain, low, 172–173
Bacterial count, in urine, A-5
Balance, 28, 251, 255–257
    aging and, 259
    dynamic, 255
    neuronal pathways for, 256–257
    static, 255
Baldness, 101, 538
Ball-and-socket joints, 142f, 143
Barbers, bloodletting by, 8
Barbiturates, A-2
Bariatric surgery, 490
Baroreceptor(s), 333, 368, 511, 512f
Baroreceptor reflexes
    in blood pressure regulation, 333, 334f,
        368–369, 368f, 370f, 373
    in cardiac regulation, 333, 334f, 337f
    in circulatory shock, 369
Barrier methods, of birth control, 550
Basal cell carcinoma, 105, 105f
Basal metabolic rate (BMR), 490
Basal nuclei, 218
    disorders of, 218
    functions of, 218
    structure of, 218, 219f
Basal surface, of epithelial tissue, 72
Base (chemical), 30–31
Base (nucleic acid), 38, 39f, 60, 61f
Base, of heart, 313–314, 315f
Basement membrane, 72
    of blood vessel, 346f
    epithelial tissue and, 72, 73f
        pseudostratified columnar, 75t
        simple columnar, 75t
        simple cuboidal, 74t
        simple squamous, 74t
        stratified squamous, 76, 77t
        transitional, 77t
Basic solutions, 30, 30f
Basilar artery, 353, 353f, 354f
Basilar membrane, 253
    anatomy of, 253, 253f
    auditory function of, 254–255, 254f
Basilic veins, 356, 357f, 358f
Basophil(s), 298–299
    function of, 299
    in inflammatory response, 299, 387
    percentage of WBCs as, 293f, 307
    production of, 295f
    structure of, 299
B cell(s), 384, 385f
    activation of, 390–392
    in adaptive immunity, 389–392
    aging and, 401
    in antibody-mediated immunity, 393–395
    antigen recognition by, 390–391
    costimulation of, 391, 391f
    development of, 389, 389f
    memory, 395
    origin of, 389, 389f
    proliferation of, 391–392, 391f
B-cell receptors, 390
Bedsores, 108t
Behavioral methods, of birth control, 550–551
Belly, of muscle, 167
Benign prostatic hyperplasia (BPH), 552
Benign tumors, 65
Beta-adrenergic blocking agents, 336
Beta carotene, 477, 479, 578
Beta cells, of pancreatic islets, 281, 283, 283f

Beta waves, 220f, 221
Bevacizumab, 65
Bicarbonate, 23t
    and acid-base balance, 519, 522f
    in blood, regulation of, 497
    carbon dioxide transport as, 296–297,
        424, 425f
    functions of, 443t
    pancreatic secretion of, 458
    renal reabsorption of, 506
    in saliva, 443
    stomach secretion of, 443t
    in urine, 503t
Biceps brachii muscle, 5f, 143f, 152f, 168f,
    178f, 179–180, 179f, 182f
    action of, 167, 180, 180t
    belly of, 167f
    heads of, 143f
    insertion of, 167f, 180t
    nomenclature for, 167
    origins of, 167, 167f, 180t
Biceps brachii tendon, 143f
Biceps femoris muscle, 169f, 183f
Biceps femoris tendon, 184f
Bicuspid valve, 319, 319f
    in blood flow, 320f, 321
    incompetent, 331
    location of, 331f
    stenosis of, 331
Bile, 455
    composition of, 455
    flow through liver, 455, 455f
    functions of, 455
    production of, 455
    reference laboratory values for, A-5
    secretion of
        daily amount of, 455
        regulation of, 456, 456f
Bile canaliculus, 454f, 455
Bile ducts
    common, 451, 455, 455f, 457f
    epithelial tissue of, 75t
Bile salts, 455
    functions of, 443t, 455–456, 456f, 461f, 463,
        463f, 476
    secretion of, 443t
    structure of, 35f
    synthesis of, 35, 476
Bilirubin, 298, 298f, 309, 455, A-2, A-5
Biochemical pathways, in metabolism,
    482–483, 482f
Biofeedback, 229, 282
Biotin
    deficiency of, 478t
    dietary guidelines for, 478t
    dietary sources of, 478t
    functions of, 477, 478t
    solubility of, 478t
Bipolar cells, of retina, 246, 247f
Bipolar neurons, 195, 196f, 196t
Birth, 572–574, 572f
    anesthetics injections in, 130
    first year following, 577
    hormones in, 573–574, 573f
    positive feedback during, 10
Birth control, 550–552, 551f
Birth control pills, 266, 550–552, 551f
Birthmarks, 100
Bismuth subsalicylate, 467
Bisphosphonates, 147
Bitter taste, 242
Bladder. See Gallbladder; Urinary bladder
Blastocele, 561, 561f, 564f
Blastocyst
    formation of, 548, 561, 561f
    implantation of, 548, 561–563, 561f
Blindness, 258t
    color, 251, 251f, 258t
    night, 246
Blind spot, 246, 248f
Blinking, 243
Blood, 291–309
    carbon dioxide transport in, 292, 296–297,
        424, 425f
    chemistry of, 308–309
    circulation of. See Circulation
    clotting of. See Blood clot
    color of, 296
    composition of, 292, 293f
    as connective tissue, 84, 85t, 292
    cultural beliefs about, 292
    diseases and disorders of, 309t
    formed elements of, 292, 293f, 294–299,
        294t, 295f
    functions of, 80, 292

liquid matrix of, 292, 293f. See also Plasma
    osmolality of, 292, A-3
    oxygen transport in, 292, 296–297, 424
    pH of, 292, A-3
        carbon dioxide and, 424, 427–428, 428f
        regulation of, 407, 428f
        and ventilation, 427–428, 428f
    protection by, 292
    reference laboratory values for, A-2–A-4
    solutes in, regulation of, 497
    temperature regulation by, 292
Blood–brain barrier, 195
Blood cells. See also specific types
    production of, 112, 294, 295f, 297, 297f,
        389, 389f
Blood clot, 301–302
    control of, 302–303
    definition of, 301
    dissolution of, 302, 303f
    fatty acids and, 476
    formation of, 292, 301–302, 301f
    reference laboratory values for, A-4
    retraction of, 302
    unwanted, dangers of, 302, 578
    vitamin K and, 477
Blood flow
    adequate, maintenance of, 361–362
    aging and, 578
    atherosclerosis and, 375
    control of, 366–368
    hormonal control of, 367–368, 367t
    local control of, 366, 366f
    nervous control of, 367–368, 367f, 367t
    in newborn, 574–575, 574f, 575f
    one-way, 313
    resistance to, 363–364, 368, 375
    in skin, and skin color, 99–100
    through heart, 313, 320–321, 320f
    through kidneys, 502, 502f, 503
    through liver, 454f, 455
    through pulmonary circulation, 349f
Blood grouping, 303–306
    ABO, 581
    Rh, 305–306, 305f
Blood islands, 567
Bloodletting, 8
Blood loss, prevention of, 299–302
Blood poisoning, 309t, 369
Blood pressure, 362–366
    arterial, 363–364, 363f, 368–373
    and capillary exchange, 365–366, 365f
    definition of, 362
    diastolic, 362–363, 362f
    elevated (hypertension), 336, 363, 584t
    generation of, 313
    guidelines for, 363
    measurement of
        auscultatory method of, 362, 362f
        brachial artery, 353–354, 362, 362f
        units for, 362
    normal, 363
    regulation of
        adrenal medullary mechanism in, 279,
            371–373, 371f, 373
        aldosterone and, 280, 369
        antidiuretic hormone mechanism in,
            275, 373, 373f, 374f
        atrial natriuretic mechanism in, 373, 374f
        autonomic, 229
        baroreceptor reflexes in, 333, 334f,
            368–369, 368f, 370f, 373
        brainstem in, 211
        chemoreceptor reflexes in, 371f, 373
        circulation and, 345
        endocrine, 263
        hormonal mechanisms of, 371–373, 374f
        long-term, 374f
        negative feedback and, 7–8, 9f
        positive feedback and, 8, 10f
        renal, 497, 510
        renin-angiotensin-aldosterone
            mechanism in, 280, 372f, 373, 374f
    resistance and, 363–364, 368
    systolic, 362–363, 362f
Blood smear, 299, 299f
Blood supply. See also specific organs and vessels
    to brain, 349f
    to heart, 314–316, 321–322, 321f
    regulation by heart, 313
Blood tests, diagnostic, 306–309
Blood transfusion
    blood grouping and, 303–306
    definition of, 303
    typing and crossmatching for, 306
Blood urea nitrogen (BUN), 308–309

Blood vessels. See also Arteries; Capillaries;
        Veins
    aging and, 345, 373–375, 578
    and body temperature, 491
    constriction of, 152, 491
    dilation of, 491
    prenatal development of, 565t, 567, 571f
    pulmonary, 345, 348, 349f
    resistance in, and blood pressure, 363–364
    structure of, 345–348, 346f
    systemic, 345, 348–361
Blood volume
    reference laboratory values for, A-2
    renal regulation of, 497
Blushing, 99
Body(ies)
    of epididymis, 533f
    of femur, 138f
    of pancreas, 457f
    of sternum, 130–131, 131f
    of stomach, 446, 447f
    of uterus, 540f, 543
    of vertebrae, 128, 128f, 129
Bodybuilding, 167, 170f
Body cavities, 15–16, 15f
Body fluids
    compartments of, 514–516
    exchange of, 516
    volumes of, 515t
Body mass index (BMI), 489
Body organization, 2–6
Body parts, names of, 10–12, 12f–13f
Body plan, 10–17
Body planes, 12–15, 14f
Body positions, 10, 11f
Body regions, names of, 10–12, 12f–13f
Body snatchers, 4
Body temperature
    and cardiac function, 340
    disorders of, 493
    normal, 102, 491
    oral vs. rectal, 491
    regulation of, 7, 7f, 491–492, 491f, 492f
        blood in, 292, 491
        hypothalamus in, 491–492, 492f
        importance of, 102
        integumentary system in, 96, 102, 103f
        muscles in, 152, 159–160
    set point for, 7, 7f
    shivering and, 160, 492
    testicular, 529–530
    water and, 31–32
Bolus, 444, 445f
Bond(s)
    carbon, 32
    chemical, 22–24
    covalent, 24, 24f
    hydrogen, 24–25, 25f
    ionic, 22–23, 23f
    peptide, 61, 62f
Bone(s). See also specific bones
    aging and, 92, 145–148
    calcium homeostasis and, 119–121, 120f
    cells of, 65f. See also specific types
    as connective tissue, 84
    diseases and disorders of, 148t
    extracellular matrix of, 112–113
    features of, terms for, 121–122, 122t
    fractures
        aging and, 145
        classification of, 120, 120f
        healing of, 119, 119f
        immobilization of, 119
        in osteogenesis imperfecta, 113
        osteoporosis and, 119
    functions of, 80, 85t, 112
    general considerations of, 121–122
    general features of, 113–119
    growth of, 116–117
        appositional, 116
        at epiphyseal plate, 116–117, 118f
        in length, 116–117, 118f
        vitamin D and, 102
        in width, 116
    histology of, 113–115, 115f
    mineral storage in, 112
    muscle attachment to, 166, 167f
    number of, 121, 121t
    ossification of, 115–116
    prenatal development of, 565t
    remodeling of, 117–119
    repair of, 119, 119f
    shapes of, 113
    structure of, 84, 85t, 113, 114f
    testosterone and, 538t

Bone collar, 116, 117f
Bone density, 549
Bone marrow
  in children vs. adults, 113
  in long bones, 113, 114f
  red, 112–113, 114f, 297, 297f, 389, 389f
  yellow, 113, 114f
Bone marrow transplant, 296
Bone mass
  age and, 145
  race and, 145
Bone matrix
  composition of, 112–113
  production of
    in endochondral ossification, 115–116, 117f
    in intramembranous ossification, 115, 116f
Bony labyrinth, 252, 253f
Botulism, 203
Bowman's capsule, 498, 500f, 502f, 503–504, 504f
Brachial arteries, 6f, 350f, 353–354, 354f
  in blood pressure measurement, 353–354, 362, 362f
  in pulse monitoring, 364f
Brachialis muscle, 179–180, 179f, 182f
  action of, 167, 180, 180t
  insertion of, 180t
  origin of, 180t
Brachial plexus, 208, 210, 210f, 210t
Brachial region, 12f
Brachial veins, 356, 358f
Brachiocephalic arteries, 348, 350f, 351f, 352, 352f, 353f, 354f
Brachiocephalic veins, 356
  left, 356, 357f, 358f
  right, 356, 358f
Brachioradialis muscle, 168f, 179f, 180, 181f, 182f
  action of, 180, 181t
  insertion of, 181t
  nomenclature for, 167
  origin of, 181t
Bradycardia, 327t
Brain, 4f, 6f, 193, 193f, 211–215
  blood supply to, 349f
  epithelial tissue of, 74t, 75t
  functions of, 193, 215–222
    motor, 216–219
    sensory, 215–216
  infant, 577
  in memory, 221
  meninges of, 222, 223f
  nervous tissue of, 87, 88t
  prenatal development of, 565t
  in speech, 220
  stem cells in, 90–91
  structure of, 211–215, 211f, 214f
  ventricles of, 222, 224f
    cerebrospinal fluid flow in, 222–224, 225f
    fourth, 222, 224f, 225f
    lateral, 222, 224f, 225f
    third, 222, 224f, 225f
Braincase, 121t, 122, 206
Brain death, 578–579
Brainstem, 14f, 211–213
  components of, 211
  functions of, 211
  respiratory areas in, 211, 424, 426f
  structure of, 211, 211f, 212f
Brain waves, 220–221, 220f
Breast(s)
  aging and, 554
  anatomy of, 544–545, 545f
  cancer of, 544, 549, 554
  imaging of, 27, 544, 554
  lactating, 545, 545f, 576–577, 576f
  male, 545
  nonlactating, 545f
  prepubescent, 544
  puberty and, 545
Breastbone. See Sternum
Breast-feeding, 576–577
  and birth control, 550
  breast changes in, 545, 545f
  endocrine regulation of, 263, 274, 275
Breath-holding, 426
Breathing, 407, 416–424. See also Ventilation
Brittle bone disease, 113
Broad ligament, 540f, 543
Broca area, 216f, 220
Bronchi. See Bronchus
Bronchial artery, 414f
Bronchial nerve, 414f
Bronchial vein, 414f
Bronchioles, 412f, 413–414, 414f
  epithelial tissue of, 413–414

respiratory, 413, 414f
  terminal, 412f, 413–414, 414f
Bronchitis, 432t, 433
Bronchopulmonary segments, 413
Bronchus (pl., bronchi), 5f, 407f, 412
  disorders of, 432t
  epithelial tissue of, 75t, 76, 413
  lobar or secondary, 412f, 413, 413f
  main or primary, 412, 412f, 413f, 416f
  segmental or tertiary, 412f, 413, 413f
B type blood, 303–304, 303f
Bubonic plague, 402t
Buccal region, 12f
Buccinator muscle, 167, 171f, 172t, 440, 442f
Buffers, 31, 31f, 518–519, 522f
Buffy coat, 306, 307f
Bulb, of hair, 100, 100f
Bulbospongiosus muscle, 177f, 177t
Bulbourethral glands, 528, 530f, 533f, 534, 535
Bulk, in diet, 473, 474
Bundle branch
  left, 325, 325f
  right, 325, 325f
Bunions, 148t
Burke, William, 4
Burns, 104–105
  artificial graft for, 105
  classification of, 104–105, 104f
  complications of, 106–107
  debridement of, 107
  depth of, 104, 104f
  first-degree, 104f, 105
  fluid therapy for, 106
  full-thickness, 104f, 105, 106, 106f
  healing of, 105
  immune effects of, 107
  metabolic effects of, 106–107
  partial-thickness, 104–105, 104f, 106, 106f
  pathology of, 106–107
  second-degree, 104f, 105
  skin graft for, 105, 107
  systemic effects of, 106–107, 107t
  third-degree, 104f, 105
Burn unit, 106f
Bursa, 141, 141f
Bursitis, 141, 148t

Cadavers, 4
Caffeine, as diuretic, 510
Calcaneal region, 13f
Calcaneal (Achilles) tendon, 169f, 184f, 185, 186f
Calcaneus (heel), 138, 139f, 140f
Calcitonin, 276
  action of, 271t, 276
  and calcium levels, 120f, 121, 276, 517
  reference laboratory values for, A-6
  secretion of, 271t, 276
  target tissues of, 271t
Calcium
  in action potential, 158
  and aging, 148
  atomic number of, 21t
  in bone matrix, 113
  calcitonin and, 517
  and cardiac function, 333
  in coagulation, 301f, 302
  deficiency of, 480t
  dietary guidelines for, 480t
  digestion of, 465
  functions of, 480t
  mass number of, 21t
  and osteoporosis, 146–147
  parathyroid hormone and, 120f, 121, 517
  percent in body, 21t
  reference laboratory values for, A-2, A-5
  regulation of, 119–121, 120f, 276, 277–278, 278f, 517
  storage in bones, 112, 119
  symbol for, 21t, 22
  vitamin D and, 99, 102, 120f, 121, 517
Calcium channel(s)
  in action potentials, 324–325, 324f
  in synaptic transmission, 201–203, 203f
Calcium channel blockers, 336
Calcium citrate, for osteoporosis, 147
Calcium homeostasis, 119–121, 120f, 276, 277–278, 278f, 517
Calcium ions, 23t
  in extracellular fluid, 517
  in intracellular fluid, 516
  regulation of, 497, 517
Calendar method, of birth control, 550
Callus
  in bone repair, 119, 119f
  in skin, 96

Calorie (cal), 474
Calyx (pl., calyces), 498, 499f
cAMP (cyclic adenosine monophosphate), 267, 267f
Canal(s). See also specific canals
  definition of, 121–122, 122t
Canaliculus (pl., canaliculi)
  bile, 454f, 455
  in bone matrix, 113, 115f
  in cancellous bone, 115, 116f
  in compact bone, 113, 115f
  in lacrimal apparatus, 244, 244f
Cancellous bone, 113, 115
  in bone repair, 119, 119f
  structure of, 114f, 115, 116f
Cancer, 65
  aging and, 402
  breast, 544, 549, 554
  cervical, 543, 554
  colorectal, 467, 549
  definition of, 65
  digestive system, 467
  epithelial damage in, 78
  immunotherapy for, 65
  lung, 78, 416, 432t
  penile, 534
  prostate, 534, 552
  radiation therapy for, 27
  skin, 99, 105, 105f
  treatment of, 65
  uterine, 549
Cancer therapy
  immunotherapy in, 398–400
  lymphedema with, 402t
  stem cells in, 296
Canine teeth, 440, 441f, 444
Capacitation, of sperm cells, 532, 559
Capacity, measures of, A-1
Capillaries
  aging and, 375
  blood flow in, 345, 347, 366, 366f
  blood pressure in, 363–364, 363f
  glomerular, 498, 502
  lymphatic, 380, 381f, 384, 385f
  peritubular, 502, 502f, 508–509
  pulmonary, 414f, 421–424, 423f
  structure of, 345, 347, 347f
Capillary exchange, 365–366, 365f
Capillary network, 347, 347f
Capitate bone, 134, 135f
Capitulum, 133f
Capsular pressure (CP), 503–504, 504f
Capsule
  of lymph node, 382, 383f
  of spleen, 383, 383f
  of thymus, 384, 384f
Carbohydrate(s)
  in absorptive state, 487, 488f
  in cell membrane, 44–45, 45f
  chemistry of, 32–33, 32t
  complex, 474
  Daily Value for, 479
  dietary amounts, recommended, 474
  dietary sources of, 474
  digestion of, 443, 460, 460f, 461–462, 461f, 473
  metabolism of, 483–486
  as nutrients, 473, 474
  uses in body, 32t, 474
Carbohydrate disorders, 56
Carbon
  atomic number of, 21t
  atomic structure of, 22, 22f
  mass number of, 21t, 22
  percent in body, 21t
  symbol for, 21t, 22
Carbon bonds, 32
Carbon dioxide
  and acid-base balance, 519, 522f
  in aerobic respiration, 484–486, 485f
  and blood pH, 424, 427, 428f
  and cardiac function, 333, 335f
  chemistry of, 31
  in citric acid cycle, 484f, 485
  exchange in lungs, 407
  exercise and levels of, 429–430
  partial pressure of, 422–424, 422t, A-3
  in plasma, 292, 293t
  reference laboratory values for, A-2
  respiratory exchange of, 407, 415f, 422–424, 423f
  transport of
    bicarbonate, 296–297, 424, 425f
    in blood, 292, 296–297, 407, 424, 425f
    in ventilation regulation, 427, 428f

Carbonic anhydrase, in carbon dioxide transport, 297, 424, 425f
Carbon monoxide
  effect on oxygen transport, 296, 432t
  reference laboratory values for, A-2
Carboxyl group
  in amino acids, 35–37, 36f
  in fatty acids, 33
Carboxypeptidase, 458
  functions of, 443t, 458, 461f, 464
  secretion of, 443t, 458
Cardiac arrhythmias, 327t
Cardiac cycle, 328, 329f
  definition of, 328
  ECG in, 328, 330f
  events of, 328, 330f
  pressure graph in, 328, 330f
  sound graph in, 328, 330f
  volume graph in, 328, 330f
Cardiac muscle, 87, 165–166, 322–324
  action potentials of, 166, 323–326, 324f, 325f
  aging and, 340, 578
  autorhythmicity of, 166, 166t
  cells of, 323, 323f
  contraction of, 323–324
  functions of, 86t, 87, 152
  intercalated disks of, 166, 323, 323f
  location of, 86t, 87
  overview of, 152f
  reduced blood flow to, 339t
  refractory period of, 325
  striations of, 86t, 87, 323, 323f
  structure of, 86t, 87, 166, 166t, 322–323, 323f
Cardiac opening, 446, 447f
Cardiac output (CO), 332
  aging and, 340
  and blood pressure, 368
  exercise and, 332
Cardiac region, of stomach, 446, 447f
Cardiac skeleton, 319–320, 319f
Cardiac sphincter, 444
Cardiac tamponade, 316
Cardiac veins, 322
  great, 316f–317f, 321f, 357f
  middle, 317f, 321f
  small, 317f, 321f, 357f
Cardiogenic shock, 369
Cardiomyopathy, 339t
Cardiopulmonary resuscitation (CPR), 131, 314, 328
Cardioregulatory center, 333, 337f
Cardiovascular disease. See also Atherosclerosis; Myocardial infarction
  dietary fats and, 477
Cardiovascular shock, 504
Cardiovascular system, 3, 6f, 313. See also Blood; Blood vessels; Heart
  aging and, 340, 578
  components of, 6f, 313, 313f
  functions of, 6f
  in postmenopausal women, 549t
  prenatal development of, 568t–569t
Caries, dental, 442
Carotene, and skin color, 100
Carotid arteries, 4f, 6f
  common, 352f
    left, 348, 350f, 351f, 352, 353f
    in pulse monitoring, 364f
    right, 351f, 352, 353f, 354f
  external, 350f, 352, 352f, 353f
  internal, 125, 350f, 352, 352f, 353f, 354f
Carotid bodies
  and blood pressure, 369
  and respiration, 427f, 428–429
Carotid canals, 126f, 127f
Carotid sinus, 352, 352f
Carpal bones, 121t, 123f, 134, 135f
Carpal region, 12f
Carpal tunnel, 135
Carpal tunnel syndrome, 135, 180
Carpometacarpal joint, 142f
Carpometacarpal pollicis, 142f
Carrier-mediated transport, 46t, 49–51
Carrier molecules, 49, 483
Cartilage, 83–84
  classification of, 80t
  composition of, 83
  extracellular matrix of, 112–113
  functions of, 83, 83t–84t, 112
  location of, 83, 83t–84t
  types of, 83–84, 83t–84t
Cartilage model, 115–117, 116f, 117f
Cartilaginous joints, 140–141
Casts for fractures, 119

Casts in urine, A-5
Catabolism, 481, 481f, 491
Catalyst(s)
    and chemical reaction rate, 29–30
    enzymes as, 37
Cataracts, 258t, 259
Cauda equina, 6f, 206, 206f
Caudate lobe, of liver, 453, 454f
Cavities, body, 15–16, 15f
C cells, 275, 276f
CD4 cells, 390f, 391
CD8 cells, 391, 396, 396f
Cecum, 451f, 459, 459f
Celiac disease, 468t
Celiac ganglion, 228f
Celiac trunk, 350f, 351f, 354–355, 355f
Cell(s), 2–3. See also specific types
    aging and, 67, 92
    characteristics of, 2–3
    communication of, 44
    death of (apoptosis), 66
    differentiation of, 66
    diversity of, 63, 65f
    energy currency of, 38, 481
    functions of, 44
    life cycle of, 61–63, 64f
    metabolism of, 44, 481
    structure–function relationships of, 59
    structure of, 43, 43f
    whole-cell activity of, 59–66
Cell body(ies), of neurons, 87, 88t, 194, 195f
Cell cycle, 61–63, 64f
Cell division, 62–63
    aging and, 577
    definition of, 62
    early, in prenatal development, 559–560
    in mitosis, 63, 64f
    period between (interphase), 61–62, 63f, 64f
Cell level of organization, 2–3, 3f
Cell-mediated immunity, 389, 395–396, 396f, 397f, 399f
Cell membrane, 43f, 44–45
    fluid-mosaic model of, 44–45, 45f
    functions of, 44
    lipid bilayer of, 44
    lipids in, 44–45, 45f
    movement through, 45–52, 46t
        by active transport, 46t, 50–51, 50f, 461
        carrier-mediated, 46t, 49–51
        by diffusion, 45–47
        by endocytosis, 46t, 51–52, 51f
        by exocytosis, 46t, 51–52
        by facilitated diffusion, 49–50, 49f
        by osmosis, 46t, 47–49, 48f
        by secondary active transport, 46t, 51, 51f
    polarization of, 153–155, 156f, 198
    potential difference across, 153–158
    resting membrane potential of, 155, 166, 198–199
    selective permeability of, 45, 516
Cellular clock, 67
Cellular physiology, 2
Cellulose, 33, 443, 444, 461–462, 474
Celsius scale, A-1
Cementum, 442, 442f
Central canal
    of compact bone, 113–115, 114f, 115f
    of spinal cord, 206, 207f, 222, 224f, 225f
Central incisor, 441f
Central nervous system (CNS), 193, 193f, 194f, 206–224. See also Brain; Spinal cord
    aging and, 230–234
    components of, 193, 193f, 194f, 206
    diseases and disorders of, 231t
    functions of, 215–222
        motor, 216–219
        sensory, 215–216
    organization of, 196
Central sulcus, 214f, 215, 216f
Central vein, of liver, 454f, 455
Centriole(s), 43f, 57, 528, 529f, 531f
    functions of, 44t
    location of, 44t
    structure of, 57, 58f
Centromere, 63, 64f
Centrosomes, 43f, 57
Cephalic phase, of stomach secretion, 448, 449f
Cephalic veins, 356, 357f, 358f
Cerebellar peduncles, 212f, 213
Cerebellum, 14f, 213
    as comparator, 218, 219f
    functions of, 218, 219f
    structure of, 211f, 213, 213f, 218
Cerebral aqueduct, 222, 224f, 225f
Cerebral arterial circle, 352, 353f, 354f

Cerebral arteries
    anterior, 354f
    middle, 354f
    posterior, 354f
Cerebral cortex, 196
    motor areas of, 216f, 217
    sensory areas of, 216, 216f
Cerebral hemispheres, 214–215
    communication between, 220
    left, 214, 214f
    right, 214, 214f
Cerebral palsy, 218, 231t
Cerebrospinal fluid (CSF), 222–224
    accumulation of (hydrocephalus), 224
    flow of, 222–224, 225f
    production of, 222, 225f
    spinal tap of, 222
Cerebrovascular accident (CVA), 231t, 232–233. See also Stroke
Cerebrum, 14f, 214–215
    aging and, 234
    structure of, 211f, 214–215, 214f
Cerumen, 251
Ceruminous glands, 251
Cervical canal, 540f
Cervical cap, 550, 551f
Cervical lymph node, 5f, 381f, 382, 442
Cervical nerves, 208–210, 210f
Cervical plexus, 208–210, 210f, 210t
Cervical region, 12f
Cervical vertebrae, 128f
    curvature of, 128, 128f
    fifth, 128f
    first (atlas), 128f, 129, 129f
    number of, 121t, 128
    second (axis), 128f, 129, 129f
    seventh, 128f, 169f, 178f
    structure of, 129, 129f
    surface view of, 133f
    third, 175f
Cervix, 539f, 540f, 543
    cancer of, 543, 554
    dilation, in childbirth, 573
    opening of, 540f
Chambers
    of eye, 244, 245f, 246–249
    of heart, 314, 316–320, 316f–317f
Cheek, 440, 441f
Chemical bonding, 22–24
    covalent, 24, 24f
    ionic, 22–23, 23f
Chemical digestion, 460
Chemical energy, 28
Chemical level of organization, 2, 3f
Chemically gated channels, 199
Chemical mediators
    in inflammation, 89, 90f, 387, 388f
    in innate immunity, 386
Chemical methods, of birth control, 550–552
Chemical reactions, 26–30
    classification of, 26–28
    decomposition, 26
    energy in, 28–29, 29f
    enzymes and, 37–38
    exchange, 28
    rate of, 29–30
    reversible, 28
    synthesis, 26
    water in, role of, 32
Chemical signals, 263–265. See also Hormone(s)
    autocrine, 263, 264t, 265
    definition of, 263
    intercellular, 263–265, 264t
    intracellular, 263
    paracrine, 263, 264t, 265
Chemistry
    basic, 21–25
    definition of, 21
    inorganic, 31–32
    organic, 31, 32–38
Chemoreceptor(s), 239
    and blood pressure, 369
    and ventilation, 427–429, 427f
Chemoreceptor reflexes
    in blood pressure regulation, 369–371, 371f, 373
    in cardiac regulation, 333, 335f, 337f
Chemotaxis, 386
Chest. See Thorax
Chickenpox, 108t, 402
Chief cells, of stomach, 446, 447f, 449f
Child
    definition of, 577
    development of, 577

Childbirth. See Birth
Chlamydial conjunctivitis, 584t
Chloride, reference laboratory values for, A-2, A-5
Chloride channels, 47
    in cystic fibrosis, 66–67
    in synaptic transmission, 201–203, 203f
Chloride ions, 23, 23f, 23t
    regulation of, 197
    renal reabsorption of, 506–508
    in urine, 503t
Chlorine
    atomic number of, 21t
    deficiency of, 480t
    dietary guidelines for, 480t
    functions of, 480t
    mass number of, 21t
    percent in body, 21t
    symbol for, 21t
Choanae, 407, 408f
Choking, 411
Cholecalciferol. See Vitamin D
Cholecystokinin, 448
    and bile secretion, 456, 456f
    functions of, 448, 448t, 451
    and pancreatic secretion, 458, 458f
    secretion of, 448, 448t, 449f
Cholera, 468t
Cholesterol, 35
    blood levels of, 309
    and cardiovascular disease, 578
    in cell membrane, 44–45, 45f
    dietary amounts, recommended, 476–477, 479–480
    dietary sources of, 476
    reference laboratory values for, A-3
    structure of, 35f
    uses in body, 35, 476
Chondrocytes, 83, 83t–84t
    in bone growth, 117, 118f
    in endochondral ossification, 116
Chordae tendineae, 318f, 319, 319f
Chorda tympani, 252f
Chorion, 561, 562f, 564f
Chorionic villi, 561, 562f
Choroid, 244, 245f, 247f
Choroid plexuses, 222, 225f
Chromatids, 63, 64f, 528, 529f
Chromatin, 52, 52f, 53f, 62, 64f
Chromium, 480t
Chromosomal disorders, 582
Chromosome(s), 52, 62–63, 579, 579f
    autosomal, 62–63, 579, 579f
    diploid number of, 62
    formation, in mitosis, 63, 64f
    homologous, 579–580
    mapping of, 583, 583f
    mitotic, 579
    number of, 62
    in oogenesis, 540–541
    sex, 62–63, 579, 579f, 580f
    in sex cells (gametes), 528, 529f
    in spermatocytes, 530
    structure of, 52, 52f
    X, 62, 579, 579f, 580f
    Y, 62, 579, 579f, 580f
Chyle, 380, 385f, 464
Chylomicrons, 463f, 464, 464f
Chyme, 448
    in colon, 460
    in small intestine, 452
    in stomach, 448–450, 449f
Chymotrypsin, 458
    functions of, 443t, 458, 461f, 464
    secretion of, 443t, 458
Cialis (tadalafil), 539
Cilia (sing., cilium), 43f, 57
    on epithelial cells, 78
    functions of, 44t, 57
    location of, 44t
    number of, 57
    structure of, 57
Ciliary body, 245, 245f
Ciliary ganglion, 228f
Ciliary muscles, 245, 245f
Cimetidine (Tagamet), 450
Cingulate gyrus, 221f
Circular folds, of small intestine, 451, 452f
Circular muscle
    of digestive tract, 438, 439f
    of small intestine, 452f
    of stomach, 446, 447f
Circulation, 344–375
    aging and, 373–375, 578
    fetal, 561–563, 562f, 574–575, 574f

function of, 345, 361
    neonatal, 574–575, 574f, 575f
    physiology of, 361–366
    prenatal development of, 567, 571f
    pulmonary, 313, 314f, 348, 349f
    systemic, 313, 314f, 348–361
Circulatory shock, 369
Circulatory system, 345. See also Circulation
Circumcision, 534
Circumduction, 144f, 145
Circumflex artery, 321, 321f
Cirrhosis, of liver, 468t
Citracal, 147
Citric acid cycle, 484f, 485, 486
Clavicle, 5f, 121t, 123f, 131–132
    fractures of, 132
    ossification of, 132
    structure of, 131–132, 131f, 132f
    surface anatomy of, 133f
Clavicular region, 12f
Cleavage lines, 97, 98f
Cleft lip, 566
Cleft palate, 566
Climacteric, 548–549
Climax
    female, 549
    male, 538
Clinical age, 559
Clinical perineum, 544, 544f
Clitoris, 539f, 544, 544f, 549
Clones, in lymphocyte development, 389
Clopidogrel bisulfate (Plavix), 300
Closed fracture, 120
Clostridium botulinum, 203
Clot
    blood. See Blood clot
    in bone repair, 119, 119f
    in tissue repair, 91–92, 91f
Clotting factors, 301–302, 301f, 457
Clotting time, A-4
CNS. See Central nervous system
Coagulation, 301–302
    disorders of, 309t
    platelet activation and, 300
    reference laboratory values for, A-4
    testing for, 307–308
Cobalamin. See Vitamin B$_{12}$
Cobalt
    deficiency of, 480t
    dietary guidelines for, 480t
    functions of, 480t
Cocaine, 203
Coccygeal bone. See Coccyx
Coccygeal nerves, 206f, 208, 210f
Coccygeal plexus, 208, 210f, 210t
Coccyx, 121t, 123f, 128, 128f, 177f
    male vs. female, 137f
    in pelvic girdle, 135, 136f
    structure of, 130, 130f
Cochlea, 252–254
    aging and, 259
    hair cells of, 253–254, 253f, 259
    structure of, 252–254, 252f, 253f
Cochlear duct, 253, 253f, 254f
Cochlear ganglion, 253f, 254
Cochlear nerve, 252f, 253f, 254f, 255, 256
Cochlear nucleus, 255
Codominance, 581
Codons, 60–61, 62f
Coenzymes, vitamins as, 477
Cognition, 193, 234
Coitus interruptus, 550
Cold, common, 432t
Cold receptors, 240
Cold sensation, 102
Cold sores, 108t
Colic flexures
    left, 459f
    right, 459f
Collagen/collagen fibers
    aging and, 2, 92, 577–578
    in bone matrix, 112–113
    in cartilage, 112–113
    in connective tissue, 80
        adipose, 80, 81t
        cartilage, 83–84, 83t–84t
        dense collagenous, 80–81, 80t, 82t
        dense elastic, 81–83, 82t
        loose, 80, 81t
    in platelet plug formation, 300, 301f
    in skin, 97, 98f
    structure of, 2
    uses in body, 477
Collagenous connective tissue, dense, 80–81, 80t, 82t

Collar bone. See Clavicle
Collateral axons, 195, 195f
Collateral ganglia, 227
Collecting duct, 498, 500f
    reabsorption in, 504, 505f, 506–508, 509f
    secretion in, 509
Colliculus (pl., colliculi), 212
    inferior, 212, 212f, 255
    superior, 212, 212f, 255
Colloid osmotic pressure, 503–504, 504f
Colon, 459
    aging and, 578
    ascending, 451f, 459, 459f
    descending, 459, 459f
    functions of, 460
    microorganisms in, 460
    movement of, 460
    sigmoid, 459, 459f
    smooth muscle of, 87t
    transverse, 440f, 459, 459f
Color blindness, 251, 251f, 258t
Colorectal cancer, 467, 549
Color vision, 246
Colostrum, 576
Column(s), of spinal cord, 206, 207f
Columnar epithelial tissue, 72, 73t
    pseudostratified, 75t, 76
    simple, 75t, 76
    stratified, 76
Comminuted fracture, 120, 120f
Commissures, 220
Common bile duct, 451, 455, 455f, 457f
Common carotid arteries, 352f
    left, 348, 350f, 351f, 352, 353f
    in pulse monitoring, 364f
    right, 351f, 352, 353f, 354f
Common cold, 432t
Common fibular (peroneal) nerve, 211
Common hepatic duct, 455
Common iliac arteries, 348–352, 350f, 356, 356f, 498f
    left, 351f, 355f
    right, 355f
Common iliac veins, 357f, 359f, 360, 498f
Communicating arteries
    anterior, 354f
    posterior, 354f
Communication
    cellular, 44
    muscles involved in, 152
Compact bone, 84, 113–115
    structure of, 113–115, 114f, 115f
Comparator function, cerebellar, 218, 219f
Complement, 386
    activation of, 386, 394f
    in inflammatory response, 387–388
    in innate immunity, 386
Complete blood count (CBC), 306–307
Complete fracture, 120, 120f
Complete protein, 477
Complex carbohydrates, 474
Compound(s)
    covalent, 25
    definition of, 25
    dissociation of, 25, 26f
    ionic, 25
    vs. molecule, 25
    symbols for, 25
Compound exocrine glands, 78, 79f
Compound fracture, 120
Computed tomography (CT), 27, 27f
Concentration
    and chemical reaction rate, 29
    of solutions, A-7
Concentration gradient, in diffusion, 46, 46f
Concentric contractions, 164
Conchae, nasal, 408
    inferior, 121t, 124, 125f, 408f
    middle, 124, 125f, 408f
    superior, 124, 408f
Condom
    male, 550, 551f
    vaginal, 550, 551f
Conduction, in body temperature regulation, 491, 491f
Conduction, of action potential
    continuous, 201, 202f
    saltatory, 201, 202f
Conduction deafness, 255, 258t
Conduction pathways, 196
Conduction system, of heart, 325–326, 325f
Condyle(s)
    definition of, 122, 122t
    of femur, 136–138, 138f, 139f
    of humerus, 133f, 134

Condyloid (ellipsoid) joints, 142f, 143
Cones
    functions of, 246
    pigments of, 246
    structure of, 246, 247f
    types of, 246
Congenital heart disease, 339t, 567
Congenital immunodeficiency, 402t
Conjunctiva, 243, 243f, 245f
Conjunctivitis, 243, 258t
Conjunctivitis, chlamydial, 584t
Connecting stalk, 563, 564f
Connective tissue, 3, 79–84. See also Blood; Bone(s); Cartilage
    adipose, 80, 81t
    aging and, 92
    blood as, 84, 85t, 292
    bone as, 84, 85t
    cells of, 80
    classification of, 80–84, 80t
    dense, 80–83, 82t
    functions of, 80
    loose, 80, 81t
    in skeletal muscle, 153, 154f
    structure of, 79–80
Consciousness, brain waves in, 220–221, 220f
Conservation of energy, law of, 28
Constant region, of antibody, 392, 392f
Constipation, 468t
Continuous conduction, 201, 202f
Contraception, 550–552, 551f
Contractility, of muscle, 84, 152–153
Contraction(s)
    cardiac muscle, 323–324
    peristaltic, 446, 446f, 450–451, 450f, 453
    skeletal muscle, 158–165
    smooth muscle, 166
    uterine, 263, 559, 573
Contraction phase, of muscle twitch, 162, 162f
Contracture
    physiologic, 164
    wound, 92
Control center, in negative feedback, 7–8
Convection, in body temperature regulation, 491, 491f
Converging pathways, 204, 205f, 206, 207f
Convoluted tubules. See Distal convoluted tubule; Proximal convoluted tubule
Copper
    deficiency of, 480t
    dietary guidelines for, 480t
    functions of, 480t
Coracoid process, 132, 132f, 178f
Corn(s), 96
Cornea, 243f, 244, 245f
    epithelial tissue of, 77t
    in focusing system, 249, 249f
    function of, 249
    prenatal development of, 565t
    structure of, 244
    transplantation of, 245
Corniculate cartilage, 410, 410f, 411f
Coronal plane, 14f, 15
Coronal suture, 122, 125f, 140f
Coronary arteries, 321–322, 348, 353f
    aging and, 373
    disorders of, 322, 338–339, 339t, 340
    left, 318f, 321, 321f, 348, 350f
    right, 317f, 318f, 321, 321f, 348, 350f
Coronary bypass, 322, 361
Coronary heart disease, 339t
Coronary sinus, 317, 317f, 318f, 320, 320f, 321f, 322
Coronary sulcus, 314, 316f
Coronary thrombosis, 322, 339t. See also Myocardial infarction
Coronoid process, 124f, 134, 134f, 143f
Corpora cavernosa, male, 533f, 534
Corpus albicans, 542f
Corpus callosum, 211f, 213f, 220, 221f
Corpus luteum, 541f, 542f, 543, 546f, 549
Corpus spongiosum, 533f, 534
Cortex
    of adrenal glands. See Adrenal glands, cortex of
    of brain. See Cerebral cortex
    of hair, 100, 100f
    of kidney, 498, 499f
    of thymus, 384, 384f
Corti, organ of, 253, 253f
Cortical nephrons, 498, 500f
Corticospinal tract, 217–218
    anterior, 217, 217f, 217t
    lateral, 217–218, 217f, 217t, 218f

Cortisol, 280–281
    action of, 268t, 271t, 280
    and fetal lung maturity, 419
    and glucose levels, 285
    reference laboratory values for, A-6
    replacement of, 274–275
    secretion of, 268t, 271t, 273–274, 280–281, 282f
    stress and, 281
    target tissues of, 268t, 271t, 280
Cortisone, 280
Costal cartilage, 83, 83t, 130, 131f
Costimulation, in immunity, 390f, 391, 391f, 396, 396f
Costovertebral joint, 142f
Cotransport, 51, 461
Cough reflex, 411–412
Coumadin (warfarin), 302
Counseling, genetic, 582
Countertransport, 51
Covalent bonds, 24, 24f
    double, 24
    nonpolar, 24
    polar, 24, 24f
    single, 24
Covalent compounds, 25
Cowper glands. See Bulbourethral glands
Coxae. See Coxal bones
Coxal bones, 14f, 121f, 123f, 135–136, 136f, 137f
Coxal joint. See Hip (hip joint)
Coxal region, 12f
CPR, 131, 314, 328
Cramps
    menstrual, 547
    muscle, 189t
Cranial cavity
    interior of, 124–125, 126f
    superior view of, 126f
Cranial fossa
    anterior, 124–125, 126f
    middle, 124–125, 126f
    posterior, 124–125, 126f
Cranial nerves, 206, 224. See also specific nerves
    functions of, 224, 226t
    motor, 224, 226t
    parasympathetic, 224, 226t, 227–229, 227t
    sensory, 224, 226t
    somatic motor, 224, 226t
    nomenclature for, 224
    origin of, 226f
Cranial region, 13f
Cranium. See Skull
Creatine phosphate, in ATP synthesis, 163–164, 163f
Creatinine
    reference laboratory values for, A-2
    tubular secretion of, 509
    in urine, 503t
Creatinine clearance, A-5
Cremaster muscle, 528–530, 532, 533f
Crenation, 49, 49f
Cretinism, 275
Cribriform plate, 126f, 408f
Cricoid cartilage, 408f, 410, 410f
Cricothyrotomy, 411
Crista ampullaris, 255–256, 257f
Cristae, 56, 57f
Crista galli, 126f
Critical Thinking Questions, answers to, A-8–A-17
Cross-bridge, actin-myosin, 159, 161f
Crossing over, 528, 529f, 580
Crossmatch, 306
Cross section, 15, 15f
Coronary thrombosis (see above)
Crossmatch (see above)
Crown, of tooth, 442, 442f
"Crow's feet," 167
Crural region, 12f
Crutch paralysis, 211
Crypt(s), of colon, 459, 459f
Cryptorchidism, 528, 531
CT scans, 27, 27f
Cubital fossa, 356
Cubital joints, 142f
Cubital vein, median, 356, 357f, 358f
Cuboidal epithelial tissue, 72, 73t
    simple, 73, 74t
    stratified, 76
Cuboid bone, 138, 139f
Cumulus cells, 541, 541f, 542f, 559
Cuneiform cartilage, 410, 410f, 411f
Cuneiforms, 138, 139f
Cupula, of ear, 255–256, 257f
Curare, 159

Cutaneous membrane, 87–88. See also Skin
Cuticle
    of hair, 100, 100f
    of nail, 101, 102f
Cyanide poisoning, 38
Cyanosis, 100, 104, 339t
Cyclic adenosine monophosphate (cAMP)
    and G proteins, 267, 267f
    as second messenger, 267
Cystic fibrosis, 51, 66–67, 432t, 582
Cystic vein, 360f
Cystitis, 514
Cytokine(s)
    in immunity, 391, 396
    in immunotherapy, 398
Cytoplasm
    functions of, 44t
    location of, 43f, 44t
    structure of, 43, 43f
Cytosine
    in DNA, 38, 39f, 60
    in RNA, 38, 60
Cytoskeleton, 57
    composition of, 57
    functions of, 57
    structure of, 57, 58f
Cytotoxic T cells, 389, 391, 395–396, 396f, 397f

Daily Reference Values (DRVs), 479
Daily Values, 479–481
Dartos muscle, 528–530, 533f
Daughter cells, 62, 63, 64f, 528, 529f, 532f, 542f
Dead space, 422, 433
Deafness
    conduction, 255, 258t
    sensorineural, 255, 258t, 259
Death, 578–579
    brain, 578–579
    cell, 66
    definitions of, 578–579
    neocortical, 579
Death genes, 67
Debridement of burns, 107
Deciduous teeth, 440, 441f
Declarative memory, 221
Decomposition reactions, 26
Decubitus ulcers, 108t
Deep, 10, 11t
Deep back muscles, 172–173, 174f, 174t
Deep femoral artery, 350f
Deep infrapatellar bursa, 143f
Deep lymphatic vessels, 414f, 415
Deep neck muscles, 170–172, 174f, 174t, 178f
Deep palmar arch, 354f
Defecation, 176, 229, 460
Defecation reflex, 460
Defibrillation, 326
Degenerative joint disease, 148t
Deglutition. See Swallowing
7-Dehydrocholesterol, 477
Delayed allergic reactions, 402t
Delta waves, 220f, 221
Deltoid injections, 179
Deltoid muscle, 168f, 169f, 178f, 179, 179f, 182f
    action of, 167, 179, 180t
    insertion of, 180t
    origin of, 180t
Deltoid tuberosity, 133f, 134
Denaturation, 37
Dendrites, 87, 194
    functions of, 87
    structure of, 87, 88t, 194, 195f
Dens, 129, 129f
Dense connective tissue, 80–83
    collagenous, 80–81, 80t, 82t
    elastic, 80t, 81–83, 82t
    structure of, 80
Dental caries, 442
Dentate gyrus, 221f
Dentin, 442, 442f, 467
Deoxyribonucleic acid (DNA), 38, 579
    aging and, 67
    composition of, 38
    mitochondrial, mutations in, 578
    in protein synthesis, 59–61, 60f
    replication of, 61–62, 63f, 64f
    sequences, mapping of, 583, 583f
    structure of, 38, 39f
    transcription of, 60–61, 60f, 61f
    in translation, 60f, 61, 62f
Deoxyribose, 38, 39f
Depolarization phase, of action potentials, 200–201, 200f, 201f
    in cardiac muscle, 324, 324f

in skeletal muscle, 156f, 157, 324f
in smooth muscle, 166
Depo-Provera, 550–552
Depression (movement), 145
Depressor anguli oris muscle, 170, 171f
action of, 170, 172t
insertion of, 172t
origin of, 172t
Dermal papilla, 97, 98f
Dermatitis, 108t
Dermis, 96
cleavage lines in, 97, 98f
functions of, 96
prenatal development of, 565t
strength of, 96, 97
structure of, 97, 97f
thickness of, 96
Descending aorta, 348–352, 351f, 353f, 355f
Descending colon, 459, 459f
Descending tracts, 206, 217–218, 217f, 217t
direct, 217, 217t, 218f
indirect, 217, 217t
Desmosomes, 78, 78f, 96
Detoxification, in liver, 456t, 457
Deuterium, 27
Development
adolescent, 577
child, 577
definition of, 7
disorders of, 148t
in first year following birth, 577
prenatal, 559–569
Developmental age, 559
Deviated nasal septum, 407
Diabetes insipidus, 275, 511
Diabetes mellitus, 263, 284–285
aging and, 288
renal reabsorption in, 508
type 1, 284–285
type 2, 284–285
Diabetic retinopathy, 258t, 259
Diacylglycerol (DAG), and G proteins, 267
Dialysis, 520–521, 520f
Diaphragm (contraception), 550, 551f
Diaphragm (muscle), 4f, 173, 412f
action of, 173, 175t
central tendon of, 175f
innervation of, 208–210
insertion of, 175t
origin of, 175t
in respiration, 416, 417f, 426f
Diaphysis
in endochondral ossification, 116, 117f
growth of, 117, 118f
of long bone, 113, 114f
Diarrhea, 466–467, 468t
systemic effects of, 467
traveler's, 466, 466t
treatment of, 462, 466
Diarthrosis, 140
Diastole
atrial, 328
ventricular, 328, 331
Diastolic pressure, 362–363, 362f
Diencephalon, 213–214
components of, 213
structure of, 211f, 212f, 213, 213f
Dietary fiber, 444, 473, 474
Dietary Guidelines for America, 473
Differential white blood count, 307, A-4
Differentiation, 7, 63–66
Diffusion, 45–47, 46f, 46t, 47f
concentration gradient in, 46, 46f
definition of, 45
in digestive system, 460–461
facilitated, 49–50, 49f, 461
of water (osmosis), 46t, 47–49, 48f
Digestion, 460–465, 460f
autonomic regulation of, 229
chemical, 460
definition of, 460
mastication and, 444
mechanical, 460
in neonate, 576
Digestive system, 3, 5f, 437–467
absorption in, 453, 460–465
accessory organs of, 438, 438f
aging and, 465–467
anatomy of, 438–440, 438f
blood flow to, 349f
cancer of, 467
components of, 5f, 438, 438f
diseases and disorders of, 468t
epithelial tissue of, 438, 439f
fluid volumes in, 465, 465f

functions of, 3–5, 5f, 438
histology of, 438, 439f
hormones of, 286
layers or tunics of, 438, 439f
of newborn, 576
prenatal development of, 565t, 567,
    568t–569t, 570f
secretions of, functions of, 443t
transport in, 460–465
Digestive tract, 438
Digit(s), of hand, 134–135, 135f
Digital arteries, 354f
Digitalis, 336
Digital regions, 12f
Dilation stage, of labor, 573
Diphtheria, 432t
Diploid number of chromosomes, 62
Diplopia, 251, 258t
Directional terms, 10, 11f, 11t
Direct tracts, 217, 217t, 218f
Disaccharidases, 452, 461f, 462
Disaccharides, 33, 33f
dietary sources of, 474
digestion of, 461f, 462
uses in body, 474
Disequilibrium, 259
Dislocation, 145
Disseminated intravascular coagulation
    (DIC), 309t
Dissociation, chemical process of, 25, 26f
Distal, 10, 11f, 11t
Distal convoluted tubule, 498, 500f, 502f
and acid-base balance, 519, 522f
reabsorption in, 504, 505f, 506–508, 509f
secretion in, 509
Distal phalanx
of finger, 135, 135f
of thumb, 135, 135f
of toe, 139f
Distant vision, 249f
Distributing arteries, 345
Diuretics, 510
Diverging pathways, 204–205, 205f,
    206–208, 207f
Dizygotic twins, 560
DNA. See Deoxyribonucleic acid
Dominance, incomplete, 581
Dominant genes, 580–581
Donor, of blood, 304
Dopamine, as neurotransmitter, 203, 204t
Dorsal, 10, 11f, 11t
Dorsal artery, of penis, 533f
Dorsal column, 206, 207f, 215, 215f, 215t
Dorsalis pedis artery, 350f, 356, 356f, 364f
Dorsal nerve, of penis, 533f
Dorsal region, 13f
Dorsal respiratory groups, 424, 426f
Dorsal root(s), 206, 207f
Dorsal root ganglion, 206, 207f
Dorsal veins
of foot, 361f
of penis, 533f
Dorsiflexion, 143, 185
Dorsum region, 12f, 13f
Double covalent bond, 24
Double helix, of DNA, 38, 39f
Double vision, 251, 258t
Down syndrome, 580
Dual-energy x-ray absorptiometry
    (DEXA), 147
Dual innervation, 229
Duchenne muscular dystrophy (DMD),
    188–189, 188f, 189t
Ductus arteriosus, 574f, 575, 575f
Ductus deferens, 6f, 527f, 528, 530f, 531f,
    532, 533f
Ductus venosus, 574f, 575, 575f
Duodenal glands, 451, 452f, 453
Duodenal papilla, 455, 455f, 457f
Duodenum, 440f, 451, 457f
absorption in, 460–461
anatomy of, 451, 451f, 452f
histology of, 451, 452f
Dupp sound, 331
Dura mater
of brain, 222, 223f, 225f
of spinal cord, 222, 223f
Dust cells, 387
Dwarfism, 148f
pituitary, 272
Dynamic equilibrium, 255
Dysautonomia, 230
Dysentery, 468t
Dyskinesias, 231t
Dystrophin, 188

Ear(s)
aging and, 259
auditory function of, 251–255
balance function of, 255–257
diseases and disorders of, 258t
external, 84, 84t, 251, 252f
inner, structure of, 252–254, 252f, 253f
middle, 252, 252f
prenatal development of, 565t
structure of, 251–254, 252f
Eardrum. See Tympanic membrane
Earwax, 251
Eccentric contractions, 164
Eccrine sweat glands, 101, 101f
ECG. See Electrocardiogram
Ectoderm
derivatives of, 565t, 568t–569t
development of, 564, 564f
Ectopic beat, 326
Ectopic pregnancy, 548, 584t
Eczema, 108t
Edema, 89, 105, 348, 365
Effector, in negative feedback, 8
Effector organs, in reflex arc, 204, 205f
Efferent arteriole, 502, 502f
Efferent ductules, 531–532, 531f
Efferent lymphatic vessels, 382, 383f
Efferent neurons. See Motor neurons
Egg cells. See Oocytes
Eicosanoids, 34–35, 266, 476
Eicosapentaenoic acid (EPA), and blood
    clots, 476
Ejaculation, 229, 538–539
Ejaculatory duct, 530f, 532, 533f
EKG. See Electrocardiogram
Elastic arteries, 345, 346f, 373
Elastic cartilage, 83–84
functions of, 83–84, 84t
location of, 84, 84t
structure of, 83, 84f
Elastic connective tissue, dense, 80t, 81–83, 82t
Elastic fibers
aging and, 92
in connective tissue, 80
adipose, 80, 81t
dense elastic, 80t, 81–83, 82t
loose, 80, 81t
in lung, 414, 414f
in skin, 97
Elasticity, of muscle, 153
Elbow (elbow joint), 143f
extension of, 144f
flexion of, 144f
as hinge joint, 141, 142f
olecranon process as, 134, 134f, 135f
Elderly, 577. See also Aging, effects of
Electricity, ions and, 25
Electric signals, 198–205. See also Action
    potential(s)
Electrocardiogram (ECG, EKG), 25, 314,
    326–328, 326f, 579
cardiac cycle in, 328, 330f
as diagnostic tool, 327–328
in myocardial infarction, 339
Electroencephalogram (EEG), 220–221, 220f
Electrolytes, 25
Electron(s), 22, 22f
bonding of, 22–24
covalent, 24, 24f
ionic, 22–23, 23f
clinical relevance of, 27
mass of, 22
number of, 22
Electron cloud, 22, 22f
Electron-transport chain, 484f, 485–486, 485f
Elements, 21–22
common in human body, 21t
symbols for, 21t, 22
Elephantiasis, 402f
Elevation (movement), 145
Ellipsoid joints, 142f, 143
Embolus, 302, 348, 375, 578
Embryo, 559, 577
35 days after fertilization, 566f
50 days after fertilization, 572f
development of, 564–567
implantation of, 561–563
Embryonic disk, 564, 564f
Embryonic period, 567
Embryo transfer, 562
Emission, in male sex act, 538–539
Emotional sweating, 101
Emphysema, 429, 432t
Emulsification, 463, 476
Enamel, tooth, 442, 442f

Encephalitis, 231t
Endocarditis, 339t
Endocardium, 322, 323f, 416f
Endochondral ossification, 115–117, 116f,
    117f, 118f
Endocrine, derivation of term, 265
Endocrine cells
of small intestine, 451, 452f
of stomach, 446, 447f, 448, 449f
Endocrine glands, 79, 265, 269, 270f. See also
    specific glands
Endocrine system, 3, 6f, 262–288
aging and, 288
components of, 6f, 265, 269, 270f
diseases and disorders of, 263
functions of, 6f, 263, 269
hormone secretion in, 269–286, 270t–271t
hypothalamic regulation in, 270–272,
    272f, 273f, 274f
negative-feedback regulation of, 268–269
vs. nervous system, 263
prenatal development of, 568t–569t
secretions of, 269–286. See also
    Hormone(s)
Endocytosis, 46t, 51–52
energy required for, 46t, 52
phagocytosis in, 51–52
pinocytosis in, 52
receptor-mediated, 51, 51f
specificity in, 51
Endoderm
derivatives of, 565t, 568t–569t
development of, 564, 564f
Endolymph, 252, 253f
auditory function of, 254, 254f
balance function of, 256, 257f
Endometriosis, 552
Endometrium, 540f, 543
aging and, 554
implantation in, 543, 561–563, 561f
in menstrual cycle, 546–548, 546f
Endomysium, 153, 154f
Endoplasmic reticulum
functions of, 54
rough, 43f, 44t, 54, 55f, 194, 195f
smooth, 43f, 44t, 54, 55f
structure of, 54, 55f
Endorphins, 203, 204t
Endosteum
in compact bones, 113–115
in long bones, 113, 114f
Endothelin, in vascular spasm, 300
Endothelium
of blood vessels, 346f
of capillaries, 347, 347f
Energy, 28–29
activation, 37, 37f
cellular use of, 44
chemical, 28
conservation of, law of, 28
conversion of, 28–29
definition of, 28
free, 491
kinetic, 28
mechanical, 28–29
potential, 28
storage of, 80
Energy currency, ATP as, 38
Enteric nervous system (ENS), 194, 230, 438
Enteric plexus, 448
Enzyme(s), 37–38
action of, 37, 38f
product control of, 483
receptor-mediated, 483
activation energy and, 37, 37f
alteration, by membrane-bound
    receptors, 267–268, 268f, 268t
definition of, 37
and disease, 482
lock-and-key model of, 37, 38f
lysosomal, 54
pancreatic, 458
synthesis of, 482
transport in blood, 292, 345
uses in body, 477
Eosinophil(s), 298–299
function of, 294t, 299
in inflammatory response, 299, 387
percentage of WBCs as, 293f, 307
production of, 295f
structure of, 294t, 299
Ependymal cells, 195, 225f
functions of, 195, 196t
structure of, 195, 196t, 197f
Epicardium, 314, 315f, 322, 323f

Epicondyle(s)
  definition of, 122t
  of femur
    lateral, 138, 138f, 140f
    medial, 138, 138f, 140f
  of humerus
    lateral, 133f, 134
    medial, 133f, 134
Epicranial aponeurosis, 167, 171f
Epidermis, 96
  cells of, production of, 96
  functions of, 96
  keratinization in, 96
  melanocytes of, 98, 99f
  prenatal development of, 565t
  for skin graft, 105
  strata of, 96, 98f
  structure of, 96, 97f, 98f
  thickness of, 96
Epididymis, 6f, 527f, 528, 530f, 531–532, 531f, 533f
Epidural space
  of brain, 222, 223f
  of spinal cord, 222, 223f
Epigastric pain, 12
Epigastric region, 12, 13f
Epiglottis, 408f, 410, 410f, 411f, 444, 445f
Epilepsy, 231t
Epimysium, 153, 154f
Epinephrine
  action of, 271t, 279
  and blood flow, 367–368
  and blood pressure, 371–373, 371f
  in fight-or-flight response, 279
  and glucose levels, 285
  in heart regulation, 333, 337f
  secretion of, 271t, 278–279, 279f, 280f
  target tissues of, 271t, 280f
Epiphyseal line, 113, 114f
Epiphyseal plate
  growth at, 116–117, 118f
  in long bones, 113, 114f
Epiphysis
  in endochondral ossification, 116
  of long bone, 113, 114f
Episiotomy, 544
Epithalamus, 213
  functions of, 213
  structure of, 213, 213f
Epithelial cells
  connections among, 78, 78f
  layers of, 73t, 76–78
  shapes of, 73t, 76–78
Epithelial tissue, 3, 72–79
  absorption by, 72
  barrier function of, 72
  cell layers and shapes in, 73t, 76–78
  characteristics of, 72, 73f
  classification of, 72–76, 73t, 74t–75t, 77t
  damage to, smoking and, 76–78
  of digestive system, 438, 439f
  functions of, 72
  of kidneys, 498–502
  passage or transport via, 72
  protection by, 72, 76
  secretion by, 72
  structure–function relationships of, 76–78
  surfaces of, 72, 78
  of tracheobronchial tree, 413–414
  of ureters, 513
  of urinary bladder, 513
Epithelium. See Epithelial tissue
Equilibrium, 28, 251, 255–257
  aging and, 259
  dynamic, 255
  neuronal pathways for, 256–257
  static, 255
Equivalents, A-7
Erectile dysfunction (ED), 539, 552–554
Erectile tissue
  female, 545, 549
  male, 532–534, 538–539
Erection, 229, 532, 538–539
Erector spinae muscles, 172, 174f
  action of, 172, 174f
  insertion of, 174t
  origin of, 174t
Ergosterol. See Vitamin D
Erythroblastosis fetalis, 305–306, 305f
Erythrocytes. See Red blood cells
Erythrocytosis, 306, 307, 309t
  primary, 307, 309t
  relative, 307, 309t
  secondary, 307, 309t

Erythropoietin, 287, 297
  abnormal levels of, 307
  action of, 287, 297, 297f
  secretion of, 287
Escherichia coli, 514
Esomeprazole (Nexium), 450
Esophageal phase, of swallowing, 445f, 446
Esophageal sphincter
  lower, 444, 447f
  upper, 444, 445f
Esophagus, 4f, 5f, 15f, 408f, 416f, 438, 438f, 444
  epithelial tissue of, 77t, 438
  in swallowing, 444–446, 445f
Essential amino acids, 477
Essential fatty acids, 477
Essential nutrients, 473
Essential vitamins, 477
Estrogen(s), 285
  action of, 268t, 271t, 285, 536t
  chemistry of, 35f
  in female puberty, 545–546
  in female sexual behavior, 549
  in labor (childbirth), 573, 573f
  in lactation, 576
  in menopause, 549, 554
  in menstrual cycle, 546f, 547–548, 547t
  in osteoporosis, 146–147
  placental, 287–288
  in pregnancy, 563, 563f
  secretion of, 268t, 271t, 285, 536t
  synthesis of, 35, 476
  synthetic, 266, 550–552
  target tissue of, 268t, 271t, 536t
Estrogen replacement therapy, 147
Ethanol. See Alcohol
Ethmoidal sinus, 124, 126f
Ethmoid bone, 121t
  in cranial cavity, 125, 126f
  cribriform plate of, 126f, 408f
  frontal view of, 124, 125f
  intramembranous ossification of, 116f
  lateral view of, 124f
  perpendicular plate of, 124, 125f
Etymology, 10
Eustachian tube, 252, 252f, 254
Evaporation, in body temperature regulation, 491, 491f
Eversion, of foot, 145, 185
Exchange reaction, 28
Excitability
  of muscle, 153–157, 199–200
  of neurons, 199–200
Excretion
  definition of, 104
  by skin, 104
  by urinary system, 497
Excursion, 145
Exercise
  and amenorrhea, 548
  and asthma, 430, 430f
  cardiac effects of, 332, 339, 340
  energy sources during, 163–164
  and muscle fibers, 164
  muscle soreness caused by, 163
  oxygen deficit in, 164
  and pain management, 240
  recovery oxygen consumption in, 164
  respiratory adaptations to, 430–433
  and stroke volume, 332
  and ventilation, 429–433
Exhalation. See Expiration
Exocrine glands, 78–79, 265
  acinar, 78, 79f
  alveolar, 78, 79f
  compound, 78, 79f
  simple, 78, 79f
  structure of, 78, 79f
  tubular, 78, 79f
  unicellular, 78, 79f
Exocytosis, 46t, 51–52, 53f
Exophthalmia, 275, 286f
Expiration, 416
  alveolar pressure changes in, 419–420
  lung recoil and, 419
  muscles of, 173, 416, 417f
  pleural pressure and, 419–420
  pressure changes during, 417, 418f
Expiratory reserve volume, 420, 421f
Explicit memory, 221
Expulsion stage, of labor, 573
Extensibility, of muscle, 153
Extension, 143, 144f
Extensor carpi radialis brevis muscle, 181f, 182
Extensor carpi radialis longus muscle, 181f, 181t, 182

Extensor carpi radialis muscle, 182, 182f
Extensor carpi ulnaris muscle, 181f, 182f
Extensor digitorum longus muscle, 168f, 186f, 187t
Extensor digitorum muscle, 181f, 181t, 182, 182f
Extensor digitorum tendons, 181f, 182f
Extensor hallucis longus muscle, 186f, 187t
External abdominal oblique muscle, 168f, 169f, 176, 176f, 178f
External anal sphincter, 177f, 177t, 459f, 460
External auditory canal, 122, 124f, 127f, 251, 252f, 254f
External carotid artery, 350f, 352, 352f, 353f
External ear, 84, 84t, 251, 252f
External iliac artery, 350f, 351f, 355f, 356, 356f
External iliac vein, 357f, 359f, 360
External intercostal muscle, 173, 175f, 178f
  action of, 173, 175f
  insertion of, 175t
  origin of, 175t
  in respiration, 173, 416, 417f, 426f
External jugular vein, 356, 357f, 358f
External nose, 5f, 407, 407f
External urinary sphincter, 513f, 514
Extracellular fluid
  composition of, 516–518
  exchange or movement in, 44–52
  ion concentration in, 517–518
  pH of, renal regulation of, 497
  regulation of, 516–518
  thirst regulation and, 516–517, 516f
  volume of, 515t
Extracellular fluid compartment, 516
Extracellular matrix, of connective tissue, 79–80
  aging and, 92
  composition of, 79
  in skeletal system, 112–113
  structure of, 79–80
Extracellular substances, 44
Extrinsic regulation, of heart, 332–340, 337f
Eye(s), 242–251
  accessory structures of, 243–244, 243f
  aging and, 259, 578
  anatomy of, 244–249, 245f
  chambers of, 244, 245f, 246–249
  color of, 581–582
  diseases and disorders of, 258t
  extrinsic muscles of, 244, 244f
  focusing system of, 249, 249f
  functions of, 249
  prenatal development of, 565t
  tunics of, 244, 245f
Eyebrows, 243, 243f
Eyelashes, 243f
Eyelids, 243, 243f

Face
  bones of. See Facial bones
  bony landmarks on, 122, 125f
  muscles of, 167–170, 168f, 171f, 172t
  prenatal development of, 566, 567f
Facial artery, 352f, 364f
Facial bones, 121t, 122
Facial expression, muscles of, 167–170, 168f, 171f, 172t
Facial (CN VII) nerve, 224, 252f
  functions of, 224, 226t
  origin of, 226f
  and parasympathetic axons, 227–229, 227t
  and taste, 242
Facial veins, 357f, 358f
Facilitated diffusion, 49–50, 49f, 461
FADH₂
  in citric acid cycle, 484f, 485
  in electron-transport chain, 484f, 485–486, 485f
Fahrenheit scale, A-1
Falciform ligament, 453, 454f
Fallopian tubes. See Uterine tubes
Falls, among elderly, 259
False ribs, 130, 131f
False vocal cords, 82t, 410, 410f, 411f
Family tree (pedigree), 582, 582f
Fascia, muscular, 153, 154f
Fasciculus (pl., fasciculi), muscle, 153, 154f
Fast-twitch muscle fibers, 164–165, 165t, 188
Fat(s), 33
  absorption of, lymphatic system and, 380
  body, and obesity, 489–490
  dietary, 474–477
    Daily Values for, 479–480
    recommended amounts of, 476–477, 479–481
    sources of, 474–476
  digestion of, 462–464

  metabolism of, 486
    in absorptive state, 487, 488f
    in postabsorptive state, 487, 488f
  monounsaturated, 34, 476
  polyunsaturated, 34, 476
  saturated, 34, 34f, 462–463, 476
  total body, estimation of, 100
  trans, 476–477
  unsaturated, 34, 34f, 462–463, 476
  uses in body, 476
Fatigue
  definition of, 164
  muscular, 164
  psychological, 164
Fat-soluble vitamins, 477
Fatty acids, 33–35
  and blood clots, 476
  as energy source, 164
  essential, 477
  lipid breakdown to, 460, 460f, 461f, 462–463, 463f
  metabolism of, 486, 487f
  in phospholipid formation, 34
  saturated, 34, 34f, 462–463, 476
  in starvation, 489
  structure of, 33–34, 34f
  in triglyceride formation, 33, 34f, 464
  unsaturated, 34, 34f, 462–463, 476
Feces, 460
  formation of, 460
  pigments in, 298, 298f
Female genitalia
  external, 539, 539f
  in postmenopausal women, 549t
Female infertility, 549
Female pelvis, 136, 136t, 137f
Female puberty, 545–546, 577
Female reproductive hormones, 536t, 545–549
  in labor (childbirth), 573–574, 573f
  in lactation, 550
  in menopause, 547–548
  in menstrual cycle, 546–548, 546f, 567t
  in pregnancy, 563, 563f
  in puberty, 545–546
Female reproductive system, 6f, 527f, 539–552
  aging and, 552–554
  anatomy of, 539–545, 539f, 540f
  components of, 6f
  diseases and disorders of, 554t
  functions of, 6f
  hormones of, 536t, 545–549
  physiology of, 545–552
Femoral arteries, 6f, 350f, 356, 356f
  deep, 350f
  in pulse monitoring, 364f
Femoral neck, 138, 138f, 143f
Femoral neck fracture, 138, 146, 146f
Femoral nerve, 210–211
Femoral region, 12f
Femoral triangle, 356
Femoral vein(s), 6f, 357f
Femur, 5f, 14f, 121f, 123f, 136–138, 136f, 138f, 140f
  at hip, 143f
  at knee, 143f
  surface anatomy of, 140f
Fertility
  female, 549, 549t
  male, 531, 539
Fertilization, 527, 530, 540–541, 542f, 548, 559, 560f
Fetal alcohol syndrome (FAS), 584t
Fetal hemoglobin, A-4
Fetal period, 567–568
Fetus, 559, 577
  bone development in, 115, 116f
  circulation in, 561–563, 562f, 574–575, 574f
  growth of, 567–569, 572f
  hormones of, and birth, 573, 573f
  skull of, 122, 124f, 140, 140f
Fever, 388
Fever blisters, 108t
Fiber, dietary, 444, 473, 474
Fibrillation, 326
Fibrin
  in clot formation, 91, 301–302, 301f
  in plasma, 292, 293f, 293t
  in tissue repair, 91–92
Fibrinogen
  in coagulation, 301f, 302
  hepatic synthesis of, 457
  in plasma, 292, 293f, 293t
  in platelet plug formation, 300, 301f
Fibrinogen receptors, 300, 301f
Fibrinolysis, 302, 303f

Fibroblast(s)
  in connective tissue, 80–83, 81t, 82t
  naming convention for, 80
Fibrocartilage, 83, 141
  function of, 83, 84t
  location of, 83, 84t
  structure of, 83, 84t
Fibrocytes, naming convention for, 80
Fibroid tumors, uterine, 553, 553f
Fibromyalgia, 189t
Fibrous joints, 140
Fibrous pericardium, 314, 315f, 416f
Fibrous tunic, of eye, 244, 245f
Fibula, 5f, 121t, 123f, 136f, 138, 139f
  at foot, 139f
  surface anatomy of, 140f
Fibular artery, 350f, 356, 356f
Fibularis (peroneus) brevis muscle, 168f, 169f,
    185, 186f, 187t
Fibularis (peroneus) longus muscle, 168f,
    169f, 185, 186f, 187t
Fibularis (peroneus) longus tendons, 186f
Fibularis (peroneus) tertius muscle, 186f, 187t
Fibular (peroneal) nerve, common, 211
Fibular vein, 357f
Fight-or-flight response, 227t, 229, 279
Filtrate, 498
Filtration
  cardiovascular shock and, 504
  definition of, 503
  regulation of, 504
  in urine production, 502–504, 503f
Filtration membrane, 498, 502f
Filtration pressure, 503–504, 504f
Filtration slits, 502f
Fimbriae, of uterine tubes, 540f, 543
Finger(s)
  abduction of, 144f
  adduction of, 144f
  bones of, 134–135, 135f
  extension of, 180
  extensor muscles of, 169f
  flexion of, 180, 182
  opposition of, 145
  prenatal development of, 566, 566f
Fingernails. See Nail(s)
First-degree burns, 104f, 105
Fish oil, 476
Fissure(s)
  on bone, 122t
  lateral, 214f, 215
  longitudinal, 214, 214f
  oblique, of lung, 413f
  orbital
    inferior, 123, 125f, 127f
    superior, 123, 125f
Fixators, 167
Flaccid paralysis, 159
Flagellum, 59
  functions of, 44t
  location of, 44t
  of sperm cells, 59
  structure of, 59
Flat bones, 113
Flavin adenine dinucleotide (FADH$_2$)
  in citric acid cycle, 484f, 485
  in electron-transport chain, 484f, 485–486, 485f
Flexion, 143, 144f
Flexor carpi muscle, 182
Flexor carpi radialis muscle, 181f, 181t
Flexor carpi radialis tendon, 182, 182f
Flexor carpi ulnaris muscle, 181f, 181t
Flexor digitorum longus muscle, 187t
Flexor digitorum muscle, 181f, 182
Flexor digitorum profundus muscle, 181t
Flexor digitorum superficialis muscle, 181t
Flexor hallucis longus muscle, 187t
Flexor (withdrawal) reflex, 208, 209f
Floating ribs, 130, 131f
Flu, 432t
Fluid(s), body. See also specific types
  compartments of, 514–516
  extracellular, 516-518
  intracellular, 514–516
Fluid balance, lymphatic system and, 380
Fluid compartments, 514–516
  exchange between, 516
  extracellular, 516
  intracellular, 514–516
  volumes of, 515t
Fluid-mosaic model, 44–45, 45f
Fluorine
  deficiency of, 480t
  dietary guidelines for, 480t
  functions of, 480t

Flu vaccine, 400–402
Focal point (FP), 249
Focusing, of light, 249
Focusing system of eye, 249, 249f
Folate (folic acid)
  deficiency of, 308
  functions of, 477
  and neural tube defects, 565
  and red blood cell production, 297
Follicle(s)
  hair, 97f, 100–101, 100f, 538t
  ovarian, 540, 541–543, 541f
  thyroid, 275, 276f
Follicle-stimulating hormone (FSH), 274
  action of, 270t, 274, 285, 536t
  in females, 536t
    and infertility, 549
    and menopause, 548
    and menstrual cycle, 546f, 547–548, 547f
    oral contraceptives and, 550
    and puberty, 545–546
  in males, 531, 535f, 536t, 537
    and puberty, 537
  secretion of, 270t, 274, 536t
  target tissues of, 270t, 536t
Fontanels, 140, 140f
Food
  chewing of (mastication), 440, 444
  consumption by quantity and type, 475t
  Daily Values of, 479–481
  digestion of, 460–465, 460f
  metabolism of, 481–490, 487f
  MyPyramid of, 473, 473f
  nutritional value of, 473
Food and Drug Administration (FDA), 479–481
Food labels, 479–481, 481f
Food poisoning, 468t
Food pyramid, 473, 473f
Foot, 12
  arches of, 140
  bones of, 136f, 139f, 140
  dorsiflexion of, 143, 185
  eversion of, 145, 185
  intrinsic muscles of, 185
  inversion of, 145, 185
  movements of, 143, 145
  plantar flexion of, 143, 185
  prenatal development of, 566, 566f
Foramen (pl., foramina)
  definition of, 121, 122t
  of skull, 125–127, 126f, 127f
Foramen magnum, 125–126, 126f, 127f
Foramen ovale, 125, 126f, 127f, 567, 571f,
    574f, 575f
Foramen rotundum, 125, 126f
Foramen spinosum, 125, 126f, 127f
Forced expiratory vital capacity, 421
Forearm
  bones of, 134, 134f
  definition of, 12, 134
  movements of
    muscles of, 179–180, 181f, 181t, 182f
    pronation and supination, 144f, 145, 180
Foreign antigens, 388–389
Foreskin (prepuce), 530f, 534
Formed elements, of blood, 292, 294–299
  descriptions of, 294t
  functions of, 294t
  number per cubic mm, 293f
  percentage in blood composition, 293f, 294
  production of, 294, 295f
  types of, 294, 294t
Fornix, 221f
Fosamax, 147
Fossa, definition of, 122, 122t
Fossa ovalis, 575f
Fourth ventricle, 222, 224f, 225f
Fovea centralis, 246, 248f
Fractures
  aging and, 145
  classification of, 120, 120f
  closed, 120
  comminuted, 120, 120f
  complete, 120, 120f
  compound, 120
  greenstick, 120
  immobilization of, 119
  impacted, 120, 120f
  incomplete, 120, 120f
  linear, 120
  oblique, 120, 120f
  open, 120
  in osteogenesis imperfecta, 113
  osteoporosis and, 119
  repair of, 119, 119f

simple, 120
spiral, 120, 120f
transverse, 120, 120f
Fraternal twins, 560
Freckles, 98, 108
Free energy, 491
Free nerve endings, 239–240, 239f, 258
Free radicals, 67, 479
Free radical theory of aging, 578
Free ribosomes, 43f, 52–54
Free surface, of epithelial tissue, 72, 78
Frenulum, 440, 441f
Frontal bone, 121t, 122
  in cranial cavity, 125, 126f
  in fetal skull, 140f
  frontal view of, 122, 125f
  intramembranous ossification of, 116f
  lateral view of, 122, 124f
Frontal fontanel, 140f
Frontal lobe, 214–215, 214f
Frontal plane, 14f, 15
Frontal region, 12f
Frontal sinus, 124, 126f, 408f
Frontonasal process, 566, 567f
Frostbite, 493
Frowning, muscles in, 170
Fructose
  chemistry of, 32–33, 33f
  dietary sources of, 474
  digestion of, 461–462
  uses in body, 32, 474
FSH. See Follicle-stimulating hormone
Full-thickness burns, 104f, 105, 106, 106f
Functional residual capacity, 420, 421f
Fundus
  of stomach, 446, 447f
  of uterus, 540f, 543
Fungal diseases, respiratory, 432t
Funny bone, 210

GABA, 203, 204t
Galactose
  digestion of, 462
  transport of, 462
Galea (epicranial) aponeurosis, 167, 171f
Gallbladder, 4f, 5f, 438f, 454f, 455, 455f
  bile storage in, 455
  epithelial tissue of, 75t
Gallstones, 468t
Gametes (sex cells), 527–528, 529f, 579
Gamma-aminobutyric acid (GABA), 203, 204t
Gamma globulins, 392. See also Antibodies
Ganglion (pl., ganglia), 88t, 193, 193f, 196. See
    also specific ganglia
Ganglion cells, of retina, 247f
Gangrene, 348
Gap junctions, 78, 78f, 323
Gardasil (HPV vaccine), 543
Gases
  in blood
    plasma, 292, 293f, 293t
    transport of, 292, 345
  partial pressure of, 422–424, 422t
  respiratory exchange of. See Gas exchange
  transport in blood, 424
Gas exchange, 407, 421–424
  aging and, 433
  circulatory system and, 345
  diffusion in lungs, 422–423
  diffusion in tissues, 423–424
  diffusion through respiratory membrane,
    414, 415f
  exercise and, 430–433
  partial pressure and, 422–424, 422t
Gastric bypass surgery, 490
Gastric glands, 446, 447f, 448, 466
Gastric phase, of stomach secretion, 448–450,
    449f
Gastric pits, 446, 447f
Gastric veins, 359f, 360f
Gastrin, 448
  functions of, 448t
  stomach secretion of, 448t, 449f
Gastritis, 450
Gastrocnemius muscle, 5f, 152f, 168f, 169f,
    184f, 185
  action of, 185, 187t
  heads of, 186f
  insertion of, 187t
  origin of, 187t
Gastroesophageal reflux disease (GERD), 12, 467
Gastrointestinal tract, 438. See also Digestive
    system
Gastroomental veins, 360f
Gastrulation, 564

Gate control theory, of pain, 240
Gated ion channels, 47, 47f, 198–199
GDP (guanosine diphosphate), 267, 267f
Gene(s), 579–582
  composition of, 579–580
  definition of, 59
  dominant, 580–581
  mapping of, 583, 583f
  random distribution of, 580
  recessive, 580–581, 581f
Gene expression, 59–61, 60f, 581–582
General anesthesia, 240
General senses, 239–241
General sensory area, 216, 216f
Genetic(s), 579–582
Genetic counseling, 582
Genetic disorders, 580, 582
Geniohyoid muscle, 173f, 173t
Genital herpes, 108t, 554t
Genitalia (genitals)
  female, 539, 539f, 549t
  male, 528–530, 530f, 533–534, 533f
Genital warts, 554t
Genome, 580, 583
Genomic map, 583, 583f
Genotype, 580, 582
GERD, 12, 467
German measles, 108t, 584t
Germ cells, 530
Germinal center, of lymph node, 382, 383f
Germinal period, 577
Germ layers
  derivatives of, 565t
  formation of, 559, 564–565, 564f
Gestation period, 572
GH. See Growth hormone
Giantism, 274
Giardiasis, 468t
Gigantism, 148t
Gingiva, 441f, 442, 442f
Gingivitis, 442
Glands, 78–79. See also specific glands
  connective tissue in, 80, 81t
  definition of, 78
  endocrine, 79, 265, 269, 270f
  exocrine, 78–79, 79f, 265
  multicellular, 78
  unicellular, 78, 79f
Glans penis, 530f, 533f, 534
Glaucoma, 248, 258t, 259
Glenoid cavity, 132, 132f, 143f
Gliding joints, 141, 142f
Globin, 296
Globulins
  hepatic synthesis of, 457
  in plasma, 292, 293f, 293t
Glomerular capillary(ies), 498, 502f
Glomerular capillary pressure (GCP),
    503–504, 504f
Glomerulonephritis, 523t
Glomerulus, 498, 500f, 502f
Glossary, G-1–G-24
Glossopharyngeal (CN IX) nerve, 224
  functions of, 224, 226t
  origin of, 226f
  and parasympathetic axons,
    227–229, 227t
  and taste, 242
Glottis, 411f
Glucagon, 285
  action of, 271t, 283t
  pancreatic production of, 457–458, 457f
  secretion of, 271t, 281, 285
  target tissues of, 271t, 283t
Glucocorticoids, 280–281
  action of, 271t, 280
  secretion of, 271t, 279f, 280–281
  stress and, 282
  target tissues of, 271t
Glucose
  blood, 308
    aging and, 288
    elevated, 284–285
    endocrine regulation of, 263, 281–285, 284f
    glucagon and, 285
    insulin and, 269, 282–285, 284f
  chemistry of, 32–33, 33f
  dietary sources of, 474
  digestion of, 461–462
  metabolism of, 483–486
    in absorptive state, 486–487, 488f
    in aerobic respiration, 163–164, 163f,
      484–486, 484f
    in anaerobic respiration, 163–164, 163f,
      483–484, 483f, 486

Glucose—*Cont.*
ATP production in, 163–164, 163f, 483–486, 483f, 484f, 485f
in citric acid cycle, 484f, 485, 486
in electron-transport chain, 484f, 485–486, 485f
in glycolysis, 483, 483f, 486
in postabsorptive state, 487, 488f
in starvation, 489
reference laboratory values for, A-2, A-5
renal reabsorption of, 508
transport of, 462, 462f
in urine, 503t, A-5
uses in body, 32, 474
Gluteal injections, 182
Gluteal muscles, 182
Gluteal region, 13f
Gluten enteropathy, 468t
Gluteus maximus muscle, 169f, 177f, 182, 183f, 184f, 184t
Gluteus medius muscle, 169f, 182, 183f, 184f, 184t
Gluteus minimus muscle, 184t
Glycerol
structure of, 33, 34f
in triglyceride formation, 33, 34f
in triglyceride metabolism, 486, 487f
Glycine, as neurotransmitter, 203, 204t
Glycogen
in absorptive state, 487, 488f
chemistry of, 33, 33f
dietary sources of, 474
digestion of, 461–462
hepatic storage of, 457
in postabsorptive state, 487, 488f
in skeletal muscle fibers, 165t
in starvation, 489
uses in body, 474
Glycolysis, 483, 483f
in aerobic respiration, 484, 484f, 486
in anaerobic respiration, 483f, 484, 486
GnRH. *See* Gonadotropin-releasing hormone
Goblet cells
aging and, 465–466
in colon, 459
in epithelial tissue, 78
in nasal cavity, 409
in small intestine, 451, 452f, 453
in stomach, 447f
as unicellular glands, 78
Goiter, 276
Golgi apparatus, 43f, 54, 194, 195f
functions of, 44t, 54
location of, 44t
structure of, 54, 55f
Gomphoses, 140
Gonadal arteries, 355f
Gonadal veins
left, 359f
right, 359f
Gonadotropin(s), 274
Gonadotropin-releasing hormone (GnRH)
action of, 536t
in females, 536t
lactation and, 550
and menstrual cycle, 546f, 547–548, 547t
and puberty, 545–546
in males, 535–537, 535f
secretion of, 535, 536t
target tissues of, 536t
Gonorrhea, 554t
Gonorrheal ophthalmia, neonatal, 584t
Gout, 148t
G protein activation, 267, 267f, 268t
Graafian follicle, 541–543, 541f, 542f
Gracilis muscle, 168f, 169f, 183f
Gram (g), 21
Granular cells, of small intestine, 451, 452f
Granulation tissue, 91f, 92
Granulocytes, 298–299
function of, 294t
production of, 295f
structure of, 294t
Granulosa cells, 541, 541f, 542f
Graves disease, 275, 286–287, 286f
Gravity, and swallowing, 446
Gray matter, 196
of brain, 196, 218
of spinal cord, 206, 207f
Great cardiac vein, 316f–317f, 321f, 357f
Greater curvature, of stomach, 446, 447f
Greater omentum, 439, 440f
Greater sciatic notch, 137f
Greater splanchnic nerve, 228f

Greater trochanter, 138, 138f, 143f
Greater tubercle, of humerus, 133f, 134
Greater vestibular glands, 544
Great saphenous vein, 357f, 360–361, 361f
Great toe, bones of, 139f, 140
Greenstick fracture, 120
Groin pull, 187
Ground substance, 79, 80
Growth
bone, 116–117, 118f
child and adolescent, 577
definition of, 7
disorders of, 148t
endocrine regulation of, 263
fetal, 567–569, 572f
puberty and, 577
Growth factors
in cancer therapy, 296
in hematopoiesis, 294, 295f
in regeneration, 91
Growth hormone (GH), 272–273
actions of, 270t, 273
aging and, 288
and amino acid transport, 464–465, 465f
disorders of, 272–273
and glucose levels, 285
reference laboratory values for, A-6
secretion of, 270t, 274
supplementation of, 273, 274
target tissue of, 270t
Growth (epiphyseal) plate, 113, 114f, 116–117, 118f
Guanine
in DNA, 38, 39f, 60
in RNA, 38, 60
Guanosine diphosphate (GDP), 267, 267f
Gynecomastia, 544
Gyrus (pl., gyri), 214, 214f, 218

Habenula, 221f
Hair(s), 5f, 100–101
aging and, 101, 108
color of, 101
as diagnostic aid, 104
distribution of, 100
functions of, 102
graying of, 101, 108
growth of, 100–101, 100f
loss of, 101
nasal, 408–409
protection provided by, 102
structure of, 100, 100f
taste, 242, 242f
Hair bulb, 100, 100f
Hair cells
of cochlea, 253–254, 253f, 259
of semicircular canals, 256, 257f
Hair follicle, 97f, 100–101, 100f, 538t
Hair follicle receptors, 239f, 240, 258
Hair root, 100, 100f
Hair shaft, 100, 100f
Hamate bone, 134, 135f
Hamstring muscles, 169f, 183f, 185
Hand
bones of, 134–135, 135f
intrinsic muscles of, 181f, 182, 182f
movements of, muscles of, 180–182, 181f, 181t
prenatal development of, 566, 566f
pronation and supination of, 144f
Hard palate, 127, 127f, 407, 408f, 441f, 442, 444, 445f
Hare, William, 4
Haversian canal, 113–115, 115f
Haversian system, 113, 114f, 115f
Hay fever, 402t
hCG. *See* Human chorionic gonadotropin
Head
of body, 10, 12f
arteries of, 350f, 352–354, 352f, 353f, 354f
blood supply to, 349f
muscles of, 167–172
sagittal section of, 14f
veins of, 356, 357f, 358f
of epididymis, 533f
of femur, 136, 138f, 143f
of fibula, 138, 139f, 140
of humerus, 132, 133f
of muscle, 167, 167f
of pancreas, 457, 457f
of radius, 134, 134f
of sperm cell, 530, 531f, 560f
of ulna, 134, 134f
Headaches, 231t
Healing, wound, 91–92, 91f

Healthy Eating index, 473
Hearing, 251–255
aging and, 259
auditory function in, 254–255, 254f
auditory structures in, 251–255
neuronal pathways for, 255
Hearing impairment, 255, 258t
aging and, 259
types of, 255
Heart, 4f, 6f, 15f, 312–340
aging and, 340, 578
anatomy of, 314–322
external, 314–316, 316f–317f
internal, 316–320, 318f
apex of, 313–314, 315f, 317f
artificial, 337
base of, 313–314, 315f
blood flow through, 313, 320–321, 320f
blood supply to, 314–316, 321–322, 321f
chambers of, 314, 316–320, 316f–317f. *See also* Atria; Ventricle(s), cardiac
conduction system of, 325–326, 325f
electrical activity of, 324–328
epithelial tissue of, 74f
exercise and, 332, 339, 340
fibrillation of, 326
functions of, 313
histology of, 322–324
location of, 313–314, 315f
neonatal, 574–575, 574f, 575f
prenatal development of, 565t, 567, 571f
regulation of, 332–340
body temperature and, 340
chemical, 333–340, 335f, 337f
extrinsic, 332–340, 337f
intrinsic, 332
nervous, 333, 334f, 337f
as seat of emotions, 313
shape of, 313
size of, 313
Starling's law of, 332
valves of, 318–320, 318f
aging and, 340
in cardiac cycle, 328, 329f, 330f
incompetent, 331, 340
replacement/repair of, 336
sounds of, 328, 330f, 331, 331f
stenosed, 331, 339t, 340
wall of, 322, 323f
Heart block, 327t
Heartburn, 450
Heart disease, 336–337, 339t
congenital, 339t, 567
prevention of, 337
treatment of, 336–337
Heart failure
aging and, 340
consequences of, 333
left, 333
right, 333
Heart-lung machine, 336
Heart murmurs, 331, 567
Heart rate (HR), 332
aging and, 340
and blood pressure, 368
exercise and, 332
regulation of, 332–340, 337f
adrenal hormones and, 279
autonomic, 229
baroreceptor reflex in, 333, 334f, 337f
brainstem in, 211
chemoreceptor reflexes in, 333, 335f, 337f
endocrine, 263
Heart sounds, 314, 328, 330f, 331, 331f
Heart transplant, 336–337
Heat, and body temperature, 491–492, 491f
Heat exhaustion, 493
Heat sensation, 102
Heat stroke, 493
Heel. *See* Calcaneus
Heimlich maneuver, 411
Helper T cells, 389, 390f, 391, 396
Hematocrit, 306–307, 307f, A-4
Hematoma, subdural, 222
Hematopoiesis, 294, 295f, 297, 297f, 389, 389f
Heme, 296
Hemiazygos vein, 358f, 359
Hemidesmosomes, 78, 78f
Hemispheres, cerebral, 214–215
communication between, 220
left, 214, 214f
right, 214, 214f
Hemocytoblasts. *See* Stem cells
Hemodialysis, 520–521, 520f
Hemoglobin, 296–297

breakdown of, 298, 298f
deficiency of, 306, 308
fetal, A-4
function of, 296–297
measurement of, 306
oxygen binding/transport by, 296–297
production of, 297
in sickle-cell anemia, 581
structure of, 296
Hemoglobin-based oxygen carriers (HBOCs), 297
HemoLink, 297
Hemolysis, 303
Hemolytic anemia, 308
Hemolytic disease of the newborn (HDN), 305–306, 305f
Hemophilia, 309t, 581, 581f
HemoPure, 297
Hemorrhagic anemia, 308
Hemorrhagic shock, 369
Hemorrhagic stroke, 232
Hemorrhoids, 375, 460
Henle, loop of
anatomy of, 498–502, 500f
reabsorption in, 504, 505f, 506, 507f, 508f
Heparin, 299, 302
Hepatic artery, 454f, 455
Hepatic cords, 454f, 455
Hepatic ducts, 454f, 455
Hepatic portal system, 359f, 360, 360f
Hepatic portal vein, 357f, 360, 360f, 454f, 455
Hepatic sinusoids, 454f, 455
Hepatic veins, 357f, 359f, 360, 360f, 455
Hepatitis, 104, 468t
Hepatocytes, 454f, 455
Herceptin, 400
Hering-Breuer reflex, 426, 427f
Hernia
hiatal, 467
inguinal, 528
Herpes, 108t, 231t, 554t
Herpes zoster, 402
Heterozygous, 580–581
Hiatal hernia, 467
High-density lipoproteins (HDL), 464, 464f
Hilum, of kidney, 497–498, 499f
Hinge joints, 141, 142f
Hip (hip joint), 142f, 143f
bones of, 135–136, 136f
extension of, 182, 184
flexion of, 182, 184
fracture of, 138, 146, 146f
frontal section through, 14f
muscles of, 182–184, 183f, 184t
Hip replacement, 146
Hippocampus, dentate gyrus of, 221f
Hispanics, bone mass in, 145
Histamine
basophil release of, 299
in inflammatory response, 89, 387
in innate immunity, 386
stomach secretion of, 448, 449f
tubular secretion of, 509
Histamine blockers, 450
Histidine, 477
Histology, definition of, 72
HIV, 554t, 584t
Hives (urticaria), 402t
Homeostasis, 7–10
aging and, 578
definition of, 7
extracellular fluid and, 516
humors and, 8
negative feedback and, 7–8
nervous system in, 193
positive feedback and, 8–10
Homologous chromosomes, 579–580
Homozygous, 580–581
Horizontal cells, of retina, 246, 247f
Horizontal plane, 14f, 15
Horizontal plate, of palatine bone, 127f
Hormone(s), 265–288. *See also specific hormones*
of adrenal cortex, 271t, 279f, 280–281, 281f
of adrenal medulla, 271t, 277–279, 279f, 280f
aging and, 288
and blood flow, 367–368, 367t
and blood pressure, 371–373, 374f
characteristics of, 265
as chemical signals, 263, 264t
chemistry of, 265–266
definition of, 263, 265
of digestive system, 286, 448t
in labor (childbirth), 573–574, 573f
in lactation, 274, 275, 576–577, 576f

lipid-soluble, 265–266
medical uses of, 266
negative feedback and, 268–269
of ovaries, 271t, 285
of pancreas, 271t, 281–285, 457–458
of parathyroid gland, 271t, 276–278
of pineal gland, 271t, 285–286
of pituitary gland, 270–275, 271t, 272f
in pregnancy, 563, 563f
receptors for, 266–268, 268t
reference laboratory values for, A-6
reproductive, 269
secretion of, 79, 268–269
blood composition and, 269
hormonal control of, 269
neural control of, 269
stress and, 282
target cell response to, 265, 265f
target tissues of, 265
of testes, 271t, 285
of thymus, 271t, 285
of thyroid gland, 271t, 275–276
transport of, 292, 345
in urine regulation, 510–513, 512f
water-soluble, 265
Hormone replacement therapy (HRT), 147, 549
Horn(s)
anterior, 206, 207f
lateral, 206, 207f
posterior, 206, 207f
Hot flashes, 549, 549t
HPV, 543
Human chorionic gonadotropin (hCG), 287–288
action of, 536t
and pregnancy, 543, 563, 563f
secretion of, 536t
target tissue of, 536t
Human genome, 583
Human Genome Project, 583
Human immunodeficiency virus (HIV), 554t, 584t
Human papillomavirus (HPV), 543
Human physiology, 2
Humeral joint, 142f
Humerus, 5f, 121t, 123f
anatomical neck of, 132–134, 133f
at elbow, 143f
fracture of, 211
head of, 132, 133f
lateral epicondyle of, 133f, 134
medial epicondyle of, 133f, 134
at shoulder, 143f
structure of, 132–133, 133f
surgical neck of, 133f, 134
Humors, 8
Hunger pangs, 451
Huntington disease, 218, 231t, 580, 582
Hyaline cartilage, 83
functions of, 83, 83t
location of, 83, 83t
structure of, 83, 83t
Hyaline membrane disease, 419
Hydrocephalus, 224
Hydrochloric acid, 448
aging and, 467
diarrhea and, 467
functions of, 443t, 448
stomach secretion of, 443t, 448
Hydrocortisone. See Cortisol
Hydrogen
atomic number of, 21t
atomic structure of, 22, 22f
isotopes of, 27
mass number of, 21t
percent in body, 21t
symbol for, 21t, 22
Hydrogenated oil, 476
Hydrogen bonds, 24–25
of amino acids, 36f, 37
of water molecule, 24–25, 25f, 31
Hydrogen ions, 23t, 30
concentration of. See pH
tubular secretion of, 509
Hydrophilic region, of lipids, 34, 35f, 44
Hydrophobic region, of lipids, 34, 35f, 44
Hydrostatic pressure, 47, 48f
Hydroxide ion, 23t
Hydroxyapatite, 113, 147
Hydroxyl group, 33
Hymen, 543–544
Hyoid bone, 121t, 123f, 127, 127f
Hyoid muscles, 170, 173f, 173t
Hypercapnia, 427
Hyperextension, 145

Hyperglycemia, 284–285
Hyperopia, 258t
Hyperparathyroidism, 277
Hyperplastic obesity, 490
Hyperpolarization, of postsynaptic cells, 203
Hypertension, 363
pregnancy-induced, 584t
treatment of, 336
Hyperthermia, 493
Hyperthyroidism, 275, 286–287, 286f
Hypertonic solutions, 48–49, 49f
Hypertrophic obesity, 489–490
Hypertrophic pyloric stenosis, 446
Hypertrophy, of muscle, 165, 189t
Hypochondriac region, right and left, 13f
Hypodermis, 96, 97f, 100
Hypogastric ganglion, 228f
Hypogastric region, 13f
Hypoglossal canal, 126f
Hypoglossal (CN XII) nerve, 224
functions of, 224, 226t
origin of, 226f
Hypoparathyroidism, 277–278
Hypophysis. See Pituitary gland
Hypopituitarism, 274–275
Hypothalamic-pituitary portal system, 271, 277f, 282f
Hypothalamus, 6f, 213–214
adrenal regulation by, 280f, 282f
and body temperature, 491–492, 492f
in female puberty, 545–546
fetal, in labor (childbirth), 573, 573f
functions of, 213–214
in lactation, 576–577, 576f
and male reproductive system, 535, 535f, 537
and menstrual cycle, 546f, 547–548
in pituitary regulation, 270–272, 272f, 273f, 274f
satiety center in, 284
structure of, 211f, 213, 213f
thirst center of, 516, 516f
in thyroid regulation, 275, 277f
Hypothermia, 493, 579
Hypothyroidism, 275
Hypotonic solutions, 48, 49f
Hypovolemic shock, 369
Hypoxia, 429
Hysterectomy, 553
H zone, 153, 155f, 158, 160f

I band, 153, 155f, 158, 160f
Ibuprofen, as anti-inflammatory, 89
Identical twins, 581
Idiopathic thrombocytopenic purpura (ITP), 300
Ileocecal junction, 451f, 452
Ileocecal sphincter, 452
Ileocecal valve, 452, 459f
Ileum, 451–452, 451f, 452f, 459f, 460–461
Iliac arteries
common, 348–352, 350f, 356, 356f, 498f
left, 351f, 355f
right, 355f
external, 350f, 351f, 355f, 356, 356f
internal, 350f, 351f, 355f, 356, 356f
Iliac crest, 135, 136f, 137f
Iliac fossa, 137f
Iliac region, right and left, 13f
Iliac spine
anterior inferior, 136f
anterior superior, 135
Iliac vein(s)
common, 357f, 359f, 360, 498f
external, 357f, 359f, 360
internal, 357f, 359f, 360
Iliocostalis muscle(s), 174f, 174t
Iliopsoas muscle, 182, 183f, 184t
Iliotibial tract, 169f, 183f
Ilium, 135, 136f, 136t, 137f
Imaging, anatomical, definition of, 2
Immediate allergic reactions, 402t
Immune, definition of, 386
Immune system. See also Immunity
aging and, 578
burns and, 107
endocrine regulation of, 263
Immunity, 386–398
acquired, 397–398, 397f
active, 397–398, 397f
active artificial, 397–398, 397f
active natural, 397–398, 397f
adaptive, 386, 388–396, 399f
aging and, 400–403
antibody-mediated, 389, 392–395, 399f
cell-mediated, 389, 395–396, 396f, 397f, 399f

definition of, 386
disorders of, 402t
innate, 386–388, 399f, 407
interactions in, 398, 399f
passive, 397–398, 397f
passive artificial, 397–398, 397f
passive natural, 397–398, 397f
respiration and, 407
Immunodeficiency, 402t
acquired. See Acquired immunodeficiency syndrome
congenital, 402t
severe combined, 402t
Immunoglobulins (Ig), 392, 393t. See also Antibodies
IgA, 392, 393t
IgD, 392, 393t
IgE, 392, 393t
IgG, 392, 393t
IgM, 392, 393t
Immunotherapy, 65, 398–400
Impacted fracture, 120, 120f
Impetigo, 108t
Implantation, 543, 548, 561f
Implicit memory, 221
Impotence, 539, 552–554
Incisive fossa, 127f
Incisors, 440, 441f, 444
Incompetent heart valves, 331, 340
Incomplete dominance, 581
Incomplete fracture, 120, 120f
Incomplete protein, 477
Incus, 121t, 252, 252f, 254f
Indirect tracts, 217, 217t
Infant(s)
definition of, 577
development of, 577
Infant respiratory distress syndrome (IRDS), 419, 574
Inferior, 10, 11f, 11t
Inferior colliculus, 212, 212f, 255
Inferior concha, 408f
Inferior lobe, of lungs, 413, 413f
Inferior meatus, 408f
Inferior mesenteric artery, 350f, 351f, 354–355, 355f
Inferior mesenteric vein, 357f, 359f, 360, 360f
Inferior nasal concha, 121t
Inferior oblique muscle, 243f, 244, 244f
Inferior rectus muscle, 243f, 244, 244f
Inferior vena cava, 6f, 314, 316f–318f, 317, 320, 320f, 356, 357f–360f
Infertility
female, 549
male, 531, 539
Inflammation, 89, 387–388, 388f
benefits of, 387
cells of, 387
chemical mediators of, 387, 388f
chronic, 89
local, 388
prostaglandins in, 287
stages of, 89, 90f
symptoms of, 89, 388
systemic, 388
in tissue repair, 91f, 92
white blood cells in, 299
Inflammatory bowel disease (IBD), 468t
Inflammatory response. See Inflammation
Influenza, 432t
Influenza vaccine, 400–402
Infrahyoid muscles, 170, 173f, 173t
Infraorbital foramen, 125f
Infrapatellar bursa, deep, 143f
Infraspinatus muscle, 169f, 178f, 180t
Infraspinous fossa, 132f
Infundibulum, 212f, 213, 213f, 270, 272f
Infusion, definition of, 303
Inguinal canal, 528, 532, 533f
Inguinal hernia, 528
Inguinal lymph node, 5f, 381f, 382
Inguinal region, 12f
Inhalation. See Inspiration
Inheritance, cellular function in, 44
Inhibin, 535f, 536t, 537
Injections
intradermal, 97
intramuscular, 97
subcutaneous, 97
Innate immunity, 386–388, 399f
vs. adaptive immunity, 388t
chemical mediators of, 386
complement in, 386
interferons in, 386
physical barriers in, 386

respiration and, 407
white blood cells in, 386
Inner cell mass, 561, 561f
Inner ear, structure of, 252–254, 252f, 253f
Inner ear infection, 258t
Inorganic chemistry, 31–32
Inositol triphosphate (IP3), and G proteins, 267
Insecticides, 159
Insemination, artificial, 531
Insertion, of muscle, 167, 167f
Inspiration, 416
alveolar pressure changes in, 419–420
muscles of, 173, 416, 417f
pleural pressure and, 419–420
pressure changes during, 417, 418f
start and stop of, 426
Inspiratory capacity, 421, 421f
Inspiratory reserve volume, 420, 421f
Insulin, 282–285
action of, 271t, 283t, 284
and amino acid transport, 464–465, 465f
in diabetes mellitus, 284–285
and glucose transport, 462
pancreatic production of, 457–458, 457f
reference laboratory values for, A-6
secretion of, 271t, 281, 283
target tissues of, 271t, 283t, 284
Insulin-like growth factors (IGFs), 273
Insulin shock, 285
Integument, definition of, 96
Integumentary system, 3, 5f, 95–108. See also Hair(s); Nail(s); Skin
aging and, 2, 108
components of, 5f, 96
as diagnostic aid, 104
diseases and disorders of, 108t
excretion by, 96, 104
functions of, 5f, 96, 102–104
physiology of, 102–104
prenatal development of, 568t–569t
protection by, 96, 102
sensation in, 96, 102
temperature regulation by, 96, 102, 103f
vitamin D production in, 96, 102
Interatrial septum, 317, 567, 571f
Intercalated disks, of cardiac muscle, 86t, 87, 166, 323, 323f
Intercarpal joint, 142f
Intercellular chemical signals, 263–265, 264t
Intercostal arteries
anterior, 351f, 354
posterior, 351f, 354
Intercostal muscles
external, 173, 175f, 175t, 416, 417f, 426f
internal, 173, 175f, 175t, 416, 417f, 426f
in respiration, 173, 416, 417f, 426f
Intercostal veins
anterior, 356–359
posterior, 358f, 359
Interferon(s), 386
in immunotherapy, 398
in innate immunity, 386
Interleukin-1, 391
Interleukin-2, 390f, 391
Interleukin(s), in adaptive immunity, 390f, 391
Interleukin-2 receptors, 390f, 391
Interlobar arteries, 502, 502f
Interlobular arteries, 502, 502f
Intermediate cuneiform, 139f
Intermediate filaments, 57, 58f
Intermetatarsal joint, 142f
Internal abdominal oblique muscle, 176, 176f, 176t, 178f
Internal anal sphincter, 459–460, 459f
Internal auditory canal, 126f
Internal carotid arteries, 125, 350f, 352, 352f, 353f, 354f
Internal iliac artery, 350f, 351f, 355f, 356, 356f
Internal iliac vein, 357f, 359f, 360
Internal intercostal muscle, 173, 175f, 178f
action of, 173, 175t
insertion of, 175t
origin of, 175t
in respiration, 173, 416, 417f, 426f
Internal jugular vein, 125, 356, 357f, 358f
Internal urinary sphincter, 513–514, 532
International Bureau of Weights and Measurements, 21
International Normalized Ratio (INR), 308
Interneurons, in reflex arc, 204, 205f
Interossei muscle, 182
Interphalangeal joint, 142f
Interphase, 61–62, 63f, 64f
Interstitial cells, 530, 531f
Interstitial cell-stimulating hormone (ICSH), 274

Interstitial fluid, volume of, 515t
Interstitial fluid loss shock, 369
Intertarsal joint, 142f
Interthalamic adhesion, 213, 213f
Interventricular arteries
    anterior, 316f–317f, 321, 321f
    posterior, 317f, 321f
Interventricular septum, 317–318, 318, 318f, 567, 571f
Interventricular sulcus
    anterior, 314, 316f–317f
    posterior, 314, 317f
Intervertebral disks, 83, 84t, 128, 128f
Intervertebral foramen, 128f, 129
Intervertebral joint, 142f
Intestinal glands, 451, 452f
Intestinal parasites, 468t
Intestinal phase, of stomach secretion, 448–450, 449f
Intestines. See also Large intestine; Small intestine
    epithelial tissue of, 75t, 76
    muscles of, 152f
    prenatal development of, 570f
Intracellular chemical signals, 263
Intracellular fluid
    composition of, 516
    volume of, 515t
Intracellular fluid compartment, 514–516
Intracellular receptors, 266, 267f, 268, 268t
Intracellular substances, 44
Intradermal injection, 97
Intramembranous ossification, 115, 116f
Intramuscular injection, 97
Intrauterine devices (IUDs), 550, 551f
Intrinsic factor, 308, 443t, 448
Intrinsic regulation, of heart, 332
Intubation, 411
Inversion, of foot, 145, 185
In vitro fertilization, 562
Involuntary movements, 216
Iodide ion, 23t
Iodine
    atomic number of, 21t
    deficiency of, 480t
    dietary guidelines for, 480t
    functions of, 480t
    mass number of, 21t
    percent in body, 21t
    radioactive, diagnostic use of, 27
    symbol for, 21t
    and thyroid hormones, 275–276
Ion(s), 22–23. See also specific types
    and electricity, 25
    important, 23t
    intracellular-extracellular exchange of, 516
    in plasma, 292, 293f, 293t
    renal reabsorption of, 504–508, 505f–509f
    in urine, 503t
Ion channels, 47, 47f
    and action potential, 199–201, 200f
    chemically gated, 199
    gated, 47, 47f, 198–199
    leak, 47, 47f, 198–199
    and resting membrane potential, 155, 156f, 198–199, 198f, 199f
    voltage-gated, 199–201, 200f
Ionic bonds, 22–23, 23f
Ionic compounds, 25
Ionized substances, equivalents of, A-7
Iris, 243f, 245, 245f, 246f, 249
Iron
    atomic number of, 21t
    deficiency of, 104, 306–307, 308, 480t
    dietary guidelines for, 479, 480t
    functions of, 480t
    in hemoglobin, 296–297, 306
    mass number of, 21t
    percent in body, 21t
    recycling of, 298, 298f
    reference laboratory values for, A-2
    symbol for, 21t
Iron-deficiency anemia, 104, 308
Iron ion, 23t
Irregular bones, 113
Irritable bowel syndrome (IBS), 468t
Ischemic stroke, 232
Ischial spine, 136t, 137f
Ischial tuberosity, 136t, 137f, 183f
Ischiocavernosus muscle, 177f, 177t
Ischium, 135, 136f, 137f
Islets of Langerhans, 281, 283f, 457–458, 457f
Isoleucine, 477
Isometric contractions, 164
Isotonic contractions, 164

Isotonic solutions, 48, 49f
Isotopes, 27
    definition of, 27
    radioactive, 27
Isthmus, of thyroid gland, 275, 276f
IUDs, 550, 551f

Jaundice, 104, 298
Jaw. See Mandible; Maxilla
Jejunum, 451, 451f, 452f, 457f, 460–461
Joint(s), 140–145
    aging and, 145–148, 578
    ball-and-socket, 142f, 143
    cartilaginous, 140–141
    classification of, 140
    definition of, 112, 140
    diseases and disorders of, 148t
    ellipsoid, 142f, 143
    fibrous, 140
    functions of, 112
    gliding, 141, 142f
    hinge, 141, 142f
    injuries of, 145
    movements of, 112, 140, 143–145, 144f
    pivot, 141, 142f
    plane, 141, 142f
    replacement of, 148t
    saddle, 141, 142f
    synovial, 141–143
        aging and, 145–148
        examples of, 143f
        structure of, 141, 141f
        types of, 141–143, 142f
Joint capsule, 141, 141f, 143f
Joint cavity, 141, 141f, 143f
Jugular foramen, 125, 126f, 127f
Jugular notch, 131, 131f, 133f
Jugular vein, 6f
    external, 356, 357f, 358f
    internal, 125, 356, 357f, 358f
Juxtaglomerular apparatus, 502, 502f
Juxtamedullary nephrons, 498, 500f

Karyotype, 579, 579f, 582
Keratin
    in epidermis, 96
    in nails, 101
    production of, 96
Keratinization
    in epidermis, 96
    in hair, 101
Keratinized stratified squamous epithelial tissue, 76, 77t
Ketone bodies, 486, 487, 487f, 488f, 489, A-4
Ketones, 283
Ketosis, 486
Kidney(s), 4f, 6f, 14f, 497, 497f
    and acid-base balance, 519, 522f
    aging and, 578
    anatomy of, 497–502, 498f, 499f
    anterior view of, 498f
    blood flow through, 349f, 502, 502f, 503
    and blood pressure, 497
    epithelial tissue of, 73, 74t, 76, 498–502
    filtration in, 502–504, 503f
    functional unit of, 498–502, 500f
    functions of, 497
    hilum of, 497–498, 499f
    hormone secretion in, 287
    inferior view of, 498f
    inflammation of, 523t
    left, 498f
    location of, 497, 498f
    longitudinal section of, 499f
    prenatal development of, 565t, 567, 570f
    right, 498f
    testosterone and, 538t
    tubular reabsorption in, 502–508, 503f, 505f
    tubular secretion in, 502–503, 503f, 508–509
    urine production in, 502–509, 503f
Kidney failure, 520–521, 523t
Kidney stones, 514
Kidney transplants, 521
Kilocalories (kcal), 474
Kilogram (kg), 21
Kinetic energy, 28
Kinins, in inflammatory response, 387
Kissing muscles, 167
Knee(s) (knee joint)
    extension of, 143, 184–185
    fibrocartilage in, 83, 84t
    flexion of, 143, 185
    as hinge joint, 141, 142f
Knee-jerk reflex, 184–185, 208, 208f
Korotkoff sounds, 362, 362f

Krebs, Sir Hans, 485
Krebs cycle, 484f, 485, 486
Kupffer cells, 387
Kyphosis, 128

Labia majora, 539f, 544, 544f
Labia minora, 539f, 544, 544f
Labor, 572–574
    definition of, 572
    first (dilation) stage of, 573
    hormones in, 573–574, 573f
    second (expulsion) stage of, 573
    third (placental) stage of, 573
Laboratory values, reference, A-2–A-6
Lacrimal apparatus, 244, 244f
Lacrimal bone, 121t, 124
    frontal view of, 124, 125f
    lateral view of, 124f
Lacrimal canaliculi, 244, 244f
Lacrimal ducts, 244, 244f
Lacrimal glands, 244, 244f
Lacrimal sac, 244, 244f
Lactase, 453
    functions of, 443t
    in newborn, 576
    secretion of, 443t
Lactation, 576–577
    and birth control, 550
    breast changes in, 545, 545f, 576
    endocrine regulation of, 263, 274, 275
    hormonal control of, 274, 275, 576–577, 576f
Lacteals, 380, 381f, 385f, 451, 452f
Lactic acid
    in anaerobic respiration, 163, 429–430, 483–484, 483f
    in muscle contraction, 163–164
    reference laboratory values for, A-2
    transport in blood, 292
Lactic dehydrogenase, A-2
Lactiferous duct, 544–545, 545f
Lactose
    in breast milk, 576
    dietary sources of, 474
    digestion in newborn, 576
    digestion of, 461–462
Lactose intolerance, 453
Lacunae
    in bone, 84, 85t, 113, 115f
    in cartilage, 83, 84t–85t
    in placenta, 561, 562f
Lag phase, of muscle twitch, 162, 162f
Lambda(e), 113, 115f
    in bone growth, 116–117
    in cancellous bone, 116f
    in compact bone, 113, 115f
Laminae, 128–129, 128f
    of cervical vertebrae, 129f
    of lumbar vertebrae, 129f
    of thoracic vertebrae, 129f
Lamina propria
    of digestive tract, 438, 439f
    of stomach, 447f
Langerhans, islets of, 281, 283f, 457–458, 457f
Lanugo, 568
Laparoscopy, 552
Large intestine, 4f, 5f, 14f, 438, 438f, 459–460
    anatomy of, 459–460, 459f
    components of, 459, 459f
    diseases and disorders of, 468t
    epithelial tissue of, 438
    functions of, 460
    movement in, 460
Large veins, 345, 346f, 347
Laryngitis, 410
Laryngopharynx, 408f, 409, 444
Laryngoscope, 411f
Larynx, 4f, 5f, 407f, 409–410, 412f
    anatomy of, 408f, 409–410, 410f
    epithelial tissue of, 76, 77t
    movement of, muscles of, 170, 173f, 173t
    during swallowing, 170, 410, 444, 445f
    testosterone and, 538t
Last menstrual period (LMP), 559, 572
Latent phase, of muscle twitch, 162, 162f
Lateral, 10, 11f, 11t
Lateral column, 206, 207f
Lateral corticospinal tract, 217–218, 217f, 217t, 218f
Lateral cuneiform, 139f
Lateral epicondyle
    of femur, 138, 138f, 140f
    of humerus, 133f
Lateral fissure, 214f, 215
Lateral horn, 206, 207f

Lateral incisor, 441f
Lateral malleolus, 138, 139f, 140f
Lateral medullary infarction, 233
Lateral pterygoid muscle
    action of, 172t
    insertion of, 172t
    origin of, 172t
Lateral rectus muscle, 244, 244f
Lateral rotation, 144f, 145
Lateral ventricle, 222, 224f, 225f
Latissimus dorsi muscle, 169f, 177, 178f, 179f, 180t
Leak channels, 47, 47f, 198–199
Lecithin, 476
Leeches, 8
Left, as directional term, 10, 11f, 11t
Left atrium, 314, 315f–318f, 316–317, 320f, 321
Left bundle branch, 325, 325f
Left coronary artery, 318f, 321, 321f, 348, 350f
Left heart failure, 333
Left hemisphere, of brain, 214, 214f, 220
Left hypochondriac region, 13f
Left iliac region, 13f
Left lumbar region, 13f
Left ventricle, 314, 315f, 316f–317f, 317–318, 318f
    blood flow through, 320f, 321
    contraction of, 318
    hypertrophy of, 340
    wall of, 318
Leg. See also Lower limb
    abduction of, 145
    adduction of, 145
    anterior compartment, muscles of, 184–185, 185t, 187t
    bones of, 136f, 138, 139f
    definition of, 12, 138
    movements of, muscles of, 184–185, 185t, 186f, 187t
    posterior compartment, muscles of
        deep, 185, 187t
        superficial, 185, 187t
    prenatal development of, 566, 566f
Leiomyomas, 553, 553f
Length, measures of, A-1
Lens, 245, 245f, 249
    aging and, 259, 578
    in focusing system, 249, 249f
    prenatal development of, 565t
Leprosy, 231t
Lesser curvature, of stomach, 446, 447f
Lesser omentum, 439, 440f, 454f
Lesser splanchnic nerve, 228t
Lesser trochanter, 138, 138f, 143f
Lesser tubercle, of humerus, 133f, 134
Leukemia, 307
Leukocytes. See White blood cells
Leukocytosis, 307
Leukopenia, 307
Leukotrienes, 266
    in inflammatory response, 387
    in innate immunity, 386
Levator ani muscle, 176, 177f, 177t
Levator labii superioris muscle, 167–170, 171f, 172t
Levator scapulae muscle, 176–177, 178f, 178t
Levitra (vardenafil), 539
Leydig cells, 530, 531f
LH. See Luteinizing hormone
Libido, menopause and, 549t
Lie detector tests, 101
Life
    characteristics of, 6–7
    stages of, 559, 577
Life cycle, of cell, 61–63, 64f
Life expectancy, 559
Life span, 559
Ligament(s). See also specific ligaments
    aging and, 92, 148
    extracellular matrix of, 113
    functions of, 80, 112
    structure of, 82t
Ligamentum teres, 143f
Ligamentum venosum, 575f
Ligand(s), 263. See also Chemical signals
Light refraction, 249, 249f
Limb buds, development of, 566, 566f
Limbic system, 221–222
    functions of, 221–222
    structure of, 221, 221f
Linea alba, 168f, 175, 176f
Linear fracture, 120
Lingual arteries, 352f
Lingual tonsils, 382, 382f, 408f, 409

Lingual veins, 358f
Linoleic acid, 477
Linolenic acid, 34f
Linolenic acid, alpha, 477
Lip(s), 440, 441f
  cleft, 566
  color of, 440
  prenatal development of, 566, 567f
Lipase, 458
  functions of, 443t, 458, 461f, 463
  secretion of, 443t, 458
Lipid(s)
  in cell membrane, 44–45, 45f
  chemistry of, 32t, 33–35
  composition of, 33
  dietary amounts, recommended, 476–477
  dietary sources of, 474–476
  digestion of, 460, 462–464, 473
  emulsification of, 463, 476
  metabolism of, 486, 487f
  as nutrients, 473, 474–477
  reference laboratory values for, A-3
  in sebum, 101
  solubility of, 462
  transport of, 463, 463f
  uses in body, 32t, 476
Lipid-soluble hormones, 265–266
Lipid-storage disorders, 56
Lipoproteins, 464, 464f
Lithium, reference laboratory values for, A-3
Liver, 4f, 5f, 14f, 438f, 440f, 453–457
  aging and, 466, 578
  anatomy of, 453–455, 454f
  bile flow through, 455, 455f
  blood flow through, 349f, 454f, 455
  detoxification in, 456t, 457
  diseases and disorders of, 468t
  functions of, 455–457, 456t
  glycogen in, 457
  histology of, 454, 455f
  lobes of, 453, 454f
  lobules of, 454f, 455
  location of, 453
  prenatal development of, 565t, 567, 570f
  secretions of, 452, 453
  storage in, 456t, 457
  synthesis in, 456t, 457
  weight of, 453
Lobar bronchi, 412f, 413, 413f
Lobes
  of brain, 214–215, 214f
  of breasts, 544–545, 545f
  of liver, 453, 454f
  of lungs, 413, 413f
Lobules
  of breasts, 545, 545f
  of liver, 454f, 455
  of seminiferous tubules, 531f
  of thymus, 384, 384f
Local anesthesia, 240
Local control, of blood flow, 366, 366f, 367f
Local current, 200–201
Local inflammation, 388
Local potentials, 201, 201f
Lock-and-key model, of enzyme action, 37, 38f
Long bones, 113
  in children, vs. adults, 113, 114f
  ossification of, 117f
  structure of, 113, 114f
Longissimus muscles, 174f, 174t
Longitudinal fissure, 214, 214f
Longitudinal muscle layer
  of small intestine, 452f
  of stomach, 446, 447f
Longitudinal section, 15, 15f
Longitudinal smooth muscle, 438, 439f
Long-term memory, 221, 234
Loop of Henle
  anatomy of, 498–502, 500f
  reabsorption in, 504, 505f, 506, 507f, 508f
Loose connective tissue, 80
  functions of, 80, 81t
  location of, 80, 81t
  structure of, 80, 81t
Loperamide, 467
Lordosis, 128
Low back pain, 172–173
Low-density lipoprotein(s) (LDL), 464, 464f
Lower esophageal sphincter, 444, 447f
Lower limb, 12, 12f, 13f
  arteries of, 350f, 356, 356f
  blood flow to, 349f
  bones of, 121t, 136–140, 136f, 140f
  muscles of, 182–185, 184f
  prenatal development of, 566, 566f

surface anatomy of, 140f, 184f
  veins of, 357f, 360–361, 361f
Lower motor neurons, 217
Lower respiratory tract, 407, 407f
Lower respiratory tract infections, 432t
Lubb sound, 331
Lumbar arteries, 351f, 355, 355f
Lumbar nerves, 208, 210–211, 210f
Lumbar region, 13f
  left, 13f
  right, 13f
Lumbar veins, ascending, 358f, 359, 359f
Lumbar vertebrae, 128f
  curvature of, 128, 128f
  fifth, 128f
  first, 128f
  number of, 121t, 128
  stability of, 130
  structure of, 129f, 130
  surface view of, 133f
Lumbosacral plexus, 208, 210–211, 210f, 210t
Lunate bone, 134, 135f
Lung(s), 4f, 5f, 407f, 413–414
  aging and, 578
  anatomy of, 412f, 413–414, 413f
  cancer of, 78, 416, 432t
  disorders of, 432t
  elasticity of, 414
  epithelial tissue of, 72–73, 74t, 75t, 76
  gas diffusion in, 422–423
  left, 315f, 416f
  lobes of, 413, 413f
  lymphatic supply to, 414f, 415–416
  neonatal, 575–576
  phagocytosis in, 416
  prenatal development of, 565t, 567, 570f
  right, 315f, 416f
  surface tension of, 419
Lung capacities, 420–421, 421f
Lung recoil, 419
Lung volumes, 420–421, 421f
Lunula, 101, 102f
Lupus erythematosus, systemic, 400–401, 400f
Luteinizing hormone (LH), 274
  action of, 268t, 270t, 274, 285, 536t
  in females, 536t
    and fertility/infertility, 549, 562
    lactation and, 550
    and menopause, 548
    and menstrual cycle, 546f, 547–548, 547t
    oral contraceptives and, 550
    and puberty, 545–546
  in males, 531, 535f, 536t, 537
  reference laboratory values for, A-6
  secretion of, 268t, 270t, 274, 536t
  target tissues of, 268t, 270t, 536t
Lymph, 380
  formation of, 380, 381f
  movement of, 380, 381f
Lymphadenitis, 402t
Lymphangitis, 402t
Lymphatic capillaries, 380, 381f, 384, 385f
Lymphatic ducts, right, 380, 381f, 385f
Lymphatic nodules, 382, 383f
Lymphatic organs, 380–384
Lymphatic sinuses, 382, 383f
Lymphatic system, 3, 5f, 380–385
  aging and, 400–403
  anatomy of, 380–384, 381f
  components of, 5f
  diseases and disorders of, 402t
  functions of, 5f, 380
  in lungs, 414f, 415–416
  overview of, 384, 385f
  prenatal development of, 568t–569t
Lymphatic tissue, 380–382
Lymphatic vessels, 5f, 380, 381f, 384, 385f
  afferent, 382, 383f
  compression of, 380
  deep, 414f, 415
  drainage of, 380, 381f
  efferent, 382, 383f
  superficial, 414f, 415
Lymphedema, 402t
Lymph node(s), 382, 414f
  axillary, 5f, 381f, 382
  capsule of, 382, 383f
  cervical, 5f, 381f, 382, 442
  distribution of, 382
  function of, 382, 384, 385f
  germinal centers of, 382, 383f
  inguinal, 5f, 381f, 382
  structure of, 382, 383f
  toothache and, 442
  trabeculae of, 382, 383f

Lymphocytes
  activation of, 390–392
  in adaptive immunity, 389–392
  aging and, 401, 578
  antigen recognition by, 390–391
  costimulation of, 390f, 391, 391f
  development of, 389, 389f
  function of, 294t, 299
  in lymphatic tissue, 380–382
  maturation of, 384
  origin of, 389, 389f
  percentage of WBCs as, 293f, 307
  production of, 295f, 380
  proliferation of, 390f, 391–392, 391f
  staining properties of, 299f
  structure of, 294t, 299
  types of, 384, 385f
Lymphoma, 402t
Lysine, 477
Lysosomes, 43f, 44t, 54, 56f
Lysozymes, 386, 443t, 444

Macrophages, 299
  in connective tissue, 80
  definition of, 387
  functions of, 299, 387
  in hemoglobin breakdown, 298, 298f
  in innate immunity, 386–387
  in phagocytosis, 299, 386–387
Macula (ear), 255, 255f, 256f
Macula (eye), 246, 248f
Macular degeneration, 258t
Magnesium
  deficiency of, 480t
  dietary guidelines for, 480t
  digestion of, 465
  functions of, 480t
  in intracellular fluid, 516
Magnesium ion, 23t
Magnetic resonance imaging (MRI), 27, 27f
Main bronchus, 412–413, 412f, 413f, 416f
Major histocompatibility complex (MHC)
    molecules
  and adaptive immunity, 390–392, 390f, 391f
  and cell-mediated immunity, 396, 396f
  class I, 396, 396f
  class II, 391–392
Malaria, 309t
Male infertility, 531, 539
Male pattern baldness, 101, 538
Male pelvis, 136, 136t, 137f
Male puberty, 537, 577
Male reproductive hormones, 535–537, 535f, 536t
Male reproductive system, 6f, 527f, 528–539
  accessory glands of, 528, 534–535
  aging and, 552–554
  anatomy of, 528–535, 530f, 533f
  components of, 6f
  diseases and disorders of, 554t
  ducts of, 528, 531–532
  functions of, 6f
  hormones of, 535–537, 535f, 536t
  physiology of, 535–539
  puberty in, 537
  secretions of, 535
  sensory impulses and integration in, 538–539
  supporting structures of, 528
Malignant tumors, 65
Malleolus
  lateral, 138, 139f, 140f
  medial, 138, 139f, 140f
Malleus, 121t, 252, 252f, 254f
Maltase, 443t
Maltose, dietary sources of, 474
Mammary glands, 6f, 76, 527f, 539, 544–545, 545f
Mammary plexus, 5f, 381f
Mammary region, 12f
Mammillary body, 213f, 214, 221f
Mammography, 27, 544, 554
Mandible, 121t, 122, 123f
  elevation of, 145
  excursion of, 145
  frontal view of, 122, 125f
  intramembranous ossification of, 116f
  lateral view of, 122, 124f
  prenatal development of, 566, 567f
  protraction of, 145
Mandibular alveolar process, 441f
Mandibular condyle, 124f
Mandibular fossa, 126, 127f
Mandibular processes, 566, 567f
Manganese, 480t
Manual region, 12f
Manubrium, 130–131, 131f

Marfan syndrome, 83
Marginal arteries
  left, 321, 321f
  right, 317f, 321, 321f
Marrow. See Bone marrow
Mask of pregnancy, 99
Mass
  definition of, 21
  international unit for, 21
  measures of, A-1
Masseter muscle, 152f, 170, 171f, 172t, 442f
Mass movements, in large intestine, 460
Mass number, 21t, 22
Mast cells
  in connective tissue, 80
  in inflammatory response, 387
Mastication, 440, 444
  aging and, 467
  muscles of, 170, 171f, 172t
Mastoid air cells, 124
Mastoid fontanel, 140f
Mastoid process, 122, 124, 124f, 127f
Matrix, 79–80. See also specific types
Matter, 21
Mature follicle, 541–543, 541f, 542f
Maxilla, 121t, 122
  frontal view of, 122, 125f
  inferior view of, 127f
  intramembranous ossification of, 116f
  lateral view of, 122, 124f
  palatine process of, 127f
  prenatal development of, 566, 567f
Maxillary artery, 352f
Maxillary processes, 566, 567f
Maxillary sinus, 124, 126f
McBurney point, 458
Mean arterial blood pressure (MAP), 368
Measles, 108t
Measurements, table of, A-1
Meatus
  definition of, 121–122, 122t
  inferior, 408f
  middle, 408f
  superior, 408f
Mechanical digestion, 460
Mechanical energy, 28–29
Mechanoreceptors, 239
Meconium, 576
Medial, 10, 11f, 11t
Medial cuneiform, 139f
Medial epicondyle
  of femur, 138, 138f, 140f
  of humerus, 133f
Medial malleolus, 138, 139f, 140f
Medial pterygoid muscle
  action of, 172t
  insertion of, 172t
  origin of, 172t
Medial rectus muscle, 244, 244f
Medial rotation, 144f, 145
Median cubital vein, 356, 357f, 358f
Median nerve, 210
Median plane, 15
Median sacral artery, 351f, 355, 355f
Median sacral crest, 130, 130f
Mediastinum, 15, 15f, 313, 315f
Mediated transport, 46t, 49–51
Medication, 229, 282
Medium arteries, 345, 347, 347f
Medium veins, 345, 346f, 347f, 348
Medroxyprogesterone, 550–552
Medulla
  adrenal. See Adrenal glands, medulla of
  of hair, 100, 100f
  of kidneys, 498, 499f
  of thymus, 384, 384f
Medulla oblongata, 211–212
  and blood pressure, 367, 367f, 368–371, 368f, 370f, 371f
  cardioregulatory center of, 333, 337f
  chemoreceptors in, 333, 335f, 337f, 369–371, 427–429, 427f
  functions of, 211–212
  and respiration, 424, 426, 426f, 427f
  and stomach secretions, 448
  structure of, 211–212, 211f, 212f
  vasomotor center of, 367, 367f
Medullary cavity, 113, 114f, 116
Medullary infarction, 233
Medullary respiratory center, 424, 426, 426f
Megakaryocytes, 299
Meiosis, 527–528, 529f, 579
  in fertilization, 559, 560f
  in oocyte development, 540–541, 542f
  in spermatogenesis, 530, 532f

Meissner corpuscles, 239f, 240, 258
Melanin
  absence or deficiency of, 98
  adaptive advantages of, 99
  aging and, 108
  functions of, 98, 102
  genetics of, 98
  in hair, 101
  location of, 98
  during pregnancy, 98–99
  production of, 98, 101
  in skin, 98–99
Melanocyte(s)
  in epidermis, 98, 99f
  in hair, 101
  melanin production by, 98, 99f
  prenatal development of, 565t
Melanocyte-stimulating hormone (MSH),
  270t, 274
Melanoma, 105, 105f
Melanosomes, 98, 99f
Melatonin, 285–286
  action of, 271t, 285–286
  aging and, 288
  secretion of, 271t, 285–286
  target tissues of, 271t
Membrane(s), 87–88. See also specific types
  cutaneous, 87–88
  definition of, 87
  mucous, 88, 89f
  serous, 16–17, 16f, 17f, 88, 89f
  synovial, 88, 89f
Membrane-bound receptors, 266, 267f
  and activation of G proteins, 267, 267f, 268f
  and alteration of enzymes, 267–268, 268f, 268t
  responses of, 266–267
Membrane channels, 45, 45f, 47, 47f
Membrane potential(s), 198–199. See also
  Action potential(s)
Membranous labyrinth, 252, 253f
Membranous urethra, 532, 533f
Memory, 221
  of adaptive immunity, 386, 388
  aging and, 234
  consolidation of, 221
  declarative, 221
  encoding of, 221
  explicit, 221
  implicit, 221
  long-term, 221, 234
  procedural, 221
  retrieval of, 221
  sensory, 221
  short-term, 221, 234
  storage of, 221
  working, 221
Memory B cells, 395
Memory engram, 221
Memory T cells, 396, 397f
Memory traces, 221
Menarche, 545
Meniere disease, 258t
Meninges, 222, 223f
Meningitis, 222, 231t
Meniscus (pl., menisci), 141, 143f
Menopause, 548–549, 549t, 554
Menses, 546, 546f, 547t
Menstrual cramps, 547
Menstrual cycle, 546–548, 546f
  absence of, 548
  cessation of, 548–549, 549t
  duration of, 546
  estrogen and, 285
  lactation and, 550
  proliferative phase of, 546f, 547–548
  secretory phase of, 546f, 548
Menstrual period
  first (menarche), 565
  last, in pregnancy calculation, 559, 572
Menstruation, first (menarche), 545
Mental activities, 193
Mental foramen, 124f, 125f
Mental region, 12f
Mercury, millimeters of (mm Hg), 362
Merkel disks, 239f, 240
Mesenteric arteries
  inferior, 350f, 351f, 354–355, 355f
  superior, 350f, 351f, 354–355, 355f
Mesenteric ganglion
  inferior, 228f
  superior, 228f
Mesenteric veins
  inferior, 357f, 359f, 360, 360f
  superior, 357f, 359f, 360, 360f
Mesenteries, 16–17, 17f, 439, 439f, 440f, 451f

Mesentery proper, 439, 440f
Mesoderm
  derivatives of, 565t, 568t–569t
  development of, 564, 564f
Mesovarium, 540, 540f, 541f
Messenger ribonucleic acid (mRNA)
  intracellular receptors and, 268
  synthesis (transcription) of, 60–61, 60f, 61f
  translation of, 60f, 61, 62f
Metabolic acidosis, 519
Metabolic alkalosis, 523
Metabolic rate, 487–490
  basal, 490
  burn injuries and, 106–107
Metabolic states, 486–487, 488f
Metabolism, 481–490
  absorptive state of, 486–487, 488f
  biochemical pathway in, 482–483, 482f
  burns and, 106–107
  carbohydrate, 483–486
  cellular, 44, 481
  definition of, 7, 481
  endocrine regulation of, 263
  lipid, 486, 487f
  postabsorptive state of, 487, 488f
  protein, 486, 487f
  regulation of, 482–483
Metacarpals, 121t, 123f, 134, 135f
Metacarpophalangeal joint, 142f
Metaphase
  of meiosis, 528, 529f
  of mitosis, 63, 64f
Metastasis, 65
Metatarsals, 121t, 123f, 136f, 139f, 140
Metatarsophalangeal joint, 142f
Methionine, 477
Metric measurements, A-1
MHC. See Major histocompatibility complex
Micelles, 463, 463f
Microfilaments, 57, 58f
Microglia, 195, 196t, 197f, 387
Microtubules, 43f, 57
  functions of, 44t, 57
  location of, 44t
  structure of, 57, 58f
Microvilli, 43f, 59
  of epithelial cells, 78
  functions of, 44t
  location of, 44t
  of small intestine, 451, 452f
  structure of, 59
Micturition reflex, 514, 515f
Midbrain, 211
  functions of, 212
  structure of, 211f, 212, 212f
Middle age, 577
Middle cardiac vein, 317f, 321f
Middle concha, 408f
Middle ear, 252, 252f
Middle ear infection, 258t
Middle lobe, of lungs, 413, 413f
Middle meatus, 408f
Middle phalanx
  of finger, 135, 135f
  of toe, 139f
Midpiece, of sperm cell, 531f
Mifepristone (RU486), 552
Milk letdown, 275, 576f, 577
Milk production. See Lactation
Milk teeth, 440
Mineral(s)
  deficiencies of, 480t
  dietary guidelines for, 479, 480t
  digestion of, 465
  functions of, 480t
  as nutrients, 473, 479
Mineralocorticoids, 280
  action of, 271t
  secretion of, 271t, 280
  target tissues of, 271t
Mini-pill, 550
Minoxidil, 538
Minute ventilation, 420
  aging and, 433
  exercise and, 431–433
Miscarriage, 584t
Mitochondrial DNA mutations, 578
Mitochondrion (pl. mitochondria), 43f, 55f,
  56–57
  ATP production in, 2, 56–57
  damage to, 67
  functions of, 2, 44t, 56–57
  location of, 44t
  of muscle fibers, 165, 165t
  of neurons, 194, 195f

of sperm cells, 531f
  structure of, 56, 57f
Mitosis, 63
  in oocyte development, 542f
  phases of, 63, 64f
  in spermatogenesis, 530, 532f
Mitotic chromosomes, 579
Mitral (bicuspid) valve, 318f, 319, 319f
  in blood flow, 320f, 321
  incompetent, 331
  stenosis of, 331
Mixed nerves, 208, 224
Mixing waves, in stomach, 450–451, 450f
M line, 153, 155f
Molality, A-7
Molarity, A-7
Molars, 440, 441f
Mole(s), 98, 105
Molecule(s)
  cellular synthesis of, 44
  vs. compounds, 25
  definition of, 24
  formation of, 24, 24f, 25
  nonpolar, 24
  organic, 32, 32t
  polar, 24
  symbols for, 25
Molybdenum
  deficiency of, 480t
  dietary guidelines for, 480t
  functions of, 480t
Monoclonal antibodies, 393, 398–400
Monocytes, 299
  function of, 294t, 299
  in innate immunity, 387
  percentage of WBCs as, 293f, 307
  production of, 295f
  structure of, 294t, 299
Monoglycerides
  lipid breakdown to, 460, 460f, 461f, 462–464
  in triglyceride formation, 464
Mononuclear phagocytic system, 387
Mononucleosis, infectious, 309t
Monosaccharides, 32–33, 33f
  carbohydrate breakdown to, 460, 460f,
    461–462, 461f
  dietary sources of, 474
  transport of, 462, 462f
  uses in body, 32–33, 474
Monosomy X, 580
Monounsaturated fats, 34, 476
Monozygotic twins, 560
Mons pubis, 539f, 544, 544f
Morning-after pills, 552
Morula, 561, 561f
Motion sickness, 256–257, 258t
Motor cortex, primary, 216f, 217
Motor neurons, 194, 195f
  aging and, 233–234
  lower, 217
  in motor unit, 157
  prenatal development of, 565t
  in reflex arc, 204, 205f
  upper, 217
Motor speech area, 216f, 220
Motor unit(s), 157
  composition of, 157
  recruitment of, 162–163
Mouth, 5f, 77t. See also Oral cavity
Movement(s)
  involuntary, 216
  joint, 143–145, 144f
  skeletal system and, 112
  voluntary, 216–217
MRI, 27, 27f
MSH. See Melanocyte-stimulating hormone
Mucin, 444
Mucosa
  of digestive tract, 438, 439f
  of small intestine, 451, 452f
  of stomach, 446, 447f
Mucous epithelium, of digestive tract, 438, 439f
Mucous fluid, 442–443
Mucous membranes, 88
  functions of, 88
  location of, 88, 89f
  structure of, 88
Mucous neck cells, of stomach, 446, 447f
Mucus
  aging and, 433
  functions of, 76, 78, 443t, 448
  in innate immunity, 386
  in nasal cavity, 409
  in respiratory system, 433
  secretion of

by goblet cells, 78, 453
  by simple columnar epithelium, 76
  by small intestine, 443t, 453
  by stomach, 443t, 448
Multicellular glands, 78
Multiple sclerosis, 231t
Multiple-wave summation, 162, 162f
Multipolar neurons, 195, 196f, 196t
Mumps, 443
Murmurs, heart, 331, 567
Muscle(s). See also Cardiac muscle; Skeletal
  muscle(s); Smooth muscle(s); specific
  muscles
  aging and, 92, 187–188, 578
  agonist, 167
  antagonist, 167
  attachment of, 166–167, 167f
  belly of, 167
  cells of, 65f, 84–85
  connective tissue in, 80, 81t
  contractility of, 84, 152–153
  contraction of. See Muscle contractions
  elasticity of, 153
  excitability of, 153–157, 199–200
  of expiration, 416, 417f
  extensibility of, 153
  fatigue of, 164
  fixator, 167
  functional groups of, 167
  general principles of, 166–167
  hypertrophy of, 165, 189t
  insertion of, 167, 167f
  of inspiration, 416, 417f
  nomenclature for, 167
  origin (head) of, 167, 167f
  prenatal development of, 565t
  prime mover, 167
  relaxation of, 160
  of respiration, 152, 173, 416, 417f, 432t
  superficial, anatomic overview of, 167,
    168f–169f, 170f
  synergist, 167
  testosterone and, 538t
  types of, 85–87, 86t–87t, 152
Muscle contractions, 84, 158–165
  concentric, 164
  cross-bridge movement in, 159, 161f
  eccentric, 164
  energy requirements for, 159, 161f,
    163–164, 163f
  heat production in, 152, 159–160
  isometric, 164
  isotonic, 164
  sliding filament model of, 158–159, 160f
  strength of, 162
  summary of, 159t
  summation of, 162, 162f
  types of, 164
Muscle fasciculi, 153, 154f
Muscle fibers, 85
  aging and, 187–188
  fast-twitch, 164–165, 165t, 188
  ratio of types, 165
  slow-twitch, 164–165, 165t
  structure of, 153, 154f
  type I, 164–165, 165t
  type IIa, 164–165, 165t
  type IIb, 164–165, 165t
Muscle tissue, 3, 84–87
  classification of, 85–87, 86t–87t, 152
  main characteristic of, 84
Muscle tone, 164
Muscle twitch, 162–163
  contraction phase of, 162, 162f
  definition of, 162
  lag (latent) phase of, 162, 162f
  relaxation phase of, 162, 162f
Muscular arteries, 345–347, 346f, 364
Muscular dystrophy, 189t
  Duchenne, 188–189, 188f, 189t
  myotonic, 189t, 231t
Muscular fascia, 153, 154f
Muscular fatigue, 164
Muscularis
  of digestive tract, 438, 439f
  of stomach, 446, 447f
Muscularis mucosae
  of digestive tract, 438, 439f
  of stomach, 447f
Muscular system, 3, 5f, 151–189. See also
  specific muscles and muscle types
  diseases and disorders of, 189t
  functions of, 5f, 152
  overview of, 152f
  prenatal development of, 568t–569t

Musculocutaneous nerve, 210
Mutation, genetic, 582
Myasthenia gravis, 158, 189t, 231t
Myelinated axons, 195
  action potentials in, 201, 202f
  vs. unmyelinated axons, 195, 198f
Myelin sheaths, 195–196
  function of, 195
  structure of, 195–196, 195f, 197f, 198f
Myocardial infarct, 322
Myocardial infarction (heart attack), 322,
    338–339, 339f
  angiogram in, 338, 338f
  aspirin use in, 300, 322
  dissolving blood clots in, 302, 338
  platelets in, 300
  referred pain in, 241
  systemic effects of, 339f
Myocardium, 322, 323f
Myoepithelial cells, 545, 545f
Myofibrils, 153, 154f
Myoglobin, in skeletal muscle fibers, 165, 165t
Myometrium, 540f, 543
Myopia, 258t
Myosin
  in blood clot retraction, 302
  in fast-twitch muscle fibers, 164–165, 165t
  in slow-twitch muscle fibers, 164–165, 165t
  uses in body, 477
Myosin heads, 153, 154f
Myosin myofilaments
  in cardiac muscle, 322–323
  in muscle contraction, 158–159, 160f
  in skeletal muscle, 153, 154f
  in smooth muscle, 165–166
  structure of, 153, 154f
Myotonic muscular dystrophy, 189t, 231t
MyPyramid, 473, 473f
Myxedema, 275

NADH. See Nicotinamide adenine
    dinucleotide
Nail(s), 101
  as diagnostic aid, 104
  functions of, 102
  growth of, 101
  structure of, 101, 102f
Nail bed, 101, 102f
Nail body, 101, 102f
Nail matrix, 101, 102f
Nail root, 101, 102f
Na⁺–K⁺ pump. See Sodium–potassium pump
Nares, 407
Naris, 408f
Nasal bone, 121t, 124
  frontal view of, 124, 125f
  intramembranous ossification of, 116f
  lateral view of, 124f
Nasal cartilages, 83, 83t
Nasal cavity, 5f, 14f, 122, 124, 125f,
    407–409, 407f
  air flow through, 407–409
  epithelial tissue of, 75t, 76
  prenatal development of, 565t
Nasal concha, 124
  inferior, 121t, 124, 125f
  middle, 124, 125f
  superior, 124
Nasal placodes, 567f
Nasal region, 12f
Nasal septum, 124, 407
  deviated, 407
  frontal view of, 124
Nasal sinuses, epithelial tissue of, 75t, 76
Nasolacrimal canal, 124, 124f
Nasolacrimal duct, 244, 244f, 408
Nasopharynx, 408f, 409, 444
Natural exposure, 397
Natural immunity
  active, 397–398, 397f
  passive, 397–398, 397f
Natural killer (NK) cells, in innate
    immunity, 387
Navicular, 138, 139f
Near vision, 249f
Neck
  of body, 10, 12f
    arteries of, 352–354, 352f
    extension of, 144f
    flexion of, 144f
    muscles of, 167, 170–172, 174f, 174t, 178f
    veins of, 356, 358f
  of femur, 138, 138f, 143f
  of humerus, 132–134, 133f
  of tooth, 442, 442f

Negative, definition of, 7
Negative feedback, 7–8
  in blood pressure maintenance, 7–8, 9f
  in body temperature regulation, 491–492
  in endocrine system, 268–269
  inadequate, and positive feedback, 8–10
  in stomach secretion, 448–450, 449f
Negative-feedback mechanisms, 7–8
Neocortical death, 579
Neonatal gonorrheal ophthalmia, 584t
Neonate, 574–576
  definition of, 577
Nephron(s), 498–502, 500f
  and acid-base balance, 519
  cortical, 498, 500f
  filtration in, 503–504, 504f
  juxtamedullary, 498, 500f
Nerve(s), 6f, 193, 193f, 196. See also specific
    nerves and nerve types
  connective tissue in, 80, 81t
  mixed, 208, 224
Nerve cells. See Neuron(s)
Nerve endings, free, 239–240, 239f, 258
Nerve fibers. See Axon(s)
Nerve tracts, 196
Nervous system, 3, 6f, 192–234. See also
    Autonomic nervous system; Central
    nervous system; Peripheral nervous
    system
  aging and, 230–234, 577–578
  cells of, 194–198
  components of, 6f
  diseases and disorders of, 231t
  divisions of, 193–194, 193f, 194f
  electrical signals in, 198–205
  vs. endocrine system, 263
  functions of, 6f, 193
  prenatal development of, 568t–569t
Nervous tissue, 3, 87
  action potentials in, 87
  functions of, 87, 88t
  location of, 87, 88t
  organization of, 196
  structure of, 87, 88t
Nervous tunic, of eye, 244, 245f, 246
Neural crest
  and digestive system, 567, 570f
  formation of, 565, 566f
Neural crest cells, 565, 565t, 566f
Neural folds, 565, 566f
Neural groove, 565, 566f
Neuralgia, 231t
Neural plate, 565, 566f
Neural tube
  defects in, 565
  and digestive system, 567, 570f
  formation of, 565, 566f
Neuritis, 231t
Neuroectoderm, 565, 565t
Neurofibromatosis, 231t
Neurogenic shock, 369
Neuroglia, 87, 88t, 195
  of central nervous system, 195
  of peripheral nervous system, 195
  types of, 195, 196t, 197f
Neurohormones, 263, 264t, 265. See also
    Hormone(s)
Neurohypophysis. See Posterior pituitary gland
Neuromodulators, 263, 264t
Neuromuscular junction, 157–158, 157f, 158f
Neuron(s), 87, 194–195
  aging and, 92, 230–234, 577–578
  bipolar, 195, 196f, 196t
  composition of, 87
  excitability of, 199–200
  functions of, 87
  motor or efferent, 194, 195f
    aging and, 233–234
    lower, 217
    in motor unit, 157
    prenatal development of, 565t
    in reflex arc, 204, 205f
    upper, 217
  multipolar, 195, 196f, 196t
  olfactory, 241, 241f
  postganglionic, 227, 227f, 228f
  preganglionic, 227, 227f, 228f
  prenatal development of, 565t
  proprioceptive, 218, 219f
  pseudo-unipolar, 195, 196f, 196t, 206
  regeneration potential of, 91
  sensory or afferent, 193–194
    aging and, 230–233
    prenatal development of, 565t
    in reflex arc, 204, 205f

structure of, 87, 88t, 194–195, 195f
  types of, 195, 196f, 196t
Neuronal pathways, 204–205. See also specific
    pathways
  converging, 204, 205f
  diverging, 204–205, 205f
Neurotransmitters, 158, 201, 263, 264t
  in action potentials, 201–203, 203f
  aging and, 233–234
  autonomic, 229
  drugs modulating, 203
  examples of, 203, 204t
  release of, 54
Neutral solution, 30, 30f
Neutron(s), 22, 22f
  clinical relevance of, 27
  mass of, 22
Neutrophil(s), 298–299
  chemical signals released by, 387
  function of, 294t
  in inflammation, 89, 90f
  in innate immunity, 386–387
  percentage of WBCs as, 293f, 307
  production of, 295f
  staining properties of, 298–299, 299f
  structure of, 294t, 298–299
Newborn, 574–576
  circulatory system of, 574–575, 574f, 575f
  digestive system of, 576
  respiratory system of, 574–575
Nexium, 450
Niacin. See Vitamin B₃
Nicotinamide adenine dinucleotide (NADH)
  in aerobic respiration, 484–486, 484f, 486
  in amino acid metabolism, 486, 487f
  in citric acid cycle, 484f, 485
  in electron-transport chain, 484f,
    485–486, 485f
  in glycolysis, 483, 483f, 484, 486
Night blindness, 246
Nipple, 544–545, 545f
Nitrogen
  atomic number of, 21t
  mass number of, 21t
  partial pressure at sea level, 422t
  percent in body, 21t
  symbol for, 21t
Nitrogenous bases, 38, 39f
Nitroglycerin, 336
Nociceptors, 239, 427, 427f
Node of Ranvier, 195–196, 195f, 197f,
    201, 202f
Nonessential amino acids, 477
Nongonococcal urethritis, 554t
Nonpolar covalent bonds, 24
Nonpolar molecules, 24
Nonspecific resistance. See Innate immunity
Noradrenaline. See Norepinephrine
Norepinephrine
  action of, 279
  and blood flow, 367–368
  and blood pressure, 371–373, 371f
  in fight-or-flight response, 279
  in heart regulation, 333, 335f, 337f
  as hormone, 265
  as neurotransmitter, 203, 204t, 265
  secretion of, 278–279, 279f, 280f
  target tissues of, 280f
Normal range, 7
Nose
  anatomy of, 407–409, 408f
  external, 5f, 407, 407f
  prenatal development of, 566, 567f
Nostrils, 407
Notochord, 564f, 565, 566f
Nuchal lines, 127f
Nuchal region, 13f
Nuclear envelope, 43f, 52, 53f, 55f
Nuclear pores, 52, 53f, 55f
Nucleases, 458
  functions of, 443t, 458
  secretion of, 443t, 458
Nucleic acid(s), 32t, 38. See also
    Deoxyribonucleic acid; Ribonucleic
    acid
Nucleolus, 43f, 52, 53f
  functions of, 44t
  location of, 44t
Nucleotides, 38, 60–61, 61f
Nucleus (atom), 22, 22f
Nucleus (cell), 43
  functions of, 44t
  location of, 43f, 44t
  of sperm cell, 530, 531f
  structure of, 43, 43f, 52, 53f

Nutrients, 473
  absorption of, 438, 453
  conversion of, hepatic, 456t, 457
  Daily Values for, 479–481
  definition of, 473
  digestion of, 438
  essential, 473
  exchange, in circulatory system, 345
  in plasma, 292, 293f, 293f
  storage of, hepatic, 456t, 457
Nutrition, 473–481
Nuva Ring, 552

Obesity, 489–490
  definition of, 489
  fat distribution in, 489
  hyperplastic, 490
  hypertrophic, 489–490
  surgery for, 490
Oblique fissure, of lung, 413f
Oblique fracture, 120, 120f
Oblique muscle(s)
  inferior, 243f, 244, 244f
  superior, 243f, 244, 244f
Oblique muscle layer, of stomach, 446, 447f
Oblique section, 15, 15f
Obturator foramen, 135, 136f, 137f
Obturator nerve, 210
Occipital artery, 352f
Occipital bone, 121t
  in cranial cavity, 125, 126f
  in fetal skull, 140f
  inferior view of, 127f
  intramembranous ossification of, 116f
  lateral view of, 124f
Occipital condyles, 126, 127f
Occipital fontanel, 140f
Occipital lobe, 214f, 215
Occipital region, 13f
Occipitofrontalis muscle, 167, 171f
  action of, 167, 172t
  insertion of, 172t
  origin of, 172t
Oculomotor (CN III) nerve, 224
  functions of, 224, 226t
  origin of, 226f
  and parasympathetic axons,
    227–229, 227t
Odorants, 241
Older adult, 577. See also Aging, effects of
Olecranon, 13f
Olecranon bursa, 143f
Olecranon fossa, 133f
Olecranon process, 134, 134f, 135f
Olfaction, 239, 241–242
  adaptation in, 242
  aging and, 233
  neuronal pathways for, 241–242, 241f
  respiration and, 407
Olfactory bulb, 221f, 241–242, 241f
Olfactory cortex, 221f, 241–242, 241f
Olfactory epithelium, 241, 241f
Olfactory (CN I) nerve, 224
  functions of, 224, 226t
  origin of, 226f
Olfactory neurons, 241, 241f
Olfactory receptors, 241
Olfactory tracts, 241–242, 241f
Oligodendrocytes, 195
  in myelin sheaths, 195–196
  structure of, 195, 196t, 197f
Omental bursa, 439, 440f
Omentum
  greater, 439, 440f
  lesser, 439, 440f, 454f
Oocytes, 540
  fertilization of, 527, 530, 540–541, 542f,
    559, 560f
  formation of, 540, 542f
  implantation of, 543
  primary, 540, 541f, 542f
  secondary, 540–541, 542f, 559, 560f
Oogenesis, 540, 542f
Oogonia, 540, 542f
Ophthalmoscope, 246, 248f
Opposition (movement), 145
Opsin, 246, 247f, 248f
Optic chiasm, 213f, 249, 250f
Optic disc, 246, 247f, 248f
Optic foramen, 122, 124, 125f, 126f
Optic (CN II) nerve, 123–124, 224, 244f,
    245f, 246, 247f, 249, 250f
  functions of, 224, 226t
  origin of, 226f
Optic radiations, 250, 250f

Optic tracts, 249–250, 250f
Oral cavity, 5f, 408f, 438f, 440–443, 441f
Oral contraceptives, 266, 550–552, 551f
Oral region, 12f
Oral rehydration therapy, 462
Oral temperature, 491
Orbicularis oculi muscle, 167, 171f, 172t, 243f
Orbicularis oris muscle, 167, 171f, 172t, 440
Orbit(s), 122–124, 125f, 242
Orbital fissures
    inferior, 123, 125f, 127f
    superior, 123, 125f
Orbital region, 12f
Organ(s), 3, 4f
    constriction of, 152
    definition of, 3
    planes of section through, 14f, 15, 15f
    prenatal development of, 559, 566–567,
        568t–569t
Organelles, 43, 52–59
    definition of, 2
    functions of, 43, 44t
    locations of, 43, 43f, 44t
    number and types of, 43
    structure of, 43, 43f
Organic chemistry, 31, 32–38
Organic molecules, 32, 32t
Organism level of organization, 3f
Organization
    cell level of, 2–3, 3f
    chemical level of, 2, 3f
    definition of, 7
    organism level of, 3f
    organ level of, 3, 3f
    organ system level of, 3–5, 3f
    structural and functional, 2–6
    tissue level of, 3, 3f
Organ level of organization, 3, 3f
Organogenesis, 566–567
Organophosphate poisoning, 159
Organ system(s), 3, 5f–6f
    coordinated activity of, 3–5
    definition of, 3
    prenatal development of, 559, 566–567,
        568t–569t
Organ system level of organization, 3–5, 3f
Orgasm
    female, 549
    male, 538
Origin, of muscle, 167, 167f
Oropharynx, 408f, 409, 444
Ortho Evra, 552
Osmolality
    of blood, A-3
    of solution, A-7
    of urine, A-5
Osmosis
    blood and, 292
    in capillary exchange, 365–366, 365f
    definition of, 47
    in intracellular-extracellular exchange, 516
    in renal filtration, 503–504, 504f
    in renal reabsorption, 504–508, 505f–509f
    in transport across cell membrane, 46t,
        47–49, 48f
Osmotic pressure, 47, 48f, 503–504, 504f
Ossification, 115–116
    centers of, 115–116, 116f
        primary, 116, 117f
        secondary, 116, 117f
    definition of, 115
    endochondral, 115–117, 116f, 117f, 118f
    intramembranous, 115, 116f
Osteoarthritis, 148t
Osteoblasts, 113
    aging and, 145
    in bone growth, 117
    in bone remodeling, 117
    in bone repair, 119, 119f
    in calcium homeostasis, 120f, 121
    in cancellous bone, 116f
    naming convention for, 80
    in ossification, 115–116
Osteoclasts
    aging and, 145
    in bone remodeling, 117
    in calcium homeostasis, 120f, 121
    in cancellous bone, 116f
    in endochondral ossification, 116
    naming convention for, 80
Osteocytes, 113
    in cancellous bones, 115, 116f
    in compact bones, 113–115, 115f
    in ossification, 115–116
Osteogenesis imperfecta, 113, 148t

Osteomalacia, 148t
Osteomyelitis, 148t
Osteon(s), 113, 114f, 115f
Osteoporosis, 119, 146–147, 148t
    diagnosis of, 147
    etiology of, 146–147
    photomicrograph of, 146f
    systemic effects of, 147t
    treatments for, 147
Otic ganglion, 228f
Otic region, 12f
Otolith(s), 255–256, 255f, 259
O type blood, 303–304, 303f
Oval window, 252
    anatomy of, 252, 252f, 253f
    auditory function of, 254, 254f
Ovarian arteries, 351f, 355
Ovarian follicles, 540, 541f
    degenerating, 541f, 543
    development of, 541–543
    mature or graafian, 541–543, 541f, 542f
    primary, 541, 541f, 542f
    primordial, 541, 541f, 542f
    secondary, 541, 541f, 542f
Ovarian ligament, 540, 540f
Ovarian veins, 360
Ovaries, 6f, 270f, 527f, 539–540, 539f, 540f
    anatomy of, 540, 541f
    epithelial tissue of, 74t, 76
    histology of, 540, 541f
    hormones of, 269, 271t, 285
    in infertility, 549
    in menopause, 548–549, 549f
    in menstrual cycle, 546f, 547–548, 547t
    oocyte development in, 540, 542f
    ovulation by, 540, 541–543, 542f, 546f,
        547–548, 547t
    sex cell production in, 527–528
Overweight, definition of, 489
Oviducts. See Uterine tubes
Ovulation, 540, 541–543, 542f, 546f,
    547–548, 547t
Oxygen
    in aerobic respiration, 484–486, 485f
    atomic number of, 21t
    chemistry of, 22, 22f, 31
    exercise and levels of, 429–430
    hemoglobin binding/transport of,
        296–297, 424
    mass number of, 21t
    partial pressure of, 422–424, 422t, A-3
    percent in body, 21t
    in plasma, 292, 293t
    and red blood cell production, 297, 297f
    respiratory exchange of, 407, 415f,
        422–424, 423f
    symbol for, 21t
    transport in blood, 292, 296–297, 407, 424
Oxygen consumption, recovery, 164
Oxygen deficit (debt), 164, 484
Oxygen saturation, A-3
Oxytocin, 275
    action of, 268t, 270t, 275, 536t
    in fertilization, 559
    in labor (childbirth), 573–574, 573f
    in lactation, 275, 576–577, 576f
    secretion of, 268t, 270t, 275, 536t
    target tissues of, 268t, 270t, 536t

Pacemaker
    artificial, 336
    sinoatrial node as, 325, 326
Pacinian corpuscles, 239f, 240, 258
Pain, 240–241
    epigastric, 12
    gate control theory of, 240
    low back, 172–173
    management of, 240
    referred, 240–241, 240f
    types of, 240
Pain receptors (nociceptors), 239, 427, 427f
Palate, 442
    cleft, 566
    hard, 127, 127f, 407, 408f, 441f, 442, 444, 445f
    prenatal development of, 566, 567f
    soft, 127, 170, 408f, 409, 441f, 442
Palatine bone, 121t, 127, 127f
Palatine process, of maxilla, 127f
Palatine tonsils, 382, 382f, 408f, 409
Palmar arch
    deep, 354f
    superficial, 354f
Palmaris longus muscle, 181f, 181t
Palmaris longus tendon, 182f
Palmar region, 12f

Palmitic acid, 34f
Pancreas, 4f, 5f, 6f, 270f, 438f, 440f, 457–458
    aging and, 466, 578
    anatomy of, 457–458, 457f
    body of, 457f
    endocrine, 269, 281–285, 457–458
    exocrine, 269, 281, 283f, 458
    functions of, 458
    head of, 457, 457f
    histology of, 457–458, 457f
    hormones of, 271t, 281–285, 457–458
    location of, 457
    neonatal, 576
    prenatal development of, 565t, 567, 570f
    secretions of, 452, 453, 458
        aqueous component of, 458, 458f
        enzymatic component of, 458, 458f
        regulation of, 458, 458f
    structure of, 281, 283f
    tail of, 457, 457f
Pancreatic amylase
    functions of, 443t, 458, 461f, 462
    in newborn, 576
    secretion of, 443t, 458
Pancreatic ducts, 283f, 451, 458
Pancreatic islets, 281, 283f, 457–458, 457f
Pantothenic acid
    deficiency of, 478t
    dietary guidelines for, 478t
    dietary sources of, 478t
    functions of, 477, 478t
    solubility of, 478t
Papilla(e)
    dermal, 97, 98f
    duodenal, 455, 455f, 457f
    renal, 499f
    tongue, 242, 242f
Papillary duct, 498, 500f
Papillary muscles, 318f, 319, 319f, 320f
Pap smear, 543, 554
Paracrine chemical signals, 263, 264t, 265
Parafollicular cells, of thyroid gland, 275, 276f
Parahippocampal gyrus, 221f
Paralysis
    crutch, 211
    flaccid, 159
    respiratory muscles, 432t
    spastic, 159
Paranasal sinuses, 124, 126f, 408, 408f
Parasites, intestinal, 159
Parasympathetic nervous system
    anatomy of, 227–229
    functions of, 227t, 229, 230t
    innervation by, 227, 228f
Parathyroid glands, 6f, 270f
    hormones of, 271t, 276–278
    prenatal development of, 565t, 567
    structure of, 276, 276f
Parathyroid hormone (PTH)
    action of, 271t, 277–278
    aging and, 288
    and calcium regulation, 120f, 121, 277–278,
        278f, 517
    deficiency of, 277–278
    excess of, 277
    reference laboratory values for, A-6
    secretion of, 271t, 276–278
    target tissues of, 271t
Parent cell, 62, 63, 64f, 528, 529f
Parietal arteries, 353f, 354
Parietal bones, 121t
    in cranial cavity, 126f
    in fetal skull, 140f
    frontal view of, 125f
    intramembranous ossification of, 116f
    lateral view of, 122, 124f
Parietal cells, of stomach, 446, 447f, 449f
Parietal lobe, 214f, 215
Parietal pericardium, 16, 17f, 314, 315f, 416f
Parietal peritoneum, 16, 17f, 439, 440f
Parietal pleura, 16, 17f, 315f, 412f, 414, 414f, 416f
Parietal serous membranes, 16, 16f
Parkinson disease, 218, 231t
Parotid duct, 442f, 443
Parotid gland, 442–443, 442f
Paroxysmal atrial tachycardia, 327t, 332
Partial pressure, 422–424, 422t, A-3
Partial-thickness burns, 104–105, 104f, 106, 106f
Parturition, 572–574, 572f, 573f
Passive immunity, 397, 397f
    artificial, 397–398, 397f
    natural, 397–398, 397f
Patch, birth control, 552
Patella, 121t, 123f, 136f, 138, 140f, 143f
Patellar groove, 138f

Patellar ligament, 143f, 183f, 184–185
Patellar reflex, 184–185, 208, 208f
Patellar region, 12f
Patellar tendon, 168f, 183f, 184
Patent ductus arteriosus, 339t
Pattern baldness, 101, 538
Pectoral girdle
    bones of, 121t, 131–132, 131f, 132f
    surface anatomy of, 133f, 135f
Pectoralis major muscle, 5f, 152f, 168f,
    177–178, 178f, 179f
    action of, 167, 177, 180t
    insertion of, 180t
    origin of, 180t
Pectoralis minor muscle, 176–177, 178f
    action of, 177, 178t
    insertion of, 178t
    origin of, 178t
    in respiration, 417f
Pectoral region, 12f
Pedal region, 12f
Pedicles
    of cervical vertebrae, 129f
    of lumbar vertebrae, 129f
    of thoracic vertebrae, 129f
    of vertebrae, 128–129, 128f
Pedigree, 582, 582f
Pelvic brim, 137f
Pelvic cavity, 15f, 16
Pelvic diaphragm. See Pelvic floor
Pelvic floor, muscles of, 176, 177f, 177t
Pelvic girdle, bones of, 121t, 135–136, 136f
Pelvic inflammatory disease (PID), 549, 554t
Pelvic inlet, 136
    and birth, 136
    male vs. female, 136t, 137f
Pelvic outlet, 136
    and birth, 136
    male vs. female, 136t, 137f
Pelvic region, 12f
Pelvis, 5f, 10–12, 135–136, 136f
    arteries of, 355f, 356
    female, 136, 136t, 137f, 539f
    male, 136, 136t, 137f
    veins of, 359–360, 359f
Pelvis, renal, 498, 499f
Penicillin(s)
    allergy to, 104
    tubular secretion of, 509
Penis, 6f, 527f, 528, 530f, 532–534, 533f
    cancer of, 534
    erectile tissue of, 532–534, 538–539
    in male sex act, 538–539
    testosterone and, 538t
Pepsin, 446, 448
    diarrhea and, 467
    functions of, 443t, 448, 461f, 464
    stomach secretion of, 443t, 448
Pepsinogen, 446
Peptic ulcers, 468t
Peptidases, 452
    functions of, 443t, 452, 461f, 464
    secretion of, 443t, 452
Peptide(s)
    digestion of, 461f, 464–465
    hormones, 265
Peptide bonds, 61, 62f
Percent concentrations, A-7
Percent Daily Value, 480, 481f
Perception, 239. See also Sense(s)
Pericardial cavity, 16, 17f, 313, 314, 315f, 416f
    epithelial tissue of, 73, 74t
    serous membranes of, 16, 17f, 88, 89f
Pericardial fluid, 16, 17f, 314, 315f
Pericardial membranes, 16, 17f, 88, 89f
Pericardial sac. See Pericardial cavity;
        Pericardium
Pericarditis, 16, 88, 316, 339t
Pericardium, 314, 317f
    disorders of, 316
    fibrous, 314, 315f, 416f
    inflammation of, 316, 339t
    parietal, 16, 17f, 314, 315f, 416f
    serous, 314, 315f
    visceral, 16, 17f, 314, 315f, 322, 323f, 416f
Perichondrium, in endochondral ossification,
        116, 117f
Perilymph, 252, 253f, 254, 254f
Perimetrium, 540f, 543
Perimysium, 153, 154f
Perineal region, 13f
Perineum
    central tendon of, 177f
    clinical, 544, 544f
    muscles of, 176, 177f, 177t

Periodontal disease, 442
Periodontal ligaments, 442, 442f
Periosteum
  on compact bones, 113–115, 115f
  in endochondral ossification, 117f
  on long bones, 113, 114f
Peripheral nervous system (PNS), 193–194,
    193f, 194f, 206, 224–230
  components of, 193–194, 193f, 194f, 206
  diseases and disorders of, 231t
  organization of, 196–197
Peripheral resistance (PR), 368
Peristaltic waves (contractions)
  in small intestine, 453
  of stomach, 450–451, 450f
  of swallowing, 446, 446f
Peritoneal cavity, 16, 17f, 440f
  epithelial tissue of, 73, 74t
  serous membrane of, 16, 17f, 88, 89f
Peritoneal dialysis, 521
Peritoneal fluid, 16, 17f
Peritoneal membranes, 16, 17f, 88, 89f
Peritoneum, 439–440
  inflammation of, 439, 459
  parietal, 16, 17f, 439, 440f
  visceral, 16, 17f, 439, 440f
Peritonitis, 16, 88, 439, 459
Peritubular capillaries, 502, 502f, 508–509
Permanent teeth, 440, 441f
Permeability, selective, 516
Pernicious anemia, 307–308, 308
Peroneal artery, 350, 356, 356f
Peroneal (fibularis) nerve, common, 211
Peroneal veins, 357f
Peroneus (fibularis) brevis muscle, 168f, 169f,
    185, 186f, 187t
Peroneus (fibularis) longus muscle, 168f,
    169f, 185, 186f, 187t
Peroneus (fibularis) longus tendons, 186f
Peroneus (fibularis) tertius muscle, 186f, 187t
Peroxisomes, 43f, 54
Perpendicular plate, of ethmoid bone, 124, 125f
Pertussis, 432t
Peyer patches, 451
pH
  in acid-base balance, 518–523
  of blood, 292, A-3
    carbon dioxide and, 424, 427–428, 428f
    regulation of, 407, 428f
    and ventilation, 427–428, 428f
  buffers and, 518–519
  and cardiac function, 333, 335f
  of extracellular fluid, 497
  reference laboratory values for, A-3, A-5
  and sperm cells, 535
  of stomach, 448
  of urine, 503t, A-5
  of vagina, 535
Phagocytes, 386–387
Phagocytic vesicles, 43f
Phagocytosis, 51–52, 299, 386–387
  antibodies in, 394f
  definition of, 386
  in inflammatory response, 89, 90f, 387–388
  in lungs, 416
  in tissue repair, 92
Phalanges (sing., phalanx)
  of foot, 121t, 123f, 136f, 139f, 140
  of hand, 121t, 123f, 134–135, 135f
Pharyngeal constrictor muscles, 170, 173f,
    173t, 444, 445f
Pharyngeal elevators, 170, 173t
Pharyngeal phase, of swallowing, 444, 445f
Pharyngeal tonsils, 382, 382f, 408f, 409
Pharynx, 5f, 14f, 409, 444
  anatomy of, 444
  in digestive system, 438, 438f, 444
  epithelial tissue of, 75t, 76, 77t
  movement of, muscles of, 170, 173f, 173t,
    444, 445f
  prenatal development of, 570f
  in respiratory system, 407f, 408f, 409
  in swallowing, 170, 173t, 409, 444, 445f
Phenotype, 580, 582
Phenylalanine, 477, 582
Phenylketonuria (PKU), 482, 582
Pheromones, 263, 264t
Phlebitis, 348
Phlebotomy, 8
Phosphatase, A-3
Phosphate
  in bone matrix, 113
  digestion of, 465
  in intracellular fluid, 516
  vitamin D and, 102

Phosphate buffer system, 519, 522f
Phosphate ions, 23t, 517–518
Phospholipids, 34
  in cell membrane, 44–45, 45f
  dietary sources of, 476
  hydrophilic region of, 34, 35f, 44
  hydrophobic region of, 34, 35f, 44
  structure of, 34, 35f
  uses in body, 476
Phosphorus
  atomic number of, 21t
  deficiency of, 480t
  dietary guidelines for, 480t
  functions of, 480t
  mass number of, 21t
  percent in body, 21t
  in phospholipid formation, 34, 35f
  reference laboratory values for, A-3
  storage in bones, 112
  symbol for, 21t
Photic sneeze reflex, 409
Photoreceptor(s), 239, 246, 247f
Phrenic arteries
  inferior, 351f, 355, 355f
  superior, 351f, 354
Phrenic nerve, 208–210, 426f
pH scale, 30, 30f
Phylloquinone. See Vitamin K
Physical barriers, in innate immunity, 386
Physiologic contracture, of muscle, 164
Physiology
  cellular, 2
  definition of, 2
  human, 2
  major goals of, 2
  systemic, 2
Pia mater
  of brain, 222, 223f, 225f
  of spinal cord, 222, 223f
Pigmentation, of skin, 98–100
Pigmented retina, 246, 247f
Pineal body, 212f
Pineal gland, 6f, 213, 213f, 270f
  hormones of, 271t, 285–286
  tumors of, 286
Pinocytosis, 52
Pisiform bone, 134f, 135
Pitch, of sound, 255
Pitocin, 275
Pituitary dwarfism, 272
Pituitary gland, 6f, 269–275, 270f
  anterior, 270
    hormones of, 270t, 272–275, 272f
    hypothalamic regulation of, 271, 272f, 273f
    in labor (childbirth), 573, 573f
    prenatal development of, 565t, 567, 570f
  in female puberty, 545–546
  in male puberty, 537
  as master gland, 270
  in menopause, 549t
  and menstrual cycle, 546f, 547–548, 547t
  posterior, 270
    hormones of, 270t, 272f, 275
    hypothalamic regulation of, 271–272,
      272f, 274f
    prenatal development of, 565t
  structure of, 213f, 270
  in thyroid regulation, 275, 277f
Pivot joints, 141, 142f
Placenta
  circulation in, 561–563, 562f, 574f
  definition of, 561
  development of, 561–563
  hormone secretion by, 287–288, 563, 563f
Placental stage, of labor, 573
Placenta previa, 584t
Plague, 402t
Plane(s)
  anatomical, 12–15, 14f
  of section, through organ, 14f, 15, 15f
Plane joints, 141, 142f
Plantar flexion, 143, 185
Plantar region, 13f
Plasma, 292–294
  antibodies in, 303
  composition of, 292, 293f, 293t
  reference laboratory values for, A-2–A-4
  volume of, 292–294, 515t
Plasma cells, 395
Plasma loss shock, 369
Plasma membrane. See Cell membrane
Plasmin, 302, 303f
Plasminogen, 302, 303f
Plateau phase, of cardiac action potential,
    324, 324f

Platelet(s), 294, 299
  activation of, importance of, 300
  in blood clot formation, 301–302, 301f
  in blood clot retraction, 302
  disorders of, 300, 307–308
  function of, 294t, 299
  number of, 293f, 294, 307–308
  production of, 294, 295f, 299
  staining properties of, 299f
  structure of, 294t, 299
Platelet adhesion, 300, 301f
Platelet aggregation, 300, 301f
Platelet count, 307–308, A-4
Platelet plug
  definition of, 300
  formation of, 300, 301f
Platelet release reaction, 300, 301f
Plavix (clopidogrel bisulfate), 300
Pleura, 414
  parietal, 16, 17f, 315f, 412f, 414, 414f, 416f
  visceral, 16, 17f, 315f, 412f, 414, 414f, 416f
Pleural cavities, 16, 17f, 315f, 412f, 414–415,
    414f, 416f
  epithelial tissue of, 73, 74t
  serous membranes of, 16, 17f, 88, 89f
Pleural fluid, 16, 17f, 414–415
Pleural membranes, 16, 17f, 88, 89f
Pleural pressure, 419–420
Pleurisy, 16, 88
Plexuses
  of digestive system, 438, 439f
  of spinal nerves, 208–211, 210f, 210t
Pneumonia, 432t
Pneumothorax, 420, 422
PNS. See Peripheral nervous system
Podocytes, 498, 502f
Poison ivy, 402t
Poison oak, 402t
Polar body, 540, 542f
Polar covalent bonds, 24, 24f
Polarization, of cell membranes, 153–155,
    156f, 198
Polar molecules, 24
Poliomyelitis, 231t
Polycystic ovarian disease, 548
Polycythemia, 8
Polycythemia vera, 307, 309t
Polygenic traits, 581–582, 582f
Polygraph tests, 101
PolyHeme, 297
Polypeptide(s), digestion of, 461f, 464
Polypeptide chain, 61, 62f
Polysaccharides, 33, 33f
  dietary sources of, 474
  digestion of, 461–462, 461f
  uses in body, 474
Polyunsaturated fats, 34, 476
Pompe disease, 56
Pons, 211, 212
  functions of, 212
  and respiration, 424, 426f
  structure of, 211f, 212, 212f, 213f
Pontine respiratory group, 424, 426f
Popliteal artery, 350f, 356, 356f
  in pulse monitoring, 364f
Popliteal region, 13f
Popliteal vein, 357f, 361, 361f
Porta, of liver, 453–455, 454f
Portal system(s)
  definition of, 360
  hepatic, 359f, 360, 360f
Portal triad, of liver, 454f, 455
Positions, body, 10, 11f
Positive, definition of, 8
Positive feedback, 8–10
  during birth, 10
  and blood pressure, 8, 10f
  harmful, 8, 10f
  and labor (childbirth), 574
  normal, 10
Positive-feedback mechanisms, 8–10
Postabsorptive state, metabolic, 487, 488f
Posterior, 10, 11f, 11t
Posterior chamber, of eye, 245f, 246–248
Posterior communicating artery, 354f
Posterior fontanel, 140f
Posterior horn, 206, 207f
Posterior interventricular artery, 317f, 321f
Posterior interventricular sulcus, 314, 317f
Posterior pituitary gland, 270
  hormones of, 270t, 272f, 275
  hypothalamic regulation of, 271–272, 272f,
    274f
  prenatal development of, 565t
Postganglionic neurons, 227, 227f, 228f

Postmenopause, 549t
Postsynaptic membrane, 158, 158f, 201, 203f
Posture, muscles of, 152, 172
Potassium
  atomic
    number of, 21t
    and cardiac function, 333
  deficiency of, 480t
  dietary guidelines for, 480t
  digestion of, 465
  functions of, 480t
  mass number of, 21t
  percent in body, 21t
  reference laboratory values for, A-3, A-5
  symbol for, 21t
Potassium channels, 47, 47f
  in action potential, 155–157, 156f, 200–201,
    200f, 324–325, 324f
  and resting membrane potential, 155, 156f,
    198–199, 198f, 199f
  in synaptic transmission, 201–203, 203f
Potassium ions, 23t
  in extracellular fluid, regulation of, 517
  in intracellular fluid, 516
  regulation of, 497
  renal reabsorption of, 506–508
  tubular secretion of, 509
  in urine, 503t
Potential energy, 28
Pouting, muscles in, 170
PQ interval, of ECG, 326–327, 326f
Pre-Bötzinger complex, 424
Precapillary sphincter, 347, 347f, 366, 366f
Predict Questions, answers to, A-18–A-26
Prednisone, 266
Prefrontal area, 216f, 217
Preganglionic neurons, 227, 227f, 228f
Pregnancy
  control of, 550–551
  diseases and disorders of, 584t
  ectopic, 584t
  hormones in, 563, 563f
  melanin production in, 98–99
  stretch marks in, 97
  tests for, 563
Preload, cardiac, 332
Premature atrial contractions, 327t
Premature ventricular contractions (PVCs),
    327t
Premolars, 440, 441f
Premotor area, 216f, 217
Prenatal development, 559–569
  blastocyst, 561, 561f
  of cardiovascular system, 568t–569t
  of digestive system, 565t, 567, 568t–569t,
    570f
  early cell division, 559–560
  of endocrine system, 568t–569t
  of face, 566, 567f
  fertilization, 559, 560f
  general body structure, 565
  germ layer formation, 564–565, 564f
  of heart, 567, 571f
  implantation, 561–563
  of integumentary system, 568t–569t
  of limb buds, 566, 566f
  of lymphatic system, 568t–569t
  of muscular system, 568t–569t
  of nervous system, 568t–569t
  neural crest formation, 565, 566f
  neural tube formation, 565, 566f
  of organ systems, 566–567, 568t–569t
  placental development, 561–563
  of reproductive system, 568t–569t
  of respiratory system, 565t, 568t–569t
  of skeletal system, 568t–569t
  of urinary system, 565t, 568t–569t
Prenatal period, 559
Prepatellar bursa, subcutaneous, 143f
Prepuce
  female, 544, 544f
  male, 530f, 534
Presbyacusis, 259
Presbyopia, 258t, 259
Pressure. See specific types and measures
Pressure sensation, 102
Pressure sores, 108t
Presynaptic terminals, 158, 158f, 201, 203f
Primary amenorrhea, 548
Primary auditory cortex, 216f
Primary bronchus, 412–413, 412f, 413f, 416f
Primary erythrocytosis, 307, 309t
Primary follicle, 541, 541f, 542f
Primary motor cortex, 216f, 217
Primary oocyte, 540, 541f, 542f
Primary ossification center, 116, 117f

Primary response, in antibody production, 393–395, 395f
Primary sensory areas, 216, 216f
Primary somatic sensory cortex, 216, 216f
Primary spermatocytes, 530, 531f, 532f
Primary teeth, 440, 441f
Primer mover, 167
Primitive streak, 564–565, 564f
Primordial follicle, 541, 541f, 542f
PR interval, of ECG, 326–327, 326f
Procedural memory, 221
Process(es)
    definition of, 122, 122t
    in prenatal development, 566, 567f
    on vertebrae, 128–129, 128f
Products, of chemical reactions, 26
Proerythroblasts, 295f, 297
Progesterone, 285
    action of, 271t, 285, 536t
    in female puberty, 545–546
    in labor (childbirth), 573, 573f
    in lactation, 576
    in menopause, 549, 554
    in menstrual cycle, 546f, 547–548, 547t
    placental, 287–288
    in pregnancy, 563, 563f
    secretion of, 271t, 285, 536t
    synthesis of, 35, 476
    synthetic, 266
    synthetic, in oral contraceptives, 550–552
    target tissue of, 536t
    target tissues of, 271t
Programmed cell death, 66
Prolactin, 274
    action of, 270t, 274
    and lactation, 274, 576–577, 576f
    reference laboratory values for, A-6
    secretion of, 270t, 274
    target tissue of, 270t
Prolapsed uterus, 543
Proliferative phase, of menstrual cycle, 546f, 547–548
Pronation, 144f, 145, 180
Pronator muscles, 180, 181f, 181t
Pronator quadratus muscle, 181f, 181t
Pronator teres muscle, 181f, 181t
Prone, definition of, 10
Prophase
    of meiosis, 528, 529f
    of mitosis, 63, 64f
Propionibacterium acnes, 103
Proprioception, 256–257
Proprioceptive neurons, 218, 219f
Proprioceptors
    and balance, 256
    and ventilation, 427, 427f
Prostacyclin, 266
Prostaglandin(s), 35, 266, 286–287
    action of, 286–287
    and blood clots, 476
    in fertilization, 559
    functions of, 265, 266, 271t
    in inflammatory response, 89, 287, 387
    in innate immunity, 386
    in labor (childbirth), 573, 573f
    in menstrual cycle, 547
    production of, 266
    secretion of, 271t
    in semen, 535
    target tissues of, 271t
Prostate gland, 6f, 527f, 528, 530f, 533f, 534
    aging and, 552–554
    cancer of, 534, 552
    secretions of, 534, 535
Prostate-specific antigen (PSA), 534
Prostatic hyperplasia, benign, 552
Prostatic urethra, 532, 533f, 534, 552
Protection
    by blood, 292
    by connective tissue, 80
    by epithelial tissue, 72, 76
    by hair, 102
    by lymphatic system, 380
    by mucous membranes, 88
    by skeleton, 112
    by skin, 96, 102
    by thoracic cage, 130
    by vertebral column, 128
    by water, 32
Protein(s)
    and acid-base balance, 519
    in blood, A-3
    chemistry of, 32t, 35–38, 36f
    complete, 477
    denaturation of, 37

dietary amounts, recommended, 477, 479
dietary sources of, 477
digestion of, 460, 460f, 464–465, 473
    enzymes as, 37–38
    functions of, 37
    hormones, 265, 266
    incomplete, 477
    metabolism of, 486, 487f
    as nutrients, 473, 477
    in plasma, 293f, 293t
    reference laboratory values for, A-3
    synthesis of
        DNA regulation of, 59–61, 60f
        in ribosomes, 61, 62f
        transcription in, 60–61, 60f, 61f
        translation in, 60f, 61, 62f
    in urine, 503t
    uses in body, 32t, 37, 477
Proteoglycan(s)
    in bone matrix, 112–113
    in cartilage, 83, 112–113
    in extracellular matrix, 80, 112–113
Prothrombin, in coagulation, 301f, 302
Prothrombin activator (prothrombinase), 301f, 302
Prothrombinase, in coagulation, 301f, 302
Prothrombin time, A-4
Prothrombin time measurement, 308
Proton(s), 22, 22f
    clinical relevance of, 27
    mass of, 22
    number of, 22
Proton acceptor, bases as, 30
Proton donor, acids as, 30
Proton pump inhibitors, 450
Protraction, 145
Provitamin(s), 477
Proximal, 10, 11f, 11t
Proximal convoluted tubule
    anatomy of, 498, 500f, 502f
    reabsorption in, 504, 505f, 506, 506f
Proximal phalanx
    of finger, 135, 135f
    of thumb, 135, 135f
    of toe, 139f
Pseudostratified columnar epithelial tissue, 76
    functions of, 75t, 76
    location of, 75t, 76
    structure of, 75t, 76
Pseudo-unipolar neurons, 195, 196f, 196t, 206
Psoriasis, 108t
Psychological fatigue, 164
Pterygoid muscles
    lateral, 172t
    medial, 172t
Pterygopalatine ganglion, 228f
PTH. See Parathyroid hormone
Puberty
    age of onset, 577
    definition of, 537, 577
    female, 545–546, 577
    male, 537, 577
    skin during, 103
Pubic region, 12f
Pubic symphysis, 15f, 135, 136f, 137f, 539f
Pubis, 135, 136f, 137f
Pudendal cleft, 544, 544f
Pudendum, 544
Pulmonary arteries, 314–316, 349f
    left, 316f–317f, 318f, 320f, 348
    right, 316f–317f, 318f, 320f, 348, 416f
    thrombosis of, 432t
Pulmonary capacities, 420–421, 421f
Pulmonary capillaries, 414f
Pulmonary circulation, 313, 314f, 348
    blood vessels in, 345, 348
    definition of, 348
    path of, 349f
Pulmonary fibrosis, 432t
Pulmonary semilunar valve, 318f, 319, 319f, 320f, 331f
Pulmonary trunk, 6f, 314–316, 315f–318f, 320f–321f, 321, 348, 349f–350f, 416f
Pulmonary veins, 314, 319f, 320f, 348, 349f, 414f
    left, 316f–317f, 318f
    right, 316f–317f, 318f, 357f, 416f
Pulmonary vessels, 345, 348, 349f
Pulmonary volume, 420–421, 421f
Pulp, of tooth, 442, 442f
Pulse, 364f, 365
Pulse pressure, 364–365
Punnett square, 581, 581f
Pupil, 243f, 245, 245f
    constriction of, 245, 246f
    dilation of, 245, 246f

Purkinje fibers, 325, 325f
Pus, 89, 299, 387
P wave, of ECG, 326, 326f
Pyloric opening, 446, 447f
Pyloric region, of stomach, 446, 447f
Pyloric sphincter, 446, 447f, 450f, 451
Pyloric stenosis, hypertrophic, 446
Pyramid(s)
    of medulla oblongata, 212, 212f
    renal, 498, 499f, 500f
Pyridoxine. See Vitamin B₆
Pyrogens, 388
Pyruvic acid
    in aerobic respiration, 484–486, 484f
    in amino acid metabolism, 486
    in anaerobic respiration, 483–484, 483f
    in glycolysis, 483, 483f, 484, 484f

QRS complex, of ECG, 326, 326f
QT interval, of ECG, 326f, 327
Quadrants, of abdomen, 12, 13f
Quadrate lobe, of liver, 453, 454f
Quadriceps femoris muscle, 5f, 152f, 168f, 183f, 184–185, 184f
    action of, 184–185
    insertion of, 184
Quadriceps femoris tendons, 143f

Rabies, 231t
Radial arteries, 350f, 354, 354f, 364f
Radial nerve, 210–211
Radial tuberosity, 134, 134f
Radiation, in body temperature regulation, 491, 491f
Radiation therapy, for cancer, 27
Radioactive isotopes, 27
Radiocarpal joint, 142f
Radioulnar joint
    distal, 142f
    proximal, 142f
Radius, 5f, 121f, 123f, 134, 134f
Ranitidine (Zantac), 450
Ranvier, node of, 195–196, 195f, 197f, 201, 202f
Rashes, as diagnostic aid, 104
Rate of chemical reactions, 29–30
Raynaud disease, 230
Reactants, in chemical reactions, 26, 29
Receptor(s). See also specific types
    hormone, 266–268
    intracellular, 266, 267f, 268, 268t
    membrane-bound, 266–268, 267f, 268t
    in negative feedback, 7–8
    responses of, 266–268
    specificity of, 266, 266f
Receptor-mediated endocytosis, 51, 51f
Receptor-mediated enzyme activity, 483
Receptor molecules, 45, 45f, 266
Receptor sites, 266
Recessive genes, 580–581, 581f
Recipient, of blood, 304
Recommended Dietary Allowances (RDA), 477–479, 478t
Recovery oxygen consumption, 164
Recruitment, of motor units, 162–163
Rectal temperature, 491
Rectum, 5f, 438f, 440f, 459, 459f
    female, 539f
    male, 530f
    prenatal development of, 570f
Rectus abdominis muscle, 5f, 168f, 175–176, 176f, 176t
Rectus femoris muscle, 168f, 183f, 184f
Rectus muscle
    inferior, 243f, 244, 244f
    lateral, 244, 244f
    medial, 244, 244f
    superior, 243f, 244, 244f
Red blood cells, 65f, 294–298
    aging and, 92
    antigens on, 303
    disorders of, 306–307, 308, 309t
    function of, 294t, 296–297
    in hypertonic solution, 49f
    in hypotonic solution, 49f
    in isotonic solution, 49f
    life history of, 297–298
    life span of, 296
    number of, 293t, 294, 306
    production of, 294, 295f, 297, 297f
    removal of, 298, 298f
    renal regulation of, 497
    rupture of, 303
    shape of, 294–296, 294t, 296f
    staining properties of, 299f

structure of, 294–296, 294t, 296f
testosterone and, 538t
in urine, A-5
Red blood count (RBC), 306, A-4
Red bone marrow
    blood cell formation in, 112, 297, 297f, 299, 389, 389f
    donation of, 296
    in inflammatory response, 388
    in long bones, 113, 114f
Red pulp, of spleen, 383, 383f
Reference Daily Intakes (RDIs), 479
Referred pain, 240–241, 240f
Reflex(es), 203–204
    aging and, 233–234
    baroreceptor, 333, 334f, 337f, 368–369, 368f, 370f, 373
    blink, 243
    chemoreceptor, 333, 335f, 337f, 369–371, 371f, 373
    cough, 411–412
    defecation, 460
    definition of, 203
    Hering-Breuer, 426, 427f
    knee-jerk (patellar), 184–185, 208, 208f
    micturition, 514, 515f
    sneeze, 409
    spinal, 208
    stretch, 208
    withdrawal (flexor), 208, 209f
Reflex arc, 203–204, 205f
Refraction, of light, 249, 249f
Refractory period, 325
Regeneration
    in spinal cord, 209
    tissue repair by, 90
Region(s)
    of abdomen, 12, 13f
    of body, 10–12, 12f–13f
Regional anatomy, 2
Regulatory molecules, transport in blood, 292
Regulatory substances, in plasma, 292, 293f, 293t
Rejection, transplant, 402t
Relative erythrocytosis, 307, 309t
Relaxation phase, of muscle twitch, 162, 162f
Remodeling, bone, 117–119
Renal arteries, 350f, 351f, 355, 355f, 498f, 499f, 502, 502f
Renal calculi, 514
Renal capsule, 497, 498f, 499f
Renal corpuscle, 498, 500f, 501f
Renal failure, 520–521, 523t
Renal papillae, 499f
Renal pelvis, 498, 499f
Renal pyramid, 498, 499f, 500f
Renal sinus, 498, 499f
Renal veins, 360, 498f, 499f
    left, 358f, 359f
    right, 359f
Renin, 280
    aging and, 288
    and blood pressure, 369, 372f, 373
    reference laboratory values for, A-6
    and urine regulation, 510, 510f, 512f
Renin-angiotensin-aldosterone mechanism, 280, 372f, 373, 374f
    and extracellular fluid volume, 517
    and urine regulation, 510, 510f, 512f
Repair, tissue, 90–92
Replacement, tissue repair by, 90–91
Replication, DNA, 61–62, 63f, 64f
Repolarization phase, of action potentials, 200f, 201, 201f
    in cardiac muscle, 324, 324f
    in skeletal muscle, 156f, 157, 324f
    in smooth muscle, 166
Reposition (movement), 145
Reproduction
    cellular function in, 44
    definition of, 7
    endocrine regulation of, 263
Reproductive hormones, 271t, 536t
    actions of, 271t, 536t
    female, 536t, 545–549
        in labor (childbirth), 573–574, 573f
        in lactation, 550
        in menopause, 547–548
        in menstrual cycle, 546–548, 546f, 567t
        in pregnancy, 563, 563f
        in puberty, 545–546
    male, 535–537, 535f, 536t
    secretion of, 271t, 536t
    synthesis of, 476, 527
    target tissues of, 271t, 536t

Reproductive system, 3, 526–554
  female, 6f, 527f, 539–552
    aging and, 552–554
    anatomy of, 539–545, 539f, 540f
    diseases and disorders of, 554t
    hormones of, 536t, 545–549
    physiology of, 545–552
  functions of, 527
  male, 527f, 528–539
    accessory glands of, 528, 534–535
    aging and, 552–554
    anatomy of, 528–535, 530f, 533f
    diseases and disorders of, 554t
    ducts of, 528, 531–532
    hormones of, 535–537, 535f, 536t
    physiology of, 535–539
    puberty in, 537
    secretions of, 535
    sensory impulses and integration in, 538–539
    supporting structures of, 528
  prenatal development of, 568t–569t
Residual volume, 420, 421f
  aging and, 433
  exercise and, 431
Resistance
  to blood flow, 363–364, 368, 375
  peripheral, 368
Resolution
  of female orgasm, 549
  of male orgasm, 538
Respiration, 407, 416–424
  aerobic, 56–57, 484–486, 484f
  aging and, 433
  altitude and, 429
  alveolar volume changes in, 419–420
  anaerobic, 429–430, 483–484, 483f, 486
  brainstem and, 211, 424, 426f
  carbon dioxide levels and, 427, 428f
  chemical control of, 427–429, 427f
  exercise and, 429–433
  Hering-Breuer reflex and, 426, 427f
  lung recoil and, 419
  muscle paralysis and, 432t
  muscles of, 152, 173, 416, 417f
  nervous control of, 426–427, 427f
  normal rate of, 424
  normal vs. labored, 416, 417f
  oxygen levels and, 427
  pH and, 427–428, 428f
  phases of, 416
  pleural pressure and, 419–420
  pressure in, 417, 418f
  processes of, 407
  rhythmic, generation of, 424–430
  thoracic volume and, 416, 417f, 418f
Respiratory acidosis, 519
Respiratory alkalosis, 523
Respiratory bronchioles, 413, 414f
Respiratory capacities, 420–421, 421f
Respiratory center, medullary, 424, 426f
Respiratory distress syndrome
  adult, 432t
  infant, 419, 574
Respiratory groups
  dorsal, 424, 426f
  ventral, 424, 426f
Respiratory membrane, 414, 415f
  diffusion of gases through, 414, 415f, 422–424, 423f
  surface area of, 422
  thickness of, 422
Respiratory rate, 420, 424, 429, 431
Respiratory system, 3, 5f, 406–433
  and acid-base balance, 519, 522f
  adaptations to exercise, 430–433
  aging and, 433
  anatomy of, 407–416, 407f
  cartilage rings of, 83, 83t
  components of, 5f, 407, 407f
  diseases and disorders of, 432t
  functions of, 5f, 407
  lower, 407, 407f
  of newborn, 574–575
  prenatal development of, 565t, 568t–569t
  upper, 407, 407f
Respiratory tract. See Respiratory system
Respiratory volumes, 420–421, 421f
Responsiveness, definition of, 7
Resting membrane potential, 198–199
  changes in. See Action potential(s)
  in skeletal muscle, 155, 156f
  in smooth muscle, 166
Resting tremor, 218
Resurrectionists, 4

Rete testis, 531–532, 531f
Reticular activating system (RAS), 213
Reticular fibers, in connective tissue, 80
Reticular formation, 212–213
  functions of, 212–213
  structure of, 212
Reticulocyte count, A-4
Retina, 244, 245f, 246
  cells of, 246, 247f
  focusing of images on, 249, 249f
  layers of, 246
  light effect on, 246, 248f
  ophthalmoscopy of, 246, 248f
  photoreceptor layer of, 246, 247f
  pigmented, 246, 247f
  sensory, 246, 247f
  structure of, 246, 247f
Retinaculum, 166, 168f, 180, 181f
Retinal, 246, 247f, 248f
Retinal detachment, 246, 258t
Retinol. See Vitamin A
Retraction, 145
Retromandibular vein, 358f
Retroperitoneal organs, 17, 17f, 440, 440f, 497
Reversible reactions, 28
Rh blood group, 305–306, 305f
Rheumatic heart disease, 339t
Rheumatoid arthritis, 148t
Rhodopsin, 246, 247f, 248f
RhoGAM, 306
Rho(D) immune globulin, 306
Rhomboid major muscle, 176–177, 177, 178f, 178t
Rhomboid minor muscle, 176–177, 177, 178f, 178t
Rhythmic ventilation, generation of, 424–430
Rib(s), 5f, 123f, 130
  costal cartilage of, 83, 83t
  counting of, 131
  false, 130, 131f
  floating, 130, 131f
  fractures of, 130
  number of, 121t
  structure of, 130, 131f
  true, 130, 131f
Rib cage. See Thoracic cage
Riboflavin. See Vitamin B₂
Ribonucleic acid (RNA), 38
  composition of, 38
  messenger (mRNA), 60–61, 60f, 61, 61f, 62f, 268
  ribosomal (rRNA), 52, 54f, 61, 62f
  structure of, 38
  synthesis (transcription) of, 60–61, 60f, 61f
  transfer (tRNA), 60f, 61
  translation of, 60f, 61, 62f
Ribose, 38
Ribosomal ribonucleic acid (rRNA), 52, 54f, 61, 62f
Ribosomes, 43f, 52–54
  free, 43f, 52–54
  functions of, 44t, 52
  location of, 44t
  production of, 52, 54f
  protein synthesis in, 52, 61, 62f
  structure of, 53f, 61
Rickets, 99, 148t
Right, as directional term, 10, 11f, 11t
Right atrium, 314, 315f, 316–317, 316f–317f, 318f
  blood flow through, 320–321, 320f
  contraction of, 320
Right bundle branch, 325, 325f
Right coronary artery, 317f, 318f, 321, 321f, 348, 350f
Right heart failure, 333
Right hemisphere, of brain, 214, 214f, 220
Right hypochondriac region, 13f
Right iliac region, 13f
Right lumbar region, 13f
Right lymphatic duct, 380, 381f, 385f
Right ventricle, 314, 315f, 316f–317f, 317–318, 318f
  blood flow through, 320–321, 320f
  contraction of, 320–321
Rigor mortis, 159
Ringworm, 108t
RNA. See Ribonucleic acid
Rods
  function of, 246
  pigments of, 246, 247f, 248f
  structure of, 246, 247f
Rogaine (minoxidil), 538
Root(s)
  of hair, 100, 100f
  of nail, 101, 102f
  of tooth, 442, 442f

Rotation, 144f, 145
Rotator cuff muscles, 178f, 179, 180t
Rough endoplasmic reticulum, 43f, 54, 194, 195f
  functions of, 44t, 54
  location of, 44t
  structure of, 54, 55f
Round ligament
  of liver, 575f
  of uterus, 540f, 543
Round window, 252
  anatomy of, 252, 252f, 253f
  auditory function of, 254, 254f
Roux-en-Y gastric bypass, 490
RU486, 552
Rubella, 108t, 584t
Rubeola, 108t
Rubrospinal tract, 217f, 217t
Ruffini corpuscles, 239f, 240
Rugae, 446, 447f

Saccule, 255, 255f, 259
Sacral arteries, median, 351f, 355, 355f
Sacral bone. See Sacrum
Sacral crest, median, 130, 130f
Sacral foramen, posterior, 130f
Sacral hiatus, 130, 130f
Sacral nerves, 208, 210–211, 210f
Sacral promontory, 129f, 130, 130f, 136f, 137f
Sacral region, 13f
Sacral vertebrae, 128, 130
Sacroiliac joint, 135, 136f, 142f
Sacrum, 121t, 123f, 128, 128f
  male vs. female, 136t
  in pelvic girdle, 135, 136f
  structure of, 130, 130f
Saddle joints, 141, 142f
Sagittal plane, 12–15, 14f
Sagittal sinuses, superior, 225f, 357f
Salicylates, reference laboratory values for, A-3
Saliva, 443
  composition of, 442–443
  functions of, 443–444, 443t
  and innate immunity, 386
  secretion of, 442–443
Salivary amylase, 443
  functions of, 443, 443t, 461–462, 461f
  in newborn, 576
  secretion of, 443, 443t
Salivary glands, 5f, 438, 438f, 442–443
  autonomic regulation of, 444
  epithelial tissue in, 76
  secretions of, 442–443
Salt(s)
  definition of, 31
  diffusion of, 46f
  taste of, 242
Saltatory conduction, 201, 202f
Saphenous veins
  graft from, 361
  great, 357f, 360–361, 361f
  small, 357f, 360–361, 361f
Sarcolemma, 153
  in muscle contraction, 158
  structure of, 153, 154f
Sarcomeres, 153
  in cardiac muscle, 322–323, 323f
  in muscle contraction, 160f
  in muscle relaxation, 158
  structure of, 153, 154f, 155f
Sarcoplasm, of skeletal muscle, 153, 154f
Sarcoplasmic reticulum, in skeletal muscle, 153, 154f
Sartorius muscle, 5f, 152f, 168f, 183f, 184–185, 184f
  action of, 184–185
Satellite cells, 165
Satiety center, 284
Saturated fats or oils, 34, 34f, 462–463, 476
Saturated fatty acids, 34, 34f
Scab, in tissue repair, 91–92, 91f
Scala tympani, 253, 253f, 254, 254f
Scala vestibuli, 253, 253f, 254, 254f
Scalene muscles, 175f, 175t, 417f
Scaphoid bone, 134, 135f
Scapula, 121t, 123f, 131–132
  inferior angle of, 133f
  movement of, muscles of, 167, 176–177, 178f, 178t
  at shoulder, 143f
  spine of, 132, 132f, 133f, 179f
  structure of, 131–132, 131f, 132f
  surface anatomy of, 133f, 135f
Scapular region, 13f
Scarlet fever, 104

Schwann cells, 195–196, 195f, 196t, 197f
Sciatic nerve, 211
Sciatic notch, greater, 137f
Sclera, 244, 245f
Scoliosis, 128
Scrotum, 528–530, 530f, 533f
  temperature regulation in, 529–530
  testosterone and, 538
Seasickness, 257
Sebaceous glands, 97f, 101, 101f, 243f
Sebum, 101, 103
Secondary active transport, 46t, 51, 51f
Secondary amenorrhea, 548, 549
Secondary bronchi, 412f, 413, 413f
Secondary erythrocytosis, 307, 309t
Secondary follicle, 541, 541f, 542f
Secondary oocyte, 540–541, 542f, 559, 560f
Secondary ossification centers, 116, 117f
Secondary response, in antibody production, 395, 395f
Secondary sexual characteristics
  female, 545, 577
  male, 537, 577
Secondary spermatocytes, 530, 531f, 532f
Secondary teeth, 440, 441f
Second-degree burns, 104f, 105
Second messengers, 267–268
Secretin, 448
  and bile secretion, 456, 456f
  functions of, 448, 448t
  and pancreatic secretion, 458, 458f
  secretion of, 448, 448t, 449f, 453
Secretory phase, of menstrual cycle, 546f, 548
Secretory vesicles, 43f, 54
  in epithelial cells, 76
  in exocytosis, 52
  functions of, 44t, 54
  location of, 44t
  structure of, 54, 55f
Segmental bronchi, 412f, 413, 413f
Segmental contractions, in small intestine, 453, 453f
Selective permeability, 45, 516
Selenium, 480t
Self-antigens, 388–389
Sella turcica, 125, 126f
Semen, 535
Semicircular canals, 252
  aging and, 259
  anatomy of, 252, 252f, 253f, 255–256, 257f
  balance function of, 255–256, 257f
Semilunar valves
  anatomy of, 319
  aortic, 318f, 319, 319f, 320f, 321f, 331f
  in cardiac cycle, 328, 329f, 330f
  function of, 319, 319f
  location of, 331f
  pulmonary, 318f, 319, 319f, 320f, 331f
  sounds of, 328, 330f, 331
Semimembranosus muscle, 169f, 183f
Semimembranosus tendon, 184f
Seminal vesicles, 6f, 527f, 528, 530f, 532, 533f
  duct of, 533f
  secretions of, 535
Seminiferous tubules, 530, 531f
Semitendinosus muscle, 169f, 183f
Semitendinosus tendon, 184f
Senior citizen, 577. See also Aging, effects of
Sensation
  definition of, 239
  in integumentary system, 96, 102
Sense(s), 238–259. See also specific senses
  aging and, 258–259, 578
  definition of, 239
  diseases and disorders of, 258t
  general, 239–241
  somatic, 239
  special, 239, 241
  visceral, 239
Sensorineural hearing loss, 255, 258t, 259
Sensory division, of peripheral nervous system, 193–194, 194f
Sensory memory, 221
Sensory (afferent) neurons, 193–194
  aging and, 230–233
  prenatal development of, 565t
  in reflex arc, 204, 205f
Sensory receptor(s)
  aging and, 258
  definition of, 239
  functions of, 193
  in reflex arc, 204, 205f
  in skin, 96, 102, 239–240, 239f
  types of, 239
  and ventilation, 427, 427f

Sensory retina, 246, 247f
Sensory speech area, 216f, 220
Separation, 145
Septal area, 221f
Septal defects, 339t, 567
Septicemia, 309t
Septic shock, 369
Serosa
 of digestive tract, 438, 439f
 of small intestine, 452f
 of stomach, 447f
Serotonin, as neurotransmitter, 203, 204t
Serous fluid, 442–443
Serous membranes, 16–17, 88
 balloon comparison to, 16, 16f
 functions of, 88
 inflammation of, 16, 88
 location of, 16, 17f, 88, 89f
 parietal, 16, 16f
 structure of, 88
 visceral, 16, 16f
Serous pericardium, 314, 315f
Serratus anterior muscles, 168f, 176–177, 177,
  178f, 178t, 179f
Sertoli cells, 530
Serum
 definition of, 292
 reference laboratory values for, A-2–A-4
Set point, 7, 7f
Severe combined immunodeficiency disease
  (SCID), 402t
Sex act
 female, 549
 male, 537–539
Sex cells (gametes), 527–528, 529f, 579
Sex chromosomes, 62–63, 579, 579f, 580f
Sex hormones, 271t, 536t
 actions of, 271t, 536t
 female, 536t, 545–549
  in labor (childbirth), 573–574, 573f
  in lactation, 550
  in menopause, 547–548
  in menstrual cycle, 546–548, 546f, 567t
  in pregnancy, 563, 563f
  in puberty, 545–546
 male, 535–537, 535f, 536t
 secretion of, 271t, 536t
 synthesis of, 476, 527
 target tissues of, 271t, 536t
Sex-linked traits, 581, 581f
Sexual behavior
 female, 549
 male, 537–539
Sexual characteristics
 female secondary, 545, 577
 male secondary, 537, 577
Sexually transmitted diseases, 554t
Shaft, of hair, 100, 100f
Shingles, 108t, 402
Shivering, 160, 492
Shock
 anaphylactic, 369
 cardiogenic, 369
 cardiovascular, 504
 circulatory, 369
 hemorrhagic, 369
 hypovolemic, 369
 insulin, 285
 interstitial fluid loss, 369
 neurogenic, 369
 plasma loss, 369
 septic, 369
Short bones, 113
Short-term memory, 221, 234
Shoulder (shoulder joint)
 as ball-and-socket joint, 143f
 dislocation of, 145
 muscles of, 176–177, 178f, 178t
Shoulder blade. See Scapula
Shoulder girdle. See Pectoral girdle
Sickle-cell anemia, 308, 581
Sickle-cell trait, 581
SIDS (sudden infant death syndrome), 432t
Sight, 242–251. See also Vision
Sigmoid colon, 459, 459f
Sildenafil (Viagra), 539
Simple epithelial tissue, 72–76, 73t
 columnar, 76
  functions of, 75t, 76
  location of, 75t, 76
  structure of, 75t, 76
 cuboidal, 73
  functions of, 73, 74t
  location of, 73, 74t
  structure of, 73, 74t

squamous, 72–73
 functions of, 72–73, 74t
 location of, 72–73, 74t
 structure of, 72, 74t
Simple exocrine glands, 78, 79f
Simple fracture, 120
Simple sugars, 32–33, 33f
Single covalent bond, 24
Sinoatrial (SA) node, 325–326, 325f
 action potentials in, 325, 325f
 block of, 327t
 location of, 325
Sinus(es). See also specific types
 definition of, 122t
 lymphatic, 382, 383f
 paranasal, 124, 126f, 408, 408f
Sinus arrhythmia, 327t
Sinusitis, 409
Sinusoid(s), in male erectile tissue, 539
Sjogren syndrome, 396–397
Skeletal muscle(s), 85. See also specific muscles
 action potentials of, 153–159, 156f, 324f
 aging and, 187–188, 578
 agonist, 167
 anatomy of, 166–186
 antagonist, 167
 attachment of, 166–167, 167f
 belly of, 167
 cells of, 85, 86t
 characteristics of, 152–165
 connective tissue in, 153, 154f
 contractility of, 152–153
 contractions of, 158–165
  energy requirements of, 159, 161f,
   163–164, 163f
  sliding filament model of, 158–159, 160f
  summary of, 159t
 elasticity of, 153
 excitability of, 153–157
 extensibility of, 153
 fatigue of, 164
 fixator, 167
 functional groups of, 167
 functions of, 85, 86t, 152
 general principles of, 166–167
 hypertrophy of, 165
 insertion of, 167, 167f
 nerve supply of, 157–158, 157f, 158f
 nomenclature for, 167
 origin (head) of, 167, 167f
 overview of, 152
 as percentage of body weight, 152
 prenatal development of, 565t
 prime mover, 167
 relaxation of, 160
 striations of, 85, 86t, 152
 structure of, 85, 86t, 153, 154f
 superficial, anatomic overview of, 167,
  168f–169f, 170f
 synergist, 167
 testosterone and, 538t
Skeletal system, 3, 5f, 111–150. See also
  Bone(s); Joint(s)
 aging and, 145–148
 components of, 5f
 diseases and disorders of, 148t
 functions of, 5f, 112
 prenatal development of, 568t–569t
Skeleton
 appendicular, 121t, 122, 123f, 131–140
 axial, 121t, 122–131, 123f
 cardiac, 319–320, 319f
 number of bones in, 121, 121t
 origin of term, 112
Skin, 5f, 96–101
 accessory structures of, 100–101
 aging and, 2, 92, 108
 cancer of, 99, 105
  sites of development, 105
  types of, 105, 105f
 cleavage lines in, 97, 98f
 color of, 98–100
  adaptive advantages of, 99
  blood flow and, 99–100
  genetics and, 98, 581–582, 582f
  during pregnancy, 98–99
 connective tissue and, 80, 81t
 as diagnostic aid, 104
 diseases and disorders of, 108t
 epithelial tissue of, 76, 77t
 excretion by, 96, 104
 functions of, 96
 glands in, 101, 101f
 and innate immunity, 386
 as membrane, 87–88

muscle attachment to, 166, 167f
 in postmenopausal women, 549t
 prenatal development of, 565t
 protection provided by, 96, 102
 sensation in, 96, 102, 239–240, 239f
 structure of, 96–100, 97f
 temperature regulation by, 96, 102, 103f
 testosterone and, 538t
 vitamin D production in, 98, 99, 102
 wound repair in, 91–92, 91f
 wrinkles in, 108
Skin-fold measurement, 100
Skin grafts, for burns, 105, 107
Skull, 5f, 122–127, 123f
 base of, 125–127
 foramina of, 125–127, 126f, 127f
 functions of, 112
 inferior view of, 127f
 lateral view of, 122, 124f
 ossification in, 115
 sutures of, 122, 124f, 140, 140f
Sleep
 aging and, 234
 ejaculation during, 538
 reticular activating system and, 213
Sliding filament model, of muscle
  contraction, 158–159, 160f
Slow Fluoride, 147
Slow-twitch muscle fibers, 164–165, 165t
Small arteries, 345, 347
Small cardiac vein, 317f, 321f, 357f
Small intestine, 4f, 5f, 438, 438f, 440f, 451–453
 absorption in, 453, 460–461
 anatomy of, 451–452, 451f, 452f
 diseases and disorders of, 468t
 epithelial tissue of, 75t, 76, 438, 451
 functions of, 451
 histology of, 451, 452f
 movement in, 453, 453f
 secretions of, 443t, 452–453
 smooth muscle of, 87
Small saphenous vein, 357f, 360–361, 361f
Small veins, 345, 347–348
Smell, sense of. See Olfaction
Smiling muscles, 167
Smoker's cough, 412
Smoking
 and asthma, 430
 and atherosclerosis, 375
 and epithelial cells, 76–78
 and heart disease, 337
Smooth endoplasmic reticulum, 43f, 54
 functions of, 44t, 54
 location of, 44t
 structure of, 54, 55f
Smooth muscle(s), 87, 165–166
 action potentials of, 166
 of anal canal, 459–460
 autorhythmicity of, 166, 166t
 circular, 438, 439f, 446, 447f
 contraction of, 166
 of digestive tract, 438, 439f
 functions of, 87, 87t, 152
 location of, 87, 87t
 longitudinal, 438, 439f, 446, 447f
 overview of, 152
 of stomach, 446, 447f
 structure of, 87, 87t, 165–166, 166f
Sneering, muscles in, 167–170
Sneeze reflex, 409
Sobriety tests, 257
Sodium, 23, 23f, 23t
 atomic number of, 21t
 deficiency of, 480t
 dietary guidelines for, 480t
 digestion of, 465
 functions of, 480t
 mass number of, 21t
 percent in body, 21t
 reference laboratory values for, A-3, A-5
 symbol for, 21t, 22
Sodium channels, 47, 47f
 in action potential, 155–157, 156f, 200–201,
  200f, 324–325, 324f
 and resting membrane potential, 155, 156f
 in synaptic transmission, 201–203, 203f
Sodium chloride
 dissociation of, 25, 26f
 formation of, molecular, 23, 23f, 25
Sodium fluoride, slow-releasing, 147
Sodium ion(s)
 and cardiac function, 333
 in extracellular fluid, 517
 in intracellular fluid, 516
 regulation of, 497, 517

renal excretion of, 517
 renal reabsorption of, 506–508, 517
 in urine, 503t
Sodium–potassium pump
 in action potentials, 155–157
 in active transport, 50–51, 50f
 in glucose transport, 462, 462f
 and resting membrane potential, 155, 199,
  199f
 in secondary active transport, 51, 51f
Soft palate, 127, 170, 408f, 409, 441f, 442
Soft spots (fontanels), 140, 140f
Soleus muscle, 168f, 169f, 184f, 185, 186f
Solutes, 45
 in blood, regulation of, 497
 and diffusion, 45–46
 and osmosis, 47
 renal reabsorption of, 504–508, 505f
Solutions
 acidic, 30, 30f
 basic (alkaline), 30, 30f
 concentrations of, A-7
  and diffusion, 46, 46f
  and osmosis, 47–48, 48f
 definition of, 45
 hypertonic, 48–49, 49f
 hypotonic, 48, 49f
 isotonic, 48, 49f
 neutral, 30, 30f
Solvents, 45
Somatic nervous system, 194, 194f
Somatic senses, 239
Somatic sensory association area, 216f
Somatic sensory cortex, 215f, 216, 216f
Somatomedins, 273
Somatotropin. See Growth hormone
Sound waves, 254–255, 254f
Sour taste, 242
Spastic paralysis, 159
Spatial summation, 205
Special senses, 239, 241. See also specific senses
Specific gravity, of urine, 503t
Specific immunity. See Adaptive immunity
Specificity
 in adaptive immunity, 386, 388
 in carrier-mediated transport, 49
 in endocytosis, 51
 in receptor sites, 266, 266f
Speech
 brain activity in, 220
 cheek's role in, 440
 tongue's role in, 440
Speech area
 motor, 216f, 220
 sensory, 216f, 220
Spermatic cord, 532, 533f
Spermatids, 530, 531f, 532f
Spermatocytes
 primary, 530, 531f, 532f
 secondary, 530, 531f, 532f
Spermatogenesis, 530–531, 532f
Spermatogonia, 530, 531f, 532f
Spermatozoon, 530. See also Sperm cell(s)
Sperm cell(s), 530–531, 531f
 acrosome of, 530, 531f
 capacitation of, 532, 559
 count of, 531
 fertilization by, 527, 530, 540–541, 542f,
  559, 560f
 flagella of, 59
 formation of, 530–531, 532f
 head of, 530, 531f, 560f
 maturation of, 532
 midpiece of, 531f
 movement of, 531, 559
 nucleus of, 530, 531f
 pH and, 535
 tail of, 531f
 temperature and, 528
 viability, in female reproductive tract, 559
Spermicidal agents, 550, 551f
Sphenoidal fontanel, 140f
Sphenoidal sinus, 124, 126f, 408f
Sphenoid bone, 121t, 122
 in cranial cavity, 125, 126f
 frontal view of, 125f
 inferior view of, 127f
 intramembranous ossification of, 116f
 lateral view of, 122, 124f
Sphygmomanometer, 362
Spina bifida, 565
Spinal anesthesia, 222
Spinal block, 222
Spinal cord, 4f, 6f, 14f, 193, 193f, 206–211
 ascending tracts of, 206, 215–216, 215f, 215t

converging pathways of, 206, 207f
cross section of, 206, 207f
descending tracts of, 206, 217–218, 217f, 217t
diverging pathways of, 206–208, 207f
functions of, 193
    motor, 216–219
    sensory, 215–216
injury to, 209
    mechanism of, 209
    primary damage in, 209
    recovery from, 217
    secondary damage in, 209
    treatment of, 209
meninges of, 222, 223f
nervous tissue of, 87, 88t
prenatal development of, 565t
protection of, 128
regeneration in, 209
structure of, 206, 206f, 207f
Spinalis muscles, 174f, 174t
Spinal nerves, 206, 206f, 208–211
nomenclature of, 208
plexuses of, 208–211, 210f, 210t
structure of, 208
Spinal reflexes, 208
Spinal tap, 222
Spine
of ilium, 135, 136f
of ischium, 137f
of scapula, 132, 132f, 133f, 179f
Spinocerebellar tract, 215f, 215t, 216
Spinothalamic tract, 215, 215f, 215t, 240
Spinous processes, 128f, 129
of cervical vertebrae, 129, 129f, 133f
of lumbar vertebrae, 128f, 129f, 130, 133f
of sacral vertebrae, 130
of thoracic vertebrae, 129, 129f
Spiral fracture, 120, 120f
Spiral ganglion, 253f, 254
Spiral lamina, 252–253, 253f
Spiral organ, 253, 253f
Spirometer, 420
Spirometry, 420
Splanchnic nerves, 227
greater, 228f
lesser, 228f
lumbar, 228f
sacral, 228f
Spleen, 4f, 5f, 14f, 381f, 383
function of, 383, 384, 385f
removal of, 384
rupture of, 384
structure of, 383, 383f
Splenectomy, 384
Splenic artery, 350f, 383f
Splenic vein, 357f, 359f, 360, 360f, 383f
Spongy bone. See Cancellous bone
Spongy urethra, 532, 533f
Sprain, 145
Squamous cell carcinoma, 105, 105f
Squamous epithelial tissue, 72, 73t
in epidermis, 96
keratinized, 76, 77t
nonkeratinized, 76
simple, 72–73, 74t
stratified, 76, 77t, 96
Squamous suture, 122, 124f, 140f
Stages of life, 559, 577
Stapes, 121t, 252, 252f, 254, 254f
Starches
dietary sources of, 474
digestion of, 461–462
Starling's law of the heart, 332
Starvation, 489
Static equilibrium, 255
Stem cells, 90–91
adult, 90–91
and cancer therapy, 296
in hematopoiesis, 294, 295f, 297, 297f, 389
Stenosed heart valves, 331, 339t, 340
Stents, for blocked coronary arteries, 322
Sternal angle, 131, 131f
Sternal region, 12f
Sternoclavicular joint, 142f
Sternocleidomastoid muscle, 152f, 168f, 169f, 171f, 172, 178f
action of, 172, 174t
insertion of, 174t
origin of, 174t
in respiration, 417f
Sternum, 5f, 121t, 123f, 130–131, 175f
structure of, 130–131, 131f
surface anatomy of, 133f
Steroid(s), 35, 265
anabolic, 537

chemistry of, 35, 35f
secretion of, 280–281
synthesis of, 476
Stethoscope
listening to heart sounds with, 314, 331, 331f
measuring blood pressure with, 362, 362f
Stomach, 5f, 14f, 438, 438f, 440f, 446–451
absorption in, 460
aging and, 578
anatomy of, 446, 447f
body of, 446, 447f
cardiac region of, 446, 447f
diseases and disorders of, 468t
emptying of, regulation of, 451
epithelial tissue of, 75t, 438, 446, 447f
fundus of, 446, 447f
greater curvature of, 446, 447f
histology of, 446, 447f
movements in, 450–451, 450f
muscular layer of, 87, 87t, 446, 447f
pH of, 448
prenatal development of, 570f
pyloric region of, 446, 447f
secretions of, 448
    cephalic phase of, 448, 449f
    daily production of, 448
    functions of, 443t
    gastric phase of, 448–450, 449f
    intestinal phase of, 448–450, 449f
    regulation of, 448–450, 449f
Stop codon, 61
Strabismus, 258t
Stratified epithelial tissue, 72, 73t, 76
columnar, 76
    functions of, 76
    location of, 76
    structure of, 76
cuboidal, 76
    functions of, 76
    location of, 76
    structure of, 76
squamous, 76
    epidermis as, 96
    functions of, 76, 77t
    keratinized, 76, 77t
    location of, 76, 77t
    nonkeratinized, 76
    structure of, 76, 77t
Stratum basale, 96, 98f
Stratum corneum, 96, 98f
Strep throat, 432t
Streptokinase, 302, 322
Stress
and adrenal medulla, 279
adverse effects of, 282
aging and, 578
and cortisol secretion, 281
and glucocorticoids, 282
and heart disease, 337
and hormones, 282
Stretch marks, 97
Stretch reflex, 208
Striated muscle
cardiac, 86t, 87, 323, 323f
skeletal, 85, 86t, 152
Stroke, 231t, 232–233
hemorrhagic, 232
imaging of, 232f
ischemic, 232
platelets in, 300
systemic effects of, 233
Stroke volume (SV), 332
and blood pressure, 368
exercise and, 332
regulation of, 332–340, 337f
    baroreceptor reflex in, 333, 334f, 337f
    chemoreceptor reflexes in, 333, 335f, 337f
Stye, 258t
Styloid process
of radius, 134, 134f
of temporal bone, 126
of ulna, 134, 134f
Subacromial bursa, 143f
Subarachnoid space
of brain, 222, 223f, 225f
of spinal cord, 222, 223f
Subclavian arteries, 350f, 352f
left, 348, 351f, 352, 353f
right, 351f, 352, 353, 353f
Subclavian veins, 356, 357f, 358f, 381f
Subcutaneous injection, 97
Subcutaneous prepatellar bursa, 143f
Subcutaneous tissue, 96, 100
Subdural hematoma, 222

Subdural space
of brain, 222, 223f
of spinal cord, 222, 223f
Sublingual duct, 442f, 443
Sublingual gland, 228f, 442–443, 442f, 444
Submandibular duct, 441f, 442f, 443
Submandibular ganglion, 228f
Submandibular gland, 228f, 442–443, 442f, 444
Submucosa
of digestive tract, 438, 439f
of small intestine, 451, 452f
of stomach, 446, 447f
Subpubic angle, 136, 136f, 136t, 137f
Subscapular fossa, 132f
Subscapularis muscle, 178f, 180t
Substantia nigra, 212, 218
Sucrase
functions of, 443t
secretion of, 443t
Sucrose, 33, 33f
dietary sources of, 474
digestion of, 461–462
Sudden infant death syndrome (SIDS), 432t
Sugars
dietary sources of, 474
uses in body, 474
Sulcus (pl., sulci), 214, 214f, 218
Sulfate ions
in extracellular fluid, regulation of, 517–518
in intracellular fluid, 516
Sulfonamide, reference laboratory values for, A-3
Sulfur
atomic number of, 21t
deficiency of, 480t
dietary guidelines for, 480t
functions of, 480t
mass number of, 21t
percent in body, 21t
symbol for, 21t
Summation
multiple-wave, 162, 162f
of muscle contractions, 162, 162f
spatial, 205
temporal, 205
Sun exposure
and signs of aging, 108
and skin cancer, 99, 105
and vitamin D production, 99, 102
Sunscreens, 105, 108
Suntan, 98
Superficial, 10, 11t
Superficial lymphatic vessels, 414f, 415
Superficial palmar arch, 354f
Superficial temporal artery, 364f
Superficial temporal vein, 358f
Superior, 10, 11f, 11t
Superior colliculus, 212, 212f, 255
Superior concha, 408f
Superior lobe, of lungs, 413, 413f
Superior meatus, 408f
Superior mesenteric artery, 350f, 351f, 354–355, 355f
Superior mesenteric vein, 357f, 359f, 360, 360f
Superior oblique muscle, 244, 244f
Superior rectus muscle, 243f, 244, 244f
Superior sagittal sinus, 225f, 357f
Superior vena cava, 6f, 314, 315f–318f, 317, 320, 320f–321f, 356, 357f–358f
Supination, 144f, 145, 180
Supinator muscle, 180, 181f, 181t
Supine, definition of, 10
Supporting cells, of nervous system, 87
Suprahyoid muscles, 170, 173f, 173t
Supraorbital foramen or notch, 125f
Suprapatellar bursa, 143f
Suprarenal arteries, 351f, 355, 355f
Suprarenal glands. See Adrenal glands
Suprarenal veins, 359f, 360
Supraspinatus muscle, 178f, 180t
Supraspinatus tendon, 178f
Supraspinous fossa, 132f
Sural region, 13f
Surface anatomy, 2
Surface mucous cells, of stomach, 446, 447f
Surface tension, of lungs, 419
Surfactant, 414, 415f, 419, 574
Surgical methods, of birth control, 552
Suspensory ligaments
of breast, 545, 545f
of eye, 245, 245f
of ovaries, 540, 540f
Sustentacular cells, 530, 531f, 535f

Sutures, 122, 124f, 140, 140f
Swallowing, 444–446
aging and, 467
esophageal phase of, 445f, 446
gravity and, 446
larynx in, 410
muscles of, 170, 173f, 173t
peristaltic waves in, 446, 446f
pharyngeal phase of, 409, 444, 445f
voluntary phase of, 444, 445f
Sweat
emotional, 101
temperature regulation by, 31–32, 102
Sweat glands, 97f, 98f, 101
aging and, 108
apocrine, 101, 101f
eccrine, 101, 101f
epithelial tissue of, 76
excretion by, 104
Sweet taste, 242, 443
Swelling. See Edema
Sympathetic chain ganglion, 227, 228f
Sympathetic nerves, 228f
Sympathetic nervous system, 227
anatomy of, 227
functions of, 227t, 229, 230t
innervation by, 227, 228f
Synapse(s), 201–203
action potentials in, 201–203, 203f
definition of, 201
in skeletal muscle, 157–158, 157f
structure of, 201, 203f
Synapsis, in meiosis, 528, 529f
Synaptic cleft, 158, 158f, 201, 203f
Synaptic vesicles, 158
Synarthrosis, 140
Syndesmoses, 140
Syndrome of inappropriate ADH secretion, 275
Synergists, 167
Synovial fluid, 88, 141, 141f
Synovial joints, 141–143
aging and, 145–148
examples of, 143f
structure of, 141, 141f
types of, 141–143, 142f
Synovial membranes, 88, 141
functions of, 88, 141
location of, 88, 89f
structure of, 88, 141, 141f, 143f
Synthesis reactions, 26
Syphilis, 554f
Systemic anatomy, 2
Systemic circulation, 313, 314f, 345, 348–361
arteries in, 348–356
definition of, 348
veins in, 356–361
Systemic inflammation, 388
Systemic lupus erythematosus (SLE), 400–401, 400f
Systemic physiology, 2
Systemic vessels, 345, 348–361
Systole
atrial, 328
ventricular, 328, 331
Systolic pressure, 362–363, 362f

T&A (tonsillectomy and adenoidectomy), 382
Tachycardia, 327t, 332
Tadalafil (Cialis), 539
Tagamet, 450
Tail
of epididymis, 533f
of pancreas, 457, 457f
of sperm cell, 531f
Tailbone. See Coccyx
Tailor's muscle, 185
Talocrural joint, 142f
Talus, 138, 139f
Tanning, of skin, 98
Target tissues, of hormones, 265
Tarsal bones, 121t, 123f, 136f, 138, 139f
Tarsometatarsal joint, 142f
Taste(s), 242
aging and, 259
loss of, 243
neuronal pathways for, 242
types of, 242
Taste area, 216f
Taste buds, 242, 242f
Taste cells, 242, 242f
Taste hairs, 242, 242f
Tay-Sachs disease, 56, 231t, 482
TBG. See Thyroxine-binding globulin

T cell(s), 384, 385f
  activation of, 390–392
  in adaptive immunity, 389–392
  aging and, 401, 578
  antigen recognition by, 390–391
  in cell-mediated immunity, 395–396, 397f
  costimulation of, 390f, 391, 396, 396f
  cytotoxic, 389, 391, 395–396, 396f, 397f
  development of, 389, 389f
  effects of, 396, 397f
  helper, 389, 390f, 391, 396
  memory, 396, 397f
  origin of, 389, 389f
  proliferation of, 390f, 391–392, 396, 396f
T-cell receptors, 390
Tears, 244, 386
Tectorial membrane, 253–254, 253f, 254f
Tectospinal tract, 217f, 217t
Teeth, 440–442
  aging and, 467
  decay of, 442
  infection in, and lymph nodes, 442
  mastication by, 444
  prenatal development of, 565t
  primary or deciduous, 440, 441f
  quadrants of, 440
  secondary or permanent, 440, 441f
  structure of, 442, 442f
Telophase
  of meiosis, 529f
  of mitosis, 63, 64f
Temperature
  body. See Body temperature
  and chemical reaction rate, 29
  receptors for, 239–240
Temperature scales, A-1
Temperature sensation, 102
Temporal, definition of, 122
Temporal artery, superficial, 352f, 364f
Temporal bone, 121t
  in cranial cavity, 125, 126f
  in fetal skull, 140f
  frontal view of, 125f
  inferior view of, 127f
  intramembranous ossification of, 116f
  lateral view of, 122, 124f
  styloid process of, 126
Temporalis muscle, 5f, 152f, 170, 171f, 172t
Temporal lobe, 214f, 215
Temporal summation, 205
Temporal vein, superficial, 358f
Temporomandibular joint (TMJ), 83, 84t, 142f
Tendinitis, 189t
Tendinous intersections, 175–176, 176f
Tendon(s)
  aging and, 92, 148
  extracellular matrix of, 113
  functions of, 80, 112
  muscle attachment to, 166, 167f
  structure of, 82t
Tendon sheaths, 141, 141f, 143f
Teniae coli, 459, 459f
Tennis elbow, 182
Tension lines, 97, 98f
Tensor fasciae latae muscle, 168f, 182, 183f, 184f, 184t
Teratogens, 584t
Teres major muscle, 169f, 178f, 180t
Teres minor muscle, 169f, 178f, 180t
Terminal bronchioles, 412f, 413, 414f
Terminal ganglia, 229
Terminology, 10–17
Tertiary bronchi, 412f, 413, 413f
Testes, 6f, 270f, 527f, 528, 530, 530f, 533f
  anatomy of, 530, 531f
  descent of, 528
  histology of, 530, 531f
  hormones of, 269, 271t, 285
  secretions of, 535, 537
  sex cell production in, 527–528
  spermatogenesis in, 530–531
  temperature regulation in, 528–530
  undescended, 528, 531
Testicular arteries, 351f, 355, 532, 533f
Testicular nerves, 532, 533f
Testicular veins, 360, 532, 533f
Testosterone, 285
  and acne, 103
  action of, 268t, 271t, 285, 536t, 537, 538t
  aging and, 537
  chemistry of, 35f
  and male puberty, 537
  and male sexual behavior, 537
  and osteoporosis, 146
  reference laboratory values for, A-6

regulation of, 535f
  secretion of, 268t, 271t, 285
  and sperm cell count, 531
  synthesis of, 35, 476
  target tissues of, 268t, 271t, 538t
Tetanus, 162–163, 162f, 231t
Tetraiodothyronine (T$_4$), 275–276, A-6
  action of, 271t
  secretion of, 271t, 275–276, 277f
  structure of, 275
  target tissues of, 271t
Thalamic nuclei, 213f, 221f
Thalamus, 213
  functions of, 213
  structure of, 211f, 212f, 213, 213f
  in visual pathway, 249–250, 250f
Thalassemia, 308
Thalidomide, 65
Theca, 541, 541f, 542f
Theophylline, for asthma, 431
Thermoreceptors, 239
Theta waves, 220f, 221
Thiamine. See Vitamin B$_1$
Thick filaments. See Myosin myofilaments
Thigh
  abduction of, 182
  adduction of, 184, 185
  adductor muscles of, 184f, 185
  anterior muscles of, 184
  bones of, 136–138, 136f, 138f
  definition of, 12
  medial muscles of, 184
  movements of, muscles of, 182–184, 183f, 184t
  posterior muscles of, 184
  rotation of, 182
Thin filaments. See Actin myofilaments
Third-degree burns, 104f, 105
Third ventricle, 222, 224f, 225f
Thirst center, 516, 516f
Thirst regulation, 516–517, 516f
Thoracic aorta, 348, 350f, 351f, 353f, 354
Thoracic arteries, internal, 351f, 352f, 353f, 354, 361
Thoracic cage, 121t, 130–131
  functions of, 130
  structure of, 130–131, 131f
  surface anatomy of, 133f
Thoracic cavity, 15, 15f
  heart location in, 314, 315f
  mediastinum and, 15, 15f
  serous membranes in, 16, 17f, 88, 89f
Thoracic duct, 5f, 380, 381f, 385f
Thoracic nerve(s), 208, 210, 210f
Thoracic region, 12f
Thoracic veins, internal, 359
Thoracic vertebrae, 128f
  curvature of, 128, 128f
  first, 175f
  number of, 121t, 128
  structure of, 129, 129f
  twelfth, 128f
Thoracic volume, 416, 417, 418f
Thoracic wall, disorders of, 432t
Thorax, 10, 12f
  arteries of, 353f, 354
  muscles of, 173, 175f, 175t
  veins of, 356–359, 358f
Threonine, 477
Threshold, for action potential, 200f, 201, 201f
Thrombin, in coagulation, 301f, 302
Thrombocytes. See Platelet(s)
Thrombocytopenia, 307–308
Thromboplastin, 301–302, 308
Thrombosis, 578
  coronary, 322, 339t. See also Myocardial infarction
  pulmonary artery, 432t
  venous, 348, 361
    burns and, 106, 107
Thromboxanes, 266
  in platelet activation, 300
  in platelet release reaction, 300, 301f
  in vascular spasm, 300
Thrombus, 302, 322, 375
Thumb
  bones of, 134–135, 135f
  muscles of, 182
  opposition of, 145
Thymine, in DNA, 38, 39f, 60
Thymosin, 285
  action of, 271t, 285
  aging and, 288
  secretion of, 271t, 285
  target tissues of, 271t

Thymus gland, 5f, 6f, 15f, 270f, 381f, 384
  aging and, 401
  function of, 384
  histology of, 384, 384f
  hormones of, 271t, 285
  lymphocyte origins in, 389, 389f
  prenatal development of, 565t, 567, 570f
  structure of, 285, 384, 384f
Thyrocervical trunk, 352f
Thyroglobulins, 275, 276f
Thyroid artery, superior, 352f
Thyroid cartilage, 408f, 409–410, 410f
Thyroid follicle, 275, 276f
Thyroid gland, 6f, 270f, 275–276
  anatomy of, 275, 276f
  disorders of, 275–276
  prenatal development of, 565t
Thyroid hormones, 265, 271t, 275–276
  aging and, 288
  deficiency of, 275
  effects of, 271t, 275
  excess of, 275, 286–287, 286f
  replacement of, 274
  secretion of, 271t, 275–276, 277f
  target tissues of, 271t
Thyroid-stimulating hormone (TSH), 273, 275–276
  action of, 268t, 270t, 273, 275–276
  disorders of, 273
  reference laboratory values for, A-6
  secretion of, 268t, 270t, 273, 275–276
  target tissues of, 268t, 270t
Thyroid vein, superior, 358f
Thyroxine. See Tetraiodothyronine
Thyroxine-binding globulin (TBG), A-6
Tibia, 5f, 121t, 123f, 136f, 138, 139f
  at foot, 139f
  at knee, 143f
  surface anatomy of, 140f
Tibial arteries
  anterior, 350f, 356, 356f
  posterior, 350f, 356, 356f, 364f
Tibialis anterior muscle, 168f, 186f, 187t
Tibialis posterior muscle, 187t
Tibial nerve, 211
Tibial tuberosity, 138, 139f, 140f
Tibial veins
  anterior, 357f
  posterior, 357f
Tidal volume, 420, 421f
  aging and, 433
  exercise and, 431
Tight junctions, 78, 78f
Tinnitus, 258t
Tissue(s). See also specific types
  aging and, 92
  blood flow to, 345, 366–368
  connective, 3, 79–84
  damage to, 89
  definition of, 3, 72
  epithelial, 3, 72–79
  maturation of, 263
  microscopic study of, 72
  muscle, 3
  nervous, 3
Tissue level of organization, 3, 3f
Tissue plasminogen activator (t-PA), 302, 303f, 322, 338
Tissue repair, 90–92
  aging and, 92
  definition of, 90
  inflammatory response in, 91f, 92
  by regeneration, 90
  by replacement, 90–91
Tocopherol. See Vitamin E
Tocotrienols. See Vitamin E
Toe(s)
  bones of, 139f, 140
  movements of, muscles of, 185, 186f, 187t
  prenatal development of, 566, 566f
Toenails. See Nail(s)
Tone
  muscle, 164
  vasomotor, 367
Tongue, 14f, 408f, 440, 441f
  extrinsic muscles of, 170, 173t
  functions of, 440
  intrinsic muscles of, 170, 173t
  movements of, muscles of, 170, 173f, 173t
  papillae of, 242, 242f
  in swallowing, 444, 445f
Tonsil(s), 5f, 381f, 382, 442
  function of, 382
  lingual, 382, 382f, 408f, 409
  palatine, 382, 382f, 408f, 409

pharyngeal, 382, 382f, 408f, 409
  prenatal development of, 565t, 567, 570f
Tonsillectomy, 382
Tooth. See Teeth
Torticollis, 172
Total lung capacity, 421, 421f
Touch receptors, 239f, 240, 427, 427f
Touch sensation, 102
Toxemia, 584t
Trabeculae
  of bone, 114f, 115, 116f
    in endochondral ossification, 116
    in intramembranous ossification, 115
  of lymph node, 382, 383f
  of spleen, 383, 383f
  of thymus, 384, 384f
Trabeculae carneae, 322, 323f
Trace elements, 473
Trachea, 4f, 5f, 14f, 15f, 407f, 410–412, 412f
  adult, 410–411
  anatomy of, 408f, 410–412, 412f
  during cough, 411–412
  epithelial tissue of, 75t, 76, 413
  foreign objects in, 412
  membranous part of, 410f
  smoking and, 412
Tracheal cartilage, 410f
Tracheobronchial tree, 412f, 413
Tracheostomy, 411
Tracheotomy, 411
Trachoma, 258t
Transcription, 60–61, 60f, 61f
Trans fats, 476–477
Transfer ribonucleic acid (tRNA), 60f, 61
Transfusion, blood
  blood grouping and, 303–306
  definition of, 303
  typing and crossmatching for, 306
Transfusion reactions, 303
Transitional epithelial tissue, 76
  functions of, 76, 77t
  location of, 76, 77t
  structure of, 76, 77t
Translation, 60f, 61, 62f
Transplantation
  bone marrow, 296
  corneal, 245
  heart, 336–337
  kidney, 521
  rejection of, 402t
Transport. See also specific mechanisms and substances
  active, 46t, 50–51, 50f, 461
  by blood, 80
  carrier-mediated, 46t, 49–51
  in digestive system, 460–465
  by water, 32
Transverse colon, 440f, 459, 459f
Transverse foramen, of cervical vertebrae, 129f
Transverse fracture, 120, 120f
Transverse perineal muscles
  deep, 177f, 177t
  superficial, 177f, 177t
Transverse plane, 14f, 15
Transverse processes, 128f, 129
  of cervical vertebrae, 129, 129f
  of lumbar vertebrae, 128f, 129f, 130
  of thoracic vertebrae, 129f
Transverse section, 15, 15f
Transverse tubules, 153, 154f
Transversus abdominis muscle, 176, 176f, 176t
Trapezius muscle, 168f, 169f, 171f, 176–177, 178f
  action of, 174t, 176–177, 178t
  insertion of, 174t, 178t
  origin of, 174t, 178t
Trapezoid bone, 134, 135f
Traveler's diarrhea, 466, 466f
Tremor, resting, 218
Triacylglycerols. See Triglycerides
Tricarboxylic acid cycle, 484f, 485, 486
Triceps brachii muscle, 169f, 179, 182f
  action of, 167, 179, 180t
  heads of, 179f, 182f
  insertion of, 167f, 180t
  origins of, 167, 167f, 180t
Trichomoniasis, 554t
Tricuspid valve, 318–319, 318f, 320f, 331f
Trigeminal (CN V) nerve, 224
  functions of, 224, 226t
  origin of, 226f
Triglycerides
  dietary sources of, 474–476
  digestion of, 462–464
  formation of, 33, 34f, 464
  metabolism of, 486

in absorptive state, 488*f*
    in postabsorptive state, 487, 488*f*
reference laboratory values for, A-3
structure of, 462
uses in body, 476
Trigone, 513, 513*f*
Triiodothyronine (T$_3$), 275–276
    action of, 268*t*, 271*t*
    secretion of, 268*t*, 271*t*, 275–276, 277*f*
    structure of, 275
    target tissues of, 268*t*, 271*t*
Triquetrum bone, 134, 135*f*
Trisomy 21, 580
Trochanter
    definition of, 122*t*
    greater, 138, 138*f*, 143*f*
    lesser, 138, 138*f*, 143*f*
Trochlea, 133*f*
Trochlear (CN IV) nerve, 224
    functions of, 224, 226*t*
    origin of, 226*f*
Trochlear notch, 134, 134*f*, 143*f*
Trophoblast, 561
Tropomyosin, 153, 154*f*
Troponin, 153, 154*f*, 161*f*
True ribs, 130, 131*f*
True vocal cords, 82*t*, 410, 410*f*, 411*f*
Trunk, of body, 10, 12*f*, 13*f*
    arteries of, 350*f*
    muscles of, 172–176
    veins of, 357*f*
Trunk cavities, 15–16, 15*f*
Trypsin, 458
    functions of, 443*t*, 458, 461*f*, 464
    secretion of, 443*t*, 458
Tryptophan, 477
TSH. *See* Thyroid-stimulating hormone
T tubules, 153, 154*f*
Tubal ligation, 551*f*, 552
Tubercle
    definition of, 122, 122*t*
    of humerus, 133*f*, 134
Tuberculosis, 432*t*
    bone infection, 148*t*
Tuberosity. *See also* specific types
    definition of, 122, 122*t*
Tubular glands, 78, 79*f*
Tubular reabsorption, 502–508, 503*f*, 505*f*
    in collecting duct, 504, 505*f*, 506–508, 509*f*
    definition of, 503
    diabetes mellitus and, 508
    in distal convoluted tubule, 504, 505*f*, 506–508, 509*f*
    in loop of Henle, 504, 505*f*, 506, 507*f*, 509*f*
    in proximal convoluted tubule, 504, 505*f*, 506, 506*f*
Tubular secretion
    definition of, 503
    in urine production, 502–503, 503*f*, 508–509
Tubules, renal. *See* Distal convoluted tubule; Proximal convoluted tubule
Tumor(s)
    benign, 65
    definition of, 65
    malignant, 65
    uterine, 553, 553*f*
Tums, 450
Tunic(s)
    of blood vessels, 345, 346*f*, 347*f*
    of digestive tract, 438, 439*f*
    of eye, 244, 245*f*
Tunica adventitia, 345, 346*f*, 347*f*
Tunica intima, 345, 346*f*, 347*f*
Tunica media, 345, 346*f*, 347*f*
Turner syndrome, 580
T wave, of ECG, 326, 326*f*
Twins, 560
Twitch, muscle. *See* Muscle twitch
Two-point discrimination, 258
Tympanic membrane, 251
    auditory function of, 254, 254*f*
    distortion of, 252
    function of, 251
    structure of, 251, 252*f*
Typhoid fever, 468*t*
Typing, of blood, 306

Ulcer(s)
    decubitus, 108*t*
    peptic, 468*t*
Ulna, 5*f*, 121*t*, 123*f*, 134
    at elbow, 143*f*
    head of, 135*f*
    structure of, 134, 134*f*
Ulnar arteries, 350*f*, 354, 354*f*

Ulnar nerve, 210
Ultraviolet-A (UVA), 105
Ultraviolet-B (UVB), 105
Ultraviolet light
    and melanin, 98, 99
    protection from, 96, 98, 105
    and skin cancer, 105
    and vitamin D production, 96, 99, 102, 277
Umami, 242
Umbilical arteries, 562*f*, 574*f*, 575, 575*f*
Umbilical cord, 562*f*, 563
Umbilical cord blood, 296
Umbilical region, 12*f*, 13*f*
Umbilical vein, 562*f*, 574*f*, 575, 575*f*
Umbilicus, 176*f*
Unicellular glands, 78, 79*f*
Uniform Anatomical Gift Act, 4
United States Preventive Services Task Force (USPSTF), 300
U.S. measurements, A-1
Universal donor, of blood, 304
Unmyelinated axons, 195
    action potentials in, 201, 202*f*
    *vs.* myelinated axons, 195, 198*f*
Unsaturated fats or oils, 34, 34*f*, 462–463, 476
Unsaturated fatty acids, 34, 34*f*
Upper esophageal sphincter, 444, 445*f*
Upper limb, 12, 12*f*, 13*f*
    arteries of, 350*f*, 353–354, 354*f*
    bones of, 121*t*, 131*f*, 132–135, 135*f*
    muscles of, 176–182, 182*f*
    prenatal development of, 566, 566*f*
    surface anatomy of, 135*f*
    veins of, 356, 357*f*, 358*f*
Upper motor neurons, 217
Upper respiratory tract, 407, 407*f*
Upper respiratory tract infections, 432*t*
Uracil, 38, 60
Urea
    ammonia conversion to, 457, 486, 487*f*, 503
    in urine, 503, 503*t*
Urea clearance, A-5
Urea nitrogen, A-3
Ureter(s), 4*f*, 6*f*, 497, 498, 498*f*, 499*f*
    anatomy of, 513–514, 513*f*
    epithelial tissue of, 77*t*, 513
    female, 497*f*
    male, 530*f*, 533*f*
Urethra, 4*f*, 6*f*, 497
    aging and, 552
    anatomy of, 513–514
    epithelial tissue of, 76, 77*t*
    female, 497*f*, 514, 539*f*, 544, 544*f*
    male, 514, 528, 530*f*, 532, 533*f*, 535, 538–539, 552
    membranous, 532, 533*f*
    prostatic, 532, 533*f*, 534, 552
    spongy, 532, 533*f*
Urethral orifice
    external male, 533*f*, 534
    female, 539*f*
Urethritis, nongonococcal, 554*t*
Uric acid
    in blood, A-3
    reference laboratory values for, A-3, A-5
    in urine, 503*t*, A-5
Urinary bladder, 4*f*, 6*f*, 440*f*, 497, 498*f*
    anatomy of, 513–514, 513*f*
    contractions of, 513
    epithelial tissue of, 76, 77*t*, 513
    female, 497*f*, 539*f*
    inflammation of, 514
    male, 530*f*, 533*f*
    and micturition reflex, 514, 515*f*
    prenatal development of, 565*f*, 567, 570*f*
    storage function of, 513
    volume changes in, 513
Urinary sphincter
    external, 513*f*, 514
    internal, 513–514, 532
Urinary system, 3–5, 6*f*, 496–523
    components of, 3, 6*f*, 497, 497*f*
    diseases and disorders of, 523*t*
    functions of, 5, 6*f*, 497
    prenatal development of, 565*t*, 568*t*–569*t*
Urination
    autonomic regulation of, 229
    muscles of, 176
Urine
    color of, A-5
    composition of, 502, 503*t*
    concentration of, 505, 505*f*, 509–514
    hormonal regulation of, 510–513, 512*f*
    and innate immunity, 386
    movement of, 514

odor of, A-5
osmolality of, A-5
pH of, 503*t*, A-5
pigments in, 298, 298*f*
production of. *See* Urine production
reference laboratory values for, A-5
specific gravity of, 503*t*
volume of, 509–514
Urine production, 502–509, 503*f*
    and antidiuretic hormones, 505*f*
    filtration in, 502–504, 503*f*
    tubular reabsorption in, 502–503, 503*f*, 504–508, 505*f*
    tubular secretion in, 502–503, 503*f*, 508–509
Urticaria, 402*t*
Uterine cavity, 543
Uterine contractions, 573
    endocrine regulation of, 263
    and fertilization, 559
Uterine tubes, 6*f*, 527*f*, 539, 539*f*, 540*f*, 543
    ampulla of, 540*f*, 543
    epithelial tissue of, 75*t*
    fimbriae of, 540*f*, 543
    implantation in, 548
    ligation of, 551*f*, 552
Uterus, 6*f*, 527*f*, 539–540, 539*f*, 540*f*, 543
    aging and, 554
    benign tumors of, 553, 553*f*
    cancer of, 549
    epithelial tissue of, 75*t*
    hormones and, 271*f*
    implantation in, 543, 561–563, 561*f*
    in infertility, 549
    in labor and birth, 572–574, 572*f*
    in menstrual cycle, 546–548, 546*f*, 547*t*
    in postmenopausal women, 549*t*, 554
    prolapsed, 543
    wall of, layers of, 540*f*, 543
Utricle, 255, 255*f*, 256*f*, 259
Uvula, 408*f*, 409, 441*f*, 442

Vaccinations, 398, 400–402
Vaccine, 398
Vagina, 6*f*, 539–540, 539*f*, 540*f*, 543–544, 544, 544*f*
    aging and, 554
    epithelial tissue of, 76, 77*t*
    pH of, 535
    in postmenopausal women, 549*t*
    in sex act, 549
Vaginal condom, 550, 551*f*
Vaginal contraceptive ring, 552
Vaginal orifice, 539*f*
Vagus (CN X) nerve, 224
    functions of, 224, 226*t*
    origin of, 226*f*
    and parasympathetic nervous system, 227–229, 227*t*, 229
    and small intestine secretions, 453
    and stomach secretions, 448, 449*f*
    and taste, 242
Valine, 477
Valves, heart. *See* Heart, valves of
Vardenafil (Levitra), 539
Variable region, of antibody, 392, 392*f*
Variables, 7
Varicose veins, 348, 375
Vasa recta, 502, 502*f*, 506–508, 507*f*
Vascular spasm, 299–300
Vascular tunic, of eye, 244–245, 245*f*
Vas deferens, 6*f*, 527*f*, 528, 530*f*, 531*f*, 532, 533*f*
Vasectomy, 551*f*, 552
Vasoconstriction, 345
Vasodilation, 347, 387
Vasomotor center, 367, 367*f*, 368*f*
Vasomotor instability, menopause and, 549, 549*t*
Vasomotor tone, 367
Vasopressin. *See* Antidiuretic hormone (ADH)
Vastus intermedius muscle, 168*f*, 183*f*
Vastus lateralis muscle, 168*f*, 183*f*, 184*f*, 185
Vastus medialis muscle, 168*f*, 183*f*, 184*f*
Vegetarian diet, 477, 482
Veins
    of abdomen, 359–360, 359*f*
    aging and, 375
    blood flow in, 345
    blood pressure in, 363–364, 363*f*
    of head, 356, 357*f*, 358*f*
    of kidneys, 502, 502*f*
    large, 345, 346*f*, 347
    of lower limb, 357*f*, 360–361, 361*f*

major, 357*f*
medium, 345, 346*f*, 347*f*, 348
of neck, 356, 358*f*
of pelvis, 359–360, 359*f*
photomicrograph of, 347*f*
in pulmonary circulation, 348, 349*f*
small, 345, 346*f*, 347–348
structure of, 345, 346*f*, 347–348
in systemic circulation, 356–361
of thorax, 356–359, 358*f*
of trunk, 357*f*
of upper limb, 356, 357*f*, 358*f*
varicose, 348, 375
Vena cava
    blood pressure in, 363*f*
    inferior, 6*f*, 314, 316*f*–318*f*, 317, 320, 320*f*, 356, 357*f*–360*f*
    superior, 6*f*, 314, 315*f*–318*f*, 317, 320, 320*f*–321*f*, 356, 357*f*–358*f*
Venous return, 332
Venous thrombosis, 348, 361
    burns and, 106, 107
Ventilation, 407, 416–424
    aging and, 433
    altitude and, 429
    alveolar volume changes in, 419–420
    brainstem and, 211, 424, 426*f*
    carbon dioxide levels and, 427, 428*f*
    chemical control of, 427–429, 427*f*
    exercise and, 429–433
    Hering-Breuer reflex and, 426, 427*f*
    lung recoil and, 419
    minute, 420, 431–433
    nervous control of, 426–427, 427*f*
    normal rate of, 424
    normal *vs.* labored, 416, 417*f*
    oxygen levels and, 427
    pH and, 427–428, 428*f*
    phases of, 416
    pleural pressure and, 419–420
    pressure in, 417, 418*f*
    rhythmic, generation of, 424–430
    thoracic volume and, 416, 417, 418*f*
Ventral, 10, 11*f*, 11*t*
Ventral column, 206, 207*f*
Ventral respiratory groups, 424, 426*f*
Ventral root, 206, 207*f*
Ventricle(s)
    brain, 222, 224*f*
        cerebrospinal fluid flow in, 222–224, 225*f*
        fourth, 222, 224*f*, 225*f*
        lateral, 222, 224*f*, 225*f*
        third, 222, 224*f*, 225*f*
    cardiac, 314, 317–318
        aging and, 340
        blood flow through, 320–321, 320*f*
        contraction of, 318, 320–321, 328, 329*f*
        function of, 317, 328
        left, 314, 315*f*, 316*f*–317*f*, 317–318, 318*f*
        neonatal, 574–575, 574*f*, 575*f*
        prenatal development of, 567, 571*f*
        relaxation of, 328, 329*f*
        right, 314, 315*f*, 316*f*–317*f*, 317–318, 318*f*
        wall of, 318
Ventricular contractions, premature, 327*t*
Ventricular diastole, 328, 331
Ventricular systole, 328, 331
Ventricular tachycardia, 327*t*
Venules, 345, 347
    blood pressure in, 363*f*
    maternal and fetal, 562*f*
Vernix caseosa, 568
Vertebrae. *See also* specific types
    dislocations of, 129
    fibrocartilage in, 83, 84*t*
    fractures of, 129, 145
    intramembranous ossification of, 116*f*
    number of, 121*t*, 128
    regional differences in, 129–130, 129*f*
    structure of, 128–130, 128*f*, 129*f*
Vertebral arch, 128, 128*f*
Vertebral arteries, 352–353, 352*f*, 353*f*, 354*f*
Vertebral canal, 128
Vertebral column, 5*f*, 14*f*, 123*f*, 127–130, 206
    curvatures in
        abnormal, 128
        normal, 128, 128*f*
    functions of, 128
    muscles moving, 172–173, 174*f*, 174*t*
    number of bones in, 121*t*, 128
    regions of, 128, 128*f*
Vertebral foramen, 128, 129*f*
Vertebral region, 13*f*
Vertigo, 259

Vesicles. *See also specific types*
  definition of, 51, 54
  membrane transport in, 51–52, 51*f*
  in ovarian follicles, 541, 541*f*
Vestibular folds, 82*t*, 408*f*, 410, 410*f*, 411*f*, 444, 445*f*
Vestibular membrane, 253, 253*f*, 254*f*
Vestibular nerve, 252*f*, 256
Vestibule
  of ear, 252, 252*f*, 253*f*, 255, 255*f*
  of nose, 408*f*
  of vagina, 544, 544*f*, 549
Vestibulocochlear (CN VIII) nerve, 224, 254
  and balance, 255, 256
  functions of, 224, 226*f*
  and hearing, 255
  origin of, 226*f*
Vestibulospinal tract, 217*f*, 217*t*
Viagra (sildenafil), 539
Villi
  chorionic, 561
  of small intestine, 451, 452*f*
Visceral arteries, 351*f*, 353*f*, 354
Visceral pericardium, 16, 17*f*, 314, 315*f*, 322, 323*f*, 416*f*
Visceral peritoneum, 16, 17*f*, 439, 440*f*
Visceral pleura, 16, 17*f*, 315*f*, 412*f*, 414, 414*f*, 416*f*
Visceral senses, 239
Visceral serous membranes, 16, 16*f*
Vision, 242–251
  aging and, 259
  anatomical structures in, 242–249
  color, 246
  disorders of, 258*t*
  distant, 249*f*
  double, 251, 258*t*
  near, 249*f*
  neuronal pathways for, 249–250, 250*f*
  testing of, 249
Visual association area, 216*f*
Visual cortex, 216*f*, 250, 250*f*
Visual field(s), 250, 250*f*
Vital capacity, 421, 421*f*
  aging and, 433
  exercise and, 431
Vitamin(s)
  as coenzymes, 477
  deficiencies of, 478*t*
  dietary guidelines for, 477–479, 478*t*
  dietary sources of, 478*t*
  discovery of, 477
  essential, 477
  excess consumption of, 479
  fat-soluble, 477, 478*t*
  functions of, 477, 478*t*
  as nutrients, 473, 477–479
  water-soluble, 477, 478*t*
Vitamin A
  deficiency of, 104, 246, 478*t*
  dietary guidelines for, 478*t*
  dietary sources of, 478*t*
  excess of, 479
  functions of, 246, 477, 478*t*
  solubility of, 477, 478*t*
  synthesis of, 477
Vitamin B₁
  deficiency of, 478*t*
  dietary guidelines for, 478*t*

dietary sources of, 478*t*
  functions of, 477, 478*t*
  solubility of, 477, 478*t*
Vitamin B₂
  deficiency of, 478*t*
  dietary guidelines for, 478*t*
  dietary sources of, 478*t*
  functions of, 477, 478*t*
  solubility of, 477, 478*t*
Vitamin B₃
  deficiency of, 478*t*
  dietary guidelines for, 478*t*
  dietary sources of, 478*t*
  functions of, 477, 478*t*
  solubility of, 477, 478*t*
Vitamin B₆
  deficiency of, 478*t*
  dietary guidelines for, 478*t*
  dietary sources of, 478*t*
  functions of, 477, 478*t*
  solubility of, 477, 478*t*
Vitamin B₁₂
  deficiency of, 307–308
  functions of, 477
  intrinsic factor and, 308, 448
  and red blood cell production, 297
Vitamin C
  and aging, 578
  as antioxidant, 479, 578
  deficiency of, 478*t*
  dietary guidelines for, 478*t*
  dietary sources of, 478*t*
  excess of, 479
  functions of, 477, 478*t*
  and osteoporosis, 146–147
  solubility of, 477, 478*t*
Vitamin D
  and aging, 148
  and bone growth, 102
  and calcium, 99, 102, 120*f*, 121, 277, 517
  deficiency of, 99, 478*t*
  dietary guidelines for, 478*t*
  dietary sources of, 102, 478*t*
  digestion of, 465
  excess of, 479
  functions of, 99, 102, 477, 478*t*
  and muscle function, 102
  and osteoporosis, 146–147
  production of, 99, 102, 277, 477
    PTH and, 277
    renal regulation of, 497
    in skin, 96, 99
  solubility of, 477, 478*t*
  transport in blood, 292
Vitamin E
  and aging, 578
  as antioxidant, 479, 578
  deficiency of, 478*t*
  dietary guidelines for, 478*t*
  dietary sources of, 478*t*
  functions of, 478*t*
  solubility of, 477, 478*t*
Vitamin K
  and coagulation, 302, 308, 477
  deficiency of, 308, 478*t*
  dietary guidelines for, 478*t*
  dietary sources of, 478*t*
  functions of, 477, 478*t*
  intestinal synthesis of, 460, 477

solubility of, 477, 478*t*
  sources of, 302
  supplementation of, 302
Vitreous chamber, 245*f*, 246–249
Vitreous humor, 245*f*, 248–249
Vocal cords
  false, 82*t*, 410, 410*f*, 411*f*
  true, 82*t*, 410, 410*f*, 411*f*
Vocal folds, 82*t*, 408*f*, 410, 410*f*, 411*f*, 444, 445*f*
Voice production, 407, 410
Voltage-gated ion channels, 199–201, 200*f*
Volume, measures of, A-1
Volume, of sound, 255
Voluntary movements, 216–217
Voluntary phase, of swallowing, 444, 445*f*
Vomer, 121*t*, 124
  frontal view of, 124, 125*f*
  inferior view of, 127*f*
Vomiting, 468*t*
von Willebrand disease, 309*t*
von Willebrand factor (VWF), 300, 301*f*
Vulva, 544

Warfarin (Coumadin), 302
Warm receptors, 240
Warts, genital, 554*t*
Waste products
  exchange, in circulatory system, 345
  in plasma, 292, 293*f*, 293*t*
  in urine, 502, 503*t*
Water
  in aerobic respiration, 484–486, 485*f*
  biological functions of, 31–32
  and body temperature, 31–32
  in chemical reactions, 32
  chemistry of, 31–32
  diffusion of (osmosis), 46*t*, 47–49, 48*f*
  digestion of, 465
  in digestive system, 465, 465*f*
  hydrogen bonds in, 24–25, 25*f*, 31
  intake of, thirst and, 516–517, 516*f*
  intracellular-extracellular exchange of, 516
  as nutrient, 473
  percent of body weight composed of, 514
  in plasma, 292, 293*f*, 293*t*
  protection by, 32
  renal reabsorption of, 504–508, 505*f*
  total body, 515*t*
  transport by, 32
Water balance, 263
"Water breaking," 573
Water molecule, 24–25, 24*f*, 25*f*, 31–32
Water-soluble hormones, 265
Water-soluble vitamins, 477
Water vapor, partial pressure of, 422*t*
Weight
  control of, 490
  definition of, 21, 489
  desirable, 489
  water as percentage of total body, 514
Wernicke area, 216*f*, 220
White blood cells, 294, 298–299
  ameboid movement of, 298
  chemotaxis by, 386
  disorders of, 307
  function of, 294*t*, 298
  identification of, 299*f*
  in inflammatory response, 89, 90*f*, 299

in innate immunity, 386
  nomenclature for, 298
  number of, 293*f*, 294, 307
  production of, 294, 295*f*
  staining properties of, 298–299, 299*f*
  structure of, 294*t*, 296*f*, 298
  types of, 298–299
  in urine, A-5
White blood count (WBC), 307, A-4
White matter, 196
  of brain, 218
  of spinal cord, 206, 207*f*
White of eye, 244
White pulp, of spleen, 383, 383*f*
Whole-cell activity, 59–66
Whooping cough, 432*t*
Windpipe. *See* Trachea
Wisdom teeth, 440, 441*f*
Withdrawal, in sex act, 550
Withdrawal (flexor) reflex, 208, 209*f*
Word origin (etymology), 10
Work, energy as capacity for, 28
Working memory, 221
Wound contracture, 92
Wound repair, 91–92, 91*f*
Wrinkles, 108
Wrist
  bones of, 134, 135*f*
  definition of, 134
  extension of, 180, 182
  extensor muscles of, 169*f*
  flexion of, 180
  flexor muscles of, 168*f*
  movements of, muscles of, 180–182, 181*f*, 181*t*
Wry neck, 172

X chromosomes, 62, 579, 579*f*, 580*f*
Xiphoid process, 130–131, 131*f*
X-linked traits, 581, 581*f*
X-rays, 27

Y chromosomes, 62, 579, 579*f*, 580*f*
Yellow bone marrow, 113, 114*f*
Y-linked traits, 581
Yolk sac, 564, 564*f*, 567, 570*f*
Young adult, 577

Zantac, 450
Z disks, 153, 154*f*, 155*f*, 160*f*
Zinc, 480*t*
Zona pellucida, 541, 541*f*, 542*f*, 559
Zygomatic arch, 122, 124*f*
Zygomatic bone, 121*t*, 122
  frontal view of, 122, 125*f*
  inferior view of, 127*f*
  intramembranous ossification of, 116*f*
  lateral view of, 122, 124*f*
Zygomaticus major muscle, 171*f*
  action of, 172*t*
  insertion of, 172*t*
  origin of, 172*t*
Zygomaticus minor muscle, 171*f*
  action of, 172*t*
  insertion of, 172*t*
  origin of, 172*t*
Zygomaticus muscles, 167
Zygote, 528, 541, 542*f*, 548, 559, 560*f*

# Prefixes, Suffixes, and Combining Forms

The ability to break down medical terms into separate components or to recognize a complete word depends on mastery of the combining forms (roots or stems) and the prefixes and suffixes that alter or modify their meanings. Common prefixes, suffixes, and combining forms are listed below in boldface type, followed by the meaning of each form and an example illustrating its use.

**a-, an-** without, lack of: *a*phasia (lack of speech), *an*aerobic (without oxygen)

**ab-** away from: *ab*ductor (leading away from)

**-able** capable: vi*able* (capable of living)

**acou-** hearing: *acou*stics (science of sound)

**acr-** extremity: *acr*omegaly (large extremities)

**ad-** to, toward, near to: *ad*renal (near the kidney)

**adeno-** gland: *adeno*ma (glandular tumor)

**-al** expressing relationship: neur*al* (referring to nerves)

**-algia** pain: gastr*algia* (stomach pain)

**angio-** vessel: *angio*graphy (radiography of blood vessels)

**ante-** before, forward: *ante*cubital (before elbow)

**anti-** against, reversed: *anti*peristalsis (reversed peristalsis)

**arthr-** joint: *arthr*itis (inflammation of a joint)

**-ary** associated with: urin*ary* (associated with urine)

**-asis** condition, state of: homeost*asis* (state of staying the same)

**auto-** self: *auto*lysis (self breakdown)

**bi-** twice, double: *bi*cuspid (two cusps)

**bio-** live: *bio*logy (study of living)

**-blast** bud, germ: fibro*blast* (fiber-producing cell)

**brady-** slow: *brady*cardia (slow heart rate)

**-c** expressing relationship: cardia*c* (referring to heart)

**carcin-** cancer: *carcin*ogenic (causing cancer)

**cardio-** heart: *cardio*pathy (heart disease)

**cata-** down, according to: *cata*bolism (breaking down)

**cephal-** head: *cephal*ic (toward the head)

**-cele** hollow: blasto*cele* (hollow cavity inside a blastocyst)

**cerebro-** brain: *cerebro*spinal (referring to brain and spinal cord)

**chol-** bile: a*chol*ic (without bile)

**cholecyst-** gallbladder: *cholecyst*okinin (hormone causing the gallbladder to contract)

**chondr-** cartilage: *chondr*ocyte (cartilage cell)

**-cide** kill: bacteri*cide* (agent that kills bacteria)

**circum-** around, about: *circum*duction (circular movement)

**-clast** smash, break: osteo*clast* (cell that breaks down bone)

**co-, com-, con-** with, together: *co*enzyme (molecule that functions with an enzyme), *com*misure (coming together), *con*vergence (to incline together)

**contra-** against, opposite: *contra*lateral (opposite side)

**crypto-** hidden: *crypt*orchidism (undescended or hidden testes)

**cysto-** bladder, sac: *cysto*cele (hernia of a bladder)

**-cyte-, cyto-** cell: erythro*cyte* (red blood cell), *cyto*skeleton (supportive fibers inside a cell)

**de-** away from: *de*hydrate (remove water)

**derm-** skin: *derm*atology (study of the skin)

**di-** two: *di*ploid (two sets of chromosomes)

**dia-** through, apart, across: *dia*pedesis (ooze through)

**dis-** reversal, apart from: *dis*sect (cut apart)

**-duct** leading, drawing: ab*duct* (lead away from)

**-dynia** pain: masto*dynia* (breast pain)

**dys-** difficult, bad: *dys*mentia (bad mind)

**e-** out, away from: *e*viscerate (take out viscera)

**ec-** out from: *ec*topic (out of place)

**ecto-** on outer side: *ecto*derm (outer skin)

**-ectomy** cut out: append*ectomy* (cut out the appendix)

**-edem-** swell: myo*edem*a (swelling of a muscle)

**em-, en-** in: *em*pyema (pus in), *en*cephalon (in the brain)

**-emia** blood: an*emia* (deficiency of blood)

**endo-** within: *endo*metrium (within the uterus)

**entero-** intestine: *entero*itis (inflammation of the intestine)

**epi-** upon, on: *epi*dermis (on the skin)

**erythro-** red: *erythro*cyte (red blood cell)

**eu-** well, good: *eu*phoria (well-being)

**ex-** out, away from: *ex*halation (breathe out)

**exo-** outside, on outer side: *exo*genous (originating outside)

**extra-** outside: *extra*cellular (outside the cell)

**-ferent** carry: af*ferent* (carrying to the central nervous system)

**-form** expressing resemblance: fusi*form* (resembling a fusion)

**gastro-** stomach: *gastro*dynia (stomach ache)

**-genesis** produce, origin: patho*genesis* (origin of disease)

**gloss-** tongue: hypo*gloss*al (under the tongue)

**glyco-** sugar, sweet: *glyco*lysis (breakdown of sugar)

**-gram** a drawing: myo*gram* (drawing of a muscle contraction)

**-graph** instrument that records: myo*graph* (instrument for measuring muscle contraction)

**hem-** blood: *hem*opoiesis (formation of blood)

**hemi-** half: *hemi*plegia (paralysis of half of the body)

**hepato-** liver: *hepato*itis (inflammation of the liver)

**hetero-** different, other: *hetero*zygous (different genes for a trait)

**hist-** tissue: *hist*ology (study of tissues)

**homeo-, homo-** same: *homeo*stasis (state of staying the same), *homo*logous (alike in structure or origin)

**hydro-** wet, water: *hydro*cephalus (fluid within the head)

**hyper-** over, above, excessive: *hyper*trophy (overgrowth)

**hypo-** under, below, deficient: *hypo*tension (low blood pressure)

**-ia, -id** expressing condition: neuralg*ia* (pain in nerve), flacc*id* (state of being weak)

**-iatr-** treat, cure: ped*iatr*ics (treatment of children)

**-im** not: *im*permeable (not permeable)

**in-** in, into: *in*jection (forcing fluid into)

**infra-** below, beneath: *infra*orbital (below the eye)

**inter-** between: *inter*costal (between the ribs)

**intra-** within: *intra*ocular (within the eye)

**-ism** condition, state of: dimorph*ism* (condition of two forms)